扩展有限元法
理论与应用

Extended Finite Element Method
Theory and Applications

Amir R. Khoei　　［伊朗］阿米尔-雷扎-科伊　著

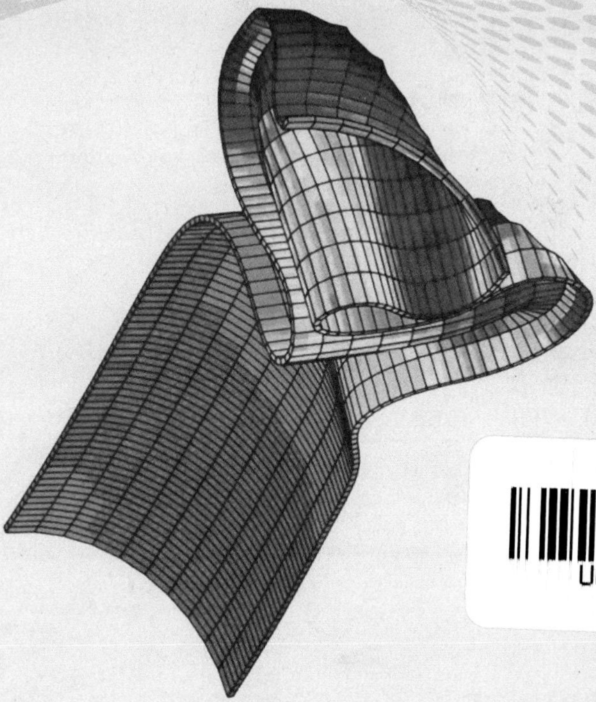

西北工業大學出版社
西　安
WILEY

陕西省版权局著作权合同登记 图字:25-2019-211

图书在版编目(CIP)数据

扩展有限元法:理论与应用 = Extended Finite Element Method: Theory and Applications:英文/(伊朗)阿米尔-雷扎-科伊著.—西安:西北工业大学出版社,2019.3

ISBN 978-7-5612-6433-1

Ⅰ.①扩… Ⅱ.①阿… Ⅲ.①有限元法-英文 Ⅳ.①O241.82-39

中国版本图书馆 CIP 数据核字(2019)第 017420 号

KUOZHAN YOUXIANYUANFA: LILUN YU YINGYONG

扩 展 有 限 元 法 : 理 论 与 应 用

(Extended Finite Element Method: Theory and Applications)

责任编辑: 卢颖慧 **策划编辑:** 杨 军
责任校对: 李阿盟 **装帧设计:** 李 飞
出版发行: 西北工业大学出版社
通信地址: 西安市友谊西路 127 号 邮编:710072
电 话: (029)88491757,88493844
网 址: www.nwpup.com
印 刷 者: 陕西天意印务有限公司
开 本: 710 mm×1 000 mm 1/16
印 张: 36.75 彩插:8
字 数: 1 233 千字
版 次: 2019 年 3 月第 1 版 2019 年 3 月第 1 次印刷
定 价: 138.00 元

如有印装问题请与出版社联系调换

影印版出版说明

本书系 WILEY 所出版的原著的影印版引进图书。原著所用科技符号的格式与我国国标有所差异，为避免可能因为修改引入的错误，结合原版图书相关著作权要求，在不影响读者理解的前提下，影印版图书保留原著格式，不予修改。

西北工业大学出版社

2019 年 2 月

EXTENDED FINITE ELEMENT METHOD

WILEY SERIES IN COMPUTATIONAL MECHANICS

Series Advisors:

René de Borst
Perumal Nithiarasu
Tayfun E. Tezduyar
Genki Yagawa
Tarek Zohdi

Extended Finite Element Method: Theory and Applications	Khoei	December 2014
Computational Fluid-Structure Interaction: Methods and Applications	Bazilevs, Takizawa and Tezduyar	January 2013
Introduction to Finite Strain Theory for Continuum Elasto-Plasticity	Hashiguchi and Yamakawa	November 2012
Nonlinear Finite Element Analysis of Solids and Structures, Second Edition	De Borst, Crisfield, Remmers and Verhoosel	August 2012
An Introduction to Mathematical Modeling: A Course in Mechanics	Oden	November 2011
Computational Mechanics of Discontinua	Munjiza, Knight, and Rougier	November 2011
Introduction to Finite Element Analysis: Formulation, Verification, and Validation	Szabó and Babuška	March 2011

EXTENDED FINITE ELEMENT METHOD
THEORY AND APPLICATIONS

Amir R. Khoei
Sharif University of Technology, Iran

WILEY

This edition first published 2015

Registered office
John Wiley & Sons, Ltd, The Atrium, Southern Gate, Chichester, West Sussex, PO19 8SQ, United Kingdom

For details of our global editorial offices, for customer services and for information about how to apply for permission to reuse the copyright material in this book please see our website at www.wiley.com.

Library of Congress Cataloging-in-Publication Data

Khoei, Amir R.

Extended finite element method : theory and applications / Amir R. Khoei.
pages cm
Includes bibliographical references and index.
ISBN 978-1-118-45768-9 (cloth)
1. Finite element method. 2. Numerical analysis. I. Title. II. Title: Finite element method.
TA347.F5K545 2015
620.1'1260151825–dc23

2014029615

Set in 9/11pt Times by SPi Publisher Services, Pondicherry, India

1 2015

To Azadeh and Arsalan

Contents

Series Preface

The series on *Computational Mechanics* is a conveniently identifiable set of books covering interrelated subjects that have received much attention in recent years, and need a place in senior undergraduate and graduate school curricula and engineering practice. The subjects of titles in the series cover applications and methods. They range from biomechanics to fluid-structure interactions to multiscale mechanics, and from computational geometry to meshfree techniques to parallel and iterative computing methods. Application areas are across the board in a wide range of industries, including civil, mechanical, aerospace, automotive, environmental, and biomedical engineering. Practicing engineers, researchers, and software developers at universities, in industry, government laboratories, and graduate students will find this series to be an indispensable source of new engineering approaches, interdisciplinary research, and a comprehensive learning experience in computational mechanics.

Since its conception in the late 1990s by Ted Belytschko, the eXtended Finite Element Method (XFEM) has become one of the most widely used numerical methods for simulating fracture. The method is highly versatile and has been applied to a variety of crack models, including linear elastic fracture mechanics and cohesive zone approaches, to shear banding and dislocations, as well as to other problems that involve discontinuities. *Extended Finite Element Method: Theory and Applications*, written by a leading expert in the field, is the most comprehensive book written to date on this important subject in computational mechanics. The book covers many aspects and application areas of the XFEM. It comes with detailed derivations and explanations, and an exhaustive bibliography that guides the reader into further developments in the field. Its engineering approach and standard notation make the book easy to read.

Preface

The finite element method is one of the most common numerical tools for obtaining approximate solutions of partial differential equations; the technique has been applied successfully in many areas of engineering sciences to study, model, and predict the behavior of structures. The area ranges from aeronautical and aerospace engineering, the automobile industry, mechanical engineering, civil engineering, biomechanics, geomechanics, material sciences, and many more. Despite its popularity, the finite element method suffers from certain drawbacks when the solution contains a non-smooth behavior, such as high gradients or singularities in the stress and strain fields, and/or strong discontinuities in the displacement field; then it becomes computationally expensive to get optimal convergence. In order to overcome such difficulties, the extended finite element method (X-FEM) has been developed to facilitate the modeling of arbitrary discontinuities such as jumps, kinks, singularities, and other non-smooth features within elements. The technique provides a powerful tool for enriching solution spaces with information from asymptotic solutions and other knowledge of the physics of the problem. The main purpose of this book is to present the theory and applications of the X-FEM in linear and nonlinear problems of continua, structures, and geomechanics.

There are a number of excellent books published on the finite element method, however, there are only three books released on the X-FEM that are geared to a specific audience. This book is aimed to provide a comprehensive study on the extended finite element modeling of continua, structures, and geomechanics that should appeal to a relatively wide audience. During the last two decades, the X-FEM has moved from purely research topic into mainstream day-to-day analysis in engineering problems. It is therefore necessary for both practicing engineers and students to become familiar with the subject. Since there is no comprehensive book explaining the X-FEM in various engineering problems, this book aims to rectify this situation and bring a comprehensive easy to follow introduction to the subject to researchers in the fields of civil, mechanical, materials, and aerospace engineering.

The book begins with an overview of the extended finite element method in Chapter 1, in which an emphasis is given on various applications of the technique in materials modeling problems. The mathematical formulation of the X-FEM is presented in Chapter 2 with special reference to solid mechanics problems. It includes the introduction of partition of unity method, enrichment functions, blending elements, the X-FEM discretization, and the numerical integration of X-FEM formulation. In this chapter, numerical implementation is presented for the linear and higher order quadrilateral elements in X-FEM modeling of linear and curved interfaces. Chapter 3 presents an overview of various X-FEM enrichment functions used in a wide variety of problems, such as bimaterials, cracks, dislocations, fluid-structure interactions, shear bands, convection-diffusion, thermo-mechanical, deformable porous media, piezoelectric, magneto-electro-elastic, topology optimization, rigid particles in Stokes flow, solidification, and so on. In Chapter 4, the problems of convergence rate and condition number within the X-FEM are discussed, and various remedies that are available in the literature are introduced for these issues. In Chapter 5, the X-FEM is developed for nonlinear behavior of materials in large deformations; it is first presented in the framework of a Lagrangian large plasticity deformation formulation, and is then described in the framework of an arbitrary Lagrangian–Eulerian method. In Chapter 6, the X-FEM method is presented for modeling frictional contact problems on the basis of the penalty method, Lagrange multipliers technique, and augmented Lagrange multipliers approach.

The implementation of X-FEM technique in linear elastic fracture mechanics is presented in Chapter 7. The basis of linear elastic fracture mechanics is first introduced by defining the stress and displacement distributions around the crack tip and the stress intensity factors for different loading modes. The governing equation of a cracked body is then derived in the framework of an X-FEM. In Chapter 8, the X-FEM technique is utilized to simulate a cracked body combined with the cohesive crack model. Various cohesive crack growths are demonstrated in the framework of extended-FEM technique based on the stress criterion, the stress intensity factor criterion, and the cohesive segments method. In Chapter 9, the X-FEM technique is presented for crack growth simulation in ductile fracture problems. A non-local damage-plasticity model is employed to capture the fracture process zone within the X-FEM technique. The Lagrangian X-FEM formulation is utilized to model large deformation crack propagation and, the process of failure and crack propagation in dynamic and cyclic loading conditions is performed using dynamic large deformation X-FEM formulation. In Chapter 10, the X-FEM is developed to model the deformable porous media with weak and strong discontinuities. The fluid phase mass balance equation is applied together with the momentum balance of bulk and fluid phases to model hydraulic fracture propagation in porous media on the basis of a *u–p* X-FEM formulation. In Chapter 11, the X-FEM is proposed for the fully coupled hydro-mechanical analysis of deformable, progressively fracturing porous media interacting with the flow of two immiscible, compressible wetting and non-wetting pore fluids. The fluid flow within the crack is simulated using Darcy's Law in which the permeability variation with porosity due to the cracking of the solid skeleton is accounted. The cohesive crack model is integrated into the numerical modeling, in which the nonlinear fracture processes occurring along the fracture process zone are simulated. Finally, Chapter 12 is devoted to the implementation of the X-FEM technique in thermo-hydro-mechanical modeling of saturated porous media. The thermo-hydro-mechanical governing equations are derived by utilizing the momentum equilibrium equation, mass balance equation, and the energy conservation relation within the X-FEM framework.

Basically, the material presented in this book is a part of established X-FEM research articles; however, for the most parts of the book, the detailed derivations have not been reported in a single source. Thus, I am indebted to the authors of all books and journal papers listed in the bibliography. I wish to express my sincere gratitude to the pioneers of the X-FEM method, in particular Ted Belytschko, John Dolbow, Nicolas Moës, and Natarajan Sukumar, whose work formed the basis of new development reported here. I would like to express my sincere gratitude to my friend and colleague, Soheil Mohammadi, for the fruitful discussions held on many occasions over a long period of time. I wish to thank my former graduate students in the Department of Civil Engineering at Sharif University of Technology, who have contributed to the advances in the application of the X-FEM; M. Anahid, S.O.R. Biabanaki, T. Mohamadnejad, P. Broumand, M. Vahab, M. Hirmand, E. Haghighat, S. Moallemi, A. Shamloo, M. Nikbakht, K. Karimi, K. Shahim, S.M.T. Mousavi, M.R. Hajiabadi, N. Hosseini, H. Akhondzadeh, and E. Abedian. Moreover, I would like to express a special thank to my students who have had a major contribution in the preparation of this manuscript; in particular, M. Vahab in the first three chapters, P. Broumand in Chapters 4 and 9, M. Anahid and S.O.R. Biabanaki in Chapter 5, S. Moallemi in Chapters 7 and 12, T. Mohamadnejad in Chapters 8 and 11, E. Haghighat in Chapter 10 and M. Hirmand in Chapter 6 and the worked examples given in a companion website of the book.

I would like to acknowledge the Iran National Science Foundation (INSF), which supported my research works on the X-FEM method through different projects over the years. I would also like to extend my acknowledgement to John Wiley & Sons, Ltd for facilitating the publication of this book; in particular Anne Hunt and Liz Wingett throughout various stages of the work, Tom Carter, who has been patient in the long process of completing this manuscript and Wahidah Abdul Wahid, Diba Lingasamy, and Lynette Woodward during the production process of the book.

Finally, I want to thank my wife, Azadeh, and my son, Arsalan, for their love and support, when instead of spending my time and attention, I disappeared for long stretches of time to work on the book.

Amir R. Khoei
Tehran, April 2014

1

Introduction

1.1 Introduction

The finite element method (FEM) is one of the most common numerical tools for obtaining the approximate solutions of partial differential equations. It has been applied successfully in many areas of engineering sciences to study, model, and predict the behavior of structures. The area ranges across aeronautical and aerospace engineering, the automobile industry, mechanical engineering, civil engineering, biomechanics, geomechanics, material sciences, and many more. The FEM does not operate on differential equations; instead, continuous boundary and initial value problems are reformulated into equivalent variational forms. The FEM requires the domain to be subdivided into non-overlapping regions, called the *elements*. In the FEM, individual elements are connected together by a topological map, called a *mesh*, and local polynomial representation is used for the fields within the element. The solution obtained is a function of the quality of mesh and the fundamental requirement is that the mesh has to conform to the geometry. The main advantage of the FEM is that it can handle complex boundaries without much difficulty. Despite its popularity, the FEM suffers from certain drawbacks. There are number of instances where the FEM poses restrictions to an efficient application of the method. The FEM relies on the approximation properties of polynomials; hence, they often require smooth solutions in order to obtain optimal accuracy. However, if the solution contains a non-smooth behavior, like high gradients or singularities in the stress and strain fields, or strong discontinuities in the displacement field as in the case of cracked bodies, then the FEM becomes computationally expensive to get optimal convergence.

One of the most significant interests in solid mechanics problems is the simulation of fracture and damage phenomena (Figure 1.1). Engineering structures, when subjected to high loading, may result in stresses in the body exceeding the material strength and thus, in progressive failure. These material failure processes manifest themselves in various failure mechanisms such as the fracture process zone (FPZ) in rocks and concrete, the shear band localization in ductile metals, or the discrete crack discontinuity in brittle materials. The accurate modeling and evolution of smeared and discrete discontinuities have been a topic of growing interest over the past few decades, with quite a few notable developments in computational techniques over the past few years. Early numerical techniques for modeling discontinuities in finite elements can be seen in the work of Ortiz, Leroy, and Needleman (1987) and Belytschko, Fish, and Englemann (1988). They modeled the shear band localization as a "weak" (strain) discontinuity that could pass through the finite element mesh using a multi-field variational principle. Dvorkin, Cuitiño, and Gioia (1990) considered a "strong" (displacement) discontinuity by modifying the principle of virtual work statement. A unified framework for modeling the strong discontinuity by taking into account the softening constitutive law and the interface traction–displacement relation was proposed by Simo, Oliver, and

Extended Finite Element Method: Theory and Applications, First Edition. Amir R. Khoei.

Figure 1.1 Building destroyed by a 8.8 magnitude earthquake on Saturday, February 27, 2010, with intense shaking lasting for about 3 minutes, which occurred off the coast of central Chile. (Source: Vladimir Platonow (Agência Brasil) [CC-BY-3.0-br (http://creativecommons.org/licenses/by/3.0/br/deed.en)], via Wikimedia Commons; http://commons.wikimedia.org/wiki/File:Terremoto_no_Chile_2010.JPG)

Armero (1993). In the strong discontinuity approach, the displacement consists of regular and enhanced components, where the enhanced component yields a jump across the discontinuity surface. An assumed enhanced strain variational formulation is used, and the enriched degrees of freedom (DOF) are statically condensed on an element level to obtain the tangent stiffness matrix for the element. An alternative approach for modeling fracture phenomena was introduced by Xu and Needleman (1994) based on the cohesive surface formulation, which was used later by Camacho and Ortiz (1996) to model the damage in brittle materials. The cohesive surface formulation is a phenomenological framework in which the fracture characteristics of the material are embedded in a cohesive surface traction–displacement relation. Based on this approach, an inherent length scale is introduced into the model, and in addition, no fracture criterion is required so the crack growth and the crack path are outcomes of the analysis.

In the FEM, the non-smooth displacement near the crack tip is basically captured by refining the mesh locally. The number of DOF may drastically increase, especially in three-dimensional applications. Moreover, the incremental computation of a crack growth needs frequent remeshings. Reprojecting the solution on the updated mesh is not only a costly operation but also it may have a troublesome impact on the quality of results. The classical FEM has achieved its limited ability for solving fracture mechanics problems. To avoid these computational difficulties, a new approach to the problem consists in taking into account the *a priori* knowledge of the exact solution. Applying the asymptotic crack tip displacement solution to the finite element basis seems to have been a somewhat early idea. A significant improvement in crack modeling was presented with the development of a partition of unity (PU) based enrichment method for discontinuous fields in the PhD dissertation by Dolbow (1999), which was referred to as the extended FEM (X-FEM). In the X-FEM, special functions are added to the finite element approximation using the framework of PU. For crack modeling, a discontinuous function such as the Heaviside step function and the two-dimensional linear elastic asymptotic crack tip displacement fields, are used to account for the crack. This enables the domain to be modeled by finite elements without explicitly meshing the crack surfaces. The location of the crack discontinuity can be arbitrary with respect to the underlying finite element mesh, and the crack propagation simulation can be performed without the need to remesh as the crack advances. A particularly appealing feature is that the finite element framework and its properties, such as the sparsity and symmetry, are retained and a single-field (displacement) variational principle is used to obtain the discrete equations. This technique provides an accurate and robust numerical method to model strong (displacement) discontinuities.

The original research articles on the X-FEM were presented by Belytschko and Black (1999) and Moës, Dolbow, and Belytschko (1999) for elastic fracture propagation on the topic of "A FEM for crack growth without remeshing". They presented a minimal remeshing FEM for crack growth by including the discontinuous enrichment functions to the finite element approximation in order to account for the presence of the crack. The essential idea was based on adding enrichment functions to the approximation space that contains a discontinuous displacement field. Hence, the method allows the crack to be arbitrarily aligned within the mesh. The same span of functions was earlier developed by Fleming *et al.* (1997) for the enrichment of the element-free Galerkin method. The method exploits the PU property of finite elements that was noted by Melenk and Babuška (1996), namely that the sum of the shape functions must be unity. This property has long been known, since it corresponds to the ability of the shape functions to reproduce a constant that represents translation, which is crucial for convergence.

The X-FEM provides a powerful tool for enriching solution spaces with information from asymptotic solutions and other knowledge of the physics of the problem. This has proved very useful for cracks and dislocations where near-field solutions can be embedded by the PU method to tremendously increase the accuracy of relatively coarse meshes. The technique offers possibilities in treating phenomena such as surface effects in nano-mechanics, void growth, subscale models of interface behavior, and so on. Thus, the X-FEM method has greatly enhanced the power of the FEM for many of the problems of interest in mechanics of materials. The aim of this chapter is to provide an overview of the X-FEM with an emphasis on various applications of the technique to materials modeling problems, including linear elastic fracture mechanics (LEFM); cohesive fracture mechanics; composite materials and material inhomogeneities; plasticity, damage and fatigue problems; shear band localization; fluid–structure interaction; fluid flow in fractured porous media; fluid flow and fluid mechanics problems; phase transition and solidification; thermal and thermo-mechanical problems; plates and shells; contact problems; topology optimization; piezoelectric and magneto-electroelastic problems; and multi-scale modeling.

1.2 An Enriched Finite Element Method

The FEM is widely used in industrial design applications, and many different software packages based on FEM techniques have been developed. It has undoubtedly become the most popular and powerful analytical tool for studying the behavior of a wide range of engineering and physical problems. Its applications have been developed from basic mechanical problems to fracture mechanics, fluid dynamics, nano-structures, electricity, chemistry, civil engineering, and material science (Figure 1.2). The FEM has proved to be very well suited to the study of fracture mechanics. However, modeling the propagation of a crack through a finite element mesh turns out to be difficult because of the modification of mesh topology. To accurately model discontinuities with FEMs, it is necessary to conform the discretization to the discontinuity. This becomes a major difficulty when treating problems with evolving discontinuities where the mesh must be regenerated at each step. Reprojecting the solution on the updated mesh is not only a costly operation but also it may have a troublesome impact on the quality of results.

Modeling moving discontinuities within the classical finite element is quite cumbersome due to the necessity of the mesh to conform to discontinuity surfaces. Mesh generation of complex geometries can be very time consuming with a classical finite element analysis. The main difficulty arises from the necessity of the mesh to conform to physical surfaces. Discontinuities such as holes, cracks, and material interfaces may not cross mesh elements. Moreover, local refinements close to discontinuities and mesh modification to track the geometrical and topological changes in crack propagation problems for example,

Figure 1.2 Bridge damage in Shaharah, Yemen, August 1986. The failure of bridges is of special concern to structural engineers in trying to learn lessons vital to bridge design, construction, and maintenance. (Source: Bernard Gagnon [CC-BY-3.0-br (http://creativecommons.org/licenses/by/3.0.en)], via Wikimedia Commons; http://en.wikipedia.org/wiki/File:Shehara_02.jpg)

can be difficult. Also, when geometries evolve and history dependent models are used, robust methods to transfer the solution to the new mesh are needed. This issue is particularly significant, since computed fields defined on these discontinuities are often the most important ones. In order to overcome these mesh-dependent difficulties, the generalized finite element method (G-FEM) and the X-FEM have been developed to facilitate the modeling of arbitrary moving discontinuities through the partition of unity enrichment of finite elements (PUM), in which the main idea is to extend a classical approximate solution basis by a set of locally supported enrichment functions that carry information about the character of the solution, for example, singularity, discontinuity, and boundary layer. As it permits arbitrary functions to be locally incorporated in the FEM or the meshfree approximation, the PUM gives flexibility in modeling moving discontinuities without changing the underlying mesh, while the set of enrichment functions evolve (and/or their supports) with the interface geometry. In addition to facilitating the modeling of moving discontinuities, enrichment also increases the local approximation power of the solution space by allowing the introduction of arbitrary functions within the basis. This is particularly useful for problems with singularities or boundary layers.

Basically, the G-FEM and the X-FEM are versatile tools for the analysis of problems characterized by discontinuities, singularities, localized deformations, and complex geometries. These methods can dramatically simplify the solution of many problems in material modeling, such as the propagation of cracks, the evolution of dislocations, the modeling of grain boundaries, and the evolution of phase boundaries. The advantage of these methods is that the finite element mesh can be completely independent of the morphology of these entities. The G-FEM and the X-FEM incorporate the analytically known or numerically computed handbook functions within some range of their applicability into the traditional FE (finite element) approximation with the PU (partition of unity) method to enhance the local and global accuracy of the computed solution. Both the X-FEM and G-FEM meshes need not conform to the boundaries of the problem. The FEM is used as the building block in the X-FEM and the G-FEM; hence, much of the theoretical and numerical developments in FEs can be readily extended and applied. Moreover, the X-FEM and G-FEM make possible an accurate solution of engineering problems in complex domains that may be practically impossible to solve using the FE method. The X-FEM and G-FEM are basically identical methods; the X-FEM was originally developed for discontinuities, such as cracks, and used local enrichments while the G-FEM was first involved with global enrichments of the approximation space. The X-FEM and G-FEM can be used with both structured and unstructured meshes. The structured meshes are appealing for many studies in materials science, where the objective is to determine the properties of a unit cell of the material. However, the unstructured meshes tend to be widely used for the analysis of engineering structures and components since it is often desirable to conform the mesh to the external boundaries of the component, although some methods under development today are able to treat even complicated geometries with structured meshes (Belytschko, Gracie, and Ventura 2009). The G-FEM allows for p–adaptivity and provides accurate numerical solutions with coarse or practically acceptable meshes by augmenting the FE space with the analytical or numerically generated solution of a given boundary value problem. The X-FEM on the other hand pays most attention to the requisite enrichment of nodes to model the internal boundary (crack or inclusion) of interest. Hence, the X-FEM is less dependent on known closed form solutions and affords greater flexibility.

1.3 A Review on X-FEM: Development and Applications

The X-FEM has gained a lot of attention in the last decade for its advantages in replicating the discontinuity of the displacement field across the crack surface and singularity at the crack front without the need for remeshing. The X-FEM enables the accurate approximation of fields that involve jumps, kinks, singularities, and other non-smooth features within elements (Karihaloo and Xiao, 2003). This is achieved by adding additional terms, that is, the enrichments, to classical FE approximations. These terms enable the approximation to capture the non-smooth features independently of the mesh. The X-FEM has shown its full potential for application in fracture mechanics (Fries and Belytschko, 2010). Applications with cracks involve discontinuities across the crack surface and singularities, or general steep gradients, at the crack front. In the classical FEM, a suitable mesh that accounts for these features has to be provided and maintained; this is particularly difficult for crack propagation in three dimensions. The X-FEM, however, can treat these types of problems on fixed meshes and considers crack propagation by a dynamic enrichment of the approximation.

Crack propagation using an enriched FEM technique was first introduced by Belytschko and Black (1999) that encompasses three major topics; the crack description, the discretized formulation, and the criteria for the crack update. In this method, the meshing task is reduced by enriching the elements near the crack tip and along the crack faces with the leading singular crack tip asymptotic displacement fields using the PUM to account for the presence of the crack. In the case that multiple crack segments need to be enriched using the near-tip fields; a mapping algorithm is used to align the discontinuity with the crack geometry. It was also shown that the use of discontinuous displacements along the crack produces a solution with zero traction along the crack faces. Moës, Dolbow, and Belytschko (1999) introduced a far more elegant and straightforward procedure to introduce a discontinuous field across the crack faces away from the crack-tip by adapting the generalized Heaviside function, and developed simple rules for the introduction of discontinuous and crack tip enrichments. Daux *et al.* (2000) introduced the junction function concept to account for multiple branched cracks and called their method the *extended finite element method* (X-FEM). They have employed this method for modeling complicated geometries such as multiple branched cracks, voids, and cracks emanating from holes without the need for the geometric entities to be meshed. The X-FEM is promising since it avoids using a mesh that conforms to the cracks, voids, or inhomogeneities as is the case with the traditional FEM. In X-FEM, a standard FE mesh for the problem is first created without accounting for the geometric entity. The presence of cracks, voids, or inhomogeneities is then represented independently of the mesh by enriching the standard displacement approximation with additional functions. For crack modeling, both discontinuous displacement fields along the crack faces and the leading singular crack tip asymptotic displacement fields are added to the displacement based FE approximation through the PUM. The additional coefficients at each enriched node are independent. Moreover, the X-FEM provides a seamless means to use higher order elements or special

FEs without significant changes in the formulation. The X-FEM improves the accuracy in problems where some aspects of the functional behavior of the solution field is known *a priori* and relevant enrichment functions can be used.

Advances in the X-FEM have been led to applications in various fields of computational mechanics and physics. The open source X-FEM codes were released by Bordas and Legay (2005), and numerical implementation and efficiency aspects of X-FEM were studied by Dunant *et al.* (2007). The X-FEM is a robust and popular method that has been used for industrial problems and is implemented by leading computational software companies. These applications have reached a high degree of robustness and are now being incorporated into the general purpose codes such as LS-DYNA and ABAQUS. Many of the techniques that are used in the X-FEM are directly related to techniques previously developed in mesh-free methods. An overview of the X-FEM has been reported by Karihaloo and Xiao (2003), Bordas and Legay (2005), Rabczuk and Wall (2006), Abdelaziz and Hamouine (2008), Belytschko, Gracie, and Ventura (2009), Rabczuk, Bordas, and Zi (2010), and Fries and Belytschko (2010). There are also three recent published books on the X-FEM that have focused on fracture mechanics problems by Mohammadi (2008, 2012) and Pommier *et al.* (2011). In what follows, a comprehensive overview is presented on various achievements of the X-FEM.

1.3.1 Coupling X-FEM with the Level-Set Method

In the context of the X-FEM, the location of non-smooth features is often defined implicitly by means of the level set method (LSM) (Osher and Sethian, 1988). The LSM complements the X-FEM extremely well as it provides the information *where* and *how* to enrich. The extension of the LSM to the description of crack *paths* in two dimensions was proposed by Stolarska *et al.* (2001) and Stolarska and Chopp (2003), and the description of crack *surfaces* in three dimensions was performed by Moës, Gravouil, and Belytschko (2002), Gravouil, Moës, and Belytschko (2002) and Sukumar, Chopp, and Moran (2003a). For crack problems, one enrichment is typically needed at the crack surface and additional enrichments are required at the crack front where both types of information, including the crack surface and the crack front, can be extracted directly from the level set functions. The discontinuous enrichment function that captures the jump in the displacement field across the crack surface depends directly on the level set function that stores the (signed) distance to the surface. The enrichment functions that capture the high gradients at the crack front depend on the level set functions indirectly; there, the level set functions imply a coordinate system in which the enrichment functions are evaluated. Thus, it can be seen that the LSM has important advantages in the context of the X-FEM. On the other hand, the X-FEM is only *one* step in the simulation of crack propagation that leads to an accurate approximation of the displacement, stress, and strain fields. The next step involves a characterization of the situation at the crack tip from which the crack increment is deduced. In fact, on the basis of fracture parameter information, such as stress intensity factors (SIFs), configurational forces, the *J*–integral, local maximum stress and strain measures, and so on, the direction and length of the increment at the crack tip in two dimensions, or at selected points on the crack front in three dimensions, can be modeled. The third and last step involves an update of the crack description such that the increments are considered appropriately (Fries and Baydoun, 2012).

Stolarska *et al.* (2001) presented the first implementation of LSM for modeling of crack propagation within the extended FE framework where the interface evolution was successfully performed by the LSM. Sukumar *et al.* (2001) employed the LSM for modeling holes and inclusions where the level set function was used to represent the local enrichment for material interfaces. Moës, Gravouil, and Belytschko (2002) and Gravouil, Moës, and Belytschko (2002) performed a combined X-FEM and the LSM to construct arbitrary discontinuities in the three-dimensional analysis of crack problems. Ventura, Xu, and Belytschko (2002) introduced the vector LSM for description of propagating crack in the element-free Galerkin method. Ji, Chopp, and Dolbow (2002) presented a hybrid X-FEM–LSM for modeling the evolution of sharp phase interfaces on fixed grids with reference to solidification problems to represent the jump in the temperature gradient that governs the velocity of the phase boundary. Stolarska and Chopp

(2003) presented an algorithm that couples the LSM with the X-FEM to investigate the effects of the proximity of multiple interconnect lines, multiple cracks, interconnect material, and integrated circuit boundaries on the growth of cracks duc to fatigue from thermal cycling. Chessa and Belytschko (2003a, b) presented a combined X-FEM and LSM for two-phase flow with surface tension effects, where the velocity was enriched by the signed distance function. They also employed the X-FEM to model arbitrary discontinuities in space–time along a moving hyper-surface using the LSM (Chessa and Belytschko, 2004). Legay, Chessa, and Belytschko (2006) proposed an Eulerian–Lagrangian method for fluid-structure interaction based on the LSM, where the level set description of the interface leads to the formulation of the fluid–structure interaction problem.

An extension of the LSM was introduced by Sethian (1996) based on the *fast marching method*. This technique prevents the need to represent the geometry of the interface during its evolution; the method is computationally attractive for monotonically advancing fronts. Sukumar, Chopp, and Moran (2003a) presented an implementation of the combined X-FEM and fast marching method for modeling planar three-dimensional fatigue crack propagation, where the fast marching method was used to handle its evolution under fatigue growth conditions. Chopp and Sukumar (2003) employed the technique for modeling fatigue crack propagation of multiple coplanar cracks based on coupling the X-FEM with the fast marching method. Sukumar *et al.* (2008) proposed a numerical technique for non-planar three-dimensional linear elastic crack growth simulation based on a coupled X-FEM and the fast marching method.

1.3.2 Linear Elastic Fracture Mechanics (LEFM)

Modeling of crack propagation with the FEM is cumbersome due to the need to update the mesh to conform the geometry of the crack surface. Several FE techniques have been developed to model cracks and crack growth without remeshing. The X-FEM is one of the most powerful techniques developed based on an enrichment strategy for finite elements on the basis of a PU. Belytschko and Black (1999) originally introduced a minimal remeshing FEM for crack growth, where discontinuous enrichment functions were added into the FE approximation to account for the presence of the crack. Moës, Dolbow, and Belytschko (1999) improved the method by incorporating a discontinuous field across the crack faces away from the crack tip for modeling crack growth, where the standard displacement based approximation was enriched near a crack by employing both discontinuous fields and near-tip asymptotic fields through a PUM. Daux *et al.* (2000) extended the X-FEM to model crack problems with multiple branches, multiple holes, and cracks emanating from holes. Sukumar *et al.* (2000) employed the X-FEM in three-dimensional fracture mechanics, where a discontinuous function and the two-dimensional asymptotic crack tip displacement fields were added to the FE approximation to account for the crack using the notion of a PU. Stolarska *et al.* (2001) introduced an algorithm that couples the LSM with the X-FEM to model crack growth, in which the LSM was used to represent the crack location, including the location of crack-tips. Moës, Gravouil, and Belytschko (2002) extended the X-FEM to handle arbitrary non-planar cracks in three dimensions by describing the crack geometry in terms of two signed distance functions that were able to construct a near tip asymptotic field with a discontinuity that conforms to the crack, even when it is curved or kinked near a tip. Ayhan and Nied (2002) proposed an enriched FE approach for obtaining the SIFs for general three-dimensional crack problems. Sukumar and Prevost (2003) presented the X-FEM for two-dimensional crack modeling in isotropic and bimaterial media within the finite element program Dynaflow^{TM}, which was later used by Huang, Sukumar, and Prévost (2003b) to demonstrate the numerical modeling of SIFs in crack problems, including crack growth simulation. Stazi *et al.* (2003) presented a method for LEFM using enriched quadratic interpolations, in which the geometry of the crack was represented by a level set function interpolated on the same quadratic FE discretization. Lee *et al.* (2004) presented a combination of the X-FEM and the mesh superposition method (*s*–version FEM) for modeling of stationary and growing cracks, in which the near-tip field was modeled by superimposed quarter point elements on an overlaid mesh and the rest of the discontinuity was implicitly described by a step function on the PU, where the two displacement fields were matched through a transition region.

Budyn *et al.* (2004) developed the X-FEM for multiple crack growth considering the junction of cracks in both homogeneous and inhomogeneous brittle materials, which does not require remeshing as the cracks grow. A similar approach was proposed by Zi *et al.* (2004) for modeling the growth and the coalescence of cracks in a quasi-brittle cell containing multiple cracks.

Advanced issues in LEFM have been conducted by researchers in more recent studies. An application of the X-FEM method to large strain fracture mechanics was presented by Legrain, Moës, and Verron (2005) with a special reference to the fracture of rubber-like materials. Moës, Béchet, and Tourbier (2006) introduced a strategy to impose the Dirichlet boundary conditions within the X-FEM by construction of a corrected Lagrange multiplier space on the boundary that preserves the optimal rate of convergence. Ventura (2006) introduced a method for eliminating the introduction of quadrature subcells when using discontinuous/non-differentiable enrichment functions in the X-FEM by replacing the discontinuous/non-differentiable functions with equivalent polynomials. Asadpoure, Mohammadi, and Vafai (2006) proposed an X-FEM for modeling cracks in orthotropic media based on a discontinuous function and two-dimensional asymptotic crack tip displacement fields. Asadpoure and Mohammadi (2007) modified their previous model by adding new enrichment functions to simulate the orthotropic cracked media, where the required near tip enrichment functions were obtained by extracting basic terms from the complex solutions in the vicinity of the crack tip. Loehnert and Belytschko (2007) employed the X-FEM to investigate the effect of crack shielding and amplification of various arrangements of micro-cracks on the SIFs of a macro-crack, including large numbers of arbitrarily aligned micro-cracks. Sukumar *et al.* (2008) proposed a numerical technique for non-planar three-dimensional elastic crack growth simulations by combining the X-FEM with the fast marching method. Tabarraei and Sukumar (2008) employed the X-FEM on polygonal and quadtree FE meshes for two-dimensional crack growth modeling, where the Laplace interpolant was used to construct basis functions on convex polygonal meshes, and the mean value coordinates were adopted for non-convex elements.

One of the main issues in the X-FEM method is the blending elements, which are constructed between the enriched and standard elements; they are often crucial for a good performance of the local partition of unity enrichments. Chessa, Wang, and Belytschko (2003) employed the enhanced strain method in blending elements to improve the performance of local PU enrichments. Laborde *et al.* (2005) modified the standard X-FEM to circumvent problems in blending elements for the case of crack problems by enriching a whole fixed area around the crack tip. Legay, Wang, and Belytschko (2005) employed the X-FEM within the spectral finite elements for modeling discontinuities in the gradients, where there was no need to implement the blending elements for high-order spectral elements. Fries and Belytschko (2006) developed an intrinsic X-FEM method without blending elements for treating arbitrary discontinuities in a FE context, where no additional unknowns were introduced at the nodes whose supports are crossed by discontinuities. Fries (2008) introduced a corrected X-FEM method without problems in blending elements based on a linearly decreasing weight function for enrichment in the blending elements. Gracie *et al.* (2008b) developed a discontinuous Galerkin formulation without blending elements that decomposes the domain into an enriched and unenriched sub-domains, where the continuity was enforced with an internal penalty method. Benvenuti, Tralli, and Ventura (2008) introduced a regularized X-FEM model for the transition from continuous to discontinuous displacements, where the emerging strain and stress fields were modeled independently using specific constitutive assumptions. Ventura, Gracie, and Belytschko (2009) introduced a weight function blending, where the enrichment function was premultiplied by a smooth weight function with a compact support to allow for a completely smooth transition between the enriched and unenriched sub-domains. Tarancon *et al.* (2009) employed the higher-order hierarchical shape functions to reduce unwanted effects of the partial enrichment in the blending elements. Shibanuma and Utsunomiya (2009) presented an alternative formulation for the X-FEM based on the concept of the PU FEM for solving the problem of blending elements, which assures the numerical accuracy in the entire domain. Loehnert, Mueller-Hoeppe, and Wriggers (2011) extended the originally corrected X-FEM method presented by Fries to the three-dimensional case with its extension to finite deformation theory. Chahine, Laborde, and Renard (2011) presented a non-conformal approximation method based on the integral matching X-FEM, in which the transition layer between the singular enrichment area and the rest of the domain was replaced by an interface associated with an integral matching condition of mortar type. Menk and Bordas (2011) presented a procedure to obtain stiffness matrices whose condition number is close to the one of the FE matrices without any enrichment using a domain decomposition technique. Chen *et al.* (2012) presented a strain smoothing procedure within the X-FEM framework for LEFM to outperform the standard X-FEM, where the edge-based smoothing was used to produce a softening effect leading to a close-to-exact stiffness, "super-convergence", and "ultra-accurate" solutions.

The implementation of the X-FEM in dynamic fracture has been mostly focused on simulation of the dynamic crack propagation and estimation of the dynamic SIFs for arbitrary two- and three-dimensional cracks. Réthoré, Gravouil, and Combescure (2005b) proposed an energy-conserving scheme in the framework of the X-FEM to model the dynamic fracture and time-dependent problems that give proof of stability of the numerical scheme in linear fracture mechanics. Menouillard *et al.* (2006, 2008) presented an explicit time stepping method based on a mass matrix lumping technique for enriched elements, and demonstrated that the critical time step of an enriched element is of the same order as that of the corresponding element without extended DOF. Elguedj, Gravouil, and Maigre (2009) presented a generalized mass lumping technique for explicit dynamics simulation using the X-FEM with arbitrary enrichment functions that was based on an exact representation of the kinetic energy of rigid body modes and enrichment modes. Gravouil, Elguedj, and Maigre (2009) presented a general explicit time integration technique for X-FEM dynamics simulations with a standard critical time step by developing a classical element-by-element strategy that couples the standard central difference scheme with the unconditionally stable-explicit scheme. Fries and Zilian (2009) studied the convergence properties of different time integration methods in the framework of X-FEM for moving interfaces, including one-step time-stepping schemes, the implicit Euler method, the trapezoidal rule, and the implicit midpoint rule. Menouillard and Belytschko (2010a) presented a method to enrich the X-FEM using the meshless approximation for dynamic fracture problems, where the mesh-free approximation was used to smooth the stress state near the crack tip during the propagation, and decreasing unphysical oscillations in the stress due to the propagation of the discontinuity. In a later work, Menouillard and Belytschko (2010b) proposed a method based on enforcing the continuity of forces corresponding to the enriched DOF to smoothly release the tip element while the crack tip travels through the element. Menouillard *et al.* (2010) proposed a new enrichment method with a time-dependent enrichment function for dynamic crack propagation in the context of the X-FEM and studied the effect of different directional criteria on the crack path. Motamedi and Mohammadi (2010a, b) presented a dynamic crack analysis for composites based on the orthotropic enrichment functions within the X-FEM framework by evaluating the dynamic SIFs using the domain separation integral method. Esna Ashari and Mohammadi (2012) proposed the X-FEM for fracture analysis of delamination problems in fiber-reinforced polymer reinforced beams, where the stress singularities near the debonding crack-tip were modeled by orthotropic bimaterial enrichment functions. Liu, Menouillard, and Belytschko (2011) developed a higher-order X-FEM method based on the spectral element method for the simulation of dynamic fracture, where the numerical oscillations were effectively suppressed and the accuracy of computed SIFs and crack path were appropriately improved. Motamedi and Mohammadi (2012) introduced the time-independent orthotropic enrichment functions for dynamic crack propagation of moving cracks in composites based on the X-FEM, where the enrichment functions were derived from the analytical solutions of a moving/propagating crack in orthotropic media.

The importance of error estimation in the X-FEM numerical analysis has been investigated by various researchers. Chahine, Laborde, and Renard (2006) performed a convergence study for a variant of the X-FEM on cracked domains by using a cut-off function to localize the singular enrichment area, and illustrated that the convergence error of the proposed variant is of order h for a linear FEM. Ródenas *et al.* (2008) presented a stress recovery procedure that provides accurate estimations of the discretization error for LEFM problems based on the superconvergent patch recovery (SPR) technique for the X-FEM framework. Panetier, Ladeveze, and Chamoin (2010) presented a method to obtain the local error bounds in the context of fracture mechanics by evaluating the discretization error for quantities of interest computed in the X-FEM using the concept of constitutive relation error. Ródenas *et al.* (2010) introduced a recovery-type error estimator yielding upper bounds of the error in energy norm for LEFM problems using the X-FEM that yields equilibrium at a local level. Shen and Lew (2010a, b) introduced an optimally convergent discontinuous Galerkin-based X-FEM for fracture mechanics problems, in which an optimal order of convergence was obtained in comparison with other variants of X-FEM technique. Nicaise, Renard, and Chahine (2011) performed an *a priori* error estimate on the standard X-FEM with a fixed enrichment area and the X-FEM with a cut-off function by estimating the error on the SIFs. Prange, Loehnert, and Wriggers (2012) presented a simple recovery based error estimator for the discretization error in the X-FEM analysis of crack problems, where enhanced smoothed stresses were recovered to enable the error estimation for arbitrary distributed cracks. Byfut and Schröder (2012) presented a higher-order X-FEM method by combining the standard X-FEM with higher-order FEM method based on the Lagrange-type and hierarchical tensor product shape functions, and demonstrated the methodological aspects that are necessary in the *hp*-adaptivity of X-FEM for obtaining the exponential convergence rate. González-Albuixech *et al.* (2013a) investigated the convergence rate of solution obtained from the domain energy integral for computation of the SIFs in the solution of two-dimensional curved crack problems using the X-FEM. Ródenas *et al.* (2013) presented a technique to obtain an accurate estimate of the error in energy norm using a moving least squares (MLS) recovery-based procedure for X-FEM problems. Rüter, Gerasimov, and Stein (2013) proposed a goal-oriented *a posteriori* error estimator for X-FEM approximations in LEFM problems to compute upper bounds on the error of the *J*–integral.

More advanced concepts have been studied by researchers in the X-FEM analysis of elastic linear fracture mechanics. Park *et al.* (2009) developed a mapping method to integrate weak singularities that result from enrichment functions in the G-FEM/X-FEM and is applicable to two- and three-dimensional problems including arbitrarily shaped triangles and tetrahedra. Mousavi and Sukumar (2010) presented an alternative Gaussian integration scheme to construct the Gauss quadrature rule over arbitrarily-shaped elements in two dimensions without the need for partitioning that was efficient and accurate in evaluation of weak form integrals. Bordas *et al.* (2010, 2011) investigated the accuracy and convergence of enriched finite element approximations by employing the strain smoothing to higher order elements, and highlighted that the strain smoothing in enriched approximations are beneficial when the enrichment functions are polynomial. Mousavi, Grinspun, and Sukumar (2011a, b) presented a higher order X-FEM with harmonic enrichment functions for complex crack problems, in which the numerically computed enrichment functions for the crack were obtained via the solution of the Laplace equation with Dirichlet and vanishing Neumann boundary conditions. Legrain, Allais, and Cartraud (2011) employed the X-FEM in the context of quadtree/octree meshes, where particular attention was paid to the enrichment of hanging nodes that inevitably arise with these meshes, and an approach was proposed for enforcing displacement continuity along hanging edges and faces. Richardson *et al.* (2011) presented a method for simulating quasi-static crack propagation that combines the X-FEM with a general algorithm for cutting triangulated domains, and introduced a simple and flexible quadrature rule based on the same geometric algorithm. Shibanuma and Utsunomiya (2011) studied the reproductions of *a priori* knowledge in the original X-FEM and the PU-FEM based X-FEM for the crack analysis, and showed that there is a serious lack of the reproduction of *a priori* knowledge in the local enrichment area close to the crack tip in case of the original X-FEM; however, *a priori* knowledge can be accurately reproduced over the entire enrichment in the PUFEM based X-FEM. Fries and Baydoun (2012) and Baydoun and Fries (2012) presented a method for two- and three-dimensional crack propagation that combines the advantages of explicit and implicit crack descriptions, and described a propagation criterion for three-dimensional fracture mechanics using the proposed hybrid explicit–implicit approach. Minnebo (2012) introduced a three-dimensional integral strategy for numerical integration of singular functions in the computation of stiffness matrix and SIFs using the interaction integral method produced by the X-FEM in LEFM. Benvenuti (2012a) proposed the Gauss quadrature of integrals of discontinuous and singular functions in the three-dimensional X-FEM analysis of regularized interfaces. González-Albuixech *et al.* (2013b) introduced a curvilinear gradient correction based on the level set information used for the crack description within the X-FEM framework to compute the SIFs in curved and non-planar cracks. Amiri *et al.* (2014) presented a method based on local maximum entropy shape functions together with enrichment functions used in

PUMs to discretize problems in LEFM. Pathak *et al.* (2013) presented a simple and efficient X-FEM approach for modeling three-dimensional crack problems, in which the crack front was divided into a number of piecewise curve segments and the level set functions were approximated using the higher order shape functions.

The implementation of X-FEM in FE programs has been performed by various researchers to model the LEFM in practical engineering problems. Rabinovich, Givoli, and Vigdergauz (2007, 2009) presented a computational tool based on a combination of X-FEM and genetic algorithm for the detection and identification of cracks in structures that was used in conjunction with non-destructive testing of specimens. Xiao and Karihaloo (2007) combined the hybrid crack element (HCE) with an X-FEM/G-FEM and incorporated it into a commercial FE package, where the HCE was used for the crack tip region and the X-FEM was employed for modeling crack faces outside the HCE, independent of the mesh with jump functions. Ahyan (2007) presented a three-dimensional enriched FE methodology within the FE program FRAC3D to compute the SIFs for cracks contained in functionally graded materials. Nistor, Pantale, and Caperaa (2008) presented the implementation of X-FEM in their home made explicit dynamic FEM code, DynELA, to simulate the crack propagation in structural problems under dynamic loading. Dhia and Jamond (2010) employed one of the key features of the X-FEM, that is, the Heaviside enrichment function, within a generic numerical method based on the Arlequin framework to reduce the costs of crack propagation simulations. Legrain (2013) proposed a NURBS (non-uniform rational b-spline) enhanced extended FE approach for the unfitted simulation of structures defined by means of CAD (Computer Aided Design) parametric surfaces, in which the geometry of the computational domain was defined using an exact CAD description. Holl *et al.* (2014) presented a multi-scale technique to investigate advancing cracks in three-dimensional spaces with a reference to gas turbine blades, which was able to capture crack growth taking localization effects from the fine scale into account.

1.3.3 Cohesive Fracture Mechanics

LEFM is only applicable when the size of the fracture process zone (FPZ) at the crack tip is small compared to the size of the crack and the size of the specimen (Bazant and Planas, 1998). Hence, alternative models must be chosen to take into account the FPZ. The cohesive crack model is one of the simplest ones that can be represented by a traction–displacement relation across the crack faces near the tip. This model was introduced in the early 1960s for metals by Dugdale (1960) and Barenblatt (1962), and then developed by Hillerborg, Modéer, and Petersson (1976) by introducing the concept of fracture energy into the cohesive crack model and establishing a number of traction–displacement relationships for concrete. The first implementation of an enriched FEM into cohesive fracture mechanics was proposed by Wells and Sluys (2001) by applying the displacement jump into the conventional FEM, in which the path of the discontinuity was completely independent of the mesh structure and the jump function was used as an enrichment function for the whole cohesive crack. Moës and Belytschko (2002a, b) developed the cohesive crack model within the X-FEM framework for modeling the growth of arbitrary cohesive cracks, where the growth of the cohesive zone was governed by requiring the SIFs at the tip of the cohesive zone to vanish. Crisfield and Alfano (2002) presented an enriched FEM for modeling the delamination in fiber-reinforced composite structures with the aid of a decohesive zone model and interface elements, in which the elements around the softening process zone were enriched using the hierarchical polynomial functions. Zi and Belytschko (2003) developed an X-FEM for the cohesive crack model with a new formulation for elements containing crack tips, in which the entire crack was modeled with only one type of enrichment function, that is, the signed distance function, including the elements containing the crack tip so that no blending of the local PU was required. Remmers, de Borst, and Needleman (2003) presented a method for modeling crack nucleation and discontinuous crack growth irrespective of the structure of the finite element mesh, in which the crack was modeled by a collection of cohesive segments with a finite length and the segments were added to finite elements by using the partition of unity property of the finite element shape functions. Mariani and Perego (2003) presented a methodology for the simulation of quasi-static cohesive crack propagation in quasi-brittle materials, where a cubic displacement discontinuity was employed to reproduce the typical cusp-like shape of the process zone at the tip of a cohesive crack. Belytschko *et al.* (2003b) proposed the X-FEM for modeling dynamic crack propagation based on switching from a continuum to a discrete discontinuity, where the loss of hyperbolicity was modeled by a hyperbolicity indicator that enables to determine both the crack speed and crack direction for a given material model. Larsson and Fagerström (2005) presented a theoretical and computational framework for linear and nonlinear fracture behaviors on the basis of the inverse deformation problem with an applied discontinuous deformation separated from the continuous deformation using the X-FEM technique.

Basically, there is a relation between strain softening and fracture mechanics. One motivation for this interest is that strain softening stems from damage and is often a prelude to fracture (Figure 1.3). In fact, it is a manifestation of progressive energy release during microscopic decohesion before a macroscopic crack is apparent. Areias and Belytschko (2005a) presented a numerical procedure for the quasi-static analysis of three-dimensional crack propagation in brittle and quasi-brittle solids, in which a viscosity-regularized continuum damage constitutive model was coupled with the X-FEM formulation resulting in a regularized "crack-band" version of X-FEM. Xiao, Karihaloo, and Liu (2007) proposed an incremental-secant modulus iteration scheme using the X-FEM/G-FEM for simulation of cracking process in quasi-brittle materials described by cohesive crack models whose softening law was composed of linear segments. Asferg, Poulsen, and Nielsen (2007) developed a partly cracked X-FEM element for cohesive crack growth based on additional enrichment of the cracked elements with the capability of modeling variations in the discontinuous displacement field on both sides of the discontinuity to obtain a better stress distribution on crack faces. Benvenuti (2008) and Benvenuti, Tralli, and Ventura (2008) introduced a regularized X-FEM model for the transition from continuous to discontinuous displacements, where the emerging strain and stress fields were modeled independently using specific constitutive assumptions that can address cohesive interfaces with vanishing and finite thickness in a unified way. Mougaard, Poulsen, and Nielsen (2011) developed a cohesive crack tip element together with a coherent fully cracked element within the X-FEM framework based on a double enriched displacement field of *linear strain triangle* type to symmetrize the elements crack opening and reproduce equal stresses at both sides of the crack at the tip. Zamani, Gracie, and Eslami (2012) performed a comprehensive study on the use of higher-order terms of the crack tip asymptotic fields as enriching functions of the X-FEM for both cohesive and traction-free cracks, where two widely used criteria, that is, the SIF criterion and the stress criterion, were used with both linear and nonlinear cohesive laws. Mougaard, Poulsen, and Nielsen (2013) presented a complete tangent stiffness for modeling crack growth in the X-FEM by including the crack growth parameters in an incremental form of the virtual work together with the constitutive conditions in front of the crack tip as direct unknowns in the FEM equations.

The X-FEM has been extensively used for crack growth simulation in concrete structures and rock mechanics problems, where the failure is accompanied by the formation of discrete cracks and zones of local damage (Figure 1.4). Unger, Eckardt, and Könke (2007) employed the X-FEM for a discrete crack simulation of concrete using an adaptive crack growth algorithm, in which different criteria were applied to predicting the direction of the extension of a cohesive crack. Deb and Das (2010) proposed the X-FEM for modeling cohesive discontinuities in rock masses, where the displacement discontinuities were modeled using the three- and six-nodded triangular elements. Xu and Yuan (2011) introduced a cohesive zone model with a threshold in combination with the X-FEM to study the effects of fracture criteria in cohesive zone models for mixed-mode cracks. Campilho *et al.* (2011a, b) employed the X-FEM to model crack propagation and to predict the fracture behavior of a thin layer of two structural epoxy adhesives under varying restraining conditions; the stiff and compliant adherends. Benvenuti and Tralli (2012) proposed a regularized X-FEM approach that can tackle in a unified and smooth way the whole process from strain localization to crack inception and propagation and can simulate both the formation of a process zone with finite width and its subsequent collapse into a macro-crack in concrete-like materials. Golewski, Golewski, and Sadowski (2012) employed the X-FEM for three-dimensional numerical modeling of compact shear specimens used for experimental testing of the mode II fracture to estimate the fracture toughness for a mode II fictitious crack. Olesen and Poulsen (2013) presented a simple element for modeling

Figure 1.3 Road damage following a 6.8 magnitude earthquake in Chūetsu on October 23, 2004, which occurred in Niigata Prefecture, Japan. The road and other routes suffered heavy damage due to landslides and faulting that resulted from liquefaction. (Source: Tubbi [CC-BY-3.0-br (http://creativecommons.org/licenses/by-sa/3.0/deed.en)], via Wikimedia Commons; http://commons.wikimedia.org/wiki/File:Chuetsu_earthquake-Yamabe_Bridge.jpg)

Figure 1.4 Teton Dam collapse on June 5, 1976: The collapse of an earthen dam sent a wall of water toward the Idaho Falls. The dam, located in Idaho, USA on the Teton River in the eastern part of the state, between Fremont and Madison counties, suffered a catastrophic failure as it was filling for the first time. (Source: US Government; http://commons.wikimedia.org/wiki/File:Teton_Dam_failure.jpg)

cohesive fracture processes in quasi-brittle materials based on the CST (constant strain triangle) element, where the crack was embedded in the element and a special shape function was introduced for the discontinuous displacements. Zhang, Wang, and Yu (2013) presented a numerical scheme based on the X-FEM for a seismic analysis of crack growth in concrete gravity dams with special reference to the dynamic analysis of the Koyna Dam during the 1967 Koyna earthquake.

1.3.4 Composite Materials and Material Inhomogeneities

Composite structures have been of great concern for possessing advantages of multiple materials resulting in substantial economic benefits. Material inhomogeneities result in a discontinuous displacement gradient that is referred to as *weak discontinuities*. Modeling the deformation and failure mechanisms in composite and polycrystalline materials is critical for improvements to the development and application of advanced structural materials. The material microstructure plays a pivotal role in dictating the modes of fracture and failure, and the macroscopic response of real materials. Sukumar *et al.* (2001) presented the first implementation of the X-FEM for modeling arbitrary holes, inclusions, and material interfaces without meshing the internal boundaries, where the X-FEM was coupled with the LSM to represent the location of holes, inclusions, and material interfaces. Moës *et al.* (2003) presented a modified level set function for problems that involve discontinuities in the gradients of the field to model the material interfaces in micro-structures with complex geometries, and demonstrated the capability of model for the homogenization of periodic basic cells. Sukumar *et al.* (2003b) proposed the X-FEM for crack propagation simulation through a polycrystalline material microstructure with an aim towards the understanding of toughening mechanisms in polycrystalline materials such as ceramics. Sukumar *et al.* (2004) extended the X-FEM to the analysis of cracks that lie at the interface of two elastically homogeneous isotropic materials, where the new crack tip enrichment functions were introduced to span the asymptotic displacement fields for an interfacial crack. Liu, Xiao, and Karihaloo (2004) developed an X-FEM formulation to directly evaluate the mixed mode SIFs without extra post-processing for homogeneous materials and bimaterials. Hettich and Ramm (2006) described the material distribution within a microstructure that consists of different solid phases, including material interfaces by means of the level set and X-FEM methods. Yan and Park (2008) applied the X-FEM combined with the LSM for the simulation of near-interfacial crack propagation in a metal-ceramic layered structure. Yvonnet, Le Quang, and He (2008) presented a numerical approach for modeling the interface effects described by the coherent interface model and determining the size-dependent effective elastic moduli of nano-composites.

There are more recent studies reported in the literature for the X-FEM analysis of composite materials that are commonly used in engineering practice to provide a desired mechanical behavior in fiber-reinforced materials, metal composites, concrete, ceramics, and so on. Huynh and Belytschko (2009) presented a method for treating fracture in composite material by the X-FEM with meshes independent of matrix/fiber interfaces, where the level sets were used to describe the geometry of the interfaces and 12-asymptotic functions were employed for bimaterial crack tips. Zhang and Li (2009) studied the mechanical response of viscoelastic materials with inclusions by using a full integration scheme for the low-Poisson ratio problem and the selective integration scheme treating the volumetric locking for the high-Poisson ratio problem encountered in viscoelastic materials. Dréau, Chevaugeon, and Moës (2010) proposed an approach to improve the geometrical representation of surfaces with the X-FEM, in which the surfaces were represented implicitly using the LSM with higher orders. Aragon, Duarte, and Geubelle (2010) presented a GFEM/XFEM formulation for modeling two-dimensional problems with gradient jumps along discrete lines, such as those of thermal and structural analysis of heterogeneous materials, by introducing new enrichment functions for the solution of multiple intersecting discontinuity lines, such as those of triple junctions in polycrystalline materials. Nouy and Clement (2010) proposed an extended stochastic FEM for numerical simulation of heterogeneous materials with random material interfaces based on coupled X-FEM and spectral stochastic methods, in which the random geometry of material interfaces was described implicitly by using random level set functions. Singh, Mishra, and Bhattacharya (2011) presented a numerical simulation for inhomogeneities/discontinuities, such as cracks, holes, and inclusions, in functionally graded materials based on the X-FEM method. Hiriyur, Waisman, and Deodatis (2011) proposed the X-FEM coupled with a Monte Carlo approach to quantify the uncertainty in the homogenized effective elastic properties of multi-phase materials that allows for an arbitrary number, aspect ratio, location, and orientation of elliptic inclusions within a matrix. Tran *et al.* (2011) developed a modified X-FEM/LSM for numerical simulation of microstructures containing high volume fractions of inclusions to prevent the numerical artifacts in complex microstructures with nearby inclusions. Curiel Sosa and Karapurath (2012) proposed the X-FEM for numerical simulation of delamination in fiber metal laminates, where the orthotropic enrichment functions were used to model the propagation of delamination in composites. Benvenuti, Vitarelli, and Tralli (2012b) presented an approach for modeling of delamination in FRP (fiber reinforced polymer)-reinforced concrete by means of a regularized X-FEM formulation, which takes into consideration the mechanical properties of concrete, adhesive, and FRP. Pathak, Singh, and Singh (2012) performed a numerical analysis of bimaterial interfacial cracks based on the element free Galerkin method and the X-FEM under mixed mode loading conditions that was applicable to dissimilar or layered materials such as ceramic-metal and composite-metal. Wang *et al.* (2012) proposed a numerical simulation of crack growth in brittle matrix of particle reinforced composites. Jiang *et al.* (2013) presented an edge-based smoothing technique combined with the X-FEM method for fracture analysis of composite materials that took the advantages of the X-FEM and ES-FEM methods. Zhao, Bordas, and Qu (2013) presented a hybrid smoothed X-FEM–LSM for modeling nanoscale inhomogeneities with interfacial energy effect, in which the Gurtin–Murdoch surface elasticity model was used to account for the interface stress effect and the Wachspress interpolants were employed to construct the shape functions in the smoothed X-FEM method.

1.3.5 Plasticity, Damage, and Fatigue Problems

Modeling of fatigue crack growth is a challenging issue in computational fracture mechanics. Fatigue plays an important role in mechanical failure of structures subjected to cyclic loadings. Sukumar, Chopp, and Moran (2003a) and Chopp and Sukumar (2003) proposed a numerical technique for three-dimensional modeling of planar and multiple coplanar fatigue crack growth simulations based on a coupled X-FEM–fast marching method. Stolarska and Chopp (2003) presented an algorithm for modeling the growth of fatigue cracks due to thermal cycling based on a coupled X-FEM–LSM. Ferrié *et al.* (2006) investigated the fatigue crack propagation for a semi-elliptical crack in the bulk of an ultrafine-grained Al–Li alloy based on the synchrotron radiation X-ray microtomography and three-dimensional X-FEM simulation. Giner *et al.* (2008a) employed the X-FEM for analysis of fretting fatigue problems with the FE software ABAQUS™. Xu and Yuan (2009) performed an X-FEM analysis combined with a cyclic cohesive model for the mixed-mode fatigue crack nucleation and propagation in quasi-brittle materials with the FE software ABAQUS™. Shi *et al.* (2010) presented a combined X-FEM–narrow band fast marching method for three-dimensional curvilinear fatigue crack growth and life prediction analysis of metallic structures with the FE software ABAQUS™. Ayhan (2011) employed a three-dimensional X-FEM analysis for fatigue crack propagation of a mode I surface crack and incremented crack front profiles for inclined and deflected surface cracks. Giner *et al.* (2011) used the X-FEM to predict fatigue lives in fretting fatigue problems by means of a combined initiation–propagation model. Singh, Mishra, and Bhattacharya (2011) and Singh *et al.* (2012) investigated the fatigue life of a homogeneous plate containing multiple discontinuities under cyclic loading conditions, where the multiple discontinuities of arbitrary size were randomly distributed in the plate. Ye, Shi, and Cheng (2012) proposed the X-FEM with ABAQUS™ software to simulate crack propagation and to predict the effect of reinforcing particles to the crack propagation behavior of composite materials. Réthoré *et al.* (2012) presented a three-dimensional fatigue crack propagation using advanced experimental, imaging, measurement, and numerical simulation based on the X-FEM technique. Bhattacharya, Singh, and Mishra (2012, 2013) presented an X-FEM numerical simulation for the fatigue crack growth of interfacial cracks in bilayered materials, and investigated the fracture fatigue behavior of center and edge cracked functionally graded materials in the presence of multiple inhomogeneities, such as holes/voids, inclusions, and minor cracks. Baietto *et al.* (2013) proposed a combined experimental–numerical tool based on the X-FEM for modeling two- and three-dimensional fretting fatigue crack growth simulation.

The need for an accurate tool to study crack growth scenarios and assess the damage tolerance is crucial for various industrial structures from the aeronautical industry to the automotive industry. Bordas and Moran (2006) employed the X-FEM together with the level sets for damage tolerance assessment of complex structures for solution of the complex three-dimensional industrial fracture mechanics problems. Xu *et al.* (2010) performed an X-FEM analysis to study the low-speed impact-induced cracking in brittle plates by evaluation of the interaction between the stress field and the initiation and propagation of the radial and circumferential cracks. Chatzi *et al.* (2011) proposed a combined X-FEM–genetic algorithm technique for detection of any type of flaw (cracks or holes) of any shape in structures. Vajragupta *et al.* (2012) presented a micro-mechanical damage simulation of dual phase steels using the X-FEM to study the interaction between failure modes in DP (dual phase) steels. Jung, Jeong, and Taciroglu (2013) performed a dynamic X-FEM analysis for identifying a scatterer embedded in elastic heterogeneous media, such as a crack, a void, or an inclusion with properties that have detectable contrasts from those of the host medium. Feerick, Liu, and McGarry (2013) presented anisotropic damage initiation criteria for the X-FEM prediction of crack initiation and propagation in cortical bone.

Ductile fracture has been the object of numerous works in engineering, which is characterized by the presence of moderate to large plastic deformations prior to the degradation mechanisms (Figure 1.5).

Figure 1.5 The *New Carissa* was a freighter that ran aground on a beach near Coos Bay, Oregon, USA, during a storm on February 4, 1999, and subsequently broke apart. (Source: Erin from Oregon City, OR (Old Carissa) [CC-BY-2.0 (http://creativecommons.org/licenses/by/2.0)], via Wikimedia Commons; http://commons.wikimedia.org/wiki/File:Wreck_off_Coos_Bay,_Oregon.jpg)

The hypothesis of confined plasticity, that is, the plasticity that is confined to the region near the crack tip, is important for prediction of the fatigue crack growth. The first implementation of the X-FEM for fatigue fracture analysis of cracks in homogeneous, isotropic, elastic–plastic two-dimensional solids subject to mixed mode in the case of confined plasticity was presented by Elguedj, Gravouil, and Combescure (2006), where the well-known Hutchinson–Rice–Rosengren (HRR) fields were used to present the singularities in elasto-plastic fracture mechanics. In a later work, Elguedj, Gravouil, and Combescure (2007) proposed a mixed augmented Lagrangian–X-FEM method for modeling of elastic–plastic fatigue crack growth with unilateral contact on the crack faces. Prabel *et al.* (2007) presented a numerical scheme based on the X-FEM for dynamic crack propagation in elastic-plastic media using an efficient level set update on non-matching meshes. Zhang, Rong, and Li (2010) investigated the crack problem in a linear viscoelastic material using an incremental X-FEM method while ignoring the nonlinear effects caused by subcritical crack growths in the crack-tip failure zone, in which the basis functions were extracted from the viscoelastic asymptotic fields at a crack tip. Kroon (2012) presented a computational framework based on the X-FEM for high-speed crack growth in rubber-like solids for dynamic steady-state condition considering the effects of inertia, viscoelasticity, and finite strains. Pourmodheji and Mashayekhi (2012) proposed a combined X-FEM–continuum damage mechanics model to validate the ductile damage evolution experimentally measured in A533B1 steel. Broumand and Khoei (2013) presented an enriched-FEM technique for crack growth simulation in large deformation ductile fracture problems using a non-local damage-plasticity model within the X-FEM framework based on the Lemaitre damage-plasticity model. Seabra *et al.* (2013) employed an X-FEM method combined with the Lemaitre ductile damage model to simulate the crack initiation and propagation by evolution of the damage in the framework of a finite strain and a non-local integral formulation. Wang *et al.* (2013) performed a numerical simulation for rubber-modified asphalt mixtures crack growth using the X-FEM in a notched semi-circular bending test. Liu and Borja (2013) presented an explicit X-FEM framework for fault rupture dynamics accommodating bulk plasticity near the fault by comparing the performance of various plasticity models for a rock medium hosting a randomly propagating fault.

The study of fundamental phenomena in nonlinear material behavior and plasticity involves the development of computationally efficient dislocation models. The first implementation of an enriched FEM technique into a dislocation problem was performed by Ventura, Moran, and Belytschko (2005), where the regular FE solution was enriched by the closed form solution for an edge dislocation. Gracie, Ventura, and Belytschko (2007) proposed an alternative X-FEM method for modeling dislocation problems, where the tangential enrichment was used to model edge dislocations as interior discontinuities. Belytschko and Gracie (2007) presented the X-FEM for modeling dislocations in problems with multiple arbitrary material interfaces, where Peach–Koehler forces were computed using the *J*–integral method. Ventura (2008) presented an X-FEM technique based on mapping the enrichment functions onto polynomials with applications to cracks and dislocations that allows one to perform the standard Gauss quadrature in the enriched elements. Gracie, Oswald, and Belytschko (2008a) improved the X-FEM solution based on enrichments in the neighborhood of a dislocation core by applying a discontinuous jump enrichment and a singular enrichment based on closed-form, infinite-domain solutions for the dislocation core. Ventura, Gracie, and Belytschko (2009) performed a similar approach by employing a singular enrichment based on closed-form solutions for an edge dislocation around the core, and step function enrichment for the remainder of the edge dislocation. Oswald *et al.* (2009) presented further developments in the X-FEM modeling of dislocations with emphasis on problems with complex geometry, such as carbon nanotubes and thin films. Gracie and Belytschko (2009) presented a coupled atomistic–continuum method for modeling dislocations and cracks that combines the X-FEM with the bridging domain method. Oswald *et al.* (2011) presented a higher-order X-FEM method for modeling dislocations that was applicable to complex geometries, interfaces with lattice mismatch strains, and both anisotropic and spatially non-uniform material properties.

The X-FEM has been also implemented for modeling of elasto-plasticity problems in solid mechanics. Dolbow and Devan (2004) presented a geometrically nonlinear enhanced assumed strain method with discontinuous enrichment of the displacement field that exhibits locking-free response in the incompressible limit. Shamloo, Azami, and Khoei (2005) presented a computational technique based on the X-FEM in plasticity behavior of pressure-sensitive materials, where a cap plasticity model was employed in numerical simulation of powder die-pressing. Khoei, Shamloo, and Azami (2006b) employed double-surface cap plasticity together with a frictional contact model within the X-FEM framework to capture the response of initially loose metal powders to complex deformation histories encountered in the compaction process of powder. Anahid and Khoei (2008) developed an X-FEM Lagrangian formulation for modeling arbitrary interfaces in large plasticity deformations. In a later work, Khoei, Biabanaki, and Anahid (2008d) extended the X-FEM into three-dimensional Lagrangian formulation to model the arbitrary interfaces in large elasto-plastic deformations of three-dimensional solid mechanics problems. Khoei, Anahid, and Shahim (2008a) developed a computational technique based on an extended arbitrary Lagrangian–Eulerian FEM for large deformation in solid mechanic problems, in which an arbitrary Lagrangian–Eulerian technique was employed to capture the advantages of both Lagrangian and Eulerian methods and alleviate the drawbacks of the mesh distortion in Lagrangian formulation. Khoei, Biabanaki, and Anahid (2009a) presented a Lagrangian–X-FEM method for large two- and three-dimensional deformation of plasticity and contact problems. Anahid and Khoei (2010) presented an enriched arbitrary Lagrangian–Eulerian FE method for modeling of moving boundaries in large plasticity deformations. Xu, Lee, and Tan (2012) presented a two-dimensional co-rotational beam element within the X-FEM formulation for simulation of pin connections and plastic hinges by enriching both the rotation and the deflection approximations to capture the non-smoothness in both small and large deformations. Bonfils, Chevaugeon, and Moës (2012) introduced a method for treating the volumetric inequality constraint in a continuum media with a coupled X-FEM/level-set strategy to represent moving interfaces in a domain.

1.3.6 Shear Band Localization

The shear band localization refers to the process in which a smoothly varying deformation field suddenly gives rise to one with deformations that are highly localized in narrow bands; this phenomenon has been observed in many engineering materials including metals, concrete, soils, and rocks. Due to an excessive large aspect ratio of shear bands, numerical modeling of such problems is always challenging. Mariano and Stazi (2004) presented the interaction between a macro-crack and a population of micro-cracks using the X-FEM to a multi-field model of micro-cracked bodies by evaluating the strain localization effects around the tip of the macroscopic crack. Samaniego and Belytschko (2005) proposed the X-FEM in shear band localization, in which the transition from continuum to discontinuum was governed by the loss of hyperbolicity and the post-localization behavior of material was modeled by means of a traction–separation law obtained from a continuum J_2 flow plasticity model. Areias and Belytschko (2006b) presented a methodology based on the X-FEM to model the shear band evolution in quasi-static regime, where the FE polynomial displacement field was enriched with a fine scale function to model the high displacement gradient in a shear band. Song *et al.* (2006a) presented a computational algorithm based on the X-FEM for modeling of arbitrary dynamic crack and shear band propagation, where the discontinuity was modeled by superposed elements and phantom nodes. Réthoré, Hild, and Roux (2007c) proposed the X-FEM in conjunction with digital image correlation to capture experimentally the tangential discontinuities in shear band localization. Khoei and Karimi (2008) employed an enriched FEM technique within a higher-order continuum model based on the Cosserat continuum theory to simulate shear band localization. Sandborn and Prevost (2011) presented a numerical strategy for detecting instabilities in elasto-plastic solids by inserting a discontinuity at these instabilities, and prescribing a frictional behavior along the discontinuity. Daneshyar and Mohammadi (2013) presented a method for modeling the shear band localization with strong tangential discontinuity by means of cohesive surfaces within the X-FEM, in which a rate-independent non-associated plasticity model was incorporated along the strong discontinuity to capture the highly localized regions.

1.3.7 Fluid–Structure Interaction

Fluid–structure interaction is of great relevance in many fields of engineering as well as in the applied sciences. Hence, the evaluation of fluid–structure interaction effects and the investigation of governing physical phenomena associated with coupled systems are always challenges in problems arising in the fields of aero- and hydro-elasticity, life sciences, or bio-engineering. Legay, Chessa, and Belytschko (2006) presented an X-FEM for fluid–structure interaction with the interface and free surfaces defined by level sets, in which the fluid was treated by an Eulerian mesh and the solid by a Lagrangian mesh, and the Lagrange multiplier method and penalty method were used to couple the fluid and structure. Wang *et al.* (2008) proposed an implementation of the fluid-structure interaction using the immersed/fictitious element method for compressible fluids, in which the fictitious fluid was treated by a Lagrangian description, and for thin elements an enrichment was added in the fluid regions around the structural elements. Gerstenberger and Wall (2008a) presented a fixed grid fluid–structure interaction scheme based on the X-FEM that can be applied to the interaction of most general structures with incompressible flow, in which the extended Eulerian fluid field and the Lagrangian structural field were partitioned and iteratively coupled using a Lagrange multiplier technique for non-matching grids. In a later work, Gerstenberger and Wall (2008b) proposed two enhancements of fixed-grid methods based on a local adaptivity and a hybrid method that combines ideas from fixed-grid methods and arbitrary Lagrangian–Eulerian formulations to improve the solutions to complex fluid–structure interaction problems. Zilian and Legay (2008) introduced a weighted residual-based approach for the enriched space–time FE simulation of the interaction of fluid flow and thin flexible structures to model the coupled systems involving large structural motion and deformation of multiple-flow-immersed solid objects. Massimi, Tezaur, and Farhat (2008) proposed a discontinuous enrichment method to three-dimensional evanescent wave problems by enriching the elements with free-space solutions of evanescent wave problems to obtain the required accuracy at practical mesh resolution for fluid–fluid and fluid–solid applications.

For large deformations of solids in fluid–structure interaction, or fluid–solid interaction problems, the Eulerian description for the fluid and Lagrangian description for the solid are often preferable. Wang and Belytschko (2009) developed a discontinuous-Galerkin method for large deformation fluid–structure interaction problems, where the fluid–structure interface was arbitrarily aligned relative to the fluid grid. Mayer, Gerstenberger, and Wall (2009) presented a three-dimensional higher-order X-FEM method for modeling moving interfaces in fluid–structure interaction problems, which provides a method for localization of a higher order interface FE mesh in an underlying three-dimensional higher order FE mesh. Mayer *et al.* (2010) proposed a three-dimensional numerical technique to tackle the finite deformation contact of flexible solids embedded in fluid flows based on a combined X-FEM–fluid–structure-contact interaction method to compute the contact of arbitrarily moving and deforming structures embedded in an arbitrary flow field. Zilian and Netuzhylov (2010) presented a mixed-hybrid velocity-based formulation of both fluid and structure discretized by a stabilized time-discontinuous space–time FEM for analysis of thin-walled structures immersed in generalized Newtonian fluids. Wall *et al.* (2010) presented an overview on recent research activities on a fixed grid fluid-structure interaction scheme based on an X-FEM for moving interfaces on fixed Eulerian fluid grids that can be applied to the interaction of most general structures with incompressible flow. Shahmiri, Gerstenberger, and Wolfgang (2011) presented a FE embedding mesh technique based on a non-overlapping domain decomposition method to embed arbitrary fluid mesh patches into an unstructured background fluid grid. Legay (2013) employed the X-FEM for structural-acoustic problems involving immersed structures at arbitrary positions, in which the immersed structures were supposed to be thin shells and were localized in the fluid domain by a signed distance level-set.

1.3.8 Fluid Flow in Fractured Porous Media

Flow of fluids in deformable porous media has been a topic of attention in engineering science, and has been crucial for understanding and predicting the physical behavior of many problems of interest in geotechnical and petroleum engineering (Figure 1.6). The first implementation of an enriched FEM was presented by de Borst, Réthoré, and Abellan (2006) based on the PU property of FE shape functions to analyze the stress evolution and fluid flow in two-phase fluid-saturated media for a biaxial plane-strain specimen with a propagating discontinuity, for example, a crack or a shear band. Réthoré, de Borst, and Abellan (2007a) proposed this formulation for modeling dynamic shear-band propagation in a fluid-saturated medium. In a later work, Réthoré, de Borst, and Abellan (2007b) presented a two-scale approach by exploiting the PU property of the FE shape functions for fluid flow within the fractured porous media, in which the flow in the cavity of a fracture was modeled as a viscous fluid at the microscopic level. Réthoré, de Borst, and Abellan (2008) extended their model to an unsaturated porous medium by developing a two-scale model for fluid flow within a deforming unsaturated and progressively fracturing porous medium, where the flow in the cohesive crack was modeled using Darcy's Law that takes into account changes in the permeability due to the progressive damage evolution inside the cohesive zone. Lecampion (2009) presented an X-FEM formulation for the solution of hydraulic fracture problems by introducing special tip functions encapsulating tip asymptotic functions typically encountered in hydraulic fractures. QingWen, YuWen, and TianTang (2009) proposed the X-FEM for numerical modeling of hydraulic fracturing in a gravity dam. Gracie and Craig (2010) employed the X-FEM for predicting the steady-state leakage from layered sedimentary aquifer systems perforated by abandoned wells, where the leakage of fluid between aquifers occurred through the aquitards and abandoned wells. Huang *et al.* (2011) proposed an enrichment scheme to model fractures and conduits in porous media flow problems that was able to capture effects of local heterogeneity introduced by subsurface features on the pressure solution.

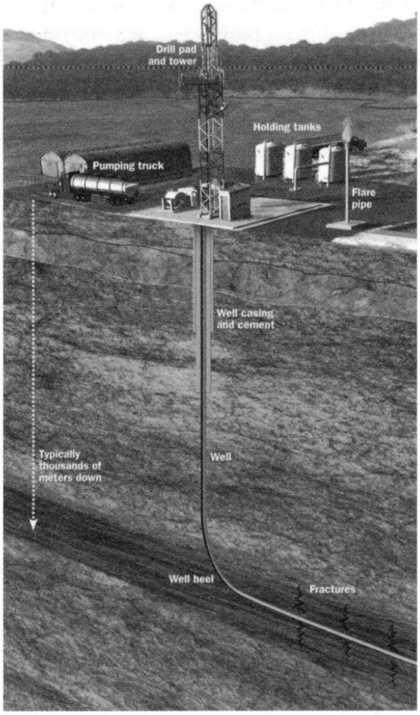

Figure 1.6 Hydraulic fracturing is the process of drilling a bore hole into a geological formation that contains *microscopic pockets* of natural gas and oil, and then fracturing the formation so that the gas and oil can escape into the bore hole and be drawn to the surface. It is a technique in which typically water is mixed with sand and chemicals, and the mixture is injected at high pressure into a wellbore to create small fractures, along which fluids such as gas, petroleum, uranium-bearing solution, and brine water may migrate to the well

Modeling of hydraulic fracture propagation has attracted considerable attention since the early 1950s because of its practical applications in a broad range of engineering areas. Hydraulic fracturing is the most commonly used stimulation technique of reservoirs, which is widely used in the petroleum industry to enhance the recovery of hydrocarbons, such as gas and oil, from underground reservoirs. Khoei and Haghighat (2011) presented an enriched FEM based on an updated Lagrangian framework combined with a generalized Newmark scheme for the time domain discretization in order to numerically simulate the saturated porous media with arbitrary material interfaces. In a later work, Khoei, Moallemi, and Haghighat (2012) presented an X-FEM technique for thermo-hydro-mechanical modeling of impermeable discontinuities in saturated porous media, where the displacement field was enriched using the Heaviside and crack tip asymptotic functions, and the pressure and temperature fields were enriched by the Heaviside and appropriate asymptotic functions. Watanabe *et al.* (2012) developed a method based on local enrichment approximations by employing the lower-dimensional interface elements to present preexisting fractures in rock material, focusing on FE analysis of coupled hydromechanical problems in discrete fracture-porous media systems. Gordeliy and Peirce (2013) presented an X-FEM method that was coupled with the elasto-hydrodynamic equations to solve elastic crack propagation due to hydraulic fracturing in an elastic medium. Mohammadnejad and Khoei (2013a) presented an X-FEM for the fully coupled analysis of deforming porous media containing weak discontinuities that interact with the flow of two immiscible, compressible wetting and non-wetting pore fluids. In a later work, Mohammadnejad and Khoei (2013b) developed a numerical model based on an extended FEM to simulate the flow of wetting and non-wetting pore fluids in progressively fracturing, partially saturated porous media, in which the mechanical and the mass transfer coupling between the crack and the porous medium surrounding the crack were taken into account. Mohammadnejad and Khoei (2013c) proposed a numerical model for the fully coupled hydro-mechanical analysis of deformable, progressively fracturing porous media using the X-FEM in conjunction with the cohesive crack model. Khoei and Vahab (2014) presented a coupled hydromechanical formulation in the framework of X-FEM method for deformable porous media subjected to crack interfaces to simulate the dynamic hydromechanical behavior of fractured porous media with opening and closing modes. Khoei *et al.* (2014b) proposed the X-FEM for mixed-mode crack growth simulation in saturated porous media using two computational algorithms to compute the interfacial forces of fluid pressure exerted on the fracture faces based on a "partitioned solution algorithm" and a "time-dependent constant pressure algorithm".

1.3.9 Fluid Flow and Fluid Mechanics Problems

Dynamic flows encompass a wide array of flows including drops, jets, plumes, thin films, and bubbles that exhibit such diverse phenomena as capillary instabilities, Homman flow, fluid bridges, particulate and bubble flows, and so on. Moreover, the interaction of different fluids is frequently observed in real problems, its numerical simulation has a long tradition and still remains an active research field. The first implementation of X-FEM in fluid flow problems was performed by Wagner *et al.* (2001) for simulation of rigid particles in Stokes flow, in which the surfaces of moving particles were not conformed to FE boundaries, and the near field form of the fluid flow about each particle was built into the FE basis using a PU enrichment. Chessa and Belytschko (2003a) presented an X-FEM method for axisymmetric two-phase flow problems with surface tension, in which the interface can move arbitrarily through the mesh and the discontinuity in the velocity gradient at the interface was modeled by a local PU. In a later work, Chessa and Belytschko (2003b) employed the X-FEM with arbitrary discontinuous gradients to two-phase immiscible flow problems, where the phase interfaces were tracked by level set functions using the same FE mesh and were updated via a stabilized conservation law. Groß and Reusken (2007) presented an extended pressure FEM for incompressible two-phase flows in which a localized force at the interface describes the effect of surface tension. Dolbow *et al.* (2008) employed the patterned-interface-reconstruction algorithm in conjunction with the X-FEM for general multi-material problems that exhibits the advantages of local, element-based interface representation, in particular the ability to enforce strict volume conservation. Fries (2009) presented an intrinsic X-FEM model for simulation of incompressible two-fluid flows, where the jumps and kinks along the interfaces in the velocity and pressure fields were modeled using an enriched PUM. Van der Bos and Gravemeier (2009) presented an enriched FEM model for simulation of the premixed combustion based on the *G*-function approach, in which a level-set or *G*–function was used to define the flame interface separating burned gas from unburned gas. Esser, Grande, and Reusken (2010) combined the X-FEM with a level set interface capturing technique for physically realistic levitated droplet problem, and applied the technique to three-dimensional simulation of a physically realistic two-phase flow problem. Abbas, Alizada, and Fries (2010) proposed an enriched FEM model for high-gradient solutions in convection-dominated problems that enables highly accurate approximations of convection-dominated problems without stabilization or mesh refinement. Sauerland and Fries (2011) investigated different enrichment schemes and time-integration schemes within the X-FEM for immiscible two-phase and free-surface flows, where the jumps and kinks in the velocity and pressure fields were captured with no restrictions on the interface topology. Cheng and Fries (2012) developed a *h*-version of the X-FEM based on a multi-level adaptive mesh refinement realized via hanging nodes on one-irregular non-conforming meshes for simulation of two-fluid incompressible flow in two and three dimensions. Liao and Zhuang (2012) presented a consistent projection-based streamline upwind/pressure stabilizing Petrov–Galerkin X-FEM to model incompressible immiscible two-phase flows, in which projections of convection and pressure gradient terms were constructed and incorporated into the stabilization formulation. Sauerland and Fries (2013) studied the issue of ill-conditioning in the X-FEM for two-phase flows, concentrating on the stable X-FEM and the application of iterative solvers.

Particulate flows arise in a wide class of research areas and industrial processes, such as fluidized suspensions, slurry transport, materials separation, rate of mixing enhancement, and so on. In fluid-rigid body problems, the field variables such as pressure and stress are discontinuous over the interface since no flow occurs inside the rigid body. Hence, modeling of the fluid flow around a stationary rigid body is crucial. Choi, Hulsen, and Meijer (2010) developed an X-FEM formulation for the direct numerical simulation of the flow of viscoelastic fluids with suspended particles, in which an arbitrary Lagrangian–Eulerian scheme was devised for moving particle problems that defines the mapping of field variables at previous time levels onto the computational mesh at the current time level. In a later work, Choi, Hulsen, and Meijer (2012) used their original formulation for simulation of the flow of a viscoelastic fluid around a stationary cylinder based on the X-FEM. Sarhangi Fard, Hulsen, and Anderson (2012a) employed the X-FEM to simulate the three-dimensional Stokes flow in complex geometries with internal moving parts and narrow gaps, in which the kinematics of the internal rigid body was enforced using a constraint Lagrangian multiplier. Sarhangi Fard *et al.* (2012b) performed a non-Newtonian viscous flow analysis in complex geometries with internal moving parts and narrow gaps based on a non-conforming mesh refinement approach and the X-FEM method. Court, Fournié, and Lozinski (2014) proposed a fictitious domain approach inspired by the X-FEM for the Stokes problem, in which the interface between the fluid and the structure was localized by a level-set function, and the Dirichlet boundary conditions were taken into account using the Lagrange multiplier.

1.3.10 Phase Transition and Solidification

Numerical modeling of melting/solidification or dissolution/precipitation is significant in the evaluation of operating conditions and designs of equipment configuration in these technologies. However, modeling the moving solid–liquid interface in such phase transition is a challenging problem. Phase transition represent diffuse material interfaces across which several fields may exhibit sharp gradients, and even discontinuities, as the interface thickness vanishes and becomes "sharp". Hence, the evaluation and enforcement of interfacial conditions is an important consideration for generating accurate approximations to solutions of boundary value problems involving the evolution of sharp interfaces. Merle and Dolbow (2002) proposed the X-FEM for modeling thermal problems with moving heat sources and phase boundaries to capture the highly localized, transient solution in the vicinity of a heat source or material interface. Chessa, Smolinski, and Belytschko (2002) presented an enriched FEM for the numerical solution of phase change problems, where the phase interface was evolved by the use of artificial heat capacity technique that was described by a level set function. Ji, Chopp, and Dolbow (2002) presented a hybrid numerical method based on the X-FEM for modeling the evolution of sharp phase interfaces on fixed grids with reference to solidification problems, where the temperature field was evolved according to the classical heat conduction in two sub-domains separated by a moving freezing front. Ji and Dolbow (2004) considered a problem stemming from phase transitions in stimulus-responsive hydrogels, wherein a sharp interface separates swelled and collapsed phases, and presented that as the reciprocal interfacial mobility vanishes, it plays the role of a penalty parameter enforcing a pure Dirichlet constraint, eventually triggering oscillations in the interfacial velocity. Zabaras, Ganapathysubramanian, and Tan (2006) studied dendritic solidification of pure materials from an under-cooled melt by using a coupled X-FEM–LSM for modeling the thermal problem and a volume-averaged stabilized formulation for modeling fluid flow.

In the solidification process, complications are described as isothermal phase change problems that are characterized mainly by the material melting temperature and its latent heat, in which an inherent difficulty with these problems is the discontinuity in the temperature gradient at the solidification front. Uchibori and Ohshima (2012) presented a numerical analysis based on a moving boundary X-FEM technique for modeling melting/solidification problems to simulate the discontinuous temperature gradient in the element crossed by the solid–liquid interface. Zhou and Qi (2010) proposed the X-FEM within the ABAQUS™ software to simulate the discontinuous interface in the liquid–solid forming process, where the discontinuous interface in the liquid–solid forming processes was handled using the LSM.

Duddu *et al.* (2011) presented a sharp-interface numerical formulation based on the X-FEM–LSM using an Eulerian description for modeling diffusional evolution of precipitates produced by phase transformations in elastic media. Skrzypczak (2012) presented a mathematical model for modeling sharp interfaces in solidification process of pure metals, where the interface motion was described by the level set function. Ghoneim, Hunedy, and Ojo (2013) proposed a numerical simulation based on an interface-enriched extended FE–LSM to study the solute-induced melting of additive powder particles during transient liquid phase bonding. Cosimo, Fachinotti, and Cardona (2013) presented an enriched FE formulation for solving isothermal phase change problems, where the discontinuity in the temperature gradient at the solidification front was modeled by enriching the FE space through a function whose definition includes a gradient discontinuity.

1.3.11 Thermal and Thermo-Mechanical Problems

Most of the critical engineering components are generally exposed to both thermal and mechanical loading during their service life, such as combustion chambers of internal combustion engines, nuclear reactor components, spacecraft, blade casing in thermal power plants, and so on. There are a number of thermal and thermo-mechanical problems in the engineering and materials science communities that are characterized by the presence of a highly localized moving heat source. Michlik and Berndt (2006) proposed a thermo-mechanical X-FEM analysis for modeling thermal barrier coatings in order to provide useful versatility in prediction of effective thermal conductivity and the in-plane Young's modulus of multi-layered thermal barrier coatings. Duflot (2008) employed the X-FEM for the analysis of steady-state thermo-elastic problems in cracked structures, where both thermal and mechanical fields are enriched in order to represent the discontinuous temperature, heat flux, displacement, and traction across the isothermal crack surface. Fagerstorm and Larsson (2008) presented a thermo-mechanically coupled interface fracture formulation based on discontinuous representation for temperature and displacements fields pertinent to the FPZ into a cohesive zone. Zamani, Gracie, and Eslami (2010) proposed the X-FEM to predict the SIFs for thermo-elastic crack problems using higher-order terms of the thermo-elastic asymptotic crack-tip fields for the enrichment of temperature and displacement fields. In a later work, Zamani and Eslami (2010) extended their model in simulation of problems with stationary cracks under dynamic thermo-mechanical loading to model the effect of mechanical and thermal shocks on a body with stationary cracks. Lee, Yang, and Maute (2011) presented an X-FEM method for the analysis of heat conduction at submicron scales of geometrically complex nano-structured materials. Yvonnet *et al.* (2011) proposed a computational procedure based on a coupled X-FEM–LSM for modeling the Kapitza thermal resistance at an arbitrarily shaped interface. Fan *et al.* (2012) employed the X-FEM to investigate the effect of thermally grown oxide on multiple surface cracking behavior in an air plasma sprayed thermal barrier coating system. Hosseini, Bayesteh, and Mohammadi (2013) proposed a computational method based on the X-FEM for fracture analysis of isotropic and orthotropic functionally graded materials under mechanical and steady-state thermal loadings. Yu and Gong (2013) employed the X-FEM for modeling the temperature field in heterogeneous materials, where the temperature field was enriched by incorporating the level-set enrichment function for the element containing material interfaces. Macri and Littlefield (2013) presented a multi-scale technique based on an enriched partition of unity approach that incorporates the thermal effects occurring on the micro-structure into the global model for simulation of heterogeneous materials undergoing substantial thermal stresses.

1.3.12 Plates and Shells

Plates and shells are widely used in thin-walled structures such as aircraft fuselages subjected to bending and pressure loadings (Figure 1.7). Through-the-thickness cracks may develop when these structures are subjected to cyclic loading, and the determination of mixed-mode SIFs is critical to the modeling of

Figure 1.7 The Impact Dynamics Research Facility is used by NASA to conduct crash testing of full-scale aircraft under controlled conditions. The aircraft are swung by cables from an A-frame structure that is approximately 400 ft long and 230 ft high. (Source: NASA; http://commons.wikimedia.org/wiki/File:Impact_Landing_Dynamics_Facility_Crash_Test_-_GPN-2000-001907.jpg)

fatigue crack propagation. Despite the practical importance, relatively little research has focused on developing robust numerical methods to determine fracture parameters and simulate crack growth in thin plates. The first implementation of an X-FEM for modeling cracks and crack growth in plates was performed by Dolbow, Moës, and Belytschko (2000b) in the context of the Mindlin–Reissner plate theory. Liang *et al.* (2003) developed an X-FEM model to evolve patterns of multiple cracks in a brittle thin film bonded to an elastic substrate, in which the film was susceptible to subcritical cracking, obeying a kinetic law that relates the velocity of each crack to its energy release rate. Huang *et al.* (2003a) employed the X-FEM to compute the steady-state energy release rate of channeling cracks in thin films, where the driving force for channeling cracks was obtained as a function of elastic mismatch, crack spacing, and the thickness ratio between the substrate and the film. Bachene, Tiberkak, and Rechak (2009a) and Bachene *et al.* (2009b) proposed the X-FEM to model vibrations of cracked plates based on Mindlin plate theory, where the effects of shear deformation and rotatory inertia were included in the development of the model. Lasry *et al.* (2010) presented an X-FEM for simulation of thin cracked plates under bending loads based on the Kirchhoff–Love theory that is well suited to very thin plates commonly used, for instance, in aircraft structures. Fan *et al.* (2011) investigated multiple cracking behavior in a thin elastic film bonded to a thick elastic substrate using an X-FEM method, in which the SIFs were obtained using a periodic FEM for the cracked film/substrate system. Natarajan *et al.* (2011) applied the X-FEM to model the linear free flexural vibrations of cracked functionally graded material plates based on the first-order shear deformation theory, and performed a parametric study on the influence of gradient index, crack location, crack length, crack orientation, and thickness on the natural frequencies of FGM plate. Lasry, Renard, and Salaün (2012) proposed the X-FEM for modeling bending plates with through-the-thickness cracks based on the Kirchhoff–Love plate model, where the reduced Hsieh–Clough–Tocher triangles and reduced Fraejis de Veubeke–Sanders quadrilaterals were used for the numerical discretization. Xu, Lee, and Tan (2013a, b) presented the X-FEM for modeling of a plate element with high gradient zones in both rotation and deflection displacement fields in the vicinity of a yield line in a plate structure with elasto-plastic material.

A large number of industrial and engineering structures, such as aircraft fuselages, storage tanks, ship hulls, and pipes, are made of plates and shells. Existence of defects and cracks in such structures may lead to substantial decrease in load capacity, fatigue crack propagation, leak before breakage, and even structural collapse. The out-of-plane effects in large thin walled structures frequently require the use of a robust dedicated method in order to properly capture the details of the three-dimensional crack tip conditions. Areias and Belytschko (2005b) developed a numerical procedure for the analysis of arbitrary crack propagation in sandwich shells, in which a new enrichment of the rotation field was proposed that satisfies the director inextensibility condition. Wyart *et al.* (2007, 2009) employed the sub-structured FE/X-FEM approach to compute the SIFs in large aircraft thin walled structures containing cracks. Bayesteh and Mohammadi (2011) studied the effect of crack tip enrichment functions in the X-FEM analysis of shells by evaluating fracture mechanics parameters such as the SIF, crack-tip opening displacement, and crack-tip opening angle. Larsson, Mediavilla, and Fagerström (2011) presented a formulation based on the cohesive zone concept applied to a kinematically consistent shell model enhanced with an X-FEM based discontinuous kinematical representation. Liu, Zhang, and Zheng (2012) investigated the plastic collapse and crack behavior of steel pressure vessels and piping using the X-FEM to provide a fundamental support for safety evaluation and life prediction of pressurized structures.

1.3.13 Contact Problems

Numerical modeling of engineering contact problems is one of the most difficult and demanding tasks in computational mechanics. Frictional contact can be observed in many problems; such as: crack propagation, metal forming operation, drilling pile, and so on (Figure 1.8). Hence, much attention must be given to the numerical research aspects of this complex problem. The first implementation of an extended finite element framework in a contact problem was performed by Dolbow, Moës, and Belytschko (2001) for modeling of crack growth with frictional contact on the crack faces. Khoei and Nikbakht (2006, 2007) developed an enriched FEM for modeling of frictional contact problems based on the penalty method in order to simulate the frictional behavior of contact between two bodies. Khoei, Shamloo, and Azami (2006b) employed a cap plasticity model in conjunction with a frictional contact model within the X-FEM framework to simulate the compaction of powder die-pressing. Vitali and Benson (2006) presented an X-FEM formulation for contact in multi-material arbitrary Lagrangian–Eulerian calculations. Liu and Borja (2008) presented an incremental quasi-static contact algorithm for path-dependent frictional crack propagation in the framework of the X-FEM, where the contact constraint was embedded within a localized element by penalty method. Giner *et al.* (2008b) performed a numerical analysis of complete sliding contact and its associated singularity based on the partition of unity FEM, in which the enriched functions were derived from the analytical expression of the asymptotic displacement field in the vicinity of the contact corner. Khoei, Biabanaki, and Anahid (2009a) presented a Lagrangian–X-FEM method for three-dimensional modeling of large-plasticity deformations and contact problems. Nistor *et al.* (2009) proposed an approach to couple the X-FEM with the Lagrangian large sliding frictionless contact algorithm, in which a hybrid X-FEM contact element was introduced in the framework of a contact search algorithm allowing for an update of contacting surfaces pairing.

The contact problem generally suffers from a numerical instability similar to that encountered in incompressible elasticity, in which the normal contact pressure exhibits spurious oscillation. This oscillation does not go away with mesh refinement, and in some cases it even gets worse as the mesh is refined. There are several stabilized approaches introduced in the literature to address the problem of contact pressure oscillation. Béchet, Moës, and Wohlmuth (2009a) introduced a stable Lagrange multiplier space to impose stiff interface conditions within the context of the X-FEM. Becker, Burman, and Hansbo (2009) presented a Nitsche X-FEM for incompressible elasticity with discontinuous elasticity modulus that satisfies the inf–sup condition for stabilized methods related to the non-mixed constant-strain method. Zilian and Fries (2009) proposed an approach for the imposition of constraints along moving or fixed immersed interfaces in the context of the X-FEM, in which the use of Lagrange multipliers or penalty

Figure 1.8 A crash test is a form of destructive testing usually performed in order to ensure safe design standards in crash worthiness and crash compatibility for various modes of transportation or related systems and components. There are various types of crash test, such as frontal-impact test (in the photograph), offset crash test, side-impact crash test, and roll-over crash test. (Source: StaraBlazkova at the Czech language Wikipedia [GFDL (www.gnu.org/copyleft/fdl.html) or CC-BY-SA-3.0 (http://creativecommons.org/licenses/by-sa/3.0/)], from Wikimedia Commons; http://commons.wikimedia.org/wiki/File:Muzeum_of_trans_Gla_crash_test.jpg)

methods was circumvented by a localized mixed hybrid formulation of the model equations. Vitali and Benson (2009) presented an approach for classical kinetic friction laws based on a multi-material arbitrary Lagrangian–Eulerian formulation, where the nodal tangent relative velocities between two materials containing more than one material in their support were updated through nodal accelerations to account for kinetic friction effects. Liu and Borja (2009) presented an X-FEM approach for the simulation of slow-rate frictional faulting in geologic media incorporating bulk plasticity and variable friction, in which the bulk plasticity was included in the formulation to trigger fault rupture and its extension as plastic yielding was localized at the fault tip. In a later work, Liu and Borja (2010a) proposed an X-FEM framework for frictional crack problems to accommodate finite deformation and bulk inelasticity, in which conditions for contact and frictional sliding were imposed in the current configuration, and integration on all types of nonlinearity, including material, geometric, and contact search, were performed implicitly. Liu and Borja (2010b) also employed a stabilized FE formulation based on the polynomial pressure projection technique that was used successfully for the Stokes equation and for coupled solid-deformation-fluid diffusion using low-order mixed finite elements.

The enforcement of the contact condition is still a challenging aspect of the contact problem, whether it be in the context of classical nonlinear contact mechanics in which element sides are aligned to the contact faces, or in the framework of the assumed enhanced strain, or X-FEM in which contact faces are allowed to pass through and cut the interior of finite elements. Khoei and Mousavi (2010) presented a node-to-segment contact algorithm based on the X-FEM to model the large deformation/large sliding contact problem using the penalty approach. Pierrès, Baietto, and Gravouil (2010) developed an X-FEM for three-dimensional modeling of the cracked structure and the crack interface as two independent global and local problems characterized by different length scales and different behaviors, in which the interface was linked to the global problem in a weak sense in order to avoid instabilities in the contact solution. Mueller-Hoeppe, Wriggers, and Loehnert (2012) proposed an X-FEM formulation for crack face contact in terms of a penalty formulation for normal contact, in which the contact surface discretization was based on tetrahedrization according to the level set field. Laursen *et al.* (2012) proposed a mortar contact formulation for deformable-deformable contact that presents promising results for development of a numerical method allowing imposition of contact constraints with the X-FEM method. Khoei, Hirmand, and Vahab (2014a) presented an augmented Lagrangian formulation for modeling frictional discontinuities in the framework of X-FEM method, in which the nonlinear contact behavior was modeled based on an active set strategy to fulfill the Kuhn–Tucker inequalities within the iterations of Newton–Raphson procedure. Khoei and Vahab (2014) presented an approach to model the contact behavior in hydromechanical deformable porous media subjected to crack interfaces in the framework of an X-FEM.

1.3.14 Topology Optimization

Continuum structural optimization methods are of great interest in engineering applications for their benefits in helping engineers achieve better designs and save time. Given the significance of the shape representation in structural shape and topology optimization problems, the LSM can be efficiently implemented in structural optimization problems. Belytschko, Xiao, and Parimi (2003c) presented an approach for topology optimization based on an implicit function description of the surface of the design, in which the implicit function was constrained by upper and lower bounds, so that only a band of nodal variables needs to be considered in each step of the optimization. Sy and Renard (2010) proposed a method based on a fictitious domain approach for structural optimization with a coupling between shape and topological gradients, in which the fictitious domain approach was inspired by the X-FEM. Wei, Wang, and Xing (2010) employed the X-FEM in conjunction with the LSM to solve structural shape and topology optimization problems, and showed that the X-FEM leads to more accurate results compared to the FEM solution without increasing the mesh density and DOF. Edke and Chang (2011) performed a design process that supports the shape optimization of structural components under a two-dimensional mixed mode fracture for maximum service life, in which the process incorporates the X-FEM–LSM for crack growth modeling without remeshing. Kreissl and Maute (2012) presented an optimal design of fluidic devices subjected to incompressible flow at low Reynolds numbers using a LSM to describe the fluid–solid interface geometry, in which the fluid flow was modeled by the incompressible Navier–Stokes equations within the X-FEM. Li, Wang, and Wei (2012) proposed the X-FEM for LSM structural optimization with a partition integral method, in which the X-FEM integral scheme without quadrature subcells and a higher order element X-FEM scheme were employed.

1.3.15 Piezoelectric and Magneto-Electroelastic Problems

Materials with a strong piezoelectric effect are generally used in various applications, as sensors, actuators, or transducers. These applications range from sub-millimeter length scales in micro-electromechanical systems up to large scales in the design of smart electromechanical structures, such as wings in the aerospace industry. As for regular materials subjected to high mechanical stresses, the knowledge of fracture behavior for these smart materials is often crucial for design of parts under high electrical and mechanical loading. Béchet, Scherzer, and Kuna (2009b) proposed an application of the X-FEM to the analysis of fracture in piezoelectric materials by introducing new enrichment functions suitable for cracks in piezoelectric structures. Verhoosel, Remmers, and Gutiérrez (2010) employed a PU-based cohesive zone FEM to mimic crack nucleation and propagation in a piezoelectric continuum, in which a multi-scale framework was proposed to appropriately represent the influence of the microstructure on the response of a miniaturized component. Rochus *et al.* (2011) investigated various X-FEM techniques to solve the electrostatic problem when the electrostatic domain was bounded by a conducting material in order to accurately evaluate the electrostatic forces acting on the devices. Rojas-Díaz *et al.* (2011) studied the static fracture analysis of two-dimensional linear magneto-electroelastic solids within the X-FEM framework considering the media possessing fully coupled piezoelectric, piezomagnetic, and magneto-electric effects. Bhargava and Sharma (2011) proposed the X-FEM in a two-dimensional finite piezoelectric media weakened by a crack, where the four-fold standard enrichment functions were taken in conjugation with the interaction integral to evaluate the intensity factors. Bhargava and Sharma (2012) introduced a generalized set of crack tip enrichment functions by redefining the existing basis functions, and presented the efficiency of these enrichment functions by validation of the Griffith crack in an infinite domain with the energy norm and the convergence of SIFs. Nguyen-Vinh *et al.* (2012) presented an X-FEM formulation for dynamic fracture of piezoelectric materials that was applied to mode I and mixed mode fractures for quasi-steady cracks. Kästner *et al.* (2013) presented an application of bilinear and biquadratic X-FEM formulations for modeling weak discontinuities in magnetic and coupled magneto-mechanical boundary value problems, where the level set representation of curved interfaces was used to resolve the location of curved interfaces and the discontinuous physical behavior.

1.3.16 Multi-Scale Modeling

Simulations with atomistic resolution of dislocation cores and crack fronts are critical to a more fundamental understanding of the physics of plasticity and failure. However, even the treatment of submicron cracks and dislocation loops by atomistic methods is generally not feasible because of the large number of atoms required. While concurrent models can deal with defects of moderate size, in the order of hundreds of Angstroms, this does not suffice for many dislocation and crack problems of interest. A key need is for methods that can apply atomistic models where needed and apply continuum models to the remainder of the domain with the capability to model the discontinuities associated with cracks and dislocations. Guidault *et al.* (2008) proposed a multi-scale technique for crack propagation using two strategies; a micro–macro approach using a domain decomposition method to account for the efficient treatment of the global and local effects due to the crack, and a local enrichment method on the basis of an X-FEM technique to describe the geometry of the crack independently of the mesh. Gracie and Belytschko (2009) presented a coupled atomistic–continuum method for modeling dislocations and cracks that combines the X-FEM with the bridging domain method, where the multi-scale strategy was used to model the crack surfaces and slip planes in the continuum, and the bridging domain method was employed to link the atomistic model with the continuum. Aubertin, Réthoré, and de Borst (2010) presented a multi-scale method that couples a molecular dynamics approach for describing fracture at the crack-tip with an X-FEM for discretizing the remainder of the domain to simulate dynamic fracture in an efficient manner on basis of elementary physical principles. Kästner, Haasemann, and Ulbricht (2011) proposed a multi-scale simulation of fiber-reinforced polymers, where the heterogeneous material structure in a representative volume element was modeled by the X-FEM. Loehnert, Prange, and Wriggers (2012) presented a discretization error controlled adaptive multi-scale technique for an accurate simulation of micro-structural effects within a macroscopic component using the corrected X-FEM, in which the incorporation of micro-structural features such as micro-cracks was achieved by means of the multi-scale projection method. Macri and Littlefield (2013) presented a multi-scale technique for modeling heterogeneous materials undergoing substantial thermal stresses based on an enriched PU approach that incorporates the thermal effects occurring on the microstructure into the global model.

2

Extended Finite Element Formulation

2.1 Introduction

The finite element method (FEM) approximation is a piecewise differentiable polynomial approximation, which is ill-suited to representing problems with discontinuities (either in the unknown field or its gradient), singularities, and boundary layers. To accurately model discontinuities with the FEM, it is necessary to conform the discretization (mesh) to the line or surface of discontinuity. This becomes a major difficulty when treating problems with evolving discontinuities where the mesh must be regenerated at each step. In the standard FEM, singularities or boundary layers are resolved by requiring significant mesh refinement in the regions where the gradients of the fields are large. In fact, modeling of discontinuities with the FEM is cumbersome due to the need to update the mesh topology to conform the geometry of the discontinuity (Figure 2.1). The discontinuity in field variables of a system can be frequently observed, such as the presence of cracks, shear bands, and inclusions in structural problems. In fluid mechanics problems, the discontinuous fields can occur between two different fluids. Discontinuities can be generally classified into a strong and a weak discontinuity, in which the former involves the discontinuity in the field variable of a model, whereas the latter describes the discontinuity in the gradient of the field variable. In structural problems, strong and weak discontinuities can be respectively referred to cracks and interfaces between different materials. In order to obtain a good convergence of the solution, a proper treatment of the interface is necessary in approximation of the discontinuous field. For example, the solution may exhibit a characteristic behavior at the interface, such as strong discontinuities or singular derivatives, as can be observed at the crack tip area. In such cases, the numerical method involved in the approximation may be enhanced to represent these characteristics of the solution. The essential feature is the incorporation of enrichment functions that contain a discontinuous field. The classical FEM relies on the local approximation properties of polynomials (Melenk and Babuška, 1996). However, for jumps, kinks, or singularities in the solution within elements, polynomials have poor approximation properties. Consequently, the accuracy of a standard FE (finite element) analysis is, in general, quite poor for problems involving arbitrary discontinuities. Several methods have been developed with the ability to introduce special characteristics of the solution into the approximation space, such as the partition of unity finite element method (PU-FEM) by Babuška and Melenk (1997), the generalized FEM method (G-FEM) by Strouboulis, Babuška, and Copps (2000) and Strouboulis, Copps, and Babuška (2001), and the extended FEM (X-FEM) by Belytschko and Black (1999) and Moës, Dolbow, and Belytschko (1999). In these methods, the partition of unity method (PUM) is considered an essential concept where special enrichment functions are added to the standard approximation space.

Extended Finite Element Method: Theory and Applications, First Edition. Amir R. Khoei.

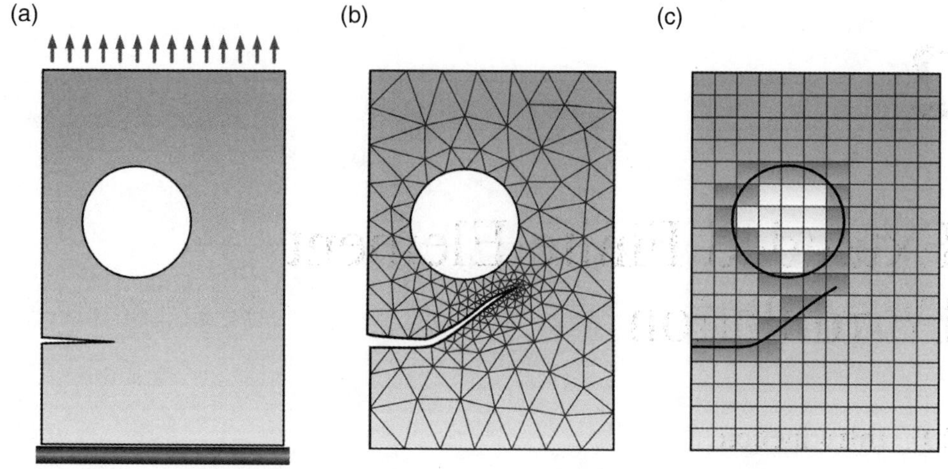

Figure 2.1 Modeling of weak and strong discontinuities in the standard-FEM and enriched-FEM techniques: (a) Crack propagation in a plate with a hole: (b) The standard-FEM using an adaptive mesh refinement in which the mesh conforms to the geometry of interfaces; (c) The enriched-FEM technique using a uniform mesh in which the elements cut by the interfaces are enriched

The enriched approximation field can be generally written as

$$\mathbf{u}(\boldsymbol{x}) = \sum_{i=1}^{\mathcal{N}} N_i(\boldsymbol{x})\,\bar{\mathbf{u}}_i + \text{enrichment terms} \tag{2.1}$$

where $\mathcal{N}$ is the set of all nodal points and $N_i(\boldsymbol{x})$ denotes the standard shape functions. This approximation is *extrinsically* enriched. Based on the PU concept, the arbitrary discontinuities may be treated on a fixed mesh, however, additional unknowns may increase the computational burden significantly. In contrast to the standard PUM, the X-FEM only uses a *local* extrinsic enrichment of the approximation space, and as a result the number of additional unknowns can be kept at a minimum. An alternative method that can be used for arbitrary discontinuities is based on the following approximation field as

$$\mathbf{u}(\boldsymbol{x}) = \sum_{i=1}^{\mathcal{N}} \hat{N}_i(\boldsymbol{x})\,\bar{\mathbf{u}}_i \tag{2.2}$$

in which no *extrinsic* enrichment terms and, consequently, no additional unknowns are involved. In this relation, the shape functions $\hat{N}_i(\boldsymbol{x})$ are set to the standard FE shape functions in those parts of the domain where a polynomial approximation space is locally adequate. However, in the vicinity of discontinuities, special shape functions are employed that are specifically designed so that they are able to capture discontinuities. Similarly, the method can be used to enrich the solution with known characteristics, such as the singular fields at the crack tip. The resulting approximation is *intrinsically* enriched, since the enrichment is implicitly incorporated in the basis (Belytschko *et al.*, 1996). Similar to the *extrinsic* enrichment of approximation space, *intrinsic* enrichment can be performed locally where it is needed, the same functions as *extrinsic* enrichment may be used for *intrinsic* enrichment, and do not show any inconsistencies resulting from the use of a local enrichment. The most important advantage of *intrinsic* enrichment is that it does not need additional unknowns, and the condition number of the final system matrix is in the same order over the element size as the standard FEM for arbitrary enrichments, while in *extrinsic* enrichment, the condition number may significantly increase with refinement, which depends on the particular enrichment (Laborde *et al.*, 2005). In order to construct the *intrinsic* enriched shape functions, the moving least-squares (MLS) method that is frequently used in the context of meshfree methods, can be used to combine a set of functions in the *intrinsic* basis (Lancaster and Salkauskas, 1981). The set of MLS functions can be obtained by minimizing a weighted least-squares error functional. The construction of suitable weight functions is crucial for the success of this procedure, and is an important ingredient of the proposed method. The same enrichment functions that are added *extrinsically* to the approximation space can be used in the *intrinsic* enrichment. In order to enable individual enrichments in different parts of the domain according to the locally present characteristics of the solution, the domain is decomposed into overlapping sub-domains. In each sub-domain, a set of shape functions is constructed based on the corresponding nodal subset by means of the MLS technique or by using the standard FE shape functions (Fries and Belytschko, 2006). In both cases, the shape functions build a PU over the sub-domain.

The *intrinsic* enrichment introduces enrichment terms as additional functions to the polynomial basis. However, the major drawback of this approach is higher computational cost as the enrichment needs to be employed everywhere to avoid artificial discontinuities. The techniques of blending "enriched" approximations to "standard" approximations have been used but they introduce additional complexity. Duflot and Hung (2004) proposed an alternative intrinsic enrichment by enriching the kernel functions in which it does not need the blending techniques. Although the technique was not computationally expensive, the accuracy of computation was poor. Such evidence widely draws attention to extrinsic enrichments. The extrinsic enrichment can be basically categorized into the extrinsic MLS enrichment and extrinsic PU enrichment. Extrinsic MLS enrichment introduces enrichment parameters that can be interpreted as stress intensity factors (SIFs) for the crack problem (Rabczuk and Wall, 2006). As a result, the SIFs can be obtained directly from the numerical analysis without computing the *J*–integral, or interaction integral. Despite the fact such properties are desirable, it was shown that this approach is less accurate. Hence, the extrinsic PU enrichment that can be compared to the standard X-FEM enrichment has been proposed as the best choice in terms of accuracy, efficiency, and flexibility.

In this chapter, the mathematical formulation of the X-FEM is presented with special reference to solid mechanics problems. It includes the introduction of PUM, enrichment functions, blending elements, the X-FEM discretization, and the numerical integration of X-FEM formulation. The numerical implementation of the method is presented for linear and higher-order quadrilateral elements in X-FEM modeling of linear and curved interfaces and the results are compared with those of classical FEM.

2.2 The Partition of Unity Finite Element Method

The PU is defined based on the definition of *clouds* (Duarte and Oden, 1996), which are overlapping open sets Ω_i of arbitrary shape centered in $\boldsymbol{x}_i$, covering the solution domain Ω^{PU} of a boundary value problem with $\Omega^{\mathrm{PU}} \subset \cup_{i=1}^{\mathcal{N}} \Omega_i$. A PU can be defined as a collection of global functions $f_i(\boldsymbol{x})$ whose support is contained in a cloud and whose value sum to unity at each point $\boldsymbol{x}$ in the solution domain as

$$\sum_{i=1}^{\mathcal{N}} f_i(\boldsymbol{x}) = 1 \qquad \forall \boldsymbol{x} \in \Omega^{\mathrm{PU}} \tag{2.3}$$

By choosing an arbitrary function $\psi(\boldsymbol{x})$ defined on Ω^{PU}, the following property can be observed as

$$\sum_{i=1}^{\mathcal{N}} f_i(\boldsymbol{x})\, \psi(\boldsymbol{x}_i) = \psi(\boldsymbol{x}) \tag{2.4}$$

which is equivalent to the definition of completeness. In the FE approach, the collection of shape functions is usually a PU. Based on the concept of the PU, the solution field $\mathbf{u}(\boldsymbol{x})$ can be discretized over the domain of the problem by taking $f_i(\boldsymbol{x}) \equiv N_i(\boldsymbol{x})$ as

$$\mathbf{u}(\boldsymbol{x}) = \sum_{i=1}^{\mathcal{N}} N_i(\boldsymbol{x})\, \bar{\mathbf{u}}_i \tag{2.5}$$

where $\mathcal{N}$ is the number of nodal points for each finite element. In the classical FEM, the standard shape functions $N_i(\boldsymbol{x})$ are used based on the polynomials. However, the interpolation field can be improved by adding expressions in accordance with the analytical solution. The concept of PU can be used to provide a mathematical framework for the development of an enriched solution. The enrichment is an act of improving the approximation of discretization based on the properties of the problem. In this manner, the approximation space employed to solve the problem is enhanced by incorporating the behavior of the undertaken phenomenon. The enriched solution can be basically obtained by increasing the order of completeness that results in higher accuracy of the approximation by including the information obtained from the analytical solution. The choice of enriched terms depends on the *a priori* solution of the problem; for example, in fracture mechanics analysis this is equivalent to an increase in accuracy of the approximation where analytical crack tip solutions are included in the enrichment terms. If the singular behavior of the solution is known, the enrichment can be defined by incorporating this knowledge directly into the finite element space. In contrast, the classical FEM has to use very small mesh sizes in order to resolve the singular behavior of the solution. There are various set of interpolation functions that can be used to approximate the solution of problem with higher accuracy compared to the spaces of polynomials. A few examples of interpolation functions that have been used with proper approximation properties for the solution of the Laplace equation, the Helmholtz equation, and the elasticity equation, were introduced by Melenk and Babuška (1996). For a certain type of equation, interpolation functions can be constructed using the structure of differential equation.

The incorporation of local enrichment into an approximation space was originally introduced by Melenk and Babuška (1996) through the PU FEM. The essential feature is based on the multiplication of enrichment functions by nodal shape functions. The enrichment is able to take on a local form by only enriching those nodes whose supports intersect a region of interest. The FE approximation of enriched domain can therefore be defined as

$$\mathbf{u}(\boldsymbol{x}) = \sum_{i=1}^{\mathcal{N}} N_i(\boldsymbol{x}) \left(\bar{\mathbf{u}}_i + \sum_{j=1}^{\mathcal{M}} p_j(\boldsymbol{x})\, \bar{\mathbf{a}}_{ij} \right) \tag{2.6}$$

or

$$\mathbf{u}(\boldsymbol{x}) = \underbrace{\sum_{i=1}^{\mathcal{N}} N_i(\boldsymbol{x})\, \bar{\mathbf{u}}_i}_{\text{standard interpolation}} + \underbrace{\sum_{i=1}^{\mathcal{N}} N_i(\boldsymbol{x}) \left(\sum_{j=1}^{\mathcal{M}} p_j(\boldsymbol{x})\, \bar{\mathbf{a}}_{ij} \right)}_{\text{enhanced interpolation}} \tag{2.7}$$

where $\bar{\mathbf{u}}_i$ are the standard nodal degrees of freedom related to the basis $N_i(\boldsymbol{x})$, and $\bar{\mathbf{a}}_{ij}$ are the enhanced degrees of freedom (DOF) related to the basis $p_j(\boldsymbol{x})$, with $\mathcal{M}$ denoting the number of enrichment functions for node i. The terms "standard" and "enhanced" refer to the fact that the "standard" interpolation field is considered as the background field upon which the "enhanced" interpolation field is superimposed. Since the FE shape functions form a partition of unity, the interpolation of the displacement field $\mathbf{u}(\boldsymbol{x})$ in Eq. (2.7) can be expressed as the combination of the standard FE interpolation field and an enhanced interpolation field, in which the enriched terms are used to improve the standard interpolation.

In the G-FEM, Eq. (2.7) is defined with different shape functions for the standard and enriched parts of the approximation as

$$\mathbf{u}(\boldsymbol{x}) = \sum_{i=1}^{\mathcal{N}} N_i(\boldsymbol{x})\, \bar{\mathbf{u}}_i + \sum_{i=1}^{\mathcal{N}} \bar{N}_i(\boldsymbol{x}) \left(\sum_{j=1}^{\mathcal{M}} p_j(\boldsymbol{x})\, \bar{\mathbf{a}}_{ij} \right) \tag{2.8}$$

where $\bar{N}_i(\boldsymbol{x})$ are the new set of shape functions associated with the enrichment part of the approximation.

2.3 The Enrichment of Approximation Space

The standard FEM is based on the approximation properties of polynomials. If the solution shows a pronounced non-polynomial behavior, such as weak or strong discontinuities, the standard FEM approximation may represent very poor performance. In fact, the standard FEM is not able to adequately represent the discontinuity or singularity in a suitable way, for example, at the crack tip area. A number of methods have been developed to overcome these difficulties; however, the enrichment of approximation space is one of the most efficient techniques that can be used to capture the weak or strong discontinuities. The enrichment can be attributed to the degree of consistency of the approximation, or to the capability of approximation to reproduce a given complex field of interest. The principal of enrichment is basically equivalent to the principal of increasing the order of completeness that can be achieved *intrinsically* or *extrinsically*. However, the enrichment aims to increase the accuracy of the approximation by including information of the analytical solution. There are basically two ways of enriching an approximation space; enriching the basic vector known as the "*intrinsic enrichment*" and enriching the approximation known as the "*extrinsic enrichment*".

2.3.1 *Intrinsic Enrichment*

Intrinsic enrichment is an idea to enhance the approximation space $\mathbf{u}(\boldsymbol{x})$ defined in (2.5) by including the new basis functions in order to capture a certain condition of a complex field, such as discontinuity or singularity. This equation can be written in a generalized form in terms of $\mathcal{M}$ basis function $\hat{N}_i(\boldsymbol{x}) = \left\{ \hat{N}_i^1, \hat{N}_i^2, \ldots, \hat{N}_i^{\mathcal{M}} \right\}$ as

$$\mathbf{u}(\boldsymbol{x}) = \sum_{i=1}^{\mathcal{P}} \hat{N}_i(\boldsymbol{x})\, \bar{\mathbf{a}}_i \equiv \hat{\mathbf{N}}^T(\boldsymbol{x})\, \bar{\mathbf{a}} \tag{2.9}$$

in which the previous definition can be considered a compact form of the PU FEM defined in (2.7), where the term corresponding to $\mathcal{M} = 1$ is extracted and referred to the standard polynomial interpolation function. In fact, the basis function $\hat{\mathbf{N}}(\boldsymbol{x})$ is expressed in the form of $\hat{\mathbf{N}}(\boldsymbol{x}) = \langle \mathbf{N}^{std}(\boldsymbol{x}), \mathbf{N}^{enr}(\boldsymbol{x}) \rangle$, in which $\mathbf{N}^{std}(\boldsymbol{x})$ refers to the standard polynomial functions $N_i(\boldsymbol{x})$, and $\mathbf{N}^{enr}(\boldsymbol{x})$ refers to the enriched shape functions obtained from the multiplication of enrichment functions by nodal shape functions as $N_i(\boldsymbol{x})\, p_j(\boldsymbol{x})$. In this relation, $\bar{\mathbf{a}}$ is a vector of coefficients obtained from one of the least-squares techniques. On the basis of a weighted least-squares discrete L_2 error norm, the coefficient vector $\bar{\mathbf{a}}$ can be evaluated by minimizing the following expression as

$$\Im(\bar{\mathbf{a}}) = \sum_{k=1}^{\mathcal{P}} w_k(\boldsymbol{x}) \left(\hat{\mathbf{N}}^T(\boldsymbol{x}_k)\, \bar{\mathbf{a}} - \bar{\mathbf{u}}_k \right)^2 \tag{2.10}$$

in which $\boldsymbol{x}_k$ refers to the position of kth node within the domain, $w_k(\boldsymbol{x})$ is the weight function, and $\mathcal{P}$ is the number of nodes around the corresponding node. Based on this minimization process, a relation can be performed between the unknowns coefficients $\bar{\mathbf{a}}$ and the nodal values $\bar{\mathbf{u}}_k$. A mesh-independent definition of weight functions $w_k(\boldsymbol{x})$ leads to the class of meshfree methods, where the MLS method is used to construct the meshfree shape functions (Fries and Belytschko, 2006). The minimization of previously mentioned L_2 error norm can be carried out by setting the derivative of function $\Im(\bar{\mathbf{a}})$ with respect to $\bar{\mathbf{a}}$ equal to zero, that is,

$$\frac{\partial}{\partial \bar{\mathbf{a}}}(\Im(\bar{\mathbf{a}})) = \sum_{k=1}^{\mathcal{P}} w_k(\boldsymbol{x}) \left[2\hat{\mathbf{N}}(\boldsymbol{x}_k) \left(\hat{\mathbf{N}}^T(\boldsymbol{x}_k)\, \bar{\mathbf{a}} - \bar{\mathbf{u}}_k \right) \right] = \mathbf{0} \tag{2.11}$$

Solving this equation for $\bar{\mathbf{a}}$ leads to

$$\bar{\mathbf{a}} = \left(\sum_{k=1}^{\mathcal{P}} w_k(\boldsymbol{x})\, \hat{\mathbf{N}}(\boldsymbol{x}_k)\, \hat{\mathbf{N}}^T(\boldsymbol{x}_k) \right)^{-1} \sum_{k=1}^{\mathcal{P}} w_k(\boldsymbol{x})\, \hat{\mathbf{N}}(\boldsymbol{x}_k)\, \bar{\mathbf{u}}_k \tag{2.12}$$

The approximation space $\mathbf{u}(\boldsymbol{x})$ can therefore be obtained by substituting Eq. (2.12) into (2.9) as

$$\mathbf{u}(\boldsymbol{x}) = \hat{\mathbf{N}}^T(\boldsymbol{x}) \left(\sum_{k=1}^{\mathcal{P}} w_k(\boldsymbol{x})\, \hat{\mathbf{N}}(\boldsymbol{x}_k)\, \hat{\mathbf{N}}^T(\boldsymbol{x}_k) \right)^{-1} \sum_{k=1}^{\mathcal{P}} w_k(\boldsymbol{x})\, \hat{\mathbf{N}}(\boldsymbol{x}_k)\, \bar{\mathbf{u}}_k \tag{2.13}$$

Based on these MLS functions in the approximation of $\mathbf{u}(\boldsymbol{x})$, the specific shape functions $\tilde{N}_i(\boldsymbol{x})$ can be defined from (2.13) in the form of $\mathbf{u}(\boldsymbol{x}) = \sum_{i=1}^{\mathcal{P}} \tilde{N}_i(\boldsymbol{x})\, \bar{\mathbf{u}}_i$ as

$$\tilde{N}_i(\boldsymbol{x}) = \hat{\mathbf{N}}^T(\boldsymbol{x}) \left(\sum_{k=1}^{\mathcal{P}} w_k(\boldsymbol{x})\, \hat{\mathbf{N}}(\boldsymbol{x}_k)\, \hat{\mathbf{N}}^T(\boldsymbol{x}_k) \right)^{-1} w_i(\boldsymbol{x})\, \hat{\mathbf{N}}(\boldsymbol{x}_i) \tag{2.14}$$

in which the set of $\mathcal{P}$ MLS functions $\tilde{N}_i(\boldsymbol{x})$ builds the PU over the domain. In fact, based on these MLS functions $\tilde{N}_i(\boldsymbol{x})$, all linear combinations of basis functions $\hat{\mathbf{N}}(\boldsymbol{x})$ can be reproduced exactly in the domain. In the absence of enrichment, the vector $\hat{\mathbf{N}}(\boldsymbol{x})$ consists of monomials only. The monomial basis may easily be enriched by other functions when a polynomial approximation cannot be expected to give satisfactory results. In fact, all enrichment functions used in the standard X-FEM may be used in the intrinsic basis vector $\hat{\mathbf{N}}(\boldsymbol{x})$, and these functions will be reproduced exactly by the resulting MLS shape functions.

2.3.2 *Extrinsic Enrichment*

The PUM, introduced in Section 2.2, is a concept for enriching the approximation *extrinsically* by adding the enrichment functions to the standard approximation, as defined by relation (2.7). In this definition, the set of functions $p_j(\boldsymbol{x})$ represents the extrinsic basis, in contrast to the intrinsic basis vector $\hat{N}_i(\boldsymbol{x})$ defined in preceding section for the implementation of MLS method in the construction of shape functions $\tilde{N}_i(\boldsymbol{x})$. It must be noted that each enrichment function in the standard PUM is required over the entire domain, and is therefore expected to increase the computational cost significantly. There are various techniques that can be applied as special versions of the partition of unity concept, including the PU-FEM, the G-FEM, and the X-FEM.

In contrast to the standard PUM, the X-FEM uses a *local* extrinsic enrichment of the approximation. Since the effects of discontinuities are generally local, the enrichment can be confined to specific zones instead of enriching the whole domain of the solution. This approach significantly improves the numerical solution by saving the significant amount of calculating time, memory storage, and other issues. The X-FEM enrichment enables elements to model the internal interfaces of both strong, and weak discontinuities, where the former refers mostly to discontinuities exerting to the main variables, for example, the displacement in crack surfaces, while the latter refers to discontinuities in gradients, for example, the displacement gradient, or strain, in the boundaries of material changes known as the *bimaterial problems*. Generally, the enhanced solution field in the X-FEM can be represented as

$$\mathbf{u}(\boldsymbol{x}) = \sum_{i=1}^{\mathcal{N}} N_i(\boldsymbol{x})\, \bar{\mathbf{u}}_i + \sum_{k=1}^{\mathcal{P}} \sum_{j=1}^{\mathcal{M}_k} \bar{N}_j(\boldsymbol{x})\, \psi_k(\boldsymbol{x})\, \bar{\mathbf{a}}_{kj} \tag{2.15}$$

or

$$\mathbf{u}(\boldsymbol{x}) = \sum_{i=1}^{\mathcal{N}} N_i(\boldsymbol{x})\, \bar{\mathbf{u}}_i + \sum_{j=1}^{\mathcal{M}_1} \bar{N}_j(\boldsymbol{x})\, \psi_1(\boldsymbol{x})\, \bar{\mathbf{a}}_{1j} + \sum_{j=1}^{\mathcal{M}_2} \bar{N}_j(\boldsymbol{x})\, \psi_2(\boldsymbol{x})\, \bar{\mathbf{a}}_{2j} + \cdots \tag{2.16}$$

where $\mathcal{M}_k \subseteq \mathcal{N}$ are the sets of nodal points, which are enriched by the functions $\psi_k(\boldsymbol{x})$, respectively. The enhanced terms are obtained by multiplication of the enriched functions $\psi_k(\boldsymbol{x})$ with the standard FE shape functions $\bar{N}_j(\boldsymbol{x})$, which do not necessarily have to be identical to the FE shape functions $N_i(\boldsymbol{x})$. It is noteworthy that, although it is most common that the set of shape functions $\bar{N}_j(\boldsymbol{x})$ utilized in the extended FE approximation are the same as standard ones, it is recommended to use the linear shape functions for higher order elements to provide continuous approximation over the domain (Stazi *et al.*, 2003). It must be noted that this procedure is not fully consistent with the standard PU concept, because $\sum_{j=1}^{\mathcal{M}_k} \bar{N}_j(\boldsymbol{x})$ is only a PU in elements for which *all* nodes are enriched. In fact, only in these elements, the enriched approximation is able to represent the functions $\psi_k(\boldsymbol{x})$ exactly. However, in elements for which only some of the nodes are enriched, called the *blending elements*, problems sometimes arise for general enrichment functions (Chessa, Wang, and Belytschko, 2003; Laborde *et al.*, 2005). Special treatments of these elements have been proposed in Chapter 4, but some of the versatility of X-FEM is lost. In fact, it must be noted that the *local* extrinsic enrichment of the X-FEM shows a systematical error in partially enriched elements.

2.4 The Basis of X-FEM Approximation

The X-FEM is one of the enriched PUM techniques that has been extensively employed in numerical modeling of discontinuous problems. The technique is a powerful and an accurate approach used to model weak and strong discontinuities without considering their geometries. In this method, the discontinuities are not considered in the mesh generation operation and special functions that depend on the nature of discontinuity are included into the finite element approximation. The aim of the technique is to simulate the weak and strong discontinuities with minimum enrichment. In X-FEM, the external boundaries are only considered in mesh generation operation and internal boundaries, such as cracks, voids, contact surfaces, and so on, have no effect on mesh configurations. This method has proper applications in problems with moving discontinuities, such as punching, phase changing, crack propagation, and shear banding.

In order to introduce the concept of discontinuous enrichment, consider that Γ_d is a discontinuity in domain Ω, as shown in Figure 2.2a. The aim is to construct a FE approximation to the field $\mathbf{u} \in \Omega$ that can be discontinuous along the discontinuity Γ_d. The traditional approach is to generate the mesh to conform to the line of discontinuity as shown in Figure 2.1, in which the element edges align with Γ_d. However, this strategy certainly creates a discontinuity in the approximation, it is cumbersome if the discontinuity Γ_d evolves in time, or if several different configurations for Γ_d are to be considered. In the X-FEM, the discontinuity along Γ_d is modeled with enrichment functions, in which the uniform mesh

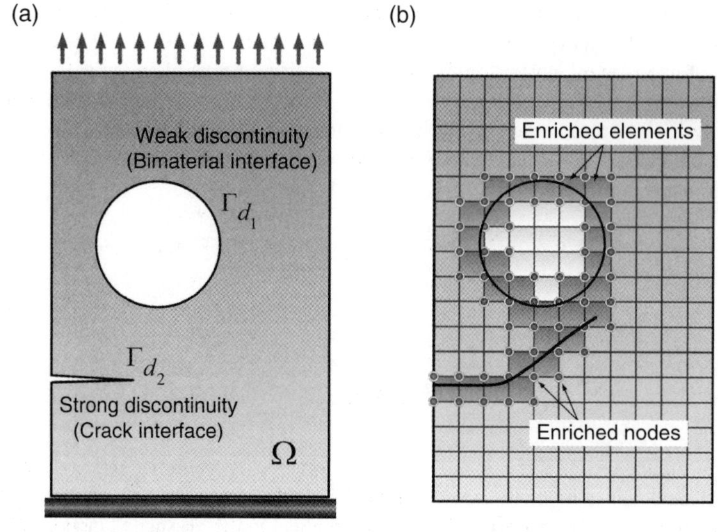

Figure 2.2 Modeling of weak and strong discontinuities in an enriched-FEM technique: (a) Definition of internal interfaces, including: the *bimaterial interface* and *crack interface*; (b) A uniform mesh in which the circled nodes have additional DOF and enrichment functions

of Figure 2.2b is capable of modeling the discontinuity in $\mathbf{u} \in \Omega$ when the circled nodes are enriched with functions, which are discontinuous across Γ_d.

Consider the enriched approximation field defined in (2.15), the first sum of the second term of the right hand side of Eq. (2.15) can be omitted (as well as the super-imposed k) if a single interface Γ_d is modeled. Hence, the enriched approximation for a single interface Γ_d can be written as

$$\mathbf{u}(\boldsymbol{x}) = \sum_{i=1}^{\mathcal{N}} N_i(\boldsymbol{x})\, \bar{\mathbf{u}}_i + \sum_{j=1}^{\mathcal{M}} N_j(\boldsymbol{x})\, \psi(\boldsymbol{x})\, \bar{\mathbf{a}}_j \tag{2.17}$$

in which the shape functions of enriched part $\bar{N}_j(\boldsymbol{x})$ are chosen similar to the FE shape functions $N_i(\boldsymbol{x})$. In this equation, $\bar{\mathbf{u}}_i$ is the standard nodal displacement, $\bar{\mathbf{a}}_j$ is the nodal DOF corresponding to the enrichment function, $\psi(\boldsymbol{x})$ is the enrichment function, and $N(\boldsymbol{x})$ is the standard shape function. In Eq. (2.17), $\mathcal{N}$ is the set of all nodal points of domain, and $\mathcal{M}$ is the set of nodes of elements located on the discontinuity Γ_d.

In the X-FEM, two approaches are simultaneously applied to the elements located on discontinuity; the PUM and the *enrichment of displacement field.* The PUM is used to enhance the approximation by adding the enrichment functions to the standard approximation. The enrichment of displacement field is applied to correct the standard displacement based approximation by incorporating discontinuous fields through a PUM. It must be noted that the enrichment varies from node to node and many nodes require no enrichment, which is an application of the PU concept. In X-FEM, the diversity in types of problems requires proposing appropriate enrichment functions. Different techniques may be used for the enrichment function; these functions are related to the type of discontinuity and its influences on the form of solution. These techniques are based on the signed distance function, level set function, branch function, Heaviside jump function, and so on. The choice of enrichment functions in displacement approximation is dependent on the conditions of problem. If the discontinuity is as a result of different types of material properties (Figure 2.3a), the level set function can be proposed as an enrichment function, however, if the discontinuity is due to different displacement fields on either sides of the discontinuity (Figure 2.3b), the Heaviside function is appropriate.

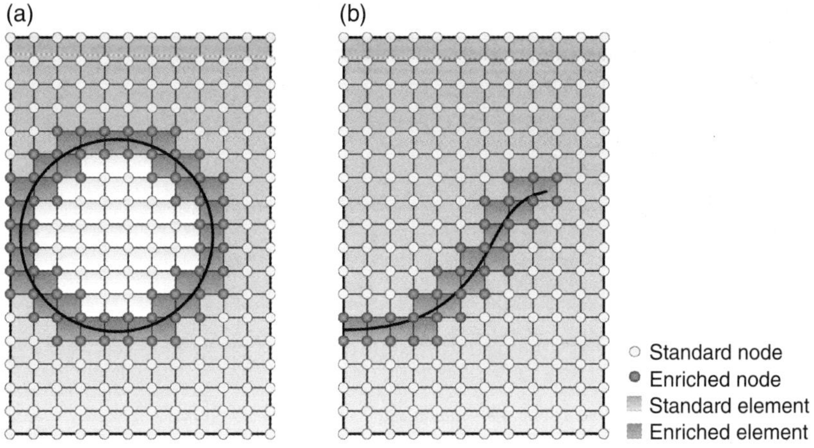

Figure 2.3 Modeling of discontinuity with the extended FEM: (a) The weak discontinuity as a *bimaterial interface*; (b) The strong discontinuity as a "*crack interface*"

The *Heaviside jump function* is applicable to a crack problem that is discontinuous across the crack line (Belytschko *et al.*, 2001). The function can be viewed as an enrichment with a windowed step function, where $N(\boldsymbol{x})$ is the window function. The window function localizes the enrichment so that the discrete equations will be sparse. For linear fracture mechanics, the crack tip singularity can be captured with the near tip asymptotic functions, or *branch functions* (Dolbow, Moës, and Belytschko, 2001). These functions span the near tip asymptotic solution for a crack tip, and provide very good accuracy for such problems. The *level set method* is a numerical scheme developed by Sethian (1996) for tracking the motion of interfaces. In this technique, the interface is represented as the zero level set of a function of one higher dimension. This method, which is used for predicting the geometry of boundaries, is distinguished as a good option for inhomogeneous fields and its application in enriching the domains that have discontinuity in the strain field is considerable. The technique of the *fast marching method*, which was first introduced by Sethian (1999), was coupled with the X-FEM to model crack growth (Chopp and Sukumar, 2003; Sukumar, Chopp, and Moran, 2003a). The method computes the crossing time map for a monotonically advancing front in an arbitrary number of spatial dimensions. In what follows, the implementation of the level set function in modeling the discontinuity due to different material properties and the Heaviside function in simulation of different displacement fields on either sides of the discontinuity are discussed, and an overview of different techniques used enrichment functions will follow in the next chapter.

2.4.1 The Signed Distance Function

The *level-set* method is a numerical technique for the tracking of moving interfaces, which is used for the description of interfaces in the domain (Sethian, 1999). In this method, the interface is represented as the zero level set of a function that is one dimension higher than the dimension of the interface and is evolved by solving the hyperbolic conservation laws. In the level set method, interfaces are modeled using *implicit* functions, allowing for a natural treatment of merging interfaces, intersection with boundaries, and so on. This is a particularly attractive feature since *no explicit* geometric treatment of the interfaces is necessary.

Consider a domain Ω divided into two domains Ω_A and Ω_B. The interface, or surface of discontinuity, between these two domains is denoted by Γ_d. Generally, the interface Γ_d can be distinguished between the open and closed interfaces, as shown in Figure 2.4. A closed interface goes fully through the domain Ω and is relevant to different materials in the domain. The open interfaces cut the domain only partially, that is, at least one end of the interface is inside the domain, such as in the case of the cracked domain. The most

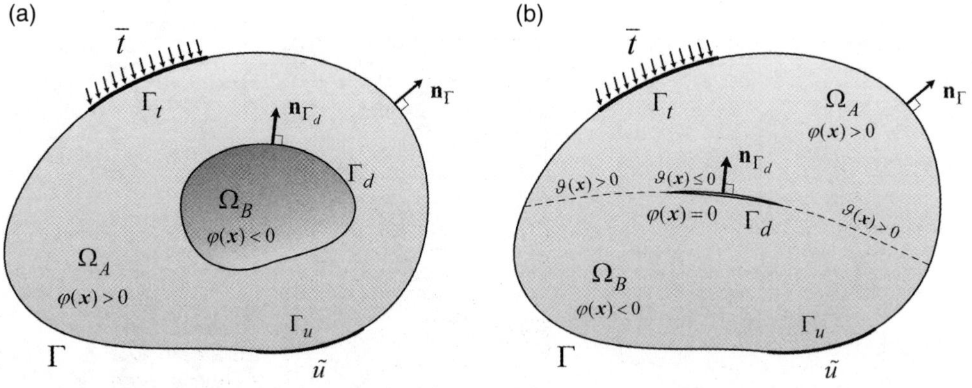

Figure 2.4 Problem with weak and strong discontinuities: (a) The weak discontinuity (closed discontinuity) as a *bimaterial interface*: (b) The strong discontinuity (open discontinuity) as a *crack interface* described by the level-set functions $\varphi(\boldsymbol{x})$ and $\vartheta(\boldsymbol{x})$

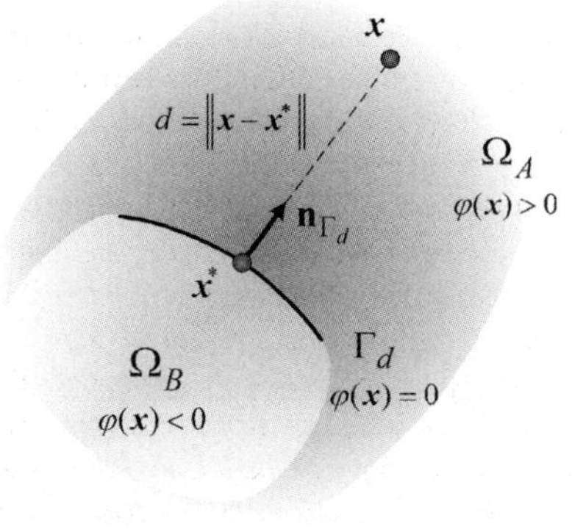

Figure 2.5 The signed distance function

common level set function is the signed distance function, which is defined for the representation of the interface position as

$$\varphi(\boldsymbol{x}) = \|\boldsymbol{x}-\boldsymbol{x}^*\| \operatorname{sign}(\mathbf{n}_{\Gamma_d}\cdot(\boldsymbol{x}-\boldsymbol{x}^*)) \tag{2.18}$$

where $\boldsymbol{x}^*$ is the closest point projection of $\boldsymbol{x}$ onto the discontinuity Γ_d, and $\mathbf{n}_{\Gamma_d}$ is the normal vector to the interface at point $\boldsymbol{x}^*$. In this definition, $\|\ \|$ denotes the Euclidean norm, where $\|\boldsymbol{x}-\boldsymbol{x}^*\|$ specifies the distance of point $\boldsymbol{x}$ to the discontinuity Γ_d (Figure 2.5).

It can be seen from the definition (2.18) that the sign is different on the two sides of a closed interface. Through this definition, the discontinuity can be represented implicitly as the zero iso-contour of the level set function as

$$\varphi(\boldsymbol{x}) \begin{cases} >0 & \text{if } \boldsymbol{x}\in\Omega_A \\ =0 & \text{if } \boldsymbol{x}\in\Gamma_d \\ <0 & \text{if } \boldsymbol{x}\in\Omega_B \end{cases} \tag{2.19}$$

A schematic view of a closed interface is presented in Figure 2.4a. In the case of a *weak discontinuity* across the interface, the level set function is useful as an enrichment function, for example, in the domain with different materials, where the displacement is continuous across the interface, but the derivatives (strains) are not. It can be shown that the norm of the gradient of the signed distance level set is equal to unity, that is, $\|\nabla\varphi\| = 1$. Obviously, it is clear that the gradient of the signed distance function at the discontinuity is indeed the unit normal $\mathbf{n}_{\Gamma_d}$ oriented to Ω_A, where $\varphi(\boldsymbol{x}) > 0$. That is, the $\nabla\varphi = \mathbf{n}_{\Gamma_d}$ equality holds for the signed distance function at the discontinuity. Note that this function could be the time dependent due to changes of the interface position during the solution that will be discussed in more detail in the next chapter.

In the case of an open discontinuity, for example, crack interfaces, it is necessary to construct another level set function such that $\vartheta(\boldsymbol{x}) \le 0$ on the cut part, and $\vartheta(\boldsymbol{x}) > 0$ on the uncut part, as shown in Figure 2.4b. In this case, the interface can be extended through the body in order to divide the domain into multiple zones. Note that this type of interface is very common, especially where the crack propagation problem is considered.

There is another choice for enrichment functions used extensively in the extended FE framework, called the *ramp* function. This function is introduced based on the absolute value of level set function that indeed has a discontinuous first derivative on the interface defined as

$$|\varphi(\boldsymbol{x})| = \begin{cases} -\varphi(\boldsymbol{x}) & \text{if } \varphi(\boldsymbol{x}) < 0 \\ +\varphi(\boldsymbol{x}) & \text{if } \varphi(\boldsymbol{x}) \ge 0 \end{cases} \tag{2.20}$$

The ramp function provides the elements with the capability of possessing independent derivatives over the edges of the interface. This type of discontinuity is referred to as a *weak discontinuity*. A schematic representation of the ramp enrichment function is presented in Figure 2.6.

On the basis of absolute value of level set function $|\varphi(\boldsymbol{x})|$, the approximation field (2.17) can be written as

$$\mathbf{u}(\boldsymbol{x}) = \sum_{i=1}^{\mathcal{N}} N_i(\boldsymbol{x})\,\bar{\mathbf{u}}_i + \sum_{j=1}^{\mathcal{M}} N_j(\boldsymbol{x})\,|\varphi(\boldsymbol{x})|\,\bar{\mathbf{a}}_j \tag{2.21}$$

It must be noted that the values of level-set function are only computed at the nodal points of FE mesh, that is, $\varphi_i = \varphi(\boldsymbol{x}_i)$, and the value of level set function $\varphi(\boldsymbol{x})$ can be approximated based on the values of nodal points using the interpolation functions $N_i(\boldsymbol{x})$ as $\varphi(\boldsymbol{x}) = \sum_{i=1}^{\mathcal{N}} N_i(\boldsymbol{x})\,\varphi_i$. Hence, the representation of the discontinuity as the zero-level of $\varphi(\boldsymbol{x})$ is only an approximation of the real interface position, which improves with mesh refinement.

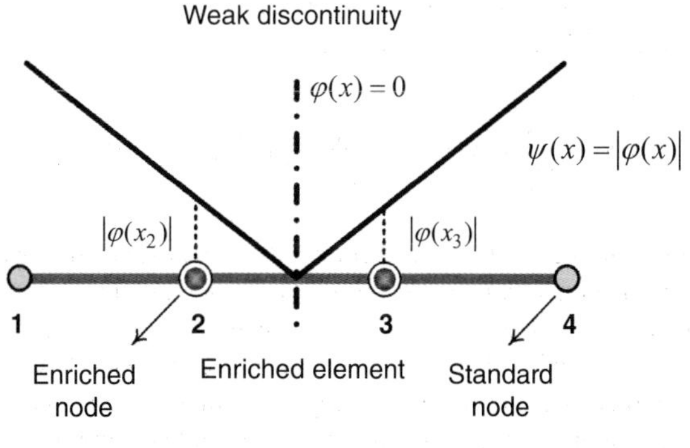

Figure 2.6 One-dimensional representation of the *ramp* enrichment function for a weak discontinuity

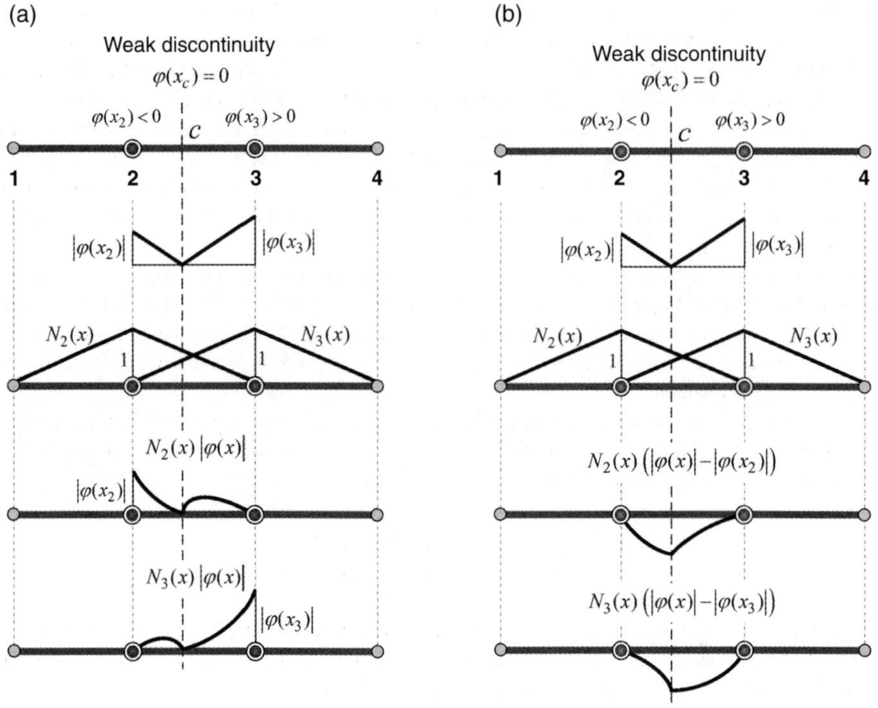

Figure 2.7 The principle of X-FEM for a weak discontinuity in an axial bar using: (a) the ramp enrichment function $N_j(x)\ |\varphi(x)|$; (b) the shifted ramp function $N_j(x)\ (|\varphi(x)| - |\varphi(x_j)|)$

In order to illustrate the effect of ramp enrichment function in the approximation field (2.21), consider an axial one-dimensional bar with three linear elements, as shown in Figure 2.7. The linear shape functions are employed for each element as $N_1(\xi) = (1 - \xi)$ and $N_2(\xi) = \xi$, where $\xi = x/\ell$ is the local natural coordinate. The middle element is assumed to have a weak discontinuity, such as bimaterial interface, at an arbitrary location x_c between nodes 2 and 3 with $\varphi(x_c) = 0$. In order to enrich the nodal points 2 and 3, the ramp enrichment function $|\varphi(\boldsymbol{x})|$ is employed according to (2.20), in which $\varphi(x_2) < 0$ at the left side of the interface and $\varphi(x_3) > 0$ at the right side of the interface, as shown in Figure 2.7a. In this figure, the enriched ramp shape functions, that is, $N_j(x)\ |\varphi(x)|$, are represented for nodal points 2 and 3. Obviously, the value of displacement $u(x)$ at an enriched node k can be obtained from (2.21) as $u(x_k) = \bar{u}_k + |\varphi(x_k)|\,\bar{a}_k$. Since $|\varphi(x_k)|$ is not necessarily zero, this expression is not equal to the real nodal value $\bar{u}_k$. Thus, the enriched displacement field (2.21) can be corrected as

$$\mathbf{u}(\boldsymbol{x}) = \sum_{i=1}^{\mathcal{N}} N_i(\boldsymbol{x})\,\bar{\mathbf{u}}_i + \sum_{j=1}^{\mathcal{M}} N_j(\boldsymbol{x})\left(|\varphi(\boldsymbol{x})| - |\varphi(\boldsymbol{x}_j)|\right)\bar{\mathbf{a}}_j \tag{2.22}$$

Based on the new definition in the last term of relation (2.22), the expected property can be obtained as $u(x_k) = \bar{u}_k$. In Figure 2.7b, the enriched fields of shifted ramp function, that is, $N_j(x)(|\varphi(x)| - |\varphi(x_j)|)$, are plotted for nodal points 2 and 3. Evidently, the continuity is hold in the displacement field but not in its derivative. Clearly, a kink is introduced that causes the jump in the gradient of the enriched fields of shifted ramp function. It is noteworthy to mention that the ramp enrichment function is commonly used in bimaterial problems. In such problems, although the displacement field is continuous along the interface, the strain field is discontinuous.

The gradient of enriched displacement field can be obtained by taking the derivation from Eq. (2.22) as

$$\nabla \mathbf{u}(\boldsymbol{x})=\sum_{i=1}^{\mathcal{N}} \nabla N_i(\boldsymbol{x})\,\bar{\mathbf{u}}_i+\sum_{j=1}^{\mathcal{M}}\left[\nabla N_j(\boldsymbol{x})\left(|\varphi(\boldsymbol{x})|-|\varphi(\boldsymbol{x}_j)|\right)+N_j(\boldsymbol{x})\,\nabla\left(|\varphi(\boldsymbol{x})|-|\varphi(\boldsymbol{x}_j)|\right)\right]\bar{\mathbf{a}}_j \tag{2.23}$$

with

$$\nabla\left(|\varphi(\boldsymbol{x})|-|\varphi(\boldsymbol{x}_j)|\right)=\operatorname{sign}(\varphi(\boldsymbol{x}))\,\nabla\varphi(\boldsymbol{x})=\operatorname{sign}(\varphi(\boldsymbol{x}))\,\mathbf{n}_{\Gamma_d} \tag{2.24}$$

which causes the jump in the gradient of the enriched field. The jump in the displacement gradient can be obtained as

$$\begin{aligned}
[\![\nabla \mathbf{u}(\boldsymbol{x})]\!] &= \nabla \mathbf{u}(\boldsymbol{x}^+)-\nabla \mathbf{u}(\boldsymbol{x}^-)\\
&=\sum_{i=1}^{\mathcal{N}} \nabla N_i(\boldsymbol{x}^+)\,\bar{\mathbf{u}}_i+\sum_{j=1}^{\mathcal{M}}\left[\nabla N_j(\boldsymbol{x}^+)\left(|\varphi(\boldsymbol{x}^+)|-|\varphi(\boldsymbol{x}_j)|\right)+N_j(\boldsymbol{x}^+)\operatorname{sign}(\varphi(\boldsymbol{x}^+))\,\mathbf{n}_{\Gamma_d}\right]\bar{\mathbf{a}}_j\\
&\quad-\sum_{i=1}^{\mathcal{N}} \nabla N_i(\boldsymbol{x}^-)\,\bar{\mathbf{u}}_i-\sum_{j=1}^{\mathcal{M}}\left[\nabla N_j(\boldsymbol{x}^-)\left(|\varphi(\boldsymbol{x}^-)|-|\varphi(\boldsymbol{x}_j)|\right)+N_j(\boldsymbol{x}^-)\operatorname{sign}(\varphi(\boldsymbol{x}^-))\,\mathbf{n}_{\Gamma_d}\right]\bar{\mathbf{a}}_j\\
&=\sum_{j=1}^{\mathcal{M}} N_j(\boldsymbol{x})\left(\operatorname{sign}(\varphi(\boldsymbol{x}^+))-\operatorname{sign}(\varphi(\boldsymbol{x}^-))\right)\mathbf{n}_{\Gamma_d}\,\bar{\mathbf{a}}_j\\
&=2\sum_{j=1}^{\mathcal{M}} N_j(\boldsymbol{x})\,\mathbf{n}_{\Gamma_d}\,\bar{\mathbf{a}}_j
\end{aligned} \tag{2.25}$$

or the jump normal to the interface can be computed as

$$[\![\nabla \mathbf{u}(\boldsymbol{x})\,\mathbf{n}_{\Gamma_d}]\!]=2\sum_{j=1}^{\mathcal{M}} N_j(\boldsymbol{x})\,\bar{\mathbf{a}}_j \tag{2.26}$$

In these relations, the continuity of the shape functions is used across the interface, that is, $N_j(\boldsymbol{x}^+)=N_j(\boldsymbol{x}^-)$.

2.4.2 *The Heaviside Function*

A jump in the displacement field is referred to as the strong discontinuity, which can be typically observed in a crack problem. The discontinuity in the displacement occurs where the displacement of one side of the crack is completely different from the displacement field of the opposite side of the crack. In such cases, the kinematics of the strong discontinuity can be defined based on the Heaviside function. The Heaviside function is one of the most popular functions used to model the crack discontinuity in the extended FE formulation. There are two ways proposed to define the Heaviside function, one as

$$H(\boldsymbol{x})=\begin{cases}0 & \text{if } \varphi(\boldsymbol{x})<0\\ 1 & \text{if } \varphi(\boldsymbol{x})>0\end{cases} \tag{2.27}$$

and the other

$$H(\boldsymbol{x})=\begin{cases}-1 & \text{if } \varphi(\boldsymbol{x})<0\\ +1 & \text{if } \varphi(\boldsymbol{x})>0\end{cases} \tag{2.28}$$

in which $\varphi(\boldsymbol{x})$ is the signed distance function, defined in (2.18). The definition (2.27) is known primarily as the common Heaviside *step* function, which is referred to be the original Heaviside function, while the definition (2.28) is generally referred to as the Heaviside *sign* function in mathematical formulation. Obviously, the Heaviside enrichment function is discontinuous at the interface. It must be noted that it may be necessary to use smoothed functions to avoid numerical problems such as instabilities in the numerical solution. The most common smoothed Heaviside functions are defined as

$$H(\boldsymbol{x})=\begin{cases} 0 & \text{if } -\varepsilon \geq \varphi(\boldsymbol{x}) \\ \dfrac{1}{2}+\dfrac{\varphi}{2\varepsilon}+\dfrac{1}{2\pi}\sin\dfrac{\pi\varphi}{\varepsilon} & \text{if } -\varepsilon<\varphi(\boldsymbol{x})<\varepsilon \\ 1 & \text{if } \quad \varphi(\boldsymbol{x}) \geq \varepsilon \end{cases} \tag{2.29}$$

and

$$H(\boldsymbol{x})=\begin{cases} 0 & \text{if } -\varepsilon \geq \varphi(\boldsymbol{x}) \\ \dfrac{1}{2}+\dfrac{3}{4}\left(\dfrac{\varphi}{\varepsilon}-\dfrac{1}{3}\left(\dfrac{\varphi}{\varepsilon}\right)^{3}\right) & \text{if } -\varepsilon<\varphi(\boldsymbol{x})<\varepsilon \\ 1 & \text{if } \quad \varphi(\boldsymbol{x}) \geq \varepsilon \end{cases} \tag{2.30}$$

where ε is a small value with respect to element size. A schematic representation of the smoothed Heaviside function (2.29), which is similar to (2.30), is presented in Figure 2.8. In numerical computation, the derivative of Heaviside function can be used extensively, which is known as the Dirac delta function. According to relations (2.29) and (2.30), the smoothed Dirac delta functions can be respectively defined as

$$\delta(\boldsymbol{x})=\begin{cases} \dfrac{1}{2\varepsilon}\left(1+\cos\dfrac{\pi\varphi}{\varepsilon}\right) & -\varepsilon<\varphi(\boldsymbol{x})<\varepsilon \\ 0 & \text{elsewhere} \end{cases} \tag{2.31}$$

and

$$\delta(\boldsymbol{x})=\begin{cases} \dfrac{3}{4\varepsilon}\left(1-\left(\dfrac{\varphi}{\varepsilon}\right)^{2}\right) & -\varepsilon<\varphi(\boldsymbol{x})<\varepsilon \\ 0 & \text{elsewhere} \end{cases} \tag{2.32}$$

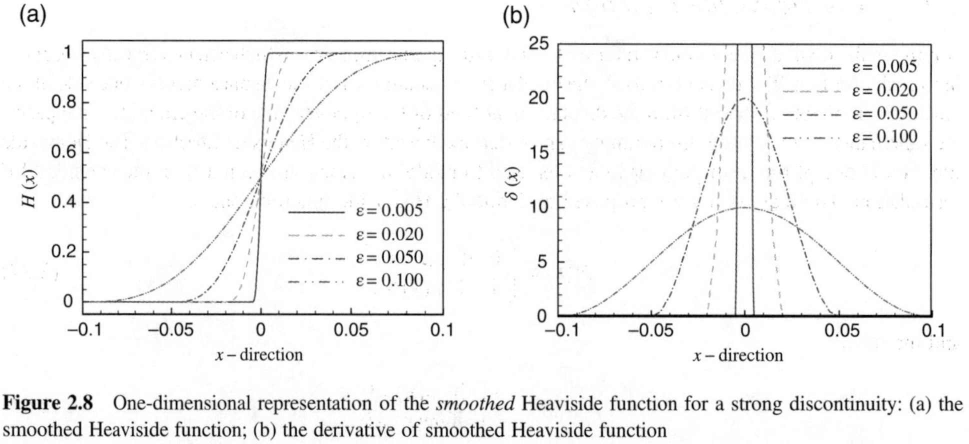

Figure 2.8 One-dimensional representation of the *smoothed* Heaviside function for a strong discontinuity: (a) the smoothed Heaviside function; (b) the derivative of smoothed Heaviside function

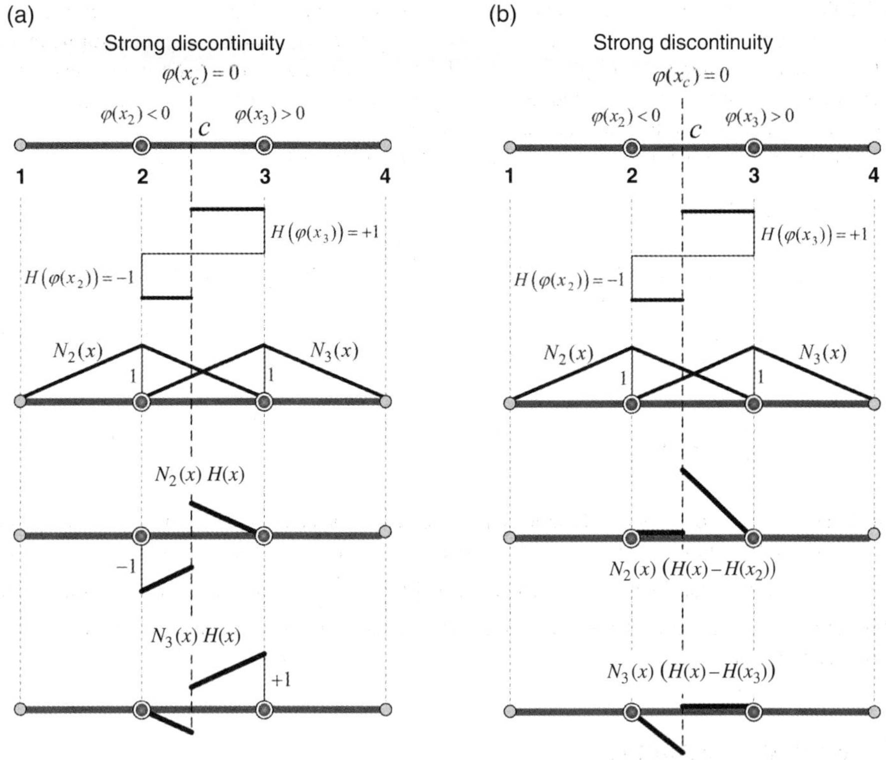

Figure 2.9 The principle of X-FEM with a strong discontinuity for an axial bar using: (a) the Heaviside function $N_j(x)H(x)$; (b) the shifted Heaviside function $N_j(x)(H(x)-H(x_j))$

On the basis of Heaviside enrichment function $H(\boldsymbol{x})$, the approximation field (2.17) can be written as

$$\mathbf{u}(\boldsymbol{x})=\sum_{i=1}^{\mathcal{N}} N_i(\boldsymbol{x})\,\bar{\mathbf{u}}_i+\sum_{j=1}^{\mathcal{M}} N_j(\boldsymbol{x})\,H(\boldsymbol{x})\,\bar{\mathbf{a}}_j \tag{2.33}$$

In order to illustrate the effect of Heaviside enrichment function in the approximation field (2.33), consider once again the axial one-dimensional bar with three linear elements, as shown in Figure 2.9. The linear shape functions are employed for each element as $N_1(\xi)=(1-\xi)$ and $N_2(\xi)=\xi$. The middle element is assumed to have a strong discontinuity, such as a crack, at an arbitrary location x_c between nodes 2 and 3 with $\varphi(x_c)=0$. In order to enrich the nodal points 2 and 3, the Heaviside sign function $H(x)$ is employed according to (2.28), in which $H(\varphi(x_2))=-1$ for the left side of the crack and $H(\varphi(x_3))=+1$ for the right side of the crack, as shown in Figure 2.9a. In this figure, the enriched Heaviside shape functions, that is, $N_j(x)\,H(x)$, are represented for nodal points 2 and 3. Obviously, the value of displacement $u(x)$ at an enriched node k can be obtained from (2.33) as $u(x_k)=\bar{u}_k+H(\varphi(x_k))\,\bar{a}_k$. Since $H(\varphi(x_k))$ is not necessarily zero, this expression is not equal to the real nodal value $\bar{u}_k$. Thus, the enriched displacement field (2.33) can be corrected to

$$\mathbf{u}(\boldsymbol{x})=\sum_{i=1}^{\mathcal{N}} N_i(\boldsymbol{x})\,\bar{\mathbf{u}}_i+\sum_{j=1}^{\mathcal{M}} N_j(\boldsymbol{x})\left(H(\boldsymbol{x})-H(\boldsymbol{x}_j)\right)\bar{\mathbf{a}}_j \tag{2.34}$$

in which the last term of relation (2.34) results in the expected property, that is, $u(x_k) = \bar{u}_k$. In Figure 2.9b, the enriched fields of shifted Heaviside function, that is, $N_j(x)\,(H(x) - H(x_j))$, are plotted for nodal points 2 and 3. Obviously, a jump in the displacement field takes place using the Heaviside enrichment function. In fact, the Heaviside enrichment function clearly presents the independence displacement fields from one side of the interface to the other side of the interface.

The jump in the displacement field can be obtained at the discontinuity interface as

$$\begin{aligned}
[\![\mathbf{u}(\boldsymbol{x})]\!] &= \mathbf{u}(\boldsymbol{x}^+) - \mathbf{u}(\boldsymbol{x}^-) \\
&= \sum_{i=1}^{\mathcal{N}} N_i(\boldsymbol{x}^+)\,\bar{\mathbf{u}}_i + \sum_{j=1}^{\mathcal{M}} N_j(\boldsymbol{x}^+)\,H(\boldsymbol{x}^+)\,\bar{\mathbf{a}}_j - \sum_{i=1}^{\mathcal{N}} N_i(\boldsymbol{x}^-)\,\bar{\mathbf{u}}_i - \sum_{j=1}^{\mathcal{M}} N_j(\boldsymbol{x}^-)\,H(\boldsymbol{x}^-)\,\bar{\mathbf{a}}_j \\
&= \sum_{j=1}^{\mathcal{M}} N_j(\boldsymbol{x})\,(H(\boldsymbol{x}^+) - H(\boldsymbol{x}^-))\,\bar{\mathbf{a}}_j \\
&= 2\sum_{j=1}^{\mathcal{M}} N_j(\boldsymbol{x})\,\bar{\mathbf{a}}_j
\end{aligned} \tag{2.35}$$

in which the continuity of the shape functions are used across the crack interface, that is, $N_j(\boldsymbol{x}^+) = N_j(\boldsymbol{x}^-)$. It is noteworthy to highlight that the implementation of the Heaviside step function (2.27) results in a half value of the previous displacement jump, that is, $[\![\mathbf{u}(\boldsymbol{x})]\!] = \sum_{j=1}^{\mathcal{M}} N_j(\boldsymbol{x})\,\bar{\mathbf{a}}_j$; however, it must be noted that the choice of the jump in the enrichment function is not significant since the nodal values will be adjusted automatically.

2.5 Blending Elements

In the X-FEM, the elements are categorized into two groups of enriched and standard elements; while in the former all nodal points are enriched with DOF, in the latter no extra DOF exist. Evidently, there are elements inter-connecting the standard and enriched elements, where both type of nodes with extra DOF and nodes without any coexist. These elements are generally referred to as the "blending", or "partially enriched" elements. In fact, the transition elements between the enriched and standard elements are called the blending elements, as shown in Figure 2.10. An appropriate construction of the elements in the

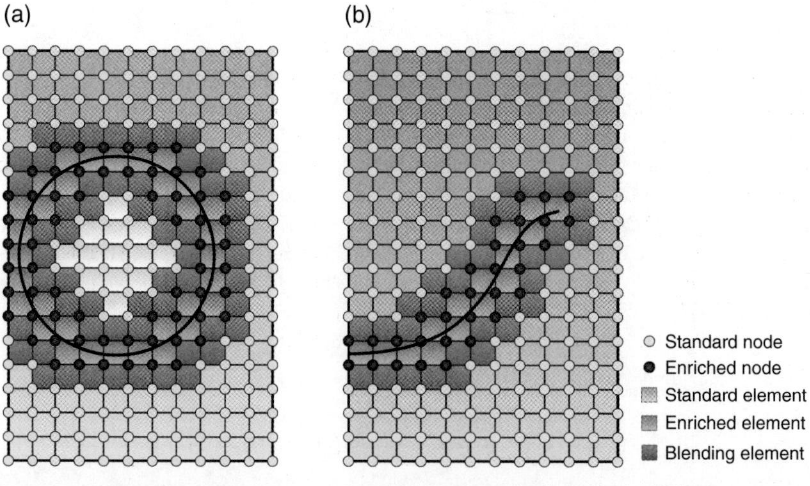

Figure 2.10 Blending elements in the X-FEM: (a) A weak discontinuity; (b) A strong discontinuity

blending area, the region where the enriched elements blend to unenriched elements, is often crucial for good performance of the local PU enrichments. The approximation field in a blending element can be written using Eq. (2.17) as

$$\mathbf{u}(\boldsymbol{x}) = \sum_{i=1}^{\mathcal{N}} N_i(\boldsymbol{x})\,\bar{\mathbf{u}}_i + \sum_{j=1}^{\mathcal{M}^{\mathcal{B}}} N_j(\boldsymbol{x})\,\psi(\boldsymbol{x})\,\bar{\mathbf{a}}_j \tag{2.36}$$

where $\mathcal{M}^{\mathcal{B}}$ is the set of nodes enriched in the blending element $\left(\mathcal{M}^{\mathcal{B}} < \mathcal{M}\right)$. In this element, the PU property of the enriched DOF is violated since all nodal points are not enriched, that is,

$$\sum_{j=1}^{\mathcal{M}^{\mathcal{B}}} N_j(\boldsymbol{x}) \neq 1 \tag{2.37}$$

and the function $\psi(\boldsymbol{x})$ cannot be recovered, that is,

$$\sum_{j=1}^{\mathcal{M}^{\mathcal{B}}} N_j(\boldsymbol{x})\,\psi(\boldsymbol{x}_j) \neq \psi(\boldsymbol{x}) \tag{2.38}$$

In fact, the definition of completeness cannot be satisfied in blending elements, which results in a kind of non-smoothness observed in the approximation field. On the other hand, blending elements introduce spurious terms in the approximation field, which produce an error in the solution of the problem. The spurious terms can be automatically corrected by the standard part of the approximation field if the order of the standard shape function $N_i(\boldsymbol{x})$ is equal or more than the order of the enriched shape function $\bar{N}_j(\boldsymbol{x})$ multiplied by the enrichment function $\psi(\boldsymbol{x})$, that is, $\bar{N}_j(\boldsymbol{x})\,\psi(\boldsymbol{x})$. This problem could be more noticeable for some special enrichment functions such as the ramp and crack tip asymptotic functions, while the values of the Heaviside enrichment function can vanish in blending elements, as presented in Table 2.1.

In order to illustrate how the blending elements lead to a lower convergence rate of the X-FEM, consider a one-dimensional axial problem with a weak discontinuity such as a bimaterial interface at its center, as shown in Figure 2.11. The X-FEM analysis is carried out using three 2-noded elements, in which the material interface is placed at the mid-point of second element. In order to model the material interface in the bar, the ramp function is used considering the enriched DOF at nodal points 2 and 3. The second element is known as an enriched element, and the two-side elements as blending elements. This example presents how a locally enriched FEM fails to reproduce a linear field when the enrichment is non-zero (Chessa, Wang, and Belytschko, 2003). The desired piecewise linear field together with the standard part $u_{(x)}^{std} = \sum_{i=1}^{\mathcal{N}} N_i(x)\,\bar{u}_i$, the enriched part $u_{(x)}^{enr} = \sum_{j=1}^{\mathcal{M}} N_j(x)\,\left(|\varphi(x)| - |\varphi(x_j)|\right)\bar{a}_j$, and the total approximation field $u_{(x)}^{total} = u_{(x)}^{std} + u_{(x)}^{enr}$ are presented in Figure 2.11. The displacement approximation in the blending element of left hand side can be written as

$$u^h(x) = \sum_{I=1}^{2} N_I \bar{u}_I + N_2\left(|\varphi(x)| - |\varphi(x_2)|\right)\bar{a}_2 \tag{2.39}$$

Table 2.1 A combination of different polynomials for the standard and enriched shape functions

The standard shape function $N_i(\boldsymbol{x})$	The enriched shape function $\bar{N}_j(\boldsymbol{x})$	The enrichment function $\psi(\boldsymbol{x})$	The order of $\bar{N}_j(\boldsymbol{x})\,\psi(\boldsymbol{x})$	Spurious terms
4-noded element (linear)	4-noded element (linear)	Heaviside	O(1) – linear	No
4-noded element (linear)	4-noded element (linear)	Ramp	O(2) – quadratic	Yes
9-noded element (quadratic)	4-noded element (linear)	Heaviside	O(1) – linear	No
9-noded element (quadratic)	4-noded element (linear)	Ramp	O(2) – quadratic	No

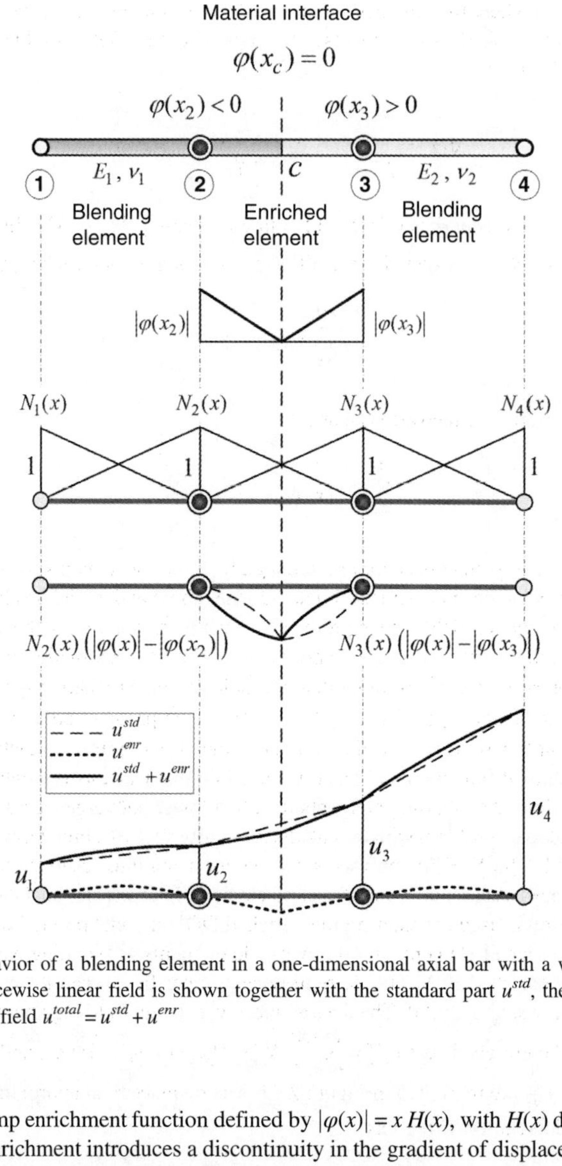

Figure 2.11 The behavior of a blending element in a one-dimensional axial bar with a weak discontinuity at its center: The desired piecewise linear field is shown together with the standard part u^{std}, the enriched part u^{enr}, and the total approximation field $u^{total}=u^{std}+u^{enr}$

where $|\varphi(x)|$ is the ramp enrichment function defined by $|\varphi(x)|=x\,H(x)$, with $H(x)$ denoting the Heaviside sign function. This enrichment introduces a discontinuity in the gradient of displacement at x_c. In relation (2.39), a linear shape function is used for both the standard approximation and the partition of unity. If ℓ is assumed as the length of the blending element, relation (2.39) can be rewritten as

$$u^h(x)=(1-\xi)\,\bar{u}_1+(\xi)\,\bar{u}_2+(\xi)\ell(1-\xi)\,\bar{a}_2 \tag{2.40}$$

where $\xi=(x-x_1)/\ell$. It can be seen from this relation that the enrichment part produces a quadratic term in the blending element. It means that if the solution consists of a piecewise linear field on each side of the discontinuity, the approximation in the blending element always produces an unwanted quadratic term.

In order to obtain the error of interpolation in the displacement field of blending element, consider that the solution of FE interpolation is denoted by $u(x)$, so the error of interpolation can be obtained as $e_u(x) \equiv u - u^h$, in which the maximum error at a point $\tilde{x}$, where $x_1 \le \tilde{x} \le x_2$, is defined by $\frac{d}{dx} e_u(x)|_{x=\tilde{x}} = 0$. Applying the Taylor series expansion about $\tilde{x}$, the error $e_u(x)$ can be written as

$$e_u(x) = e_u(\tilde{x}) + \frac{d}{dx} e_u(x)\bigg|_{x=\tilde{x}} (x-\tilde{x}) + \frac{1}{2} \frac{d^2}{dx^2} e_u(x)\bigg|_{x=\tilde{x}} (x-\tilde{x})^2 + O(h^3) \tag{2.41}$$

By neglecting the error of $O(h^3)$ and removing the second term to obtain the maximum error, it results in

$$e_u(x) = e_u(\tilde{x}) + \frac{1}{2} \frac{d^2}{dx^2} e_u(x)\bigg|_{x=\tilde{x}} (x-\tilde{x})^2 \tag{2.42}$$

Since the solution of interpolation $u^h(x)$ at node I is defined as $u^h(x_I) = \bar{u}_I$, the error of interpolation at point x_2 is equal to zero, that is, $e_u(x_2) = 0$. As a result, the maximum error of interpolation in the blending element can be obtained from (2.42) by setting $x = x_2$ as

$$e_u(\tilde{x}) \le \left| \frac{1}{2} \frac{d^2}{dx^2} e_u(x)\bigg|_{x=\tilde{x}} (x_2-\tilde{x})^2 \right| \tag{2.43}$$

It can be seen from (2.40) that the second derivative of the error is $\frac{d^2}{dx^2} e_u(x) = \frac{d^2 u}{dx^2} + \frac{2\bar{a}_2}{\ell}$, and from the problem $\frac{1}{2}(x_2-\tilde{x})^2 \le \frac{1}{8}\ell^2$. Hence, relation (2.43) results in

$$e_u(\tilde{x}) \le \frac{1}{8} \ell^2 \left| \frac{d^2 u}{dx^2} + \frac{2\bar{a}_2}{\ell} \right| \tag{2.44}$$

In fact, the last term of this relation, that is, $2\bar{a}_2/\ell$, appeared because of the enriched part of the displacement approximation and increases the error of interpolation in the blending element from $O(h^2)$ to $O(h)$. Although this error occurs in only a few elements, the effect is to reduce the rate of convergence of the entire solution. In the case that an enrichment of a higher order polynomial is used, the interpolation error of the blending element can be increased even further.

In order to overcome these difficulties, a number of research studies have been carried out on the issue of blending elements. One of the primary studies was performed by Chessa, Wang, and Belytschko (2003), in which the hierarchical shape functions were added to the standard FE part of the approximation field of blending elements to compensate for the parasitic terms. An enhanced strain technique was proposed by Chessa, Wang, and Belytschko (2003) and Gracie, Wang, and Belytschko (2008b) to remove the parasitic terms in the local PU finite elements, which leads to a good performance of local PU enrichments, and results in an optimal rate of convergence. A modified X-FEM technique was employed by Fries (2008), in which the enrichment functions have zero values in the standard elements, unchanged in the elements with all their nodes being enriched, and vary continuously in the blending elements. A discontinuous Galerkin (DG) method was developed by Gracie, Wang, and Belytschko (2008b), which decomposes the domain of solution into the enriched and unenriched sub-domains, where the continuity between patches is enforced with an internal penalty parameter. A comprehensive overview of the techniques proposed to circumvent the errors caused by parasitic terms in the approximation space of the blending elements, is given in Chapter 4.

2.6 Governing Equation of a Body with Discontinuity

In order to numerically model a continuum body with a weak or strong discontinuity, the X-FEM is employed by adding appropriate enrichment functions to the standard FE approximation. It must be noted that in the derivation of X-FEM formulation, it is implicitly assumed that the displacement field is continuous over the domain of problem; however, necessary modifications must be applied in the variational formulation of X-FEM solution to model the discontinuous displacement field. In what follows, the *Divergence theorem* is first presented for discontinuous problems to provide a mathematical description for the variational formulation of the X-FEM solution. An appropriate term is then obtained to incorporate the discontinuity into the X-FEM formulation. The discrete system of X-FEM equation is finally obtained by substituting the trial and test functions into the weak form of equilibrium equation.

2.6.1 The Divergence Theorem for Discontinuous Problems

The *Divergence theorem*, which is also known as the *Gauss–Green theorem*, is extensively applied in continuum mechanics problems to provide a mathematical description for the variational formulation of the FEM. Based on the *Divergence theorem*, the integration over the domain Ω can be converted to the integration over the boundary Γ for a continuous function $\mathbf{F}$ as

$$\int_{\Omega} \operatorname{div} \mathbf{F} \, d\Omega = \int_{\Gamma} \mathbf{F} \cdot \mathbf{n}_{\Gamma} \, d\Gamma \tag{2.45}$$

where $\mathbf{n}_{\Gamma}$ is the outward unit normal vector to the domain Ω.

In order to modify the *Divergence theorem* for discontinuous continuum problems, an appropriate term is derived to incorporate the discontinuity into the extended finite element formulation. Consider a two-dimensional domain Ω that is divided into two distinct parts Ω^+ and Ω^- through an open discontinuity interface Γ_d, as shown in Figure 2.12. The external boundary of domain Ω^+ is denoted by Γ^+ with the outward unit normal vector $\mathbf{n}_{\Gamma^+}$, and the external boundary of domain Ω^- is denoted by Γ^- with the outward unit normal vector $\mathbf{n}_{\Gamma^-}$. Moreover, the discontinuity interface Γ_d is assumed with the unit normal vector $\mathbf{n}_{\Gamma_d}$ oriented to Ω^+, which consists of the cut part Γ_{d_1} with the unit normal vector $\mathbf{n}_{\Gamma_{d_1}}$ and the uncut part Γ_{d_2} with the unit normal vector $\mathbf{n}_{\Gamma_{d_2}}$ oriented to Ω^+ (or $\bar{\mathbf{n}}_{\Gamma_{d_1}} = -\mathbf{n}_{\Gamma_{d_1}}$ and $\bar{\mathbf{n}}_{\Gamma_{d_2}} = -\mathbf{n}_{\Gamma_{d_2}}$ oriented to Ω^-). The domain integration in the left hand side of Eq. (2.45) can be decomposed into two parts as

$$\int_{\Omega} \operatorname{div} \mathbf{F} \, d\Omega = \int_{\Omega^+} \operatorname{div} \mathbf{F} \, d\Omega + \int_{\Omega^-} \operatorname{div} \mathbf{F} \, d\Omega \tag{2.46}$$

Based on the *Divergence theorem*, the integrations at the right hand side of this equation can be respectively written for two sub-domains Ω^+ and Ω^- as

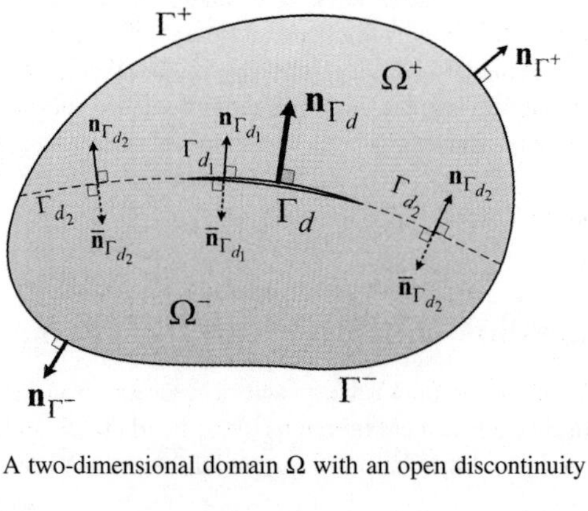

Figure 2.12 A two-dimensional domain Ω with an open discontinuity interface Γ_d

$$\begin{aligned}\int_{\Omega^+} \operatorname{div}\mathbf{F}\,d\Omega &= \int_{\Gamma^+} \mathbf{F}\cdot\mathbf{n}_{\Gamma^+}\,d\Gamma + \int_{\Gamma_{d_1}^+} \mathbf{F}\cdot\bar{\mathbf{n}}_{\Gamma_{d_1}}\,d\Gamma + \int_{\Gamma_{d_2}^+} \mathbf{F}\cdot\bar{\mathbf{n}}_{\Gamma_{d_2}}\,d\Gamma \\ &= \int_{\Gamma^+} \mathbf{F}\cdot\mathbf{n}_{\Gamma^+}\,d\Gamma + \int_{\Gamma_{d_1}^+} \mathbf{F}\cdot\left(-\mathbf{n}_{\Gamma_{d_1}}\right)d\Gamma + \int_{\Gamma_{d_2}^+} \mathbf{F}\cdot\left(-\mathbf{n}_{\Gamma_{d_2}}\right)d\Gamma \\ &= \int_{\Gamma^+} \mathbf{F}\cdot\mathbf{n}_{\Gamma^+}\,d\Gamma - \int_{\Gamma_d^+} \mathbf{F}\cdot\mathbf{n}_{\Gamma_d}\,d\Gamma\end{aligned} \tag{2.47}$$

and

$$\begin{aligned}\int_{\Omega^-} \operatorname{div}\mathbf{F}\,d\Omega &= \int_{\Gamma^-} \mathbf{F}\cdot\mathbf{n}_{\Gamma^-}\,d\Gamma + \int_{\Gamma_{d_1}^-} \mathbf{F}\cdot\mathbf{n}_{\Gamma_{d_1}}\,d\Gamma + \int_{\Gamma_{d_2}^-} \mathbf{F}\cdot\mathbf{n}_{\Gamma_{d_2}}\,d\Gamma \\ &= \int_{\Gamma^-} \mathbf{F}\cdot\mathbf{n}_{\Gamma^-}\,d\Gamma + \int_{\Gamma_d^-} \mathbf{F}\cdot\mathbf{n}_{\Gamma_d}\,d\Gamma\end{aligned} \tag{2.48}$$

in which $\Gamma = \Gamma^+ \cup \Gamma^-$. Substituting Eqs. (2.47) and (2.48) into (2.46), it yields to

$$\int_{\Omega} \operatorname{div}\mathbf{F}\,d\Omega = \int_{\Gamma} \mathbf{F}\cdot\mathbf{n}_{\Gamma}\,d\Gamma - \int_{\Gamma_d^+} \mathbf{F}\cdot\mathbf{n}_{\Gamma_d}\,d\Gamma + \int_{\Gamma_d^-} \mathbf{F}\cdot\mathbf{n}_{\Gamma_d}\,d\Gamma \tag{2.49}$$

Assuming that the value of $\mathbf{F}$ on Γ_d^+ and Γ_d^- are denoted by $\mathbf{F}^+$ and $\mathbf{F}^-$, respectively, Eq. (2.49) can be written as

$$\int_{\Omega} \operatorname{div}\mathbf{F}\,d\Omega = \int_{\Gamma} \mathbf{F}\cdot\mathbf{n}_{\Gamma}\,d\Gamma - \int_{\Gamma_d} (\mathbf{F}^+ - \mathbf{F}^-)\cdot\mathbf{n}_{\Gamma_d}\,d\Gamma \tag{2.50}$$

By defining the jump of $\mathbf{F}$ as $[\![\mathbf{F}]\!] = \mathbf{F}^+ - \mathbf{F}^-$, the Divergence theorem can therefore be expressed for discontinuous problems as

$$\int_{\Omega} \operatorname{div}\mathbf{F}\,d\Omega = \int_{\Gamma} \mathbf{F}\cdot\mathbf{n}_{\Gamma}\,d\Gamma - \int_{\Gamma_d} [\![\mathbf{F}]\!]\cdot\mathbf{n}_{\Gamma_d}\,d\Gamma \tag{2.51}$$

For problems including more than one discontinuity in the domain, Eq. (2.51) can be written as

$$\int_{\Omega} \operatorname{div}\mathbf{F}\,d\Omega = \int_{\Gamma} \mathbf{F}\cdot\mathbf{n}_{\Gamma}\,d\Gamma - \sum_{i=1}^{N^d} \int_{\Gamma_{d_i}} [\![\mathbf{F}_i]\!]\cdot\mathbf{n}_{\Gamma_{d_i}}\,d\Gamma \tag{2.52}$$

where N^d is the number of discontinuities in the domain. This equation can be considered as an extension of the Divergence theorem that provides a mathematical basis for the variational formulation of the X-FEM solution in discontinuous problems.

2.6.2 The Weak form of Governing Equation

In order to derive the extended FE formulation for a continuum body with a weak or strong discontinuity, consider a two-dimensional body Ω with an internal discontinuous boundary denoted by Γ_d, as shown in Figure 2.4, the static equilibrium equation of the body can be written as

$$\nabla\cdot\boldsymbol{\sigma} + \mathbf{b} = 0 \tag{2.53}$$

where ∇ is the gradient operator, $\boldsymbol{\sigma}$ is the Cauchy stress tensor that can be related to the strain tensor $\boldsymbol{\varepsilon}$ by an appropriate constitutive relation, and $\mathbf{b}$ is the body force vector. The behavior of the bulk material is assumed to be the linear elastic, so its constitutive relation can be defined as $\boldsymbol{\sigma} = \mathbf{D}\boldsymbol{\varepsilon}$, with $\mathbf{D}$ denoting the tangential constitutive matrix of the bulk material.

The essential and natural boundary conditions are defined as follows; the displacement (Dirichlet) boundary condition as $\mathbf{u} = \tilde{\mathbf{u}}$ on $\Gamma_u \subset \Gamma$ where $\tilde{\mathbf{u}}$ is the prescribed displacement on the boundary Γ_u, and the traction (Neumann) boundary condition as $\boldsymbol{\sigma} \cdot \mathbf{n}_\Gamma = \bar{\mathbf{t}}$ on $\Gamma_t \subset \Gamma$, where $\bar{\mathbf{t}}$ is the prescribed traction applied on the boundary Γ_t, and $\mathbf{n}_\Gamma$ is the unit outward normal vector to the external boundary Γ. It is assumed that $\Gamma_u \cup \Gamma_t = \Gamma$ and $\Gamma_u \cap \Gamma_t = \emptyset$. Moreover, the internal boundary condition is defined as $\boldsymbol{\sigma} \cdot \mathbf{n}_{\Gamma_d} = \bar{\mathbf{t}}_d$ on the discontinuity Γ_d, in which $\mathbf{n}_{\Gamma_d}$ is the unit normal vector to the discontinuity Γ_d pointing to Ω^+. In a domain with the weak discontinuity, such as bimaterial problems, $\bar{\mathbf{t}}_d$ is the traction transferred across the discontinuity Γ_d. While in a domain with the strong discontinuity, such as crack interfaces, the traction free condition is assumed as $\boldsymbol{\sigma} \cdot \mathbf{n}_{\Gamma_d} = 0$.

In order to derive the weak form of the equilibrium equation (2.53), the FE Galerkin discretization technique is employed by integrating the product of the equilibrium equation multiplied by admissible test function over the domain. The trial function $\mathbf{u}(\boldsymbol{x}, t)$ is required to satisfy all essential boundary conditions and to be smooth enough to define the derivatives of equations. In addition, the test function $\delta\mathbf{u}(\boldsymbol{x}, t)$ is defined in the same approximating space to have the properties of trial function. The weak form of equilibrium equation (2.53) can be obtained multiplying (2.53) by the test functions $\delta\mathbf{u}(\boldsymbol{x}, t)$ and integrating over the domain Ω as

$$\int_\Omega \delta\mathbf{u}(\boldsymbol{x}, t)\,(\nabla \cdot \boldsymbol{\sigma} + \mathbf{b})\,d\Omega = 0 \tag{2.54}$$

Applying the Divergence theorem (2.51) for discontinuous problems, Eq. (2.54) can be written as

$$\int_\Omega \nabla\delta\mathbf{u} : \boldsymbol{\sigma}\,d\Omega + \int_{\Gamma_d} [\![\delta\mathbf{u} \cdot \boldsymbol{\sigma}]\!]\mathbf{n}_{\Gamma_d} d\Gamma - \int_{\Gamma_t} \delta\mathbf{u} \cdot \bar{\mathbf{t}}\,d\Gamma - \int_\Omega \delta\mathbf{u} \cdot \mathbf{b}\,d\Omega = 0 \tag{2.55}$$

in which the second integral in this equation can be eliminated from the integral equation if a weak or strong discontinuity is used. In fact, in the case of weak discontinuity, for example, bimaterial interfaces, the second integral can be computed by assigning the positive and negative sides to Γ_d and imposing the internal boundary condition $(\delta\mathbf{u} \cdot \bar{\mathbf{t}}_d)^+ = (\delta\mathbf{u} \cdot \bar{\mathbf{t}}_d)^-$ as

$$\int_{\Gamma_d} [\![\delta\mathbf{u} \cdot \boldsymbol{\sigma}]\!]\mathbf{n}_{\Gamma_d} d\Gamma = \int_{\Gamma_d} [\![\delta\mathbf{u} \cdot \bar{\mathbf{t}}_d]\!] d\Gamma = \int_{\Gamma_d} (\delta\mathbf{u}^+ \bar{\mathbf{t}}_d^+ - \delta\mathbf{u}^- \bar{\mathbf{t}}_d^-)\,d\Gamma = 0 \tag{2.56}$$

In the case of strong discontinuity, for example, crack interfaces, the second integral can be evaluated by imposing the traction free boundary condition as

$$\int_{\Gamma_d} [\![\delta\mathbf{u} \cdot \boldsymbol{\sigma}]\!]\mathbf{n}_{\Gamma_d} d\Gamma = \int_{\Gamma_d} [\![\delta\mathbf{u} \cdot \bar{\mathbf{t}}_d]\!] d\Gamma = \int_{\Gamma_d} (\delta\mathbf{u}^+ - \delta\mathbf{u}^-)\bar{\mathbf{t}}_d\,d\Gamma = 0 \tag{2.57}$$

It must be noted that the value of second integral in (2.55) is not generally zero, as will be discussed in next chapters for the cohesive crack problem, contact problem, and so on. Finally, the integral equation (2.55) can be simplified to

$$\int_\Omega \nabla\delta\mathbf{u} : \boldsymbol{\sigma}\,d\Omega - \int_{\Gamma_t} \delta\mathbf{u} \cdot \bar{\mathbf{t}}\,d\Gamma - \int_\Omega \delta\mathbf{u} \cdot \mathbf{b}\,d\Omega = 0 \tag{2.58}$$

This equation is an expression for the weak form of equilibrium equation that can be solved by substituting the trial function $\delta\mathbf{u}(\boldsymbol{x}, t)$ into (2.58) to obtain the discretized form of extended FE formulation.

2.7 The X-FEM Discretization of Governing Equation

In order to discretize the integral equation (2.58), the extended FE discretization method is applied employing the enhanced approximation field (2.17). In the X-FEM, the conventional FE shape functions are enriched by adding enrichment functions that are achieved based on the type of the discontinuity. Various enrichment functions are used to model the discontinuity in an element; the most important ones are the signed distance function and the Heaviside step function, as mentioned in preceding sections. Consider the displacement field of an enriched element $\mathbf{u}(\boldsymbol{x}, t)$ given in (2.17), the enhanced approximation field of the shifted enrichment function $\psi(\boldsymbol{x})$ can be defined as

$$\begin{aligned} \mathbf{u}(\boldsymbol{x}, t) &= \sum_{i=1}^{\mathcal{N}} N_i(\boldsymbol{x})\, \bar{\mathbf{u}}_i + \sum_{j=1}^{\mathcal{M}} N_j(\boldsymbol{x})\, \left(\psi(\boldsymbol{x}) - \psi(\boldsymbol{x}_j)\right) \bar{\mathbf{a}}_j \\ &\equiv \mathbf{N}^{std}(\boldsymbol{x})\, \bar{\mathbf{u}} + \mathbf{N}^{enr}(\boldsymbol{x})\, \bar{\mathbf{a}} \end{aligned} \tag{2.59}$$

in which the last term of this relation is added to satisfy the expected value of displacement $\mathbf{u}(\boldsymbol{x})$ on an enriched node k, that is, $\mathbf{u}(\boldsymbol{x}_k) = \bar{\mathbf{u}}_k$. Similarly, the test function $\delta\mathbf{u}(\boldsymbol{x}, t)$ can be defined in the same approximating space as the displacement field $\mathbf{u}(\boldsymbol{x}, t)$ as

$$\begin{aligned} \delta\mathbf{u}(\boldsymbol{x}, t) &= \sum_{i=1}^{\mathcal{N}} N_i(\boldsymbol{x})\, \delta\bar{\mathbf{u}}_i + \sum_{j=1}^{\mathcal{M}} N_j(\boldsymbol{x})\, \left(\psi(\boldsymbol{x}) - \psi(\boldsymbol{x}_j)\right) \delta\bar{\mathbf{a}}_j \\ &\equiv \mathbf{N}^{std}(\boldsymbol{x})\, \delta\bar{\mathbf{u}} + \mathbf{N}^{enr}(\boldsymbol{x})\, \delta\bar{\mathbf{a}} \end{aligned} \tag{2.60}$$

In relations (2.59) and (2.60), $N_i^{std}(\boldsymbol{x}) \equiv N_i(\boldsymbol{x})$ and $N_j^{enr}(\boldsymbol{x}) \equiv N_j(\boldsymbol{x})\left(\psi(\boldsymbol{x}) - \psi(\boldsymbol{x}_j)\right)$, which is defined for a two-dimensional problem as

$$\begin{aligned} \mathbf{N}_i^{std} &= \begin{bmatrix} N_i(\boldsymbol{x}) & 0 \\ 0 & N_i(\boldsymbol{x}) \end{bmatrix} \\ \mathbf{N}_j^{enr} &= \begin{bmatrix} N_j(\boldsymbol{x})\, \left(\psi(\boldsymbol{x}) - \psi(\boldsymbol{x}_j)\right) & 0 \\ 0 & N_j(\boldsymbol{x})\, \left(\psi(\boldsymbol{x}) - \psi(\boldsymbol{x}_j)\right) \end{bmatrix} \end{aligned} \tag{2.61}$$

The strain vector $\boldsymbol{\varepsilon}(\boldsymbol{x}, t)$ can be accordingly defined corresponding to the approximate displacement field (2.59) in terms of the standard and enriched nodal values as

$$\begin{aligned} \boldsymbol{\varepsilon}(\boldsymbol{x}, t) &= \sum_{i=1}^{\mathcal{N}} \frac{\partial N_i}{\partial \boldsymbol{x}}\, \bar{\mathbf{u}}_i + \sum_{j=1}^{\mathcal{M}} \left[\frac{\partial N_j}{\partial \boldsymbol{x}} \left(\psi(\boldsymbol{x}) - \psi(\boldsymbol{x}_j)\right) + N_j(\boldsymbol{x}) \frac{\partial}{\partial \boldsymbol{x}} \left(\psi(\boldsymbol{x}) - \psi(\boldsymbol{x}_j)\right) \right] \bar{\mathbf{a}}_j \\ &\equiv \mathbf{B}^{std}(\boldsymbol{x})\, \bar{\mathbf{u}} + \mathbf{B}^{enr}(\boldsymbol{x})\, \bar{\mathbf{a}} \end{aligned} \tag{2.62}$$

Similarly, the variation of strain field $\delta\boldsymbol{\varepsilon}(\boldsymbol{x}, t)$ required for the weak form of equilibrium equation (2.58) can be obtained as

$$\begin{aligned} \delta\boldsymbol{\varepsilon}(\boldsymbol{x}, t) &= \sum_{i=1}^{\mathcal{N}} \frac{\partial N_i}{\partial \boldsymbol{x}}\, \delta\bar{\mathbf{u}}_i + \sum_{j=1}^{\mathcal{M}} \left[\frac{\partial N_j}{\partial \boldsymbol{x}} \left(\psi(\boldsymbol{x}) - \psi(\boldsymbol{x}_j)\right) + N_j(\boldsymbol{x}) \frac{\partial}{\partial \boldsymbol{x}} \left(\psi(\boldsymbol{x}) - \psi(\boldsymbol{x}_j)\right) \right] \delta\bar{\mathbf{a}}_j \\ &\equiv \mathbf{B}^{std}(\boldsymbol{x})\, \delta\bar{\mathbf{u}} + \mathbf{B}^{enr}(\boldsymbol{x})\, \delta\bar{\mathbf{a}} \end{aligned} \tag{2.63}$$

In relations (2.62) and (2.63), $B_i^{std}(\boldsymbol{x}) \equiv \partial N_i/\partial \boldsymbol{x}$ and $B_j^{enr}(\boldsymbol{x}) \equiv \frac{\partial}{\partial \boldsymbol{x}}\left[N_j(\boldsymbol{x})\left(\psi(\boldsymbol{x})-\psi(\boldsymbol{x}_j)\right)\right]$, which is defined for a two-dimensional problem as

$$\mathbf{B}_i^{std} = \begin{bmatrix} \partial N_i/\partial x & 0 \\ 0 & \partial N_i/\partial y \\ \partial N_i/\partial y & \partial N_i/\partial x \end{bmatrix} \tag{2.64a}$$

$$\mathbf{B}_j^{enr} = \begin{bmatrix} \partial\left[N_j(\boldsymbol{x})\left(\psi(\boldsymbol{x})-\psi(\boldsymbol{x}_j)\right)\right]/\partial x & 0 \\ 0 & \partial\left[N_j(\boldsymbol{x})\left(\psi(\boldsymbol{x})-\psi(\boldsymbol{x}_j)\right)\right]/\partial y \\ \partial\left[N_j(\boldsymbol{x})\left(\psi(\boldsymbol{x})-\psi(\boldsymbol{x}_j)\right)\right]/\partial y & \partial\left[N_j(\boldsymbol{x})\left(\psi(\boldsymbol{x})-\psi(\boldsymbol{x}_j)\right)\right]/\partial x \end{bmatrix} \tag{2.64b}$$

The discretized form of the extended FE formulation can be obtained by substituting the test functions $\delta\mathbf{u}(\boldsymbol{x}, t)$ and $\delta\boldsymbol{\varepsilon}(\boldsymbol{x}, t)$ into Eq. (2.58) as

$$\begin{aligned} &\int_\Omega \left(\mathbf{B}^{std}\,\delta\bar{\mathbf{u}} + \mathbf{B}^{enr}\,\delta\bar{\mathbf{a}}\right)^T \boldsymbol{\sigma}\, d\Omega \\ &-\int_{\Gamma_t} \left(\mathbf{N}^{std}\,\delta\bar{\mathbf{u}} + \mathbf{N}^{enr}\,\delta\bar{\mathbf{a}}\right)^T \bar{\mathbf{t}}\, d\Gamma \\ &-\int_\Omega \left(\mathbf{N}^{std}\,\delta\bar{\mathbf{u}} + \mathbf{N}^{enr}\,\delta\bar{\mathbf{a}}\right)^T \mathbf{b}\, d\Omega = 0 \end{aligned} \tag{2.65}$$

This equation can be rewritten as

$$\begin{aligned} &\delta\bar{\mathbf{u}}^T \left\{ \int_\Omega (\mathbf{B}^{std})^T \boldsymbol{\sigma}\, d\Omega - \int_{\Gamma_t} (\mathbf{N}^{std})^T \bar{\mathbf{t}}\, d\Gamma - \int_\Omega (\mathbf{N}^{std})^T \mathbf{b}\, d\Omega \right\} \\ &+ \delta\bar{\mathbf{a}}^T \left\{ \int_\Omega (\mathbf{B}^{enr})^T \boldsymbol{\sigma}\, d\Omega - \int_{\Gamma_t} (\mathbf{N}^{enr})^T \bar{\mathbf{t}}\, d\Gamma - \int_\Omega (\mathbf{N}^{enr})^T \mathbf{b}\, d\Omega \right\} = 0 \end{aligned} \tag{2.66}$$

The discrete system of X-FEM equations can be obtained from (2.66) as $\mathbf{K}\,\bar{\mathbf{U}} - \mathbf{F} = 0$, where $\bar{\mathbf{U}}^T = [\bar{\mathbf{u}}^T, \bar{\mathbf{a}}^T]$ is the vector of unknowns at the nodal points, $\mathbf{K}$ is the total stiffness matrix, and $\mathbf{F}$ is the external force vector. The discrete system of equations can be finally obtained as

$$\begin{bmatrix} \mathbf{K}_{uu} & \mathbf{K}_{ua} \\ \mathbf{K}_{au} & \mathbf{K}_{aa} \end{bmatrix} \begin{Bmatrix} \bar{\mathbf{u}} \\ \bar{\mathbf{a}} \end{Bmatrix} = \begin{Bmatrix} \mathbf{F}_u \\ \mathbf{F}_a \end{Bmatrix} \tag{2.67}$$

where

$$\mathbf{K} = \begin{bmatrix} \int_\Omega (\mathbf{B}^{std})^T \mathbf{D}\, \mathbf{B}^{std} d\Omega & \int_\Omega (\mathbf{B}^{std})^T \mathbf{D}\, \mathbf{B}^{enr} d\Omega \\ \int_\Omega (\mathbf{B}^{enr})^T \mathbf{D}\, \mathbf{B}^{std} d\Omega & \int_\Omega (\mathbf{B}^{enr})^T \mathbf{D}\, \mathbf{B}^{enr} d\Omega \end{bmatrix} \tag{2.68}$$

and

$$\mathbf{F} = \begin{Bmatrix} \int_{\Gamma_t} (\mathbf{N}^{std})^T \bar{\mathbf{t}}\, d\Gamma + \int_\Omega (\mathbf{N}^{std})^T \mathbf{b}\, d\Omega \\ \int_{\Gamma_t} (\mathbf{N}^{enr})^T \bar{\mathbf{t}}\, d\Gamma + \int_\Omega (\mathbf{N}^{enr})^T \mathbf{b}\, d\Omega \end{Bmatrix} \tag{2.69}$$

2.7.1 Numerical Implementation of X-FEM Formulation

The X-FEM discretization of equilibrium equation for a body with the weak or strong discontinuity is presented in previous section by introducing the enhanced stiffness matrix and load vector. In this section, the numerical implementation of the X-FEM formulation is described in the framework of an isoparametric element similar to the concept of standard FE model. In an isoparametric element, the natural coordinate system is used to map an arbitrary quadrilateral element into the parent element of the same order. Consider an arbitrary linear quadrilateral element in the global Cartesian coordinates, as shown in Figure 2.13. In order to derive the stiffness matrix of an isoparametric element, a local natural coordinate system (ξ, η) whose origin is at the centroid of the quadrilateral element is assumed. The shape functions of parent element in the natural coordinate system are defined as $N_i(\xi,\eta) = \frac{1}{4}(1+\xi_i\xi)(1+\eta_i\eta)$.

The standard FE approximation of displacement field can be obtained as $\mathbf{u} = \mathbf{N}^{std}\,\bar{\mathbf{u}}$, in which $\mathbf{N}^{std}$ and $\bar{\mathbf{u}}$ are defined as

$$\mathbf{N}^{std} = \begin{bmatrix} N_1 & 0 & N_2 & 0 & N_3 & 0 & N_4 & 0 \\ 0 & N_1 & 0 & N_2 & 0 & N_3 & 0 & N_4 \end{bmatrix} \tag{2.70a}$$

$$\bar{\mathbf{u}}^T = \begin{bmatrix} u_{x_1} & u_{y_1} & u_{x_2} & u_{y_2} & u_{x_3} & u_{y_3} & u_{x_4} & u_{y_4} \end{bmatrix} \tag{2.70b}$$

The enhanced FE approximation of a displacement field can be obtained based on the standard and enriched parts using $\mathbf{u} = \mathbf{N}^{enh}\,\bar{\mathbf{U}}$, in which the enhanced shape functions $\mathbf{N}^{enh} = [\mathbf{N}^{std}, \mathbf{N}^{enr}]$ and the vector of unknown nodal points $\bar{\mathbf{U}}^T = [\bar{\mathbf{u}}^T, \bar{\mathbf{a}}^T]$ are defined as

$$\mathbf{N}^{enh} = \left[\begin{matrix} N_1 & 0 & N_2 & 0 & N_3 & 0 & N_4 & 0 \\ 0 & N_1 & 0 & N_2 & 0 & N_3 & 0 & N_4 \end{matrix} \quad \cdots \right.$$
$$\left. \cdots \quad \begin{matrix} N_1\bar{\psi}_1 & 0 & N_2\bar{\psi}_2 & 0 & N_3\bar{\psi}_3 & 0 & N_4\bar{\psi}_4 & 0 \\ 0 & N_1\bar{\psi}_1 & 0 & N_2\bar{\psi}_2 & 0 & N_3\bar{\psi}_3 & 0 & N_4\bar{\psi}_4 \end{matrix} \right] \tag{2.71a}$$

$$\bar{\mathbf{U}}^T = \left[u_{x_1} \; u_{y_1} \; u_{x_2} \; u_{y_2} \; u_{x_3} \; u_{y_3} \; u_{x_4} \; u_{y_4} \; \cdots \right.$$
$$\left. \cdots \; a_{x_1} \; a_{y_1} \; a_{x_2} \; a_{y_2} \; a_{x_3} \; a_{y_3} \; a_{x_4} \; a_{y_4} \right] \tag{2.71b}$$

In this expression, it is assumed that the four nodes of the isoparametric element are enriched using a single enrichment function $\psi(\boldsymbol{x})$, where $\bar{\psi}_i$ denotes the shifted enrichment function, that is, $\bar{\psi}_i(\boldsymbol{x}) = \psi(\boldsymbol{x}) - \psi(\boldsymbol{x}_i)$.

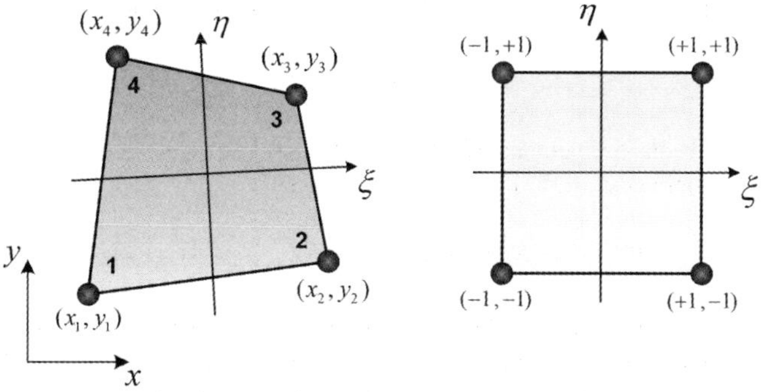

Figure 2.13 An isoparametric Q4 element: The arbitrary (physical) and the parent elements

Accordingly, the strain vector corresponding to the standard approximation of displacement field $\mathbf{u}=\mathbf{N}^{std}\,\bar{\mathbf{u}}$ can be obtained as $\boldsymbol{\varepsilon}=\mathbf{B}^{std}\,\bar{\mathbf{u}}$, in which $\mathbf{B}^{std}$ is the standard discretized gradient operator defined as

$$\mathbf{B}^{std}=\begin{bmatrix} N_{1,x} & 0 & N_{2,x} & 0 & N_{3,x} & 0 & N_{4,x} & 0 \\ 0 & N_{1,y} & 0 & N_{2,y} & 0 & N_{3,y} & 0 & N_{4,y} \\ N_{1,y} & N_{1,x} & N_{2,y} & N_{2,x} & N_{3,y} & N_{3,x} & N_{4,y} & N_{4,x} \end{bmatrix} \tag{2.72}$$

The enhanced strain field can also be obtained using the standard and enriched parts of the enhanced displacement field $\mathbf{u}=\mathbf{N}^{enh}\,\bar{\mathbf{U}}$ by the definition of $\boldsymbol{\varepsilon}=\mathbf{B}^{enh}\,\bar{\mathbf{U}}$, where the enhanced discretized gradient operator $\mathbf{B}^{enh}=[\mathbf{B}^{std},\ \mathbf{B}^{enr}]$ is defined as

$$\mathbf{B}^{enh}=\begin{bmatrix} N_{1,x} & 0 & N_{2,x} & 0 & N_{3,x} & 0 & N_{4,x} & 0 \\ 0 & N_{1,y} & 0 & N_{2,y} & 0 & N_{3,y} & 0 & N_{4,y} & \cdots \\ N_{1,y} & N_{1,x} & N_{2,y} & N_{2,x} & N_{3,y} & N_{3,x} & N_{4,y} & N_{4,x} \end{bmatrix}$$
$$\begin{matrix} (N_1\bar{\psi}_1)_{,x} & 0 & (N_2\bar{\psi}_2)_{,x} & 0 & (N_3\bar{\psi}_3)_{,x} & 0 & (N_4\bar{\psi}_4)_{,x} & 0 \\ \cdots\ 0 & (N_1\bar{\psi}_1)_{,y} & 0 & (N_2\bar{\psi}_2)_{,y} & 0 & (N_3\bar{\psi}_3)_{,y} & 0 & (N_4\bar{\psi}_4)_{,y} \\ (N_1\bar{\psi}_1)_{,y} & (N_1\bar{\psi}_1)_{,x} & (N_2\bar{\psi}_2)_{,y} & (N_2\bar{\psi}_2)_{,x} & (N_3\bar{\psi}_3)_{,y} & (N_3\bar{\psi}_3)_{,x} & (N_4\bar{\psi}_4)_{,y} & (N_4\bar{\psi}_4)_{,x} \end{matrix}\Bigg] \tag{2.73}$$

In the case of a *weak* discontinuity, where the ramp enrichment function is used, that is, $\bar{\psi}_i(\boldsymbol{x})=|\varphi(\boldsymbol{x})|-|\varphi(\boldsymbol{x}_i)|$, the derivative of enrichment function $\bar{\psi}_i(\boldsymbol{x})$ can be obtained as

$$\bar{\psi}_{i,x}(\boldsymbol{x})=\mathrm{sign}(\varphi(\boldsymbol{x}))\,\varphi_{i,x}(\boldsymbol{x}) \tag{2.74}$$

in which the value of level set function $\varphi(\boldsymbol{x})$ is obtained using the values of nodal points φ_i and applying the interpolation functions $N_i(\boldsymbol{x})$ as $\varphi(\boldsymbol{x})=\sum_{i=1}^{\mathcal{N}} N_i(\boldsymbol{x})\,\varphi_i$. As a result, the derivative of $\varphi(\boldsymbol{x})$ can be obtained by $\varphi_{i,x}(\boldsymbol{x})=\sum_{i=1}^{\mathcal{N}} N_{i,x}(\boldsymbol{x})\,\varphi_i$. Hence, the enriched discretized gradient operator B_i^{enr} defined in (2.73) can be computed as

$$\begin{aligned} B_i^{enr}(\boldsymbol{x}) &= (N_i(\boldsymbol{x})\,\bar{\psi}_i(\boldsymbol{x}))_{,x} \\ &= N_{i,x}(\boldsymbol{x})\,(|\varphi(\boldsymbol{x})|-|\varphi(\boldsymbol{x}_i)|)+N_i(\boldsymbol{x})\,\mathrm{sign}(\varphi(\boldsymbol{x}))\,\varphi_{i,x}(\boldsymbol{x}) \end{aligned} \tag{2.75}$$

In the case of a *strong* discontinuity, where the Heaviside enrichment function is used $\bar{\psi}_i(\boldsymbol{x})=H(\boldsymbol{x})-H(\boldsymbol{x}_i)$, the derivative of Heaviside function is the Dirac delta function, that is,

$$\bar{\psi}_{i,x}(\boldsymbol{x})=H_{,x}(\boldsymbol{x})=\delta \tag{2.76}$$

in which $H_{,x}(\boldsymbol{x})=1$ at the discontinuity interface and $H_{,x}(\boldsymbol{x})=0$ otherwise. Hence, the enriched discretized gradient operator B_i^{enr} defined in (2.73) can be computed as

$$B_i^{enr}(\boldsymbol{x})=(N_i(\boldsymbol{x})\,\bar{\psi}_i(\boldsymbol{x}))_{,x}=N_{i,x}(\boldsymbol{x})\,(H(\boldsymbol{x})-H(\boldsymbol{x}_i)) \tag{2.77}$$

Finally, the total stiffness matrix can be obtained by substituting the enhanced discretized gradient operator $\mathbf{B}^{enh}$ from (2.73) into (2.68). Since the derivatives of the standard and enriched shape functions are defined in terms of Cartesian coordinates, a relation between the derivatives in the parent and Cartesian coordinates can be performed using the Jacobian matrix as

$$\begin{Bmatrix} \dfrac{\partial}{\partial x} \\ \dfrac{\partial}{\partial y} \end{Bmatrix} = \mathbf{J}^{-1} \begin{bmatrix} \dfrac{\partial}{\partial \xi} \\ \dfrac{\partial}{\partial \eta} \end{bmatrix} \tag{2.78}$$

in which a map between the parent and current element is given for the Q4 element as $\boldsymbol{x} = \sum_{i=1}^{4} N_i(\xi, \eta)\, \boldsymbol{x}_i$. In this equation, the Jacobian matrix $\mathbf{J}$ is defined as

$$\mathbf{J} = \begin{bmatrix} \dfrac{\partial x}{\partial \xi} & \dfrac{\partial y}{\partial \xi} \\ \dfrac{\partial x}{\partial \eta} & \dfrac{\partial y}{\partial \eta} \end{bmatrix} = \begin{bmatrix} \sum_{i=1}^{4} \dfrac{\partial N_i}{\partial \xi} x_i & \sum_{i=1}^{4} \dfrac{\partial N_i}{\partial \xi} y_i \\ \sum_{i=1}^{4} \dfrac{\partial N_i}{\partial \eta} x_i & \sum_{i=1}^{4} \dfrac{\partial N_i}{\partial \eta} y_i \end{bmatrix} \tag{2.79}$$

Thus, the total stiffness matrix can be obtained from (2.68) as

$$\begin{aligned} \mathbf{K}_{ij}^{\alpha\beta} &= \int_{\Omega} (\mathbf{B}_i^{\alpha})^T \mathbf{D}\, \mathbf{B}_j^{\beta}\, d\Omega \\ &= \int_{-1}^{+1} \int_{-1}^{+1} \left(\mathbf{B}_{i\,(\xi,\eta)}^{\alpha}\right)^T \mathbf{D}\, \mathbf{B}_{j\,(\xi,\eta)}^{\beta} \det\mathbf{J}\, d\xi d\eta \qquad (\alpha, \beta = std, enr) \end{aligned} \tag{2.80}$$

in which $\mathbf{K}$ is a 8×8 matrix for the Q4 standard element and 16×16 matrix for the Q4 enriched element. The above stiffness matrix must be evaluated numerically using the numerical integration algorithm.

2.7.2 *Numerical Integration Algorithm*

In the classical FEM, the standard shape functions are in terms of the polynomial order and the Gauss integration rule can be used efficiently to evaluate the integral of stiffness matrix. However, in the X-FEM the enhanced shape functions may be obtained in terms of non-polynomial order. Moreover, the enrichment functions may not be smooth over an enriched element due to presence of the weak or strong discontinuity inside the element. Hence, the standard Gauss quadrature rule cannot be used if the element is crossed by a discontinuity, and necessary modifications are necessary for numerical integration over an enriched element. The primary approach may be suggested based on the increase of the number of Gauss integration points, as shown in Figure 2.14a, however, this approach may result in a substantial loss of accuracy. In order to overcome these difficulties, two techniques are introduced for the numerical integration of an enriched element; the triangular/quadrilateral partitioning method and the rectangular sub-grids method.

In the triangular/quadrilateral partitioning method, the element bisected by the interface, is divided into triangular/quadrilateral sub-elements, as shown in Figure 2.14b, and the Gauss integration rule is performed over each sub-polygons. The sub-polygons do not have to be conforming and no new unknowns are created from this decomposition. It is important to note that the interface is generally described by a discretized level set function that is interpolated by the standard FE shape functions, so the zero level set is defined by $\varphi(\boldsymbol{x}) = \sum_{i=1}^{\mathcal{N}} N_i(\boldsymbol{x})\, \varphi_i = 0$. The interface which is the zero-level of $\varphi(\boldsymbol{x})$, is, in general, curved, and results from obtaining the roots in the *reference* element and projecting these points into the *real* element geometry by an isoparametric mapping. In fact, the interface is planar inside the elements only for linear interpolants with 3-noded triangular elements. For quadrilateral elements, it can be justified to

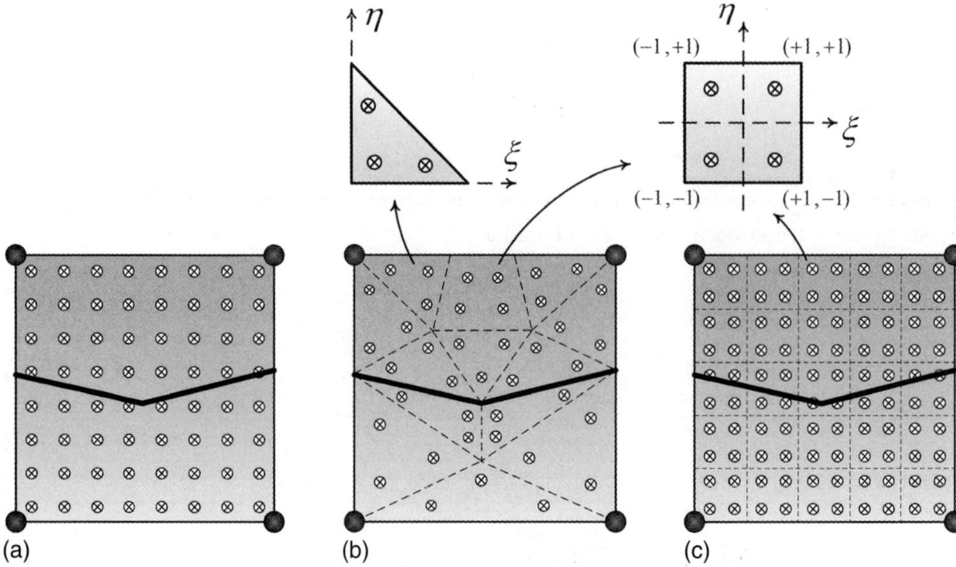

Figure 2.14 Numerical integration of an enriched element: (a) the standard Gauss integration points; (b) the triangular/quadrilateral partitioning method; (c) the rectangular sub-grids method

replace the curved interface by a straight line, which is determined from the intersections of the interface with the element edges. In this case, it is difficult to construct the exact interpolation functions of $\varphi(\boldsymbol{x})$ because they embody a constraint. Thus, it is often preferable to decompose the cut quadrilateral reference elements into sub-triangles and use the linear interpolation in each sub-triangle; so the interface is then always piecewise straight, and triangular sub-elements for integration purposes are easily obtained. However, if an element with linear shape functions is cut by a curved interface, in which the interface is discretized by the level set function, then the level set cannot capture the curvature of the interface correctly. In such a case, additional points can be introduced along the interface to increase the number of sub-polygons and to maintain the accuracy of the integration.

Consider an element cut by an interface into two distinct parts, Ω^+ and Ω^-; the integration of stiffness matrix (2.80) over Ω can be performed as

$$\begin{aligned}
\mathbf{K}_{ij}^{\alpha\beta} &= \int_{\Omega} \left(\mathbf{B}_{i\,(\xi,\eta)}^{\alpha}\right)^T \mathbf{D}\,\mathbf{B}_{j\,(\xi,\eta)}^{\beta}\, d\Omega \\
&= \int_{\Omega^+} \left(\mathbf{B}_{i\,(\xi,\eta)}^{\alpha}\right)^T \mathbf{D}\,\mathbf{B}_{j\,(\xi,\eta)}^{\beta}\, d\Omega + \int_{\Omega^-} \left(\mathbf{B}_{i\,(\xi,\eta)}^{\alpha}\right)^T \mathbf{D}\,\mathbf{B}_{j\,(\xi,\eta)}^{\beta}\, d\Omega \\
&= \sum_{\ell=1}^{\mathcal{N}_{Sub}^{+}} \left(\sum_{k=1}^{\mathcal{N}_{GP}^{+}} \left(\mathbf{B}_{i\,(\xi_k,\eta_k)}^{\alpha}\right)^T \mathbf{D}\,\mathbf{B}_{j\,(\xi_k,\eta_k)}^{\beta} w_k \right)_{\ell} + \sum_{\ell=1}^{\mathcal{N}_{Sub}^{-}} \left(\sum_{k=1}^{\mathcal{N}_{GP}^{-}} \left(\mathbf{B}_{i\,(\xi_k,\eta_k)}^{\alpha}\right)^T \mathbf{D}\,\mathbf{B}_{j\,(\xi_k,\eta_k)}^{\beta} w_k \right)_{\ell}
\end{aligned} \tag{2.81}$$

where $\mathcal{N}_{Sub}^{+}$ and $\mathcal{N}_{Sub}^{-}$ are the number of sub-polygons in Ω^+ and Ω^-, and $\mathcal{N}_{GP}^{+}$ and $\mathcal{N}_{GP}^{-}$ are the number of Gauss points at each sub-polygon in Ω^+ and Ω^-, respectively. In this relation, w_k is the weight of quadrature point. Although special integration formulas are available for polygons with n^- edges, such as those given by Natarajan, Bordas, and Mahapatra (2009), Mousavi and Sukumar (2010a), and Mousavi, Xiao, and Sukumar (2010b), a further decomposition into quadrilaterals and triangles is often useful (Figure 2.14b). Hence, the standard Gauss quadrature rule can be employed, which enables the exact

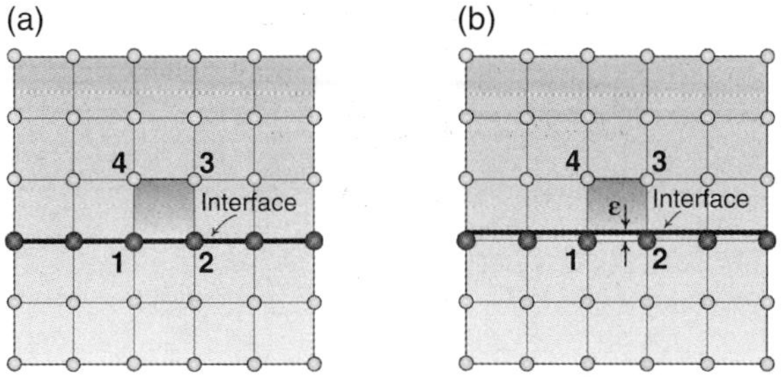

Figure 2.15 Effects of crack near edge: (a) the crack is aligned with a mesh where nodes 1 and 2 are enriched and the nodes 3 and 4 are not enriched; (b) the crack is almost aligned with a mesh where nodes 3 and 4 should not be enriched to avoid singular stiffness matrix

integration of polynomials up to a certain order in the reference element. However, this property is only maintained for the cut reference elements that decompose into triangular sub-elements. For quadrilateral sub-elements, it requires an isoparametric mapping of the Gauss points into the sub-elements, and, although very good results may be obtained, the "exact integration" property for the reference element can be lost.

In an alternative technique, the element cut by the interface can be divided into rectangular sub-grids, as shown in Figure 2.14c. In this technique, it is not necessary to conform the sub-quadrilaterals to the geometry of interface; however, enough sub-divisions are needed to reduce the error of numerical integration. Based on the rectangular sub-grids integration, the natural coordinates of Gauss quadrature points are independent at each sub-quadrilateral. Consider the element cut by the interface contains 5 × 5 rectangular sub-grids with four Gauss points at each sub-quadrilateral; the total number of Gauss points is 100 for an enriched element. It must be noted that, since the interface does not coincide with the rectangular edges, it may cause some approximation in the numerical simulation; as a result, an accurate solution can be achieved by increasing the number of sub-quadrilaterals.

Although both techniques are quite accurate and are widely used in the extended FE framework, they each have their own advantages and disadvantages. In the triangular-partitioning method, the integration could be much more accurate since the domain is divided into smooth sub-domains; however, it would be quite cumbersome to develop partitioning for various configurations of interface. Moreover, the probability of updating the interface due to its evolution in time, particularly due to crack propagation, may cause the position of Gauss points to vary during the solution, and as a result leads to the need for data transferring between the old and new Gauss points. On the contrary in the rectangular sub-grids approach, the data transfer is not necessary as the sub-grids are independent from the configuration of the interface. It is manifest that the implementation of rectangular sub-grid integration is much easier than the triangular-partitioning algorithm; however, the accuracy cannot be guaranteed in those rectangular sub-grids cut by the interface. In order to avoid the inaccuracy involved in this method, the finer sub-grids can commonly be utilized to confine this event. Practically, it is observed that the rectangular sub-grids scheme results in an adequate accuracy.

One of the most important issues in numerical integration of X-FEM formulation is incompatible enrichments. In the X-FEM, the condition number of stiffness matrix is generally much higher than that of the standard FEM. This is mainly due to the fact that the values of enriched shape functions are less than the standard ones, particularly at the Gauss points that are very close to the interface. Moreover, the enrichment terms related to extra DOF usually have greater non-zero values in the opposite side of the interface when using the Heaviside enrichment function. This may result in increasing the order of differences at the extra DOF, and causes a bad condition number of stiffness matrix that leads to ill conditioning, or even singularity because of minor values corresponding to these extra DOF. In order to clearly demonstrate this issue, consider that the interface passes through an element edge, as shown in Figure 2.15a, where nodes (1) and (2) are enriched by the Heaviside function and nodes (3) and (4) are not enriched since their

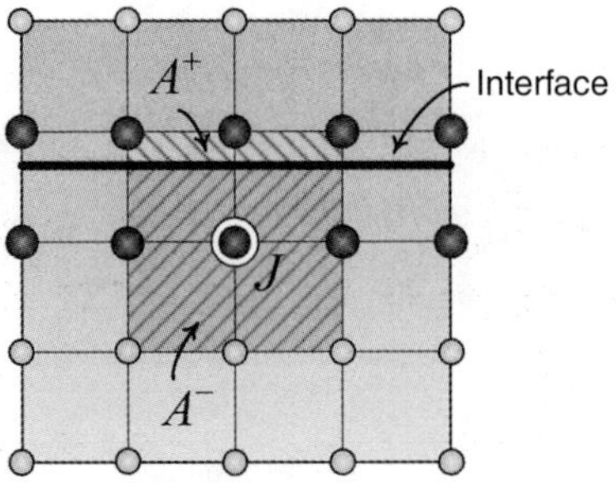

Figure 2.16 The criterion for node enrichment based on the definition of area of the influence domain of a node J

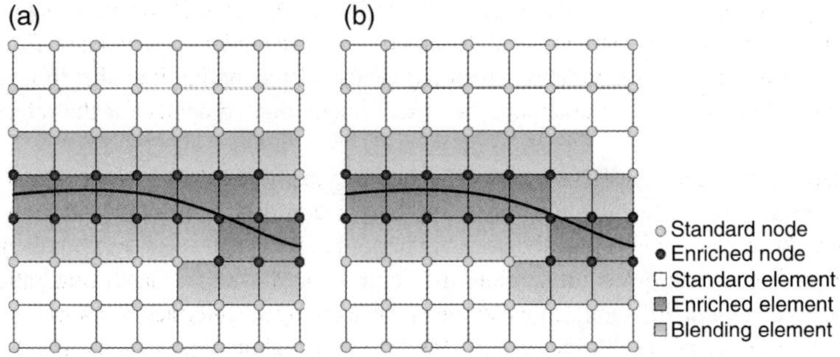

Figure 2.17 The modification of an enriched element when the interface passes through a nodal point: (a) before modification; (b) after modification

supports are not cut by the interface. Now, if the interface passes in a very thin band of width ε along the element edge, as shown in Figure 2.15b, the nodes (3) and (4) must be enriched on the basis of support criterion, which leads to ill-conditioned stiffness matrix since the resulting basis functions are almost identical. In such particular circumstance, nodes (3) and (4) must not be enriched by the Heaviside function. To overcome this situation, a criterion can be defined in which, for a certain node J, if the values of $A^+/(A^+ + A^-)$ or $A^-/(A^+ + A^-)$, where A^+ and A^- are, respectively, the area of the influence domain of that node (Figure 2.16), are smaller than the allowable tolerance value of 10^{-4}, the node must not be enriched. Figure 2.17 represents this methodology to remove this issue by modification of an enriched element.

2.8 Application of X-FEM in Weak and Strong Discontinuities

In order to demonstrate the accuracy and versatility of the X-FEM in weak and strong discontinuities, four examples are modeled numerically. The examples consist of an elastic one-dimensional bar and an elastic two-dimensional plate, which are in tension with a "*crack interface*" and a "*material interface*" at their centers. The examples are solved using both FEM and X-FEM techniques, and the results are compared. In X-FEM analysis, the FE mesh is defined independent of the shape of discontinuity, while the FEM analysis is performed by conforming the mesh to the geometry of the interface. The strong discontinuity of crack interface is modeled using the Heaviside enrichment function, and the weak discontinuity of bimaterial interface is modeled using the ramp enrichment function. Finally, the evolutions of standard, enriched, and total displacement fields are presented in the X-FEM solutions (see Companion Website).

2.8.1 Modeling an Elastic Bar with a Strong Discontinuity

In order to illustrate the performance of X-FEM formulation described in preceding sections, an elastic one-dimensional bar with a discontinuity at its center, such as a crack interface, is modeled numerically. The geometry and boundary conditions together with the FEM and X-FEM meshes of the problem are shown in Figure 2.18.

The bar is restrained at one end and is subjected to a prescribed displacement u_0 at the free end. The X-FEM analysis is performed using three 2-noded elements, in which the discontinuity is placed at the mid-point of element (2), as shown in Figure 2.18. In order to model the discontinuity in the bar, the Heaviside function is employed considering the enriched degrees of freedom at nodal points 2 and 3. The results of X-FEM model are compared with the FEM analysis carried out using four 2-noded elements. In the FEM analysis, the stiffness matrix is obtained for each bar element as

$$ {}^{(1)}\mathbf{K} = {}^{(3)}\mathbf{K} = \frac{AE}{\ell}\begin{bmatrix} +1 & -1 \\ -1 & +1 \end{bmatrix} \text{ and } {}^{(2)}\mathbf{K} = \frac{AE}{\ell/2}\begin{bmatrix} +1 & -1 \\ -1 & +1 \end{bmatrix} \tag{2.82} $$

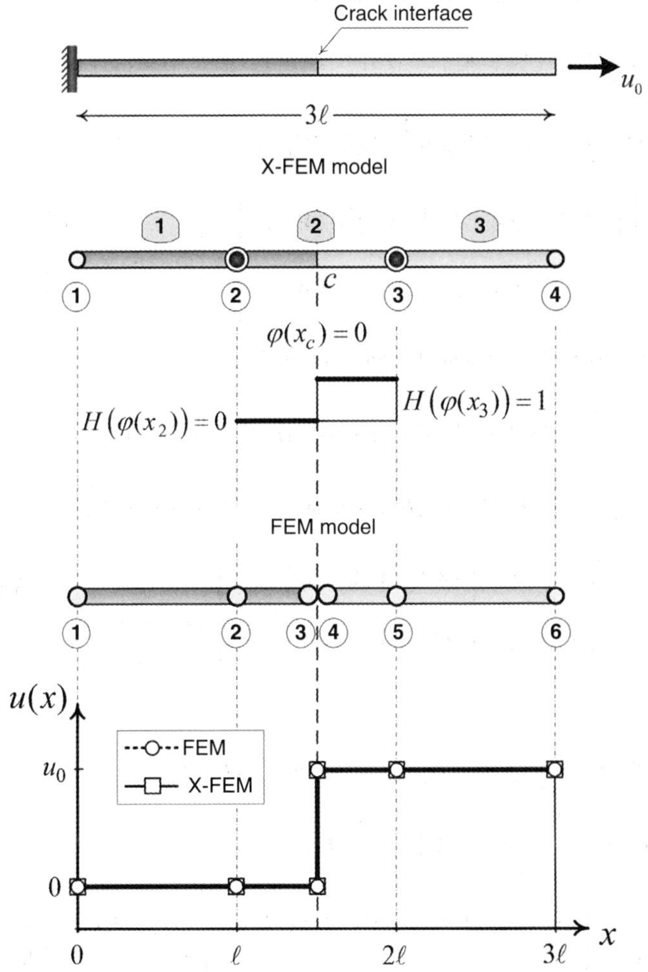

Figure 2.18 Modeling of an elastic one-dimensional bar with a discontinuity at its center: A comparison between the X-FEM and FEM techniques

and the total stiffness matrix of FEM model can be obtained as

$$\mathbf{K}^{\mathrm{FEM}} = \frac{AE}{\ell}\begin{bmatrix} 1 & -1 & 0 & 0 & 0 & 0 \\ -1 & 3 & -2 & 0 & 0 & 0 \\ 0 & -2 & 2 & 0 & 0 & 0 \\ 0 & 0 & 0 & 2 & -2 & 0 \\ 0 & 0 & 0 & -2 & 3 & -1 \\ 0 & 0 & 0 & 0 & -1 & 1 \end{bmatrix} \begin{matrix} u_1 \\ u_2 \\ u_3 \\ u_4 \\ u_5 \\ u_6 \end{matrix} \tag{2.83}$$

The linear system of FEM equations can be solved by applying the prescribed displacements as $u_1 = 0$ and $u_6 = u_0$ that result in the nodal displacements of $u_2 = u_3 = 0$ and $u_4 = u_5 = u_0$.

In X-FEM analysis, the discontinuity is modeled using the Heaviside step function (2.27) by enrichment of the nodal points 2 and 3, in which $H(\varphi(x_2)) = 0$ and $H(\varphi(x_3)) = 1$. The standard and enriched parts of the shape function $\mathbf{N}(x)$, the vector of Cartesian shape function derivatives $\mathbf{B}(x)$, and the stiffness matrix $\mathbf{K}$ for these three elements are as follows; for element (1)

$$\begin{aligned} &{}^{(1)}\mathbf{N}^{std}_{(x)} = \left\langle 1-\frac{x}{\ell} \;\; \frac{x}{\ell} \right\rangle, \quad {}^{(1)}\mathbf{N}^{enr}_{(x)} = \frac{x}{\ell}\left(H(\varphi)-H(\varphi_2)\right) \\ &{}^{(1)}\mathbf{B}^{std}_{(x)} = \left\langle -\frac{1}{\ell} \;\; \frac{1}{\ell} \right\rangle, \quad {}^{(1)}\mathbf{B}^{enr}_{(x)} = \frac{1}{\ell}\left(H(\varphi)-H(\varphi_2)\right) \\ &{}^{(1)}\mathbf{K} = \begin{bmatrix} \int_\ell (\mathbf{B}^{std})^T AE\,\mathbf{B}^{std} dx & \int_\ell (\mathbf{B}^{std})^T AE\,\mathbf{B}^{enr} dx \\ \int_\ell (\mathbf{B}^{enr})^T AE\,\mathbf{B}^{std} dx & \int_\ell (\mathbf{B}^{enr})^T AE\,\mathbf{B}^{enr} dx \end{bmatrix} = \frac{AE}{\ell}\begin{bmatrix} +1 & -1 & 0 \\ -1 & +1 & 0 \\ 0 & 0 & 0 \end{bmatrix} \begin{matrix} u_1 \\ u_2 \\ a_2 \end{matrix} \end{aligned} \tag{2.84a}$$

in which $H(\varphi) - H(\varphi_2) = 0$ for element (1). For element (2)

$$\begin{aligned} &{}^{(2)}\mathbf{N}^{std}_{(x)} = \left\langle 1-\frac{x}{\ell} \;\; \frac{x}{\ell} \right\rangle, \quad {}^{(2)}\mathbf{N}^{enr}_{(x)} = \left\langle \left(1-\frac{x}{\ell}\right)\left(H(\varphi)-H(\varphi_2)\right) \;\; \frac{x}{\ell}\left(H(\varphi)-H(\varphi_3)\right) \right\rangle \\ &{}^{(2)}\mathbf{B}^{std}_{(x)} = \left\langle -\frac{1}{\ell} \;\; \frac{1}{\ell} \right\rangle, \quad {}^{(2)}\mathbf{B}^{enr}_{(x)} = \left\langle -\frac{1}{\ell}\left(H(\varphi)-H(\varphi_2)\right) \;\; \frac{1}{\ell}\left(H(\varphi)-H(\varphi_3)\right) \right\rangle \\ &{}^{(2)}\mathbf{K} = \begin{bmatrix} \int_\ell (\mathbf{B}^{std})^T AE\,\mathbf{B}^{std} dx & \int_\ell (\mathbf{B}^{std})^T AE\,\mathbf{B}^{enr} dx \\ \int_\ell (\mathbf{B}^{enr})^T AE\,\mathbf{B}^{std} dx & \int_\ell (\mathbf{B}^{enr})^T AE\,\mathbf{B}^{enr} dx \end{bmatrix} = \frac{AE}{\ell}\begin{bmatrix} 1 & -1 & 0.5 & 0.5 \\ -1 & +1 & -0.5 & -0.5 \\ 0.5 & -0.5 & 0.5 & 0 \\ 0.5 & -0.5 & 0 & 0.5 \end{bmatrix} \begin{matrix} u_2 \\ u_3 \\ a_2 \\ a_3 \end{matrix} \end{aligned} \tag{2.84b}$$

and for element (3)

$$\begin{aligned} &{}^{(3)}\mathbf{N}^{std}_{(x)} = \left\langle 1-\frac{x}{\ell} \;\; \frac{x}{\ell} \right\rangle, \quad {}^{(3)}\mathbf{N}^{enr}_{(x)} = \left(1-\frac{x}{\ell}\right)\left(H(\varphi)-H(\varphi_3)\right) \\ &{}^{(3)}\mathbf{B}^{std}_{(x)} = \left\langle -\frac{1}{\ell} \;\; \frac{1}{\ell} \right\rangle, \quad {}^{(3)}\mathbf{B}^{enr}_{(x)} = -\frac{1}{\ell}\left(H(\varphi)-H(\varphi_3)\right) \\ &{}^{(3)}\mathbf{K} = \begin{bmatrix} \int_\ell (\mathbf{B}^{std})^T AE\,\mathbf{B}^{std} dx & \int_\ell (\mathbf{B}^{std})^T AE\,\mathbf{B}^{enr} dx \\ \int_\ell (\mathbf{B}^{enr})^T AE\,\mathbf{B}^{std} dx & \int_\ell (\mathbf{B}^{enr})^T AE\,\mathbf{B}^{enr} dx \end{bmatrix} = \frac{AE}{\ell}\begin{bmatrix} +1 & -1 & 0 \\ -1 & +1 & 0 \\ 0 & 0 & 0 \end{bmatrix} \begin{matrix} u_3 \\ u_4 \\ a_3 \end{matrix} \end{aligned} \tag{2.84c}$$

Thus, the total stiffness matrix of X-FEM model can be obtained as

$$\mathbf{K}^{\mathrm{XFEM}} = \frac{AE}{\ell}\begin{bmatrix} 1 & -1 & 0 & 0 & 0 & 0 \\ -1 & 2 & -1 & 0 & 0.5 & 0.5 \\ 0 & -1 & 2 & -1 & -0.5 & -0.5 \\ 0 & 0 & -1 & 1 & 0 & 0 \\ 0 & 0.5 & -0.5 & 0 & 0.5 & 0 \\ 0 & 0.5 & -0.5 & 0 & 0 & 0.5 \end{bmatrix}\begin{matrix} u_1 \\ u_2 \\ u_3 \\ u_4 \\ a_2 \\ a_3 \end{matrix} \tag{2.85}$$

The solution of linear system of X-FEM equations results in the following standard and enriched nodal displacements as $u_1 = u_2 = 0$, $u_3 = u_4 = u_0$, and $a_2 = a_3 = u_0$. In Figure 2.18, the evolutions of axial displacement are plotted along the bar for the FEM and X-FEM analyses. Obviously, the results of displacement are identical along the bar representing a great performance of the X-FEM with the Heaviside enrichment function.

2.8.2 *Modeling an Elastic Bar with a Weak Discontinuity*

The next example is chosen to illustrate the performance of X-FEM formulation for a bimaterial problem, in which an axial bar is modeled with a weak discontinuity at its center, as shown in Figure 2.19. The geometry and boundary conditions of the problem together with the FEM and X-FEM meshes are presented in this figure. The material properties of the left part are E_1 and ν_1, and those of the right part are E_2 and ν_2, where $E_1 = 2E_2 = E$ and $\nu_1 = \nu_2 = \nu$. The bar is restrained at one end and subjected to an axial load P at the free end. Similar to previous example, the X-FEM analysis is performed using three 2-noded elements, in which the material interface is placed at the mid-point of element (2). In order to model the material interface in the bar, the ramp function is employed considering the enriched degrees of freedom at nodal points 2 and 3. The results of X-FEM model are also compared with the FEM analysis carried out using four 2-noded elements.

The FEM analysis is performed using the total stiffness matrix calculated as

$$\mathbf{K}^{\mathrm{FEM}} = \frac{AE}{\ell}\begin{bmatrix} 1 & -1 & 0 & 0 & 0 \\ -1 & 3 & -2 & 0 & 0 \\ 0 & -2 & 3 & -1 & 0 \\ 0 & 0 & -1 & 1.5 & -0.5 \\ 0 & 0 & 0 & -0.5 & 0.5 \end{bmatrix}\begin{matrix} u_1 \\ u_2 \\ u_3 \\ u_4 \\ u_5 \end{matrix} \tag{2.86}$$

The solution of linear system of FE equations results in the following nodal displacements as

$$\bar{\mathbf{U}}^T = \langle u_1 \quad u_2 \quad u_3 \quad u_4 \quad u_5 \rangle = \frac{P\ell}{AE}\langle 0 \quad 1 \quad 1.5 \quad 2.5 \quad 4.5 \rangle \tag{2.87}$$

In the X-FEM analysis, the material interface is modeled using the ramp function by enrichment of the nodal points 2 and 3, in which $\varphi(x_2) < 0$ at the left side of the interface and $\varphi(x_3) > 0$ at the right side of the interface. The standard and enriched parts of the shape function $\mathbf{N}(x)$, the vector of Cartesian shape function derivatives $\mathbf{B}(x)$, and the stiffness matrix $\mathbf{K}$ for these three elements are obtained in a similar way to the previous example. The level set function of element (1) is defined using a local coordinate system as $\varphi(x) = x - 1.5\ell$, hence

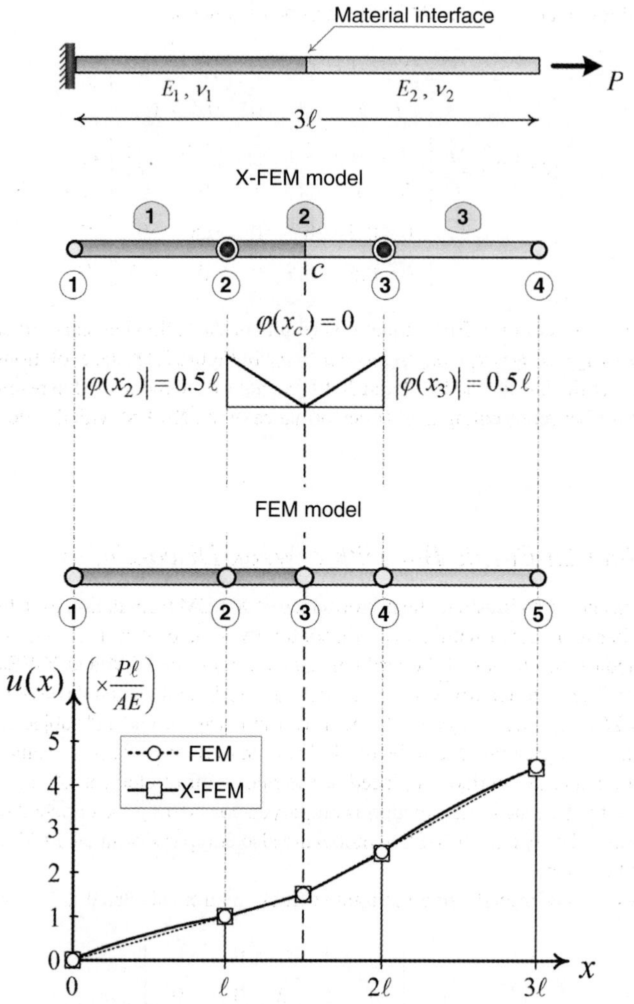

Figure 2.19 Modeling of an elastic one-dimensional bar with a material interface at its center: A comparison between the X-FEM and FEM techniques

$$
{}^{(1)}\mathbf{N}^{std}_{(x)} = \left\langle 1-\frac{x}{\ell}\;\; \frac{x}{\ell} \right\rangle,\quad {}^{(1)}\mathbf{N}^{enr}_{(x)} = \frac{x}{\ell}(|\varphi(x)|-|\varphi(x_2)|) = x-\frac{x^2}{\ell}
$$

$$
{}^{(1)}\mathbf{B}^{std}_{(x)} = \left\langle -\frac{1}{\ell}\;\; \frac{1}{\ell} \right\rangle,\quad {}^{(1)}\mathbf{B}^{enr}_{(x)} = \frac{1}{\ell}(|\varphi(x)|-|\varphi(x_2)|) + \frac{x}{\ell}\,\mathrm{sign}(\varphi(x)) = 1-\frac{2x}{\ell} \tag{2.88a}
$$

$$
{}^{(1)}\mathbf{K} = \frac{AE}{\ell}\begin{bmatrix} +1 & -1 & 0 \\ -1 & +1 & 0 \\ 0 & 0 & \ell^2/3 \end{bmatrix} \begin{matrix} u_1 \\ u_2 \\ a_2 \end{matrix}
$$

By defining the level set function of element (2) in a local coordinate system as $\varphi(x) = x - 0.5\ell$, it results

$$
\begin{aligned}
{}^{(2)}\mathbf{N}_{(x)} &= \left\langle {}^{(2)}\mathbf{N}^{std}_{(x)} \quad {}^{(2)}\mathbf{N}^{enr}_{(x)} \right\rangle \\
&= \left\langle 1-\frac{x}{\ell} \quad \frac{x}{\ell} \quad \left(1-\frac{x}{\ell}\right)(|\varphi(x)|-|\varphi(x_2)|) \quad \frac{x}{\ell}(|\varphi(x)|-|\varphi(x_3)|) \right\rangle \\
&= \left\langle 1-\frac{x}{\ell} \quad \frac{x}{\ell} \quad \left(1-\frac{x}{\ell}\right)\left(\left|x-\frac{\ell}{2}\right|-\frac{\ell}{2}\right) \quad \frac{x}{\ell}\left(\left|x-\frac{\ell}{2}\right|-\frac{\ell}{2}\right) \right\rangle
\end{aligned}
$$

$$
\begin{aligned}
{}^{(2)}\mathbf{B}_{(x)} &= \left\langle {}^{(2)}\mathbf{B}^{std}_{(x)} \quad {}^{(2)}\mathbf{B}^{enr}_{(x)} \right\rangle \\
&= \left\langle -\frac{1}{\ell} \quad \frac{1}{\ell} \quad -\frac{1}{\ell}(|\varphi(x)|-|\varphi(x_2)|)+\left(1-\frac{x}{\ell}\right)\operatorname{sign}(\varphi(x)) \quad \frac{1}{\ell}(|\varphi(x)|-|\varphi(x_3)|)+\frac{x}{\ell}\operatorname{sign}(\varphi(x)) \right\rangle \\
&= \left\langle -\frac{1}{\ell} \quad \frac{1}{\ell} \quad -\frac{1}{\ell}\left(\left|x-\frac{\ell}{2}\right|-\frac{\ell}{2}\right)+\left(1-\frac{x}{\ell}\right)\left(x-\frac{\ell}{2}\right)\Big/\left|x-\frac{\ell}{2}\right| \quad \frac{1}{\ell}\left(\left|x-\frac{\ell}{2}\right|-\frac{\ell}{2}\right)+\frac{x}{\ell}\left(x-\frac{\ell}{2}\right)\Big/\left|x-\frac{\ell}{2}\right| \right\rangle
\end{aligned}
$$

$$
{}^{(2)}\mathbf{K} = \frac{AE}{\ell}\begin{bmatrix} 3/4 & -3/4 & \ell/8 & \ell/8 \\ -3/4 & 3/4 & -\ell/8 & -\ell/8 \\ \ell/8 & -\ell/8 & \ell^2/4 & \ell^2/8 \\ \ell/8 & -\ell/8 & \ell^2/8 & \ell^2/4 \end{bmatrix}\begin{matrix} u_2 \\ u_3 \\ a_2 \\ a_3 \end{matrix} \tag{2.88b}
$$

and for element (3), the level set function in a local coordinate system is defined as $\varphi(x)=x+0.5\ell$, hence

$$
{}^{(3)}\mathbf{N}^{std}_{(x)} = \left\langle 1-\frac{x}{\ell} \quad \frac{x}{\ell} \right\rangle, \quad {}^{(3)}\mathbf{N}^{enr}_{(x)} = \left(1-\frac{x}{\ell}\right)(|\varphi(x)|-|\varphi(x_3)|) = x-\frac{x^2}{\ell}
$$

$$
{}^{(3)}\mathbf{B}^{std}_{(x)} = \left\langle -\frac{1}{\ell} \quad \frac{1}{\ell} \right\rangle, \quad {}^{(3)}\mathbf{B}^{enr}_{(x)} = -\frac{1}{\ell}(|\varphi(x)|-|\varphi(x_3)|)+\left(1-\frac{x}{\ell}\right)\operatorname{sign}(\varphi(x)) = 1-\frac{2x}{\ell}
$$

$$
{}^{(3)}\mathbf{K} = \frac{AE}{2\ell}\begin{bmatrix} +1 & -1 & 0 \\ -1 & +1 & 0 \\ 0 & 0 & \ell^2/3 \end{bmatrix}\begin{matrix} u_3 \\ u_4 \\ a_3 \end{matrix} \tag{2.88c}
$$

Thus, the total stiffness matrix of the X-FEM model can be obtained as

$$
\mathbf{K}^{\text{XFEM}} = \frac{AE}{\ell}\begin{bmatrix} 1 & -1 & 0 & 0 & 0 & 0 \\ -1 & 7/4 & -3/4 & 0 & \ell/8 & \ell/8 \\ 0 & -3/4 & 5/4 & -1/2 & -\ell/8 & -\ell/8 \\ 0 & 0 & -1/2 & 1/2 & 0 & 0 \\ 0 & \ell/8 & -\ell/8 & 0 & 7\ell^2/12 & \ell^2/8 \\ 0 & \ell/8 & -\ell/8 & 0 & \ell^2/8 & 5\ell^2/12 \end{bmatrix}\begin{matrix} u_1 \\ u_2 \\ u_3 \\ u_4 \\ a_2 \\ a_3 \end{matrix} \tag{2.89}
$$

The solution of linear system of X-FEM equations results in the following standard and enriched nodal displacements as

$$
\begin{aligned}
\bar{\mathbf{U}}^T &= \langle u_1 \quad u_2 \quad u_3 \quad u_4 \quad a_2 \quad a_3 \rangle \\
&= \frac{P\ell}{AE}\langle 0 \quad 1 \quad 2.4317 \quad 4.4317 \quad 0.2295/\ell \quad 0.3607/\ell \rangle
\end{aligned} \tag{2.90}
$$

In Figure 2.19, the evolutions of axial displacement are plotted along the bar for the FEM and X-FEM analyses. It is noteworthy to highlight that a slight difference between the FEM and X-FEM solutions is due to loss of the PU in blending elements, that is, elements (1) and (3), as noted in Section 2.5. However, this error can be reduced using various techniques proposed for blending elements in Chapter 4.

2.8.3 Modeling an Elastic Plate with a Crack Interface at its Center

The next example is of an elastic plate with a strong discontinuity at its center, for example, a crack interface, chosen to illustrate the performance of X-FEM formulation for a two-dimensional problem. The geometry, boundary conditions, and X-FEM mesh of the problem are shown in Figure 2.20. The material properties of the plate are chosen as follows; $E = 2 \times 10^6$ kg/cm^2 and $\nu = 0.3$. The plate is in the plane strain condition, which is fixed at the left edge and subjected to a prescribed displacement $u_0 = 1$ cm at the right edge. In order to model the discontinuity in the plate, the Heaviside sign function is employed considering the enriched degrees of freedom at nodal points of element (2) as

$$H(x) = \begin{cases} +1 & x > x_c \\ -1 & x < x_c \end{cases} \tag{2.91}$$

The problem is modeled by three 4-noded elements using the standard shape functions $N^{std}_{I(\xi,\eta)} = \frac{1}{4}(1+\xi_I\xi)(1+\eta_I\eta)$ and the enrichment shape functions $N^{enr}_{J(\xi,\eta)} = N^{std}_{J(\xi,\eta)}(H(\boldsymbol{x}) - H(\boldsymbol{x}_J))$. The numerical integration is performed with four Gauss integration points in elements (1) and (3), and employing the rectangular sub-grids method for element (2), where the cut element is divided into four sub-grids with four Gauss points at each sub-rectangle. The stiffness matrices for these three elements can be obtained according to (2.68) as follows

$$^{(1)}\mathbf{K} = {}^{(3)}\mathbf{K} = \int_{-1}^{+1}\int_{-1}^{+1} \mathbf{B}^T_{(\xi,\eta)} \mathbf{D}\, \mathbf{B}_{(\xi,\eta)} \det[\mathbf{J}] d\xi\, d\eta \equiv \left({}^{(1)}\mathbf{K}_{uu} = {}^{(3)}\mathbf{K}_{uu}\right)$$

$$= 1.0 \times 10^6 \begin{array}{c} \begin{array}{cccccccc} u_{x_1} & u_{y_1} & u_{x_2} & u_{y_2} & u_{x_3} & u_{y_3} & u_{x_4} & u_{y_4} \end{array} \\ \left[\begin{array}{cccccccc} 1.154 & 0.481 & -0.769 & 0.096 & -0.577 & -0.481 & 0.192 & -0.096 \\ 0.481 & 1.154 & -0.096 & 0.192 & -0.481 & -0.577 & 0.096 & -0.769 \\ -0.769 & -0.096 & 1.154 & -0.481 & 0.192 & 0.096 & -0.577 & 0.481 \\ 0.096 & 0.192 & -0.481 & 1.154 & -0.096 & -0.769 & 0.481 & -0.577 \\ -0.577 & -0.481 & 0.192 & -0.096 & 1.154 & 0.481 & -0.769 & 0.096 \\ -0.481 & -0.577 & 0.096 & -0.769 & 0.481 & 1.154 & -0.096 & 0.192 \\ 0.192 & 0.096 & -0.577 & 0.481 & -0.769 & -0.096 & 1.154 & -0.481 \\ -0.096 & -0.769 & 0.481 & -0.577 & 0.096 & 0.192 & -0.481 & 1.154 \end{array}\right] \end{array} \begin{array}{c} u_{x_1} \\ u_{y_1} \\ u_{x_2} \\ u_{y_2} \\ u_{x_3} \\ u_{y_3} \\ u_{x_4} \\ u_{y_4} \end{array} \tag{2.92a}$$

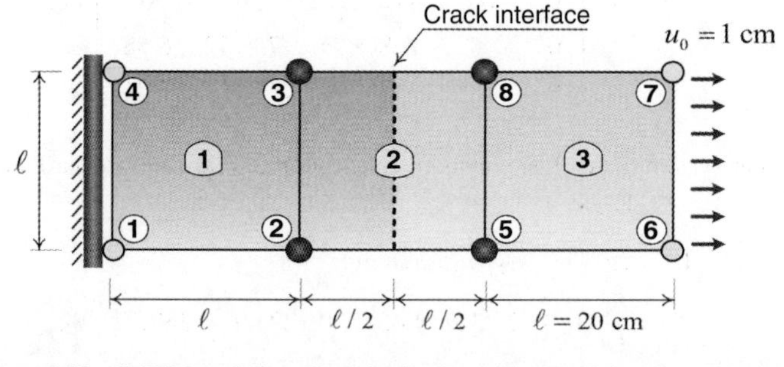

Figure 2.20 X-FEM modeling of an elastic plate with a crack interface at its center

in which the stiffness matrices of elements (1) and (3) corresponding to enriched degrees of freedom are zero, that is, $\mathbf{K}_{ua} = \mathbf{K}_{au} = \mathbf{K}_{aa} = \mathbf{0}$. The stiffness matrix of element (2) that contains the crack interface can be obtained using (2.68) as

$$
{}^{(2)}\mathbf{K} = \begin{bmatrix} \mathbf{K}_{uu} & \mathbf{K}_{ua} \\ \mathbf{K}_{au} & \mathbf{K}_{aa} \end{bmatrix} = \begin{bmatrix} \int_{-1}^{+1}\int_{-1}^{+1} \mathbf{B}^{std\,T}_{(\xi,\eta)} \mathbf{D}\, \mathbf{B}^{std}_{(\xi,\eta)} \det[\mathbf{J}]\, d\xi\, d\eta & \int_{-1}^{+1}\int_{-1}^{+1} \mathbf{B}^{std\,T}_{(\xi,\eta)} \mathbf{D}\, \mathbf{B}^{enr}_{(\xi,\eta)} \det[\mathbf{J}]\, d\xi\, d\eta \\ \int_{-1}^{+1}\int_{-1}^{+1} \mathbf{B}^{enr\,T}_{(\xi,\eta)} \mathbf{D}\, \mathbf{B}^{std}_{(\xi,\eta)} \det[\mathbf{J}]\, d\xi\, d\eta & \int_{-1}^{+1}\int_{-1}^{+1} \mathbf{B}^{enr\,T}_{(\xi,\eta)} \mathbf{D}\, \mathbf{B}^{enr}_{(\xi,\eta)} \det[\mathbf{J}]\, d\xi\, d\eta \end{bmatrix}
$$

$$
\begin{array}{c}
\begin{array}{cccccccccccccccc}
u_{x_2} & u_{y_2} & u_{x_5} & u_{y_5} & u_{x_8} & u_{y_8} & u_{x_3} & u_{y_3} & a_{x_2} & a_{y_2} & a_{x_5} & a_{y_5} & a_{x_8} & a_{y_8} & a_{x_3} & a_{y_3}
\end{array} \\
= 1.0\times 10^{6} \left[\begin{array}{rrrrrrrrrrrrrrrr}
1.154 & 0.481 & -0.769 & 0.096 & -0.577 & -0.481 & 0.192 & -0.096 & 0.962 & 0.24 & 0.769 & 0.144 & 0.577 & 0.433 & 0.385 & -0.048 \\
0.481 & 1.154 & -0.096 & 0.192 & -0.481 & -0.577 & 0.096 & -0.769 & 0.24 & 0.481 & 0.337 & -0.192 & 0.529 & 0.577 & 0.048 & -0.096 \\
-0.769 & -0.096 & 1.154 & -0.481 & 0.192 & 0.096 & -0.577 & 0.481 & -0.769 & 0.144 & -0.962 & 0.24 & -0.385 & -0.048 & -0.577 & 0.433 \\
0.096 & 0.192 & -0.481 & 1.154 & -0.096 & -0.769 & 0.481 & -0.577 & 0.337 & 0.192 & 0.24 & -0.481 & 0.048 & 0.096 & 0.529 & -0.577 \\
-0.577 & -0.481 & 0.192 & -0.096 & 1.154 & 0.481 & -0.769 & 0.096 & -0.577 & -0.433 & -0.385 & 0.048 & -0.962 & -0.24 & -0.769 & -0.144 \\
-0.481 & -0.577 & 0.096 & -0.769 & 0.481 & 1.154 & -0.096 & 0.192 & -0.529 & -0.577 & -0.048 & 0.096 & -0.24 & -0.481 & -0.337 & 0.192 \\
0.192 & 0.096 & -0.577 & 0.481 & -0.769 & -0.096 & 1.154 & -0.481 & 0.385 & 0.048 & 0.577 & -0.433 & 0.769 & -0.144 & 0.962 & -0.24 \\
-0.096 & -0.769 & 0.481 & -0.577 & 0.096 & 0.192 & -0.481 & 1.154 & -0.048 & -0.096 & -0.529 & 0.577 & -0.337 & -0.192 & -0.24 & 0.481 \\
0.962 & 0.24 & -0.769 & 0.337 & -0.577 & -0.529 & 0.385 & -0.048 & 1.923 & 0.481 & 0 & 0 & 0 & 0 & 0.769 & -0.096 \\
0.24 & 0.481 & 0.144 & 0.192 & -0.433 & -0.577 & 0.048 & -0.096 & 0.481 & 0.962 & 0 & 0 & 0 & 0 & 0.096 & -0.192 \\
0.769 & 0.337 & -0.962 & 0.24 & -0.385 & -0.048 & 0.577 & -0.529 & 0 & 0 & 1.923 & -0.481 & 0.769 & 0.096 & 0 & 0 \\
0.144 & -0.192 & 0.24 & -0.481 & 0.048 & 0.096 & -0.433 & 0.577 & 0 & 0 & -0.481 & 0.962 & -0.096 & -0.192 & 0 & 0 \\
0.577 & 0.529 & -0.385 & 0.048 & -0.962 & -0.24 & 0.769 & -0.337 & 0 & 0 & 0.769 & -0.096 & 1.923 & 0.481 & 0 & 0 \\
0.433 & 0.577 & -0.048 & 0.096 & -0.24 & -0.481 & -0.144 & -0.192 & 0 & 0 & 0.096 & -0.192 & 0.481 & 0.962 & 0 & 0 \\
0.385 & 0.048 & -0.577 & 0.529 & -0.769 & -0.337 & 0.962 & -0.24 & 0.769 & 0.096 & 0 & 0 & 0 & 0 & 1.923 & -0.481 \\
-0.048 & -0.096 & 0.433 & -0.577 & -0.144 & 0.192 & -0.24 & 0.481 & -0.096 & -0.192 & 0 & 0 & 0 & 0 & -0.481 & 0.962
\end{array}\right]
\begin{array}{c}
u_{x_2} \\ u_{y_2} \\ u_{x_5} \\ u_{y_5} \\ u_{x_8} \\ u_{y_8} \\ u_{x_3} \\ u_{y_3} \\ a_{x_2} \\ a_{y_2} \\ a_{x_5} \\ a_{y_5} \\ a_{x_8} \\ a_{y_8} \\ a_{x_3} \\ a_{y_3}
\end{array}
\end{array}
\tag{2.92b}
$$

By assembling the stiffness matrices of elements (1), (2) and (3), a set of linear system of equations can be obtained as

$$
\begin{bmatrix} {}^{(1)}\mathbf{K} + {}^{(2)}\mathbf{K}_{uu} + {}^{(3)}\mathbf{K} & {}^{(2)}\mathbf{K}_{ua} \\ {}^{(2)}\mathbf{K}_{au} & {}^{(2)}\mathbf{K}_{aa} \end{bmatrix}_{24\times 24} \begin{Bmatrix} \bar{\mathbf{u}}_{16\times 1} \\ \bar{\mathbf{a}}_{8\times 1} \end{Bmatrix} = \begin{Bmatrix} \mathbf{F}^{std}_{\text{ext}} \\ \mathbf{0} \end{Bmatrix}_{24\times 1} \tag{2.93}
$$

The solution of this linear system of X-FEM equations results in the following standard and enriched nodal displacements as

$$
\begin{aligned}
\bar{\mathbf{U}}^{T} &= \langle u_{x_1} \;\; u_{y_1} \;\; u_{x_2} \;\; u_{y_2} \;\; u_{x_3} \;\; u_{y_3} \;\; u_{x_4} \;\; u_{y_4} \;\; u_{x_5} \;\; u_{y_5} \;\; u_{x_6} \;\; u_{y_6} \;\; u_{x_7} \;\; u_{y_7} \;\; u_{x_8} \;\; u_{y_8} \rangle \\
&= \langle 0 \;\; 0 \;\; 0 \;\; 0 \;\; 0 \;\; 0 \;\; 0 \;\; 0 \;\; 1 \;\; 0 \;\; 1 \;\; 0 \;\; 1 \;\; 0 \;\; 1 \;\; 0 \rangle \\
\bar{\mathbf{a}}^{T} &= \langle a_{x_2} \;\; a_{y_2} \;\; a_{x_5} \;\; a_{y_5} \;\; a_{x_8} \;\; a_{y_8} \;\; a_{x_3} \;\; a_{y_3} \rangle \\
&= \langle 0.5 \;\; 0 \;\; 0.5 \;\; 0 \;\; 0.5 \;\; 0 \;\; 0.5 \;\; 0 \rangle
\end{aligned}
\tag{2.94}
$$

In Figure 2.21a, the evolutions of standard displacement $\mathbf{u}^{std}_{(\boldsymbol{x})} = \mathbf{N}^{std}(\boldsymbol{x})\,\bar{\mathbf{u}}$ and the enriched displacement $\mathbf{u}^{enr}_{(\boldsymbol{x})} = \mathbf{N}^{std}(\boldsymbol{x})(H(\boldsymbol{x}) - H(\boldsymbol{x}_J))\,\bar{\mathbf{a}}$ are plotted along the plate. In Figure 2.21b, the evolutions of total displacement field $\mathbf{u}^{total}_{(\boldsymbol{x})} = \mathbf{u}^{std}_{(\boldsymbol{x})} + \mathbf{u}^{enr}_{(\boldsymbol{x})}$ are plotted along the plate, which is compared with that obtained by the FEM solution. Obviously, a great performance of the X-FEM with the Heaviside enrichment function can be seen in this figure.

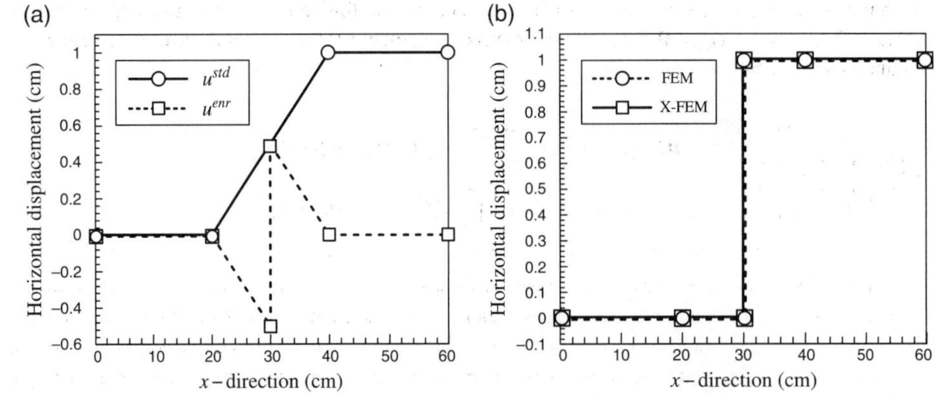

Figure 2.21 The evolutions of displacements along the plate with a crack interface at its center: (a) the standard and enriched displacement fields in X-FEM solution; (b) a comparison between the X-FEM and FEM solutions

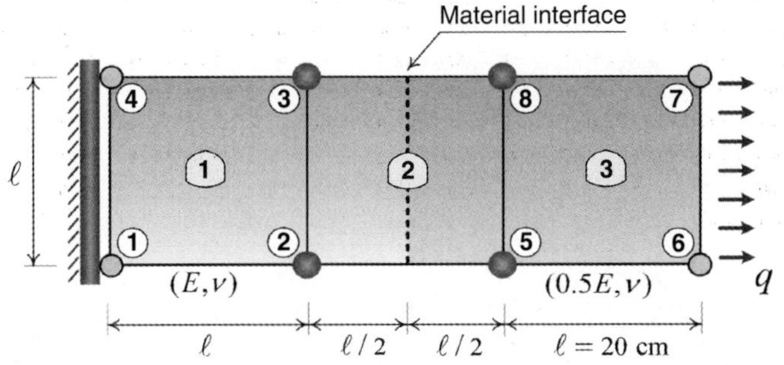

Figure 2.22 X-FEM modeling of an elastic plate with a material interface at its center

2.8.4 Modeling an Elastic Plate with a Material Interface at its Center

The last example is of a plate in tension with the material interface at its center chosen to illustrate the performance of X-FEM formulation for a bimaterial problem. The geometry, boundary conditions, and X-FEM mesh of the problem are shown in Figure 2.22. The material properties of the left part are E and ν, and those of the right part are $0.5E$ and ν, where $E = 2 \times 10^6$ kg/cm^2 and $\nu = 0.3$. The plate is in plane strain condition, which is fixed at the left edge and subjected to a tensile load $q = 2 \times 10^4$ kg/cm at the free end. In order to model the material interface in the plate, the ramp function is employed considering the enriched degrees of freedom at nodal points of element (2). The ramp function is computed using the absolute value of level set function $|\varphi(\mathbf{x})|$, in which $\varphi(x_2) = \varphi(x_3) < 0$ at the left side of the interface and $\varphi(x_5) = \varphi(x_8) > 0$ at the right side of the interface.

X-FEM analysis is carried out using three 4-noded elements with the standard shape functions $N^{std}_{I(\xi,\eta)} = \frac{1}{4}(1 + \xi_I \xi)(1 + \eta_I \eta)$ and enriched shape functions $N^{enr}_{J(\xi,\eta)} = N^{std}_{J(\xi,\eta)}(|\varphi(\mathbf{x})| - |\varphi(\mathbf{x}_J)|)$. The numerical integration is performed using 4×4 Gauss integration points for blending elements (1) and (3), and employing the rectangular sub-grids method for element (2), where the cut element is divided into four sub-grids with four Gauss points at each sub-rectangle. The stiffness matrices for these three elements can be obtained according to (2.68) as follows

$$
{}^{(1)}\mathbf{K} = 2 \times {}^{(3)}\mathbf{K} = \int_{-1}^{+1}\int_{-1}^{+1} \mathbf{B}_{(\xi,\eta)}^{T}\mathbf{D}\,\mathbf{B}_{(\xi,\eta)} \det[\mathbf{J}]d\xi\, d\eta = \begin{bmatrix} (\mathbf{K}_{uu})_{8\times 8} & (\mathbf{K}_{ua})_{8\times 4} \\ (\mathbf{K}_{au})_{4\times 8} & (\mathbf{K}_{aa})_{4\times 4} \end{bmatrix}
$$

$$
\begin{matrix} u_{x_1} & u_{y_1} & u_{x_2} & u_{y_2} & u_{x_3} & u_{y_3} & u_{x_4} & u_{y_4} & a_{x_2} & a_{y_2} & a_{x_3} & a_{y_3} \end{matrix}
$$

$$
= 1.0 \times 10^6 \begin{bmatrix}
1.154 & 0.481 & -0.769 & 0.096 & -0.577 & -0.481 & 0.192 & -0.096 & 1.282 & 0.641 & -1.282 & -3.205 \\
0.481 & 1.154 & -0.096 & 0.192 & -0.481 & -0.577 & 0.096 & -0.769 & -0.641 & 4.487 & -3.205 & -4.487 \\
-0.769 & -0.096 & 1.154 & -0.481 & 0.192 & 0.096 & -0.577 & 0.481 & 1.282 & -0.641 & -1.282 & 3.205 \\
0.096 & 0.192 & -0.481 & 1.154 & -0.096 & -0.769 & 0.481 & -0.577 & 0.641 & 4.487 & 3.205 & -4.487 \\
-0.577 & -0.481 & 0.192 & -0.096 & 1.154 & 0.481 & -0.769 & 0.096 & -1.282 & -3.205 & 1.282 & 0.641 \\
-0.481 & -0.577 & 0.096 & -0.769 & 0.481 & 1.154 & -0.096 & 0.192 & -3.205 & -4.487 & -0.641 & 4.487 \\
0.192 & 0.096 & -0.577 & 0.481 & -0.769 & -0.096 & 1.154 & -0.481 & -1.282 & 3.205 & 1.282 & -0.641 \\
-0.096 & -0.769 & 0.481 & -0.577 & 0.096 & 0.192 & -0.481 & 1.154 & 3.205 & -4.487 & 0.641 & 4.487 \\
1.282 & -0.641 & 1.282 & 0.641 & -1.282 & -3.205 & -1.282 & 3.205 & 129.915 & 0.0 & 49.573 & 0.0 \\
0.641 & 4.487 & -0.641 & 4.487 & -3.205 & -4.487 & 3.205 & -4.487 & 0.0 & 70.085 & 0.0 & -18.803 \\
-1.282 & -3.205 & -1.282 & 3.205 & 1.282 & -0.641 & 1.282 & 0.641 & 49.573 & 0.0 & 129.915 & 0.0 \\
-3.205 & -4.487 & 3.205 & -4.487 & 0.641 & 4.487 & -0.641 & 4.487 & 0.0 & -18.803 & 0.0 & 70.085
\end{bmatrix}
\begin{matrix} u_{x_1} \\ u_{y_1} \\ u_{x_2} \\ u_{y_2} \\ u_{x_3} \\ u_{y_3} \\ u_{x_4} \\ u_{y_4} \\ a_{x_2} \\ a_{y_2} \\ a_{x_3} \\ a_{y_3} \end{matrix}
\tag{2.95a}
$$

and the stiffness matrix of element (2) that contains the material interface can be obtained using (2.68) as

$$
{}^{(2)}\mathbf{K} = \begin{bmatrix} (\mathbf{K}_{uu})_{8\times 8} & (\mathbf{K}_{ua})_{8\times 8} \\ (\mathbf{K}_{au})_{8\times 8} & (\mathbf{K}_{aa})_{8\times 8} \end{bmatrix} = \begin{bmatrix} \int_{-1}^{+1}\int_{-1}^{+1} \mathbf{B}_{(\xi,\eta)}^{std\,T}\mathbf{D}\,\mathbf{B}_{(\xi,\eta)}^{std} \det[\mathbf{J}]d\xi\, d\eta & \int_{-1}^{+1}\int_{-1}^{+1} \mathbf{B}_{(\xi,\eta)}^{std\,T}\mathbf{D}\,\mathbf{B}_{(\xi,\eta)}^{enr} \det[\mathbf{J}]d\xi\, d\eta \\ \int_{-1}^{+1}\int_{-1}^{+1} \mathbf{B}_{(\xi,\eta)}^{enr\,T}\mathbf{D}\,\mathbf{B}_{(\xi,\eta)}^{std} \det[\mathbf{J}]d\xi\, d\eta & \int_{-1}^{+1}\int_{-1}^{+1} \mathbf{B}_{(\xi,\eta)}^{enr\,T}\mathbf{D}\,\mathbf{B}_{(\xi,\eta)}^{enr} \det[\mathbf{J}]d\xi\, d\eta \end{bmatrix}
$$

$$
\begin{matrix} u_{x_2} & u_{y_2} & u_{x_5} & u_{y_5} & u_{x_8} & u_{y_8} & u_{x_3} & u_{y_3} & a_{x_2} & a_{y_2} & a_{x_5} & a_{y_5} & a_{x_8} & a_{y_8} & a_{x_3} & a_{y_3} \end{matrix}
$$

$$
= 1.0 \times 10^6 \begin{bmatrix}
0.913 & 0.421 & -0.577 & 0.012 & -0.433 & -0.349 & 0.096 & -0.084 & 1.242 & 0.080 & 1.643 & 0.160 & 1.723 & 2.083 & 2.123 & 2.484 \\
0.421 & 1.034 & -0.132 & 0.144 & -0.373 & -0.433 & 0.084 & -0.745 & 1.122 & -2.865 & 1.042 & -1.462 & 2.324 & 2.424 & 2.724 & 3.826 \\
-0.577 & -0.132 & 0.817 & -0.300 & 0.192 & 0.060 & -0.433 & 0.373 & -2.845 & 0.881 & -2.925 & 0.801 & -0.441 & -1.122 & -0.521 & -1.522 \\
0.012 & 0.144 & -0.300 & 0.697 & -0.060 & -0.409 & 0.349 & -0.433 & 0.321 & -2.744 & 0.401 & -3.025 & -0.881 & 2.063 & -1.282 & 1.783 \\
-0.433 & -0.373 & 0.192 & -0.060 & 0.817 & 0.300 & -0.577 & 0.132 & -0.521 & 1.522 & -0.441 & 1.122 & -2.925 & -0.801 & -2.845 & -0.881 \\
-0.349 & -0.433 & 0.060 & -0.409 & 0.300 & 0.697 & -0.012 & 0.144 & 1.282 & 1.783 & 0.881 & 2.063 & -0.401 & -3.025 & -0.321 & -2.744 \\
0.096 & 0.084 & -0.433 & 0.349 & -0.577 & -0.012 & 0.913 & -0.421 & 2.123 & -2.484 & 1.723 & -2.083 & 1.643 & -0.160 & 1.242 & -0.080 \\
-0.084 & -0.745 & 0.373 & -0.433 & 0.132 & 0.144 & -0.421 & 1.034 & -2.724 & 3.826 & -2.324 & 2.424 & -1.042 & -1.462 & -1.122 & -2.865 \\
1.242 & 1.122 & -2.845 & 0.321 & -0.521 & 1.282 & 2.123 & -2.724 & 95.753 & -6.010 & 49.279 & -3.606 & 18.029 & -10.817 & 38.862 & 1.202 \\
0.080 & -2.865 & 0.881 & -2.744 & 1.522 & 1.783 & -2.484 & 3.826 & -6.010 & 46.675 & -8.413 & 28.245 & -13.221 & -9.014 & -1.202 & -8.213 \\
1.643 & 1.042 & -2.925 & 0.401 & -0.441 & 0.881 & 1.723 & -2.324 & 49.279 & -8.413 & 94.151 & -6.010 & 40.465 & 1.202 & 18.029 & 13.221 \\
0.160 & -1.462 & 0.801 & -3.025 & 1.122 & 2.063 & -2.083 & 2.424 & -3.606 & 28.245 & -6.010 & 41.066 & -1.202 & -2.604 & 10.817 & -9.014 \\
1.723 & 2.324 & -0.441 & -0.881 & -2.925 & -0.401 & 1.643 & -1.042 & 18.029 & -13.221 & 40.465 & -1.202 & 94.151 & 6.010 & 49.279 & 8.413 \\
2.083 & 2.424 & -1.122 & 2.063 & -0.801 & -3.025 & -0.160 & -1.462 & -10.817 & -9.014 & 1.202 & -2.604 & 6.010 & 41.066 & 3.606 & 28.245 \\
2.123 & 2.724 & -0.521 & -1.282 & -2.845 & -0.321 & 1.242 & -1.122 & 38.862 & -1.202 & 18.029 & 10.817 & 49.279 & 3.606 & 95.753 & 6.010 \\
2.484 & 3.826 & -1.522 & 1.783 & -0.881 & -2.744 & -0.080 & -2.865 & 1.202 & -8.213 & 13.221 & -9.014 & 8.413 & 28.245 & 6.010 & 46.675
\end{bmatrix}
\begin{matrix} u_{x_2} \\ u_{y_2} \\ u_{x_5} \\ u_{y_5} \\ u_{x_8} \\ u_{y_8} \\ u_{x_3} \\ u_{y_3} \\ a_{x_2} \\ a_{y_2} \\ a_{x_5} \\ a_{y_5} \\ a_{x_8} \\ a_{y_8} \\ a_{x_3} \\ a_{y_3} \end{matrix}
\tag{2.95b}
$$

By assembling the stiffness matrices of elements (1), (2), and (3), a set of linear system of equations can be obtained as $\mathbf{K}_{24\times 24}\bar{\mathbf{U}}_{24\times 1} = \mathbf{F}_{24\times 1}$. The solution of linear system of X-FEM equations results in the following standard and enriched nodal displacements as

$$
\begin{aligned}
\bar{\mathbf{U}}^T &= \langle u_{x_2} \quad u_{y_2} \quad u_{x_3} \quad u_{y_3} \quad u_{x_5} \quad u_{y_5} \quad u_{x_6} \quad u_{y_6} \quad u_{x_7} \quad u_{y_7} \quad u_{x_8} \quad u_{y_8} \rangle \\
&= \langle 0.18 \quad 0.04 \quad 0.18 \quad -0.04 \quad 0.43 \quad 0.08 \quad 0.79 \quad 0.07 \quad 0.79 \quad -0.07 \quad 0.43 \quad -0.08 \rangle \\
\bar{\mathbf{a}}^T &= \langle a_{x_2} \quad a_{y_2} \quad a_{x_5} \quad a_{y_5} \quad a_{x_8} \quad a_{y_8} \quad a_{x_3} \quad a_{y_3} \rangle \\
&= \langle 0.0013 \quad 0.0028 \quad 0.0034 \quad 0.0002 \quad 0.0034 \quad -0.0002 \quad 0.0013 \quad -0.0028 \rangle
\end{aligned}
\tag{2.96}
$$

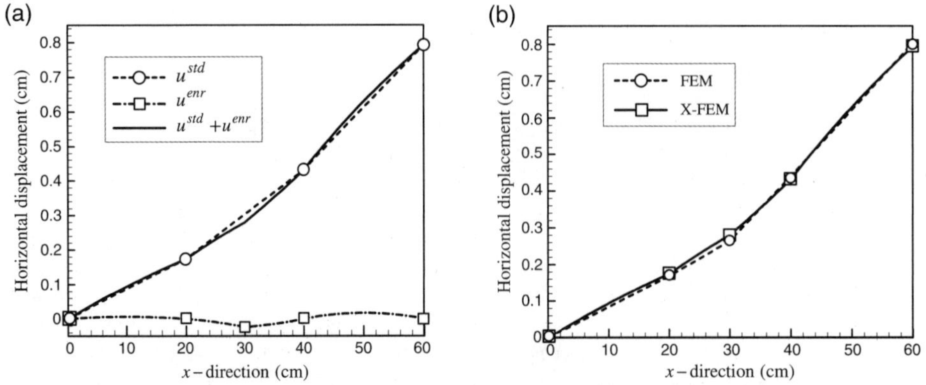

Figure 2.23 The evolutions of displacements along the plate with a material interface at its center: (a) the standard, enriched, and total displacement fields in X-FEM solution; (b) a comparison between the X-FEM and FEM solutions

In Figure 2.23a, the evolutions of standard displacement $\mathbf{u}^{std}_{(x)} = \mathbf{N}^{std}(\boldsymbol{x})\,\bar{\mathbf{u}}$, the enriched displacement $\mathbf{u}^{enr}_{(x)} = \mathbf{N}^{std}(\boldsymbol{x})(|\varphi(\boldsymbol{x})| - |\varphi(\boldsymbol{x}_J)|)\;\bar{\mathbf{a}}$, and the total displacement $\mathbf{u}^{total}_{(x)} = \mathbf{u}^{std}_{(x)} + \mathbf{u}^{enr}_{(x)}$ are plotted along the plate. In Figure 2.23b, a comparison is performed between the X-FEM and FEM displacement fields, which shows a great performance of the X-FEM with the ramp enrichment function.

2.9 Higher Order X-FEM

The implementation of higher order elements has been of great interest in the FEM. Higher order elements improve both the accuracy and rate of convergence of FEM solution. Hence, it is importance to utilize the higher order elements in the X-FEM. In Figure 2.24, a representation of curved interface discretized by the linear and higher order shape functions is demonstrated. It has been shown that the jumps and kinks can be accurately approximated by the X-FEM within an element, where the optimal convergence rate is frequently achieved for the linear elements and piecewise planar interfaces. However, the extension to *higher order convergence* for arbitrarily curved weak discontinuities leads to two major difficulties; firstly an accurate quadrature of the Galerkin weak form for the elements crossed by the curved interface, and secondly a proper enrichment formulation that prevents any problems in the blending elements.

The first study on the implementation of higher order elements in X-FEM was performed by Stazi *et al.* (2003) for a curved crack problem. They used different order of shape functions for the standard and enriched parts, where the higher order shape functions are only employed in the standard part. Moreover, the geometry of the crack was represented by a level set function interpolated on the same quadratic finite element discretization. The signed distance function is approximated by the standard shape functions as

$$\varphi(\boldsymbol{x}) = \sum_{i=1}^{\mathcal{N}} N_i^{std}(\boldsymbol{x})\,\varphi_i \tag{2.97}$$

where φ_i are the nodal values of level set function $\varphi(\boldsymbol{x})$ and $N_i^{std}(\boldsymbol{x})$ are the quadratic shape functions. This technique is able to capture the curved crack interface with a high accuracy, where no increase of enriched degrees of freedom was required. As can be seen from Figure 2.25, the higher order elements better candidates in modeling curved discontinuity than the linear elements that are capable of representing the curved interface. Based on this technique, a higher-order description of the interface is achieved by

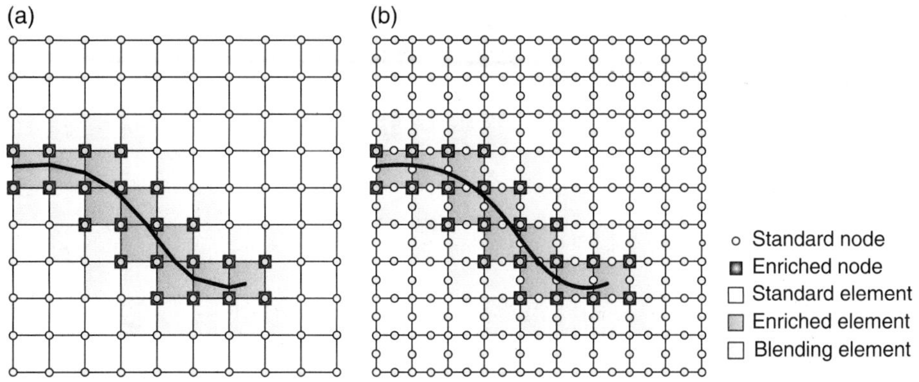

Figure 2.24 The X-FEM modeling of a curved interface with (a) linear elements, and (b) higher order elements

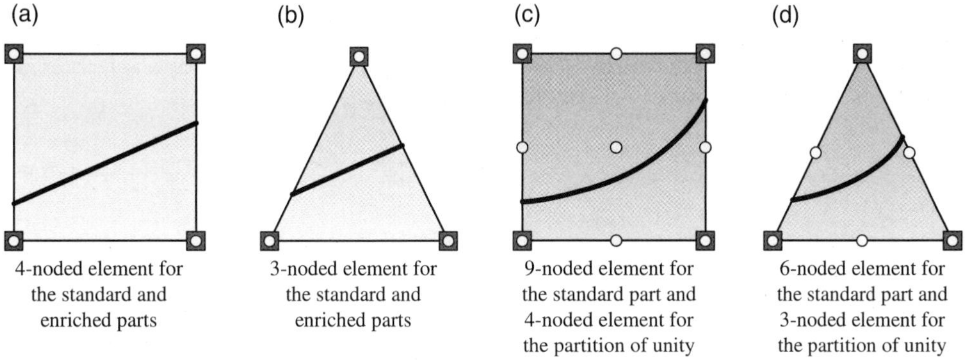

Figure 2.25 The X-FEM modeling of a curved interface using quadrilateral and triangular elements: (a and b) linear elements; (c and d) higher order elements

employing a higher-order FE interpolation of the interface using the level set method. In order to perform the numerical integration, an efficient strategy of sub-dividing the cut element into subcells was proposed by Cheng and Fries (2009). For the generated subcells, the reference elements are defined to map the integration points to the real domain, as shown in Figure 2.26. In this figure, a linear Q4 element, which is cut by the curved interface, is first divided into two triangular subcells. Assuming that each edge of the element is cut only once, the sub-triangle cut by the discontinuity is further divided into triangular and quadrilateral subcells with only *one* curved side. The standard reference elements are defined as special 5-noded triangular and 6-noded quadrilateral reference elements, where four nodes are assigned on only one side and two nodes on all other sides. The side with four nodes is mapped to the curved side of the triangular or quadrilateral subcells in the real domain. An alternative sub-division is presented in the right hand side of Figure 2.26, where the two sub-triangles cut by the curved interface are further divided into two triangular and quadrilateral subcells. It has been shown that four nodes along the edge which is mapped onto the curved interface are sufficient to achieve good accuracy for quadratic and cubic elements.

One of the main difficulties with the higher order X-FEM is the blending elements. In Section 2.5, it was illustrated through a simple one-dimensional axial bar that the blending elements result in a low convergence rate in the X-FEM. It was shown that, because of the enriched part of displacement field, the error of interpolation increases from $O(h^2)$ to $O(h)$ in blending elements when the linear shape functions are used

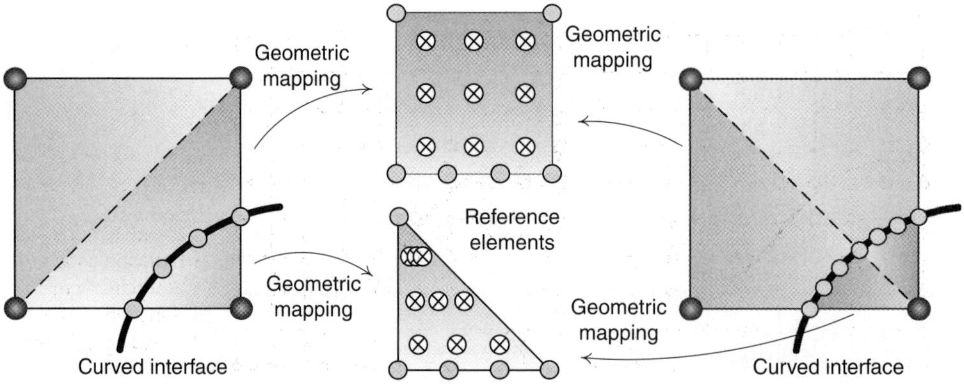

Figure 2.26 Numerical integration of an element cut by a curved interface based on the sub-division of cut element into subcells and mapping the integration points of reference elements to the real domain

for the enriched part of approximation field. In the case that enrichment is based on a polynomial of higher order n ($n > 1$), the error of interpolation can be increased even further in blending elements. On the basis of relation (2.44), the error of interpolation can be obtained as

$$e_u(\tilde{x}) \leq \frac{1}{8}\ell^2 \left| \frac{d^2u}{dx^2} + \frac{2\bar{a}_2}{\ell^n} \right| \tag{2.98}$$

It can be clearly seen from this error estimation that the term of $2\bar{a}_2/\ell^n$ increases the error of interpolation even further compared to the case where linear shape functions are used for the enriched part. Hence, the higher order shape functions must not be employed in the enriched parts, particularly when the weak discontinuity is used. Although this causes no problems for strong discontinuities, such as the crack problem based on the Heaviside enrichment function, it degrades the accuracy slightly for crack tip enrichments.

Laborde *et al.* (2005) investigated the rate of convergence for the higher order X-FEM. They highlighted that for the Heaviside enrichment function, which is used to represent the jump in the displacement field across the discontinuity, the corresponding PU must be of the same degree as the FE polynomials, so that the displacement can be approximated in a satisfactory way along the discontinuity interface. In a different work, Byfut and Schröder (2012) showed that the higher-order Lagrange type PU functions can be used to improve the overall convergence rate. Based on their numerical experiments, it was shown that the optimal algebraic and even exponential convergence rates can be obtained in the X-FEM. However, they noted that the computational cost can be increased due to the expensive integration of the enrichment functions; and even, for higher-order PU functions, the condition numbers for the resulting linear systems of equations increased drastically. In order to alleviate these problems and to continue within the framework of the PUM, it was proposed to use a PU that is of lower-order for elements cut by the discontinuity interface and of higher-order away from the discontinuity. However, using the Lagrange type shape functions, it is difficult to handle element-wise anisotropic and, in particular, varying polynomial degrees; for example, elements with lower polynomial degrees at the discontinuity interface and elements with higher polynomial degrees away from the discontinuity. To this end, it was suggested to introduce the first-order shape functions for the enriched part of approximation field, as employed by Stazi *et al.* (2003) and Laborde *et al.* (2005). However, this basically means that the concept of the PUM is left aside. Moreover, the convergence rates for this approach are clearly suboptimal. Consequently, a solution was proposed by Fries (2008) to cancel the "unwanted terms" in the transition layer between the enriched and non-enriched elements. An alternative approach was presented by Byfut and Schröder (2012) based on the definition of a global lower-order PU along with some higher order polynomial enrichment functions. In the literature, this technique is referred to as p-adaptivity in the generalized FEM, or as the hp-cloud method. It must be noted that, apart from their ability to easily allow for varying and, moreover, anisotropic polynomial degrees, these hierarchical shape functions have proven to be highly efficient in hp-FEM while significantly improving the condition number of the resulting linear system of equations compared to Lagrange-type shape functions. Note that the hierarchical shape functions can be generated for higher order FEM by tensor products of integrated Legendre or Gauss–Lobatto polynomials. Due to their tensor product structure and, in particular, their definition via a recurrence relation, these shape functions and their derivatives can be evaluated as numerically stable.

2.10 Implementation of X-FEM with Higher Order Elements

In this section, the performance of the higher order X-FEM is presented for the weak and strong discontinuous problems. According to the enriched displacement field (2.59), the displacement approximation consists of the standard and enriched parts, where the higher order shape functions must be carefully taken into the computation. On the basis of previous discussion, it must be noted do not use the higher order shape functions $N_I(\boldsymbol{x})$ as a PU support in the enriched parts. In fact, the quadratic or higher order polynomials can be used as standard shape functions, whereas the linear shape function must be employed to build the PU. As can be seen from Table 2.1, the 9-noded quadratic element is employed for the standard part and the 4-noded rectangular element for a PU. Hence, the higher order enriched displacement field (2.59) can be written as

$$\mathbf{u}(\boldsymbol{x}, t) = \sum_{i=1}^{\mathcal{N}} N_i(\boldsymbol{x})\, \bar{\mathbf{u}}_i + \sum_{j=1}^{\mathcal{M}} \widehat{N}_j(\boldsymbol{x}) \left(\psi(\boldsymbol{x}) - \psi(\boldsymbol{x}_j)\right) \bar{\mathbf{a}}_j \tag{2.99}$$

in which $N_i(\boldsymbol{x})$ are the quadratic shape functions, whereas $\widehat{N}_j(\boldsymbol{x})$ are the linear shape functions that construct the PU in enriched parts. It must be noted that the implementation of different order interpolants is crucial for the standard and enriched shape functions in achieving the optimal convergence rate. The reason can be related to the presence of elements located in the blending sub-domain in which only some nodal points of these elements are enriched. Since these elements are partially enriched, the enriched nodes of these elements do not form a PU.

2.10.1 *Higher Order X-FEM Modeling of a Plate with a Material Interface*

In order to illustrate the performance of higher-order X-FEM, an elastic plate with a weak discontinuity of bimaterial interface at its center is modeled numerically. The plate is 1×1 m with the geometry and boundary conditions presented in Figure 2.27. The material properties of the plate are chosen as follows; the top part of plate includes $E_1 = 2 \times 10^6$ kg/cm^2 and $\nu_1 = 0.3$, and the bottom part of plate consists of $E_2 = 2 \times 10^5$ kg/cm^2 and $\nu_2 = 0.3$. The plate is in the plane strain condition, which is fixed at the top edge and is subjected to a tension loading $q = 10$ kg/cm at the bottom edge. In order to present the accuracy of the proposed computational algorithm in modeling the material interface inside the plate, X-FEM analysis is carried out in three different cases; that is, one 4-noded linear element for the standard and enriched parts, one 9-noded quadratic element for the standard part together with a 4-noded element for the PU, and one 9-noded quadratic element for standard and enriched parts. In all X-FEM analyses, the material interface is placed at the middle of element, which is modeled by the ramp enrichment function. The stiffness matrix of enriched element is obtained using Eq. (2.68) by partitioning the element into eight sub-triangles with three Gauss integration points at each sub-triangle. The results of X-FEM analyses are

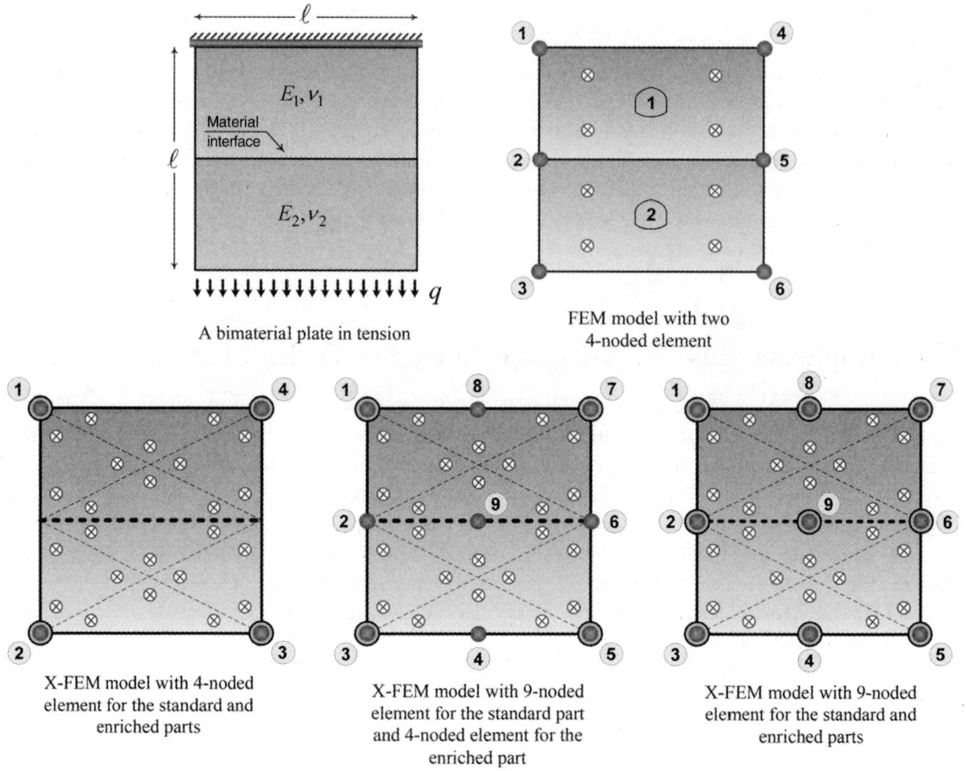

Figure 2.27 Modeling an elastic plate with a material interface at its center with linear and higher-order X-FEM: A comparison between the FEM and X-FEM techniques

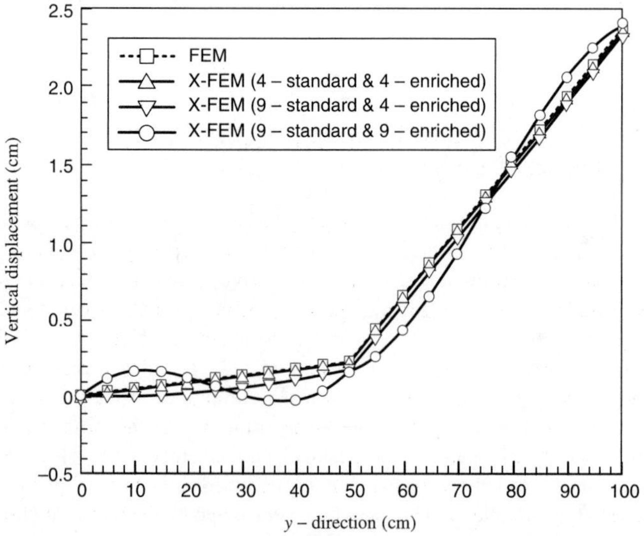

Figure 2.28 The evolutions of displacement fields along the plate for the FEM and X-FEM analyses

compared with the FEM technique carried out using two 4-noded elements. In Figure 2.28, the evolutions of displacement fields are plotted along the plate for three X-FEM solutions and the results are compared with the FEM analysis. Obviously, there is a remarkable agreement between the FEM solution and that obtained by the X-FEM model using the standard and enriched Q4 elements. It can be observed from the higher order X-FEM solutions that, although the Q9 standard element with Q4 enriched element presents a better accuracy of the solution, the X-FEM model with standard and enriched Q9 elements noticeably decreases accuracy of the solution. As noted earlier, it must be highlighted that the higher order shape functions must not be used as PU in the enriched parts when the weak discontinuity is employed (see Companion Website).

2.10.2 Higher Order X-FEM Modeling of a Plate with a Curved Crack Interface

The next example is of an elastic plate with a curved crack interface, chosen to illustrate the performance of higher order X-FEM model in a problem with strong discontinuity. The plate is 2 × 2 m with the geometry and boundary conditions presented in Figure 2.29. The material properties of the plate are chosen as follows; $E_1 = 2 \times 10^6$ kg/cm^2 and $\nu_1 = 0.3$. The plate is in the plane strain condition, which is fixed at the top edge and is subjected to a prescribed displacement $u_0 = 1$ mm at the bottom edge. In order to present the performance of a higher order X-FEM technique, the plate is modeled using one 9-noded quadratic element for the standard part together with a 4-noded linear element for the PU. In X-FEM analysis, the curved interface is modeled by the Heaviside enrichment function, where the standard Q9 shape functions are employed to represent the geometry of the crack. In fact, the higher- order Q9 shape functions are used for discretization of level set function $\varphi(\boldsymbol{x})$, while 4-noded shape functions are utilized for the PU in the enriched part. Hence, the signed distance function is approximated using the standard shape functions as $\varphi(\boldsymbol{x}) = \sum_{i=1}^{9} N_i^{std}(\boldsymbol{x})\,\varphi_i$. The stiffness matrix of the enriched element is obtained using Eq. (2.68) by partitioning the element into 6 × 6 rectangular sub-grids with one Gauss integration point at each sub-rectangle. The results of X-FEM analysis is compared with the FEM technique carried out using

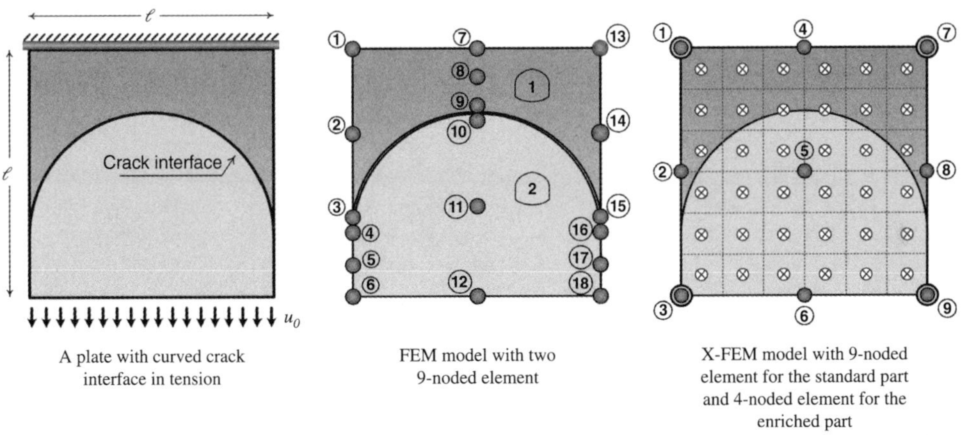

Figure 2.29 Modeling an elastic plate with a curved crack interface using the higher–order X-FEM: A comparison between the FEM and X-FEM techniques

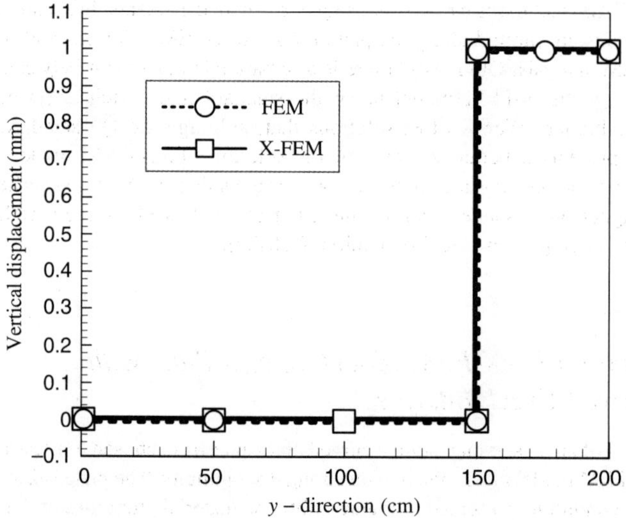

Figure 2.30 The evolutions of displacement fields along the left edge of the plate for the FEM and X-FEM analyses

two 9-noded elements. In Figure 2.30, the evolutions of displacement fields are plotted along the left edge of the plate for the X-FEM and FEM analyses. Great agreement can be seen between the FEM and higher order X-FEM technique when the Heaviside enrichment function $H(\boldsymbol{x})$ is employed, however, it must be noted that the order of standard shape function $N_i(\boldsymbol{x})$ must be equal or more than the order of enriched shape function $\widehat{N}_j(\boldsymbol{x})$, if the strong discontinuity is employed.

3

Enrichment Elements

3.1 Introduction

Problems involving irregularly shaped domains with arbitrary interfaces occur in a wide range of engineering models. These include models in computational mechanics, computational biology, fluid mechanics, solidification, and so on. Numerical solutions for these types of problems can be difficult to obtain, and thus a variety of methods have been developed to solve these problems. A common technique for the solution of problems with embedded interfaces or irregular domains is the construction of a conforming mesh. This allows the problem to be defined and solved in a straightforward manner using the finite element method (FEM). The FEM has been proven an attractive technique that has been used in a large number of applications. However, the need for a conforming mesh can be a major drawback in some cases. The mesh construction is often time consuming for difficult domain shapes, and poorly constructed meshes can produce ill-conditioned approximations. This is an even more serious drawback for problems with evolving interfaces, as the domain must be remeshed as the geometry changes (Smith, Vaughan, and Chopp, 2007).

One method that has been shown to perform well is the extended finite element method (X-FEM). The X-FEM is a *partition of unity method* (PUM) in which enrichment functions are added to the conventional finite element (FE) function space in order to capture discontinuous behavior at an internal interface. It was first utilized in combination with the *level set method* (LSM), and then employed for a variety of applications, including solidification, multiphase flow, and biofilm growth. What distinguishes the X-FEM from conventional FEMs is the ability to account for an internal interface or irregularly shaped external domain without the need for a conforming mesh; both the location of the interface and the local solution behavior are embedded in the enrichment functions. This allows the method to solve moving interface problems without the need for remeshing as well as to capture large solution gradients near an interface without the need for extremely fine grids.

Tracking evolving interfaces occur in a wide variety of problems, and have been studied by researchers for many years. The goal in tracking moving interfaces is to deal accurately and robustly with the formation of cusps and corners, topological changes in the propagating interface, and stability issues in three-dimensional (3D) problems. The LSM is among various techniques used efficiently for tracking the evolution of complex interfaces (Sethian, 1999). The technique offers remarkably powerful tool for understanding, analyzing, and computing the interface motion in a host of settings. The technique is based on the solution of an initial value partial differential equation for a level set function, in which the idea is taken from the hyperbolic conservation law. The topological changes, corner and cusp development, and accurate determination of geometric properties, such as curvature and normal direction, are naturally obtained in this technique. The

Extended Finite Element Method: Theory and Applications, First Edition. Amir R. Khoei.

methodology results in a robust, accurate, and efficient numerical algorithm for propagating interfaces in highly complex settings.

To account for an internal interface, enrichment functions must be added to the space of standard FE shape functions. If the enrichments depend only on the signed distance to the interface, as described in Section 2.4.1, it allows the X-FEM to be easily coupled to the LSM. In the X-FEM, the enrichment functions added to the standard FE approximation basically provide two main requirements; firstly, they instruct the location of the interface into the function space itself, which allows for the application of both Dirichlet and Neumann interface conditions without the need for a conforming mesh, and secondly it makes the opportunity to include information about the known asymptotic behavior of the solution near the interface, allowing the method to accurately capture boundary layer behavior without the need for a very fine mesh. Thus, the ideal enrichment function for a given application is dependent on both the type of solution behavior expected near the interface and the types of boundary conditions to be applied. For example, the Heaviside enrichment function, described in Section 2.4.1, is the simplest and most flexible enrichment function that can be used to solve problems with a jump in the solution value, the normal derivative, or both. Because the enrichment itself does not specify a jump in the solution or its derivative, two separate conditions must be applied in order to maintain well-posedness. However, for solutions that are continuous across the interface but include a jump in the normal derivative, a continuous enrichment function must be employed, such as the *ramp* function or the *absolute value* of level set function. It is important to note that, in these functions, the enrichment decays to zero at the interface, which is an important criterion for stability when a Dirichlet interface condition is applied.

Once a suitable enrichment function has been selected, a method for selecting nodes for enrichment is needed. The most common approach is *topological* enrichment, where a node is enriched if the interface cuts across its support. This method is most effective when using enrichments that do not increase the order of the FE approximation, such as the Heaviside enrichment function (see Section 2.5). However, for enrichments that increase the order of the FE approximation, such as the ramp or absolute value enrichment function, the *geometric* enrichment is needed where all nodes within a specific signed distance of the interface are enriched. It is important to note that if a boundary layer is used for the enrichment, it ensures that the enriched region captures the entire layer. Furthermore, it allows the higher order enriched approximations to blend with the standard approximation far away from the interface, which preserves the accuracy of the solution near the interface without the need for complicated blending elements.

Obviously, there are two essential resolutions that must be made when implementing the X-FEM for a specific problem. Firstly, the type of enrichment function that depends on the type of interface conditions as well as the expected solution behavior near the interface. Secondly, the method of enrichment; while the *topological* enrichment is appropriate for most problems, the *geometric* enrichment is preferable for problems that include boundary layers or require continuous enrichment functions. In this chapter, an introduction is first given for tracking evolving interfaces based on the LSM and fast marching method (FMM). It is then followed by an overview of various X-FEM enrichment functions used in a wide variety of problems, such as bimaterials, cracks, dislocations, fluid-structure interactions, shear bands, convection-diffusion, thermo-mechanical, deformable porous media, piezoelectric, magneto-electroelastic, topology optimization, rigid particles in Stokes flow, solidification, and so on.

3.2 Tracking Moving Boundaries

In a variety of problems, the goal is to track a propagation front that can form sharp corners, change topology, and break and merge as it involves. Such phenomena can occur in fluid mechanics, material science, micro-fabrication of electronic components, combustion, metrology, control theory, and image processing, as well as more mathematical problems in minimal surfaces and construction of geodesics. Computational methods for tracking moving interfaces have played an important role in numerical modeling of a wide range of these applications. In many of these applications, the solution of an elliptic equation involving irregularly shaped moving boundaries is required in order to obtain the velocity field of the

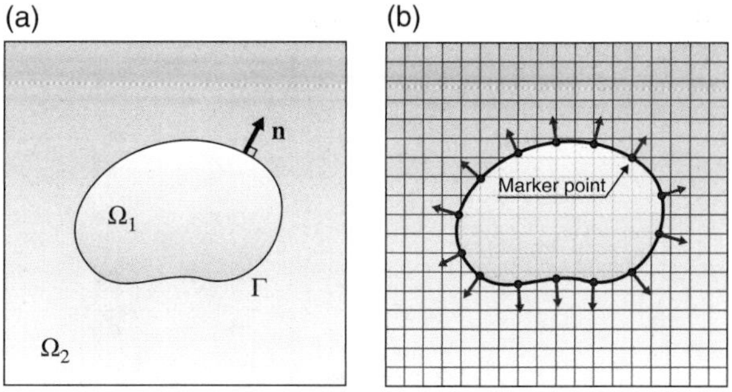

Figure 3.1 Tracking moving boundary; (a) problem definition; (b) the marker point method

interface. There are various techniques proposed in literature for computing the motion of interfaces, each with their own strengths and weaknesses making no single technique optimal for all moving interface problems. The traditional techniques can be generally categorized as the marker/string method, the cell-based method, and the characteristic methods (Sethian, 1999).

In the *marker/string method*, a discrete parameterized version of the interface boundary is employed, in which the marker particles are used for two-dimensional (2D) problem, and a nodal triangularization of the interface is used for 3D problem (Sethian, 1996). The positions of marker points are updated in time according to approximations to the equations of motions by determination of the front information about the normal and curvature from the marker representation. Such representations can be quite accurate; however, limitations exist for complex motions. If the corners and cusps develop in the evolving front, markers usually form "swallowtail" solutions that must be removed through de-looping techniques that attempt to enforce an entropy condition inherent in such motion. In addition, the topological changes are difficult to handle; when regions merge, some markers must be removed. Furthermore, significant instabilities in the front can result; since the underlying marker particle motions represent a weakly ill-posed initial value problem. Finally, extensions of such methods to three dimensions require additional work (Figure 3.1).

In the *cell-based method*, the computational domain is divided into a set of cells that contain "volume fractions" and is known as the "volume of fluid" technique, which, rather than track the boundary of the propagating front, track the motion of the interior region (Adalsteinsson and Sethian, 1995). In this algorithm, the interior is discretized by employing a grid on the domain and assigning to each cell a "volume fraction" corresponding to the amount of interior fluid currently located in that cell. The volume fractions are numbers between 0 and 1, and represent the fraction of each cell containing the physical material. At any time, the front can be reconstructed from these volume fractions. Various elaborate reconstruction techniques have been developed to include pitched slopes and curved surfaces. The accompanying accuracy depends on the sophistication of the reconstruction and the so-called "advection sweeps". The advantages of such techniques include the ability to easily handle topological changes, design adaptive mesh methods, and extending the results to three dimensions. However, determination of geometric quantities such as the normal and curvature can be inaccurate (Figure 3.2).

In *characteristic methods*, which are based on the "ray-trace" techniques, the characteristic equations for the propagating interface are used, and the entropy condition at forming corners is formally enforced by constructing the envelope of the evolving characteristics. Such methods handle the looping problems more naturally, but may be complex in three dimensions and require adaptive adding and removing rays, which can cause instabilities and/or over-smoothing (Sethian, 1999) (Figure 3.3).

On the basis of this idea, two efficient techniques of the LSM and FMM have been developed. These are computational techniques proposed for tracking propagating interfaces on the basis of finite difference schemes. They share the virtues of working in an arbitrary number of space dimensions with no change, handle topological merger and splitting with no special procedures, and accurately and efficiently

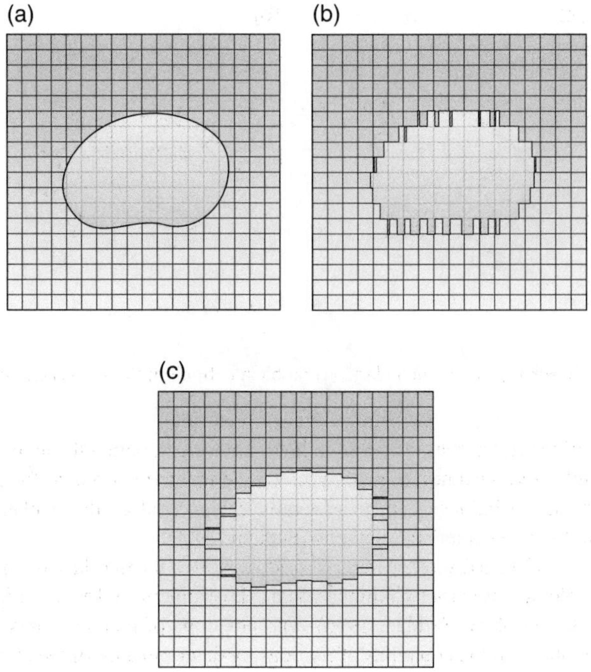

Figure 3.2 The cell-based method; (a) An interface in the domain; (b) The representation of an interface in x–pass; (c) The representation of an interface in y–pass

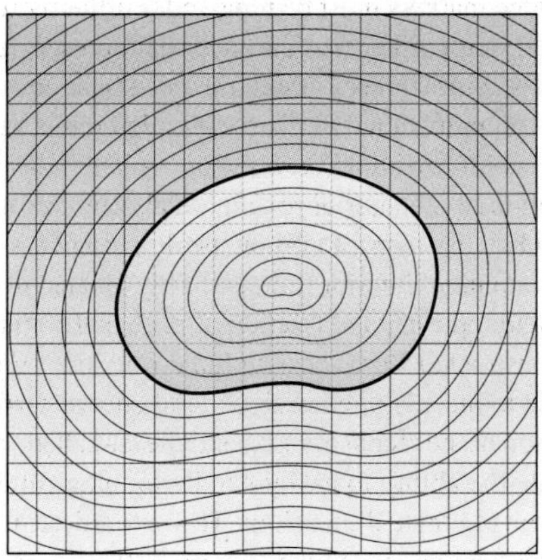

Figure 3.3 The characteristic method

compute the motion of fronts with sharp corners moving under speed laws that may include large variations in velocities. Basically, the LSM approximates the solution of an initial value partial differential equation (Osher and Sethian, 1988), while the FMM approximates the solution of boundary value partial differential equations with the view of propagating interface (Sethian, 1999). At the core, both techniques rely on viscosity solutions for Hamilton–Jacobi equations, linking finite difference upwind schemes for hyperbolic conservation laws to propagating fronts, and aspects of the theory of curve and surface evolution. They have been used in a large variety of applications, including problems in fluid interface motion, combustion, dendritic solidification, etching and deposition in semi-conductor manufacturing, robotic navigation and path planning, image segmentation in medical imaging scans, computation of seismic travel times, and aspects of computational geometry and computer vision.

3.3 Level Set Method

The general idea behind the LSM is to apply a function $\varphi(\boldsymbol{x}, t)$ to the space where the interface exists, with $\boldsymbol{x}$ denoting a point in that space and t a point in time. The function is initialized at $t = 0$, and then a scheme is used to approximate the value of $\varphi(\boldsymbol{x}, t)$ over each time step. In order to initialize the value of $\varphi(\boldsymbol{x}, t)$ at each point of the mesh, the function φ is defined as $\varphi(\boldsymbol{x}, t) = \pm\, d$ for any point $\boldsymbol{x}$ in the mesh, where d is the distance from the point $\boldsymbol{x}$ to the curve at the time $t = 0$. The positive sign is used if the point $\boldsymbol{x}$ is outside the closed curve; the negative sign is used if the point $\boldsymbol{x}$ is inside the closed curve. Thus, the LSM is known as the evolving curve corresponding to the locus of all points $\boldsymbol{x}$ at any time t_0, such that $\varphi(\boldsymbol{x}, t_0) = 0$, and that locus is a level curve of the function φ. The locus of all points $\boldsymbol{x}$ such that $\varphi(\boldsymbol{x}, t_0) = c$, contour around the original curve, where c is an arbitrary positive or negative constant. A computationally intensive algorithm can be used to set up the initial value of function φ at each point of the mesh, where the distance between each point of the mesh and each point on the curve is determined and the minimum value is set as the mesh point value of φ.

In the LSM, the real interface is replaced by a new interface of same order of the original domain; for example, for a one-dimensional domain, the extents of the interface are points in which the LSM is replaced by a one-dimensional curve. Similarly for two- and three-dimensional problems the interface is replaced by a surface and a volume function of the same order of the original domain. This approach was devised by Osher and Sethian (1988) as a simple and versatile method for computing and analyzing the motion of an interface Γ in two or three dimensions, in which Γ bounds a (possibly multiply connected) region Ω. The goal is to compute and analyze the subsequent motion of Γ under a velocity field $\boldsymbol{v}$. This velocity is dependent on the position, the time, the geometry of the interface, and the external physics. The interface is captured for a time increment as the zero level set of a smooth (at least Lipschitz continuous) function $\varphi(\boldsymbol{x}, t) = 0$ (Osher and Fedkiw, 2001). A schematic view of a level set function for a two-dimensional problem is presented in Figure 3.4.

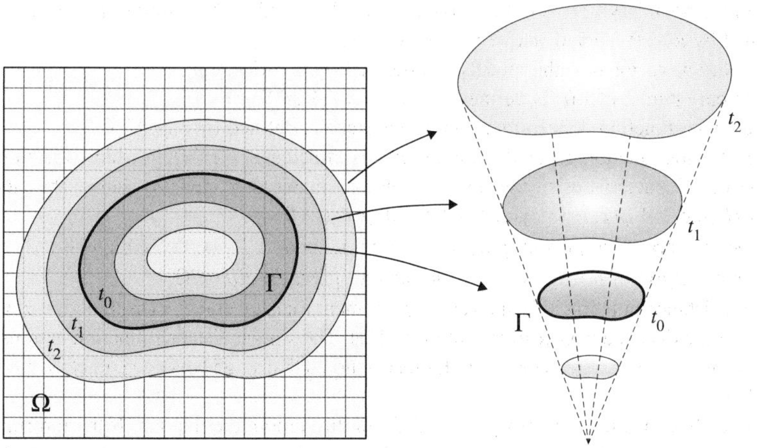

Figure 3.4 The level set function for a two-dimensional problem

Consider a domain Ω containing an interface Γ, the level set function $\varphi(\boldsymbol{x}, t)$ can be defined by dividing the domain into two separate zones Ω_A and Ω_B as

$$\begin{aligned} \varphi(\boldsymbol{x},t)>0 \quad & \text{if } x\in\Omega_A \\ \varphi(\boldsymbol{x},t)=0 \quad & \text{if } x\in\Gamma \\ \varphi(\boldsymbol{x},t)<0 \quad & \text{if } x\in\Omega_B \end{aligned} \tag{3.1}$$

in which the interface can be captured for each time increment by merely locating the set $\Gamma(t)$ for which φ vanishes. This deceptively trivial statement is of great significance for numerical computation, primarily because topological changes such as breaking and merging are well defined and performed "without emotional involvement" (Osher and Sethian, 1988). A common definition of level set function is based on the signed distance function introduced in the previous chapter as

$$\varphi(\boldsymbol{x}) = \|\boldsymbol{x}-\boldsymbol{x}_\Gamma\| \operatorname{sign}(\mathbf{n}.(\boldsymbol{x}-\boldsymbol{x}_\Gamma)) \tag{3.2}$$

where $\boldsymbol{x}_\Gamma$ is the closest point projection of $\boldsymbol{x}$ onto the interface Γ, $\mathbf{n}$ is the normal vector to the interface at point $\boldsymbol{x}_\Gamma$, and $\|\boldsymbol{x}-\boldsymbol{x}_\Gamma\|$ specifies the distance of point $\boldsymbol{x}$ to the interface Γ. The singed distance function introduces a distance and a sign to each particular position with a linear behavior that is desirable for numerical simulations.

3.3.1 Numerical Implementation of LSM

In order to numerically implement the LSM, a simple and computationally fast remedy is presented for reinitializing the function $\varphi(\boldsymbol{x}, t)$ to be signed distance to Γ, at least near the boundary, smoothly extending the velocity field off of the front Γ, and solving the equation of motion only locally near the interface Γ, that makes the cost of LSM competitive with that of the boundary integral methods. It must be emphasized that all this is easy to implement in the presence of boundary singularities and/or topological changes and in two-/three-dimensional problems. Moreover, in the case that the velocity field is a function of the direction of the unit normal, such as the kinetic crystal growth, the equation of motion becomes the first-order Hamilton–Jacobi equation.

A list of key terms and parameters involved in the LSM can be defined as (Osher and Fedkiw, 2001)

1. The interface boundary $\Gamma(t)$ is defined by $\varphi(\boldsymbol{x}, t) = 0$, in which the interior and exterior regions are determined by $\varphi(\boldsymbol{x}, t) < 0$ and $\varphi(\boldsymbol{x}, t) > 0$, respectively,
2. The unit normal vector $\mathbf{n}$ to the interface is defined by $\mathbf{n} = \nabla\varphi/|\nabla\varphi|$,
3. The mean curvature κ of $\Gamma(t)$ is defined as by $\kappa = \nabla\cdot(\nabla\varphi/|\nabla\varphi|)$,
4. The Dirac delta function concentrated on an interface is defined by $\delta(\varphi)|\nabla\varphi|$,
5. The characteristic function ψ of a domain Ω can be defined as $\psi = H(\varphi)$, in which ψ is chosen to determine specific characteristics of the domain, such as material properties. Hence, the Heaviside step function $H(\varphi) \equiv 1$ if $\varphi(\boldsymbol{x}, t) > 0$, and $H(\varphi) \equiv 0$ if $\varphi(\boldsymbol{x}, t) < 0$.
6. The surface integral of a quantity $p(\boldsymbol{x}, t)$ over Γ is defined as $\int_\Gamma p(\boldsymbol{x}, t)\, \delta(\varphi)|\nabla\varphi| d\Gamma$,
7. The domain integral of $p(\boldsymbol{x}, t)$ over Ω is defined as $\int_\Omega p(\boldsymbol{x}, t)\, H(\varphi)\, d\Omega$,
8. The singed distance function is defined as a replacement for a general level set function $\varphi(\boldsymbol{x}, t)$, which is the value of the distance from $\boldsymbol{x}$ to the interface $\Gamma(t)$. The singed distance function assures that φ does not become too flat or too steep near $\Gamma(t)$. It must be noted that singed distance function has the property of $\|\nabla\varphi\| = 1$.
9. The basic LSM concerns a function $\varphi(\boldsymbol{x}, t)$ defined throughout the domain. If the information is only necessary near the zero level set, the local LSM can be defined merely close to the interface. A local level set approach called "narrow banding" was introduced by Adalsteinsson and Sethian (1999).

Consider an interface, either a curve in two dimensions or a surface in three dimensions, separating one region from another, and imagine that this curve/surface moves in its normal direction with a known speed function. The goal is to track the motion of the interface in its normal direction as it evolves. It is important that the propagating interface is able to develop corners and discontinuities during the motion, which requires the introduction of a weak solution. The correct weak solution comes from enforcing an entropy condition for the propagating interface. Furthermore, this entropy-satisfying weak solution is the one obtained by considering the limit of smooth solutions for the problem in which the curvature plays a regularizing role. For example, consider a curved interface propagating with the normal velocity of v_N, where a corner forms due to a shock in the slope, as the front moves. If the motion of each individual point is continued, the result is the swallowtail solution (Adalsteinsson and Sethian, 1999), which is multi-valued and does not correspond to a clear interface separating two regions. However, an appropriate weak solution can be derived by assuming an associated smooth flow obtained by adding the curvature κ into the speed law, as $v_N = v_{N_0} - \varepsilon\kappa$. The limit of this smooth solution as ε goes to zero produces a proper weak solution obtained by enforcing an entropy condition.

In order to track the motion of interface $\Gamma(t)$, an appropriate algorithm is used to update the value of level set function φ in time at each point of the mesh. To provide an expression for the time derivative of φ, consider a function $\boldsymbol{x}(t)$ that describes the path of a point on the curve through the time, where $\varphi(\boldsymbol{x}(t), t) = 0$. Taking the derivation from the level set function, it yields

$$\frac{\partial\varphi}{\partial t} + \boldsymbol{v}\cdot\nabla\varphi = 0 \tag{3.3}$$

where $\boldsymbol{v}$ is the desired velocity vector on the interface. Since the normal component of the velocity is only needed for the motion of the interface, that is, $v_N = \boldsymbol{v}\cdot\mathbf{n}$ with $\mathbf{n} = \nabla\varphi/|\nabla\varphi|$ denoting the outward normal vector to the interface, Eq. (3.3) can be written as

$$\frac{\partial\varphi}{\partial t} + v_N|\nabla\varphi| = 0 \tag{3.4}$$

This scheme is known as the Hamilton–Jacobi equation of motion. In the case of closed interfaces, a curvature dependent speed is assumed for the interface front as $v_N = v_{N_0} - \varepsilon\kappa$, with κ denoting the curvature at the point $\boldsymbol{x}$. Equation (3.4) can then be rewritten as

$$\frac{\partial\varphi}{\partial t} = -v_{N_0}|\nabla\varphi| + \varepsilon\kappa|\nabla\varphi| \tag{3.5}$$

and the value of φ after a time interval Δt can then be approximated using the first-order Taylor expansion with respect to t as

$$\varphi(\boldsymbol{x}(t+\Delta t), t+\Delta t) = \varphi(\boldsymbol{x}(t), t) + \frac{\partial\varphi}{\partial t}\Delta t \tag{3.6}$$

This level set approach has been used in a large number of problems involving moving interfaces; however, an extension of this technique includes a fast technique of the solution.

3.3.2 Coupling the LSM with X-FEM

The LSM is proven as a powerful approach to model the motion of interfaces. There are many advantages of using the LSM for tracking interfaces. First, unlike many other interface tracking schemes, the motion of the interface is computed on a fixed mesh. Second, the method handles changes in the topology of the interface naturally. Third, the evolution equation is of the form (3.4) regardless of the dimension of the interface. Hence, extending the method to higher dimensions is easily accomplished. Finally, the geometric properties of the interface can be obtained from the level set function φ. The only drawback of the LSM technique is that the level set representation requires a function of a higher dimension than the original interface, potentially leading to higher storage and computational costs. However, it is not necessary to perform the level set computation for the whole domain and it can be carried out in a region surrounding the interface as the motion is needed near the interface in a predetermined region on either side of the interface, which is called the narrow band.

Coupling the LSM with the X-FEM technique presents a powerful tool and complimentary capability for numerical simulation of moving boundaries. The X-FEM allows the representation of crack discontinuities and voids independently of the mesh. In crack growth simulation, the level set representation of the crack simplifies the selection of the enriched nodes, as well as the definition of the enrichment functions. Coupling the LSM and the X-FEM has been successfully employed in modeling cracks and inhomogeneities, in which its application is widely spread in the problems with different boundary conditions. A methodology to model arbitrary holes and inclusions without meshing the internal boundaries was performed by Sukumar *et al.* (2001), in which the LSM was coupled with the X-FEM. In their study, the level set function is employed to represent the holes and inclusions, and to develop the local enrichment for material interfaces to model inclusions. An algorithm that couples the LSM with the X-FEM was proposed by Stolarska *et al.* (2001) to model the growth of fatigue crack, where the LSM is used to represent the evolution of an open curved crack, including the location of crack tips. They used the LSM to model the crack and update the crack tip at each iteration, in which the geometry of crack is represented by two zero level sets that are orthogonal to one another at the crack tip. A three-dimensional analysis of arbitrary crack problems was performed by Moës, Gravouil, and Belytschko (2002) that combines the three-dimensional X-FEM and constructs arbitrary discontinuities through a discontinuous PU with LSM. The work was extended by Gravouil, Moës, and Belytschko (2002) for treating the growth of non-planar three-dimensional cracks. In their study, the crack is defined by two orthogonal level sets (signed distance functions), in which one level set describes the crack as a two-dimensional surface in a three-dimensional space, and the other is used to describe the one-dimensional crack front, which is the intersection of the two level sets. Moreover, they used the Hamilton–Jacobi equation to update the level sets, where a velocity extension is employed to preserve the old crack surface and generate the growing surface.

A vector LSM was proposed by Ventura, Xu, and Belytschko (2002) in modeling crack propagation using the element-free Galerkin method (EFGM). They only used the nodal data to describe the crack growth; no geometrical entity is introduced for the crack trajectory, and no partial differential equations are solved to update the level sets. An enriched FEM and level sets was presented by Chessa and Belytschko (2003a) for treating two-phase flow with surface tension effects. In their study, the velocity field is enriched by the signed distance function so that the velocity gradient discontinuity at the interface within an element is accurately treated. Moreover, the position of the fluid interface is tracked via a level set function that allows for a simple construction of the enriched FEM and also simplifies the computation of the interface curvature. They used a modified velocity enrichment function, which is derived from the solution of Laplace equation for the signed distance function in the blending domain; so that the pathological terms resulting in the blending elements are reduced. An enriched FEM with arbitrary discontinuities in space–time was also presented by Chessa and Belytschko (2004), in which a local PU enrichment is used to model discontinuities along a moving hyper-surface using the LSM. An Eulerian–Lagrangian method was proposed by Legay, Chessa, and Belytschko (2006) for fluid–structure interaction based on LSM, in which the LSM is used to provide a compact framework for derivation of the weak form and development of the discretized equations. They used an Eulerian description to model the fluid domain and a Lagrangian description for the solid phase; where the level set description of the interface leads to the formulation of fluid–structure interaction problem. A vector LSM was presented by Budyn *et al.* (2004) in modeling the growth of multiple cracks for homogeneous and inhomogeneous linear elastic materials.

3.4 Fast Marching Method

The robustness of level set method in simulation of evolving interfaces has been proven through a large number of numerical computations. However, in the level set formulation, both the level set function and the speed are embedded into a higher dimension that implies computational labor through the entire grid, which is inefficient. A rough evaluation of the level set procedure assumes N grid points in each space dimension of a three-dimensional problem. For a simple problem of straightforward propagation with the normal velocity v_N, the computational cost is in the order of $O(N^3)$ (Sethian 1996). An alternative and efficient technique that considerably speed up the computational solution can be performed in a neighborhood of the zero level set; this is known as the *narrow band level set* method that was originally introduced by Adalsteinsson and Sethian (1995). It is clear that performing calculations over the entire computational domain is wasteful. Hence, an efficient modification is to perform computations only on a neighborhood of the zero level set. The narrow band approach causes that the computational costs reduce to $O(kN^2)$ in three-dimensional problems, where k is the number of cells in the narrow band. It means that the evaluation of velocities needs only to be performed over the points lying in the narrow band. A schematic view of the narrow band approach is presented in Figure 3.5. The use of narrow band approach leads to the level set front advancement algorithm that is computationally equivalent in terms of complexity to the traditional marker method and the cell technique, while maintaining the advantages of topological merger, accuracy, and easy extension to multi-dimensions. Typically, the speed associated with the narrow band method is about 10 times faster on a 160×160 grid than the full matrix method (Sethian 1985). In this technique, the entire two-dimensional grid of data is stored in a square array, in which a one-dimensional object can be used to keep track of the points in this array. In fact, only the values of φ at points located in a narrow band around the front of a user-defined width are updated. The values of φ at the grid points on the boundary of the narrow band are frozen. When the front moves near the edge of narrow band boundary, the calculation is stopped, and a new narrow band is built with the zero level set interface boundary at the center. This rebuilding process is known as the *reinitialization* procedure.

An extreme one-cell version of the narrow band approach leads to the *fast marching level set* method, which was introduced by Sethian (1996). This method is developed under some restrictions, which is known as the *stationary level set* formulation. In this technique, it is assumed that the front propagates with a normal velocity v_N, which is either always positive or always negative. However, the variations

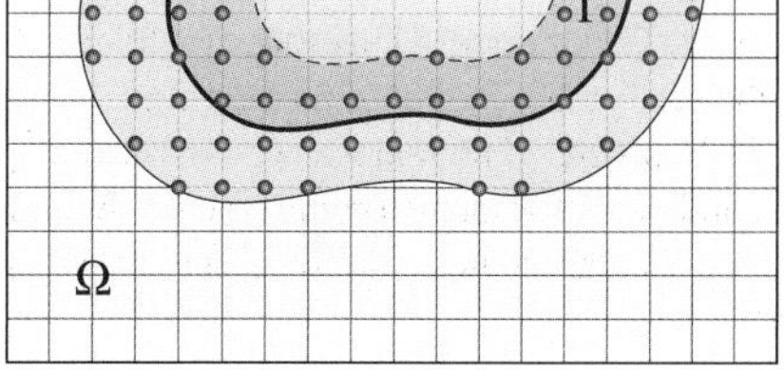

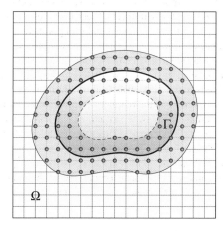

Figure 3.5 The narrow band level set method: The nodal points in the narrow band are used to update the values of the level set

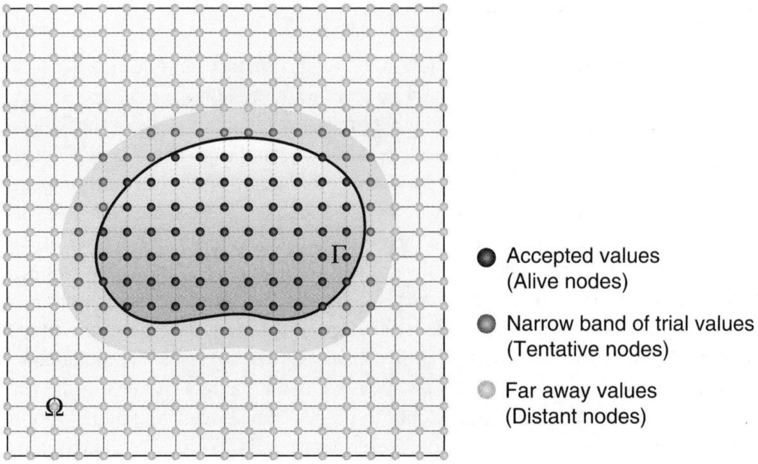

Figure 3.6 The fast marching method: The upwind construction of accepted values

may differ from one point to another. In this case, the level set formulation is converted from a time-dependent partial differential equation to a stationary one, which is a simplified form of the Hamilton-Jacobi equation (3.4), that is, $v_N|\nabla\varphi| = 1$, which is called the Eikonal equation. The idea behind the FMM is to solve the Eikonal equation by first replacing the gradient by suitable upwind operators, and then systematically advancing the front by marching outwards from the boundary data in an upwind fashion. In the upwind scheme, the information propagates one way; in fact, the fast marching algorithm is based on the solution of Eikonal equation by building its solution outward from the smallest values of φ to the largest ones. The idea is to sweep the front ahead in an upwind fashion by considering a set of points in the narrow band around the existing front and to march this narrow band forward, freezing the values of existing points and bringing new ones into the narrow-band structure. The key part of the algorithm is the selection of grid points in the narrow band that must be updated, as shown in Figure 3.6.

The efficiency and effectiveness of the upwind scheme is dependent on a fast computational algorithm in determination of the grid points in the narrow band with the smallest value for φ. In this manner, a list of narrow band points is initially sorted in a way that the smallest member can be easily located. The values of these points in the list are stored together with their indices that present the location of grid points. An additional array is defined to address the two-dimensional grid points to the location of those grid points in the list. By indicating the point with smallest value of φ, the neighbors of that point can be determined using the pointers from the grid array to the list of narrow band points. The values of the neighbors are then computed, and the results are bubbled upwards in the list until they reach their correct locations, at the same time readjusting the pointers in the grid array. In this algorithm, the computational cost is optimal and in the order of $O(N \log N)$, where N is the number of points in the narrow band (Sethian 1996).

In the FMM, all the nodes in the mesh are sorted into three disjointed sets, a set of *accepted nodes* A, a set of *tentative nodes* T, and a set of *distant nodes* D, as shown in Figure 3.6. The method systematically moves nodes from the set D to the set T and finally into the set A and terminates when all nodes are in the set A (Figure 3.7). Briefly, the set A consists of all nodes $\boldsymbol{x}$ whose value of $\varphi(\boldsymbol{x})$ has been computed, the set T consists of all nodes that are candidates for inclusion into the set A, and the set D consists of all nodes which are too far from the set A to be candidates. Based on these three sets, the algorithm proceeds as follows (Sukumar et al. 2003a);

1. Initialize a core set of nodes to be in the set A. The value of $\varphi(\boldsymbol{x})$ for $\boldsymbol{x} \in \mathsf{A}$ is determined by direct computation. Each element of the mesh through which the zero contour of $\varphi(\boldsymbol{x})$ crosses, that is, the initial front position, has each of its nodes start in the set A and the value of each node is determined by directly computing the distance from each node to the level contour in the element.

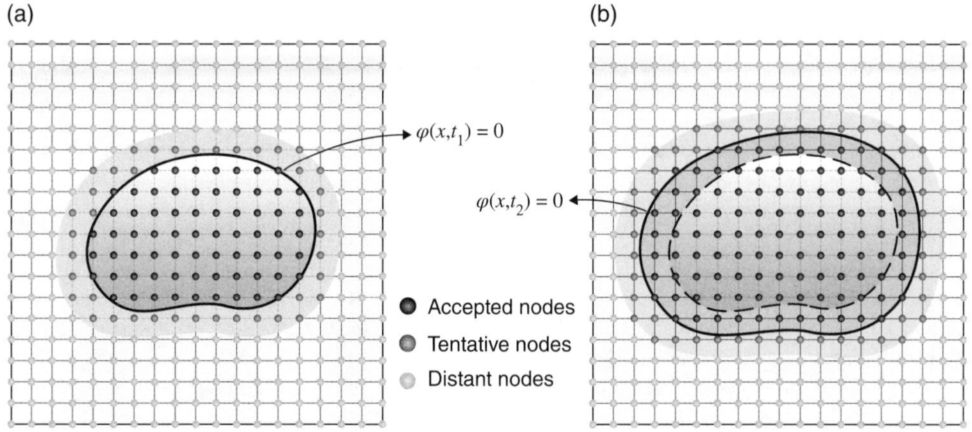

Figure 3.7 The computational procedure of the fast marching method: (a) the set of nodes at t_1 for an interface $\varphi(\boldsymbol{x}, t_1) = 0$; (b) the set of nodes at t_2 for an interface $\varphi(\boldsymbol{x}, t_2) = 0$

2. For each node $\boldsymbol{x} \in \mathsf{A}$, each neighboring node $\boldsymbol{y} \in \mathsf{A}$ connected to $\boldsymbol{x}$ is assigned a tentative value $\varphi(\boldsymbol{y})$ and placed in the set T. The tentative value is constructed by using second-order one-sided finite difference approximations for $v_N|\nabla\varphi| = 1$.
3. Start the main loop by taking the node $\boldsymbol{x} \in \mathsf{T}$ with the smallest value for $\varphi(\boldsymbol{x})$ and moves it from the set T to the set A.
4. For each node $\boldsymbol{y}$ adjacent to node $\boldsymbol{x}$ selected in step 3, and not already in A update its value $\varphi(\boldsymbol{y})$. If $\boldsymbol{y} \in \mathsf{T}$, then the set T must be re-sorted to account for the changed value of $\varphi(\boldsymbol{y})$. If $\boldsymbol{y} \in \mathsf{D}$, it is moved from the set D to T.
5. If $\mathsf{T} \neq \emptyset$, then go to step 3.

3.4.1 Coupling the FMM with X-FEM

The level set framework, and in particular the FMM provide a seamless means to evolve an interface (crack front). This technique obviates the need to represent and maintain the geometry of the interface during its evolution. The LSM is used as a numerical technique for tracking moving interfaces, while the FMM is employed as a computationally attractive alternative for strictly monotonically advancing fronts. In both methods, the evolving interface is represented as a level contour of a function of one higher dimension. In the FMM, the motion of the interface is embedded in the solution of an elliptic equation in terms of $\varphi(\boldsymbol{x}, t)$.

The first implementation of a combined X-FEM and FMM was proposed by Sukumar, Chopp, and Moran (2003a) for modeling planar three-dimensional fatigue crack propagation. They used the level set function to represent the planar crack, and the FMM to handle its evolution under fatigue growth conditions. The technique was then extended by Chopp and Sukumar (2003) for modeling fatigue crack propagation of multiple coplanar cracks based on coupling the X-FEM with the FMM. In their study, the crack geometry, including one or more cracks, is represented by a single signed distance (level set) function, and merging of distinct cracks is handled by the FMM in conjunction with the Paris crack growth law to advance the crack front. An extension of this work was performed by Stolarska and Chopp (2003) for simulation of thermal fatigue cracking in integrated circuits. They presented an algorithm that coupled the LSM with the X-FEM to investigate the effects of the proximity of multiple interconnect lines, multiple cracks, interconnect material, and integrated circuit boundaries on the growth of cracks due to fatigue from thermal cycling.

3.5 X-FEM Enrichment Functions

One strategy to improve the performance of the FEM using piecewise polynomials for representation of the singularities, high gradients, or problems with oscillatory solutions is to enrich the FE approximation basis. It is shown that the convergence rate of the FE approximation using piecewise functions without enrichment functions is much slower than when the approximation basis is enhanced with them. However, the addition of enrichment functions to approximate the displacement field near the singularity may destroy the band structure of the stiffness matrix and may lead to an ill-conditioned system when the basis functions are nearly linearly dependent. Hence, the most important step in developing the X-FEM approximation field is to define proper enrichment functions being in agreement with the phenomenon under consideration. Melenk and Babuška (1996) proposed the PU FEM to solve problems with highly oscillatory solutions. For the Helmholtz equation it was shown that if the enrichment functions have the same oscillatory behavior as the solution, the convergence of the FEM is improved.

In the X-FEM formulation, the enrichment functions added to the standard FE approximation provide two main requirements; firstly, they instruct the location of the interface into the function space itself, and secondly create the opportunity to include information about the known asymptotic behavior of the solution near the interface. Thus, the ideal enrichment function for a given application is dependent on the type of solution behavior expected near the interface. The two enrichment functions of the *Heaviside* and *ramp* functions, described in Chapter 2, are among the most popular and successful enrichment functions proven their capabilities in modeling strong and weak discontinuities. The *Heaviside* enrichment function is used to solve problems with a jump in the solution value, the normal derivative, or both, while the *ramp* function is used for problems that are continuous across the interface but include a jump in the normal derivative. However, for a wide range of applications with weak and strong discontinuities, it is necessary to define the proper enrichment functions as a primary key to derive the X-FEM approximation field. In what follows, the most applicable enrichment functions are introduced and their applications in various problems are demonstrated.

3.5.1 Bimaterials, Voids, and Inclusions

Defects in engineering problems such as pores, voids, and inclusions are important to the structural integrity and durability of components. There are various applications of weak discontinuity, known as a closed discontinuity (Figure 2.4a), in solid mechanics problems, such as voids, inclusions, bimaterial interfaces, grain boundaries, and so on, that must be modeled in order to predict the mechanical behavior of materials. Modeling of these discontinuities with FEM requires conforming the internal boundaries to the FE mesh. Although mesh generation is well established for complex geometries, meshing arbitrary number and distribution of defects and inclusions is a time-consuming and burdensome task. Hence, the X-FEM is used by taking advantages from enrichment functions and the level set concept to perform the mesh independent of the material interfaces. The computational geometry issues that are associated with the FE mesh and the internal geometric entities, such as voids or inclusions, are an important consideration in X-FEM computations. In light of this, the LSM is an appealing choice that would greatly simplify and speed-up the geometric computations in the X-FEM. In addition, the level set function can also be used to construct enrichment functions within the X-FEM framework. This provides motivation and is the seemingly natural choice to use the LSM in conjunction with the X-FEM.

The first implementation of X-FEM in modeling of voids and inclusions was performed by Sukumar *et al.* (2001), where the LSM is used to represent the location of voids, inclusions, and material interfaces and to develop the local enrichment for material interfaces. A combined X-FEM/LSM was employed by Moës *et al.* (2003) to model micro-structures with complex geometries, and to present the capability of model for the homogenization of periodic basic cells. An X-FEM/level set approach was proposed by Yvonnet, Le Quang, and He (2008) to compute the overall elastic properties of nano-materials and nano-structures with surface/interface effects using a coherent interface model, and to determine the size-dependent effective elastic moduli of nano-composites with randomly distributed nano-pores. The X-FEM technique was employed by Wu *et al.* (2010) in geomaterial problems to study the behavior of heterogeneous materials by a numerical homogenization method based on the X-FEM technique. Since the major obstacle in applying the standard FEM for estimation of effective properties of heterogeneous materials resides in difficulties constructing the mesh to conform to complexity of structures, the X-FEM is used to model the multiple inclusions exist through the medium. X-FEM numerical modeling was proposed by Singh, Mishra, and Bhattacharya (2011) for functionally graded materials with inhomogeneities/discontinuities, in which the domain contains multiple discontinuities/flaws such as cracks and/or voids/inclusions distributed all over the domain. A mathematical derivation of enrichment functions in the X-FEM was presented by Zhu (2012) for numerical modeling of weak and strong discontinuities, which consists of combining the LSM with characteristic functions as well as domain decomposition and reproduction technique.

According to relation (2.17), the X-FEM approximation field in modeling of voids and inclusions with multiple material interfaces can be written as

$$\mathbf{u}(\boldsymbol{x}) = \sum_{I \in \mathcal{N}} N_I(\boldsymbol{x})\, \bar{\mathbf{u}}_I + \sum_{\mathcal{K}} \sum_{J \in \mathcal{N}^{enr}} N_J(\boldsymbol{x}) \left[\psi_{\mathcal{K}}(\varphi(\boldsymbol{x})) - \psi_{\mathcal{K}}(\varphi(\boldsymbol{x}_J))\right] \bar{\mathbf{a}}_J^{\mathcal{K}} \tag{3.7}$$

where $\mathcal{N}$ is the set of all nodal points of domain, $\mathcal{N}^{enr}$ is the set of enriched nodes, $N(\boldsymbol{x})$ is the standard shape functions, $\psi_{\mathcal{K}}$ is the enrichment function, and $\varphi(\boldsymbol{x})$ is the signed distance function representing the interface, as defined in (3.2). The variables $\bar{\mathbf{u}}_I$ and $\bar{\mathbf{a}}_J$ are the standard and enriched degrees of freedom, respectively, and $\mathcal{K}$ is the number of material interfaces.

For problems with weak discontinuities, such as voids, inclusions, or bimaterials, which are continuous across the interface but feature a jump in the normal derivative, a continuous enrichment function can be employed as

$$\psi_{\text{ramp}}(\varphi(\boldsymbol{x})) = |\varphi(\boldsymbol{x})| = \begin{cases} -\varphi(\boldsymbol{x}) & \text{if } \varphi(\boldsymbol{x}) \le 0 \\ +\varphi(\boldsymbol{x}) & \text{if } \varphi(\boldsymbol{x}) > 0 \end{cases} \tag{3.8}$$

This enrichment function is known as the *ramp* function, which is introduced based on the absolute value of level set function. It is important to note that the enrichment decays to zero at the interface, which is an important criterion for the stability when a Dirichlet interface condition is applied. This enrichment function is able to produce a jump across the discontinuity in the stress and strain fields of bimaterial problems, or in the pressure fields of incompressible two-phase flows with surface tension, where $\nabla\psi_{\text{ramp}}(\varphi(\boldsymbol{x})) = \text{sign}(\varphi(\boldsymbol{x})) \cdot \nabla\varphi(\boldsymbol{x})$. Figure 3.8 presents this enrichment function together with the enriched shape functions for a 4-noded quadratic element.

An alternative definition of enrichment function was proposed by Moës *et al.* (2003) for the weak discontinuity problems as

$$\psi_{\text{ridge}}(\varphi(\boldsymbol{x})) = \sum_{I \in \mathcal{N}} N_I(\boldsymbol{x})\, |\varphi_I| - \left| \sum_{I \in \mathcal{N}} N_I(\boldsymbol{x})\, \varphi_I \right| \tag{3.9}$$

The main advantage of the earlier enrichment function is that this has a non-zero value only in the elements cut by the interface. This modified level-set enrichment function is shown in Figure 3.9 together with the enriched shape functions for a 4-noded quadratic element. The previous enrichment function has a ridge centered on the interface and zero value on the elements that are not crossed by the interface.

An enrichment function that is tailored to the solution behavior near the interface was proposed by Smith (2008) on the basis of an exponential function as

$$\psi_{\exp}(\varphi(\boldsymbol{x})) = \begin{cases} -\varphi(\boldsymbol{x}) & \text{if } \varphi(\boldsymbol{x}) \le 0 \\ 1 - e^{-\mu\,\varphi(\boldsymbol{x})} & \text{if } \varphi(\boldsymbol{x}) > 0 \end{cases} \tag{3.10}$$

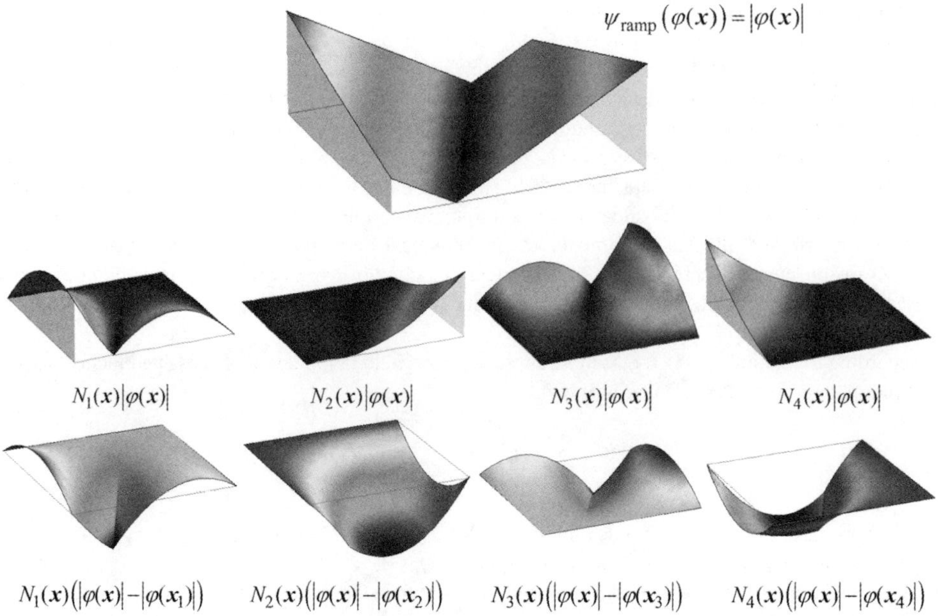

Figure 3.8 The *ramp* enrichment function together with the enriched shape functions for a 4-noded quadratic element

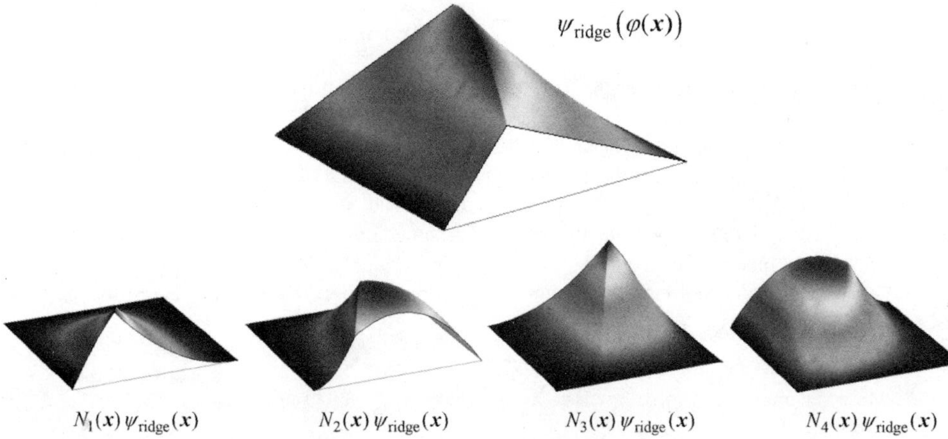

Figure 3.9 The modified level-set enrichment function together with the enriched shape functions for a 4-noded quadratic element

This exponential enrichment function can be used for problems which display exponential behavior within an interfacial boundary layer. The enrichment contains a parameter μ^{-1}, which is the thickness of the boundary layer. This thickness is problem dependent and is generally determined by asymptotic analysis. Figure 3.10 presents the exponential enrichment function together with the enriched shape functions for a 4-noded quadratic element.

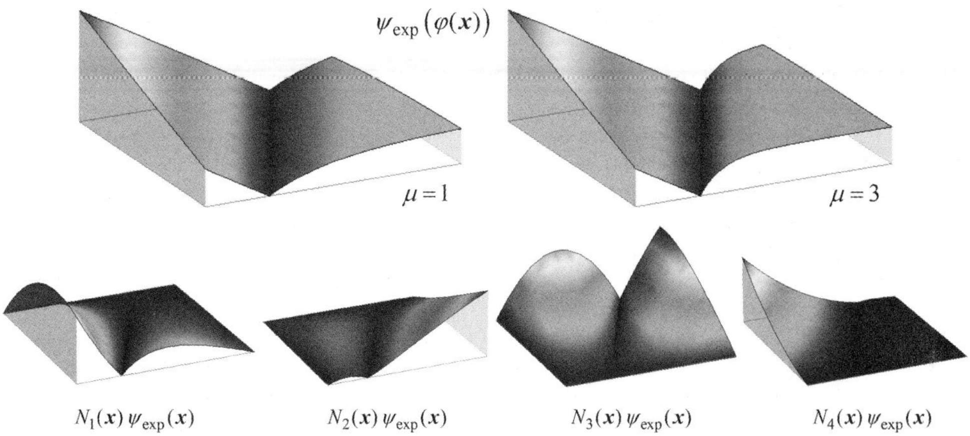

Figure 3.10 The exponential enrichment function together with the enriched shape functions for a 4-noded quadratic element at $\mu = 1$

3.5.2 Strong Discontinuities and Crack Interfaces

A jump in the displacement field is referred to as the strong discontinuity, which can be typically observed in crack problems. The discontinuity in the displacement field occurs where the displacement of one side of the interface is different from the other side of the interface. An enriched technique was originally proposed by Ortiz, Leroy, and Needleman (1987) that was able to deal successfully with the strong discontinuity in FE. The method, which is assumed in the class of *embedded elements*, captures strong discontinuities in FE to improve the resolution of shear band localization. The method was applied to study the strain localization in both compressible and nearly incompressible solid mechanics. In their method, additional DOF, due to localized deformation modes are eliminated at the element level by static condensation. Based on this idea, Belytschko, Fish, and Englemann (1988) proposed a method to model the strain localization by imposing two parallel weak discontinuity lines in a single element, so that the element was able to contain the band of localized strain. One of the distinct differences between these two methods is that the width of the localized band is smaller than the mesh size in the method proposed by Belytschko, Fish, and Englemann (1988), whereas the localization band width is of the same size as the mesh size in the method by Ortiz, Leroy, and Needleman (1987), and a very fine mesh was required to resolve the localization band. In a further work, Dvorkin, Cuitiño, and Gioia (1990) proposed a method based on displacement interpolated embedded localization lines, which was insensitive to mesh size and distortions. This method was much more flexible than schemes that allow discontinuities only at element interfaces, and was used easily to model the crack growth without remeshing. The idea was similar to the X-FEM, which was developed a decade later by Belytschko and Black (1999); in fact, the embedded elements introduce additional unknowns into the variational formulation. The X-FEM has been proposed in modeling of strong discontinuities from the earliest studies of the technique by Moës, Dolbow, and Belytschko (1999), Sukumar *et al.* (2000), and Stolarska *et al.* (2001). An innovative method was also proposed by Hansbo and Hansbo (2004) for modeling strong discontinuities, where the approximation was constructed from two different fields. In their method, the crack kinematics was obtained by overlapping elements instead of introducing additional degrees of freedom. The idea was based on the superposition of an additional element to the element cut by the discontinuity, in order to construct the enriched displacement field. Areias and Belytschko (2006a) presented that the Hansbo–Hansbo method can be derived from the standard X-FEM technique by using a linear combination of the X-FEM basis.

There are numerous approaches to track the crack path and to represent the crack growth. Most commonly, the crack interface can be approximated by the LSM where the level set is discretized with the same shape functions of the FE mesh. In such case, the crack interface is described by piecewise linear crack segments when the linear quadratic element is used. However, it is possible to describe the curved crack interface using the B-splines, or NURBS functions; or the level set can be approximated with higher-order shape functions. An advantage of tracking the crack path with level sets is that no explicit representation of the crack geometry is needed. The crack update can be performed by solving the Hamilton–Jacobi equation at each point of the "*level-set mesh*". Once the new level set functions are determined, a projection is needed to give their values on the FE mesh.

The key point in modeling of crack and strong discontinuities using the LSM is to represent the discontinuity as a zero level set function, which is defined as a signed distance function. Since the crack is a discontinuity which does not divide the domain into two distinct parts completely, known as an open discontinuity (Figure 2.4b), it is necessary to construct two level set functions; a *normal level-set* function and a *tangential level-set* function, in which both the two level set functions are defined as a signed distance functions. The *normal level-set* function $\varphi(\boldsymbol{x})$ is introduced to describe the crack path similar to that defined for the closed discontinuity, such as voids, inclusions, or biomaterial, by $\varphi(\boldsymbol{x})=0$ in previous section. The *tangential level-set* function is defined using a second level-set function $\vartheta(\boldsymbol{x})$ that defines the tangential extension of the crack tip by $\varphi(\boldsymbol{x})=0$ and $\vartheta(\boldsymbol{x})\leq 0$, as shown in Figure 3.11a. An extension of the LSM for description of a crack in the tangential direction was given by Stolarska *et al.* (2001). The tangential level-set function is constructed such that it is orthogonal to the contact interface at the crack tip. If the crack is an interior crack, two tangential level set functions are defined according to each crack tip. The construction of normal and tangential level set functions for an interior crack is presented in Figure 3.11b and c.

In the X-FEM formulation, the kinematics of the strong discontinuity can be defined based on the sign and step enrichment functions. A typical choice of the enrichment function is the sign of the level-set function defined as

$$\psi_{\text{sign}}(\varphi(\boldsymbol{x}))=\text{sign}(\varphi(\boldsymbol{x}))=\begin{cases}-1 & \text{if } \varphi(\boldsymbol{x})<0\\ 0 & \text{if } \varphi(\boldsymbol{x})=0\\ +1 & \text{if } \varphi(\boldsymbol{x})>0\end{cases} \tag{3.11}$$

An alternative discontinuous enrichment function can be defined based on the step enrichment function as

$$\psi_{\text{step}}(\varphi(\boldsymbol{x}))=H(\varphi(\boldsymbol{x}))=\begin{cases}0 & \text{if } \varphi(\boldsymbol{x})\leq 0\\ 1 & \text{if } \varphi(\boldsymbol{x})>0\end{cases} \tag{3.12}$$

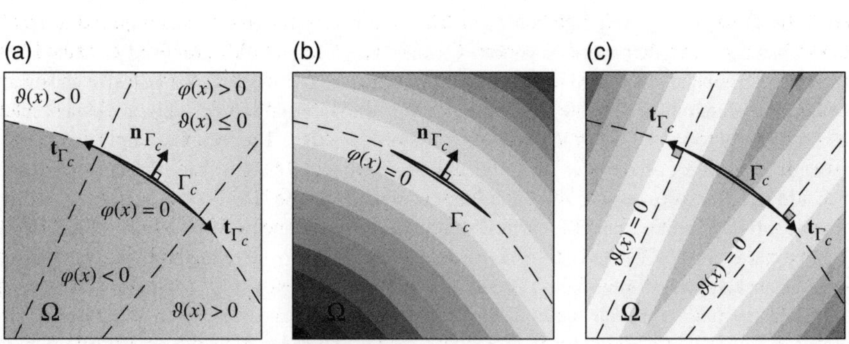

Figure 3.11 Construction of normal and tangential level set functions for an interior crack: (a) the domain Ω with a crack Γ_c; (b) the normal level-set function $\varphi(\boldsymbol{x})$ for description of the crack-path; (c) the tangential level-set function $\vartheta(\boldsymbol{x})$ for description of the crack tip

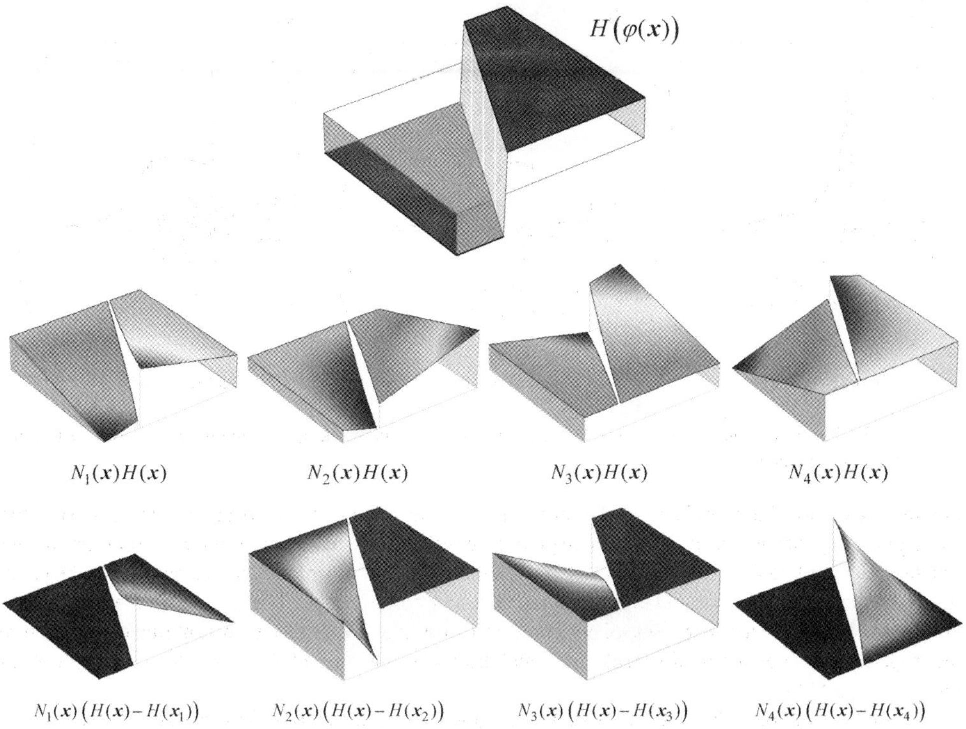

Figure 3.12 The Heaviside enrichment function together with the enriched shape functions for a 4-noded quadratic element

Thus, the enriched displacement field can be expressed as

$$\mathbf{u}(\boldsymbol{x}) = \sum_{i=1}^{\mathcal{N}} N_i(\boldsymbol{x})\,\bar{\mathbf{u}}_i + \sum_{j=1}^{\mathcal{M}} N_j(\boldsymbol{x})\,\big(H(\boldsymbol{x}) - H(\boldsymbol{x}_j)\big)\,\bar{\mathbf{a}}_j \tag{3.13}$$

The sign and Heaviside step enrichments lead to identical results as they span the same approximation space. It is important to note that the shifted step enrichments vanish outside the elements crossed by the discontinuities, so only these elements need to be modified; essentially there are no blending elements. Even for un-shifted step enrichments, blending causes no difficulties because the enrichments are constant in the blending elements. In Figure 3.12, the Heaviside function is shown together with the enriched shape functions for a 4-noded quadratic element.

3.5.3 Brittle Cracks

There are many applications that involve discontinuity along the interface that end inside the domain, such as the crack interface with a tip in two-dimensional or a front in three-dimensional problems. Moreover, in many physical problems, the high gradients are present in the field variables at the interface tip or front (Figure 3.13a). Thus, for the simulation of such problems, both the discontinuities along the interface and the high gradients at the interface tip/front must be considered appropriately. Cracks and dislocations are typical examples of such problems, where the displacement fields involve jumps across the open

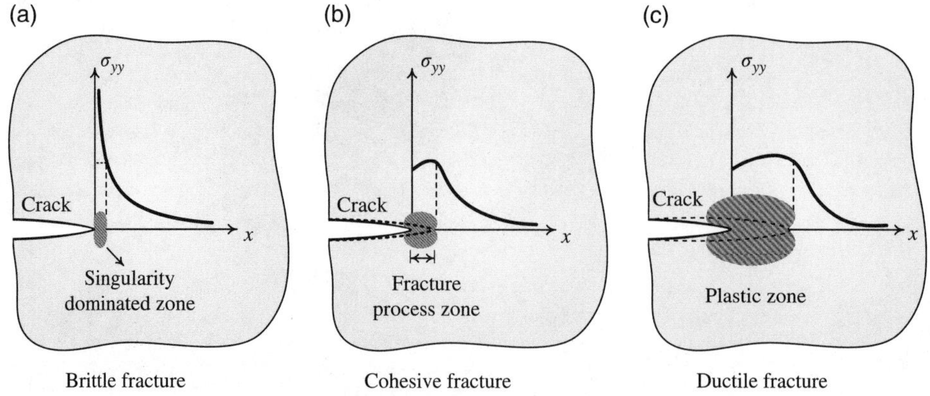

Figure 3.13 The schematic distributions of normal stress ahead of the crack tip for the brittle (a), cohesive (b), and ductile fracture (c) mechanics

interfaces and have high gradients at the crack tips or dislocation-cores, respectively. In contrast to the weak and strong discontinuities described in preceding sections where the *ramp* and *Heaviside* enrichment functions are applied for a large number of applications, the enrichment functions employed at the crack tips/fronts depend on the particular physical model under consideration. In the X-FEM, the crack simulation is performed by using two sets of enrichment functions; a *Heaviside* enrichment function to capture the displacement jump across the crack faces, and the *asymptotic branch* functions that span the asymptotic crack tip fields.

The implementation of X-FEM for modeling the crack in brittle materials was started from the original studies of X-FEM by Belytschko and Black (1999) and Moës, Dolbow, and Belytschko (1999). In the work of Belytschko and Black (1999), the asymptotic enrichment functions were employed for the complete crack path and the kinks were treated in the crack path by special mappings. However, Moës, Dolbow, and Belytschko (1999) proposed the asymptotic enrichment functions at the crack tip region and the Heaviside enrichment function along the crack path, which has been established as a standard procedure in the X-FEM technique. Daux *et al.* (2000) developed the X-FEM technique for modeling cracks with multiple branches and intersections, where a step enrichment function is introduced for crack junctions that does not rely on level-sets. The X-FEM was applied to three-dimensional crack problems by Sukumar *et al.* (2000). A local enrichment strategy was proposed by Dolbow, Moës, and Belytschko (2000a) for arbitrary discontinuities in the X-FEM framework to model crack growth in two-dimensional elasticity and Mindlin–Reissner plate. The X-FEM was used by Dolbow, Moës, and Belytschko (2001) for modeling crack growth with frictional contact on the crack faces. The curved crack simulation was presented by Stazi *et al.* (2003) using higher-order X-FEM. The 3D crack initiation and propagation was performed by Areias and Belytschko (2005a) for the quasi-static analysis of brittle and quasi-brittle solids. Xiao and Karihaloo (2006) proposed an alternative approach to enrich the FE approximation with higher order terms of the asymptotic expansion of the crack tip field as enrichment functions, and obtained the stress intensity factors directly without extra post-processing.

Modeling the crack in brittle materials can be performed in the framework of linear elastic fracture mechanics (LEFM). In that case, the stresses and strain fields at the crack tip are singular. The following four enrichment functions are based on the asymptotic solution of Williams (1957) defined for the two-dimensional isotropic media as

$$\begin{aligned}\psi_{\text{tip}}^{\text{brittle}}(r,\theta) &= \left\{\psi_{\text{tip}}^{1}, \psi_{\text{tip}}^{2}, \psi_{\text{tip}}^{3}, \psi_{\text{tip}}^{4}\right\} \\ &= \left\{\sqrt{r}\sin\frac{\theta}{2}, \sqrt{r}\cos\frac{\theta}{2}, \sqrt{r}\sin\frac{\theta}{2}\sin\theta, \sqrt{r}\cos\frac{\theta}{2}\sin\theta\right\}\end{aligned} \tag{3.14}$$

These functions are defined in a local polar coordinate system (r, θ) at the crack tip. The distributions of these four asymptotic functions are shown in Figure 3.14 together with the enriched shape functions for a 4 noded quadratic element. Obviously, only the first asymptotic function represents the discontinuity near the tip on both sides of the crack while the other three functions can be used to ameliorate the accuracy of the approximation. In three-dimensional problems, the same crack tip enrichment is often used along the crack front with θ being the angle to the tangent plane at the front. The justification for using these enrichment functions in three-dimensional problems is that the asymptotic fields near the crack front are two-dimensional in nature, except near surfaces.

For orthotropic materials, the asymptotic functions given by Eq. (3.14) have to be modified because the material property is a function of material orientation. Orthotropic materials such as composites are widely used in different branches of engineering and structural systems, including the aerospace and automobile industries, power plants, and so on. Since the ratio of strength to weight and stiffness of such materials in many cases is higher than other conventional materials, extensive applications of orthotropic materials can be observed in engineering materials. There are several analytical investigations reported on the fracture behavior of composite materials for obtaining the stress and displacement fields around a linear crack in an anisotropic medium. On the basis of asymptotic solution of an orthotropic crack problem, Asadpoure and Mohammadi (2007) and Hattori *et al.* (2012) proposed special near-tip enrichment functions for orthotropic materials as

$$\begin{aligned}\psi_{\text{tip}}^{\text{orthotropic}}(r,\theta) &= \left\{\psi_{\text{tip}}^1, \psi_{\text{tip}}^2, \psi_{\text{tip}}^3, \psi_{\text{tip}}^4\right\} \\ &= \left\{\sqrt{r}\sin\frac{\theta_1}{2}\sqrt{g_1(\theta)},\ \sqrt{r}\cos\frac{\theta_1}{2}\sqrt{g_1(\theta)},\ \sqrt{r}\sin\frac{\theta_2}{2}\sqrt{g_2(\theta)},\ \sqrt{r}\cos\frac{\theta_2}{2}\sqrt{g_2(\theta)}\right\}\end{aligned} \tag{3.15}$$

where (r, θ) represent the crack tip polar coordinate system. There are generally two types of orthotropic materials defined in the literature; the functions $g_i(\theta)$ $(i = 1, 2)$ and θ_i $(i = 1, 2)$ for these two types of orthotropic materials are respectively defined as

$$\begin{aligned} g_i(\theta) &= \left(\cos^2\theta + \frac{1}{e_i^2}\sin^2\theta\right)^{1/2} \\ \theta_i &= \arctan\left(\frac{1}{e_i}\tan\theta\right)\end{aligned} \tag{3.16a}$$

$$\begin{aligned} g_i(\theta) &= \left(\cos^2\theta + \ell^2\sin^2\theta + (-1)^i\ell^2\sin 2\theta\right)^{1/2} \\ \theta_i &= \arctan\left(\frac{\gamma_2\ell^2\sin\theta}{\cos\theta + (-1)^i\gamma_1\ell^2\sin\theta}\right)\end{aligned} \tag{3.16b}$$

where γ_i, e_i $(i = 1, 2)$, and ℓ are related to material constants, which depend on the orientation of the material. In Figure 3.15, the typical representation of near tip asymptotic fields are presented for the second type of orthotropic materials.

The enriched displacement field for a cracked problem can therefore be rewritten by employing the crack tip asymptotic functions (3.14) as

$$\begin{aligned}\mathbf{u}(\boldsymbol{x}) &= \sum_{I\in\mathcal{N}} N_I(\boldsymbol{x})\,\bar{\mathbf{u}}_I + \sum_{J\in\mathcal{N}^{dis}} N_J(\boldsymbol{x})\left(H(\boldsymbol{x}) - H(\boldsymbol{x}_J)\right)\bar{\mathbf{d}}_J \\ &+ \sum_{K\in\mathcal{N}^{tip}} N_K(\boldsymbol{x})\sum_{\alpha=1}^{4}\left(\psi_{\text{tip}}^{\alpha}(\boldsymbol{x}) - \psi_{\text{tip}}^{\alpha}(\boldsymbol{x}_K)\right)\bar{\mathbf{b}}_K^{\alpha}\end{aligned} \tag{3.17}$$

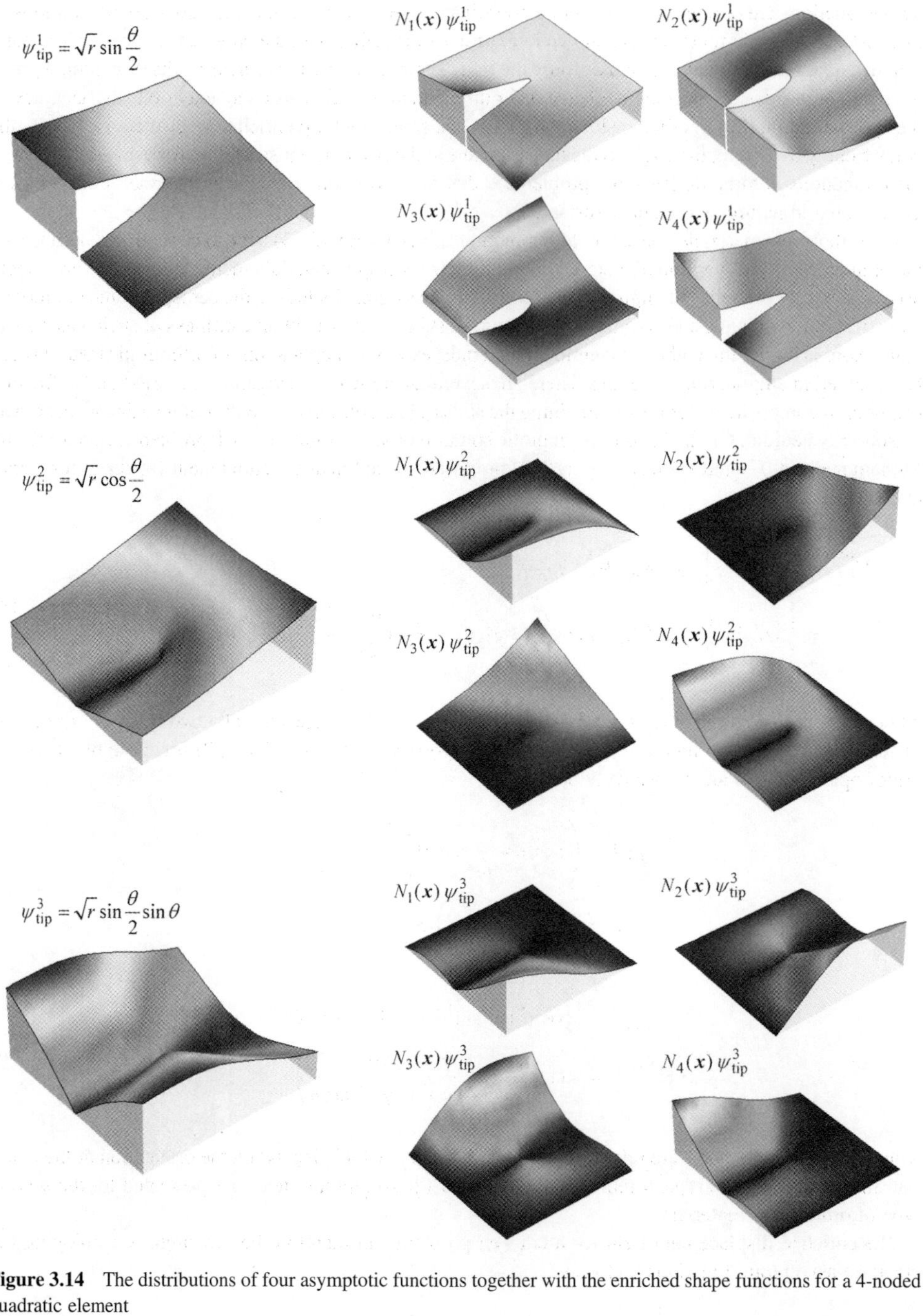

Figure 3.14 The distributions of four asymptotic functions together with the enriched shape functions for a 4-noded quadratic element

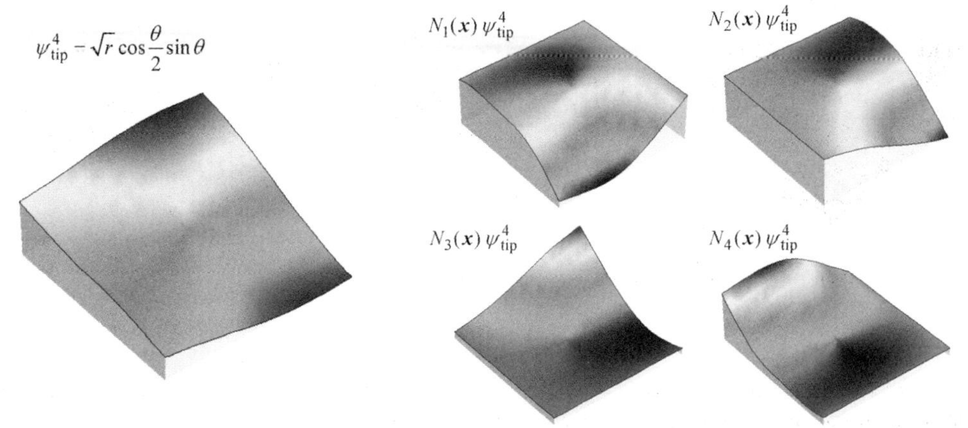

Figure 3.14 *(Continued)*

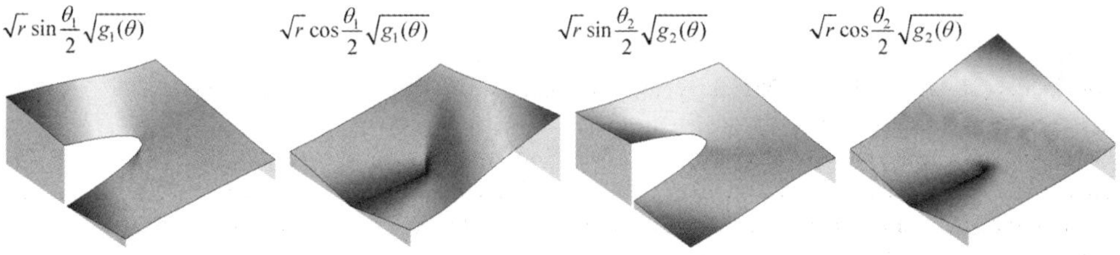

Figure 3.15 Typical representation of near tip asymptotic fields for the second type of orthotropic materials

in which the second enrichment term is used at the crack tip region, where $\mathcal{N}^{tip}$ is the set of nodes that contain the crack tip in the support of their shape functions enriched by the asymptotic functions, and $\bar{\mathbf{b}}_K^{\alpha}$ are the additional enriched nodal degrees of freedom associated with the asymptotic functions at node K.

3.5.4 Cohesive Cracks

The LEFM is only applicable when the size of the fracture process zone (FPZ) at the crack tip is small compared to the size of the crack and the size of the specimen. Based on the assumptions of LEFM, the stress at a crack tip is theoretically infinite. However, from a physical point of view no material can withstand such an infinite stress state, and a small plastic/fractured zone is formed around the crack tip (Figure 3.13b). Even if the FPZ is small, describing what is going on inside it may be convenient for the purpose of understanding the fracture mechanisms. The cohesive fracture models are among various techniques that can be used to efficiently describe the FPZ. In cohesive crack models, the stresses and strains are no longer singular at the crack tips. Cohesive crack models are most suitable for quasi-brittle and ductile materials that account for the presence of a plastic zone. In a cohesive crack model, the propagation is governed by a traction-displacement relation across the crack faces near the tip. This model was originally presented by Dugdale (1960) and Barenblatt (1962) for metals. Hillerborg, Modéer, and Petersson (1976) introduced the concept of fracture energy into the cohesive crack model, and proposed a number of traction–displacement relationships for concrete.

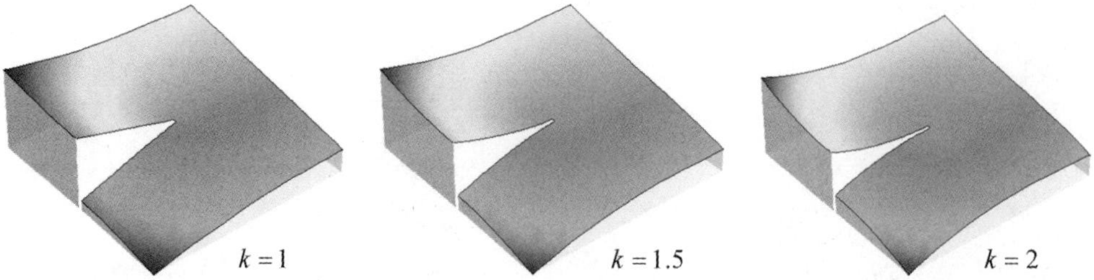

Figure 3.16 The cohesive crack tip function $\psi_{\text{tip}}^{\text{cohesive}}(r,\theta) = r^k \sin\frac{\theta}{2}$ for $k = 1$, 1.5, and 2

In order to model a cohesive crack tip numerically, alternative functions are necessary since the stresses at the tip are not singular. An asymptotic analysis of the mechanical fields in a cohesive zone for very large structure was performed by Planas and Elices (1992, 1993). On the basis of an analytical solution, the following enrichment functions are proposed for two dimensional cohesive crack tips as

$$\psi_{\text{tip}}^{\text{cohesive}}(r,\theta) = r^k \sin\frac{\theta}{2} \tag{3.18}$$

with k being either 1, 1.5, or 2. Obviously, the non-singular behavior of branch functions can be observed at the cohesive crack tip, as shown in Figure 3.16.

The first implementation of an enhanced FEM technique for modeling of cohesive crack growth was presented by Wells and Sluys (2001) based on the PUM. They employed the step enrichment function through the entire crack including the crack tip. However, it must be noted that if only the step enrichment function is applied at all nodes, the X-FEM approximation is not able to treat crack tips or fronts that lie inside elements. To overcome this difficulty, Wells and Sluys (2001) assumed that the crack can be virtually extended to the next element edge. The X-FEM was proposed by Moës and Belytschko (2002a) in the cohesive crack simulation employing the cohesive branch functions (3.18), where the growth of the cohesive zone was governed by requiring the stress intensity factors at the tip of the cohesive zone to vanish. Mariani and Perego (2003) proposed a methodology for the simulation of quasi-static cohesive crack propagation in quasi-brittle materials. In their method, a cubic displacement discontinuity was used based on the product of the step function and polynomial ramp functions, which was able to reproduce the typical cusp-like shape of the process zone at the tip of a cohesive crack. The X-FEM was developed by Zi and Belytschko (2003) in numerical simulation of the static cohesive crack, where a new formulation was proposed for elements containing crack tips. This method overcomes the deficiency produced in the work of Wells and Sluys (2001), and provides a method for crack tips within elements without using the singular elements. In their method, an element containing a crack tip is enriched by the sign of a signed distance function whose gradient is normal to the crack and a polynomial along the crack, in which the parent domain of the partially cracked tip element is divided into two parts; the cracked and the uncracked parts. A similar technique was proposed by Rabczuk *et al.* (2008), where a new crack tip element was introduced for the phantom-node method. The main idea was based on implementation of the reduced integration FE with hourglass control to model the crack tip within an element. Meschke and Dumstorff (2007) proposed the X-FEM for modeling the cohesive and cohesionless cracks by employing the linear asymptotic enrichment functions (3.14), where $\sqrt{r}$ was replaced by r. There are various applications of cohesive crack simulation within the X-FEM framework, such as modeling the quasi-brittle fracture in functionally graded materials by Comi and Mariani (2007), modeling the cohesive crack growth in concrete structures by Unger, Eckardt, and Könke (2007), and so on.

3.5.5 Plastic Fracture Mechanics

In contrast to broad applications of the extended FE in linear elastic problems, there are less numerical modeling reported in elasto-plasticity (Figure 3.13c). Among those contributions, the most application of X-FEM in plasticity problems are as follows; the implementation of X-FEM model for resolution of highly localized strains in narrow damage process zones of quasi-brittle materials by Patzák and Jirásek (2003); the X-FEM modeling of shear band localization in Cosserat plasticity theory by Khoei and Karimi (2008); the X-FEM modeling of plasticity problem with contact friction in pressure sensitive materials of powder compaction process by Khoei, Shamloo, and Azami (2006b); the X-FEM modeling of frictional contact plasticity on arbitrary interfaces by Khoei and Nikbakht (2007) and Kim, Dolbow, and Laursen (2007); the 2D/3D X-FEM modeling of large plasticity deformations on arbitrary interfaces by Anahid and Khoei (2008) and Khoei, Biabanaki, and Anahid (2008d); an arbitrary Lagrangian–Eulerian X-FEM for modeling of moving boundaries in large plasticity deformations by Khoei, Anahid, and Shahim (2008a) and Anahid and Khoei (2010); a combined Lagrangian–XFEM in modeling large plasticity deformations and contact problems by Khoei, Biabanaki, and Anahid (2009a); the X-FEM modeling of crack propagation in plastic fracture mechanics by Elguedj, Gravouil, and Combescure (2006); a mixed augmented Lagrangian–XFEM model for elasto-plastic fatigue crack growth with unilateral contact by Elguedj, Gravouil, and Combescure (2007); a level set X-FEM for non-matching meshes with application to dynamic crack propagation in elastic–plastic problems by Prabel *et al.* (2007); and an X-FEM for large deformation ductile fracture problems with a non-local damage–plasticity model by Broumand and Khoei (2013).

In elastic-plastic fracture mechanics (EPFM), the elastic strain rates are generally negligible compared to the plastic strain rates at the vicinity of crack tip; and the singularity of stresses around the crack- tip is in the order of $r^{1/(n+1)}$ for nonlinear hardening material behavior based on the Ramberg–Osgood plasticity theory (Hutchinson, 1967). Elguedj, Gravouil, and Combescure (2006) employed the well-known Hutchinson–Rice–Rosengren (HRR) fields to represent the nature of singularity for a power-law hardening material in EPFM. They performed the X-FEM analysis in the context of confined plasticity to predict the fatigue crack growth without remeshing. A Fourier analysis of the HRR fields was applied in order to extract a proper elastic–plastic enrichment basis for the X-FEM solution. On the basis of Fourier decomposition of solutions, they presented that the HRR fields can be well approximated by using a truncated Fourier expansion with only the first four non-zero harmonic terms as

$$\psi_{\text{tip}}^{\text{plastic}}(r,\theta) = r^{\frac{1}{n+1}}\left\{ \sin\frac{k\theta}{2},\ \cos\frac{k\theta}{2} \right\} \tag{3.19}$$

where $k \in [1, 3, 5, 7]$ and n is the hardening exponent, in which $n = 1$ corresponds to the high hardening with linear-elastic behavior and $n \to \infty$ corresponds to the low hardening with rigid perfectly plastic material behavior. Elguedj, Gravouil, and Combescure (2006) proposed the following three sets of basis functions to compare the numerical rank of the tip stiffness matrix as

$$\psi_{\text{tip}}^{\text{plastic}}(r,\theta) = r^{\frac{1}{n+1}}\left\{ \sin\frac{\theta}{2},\ \cos\frac{\theta}{2},\ \sin\frac{\theta}{2}\sin\theta,\ \cos\frac{\theta}{2}\sin\theta,\ \sin\frac{\theta}{2}\sin 2\theta,\ \cos\frac{\theta}{2}\sin 2\theta \right\} \tag{3.20a}$$

$$\psi_{\text{tip}}^{\text{plastic}}(r,\theta) = r^{\frac{1}{n+1}}\left\{ \sin\frac{\theta}{2},\ \cos\frac{\theta}{2},\ \sin\frac{\theta}{2}\sin\theta,\ \cos\frac{\theta}{2}\sin\theta,\ \sin\frac{\theta}{2}\sin 3\theta,\ \cos\frac{\theta}{2}\sin 3\theta \right\} \tag{3.20b}$$

$$\begin{aligned}\psi_{\text{tip}}^{\text{plastic}}(r,\theta) = r^{\frac{1}{n+1}}\Bigg\{ & \sin\frac{\theta}{2},\ \cos\frac{\theta}{2},\ \sin\frac{\theta}{2}\sin\theta,\ \cos\frac{\theta}{2}\sin\theta, \\ & \sin\frac{\theta}{2}\sin 2\theta,\ \cos\frac{\theta}{2}\sin 2\theta,\ \sin\frac{\theta}{2}\sin 3\theta,\ \cos\frac{\theta}{2}\sin 3\theta \Bigg\}\end{aligned} \tag{3.20c}$$

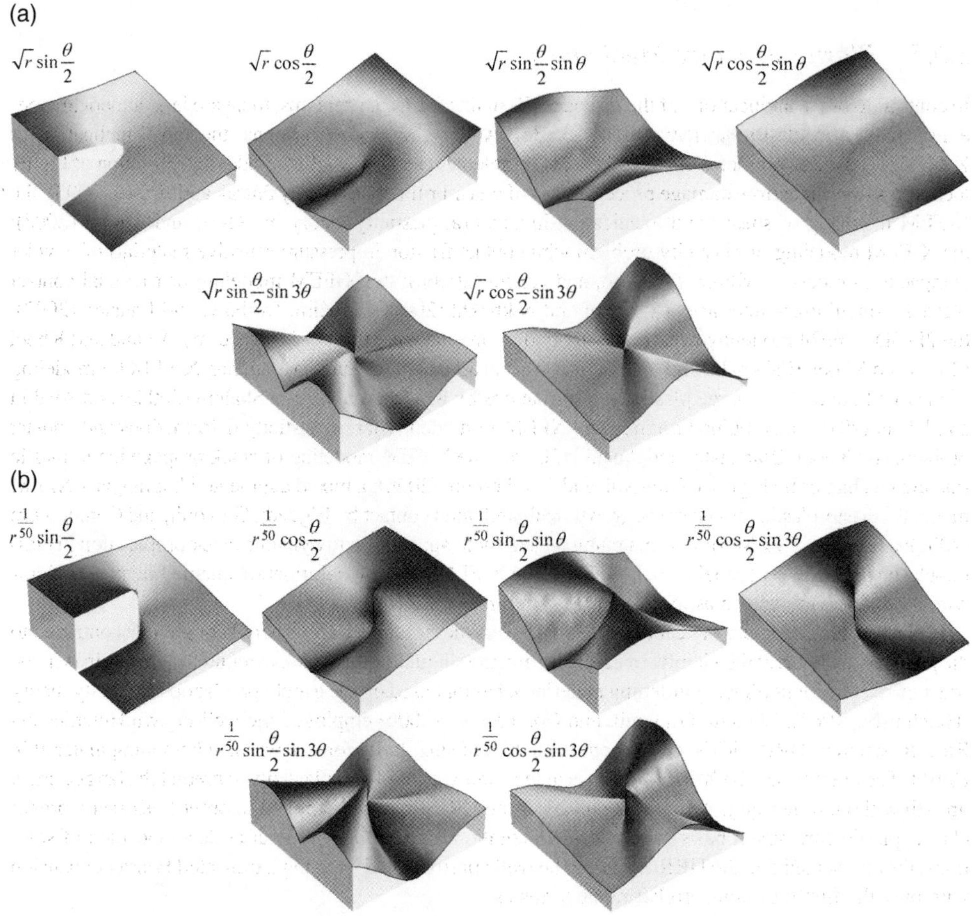

Figure 3.17 The distributions of six asymptotic elasto-plastic functions $\psi_{\text{tip}}^{\text{plastic}}(r,\theta)$ for two values of (a) $n=1$ and (b) $n=49$

It was interesting to note that the stiffness matrix with basis functions (3.20a) and (3.20c) has 8 or 12 "hourglass" modes, while it has 4 "hourglass" modes with the basis functions (3.20b). Thus, the set of asymptotic functions (3.20b) seems the proper choice for the elastic–plastic enrichment basis in EPFM, as also proposed by Rao and Rahman (2004) in the context of the EFGM. The distributions of these six asymptotic functions are shown in Figure 3.17 for two values of $n=1$ and $n=49$.

The X-FEM technique was also employed by Broumand and Khoei (2013) for the crack growth simulation in large deformation ductile fracture problems using a non-local damage-plasticity model. They used the Lemaitre damage-plasticity model to capture the material degradation effect, in which the non-locality is enforced by solving a Helmholtz type equation in combination with the governing equation of the system based on an operator-split technique. In their study, the performance of X-FEM plasticity model was investigated by studying the convergence rate of solution in ductile fracture problem. It was observed that the nonlinearity of plastic behavior has a deteriorating effect on the convergence of solution, and the sub-optimal convergence rates can be obtained for the plastic asymptotic functions. Thus, a technique that improves the convergence rate of plastic solution remains as an open question.

3.5.6 Multiple Cracks

Modeling crack problems in fracture mechanics is imperative to quantify and predict the behavior of cracked structures. Although the numerical simulation of a neat crack in the whole domain is preferable, the existence of a massive number of cracks in most practical problems cannot be ignored. Hence, an accurate modeling of cracks with multiple branches and interactions between them is required for simulation-based life-cycle design analysis. Basically, two types of cracks can be distinguished in fracture mechanics; micro-cracks and macro-cracks. Micro-cracks are generally infinite in number and variable in size, with a maximum size limited, while macro-cracks are usually few and have considerable length. The former are considered continuous cracks, called smeared cracks, where cracks are not directly entered into the numerical model and their effects are imposed onto the material behavior in fractured zone, while the latter are considered discontinuous cracks where the discrete interfaces are directly entered in the numerical modeling as major cracks.

The X-FEM is successfully employed to model the discrete crack interfaces in a continuous domain. However, multiple cracking and intersecting cracks require further modification in X-FEM formulation. The first implementation of the X-FEM for modeling arbitrary branched and intersecting cracks with multiple branches, multiple holes, and cracks emanating from holes was presented by Daux *et al.* (2000) by introducing the *junction* enrichment function. Chopp and Sukumar (2003) extended the method to multiple coplanar cracks, where the entire multiple crack geometry was represented by a single level set signed distance function and merging of distinct cracks was handled by the FMM. An X-FEM model to analyze the growth and the coalescence of cracks in a quasi-brittle material containing multiple cracks was presented by Zi *et al.* (2004), in which the method does not require a special enrichment for the junction of two cracks and the junction is automatically captured by the combination of the step enrichment functions. Budyn *et al.* (2004) proposed the X-FEM for modeling the growth of multiple cracks in linear elastic media with homogeneous and inhomogeneous materials behavior. The effect of crack shielding and amplification of various arrangements of micro-cracks on the stress intensity factors of a macro-crack, including large numbers of arbitrarily aligned micro-cracks, was investigated by Loehnert and Belytschko (2007) using the X-FEM.

In order to model branched crack geometry in the X-FEM, consider the branched crack as a main crack (solid line) and a branched secondary crack (dashed line), as shown in Figure 3.18. The procedure of enrichment can be performed as follows; the main crack is enriched in the absence of the secondary crack

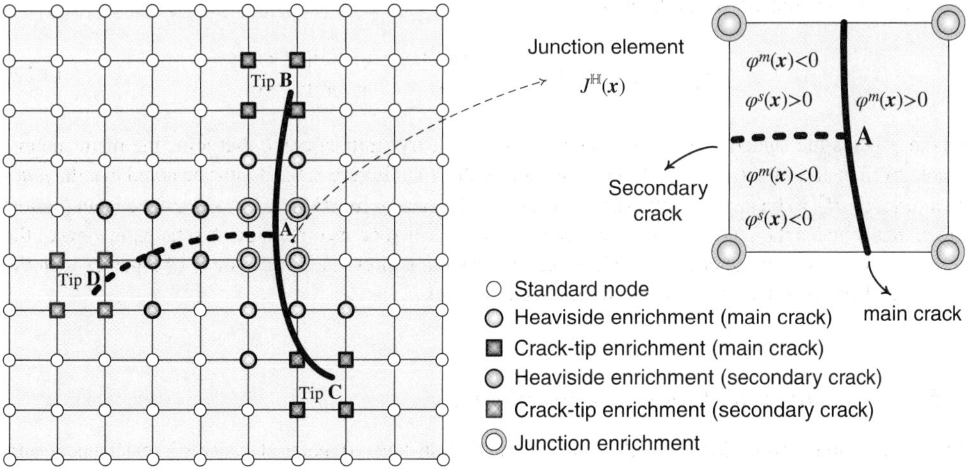

Figure 3.18 The discontinuous junction function $J^{\mathbb{H}}(x)$ for the multiple cracks

that involves the tip enrichments for the crack tips B and C, denoted by squares and the discontinuous interior enrichment denoted by circles. The secondary crack is then enriched in the absence of the main crack; the crack tip D is enriched by the tip enrichments (squares) and all nodes for which the support intersects the secondary crack but do not contain points A or D are part of the discontinuous interior enrichment (circles). Finally, the *junction enrichment* function is introduced to enrich all nodes for which the support of the FE shape functions contains the junction point A (double-circles). The procedure for modeling a multiple branched crack can be performed similarly by discretizing the branched crack into a main crack and several cracks joining the main crack. In the case that a crack tip joins with another crack, the procedure is carried out by removing the tip enrichment of the approaching crack and replacing it by a Heaviside enrichment function. In the element containing the junction of the two cracks, a "*junction*" enrichment function $J^{\mathbb{H}}(\boldsymbol{x})$ is introduced to account for the linked discontinuities.

The enriched displacement field for a branched crack, which has a main crack to which several secondary cracks are connected, can be written as

$$\begin{aligned}\mathbf{u}(\boldsymbol{x}) &= \sum_{I\in\mathcal{N}} N_I(\boldsymbol{x})\,\bar{\mathbf{u}}_I + \sum_{j=1}^{\mathcal{M}^{dis}}\sum_{J\in\mathcal{N}^{dis}} N_J(\boldsymbol{x})\,(H(\boldsymbol{x})-H(\boldsymbol{x}_J))\,\bar{\mathbf{d}}_J \\ &+\sum_{k=1}^{\mathcal{M}^{tip}}\sum_{K\in\mathcal{N}^{tip}} N_K(\boldsymbol{x})\sum_{\alpha=1}^{4}\left(\psi^{\alpha}_{\text{tip}}(\boldsymbol{x})-\psi^{\alpha}_{\text{tip}}(\boldsymbol{x}_K)\right)\bar{\mathbf{b}}^{\alpha}_K + \sum_{l=1}^{\mathcal{M}^{jun}}\sum_{L\in\mathcal{N}^{jun}} N_L(\boldsymbol{x})\,(J^{\mathbb{H}}(\boldsymbol{x})-J^{\mathbb{H}}(\boldsymbol{x}_L))\,\bar{\mathbf{c}}_L\end{aligned} \tag{3.21}$$

where $\mathcal{M}^{dis}$ is the number of cracks including the main crack and the secondary cracks, $\mathcal{M}^{tip}$ is the number of crack tips, and $\mathcal{M}^{jun}$ is the number of junctions with $\mathcal{M}^{jun}=\mathcal{M}^{tip}-1$. In the earlier relation, $\mathcal{N}^{dis}$ is the subset of nodes to enrich for the jth $\in \mathcal{M}^{dis}$ crack discontinuity and $\bar{\mathbf{d}}_J$ are the corresponding additional degrees of freedom; the nodes in $\mathcal{N}^{dis}$ are such that their support intersects the jth crack but do not contain any of its two extremities. $\mathcal{N}^{tip}$ is the subset of nodes to enrich for the kth $\in \mathcal{M}^{tip}$ crack tip and $\bar{\mathbf{b}}^{\alpha}_K$ are the corresponding additional degrees of freedom; the nodes in $\mathcal{N}^{tip}$ are such that their support contain the kth crack tip. $\mathcal{N}^{jun}$ is the subset of nodes to enrich for the lth $\in \mathcal{M}^{jun}$ junction and $\bar{\mathbf{c}}_L$ are the corresponding additional degrees of freedom; the nodes in $\mathcal{N}^{jun}$ are such that their support contain the lth junction.

In order to define a discontinuous *junction* function $J^{\mathbb{H}}(\boldsymbol{x})$, consider the main (major) crack is denoted by m and the secondary (minor) crack by s, where the approaching crack s is arrested by the main crack m. The *junction* function $J^{\mathbb{H}}(\boldsymbol{x})$ is defined as

$$\psi^{\mathbb{H}}_{\text{junction}}(\varphi(\boldsymbol{x})) = J^{\mathbb{H}}(\boldsymbol{x}) = \begin{cases} H(\varphi^s(\boldsymbol{x})) & \text{if } \varphi^m(\boldsymbol{x})\le 0 \\ 0 & \text{if } \varphi^m(\boldsymbol{x})>0 \end{cases} \tag{3.22}$$

where $\varphi^s(\boldsymbol{x})$ is the signed distance function of the secondary (minor) crack that joins the main (major) crack given by the signed distance function $\varphi^m(\boldsymbol{x})$, as shown in Figure 3.18. It must be noted that the *junction* function $J^{\mathbb{H}}(\boldsymbol{x})$ behaves exactly similar to a single Heaviside function on one side of the main (major) crack that contains the secondary (minor) crack and is vanished on its other side. A schematic view of the discontinuous *junction* function on an intersection element is presented in Figure 3.19 together with the enriched shape functions for a 4-noded quadratic element.

3.5.7 *Fracture in Bimaterial Problems*

There are various applications of multi-functional and multi-layered material systems in mechanical, aerospace, and biomedical applications. The overall mechanical behavior and response of layered systems are directly dependent on the mechanical properties and fracture behavior of the interfaces. Composite materials, such as fiber reinforced concrete, polymers, and heterogeneous soils, are typical examples of multi-phase materials. Their particular micro-structure is often characterized by a complex distribution

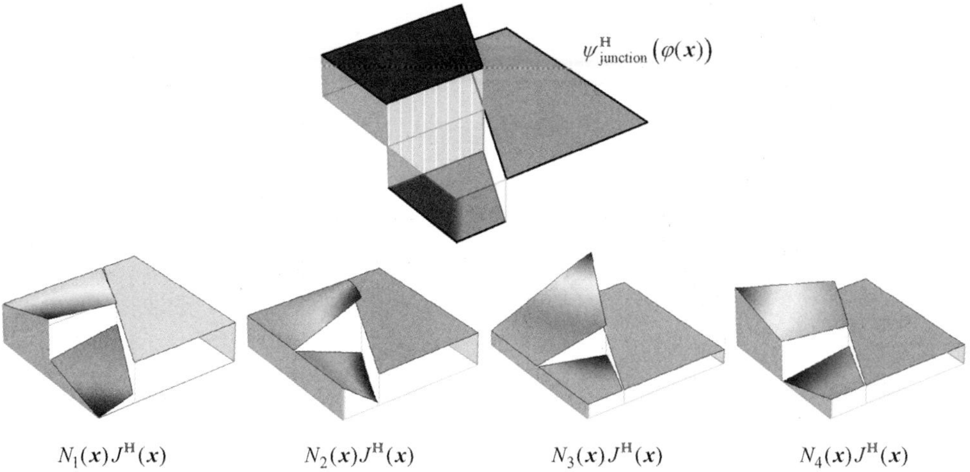

Figure 3.19 The discontinuous *junction* function together with the enriched shape functions for a 4-noded quadratic element

of material constituents, which can vary deterministically or even randomly within a macroscopic structure. The presence of weak fiber–matrix interfaces in composites provides the preferential crack paths that enhance the overall fracture toughness of the composite. Debonding in adhesive joints, composite laminates and at film-substrate interfaces, along bimaterial interface and the structural integrity of thin films are various applications that highlight the role of interfacial mechanics. Crack termination at a bimaterial interface with an oblique angle plays an important role in heterogeneous composites, where the distribution and orientation of the fillers is complex. In such composites, the multiple interaction of the crack tip with fillers can lead to the case of a crack terminating at filler/matrix interface with an oblique angle. The development of a robust technique to characterize the crack driving force and interfacial toughness in bimaterial systems can lead to a better understanding of the role and influence of the mismatch in properties and their effects on crack growth.

The numerical implementation of the X-FEM for modeling bimaterial interfacial cracks was originally proposed by Nagashima, Omoto, and Tani (2003) and Sukumar *et al.* (2004). In the work of Nagashima, Omoto, and Tani (2003), the crack was modeled between dissimilar materials by enriching the crack tip displacement field with the same basis as for a crack located in an isotropic material, while Sukumar *et al.* (2004) employed the X-FEM by applying the near tip asymptotic functions of bimaterial interfacial crack that span the asymptotic displacement fields for an interfacial crack. Liu, Xiao, and Karihaloo (2004) improved the accuracy of mixed mode stress intensity factors for a bimaterial interfacial crack by enriching the FE approximation of the nodes surrounding the crack tip with not only the first term, but also the higher order terms of the crack tip asymptotic field. In the aforementioned studies, the FE mesh had to be aligned with the material interfaces, and the enrichment functions were only used at the crack tips. To overcome this restriction, Hettich and Ramm (2006) applied the X-FEM for modeling the material interfaces that are not aligned with element edges by employing the level sets to enrich the displacement field for discontinuous displacement derivatives. A similar technique was proposed by Huynh and Belytschko (2009) for treating fractures in composite materials with meshes that are independent of matrix/fiber interfaces. In their study, the level sets were used to describe the geometry of the crack interfaces and 12 asymptotic functions were applied for bimaterial crack tips. The bimaterial interfacial crack was also modeled by Pathak, Singh, and Singh (2012) on the basis of EFGM and X-FEM, where the standard crack tip asymptotic field of isotropic material was utilized. Bouhala *et al.* (2013) employed the X-FEM for modeling the cracked bimaterial problem in the case of a crack terminating at a bimaterial interface by applying various mechanical mismatches and crack angles.

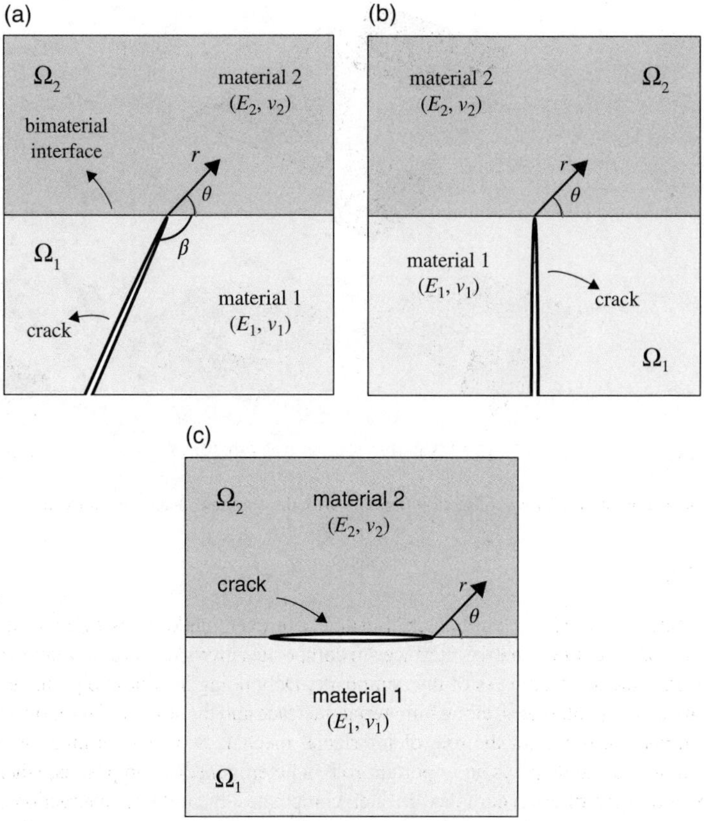

Figure 3.20 Fracture in bimaterial media: (a) a crack terminating at the bimaterial interface with an oblique angle; (b) an arbitrary crack perpendicular to the bimaterial interface; (c) an interfacial crack along the bimaterial interface

In the case of a crack terminating at a bimaterial interface (Figure 3.20a), the analytical asymptotic solution in the vicinity of the crack tip is dependent on the mechanical mismatch between the two materials (E_1/E_2), and the angle between the crack and the interface β. Bouhala *et al.* (2013) derived the enrichment functions for a crack tip terminating at a bimaterial interface following the Williams's approach (1952, 1957) based on the following 16 functions as

$$\begin{aligned}
\psi_{\text{tip}}^{\text{bimaterial}}(r,\theta) &= \left\{\psi_{\text{tip}}^{1}, \psi_{\text{tip}}^{2}, \dots, \psi_{\text{tip}}^{16}\right\} \\
&= \left\{ r^{\lambda_{\text{Re}}} e^{+\lambda_{\text{Im}}\theta} \cos(\lambda_{\text{Im}} \log r) \cos(\lambda_{\text{Re}}+1)\theta,\ r^{\lambda_{\text{Re}}} e^{+\lambda_{\text{Im}}\theta} \cos(\lambda_{\text{Im}} \log r) \sin(\lambda_{\text{Re}}+1)\theta, \right. \\
&\quad r^{\lambda_{\text{Re}}} e^{+\lambda_{\text{Im}}\theta} \sin(\lambda_{\text{Im}} \log r) \cos(\lambda_{\text{Re}}+1)\theta,\ r^{\lambda_{\text{Re}}} e^{+\lambda_{\text{Im}}\theta} \sin(\lambda_{\text{Im}} \log r) \sin(\lambda_{\text{Re}}+1)\theta, \\
&\quad r^{\lambda_{\text{Re}}} e^{+\lambda_{\text{Im}}\theta} \cos(\lambda_{\text{Im}} \log r) \cos(\lambda_{\text{Re}}-1)\theta,\ r^{\lambda_{\text{Re}}} e^{+\lambda_{\text{Im}}\theta} \cos(\lambda_{\text{Im}} \log r) \sin(\lambda_{\text{Re}}-1)\theta, \\
&\quad r^{\lambda_{\text{Re}}} e^{+\lambda_{\text{Im}}\theta} \sin(\lambda_{\text{Im}} \log r) \cos(\lambda_{\text{Re}}-1)\theta,\ r^{\lambda_{\text{Re}}} e^{+\lambda_{\text{Im}}\theta} \sin(\lambda_{\text{Im}} \log r) \sin(\lambda_{\text{Re}}-1)\theta, \\
&\quad r^{\lambda_{\text{Re}}} e^{-\lambda_{\text{Im}}\theta} \cos(\lambda_{\text{Im}} \log r) \cos(\lambda_{\text{Re}}+1)\theta,\ r^{\lambda_{\text{Re}}} e^{-\lambda_{\text{Im}}\theta} \cos(\lambda_{\text{Im}} \log r) \sin(\lambda_{\text{Re}}+1)\theta, \\
&\quad r^{\lambda_{\text{Re}}} e^{-\lambda_{\text{Im}}\theta} \sin(\lambda_{\text{Im}} \log r) \cos(\lambda_{\text{Re}}+1)\theta,\ r^{\lambda_{\text{Re}}} e^{-\lambda_{\text{Im}}\theta} \sin(\lambda_{\text{Im}} \log r) \sin(\lambda_{\text{Re}}+1)\theta, \\
&\quad r^{\lambda_{\text{Re}}} e^{-\lambda_{\text{Im}}\theta} \cos(\lambda_{\text{Im}} \log r) \cos(\lambda_{\text{Re}}-1)\theta,\ r^{\lambda_{\text{Re}}} e^{-\lambda_{\text{Im}}\theta} \cos(\lambda_{\text{Im}} \log r) \sin(\lambda_{\text{Re}}-1)\theta, \\
&\quad \left. r^{\lambda_{\text{Re}}} e^{-\lambda_{\text{Im}}\theta} \sin(\lambda_{\text{Im}} \log r) \cos(\lambda_{\text{Re}}-1)\theta,\ r^{\lambda_{\text{Re}}} e^{-\lambda_{\text{Im}}\theta} \sin(\lambda_{\text{Im}} \log r) \sin(\lambda_{\text{Re}}-1)\theta \right\}
\end{aligned} \tag{3.23}$$

where (r, θ) is the polar coordinates system. The characteristic exponent λ_k is dependent on the mismatch between the two materials (E_1/E_2) and the angle between the crack and the interface β, and can be determined by applying a parametric solution whose parameters are obtained from the problem boundary conditions. The characteristic exponent can be real single $\lambda = \lambda_1$ in the case of a crack perpendicular to the interface (Figure 3.20b), real double $\lambda_k = (\lambda_1, \lambda_2)$ if the crack is oblique (Figure 3.20a), or complex if the crack goes to zero or π (Figure 3.20c), that is, $\lambda_1 = \lambda_{\mathrm{Re}} + i\lambda_{\mathrm{Im}}$ and $\lambda_2 = \lambda_{\mathrm{Re}} - i\lambda_{\mathrm{Im}}$, where λ_{Re} and λ_{Im} are the real and imaginary components of the complex values and $i = \sqrt{-1}$. The passage from real double to complex values depends on both the crack angle β and the mechanical mismatch (E_1/E_2).

These enrichment functions are represented in two special cases; if an arbitrary crack is normal to the bimaterial interface (Figure 3.20b) and when an interfacial crack is along the bimaterial interface (Figure 3.20c). For a bimaterial crack with a crack perpendicular to the interface, the near tip stress field is of the order $r^{-\lambda}(0 < \lambda < 1)$, where the exponent λ is given by the solution of a transcendental equation (Zak and Williams, 1963). However, for a bimaterial interfacial crack, the complex stress intensity factors are proposed since the stress singularity in the vicinity of the crack tip of a bimaterial interface crack is oscillatory in nature with the presence of an inverse $\sqrt{r}$ singularity that introduces significant complexity in an element formulation (Rice, 1988).

Consider a bimaterial with a crack perpendicular to the interface where the crack terminates at the interface, as shown in Figure 3.20b. The X-FEM was proposed by Sukumar and Prévost (2003) and Huang *et al.* (2003a) and Huang, Sukumar, and Prévost (2003b) for modeling a crack normal to the bimaterial interface in quasi-static crack growth simulation. The tip asymptotic functions for a crack perpendicular to the bimaterial interface, where the resulting characteristic exponent λ is real single, are defined as

$$\begin{aligned}\psi_{\mathrm{tip}}^{\text{N-bimaterial}}(r,\theta) &= \left\{\psi_{\mathrm{tip}}^1, \psi_{\mathrm{tip}}^2, \psi_{\mathrm{tip}}^3, \psi_{\mathrm{tip}}^4\right\} \\ &= \left\{r^\lambda \cos(\lambda+1)\theta,\, r^\lambda \sin(\lambda+1)\theta,\, r^\lambda \cos(\lambda-1)\theta,\, r^\lambda \sin(\lambda-1)\theta\right\}\end{aligned} \tag{3.24}$$

where $\lambda(0 < \lambda < 1)$ is the stress singularity exponent. In the case of no mismatch, the stress singularity reduces to the classical inverse $\sqrt{r}$ stress singularity $(\lambda = 0.5)$ for homogeneous linear elastic materials. If the material (2) is stiffer than material (1), the singularity is weaker $(\lambda > 0.5)$, and if the material (2) is more compliant than material (1), the singularity is stronger $(\lambda < 0.5)$. These asymptotic functions span the asymptotic crack tip displacement field, where the second and fourth functions are discontinuous across the crack $(\theta = \pm\pi)$.

In the case of an oblique crack with the crack terminating at a bimaterial interface with the crack angle β, and the resulting characteristic exponent λ is real double $\lambda_k = (\lambda_1, \lambda_2)$, the crack tip asymptotic functions can be defined based on the following eight functions as

$$\begin{aligned}\psi_{\mathrm{tip}}^{\text{O-bimaterial}}(r,\theta) &= \left\{\psi_{\mathrm{tip}}^1, \psi_{\mathrm{tip}}^2, \ldots, \psi_{\mathrm{tip}}^8\right\} \\ &= \left\{r^{\lambda_1}\cos(\lambda_1+1)\theta,\, r^{\lambda_1}\sin(\lambda_1+1)\theta,\, r^{\lambda_1}\cos(\lambda_1-1)\theta,\, r^{\lambda_1}\sin(\lambda_1-1)\theta,\right. \\ &\qquad \left. r^{\lambda_2}\cos(\lambda_2+1)\theta,\, r^{\lambda_2}\sin(\lambda_2+1)\theta,\, r^{\lambda_2}\cos(\lambda_2-1)\theta,\, r^{\lambda_2}\sin(\lambda_2-1)\theta\right\}\end{aligned} \tag{3.25}$$

In the case that the crack angle goes to zero or π, the bimaterial interfacial crack occurs where the resulting characteristic exponents λ_1 and λ_2 become complex. For the bimaterial interfacial crack, the crack tip asymptotic functions are defined by Sukumar *et al.* (2004) based on the following 12 functions as

$$
\begin{aligned}
\psi_{\text{tip}}^{\text{I-bimaterial}}(r,\theta) &= \left\{\psi_{\text{tip}}^{1}, \psi_{\text{tip}}^{2}, \ldots, \psi_{\text{tip}}^{12}\right\} \\
&= \left\{ \sqrt{r}\cos(\varepsilon \log r)e^{-\varepsilon\theta}\sin\frac{\theta}{2},\ \sqrt{r}\cos(\varepsilon \log r)e^{-\varepsilon\theta}\cos\frac{\theta}{2}, \right. \\
&\quad \sqrt{r}\cos(\varepsilon \log r)e^{+\varepsilon\theta}\sin\frac{\theta}{2},\ \sqrt{r}\cos(\varepsilon \log r)e^{+\varepsilon\theta}\cos\frac{\theta}{2}, \\
&\quad \sqrt{r}\cos(\varepsilon \log r)e^{+\varepsilon\theta}\sin\frac{\theta}{2}\sin\theta,\ \sqrt{r}\cos(\varepsilon \log r)e^{+\varepsilon\theta}\cos\frac{\theta}{2}\sin\theta, \\
&\quad \sqrt{r}\sin(\varepsilon \log r)e^{-\varepsilon\theta}\sin\frac{\theta}{2},\ \sqrt{r}\sin(\varepsilon \log r)e^{-\varepsilon\theta}\cos\frac{\theta}{2}, \\
&\quad \sqrt{r}\sin(\varepsilon \log r)e^{+\varepsilon\theta}\sin\frac{\theta}{2},\ \sqrt{r}\sin(\varepsilon \log r)e^{+\varepsilon\theta}\cos\frac{\theta}{2}, \\
&\quad \left. \sqrt{r}\sin(\varepsilon \log r)e^{+\varepsilon\theta}\sin\frac{\theta}{2}\sin\theta,\ \sqrt{r}\sin(\varepsilon \log r)e^{+\varepsilon\theta}\cos\frac{\theta}{2}\sin\theta \right\}
\end{aligned}
\tag{3.26}
$$

where ε is the bimaterial constant that depends on the properties of both materials. If the bimaterial constant ε is set to zero (no mismatch), the previous crack tip asymptotic functions result in the span of near tip enrichment functions used for the isotropic material in (3.14). The enriched displacement field can, therefore, be written for the bimaterial interfacial crack as

$$
\begin{aligned}
\mathbf{u}(\boldsymbol{x}) &= \sum_{I\in\mathcal{N}} N_I(\boldsymbol{x})\,\bar{\mathbf{u}}_I + \sum_{J\in\mathcal{N}^{Hev}} N_J(\boldsymbol{x})\,(H(\boldsymbol{x})-H(\boldsymbol{x}_J))\,\bar{\mathbf{d}}_J \\
&\quad + \sum_{K\in\mathcal{N}^{ramp}} N_K(\boldsymbol{x})\,(\psi_{\text{ramp}}(\boldsymbol{x})-\psi_{\text{ramp}}(\boldsymbol{x}_K))\,\bar{\mathbf{a}}_K + \sum_{L\in\mathcal{N}^{tip}} N_L(\boldsymbol{x}) \sum_{\alpha=1}^{12}\left(\psi_{\text{tip}}^{\alpha}(\boldsymbol{x})-\psi_{\text{tip}}^{\alpha}(\boldsymbol{x}_L)\right)\bar{\mathbf{b}}_L^{\alpha}
\end{aligned}
\tag{3.27}
$$

in which the second term in this relation is used to model the discontinuity along the crack interface using the Heaviside function, the third term is introduced to enforce the weak discontinuity at the bimaterial interface using a ramp enrichment function in order to allow for the non-alignment of the mesh to the interface, and the last term is employed to enrich the nodal points of crack tip element using 12 asymptotic functions. The distributions of these 12 asymptotic functions are shown in Figure 3.21 for the bimaterial constant $\varepsilon = 0.076$.

3.5.8 Polycrystalline Microstructure

The failure mechanisms in brittle polycrystalline materials such as ceramics are important for improvements in the development and application of advanced structural materials. The material microstructure plays a critical role in dictating the modes of fracture and failure, and the macroscopic response of real materials. The grain morphology, elastic modulus, and the toughness of the individual microstructural constituents and interfaces are key parameters that control the failure mechanisms in polycrystalline materials. The crack propagation through a material microstructure depends on the mechanical state in the vicinity of the crack tip, and hence local differences in toughness (grain interior vs. grain boundaries) significantly influence the crack path. The grains can be oriented in different directions, which results in dissimilar anisotropic elastic properties (Figure 3.22a). The strain singularity can be occurred at the junctions formed by several grains or at the crack tips. Their shape and strength is, however, dependent

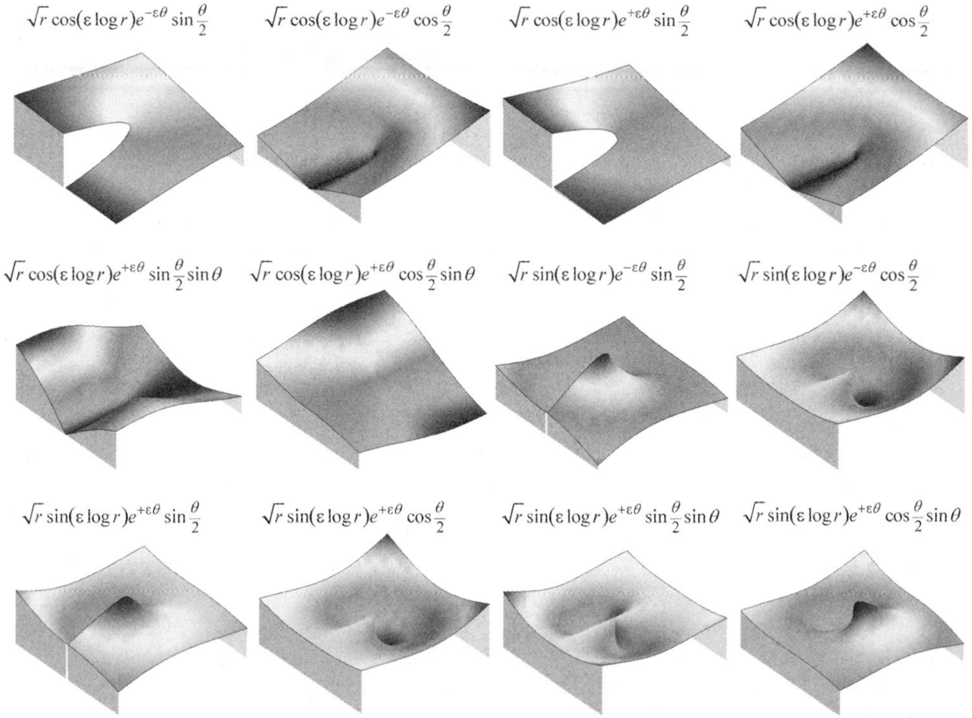

Figure 3.21 The distributions of 12 asymptotic functions for a bimaterial interfacial crack at the bimaterial constant $\varepsilon = 0.076$

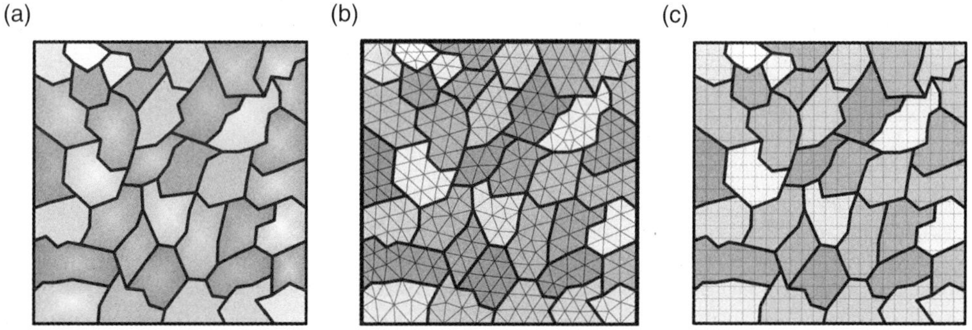

Figure 3.22 Modeling of polycrystals: (a) polycrystalline topology; (b) FEM modeling of polycrystals; (c) X-FEM modeling of polycrystals

on the orientation and the geometry of the surrounding grains, which makes it hard to find closed-form solutions. Based on the previous discussion, it is clear that a numerical fracture model is needed to model crack propagation by incorporating these microstructural features that have the potential to describe toughening mechanisms in polycrystals and provide a framework for microstructural design. The displacement discontinuities along cracks as well as the strain discontinuities along grain boundaries can further complicate the simulation of polycrystals with the FEM. The FE modeling of polycrystals relies on the use of mesh generators to design the mesh around the polycrystalline topology (Figure 3.22b); the situations becomes more complex in the case of a three-dimensional configuration, or when the grains are shaped like a wedge, in which it is not possible to achieve the desired quality. The X-FEM can be performed efficiently to model polycrystalline materials without using a mesh generator to mimic the polycrystalline; it only requires a simple background mesh on which the polycrystalline topology is superimposed, as schematically depicted in Figure 3.22c.

The first implementation of X-FEM in polycrystalline microstructures was performed by Sukumar *et al.* (2003b) with the aim of an understanding of toughening mechanisms in polycrystalline materials. In their model, the same isotropic elastic material properties were assumed for each crystal, and the displacement field was discretized by employing the Heaviside and asymptotic crack tip enrichment functions for the cracks together with the ramp enrichment function for material boundaries. Sukumar and Srolovitz (2004) employed the same computational algorithm for brittle fracture simulations in polycrystalline microstructures to evaluate the convergence rate of the X-FEM solution with the theoretical convergence rate in the presence of a singularity. Moës *et al.* (2003) employed a modified level set function to model the non-smooth behavior of the solution at material interfaces in micro-structures with complex geometries, and presented the capability of model for the homogenization of periodic basic cells. Simone, Duarte, and Giessen (2006) presented a generalized FEM based on the PU property for modeling polycrystalline microstructures with discontinuous displacement fields at the grain boundaries and grain junctions, in which the effect of free grain boundary sliding on an elasticity of polycrystals was investigated. Aragon, Duarte, and Geubelle (2010) developed a general GFEM/X-FEM formulation to model polycrystalline materials with gradient jumps along discrete lines by introducing a new enrichment function that is able to solve problems with multiple intersecting discontinuity lines. In their study, an enrichment function was introduced that addresses the problem of having multiple interfaces intersecting inside the FE; this enrichment function is continuous over the FE and has a discontinuous gradient in the direction perpendicular to the line segments that represent line loads or grain boundaries. Menk and Bordas (2010) proposed a numerical procedure to compute the enrichment functions for anisotropic polycrystals. Although their technique is computationally expensive, it gives a systematic algorithm to compute the enrichment functions when their analytical form is not known *a priori*.

The enriched displacement field for modeling the brittle fracture in polycrystalline structures can be performed based on the following three discontinuities; the displacement discontinuity along the crack, the strain discontinuity along the grain boundaries, and the strain singularity at the crack tips and grain junctions. Since the elements do not conform to the crack geometry, the first type of enrichment is crucial for obtaining a reasonable solution in the case of a cracked structure. The second and third types of enrichment are necessary for obtaining the optimal convergence rates, because the representation of an exact solution by the polynomial shape functions is very poor in the case of strain singularities or weak discontinuities. As discussed in preceding sections, the common enrichment functions for the first two types of discontinuities are the *Heaviside* and *ramp* enrichment functions, respectively. For the third type of discontinuity, the tip asymptotic functions described in preceding section can be used properly to capture the singularity at the crack tips; however, there is no general form of enrichment proposed for the singularity at the grain junctions. Li, Zhang, and Recho (2001) proposed a procedure to determine the stress singularity at the vicinity of a notch formed by different anisotropic materials. A slight modification of this approach was performed by Menk and Bordas (2010) to numerically calculate the singular fields in the interior of the structural domain as

$$\psi_{\text{tip}}^{\text{notch}}(r,\theta) = r^{\lambda}\,\Psi(\theta) \tag{3.28}$$

where the polar coordinate system is located at the junction with the radius r and angle θ. In the previous equation, $\lambda (0 < \lambda < 1)$ is the stress singularity exponent and $\Psi(\theta)$ is the angular function. Although an explicit form of the angular part is unknown, the function values can be obtained approximately by linear interpolation using the values calculated by numerical differentiation.

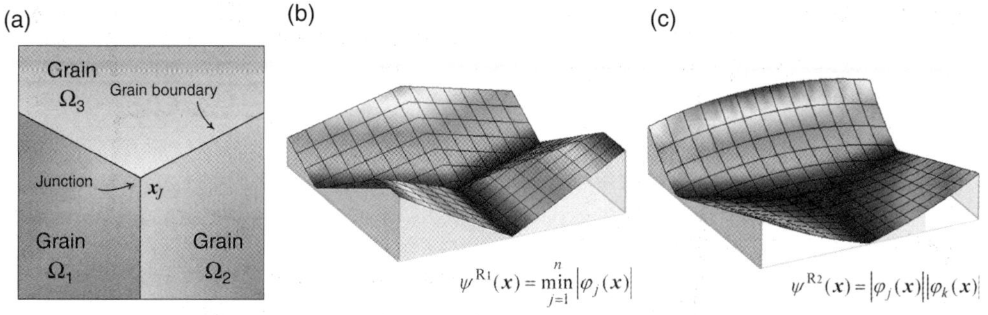

Figure 3.23 Modeling an element with multiple interfaces intersecting inside a polycrystalline material: (a) problem definition; (b) the enrichment function proposed by Aragon, Duarte, and Geubelle (2010) based on the minimum value of distance level set functions; (c) the enrichment function proposed by Khoei *et al.* (2014b) based on multiplication of the distance level set functions obtained from two line segments at each grain

In order to model the problem of multiple intersecting discontinuity lines in polycrystalline materials, Aragon, Duarte, and Geubelle (2010) introduced a *junction ramp* enrichment function. This enrichment function is continuous over the element and has a discontinuous gradient in the direction perpendicular to the line segments that represent the grain boundaries. Consider a problem of having multiple interfaces intersecting inside a FE, in which the domain of element is sub-divided into three regions Ω_i $(i = 1, 2, 3)$, as shown in Figure 3.23a. In the case of a polycrystalline microstructure, a region Ω_i represents one grain in the domain. The enrichment function for a junction $\boldsymbol{x}_J$ between m grains is defined by Aragon, Duarte, and Geubelle (2010) as

$$\psi^{R1}_{\text{Junction}}(\boldsymbol{x}) = \min_{j=1}^{n} |\varphi_j(\boldsymbol{x})| \qquad \boldsymbol{x} \in \Omega_i \tag{3.29}$$

where $|\varphi_j(\boldsymbol{x})|$ is the distance level set function to the jth line segment Γ_j, and n is the number of line segments (grain boundaries) at each grain Ω_i. In this definition, the enrichment function is obtained by computing the distance from point $\boldsymbol{x}$ to the closest line in the domain. In order to obtain a well-conditioned stiffness matrices, the authors introduced a constant unit value to the previous definition, however, it must be noted that the main problem with this enrichment function is that a discontinuous displacement gradient can be observed all along the line segments that include the grain boundaries, as shown in Figure 3.23b for an element with three regions.

In order to overcome this difficulty, the *junction ramp* enrichment function (3.29) can be modified based on multiplication of the distance level set functions obtained from two line segments Γ_j and Γ_k at each grain Ω_i as

$$\psi^{R2}_{\text{Junction}}(\boldsymbol{x}) = |\varphi_j(\boldsymbol{x})||\varphi_k(\boldsymbol{x})| \qquad \boldsymbol{x} \in \Omega_i \tag{3.30}$$

This definition clearly presents a discontinuous displacement gradient only along the grain boundaries, as shown in Figure 3.23c.

An alternative *junction ramp* enrichment function was introduced by Khoei *et al.* (2014b) to enrich an element cut by multiple interfaces, as depicted in Figure 3.24. Consider the main (major) interface is denoted by m and the secondary (minor) interface by S, where the approaching interface S is arrested by the main interface m. The *junction ramp* function $J^{\mathbb{R}}(\boldsymbol{x})$ is defined as

$$\psi^{\mathbb{R}}_{\text{junction}}(\varphi(\boldsymbol{x})) = J^{\mathbb{R}}(\boldsymbol{x}) = \begin{cases} |\varphi^m(\boldsymbol{x})||\varphi^S(\boldsymbol{x})| & \text{if } \varphi^m(\boldsymbol{x}) \le 0 \\ |\varphi^m(\boldsymbol{x})| & \text{if } \varphi^m(\boldsymbol{x}) > 0 \end{cases} \tag{3.31}$$

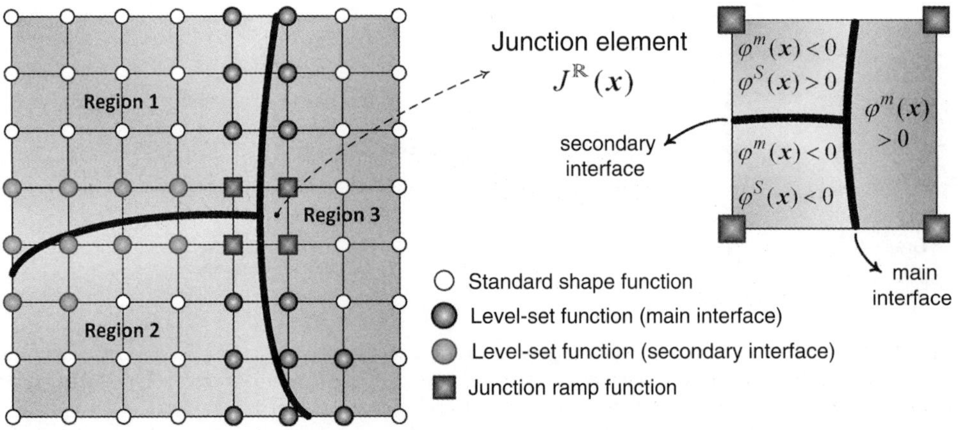

Figure 3.24 The *junction ramp* function $J^{\mathbb{R}}(x)$ for multiple interfaces intersecting inside an element

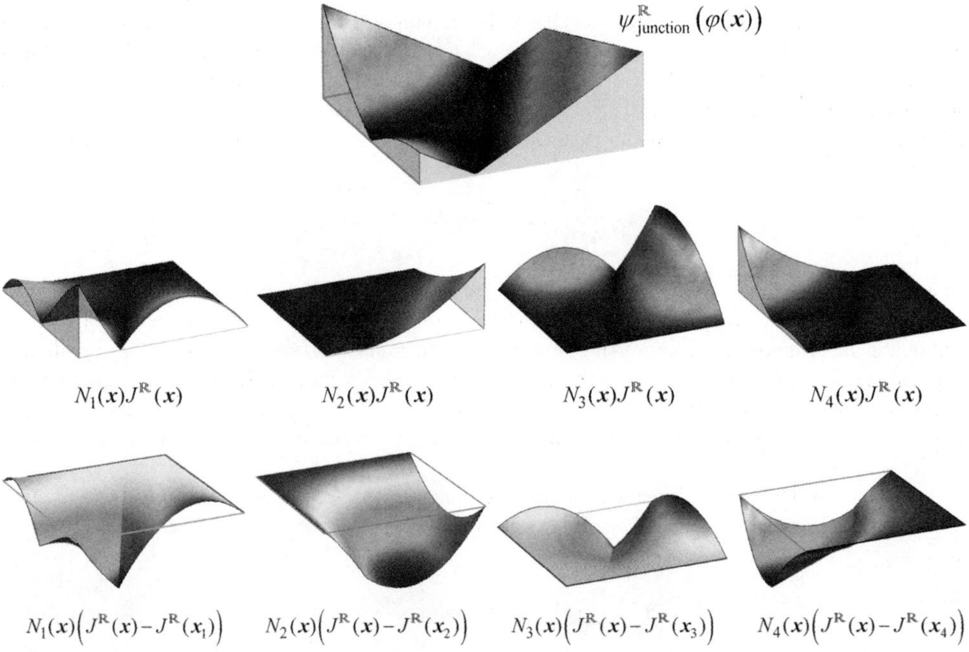

Figure 3.25 The junction ramp function together with the enriched shape functions for a 4-noded quadratic element

where $|\varphi^S(x)|$ is the signed distance function of the secondary (minor) interface that joins the main (major) interface given by the signed distance function $|\varphi^m(x)|$. Obviously, the junction enrichment function (3.31) provides the continuity in the displacement field, while there are discontinuities in the displacement derivatives across the main (major) and secondary (minor) material interfaces. A schematic view of the *junction ramp* function $J^{\mathbb{R}}(x)$ is shown in Figure 3.25 for a 4-noded quadratic element with multiple material interfaces. Note that, when there is a single line segment, such as a single material interface, the enrichment function (3.31) reduces to the standard level set function extensively used for bimaterial problems.

3.5.9 *Dislocations*

The study of fundamental phenomena in plasticity often involves the simulation of the motion of thousands of dislocations; however, the cost of these computations is often very large and a limiting factor on model size. So the development of computationally more efficient dislocation models is desirable. In the continuum-based approach, the solution of a domain with a number of dislocations is basically determined by superimposing the analytical infinite domain solution of each dislocation and an image stress field. The solution can then be obtained by a FE model with boundary tractions chosen to cancel the tractions of the infinite dislocation fields. A disadvantage of this method is that at each quadrature point on the boundary of the FE domain, a sum over all dislocations must be performed. When a dislocation core is near a boundary, an accurate integration of the tractions requires a large number of quadrature points. Furthermore, calculation of the driving stress at each dislocation requires the superposition of all infinite domain fields, which introduces a very large computational cost. So, the cost of the superposition calculations becomes quite large for many dislocations. Moreover, the standard FEM cannot capture a discontinuity in the displacement field within a single element; the slip across the glide plane is only captured in an average sense. A method is therefore desirable that does not depend on analytical solutions, does not require the computation of image stresses, or uses superposition.

The X-FEM has proved practical in continuum models of dislocations. Modeling of dislocations by a local PU in an elastic solid presents some analogies and substantial differences to crack problems. Similar to crack problems, dislocation solutions can be characterized by discontinuities and singular points. The glide plane of the dislocation can be regarded as a discontinuity surface for the displacement field, but the dislocation acts at the same time as a distortion in the medium, so that the presence of dislocations in a solid generates non-zero displacements and stress fields even though no external forces are acting on the system. The first implementation of a partition of unity method into a dislocation was performed by Ventura, Moran, and Belytschko (2005), where the regular FE solution was enriched by the closed form solution for an edge dislocation. They modeled an edge dislocation in a hollow cylinder and in an infinite medium; however, the exact solution adopted for the enrichment was computationally expensive. An alternative technique based on the X-FEM was proposed by Gracie, Ventura, and Belytschko (2007), in which the tangential enrichment was applied to model edge dislocations directly as interior discontinuities in the standard FEM. In contrast to previous research work, the new proposed technique had several advantages; for example, the superposition of infinite domain solutions was avoided, the interior discontinuities were specified on the dislocation slip surfaces, and the cost of a computation was roughly comparable to that of a standard FE model of the same size. Belytschko and Gracie (2007) presented the X-FEM for modeling dislocations in systems with multiple arbitrary material interfaces. Their method had two advantages over superposition and image field methods; firstly, it scales linearly with the number of dislocations for a given mesh since the Peach–Koehler force can be determined from local quantities and its computational complexity does not depend on the number of dislocations in the domain, and secondly it is applicable to problems of multiple arbitrary material interfaces and can be easily extended to anisotropic materials. However, modeling the core of a dislocation by a simple discontinuous model does not yield to an optimal convergence rate; so to overcome this deficiency, Gracie, Oswald, and Belytschko (2008a) improved the X-FEM solution by adding enrichments in the neighborhood of the dislocation core. In their study, two separate enrichment functions were employed by applying a discontinuous jump enrichment and a singular enrichment based on the closed-form, infinite-domain solutions for the dislocation core. Ventura, Gracie, and Belytschko (2009) proposed an alternative numerical integration of the weak form when the enrichment function is self-equilibrating, that is, when the enriched part of the displacement approximation satisfies equilibrium. They performed the integration based on transforming the domain integrals in the weak form into equivalent contour integrals, and shown that the contour form is

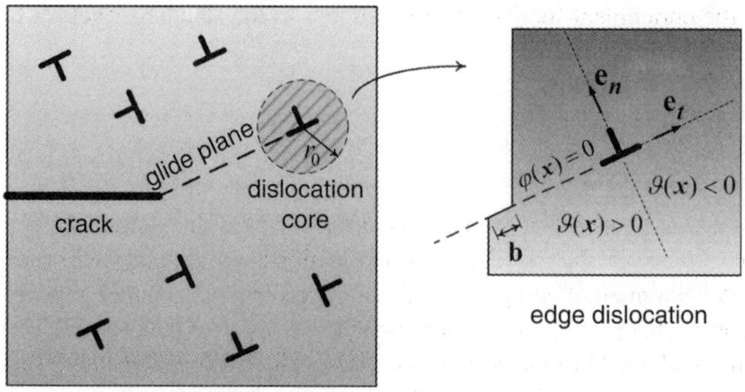

Figure 3.26 A random distribution of dislocations in a continuum body; (a) Illustration of an edge dislocation emanating from a crack tip; (b) Description of an edge dislocation by functions $\varphi(\boldsymbol{x})$ and $\vartheta(\boldsymbol{x})$

computationally more efficient than the domain form, especially when the enrichment function is singular and/or discontinuous. In the works of Gracie, Oswald, and Belytschko (2008a) and Ventura, Gracie, and Belytschko (2009), a singular enrichment based on closed-form solutions for an edge dislocation was used around the core and the remainder of the edge dislocation was modeled by a step function enrichment, however, it must be noted that closed-form solutions are only readily available for isotropic materials, and in many cases anisotropic materials are of more interest in dislocation studies. Recently, a method was presented by Gracie and Belytschko (2009) for modeling dislocations and cracks by atomistic/continuum models that combines the X-FEM with the bridging domain method. The potential of this method is that it circumvents the need for closed form solutions and can rely on more accurate atomistic solutions for the core.

In order to enrich the displacement field for a dislocation problem, the dislocation is modeled using a discontinuous jump enrichment away from the core and a singular enrichment around the core based on the closed-form solutions for dislocations. The former includes a discontinuity on a closed surface in the domain, for example, an edge dislocation enrichment introduces a tangential jump of constant magnitude across the slip plane; and the latter consists of an enrichment field defined by analytical solutions of core mechanics in the vicinity of the core. In such case, the dislocation is modeled by line or surface of discontinuity, where one surface of the discontinuity is displaced relative to the other by the Burgers vector (Figure 3.26a), as in the dislocation model of Volterra (1907), who conceptualized a dislocation as a cut, followed by a displacement and then a reattachment of the material of the two sides of the cut. The enriched displacement field for modeling the dislocation can be defined as

$$\mathbf{u}(\boldsymbol{x}) = \sum_{I\in\mathcal{N}} N_I(\boldsymbol{x})\,\bar{\mathbf{u}}_I + \sum_{\alpha\in\mathcal{M}^{disloc}} \boldsymbol{\psi}^{\alpha}_{\text{step}}(\boldsymbol{x}) + \sum_{\alpha\in\mathcal{M}^{disloc}} \boldsymbol{\psi}^{\alpha}_{\text{core}}(\boldsymbol{x}) \tag{3.32}$$

where $\mathcal{M}^{disloc}$ is the number of dislocations and, $\boldsymbol{\psi}^{\alpha}_{\text{step}}(\boldsymbol{x})$ and $\boldsymbol{\psi}^{\alpha}_{\text{core}}(\boldsymbol{x})$ are correspondingly the discontinuous jump enrichment function away from the core and the singular enrichment function around the core. The tangential jump enrichment function $\boldsymbol{\psi}^{\alpha}_{\text{step}}(\boldsymbol{x})$ across the slip plane is defined as

$$\boldsymbol{\psi}^{\alpha}_{\text{step}}(\boldsymbol{x}) = \mathbf{b}^{\alpha} \sum_{J\in\mathcal{N}^{\alpha}_{\text{step}}} N_J(\boldsymbol{x})\,[H(\varphi^{\alpha}(\boldsymbol{x})) - H(\varphi^{\alpha}(\boldsymbol{x}_J))]\,H(\vartheta^{\alpha}(\boldsymbol{x})) \tag{3.33}$$

where $\mathcal{N}^{\alpha}_{\text{step}}$ is the set of nodes belonging to all elements with at least one edge crossed by the glide plane of dislocation α, and $\mathbf{b}^{\alpha}$ is the Burgers vector that is assumed to be known. The surface of the glide plane of dislocation α is defined by $\varphi^{\alpha}(\boldsymbol{x}) = 0$ and $\vartheta^{\alpha}(\boldsymbol{x}) > 0$, where the core of the dislocation is given by the intersection of the surface $\varphi^{\alpha}(\boldsymbol{x})=0$ and $\vartheta^{\alpha}(\boldsymbol{x})=0$, as shown in Figure 3.26b. The two level-set functions $\varphi^{\alpha}(\boldsymbol{x})$ and $\vartheta^{\alpha}(\boldsymbol{x})$ define the open interface as discussed in Section 3.5.2. For an edge dislocation, the tangential jump enrichment function $\psi^{\alpha}_{\text{step}}(\boldsymbol{x})$ only introduces a jump tangent to the glide plane. It must be noted that since the Burgers vector of a dislocation is known, no enrichment unknowns are introduced. In fact, the step function enrichment is modified by a tangential regularization at the core of the dislocation, since otherwise the field is incompatible.

The local singular core enrichment function $\psi^{\alpha}_{\text{core}}(\boldsymbol{x})$ can be obtained according to the type of dislocation, which is derived from the general solution of an infinite domain problem. For the edge component of a dislocation, the singular core enrichment $\psi^{\alpha}_{\text{core}}(\boldsymbol{x})$ is defined as

$$\psi^{\alpha-\text{edge}}_{\text{core}}(\boldsymbol{x})=\sum_{J\in\mathcal{N}^{\alpha}_{\text{core}}}N_J(\boldsymbol{x})\frac{\mathbf{b}^{\alpha}\cdot\mathbf{e}_t}{2\pi}\left[\left(\tan^{-1}\left(\frac{y}{x}\right)+\frac{xy}{2(1-\nu)(x^2+y^2)}\right)\mathbf{e}_t\right.$$
$$\left.-\left(\frac{1-2\nu}{4(1-\nu)}\ln\left(x^2+y^2\right)+\frac{x^2-y^2}{4(1-\nu)(x^2+y^2)}\right)\mathbf{e}_n\right] \tag{3.34}$$

where $\mathcal{N}^{\alpha}_{\text{core}}$ is the set of nodes in a circular domain of radius r_{α} centered at dislocation core α, which is not in the set of nodes of elements cut by glide plane $\mathcal{N}^{\alpha}_{\text{step}}$. The unit vectors $\mathbf{e}_t$ and $\mathbf{e}_n$ are expressed in terms of the level set functions that describe the configuration of the dislocation as $\mathbf{e}_t=-\nabla\vartheta^{\alpha}(\boldsymbol{x})$ and $\mathbf{e}_n=-\nabla\varphi^{\alpha}(\boldsymbol{x})$.

For the screw component of a dislocation, the singular core enrichment $\psi^{\alpha}_{\text{core}}(\boldsymbol{x})$ is given by

$$\psi^{\alpha-\text{screw}}_{\text{core}}(\boldsymbol{x})=\sum_{J\in\mathcal{N}^{\alpha}_{\text{core}}}N_J(\boldsymbol{x})\left[\frac{\mathbf{b}^{\alpha}\cdot(\mathbf{e}_t\times\mathbf{e}_n)}{2\pi}\tan^{-1}\left(\frac{y}{x}\right)(\mathbf{e}_t\times\mathbf{e}_n)\right] \tag{3.35}$$

in which the accuracy of the singular core enrichment can be improved by increasing the size of the enrichment domain r_{α}. It must be noted that the jump enrichment is introduced for the nodes of elements cut by the glide plane that are sufficiently far from the dislocation core. Since the strain gradients are small far from the core, the adoption of the tangential enrichment away from the core reduces the computational cost of the singular core enrichment.

3.5.10 Shear Band Localization

Localization of deformation refers to the emergence of narrow regions in a structure where all further deformation tends to localize, in spite of the fact that the external actions continue to follow a monotonic loading program. The remaining parts of the structure usually unload and behave in an almost rigid manner. Indeed such localization is almost certain to occur if strain softening or non-associated behavior exists, though it can be triggered even when ideal plasticity is assumed. The phenomenon has a detrimental effect on the integrity of the structure and often acts as a direct precursor to structural failure. It is observed for a wide range of materials, including rocks, concrete, soils, metals, alloys, and polymers, although the scale of localization phenomena in the various materials may differ by some orders of magnitude: the bandwidth is typically less than a millimeter in metals and several meters for crystal faults in rocks. The capturing of such local phenomena in a FE analysis is difficult as discontinuities of displacement occur and these are generally precluded in a standard displacement analysis. Early efforts were devoted for obtaining the failure surfaces and their associated safety factors. Later, FE techniques allowed a more precise analysis of the stress and strain fields, using more sophisticated material models. However, the problem is no longer mathematically well posed after the onset of localization in strain-softening materials, because the local continuum allows for an infinitely small band width in shear or in front of a crack tip.

FE techniques together with suitable constitutive models provide a valuable tool to simulate how the localized zone deforms during the loading process. However, the standard approach presents some important limitations when failure surfaces are to be obtained. Various techniques have been implemented to provide a physically acceptable solution. Some impose restrictions on the constitutive moduli in the post-localization regime, while others artificially restrict the size of FE by comparison to the localization zone. The former is based on enriching the continuum with non-conventional constitutive relations in such way that an internal or characteristic length scale is introduced. Non-local theories are the Cosserat continuum (de Borst, 1991), the higher gradient theories (Triantafyllidis and Ainfantis, 1986), and the integral theory or the gradient theory (Muhlhaus and Ainfantis, 1991; Fleck and Hutchinson, 1993). The latter is based on a suitable mesh refinement using normal, continuous, approximations to all the variables (Pastor, Peraire, and Zienkiewicz, 1991; Belytschko and Tabbara, 1993; Zienkiewicz, Huang, and Pastor, 1995; Khoei and Gharehbaghi, 2003). The main function of this approach is a progressive mesh refinement, which can model well the discontinuity in the limit by using the standard continuous functions; however, the technique is computationally expensive.

In shear band localization, there is a narrow region in which the strain is highly localized, and thus, a discontinuity can be considered in the displacement field. In light of this, the use of the X-FEM for the spatial discretization of governing equations emerges as natural because of its ability to include discontinuities in the approximating fields. The first implementation of the X-FEM in shear band localization was performed by Samaniego and Belytschko (2005), in which the transition from continuum to discontinuum was governed by the loss of hyperbolicity and the post-localization behavior of material was modeled by means of a traction–separation law obtained from a continuum J_2 flow plasticity model. A methodology based on the X-FEM was proposed by Areias and Belytschko (2006b) to model the shear band evolution in quasi-static regime, in which a fine scale function was used to simulate the high displacement gradient in the shear band. In this method, a two scale method for shear bands was employed that was capable of accurately resolving the structure of shear bands for visco-plastic materials. Moreover, a criterion was introduced for the enrichment technique based on the loss of stability of the boundary value problem, as an indicator of shear band initiation point and direction. A computational algorithm was presented by Song, Areias, and Belytschko (2006a) for modeling of arbitrary dynamic crack and shear band propagation based on a rearrangement of the extended FE basis and the nodal degrees of freedom, where the discontinuity was modeled by superposed elements and phantom nodes. In this method, the shear band was treated by adding phantom degrees of freedom that was particularly suited to explicit time integration methods. An application of the X-FEM in conjunction with the digital image correlation was proposed by Réthoré, Hild, and Roux (2007c) to capture experimentally the tangential discontinuities in shear band localization. An enriched FE technique was employed by Khoei and Karimi (2008) based on the Cosserat continuum theory to simulate the shear band localization. In the Cosserat–XFEM technique, the governing equations were regularized by adding the rotational degrees of freedom to the conventional degrees of freedom and including the internal length parameter in the model. A numerical strategy based on the X-FEM was presented by Sandborn and Prevost (2011) for detecting instabilities in elasto-plastic solids, inserting a discontinuity at these instabilities, and prescribing a frictional behavior along the discontinuity.

In shear band problems, the exact structure of displacement normal to the plane of the shear band is often irrelevant to the overall response of a structure. Thus, the tangential component of the enriched displacement field (3.7) can be employed with the step enrichment (3.12) to model the shear band as

$$\mathbf{u}(\boldsymbol{x}) = \sum_{i=1}^{\mathcal{N}} N_i(\boldsymbol{x})\,\bar{\mathbf{u}}_i + \sum_{j=1}^{\mathcal{M}} N_j(\boldsymbol{x})\left(\psi(\varphi(\boldsymbol{x})) - \psi(\varphi(\boldsymbol{x}_j))\right)\mathbf{e}_t\,\bar{a}_j \tag{3.36}$$

where $\mathbf{e}_t$ is the unit tangent vector along the interface. In contrast to the situation where all components are discontinuous, only one set of X-FEM unknowns $\bar{a}_j$ is needed.

In shear band localization, the model of an ideal discontinuity where a property changes its value within an infinitesimal length scale is no longer justified. Instead, the structure and length scale of the variation normal to the interface is part of the solution. In such case, the step enrichment may not be suitable and regularized step functions or sub-scale models are often useful. Thus, the enrichment functions vary from − 1 to + 1 within a certain width across the interface. There are several examples of regularized step functions; the following two functions were accordingly introduced by Areias and Belytschko (2006b) and Benvenuti, Tralli, and Ventura (2008) in terms of the signed-distance function $\varphi(\boldsymbol{x})$ as

$$\psi_{\tanh}(\varphi(\boldsymbol{x})) = \tanh(\ell\,.\,\varphi(\boldsymbol{x})) \tag{3.37a}$$

$$\psi_{\exp}(\varphi(\boldsymbol{x})) = \mathrm{sign}(\varphi(\boldsymbol{x}))(1-\exp(-\ell\,.\,|\varphi(\boldsymbol{x})|)) \tag{3.37b}$$

in which the parameter ℓ is used to scale the gradient, and for $\ell \to \infty$ the sign enrichment function (3.11) can be recovered. It must be noted that these functions are only bounded by ± 1 so that there is only an indirect control over the width of the jump through the parameter ℓ. A function that provides direct control over the width by the characteristic length scale ε can be defined as (Benvenuti, 2008)

$$\psi_{\mathrm{scale}}(\varphi(\boldsymbol{x})) = \begin{cases} -1 & \text{if } -\varepsilon \geq \varphi(\boldsymbol{x}) \\ f(\varphi(\boldsymbol{x})) & \text{if } -\varepsilon < \varphi(\boldsymbol{x}) < \varepsilon \\ +1 & \text{if } \quad \varphi(\boldsymbol{x}) \geq \varepsilon \end{cases} \tag{3.38}$$

in which the function $f(\varphi(\boldsymbol{x}))$ can for example be chosen as follows

$$f_1(\varphi(\boldsymbol{x})) = \frac{3}{8}\left(\frac{\varphi}{\varepsilon}\right)^5 - \frac{5}{4}\left(\frac{\varphi}{\varepsilon}\right)^3 + \frac{15}{8}\left(\frac{\varphi}{\varepsilon}\right) \tag{3.39a}$$

$$f_2(\varphi(\boldsymbol{x})) = \frac{35}{128}\left(\frac{\varphi}{\varepsilon}\right)^9 - \frac{45}{32}\left(\frac{\varphi}{\varepsilon}\right)^7 + \frac{189}{64}\left(\frac{\varphi}{\varepsilon}\right)^5 - \frac{105}{32}\left(\frac{\varphi}{\varepsilon}\right)^3 + \frac{315}{128}\left(\frac{\varphi}{\varepsilon}\right) \tag{3.39b}$$

in which the function f_1 is C_2–continuous at $|\varphi(\boldsymbol{x})| = \varepsilon$, whereas f_2 is C_4–continuous at $|\varphi(\boldsymbol{x})| = \varepsilon$. The distributions of the enrichment function (3.38) are shown in Figure 3.27 together with the enriched shape

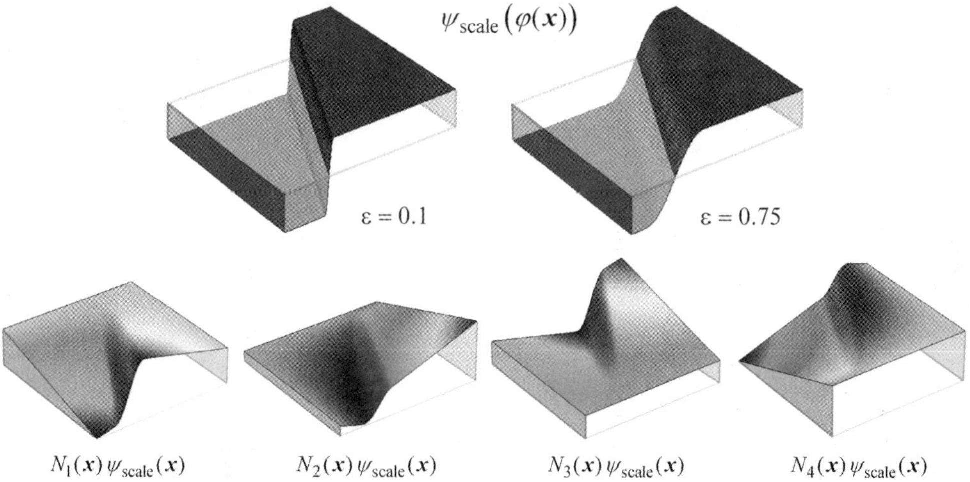

Figure 3.27 The regularized step functions together with the enriched shape functions for a 4-noded quadratic element at $\varepsilon = 0.5$

Table 3.1 A summary of various X-FEM enrichment functions for different classes of solid mechanics problems

Kind of problem	Field variable (displacement)	Gradient of field variable (strain)	X-FEM enrichment functions
Bimaterial interfaces, voids, inclusions, grain boundaries	Continuous	Discontinuous	$\psi_{\text{ramp}}(\varphi(\boldsymbol{x}))=\lvert\varphi(\boldsymbol{x})\rvert$ $\psi_{\text{ridge}}(\varphi(\boldsymbol{x}))=\sum_{I\in\mathcal{N}}N_I(\boldsymbol{x})\,\lvert\varphi_I\rvert-\left\lvert\sum_{I\in\mathcal{N}}N_I(\boldsymbol{x})\,\varphi_I\right\rvert$
Strong discontinuity, crack interfaces	Discontinuous	—	$\psi_{\text{sign}}(\varphi(\boldsymbol{x}))=\text{sign}(\varphi(\boldsymbol{x}))$ $\psi_{\text{step}}(\varphi(\boldsymbol{x}))=H(\varphi(\boldsymbol{x}))$
Brittle crack tip (isotropic material)	Discontinuous	High gradient	$\psi_{\text{tip}}^{\text{brittle}}(r,\theta)=\left\{\sqrt{r}\sin\frac{\theta}{2},\ \sqrt{r}\cos\frac{\theta}{2},\ \sqrt{r}\sin\frac{\theta}{2}\sin\theta,\ \sqrt{r}\cos\frac{\theta}{2}\sin\theta\right\}$
Brittle crack tip (orthotropic material)	Discontinuous	High gradient	$\psi_{\text{tip}}^{\text{orthotropic}}(r,\theta)=\left\{\sqrt{r}\sin\frac{\theta_1}{2}\sqrt{g_1(\theta)},\ \sqrt{r}\cos\frac{\theta_1}{2}\sqrt{g_1(\theta)},\ \sqrt{r}\sin\frac{\theta_2}{2}\sqrt{g_2(\theta)},\ \sqrt{r}\cos\frac{\theta_2}{2}\sqrt{g_2(\theta)}\right\}$
Cohesive crack tip	Discontinuous	High gradient	$\psi_{\text{tip}}^{\text{cohesive}}(r,\theta)=r^k\sin\frac{\theta}{2}\quad(k=1,\,1.5,\,2)$
Plastic crack tip	Discontinuous	High gradient	$\psi_{\text{tip}}^{\text{plastic}}(r,\theta)=r^{\frac{1}{n+1}}\left\{\sin\frac{\theta}{2},\ \cos\frac{\theta}{2},\ \sin\frac{\theta}{2}\sin\theta,\ \cos\frac{\theta}{2}\sin\theta,\ \sin\frac{\theta}{2}\sin 3\theta,\ \cos\frac{\theta}{2}\sin 3\theta\right\}$
Multiple cracks (discontinuous *junction* function)	Discontinuous	High gradient	$\psi_{\text{junction}}^{\mathbb{H}}(\varphi(\boldsymbol{x}))=J^{\mathsf{H}}(\boldsymbol{x})$
Crack tip perpendicular to bimaterial interface	Discontinuous	High gradient	$\psi_{\text{tip}}^{\text{N-bimaterial}}(r,\theta)=\left\{\psi_{\text{tip}}^1,\psi_{\text{tip}}^2,\psi_{\text{tip}}^3,\psi_{\text{tip}}^4\right\}$ $=\left\{r^\lambda\cos(\lambda+1)\theta,\ r^\lambda\sin(\lambda+1)\theta,\ r^\lambda\cos(\lambda-1)\theta,\ r^\lambda\sin(\lambda-1)\theta\right\}$
Crack tip terminating at a bimaterial interface	Discontinuous	High gradient	$\psi_{\text{tip}}^{\mathbb{O}\text{-bimaterial}}(r,\theta)=\left\{\psi_{\text{tip}}^1,\psi_{\text{tip}}^2,\ldots,\psi_{\text{tip}}^8\right\}$ $=\left\{r^{\lambda_1}\cos(\lambda_1+1)\theta,\ r^{\lambda_1}\sin(\lambda_1+1)\theta,\ r^{\lambda_1}\cos(\lambda_1-1)\theta,\ r^{\lambda_1}\sin(\lambda_1-1)\theta,\ r^{\lambda_2}\cos(\lambda_2+1)\theta,\ r^{\lambda_2}\sin(\lambda_2+1)\theta,\ r^{\lambda_2}\cos(\lambda_2-1)\theta,\ r^{\lambda_2}\sin(\lambda_2-1)\theta\right\}$

Table 3.1 (*continued*)

Kind of problem	Field variable (displacement)	Gradient of field variable (strain)	X-FEM enrichment functions
Bimaterial interfacial crack	Discontinuous	High gradient	$\psi_{\text{tip}}^{\text{I-bimaterial}}(r,\theta)=\left\{\psi_{\text{tip}}^{1},\psi_{\text{tip}}^{2},\ldots,\psi_{\text{tip}}^{12}\right\}$ $=\Big\{\sqrt{r}\cos(\varepsilon\log r)e^{-\varepsilon\theta}\sin\frac{\theta}{2},\ \sqrt{r}\cos(\varepsilon\log r)e^{-\varepsilon\theta}\cos\frac{\theta}{2},$ $\sqrt{r}\cos(\varepsilon\log r)e^{+\varepsilon\theta}\sin\frac{\theta}{2},\ \sqrt{r}\cos(\varepsilon\log r)e^{+\varepsilon\theta}\cos\frac{\theta}{2},$ $\sqrt{r}\cos(\varepsilon\log r)e^{+\varepsilon\theta}\sin\frac{\theta}{2}\sin\theta,\ \sqrt{r}\cos(\varepsilon\log r)e^{+\varepsilon\theta}\cos\frac{\theta}{2}\sin\theta,$ $\sqrt{r}\sin(\varepsilon\log r)e^{-\varepsilon\theta}\sin\frac{\theta}{2},\ \sqrt{r}\sin(\varepsilon\log r)e^{-\varepsilon\theta}\cos\frac{\theta}{2},$ $\sqrt{r}\sin(\varepsilon\log r)e^{+\varepsilon\theta}\sin\frac{\theta}{2},\ \sqrt{r}\sin(\varepsilon\log r)e^{+\varepsilon\theta}\cos\frac{\theta}{2},$ $\sqrt{r}\sin(\varepsilon\log r)e^{+\varepsilon\theta}\sin\frac{\theta}{2}\sin\theta,\ \sqrt{r}\sin(\varepsilon\log r)e^{+\varepsilon\theta}\cos\frac{\theta}{2}\sin\theta\Big\}$
Grain junctions in polycrystalline structures	Discontinuous	High gradient	$\psi_{\text{tip}}^{\text{notch}}(r,\theta)=r^{\lambda}\,\Psi(\theta)$
Multiple interfaces (*junction ramp* function)	Continuous	Discontinuous	$\psi_{\text{junction}}^{\mathbb{R}}(\varphi(\boldsymbol{x}))=J^{\mathbb{R}}(\boldsymbol{x})=\lvert\varphi_j(\boldsymbol{x})\rvert\lvert\varphi_k(\boldsymbol{x})\rvert$
Dislocation (tangential jump function)	Discontinuous	—	$\boldsymbol{\psi}_{\text{step}}^{\alpha}(\boldsymbol{x})=\mathbf{b}^{\alpha}\sum_{J\in\mathcal{N}_{\text{step}}^{\alpha}}N_J(\boldsymbol{x})\left[H(\varphi^{\alpha}(\boldsymbol{x}))-H(\varphi^{\alpha}(\boldsymbol{x}_J))\right]H(\vartheta^{\alpha}(\boldsymbol{x}))$
Dislocation (edge function)	Discontinuous	High gradient	$\boldsymbol{\psi}_{\text{core}}^{\alpha-\text{edge}}(\boldsymbol{x})=\sum_{J\in\mathcal{N}_{\text{core}}^{\alpha}}N_J(\boldsymbol{x})\frac{\mathbf{b}^{\alpha}\cdot\mathbf{e}_t}{2\pi}\left[\left(\tan^{-1}\left(\frac{y}{x}\right)+\frac{xy}{2(1-\nu)(x^2+y^2)}\right)\mathbf{e}_t-\left(\frac{1-2\nu}{4(1-\nu)}\ln(x^2+y^2)+\frac{x^2-y^2}{4(1-\nu)(x^2+y^2)}\right)\mathbf{e}_n\right]$
Dislocation (screw function)	Discontinuous	High gradient	$\boldsymbol{\psi}_{\text{core}}^{\alpha-\text{screw}}(\boldsymbol{x})=\sum_{J\in\mathcal{N}_{\text{core}}^{\alpha}}N_J(\boldsymbol{x})\left[\frac{\mathbf{b}^{\alpha}\cdot(\mathbf{e}_t\times\mathbf{e}_n)}{2\pi}\tan^{-1}\left(\frac{y}{x}\right)(\mathbf{e}_t\times\mathbf{e}_n)\right]$
Shear band localization	Discontinuous	Discontinuous	$\psi_{\tanh}(\varphi(\boldsymbol{x}))=\tanh(\ell\cdot\varphi(\boldsymbol{x}))$ $\psi_{\exp}(\varphi(\boldsymbol{x}))=\text{sign}(\varphi(\boldsymbol{x}))(1-\exp(-\ell\cdot\lvert\varphi(\boldsymbol{x})\rvert))$

functions for a 4-noded quadratic element at $\varepsilon = 0.5$. In some applications, the characteristic width over which the "jump" takes place may be known or estimated *a priori* and only one regularized step function is used for the enrichment, however, if determining the width is part of the solution several regularized step functions may be used with the aim to span the whole range of gradients between smooth and strongly discontinuous fields.

In Table 3.1, a summary of various enrichment functions employed in the X-FEM is presented for different classes of solid mechanics problems. In this table, the continuity and discontinuity conditions for both the field variable (displacement) and its gradient (strain) are indicated.

4

Blending Elements

4.1 Introduction

It has been well established that the conventional finite element method (FEM) lacks the accuracy and convergence rate when the exact solution of differential equation or its gradients have singularities or discontinuities. These issues are common in fracture mechanics, contact problems, bimaterial structures, two-phase flow, and so on. In solid mechanics, stresses and strains jump along material interfaces, displacements change discontinuously along cracks, and stresses become singular at a crack tip. In fluid mechanics, high gradients are present near shocks and boundary layers. In fact, the conventional FEM is ideally suited to the approximation of smooth solutions that rely on the approximation properties of polynomials. However, special care must be taken for approximating non-smooth solutions with the FEM. In such cases, a considerable mesh refinement, or the use of higher order elements is required that result in high computational cost and numerical difficulties. For example, it can be shown that for an elastic crack problem the convergence rate of the energy norm with first order finite elements (FEs) reduces from $O(h)$ to $O(\sqrt{h})$ (Strang and Fix, 1973). In the FEM, the construction of an appropriate mesh is crucial to the success of the approximation; the element edges have to be aligned with a discontinuity, and a mesh refinement is required where the solution is expected to have singularities or large gradients; particularly, in the case of moving discontinuities, such as crack propagation or the interface movement between two fluids, the continuation of a proper mesh is difficult or even impossible.

In contrast, the extended finite element method (X-FEM) offers the inclusion of *a priori* known solution properties into the approximation space. The simulation is generally carried out on a fixed, simple, structured mesh so that the mesh construction and continuation are reduced to a minimum. In the extended FEM, the approximation field is enhanced by adding the enrichment functions in the framework of the partition of unity finite element method (PU FEM), and is expected to improve the approximation quality and the convergence rate (Belytschko and Black, 1999). However, it was observed that although the accuracy of solution increases with higher order elements, the convergence rate becomes worse in the standard X-FEM; Stazi *et al.* (2003) employed the higher order elements to model the curved crack problem, and observed that the optimal convergence rate cannot be achieved with the X-FEM. Researchers agree that the *blending elements* in the transitional zone between the fully enriched and the standard elements are responsible for the lack of optimal convergence rate. In these elements, the PU is violated and the parasitic terms introduce the error into the solution space, which result in poorly-conditioned stiffness matrices, high condition numbers, and an increase in effort required to solve the system of equations numerically.

Basically, there are two important issues with the blending elements in the X-FEM; firstly, the enrichment function can no longer be reproduced exactly due to the lack of a PU, and secondly blending elements produce unwanted terms into the approximation, which cannot be compensated for by the

Extended Finite Element Method: Theory and Applications, First Edition. Amir R. Khoei.

standard FE part of the approximation. For example, a linear function can no longer be represented in a linear blending element if the enrichment introduces nonlinear terms. The first issue does not pose a significant problem in the X-FEM; however, the second one may significantly reduce the convergence rate for general enrichment functions. Thus, a special treatment is required in blending elements to remove the unwanted terms. There are various techniques proposed in literature to overcome those issues related to the parasitic terms in the X-FEM blending elements. Chessa, Wang, and Belytschko (2003) employed the enhanced strain method, or *p*–refinement, in the blending elements to improve the performance of local PU enrichments. The method was based on the Hu–Washizu variational principle, and was able to eliminate the parasitic terms by choosing an appropriate enhanced strain field. The enhanced strain elements were used only in the blending sub-domains for any arbitrary enrichment functions, but for each enrichment function a set of enhanced strains must be constructed. It was shown that for polynomial enrichments, such as the signed distance function that is used to model discontinuous derivatives within elements, the enhanced strain blending element recovers linear convergence in the energy. Laborde *et al.* (2005) modified the standard X-FEM to circumvent problems in blending elements for the case of crack problems by enriching a whole fixed area around the crack tip. Since in the standard X-FEM only the nodes of the crack-tip element are enriched, the support of the additional basis functions vanishes when h goes to zero. However, it was shown that the "fixed enrichment area" around the crack tip can be used efficiently to achieve the expected optimal rate of convergence. It must be noted that for the particular case of the Heaviside enrichment function, there is no problem with the blending elements. For solutions that involve a kink along an interface, the Heaviside function is often chosen, although additional constraints are needed to ensure the continuity. This may also be considered as a special technique to avoid problems in blending elements, however, other enrichment functions are more appropriate since they do not need additional constraints, but introduce problems in blending elements. All these approaches share the property that they are not easily extended to arbitrary enrichment functions, element types, and mathematical models.

A corrected X-FEM was introduced by Fries (2008), which has two important differences to the standard X-FEM; firstly, in addition to those nodes that are enriched in the standard X-FEM, all nodal points in the blending elements are enriched. As a result, a complete PU is formed in the reproducing and blending elements. Secondly, the enrichment functions of the standard X-FEM are modified except in the reproducing elements, in which they are zero in the standard elements and they vary continuously between the standard and reproducing elements in the blending elements. In the corrected X-FEM, the modified enrichment function can be reproduced exactly everywhere in the domain, and the original enrichment function in the reproducing elements. Most importantly, there are no unwanted terms in the blending elements; so the corrected X-FEM is able to achieve the optimal convergence rate. An alternative technique was developed by Gracie, Wang, and Belytschko (2008b) based on a discontinuous Galerkin (DG) X-FEM approach to circumvent the spurious behavior of the blending elements. In the DG-XFEM, the domain is decomposed into non-overlapping patches; in which the enrichments are applied over these patches, and continuity between the patches is enforced using an internal penalty method. The technique has a desirable characteristic in which the enrichment is local and all shape functions form a PU. This is in contrast to the standard X-FEM where the PU property of the shape functions that pre-multiply the enrichment functions do not satisfy the PU property everywhere in the domain. In this chapter, the problems of convergence rate and condition number are first discussed using simple one-dimensional examples; and then, various remedies that are available in the literature for these issues are introduced using some numerical examples.

4.2 Convergence Analysis in the X-FEM

In the X-FEM, the problem domain $\Omega \subset \mathbb{R}^n$ is generally decomposed into three sub-domains (Figure 4.1); the sub-domain Ω_{enr} with the set of elements in which their nodal points are fully enriched, the sub-domain Ω_{std} that includes the set of elements where their nodal points are not enriched, and the sub-domain Ω_{bld}

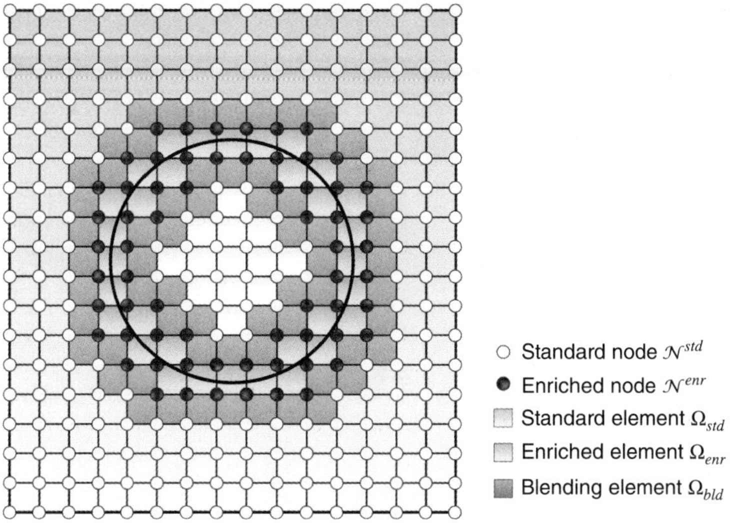

Figure 4.1 The domain of problem in X-FEM illustrating the enriched sub-domain Ω_{enr}, the standard sub-domain Ω_{std}, and the partially enriched sub-domain Ω_{bld}

with the set of elements where their nodal points are partially enriched called as the blending elements. Since these elements are "*partially enriched*" the enriched nodes in these elements do not form a PU.

For an enriched element with the shape functions $N_I(\boldsymbol{x})$ and the enrichment function $\psi(\boldsymbol{x})$, the displacement field $u^h(\boldsymbol{x})$ can be defined as

$$\begin{aligned} u^h(\boldsymbol{x}) &= u_{std}(\boldsymbol{x}) + u_{enr}(\boldsymbol{x}) \\ &= \sum_{I \in \mathcal{N}^{std}} N_I(\boldsymbol{x})\, u_I + \sum_{J \in \mathcal{N}^{enr}} N_J(\boldsymbol{x})\, \psi(\boldsymbol{x})\, a_J \end{aligned} \tag{4.1}$$

where $\mathcal{N}^{std}$ and $\mathcal{N}^{enr}$ are the set of standard and enriched nodal points respectively, where $\mathcal{N} = \mathcal{N}^{std} \cup \mathcal{N}^{enr}$ is the set of all nodal points. In this relation, u_I and a_J are the relevant nodal degrees of freedom (DOF) and $\boldsymbol{x}$ is the position vector.

In the X-FEM, there are basically two important issues with the blending elements; the first is related to the PU property of the blending element. In fact, the PU is violated in blending elements since all nodal points are not enriched. It means that the enrichment functions cannot be reproduced properly (Fries, 2008). On the other hand, the displacement field $u^h(\boldsymbol{x})$ for an enrichment function $\psi(\boldsymbol{x})$ considering the standard DOF as $u_I = 0$ and the enriched DOF as $a_J = 1$ results in

$$u^h(\boldsymbol{x}) = \begin{cases} \sum_{J \in \mathcal{N}^{enr}} N_J(\boldsymbol{x})\, \psi(\boldsymbol{x}) = \psi(\boldsymbol{x}) & \forall \boldsymbol{x} \in \Omega_{enr} \\ \sum_{J \in \mathcal{N}^{enr}} N_J(\boldsymbol{x})\, \psi(\boldsymbol{x}) \neq \psi(\boldsymbol{x}) & \forall \boldsymbol{x} \in \Omega_{bld} \\ \sum_{J \in \mathcal{N}^{enr}} N_J(\boldsymbol{x})\, \psi(\boldsymbol{x}) = 0 & \forall \boldsymbol{x} \in \Omega_{std} \end{cases} \tag{4.2}$$

It can be observed from these relations that the displacement field (4.1) can reproduce the enrichment in the sub-domain Ω_{enr} and it vanishes in the sub-domain Ω_{std} for $a_J = 1$. However, it consists of the product of a subset of the PU shape functions and the enrichment in the blending sub-domain Ω_{bld} so the enrichment cannot be reproduced.

The next apprehension regarding the blending elements is that they introduce unwanted parasitic terms into the approximation space that cannot be compensated by the standard FE part. Parasitic terms play an important role in the approximation properties and convergence of the enriched solution. The unwanted parasitic terms spoil the accuracy of the solution and convergence rate. By the convergence, it means that as the specific element size h goes to zero, the solution of FEM, in a defined norm like the energy norm, approaches to the exact solution of the governing differential equation. Chessa, Wang, and Belytschko (2003) demonstrated this issue for a one-dimensional problem by applying the ramp function as an enrichment function. The same analysis was performed by Tarancon *et al.* (2009) for an arbitrary enrichment function. Based on the earlier discussion, it can be concluded that while a local PU enrichment improves the approximation in the enriched elements, the overall improvement is often quite modest because of the impairment in the blending elements.

In order to illustrate the lower convergence rate of blending elements, consider a one-dimensional bar, as discussed earlier in Chapter 2. The bar is modeled by three linear elements, in which the nodal points 1 and 2 are enriched, as shown in Figure 4.2. The displacement field for the enriched element 1 can be written according to (4.1) as

$$u^h(x) = u_1(1-\xi_1) + u_2\,\xi_1 + a_1(1-\xi_1)\psi(x) + a_2\,\xi_1\,\psi(x) \tag{4.3}$$

and the displacement field for the blending element 2 can be written as

$$u^h(x) = u_2(1-\xi_2) + u_3\,\xi_2 + a_2(1-\xi_2)\psi(x) \tag{4.4}$$

where

$$\xi_1 = \frac{x-x_1}{h} \quad \text{and} \quad \xi_2 = \frac{x-x_2}{h} \qquad 0 \le \xi_1, \xi_2 \le 1 \tag{4.5}$$

where h is the length of element. It can be observed from (4.3) to (4.4) that by approaching to node 2 from element 1 ($\xi_1 = 1$), or from element 2 ($\xi_2 = 0$), the same solution can be achieved, which is the result of C_0 continuity in node 2. In fact, the term $a_2(1-\xi_2)\psi(x)$ in the blending element 2 is crucial for the existence of C_0 continuity that is a necessary condition for conforming finite element. However, this term cannot reproduce the enrichment function $\psi(x)$ for any value of a_2 in element 2, and for the non-zero value of a_2 due to fully enriched element 1, it is impossible to produce a linear field. This higher order term which introduces parasitic error into the solution space is called the "pathological term". In order to obtain the error bound for this issue, the Taylor series of $\psi(x)$ is written about a point x_0 as

$$\psi(x) = \psi(x_0) + \left.\frac{d\psi}{dx}\right|_{x=x_0}(x-x_0) + O(h^2) \tag{4.6}$$

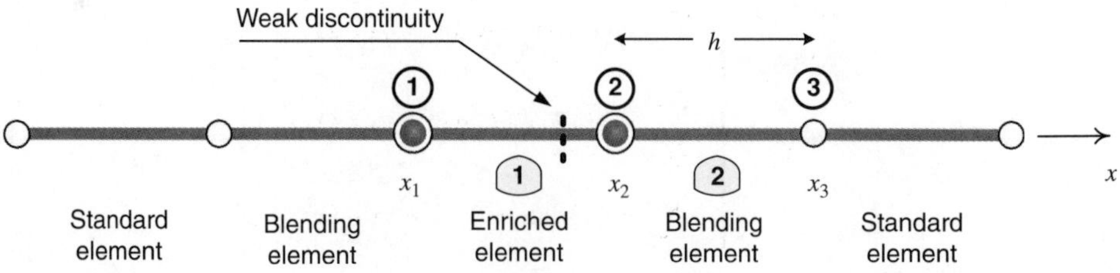

Figure 4.2 The convergence analysis of a one-dimensional bar where element 1 is fully enriched and element 2 is partially enriched

Substitution relation (4.6) in (4.4) and removing the higher order terms $O(h^2)$, it yields to

$$u^h(x) \approx u_2(1-\xi_2) + u_3\,\xi_2 + a_2(1-\xi_2)\left(\psi(x_0) + \left.\frac{d\psi}{dx}\right|_{x=x_0}(x-x_0)\right) \tag{4.7}$$

Considering the FE interpolation of the solution as $u^h(x)$, the error of the interpolation e_u can be obtained as

$$e_u(x) = u(x) - u^h(x) \tag{4.8}$$

Assuming that $\bar{x}$ is a point with the maximum error in the blending element 2, which is located in the half of the element close to node 2 such that $\bar{x}-x_2 \le h/2$. Hence, the Taylor series of the error e_u about $\bar{x}$ can be written as

$$e_u(x) = e_u(\bar{x}) + \left.\frac{de_u}{dx}\right|_{x=\bar{x}}(x-\bar{x}) + \frac{1}{2}\left.\frac{d^2e_u}{dx^2}\right|_{x=\bar{x}}(x-\bar{x})^2 \tag{4.9}$$

Since the interpolation has an exact value at the nodal points, the error of interpolation at point x_2 is equal to zero, that is, $e_u(x_2)=0$. Moreover, to obtain the maximum error at the point $\bar{x}$, the derivative of error must be zero, that is, $\left.\frac{de_u}{dx}\right|_{x=\bar{x}}=0$, so setting $x=x_2$ in Eq. (4.9) results in

$$e_u(\bar{x}) = -\frac{1}{2}\left.\frac{d^2e_u}{dx^2}\right|_{x=\bar{x}}(x_2-\bar{x})^2 \tag{4.10}$$

since $|\bar{x}-x_2| \le h/2$, it leads to

$$e_u(\bar{x}) = \left|\frac{1}{2}\left.\frac{d^2e_u}{dx^2}\right|_{x=\bar{x}}(x_2-\bar{x})^2\right| \le \frac{1}{8}h^2 \max\left|\frac{d^2e_u}{dx^2}\right|_{x=\bar{x}} \tag{4.11}$$

Taking derivation from (4.7) to (4.8) and substitution in (4.10), it results in

$$\frac{d^2e_u}{dx^2} = \frac{d^2u}{dx^2} + \frac{2a_2}{h}\left.\frac{d\psi}{dx}\right|_{x=\bar{x}} \tag{4.12}$$

Hence, the error (4.11) results in

$$\max|e_u(\bar{x})| \le \frac{1}{8}h^2 \max\left|\frac{d^2u}{dx^2} + \frac{2a_2}{h}\frac{d\psi}{dx}\right|_{x=\bar{x}} \tag{4.13}$$

Obviously, it can be seen from this expression that the term $\frac{2a_2}{h}\frac{d\psi}{dx}$ that does not appear in the standard elements, increases the error of blending element, and reduces the global convergence rate. It can be concluded from the error of (4.13) that a constant enrichment function, such as the Heaviside enrichment function, does not introduce the error into the blending elements. Furthermore, it must be noted that the pathological term cannot be simply removed from the blending elements, since it is crucial for the C_0 continuity of these elements.

It is noteworthy to highlight that for the case of polynomial enrichment function, the pathological terms can be corrected by the standard part of the approximation field if the order of standard shape function is equal or more than the order of the enriched shape function. For example, by adding a quadratic term into the blending element, the displacement field (4.4) can be written as

$$u^h(x) = u_2(1-\xi_2) + u_3\,\xi_2 + a_2(1-\xi_2)\,\psi(x) + b_2\,\xi_2(1-\xi_2) \tag{4.14}$$

where b_2 is additional DOF associated with the quadratic term. Hence, the error of (4.13) can be obtained as

$$\max|e_u(\bar{x})| \le \frac{1}{8}h^2 \max\left|\frac{d^2u}{dx^2} + \frac{2a_2}{h}\frac{d\psi}{dx} + \frac{2b_2}{h^2}\right|_{x=\bar{x}} \tag{4.15}$$

If $b_2 = -a_2 h\dfrac{d\psi}{dx}$, this expression leads to an optimal error, that is,

$$\max|e_u(\bar{x})| \le \frac{1}{8}h^2 \max\left|\frac{d^2u}{dx^2}\right| \tag{4.16}$$

4.3 Ill-Conditioning in the X-FEM Method

The extended FEM enhances the approximation properties of the finite element space by using additional enrichment functions, however, the resulting stiffness matrices can become ill-conditioned and result in a large number of iterations to obtain an acceptable solution. Of course, the high condition number of stiffness matrix can be observed in the nature of all PU based methods (Babuška and Melenk, 1997). However, in the X-FEM, even if the chosen local approximation functions, that is, enrichment functions, are linearly independent, their multiplication with PU functions may result in dependent functions that do not form a basis for a partition of unity method (PUM) and may lead to badly conditioned stiffness matrices. Béchet *et al.* (2005) proposed a pre-conditioner specially tailored to the X-FEM, which stabilizes the enrichments by applying Cholesky decompositions to certain sub-matrices of the stiffness matrix. Since the problem of almost linearly dependent enrichment functions was eliminated for each node, these sub-matrices were formed by the DOF associated with each enriched nodal point, which can be considered as a local stabilization. However, there are still situations where the enrichment functions associated with several nodal points become almost linearly dependent. A pre-conditioning method, which was originally proposed by Farhat and Roux (1991) in the finite element tearing and interconnecting method, was applied by Menk and Bordas (2011) for the X-FEM that implements the idea of domain decomposition to the sub-matrix associated with the enriched DOF. The Cholesky decompositions were employed to the stiffness matrix associated with the enriched DOF, and the continuity of the solution was guaranteed by using additional constraints.

In order to illustrate the ill-conditioned stiffness matrix, consider a system of equations given by $\mathbf{A}\boldsymbol{x} = \mathbf{b}$, where $\mathbf{A}$ is a coefficient matrix, $\boldsymbol{x}$ is the unknown vector, and $\mathbf{b}$ is the known vector. The numerical computations are generally performed with floating point numbers that have finite precision. Hence, it is important to know how the solution of the system is sensitive to the changes of the known vector $\mathbf{b}$, or coefficient matrix $\mathbf{A}$. For example, consider the following system of equations

$$\begin{bmatrix} 1 & 2 \\ 2 & 3.999 \end{bmatrix}\begin{bmatrix} x_1 \\ x_2 \end{bmatrix} = \begin{bmatrix} 4.0 \\ 7.999 \end{bmatrix}$$

The solution to this system of equations leads to

$$\begin{bmatrix} x_1 \\ x_2 \end{bmatrix} = \begin{bmatrix} 2 \\ 1 \end{bmatrix}$$

However, applying a small perturbation in the vector $\mathbf{b}$ as

$$\begin{bmatrix} 1 & 2 \\ 2 & 3.999 \end{bmatrix}\begin{bmatrix} x_1 \\ x_2 \end{bmatrix} = \begin{bmatrix} 4.001 \\ 7.999 \end{bmatrix}$$

The solution of system of equations surprisingly changes to

$$\begin{bmatrix} x_1 \\ x_2 \end{bmatrix} = \begin{bmatrix} -1.999 \\ 3.0 \end{bmatrix}$$

The reason it has happened lies in the fact that the columns of the coefficient matrix are "partially" linear dependent and the matrix is "near" singular. In such case, the small changes in the vector **b** causes the large variations in the unknown vector $\boldsymbol{x}$.

Basically, the determinant of a matrix can be considered as an obvious candidate to indicate that the matrix is "near" singular. However, it is not generally a good numerical tool since it is computationally expensive and in many cases has ambiguous values. For example, consider the matrix **A** equal to $2\mathbf{I}_{n\times n}$ where $\mathbf{I}_{n\times n}$ is an identity matrix of size n, the determinant of this matrix is 2^n which goes to infinity when n becomes a large value. Now consider the matrix **A** as $10^{-4}\mathbf{I}_{n\times n}$, the determinant of this matrix is almost zero for a large value of n. While it is obvious from these two cases that both matrices are well behaved and are not singular. The best indicator for determination of singularity of the matrix **A** is the condition number $\kappa(\mathbf{A})$ which is defined based on the ratio of the largest to smallest eigenvalues as

$$\kappa(\mathbf{A}) = \frac{\lambda_{max}(\mathbf{A})}{\lambda_{min}(\mathbf{A})} \tag{4.17}$$

Obviously the system of equations becomes ill-conditioned if the smallest eigenvalue approaches to zero. Assume that the known vector **b** has an error of $\Delta\mathbf{b}$ that results in an error of $\Delta\boldsymbol{x}$ in the solution vector $\boldsymbol{x}$, the system of equations takes the form of

$$\mathbf{A}(\boldsymbol{x} + \Delta\boldsymbol{x}) = \mathbf{b} + \Delta\mathbf{b} \tag{4.18}$$

Subtracting the existent system of equations $\mathbf{A}\boldsymbol{x} = \mathbf{b}$ from this equation, it yields to

$$\mathbf{A}(\Delta\boldsymbol{x}) = \Delta\mathbf{b} \tag{4.19}$$

For a positive definite matrix **A**, an upper bound can be defined for $\Delta\boldsymbol{x}$ based on the smallest eigenvalue λ_{min} as

$$\Delta\boldsymbol{x} = \mathbf{A}^{-1}\Delta\mathbf{b} \quad \text{with} \quad \|\Delta\boldsymbol{x}\| \le \frac{\|\Delta\mathbf{b}\|}{\lambda_{min}(\mathbf{A})} \tag{4.20}$$

This definition comes from the fact that the largest eigenvalue of $\mathbf{A}^{-1}$ is $1/\lambda_{min}(\mathbf{A})$, thus the largest value of error in $\Delta\boldsymbol{x}$ occurs when $\Delta\mathbf{b}$ is in the direction of the eigenvector corresponding to λ_{min}. Based on the definition (4.20), it is obvious that an upper bound for λ_{min} results in an upper bound for the absolute error $\|\Delta\boldsymbol{x}\|$. An upper bound for the smallest eigenvalue λ_{min} can be obtained using the Rayleigh quotient $q_{\mathbf{a}}$, which is defined for any non-zero vector $\mathbf{a} \in \mathbb{R}^n$ as (Menk and Bordas, 2011)

$$q_{\mathbf{a}} = \frac{\mathbf{a}^T\mathbf{A}\,\mathbf{a}}{\mathbf{a}^T\mathbf{a}} \ge \lambda_{min} \ge 0 \tag{4.21}$$

However, the value of λ_{min} can be misleading; for example consider the matrix **A** as $10^{-4}\mathbf{I}_{n\times n}$ with $\lambda_{min} = 10^{-4}$, it can be seen that a small value of λ_{min} does not necessarily mean the singularity of a matrix. Thus, an absolute value of the error is not enough and a relative error can be used to evaluate the matrix singularity.

For this purpose, a lower bound can first be obtained for the solution $\boldsymbol{x}$ based on the largest eigenvalue λ_{max} as

$$\boldsymbol{x} = \mathbf{A}^{-1}\mathbf{b} \quad \text{with} \quad \|\boldsymbol{x}\| \geq \frac{\|\mathbf{b}\|}{\lambda_{max}(\mathbf{A})} \tag{4.22}$$

The relative error can then be obtained by dividing the maximum absolute error (4.20) to the smallest solution of the system (4.22) as

$$\frac{\|\Delta\boldsymbol{x}\|}{\|\boldsymbol{x}\|} \leq \frac{\lambda_{max}}{\lambda_{min}} \frac{\|\Delta\mathbf{b}\|}{\|\mathbf{b}\|} \tag{4.23}$$

This expression illustrates how the condition number $\kappa(\mathbf{A}) = \lambda_{max}/\lambda_{min}$ can be used to relate the relative error of vector $\mathbf{b}$ to the relative error of solution $\boldsymbol{x}$. It is noteworthy to highlight that the condition number does not have the drawbacks of the determinant and smallest eigenvalue; for example, the condition numbers of matrices $\mathbf{I}_{n\times n}$, $2\mathbf{I}_{n\times n}$ and $10^{-4}\mathbf{I}_{n\times n}$ are all equal to one. In fact, the ideal value of condition number is one and the ill-conditioned system has a high condition number. As a rule of thumb, it can be stated that for every power of 10 in the condition number, one significant digit of the solution of $\mathbf{A}\boldsymbol{x} = \mathbf{b}$ is lost. For example, for a double precision computation with 16 significant digits, a system with the condition number of 10^{11} has only five available significant digits.

Furthermore, the high condition number has a deteriorating effect on the convergence rate of iterative solvers which are used for the solution of $\mathbf{A}\boldsymbol{x} = \mathbf{b}$. For example, it can be shown that for the Conjugate Gradient (CG) method the error of iteration i can be related to the condition number of the system by

$$\|\boldsymbol{x}^{(i)} - \boldsymbol{x}\| \leq 2\frac{\sqrt{\kappa(\mathbf{A})} - 1}{\sqrt{\kappa(\mathbf{A})} + 1} \|\boldsymbol{x}^{(0)} - \boldsymbol{x}\| \tag{4.24}$$

where $\boldsymbol{x}$ is the exact solution and $\boldsymbol{x}^{(i)}$ is the solution of ith iteration. It can be seen from this definition that the number of iterations to reach the convergence is proportional to $\sqrt{\kappa(\mathbf{A})}$ (Barrett *et al.*, 1994).

4.3.1 One-Dimensional Problem with Material Interface

In order to illustrate how the enrichments can result in an extremely ill-conditioned system, a simple one-dimensional problem with material interface is presented here, which was originally proposed by Menk and Bordas (2011). Consider a one-dimensional bar modeled by three linear elements where the material interface is placed at the distance of ε from node 2, as shown in Figure 4.3a. All four nodal points are enriched, and their enrichment functions are plotted in Figure 4.3b for the distance of $\varepsilon = 0.2h$. The enrichment function $\varphi_i(x)$ for each node i is defined as

$$\varphi_i(x) = (D(x) - D(x_i))R(x) \tag{4.25}$$

where $D(x)$ is the signed distance function with respect to the material interface Γ_d defined as

$$D(x) = \|x - x_d\| \tag{4.26}$$

and $R(x)$ is a cutoff function defined as

$$R(x) = \sum_{i\in\mathcal{N}} N_i(x) \tag{4.27}$$

where $\mathcal{N} = \{\text{supp}(N_i) \cap \Gamma_d \neq \emptyset\}$.

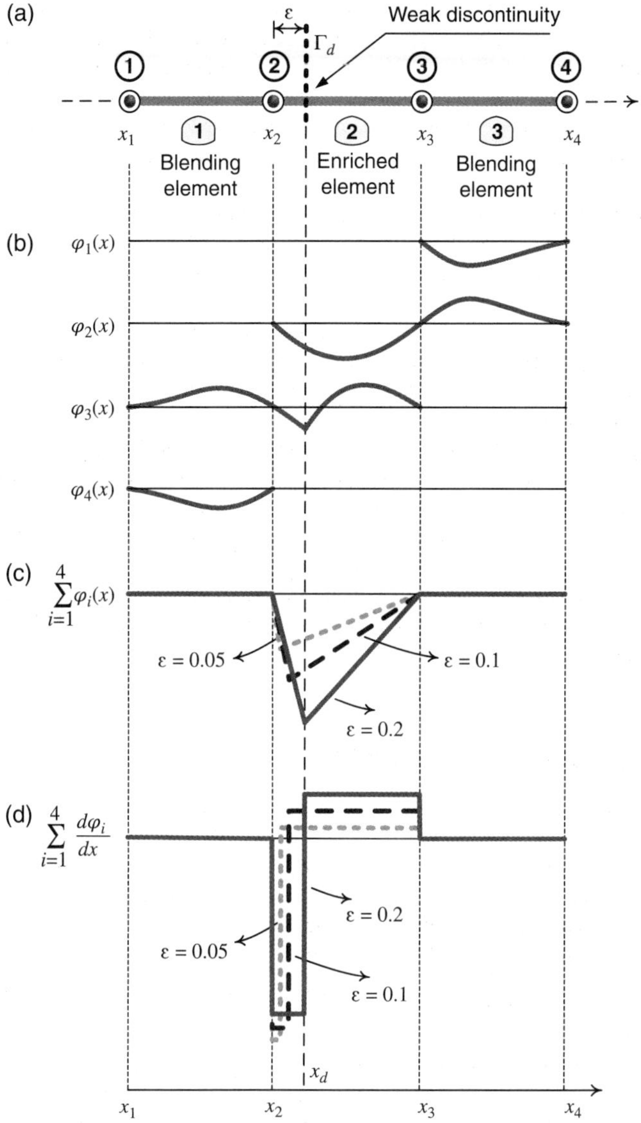

Figure 4.3 A one-dimensional problem with material interface: (a) the problem definition; (b) the enrichment functions of four nodal points for the material interface placed at $\varepsilon = 0.1$; (c) the sum of enrichment functions for different values of ε; (d) the sum of derivatives of enrichment functions for different values of ε

It can be seen from Figure 4.3c that the sum of enrichment functions may be vanished, and the sum approaches to zero for smaller values of ε. The same behavior can be observed for the sum of derivatives of enrichment functions, as shown in Figure 4.3d. In fact, a linear combination of enrichment functions has a derivative that approaches zero as $\varepsilon \rightarrow 0$. For this example, the energy term of axial bar given in (4.21) can be obtained as

$$U(\mathbf{a}) = \mathbf{a}^T \mathbf{K}\,\mathbf{a} = \int_{\ell} E \left(\frac{du^h}{dx} \right)^2 dx \tag{4.28}$$

in which the bilinear form vanishes for the sum of enrichment functions if the interface approaches to node 2 but the corresponding vector of coefficients **a** remains unchanged. The vector **a** contains the values of ones and zeros such that the standard shape functions are canceled and the enriched shape functions are summed up. Thus, it can be concluded that a linear combination of enrichment functions vanishes if $\varepsilon \to 0$, and the corresponding vector of coefficients **a** remains unchanged, however, $\mathbf{a}^T\mathbf{K}\,\mathbf{a} \to 0$ and as a result, the smallest eigenvalue λ_{min} approaches to zero according to the Rayleigh quotient (4.21), which results in an ill-conditioned equation system.

This simple one-dimension example clearly illustrates the problem of ill-conditioning in the standard enrichment techniques. In the FEM, there are several criteria used to evaluate the mesh quality; in fact, if the mesh is of a certain quality the stiffness matrix is more likely to be well-conditioned. However, it is not easy to obtain such criteria for the X-FEM, and a well-conditioned equation system cannot be guaranteed, even in a very simple case. One possible solution to circumvent this problem can be performed by postulating that only a certain number of elements are allowed inside the enrichment radius, or by avoiding the interface enrichment if the interface is too close to the element edges. However, from a more general perception, a procedure to eliminate the negative effects of the enrichments on the condition number in a computationally efficient way is desirable. A good candidate for such a procedure was presented by Menk and Bordas (2011) based on a domain decomposition in which additional constraints were used to ensure the continuity of the solution. In this manner, the problem of ill-conditioning was removed using the Cholesky decomposition employed together with the LQ (linear quotient) decomposition. They showed that the condition numbers obtained by this method are close to that of the corresponding FEM matrices without enrichment, and the convergence of the iterative solver improves significantly. However, it must be noted that the computational effort is dependent on the choice of domain decomposition.

4.4 Blending Strategies in X-FEM

The PU enrichments are attractive for problems involving arbitrary discontinuities, singularities, and features with high gradients, such as crack tips and bimaterials. Because of computational efficiency, the PU enrichments are best restricted to sub-domains where the enrichment is useful, that is, a local PU. However, it has been shown that the accuracy and rate of convergence of the method suffers if the elements in the blending domain between the enriched and standard sub-domains are not properly constructed. In the blending elements, the PU property of the enriched DOF is violated since all nodal points are not enriched, that is, $\sum_{J\in\mathcal{M}^B} N_J(\boldsymbol{x}) \neq 1$, and consequently the approximation is no longer able to represent the enrichment function $\psi(\boldsymbol{x})$ exactly, that is, $\sum_{J\in\mathcal{M}^B} N_J(\boldsymbol{x})\,\psi(\boldsymbol{x}_J) \neq \psi(\boldsymbol{x})$. This fact, however, does not pose a severe problem since the capture of *local* phenomena through the enrichment is more interested. Through the choice of the nodal subset $\mathcal{M}^B$ the local area of the domain can be prescribed directly where the enrichment function is represented exactly since functions $N_J(\boldsymbol{x})$ build a PU over the sub-domain. Hence, it is not necessary to represent the enrichment function exactly in the surrounding string of blending elements.

The next issue that has a significant problem in the X-FEM is the introduction of unwanted terms in the approximation which, in general, cannot be compensated by the standard FE part of the approximation. The appearance of unwanted terms in the blending elements is much more severe than the fact that $\psi(\boldsymbol{x})$ can no longer be represented exactly. These terms can degrade the convergence of the X-FEM significantly. It must be noted that it is not possible to simply set $N_J(\boldsymbol{x})$ to zero in the blending elements, and express that the approximation would not introduce unwanted terms in the blending elements and would still be able to represent the enrichment function exactly in the reproducing elements. In fact, the resulting local enrichment functions would be discontinuous along the element edges between the reproducing and blending elements. Thus, a special treatment is necessary in blending elements in order to remove the unwanted terms. There are various methods proposed in literature to alleviate the above problems in blending elements. These approaches that have been developed to improve the accuracy, convergence rate, and condition number in the blending elements are the *enhanced strain* method, the *hierarchical* method, the *cutoff function* method, the *discontinuous Galerkin* (DG) method, and the *modified enrichment function* method.

Chessa, Wang, and Belytschko (2003) performed an analysis of the interpolation error for a one-dimensional blending element enriched with polynomial functions to reflect a gradient discontinuity, and illustrated that the error is of higher order than in the other elements. They proposed two approaches of recovering PU in the blending elements; first by modifying the X-FEM blending formulation based on an enhanced strain method, and second by adding the higher-order polynomial functions hierarchically to blending elements. Both techniques attempt to compensate for the unwanted terms appearing in the displacement interpolation because of the partial enrichment. Legay, Wang, and Belytschko (2005) employed the X-FEM concept within the spectral finite elements for modeling discontinuities in the gradients, and highlighted that there is no need to implement special blending elements for high-order spectral elements if the shape functions used for the local PU are at least one order lower than those used for the basic approximation. They emphasized that this is because of the fact that the polynomial basis of high-order spectral elements can provide the polynomial terms needed to eliminate the parasitic terms in blending elements. However, for the first order spectral elements the convergence rate was sub-optimal and the results improved tremendously when the blending elements were enhanced by the assumed strain method. Gracie, Wang, and Belytschko (2008b) developed a DG formulation without blending elements in which the continuity was enforced with an internal penalty method. Fries (2008) proposed a modification of the X-FEM approximation in which all nodes in the blending elements were enriched with new modified enrichment functions such that the PU was satisfied everywhere in the domain. Tarancon *et al.* (2009) extended the work of Chessa, Wang, and Belytschko (2003) to reduce unwanted effects of partial enrichment in the blending elements by adding higher order hierarchical shape functions.

It is noteworthy to highlight that these techniques can be combined with a general enrichment strategy based on the "*geometrical*" enrichment. In the standard X-FEM, the enrichment is such that only one layer of elements supports the complete enrichment basis, which is called as the "*topological*" enrichment (Figure 4.4a). In the topological enrichment, only the element that contains the crack-tip (in facture problem) and/or only one layer of elements around the interface (in bimaterial problem) are enriched. One drawback of this enrichment scheme is that the accuracy depends on the position of the crack tip

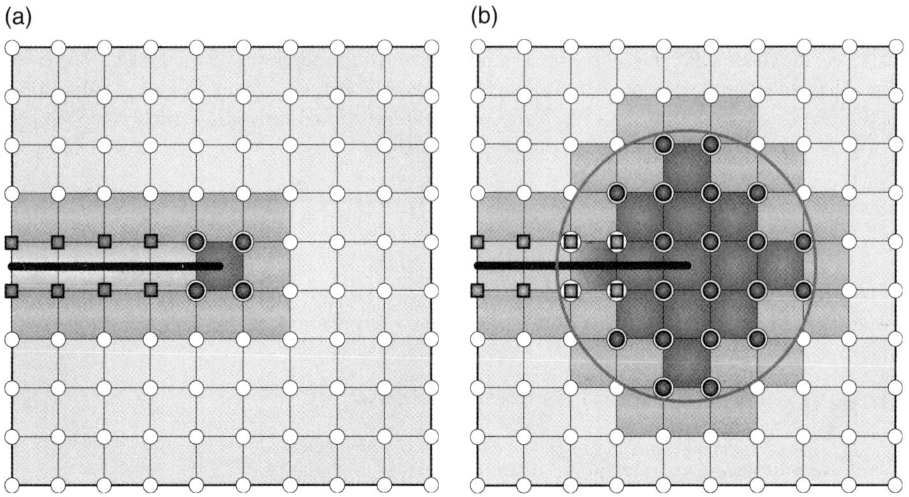

Figure 4.4 The crack tip enrichment strategies: (a) the *topological* enrichment (b) the *geometrical* enrichment

(for instance, close to a node or an edge), which leads to a poor convergence rate, since only a layer of elements are partially enriched. The characteristic of "*topological*" enrichment is that the size of the enriched area is proportional to the element size h. In contrast, the "*geometrical*" enrichment can be used to achieve the optimal rate of convergence by enriching the elements that are contained in a predefined area (Figure 4.4b). Béchet *et al.* (2005) and Laborde *et al.* (2005) illustrated that for an optimal convergence rate, crack tip enrichment must be geometrical within a prescribed geometry independent of mesh size.

4.5 Enhanced Strain Method

An approach to improve the performance of local PU enrichments in the blending elements was introduced by Chessa, Wang, and Belytschko (2003) based on an enhanced strain formulation. According to this method by properly choosing an enhanced strain field, the unwanted terms can be eliminated in the enriched displacement field. The enhanced strain method is a general approach used to overcome the locking issue in incompressible material problems within the framework of a mixed FEM formulation. In the X-FEM, the enhanced strain method is used only in the blending domain to eliminate the undesirable terms that arise from the enrichment function in the blending sub-domain. It can be used for arbitrary enrichment functions, but for each enrichment function a set of enhanced strains need to be constructed. Based on this method, the assumed strain field $\bar{\boldsymbol{\varepsilon}}$ is the sum of the symmetric displacement gradient and an enhanced strain field $\boldsymbol{\varepsilon}^{\text{enh}}$ as

$$\bar{\boldsymbol{\varepsilon}} = \nabla^s \mathbf{u}^h + \boldsymbol{\varepsilon}^{\text{enh}} \tag{4.29}$$

in which the symmetric displacement gradient is defined as

$$\nabla^s \mathbf{u}^h = \frac{1}{2}\left(\nabla \mathbf{u}^h + \left(\nabla \mathbf{u}^h\right)^T\right) \tag{4.30}$$

and the enhanced strain $\boldsymbol{\varepsilon}^{\text{enh}}$ is defined in the blending element Ω_{bld} as

$$\boldsymbol{\varepsilon}^{\text{enh}} = \mathbf{N}^{\mathcal{B}} \mathbf{b}^{\mathcal{B}} \tag{4.31}$$

where $\mathbf{N}^{\mathcal{B}}$ are the shape functions of assumed strain element whose support is limited to Ω_{bld} and $\mathbf{b}^{\mathcal{B}}$ is the corresponding enhanced vector of DOFs. It must be mentioned that this enhanced field is only applied to blended elements. Applying the Hu–Washizu multi-field variational principle, which is the variational form of potential energy in the mixed formulation, it leads to

$$\delta \Pi^{\text{HW}} = \int_\Omega \delta\bar{\boldsymbol{\varepsilon}} : \boldsymbol{\sigma}(\bar{\boldsymbol{\varepsilon}})\, d\Omega + \int_\Omega \delta\left[\bar{\boldsymbol{\sigma}} : \left(\nabla^s \mathbf{u}^h - \bar{\boldsymbol{\varepsilon}}\right)\right] d\Omega - \delta w^{\text{ext}} = 0 \tag{4.32}$$

It is assumed that $\boldsymbol{\varepsilon}^{\text{enh}}$ is orthogonal to the assumed stress field $\bar{\boldsymbol{\sigma}}$, thus the second term in the right-hand-side of (4.32) is omitted and the following two field formulation can be obtained as

$$\delta \Pi^{\text{HW}} = \int_\Omega \delta\bar{\boldsymbol{\varepsilon}} : \boldsymbol{\sigma}(\bar{\boldsymbol{\varepsilon}})\, d\Omega - \delta w^{\text{ext}} = 0 \tag{4.33}$$

The symmetric displacement gradient (4.30) can be written as

$$\nabla^s \mathbf{u}^h = \mathbf{B}\mathbf{d} \tag{4.34}$$

where the matrix $\mathbf{B}$ is defined for each node I using the standard and enriched components as

$$\mathbf{B}_I = \begin{bmatrix} \partial N_I/\partial x & 0 & \partial(N_I\psi)/\partial x & 0 \\ 0 & \partial N_I/\partial y & 0 & \partial(N_I\psi)/\partial y \\ \partial N_I/\partial y & \partial N_I/\partial x & \partial(N_I\psi)/\partial y & \partial(N_I\psi)/\partial x \end{bmatrix} \tag{4.35}$$

and the corresponding vector of DOF is defined based on the standard and enriched components as $\mathbf{d}_I = \langle u_{xI} \; u_{yI} \; a_{xI} \; a_{yI} \rangle$. Substituting Eqs. (4.29), (4.31), and (4.34) into (4.33) yields to

$$\int_\Omega \delta\left(\mathbf{Bd} + \mathbf{N}^{\mathcal{B}}\mathbf{b}^{\mathcal{B}}\right)^T \mathbf{D}\left(\mathbf{Bd} + \mathbf{N}^{\mathcal{B}}\mathbf{b}^{\mathcal{B}}\right) d\Omega - \delta w^{\text{ext}} = 0 \tag{4.36}$$

or

$$\int_\Omega \left(\delta\mathbf{d}^T\left(\mathbf{B}^T\mathbf{DBd} + \mathbf{B}^T\mathbf{D}\,\mathbf{N}^{\mathcal{B}}\mathbf{b}^{\mathcal{B}}\right) + \delta\mathbf{b}^{\mathcal{B}^T}\left(\mathbf{N}^{\mathcal{B}^T}\mathbf{DBd} + \mathbf{N}^{\mathcal{B}^T}\mathbf{D}\,\mathbf{N}^{\mathcal{B}}\mathbf{b}^{\mathcal{B}}\right)\right) d\Omega - \delta\mathbf{d}^T\mathbf{f}^{\text{ext}} = 0 \tag{4.37}$$

This integral equation must hold for all admissible $\delta\mathbf{d}$ and $\delta\mathbf{b}^{\mathcal{B}}$, hence, it results in the following system of mixed formulation as

$$\left(\int_\Omega \mathbf{B}^T\mathbf{DB}\, d\Omega\right)\mathbf{d} + \left(\int_\Omega \mathbf{B}^T\mathbf{DN}^{\mathcal{B}} d\Omega\right)\mathbf{b}^{\mathcal{B}} = \mathbf{f}^{\text{ext}} \tag{4.38a}$$

$$\left(\int_\Omega \mathbf{N}^{\mathcal{B}^T}\mathbf{DB}\, d\Omega\right)\mathbf{d} + \left(\int_\Omega \mathbf{N}^{\mathcal{B}^T}\mathbf{DN}^{\mathcal{B}} d\Omega\right)\mathbf{b}^{\mathcal{B}} = \mathbf{0} \tag{4.38b}$$

or the discrete assumed strain element equations

$$\begin{bmatrix} \mathbf{K}^{\mathcal{B}}_{dd} & \mathbf{K}^{\mathcal{B}}_{db} \\ \mathbf{K}^{\mathcal{B}}_{bd} & \mathbf{K}^{\mathcal{B}}_{bb} \end{bmatrix} \begin{Bmatrix} \mathbf{d} \\ \mathbf{b}^{\mathcal{B}} \end{Bmatrix} = \begin{Bmatrix} \mathbf{f}^{\text{ext}} \\ \mathbf{0} \end{Bmatrix} \tag{4.39}$$

where $\mathbf{D}$ is the material property matrix and $\mathbf{f}^{\text{ext}}$ is the vector of external force. It must be noted that the enhanced strain DOFs $\mathbf{b}^{\mathcal{B}}$ can be solved on the element level from (4.39) if the support of $\mathbf{N}^{\mathcal{B}}$ is restricted to a single element. Although the assumed stress field $\bar{\boldsymbol{\sigma}}$ is not explicitly needed due to the orthogonality assumption (4.33), it must at least contain piecewise constant stresses. Hence, the enhanced strain field can be constructed such that it is orthogonal to an arbitrary constant stress, that is,

$$\int_{\Omega_{bld}} \boldsymbol{\varepsilon}^{\text{enh}} d\Omega = 0 \tag{4.40}$$

in which, for arbitrary values of $\mathbf{b}^{\mathcal{B}}$, it yields to

$$\int_{\Omega_{bld}} \mathbf{N}^{\mathcal{B}} d\Omega = 0 \tag{4.41}$$

In order to avoid zero energy modes, the kernel of the enhanced strain field must be null, or in other words the set of $\mathbf{N}^{\mathcal{B}}$ must be linearly independent. As it was stated earlier, the purpose of enhanced strain method is to construct an enhanced strain field to compensate the effects of the parasitic terms in the blending elements. To this end, it is important to determine which terms are parasitic in partially enriched elements. The enriched part of the strain field in a blended element can be expressed as

$$\boldsymbol{\varepsilon}^{\text{enr}} = \sum_{I \in \mathcal{N}^{enr}} \nabla^{\text{s}}(N_I(\xi,\eta)\,\psi)\,\mathbf{a}_I \tag{4.42}$$

where (ξ, η) is the element parent coordinate space. The corresponding enriched strain space function can be given as

$$\mathcal{S}^{\text{enr}} = \text{span}\{\nabla^{\text{s}}(N_I(\xi,\eta)\,\psi)\} \tag{4.43}$$

Accordingly from equation (4.31), the enhanced strain space function is given by

$$\mathcal{S}^{\text{enh}} = \text{span}\{N_I^{\varepsilon}(\xi,\eta)\} \tag{4.44}$$

where $\text{span}\{N_I^{\varepsilon}(\xi,\eta)\}$ means the set of all linear combinations of N_I^{ε}. It can be shown that if N_I^{ε} are chosen such that $\mathcal{S}^{\text{enr}} \subseteq \mathcal{S}^{\text{enh}}$ the enhanced part is able to eliminate the negative effect of parasitic terms. The main drawback of this method is that a proper enhanced strain field must be obtained for each type of problem separately.

4.5.1 An Enhanced Strain Blending Element for the Ramp Enrichment Function

In order to illustrate how to construct an enhanced strain field for the blending elements in a problem with the discontinuity in the gradient of displacement where the ramp enrichment function is employed, the enhanced strain method described in preceding section is utilized within the Laplace equation. This example was proposed by Chessa, Wang, and Belytschko (2003) to produce a blending element for the discontinuous gradient enrichment in the Laplace equation using the enhanced strain method. Consider the two-dimensional Laplace equation on a domain Ω as

$$\nabla\cdot(k\,\nabla u(\boldsymbol{x})) = f \quad \forall \boldsymbol{x} \in \Omega \tag{4.45}$$

where k is a discontinuous coefficient defined as

$$k = \begin{cases} k_1 & \forall \boldsymbol{x} \in \Omega_1 \\ k_2 & \forall \boldsymbol{x} \in \Omega_2 \end{cases} \tag{4.46}$$

in which the domain is divided into two regions Ω_1 and Ω_2 where the interface between two sub-domains is denoted by Γ_d. Moreover, the gradient of u is defined by $\mathbf{g}(\boldsymbol{x}) = \nabla u$. Since the coefficient k is discontinuous over Ω, the gradient of u is also discontinuous on Γ_d. In order to produce the discontinuity in the gradient of u, the signed distance function is employed as an enrichment function given by $\psi_{\text{Ramp}}(\boldsymbol{x}) = |\varphi(\boldsymbol{x})|$, where $\varphi(\boldsymbol{x})$ is defined as

$$\varphi(\boldsymbol{x}) = \begin{cases} -d(\boldsymbol{x}) & \forall \boldsymbol{x} \in \Omega_1 \\ d(\boldsymbol{x}) & \forall \boldsymbol{x} \in \Omega_2 \\ 0 & \forall \boldsymbol{x} \in \Gamma_d \end{cases} \tag{4.47}$$

where d is the distance to the interface, that is,

$$d(\boldsymbol{x}) = \min \|\boldsymbol{x} - \boldsymbol{x}^*\| \tag{4.48}$$

where x^* is the closest point projection of x onto the discontinuity Γ_d. Consider that $\varphi(x)$ is interpolated based on the values of nodal points φ_I using the linear FE shape functions as

$$\psi_{\text{Ramp}}(\boldsymbol{x}) = \left| \sum_{I \in \mathcal{N}^{std}} N_I(\boldsymbol{x})\, \varphi_I \right| \tag{4.49}$$

Hence, the enriched approximation can be written according to (4.1) as

$$u^h(\boldsymbol{x}) = \sum_{I \in \mathcal{N}^{std}} N_I(\boldsymbol{x})\, u_I + \sum_{J \in \mathcal{N}^{enr}} N_J(\boldsymbol{x}) |\varphi(\boldsymbol{x})| a_J \tag{4.50}$$

In order to construct the enhanced strain field for a blending element, the gradient field $\mathbf{g}(x) = \nabla u$ can be written using relation (4.50) in the blended domain Ω_{bld} as

$$\mathbf{g}(\boldsymbol{x}) = \sum_{I \in \mathcal{N}^{std}} \frac{\partial N_I}{\partial \boldsymbol{x}} u_I + \sum_{J \in \mathcal{N}^{enr}} \left(\frac{\partial N_J}{\partial \boldsymbol{x}} |\varphi(\boldsymbol{x})| + N_J(\boldsymbol{x}) H(\varphi(\boldsymbol{x})) \frac{\partial \varphi}{\partial \boldsymbol{x}} \right) a_J \tag{4.51}$$

where $H(\varphi(x))$ is the Heaviside function defined as

$$H(\varphi(\boldsymbol{x})) = \begin{cases} -1 & \text{if } \varphi(\boldsymbol{x}) \le 0 \\ +1 & \text{if } \varphi(\boldsymbol{x}) > 0 \end{cases} \tag{4.52}$$

For elements in the blended domain Ω_{bld}, the signed distance function can be written as

$$|\varphi(\boldsymbol{x})| = \begin{cases} -\sum_{I \in \mathcal{N}^{std}} N_I(\boldsymbol{x})\, \varphi_I & \boldsymbol{x} \in \Omega_1 \\ \sum_{I \in \mathcal{N}^{std}} N_I(\boldsymbol{x})\, \varphi_I & \boldsymbol{x} \in \Omega_2 \end{cases} \tag{4.53}$$

Assuming that the Jacobian is approximately constant in the blending element, the enriched strain space $\mathcal{S}^{\text{enr}}$ in Ω_{bld} can be expressed in terms of element shape functions as

$$\mathcal{S}^{\text{enr}} \subset \text{span}\{N_J(\xi,\eta)\, \nabla_\xi N_I(\xi,\eta)\} \tag{4.54}$$

where $\nabla_\xi N_I$ is the gradient of the shape functions N_I with respect to the parent coordinates (ξ, η). For a 4-noded bilinear element, the shape functions in term of the parent coordinates are defined as $N_I(\xi,\eta) = \frac{1}{4}(1+\xi_I\xi)(1+\eta_I\eta)$. Hence, the enriched strain space (4.54) can be written as

$$\mathcal{S}^{\text{enr}} = \text{span}\{1, \xi, \eta, \xi\eta, \xi^2, \eta^2, \xi^2\eta, \xi\eta^2\} \tag{4.55}$$

In order to retain the constant strain terms and remove the unwanted terms, the enhanced strain space $\mathcal{S}^{\text{enh}}$ can be defined as

$$\mathcal{S}^{\text{enh}} = \text{span}\{\xi, \eta, \xi\eta, \xi^2 - \eta^2, \xi^2\eta, \xi\eta^2\} \tag{4.56}$$

It can be seen that the terms imposed on the enhanced strain space comply with orthogonality condition. The enhanced gradient field can therefore be written for the blended domain Ω_{bld} as

$$\mathbf{g}^{\text{enh}}_{(x)} = \xi\boldsymbol{\alpha}_1 + \eta\boldsymbol{\alpha}_2 + \xi\eta\boldsymbol{\alpha}_3 + (\xi^2 - \eta^2)\boldsymbol{\alpha}_4 + \xi^2\eta\boldsymbol{\alpha}_5 + \xi\eta^2\boldsymbol{\alpha}_6 \tag{4.57}$$

This field spans $\mathcal{S}^{\text{enh}}$, satisfies the orthogonality condition (4.41), and has no zero energy modes.

4.5.2 An Enhanced Strain Blending Element for Asymptotic Enrichment Functions

In a crack problem, each patch that contains the crack interface, or is near the crack-tip must be enriched by one of the two types of enrichments. For the patches that are crossed by the crack interface, the Heaviside enrichment function is used to enhance the solution. The Heaviside enrichment function does not involve any blending elements. However, patches near the crack tip are enriched by a set of singular enrichment functions based on the near-field asymptotic solution. These functions are often referred to as the branch functions given by

$$\boldsymbol{\psi}_{\text{Tip}}(r,\theta)=\{\mathcal{B}_\alpha(r,\theta)\}=\left\{\sqrt{r}\sin\frac{\theta}{2},\sqrt{r}\cos\frac{\theta}{2},\sqrt{r}\sin\frac{\theta}{2}\sin\theta,\sqrt{r}\cos\frac{\theta}{2}\sin\theta\right\} \tag{4.58}$$

where (r, θ) are the polar coordinates of the local coordinate system defined with an origin at the tip of crack and with basis vectors defined by the unit vector's tangent and normal to the crack at the crack tip.

The enriched approximation field by means of the X-FEM can be written as

$$\mathbf{u}^h(\boldsymbol{x})=\sum_{I\in\mathcal{N}^{std}}N_I(\boldsymbol{x})\,\mathbf{u}_I+\sum_{J\in\mathcal{N}^{enr}_{Hev}}N_J(\boldsymbol{x})\,H(\boldsymbol{x})\,\mathbf{a}_J+\sum_{K\in\mathcal{N}^{enr}_{tip}}N_K(\boldsymbol{x})\left(\sum_{\alpha=1}^{4}\mathcal{B}_\alpha(\boldsymbol{x})\,\mathbf{b}_{\alpha K}\right) \tag{4.59}$$

where $\mathcal{N}^{enr}_{Hev}$ is the set of nodes enriched by the Heaviside function $H(\boldsymbol{x})$, and $\mathcal{N}^{enr}_{tip}$ is the set of nodes enriched by the asymptotic functions $\mathcal{B}_\alpha(r, \theta)$; and $\mathbf{a}_J$ and $\mathbf{b}_{\alpha K}$ are the corresponding DOF.

For the asymptotic enrichment functions of linear elastic fracture mechanics (LEFM), the assumed strain shape functions were derived by Gracie, Wang, and Belytschko (2008b) for blending elements. It was shown that for the singular enrichment functions, the parasitic terms in the approximation space of the blending elements are spanned by the set

$$\mathcal{S}^{\text{enh}}=\text{span}\left\{\boldsymbol{\psi}_{\text{Tip}}(\boldsymbol{x}),\ \xi_i\frac{\partial\boldsymbol{\psi}_{\text{Tip}}(\boldsymbol{x})}{\partial x_j}\right\}\quad\forall\boldsymbol{x}\in\Omega_{bld} \tag{4.60}$$

where ξ_i are parent element coordinates. Moreover, it is assumed that the Jacobian between the parent and global coordinate system is constant. For a constant stress triangular element, it can be shown that the parasitic terms of the strain approximation in the blending elements are spanned by a set of functions

$$\begin{aligned}\mathcal{S}^{\text{enh}}=\text{span}\Bigg\{&\sqrt{r}\cos\frac{\theta}{2},\sqrt{r}\sin\frac{\theta}{2},\sqrt{r}\cos\frac{\theta}{2}\sin\theta,\sqrt{r}\sin\frac{\theta}{2}\sin\theta,\\&\frac{\xi_1}{\sqrt{r}}\cos\frac{\theta}{2},\ \frac{\xi_2}{\sqrt{r}}\cos\frac{\theta}{2},\ \frac{\xi_1}{\sqrt{r}}\sin\frac{\theta}{2},\ \frac{\xi_2}{\sqrt{r}}\sin\frac{\theta}{2},\\&\frac{\xi_1}{\sqrt{r}}\left(\cos\frac{5\theta}{2}-\cos\frac{\theta}{2}\right),\frac{\xi_2}{\sqrt{r}}\left(\cos\frac{5\theta}{2}-\cos\frac{\theta}{2}\right),\\&\frac{\xi_1}{\sqrt{r}}\left(3\cos\frac{5\theta}{2}+\cos\frac{\theta}{2}\right),\frac{\xi_2}{\sqrt{r}}\left(3\cos\frac{5\theta}{2}+\cos\frac{\theta}{2}\right),\\&\frac{\xi_1}{\sqrt{r}}\left(3\sin\frac{5\theta}{2}+\sin\frac{\theta}{2}\right),\frac{\xi_2}{\sqrt{r}}\left(3\sin\frac{5\theta}{2}+\sin\frac{\theta}{2}\right),\\&\frac{\xi_1}{\sqrt{r}}\left(\sin\frac{5\theta}{2}-\sin\frac{\theta}{2}\right),\frac{\xi_2}{\sqrt{r}}\left(\sin\frac{5\theta}{2}-\sin\frac{\theta}{2}\right)\Bigg\}\end{aligned} \tag{4.61}$$

Since these functions do not satisfy the orthogonality condition, they cannot be used as the enhanced shape functions. The enhanced shape functions can be obtained by orthogonalizing the span $\mathcal{S}^{\text{enh}}$ as

$$N_I^{\varepsilon} = \mathcal{S}_I^{\text{enh}} - \frac{1}{A_0}\int_{\Omega_{bld}} \mathcal{S}_I^{\text{enh}} \, d\Omega \tag{4.62}$$

where A_0 is the area of the element defined by Ω_{bld}, and I ranges over all members in $\mathcal{S}^{\text{enh}}$. It must be noted that although this approach improves the accuracy and optimum convergence rate, one significant limitation of this approach is that a linearly independent basis spanning the parasitic terms is required. For complicated enrichment functions, the determination of the basis is not trivial; as can be seen from (4.61) a space of 16 functions must be used to alleviate the blending problems that arise from the use of the elastic crack enrichments.

4.6 The Hierarchical Method

An alternative approach to improve the accuracy and convergence rate of local PU enrichments in the blending elements is based on the *hierarchical method* that utilizes different order of polynomials for the standard and enriched shape functions. It was shown in Section 4.2 for the convergence analysis of a one-dimensional bar that if the order of shape functions of the standard part is higher than the enriched part, the parasitic terms in the blending elements can be corrected by the standard part of the approximation field, and the accuracy and convergence rate of the solution can be improved. It was shown by Chessa, Wang, and Belytschko (2003) that for the polynomial enrichment functions of order p, if the shape function of standard part N^{std} is $s-$ order complete and the shape function of enriched part N^{enr} is $e-$ order complete, the parasitic terms can be removed such that $s \geq e+p$. In this case, the displacement field $\mathbf{u}^h(\boldsymbol{x})$ can reproduce a linear field in the blending elements even if the enrichment is active.

The hierarchical blending elements with the aim of compensating unwanted terms in the blending elements were introduced by Chessa, Wang, and Belytschko (2003), where the polynomial order of the approximation in the blending elements was increased. A similar approach based on the modified X-FEM technique was proposed by Fries (2008) employing additional nodes in the blending elements, where all nodes in the blending elements were enriched. There are some differences between the two approaches; the hierarchical blending elements are designed for polynomial enrichment functions, whereas the modified X-FEM is for general enrichment functions. Moreover, the placement of the additional nodes in the blending elements and the definition of their shape functions differ in the two approaches; in hierarchical blending elements, nodes and shape functions are introduced based on classical higher order finite elements, whereas in the modified X-FEM, only the existing nodes of the blending elements are used. It must also be pointed out that the two approaches have different motivations; in hierarchical blending elements, the technique eliminates the negative effects of the unwanted terms by increasing the polynomial order in the element, while the modified X-FEM avoids unwanted terms from the beginning. The hierarchical blending elements were first employed by Chessa, Wang, and Belytschko (2003) for the material interface (ramp) enrichment, and then extended by Tarancon *et al.* (2009) to the elastic crack enrichments.

4.6.1 A Hierarchical Blending Element for Discontinuous Gradient Enrichment

In order to construct the hierarchical shape functions for blending elements in a problem with the discontinuity in the gradient of displacements where the ramp enrichment function (4.49) is used, the standard shape functions of the approximation in Ω_{bld} must be one degree higher than the enriched shape functions. A blending element that satisfies the expression $s \geq e+p$ can be constructed by recognizing the fact that pathological terms in the blending elements due to the partial enrichment are quadratic primarily normal to the discontinuity. Thus, it is sufficient to use a hierarchical element with quadratic shape functions on the edges not coincident with the boundary between Ω_{bld} and Ω_{std}. Chessa, Wang, and Belytschko (2003) constructed a hierarchical blending element for a linear 3-noded triangular element by adding mid-side nodes to the sides connecting the enriched nodes to the standard nodes, as shown in Figure 4.5a. The shape functions for the standard and enriched displacement fields $\mathbf{u}^{std}(\boldsymbol{x})$ and $\mathbf{u}^{enr}(\boldsymbol{x})$ in the blending elements are defined as

$$
\begin{aligned}
&N_1^{std}(\xi,\eta)=1-\xi-\eta\\
&N_2^{std}(\xi,\eta)=\xi\\
&N_3^{std}(\xi,\eta)=\eta\\
&N_4^{std}(\xi,\eta)=4\xi(1-\xi-\eta)\\
&N_5^{std}(\xi,\eta)=4\eta(1-\xi-\eta)\\
&N_1^{enr}(\xi,\eta)=1-\xi-\eta\\
&N_3^{enr}(\xi,\eta)=\eta
\end{aligned}
\tag{4.63}
$$

where ξ and η are the parent triangular coordinates. Since the corner shape functions are the same as the shape functions in the neighboring elements, $\mathbf{u}^{std}(\boldsymbol{x})$ is linear along the element edge between Ω_{bld} and Ω_{std}, such as edge $\overline{13}$. Furthermore, the mid-side nodes are always contiguous with a similar element; therefore the approximation becomes C_0 on Ω.

Similarly, a hierarchical blending element for a linear quadrilateral element can be constructed by adding mid-side nodes to the sides connecting the enriched nodes to the standard nodes of a 4-noded element, as shown in Figure 4.5b. The shape functions for the standard and enriched displacement fields $\mathbf{u}^{std}(\boldsymbol{x})$ and $\mathbf{u}^{enr}(\boldsymbol{x})$ in the blending elements are defined as

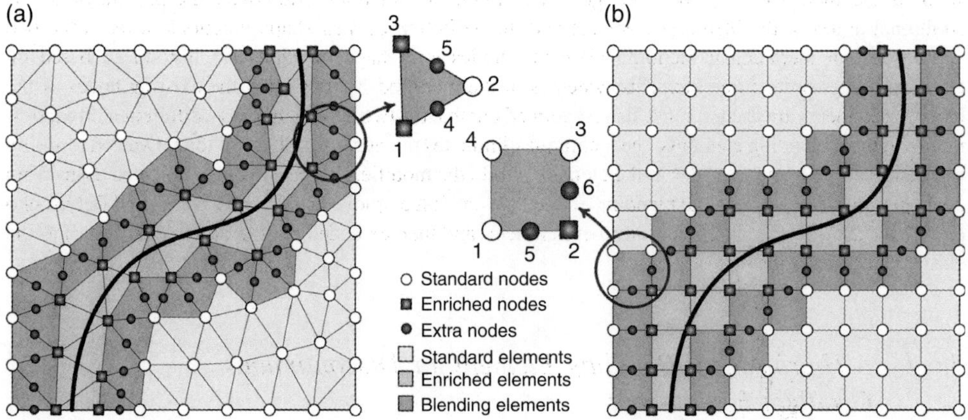

Figure 4.5 Constructing hierarchical blending elements for discontinuous gradient enrichment using: (a) the linear 3-noded triangular elements (b) the linear 4-noded quadrilateral elements

$$
\begin{aligned}
N_I^{std}(\xi,\eta) &= \frac{1}{4}(1+\xi_I\xi)(1+\eta_I\eta) \qquad (I=1,2,3,4)\\
N_5^{std}(\xi,\eta) &= \frac{1}{2}(1-\xi^2)(1-\eta)\\
N_6^{std}(\xi,\eta) &= \frac{1}{2}(1+\xi)(1-\eta^2)\\
N_2^{enr}(\xi,\eta) &= \frac{1}{4}(1+\xi)(1-\eta)
\end{aligned} \tag{4.64}
$$

where ξ and η are the parent quadrilateral coordinates.

4.6.2 *A Hierarchical Blending Element for Crack Tip Asymptotic Enrichments*

The hierarchical method was employed by Tarancon *et al.* (2009) for LEFM problems by adding appropriate hierarchical shape functions in order to compensate for the pathological terms in the blending elements caused by partial enrichment. In this method, the blending elements are enhanced in the X-FEM by increasing the polynomial order of the standard interpolation locally so as to achieve the higher accuracy at both the local and global levels. Hence, the order of standard shape functions is increased while the C_0 continuity of the approximation is also preserved. In a two-dimensional LEFM problem, the hierarchical nodes are added on the sides between blending elements; those are the element sides connecting a node enriched by the functions associated with the crack-tip asymptotic displacement field to a node without this type of enrichment; in other words, sides that connect a node belonging to $\mathcal{N}_{tip}^{enr}$ in (4.59) with another that does not belong to $\mathcal{N}_{tip}^{enr}$. If the added hierarchical shape functions are of cubic order or above, hierarchical nodes can also be added inside the blending elements as the bubble modes.

Based on the hierarchical shape functions, the enhanced displacement field (4.59) can be written for blending elements as

$$
\begin{aligned}
\mathbf{u}^h(\boldsymbol{x}) = &\sum_{I\in\mathcal{N}^{std}} N_I(\boldsymbol{x})\,\mathbf{u}_I + \sum_{J\in\mathcal{N}_{Hev}^{enr}} N_J(\boldsymbol{x})\,H(\boldsymbol{x})\,\mathbf{a}_J + \sum_{K\in\mathcal{N}_{tip}^{enr}} N_K(\boldsymbol{x})\left(\sum_{\alpha=1}^{4}\mathcal{B}_\alpha(\boldsymbol{x})\,\mathbf{b}_{\alpha K}\right)\\
&+\sum_{P\in\mathcal{N}_P^{enr}} \widehat{N}_P(\boldsymbol{x})\,\mathbf{c}_P + \sum_{Q\in\mathcal{N}_Q^{enr}} \widehat{N}_Q(\boldsymbol{x})\,H(\boldsymbol{x})\,\mathbf{d}_Q
\end{aligned} \tag{4.65}
$$

where $\mathcal{N}_P^{enr}$ is the set of added hierarchical nodes associated with sides that connect an enriched node (member of set $\mathcal{N}_{tip}^{enr}$) with a standard node (without any kind of enrichment) or that are associated with bubble modes in the blending elements not divided by the crack; and $\mathcal{N}_Q^{enr}$ is the set of the remaining added hierarchical nodes, that is, those associated with sides that connect an enriched node of $\mathcal{N}_{tip}^{enr}$ to an enriched node of $\mathcal{N}_{Hev}^{enr}$ that does not belong to $\mathcal{N}_{tip}^{enr}$ (which is only enriched with the Heaviside function), or those associated with the interior of the blending elements divided by the crack. In the earlier relation, $\widehat{N}_P$ and $\widehat{N}_Q$ are the hierarchical shape functions related to hierarchical nodes, and $\mathbf{c}_P$ and $\mathbf{d}_Q$ are corresponding DOFs. It is necessary to separate the added hierarchical nodes into $\mathcal{N}_P^{enr}$ and $\mathcal{N}_Q^{enr}$, with their respective interpolation functions, so that the enhancement of the blending elements allows for the discontinuity in the crack-divided elements. In Figure 4.6, the standard geometrical enrichment is shown together with the proposed hierarchical method. In this figure, the set of nodes $\mathcal{N}_{Hev}^{enr}$, the set of nodes $\mathcal{N}_{tip}^{enr}$, and the set of nodes belonging to both Heaviside and asymptotic functions are shown together with the set of added

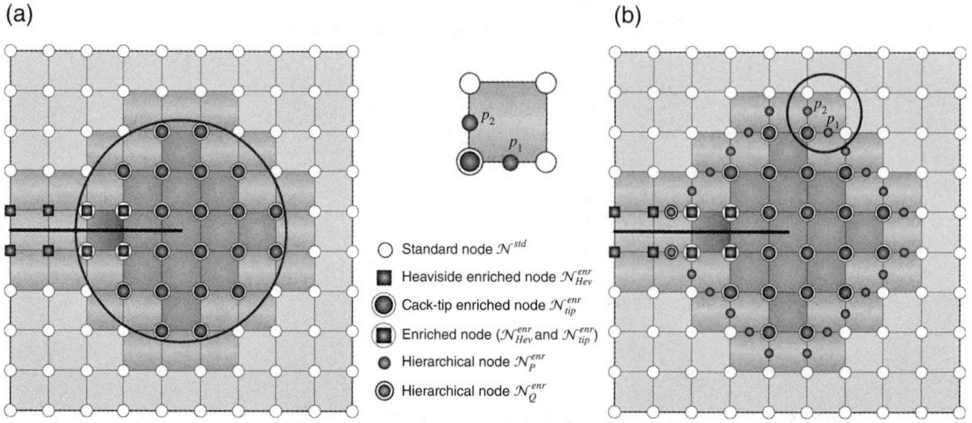

Figure 4.6 Constructing hierarchical blending elements for crack tip asymptotic enrichments with linear quadrilateral elements: (a) the standard nodal enrichments; (b) the added hierarchical nodes in the blending elements

hierarchical nodes $\mathcal{N}_P^{enr}$ associated with sides that connect an enriched node (member of $\mathcal{N}_{tip}^{enr}$) with a standard node, and the set of remaining added hierarchical nodes $\mathcal{N}_Q^{enr}$ associated with sides that connect an enriched node of $\mathcal{N}_{tip}^{enr}$ to an enriched node of $\mathcal{N}_{Hev}^{enr}$ that does not belong to $\mathcal{N}_{tip}^{enr}$.

Consider the top right linear blending element of Figure 4.6b with the hierarchical nodes p_1 and p_2, the corresponding hierarchical shape functions can be expressed as

$$\begin{aligned}\widehat{N}_{P1}(\xi,\eta) &= \frac{1}{2}(1-\eta)\varphi_2(\xi) \\ \widehat{N}_{P2}(\xi,\eta) &= \frac{1}{2}(1-\xi)\varphi_2(\eta)\end{aligned} \tag{4.66}$$

where ξ and η are parent element coordinates and φ_i is a polynomial of degree i, usually a Legendre polynomial; in the case of $i = 2$, the polynomial φ_i is given by

$$\varphi_2(\xi) = -\frac{1}{2}\sqrt{\frac{3}{2}}(1-\xi^2) \tag{4.67}$$

Similar to the enhanced assumed method, the main drawback of the hierarchical method is that it has to be tailored to every special problem to effectively improve the solution. It must be noted that although for polynomial enrichments, such as the signed distance function, it is possible to recover the order of the standard interpolation in the blending elements by use of a hierarchical element, this is more awkward since for a moving interface problem, the nodal connectivity would change during the computation. Nevertheless, the method provides the optimal rate of convergence and it is more accurate than the standard X-FEM without special blending elements or methods that treat the discontinuity without enrichment by accounting for it in the integration of the weak form.

4.7 The Cutoff Function Method

An effective and simple approach was proposed by Fries (2008) based on a linearly decreasing weight function for the enrichment in the blending elements, called as the modified X-FEM technique. This approach allows to obtain a conforming approximation and to eliminate partially enriched elements,

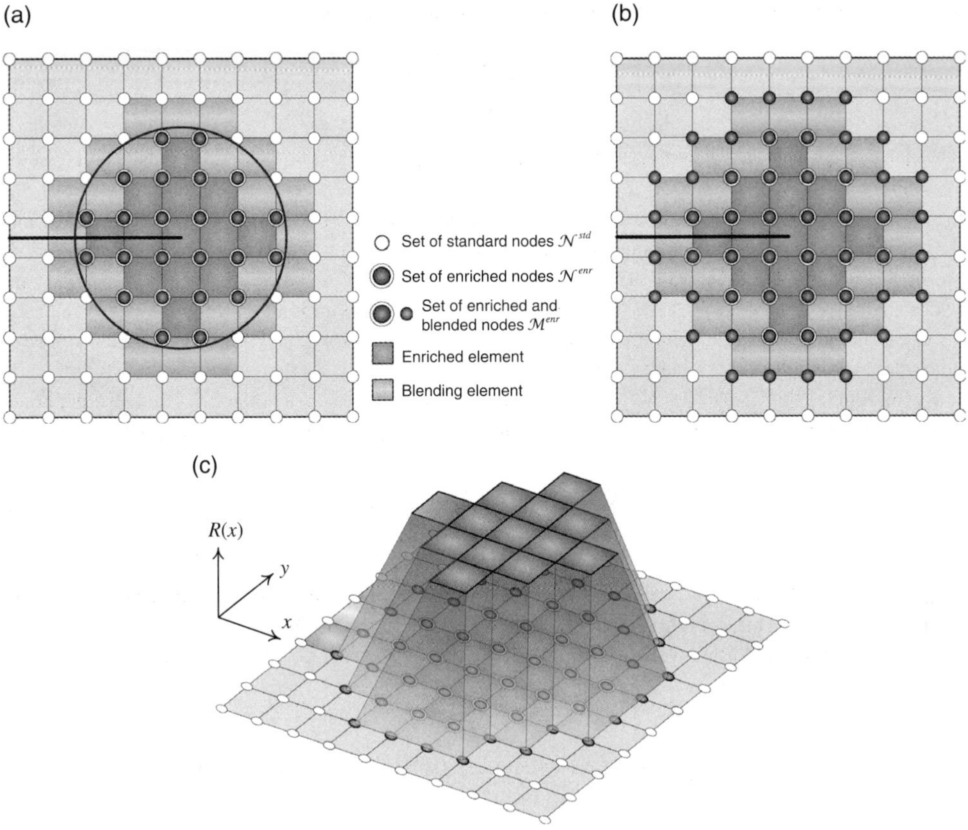

Figure 4.7 The cutoff function method for blending elements: (a) illustration of reproducing elements, blending elements, and the set of nodes $\mathcal{N}^{enr}$ for a crack-tip problem in the standard X-FEM; (b) the set of nodes $\mathcal{M}^{enr}$ and $\mathcal{N}^{enr}$ in the modified X-FEM; (c) the graphical representation of the cutoff function $R(\boldsymbol{x})$

so that the PU property can be satisfied everywhere in the domain. In this method, in addition to those nodes that are enriched in the standard X-FEM (Figure 4.7a), all nodes in the blending elements are enriched that presents a complete PU in the reproducing and blending elements (Figure 4.7b). Moreover, the enrichment functions of the standard X-FEM are modified except in the reproducing elements; they are zero in the standard elements, and vary continuously in the blending elements between the standard and reproducing elements. This modification avoids unwanted terms in the blending elements and still leads to continuous local enrichment functions as long as the enrichment function $\psi(\boldsymbol{x})$ is continuous. In fact, the problem in blending elements is now transferred to the neighboring layer of elements called as *n-layer*. In order to alleviate the negative effect of parasitic terms in *n-layer* elements, the enrichment functions in blending elements are modified by applying a cutoff function $R(\boldsymbol{x})$ which is *one* in fully enriched elements and varies continuously to *zero* in blending elements (Figure 4.7c). As a result, the global enrichment function is localized through the multiplication by a ramp function $R(\boldsymbol{x})$. Consequently, all nodes of the domain where this localized enrichment function is non-zero are enriched, and the enrichment vanishes in those elements where only some of the nodes are enriched and no difficulties occur in blending elements.

Based on the modified X-FEM technique, a modified enrichment function $\psi_{\text{Mod}}(\boldsymbol{x})$ is defined as

$$\psi_{\text{Mod}}(\boldsymbol{x}) = R(\boldsymbol{x})\,\psi(\boldsymbol{x}) \tag{4.68}$$

where the cutoff function $R(\boldsymbol{x})$ is defined based on the ramp function as

$$R(\boldsymbol{x}) = \sum_{I \in \mathcal{N}^{enr}} N_I(\boldsymbol{x}) \tag{4.69}$$

in which a graphical representation of $R(\boldsymbol{x})$ is shown in Figure 4.7c, where $\mathcal{N}^{enr}$ is the set of nodes of the elements that are fully prescribed in the enrichment radius. It can be seen from (4.68) and (4.69) that the modified enrichment function is zero in *n-layer* elements, therefore no parasitic term is introduced into the approximation space. It is obvious that $\psi_{\text{Mod}}(\boldsymbol{x}) = \psi(\boldsymbol{x})$ in the reproducing elements where their nodes are in $\mathcal{N}^{enr}$, and $\psi_{\text{Mod}}(\boldsymbol{x}) = 0$ in the standard finite elements. In the blending elements where some nodes are in $\mathcal{N}^{enr}$, the modified enrichment function $\psi_{\text{Mod}}(\boldsymbol{x})$ varies continuously between $\psi(\boldsymbol{x})$ and zero. It must be noted that due to the multiplication with $R(\boldsymbol{x})$, the order of $\psi_{\text{Mod}}(\boldsymbol{x})$ is increased when compared with $\psi(\boldsymbol{x})$, and slightly more integration points may be appropriate for the numerical integration.

By introducing the set of nodes $\mathcal{M}^{enr}$ which is constructed from the set of nodes $\mathcal{N}^{enr}$ and all other nodes of the blending elements, a modified version of the enriched approximation field (4.1) can be written as

$$\mathbf{u}^h(\boldsymbol{x}) = \sum_{I \in \mathcal{N}^{std}} N_I(\boldsymbol{x})\, \mathbf{u}_I + \sum_{J \in \mathcal{M}^{enr}} N_J(\boldsymbol{x})\, \psi_{\text{Mod}}(\boldsymbol{x})\, \mathbf{a}_J \tag{4.70}$$

in which the set of nodes $\mathcal{M}^{enr}$ consists of all element nodes of the reproducing and blending elements with $\mathcal{N}^{enr} \subset \mathcal{M}^{enr}$. This definition can easily be implemented for any kind of enrichment functions. In most general form a shifted version of the enriched approximation field with multiple enrichment functions can be written as

$$\mathbf{u}^h(\boldsymbol{x}) = \sum_{I \in \mathcal{N}^{std}} N_I(\boldsymbol{x})\, \mathbf{u}_I + \sum_{\alpha=1}^{NE} \sum_{J \in \mathcal{M}^{enr}} N_J(\boldsymbol{x}) \left(\psi^{\alpha}_{\text{Mod}}(\boldsymbol{x}) - \psi^{\alpha}_{\text{Mod}}(\boldsymbol{x}_J)\right) \mathbf{a}^{\alpha}_J \tag{4.71}$$

where NE is the number of enrichment functions.

4.7.1 The Weighted Function Blending Method

An extension of the modified X-FEM was presented by Ventura, Gracie, and Belytschko (2009) on the basis of a weighted X-FEM technique for multiple interacting enrichments. The weighted X-FEM is similar to that proposed by Fries (2008), however, it introduces a general form based on multiplication of the standard enrichment function by a monotonically decreasing weight function with compact support. The weight function is chosen in a manner to avoid inter-element discontinuities and hence has little effect on the order of quadrature required for the integration of discrete equations. The weighted X-FEM technique was originally developed for dislocation problems with the aim of introducing a weight function, where all enriched DOF are constrained to be equal. In this approach, a weight function, or a smooth function, $S(\boldsymbol{x})$ is constructed so that $S(\boldsymbol{x}) > 0$ only in the sub-domain to be enriched; hence, the node J is enriched if there exists an $\boldsymbol{x} \in \text{supp}\{N_J(\boldsymbol{x})\}$ such that $S(\boldsymbol{x}) > 0$, with supp $\{N_J(\boldsymbol{x})\}$ denoting the support of shape function of node J. In fact, the approximation field (4.70) locally enriches the standard FEM approximation by $\psi_{\text{Mod}}(\boldsymbol{x})$, and locality is ensured by the compact support of $S(\boldsymbol{x})$.

The enriched approximation field based on the weighted X-FEM can therefore be written as

$$\mathbf{u}^h(\boldsymbol{x}) = \sum_{I \in \mathcal{N}^{std}} N_I(\boldsymbol{x})\, \mathbf{u}_I + \sum_{J \in \mathcal{M}^{enr}} N_J(\boldsymbol{x})\, S(\boldsymbol{x})\, \psi(\boldsymbol{x})\, \mathbf{a}_J \tag{4.72}$$

where $\psi(\boldsymbol{x})$ is an enrichment function. In this definition, the compact support of weight function $S(\boldsymbol{x})$ ensures the locality of the enrichment. For some particular problems, such as conditioning issues and reducing the number of DOF, and/or implementation to dislocation problems, it is more convenient to constrain all enriched DOF to be equal, that is, $\mathbf{a}_J = \mathbf{a}$; hence, Eq. (4.72) can be simplified as

$$\mathbf{u}^h(\boldsymbol{x}) = \sum_{I \in \mathcal{N}^{std}} N_I(\boldsymbol{x})\ \mathbf{u}_I + S(\boldsymbol{x})\ \psi(\boldsymbol{x})\,\mathbf{a} \sum_{J \in \mathcal{M}^{enr}} N_J(\boldsymbol{x}) \tag{4.73}$$

Since the PU property of the shape functions is satisfied for an $\boldsymbol{x} \in \text{supp}\{S(\boldsymbol{x})\}$, Eq. (4.73) can be written as

$$\mathbf{u}^h(\boldsymbol{x}) = \sum_{I \in \mathcal{N}^{std}} N_I(\boldsymbol{x})\ \mathbf{u}_I + S(\boldsymbol{x})\ \psi(\boldsymbol{x})\,\mathbf{a} \tag{4.74}$$

Similar to the standard X-FEM, the performance of the weighted X-FEM can be improved and the application of essential boundary conditions can be facilitated by *shifting* the approximation field. Two different shifted approximations are possible depending on how the shifts are constructed. The first type can be defined based on the shifted product of the weight and enrichment functions as

$$\mathbf{u}^h(\boldsymbol{x}) = \sum_{I \in \mathcal{N}^{std}} N_I(\boldsymbol{x})\ \mathbf{u}_I + \sum_{J \in \mathcal{M}^{enr}} N_J(\boldsymbol{x})(S(\boldsymbol{x})\,\psi(\boldsymbol{x}) - S(\boldsymbol{x}_J)\,\psi(\boldsymbol{x}_J))\mathbf{a}_J \tag{4.75}$$

while the second type can be defined based on the shifted enrichment function as

$$\mathbf{u}^h(\boldsymbol{x}) = \sum_{I \in \mathcal{N}^{std}} N_I(\boldsymbol{x})\ \mathbf{u}_I + \sum_{J \in \mathcal{M}^{enr}} N_J(\boldsymbol{x})\,S(\boldsymbol{x})\,(\psi(\boldsymbol{x}) - \psi(\boldsymbol{x}_J))\mathbf{a}_J \tag{4.76}$$

In the case that all enriched DOF are constrained to be equal at each nodes, that is, $\mathbf{a}_J = \mathbf{a}$, Eqs. (4.75) and (4.76) can be respectively rewritten as

$$\mathbf{u}^h(\boldsymbol{x}) = \sum_{I \in \mathcal{N}^{std}} N_I(\boldsymbol{x})\ \mathbf{u}_I + \left(S(\boldsymbol{x})\,\psi(\boldsymbol{x}) - \sum_{J \in \mathcal{M}^{enr}} N_J(\boldsymbol{x})\,S(\boldsymbol{x}_J)\,\psi(\boldsymbol{x}_J) \right)\mathbf{a} \tag{4.77}$$

and

$$\mathbf{u}^h(\boldsymbol{x}) = \sum_{I \in \mathcal{N}^{std}} N_I(\boldsymbol{x})\ \mathbf{u}_I + S(\boldsymbol{x}) \left(\psi(\boldsymbol{x}) - \sum_{J \in \mathcal{M}^{enr}} N_J(\boldsymbol{x})\,\psi(\boldsymbol{x}_J) \right)\mathbf{a} \tag{4.78}$$

in which $S(\boldsymbol{x})$ is constructed based on the polynomial ramp function that smoothes the transition from enriched to unenriched elements. The non-enriched elements are those element with $S(\boldsymbol{x}) = 0$, and the enriched elements are those of $S(\boldsymbol{x}) \neq 0$. Considering the enriched domain $\widehat{\Omega}_{enr}$ is defined as the union of all the enriched elements that consists of the reproducing elements in Ω_{enr} and the blending elements in Ω_{bld}; and $\widehat{\Gamma}_{enr}$ is defined as the boundary between the enriched and unenriched sub-domains that includes the set of all edges of the enriched elements which are not inside the enriched domain, the weight function $S(\boldsymbol{x})$ can therefore be defined as

$$\left\{ S(\boldsymbol{x}) \middle| S(\boldsymbol{x}) \in C_0,\ S(\boldsymbol{x}) > 0\ \forall \boldsymbol{x} \in \widehat{\Omega}_{enr} \backslash \widehat{\Gamma}_{enr},\ S(\boldsymbol{x}) = 0\ \forall \boldsymbol{x} \in \Omega \backslash \widehat{\Omega}_{enr} \right\} \tag{4.79}$$

Since the weight function $S(\boldsymbol{x}) = 0$, $\forall \boldsymbol{x} \in \widehat{\Gamma}_{enr}$, the displacement compatibility is satisfied between the enriched and unenriched sub-domains. It is noteworthy to highlight that no partially enriched elements

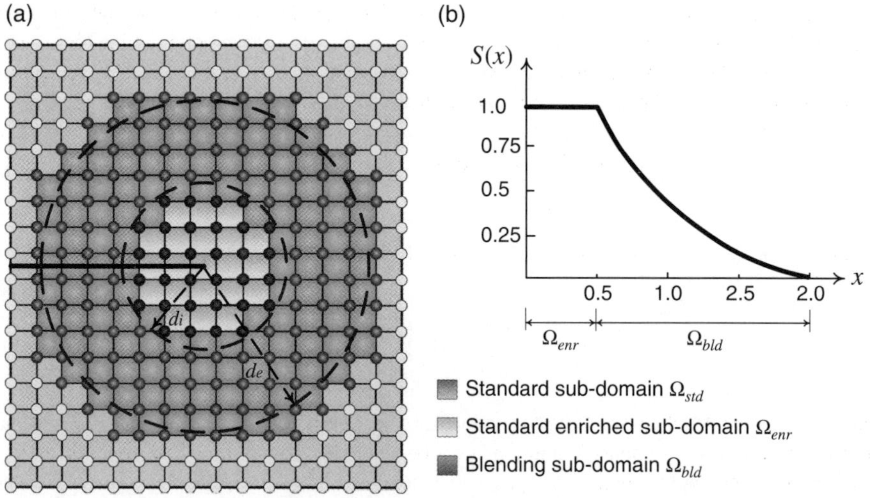

Figure 4.8 The weight function blending method: (a) Decomposition of the enrichment domain $\widehat{\Omega}_{enr}$ into the standard enriched sub-domain Ω_{enr}, where $S(\boldsymbol{x})=1$, and the blending sub-domain Ω_{bld}, where $0<S(\boldsymbol{x})<1$; (b) A typical polynomial weight function for $n=2$, $d_i=0.5$, and $d_e=3$

are allowed, and the PU property is satisfied in all elements. In this method, the enriched domain $\widehat{\Omega}_{enr}$ generally consists of the union of two sub-domains, as shown in Figure 4.8a; the standard enriched sub-domain Ω_{enr}, where $S(\boldsymbol{x})=1,\ \forall \boldsymbol{x}\in\Omega_{enr}$, and the blending sub-domain Ω_{bld}, where $0<S(\boldsymbol{x})<1,\ \forall \boldsymbol{x}\in \Omega_{bld}\backslash\,\Gamma_{bld}$ with $S(\boldsymbol{x})=0,\ \forall \boldsymbol{x}\in\widehat{\Gamma}_{enr}$ and $S(\boldsymbol{x})=1,\ \forall \boldsymbol{x}\in\Gamma_{enr}$. It is often desirable for Ω_{bld} to span several element layers.

In order to clarify the weight function $S(\boldsymbol{x})$ defined in (4.79), consider Γ_d be a discontinuity such as the crack body, or material interface. For each point $\boldsymbol{x}\in\widehat{\Omega}_{enr}$, the signed distance function $d=d(\boldsymbol{x})$ is defined from $\boldsymbol{x}$ to Γ_d. Thus, the weight function $S(d(\boldsymbol{x}))$ can be defined as

$$S(d(\boldsymbol{x}))=\begin{cases}1 & 0\le |d|\le d_i\\ (1-g)^n & d_i\le |d|<d_e\\ 0 & d_e\le |d|\end{cases}\tag{4.80}$$

where d_i and d_e are two constants which are chosen such that $S(d)=1$ for $|d|\le d_i$ and $S(d)=0$ for $|d|>d_e$; moreover, n is a positive integer and g is a linear ramp function defined as

$$g=\frac{|d|-d_i}{d_e-d_i}\tag{4.81}$$

in which a graphical representation of $S(d)$ is shown in Figure 4.8b for the values $n=2$, $d_i=0.5$, and $d_e=2$. An alternative definition of the weight function for more complex enrichments is to first define the sub-domain Ω_{enr} where $S(d)=1$; and then define the distances d, d_i, and d_e relative to the boundary of Ω_{enr} instead of Γ.

4.7.2 A Variant of the Cutoff Function Method

A similar approach to the weight function method was proposed by Chahine, Laborde, and Renard (2008) based on a variant of the cutoff function method that provides an optimal convergence rate in linear fracture problems with the X-FEM. In this method, a cutoff function is employed to localize the singular enrichments in order to reduce the number of the enrichment DOF and improve the conditioning of the linear system. Since one of the difficulties in blending elements is the mathematical analysis of the coupling between the two types of enrichment, it has been shown that an optimal convergence rate can be achieved with a lower computational cost compared to the standard X-FEM. In this method instead of employing the PU, the whole area around the crack tip is enriched using a cutoff function. The resulting enriched displacement field is defined as

$$\mathbf{u}^h(\boldsymbol{x}) = \sum_{I \in \mathcal{N}^{std}} N_I(\boldsymbol{x})\, \mathbf{u}_I + \sum_{J \in \mathcal{N}^{enr}_{Hev}} N_I(\boldsymbol{x})\, H(\boldsymbol{x})\, \mathbf{a}_J + \sum_{\alpha=1}^{4} \mathcal{B}_\alpha(\boldsymbol{x})\, \chi(\boldsymbol{x})\, \mathbf{b}_\alpha \tag{4.82}$$

where $\chi(\boldsymbol{x})$ is a cutoff function defined over the area of $0 < r_0 < r_1$ as

$$\chi(r(\boldsymbol{x})) = \begin{cases} 1 & r < r_0 \\ 0 < \chi(r) < 1 & r_0 < r < r_1 \\ 0 & r_1 < r \end{cases} \tag{4.83}$$

The proposed enrichment can be compared with the standard X-FEM, where the singular enrichment term of (4.82) is replaced by $\sum_{K \in \mathcal{N}^{enr}_{tip}} N_K(\boldsymbol{x}) \left(\sum_{\alpha=1}^{4} \mathcal{B}_\alpha(\boldsymbol{x})\, \mathbf{b}_{\alpha K} \right)$. It can be seen from (4.82) that only four DOF are added for each component of displacement field for the whole enriched area around the crack-tip. This reduces the computational cost and has a positive effect on the condition number.

4.8 A DG X-FEM Method

An alternative technique to circumvent the errors caused by parasitic terms in the approximation space of the blending elements at the edge of the enriched sub-domain is based on the DG X-FEM formulation. The DG method is a numerical technique used for solving partial differential equations that was originally introduced by Reed and Hill (1973) for hyperbolic neutron transport equations. The method has been successfully employed for hyperbolic and nearly hyperbolic problems, and then extended to purely elliptic problems. There are many variants of DG method which have different *consistency* and *stability* properties (Bassi and Rebay, 1997; Brezzi *et al.*, 1999; Arnold *et al.*, 2002). The DG method is somewhere between a finite element and a finite volume method that takes the advantages of both methods. It provides a practical framework for the development of high order accurate methods using unstructured grids. The method is well suited for large-scale time-dependent computations in which high accuracy is required. An important distinction between the DG method and the conventional FEM is that in the DG method the resulting equations are local to the generating element. The solution within each element is not reconstructed by looking to neighboring elements. Its compact formulation can be applied near boundaries without special treatment, which greatly increases the robustness and accuracy of any boundary condition implementation.

In the X-FEM, the singular enrichment functions are basically employed on a region that contains the crack tip, and are set to zero outside this region. As a result the enrichment functions are discontinuous at the interface between the two regions, and hence a DG approximation can be adopted therein to weakly enforce continuity. The idea was proposed by Gracie, Wang, and Belytschko (2008b) by decomposing the domain into enriched and unenriched sub-domains, in which the continuity between sub-domains is enforced with an internal penalty method. The method was similar to the discontinuous enrichment method (DEM) originally proposed by Farhat, Harari, and Franca (2001). In the DG-XFEM, the domain is decomposed into patches where enrichments are to be added; each patch is discretized independently, and enrichments are applied over entire patches but not over the entire domain. As a result the enrichment is local but does not require blending elements. Continuity between the patches is enforced in a weak sense using the internal penalty method. One drawback of this approach is that stabilization is required, and the stabilization parameter is problem dependent; so a few iterations are required to obtain a suitable value. Shen and Lew (2010a) employed the DG method to connect solutions on the domains with and without enrichment functions by adopting the Bassi–Rebay numerical fluxes. An important feature of this approach was that for isotropic, unstressed, linear elastic problems, the stabilization term is *problem independent.*

In order to describe the X-FEM within a framework of DG formulation, the domain Ω is decomposed into a set of non-overlapping patches Ω_β^P, where β varies between 1 and $\mathcal{M}_p$ with $\mathcal{M}_p$ denoting the number of patches. It is assumed that each patch β is enriched by the enrichment functions $\psi^{\alpha\beta}$; where $\alpha = 1$ to $\mathcal{M}_\beta^{enr}$ with $\mathcal{M}_\beta^{enr}$ denoting the number of enrichment functions for patch β. Consider that Γ_β^P is the boundary of patch β, and $\Gamma_{\beta\theta}$ denotes the intersection of the boundaries of patches β and θ. It must be noted that several patches can be enriched by the same enrichment function; this can occur when the domains of two enrichment functions overlap. In Figure 4.9, a typical example is presented that illustrates the decomposition of a domain Ω into four patches. In this figure, the enrichment functions ψ^1 and ψ^2 are applied over the sub-domains Ω_1^{enr} and Ω_2^{enr}, respectively. The boundaries of these sub-domains Ω_1^{enr} and Ω_2^{enr} decompose Ω into four patches, where patches Ω_1^P and Ω_2^P are enriched by ψ^1 and ψ^2, respectively, patch Ω_3^P is enriched by both ψ^1 and ψ^2, and patch Ω_4^P is unenriched.

Thus, the displacement approximation field on the patch β can be written as

$$\mathbf{u}^\beta(\boldsymbol{x}) = \sum_{I \in \mathcal{N}_\beta^P} N_I(\boldsymbol{x})\, \mathbf{u}_I^\beta + \sum_{I \in \mathcal{N}_\beta^P} N_I(\boldsymbol{x}) \sum_{\alpha=1}^{\mathcal{M}_\beta^{enr}} \psi^{\alpha\beta}(\boldsymbol{x})\, \mathbf{a}_I^{\alpha\beta} \tag{4.84}$$

where $\mathcal{N}_\beta^P$ is the set of all nodes in patch β. If the patch β is not enriched by any enrichment functions $\mathcal{M}_\beta^{enr} = 0$, and this equation simplifies to the standard FEM approximation. For those patches enhanced by the enrichment functions $\psi^{\alpha\beta}$, since all nodes of each patch are enriched, the PU is satisfied

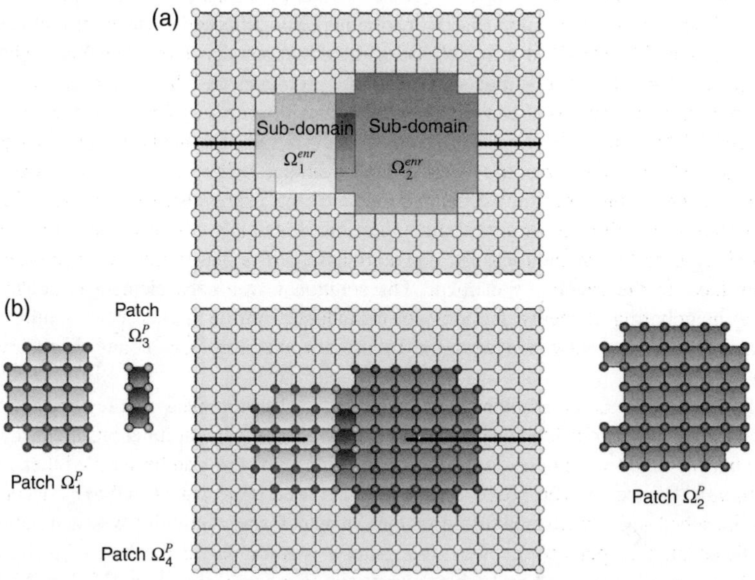

Figure 4.9 Discontinuous Galerkin X-FEM: (a) Illustration of a domain with two overlapping enrichment sub-domains; (b) Decomposition of the proposed sub-domains into four patches defined by the boundaries of the enrichment domain

everywhere on the patch. In order to enforce the continuity between the patches, the internal penalty method is employed. The displacement jump across $\Gamma_{\beta\theta}$ can be defined as

$$[\![\mathbf{u}^{\beta\theta}]\!] = \frac{1}{2}(\mathbf{u}^{\beta} - \mathbf{u}^{\theta}) \quad \beta = 1,\ldots,\mathcal{M}_p \text{ and } \theta = 1,\ldots,\mathcal{M}_p \tag{4.85}$$

and the average traction is defined by

$$\langle \mathbf{t}^{\beta\theta} \rangle = \frac{1}{2}(\mathbf{t}^{\beta} + \mathbf{t}^{\theta}) \quad \beta = 1,\ldots,\mathcal{M}_p \text{ and } \theta = 1,\ldots,\mathcal{M}_p \tag{4.86}$$

where $\mathbf{u}^{\beta}$ and $\mathbf{t}^{\beta} = \boldsymbol{\sigma}^{\beta}\cdot\mathbf{n}^{\beta}$ are the displacement and traction on the boundary of patch β, and $\mathbf{n}^{\beta}$ and $\boldsymbol{\sigma}^{\beta}$ are the normal vector and stress on the boundary of patch β.

The weak form of DG X-FEM formulation can therefore be written for all admissible trial and test function $\mathbf{u}^{\beta} \in \mathcal{U}^{\beta}$ and $\delta v^{\beta} \in \mathcal{V}^{\beta}$ as (Gracie, Wang, and Belytschko, 2008b)

$$\begin{aligned}\delta\Pi^{\mathrm{DG}} = \sum_{\beta=1}^{\mathcal{M}_p}\Bigg[&\int_{\Omega_{\beta}^{P}} \boldsymbol{\varepsilon}(\delta v^{\beta})\cdot\boldsymbol{\sigma}(\boldsymbol{\varepsilon}(\mathbf{u}^{\beta}))\,d\Omega \\ &+\sum_{\theta>\beta}^{\mathcal{M}_p}\left(\frac{\alpha}{A}\int_{\Gamma_{\beta\theta}}[\![\delta v^{\theta\beta}]\!]\,[\![\mathbf{u}^{\theta\beta}]\!]\,d\Gamma - \mu_1\int_{\Gamma_{\beta\theta}}\langle\delta\mathbf{t}^{\theta\beta}\rangle\,[\![\mathbf{u}^{\theta\beta}]\!]\,d\Gamma - \mu_2\int_{\Gamma_{\beta\theta}}[\![\delta v^{\theta\beta}]\!]\,\langle\mathbf{t}^{\theta\beta}\rangle\,d\Gamma\right)\Bigg] - \delta w^{ext} = 0\end{aligned} \tag{4.87}$$

where A is a measure of the domain of an element. In the above equation, it is assumed that $\mu_1 = \mu_2 = 1$, and α is taken as a penalty-like parameter that depends on the problem and it is multiplied into a stabilization term. The space of all admissible trial and test functions are defined as

$$\mathcal{U}^{\beta} = \left\{\mathbf{u}\,\middle|\,\mathbf{u}\in\mathcal{H}^1\left(\Omega_{\beta}^{P}\right),\ \mathbf{u} = \bar{\mathbf{u}} \text{ on } \Gamma_u\right\} \tag{4.88a}$$

$$\mathcal{V}^{\beta} = \left\{v\,\middle|\,v\in\mathcal{H}^1\left(\Omega_{\beta}^{P}\right),\ v = 0 \text{ on } \Gamma_u\right\} \tag{4.88b}$$

where $\mathcal{H}^1\left(\Omega_{\beta}^{P}\right)$ is the Hilbert space of order one defined on domain Ω_{β}^{P}.

Finally, the discrete DG X-FEM equations can be obtained for small strain elastic problems as

$$\left(\mathbf{K}^{\mathrm{XFEM}} + \mathbf{K}^{\mathrm{DG}}\right)\mathbf{d} = \mathbf{f}^{ext} \tag{4.89}$$

where $\mathbf{d}$ is the vector of nodal DOFs defined as $\mathbf{d}^T = \left\{\mathbf{d}_1^1, \mathbf{d}_2^1, \ldots, \mathbf{d}_{\mathcal{N}_1^P}^1, \ldots, \mathbf{d}_1^{\mathcal{M}_p}, \mathbf{d}_2^{\mathcal{M}_p}, \ldots, \mathbf{d}_{\mathcal{N}_{\mathcal{M}_p}^P}^{\mathcal{M}_p}\right\}$ with $\mathcal{N}_{\beta}^{P}$ denoting the number of nodes in patch β. For node I of patch β the nodal vector is given by $\left(\mathbf{d}_I^{\beta}\right)^T = \left\{\mathbf{u}_I^{\beta}, (\mathbf{a}_I^1)^{\beta}, \ldots, \left(\mathbf{a}_I^{\mathcal{M}_{\beta}^{enr}}\right)^{\beta}\right\}$. In Eq. (4.89), the stiffness matrices $\mathbf{K}^{\mathrm{XFEM}}$ and $\mathbf{K}^{\mathrm{DG}}$ are defined as

$$\mathbf{K}^{\mathrm{XFEM}} = \sum_{\beta=1}^{\mathcal{M}_p}\mathbf{K}^{P\beta} \quad \text{with} \quad \mathbf{K}_{IJ}^{P\beta} = \int_{\Omega_e}\mathbf{B}_I^T\mathbf{C}\mathbf{B}_J\,d\Omega \quad I,J\in\mathcal{N}_{\beta}^{P} \tag{4.90a}$$

$$\mathbf{K}^{\mathrm{DG}} = \sum_{\beta=1}^{\mathcal{M}_p}\sum_{\theta>\beta}^{\mathcal{M}_p}\left[\mathbf{K}_{\alpha}^{\beta\theta} - \mathbf{K}_{\mu}^{\beta\theta} - \left(\mathbf{K}_{\mu}^{\beta\theta}\right)^T\right] \tag{4.90b}$$

where

$$\mathbf{K}_{\alpha\,IJ}^{\beta\theta} = \frac{\alpha}{4A}\int_{\Gamma_{\beta\theta}} \left(\bar{\mathbf{N}}_{IJ}^{\beta\theta}\right)^T \mathbf{C}\bar{\mathbf{N}}_{IJ}^{\beta\theta}\, d\Gamma \qquad I \in \mathcal{N}_{\beta}^{P},\ J \in \mathcal{N}_{\theta}^{P} \tag{4.91a}$$

$$\mathbf{K}_{\mu\,IJ}^{\beta\theta} = \frac{1}{4}\int_{\Gamma_{\beta\theta}} \left(\bar{\mathbf{B}}_{IJ}^{\beta\theta}\right)^T \mathbf{C}\bar{\mathbf{N}}_{IJ}^{\beta\theta}\, d\Gamma \qquad I \in \mathcal{N}_{\beta}^{P},\ J \in \mathcal{N}_{\theta}^{P} \tag{4.91b}$$

$$\bar{\mathbf{B}}_{\mu\,IJ}^{\beta\theta}(\boldsymbol{x}) = \begin{cases} \left[\mathbf{B}_I^{\beta}\cdot\mathbf{n}^{\beta},\ \mathbf{B}_J^{\theta}\cdot\mathbf{n}^{\theta}\right] & \text{if } \boldsymbol{x}\in \text{supp}(I) \text{ and } \boldsymbol{x}\in\text{supp}(J) \\ 0 & \text{if } \boldsymbol{x}\notin \text{supp}(I) \text{ and } \boldsymbol{x}\notin\text{supp}(J) \end{cases} \tag{4.91c}$$

$$\bar{\mathbf{N}}_{\mu\,IJ}^{\beta\theta}(\boldsymbol{x}) = \begin{cases} \left[\mathbf{N}_I^{\beta},\ -\mathbf{N}_J^{\theta}\right] & \text{if } \boldsymbol{x}\in \text{supp}(I) \text{ and } \boldsymbol{x}\in\text{supp}(J) \\ 0 & \text{if } \boldsymbol{x}\notin \text{supp}(I) \text{ and } \boldsymbol{x}\notin\text{supp}(J) \end{cases} \tag{4.91d}$$

where supp(I) is the support of node I. The matrices $\mathbf{N}_I^{\beta}$ and $\mathbf{B}_I^{\beta}$ are defined in the standard manner for node I of patch β as $\mathbf{u}^{\beta} = \sum_{I\in\mathcal{N}_{\beta}^{P}} \mathbf{N}_I^{\beta}\mathbf{d}_I^{\beta}$ and $\boldsymbol{\varepsilon}^{\beta} = \nabla^s\mathbf{u}^{\beta} = \sum_{I\in\mathcal{N}_{\beta}^{P}} \mathbf{B}_I^{\beta}\mathbf{d}_I^{\beta}$, in which $\mathbf{N}_I^{\beta}$ and $\mathbf{B}_I^{\beta}$ can be decomposed into the standard and enriched parts as $\mathbf{N}_I^{\beta} = \left[\mathbf{N}_I^{std,\beta}, \mathbf{N}_I^{enr,\beta}\right]$ and $\mathbf{B}_I^{\beta} = \left[\mathbf{B}_I^{std,\beta}, \mathbf{B}_I^{enr,\beta}\right]$. In a two-dimensional problem, the matrices $\mathbf{N}_I^{std,\beta}$, $\mathbf{N}_I^{enr,\beta}$, $\mathbf{B}_I^{std,\beta}$, and $\mathbf{B}_I^{enr,\beta}$ are defined as

$$\mathbf{N}_I^{std,\beta} = \begin{bmatrix} N_I & 0 \\ 0 & N_I \end{bmatrix} \tag{4.92a}$$

$$\mathbf{N}_I^{enr,\beta} = \begin{bmatrix} \psi^{1\beta}N_I & 0 & \cdots & \psi^{\mathcal{M}_{\beta}^{enr}\beta}N_I & 0 \\ 0 & \psi^{1\beta}N_I & \cdots & 0 & \psi^{\mathcal{M}_{\beta}^{enr}\beta}N_I \end{bmatrix} \tag{4.92b}$$

$$\mathbf{B}_I^{std,\beta} = \begin{bmatrix} N_{I,x} & 0 \\ 0 & N_{I,y} \\ N_{I,y} & N_{I,x} \end{bmatrix} \tag{4.92c}$$

$$\mathbf{B}_I^{enr,\beta} = \begin{bmatrix} \left(\psi^{1\beta}N_I\right)_{,x} & 0 & \cdots & \left(\psi^{\mathcal{M}_{\beta}^{enr}\beta}N_I\right)_{,x} & 0 \\ 0 & \left(\psi^{1\beta}N_I\right)_{,y} & \cdots & 0 & \left(\psi^{\mathcal{M}_{\beta}^{enr}\beta}N_I\right)_{,y} \\ \left(\psi^{1\beta}N_I\right)_{,y} & \left(\psi^{1\beta}N_I\right)_{,x} & \cdots & \left(\psi^{\mathcal{M}_{\beta}^{enr}\beta}N_I\right)_{,y} & \left(\psi^{\mathcal{M}_{\beta}^{enr}\beta}N_I\right)_{,x} \end{bmatrix} \tag{4.92d}$$

in which the matrix $\mathbf{K}^{\text{XFEM}}$ is constructed by assembling the standard and enriched X-FEM stiffness matrices of each patch, and $\mathbf{K}^{\text{DG}}$ is produced using the DG terms that enforce the continuity between the patches.

This computational algorithm is known as the *patch-based* DG-XFEM. Gracie, Wang, and Belytschko (2008b) employed this technique over each element in the mesh, which is called as the *element-based* DG-XFEM. In the element-based DG-XFEM, each element in the mesh is treated as a patch as in the standard DG formulations; hence, the nodes of each element are independent and the DG terms are applied over all element edges. As noted earlier, one drawback of the patched based DG-XFEM is that it needs stabilization and the stabilization parameter is problem dependant. Shen and Lew (2010a) proposed a different version of DG-XFEM by adopting the Bassi–Rebay numerical flux in which the stabilization term is *problem independent.* In this method, the two asymptotic displacement fields are included near the crack-tip for modes I and II fracture as the enrichment functions on a region in the vicinity of the crack tip and are set to zero everywhere else. In order to handle the discontinuous displacement field between the enriched and unenriched sub-domains, the continuity between sub-domains is enforced based on the Bassi–Rebay numerical flux, which is both stable and consistent.

4.9 Implementation of Some Optimal X-FEM Type Methods

In order to illustrate the performance of some proposed optimal X-FEM type methods described in preceding sections and to have a better understanding of their robustness, three numerical examples are presented here. The examples are solved using the standard X-FEM (STD XFEM), the X-FEM with modified enrichment function introduced by Moës *et al.* (2003) (MOES), the hierarchical method by Chessa, Wang, and Belytschko (2003) (CHESSA), and the corrected X-FEM technique by Fries (2008) (FRIES), and the results are compared for three bimaterial problems. It was highlighted that the hierarchical method can be efficiently used to eliminate the parasitic terms in blending elements of polynomial order, however, it has two considerable weaknesses; the first is the insertion of extra DOF that have to be tailored to every special problem, and the next one is that it is well suited for polynomial enrichments. The corrected X-FEM technique was presented as an efficient approach that can be used successfully for any type of enrichment functions to eliminate the parasitic effects and to achieve an optimum convergence rate.

The modified enrichment function proposed by Moës *et al.* (2003) is an alternative approach used to improve the accuracy and convergence rate of the standard X-FEM while dealing with the weak discontinuities, such as holes, inclusions, and bimaterial problems, where the displacement field is continuous but its derivatives are discontinuous. For such problems, the absolute value of the level set function $\varphi(\boldsymbol{x})$ can be used as $\psi(\boldsymbol{x}) = |\varphi(\boldsymbol{x})|$ that has a discontinuous first derivative on the material interface. It must be noted that the values of level-set function are only computed at the nodal points of FE mesh, so the level-set function $\varphi(\boldsymbol{x})$ can be approximated based on the values of nodal points φ_I using the interpolation functions $N_I(\boldsymbol{x})$ as

$$\psi_1(\boldsymbol{x}) = \left| \sum_I N_I(\boldsymbol{x}) \varphi_I \right| \tag{4.93}$$

It was shown by Sukumar *et al.* (2001) and Belytschko *et al.* (2003a) that a smoothing function of $\psi_1(\boldsymbol{x})$ away from the element layer containing the interface, somewhat improves the accuracy and convergence rate. Moës *et al.* (2003) proposed an alternative definition of the absolute value of level set function $\varphi(\boldsymbol{x})$ for weak discontinuity problems that shows a better accuracy and convergence rate as

$$\psi_2(\boldsymbol{x}) = \sum_I N_I(\boldsymbol{x}) |\varphi_I| - \left| \sum_I N_I(\boldsymbol{x}) \varphi_I \right| \tag{4.94}$$

The main advantage of this enrichment function is that it has a non-zero value only in the elements cut by the interface. This enrichment function has a ridge centered on the interface and zero value on the elements that are not crossed by the interface, as shown in Figure 4.10.

The first example is a plate with a circular hole at its center under uniaxial tension chosen to demonstrate the performance of X-FEM for a bimaterial problem. In order to obtain the convergence rate of solution for various approaches, the norm of energy is compared between the numerical analysis and that obtained from the analytical solution. The next example is a plate with a material interface at its center chosen to illustrate the performance of various X-FEM approaches in a large deformation problem. The last example is chosen to illustrate the performance of X-FEM technique in a practical engineering problem of the fiber reinforced concrete. All three examples are simulated in a plane strain condition, and the acceptable tolerance of residual is set to 10^{-14}. The convergence analysis is performed for all three examples to investigate the accuracy and convergence rate of each approach. In the first example the X-FEM analysis is

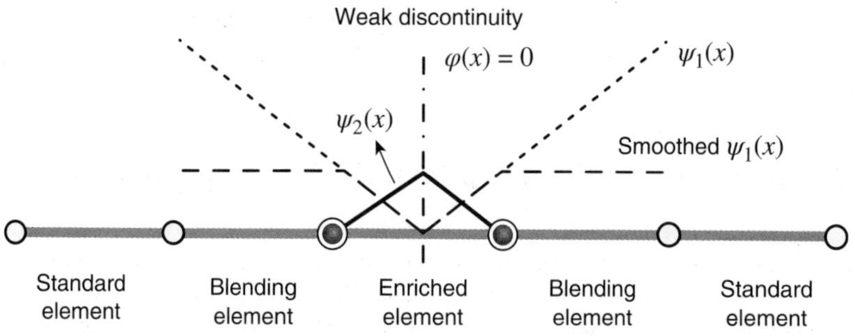

Figure 4.10 One-dimensional representation of various enrichment functions for the material interface problem

compared with the analytical solution, while for the next two examples a very fine FE mesh is carried out as the reference solution to obtain the error of the energy norm. The energy norm used for evaluation of the error is defined as

$$e_E=\left(\int_\Omega \left(\boldsymbol{\varepsilon}-\boldsymbol{\varepsilon}^h\right)^T \mathbf{D}\left(\boldsymbol{\varepsilon}-\boldsymbol{\varepsilon}^h\right)d\Omega\right)^{1/2} \tag{4.95}$$

where $\boldsymbol{\varepsilon}$ and $\boldsymbol{\varepsilon}^h$ are the exact and numerical values of strain tensor, respectively, and $\mathbf{D}$ is the material property tensor.

4.9.1 A Plate with a Circular Hole at Its Centre

The first example is a plate with a traction-free circular hole at its center that is subjected to a uniaxial tension, as shown in Figure 4.11. The analytical solution is available for this example as given by Szabó and Babuška (1991), and the X-FEM analysis was originally performed by Sukumar *et al.* (2001). In order to model an infinite plate, the X-FEM analysis is carried out using a square plate of 2 × 2 cm with a circular hole of radius $a=0.4$ cm at its center. The plate is subjected to a uniform uniaxial tension of $\sigma_0=1$ kg/cm^2 along x-direction. In order to remove the rigid body modes, the exact tractions and appropriate constraints are imposed on the boundary of the plate. The material properties of the plate are as follows; a Young modulus of $E=1\times10^5$ kg/cm^2 and a Poisson ratio of $\nu=0.3$. In order to avoid singular stiffness matrices in the X-FEM analysis, the circular hole is assumed to be filled with a soft material as $E=1\times10^2$ kg/cm^2.

The displacement fields of the plate can be obtained from the analytical solution in a polar coordinates system (r,θ) under a uniaxial tension of $\sigma_0=1$kg/cm^2 as

$$u(r,\theta)=\frac{a}{8\mu}\left[\frac{r}{a}(\kappa+1)\cos\theta+2\frac{a}{r}((1+\kappa)\cos\theta+\cos 3\theta)-2\frac{a^3}{r^3}\cos 3\theta\right] \tag{4.96a}$$

$$v(r,\theta)=\frac{a}{8\mu}\left[\frac{r}{a}(\kappa-3)\sin\theta+2\frac{a}{r}((1-\kappa)\sin\theta+\sin 3\theta)-2\frac{a^3}{r^3}\sin 3\theta\right] \tag{4.96b}$$

where μ is the shear modulus, and κ is the Kolosov constant defined as $\kappa=3-4\nu$ for plane strain and $\kappa=(3-\nu)/(1+\nu)$ for plane stress problems. The analytical stress components can be obtained as

$$\sigma_x(r,\theta)=\frac{1}{2}\left[1-\frac{a^2}{r^2}+\left(1-4\frac{a^2}{r^2}+3\frac{a^4}{r^4}\right)\cos 2\theta\right] \tag{4.97a}$$

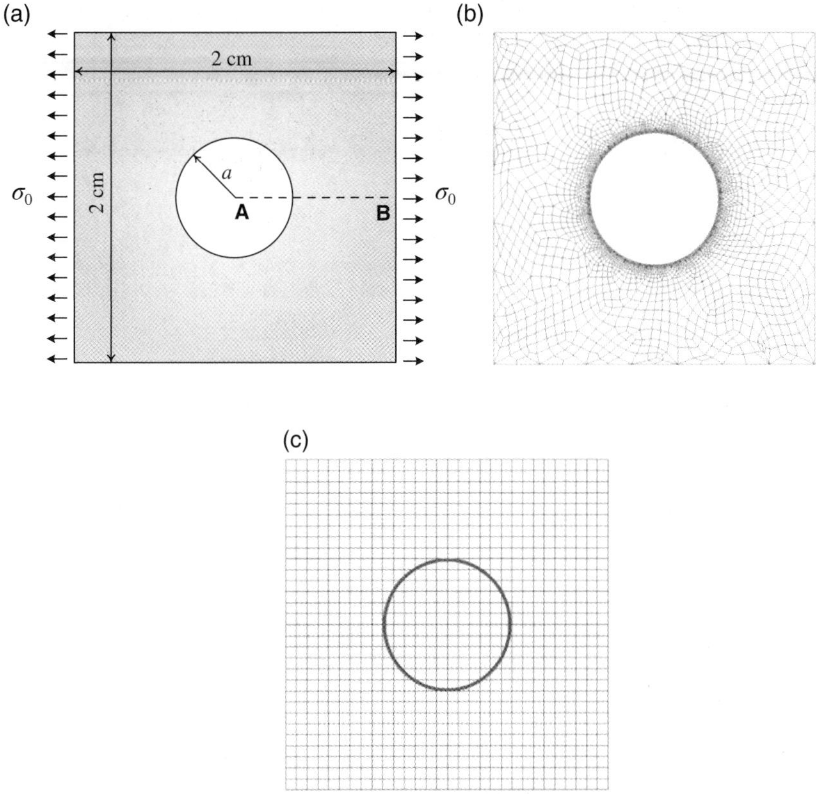

Figure 4.11 A plate with a circular hole at its center: (a) problem definition; (b) the FEM mesh; (c) the X-FEM mesh of 30×30 elements

$$\sigma_y(r,\theta) = \frac{1}{2}\left[1 + \frac{a^2}{r^2} - \left(1 + 3\frac{a^4}{r^4}\right)\cos 2\theta\right] \tag{4.97b}$$

$$\tau_{xy}(r,\theta) = \frac{1}{2}\left[-1 - 2\frac{a^2}{r^2} + 3\frac{a^4}{r^4}\right]\sin 2\theta \tag{4.97c}$$

In order to perform a convergence study, the X-FEM analyses are carried out using uniform structured meshes of 30×30, 40×40, 50×50, 60×60, 70×70, and 80×80. In Figure 4.12, the distributions of stress contours σ_x and σ_y are presented for the FEM and standard X-FEM. Obviously, a reasonable agreement can be seen between two techniques. In Figure 4.13, the errors of the energy norm are plotted on a log-log plot using various blending strategies. This graph illustrates that the induced error is negligible and all approaches result in an optimal convergence rate.

Finally, the evolutions of stress σ_x and σ_y along the horizontal line passing through the center of the plate are plotted in Figure 4.14 for various blending strategies.

4.9.2 A Plate with a Horizontal Material Interface

The next example is chosen to illustrate the performance of various blending strategies for a large deformation problem of a plate with horizontal material interface subjected to a uniaxial compression,

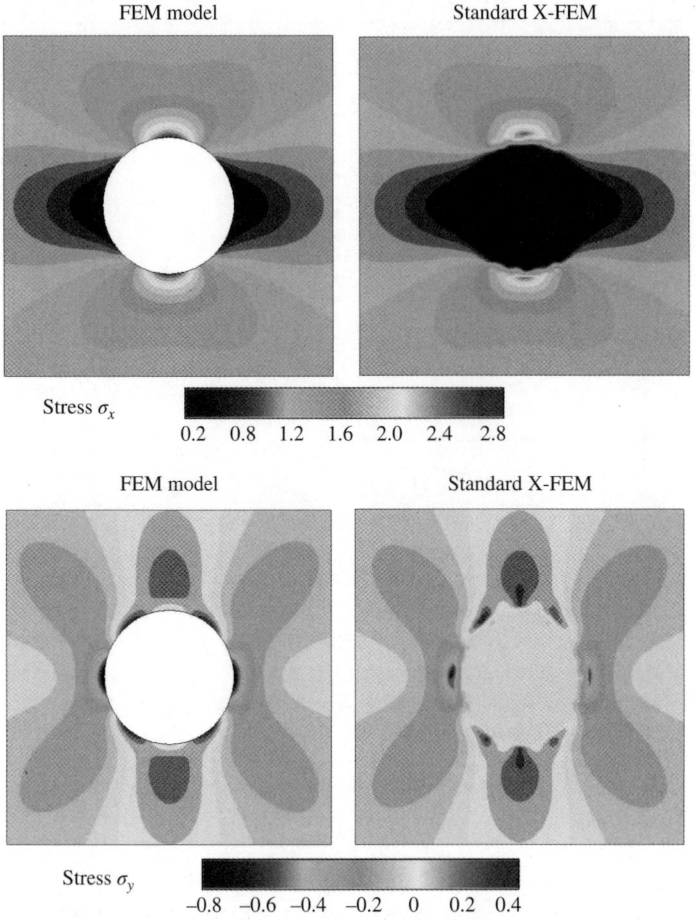

Figure 4.12 The distributions of stress σ_x and σ_y contours for a plate with a circular hole: A comparison between the FEM and standard X-FEM analyses

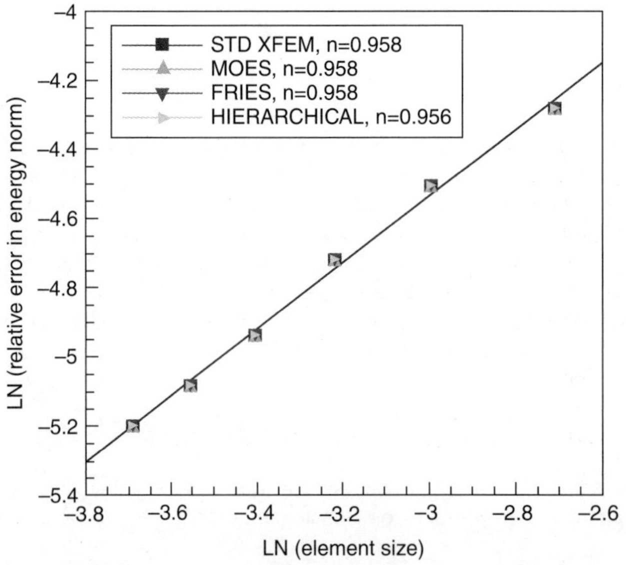

Figure 4.13 The error of the energy norm with various blending strategies for a plate with a circular hole

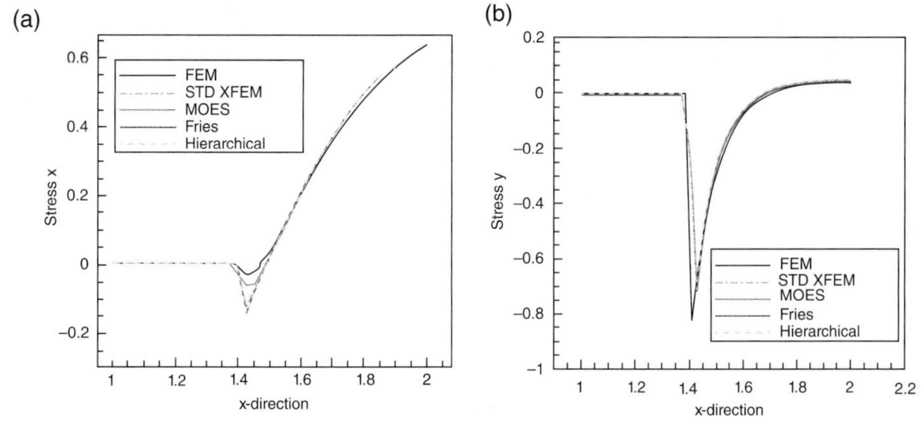

Figure 4.14 The evolutions of stress σ_x and σ_y along the center line using various blending strategies for a plate with a circular hole

as shown in Figure 4.15. A square plate of 2×2 m is modeled in plane strain condition that is composed of two materials with the Young modulus of $E_1 = 1 \times 10^5$ kN/m^2 for the upper part and $E_2 = 1 \times 10^2$ kN/m^2 for the lower part. The Poisson ratio is assumed to be the same for both materials as $\nu_1 = \nu_2 = 0.3$. The plate is restrained at the bottom edge and a uniform compaction with the intensity of $q = 55$ kN/m is applied at the top edge. A convergence study is carried out for this example by evaluation of the relative error of energy norm, and the rate of convergence is compared with an optimal convergence rate. To perform the convergence study, five uniform structured meshes of 20×20, 40×40, 50×50, 60×60, and 80×80 are employed. Since the exact solution is not available for a comparison, a finite element analysis with a fine mesh of 90×90 is carried out as a reference solution. In Figure 4.16, the distributions of the stress σ_y contour are shown for the FEM and X-FEM technique with various blending strategies. Obviously, the effect of parasitic terms in the blending layer is obvious in the standard X-FEM while other approaches are in good agreement with the FEM. In Figure 4.17, the errors of the energy norm are plotted on a log-log plot using various blending strategies. Remarkable improvement can be seen for all modified X-FEM techniques compared to the standard X-FEM with a sub-optimal convergence rate of 0.492. It is noteworthy to highlight that while all proposed approaches result in almost optimum convergence rate, the hierarchical method presents a sub-optimal convergence rate of 0.795.

4.9.3 The Fiber Reinforced Concrete in Uniaxial Tension

In the last example, a practical engineering problem is investigated by modeling the fiber reinforced composite in a uniaxial tension loading. The optimal X-FEM type methods are carried out to illustrate the performance of various blending strategies in modeling of this complex geometry. In Figure 4.18a, the geometry of a representative volume element is presented, where the circular fibers are distributed with random spatial distributions. It is assumed that the composite is made up of 20% volume of fibers embedded in an epoxy with no relative displacement on the interface of fiber and epoxy. In Figure 4.18b and c, the finite element mesh of the composite lamina is shown together with a typical structured mesh of linear quadrilateral elements used for the X-FEM analyses. An elastic square plate of 2×2 m with a Young modulus of $E = 1 \times 10^4$ kN/m^2 and a Poisson ratio of $\nu = 0.3$ is modeled in a plane strain condition that is reinforced with the circular fibers of radius $R = 0.15$ m with the Young modulus of $E = 1 \times 10^5$ kN/m^2 and Poisson ratio of $\nu = 0.3$. The plate is subjected to the uniaxial tension loading $q = 8 \times 10^3$ N/m.

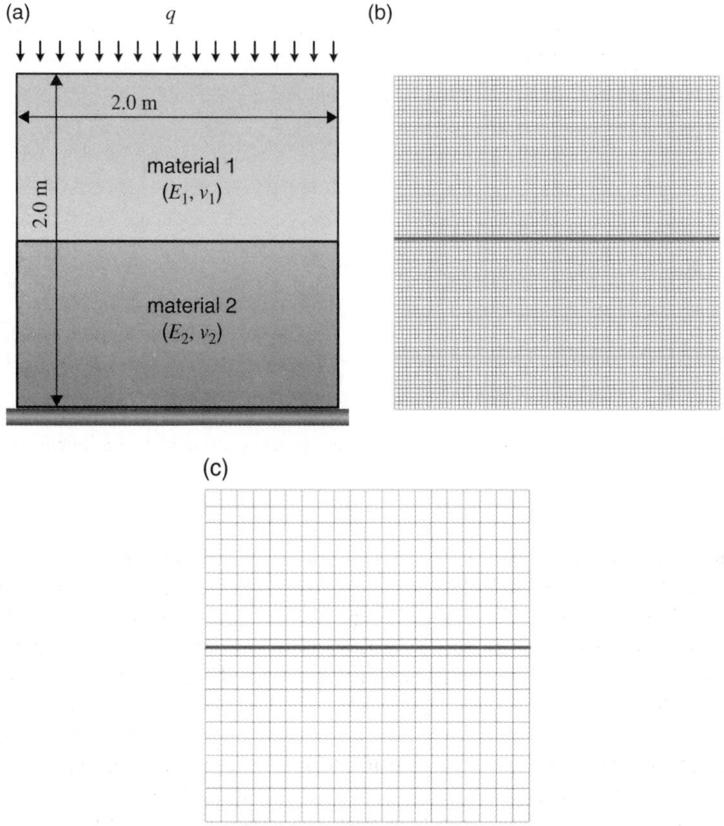

Figure 4.15 A plate with horizontal material interface: (a) Problem definition; (b) the FEM mesh of 90 × 90 elements (c) the X-FEM mesh of 20 × 20 elements

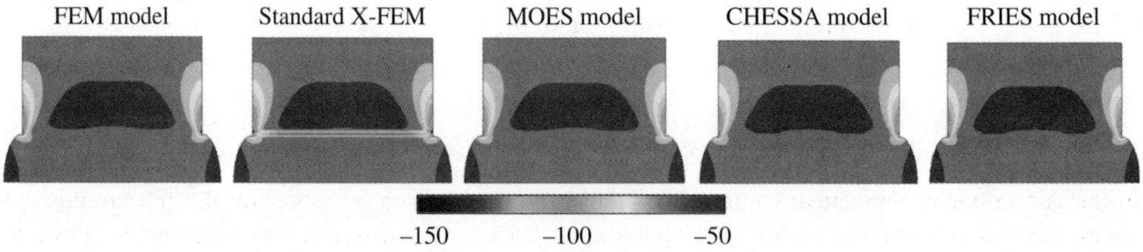

Figure 4.16 The distributions of stress σ_y contour for a plate with horizontal material interface: A comparison between the FEM and X-FEM techniques with various blending strategies

A convergence study is carried out for this example by evaluation of the relative error of energy norm, and the rate of convergence is compared with the optimal convergence rate. To perform the convergence study, four uniform structured meshes of 40 × 40, 50 × 50, 70 × 70, and 80 × 80 are employed. Since the exact solution is not available for a comparison, a FE analysis with a fine mesh of 9134 elements is carried out as a reference solution. In Figure 4.19a, the evolutions of displacement u_x are plotted for the mesh of

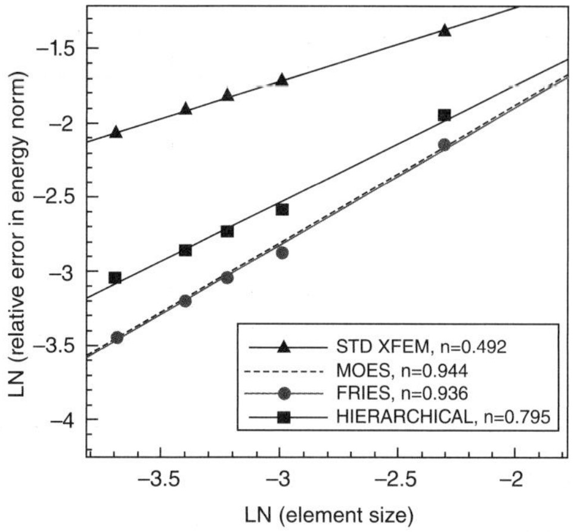

Figure 4.17 The error of the energy norm with various blending strategies for a plate with horizontal material interface

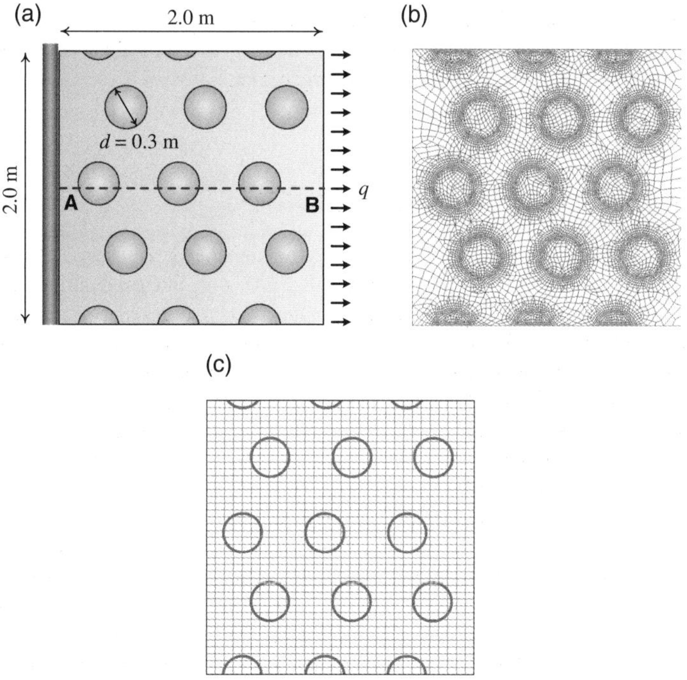

Figure 4.18 The fiber reinforced concrete in uniaxial tension: (a) Problem definition; (b) FEM mesh (c) the X-FEM mesh of 40 × 40 elements

70 × 70 along the line passing through the center of the specimen using various blending strategies. Clearly, the effect of parasitic terms in the blending layer can be observed in the standard X-FEM. In Figure 4.19b, the errors of the energy norm are plotted on a log-log plot using various blending strategies. Obviously, a sub-optimal convergence rate of 0.492 can be seen from the standard X-FEM while other

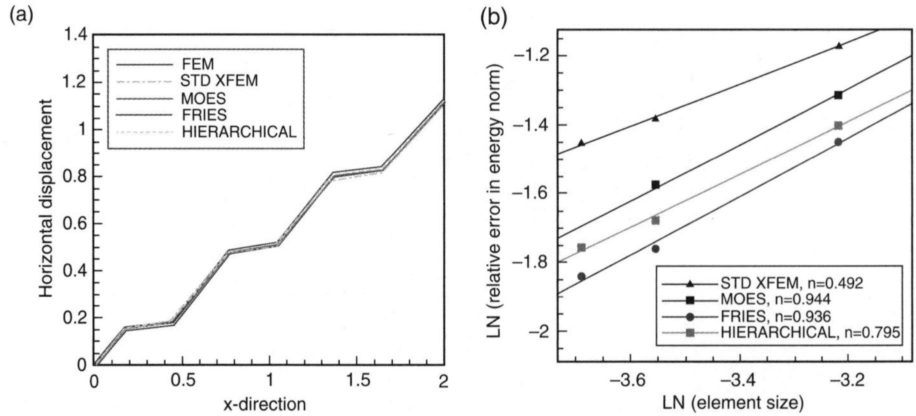

Figure 4.19 A fiber reinforced concrete in uniaxial tension: (a) The evolutions of displacement u_x along the line passing through the center of the specimen; (b) The error of the energy norm using various blending strategies

proposed approaches improve the convergence rate of the solution, however, the hierarchical approach of Chessa, Wang, and Belytschko (2003) still presents a sub-optimal convergence rate of 0.795. It can be highlighted that the modified enrichment function of Moës *et al.* (2003) and the corrected X-FEM technique of Fries (2008) result in almost optimum convergence rate.

4.10 Pre-Conditioning Strategies in X-FEM

The problem of ill-conditioning in the X-FEM was demonstrated in Section 4.3 through a one-dimensional example. It has been shown that in the X-FEM with independent enrichment functions, multiplication of the enrichment functions with the PU functions may result in dependent functions that may lead to badly conditioned stiffness matrices. One remedy to circumvent this problem is to allow only a certain number of elements inside the enrichment radius, or to avoid the interface enrichment if the interface is too close to the element edges. However, from a more general perception a procedure to eliminate the negative effects of the enrichments on the condition number in a computationally efficient way is desirable. Laborde *et al.* (2005) proposed a technique based on the *DOF gathering* method to decrease the number of unknowns and the condition number in fracture problems. They introduced the "*geometrical*" enrichment strategy to achieve the optimal convergence rate in the X-FEM solution by enriching the elements that are contained in a predefined area. However, adding singular functions to all nodes of a fixed area around the crack tip increases the number of DOF and results in enormous increase of the condition number of the system of equations. In the *DOF gathering* method, a bonding condition is introduced on the enrichment area around the crack-tip, in which for each singular shape function, the equality of the corresponding DOF is prescribed. In fact, all DOF related to each asymptotic function are assumed to be equal in the predefined area, that is, instead of eight enriched DOF for each node, eight DOF are assumed for the whole enriched area. Although the numerical tests highlighted that the condition number is significantly improved, a reduction of the convergence rate was observed in the X-FEM solution. It was shown that the lack of accuracy comes from the elements in the transition layer, at the boundary of the enriched zone. They defined the PU independently of the mesh from an overlapping domain by two patches, where the crack tip lies in the interior of one patch and showed that an optimal convergence rate can be achieved. The *DOF gathering* method is a popular technique and has been widely used in conjunction with the blending strategies to improve the X-FEM performance; for example, Tarancon *et al.* (2009) employed this method with the enhanced strain method, and Shen and Lew (2010a) applied the technique with the DG X-FEM technique.

An efficient technique that can be used to improve the conditioning issues in the X-FEM is based on the *pre-conditioning* strategy. Béchet *et al.* (2005) proposed a pre-conditioner for the X-FEM solution that stabilizes the enrichments by applying the Cholesky decompositions to certain sub-matrices of the stiffness matrix. Since the problem of almost linearly dependent enrichment functions was eliminated for each node, these sub-matrices were formed by the DOF associated with each enriched nodal point which can be considered as a local stabilization. However, there are still situations where the enrichment functions associated with several nodal points become almost linearly dependent. An alternative solution for this problem was presented by Menk and Bordas (2011) based on a domain decomposition in which additional constraints were used to ensure the continuity of the solution. In this manner, the Cholesky decomposition is employed together with the LQ decomposition. It was shown that the condition numbers obtained by this method are close to that of the corresponding FEM matrices without enrichment, and the convergence of the iterative solver improves significantly.

4.10.1 Béchet's Pre-Conditioning Scheme

A pre-conditioner matrix, in its simplest form, is a matrix that transforms a system of equations into an alternative equivalent system with a lower condition number that increases the speed of the solution. Since the implementation of a pre-conditioner in an iterative method incurs some extra cost, both from the point of constructing and applying it to each iteration, there must be a trade-off between the cost of constructing and applying the pre-conditioner, and the gain in convergence speed. For a symmetric positive definite system of equation $\mathbf{Ku} = \mathbf{f}$, a pre-conditioned system can be defined as

$$\underbrace{\mathbf{P}^T\mathbf{K}\mathbf{P}}_{\widehat{\mathbf{K}}}\underbrace{\mathbf{P}^{-1}\mathbf{u}}_{\widehat{\mathbf{u}}} = \underbrace{\mathbf{P}^T\mathbf{f}}_{\widehat{\mathbf{f}}} \tag{4.98}$$

where $\mathbf{P}$ is a pre-conditioner matrix. This preconditioning scheme retains the symmetry of the coefficient matrix $\widehat{\mathbf{K}}$. Once the system of equation (4.98) is solved, the solution $\mathbf{u}$ can be obtained from $\mathbf{u} = \mathbf{P}\,\widehat{\mathbf{u}}$. It is important to choose a proper pre-conditioner matrix $\mathbf{P}$ with a minimum computational cost such that the condition number of the transformed stiffness matrix $\widehat{\mathbf{K}}$ becomes smaller than the stiffness matrix $\mathbf{K}$. As a result, an iterative solution with a lower number of iterations can be used to obtain an acceptable result.

Béchet *et al.* (2005) proposed a specialized pre-conditioner for the X-FEM that uses a local (nodal) Cholesky based decomposition. The scheme is aimed to take advantage of the knowledge of the enrichment to produce linear system of equations that are easier to solve with "off the shelf" pre-conditioners and solvers. This type of pre-conditioner attempts to orthogonalize the finite element basis generated for each standard DOF. Consider the linear system of equations $\mathbf{Ku} = \mathbf{f}$ as

$$\begin{bmatrix} \ddots & \vdots & \vdots & \\ & a & b & \cdots \\ & b & c & \cdots \\ & & & \ddots \end{bmatrix}\mathbf{u} = \mathbf{f} \tag{4.99}$$

in which there is only one enrichment function assumed here, and thus for each enriched node, one enriched DOF is related to each standard DOF, that is, the size of the sub-matrix becomes 2×2 representing terms, a, b, and c. In order to perform the orthogonality condition between the standard and enriched terms in matrix $\mathbf{K}$, that is, the term b becomes zero, a modified linear system of equation needs to be solved as

$$\begin{bmatrix} \ddots & \vdots & \vdots & \\ & a & 0 & \cdots \\ & 0 & c & \cdots \\ & & & \ddots \end{bmatrix} \widehat{\mathbf{u}} = \widehat{\mathbf{f}} \tag{4.100}$$

To obtain this system of equations, a Cholesky decomposition can be performed on the proposed sub-matrix; this decomposition can then be used to pre- and post-multiply the appropriate terms into Eq. (4.99). The matrix is therefore completed with diagonal "ones" to match the size of the original system. Of course the implementation is not exactly as described for obvious performance issues, but it is mathematically equivalent. Let **A** be the sub-matrix needs to be orthogonalized as

$$\mathbf{A} = \begin{bmatrix} a & b \\ b & c \end{bmatrix} \tag{4.101}$$

The Cholesky decomposition of **A** can be defined as

$$\mathbf{A} = \mathbf{C}\,\mathbf{C}^T \tag{4.102}$$

where **C** is the lower triangular matrix. In order to avoid the diagonal terms with very different magnitudes, the inverse of the decomposition is scaled. Let **D** be the scaling matrix which is a diagonal matrix constructed by the square root of the diagonal values of **A** as

$$D_{ij} = \sqrt{A_{ij}\delta_{ij}} \quad \text{(no summation)} \tag{4.103}$$

then

$$\mathbf{P} = \mathbf{C}^{-1}\mathbf{D} \tag{4.104}$$

where **P** is the pre-conditioning sub-matrix for only one set of enriched DOFs. The full pre-conditioner matrix $\mathbf{P}^*$ for the whole system can then be constructed from these sub-matrices **P**. The resulting system of equation can therefore be obtained as

$$\mathbf{P}^*\mathbf{K}\,\mathbf{P}^{*T}\,\widehat{\mathbf{u}} = \mathbf{P}^*\mathbf{f} \tag{4.105}$$

in which $\mathbf{u} = \mathbf{P}^{*T}\widehat{\mathbf{u}}$. One advantage of this method is that it can be applied to any kind of enrichment function.

4.10.2 Menk–Bordas Pre-Conditioning Scheme

The Béchet pre-conditioning scheme removes the linear dependency for each node locally, however, there are situations where the enrichment functions associated with several nodal points become almost linear dependent. In order to overcome this difficulty, Menk and Bordas (2011) proposed a pre-conditioning strategy that produces the well-conditioned stiffness matrices for the X-FEM solution. In this approach, the condition number is always close to the one of the corresponding FEM stiffness matrices without any enrichment. The method is very close to the FE tearing and interconnecting method proposed by Farhat and Roux (1991). The method employs the domain decomposition only to the sub-matrices associated with the enriched DOF, in which the sub-domains are treated as independent structures. A Cholesky decomposition is applied to the stiffness matrix associated with each sub-domain, and the continuity of the solution along the sub-domain boundaries is ensured by using additional constraints. The inverse of the Cholesky factor can then be used to form a pre-conditioner. The disadvantage of such matrix decomposition algorithm is that the computation time as well as the memory consumption depends cubically on the matrix size. Although in most applications the number of enriched DOF is much smaller than the number of standard DOF, there are some applications such as polycrystal problems, multiple cracks, and geometrical enrichment, in which the number of enriched DOF is still very large. However, if the structure could be decomposed into several smaller disconnected domains, the sub-matrix associated with the enriched DOF would be a block diagonal matrix. In such case, a Cholesky decomposition can be applied only to each one of these smaller blocks that would decrease the numerical effort.

The DOF of the X-FEM stiffness matrix can be ordered such that

$$\mathbf{K} = \begin{bmatrix} \mathbf{K}_{\text{FEM,FEM}} & \mathbf{K}_{\text{XFEM,FEM}} \\ \mathbf{K}_{\text{FEM,XFEM}} & \mathbf{K}_{\text{XFEM,XFEM}} \end{bmatrix} \tag{4.106}$$

where $\mathbf{K}_{\text{FEM,FEM}}$ refers to the sub-matrix derived by the standard DOF, and $\mathbf{K}_{\text{XFEM,XFEM}}$ is the one obtained by the enriched DOF.

Consider that the domain Ω is decomposed into several non-overlapping sub-domains Ω_i, where each sub-domain is constructed by a union of elements. The matrix $\mathbf{K}_{\text{XFEM,XFEM}}$ is evaluated as if these domains are disconnected, however, the matrix $\mathbf{K}_{\text{FEM,FEM}}$ is evaluated as the standard manner where the domain Ω is a one connected domain. In Figure 4.20, a typical decomposition of the domain Ω into three sub-domains is presented, in which each sub-domain is formed by a union of elements. The number of standard DOF does not change, since the decomposition has no effect on the corresponding part of the stiffness matrix. But the number of enriched DOF increases due to the nodes at the boundary of sub-domains. Since the sub-domains are disconnected, the sub-matrix of enriched DOF can be written as

$$\mathbf{K}_{\text{XFEM,XFEM}} = \begin{bmatrix} \mathbf{K}^{\Omega_1}_{\text{XFEM,XFEM}} & 0 & \\ 0 & \mathbf{K}^{\Omega_2}_{\text{XFEM,XFEM}} & \\ & & \ddots \end{bmatrix} \tag{4.107}$$

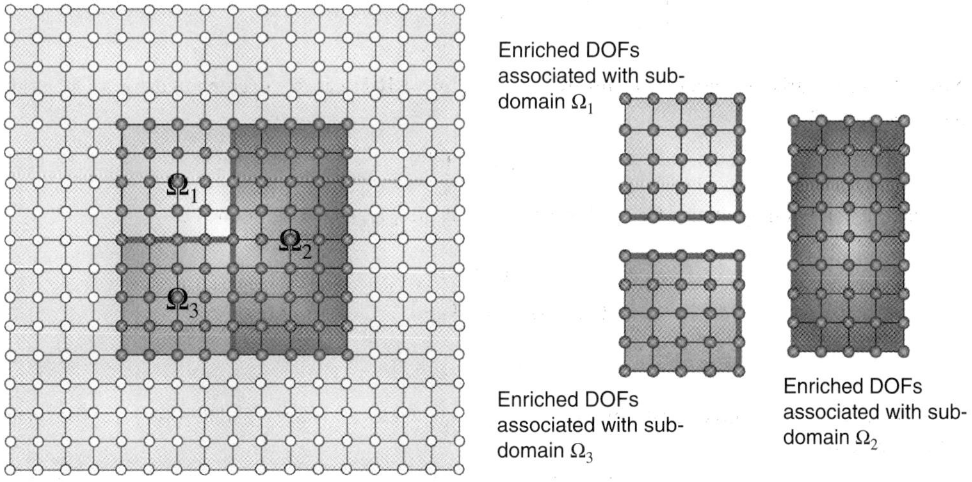

Figure 4.20 Decomposition of a domain Ω into three sub-domains with associated enriched DOF

where $\mathbf{K}_{\text{XFEM,XFEM}}^{\Omega_i}$ are the non-zero enriched DOF of all the enrichments inside the domain Ω_i. In order to ensure the continuity of the displacements along the boundaries of sub-domains, an additional constraint matrix $\mathbf{G}$ is employed into the X-FEM stiffness matrix (4.106) as

$$\mathbf{K} = \begin{bmatrix} \mathbf{K}_{\text{FEM,FEM}} & \mathbf{K}_{\text{XFEM,FEM}} & 0 \\ \mathbf{K}_{\text{FEM,XFEM}} & \mathbf{K}_{\text{XFEM,XFEM}} & \mathbf{G}^T \\ 0 & \mathbf{G} & 0 \end{bmatrix} \tag{4.108}$$

In order to construct the constraint matrix $\mathbf{G}$, consider that the node i is at the boundary of two sub-domains, for example, it belongs to Ω_1 and Ω_2 but not Ω_3 as shown in Figure 4.20. Moreover, assume that there is one enriched degree of freedom associated with this node in the standard X-FEM. However, due to the domain decomposition there are two enriched DOFs associated with this node in the stiffness matrix (4.108), one associated with Ω_1, the other with Ω_2. Consider that the vector $\mathbf{u}_{\text{XFEM}}^i$ contains the corresponding DOFs, a constraint condition can then be defined for the continuity of nodal enrichments as

$$\langle 1 \;\; -1 \rangle \mathbf{u}_{\text{XFEM}}^i = 0 \tag{4.109}$$

In the case that node i is connected to all three domains, it leads to

$$\begin{bmatrix} 1 & -1 & 0 \\ 0 & 1 & -1 \end{bmatrix} \mathbf{u}_{\text{XFEM}}^i = 0 \tag{4.110}$$

Thus, the full constraint matrix $\mathbf{G}$ for the whole system can be constructed by imposing these constraint conditions for all enriched nodes that are connected to two or more sub-domains. More precisely, for each nodal enrichment, whose support is contained in n distinct sub-domains $n-1$ rows must be added to the matrix $\mathbf{G}$. Hence, the matrix $\mathbf{G}$ can be constructed using $(n-1) \times n$ matrix as

$$\begin{bmatrix} 1 & -1 & 0 & & \dots & 0 \\ 0 & 1 & -1 & & \dots & 0 \\ \vdots & & & \vdots & & \vdots \\ 0 & \dots & & 1 & -1 & 0 \\ 0 & \dots & & 0 & 1 & -1 \end{bmatrix} \mathbf{u}_{\text{XFEM}}^i = 0 \tag{4.111}$$

Now, the sub-matrix of enriched DOFs $\mathbf{K}_{\text{XFEM,XFEM}}^{\Omega_i}$ in (4.107) can be decomposed by a Cholesky decomposition $\mathbf{C}_i$ as

$$\mathbf{K}_{\text{XFEM,XFEM}}^{\Omega_i} = \mathbf{C}_{\Omega_i}^T \mathbf{C}_{\Omega_i} \tag{4.112}$$

and the pre-conditioning matrix $\mathbf{P}_{\text{XFEM}}$ for the matrix $\mathbf{K}_{\text{XFEM,XFEM}}$ can be defined as

$$\mathbf{P}_{\text{XFEM}} = \begin{bmatrix} \mathbf{C}_{\Omega_1}^{-1} & 0 & \\ 0 & \mathbf{C}_{\Omega_2}^{-1} & \\ & & \ddots \end{bmatrix} \tag{4.113}$$

This pre-conditioning matrix transforms $\mathbf{K}_{\text{XFEM,XFEM}}$ into an identity matrix, and thus removes the small eigenvalues, that is,

$$\mathbf{P}_{\text{XFEM}}^T \mathbf{K}_{\text{XFEM,XFEM}} \mathbf{P}_{\text{XFEM}} = \mathbf{I} \tag{4.114}$$

A pre-conditioner for the whole system can then be constructed as

$$\mathbf{P} = \begin{bmatrix} \mathbf{P}_{\mathrm{FEM}} & 0 & 0 \\ 0 & \mathbf{P}_{\mathrm{XFEM}} & 0 \\ 0 & 0 & \mathbf{I} \end{bmatrix} \tag{4.115}$$

in which $\mathbf{P}_{\mathrm{FEM}}$ can be defined using any pre-conditioner for the standard DOF. It must be noted that the final system of equation can still be ill-conditioned, since the matrix $\widehat{\mathbf{K}} = \mathbf{P}^T \mathbf{K} \mathbf{P}$ contains the matrix $\mathbf{G}\mathbf{P}_{\mathrm{XFEM}}$, whose rows can be linearly dependent. Applying an LQ decomposition to $\mathbf{G}\mathbf{P}_{\mathrm{XFEM}}$, it results in

$$\mathbf{Q} = \mathbf{L}^{-1} \mathbf{G} \mathbf{P}_{\mathrm{XFEM}} \tag{4.116}$$

where $\mathbf{Q}$ is an orthogonal matrix and $\mathbf{L}$ is a lower triangular matrix. Thus, the final pre-conditioning matrix $\mathbf{P}^*$ can be defined as

$$\mathbf{P}^* = \begin{bmatrix} \mathbf{P}_{\mathrm{FEM}} & 0 & 0 \\ 0 & \mathbf{P}_{\mathrm{XFEM}} & 0 \\ 0 & 0 & \mathbf{L}^{-1} \end{bmatrix} \tag{4.117}$$

and the transformed system of equation can be finally given as

$$\widehat{\mathbf{K}} = \begin{bmatrix} \widehat{\mathbf{K}}_{\mathrm{FEM,FEM}} & \widehat{\mathbf{K}}_{\mathrm{XFEM,FEM}} & 0 \\ \widehat{\mathbf{K}}_{\mathrm{FEM,XFEM}} & \mathbf{I} & \mathbf{Q}^T \\ 0 & \mathbf{Q} & 0 \end{bmatrix} \tag{4.118}$$

It must be noted that the Cholesky decomposition is available only for positive definite matrices. If the domain decomposition is used in combination with the FEM, the stiffness matrix of a sub-domain may become indefinite, if appropriate boundary conditions are missing for this particular sub-domain. In the absence of displacement boundary conditions, the sub-structures can perform as rigid body motions which results in zero eigenvalues; thus, the Cholesky decomposition is likely to fail in that case.

5

Large X-FEM Deformation

5.1 Introduction

In computational mechanics, the finite element (FE) formulation dealing with geometric and material nonlinearities has been well developed and a significant amount of work has been accomplished in large deformation analysis. Nevertheless, standard FE approaches are still ineffective in handling extreme material distortions owing to severe mesh distortion. Therefore, the adaptive FE technique has been introduced to allow the remeshing procedure during computation. In large deformation problems with internal discontinuities or material interfaces, the opportunity for mesh adaptation in different stages of process is of great importance. The need for mesh conforming to the shape of the interface must be preserved at each stage of simulation. In numerical simulation, the requirement of mesh adaptation may consume high amounts of capacity and time. Thus, it is necessary to perform an innovative procedure to alleviate these difficulties by allowing the discontinuities to be mesh-independent. There are several new FE techniques that have been developed to model the discontinuities without remeshing. These include the implementation of the partition of unity (PU) into the FE model (Babuška and Melenk, 1997; Simone, 2004), the meshfree technique (Rabczuk and Belytschko, 2004; Duflot, 2006), and the extended finite element method (X-FEM) (Belytschko and Black, 1999; Moës, Dolbow, and Belytschko, 1999).

The X-FEM has been successfully employed in modeling cracks and inhomogeneities, in which its application is widely spread in the problems with different boundary conditions. In X-FEM, the need for remeshing and mesh adaption can be neglected if the discontinuity or crack propagation happens. In this technique, the discontinuities or interfaces are taken into account by adding appropriate functions into the standard approximation through a PU method (Melenk and Babuška, 1996). The mesh adaptation process is therefore substituted by partitioning the domain with some polygonal sub-elements whose Gauss points are used for integration of the domain of elements. The important advantage of the method is that the number of degrees of freedom remains constant and thus, we do not need to solve a greater equation. In this case, additional functions are used to enrich the displacement fields of the domain, which compensate for displacement or strain fields discontinuity.

The most application of X-FEM technique has been reported in modeling crack growth and crack propagation (Belytschko and Black, 1999; Moës, Dolbow, and Belytschko, 1999). The technique allows modeling the entire crack geometry independent of the mesh, and completely avoids the need to remesh as the crack grows. The method was used to model crack growth and arbitrary discontinuities by enriching the discontinuous approximation in terms of a signed distance and level set functions (Daux *et al.*, 2000; Belytschko *et al.*, 2001; Sukumar *et al.*, 2001). The X-FEM was applied in various aspects of crack problems, including: crack growth with frictional contact (Dolbow, Moës, and Belytschko, 2001), cohesive crack propagation (Moës and Belytschko, 2002a), quasi-static crack growth (Sukumar and Prévost, 2003),

Extended Finite Element Method: Theory and Applications, First Edition. Amir R. Khoei.

fatigue crack propagation (Chopp and Sukumar, 2003; Sukumar, Chopp, and Moran, 2003a), stationary and growing cracks (Ventura, Budyn, and Belytschko, 2003; Lee *et al.*, 2004), bimaterial interfacial cracks (Sukumar *et al.*, 2004), three-dimensional crack propagation (Areias and Belytschko, 2005a), and brittle dynamic crack propagation (Combescure *et al.*, 2008).

The most computational aspects of X-FEM have been presented in linear elastic problems. An overview of the technique has been addressed by Bordas *et al.* (2007b) in the framework of an object-oriented-enriched FE programming. On the practical side, since for a large number of industrially relevant solid mechanics problems, the mesh configuration changes continually throughout the deformation process, the introduction of a large deformation X-FEM is crucial for the solution of large scale industrial problems. Typical situations arise in metal forming simulations where element distortions additionally cause rapid solution degradation and in the majority of practical simulations prevent achievement of any solution for the range of the deformations of practical interest. There are several numerical modeling that has been reported in elasto-plasticity problems; among those contributions, the most implementation of X-FEM in plasticity problems are as follows; the crack propagation in plastic fracture mechanics by Elguedj, Gravouil, and Combescure (2006) and Prabel *et al.* (2007), a mixed augmented Lagrangian–X-FEM model in elasto-plastic fatigue crack growth by Elguedj, Gravouil, and Combescure (2007), the plasticity forming of powder compaction with contact friction by Khoei, Shamloo, and Azami (2006b), the plasticity of frictional contact on arbitrary interfaces by Khoei and Nikbakht (2007), a combined Lagrangian–X-FEM and an arbitrary Lagrangian–Eulerian (ALE)–X-FEM descriptions in large plastic deformations by Khoei, Anahid, and Shahim (2008a) and Khoei, Biabanaki, and Anahid (2009a).

In large deformation problems where displacements are also large, it is necessary to couple governing equations with an update of the calculation region mapped by the elements. Basically there are two types of description; the "Lagrangian" and "Eulerian" descriptions. The *Lagrangian* (*or material*) description describes the material behavior with respect to either the original domain in the case of a *total* Lagrangian frame or the domain at the previous calculation step in the case of an *updated* Lagrangian scheme. In the total Lagrangian description, the element geometry changes are determined by the deformation of the material itself. On the other hand, the position of a particle is obtained by comparing with its initial position. If the initial coordinate is updated at each time step by following the motion of material the description becomes updated Lagrangian, which always refers the position of a particle to its previous position in the updated coordinate. The Lagrangian approach has shown to be adequate for problems that do not exhibit large mass fluxes among different parts of the sample. But in practical problems, as those that appear in realistic design processes, the Lagrangian approach leads to highly distorted and usually useless meshes. This difficulty can be particularly observed in higher order elements. Because of severe distortion of elements, the determinant of Jacobian matrix may become negative at quadrature points, aborting the calculations or causing numerical errors. An *Eulerian* (*or spatial*) description can then be used to describe the material behavior with respect to the fixed position in space in which the physical quantities are expressed as a function of time and position vector in the geometrical space. This approach is well suited to fluid mechanics problems with high velocity flows of material. In the Eulerian method, the elements are fixed in space and the material flows through the elements. Therefore, the elements undergo no distortion even though the material experiences large deformation. However, an important disadvantage of the Eulerian method is that it is less suited to path dependent material due to the convection of material through the elements and problems with free boundary motion. Hence, it can be concluded that neither Lagrangian nor Eulerian formulation alone is well suited to simulation of processes involving large deformation and path dependent material. For this purpose, a combined technique of Lagrangian and Eulerian approaches that captures the advantages of both methods has been developed. The ALE technique has been proposed to take the advantages of both Lagrangian and Eulerian methods while minimizing the disadvantages. The term of "arbitrary" describes that the ALE method is specified based on an arbitrary motion of the mesh so that it optimizes the shape of mesh and elements under no distortion. The basic idea of the ALE formulation is the use of a referential domain for the description of motion, different from the material domain (Lagrangian description) and the spatial domain (Eulerian description). In comparison to fluid dynamics, where the ALE formulation originated, the main difficulty of ALE solid mechanics is the path dependent behavior of plasticity models. The relative motion between the mesh and the material is accounted for in the treatment of the constitutive equation.

In this chapter, the X-FEM is proposed to simulate the nonlinear behavior of materials in large deformations. The conventional FE approximation is enriched by employing additional terms based on the Heaviside and level set enrichment functions. In order to model the material interface, the level set method is used for the elements cut by the interface. To simulate the discontinuities in displacement fields, the Heaviside function is implemented as an enrichment function. An element sub-division procedure is used to subdivide the interface into triangular and quadrilateral sub-elements whose Gauss points are used for integration of the domain of elements. In what follows, the X-FEM technique is first presented in the framework of a Lagrangian large plasticity deformation formulation, and is then described in the framework of an ALE method.

5.2 Large FE Deformation

The nonlinearities in elasto-plastic analyses arise from two distinct sources; the constitutive nonlinearities and geometric nonlinearities, the latter being due to large deformations. Whether the displacements, or strains, are large or small, it is imperative that the equilibrium conditions between the internal and external forces have to be satisfied. Thus, the equilibrium equation for a continuum body Ω with a weak or strong discontinuous boundary Γ_d, as shown in Figure 5.1, can be written as

$$\nabla \cdot \mathbf{P} + \mathbf{b} = 0 \tag{5.1}$$

where ∇ is the gradient operator, $\boldsymbol{X}$ is the Lagrangian coordinate, $\mathbf{b}$ is the body force, and $\mathbf{P}$ is the first Piola–Kirchhoff stress tensor. The essential and natural boundary conditions are defined as follows; the displacement boundary condition as $\mathbf{u} = \tilde{\mathbf{u}}$ on Γ_u where $\tilde{\mathbf{u}}$ is the prescribed displacement on the boundary Γ_u, and the traction boundary condition as $\boldsymbol{\sigma} \cdot \mathbf{n}_\Gamma = \bar{\mathbf{t}}$ on Γ_t, where $\bar{\mathbf{t}}$ is the prescribed traction applied on the boundary Γ_t, and $\mathbf{n}_\Gamma$ is the unit outward normal vector to the external boundary Γ. Moreover, the internal boundary condition is defined as $\boldsymbol{\sigma} \cdot \mathbf{n}_{\Gamma_d} = \bar{\mathbf{t}}_d$ on the discontinuity Γ_d, in which $\mathbf{n}_{\Gamma_d}$ is the unit normal vector to the discontinuity Γ_d. In a domain with the weak discontinuity, such as bimaterial problems, $\bar{\mathbf{t}}_d$ is the traction transferred across the discontinuity Γ_d. While in a domain with the strong discontinuity, such as crack interfaces, the traction free condition is assumed as $\boldsymbol{\sigma} \cdot \mathbf{n}_{\Gamma_d} = 0$.

The differential Eq. (5.1) is written in the reference configuration. In large deformations, the domain of the body in the initial state is denoted by Ω_0 called the initial configuration. In order to describe the deformation of the body, a configuration is used to which the governing equations are referred and is called the

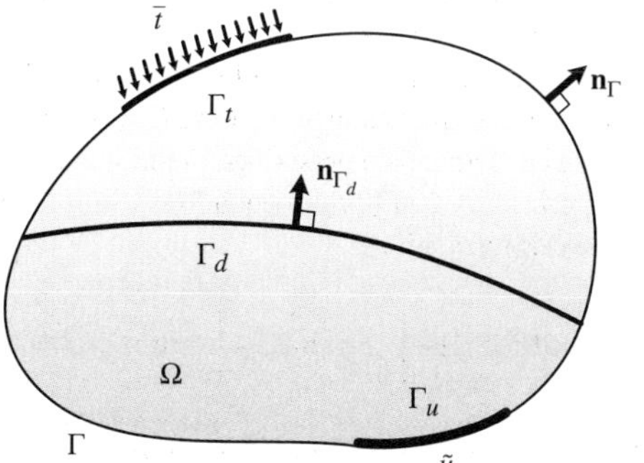

Figure 5.1 Problem definition of a continuum body Ω with a weak or strong discontinuous boundary Γ_d

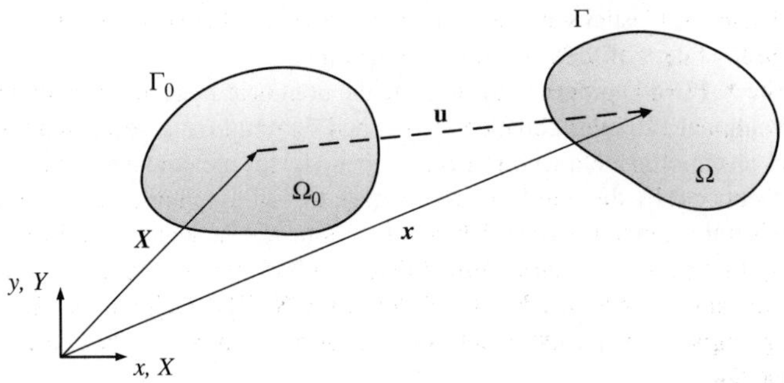

Figure 5.2 The initial (undeformed) configuration together with the current (deformed) configuration for a continuum body

reference configuration (Figure 5.2). A general definition of strains, which is valid whether the displacements or strains are large or small, was introduced by Green and St. Venant. Based on the Green strain tensor, the nonlinear strain displacement relationship can be defined in terms of the infinitesimal and large displacement components as

$$\boldsymbol{E}=\boldsymbol{E}_L+\boldsymbol{E}_{NL}=\boldsymbol{E}_L+\frac{1}{2}\boldsymbol{A}_\theta\boldsymbol{\theta} \tag{5.2}$$

where $\boldsymbol{E}_L$ and $\boldsymbol{E}_{NL}$ are the linear and nonlinear strains and are defined as

$$\boldsymbol{E}_L=\begin{Bmatrix}\dfrac{\partial u}{\partial X}\\ \dfrac{\partial v}{\partial Y}\\ \dfrac{\partial v}{\partial X}+\dfrac{\partial u}{\partial Y}\end{Bmatrix},\ \boldsymbol{E}_{NL}=\begin{Bmatrix}\dfrac{1}{2}\left(\dfrac{\partial u}{\partial X}\right)^2+\dfrac{1}{2}\left(\dfrac{\partial v}{\partial X}\right)^2\\ \dfrac{1}{2}\left(\dfrac{\partial u}{\partial Y}\right)^2+\dfrac{1}{2}\left(\dfrac{\partial v}{\partial Y}\right)^2\\ \dfrac{\partial u}{\partial X}\dfrac{\partial u}{\partial Y}+\dfrac{\partial v}{\partial X}\dfrac{\partial v}{\partial Y}\end{Bmatrix} \tag{5.3}$$

In small displacement theory, the general first order linear strain approximation is obtained by neglecting the quadratic terms. In Eq. (5.2), the nonlinear terms of strain $\boldsymbol{E}_{NL}$ is defined as $\boldsymbol{E}_{NL}=\frac{1}{2}\boldsymbol{A}_\theta\boldsymbol{\theta}$, with $\boldsymbol{\theta}$ denoting the displacement gradient and A_θ a suitably defined matrix operator that contains displacement derivatives and can be derived from the definition of $\boldsymbol{E}_{NL}$. For a set of virtual displacements, the corresponding virtual Green strains of Eq. (5.2) are given as $\delta\boldsymbol{E}=\delta\boldsymbol{E}_L+\delta\boldsymbol{E}_{NL}$.

In order to develop a FE formulation, we need to solve Eq. (5.1) numerically for spatial discretization. Following the standard procedure of the finite element method (FEM), the domain Ω is divided into elements. If the displacements within an element are prescribed in the standard manner by a finite number of nodal values, we can obtain the necessary equilibrium equations using the virtual work principle. Thus, Eq. (5.1) can be written in the weak form as

$$\int_\Omega \delta\mathbf{F}^T\mathbf{P}d\Omega-\int_\Omega \delta\mathbf{u}^T\mathbf{b}d\Omega-\int_{\Gamma_t}\delta\mathbf{u}^T\bar{\mathbf{t}}d\Gamma=0 \tag{5.4}$$

where $\bar{\mathbf{t}}$ is the traction corresponding to the first Piola–Kirchhoff stress. The deformation gradient can be defined according to $\boldsymbol{x}=\boldsymbol{X}+\mathbf{u}$ by $\mathbf{F}=\mathbf{I}+\partial\mathbf{u}/\partial\boldsymbol{X}$, or $F_{ij}=\partial x_i/\partial X_j$, and as a result the variation of deformation gradient is given by $\delta\mathbf{F}=\partial(\delta\mathbf{u})/\partial\boldsymbol{X}$, or $\delta F_{ij}=\partial(\delta u_i)/\partial X_j$. Applying the standard FE Galerkin discretization process to Eq. (5.4) with the independent approximations of $\mathbf{u}$ defined as $\mathbf{u}=\mathbf{N}^T\bar{\mathbf{u}}$, we will arrive at

$$\boldsymbol{\Psi}(\bar{\mathbf{u}}) = \int_\Omega \mathbf{B}^T\mathbf{P}\, d\Omega - \mathbf{f} = 0 \tag{5.5}$$

where $\mathbf{B}$ is the matrix of Cartesian shape function derivatives defined by $\mathbf{B}_I = \partial N_I/\partial \boldsymbol{X}$ for each node I, in which the relation between the matrix $\mathbf{B}$ and the deformation gradient $\mathbf{F}$ can be given as $F_{ij} = \partial(N_I x_{iI})/\partial X_j = x_{iI}B_{jI}$. In Eq. (5.5), the load vector $\mathbf{f}$ is defined as

$$\mathbf{f} = \int_\Omega \mathbf{N}^T\mathbf{b}\, d\Omega + \int_{\Gamma_t} \mathbf{N}^T\bar{\mathbf{t}}\, d\Gamma \tag{5.6}$$

There is of little interest to express the nodal forces in terms of the nominal stress $\mathbf{P}$ since this type of stress is not symmetric. Therefore, we will write Eq. (5.5) in terms of the second Piola–Kirchhoff (PK2) stress $\mathbf{S}$, in which $\mathbf{P} = \mathbf{F}\ \mathbf{S}$. Substituting this transformation into expression (5.5), it can then be rewritten as

$$\boldsymbol{\Psi}(\bar{\mathbf{u}}) = \int_\Omega \bar{\mathbf{B}}^T\mathbf{S}\, d\Omega - \mathbf{f} = 0 \tag{5.7}$$

where the matrix $\bar{\mathbf{B}}$ is defined based on $\mathbf{B}$ and the deformation gradient $\mathbf{F}$, as $\bar{\mathbf{B}} = \mathbf{F}^T\mathbf{B}$, that is,

$$\bar{\mathbf{B}}_I = \begin{bmatrix} \dfrac{\partial N_I}{\partial X}\dfrac{\partial x}{\partial X} & \dfrac{\partial N_I}{\partial X}\dfrac{\partial y}{\partial X} \\ \dfrac{\partial N_I}{\partial Y}\dfrac{\partial x}{\partial Y} & \dfrac{\partial N_I}{\partial Y}\dfrac{\partial y}{\partial Y} \\ \dfrac{\partial N_I}{\partial X}\dfrac{\partial x}{\partial Y}+\dfrac{\partial N_I}{\partial Y}\dfrac{\partial x}{\partial X} & \dfrac{\partial N_I}{\partial X}\dfrac{\partial y}{\partial Y}+\dfrac{\partial N_I}{\partial Y}\dfrac{\partial y}{\partial X} \end{bmatrix} \tag{5.8}$$

in which the matrix $\bar{\mathbf{B}}$ can be rewritten in terms of $\boldsymbol{A}_\theta$ and the matrix of Cartesian shape function derivatives $\mathbf{G}$, as $\bar{\mathbf{B}}_I = \mathbf{B}_I + \boldsymbol{A}_{\theta(\bar{\mathbf{u}})}\mathbf{G}_I$, that is,

$$\bar{\mathbf{B}}_I = \begin{bmatrix} \dfrac{\partial N_I}{\partial X} & 0 \\ 0 & \dfrac{\partial N_I}{\partial Y} \\ \dfrac{\partial N_I}{\partial Y} & \dfrac{\partial N_I}{\partial X} \end{bmatrix} + \begin{bmatrix} \dfrac{\partial u}{\partial X} & \dfrac{\partial v}{\partial X} & 0 & 0 \\ 0 & 0 & \dfrac{\partial u}{\partial Y} & \dfrac{\partial v}{\partial Y} \\ \dfrac{\partial u}{\partial Y} & \dfrac{\partial v}{\partial Y} & \dfrac{\partial u}{\partial X} & \dfrac{\partial v}{\partial X} \end{bmatrix} \begin{bmatrix} \dfrac{\partial N_I}{\partial X} & 0 \\ 0 & \dfrac{\partial N_I}{\partial X} \\ \dfrac{\partial N_I}{\partial Y} & 0 \\ 0 & \dfrac{\partial N_I}{\partial Y} \end{bmatrix} \tag{5.9}$$

In order to obtain the tangential stiffness matrix, the FE Galerkin discretization formulation (5.5) is appropriately taken variations with respect to $\delta\bar{\mathbf{u}}$. In this equation, the only variable that depends on the displacement is the nominal stress, thus

$$\delta\boldsymbol{\Psi} = \int_\Omega \mathbf{B}^T\delta\mathbf{P}\, d\Omega \equiv \bar{\mathbf{K}}_T\ \delta\bar{\mathbf{u}} \tag{5.10}$$

For a set of virtual displacements, the corresponding $\delta\mathbf{P}$ in Eq. (5.10) can be obtained by taking the derivative of transformation $\mathbf{P}=\mathbf{FS}$ as

$$\delta\mathbf{P}=\mathbf{F}\,\delta\mathbf{S}+\delta\mathbf{F}\,\mathbf{S} \tag{5.11}$$

Substituting relation (5.11) into Eq. (5.10) yields

$$\begin{aligned}\delta\mathbf{\Psi}&=\int_{\Omega}\mathbf{B}^T\mathbf{F}\,\delta\mathbf{S}\,d\Omega+\int_{\Omega}\mathbf{B}^T\delta\mathbf{F}\,\mathbf{S}\,d\Omega\\&\equiv\delta\mathbf{\Psi}^{\text{mat}}+\delta\mathbf{\Psi}^{\text{geo}}\end{aligned} \tag{5.12}$$

This equation shows that the stiffness matrix $\bar{\mathbf{K}}_T$ consists of two parts; the first part involves the derivative of stress $\delta\mathbf{S}$, which depends on the material response and leads to the material tangent stiffness matrix $\mathbf{K}^{\text{mat}}$, and the second part involves the current state of stress $\mathbf{S}$, which accounts for the geometric effects of the deformation (including rotation and stretching) and leads to the geometric stiffness matrix $\mathbf{K}^{\text{geo}}$ (Khoei, 2005).

In order to derive the material tangent stiffness matrix $\mathbf{K}^{\text{mat}}$ in Eq. (5.12), implement the constitutive law definition with respect to the incremental PK2 stress, that is, $\delta\mathbf{S}=\mathbf{D}_S^{ep}\,\delta E$, into Eq. (5.12), we will have

$$\delta\mathbf{\Psi}^{\text{mat}}=\int_{\Omega}\mathbf{B}^T\mathbf{F}\,\mathbf{D}_S^{ep}\,\delta E\;d\Omega\equiv\mathbf{K}^{\text{mat}}\delta\bar{\mathbf{u}} \tag{5.13}$$

where the incremental Green strain δE is defined as $\delta E=\bar{\mathbf{B}}\,\delta\bar{\mathbf{u}}$. Substituting $\bar{\mathbf{B}}=\mathbf{F}^T\mathbf{B}$ in the previous relation yields to

$$\mathbf{K}^{\text{mat}}=\int_{\Omega}\bar{\mathbf{B}}^T\,\mathbf{D}_S^{ep}\,\bar{\mathbf{B}}\;d\Omega \tag{5.14}$$

where the stress–strain relationship for an elastic isotropic Kirchhoff material is defined as $\mathbf{S}=\lambda\,\text{trace}(\boldsymbol{E})\mathbf{I}+2\mu\,\mathbf{I}$, with $\mathbf{I}$ denoting the fourth-order symmetric identity tensor, and λ and μ the Lamé constants.

In order to derive the geometric stiffness matrix $\mathbf{K}^{\text{geo}}$ in Eq. (5.12), applying the relation $\delta\mathbf{F}=\mathbf{B}\,\delta\bar{\mathbf{u}}$ into the geometric term $\delta\mathbf{\Psi}^{\text{geo}}$, after some manipulations it results in

$$\delta\mathbf{\Psi}^{\text{geo}}=\int_{\Omega}\mathbf{G}^T\mathbf{M}_S\mathbf{G}\,\delta\bar{\mathbf{u}}\;d\Omega\equiv\mathbf{K}^{\text{geo}}\delta\bar{\mathbf{u}} \tag{5.15}$$

where $\mathbf{M}_S$ is a 4×4 matrix of the three PK2 stress components, which is defined for two-dimensional problems as

$$\mathbf{M}_S=\begin{bmatrix}S_{XX}\mathbf{I}_{2\times2} & S_{XY}\mathbf{I}_{2\times2}\\ sym\cdot & S_{YY}\mathbf{I}_{2\times2}\end{bmatrix} \tag{5.16}$$

where $\mathbf{I}$ is the identity matrix. Thus, we can define the total tangential stiffness matrix $\bar{\mathbf{K}}_T$, used in Eq. (5.10), as

$$\bar{\mathbf{K}}_T=\mathbf{K}^{\text{mat}}+\mathbf{K}^{\text{geo}}=\int_{\Omega}\bar{\mathbf{B}}^T\mathbf{D}_S^{ep}\,\bar{\mathbf{B}}\;d\Omega+\int_{\Omega}\mathbf{G}^T\mathbf{M}_S\,\mathbf{G}\,d\Omega \tag{5.17}$$

All the ingredients necessary for computing large deformation problems are now available. As the first step, displacements $\bar{\mathbf{u}}^0$ are obtained according to the small displacement solution. This determines the actual strains by considering the nonlinear contribution defined by Eq. (5.4) together with the appropriate linear contributions. Corresponding stresses can be obtained by the elastic expressions and $\mathbf{\Psi}^0$ determined according to Eq. (5.7). For successive iterations, $(\bar{\mathbf{K}}_T)_n$ is obtained from Eq. (5.17). The Cauchy stress $\boldsymbol{\sigma}$ is calculated based on PK2 stress using the relation of $\boldsymbol{\sigma} = J^{-1}\mathbf{F}\,\mathbf{S}\,\mathbf{F}^T$, with J denoting the determinant of $\mathbf{F}$, that is, $J = \det(\mathbf{F})$.

5.3 The Lagrangian Large X-FEM Deformation Method

5.3.1 The Enrichment of Displacement Field

The enriched FEM is a powerful and accurate approach to model discontinuities, which are independent of the FE mesh topology. In this technique, the discontinuities, or interfaces, are not considered in the mesh generation operation and special functions that depend on the nature of discontinuity, are included into the FE approximation. The aim of this method is to simulate the discontinuity, or interface, with minimum enrichment. In X-FEM, the external boundaries are only consideration in mesh generation operation and the internal boundaries, such as cracks, voids, contact surfaces, and so on, have no effect on mesh configurations. This method has proper applications in problems with moving discontinuities, such as punching, phase changing, crack propagation, and shear banding.

In order to introduce the concept of discontinuous enrichment, consider that Γ_d be a discontinuity or interface in the domain Ω, as shown in Figure 5.3a. The aim is to construct a FE approximation to the field

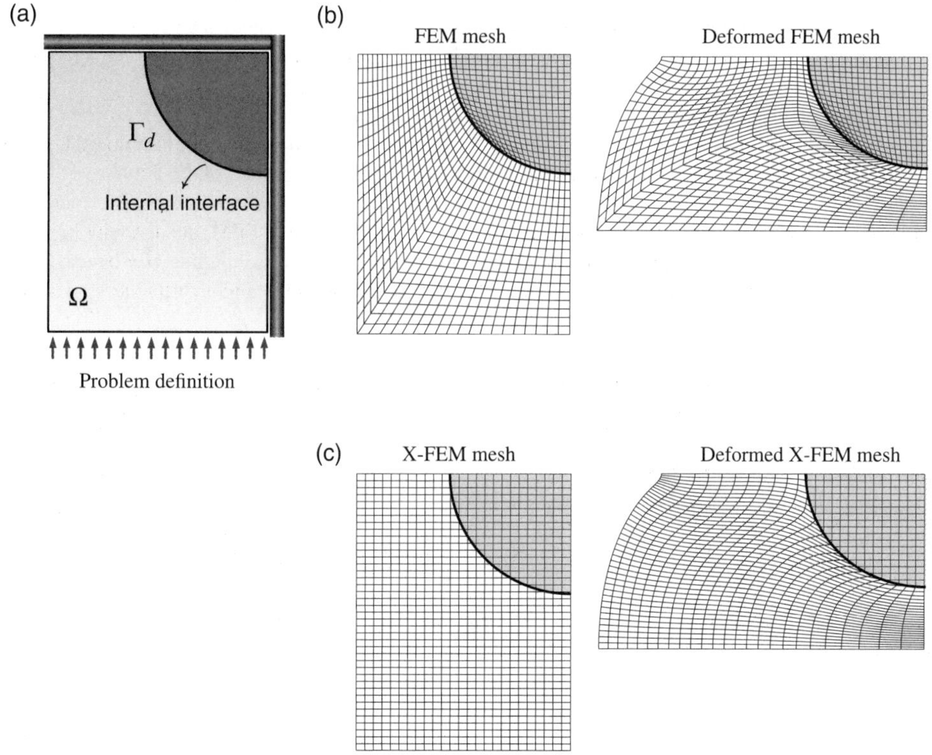

Figure 5.3 Modeling of an internal interface: (a) Problem definition; (b) The FE mesh which conforms to the geometry of interface; (c) A uniform mesh in which the elements cut by the interface

$\mathbf{u} \in \Omega$ which can be discontinuous along interface surface Γ_d. The traditional approach is to generate the mesh to conform to the line of discontinuity surface as shown in Figure 5.3b, in which the element edges align with Γ_d. While this strategy certainly creates a discontinuity in the approximation field, it is cumbersome if the line Γ_d evolves in time, or if several different configurations for Γ_d are to be considered (Khoei and Lewis, 1999). Here, the intention is to model the discontinuity Γ_d by extrinsic enrichment, in which the uniform mesh of Figure 5.3c is capable of modeling the discontinuity in **u**, or its derivatives with appropriate enrichment functions.

The standard FE approximation can be enriched with additional functions by using the notion of the partition of unity (PU) (Melenk and Babuška, 1996). The enriched displacement field can be expressed in the following form

$$\mathbf{u}(X,t)=\sum_{I=1}^{\mathcal{N}} N_I(X)\,\bar{\mathbf{u}}_I(t)+\sum_{J=1}^{\mathcal{M}} N_J(X)\,\psi(X)\,\bar{\mathbf{a}}_J(t) \quad \text{for } n_I \in \mathcal{N} \text{ and } n_J \in \mathcal{M} \tag{5.18}$$

The first term of this equation denotes the standard FE approximation and the second term indicates the enrichment function considered in the X-FEM. In this equation, $\bar{\mathbf{u}}_I$ is the classical nodal displacement, $\bar{\mathbf{a}}_J$ the nodal degrees of freedom corresponding to the enrichment functions, $\psi(X)$ the enrichment function, and $N(X)$ the standard shape function. In Eq. (5.18), $\mathcal{N}$ is the set of all nodal points of domain, and $\mathcal{M}$ is the set of nodes of elements located on discontinuity, that is,

$$\mathcal{M}=\{n_J : n_J \in \mathcal{N},\ \Upsilon_J \cap \Gamma_d \neq 0\} \tag{5.19}$$

In this equation, $\Upsilon_J = \text{supp}(n_J)$ is the support of the nodal shape function $N_J(X)$, which consists of the union of all elements with n_J as one of its vertices, or in other words the union of elements in which $N_J(X)$ is non-zero.

In this method, the displacement field is enriched for the elements intersected by the interface based on the *partition of unity method* (PUM). An element subdivision procedure is used to partition the domain with some polygonal sub-elements whose Gauss points are used for integration of the domain of elements (Figure 5.4). The enrichment of displacement field is applied to correct the standard displacement based approximation by incorporating discontinuous fields through a PUM. In X-FEM, the diversity in types of problems requires proposing appropriate enrichment functions. Different techniques may be used for the enrichment function; these functions are related to the type of discontinuity and its influences on the form

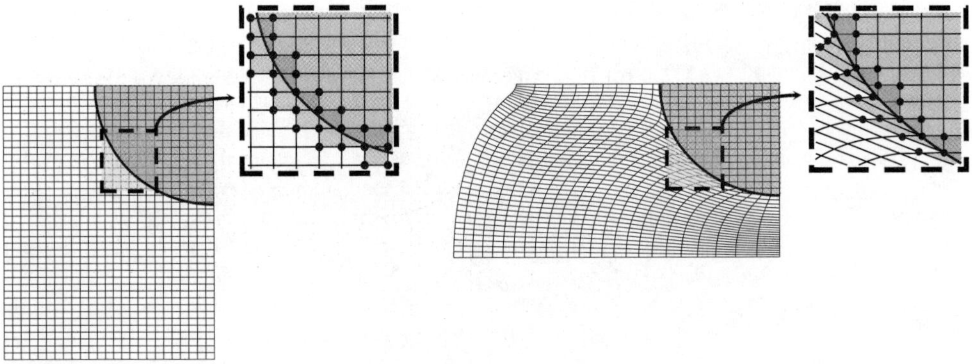

Figure 5.4 The extended finite element mesh: the circled nodes have additional degrees of freedom and enrichment functions

of solution. These techniques are based on the signed distance function, branch function, Heaviside jump function, level set function, and so on. The *level set* method is a numerical scheme developed by Sethian (1999) for tracking moving interfaces. In this technique, the interface is represented as the zero level set of a function of one higher dimension. This method, which is used for predicting the geometry of boundaries, is very suitable for bimaterial problems in which the displacement field is continuous and the strain field is discontinuous.

The choice of enrichment function for displacement approximation is dependent on the conditions of problem. If the discontinuity is as a result of different types of material properties, then the level set based function is proposed as an enrichment function; however, if the discontinuity is due to different displacement fields on either sides of the discontinuity, such as contact surface, then the Heaviside function is appropriate. In what follows, the implementation of the level set based function in modeling the material discontinuity, and the Heaviside function in simulation of different displacement fields on either side of the contact surface is presented.

In order to model the material interface due to different material properties in two intersected parts, the level set method is used. A moving interface $\Gamma_d(t)$ is formulated by the level set function $\varphi(\boldsymbol{X}, t)$ as

$$\Gamma_d(t) = \left\{ \boldsymbol{X} \in \Re^2 : \ \varphi(\boldsymbol{X}, t) = 0 \right\} \tag{5.20}$$

in which the value of level set function is zero on the border. The sign of its value is negative in one side and positive on the other side. The absolute value of level set function presents the distance from the border, that is,

$$\varphi(\boldsymbol{X}, t) = \min \|\boldsymbol{X} - \boldsymbol{X}^*\| \operatorname{sign}((\boldsymbol{X} - \boldsymbol{X}^*) \cdot \mathbf{n}_{\Gamma_d}) \tag{5.21}$$

where $\boldsymbol{X}^*$ is the point on the interface boundary which has the closest distance from the point $\boldsymbol{X}$, and $\mathbf{n}_{\Gamma_d}$ is the normal vector to the interface at point $\boldsymbol{X}^*$. These changes in the values of level set make an appropriate behavior for the corrected shape functions. The new degrees of freedom $\bar{\mathbf{a}}_J$ corresponding to the level set enrichment function are considered in Eq. (5.18) in order to attribute to the nodes that belong to the set of $\mathcal{M}$. The presence of these new degrees of freedom improves the strain discontinuity due to different types of material properties in two intersected parts of element.

In order to improve the numerical computation in X-FEM, it is preferable to have a uniform distribution of shape functions around the interface. To this goal, a technique proposed by Sukumar *et al.* (2001), is applied to smoothing the values of level set by employing the nodes that belong to the elements in the neighbor of intersected elements. Two types of level set enrichment function are employed to model the material interface. The first function is based on the absolute value of level set function proposed by Sukumar *et al.* (2001), which indeed has a discontinuous first derivative on the interface as

$$\psi^1(\boldsymbol{X}) = \left| \sum_I N_I(\boldsymbol{X}) \, \varphi_I \right| \tag{5.22}$$

where φ_I indicates the nodal value of level set function. The second enrichment function proposed here is an extension of function (5.22) introduced by Moës *et al.* (2003), that is,

$$\psi^2(\boldsymbol{X}) = \sum_I N_I(\boldsymbol{X}) \, |\varphi_I| - \left| \sum_I N_I(\boldsymbol{X}) \, \varphi_I \right| \tag{5.23}$$

This enrichment function has a ridge centered on the interface and zero value on the elements that are not crossed by the interface.

In order to model the discontinuities in displacement fields, the Heaviside enrichment function is implemented in simulation of contact surfaces, in which the function $H(X)$ takes the value of +1 in one side of contact surface, and −1 on the other side, as

$$H(X)=\begin{cases} +1 & \text{if } (\boldsymbol{X}-\boldsymbol{X}^*)\cdot \mathbf{n}_{\Gamma_d}\geq 0 \\ -1 & \text{otherwise} \end{cases} \tag{5.24}$$

It must be noted that the Heaviside function is a powerful and reliable tool in enriching the domain with strong discontinuities, such as contact surface.

5.3.2 The Large X-FEM Deformation Formulation

Considering the enriched approximation of displacement field defined by Eq. (5.18), the value of $\mathbf{u}(X, t)$ on an enriched node K in set $\mathcal{M}$ can be written as

$$\mathbf{u}(\boldsymbol{X}_K, t)=\bar{\mathbf{u}}_K(t)+\psi(\boldsymbol{X}_K)\,\bar{\mathbf{a}}_K(t) \tag{5.25}$$

Since $\psi(X_K)$ is not necessarily zero, this expression is not equal to the real nodal value $\bar{\mathbf{u}}_K$. Thus, the enriched displacement field can be corrected as

$$\mathbf{u}(\boldsymbol{X}, t)=\sum_{I=1}^{\mathcal{N}} N_I(\boldsymbol{X})\,\bar{\mathbf{u}}_I(t)+\sum_{J=1}^{\mathcal{M}} N_J(\boldsymbol{X})\,(\psi(\boldsymbol{X})-\psi(\boldsymbol{X}_J))\,\bar{\mathbf{a}}_J(t) \tag{5.26}$$

Based on the new definition in last term of relation (5.26), the expected property can be obtained as $\mathbf{u}(\boldsymbol{X}_K)=\bar{\mathbf{u}}_K$. Substituting the level set enrichment function $\psi(\varphi(\boldsymbol{X}))$ in relation (5.26), the X-FEM displacement approximation can be rewritten as

$$\mathbf{u}(\boldsymbol{X}, t)=\sum_{I=1}^{\mathcal{N}} N_I(\boldsymbol{X})\,\bar{\mathbf{u}}_I(t)+\sum_{J=1}^{\mathcal{M}} N_J(\boldsymbol{X})\left(\psi(\varphi(\boldsymbol{X}))-\psi(\varphi(\boldsymbol{X}_J))\right)\bar{\mathbf{a}}_J(t) \tag{5.27}$$

Applying the enrichment function $\psi(X)$ based on its nodal values as $\psi(X)=\sum_{I\in\mathcal{N}} N_I(X)\,\psi_I$, Eq. (5.27) can be rewritten as

$$\mathbf{u}(\boldsymbol{X}, t)=\sum_{I=1}^{\mathcal{N}} N_I(\boldsymbol{X})\,\bar{\mathbf{u}}_I(t)+\sum_{J=1}^{\mathcal{M}} N_J(\boldsymbol{X})\left(\sum_{K=1}^{\mathcal{M}} N_K(\boldsymbol{X})\,\psi_K-\psi_J\right)\bar{\mathbf{a}}_J(t) \tag{5.28}$$

Taking the derivative of displacement field (5.28), results in

$$\frac{\partial \mathbf{u}}{\partial \boldsymbol{X}}=\sum_{I=1}^{\mathcal{N}}\frac{\partial N_I}{\partial \boldsymbol{X}}\bar{\mathbf{u}}_I+\sum_{J=1}^{\mathcal{M}}\left[\frac{\partial N_J}{\partial \boldsymbol{X}}\left(\sum_{K=1}^{\mathcal{M}} N_K(\boldsymbol{X})\,\psi_K-\psi_J\right)+N_J(\boldsymbol{X})\,\frac{\partial}{\partial \boldsymbol{X}}\left(\sum_{K=1}^{\mathcal{M}} N_K(\boldsymbol{X})\,\psi_K\right)\right]\bar{\mathbf{a}}_J \tag{5.29}$$

in which the first term denotes the standard FE approximation composed by the derivatives of shape functions in whole domain, and the second term presents the derivative of enriched shape functions constructed in the domain of set $\mathcal{M}$. Thus, the well known strain matrix $\mathbf{B}_I$ relating the increments of strain and displacement for node I in the case of X-FEM formulation can be defined based on the standard and enriched parts as

$$\mathbf{B}_I^u = \begin{bmatrix} \dfrac{\partial N_I}{\partial X} & 0 \\ 0 & \dfrac{\partial N_I}{\partial Y} \\ \dfrac{\partial N_I}{\partial Y} & \dfrac{\partial N_I}{\partial X} \end{bmatrix} \tag{5.30a}$$

$$\mathbf{B}_I^a = \begin{bmatrix} \dfrac{\partial}{\partial X}\left[N_I(\boldsymbol{X})\left(\sum_{K=1}^{\mathcal{M}} N_K(\boldsymbol{X})\,\psi_K - \psi_I\right)\right] & 0 \\ 0 & \dfrac{\partial}{\partial Y}\left[N_I(\boldsymbol{X})\left(\sum_{K=1}^{\mathcal{M}} N_K(\boldsymbol{X})\,\psi_K - \psi_I\right)\right] \\ \dfrac{\partial}{\partial Y}\left[N_I(\boldsymbol{X})\left(\sum_{K=1}^{\mathcal{M}} N_K(\boldsymbol{X})\,\psi_K - \psi_I\right)\right] & \dfrac{\partial}{\partial X}\left[N_I(\boldsymbol{X})\left(\sum_{K=1}^{\mathcal{M}} N_K(\boldsymbol{X})\,\psi_K - \psi_I\right)\right] \end{bmatrix} \tag{5.30b}$$

and the nonlinear term of relation (5.8) $\bar{\mathbf{B}}_I$ can be obtained by using $\bar{\mathbf{B}}_I = \mathbf{B}_I + \boldsymbol{A}_{\theta(\bar{\mathbf{u}})}\mathbf{G}_I$, in which the standard and enriched parts of matrix $\mathbf{G}_I$ in the X-FEM formulation can be defined as

$$\mathbf{G}_I^u = \begin{bmatrix} \dfrac{\partial N_I}{\partial X} & 0 \\ 0 & \dfrac{\partial N_I}{\partial X} \\ \dfrac{\partial N_I}{\partial Y} & 0 \\ 0 & \dfrac{\partial N_I}{\partial Y} \end{bmatrix} \tag{5.31a}$$

$$\mathbf{G}_I^a = \begin{bmatrix} \dfrac{\partial}{\partial X}\left[N_I(\boldsymbol{X})\left(\sum_{K=1}^{\mathcal{M}} N_K(\boldsymbol{X})\,\psi_K - \psi_I\right)\right] & 0 \\ 0 & \dfrac{\partial}{\partial X}\left[N_I(\boldsymbol{X})\left(\sum_{K=1}^{\mathcal{M}} N_K(\boldsymbol{X})\,\psi_K - \psi_I\right)\right] \\ \dfrac{\partial}{\partial Y}\left[N_I(\boldsymbol{X})\left(\sum_{K-1}^{\mathcal{M}} N_K(\boldsymbol{X})\,\psi_K - \psi_I\right)\right] & 0 \\ 0 & \dfrac{\partial}{\partial Y}\left[N_I(\boldsymbol{X})\left(\sum_{K=1}^{\mathcal{M}} N_K(\boldsymbol{X})\,\psi_K - \psi_I\right)\right] \end{bmatrix} \tag{5.31b}$$

On substituting the trial function of Eq. (5.28) into the weak form of equilibrium equation of elasto-plasticity (5.4) and taking variations, the discrete system of equations can be obtained as $\bar{\mathbf{K}}_T\,\delta\bar{\mathbf{d}} - \delta\mathbf{f} = 0$, where $\delta\bar{\mathbf{d}}$ is the vector of unknowns of $\delta\bar{\mathbf{u}}$ and $\delta\bar{\mathbf{a}}$ at the nodal points, and $\bar{\mathbf{K}}_T$ is the tangential stiffness matrix as

$$(\bar{\mathbf{K}}_T)_{IJ} = \begin{bmatrix} \bar{\mathbf{K}}_{IJ}^{uu} & \bar{\mathbf{K}}_{IJ}^{ua} \\ \bar{\mathbf{K}}_{IJ}^{au} & \bar{\mathbf{K}}_{IJ}^{aa} \end{bmatrix} \tag{5.32}$$

where the total stiffness matrix can be obtained using Eqs. (5.17), (5.30), and (5.31) as

$$\bar{\mathbf{K}}_{IJ}^{\alpha\beta} = \int_{\Omega^e} (\bar{\mathbf{B}}_I^{\alpha})^T \mathbf{D}_S^{ep} \bar{\mathbf{B}}_J^{\beta} \, d\Omega + \int_{\Omega^e} (\mathbf{G}_I^{\alpha})^T \mathbf{M}_S \mathbf{G}_J^{\beta} \, d\Omega \quad (\alpha, \beta = u, a) \tag{5.33}$$

The external force vector can be calculated using relation (5.6) as

$$\mathbf{f}_I = \{\mathbf{f}_I^u \;\; \mathbf{f}_I^a\}^T \tag{5.34}$$

where

$$\mathbf{f}_I^{\alpha} = \int_{\Omega^e} N_I^{\alpha} \rho \, \mathbf{b} d\Omega + \int_{\Gamma^e} N_I^{\alpha} \bar{\mathbf{t}} d\Gamma \quad (\alpha = u, a) \tag{5.35}$$

where $N_I^{\alpha} \equiv N_I$ is related to the FE displacement degree of freedom (DOF), and $N_I^{\alpha} \equiv N_I \psi$ to an enriched DOF.

5.3.3 Numerical Integration Scheme

In the standard FEM, the shape functions are introduced in terms of the polynomial order and the Gauss integration rule can be used efficiently to evaluate the integral of stiffness matrix. However, in the X-FEM the enrichment functions may not be smooth over an enriched element due to presence of the weak or strong discontinuity inside of the element. Hence, the standard Gauss quadrature rule cannot be used if the element is crossed by a discontinuity, and it is necessary to modify the element quadrature points to accurately evaluate the contribution to the weak form for both sides of interface. In the standard FEM, the numerical integration can be performed by discretizing the domain as

$$\Omega = \bigcup_{e=1}^{m} \Omega_e \tag{5.36}$$

where m is the number of elements and Ω_e is the element sub-domain. In the X-FEM, the elements located on the interface boundary can be partitioned by polygonal sub-domains Ω_s with the boundaries aligned with the material interface, that is,

$$\Omega_e = \bigcup_{s=1}^{m_s} \Omega_s \tag{5.37}$$

where m_s denotes the number of sub-polygons of an element. The Gauss points of sub-polygons are used for numerical integration across the discontinuity surface. It is essential to mention that these sub-polygons only generated for numerical integration and no new degrees of freedom are added to system. In order to generate these sub-polygons, the triangular/quadrilateral partitioning method can be performed, as shown in Figure 5.5a, and the Gauss integration rule can be carried out over each sub-polygons. An alternative technique can be used by dividing the element cut by the interface into rectangular sub-grids, as shown in Figure 5.5b. In this technique, it is not necessary to conform the sub-quadrilaterals to the geometry of interface, however, it is required to have enough sub-divisions to reduce the error of numerical integration.

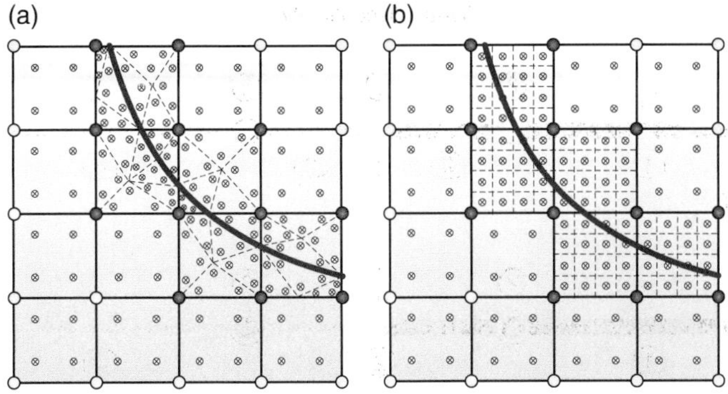

Figure 5.5 Numerical integration of enriched elements: (a) the triangular partitioning method; (b) the rectangular sub-grids method

5.4 Numerical Modeling of Large X-FEM Deformations

In order to demonstrate the accuracy and versatility of X-FEM in large deformations, two examples are modeled numerically. The examples consist of a one-dimensional bar and a two-dimensional plate, which are in tension with a "*material interface*" at their centers. The examples are solved using both FEM and X-FEM techniques, and the results are compared. In X-FEM analyses, the FE mesh is defined independent of the shape of discontinuity, while the FEM analysis is performed by conforming the mesh to the geometry of the interface. The weak discontinuity of bimaterial interface is modeled using the ramp enrichment function. Finally, the performance of blending elements is investigated for both examples (see Companion Website).

5.4.1 Modeling an Axial Bar with a Weak Discontinuity

The first example is chosen to illustrate the performance of large X-FEM deformation formulation for a bimaterial problem, in which an axial bar is modeled with a weak discontinuity at its center, as shown in Figure 5.6. The geometry and boundary conditions of the problem together with the FEM and X-FEM meshes are presented in this figure. The material properties are chosen based on a hypo-elastic constitutive rule, in which the Young's modulus of the left part is $E_1 = 2 \times 10^6$ kg/cm^2 and that of the right part is $E_2 = 7 \times 10^5$ kg/cm^2. The bar is restrained at one end and subjected to a ramp load p at the free end. The X-FEM analysis is performed using three 2-noded elements, in which the material interface is placed at the mid-point of element (2). In order to model the material interface in the bar, the ramp enrichment function $\varphi(X) = |X - 1.5\ell|$ is employed considering the enriched degrees of freedom at nodal points 2 and 3. In X-FEM, the enrichment function is defined using the shifted distance function $\psi(X) = \varphi(X) - \varphi(X_I) = |X - 1.5\ell| - 0.5\ell$, with X_I denoting the position of corresponding nodal point. The results of X-FEM model are finally compared with the FEM analysis carried out using four 2-noded elements.

In the one-dimensional problem, the only component of Green strain can be expressed as

$$E_{XX} = E_L + E_{NL} = \frac{\partial u}{\partial X} + \frac{1}{2}\left(\frac{\partial u}{\partial X}\right)^2 \tag{5.38a}$$

and the corresponding virtual Green strain is defined as

$$\delta E_{XX} = \delta E_L + \delta E_{NL} = \frac{\partial(\delta u)}{\partial X} + \left(\frac{\partial u}{\partial X}\right)\left(\frac{\partial(\delta u)}{\partial X}\right) \tag{5.38b}$$

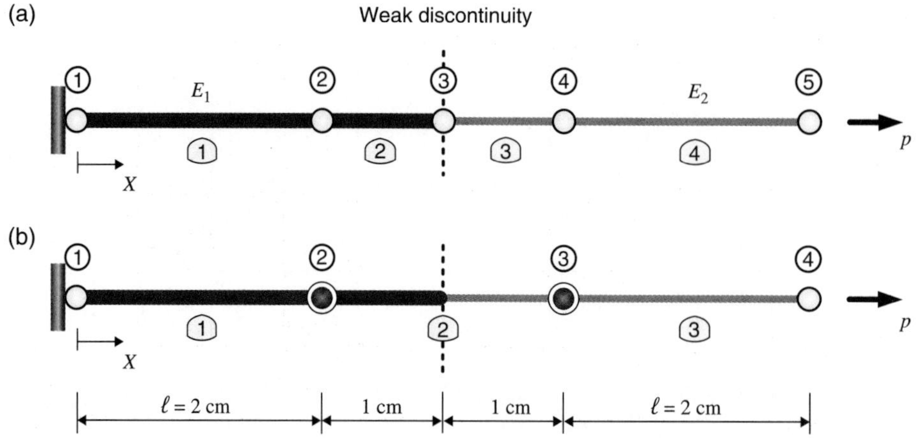

Figure 5.6 An axial bar with a material interface at its center in large deformation: (a) FEM model; (b) X-FEM model

For a one-dimensional linear element, the standard shape functions are defined as $\mathbf{N}^{std}_{(X)} = \langle 1 - \frac{X}{\ell} \ \ \frac{X}{\ell} \rangle$ and the enriched shape functions are given by $\mathbf{N}^{enr}_{(X)} = \mathbf{N}^{std}_{(X)}\,\psi(X)$. Thus, the incremental Green strain for an enriched element (2) can be obtained from (5.38b) as

$$\delta E_{XX} = \left\langle -\frac{1}{\ell} \quad \frac{1}{\ell} \quad -\frac{1}{\ell}\psi(X) + \left(1-\frac{X}{\ell}\right)\frac{\partial\psi}{\partial X} \quad \frac{1}{\ell}\psi(X) + \left(\frac{X}{\ell}\right)\frac{\partial\psi}{\partial X} \right\rangle \begin{Bmatrix} \delta u_2 \\ \delta u_3 \\ \delta a_2 \\ \delta a_3 \end{Bmatrix}$$
$$+ \left(\left\langle -\frac{1}{\ell} \quad \frac{1}{\ell} \quad -\frac{1}{\ell}\psi(X) + \left(1-\frac{X}{\ell}\right)\frac{\partial\psi}{\partial X} \quad \frac{1}{\ell}\psi(X) + \left(\frac{X}{\ell}\right)\frac{\partial\psi}{\partial X} \right\rangle \begin{Bmatrix} u_2 \\ u_3 \\ a_2 \\ a_3 \end{Bmatrix} \right)^T \tag{5.38c}$$
$$\times \left\langle -\frac{1}{\ell} \quad \frac{1}{\ell} \quad -\frac{1}{\ell}\psi(X) + \left(1-\frac{X}{\ell}\right)\frac{\partial\psi}{\partial X} \quad \frac{1}{\ell}\psi(X) + \left(\frac{X}{\ell}\right)\frac{\partial\psi}{\partial X} \right\rangle \begin{Bmatrix} \delta u_2 \\ \delta u_3 \\ \delta a_2 \\ \delta a_3 \end{Bmatrix}$$

This relation can be simplified as $\delta E_{XX} = \bar{\mathbf{B}}^{enh}_{(X,\bar{\mathbf{d}})}\,\delta\bar{\mathbf{d}}$, in which the matrix $\bar{\mathbf{B}}^{enh}$ is defined by

$$\begin{aligned}\bar{\mathbf{B}}^{enh}_{(X,\bar{\mathbf{d}})} &= \left\langle \mathbf{B}^{std}_{(X)} \ \ \mathbf{B}^{enr}_{(X)} \right\rangle + \left(\left\langle \mathbf{B}^{std}_{(X)} \ \ \mathbf{B}^{enr}_{(X)} \right\rangle \begin{Bmatrix} \bar{\mathbf{u}} \\ \bar{\mathbf{a}} \end{Bmatrix} \right)^T \left\langle \mathbf{B}^{std}_{(X)} \ \ \mathbf{B}^{enr}_{(X)} \right\rangle \\ &= \mathbf{B}^{enh}_{(X)} + \bar{\mathbf{d}}^T \mathbf{B}^{enh\,T}_{(X)} \mathbf{B}^{enh}_{(X)} \end{aligned} \tag{5.38d}$$

where $\mathbf{B}^{enh}_{(X)} = \left\langle \mathbf{B}^{std}_{(X)} \ \ \mathbf{B}^{enr}_{(X)} \right\rangle$ and $\delta\bar{\mathbf{d}} = \langle \delta\bar{\mathbf{u}} \ \ \delta\bar{\mathbf{a}} \rangle^T$. The matrix of Cartesian shape function derivatives $\mathbf{G}$ can be obtained based on the standard and enriched parts as

$$\mathbf{G}^{enh}_{(X)} = \left\langle \mathbf{G}^{std}_{(X)} \ \ \mathbf{G}^{enr}_{(X)} \right\rangle = \left\langle \frac{\partial \mathbf{N}^{std}}{\partial X} \frac{\partial \mathbf{N}^{enr}}{\partial X} \right\rangle \tag{5.38e}$$

The total stiffness matrix $\bar{\mathbf{K}}_T$ for an enriched element (2) can be derived using the material and geometric stiffness matrices as

$$\bar{\mathbf{K}}_T = \mathbf{K}^{\text{mat}} + \mathbf{K}^{\text{geo}} = \int_\ell \bar{\mathbf{B}}^{enh\,T}_{(X,\bar{\mathbf{d}})}\, E_0\, \bar{\mathbf{B}}^{enh}_{(X,\bar{\mathbf{d}})}\, dX + \int_\ell \mathbf{G}^{enh\,T}_{(X)}\, S_{XX}\, \mathbf{G}^{enh}_{(X)}\, dX \tag{5.38f}$$

in which the hypo-elastic constitutive rule is employed as $S_{XX} = E_0\, E_{XX}$, with E_0 denoting the Young's modulus of the material. The discrete system of equations can therefore be obtained as

$$\boldsymbol{\Psi}(\bar{\mathbf{d}}) = \bar{\mathbf{K}}_T(\bar{\mathbf{d}})\, \delta\bar{\mathbf{d}} - \delta\mathbf{f} = 0 \tag{5.38g}$$

This system of nonlinear equations can be solved using the Newton–Raphson procedure. In Table 5.1, the nodal displacements of free-end are given together with the residual forces $\boldsymbol{\Psi}(\bar{\mathbf{d}})$ for three values of external loading p at various iterations using the FEM and X-FEM analyses. Obviously, the results of displacement are fairly acceptable along the bar representing a good performance of the X-FEM in large deformation.

In order to illustrate the effects of blending elements in the large X-FEM deformation analysis, this example is solved once again using five 2-noded elements, as shown in Figure 5.7, in which a corrected X-FEM formulation is employed based on the Fries strategy described in Chapter 4. In the corrected X-FEM analysis, in addition to those nodes that are enriched in the standard X-FEM, all nodal points in the blending elements are enriched. As a result, nodal points 2, 3, 4, and 5 are enriched using the modified enrichment function $\psi_{\text{Mod}}(X)$ as

$$\psi_{\text{Mod}}(X) = R(X)\,\psi(X) = R(X)\left(|X-3| - |X_I - 3|\right) \tag{5.39a}$$

where X_I is the coordinate of corresponding nodal point, and $R(X)$ is the cutoff function defined based on the ramp function as

Table 5.1 An axial bar with a weak discontinuity

Loading p	Model	Iteration	Nodal displacements at the free-end	Residual norms
$p = 1 \times 10^3$ kg	FEM	1	0.0057857	2.5571
		2	0.0057754	1.4353E–005
	X-FEM	1	0.0055965	2.6334
		2	0.005587	1.5431E–005
$p = 1 \times 10^4$ kg	FEM	1	0.057857	256.92
		2	0.056862	0.12804
		3	0.056861	1E–006
	X-FEM	1	0.055965	264.67
		2	0.05504	0.14571
		3	0.05504	3.26E–008
$p = 1 \times 10^5$ kg	FEM	1	0.57857	26,902
		2	0.50272	790.56
		3	0.5009	0.58711
		4	0.50089	0.0001
	X-FEM	1	0.55965	27796
		2	0.4887	885.94
		3	0.48705	0.76509
		4	0.48705	5.1307E–007

The nodal displacements of free-end and the residual norms for three values of loading p at various iterations using the FEM and X-FEM analyses

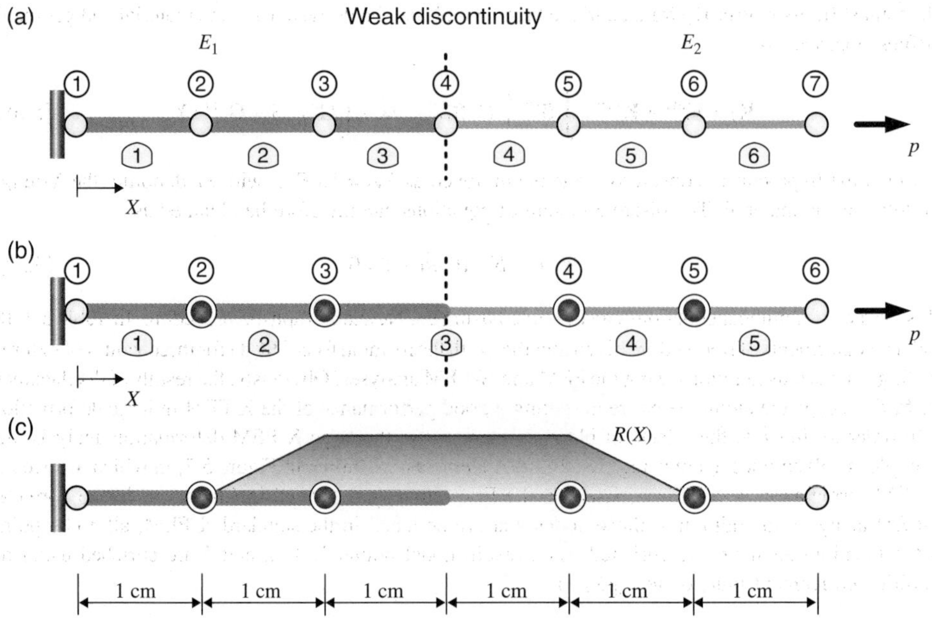

Figure 5.7 An axial bar with a material interface at its center in large deformation: (a) the FEM model; (b) the corrected X-FEM model including blending elements; (c) A graphical representation of the cutoff function

$$R(X)=\begin{cases}0 & 0\le X\le 1\\ X-1 & 1<X\le 2\\ 1 & 2<X\le 4\\ 5-X & 4<X\le 5\\ 0 & 5<X\le 6\end{cases} \tag{5.39b}$$

A graphical representation of $R(X)$ is shown in Figure 5.7c. The matrix **B** for all FEM elements as well as the standard elements (1) and (5) in the X-FEM model is equal to $\mathbf{B}_{(X)}^{std}=\langle -1+1\rangle$, while for the reproducing and blending elements are as follows

$$\begin{aligned}{}^{(2)}\mathbf{B}_{(X)}^{enh}&=\left\langle {}^{(2)}\mathbf{B}_{(X)}^{std}\quad {}^{(2)}\mathbf{B}_{(X)}^{enr}\right\rangle\\ &=\langle -1\quad +1\quad -(|X-3|-2)(X-1)-(-X+2)(X-1)+(-X+2)(|X-3|-2)\\ &\qquad (|X-3|-1)(X-1)-(X-1)(X-1)+(X-1)(|X-3|-2)\rangle\end{aligned} \tag{5.39c}$$

$$\begin{aligned}{}^{(3)}\mathbf{B}_{(X)}^{enh}&=\left\langle {}^{(3)}\mathbf{B}_{(X)}^{std}\quad {}^{(3)}\mathbf{B}_{(X)}^{enr}\right\rangle\\ &=\langle -0.5\quad +0.5\quad -0.5(|X-3|-1)-(-0.5X+2)\quad 0.5(|X-3|-1)-(0.5X-1)\rangle\end{aligned} \tag{5.39d}$$

$$\begin{aligned}{}^{(4)}\mathbf{B}_{(X)}^{enh}&=\left\langle {}^{(4)}\mathbf{B}_{(X)}^{std}\quad {}^{(4)}\mathbf{B}_{(X)}^{enr}\right\rangle\\ &=\langle -1\quad +1\quad -(|X-3|-1)(5-X)+(5-X)(5-X)-(5-X)(|X-3|-1)\\ &\quad +(5-X)(|X-3|-2)+(X-4)(5-X)-(X-4)(|X-3|-2)\rangle\end{aligned} \tag{5.39e}$$

Table 5.2 Modeling an axial bar with the X-FEM considering the blending elements

Loading p	Model	Iteration	Nodal displacements at the free-end	Residual forces $\mathbf{\Psi}(\bar{\mathbf{d}})$
$p = 1 \times 10^3$ kg	FEM	1	0.0057857	2.5571
		2	0.0057754	9.8995E–006
	X-FEM	1	0.0057857	2.5571
		2	0.0057754	1.2819E–005
$p = 1 \times 10^4$ kg	FEM	1	0.057857	256.92
		2	0.056861	0.098936
		3	0.056861	1.4142E–005
	X-FEM	1	0.057857	256.92
		2	0.056862	0.12804
		3	0.056861	5E–006
$p = 1 \times 10^5$ kg	FEM	1	0.57857	26,902
		2	0.50225	584.17
		3	0.50091	6.6716
		4	0.5009	0.07361
		5	0.50089	0.00072863
	X-FEM	1	0.57857	26,902
		2	0.50272	790.56
		3	0.5009	0.58712
		4	0.50089	4.4721E–005

A comparison between the FEM and X-FEM techniques.

Based on these $\mathbf{B}$ matrices, the matrix $\bar{\mathbf{B}}$ and the material and geometric stiffness matrices ($\mathbf{K}^{\text{mat}}$ and $\mathbf{K}^{\text{geo}}$) can be obtained. By assembling the stiffness matrices of the standard, reproducing and blending elements, a system of nonlinear equations can be obtained that can be solved using the Newton–Raphson procedure. In Table 5.2, the nodal displacements at free-end of the bar are given together with the residual forces $\mathbf{\Psi}(\bar{\mathbf{d}})$ for three values of external loading p at various iterations using the FEM and X-FEM analyses. A complete agreement can be observed between the FEM and X-FEM techniques.

5.4.2 *Modeling a Plate with the Material Interface*

The next example is of a plate in tension with the material interface at its center chosen to illustrate the performance of large X-FEM deformation formulation for a two-dimensional problem. The geometry, boundary conditions, and X-FEM mesh of the problem are shown in Figure 5.8. The hypo-elastic constitutive rule is employed with material properties given for the left part as $E_1 = 2 \times 10^6$ kg/cm^2 and $\nu_1 = 0.3$, and those of the right part as $E_2 = 7 \times 10^5$ kg/cm^2 and $\nu_2 = 0.3$. The plate is restrained at the left edge and subjected to a ramp load p at the right edge. In order to model the material interface in the plate, the ramp enrichment function $\varphi(X) = |X - 3|$ is employed considering the enriched degrees of freedom at nodal points of element (2). In the X-FEM model, the enrichment function is defined using the shifted distance function $\psi(X) = \varphi(X) - \varphi(X_I) = |X - 3| - 1$.

The X-FEM analysis is carried out using three 4-noded elements with the standard shape functions $N^{std}_{I(\xi,\eta)} = \frac{1}{4}(1 + \xi_I \xi)(1 + \eta_I \eta)$ and the enriched shape functions $N^{enr}_{J(\xi,\eta)} = N^{std}_{J(\xi,\eta)}\, \psi(X)$. The matrix $\bar{\mathbf{B}}$ for an enriched element (2) can be defined as

$$\bar{\mathbf{B}}^{enh}_{I\,(X,\bar{\mathbf{d}})} = \mathbf{B}^{enh}_{I\,(X)} + \boldsymbol{A}_{\theta\,(X,\bar{\mathbf{d}})} \mathbf{G}^{enh}_{I\,(X)} \tag{5.40a}$$

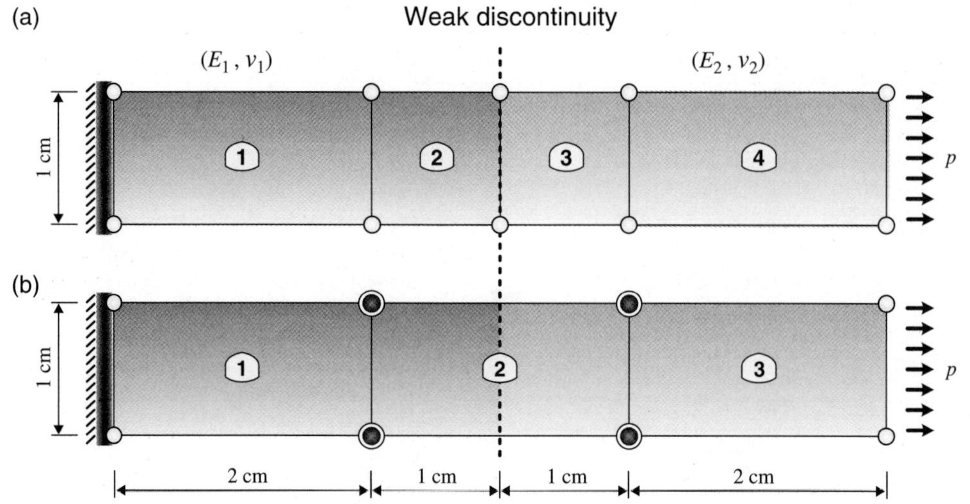

Figure 5.8 A plate with a material interface at its center in large deformation: (a) FEM model; (b) X-FEM model

where

$$\mathbf{B}_{I(\boldsymbol{X})}^{enh} = \begin{bmatrix} N_{I,X}^{std} & 0 \\ 0 & N_{I,Y}^{std} \\ N_{I,Y}^{std} & N_{I,X}^{std} \end{bmatrix} + \begin{bmatrix} N_{I,X}^{enr} & 0 \\ 0 & N_{I,Y}^{enr} \\ N_{I,Y}^{enr} & N_{I,X}^{enr} \end{bmatrix} \tag{5.40b}$$

$$\mathbf{G}_{I(\boldsymbol{X})}^{enh} = \begin{bmatrix} N_{I,X}^{std} & 0 \\ 0 & N_{I,X}^{std} \\ N_{I,Y}^{std} & 0 \\ 0 & N_{I,Y}^{std} \end{bmatrix} + \begin{bmatrix} N_{I,X}^{enr} & 0 \\ 0 & N_{I,X}^{enr} \\ N_{I,Y}^{enr} & 0 \\ 0 & N_{I,Y}^{enr} \end{bmatrix} \tag{5.40c}$$

$$\boldsymbol{A}_{\theta(\boldsymbol{X},\bar{\mathbf{d}})} = \sum_{J=1}^{4} \begin{bmatrix} N_{J,X}^{std} & N_{J,X}^{std} & 0 & 0 \\ 0 & 0 & N_{J,Y}^{std} & N_{J,Y}^{std} \\ N_{J,Y}^{std} & N_{J,Y}^{std} & N_{J,X}^{std} & N_{J,X}^{std} \end{bmatrix} \begin{Bmatrix} u_{xJ} \\ u_{yJ} \\ u_{xJ} \\ u_{yJ} \end{Bmatrix} + \sum_{J=1}^{4} \begin{bmatrix} N_{J,X}^{enr} & N_{J,X}^{enr} & 0 & 0 \\ 0 & 0 & N_{J,Y}^{enr} & N_{J,Y}^{enr} \\ N_{J,Y}^{enr} & N_{J,Y}^{enr} & N_{J,X}^{enr} & N_{J,X}^{enr} \end{bmatrix} \begin{Bmatrix} a_{xJ} \\ a_{yJ} \\ a_{xJ} \\ a_{yJ} \end{Bmatrix} \tag{5.40d}$$

It must be noted that the derivatives in relations (5.40a), (5.40b), and (5.40c) are obtained with reference to the initial configuration. The current position vector is defined using $\boldsymbol{x} = \boldsymbol{X} + \mathbf{u}$ as

$$\boldsymbol{x} = \sum_{I=1}^{4} N_I(\xi,\eta)\,\boldsymbol{x}_I + \sum_{J=1}^{4} N_J(\xi,\eta)\,(\psi(X) - \psi(X_J))\,\bar{\mathbf{a}}_J \tag{5.40e}$$

and the derivatives of the standard and enriched shape functions can be obtained as

$$\begin{aligned} \frac{\partial N_I}{\partial X} &= \frac{\partial N_I}{\partial \xi}\bar{F}_{11} + \frac{\partial N_I}{\partial \eta}\bar{F}_{12} \\ \frac{\partial N_I}{\partial Y} &= \frac{\partial N_I}{\partial \xi}\bar{F}_{21} + \frac{\partial N_I}{\partial \eta}\bar{F}_{22} \end{aligned} \tag{5.40f}$$

where $\bar{\mathbf{F}}$ is defined based on the mapping between the parent and initial configurations as

$$\bar{\mathbf{F}} = \begin{bmatrix} \dfrac{\partial \xi}{\partial X} & \dfrac{\partial \eta}{\partial X} \\ \dfrac{\partial \xi}{\partial Y} & \dfrac{\partial \eta}{\partial Y} \end{bmatrix} \tag{5.40g}$$

The Jacobian matrix $\mathbf{J}$ for the current configurations is defined as

$$\mathbf{J} = \begin{bmatrix} \dfrac{\partial x}{\partial \xi} & \dfrac{\partial y}{\partial \xi} \\ \dfrac{\partial x}{\partial \eta} & \dfrac{\partial y}{\partial \eta} \end{bmatrix} = \begin{bmatrix} \sum_{I=1}^{4} \dfrac{\partial N_I}{\partial \xi} x_I & \sum_{I=1}^{4} \dfrac{\partial N_I}{\partial \xi} y_I \\ \sum_{I=1}^{4} \dfrac{\partial N_I}{\partial \eta} x_I & \sum_{I=1}^{4} \dfrac{\partial N_I}{\partial \eta} y_I \end{bmatrix} \tag{5.40h}$$

in which the Jacobian matrix is defined between the parent and current elements, and its components can be obtained according to (5.40e) as

$$\begin{aligned} J_{11} &= \frac{\partial x}{\partial \xi} = \sum_{I=1}^{4} \frac{\partial N_I}{\partial \xi} x_I + \sum_{J=1}^{4} \frac{\partial N_J}{\partial \xi} (\psi(X) - \psi(X_J)) a_{xJ} \\ J_{12} &= \frac{\partial y}{\partial \xi} = \sum_{I=1}^{4} \frac{\partial N_I}{\partial \xi} y_I + \sum_{J=1}^{4} \frac{\partial N_J}{\partial \xi} (\psi(X) - \psi(X_J)) a_{yJ} \\ J_{21} &= \frac{\partial x}{\partial \eta} = \sum_{I=1}^{4} \frac{\partial N_I}{\partial \eta} x_I + \sum_{J=1}^{4} \frac{\partial N_J}{\partial \eta} (\psi(X) - \psi(X_J)) a_{xJ} \\ J_{22} &= \frac{\partial y}{\partial \eta} = \sum_{I=1}^{4} \frac{\partial N_I}{\partial \eta} y_I + \sum_{J=1}^{4} \frac{\partial N_J}{\partial \eta} (\psi(X) - \psi(X_J)) a_{yJ} \end{aligned} \tag{5.40i}$$

Finally, the total stiffness matrix $\bar{\mathbf{K}}_T$ for an enriched element (2) can be obtained using the material and geometric stiffness matrices as

$$\begin{aligned} (\bar{\mathbf{K}}_T)_{IJ} &= \mathbf{K}^{\text{mat}} + \mathbf{K}^{\text{geo}} \\ &= \int_{-1}^{+1}\int_{-1}^{+1} \bar{\mathbf{B}}_{I\,(\bar{\mathbf{d}})}^{enh\,T} \mathbf{D}_S^e \, \bar{\mathbf{B}}_{J\,(\bar{\mathbf{d}})}^{enh} \det[\mathbf{J}] d\xi\, d\eta + \int_{-1}^{+1}\int_{-1}^{+1} \mathbf{G}_I^{enh\,T} \mathbf{M}_S \, \mathbf{G}_J^{enh} \det[\mathbf{J}] d\xi\, d\eta \end{aligned} \tag{5.40j}$$

By assembling the stiffness matrices of the standard and enriched elements, a system of nonlinear equations $\mathbf{\Psi}(\bar{\mathbf{d}}) = \bar{\mathbf{K}}_T(\bar{\mathbf{d}})\,\delta\bar{\mathbf{d}} - \delta\mathbf{f} = 0$ can be obtained that can be solved using the Newton–Raphson procedure. In Table 5.3, the displacement at the right edge of the plate is given together with the residual force $\mathbf{\Psi}(\bar{\mathbf{d}})$ for three values of external loading p at various iterations using the FEM and X-FEM analyses. A good agreement can be seen between the FEM and X-FEM techniques.

In order to illustrate the effects of blending elements in the large X-FEM deformation analysis, the previous example is solved once again using five 4-noded elements, as shown in Figure 5.9. The corrected X-FEM formulation is employed based on the Fries strategy by enriching all nodal points in the blending and reproducing elements. In this manner, all nodal points of elements (2), (3), and (4) are enriched using the modified enrichment function $\psi_{\text{Mod}}(X) = R(X)\,\psi(X)$, in which the cutoff function $R(X)$ is defined in (5.39b). By employing the solution procedure described earlier for the standard X-FEM technique, a

Table 5.3 A plate with a weak discontinuity

Loading p	Model	Iteration	Displacements of the free-end	Residual force $\mathbf{\Psi}(\bar{\mathbf{d}})$
$p = 1 \times 10^3$ kg	FEM	1	0.005238	707.107
		2	0.005229	2.223
	X-FEM	1	0.005247	707.107
		2	0.005238	2.392
$p = 1 \times 10^4$ kg	FEM	1	0.052376	7071.068
		2	0.051561	222.620
		3	0.051561	0.080
	X-FEM	1	0.052467	7,071.068
		2	0.051647	239.425
		3	0.051647	0.084
$p = 1 \times 10^5$ kg	FEM	1	0.523757	70,710.678
		2	0.460301	22,616.054
		3	0.458879	503.564
		4	0.458878	0.329
	X-FEM	1	0.524671	70,710.678
		2	0.460904	24,215.743
		3	0.459463	527.905
		4	0.459462	0.351

The displacement of free-end and the residual force $\mathbf{\Psi}(\bar{\mathbf{d}})$ for three values of loading p at various iterations using the FEM and X-FEM analyses

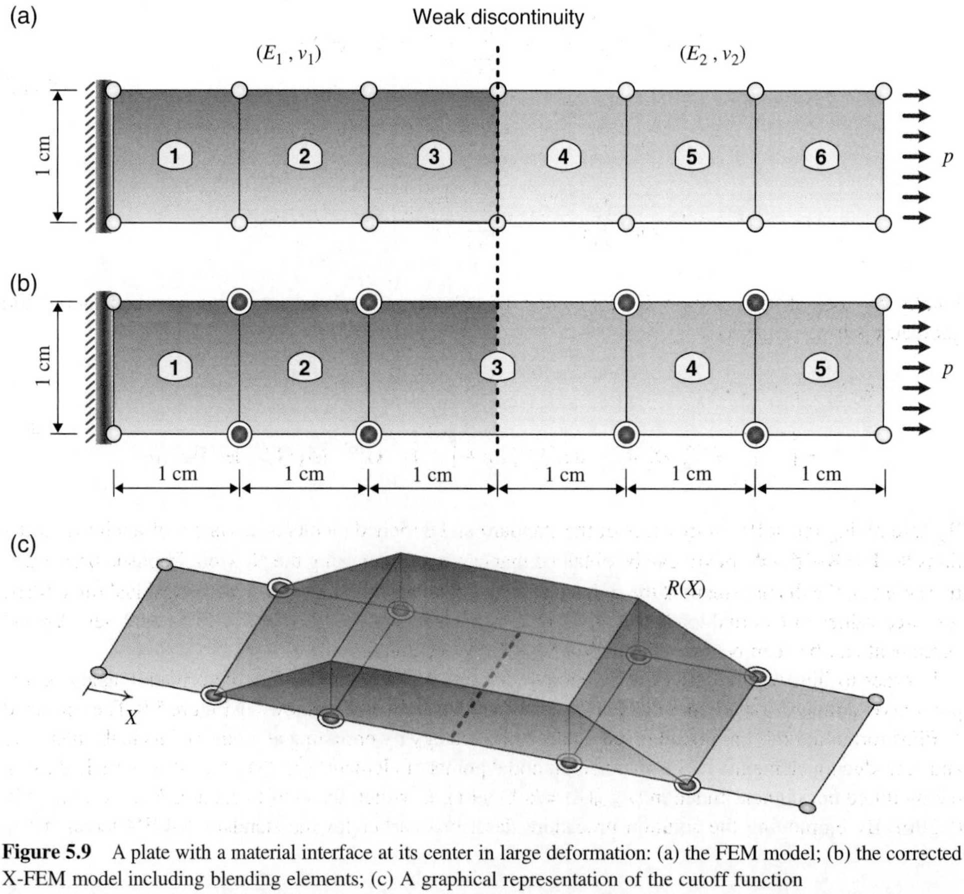

Figure 5.9 A plate with a material interface at its center in large deformation: (a) the FEM model; (b) the corrected X-FEM model including blending elements; (c) A graphical representation of the cutoff function

Table 5.4 Modeling a plate with the X-FEM considering the blending elements

Loading p	Model	Iteration	Displacements of the free-end	Residual force $\boldsymbol{\Psi}(\bar{\mathbf{d}})$
$p = 1 \times 10^3$ kg	FEM	1	0.005238	707.107
		2	0.005230	2.107
	X-FEM	1	0.005247	707.107
		2	0.005238	2.234
$p = 1 \times 10^4$ kg	FEM	1	0.052379	7071.068
		2	0.051564	211.141
		3	0.051564	0.078
	X-FEM	1	0.052469	7071.068
		2	0.051649	223.779
		3	0.051649	0.082
$p = 1 \times 10^5$ kg	FEM	1	0.523793	70,710.678
		2	0.460320	21,588.093
		3	0.458897	495.811
		4	0.458896	0.320
	X-FEM	1	0.524691	70,710.678
		2	0.460915	22,808.256
		3	0.459473	520.481
		4	0.459472	0.344

A comparison between the FEM and X-FEM techniques.

system of nonlinear equations can be obtained that can be solved using the Newton–Raphson procedure. In Table 5.4, the displacement at the right edge of the plate is given together with the residual force $\boldsymbol{\Psi}(\bar{\mathbf{d}})$ for three values of external loading p at various iterations using the FEM and X-FEM analyses. A good agreement can be seen between the FEM and X-FEM techniques.

5.5 Application of X-FEM in Large Deformation Problems

In order to illustrate the accuracy and versatility of the X-FEM in large plasticity deformations, three numerical examples are presented, including the free-die pressing with a horizontal material interface and a rigid central core, and the free-die pressing of a shaped-tablet component. The examples are solved using both FEM and X-FEM techniques, and the results are compared. In order to perform a real comparison, the number of elements in FEM and X-FEM meshes is taken to be almost equal. In all examples, two uniform X-FEM meshes of 600 and 1000 linear elements are used independent of the shape of discontinuity to assess the accuracy of discretization. While in FEM analyses, the FEM meshes need to be conformed to the geometry of interface. All examples are simulated by a plane strain representation and the convergence tolerance is set to 10^{-14}. For each example, both enrichment functions ψ^1 and ψ^2, given in Eqs. (5.22) and (5.23), are implemented to simulate the discontinuity due to different types of material properties in two intersected parts of an element. The results of X-FEM analyses obtained by the enrichment functions ψ^1 and ψ^2 are denoted by X-FEM–ψ^1 and X-FEM–ψ^2, respectively. A convergence study is performed for each example by evaluation of the relative error of energy norm, and the rate of convergence is compared with the optimal convergence rate. For this purpose, four various X-FEM meshes of 600, 1000, 2000, and 3400 quadrilateral elements are considered. Since the exact solution is not available for a comparison, a FE analysis with very fine mesh is carried out as a reference solution to illustrate the applicability of the proposed model in solving problems with arbitrary and multiple material interfaces due to closed-die pressing.

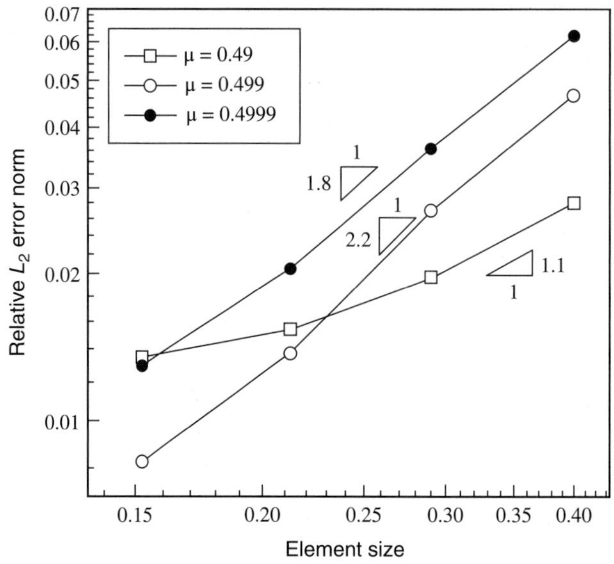

Figure 5.10 The variation of relative L_2 error norm with the element size for a unit volumetric cubic component, as the Poisson ratio ν tends to 0.5

It must be noted that although locking in incompressible limit is a major concern in the FE analysis, no volumetric locking has been observed in the analyses of X-FEM plasticity problems. In order to investigate the stability in incompressible limit, a unit volumetric cubic component is modeled using the material parameters pertinent to the von-Mises yield criterion. In Figure 5.10, the variations of relative L_2 error norm with element size are plotted, as the Poisson ratio ν tends to 0.5. As it can be seen from the figure, there is no volumetric locking even with the extremely unreasonable Poisson ratio of 0.4999. However, it must be noted that the treatment of near-incompressibility is important when significant plastic flows occurs. In this case, the well-known mixed formulation based on the mixed X-FEM could be used to prevent the locking of numerical approximation in an incompressible limit (Vidal, Villon, and Huerta, 2003; Legrain, Moës, and Huerta, 2008). The results presented in next section clearly indicate that the proposed approach can be efficiency used to model the large plasticity deformation behavior of solid mechanics problems.

5.5.1 Die-Pressing with a Horizontal Material Interface

The first example is of a free-die pressing with horizontal material interface, as shown in Figure 5.11a. The component is restrained at the top edge and a uniform compaction is imposed at the bottom up to 1.3 cm height reduction. The upper part of the component is assumed to have an elasto-plastic von-Mises behavior with a Young's modulus of 2.1×10^5 kg/cm^2, Poisson ratio 0.35, yield stress 2400 kg/cm^2, and a hardening parameter of 3.0×10^4 kg/cm^2, while the lower part is considered to be elastic with the Young's modulus of 2.1×10^6 kg/cm^2 and Poisson ratio 0.35. Four regular meshes corresponding to the FEM and X-FEM techniques are employed. In FE simulation, the FE mesh is conformed to the boundary of horizontal interface (Figure 5.11b), while in X-FEM analysis the interface passes through the elements (Figure 5.11c). Both functions ψ^1 and ψ^2 are proposed as enrichment functions due to discontinuity in different material properties. For those elements intersected by the interface, a partitioning procedure is used to generate the sub-triangles. The Gauss points of sub-triangles are then employed to evaluate the stiffness matrix of elements.

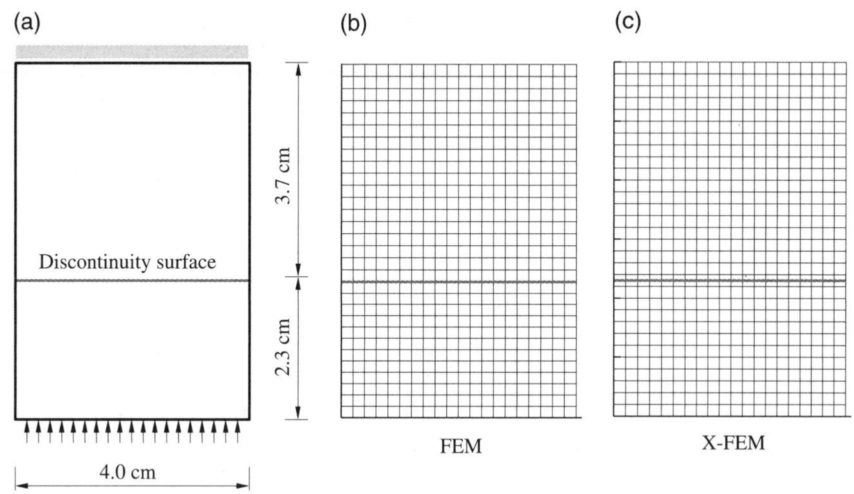

Figure 5.11 Die pressing with horizontal material interface: (a) Problem definition; (b) The coarse FEM mesh; (c) The coarse X-FEM mesh

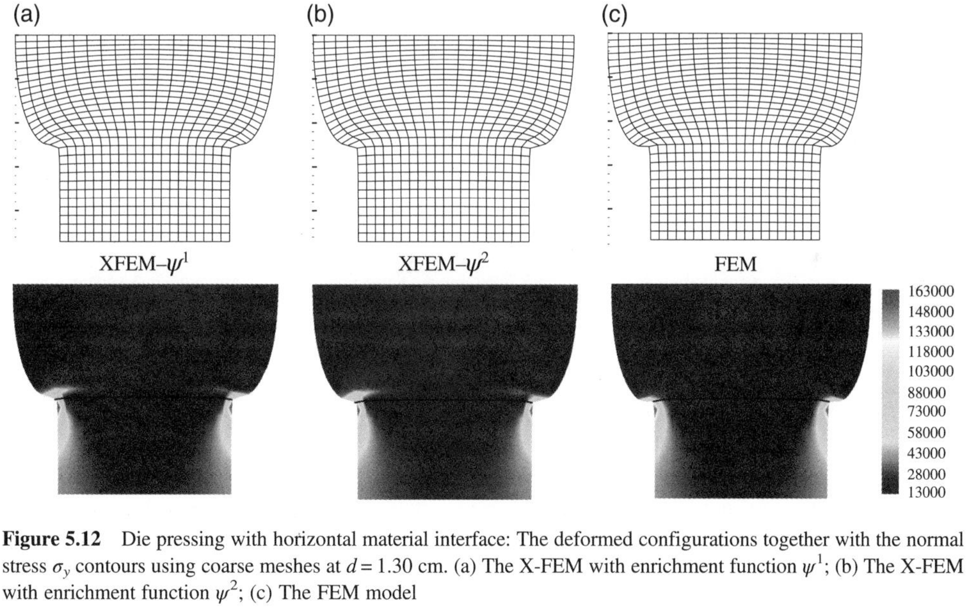

Figure 5.12 Die pressing with horizontal material interface: The deformed configurations together with the normal stress σ_y contours using coarse meshes at $d = 1.30$ cm. (a) The X-FEM with enrichment function ψ^1; (b) The X-FEM with enrichment function ψ^2; (c) The FEM model

The deformed configuration of the X-FEM and FEM techniques for 600 elements are presented in Figure 5.12 corresponding to the compaction deformation of 1.3 cm. Also presented in this figure is the distribution of normal stress σ_y contours for both X-FEM and FEM at the final die-pressing. In order to make a comparison between two techniques, the curves of normal stress σ_y are plotted in Figure 5.13 along the material interface, and perpendicular to the interface at the left hand side for both X-FEM and FEM. Also, plotted in Figure 5.14a are the evolutions of von-Mises stress along the material interface

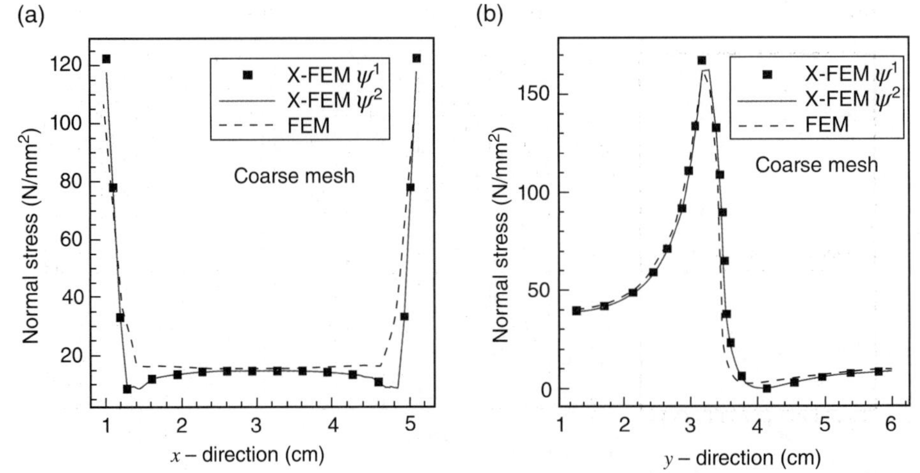

Figure 5.13 Die pressing with horizontal material interface; A comparison of the normal stress σ_y between FEM and X-FEM techniques using the enrichment functions ψ^1 and ψ^2: (a) The stress distribution along the interface; (b) The stress distribution perpendicular to the interface at the left-hand side

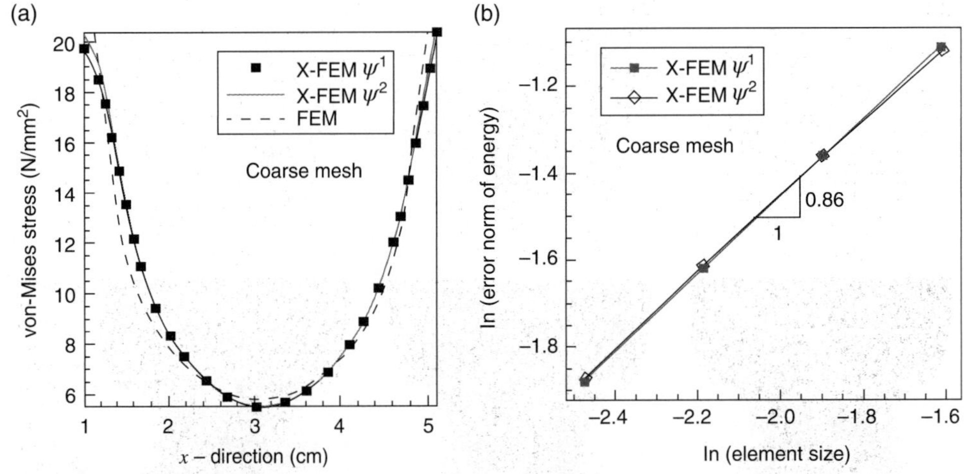

Figure 5.14 Die pressing with horizontal material interface: A comparison between the FEM and X-FEM techniques using the enrichment functions ψ^1 and ψ^2 along the interface. (a) The variation of von-Mises stress; (b) The rate of convergence in energy norm

using both techniques at the deformation of 1.3 cm. A good agreement can be observed between two numerical simulations. In order to present the rate of convergence, the error of energy norm is obtained for four X-FEM meshes of 600, 1000, 2000, and 3400 quadrilateral elements using both enrichment functions. Since the exact solution is not available for this example, a very fine FEM mesh of 3400 elements is used to obtain the reference solution. In Figure 5.14b, the variations of energy error versus the element size are plotted with respect to the reference solution on a log-log plot. The convergence rate α is equal to 0.86 for both enrichment functions, which is very close to the optimal convergence rate. In order to present the computational efforts involved in numerical simulation, the convergence rates of the Newton iterations using FEM and X-FEM models are given in Tables 5.5 and 5.6. Table 5.5 summarizes the required number of Newton iterations per load step for both X-FEM and FEM techniques. In Table 5.6, the values of residual norm during various iterations are given for a typical load step at the deformation of 1.0 cm. Obviously, the X-FEM convergence rate of the Newton iterations based on both enrichment functions ψ^1 and

Table 5.5 Die pressing with horizontal material interface

	X-FEM ψ^1 (600 elements)	X-FEM ψ^2 (600 elements)	X-FEM ψ^1 (1000 elements)	X-FEM ψ^2 (1000 elements)	FEM (600 elements)	FEM (1000 elements)
Average number of iterations	6.137	6.207	6.073	6.114	6.117	6.253

Newton iterations per load step.

Table 5.6 Die pressing with horizontal material interface

Iteration	X-FEM ψ^1 (600 elements)	X-FEM ψ^2 (600 elements)	X-FEM ψ^1 (1000 elements)	X-FEM ψ^2 (1000 elements)	FEM (600 elements)	FEM (1000 elements)
1	0.8795E–01	0.8795E–1	0.8084E–01	0.8084E–01	0.8793E–01	0.8081E–01
2	0.3302E–03	0.3315E–3	0.7661E–04	0.7657E–04	0.1405E–04	0.4231E–04
3	0.1212E–05	0.1234E–5	0.2228E–06	0.2217E–06	0.2806E–07	0.1791E–06
4	0.2834E–07	0.2857E–7	0.1839E–08	0.1826E–08	0.2055E–09	0.1693E–08
5	0.2113E–09	0.2121E–09	0.1707E–10	0.1693E–10	0.1530E–11	0.1501E–10
6	0.1623E–11	0.1625E–11	0.1602E–12	0.1587E–12	0.1141E–13	0.1325E–12
7	0.1249E–13	0.1246E–13	0.1562E–14	0.1492E–14	0.3480E–15	0.1167E–14
8	0.5313E–15	0.1445E–15				

Residual norm for a load step at $d = 1.0$ cm.

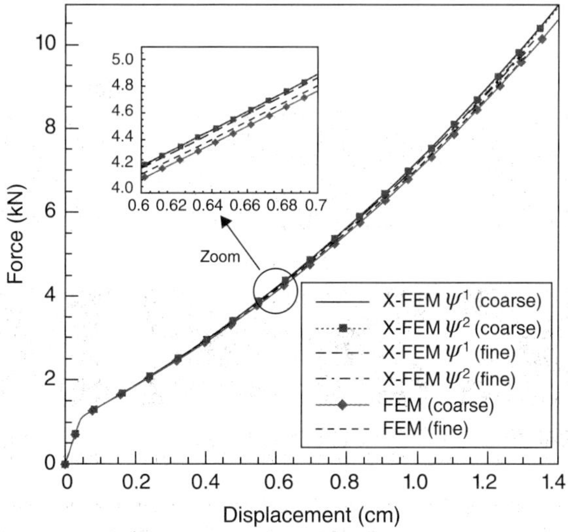

Figure 5.15 The variation of reaction force with vertical displacement for a die pressing with horizontal material interface: A comparison between FEM and X-FEM techniques

ψ^2 are almost close to the FEM solution. Finally, a comparison of the reaction force versus vertical displacements is performed in Figure 5.15 between the X-FEM and FEM approaches. The results clearly demonstrate that how the X-FEM technique can be efficiently used to model the large elasto-plastic deformations.

5.5.2 Die-Pressing with a Rigid Central Core

The second example refers to die pressing of the rectangular component with a rigid central core, as shown in Figure 5.16a. The geometry, material properties, and boundary conditions are similar to those given in previous example. The initial coarse meshes for both FEM and X-FEM techniques are presented in Figure 5.16b and c. The component is modeled for the volume reduction of 50%. In Figure 5.17, the

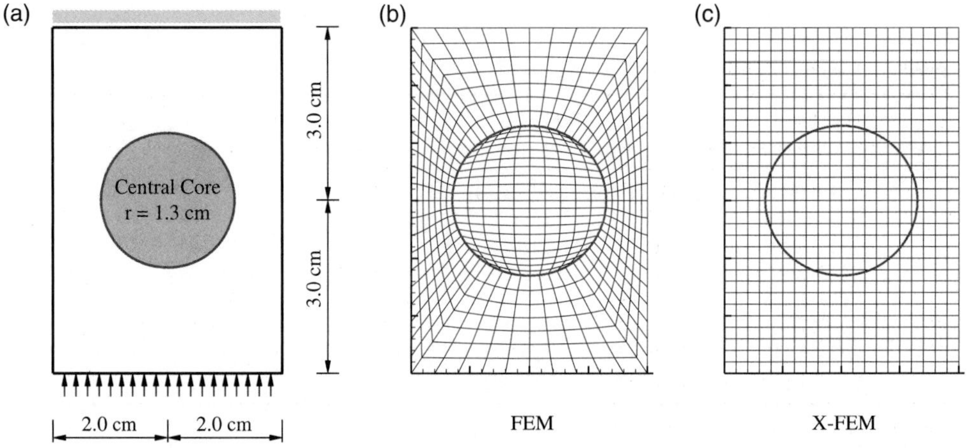

Figure 5.16 Die pressing with a rigid central core: (a) Problem definition; (b) The coarse FEM mesh; (c) The coarse X-FEM mesh

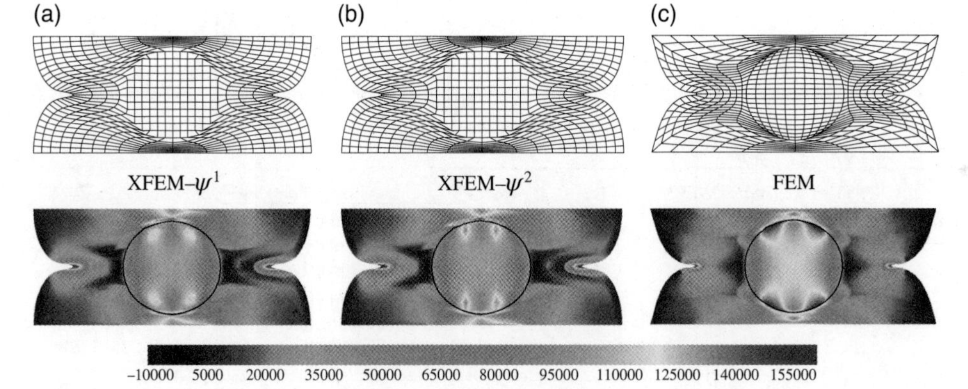

Figure 5.17 Die pressing with a rigid central core: The deformed configurations together with the normal stress σ_y contours using coarse meshes at $d = 3.0$ cm. (a) The X-FEM with enrichment function ψ^1; (b) The X-FEM with enrichment function ψ^2; (c) The FEM model

deformed configuration and the normal stress σ_y distribution of compacted component are shown at the final stage of die-pressing for the FEM and those of X-FEM using the enrichment functions ψ^1 and ψ^2. In order to compare the results of two different techniques, the force-displacement curves are plotted in Figure 5.18 using the X-FEM and FEM analyses. In Tables 5.7 and 5.8, the average required number

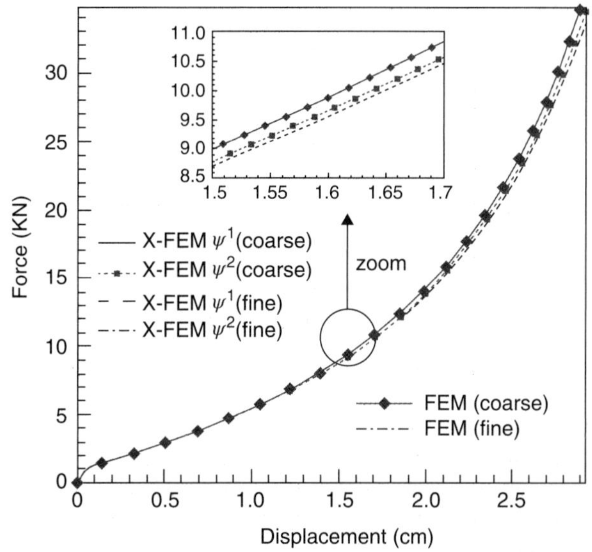

Figure 5.18 The variation of reaction force with vertical displacement for die pressing with a rigid central core: A comparison between FEM and X-FEM techniques

Table 5.7 Die pressing with a rigid central core

	X-FEM ψ^1 (600 elements)	X-FEM ψ^2 (600 elements)	X-FEM ψ^1 (1000 elements)	X-FEM ψ^2 (1000 elements)	FEM (620 elements)	FEM (1020 elements)
Average number of iterations	6.019	6.359	6.521	5.830	5.689	6.250

Newton iterations per load step.

Table 5.8 Die pressing with a rigid central core

Iteration	X-FEM ψ^1 (600 elements)	X-FEM ψ^2 (600 elements)	X-FEM ψ^1 (1000 elements)	X-FEM ψ^2 (1000 elements)	FEM (620 elements)	FEM (1020 elements)
1	0.2582E−01	0.2337E−01	0.2358E−01	0.2589E−01	0.2421E−01	0.2412E−01
2	0.3446E−04	0.9383E−04	0.5076E−04	0.6479E−03	0.4817E−04	0.3554E−04
3	0.8422E−07	0.1986E−06	0.2861E−05	0.9045E−06	0.6671E−06	0.1316E−06
4	0.1386E−09	0.3728E−09	0.6010E−08	0.1438E−08	0.1390E−08	0.1320E−09
5	0.1799E−12	0.4921E−12	0.1492E−10	0.1326E−11	0.2489E−11	0.3596E−12
6	0.9288E−15	0.1346E−14	0.2199E−13	0.3729E−14	0.2639E−14	0.5789E−15
7			0.4887E−15			

Residual norm for a load step at $d = 0.9$ cm.

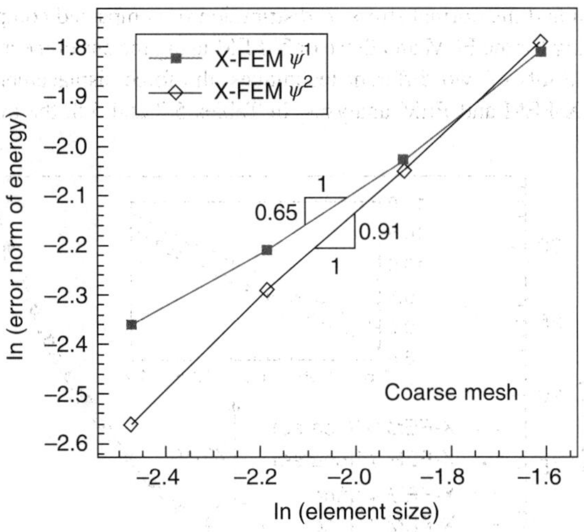

Figure 5.19 The rate of convergence in energy norm for die pressing with a rigid central core

of Newton iterations per load step and the values of residual norm at the deformation of 0.9 cm are compared for two different techniques. From these tables it can be concluded that the enrichment function ψ^2 using a fine X-FEM mesh improves the required number of Newton iterations. Furthermore, it shows that the X-FEM convergence rate of the Newton iterations using the enrichment function ψ^2 is very close to the FEM solution, however, a very fine mesh is necessary to obtain a good convergence. In order to investigate the accuracy of the computational model, a convergence study is carried out using four quadrilateral X-FEM meshes, in which a very fine FEM mesh of 3500 elements is used as a reference solution. The variations of energy error versus the element size are plotted in Figure 5.19. The convergence rate α is equal to 0.65 and 0.91 for X-FEM– ψ^1 and X-FEM– ψ^2, respectively. Obviously, the enrichment function ψ^2 improves the convergence rate in comparison to ψ^1.

5.5.3 *Closed-Die Pressing of a Shaped-Tablet Component*

The next example demonstrates the performance of X-FEM technique in modeling large elasto-plastic deformation of shaped-tablet pressing, as shown in Figure 5.20. The geometry and boundary conditions of this practical example are shown in this figure. A shaped tablet is compacted by simultaneous action of top and bottom punches. This component is a challenging example for the proposed X-FEM approach because it involves two crossing interfaces; the first is the surface between punch and tablet and the second consists of the contact interface between the die and tablet. In addition, these two interfaces cross a single element at the intersection of two different materials, as shown in Figure 5.21a. The punch and sleeve are both elastic with a Young's modulus of 2.1×10^6 kg/cm^2 and Poisson ratio of 0.35. The tablet has a non-linear elasto-plastic behavior characterized by the von-Mises constitutive model. The material parameters of tablet component are similar to the first example. On the virtue of symmetry, the process is modeled for one-quarter of component.

In Figure 5.21a and b, the X-FEM and FEM meshes are shown at the initial stage of compaction. In the X-FEM analysis, the discontinuity in homogeneity due to different material properties of tablet and punch is modeled by the enrichment functions ψ^1 and ψ^2, and the discontinuity in displacement fields due to contact interface between the die and tablet is simulated by the Heaviside enrichment function. In the FEM analysis, the FE mesh is combined with the contact elements along the contact surface. In

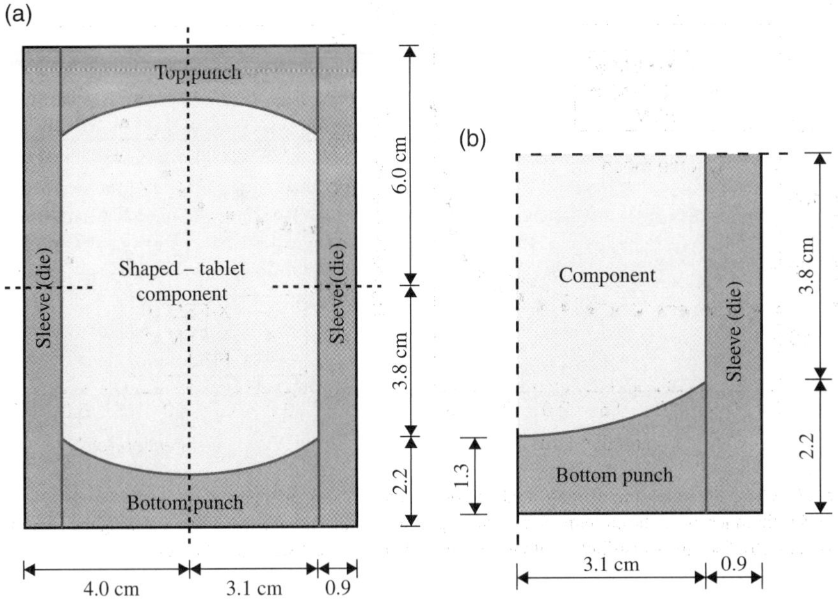

Figure 5.20 Closed-die pressing of shaped tablet component: (a) Problem definition; (b) One-quarter of specimen

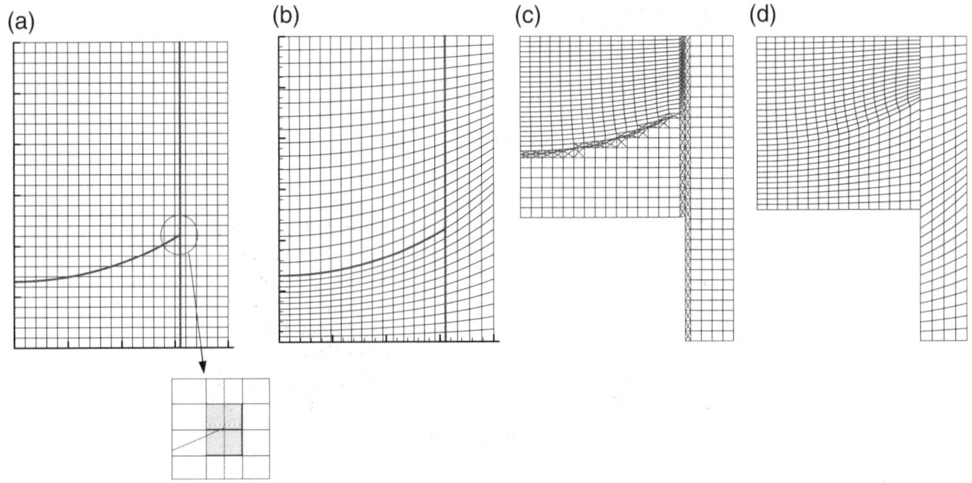

Figure 5.21 Closed-die pressing of shaped tablet component: (a and b) The initial configuration of X-FEM and FEM meshes of 600 elements, (c and d) The deformed configuration of X-FEM and FEM meshes of 600 elements

Figure 5.21c and d, the deformed X-FEM and FEM meshes corresponding to 600 elements are shown at the compaction deformation of 2.60 cm. Triangular partitioning of enriched elements in X-FEM technique can be observed in this figure. The variations of normal stresses σ_y along the curved interface and vertical contact surface are shown in Figure 5.22 at the final stage of die-pressing for both techniques. It can be seen from the figure that the enrichment function ψ^2 results in a better distribution of stress, and can be

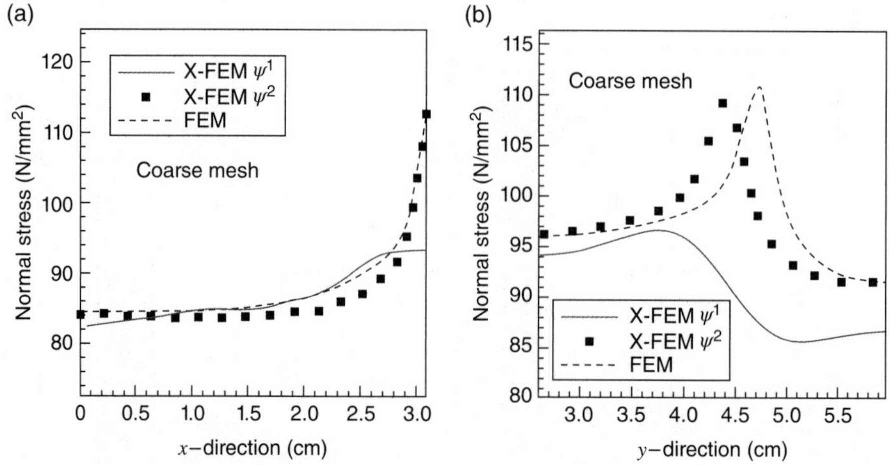

Figure 5.22 Closed-die pressing of shaped tablet component: A comparison of normal stress σ_y between the FEM and X-FEM techniques using the enrichment functions ψ^1 and ψ^2; (a) The stress distribution along the curved interface at $d = 2.60$ cm; (b) The stress distribution along the vertical contact surface at $d = 2.60$ cm

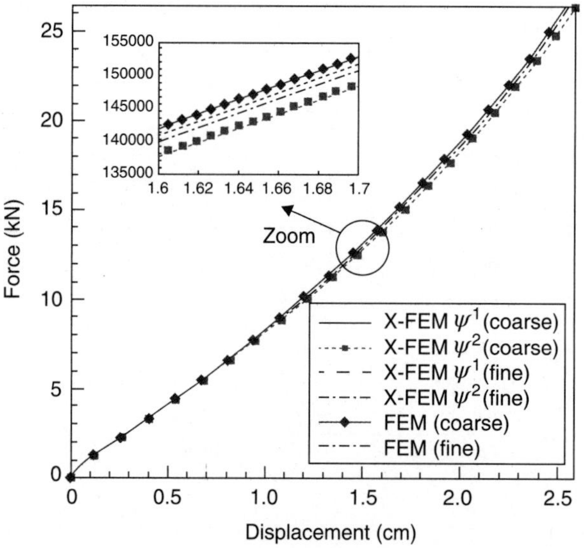

Figure 5.23 The variation of bottom punch reaction with vertical displacement for a closed-die pressing of shaped tablet component: A comparison between FEM and X-FEM techniques

compared with that obtained from the FE solution. This can be particularly observed in the intersection of two interfaces. It can be therefore concluded that the enrichment function ψ^2 can be more reasonable for elements cut by multiple interfaces.

The evolution of bottom punch reaction force with its vertical displacement is depicted in Figure 5.23. Complete agreements can be observed between the X-FEM and FEM. In Table 5.9, the average required number of iterations per load step is summarized for both X-FEM and FEM. Also given in Table 5.10 are

Table 5.9 Closed-die pressing of shaped tablet component

	X-FEM ψ^1 (600 elements)	X-FEM ψ^2 (600 elements)	X-FEM ψ^1 (1000 elements)	X-FEM ψ^2 (1000 elements)	FEM (600 elements)	FEM (1000 elements)
Average number of iterations	7.021	6.587	6.935	6.210	5.931	5.383

Newton iterations per load step.

Table 5.10 Closed-die pressing of shaped tablet component

Iteration	X-FEM ψ^1 (600 elements)	X-FEM ψ^2 (600 elements)	X-FEM ψ^1 (1000 elements)	X-FEM ψ^2 (1000 elements)	FEM (600 elements)	FEM (1000 elements)
1	0.40832–02	0.4966E–02	0.4514E–02	0.5174E–02	0.4779E–02	0.4366E–02
2	0.5383E–06	0.8647E–07	0.2849E–06	0.8585E–07	0.1611E–06	0.1446E–06
3	0.8854E–08	0.7628E–09	0.4034E–08	0.7587E–09	0.6497E–09	0.1154E–09
4	0.1572E–09	0.9073E–11	0.6280E–10	0.8955E–11	0.1251E–10	0.9691E–13
5	0.2899E–11	0.1193E–12	0.1018E–11	0.1233E–12	0.2427E–12	0.1571E–15
6	0.5499E–13	0.1614E–14	0.1703E–13	0.1660E–14	0.4717E–14	
7	0.1052E–14		0.2651E–15			

Residual norm for a load step at $d = 2.0$ cm.

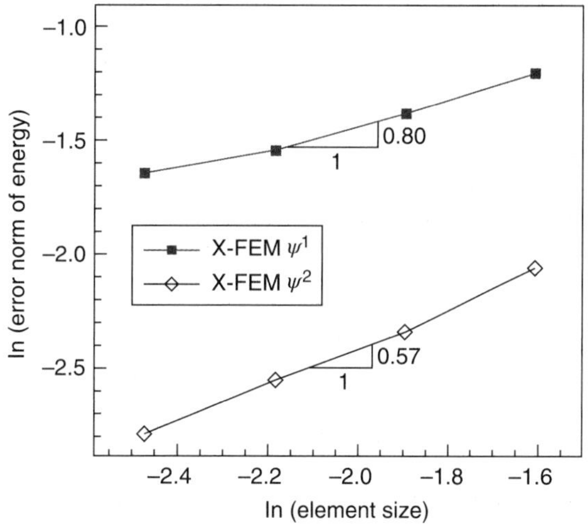

Figure 5.24 The rate of convergence in energy norm for the closed-die pressing of shaped tablet component

the values of residual norm for a typical load step at the deformation of 2.0 cm. It can be seen from these tables that the X-FEM convergence rate of the Newton iterations using the enrichment function ψ^2 is better than ψ^1 compared to the FE solution. The convergence study is performed using the simulation of a very fine FE mesh of 3400 quadrilateral elements as a reference solution. Finally, the relative error of energy norm is measured for both enrichment functions and the results are summarized in Figure 5.24.

5.6 The Extended Arbitrary Lagrangian–Eulerian FEM

In preceding sections, the X-FEM has been successfully applied into large deformation problems exhibiting discontinuities and inhomogeneities. In this technique, the discontinuities are taken into account by adding appropriate functions into the standard approximation through a PUM. The discontinuity in the displacement field is modeled by additional DOF at nodal points of elements, which have been cut by the discontinuity. This description allows releasing the computational grids from conforming to the shape of the discontinuity. However, in large deformation analysis with large mass fluxes, the conventional FE and extended FE techniques using the updated Lagrangian formulation may suffer from serious numerical difficulties when the deformation of material is significantly large. This difficulty can be particularly observed in higher order elements when severe distortion of elements may lead to singularities in an isoparametric mapping of the elements, aborting the calculations or causing numerical errors.

In order to overcome this difficulty, the ALE approach has been proposed in the literature for classical FEM (Haber, 1984; Kennedy and Belytschko, 1981). In the ALE approach, the mesh motion is taken arbitrarily from material deformation to keep element shapes optimal. There are basically two methods of solutions for the governing equations of ALE formulation. The majority of ALE analyses, whether quasi-static or dynamic, are based on the computationally convenient operator split technique (Benson, 1989; Rodriguez-Ferran, Casadei, and Huerta, 1998). In this method, the whole process can be decoupled into a regular Lagrangian phase followed by an Eulerian phase. In an uncoupled technique, the analysis is first carried out according to Lagrangian phase at each time step until the required convergence is attained. The Eulerian phase is then applied to keep the mesh configuration regular. However, from a theoretical point of view, the fully coupled ALE approach represents a true kinematic description in which a moving reference configuration is applied to describe the motion instead of material configuration in the Lagrangian formulation (Bayoumi and Gadala, 2004; Yamada and Kikuchi, 1993). In ALE description, the choice of the material, spatial, or any arbitrary configuration yields to a Lagrangian, Eulerian, or ALE description, respectively. Since the mesh and material movements are uncoupled, the convective term appears in the balance of momentum equation.

The aim of this study is to develop an ALE technique in large deformation analysis of the X-FEM based on an operator splitting technique. In the Lagrangian phase, a typical X-FEM analysis is first carried out with updated Lagrangian approach. The Eulerian phase is then applied to update the mesh, while the material interface is independent of the FE mesh. The main difficulty in ALE formulation of solid mechanics is the path dependent material behavior. The constitutive equation of ALE nonlinear mechanics contains a convective term which reflects the relative motion between the physical motion and the mesh motion. However, for the purpose of simplicity, only linear elastic behavior of materials is considered here. In linear elastic simulation, the convective effects only occur in the mass balance, momentum balance, and energy balance, but not in the constitutive equation. This uncoupled approach makes easily the extension of a pure Lagrangian extended FE code to the ALE technique and allows the use of original updated Lagrangian program to solve the relevant ALE equations. In what follows, the ALE method is first introduced in large deformation problems. The algorithm of an uncoupled ALE solution together with the mesh motion strategy and stress update procedure are then demonstrated based on the Godunov method for transferring of variables from Lagrangian mesh to relocated mesh. The extended arbitrary Lagrangian–Eulerian FEM (X-ALE-FEM) technique is finally described based on the combined ALE and X-FEM to reduce the mesh distortion occurred in conventional large X-FEM deformation by introducing a technique to update the nodal values of level set and stress components during the Eulerian phase.

5.6.1 ALE Formulation

The ALE formulation was first applied to nonlinear solid mechanics path-dependent materials with definition of the tangent stiffness matrix and consistent linearization process by Haber (1984), Liu, Belytschko, and Chang (1986), Liu *et al.* (1988), and Benson (1989). The technique was then implemented in various solid mechanics problems, including: the incompressible hyper-elasticity (Yamada and Kikuchi, 1993), metal forming simulation (Ghosh and Raju, 1996), transient dynamic analysis (Rodriguez-Ferran, Perez-Foguet, and Huerta, 2002), hyperelasto-plasticity (Armero and Love, 2003), finite strain plasticity (Bayoumi and Gadala, 2004), and pressure-sensitive materials (Khoei *et al.*, 2006a; Khoei, Anahid, and Shahim, 2008b). A key issue in the ALE formulation is an efficient mesh motion technique in order to achieve an acceptable and a reasonable result. There are various mesh relocation techniques, based on an ALE split operator (Benson, 1989), an uniform distribution of the equivalent plastic strain indicator (Ghosh and Raju, 1996), the transfinite mapping algorithm using the nodal relocation (Wang and Gadala, 1997; Gadala and Wang, 1998), and so on. These techniques are able to keep the elements with a good shape by equalizing the size of elements and by avoiding the shape distortion without changing the mesh topology.

The main difficulty in extending the ALE formulation from FEM to X-FEM is the convective term that reflects the relative motion between the physical motion and the mesh motion. The correct treatment of this convective term in the X-FEM is a key point in the X-ALE-FEM modeling of solid mechanic problems. The most popular approach to deal with the convective term is the use of a split, or fractional-step method. Each time-step is divided into a Lagrangian phase and an Eulerian phase. Convection is neglected in the material phase, which is thus identical to a time-step in a standard Lagrangian X-FEM analysis. The stress and plastic internal variables are then transferred from Lagrangian mesh to the relocated mesh in order to evaluate the relative mesh-material motion in the convection phase. Special care has to be taken with respect to the time integration of the constitutive equations, often denoted the stress update, since the stress field is usually discontinuous across the elements due to the fact that stress values are only evaluated at discrete integration points, which normally lie inside the element. To handle this, an approach called the Godunov scheme, is used here for the stress update.

5.6.1.1 Kinematics

In the ALE description, three different configurations are considered; the material domain Ω_0, spatial domain Ω, and reference domain $\hat{\Omega}$, which is called the ALE domain. The material motion is defined by $x_i^{\mathrm{m}} = f_i(X_j, t)$, with X_j denoting the material point coordinates and $f_i(X_j, t)$ a function which maps the body from the initial or material configuration Ω_0 to the current or spatial configuration Ω. The initial position of material points is denoted by x_i^{g} called the reference or ALE coordinate in which $x_i^{\mathrm{g}} = f_i(X_j, 0)$. The reference domain $\hat{\Omega}$ is defined to describe the mesh motion and is coincident with mesh points so it can be denoted by computational domain. The mesh motion is described by $x_i^{\mathrm{m}} = \hat{f}_i\left(x_j^{\mathrm{g}}, t\right)$. The material coordinate can then be related to ALE coordinate by $x_i^{\mathrm{g}} = \hat{f}_i^{-1}\left(x_j^{\mathrm{m}}, t\right)$. The mesh displacement can be defined as

$$u_i^{\mathrm{g}}\left(x_j^{\mathrm{g}}, t\right) = x_i^{\mathrm{m}} - x_i^{\mathrm{g}} = \hat{f}_i\left(x_j^{\mathrm{g}}, t\right) - x_i^{\mathrm{g}} \tag{5.41}$$

It must be noted that the mesh motion can be simply obtained from material motion replacing the material coordinate by ALE coordinate. The mesh velocity can be defined as

$$v_i^{\mathrm{g}}\left(x_j^{\mathrm{g}}, t\right) = \frac{\partial \hat{f}_i\left(x_j^{\mathrm{g}}, t\right)}{\partial t} = \left.\frac{\partial x_i^{\mathrm{m}}}{\partial t}\right|_{x_j^{\mathrm{g}}} \tag{5.42}$$

in which the ALE coordinate x_j^{g} and material coordinate X_j in material velocity are fixed. In ALE formulation, the convective velocity c_i is defined using the difference between the material and mesh velocities as

$$c_i = v_i^{\mathrm{m}} - v_i^{\mathrm{g}} = \frac{\partial x_i^{\mathrm{m}}}{\partial x_j^{\mathrm{g}}} \left.\frac{\partial x_j^{\mathrm{g}}}{\partial t}\right|_{X_k} = \frac{\partial x_i^{\mathrm{m}}}{\partial x_j^{\mathrm{g}}} w_j \tag{5.43}$$

where the material velocity $v_i^{\mathrm{m}} = \left(\partial x_i^{\mathrm{m}}/\partial t\right)_{X_j}$ can be obtained using the chain rule expression with respect to the ALE coordinate x_j^{g} and time t. In Eq. (5.43), the referential velocity w_i is defined by $w_{\mathrm{i}} = (\partial x_i^{\mathrm{g}}/\partial t)_{X_j}$. This relationship between the convective velocity c_i, material velocity v_i^{m}, mesh velocity v_i^{g}, and referential velocity w_i is frequently used in ALE formulation.

The general relationship between the material time derivatives and referential time derivatives of any scalar function f_i can then be written as

$$\left.\frac{\partial f_i}{\partial t}\right|_{X_j} = \left.\frac{\partial f_i}{\partial t}\right|_{x_j^{\mathrm{g}}} + \left.\frac{\partial f_i}{\partial x_i^{\mathrm{g}}}\frac{\partial f_i}{\partial t}\right|_{X_j} = \left.\frac{\partial f_i}{\partial t}\right|_{x_j^{\mathrm{g}}} + \frac{\partial f_i}{\partial x_j^{\mathrm{m}}} c_j \tag{5.44}$$

This equation can be used to deduce the fundamental conservation laws of continuum mechanics, that is, the momentum, mass, and constitutive equations, in nonlinear ALE description.

5.6.1.2 ALE Governing Equations

In the ALE technique, the governing equations can be derived by substituting the relationship between the material time derivatives and referential time derivatives, that is, Eq. (5.44), into the continuum mechanics governing equations. This substitution gives rise to convective terms in the ALE equations that account for the transport of material through the grid. Thus, the momentum equation in ALE formulation can be written similar to the updated Lagrangian description by consideration of the material time derivative terms as

$$\rho\, \dot{v}_i^{\mathrm{m}} = \sigma_{ji,j} + \rho b_i \tag{5.45}$$

where ρ is the density, σ the Cauchy stress, and b_i the body force. In this equation, the material time derivative of velocity $\dot{v}_i^{\mathrm{m}}$ can be obtained by specializing the general relationship (5.44) to $\dot{v}_i^{\mathrm{m}}$ as

$$\dot{v}_i^{\mathrm{m}} = \left.\frac{\partial v_i^{\mathrm{m}}}{\partial t}\right|_{x_j^{\mathrm{g}}} + \frac{\partial v_i^{\mathrm{m}}}{\partial x_j^{\mathrm{m}}} c_j \tag{5.46}$$

Substituting Eq. (5.46) into (5.45), the momentum equation can then be written as

$$\rho\left(\left.\frac{\partial v_i^{\mathrm{m}}}{\partial t}\right|_{x_j^{\mathrm{g}}} + \frac{\partial v_i^{\mathrm{m}}}{\partial x_j^{\mathrm{m}}} c_j\right) = \frac{\partial \sigma_{ij}}{\partial x_j^{\mathrm{m}}} + \rho b_i \tag{5.47}$$

The mass balance in ALE formulation can be similarly derived by specializing the general relationship (5.44) to the density ρ as

$$\left.\frac{\partial \rho}{\partial t}\right|_{x_j^{\mathrm{g}}} + \frac{\partial \rho}{\partial x_j^{\mathrm{m}}} c_j = -\rho \left.\frac{\partial v_j^{\mathrm{m}}}{\partial x_j^{\mathrm{m}}}\right|_{X_j} \tag{5.48}$$

Finally, in order to describe the constitutive equation for nonlinear ALE formulation, the general relationship (5.44) is specialized to the stress tensor, thus

$$\left.\frac{\partial \boldsymbol{\sigma}}{\partial t}\right|_{x_j^{\mathrm{g}}} + \frac{\partial \boldsymbol{\sigma}}{\partial x_j^{\mathrm{m}}} c_j = \mathbf{q} \tag{5.49}$$

where $\mathbf{q}$ accounts for both the pure straining of the material and the rotational terms that counteract the non-objectivity of the material stress rate (Rodriguez-Ferran, Casadei, and Huerta, 1998).

The basis of any mechanical initial boundary value problem in the framework of material description is the balance of momentum equation. In the framework of the referential configuration, the mass balance and the constitutive equations are also considered, which are defined as the partial differential equations of referential description. In the quasi-static problems, the inertia force $\rho\mathbf{a}$ is negligible with respect to other forces of momentum equation, hence, the equilibrium equation in ALE and Lagrangian descriptions is exactly identical. In addition, considering the constant value of density ρ, the balance of mass equation results in $\partial v_j^{\mathrm{m}}/\partial x_j^{\mathrm{m}}\big|_{X_j}=0$, which is already satisfied. Thus, the governing equations in ALE formulation can be summarized into Eqs. (5.47) and (5.49).

5.6.2 *The Weak Form of ALE Formulation*

In order to present the weak form of initial boundary value problems in ALE description, the mass balance and the balance of linear momentum can be written in the integral form over the spatial domain Ω, multiplied by the test functions $\delta\rho$, $\delta\mathbf{v}$, and $\delta\sigma$. Clearly, there must be a relationship between the strong form and the weak form of governing equations, in which these two forms are identical. The weak form of momentum equation can be obtained by multiplying the strong form of Eq. (5.47) with the test function $\delta\mathbf{v}\in\mathrm{U}^0$, where $\mathrm{U}^0=\{\delta\mathbf{v}\mid\delta\mathbf{v}\in C_0,\ \delta\mathbf{v}=\mathbf{0}\text{ on }\Gamma_v\}$ and Γ_v indicates the part of the boundary in which the velocities are prescribed. Consider $\mathbf{v}\in\mathrm{U}$ is the trial solution with $\mathrm{U}=\{\mathbf{v}\mid\mathbf{v}\in C_0,\mathbf{v}=\hat{\mathbf{v}}\text{ on }\Gamma_v\}$ and $\hat{\mathbf{v}}$ is the prescribed velocities in Γ_v, the integration over the spatial domain results in

$$\int_\Omega\delta\mathbf{v}\,\rho\left(\left.\frac{\partial\mathbf{v}^{\mathrm{m}}}{\partial t}\right|_{\boldsymbol{x}^{\mathrm{g}}}+\frac{\partial\mathbf{v}^{\mathrm{m}}}{\partial\boldsymbol{x}^{\mathrm{m}}}\mathbf{c}\right)d\Omega=\int_\Omega\delta\mathbf{v}\left(\mathrm{div}_{x^{\mathrm{m}}}\boldsymbol{\sigma}+\rho\mathbf{b}\right)d\Omega \tag{5.50}$$

or

$$\int_\Omega\delta\mathbf{v}\,\rho\left.\frac{\partial\mathbf{v}^{\mathrm{m}}}{\partial t}\right|_{\boldsymbol{x}^{\mathrm{g}}}d\Omega+\int_\Omega\delta\mathbf{v}\,\rho\frac{\partial\mathbf{v}^{\mathrm{m}}}{\partial\boldsymbol{x}^{\mathrm{m}}}\mathbf{c}\,d\Omega=\int_\Omega\delta\mathbf{v}\,\mathrm{div}_{x^{\mathrm{m}}}\boldsymbol{\sigma}\,d\Omega+\int_\Omega\delta\mathbf{v}\,\rho\mathbf{b}\,d\Omega \tag{5.51}$$

To eliminate the stress derivatives, the first term of right hand side in this equation is rewritten using the integration part by part as

$$\int_\Omega\delta\mathbf{v}\,\rho\left.\frac{\partial\mathbf{v}^{\mathrm{m}}}{\partial t}\right|_{\boldsymbol{x}^{\mathrm{g}}}d\Omega+\int_\Omega\delta\mathbf{v}\,\rho\frac{\partial\mathbf{v}^{\mathrm{m}}}{\partial\boldsymbol{x}^{\mathrm{m}}}\mathbf{c}\,d\Omega=-\int_\Omega\delta\mathbf{v}\,\mathrm{div}_{x^{\mathrm{m}}}\boldsymbol{\sigma}\,d\Omega+\int_\Omega\delta\mathbf{v}\,\rho\mathbf{b}\,d\Omega+\int_{\Gamma_t}\delta\mathbf{v}\,\hat{\mathbf{t}}\,d\Gamma \tag{5.52}$$

where Γ_t refers to the part of boundary in which the traction vector $\hat{\mathbf{t}}$ is prescribed.

Since the mass balance is enforced in the referential description as a partial differential equation, a weak form must be developed. Considering the trial solution as $\rho\in C_0$, the weak form of the balance of mass can be obtained by integration of the strong form of mass balance, given in Eq. (5.48) over the spatial domain Ω, which has been multiplied by a test function $\delta\rho\in C_0$ as

$$\int_\Omega\delta\rho\left(\left.\frac{\partial\rho}{\partial t}\right|_{\boldsymbol{x}^{\mathrm{g}}}+\frac{\partial\rho}{\partial\boldsymbol{x}^{\mathrm{m}}}\mathbf{c}+\rho\left.\frac{\partial\mathbf{v}^{\mathrm{m}}}{\partial\boldsymbol{x}^{\mathrm{m}}}\right|_X\right)d\Omega=0 \tag{5.53}$$

In this equation, the only first derivatives are appeared with respect to the mass density ρ and the velocity $\mathbf{v}$.

Similar to the momentum equation and mass balance, the weak form of constitutive equation for non-linear ALE formulation can be obtained by multiplying the strong form of Eq. (5.49) with a test function $\delta\sigma$ and integrating over the spatial domain as

$$\int_{\Omega} \delta\sigma \left(\left.\frac{\partial \sigma}{\partial t}\right|_{\boldsymbol{x}^{\mathrm{g}}} + \frac{\partial \sigma}{\partial \boldsymbol{x}^{\mathrm{m}}} \mathbf{c} \right) d\Omega = \int_{\Omega} \delta\sigma \, \mathbf{q} \, d\Omega \tag{5.54}$$

or

$$\int_{\Omega} \delta\sigma \left.\frac{\partial \sigma}{\partial t}\right|_{\boldsymbol{x}^{\mathrm{g}}} d\Omega + \int_{\Omega} \delta\sigma \frac{\partial \sigma}{\partial \boldsymbol{x}^{\mathrm{m}}} \mathbf{c} \, d\Omega = \int_{\Omega} \delta\sigma \, \mathbf{q} \, d\Omega \tag{5.55}$$

5.6.3 The ALE FE Discretization

In the FEM, the reference domain $\hat{\Omega}$ is subdivided into a number of elements, in which, for each element e, the ALE coordinates $\boldsymbol{x}^{\mathrm{g}}$ are defined as

$$\boldsymbol{x}^{\mathrm{g}}(\boldsymbol{\xi}) = \sum_{I=1}^{N^e} N_I(\boldsymbol{\xi}) \boldsymbol{x}_I^{\mathrm{g}} \tag{5.56}$$

where $\boldsymbol{\xi}$ denotes the parent element coordinates, $N_I(\boldsymbol{\xi})$ is the interpolation shape function, $\boldsymbol{x}_I^{\mathrm{g}}$ stands for the ALE coordinates of node I, and N^e is the number of nodes of element e. The mesh displacement field can then be written following Eq. (5.41) as

$$\mathbf{u}^{\mathrm{g}}(\boldsymbol{\xi}) = \boldsymbol{x}^{\mathrm{m}}(\boldsymbol{\xi}) - \boldsymbol{x}^{\mathrm{g}}(\boldsymbol{\xi}) = \sum_{I=1}^{N^e} N_I(\boldsymbol{\xi}) \mathbf{u}_I^{\mathrm{g}} \tag{5.57}$$

Thus, the mesh, material, and convective velocities can be defined as

$$\begin{aligned}
\mathbf{v}^{\mathrm{g}}(\boldsymbol{\xi}) &= \left.\frac{\partial \boldsymbol{x}^{\mathrm{m}}}{\partial t}\right|_{\boldsymbol{x}^{\mathrm{g}}} = \sum_{I=1}^{N^e} N_I(\boldsymbol{\xi}) \mathbf{v}_I^{\mathrm{g}} \\
\mathbf{v}^{\mathrm{m}}(\boldsymbol{\xi}) &= \left.\frac{\partial \boldsymbol{x}^{\mathrm{m}}}{\partial t}\right|_{X} = \sum_{I=1}^{N^e} N_I(\boldsymbol{\xi}) \mathbf{v}_I^{\mathrm{m}} \\
\mathbf{c}(\boldsymbol{\xi}) &= \mathbf{v}^{\mathrm{m}} - \mathbf{v}^{\mathrm{g}} = \sum_{I=1}^{N^e} N_I(\boldsymbol{\xi}) \left(\mathbf{v}_I^{\mathrm{m}} - \mathbf{v}_I^{\mathrm{g}} \right) = \sum_{I=1}^{N^e} N_I(\boldsymbol{\xi}^e) \mathbf{c}_I(t)
\end{aligned} \tag{5.58}$$

Furthermore, the internal variables, such as density and stresses, can be approximated in the same manner as

$$\begin{aligned}
\rho(\boldsymbol{\xi}, t) &= \sum_{I=1}^{N^e} N_I^{\rho}(\boldsymbol{\xi}^e) \rho^I(t) \\
\boldsymbol{\sigma}(\boldsymbol{\xi}, t) &= \sum_{I=1}^{N^e} N_I^{\sigma}(\boldsymbol{\xi}^e) \boldsymbol{\sigma}^I(t)
\end{aligned} \tag{5.59}$$

where N_I^ρ and N_I^σ stand for the shape functions of density and stress, respectively. These shape functions may differ from those used to approximate the displacement field N_I. It must be noted that because the convective terms appear in governing equations, the implementation of standard Galerkin FE formulation may result in numerical instabilities. This point is especially true for severe dynamic systems. One way to alleviate these difficulties is to employ the Petrov–Galerkin formulation. In this approach, different sets of shape functions are used to interpolate the trial and test functions for displacement, stress, and density.

Substituting the material and the convective velocity ($\mathbf{v}^{\mathrm{m}}$ and $\mathbf{c}$) given in Eq. (5.58), and the density and the stresses (ρ and σ) given in Eq. (5.59), into the weak form of the balance of linear momentum given in Eq. (5.52), yields to

$$\mathbf{M}\frac{d\mathbf{v}^{\mathrm{m}}}{dt}+\mathbf{L}\mathbf{v}^{\mathrm{m}}+\mathbf{f}^{\mathrm{int}}=\mathbf{f}^{\mathrm{ext}} \tag{5.60}$$

where

$$\begin{aligned}\mathbf{M}&=\int_\Omega \rho\mathbf{N}^T\mathbf{N}\,d\Omega\\ \mathbf{L}&=\int_\Omega \rho\mathbf{N}^T\mathbf{c}\frac{d\mathbf{N}}{d\boldsymbol{x}^{\mathrm{m}}}\,d\Omega\end{aligned} \tag{5.61}$$

and

$$\begin{aligned}\mathbf{f}^{\mathrm{int}}&=\int_\Omega \frac{d\mathbf{N}^T}{d\boldsymbol{x}^{\mathrm{m}}}\boldsymbol{\sigma}\,d\Omega\\ \mathbf{f}^{\mathrm{ext}}&=\int_\Omega \rho\mathbf{N}^T\mathbf{b}\,d\Omega+\int_\Gamma \mathbf{N}^T\hat{\mathbf{t}}\,d\Gamma\end{aligned} \tag{5.62}$$

As mentioned earlier, the term of inertia forces can be neglected in the quasi-static problems. Thus, the momentum Eq. (5.60) can be simplified to $\mathbf{f}^{\mathrm{int}}=\mathbf{f}^{\mathrm{ext}}$.

In a similar manner, the FE formulation for the mass balance can be obtained by substituting the material velocity from Eq. (5.58) and the density from Eq. (5.59) into Eq. (5.50) as

$$\mathbf{M}^\rho\frac{d\rho}{dt}+\mathbf{L}^\rho\rho+\mathbf{K}^\rho\rho=0 \tag{5.63}$$

where

$$\begin{aligned}\mathbf{M}^\rho&=\int_\Omega (\mathbf{N}^\rho)^T\mathbf{N}^\rho d\Omega\\ \mathbf{L}^\rho&=\int_\Omega (\mathbf{N}^\rho)^T\mathbf{c}\,\frac{d\mathbf{N}^\rho}{d\boldsymbol{x}^{\mathrm{m}}}\,d\Omega\\ \mathbf{K}^\rho&=\int_\Omega (\mathbf{N}^\rho)^T\mathrm{div}_{x^{\mathrm{m}}}\mathbf{v}^{\mathrm{m}}\,\mathbf{N}^\rho\,d\Omega\end{aligned} \tag{5.64}$$

Finally, the FE formulation for the constitutive equation can be obtained by replacing the convective velocity from Eq. (5.58) and the stresses from Eq. (5.59) into (5.55) as

$$\mathbf{M}^\sigma\frac{d\boldsymbol{\sigma}}{dt}+\mathbf{L}^T\boldsymbol{\sigma}=\mathbf{q} \tag{5.65}$$

where

$$\begin{aligned}\mathbf{M}^\sigma&=\int_\Omega (\mathbf{N}^\sigma)^T\mathbf{N}^\sigma d\Omega\\ \mathbf{L}^\sigma&=\int_\Omega (\mathbf{N}^\sigma)^T\mathbf{c}\,\frac{d\mathbf{N}^\sigma}{d\boldsymbol{x}^{m}}\,d\Omega\end{aligned} \tag{5.66}$$

5.6.4 The Uncoupled ALE Solution

There are basically two methods of solutions for the governing equations of ALE formulation (5.60), (5.63), and (5.65); the fully coupled solution and uncoupled solution (Belytschko, Liu, and Moran, 2000a). In the fully coupled solution method, no further simplifications can be considered and various terms must be calculated simultaneously. This approach was proposed by Yamada and Kikuchi (1993), Bayoumi and Gadala (2004), and Khoei *et al.* (2006a). In the uncoupled solution technique, the fully coupled equations are not employed and the whole process is decoupled into a Lagrangian phase and an Eulerian phase by employing a splitting operator. Such a technique has been employed by Benson (1989), Rodriguez-Ferran, Casadei, and Huerta (1998), and Khoei *et al.* (2008b). In an uncoupled technique, the analysis is first carried out according to Lagrangian phase at each time step until the required convergence is attained. The Eulerian phase is then applied to keep the mesh configuration regular. In this study, the uncoupled ALE solution is applied as it makes possible to upgrade the standard Lagrangian X-FEM program to X-ALE-FEM case with as little expenditure as possible. In this case, the Lagrangian computation can be included as a sub-step of the new ALE computation.

The basis of splitting operator in uncoupled solution is to separate the material (Lagrangian) phase from convective (Eulerian) phase, which is combined with a smoothing phase (Figure 5.25). In the Lagrangian phase, the convective effects are neglected, so the material body deforms from its material configuration to its spatial one. In this framework, the nodal and quadrature points may lead eventually to a high distortion of the spatial discretization after the Lagrangian step. In order to reduce this distortion a smoothing phase is then applied, which leads to the final spatial discretization. This allows the computation of mesh velocity, which leads to the convective velocity, or Eulerian phase. The advantage of ALE splitting operator is that the calculations are performed in the Lagrangian step with no convective terms to achieve the equilibrium. When the equilibrium is achieved in Lagrangian step, an Eulerian step is performed by transferring the internal variables from the Lagrangian mesh to the relocated mesh.

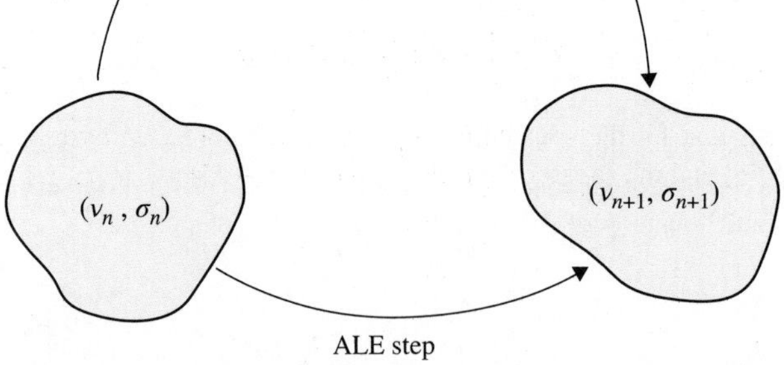

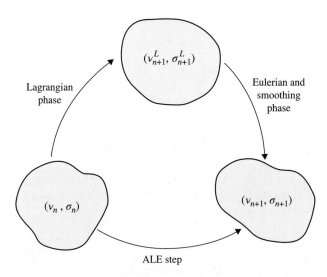

Figure 5.25 Decomposition of the ALE step into a Lagrangian phase and an Eulerian phase combined with a smoothing phase

5.6.4.1 Material (Lagrangian) Phase

In matcrial phase, the convective terms are neglected, so the momentum equation is identical to a time–step in a standard Lagrangian analysis. Thus, the momentum balance (5.60) in quasistatic analysis becomes

$$\int_{\Omega} \frac{d\mathbf{N}}{d\boldsymbol{x}^{\mathrm{m}}} \boldsymbol{\sigma}\, d\Omega = \int_{\Omega} \rho \mathbf{N}^T \mathbf{b}\, d\Omega + \int_{\Gamma} \mathbf{N}^T \hat{\mathbf{t}}\, d\Gamma \tag{5.67}$$

which is a static equilibrium equation with no time, velocity, and convective terms in ALE momentum balance.

The ALE constitutive Eq. (5.65) for the material phase can be simplified as

$$\mathbf{M}^{\sigma} \frac{d\boldsymbol{\sigma}}{dt} = \mathbf{q} \tag{5.68}$$

which needs to be integrated at each time step to update the stress from $\boldsymbol{\sigma}_n$ at time t_n to $\boldsymbol{\sigma}_{n+1}^L$ after the Lagrangian phase. It means that in the absence of convective terms, the grid points move attached to material particles. Thus, the Lagrangian phase can be performed with the same stress update algorithm used in Lagrangian simulation, which handles the constitutive equation at the Gauss point level.

5.6.4.2 Smoothing Phase

In order to reduce the mesh distortion in spatial configuration, a remeshing procedure must be applied between the Lagrangian and the Eulerian phase. The algorithm produces smoother meshes without redefining the element connectivity. Since the mesh moves independently from the material, the mesh velocity can be evaluated in order to compute the convective velocity. There are various remeshing strategies that have been proposed by researchers (Benson, 1989; Ghosh and Raju, 1996; Gadala and Wang, 1998). In this study, the simple algorithms are used based on the "Laplacian approach" and the "mid-area averaging technique". In these approaches, the mesh distortion is controlled by moving the inner nodes in an appropriate way. In addition, the boundary nodes are remained on the boundary by allowing only a tangential movement to those nodes.

The Laplacian approach is one of the most popular and simple smoothing strategy, which has been used by researchers to produce smoother meshes. In this technique, the spatial position of smoothed node $\boldsymbol{x}_i$ can be computed using the spatial position after the Lagrangian phase $\boldsymbol{x}_i^L$ as (Herrmann, 1976)

$$\boldsymbol{x}_i = \frac{1}{(2-w)N} \sum_{e=1}^{N} (\boldsymbol{x}_{e1} + \boldsymbol{x}_{e2} - w\boldsymbol{x}_{e3}) \tag{5.69}$$

where $\boldsymbol{x}_i$ presents the spatial position of node i, and N is the number of 4-node elements connecting to node i (typically $N=4$). For each element $1 \le e \le N$, $\boldsymbol{x}_{e1}$ and $\boldsymbol{x}_{e2}$ are the coordinates of the nodes of element e connected to $\boldsymbol{x}_i$ by an edge, and $\boldsymbol{x}_{e3}$ is the coordinate of the node of element e at the opposite corner of $\boldsymbol{x}_i$, as shown in Figure 5.26a. In this relation, w is the weighting factor $0 \le w \le 1$, in which for $w=0$ yields to the commonly used Laplacian scheme.

The mid-area averaging technique is a modification of the Laplacian approach. In this method, the considered node is in the centroid of all connected elements, and the area of different elements is taken into account. Thus, the spatial position of smoothed node $\boldsymbol{x}_i$ can be computed as

$$\boldsymbol{x}_i = \frac{\sum_{e=1}^{N} A_e \boldsymbol{x}_{Se}}{\sum_{e=1}^{N} A_e} \tag{5.70}$$

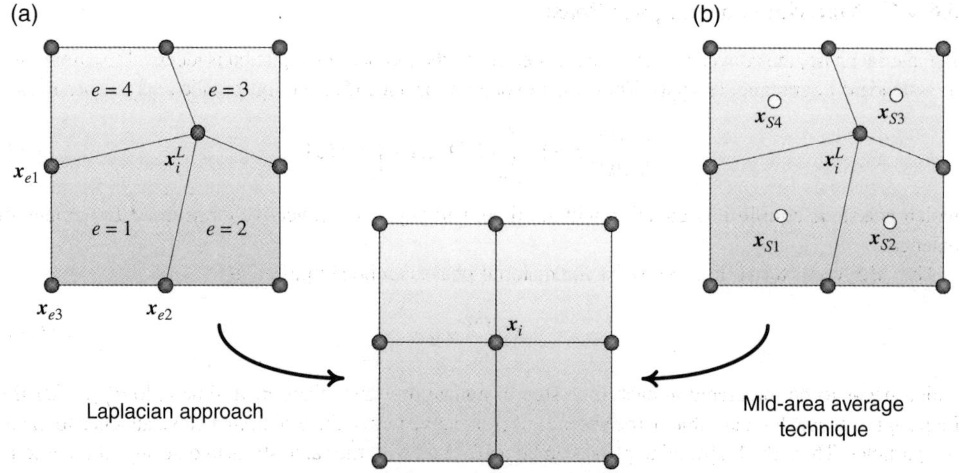

Figure 5.26 Remeshing procedure in smoothing phase: (a) Laplacian approach; (b) Mid-area averaging technique

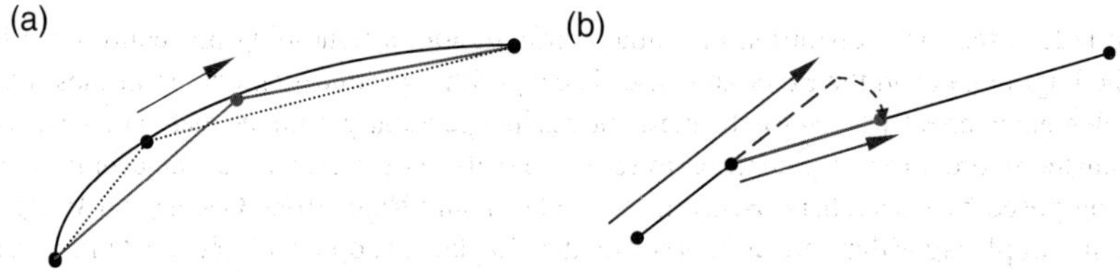

Figure 5.27 Illustration of boundaries motion: (a) Curve treatment; (b) Extension

where $\boldsymbol{x}_{Se}$ is the location of the centroid of element e as shown in Figure 5.26b, A_e is the area of element e, and N indicates the number of four-node elements connecting to node i. After smoothing the position of all nodal points $\boldsymbol{x}_{n+1}$ at time t_n, the convective term $\mathbf{c}_{n+1}$ can be computed using the material (Lagrangian) displacement $\mathbf{u}_{n+1}^L$ and mesh displacement $\mathbf{u}_{n+1}^g$ for quasi-static problems as $\mathbf{c}_{n+1} = \mathbf{u}_{n+1}^L - \mathbf{u}_{n+1}^g$.

In order to perform the smoothing procedure for boundary nodes, it is assumed that the nodal points remain on the boundary by allowing only a tangential movement to these nodes. In this case, the boundary nodes are allowed to move in a normal direction following the motion of material points. For boundaries with pre-known deformed shape, it is sufficient to assign a motion in the normal direction equal to the known material motion. To assign a desired tangential motion for nodal points, the following algorithm is performed here. First, a second order polynomial is constructed using the considered node and two connected nodes on the boundary. The position of mid-node is then corrected according to the position of two connected nodes, as shown in Figure 5.27a. This procedure will not work properly when the determinant of matrix of its solution becomes zero. To avoid this problem, the approach is modified in the manner so that the extension of mid-node lies on the next connecting line, as shown in Figure 5.27b.

5.6.4.3 Convection (Eulerian) Phase

The final part of the operator splitting technique includes data transferring of solution obtained by the Lagrangian phase onto the new relocated mesh, which was developed through the mesh smoothing algorithm. In the Eulerian (or convection) phase, the convective terms that were neglected during the

Lagrangian phase are taken into account. Since we are dealing with history-dependent materials and due to the fact that different material integration points have different histories, these quantities must be updated in order to compute the history-dependent variables in the next time step. These variables are computed at discrete integration points, which normally lie inside the element; this yields to discontinuous fields that produces some difficulties, since the spatial gradients of these variables are required. To overcome these difficulties, a smooth gradient field is obtained based on the Godunov technique that circumvents the computation of history variable gradients.

The constitutive Eq. (5.65) for the convection phase can be written as

$$\left.\frac{\partial \sigma}{\partial t}\right|_{x_j^{\mathrm{g}}} + \frac{\partial \sigma}{\partial x_j^{\mathrm{m}}} c_j = 0 \tag{5.71}$$

which needs to be integrated at each time step to update the stress from $\boldsymbol{\sigma}_{n+1}^L$ to $\boldsymbol{\sigma}_{n+1}$ at time t_{n+1}. As noted earlier, the main difficulty in Eq. (5.71) is the stress gradient, which cannot be properly computed at the element level. In order to avoid computation of the gradients of discontinuous fields, the Godunov technique is implemented here to transfer the internal variables ρ_{n+1}^L, $\bar{\varepsilon}_{n+1}^{p^L}$, and $\boldsymbol{\sigma}_{n+1}^L$ from the Lagrangian mesh to the relocated mesh.

The Godunov method assumes a piecewise constant field of the solution of internal variable after the Lagrangian phase. In the FE framework, this is the situation if one-point quadratures are employed, however, to allow for a subsequent generalization to multiple-point quadratures, the FE can be subdivided into various sub-elements, each corresponding to the influence domain of a Gauss point. Considering the scalar quantity ψ be any components of the stress $\boldsymbol{\sigma}$, or the effective plastic strain $\bar{\varepsilon}^p$, the value of internal variable ψ_{n+1} at time t_{n+1} can be obtained from the Lagrangian solution ψ_{n+1}^L as

$$\psi_{n+1} = \psi_{n+1}^L - \frac{\Delta t}{2A} \sum_{s=1}^{N_s} \{ f_s (\psi_{n+1}^{Lc} - \psi_{n+1}^L) [1 - \alpha_0 \operatorname{sign}(f_s)] \} \tag{5.72}$$

where A is the area of sub-element, N_s the number of edges of sub-element, and ψ_{n+1}^{Lc} is the value of ψ_{n+1}^L in the contiguous sub-element across edge s (Figure 5.28). The upwind parameter α_0 is in the range of

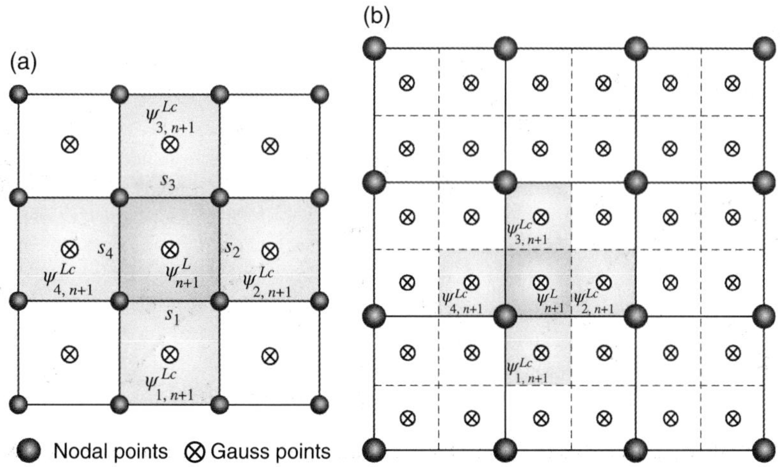

Figure 5.28 Illustration of the Godunov technique: (a) One–point–quadrature; (b) Multiple-point-quadrature

$0 \le \alpha_0 \le 1$, where $\alpha_0 = 1$ corresponds to a full-donor approximation and $\alpha_0 = 0$ is a centered approximation. In above relation, f_s is the flux of convective velocity $\mathbf{c}$ across edge s defined as

$$f_s = \int_s \mathbf{c} \cdot \mathbf{n}\, ds \tag{5.73}$$

Based on this approach, if for instance, a quadrilateral Q4 element with 2×2 integration points are employed, each element is divided into four sub-elements, as shown in Figure 5.28. In each sub-element, ψ is assumed to be constant, and represented by the Gauss point value. Thus, ψ is a piecewise constant field with respect to the mesh of sub-elements, and relation (5.72) can be employed to update the value of ψ for each sub-element.

5.6.5 The X-ALE-FEM Computational Algorithm

In the X-ALE-FEM analysis, the X-FEM is performed together with an operator splitting technique, in which each time step consists of two stages; a Lagrangian (material) phase and an Eulerian (smoothing) phase. In the material phase, the X-FEM analysis is carried out based on an updated Lagrangian approach. It means that the convective terms are neglected and only material effects are considered. The time step is then followed by an Eulerian phase in which the convective terms are considered into account. In this step, the nodal points move arbitrarily in the space so that the computational mesh has regular shape and the mesh distortion can be prevented, however, the material interface is independent of the FE mesh.

Figure 5.29 presents the mesh configuration including the material interface before and after Eulerian phase. In this figure, the position of interface has been shown in two different cases. In Figure 5.29a, the interface does not move from one element to another during smoothing phase. On the other hand, the number of elements that have been cut by the interface and partitioned by sub-polygons does not change during smoothing phase. While in Figure 5.29b, the material interface may move from one element to another. As can be seen, elements 1, 3, and 4 have been cut by the interface before mesh motion procedure, however, only element 3 is cut by the interface after Eulerian phase. Thus, the number of enriched nodes

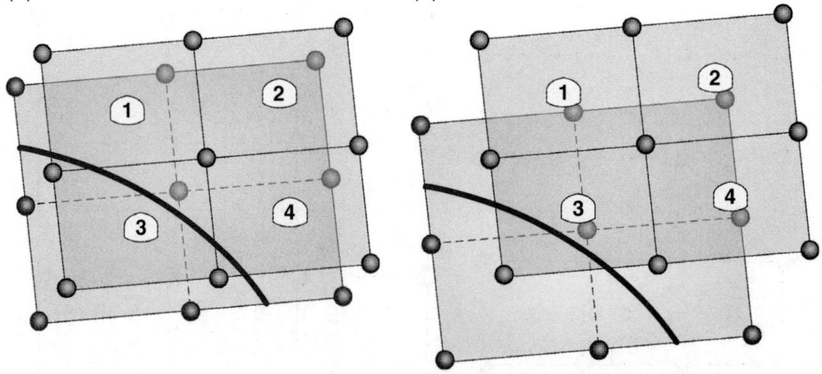

Figure 5.29 Mesh configuration before and after Eulerian phase together with the material interface in X-ALE-FEM analysis: The dashed and solid elements correspond to mesh configuration before and after mesh motion, respectively. (a) The interface does not move from one element to another; (b) The interface moves from one element to another

may be different during the X-ALE-FEM analysis, which results in different number of DOF in two successive steps. There are two main requirements, which need to be considered in the smoothing phase:

1. Due to movement of nodal points in the mesh motion process, a procedure must be applied to determine the new nodal values of level set enrichment function.
2. In the X-FEM analysis, the numerical integration of elements cut by the interface is generally performed using the Gauss quadrature points of sub-polygons obtained by partitioning procedure. However, in the case that the material interface leaves one element to another during the mesh update procedure, the number of Gauss quadrature points of an element may differ before and after mesh motion. Hence, an accurate and efficient technique must be applied into the Godunov scheme to update the stress values.

5.6.5.1 Level Set Update

During smoothing phase, the nodal points are relocated in order to keep the computational mesh in regular shape, however, the material interface is independent of the FE mesh, as shown in Figure 5.30. In this figure, the dashed lines illustrate the old mesh and the relocated elements are depicted by solid lines. The aim is to obtain the level set value for node I after Eulerian phase. For this purpose, the distance of relocated node I from the interface must be determined. The intersection of interface with the edges of old elements can be calculated at the end of Lagrangian step, called points A, B, C, and D in Figure 5.30a. The procedure used here to update the nodal values of level set is as follows; the arc ABCD is first approximated by several straight lines that connect the intersection of the interface with the edges of old elements, that is, lines AB, BC, and CD in Figure 5.30b. The support domain of node I has been indicated by elements 1, 2, 3, and 4 in Figure 5.30a. Among the elements of support domain, we define the Extended Support Domain (ESD) of node I that contains those elements cut by the interface at the end of Lagrangian phase. As can be seen from Figure 5.30b, the ESD of node I includes elements 2, 3, and 4. For each element of ESD, the distance of relocated node I is calculated from the element interface. The value of level set function for relocated node I is the minimum value of distances d_1, d_2, and d_3, as shown in Figure 5.30b. In order to determine the sign of level set value for relocated node I, this procedure is carried out once again to evaluate the value of level set for old node I. If the sign of these two level set values are similar, it means that the old and relocated nodes are in the same side of interface. Thus, the sign of level set value for relocated node I must be indicated similar to that obtained for old node I.

However, the proposed algorithm is not appropriate for determination of the level set value of nodal points, which are not enriched during Lagrangian phase, and included in an element split by the interface

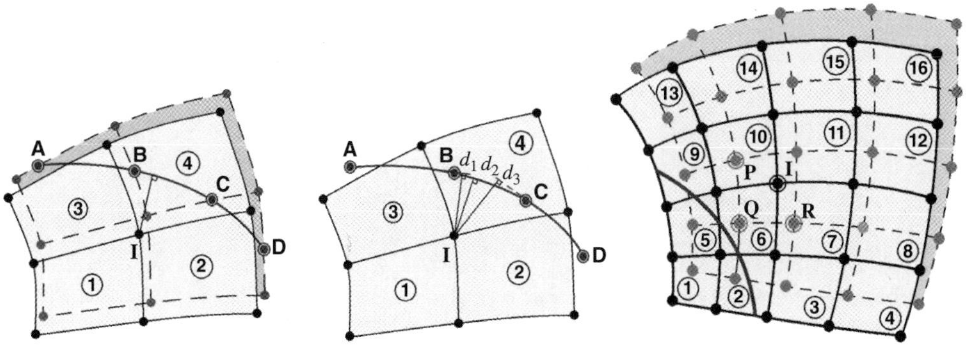

Figure 5.30 The procedure for determination of level set values after mesh motion

after mesh motion, for example, node I in Figure 5.30c. In this case, the ESD of node I in Figure 5.30c contains no elements, and the definition of ESD must be modified. In order to modify the domain of ESD for these nodal points, we first indicate the support domain of node I, for example, elements 6, 7, 10, and 11 in Figure 5.30c. All nodal points of this domain which were enriched during the Lagrangian phase are selected, that is, the circled nodes P, Q, and R in Figure 5.30c. Those elements in union of circled nodes' support domains that are cut by the interface can be considered the modified ESD of node I. The procedure will then be followed by determination of level set value for node I, as demonstrated earlier. The level set function can be finally obtained using the nodal values of level set from $\varphi(\boldsymbol{x}) = \sum_{I \in \mathcal{N}} N_I(\boldsymbol{x})\varphi_I$. The iso-zero of the level set function determines the location of the interface.

5.6.5.2 Stress Update with Sub-Triangular Numerical Integration

A key point in Godunov-like stress update procedure is that the number of Gauss points before and after mesh motion is equal. In addition, the natural coordinates of Gauss quadrature points remain constant during smoothing phase. However, these conditions may not be necessarily satisfied in the Eulerian phase of an X-ALE-FEM analysis, since the elements cut by the interface are divided into sub-triangles whose Gauss points are used for numerical integration. Based on three Gauss quadrature points per sub-element, the total number of Gauss points is 18 for each split element, for example, elements 1, 2, and 3 in Figure 5.31a. While in other elements, the standard Gauss points of a set of 4×4 points are used for numerical integration, for example, element 4 in Figure 5.31a. During smoothing phase of X-ALE-FEM analysis, four different cases may occur regarding the numbers of Gauss quadrature points:

1. The element is not cut by the interface either before or after mesh motion (e.g., element number 4 in Figure 5.31a and b. In this case, the FE standard Gauss points are used for numerical integration before and after smoothing phase. Thus, the number of Gauss points and their natural coordinates remain constant, and the Godunov scheme can be used without any modification.
2. The element, which has been cut by the interface before smoothing phase, is still included by the interface after mesh updating procedure, for example, element 1 in Figure 5.31a and b. As mentioned earlier, both the old and relocated elements contain 18 Gauss quadrature points. But the natural coordinates of Gauss points must be evaluated after the mesh update procedure due to new triangular sub-elements obtained by partitioning the element. Thus, an efficient algorithm must be applied into the

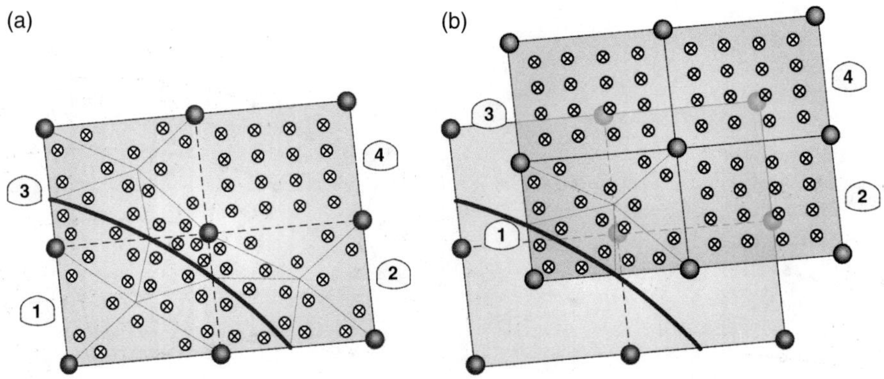

Figure 5.31 The distribution of Gauss points in X-ALE-FEM analysis: (a) Old mesh; (b) Relocated mesh

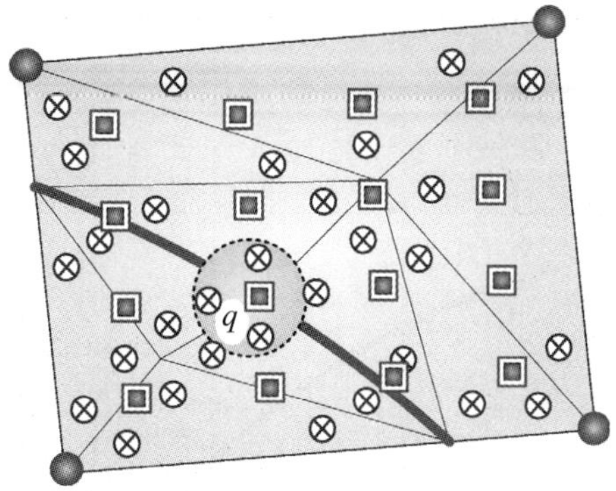

Figure 5.32 The support domain of Gauss integration point q: ▣ The Gauss points corresponding to the standard FE element. ⊗ The Gauss points corresponding to sub-triangles

Godunov technique to update the stress $\boldsymbol{\sigma}_{n+1}^{L}$ obtained from the Lagrangian phase to $\boldsymbol{\sigma}_{n+1}$ after the Eulerian phase. Consider that the internal variables such as stresses must be transferred from the triangular Gauss points to the virtual set of FE standard Gauss points in old mesh, as shown in Figure 5.32. According to this figure, for each FE Gauss point q there is a support domain which consists of three nearest triangular Gauss points. The stress value at q can be interpolated from the stress values of these three nearest Gauss points. To obtain more accurate interpolation, a set of 4×4 standard Gauss points is selected here. After obtaining the stress values at the FE standard Gauss points, the Godunov scheme can be used to update the stress values from the virtual Gauss points to relocated 4×4 Gauss points. It must be noted that the proposed algorithm is also adopted to transfer the stress values from relocated 4×4 Gauss points to Gauss quadrature points of sub-triangles. In this case, a support domain is defined for each triangular Gauss points, which consists of three nearest standard FE Gauss points.

3. The interface leaves one element to another during the smoothing phase. In this case, the element, which was split by the interface before mesh updating procedure, does not contain the interface after mesh motion, for example, elements 2 and 3 in Figure 5.31a and b. It will therefore result in different numbers of Gauss points in the old and relocated elements. Here, the proposed technique illustrated previously can be used to transfer the stress values from triangular Gauss points to a virtual set of 4×4 Gauss points in the old element. The stress values at the Gauss quadrature points of new element can then be obtained using a Godunov update algorithm.
4. The material interface moves into the relocated element during the Eulerian phase. In this case, the relocated element is cut by the interface while it was not included before the mesh motion. In such case, the stress values have been calculated at the standard FE Gauss points of old element at the end of Lagrangian phase. These values need to be updated using the Godunov algorithm to obtain the stress values at relocated 4×4 Gauss points. It must be noted that the stress values are required at the Gauss quadrature points of sub-triangles in relocated element. Hence, a support domain can be assigned for each triangular Gauss points, which consists of three nearest standard FE Gauss points, we can therefore obtain the required stress value with a simple interpolation.

5.6.5.3 Stress Update with Sub-Quadrilateral Numerical Integration

It must be noted that for the elements cut by the interface boundary, the standard Gauss quadrature points are insufficient for numerical integration, and may not adequately integrate the interface boundary. Thus, it is necessary to modify the element quadrature points to accurately evaluate the contribution

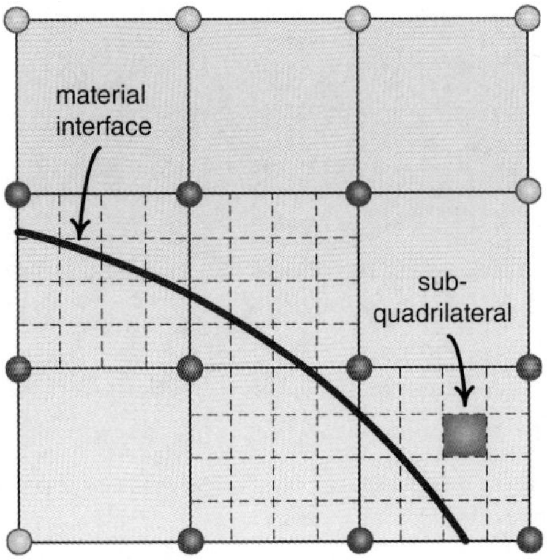

Figure 5.33 The sub-quadrilateral partitioning for numerical integration of elements cut by the interface

to the weak form for both sides of interface. In the X-FEM, the elements located on the interface boundary can be partitioned by sub-polygons based on sub-triangles and sub-quadratics, however, because of two reasons the sub-triangles described previously are not suitable in an X-ALE-FEM analysis. Firstly, due to the relative motion between the interface and elements, the natural coordinates of Gauss quadrature points may differ during smoothing phase even though the interface does not leave the element to another in this phase, and secondly, during the evolution of relative motion between the interface and nodal points, it is possible that a nodal point lies close to the interface. In this case, partitioning the bounded part between the node and the interface by sub-triangles results in serious numerical errors.

Thus, the numerical integration of elements cut by the interface is performed based on sub-quadrilaterals, as shown in Figure 5.33. In this technique, it is not necessary to conform the sub-quadrilaterals to the geometry of interface, however, it is required to have enough sub-divisions to reduce the error of numerical integration. Considering each split element contains 4×4 subdivision, the total number of Gauss points is 64 for each split element. While in standard elements, a set of 2×2 Gauss points are used for numerical integration. Based on the sub-quadrilaterals integration scheme, the natural coordinates of Gauss quadrature points are independent of the interface position and do not change during smoothing phase. Furthermore, the most important feature is that the error of numerical integration can be significantly reduced in this technique. However, the key requirement of the Godunov scheme is the equivalent of integration points before and after mesh update process that is not satisfied in the Eulerian phase of an X-ALE-FEM analysis, when the interface leaves one element to another, or moves into the relocated element during smoothing procedure. In this circumstance, four different cases may again occur regarding the numbers of Gauss quadrature points:

1. The element is not cut by the interface either before or after mesh motion, for example, elements 1, 2, 3, and 7 in Figure 5.34a and b. In this case, the standard FE Gauss points are used for numerical integration before and after smoothing phase, the number of Gauss points and their natural coordinates remain constant, and the Godunov scheme is directly used.
2. The element, which has been cut by the interface before smoothing phase, is still included by the interface after mesh updating procedure, for example, elements 4, 5, and 9 in Figure 5.34a and b. Based on the new integration scheme, both the old and relocated elements contain 64 Gauss quadrature points, the natural coordinates of Gauss points do not change, and no modification is needed for the Godunov scheme.

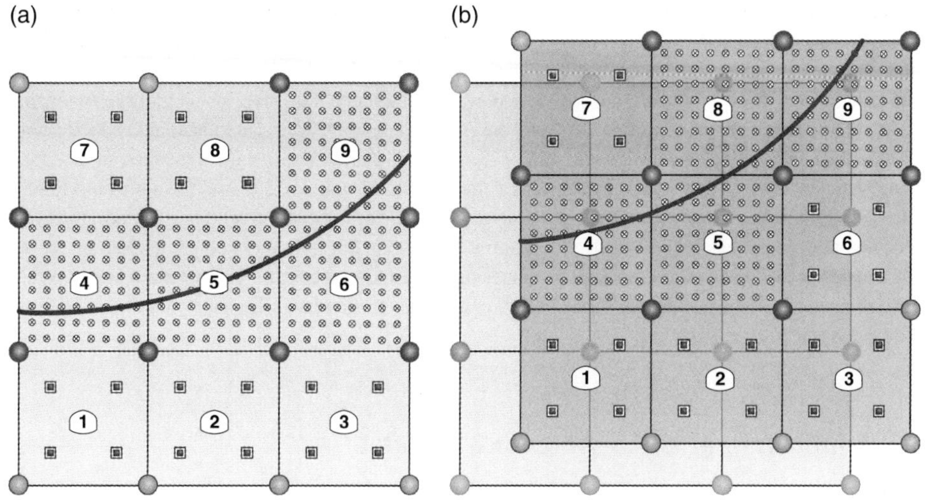

Figure 5.34 The distribution of Gauss integration points in the X-ALE-FEM analysis: (a) Old mesh; (b) Relocated mesh

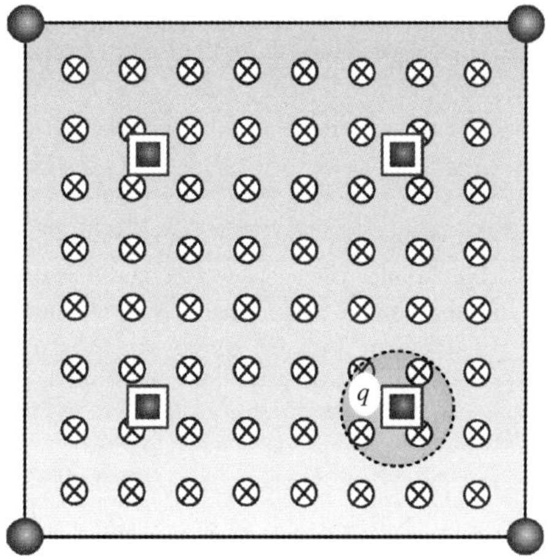

Figure 5.35 The support domain of Gauss integration point q : ▣ The 2×2 Gauss points corresponding to relocated mesh and, ⊗ The virtual Gauss points in relocated mesh

3. The interface leaves one element to another during the smoothing phase. In this case, the element, which was split by the interface before mesh updating procedure, does not contain the interface after mesh motion, for example, element 6 in Figure 5.34a and b. It therefore results in different numbers of Gauss points in the old and relocated elements. As mentioned earlier, the old and relocated elements contain 64 and 4 Gauss quadrature points respectively. Thus, a modified-Godunov technique is applied to update the stress from $\boldsymbol{\sigma}_{n+1}^{L}$ obtained from the Lagrangian phase to $\boldsymbol{\sigma}_{n+1}$ after the Eulerian phase. First, the internal variables such as stresses are updated from the 8×8 Gauss points in the old element to a virtual set of 8×8 FE standard Gauss points in the relocated mesh via an updated Godunov algorithm. The stress values are then transferred from the virtual Gauss points to relocated 2×2 Gauss points in the updated element, as shown in Figure 5.35. According to this figure, for each FE Gauss

point q there is a support domain which consists of three nearest virtual Gauss points. The stress value at q can be interpolated from the stress values of these three nearest Gauss points.

4. The material interface moves into the relocated element during the Eulerian phase. In this case, the relocated element is cut by the interface while it was not included before the mesh motion, for example, element 8 in Figure 5.34a and b. In this case, the stress values have been calculated at the 2×2 standard FE Gauss points of old element at the end of Lagrangian phase. These values must be updated using the Godunov algorithm to obtain the stress values at the virtual 2×2 Gauss points in the relocated element. It must be noted that the stress values are required at relocated 8×8 Gauss quadrature points. For this aim, the stress values are firstly computed at the nodal points of relocated element by using an averaging technique from the related values at nearest Gauss points. The required stress values at relocated 8×8 Gauss points can then be obtained using the standard FE shape functions of relocated nodal values.

5.7 Application of the X-ALE-FEM Model

In order to illustrate the applicability of proposed computational algorithm in large deformation problems, two numerical examples are presented, including the coining problem and a plate in tension. The examples are solved using FEM, X-FEM, and X-ALE-FEM techniques, and the results are compared. The initial mesh used for X-FEM and X-ALE-FEM are similar and independent of discontinuity shape, while in FEM analysis the FE mesh needs to be conformed to the geometry of discontinuity. In order to perform a real comparison, the number of elements in FEM and X-FEM analyses is almost equal. Both examples are simulated by a plane strain representation and the convergence tolerance is set to 5×10^{-14}.

5.7.1 The Coining Test

The first example illustrates the performance of X-ALE-FEM technique in large deformation simulation of a practical and challenging ALE example. The proposed technique is used to simulate the coining test by pressing a rigid component into the flexible elastic foam, as shown in Figure 5.36. In this example, a rigid component with a Young's modulus of 2.1×10^6 kg/cm^2 and Poisson ratio of 0.35, is pressed from the bottom edge into the elastic foam with a Young's modulus of 2.1×10^5 kg/cm^2 and Poisson coefficient

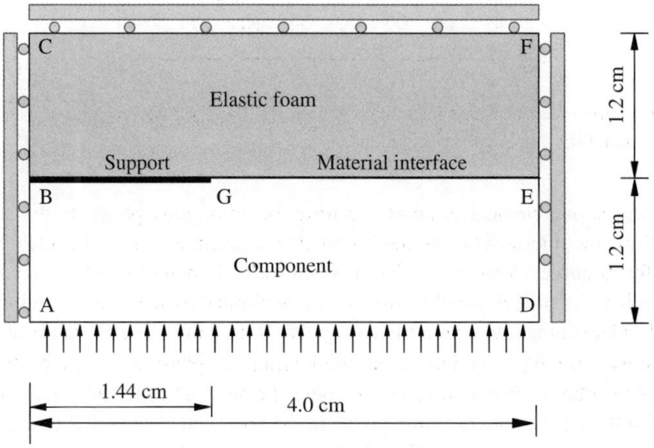

Figure 5.36 Coining test with a horizontal interface: Problem definition

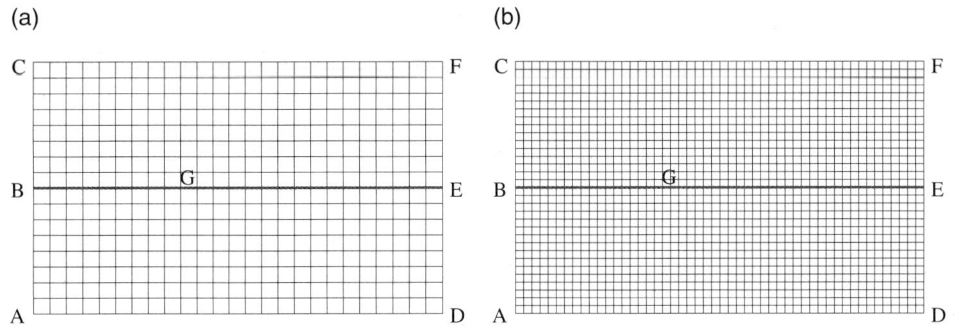

Figure 5.37 Coining test with a horizontal interface: The FEM and X-ALE-FEM meshes of (a) 400 elements; (b) 1600 elements

of 0.35. The rigid part is constrained at the top edge along line BG, as shown in Figure 5.36. The geometry and boundary conditions of the problem are shown in this figure with frictionless contact surfaces along AC, CF, DF, and BG. In Figure 5.37, the FEM and X-ALE-FEM meshes are shown together with the material interface at the initial stage of compaction. For both simulations, the material interface conforms to the edge of elements at the initial configuration in order to simplify the displacement boundary conditions along edge BG. However, the interface is separated from the edge of elements in X-ALE-FEM analysis, as the compaction proceeds. It must be noted that the X-FEM analysis leads to similar results obtained by FEM modeling, as the interface passes through the nodal points. The numerical modeling of this example poses two difficulties using the Lagrangian formulation. First, the implementation of displacement boundary conditions along line BG results in lateral displacement of material points, and leads to unrealistic analysis due to increase of the width of edge BG. This error may be overcome by implementation of the contact interface along edge BG. The second difficulty is due to highly distorted mesh particularly around point G, aborting the calculations and causing numerical errors at the compaction of 0.4 cm.

In Figure 5.38a and c, the deformed FE meshes are shown for the fine mesh at the pressing of 0.3 and 0.4 cm. By using a simple ALE remeshing strategy with the equal height and width of elements prescribed in the regions of ABED and BCFE, the X-ALE-FEM analysis can proceed until 0.58 cm height reduction. The deformed X-ALE-FEM configuration for the fine mesh is presented in Figure 5.38b, d, and e corresponding to the compaction deformations of 0.3, 0.4, and 0.58 cm.

The normal stress σ_y contours of the rigid part are shown in Figure 5.39 for both FEM and X-ALE-FEM at the height reduction of 0.4 cm. A good agreement can be seen between two different techniques. A comparison has been performed using two techniques by evolution of the vertical reaction and vertical displacement of point E with the compaction deformation in Figure 5.40a and b. Good agreement can be observed up to the compaction of 0.40 cm (the final compaction obtained by the FEM analysis) using the FEM and X-ALE-FEM techniques.

5.7.2 *A Plate in Tension*

The next example is of a rectangular plate under uniaxial extension up to an elongation of 6.4 cm, as shown in Figure 5.41. The plate has an elasto-plastic von-Mises behavior with a Young's modulus of 2.1×10^6 kg/cm^2, Poisson ratio of 0.35, yield stress of 2400 kg/cm^2, and a hardening parameter of 100 kg/cm^2. The necking may be physically occurred at each part of the specimen. In order to control the phenomenon, a geometric imperfection of 1% reduction in width is induced in the central part of plate. On the virtue of symmetry, only one-quarter of the metallic plate is modeled in a plane strain condition, as

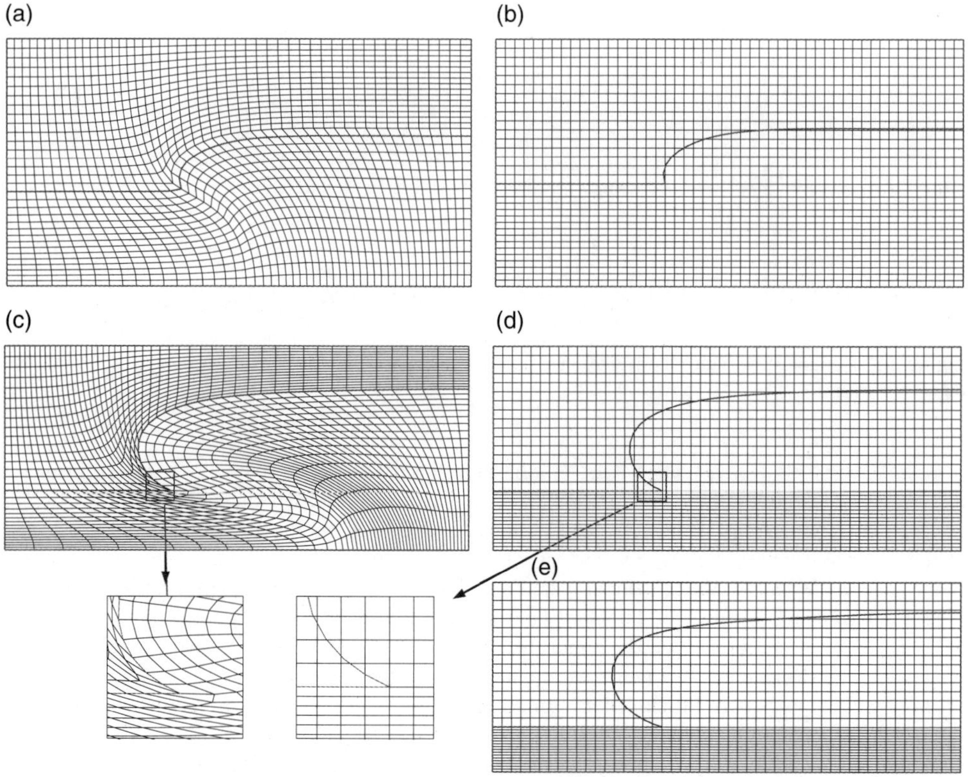

Figure 5.38 Coining test with a horizontal interface: The deformed configuration using fine meshes; (a) The FEM at $d = 0.35$ cm; (b) The X-ALE-FEM at $d = 0.35$ cm; (c) The FEM at $d = 0.7$ cm; (d) The X-ALE-FEM at $d = 0.7$ cm; (e) The X-ALE-FEM at $d = 0.83$ cm

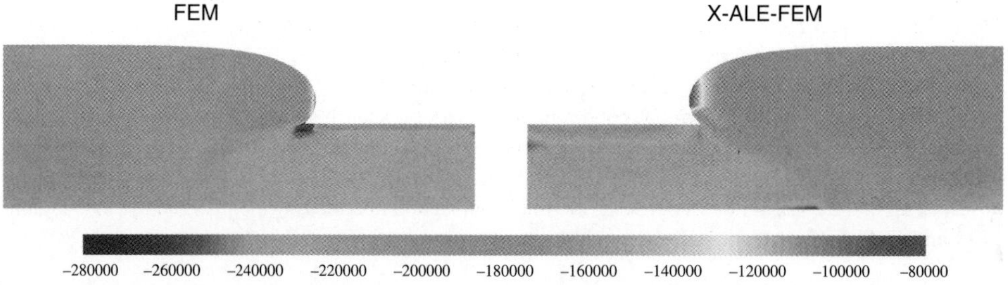

Figure 5.39 Coining test with a horizontal interface: The normal stress σ_y contours of rigid part at $d = 0.5$ cm. A comparison between FEM and X-ALE-FEM analyses

shown in Figure 5.41a. The analysis is performed using the FEM, X-FEM, and X-ALE-FEM techniques and the results are compared. In order to make a comprehensive comparison, the FEM analysis is carried out using both the Lagrangian and ALE techniques. The FE mesh has 480 linear quadrilateral elements, as shown in Figure 5.41b. In order to demonstrate the performance of X-FEM and X-ALE-FEM in modeling discontinuities due to significant difference of material properties, the numerical simulation is performed

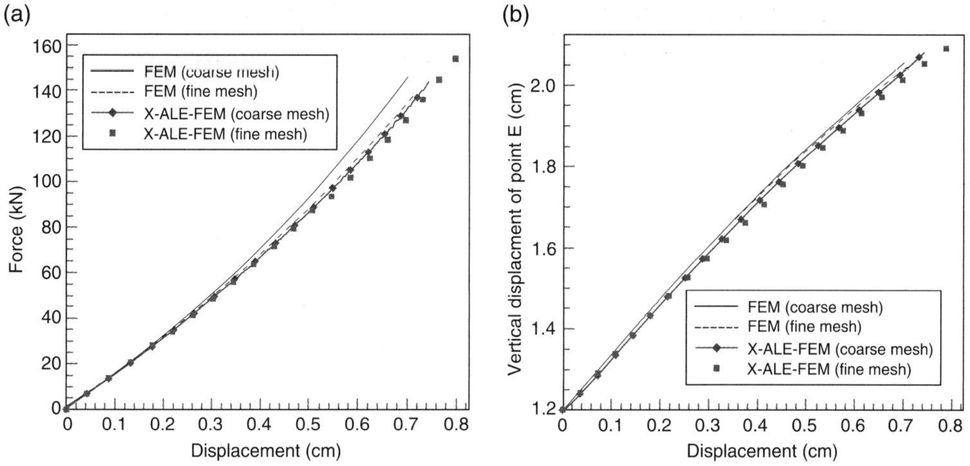

Figure 5.40 A comparison between FEM and X-ALE-FEM analyses for the coining test with a horizontal interface: (a) The variation of reaction force with vertical displacement; (b) The variation of vertical displacement of point E with displacement

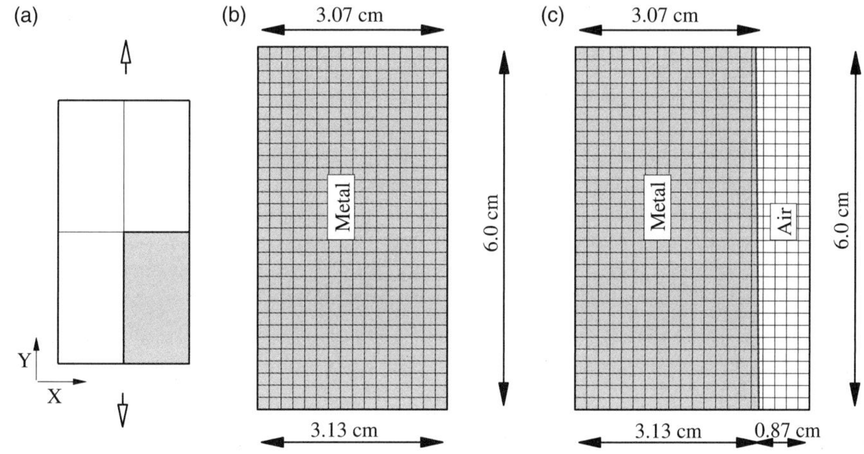

Figure 5.41 A plate in tension: (a) Problem definition; (b) The FE mesh of 480 elements; (c) The X-FEM and X-ALE-FEM meshes of 600 elements

using a metallic part whose geometry and properties are similar to FE model connected to a rectangular part with zero Young's modulus. Figure 5.41c illustrates the X-FEM and X-ALE-FEM meshes of 600 uniform elements together with the geometry of model. The boundary conditions are applied by restricting the top edge in y-direction and the left edge in x-direction. In the X-FEM and X-ALE-FEM models, all nodal points of the right edge are restrained in x-direction. A very simple remeshing strategy is applied for the ALE analysis in which the equal height of elements is prescribed in whole domain. The Laplacian approach is used in the X-ALE-FEM analysis to reduce the mesh distortion and produce a smooth mesh. Figure 5.42 presents the deformed configuration obtained by four techniques, that is, the Lagrangian, ALE, X-FEM, and X-ALE-FEM techniques, up to an elongation of 3.2 cm for one quarter of the specimen. The variations of normal stress σ_y along the vertical symmetry plane are shown in Figure 5.43 at the

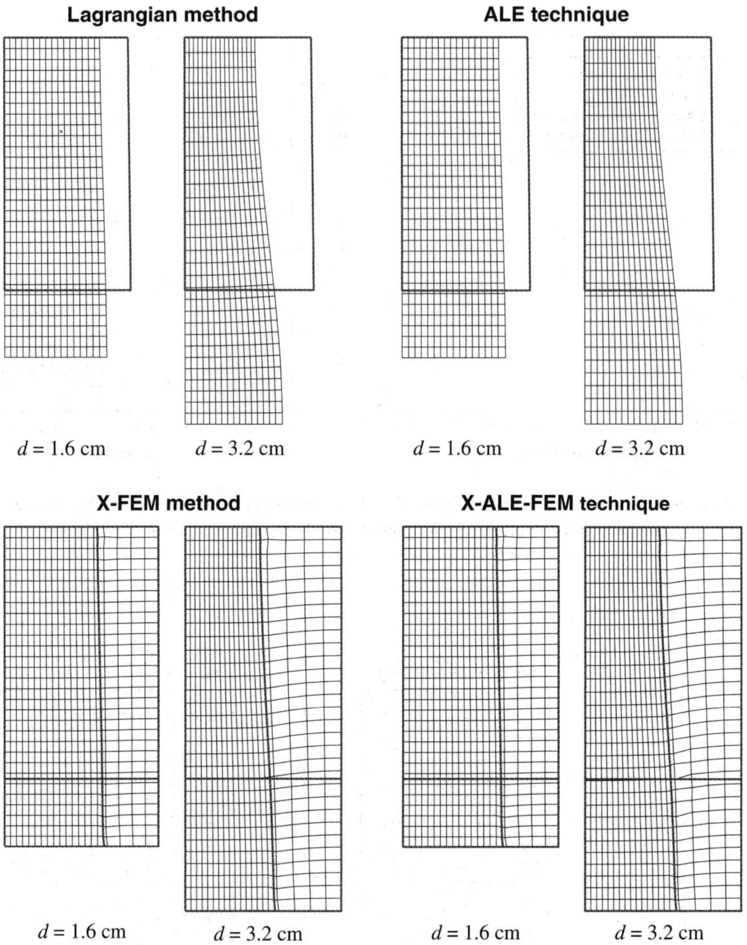

Figure 5.42 A plate in tension: The deformed mesh configurations at different bottom displacements using the Lagrangian, ALE, X-FEM, and X-ALE-FEM techniques

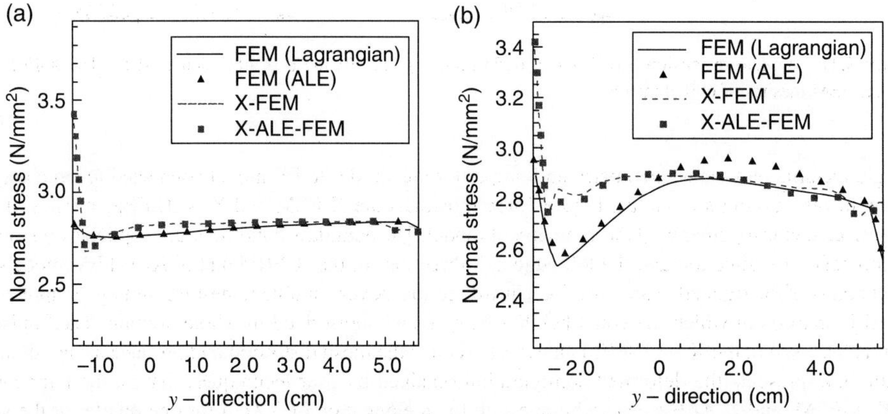

Figure 5.43 A plate in tension: A comparison of normal stress σ_y between the FEM, X-FEM, and X-ALE-FEM techniques. The stress distribution along vertical symmetry plane; (a) $d = 1.6$ cm; (b) $d = 3.2$ cm

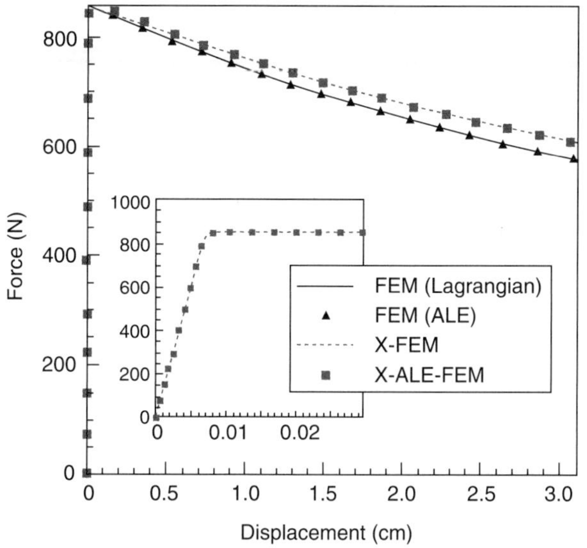

Figure 5.44 A plate in tension: The variation of reaction force with vertical displacement. A comparison between FEM, X-FEM, and X-ALE-FEM analyses

half and final stages of tensile process for four different techniques. In Figure 5.44, a quantitative comparison is performed between various methods by evolution of the vertical reaction at different vertical displacements. Obviously, it can be concluded that the proposed X-ALE-FEM can be successfully used to model the large deformation in elasto-plastic behavior.

6

Contact Friction Modeling with X-FEM

6.1 Introduction

Numerical modeling of contact between two bodies is a very important task in computational mechanics and has enormous applications in science and engineering. The core of science and technology of interacting surfaces, which is called *tribology*, is friction (Bhushan, 1999). The *friction phenomenon* occurs during many industrial processes, such as the metal forming process. In metal forming processes, friction is one of the most important factors that contribute to material deformation. During the material forming process relative movements occur in contact area due to large deformations between the tools and material. These relative movements produce normal and tangential stresses, which have an important role in metal flow and may cause serious inhomogeneities in final products. Generally, many aspects of the process are affected by friction, such as pressing forces, density distribution, final shape, tool wear, cracks, and residual stresses (Khoei, 2005).

Frictional contact may be observed in crack surfaces during mixed-mode crack propagation, or along a discontinuity surface in a body. Geological discontinuities such as natural joints, faults, flaws, and bedding planes, are usually subjected to frictional contact condition. Contact problems mostly involve the friction phenomenon. An appropriate frictional contact model is of great importance to efficiently asses the behavior of the cracked media. Frictional effects may, however, be neglected in many cases if the frictional forces on the contacting boundaries are sufficiently small. In such case, a frictionless contact problem can be employed instead of a frictional contact problem. Extensive studies have been carried out in the field of frictional/frictionless contact problems based on the experimental and numerical approaches. In this chapter, the extended finite element method (X-FEM) model is developed to model the frictional contact in embedded interfaces.

In the framework of the finite element method (FEM), distinct boundaries are considered for the bodies in contact, called the "master" and "slave" bodies, and the contact conditions are taken into account. Modeling contact problems with the FEM approach can be generally divided into three categories; the "traction boundary condition" method, in which the frictional forces are appended to the external force vector as a traction boundary force; the "contact node algorithm", where an iterative algorithm is implemented to obtain the velocity or displacement value that is dependent on the nodal forces at the interface friction boundary; and the "interface element method", which is used as an alternative contact node algorithm that utilizes the thin elements having very high aspect (length/width) ratios along the friction boundary.

Basically, two main techniques have been employed to impose the contact constraints in the FEM solution of contact problems; the method of "Lagrangian multipliers" and the "penalty method". In the *Lagrange multipliers* approach, the contact condition is introduced as a constraint in the solution

Extended Finite Element Method: Theory and Applications, First Edition. Amir R. Khoei.

of governing variational formulation of the continuum problem. In this manner, the contact forces, which are known as Lagrange multipliers, are taken as primary unknowns and the contact constraints are enforced accurately (Simo and Laursen, 1992). The discretization process is similar to unconstrained problems except that, the Lagrange multiplier field must be approximated. Consequently, the Lagrange multiplier method affects the dimension and symmetry of the formulation by introducing new unknowns. In the *penalty method*, the contact constraints are imposed by assuming certain stiffness at the contact surface between two bodies. In this manner, the penetration between two contacting boundaries is introduced and the normal contact force is related to the penetration through the contact stiffness, known as the penalty parameter (Perić and Owen, 1992). The accuracy of satisfying contact constraints highly depends on the magnitude of penalty parameter; the larger the value of the penalty parameter, the more accurate contact constraints achieved. However, very large values for the penalty parameter will result in an ill conditioned formulation when the penalty parameter is combined with finite stiffness of bodies in contact. Based on these two techniques, other constraint algorithms have been developed and successfully employed in contact problems. One of these techniques is the *augmented Lagrange multipliers* method, which is based on the combination of the penalty and Lagrange multipliers techniques (Simo and Laursen, 1992). This method eliminates the drawbacks of the penalty and Lagrange multipliers methods and attempts to achieve a predetermined tolerance for the contact constraints through an iterative procedure.

In comparison to the FEM solution to contact problems, less attention has been paid to address the contact friction problem based on the X-FEM. The most advances in the X-FEM solution of continuum problems have focused on crack propagation with opening/closing modes (Dolbow *et al.*, 2001). In the X-FEM, the discontinuity is modeled independent of the finite element (FE) mesh and the displacement field of elements cut by the interface is approximated by discontinuous enrichment functions. The X-FEM was proposed by Khoei and Nikbakht (2007) and Liu and Borja (2008) to model the frictional contact problem using the penalty method. The X-FEM technique was proposed by Liu and Borja (2010a) for embedded frictional cracks to accommodate the finite deformation, including finite stretching and rotation. An approach was developed by Nistor *et al.* (2009) to couple the X-FEM with the Lagrangian large sliding frictionless contact algorithm. In this method, a hybrid X-FEM contact element was introduced in the framework of a contact search algorithm allowing for an update of contacting surfaces pairing. Recently, a node-to-segment contact algorithm was presented by Khoei and Mousavi (2010) based on the X-FEM to model the large deformation-large sliding contact problem using the penalty approach. In this chapter, the X-FEM formulation is presented for frictional contact problems on the basis of penalty method, Lagrange multipliers technique, and augmented Lagrange multipliers approach.

6.2 Continuum Model of Contact Friction

The contact phenomenon has been subjected to extensive research work due to its importance and complexity. The nature of friction forces developed between two bodies in the contact surface is extremely complex and is affected by a long list of factors (Oden and Martins, 1985); the constitution of the interface, the time scales and frequency of the contact, the response of the interface to normal forces, inertia and thermal effects, roughness of the contacting surfaces, history of loadings, wear and general failure of the interface materials, the presence or absence of lubricants, and so on. Thus, friction is not a single phenomenon but is a collection of many complex mechanical and chemical phenomena entwined in a mosaic whose features cannot be grasped simply by means of isolated simple experiments. Successful computational methods in nonlinear contact mechanics are generally those based on a thorough familiarity with the natural phenomena being simulated and on a good understanding of the mathematical models.

The plasticity theory of friction was originally developed by Curnier (1984), and then extended to analyze the *dynamic sliding friction* in the framework of FEM by Laursen and Simo (1993). The dynamic sliding friction refers to a large and important class of truly dynamic problems, which includes such effects as frictional damping, dynamic sliding, stick-slip motion, chattering, and so on. The constitution of the material interface is essentially stable; there is no marked penetration or normal plastic deformation of the interface and, at least from a global point of view, the frictional forces developed on the contact surface appear to depend on the sliding velocity of one surface relative to another. The first and perhaps most difficult step in the analysis of dynamic friction is the development of an acceptable model for the contact interface. The plasticity theory of friction provides a theoretical description of motion at the interface of bodies in contact.

6.2.1 Contact Conditions: The Kuhn–Tucker Rule

Consider two bodies, a master and a slave body, as shown in Figure 6.1a, with initial configurations denoted by Ω^M and Ω^S, respectively. The relative displacement of the bodies from point S on the surface of the slave body Γ^S to the point M on the surface of the master body Γ^M is defined as

$$[\![\mathbf{u}]\!] = \mathbf{u}^S - \mathbf{u}^M \qquad \text{on} \qquad \Gamma = \Gamma^S \cup \Gamma^M \tag{6.1}$$

where $\mathbf{u}^S$ and $\mathbf{u}^M$ denote the displacements of the slave point S and master point M, respectively. Denoting the vector of unit normal on the surface of master body by $\mathbf{n}_{\Gamma_c}$, the normal and tangential gap functions can be defined respectively as

$$g_N = [\![\mathbf{u}]\!] \cdot \mathbf{n}_{\Gamma_c} \equiv \mathbf{n}_{\Gamma_c}^T [\![\mathbf{u}]\!] \qquad \text{on} \qquad \Gamma = \Gamma^S \cup \Gamma^M \tag{6.2}$$

and

$$\mathbf{g}_T = [\![\mathbf{u}_T]\!] = (\mathbf{I} - \mathbf{n}_{\Gamma_c} \otimes \mathbf{n}_{\Gamma_c}) [\![\mathbf{u}]\!] \qquad \text{on} \qquad \Gamma = \Gamma^S \cup \Gamma^M \tag{6.3}$$

where $\mathbf{n}_{\Gamma_c} \otimes \mathbf{n}_{\Gamma_c}$ is the projection tensor and $\mathbf{I}$ is the identity tensor. The mathematical description of non-penetration condition, which precludes the penetration of two bodies is stated $g_N \geq 0$. When $g_N = 0$, the contact condition occurs and the normal contact force p_N developed on the contact surface of master body is a non-positive value (i.e., $p_N \leq 0$), as depicted in Figure 6.1b. In this condition, the contact interface consists of the intersection of surfaces of the two bodies denoted by $\Gamma_c = \Gamma^S \cap \Gamma^M$. In fact, the conditions

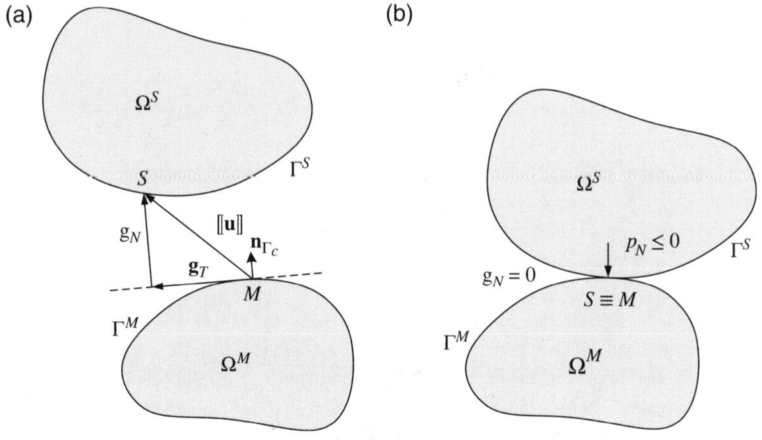

Figure 6.1 Definition of contact between two bodies: (a) the normal and tangential gap functions; (b) representation of contact condition

$g_N \geq 0$ and $p_N = 0$ hold if a gap between two bodies occurs. Hence, it leads to a mathematical description of contact conditions through the following statements

$$g_N \geq 0, \quad p_N \leq 0, \quad g_N p_N = 0 \tag{6.4}$$

These statements are known as the *Hertz–Signorini–Moreau* conditions used in the optimization theory, which are usually called the standard *Kuhn–Tucker* conditions (Wriggers, 2006). The first relation of Kuhn–Tucker relations expresses the non-penetration condition, the second one expresses that a closed gap must lead to a compressive contact pressure, and the last one ensures that the contact tractions vanishes when the gap is open. These conditions provide a basis to treat contact problems in the context of constrained optimization theory and are best suited for a variational formulation.

6.2.2 Plasticity Theory of Friction

Dynamic sliding friction can refer to large frictional relative motions between two bodies, similar to the elasto-plasticity theory of continuum mechanics. In such case, the displacement can be decomposed into an elastic deformation and sliding component. Thus, the plasticity theory of friction can be achieved by an analogy between the plasticity behavior and frictional phenomenon. In order to formulate such a theory of friction several requirements have to be considered. These requirements, which are similar to the requirements that have to be considered in the theory of elasto-plasticity, are as follows; the *stick (or adhesion) law* that is a mathematical description of the stress state under sticking (elastic) condition, the *stick–slip law* that is a theoretical description of the relationship between the stress and stick–slip (elasto-plastic) condition, the *wear and tear rule* that indicates a hardening and softening behavior during sliding, the *slip criterion* that is a yield criterion indicating the stress level at which the relative slip motion occurs, and the *slip rule* that defines a flow rule indicating the relationship between stresses and slip motion.

In order to derive the constitutive model for the contact problem with a non-linear frictional evolution law, consider two bodies in contact, a master and a slave body, as shown in Figure 6.2. The unit normal vector on the master surface is denoted by $\mathbf{n}_{\Gamma_c}$. Since there is no normal gap between the two bodies in contact condition, only a tangential relative displacement is considered. In order to perform a meaningful decomposition of the relative displacement into normal and tangential components, it is imperative that the direction of the outward normal $\mathbf{n}_{\Gamma_c}$ to the master surface remains nearly constant throughout the sliding process. During the contact and sliding of the two bodies, the normal and tangential loads $\mathbf{p}_N$ and $\mathbf{p}_T$ are defined on the contact surface of master body.

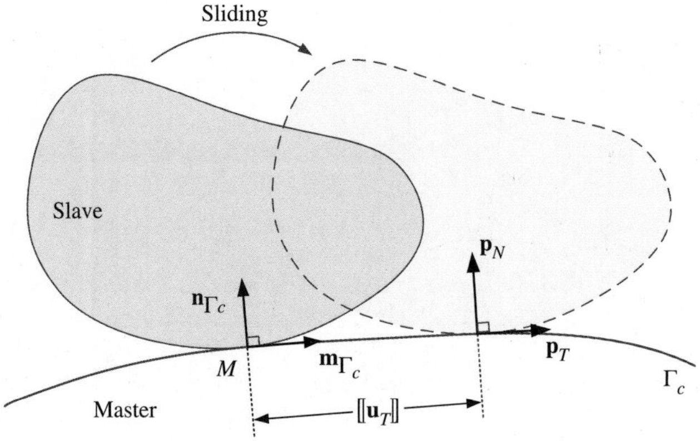

Figure 6.2 The master and slave bodies in contact due to a relative tangential displacement

A *stick–slip* theory of friction was developed by Curnier (1984) based on the decomposition of the relative displacement at the contact surface into the sum of two parts; one reversible, rather unusual, called *stick* (or *adherence*), that can be attributed to the elastic deformation of asperities, and the other irreversible, more familiar, called *slip*, that can be attributed partly to the plastic deformation of asperities. In order to decompose the relative displacement into the stick and slip, it is essential to conserve the same origin M, defined as the initial contact point of the slave on the master surface, for all subsequent measurements of the relative displacement. In this way, the tangential relative displacement consists of stick and slip components that are, in principle, analogous to the decomposition of elastic and plastic behavior in plasticity. The decomposition of the tangential relative displacement at the contact surface can be written as

$$[\![\mathbf{u}_T]\!] = [\![\mathbf{u}_T^A]\!] + [\![\mathbf{u}_T^S]\!] \equiv [\![\mathbf{u}_T^e]\!] + [\![\mathbf{u}_T^p]\!] \quad \text{on} \quad \Gamma_c \tag{6.5}$$

where $[\![\mathbf{u}_T]\!]$ is the tangential relative displacement defined by $[\![\mathbf{u}_T]\!] = (\mathbf{I} - \mathbf{n}_{\Gamma_c} \otimes \mathbf{n}_{\Gamma_c})[\![\mathbf{u}]\!]$. In this relation, the superscripts A (and e) stand for the adherence/stick (and elastic), and S (and p) indicate the slip (and plastic) parts of $[\![\mathbf{u}_T]\!]$, respectively. Note that the projection tensor has the properties $(\mathbf{n}_{\Gamma_c} \otimes \mathbf{n}_{\Gamma_c})^2 = \mathbf{n}_{\Gamma_c} \otimes \mathbf{n}_{\Gamma_c}$ and $(\mathbf{n}_{\Gamma_c} \otimes \mathbf{n}_{\Gamma_c})(\mathbf{I} - \mathbf{n}_{\Gamma_c} \otimes \mathbf{n}_{\Gamma_c}) = \mathbf{0}$.

In order to resolve the resulting unilateral contact condition, the Lagrange multipliers method, or the penalty method, can be typically employed. In the case of the Lagrange multipliers method, a large number of additional unknowns need to be included owing to incorporation of $\mathbf{p}_N$ as a new variable. On the contrary, the penalty method needs no additional variable because the impenetrability condition is approximately satisfied via embedding very stiff springs on the contact surface, and the normal contact forces $\mathbf{p}_N$ can be obtained from the multiplication of the penalty parameter k_N and the relative displacement in the normal direction $[\![\mathbf{u}_N]\!]$. Similarly, the stick (or elastic) component of the tangential contact forces $\mathbf{p}_T$ can be obtained by multiplying the tangential penalty parameter k_T and the elastic part of relative displacement in the tangential direction $[\![\mathbf{u}_T^e]\!]$. The penalty parameters k_N and k_T can be respectively considered as the normal and tangential stiffness constants at the contact surface. Based on this discussion, the constitutive law for the contact loads can be summarized as

$$\begin{aligned} \mathbf{p}_N &= \left(\mathbf{D}_f^e\right)_N [\![\mathbf{u}^e]\!] \\ \mathbf{p}_T &= \left(\mathbf{D}_f^e\right)_T [\![\mathbf{u}^e]\!] \end{aligned} \tag{6.6}$$

where $\left(\mathbf{D}_f^e\right)_N$ and $\left(\mathbf{D}_f^e\right)_T$ are the normal and tangential parts of the elastic modulus tensor for contact friction defined by

$$\begin{aligned} \left(\mathbf{D}_f^e\right)_N &= k_N (\mathbf{n}_{\Gamma_c} \otimes \mathbf{n}_{\Gamma_c}) \\ \left(\mathbf{D}_f^e\right)_T &= k_T (\mathbf{I} - \mathbf{n}_{\Gamma_c} \otimes \mathbf{n}_{\Gamma_c}) \end{aligned} \tag{6.7}$$

In order to perform the decomposition of relative displacement into the stick and slip, a slip criterion is introduced that refers to as a friction criterion. A slip criterion is a theoretical description of the relationship between the stress and stick-slip motion at the contact surface between two bodies. It is important to indicate which one of the two modes, sticking or slipping, occurs. In the stick condition, there is no relative movement allowed in the tangential direction, while in the sliding condition, a point can move in the tangential direction of contact interface. One of the most commonly used slip criteria, which has been employed for the description of frictional contact behavior, is Coulomb's friction law defined as a function of contact forces $\mathbf{p}_c$ by

$$F_f(\mathbf{p}_c, w) = \|\mathbf{p}_T\| - \mu_f(\mathbf{p}_N, w)\,\|\mathbf{p}_N\| \quad \begin{cases} = 0 & \text{slip} \\ < 0 & \text{stick} \end{cases} \tag{6.8}$$

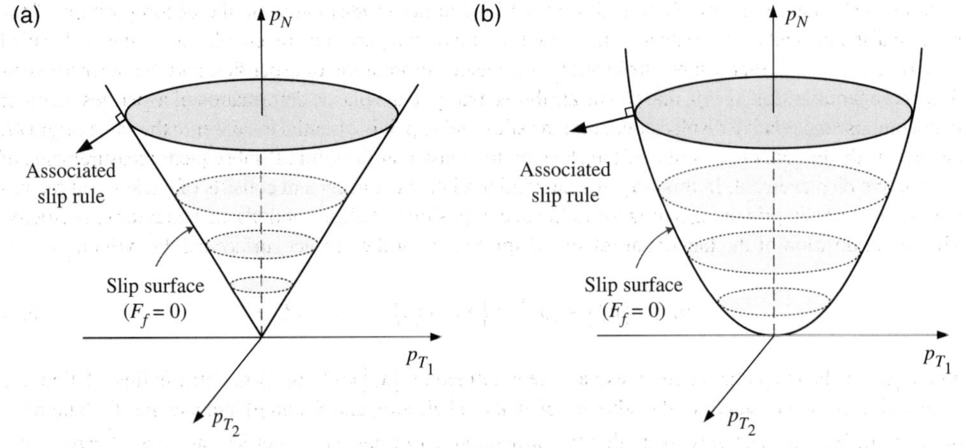

Figure 6.3 The slip criteria and associated slip rule: (a) linear slip criterion; (b) nonlinear slip criterion

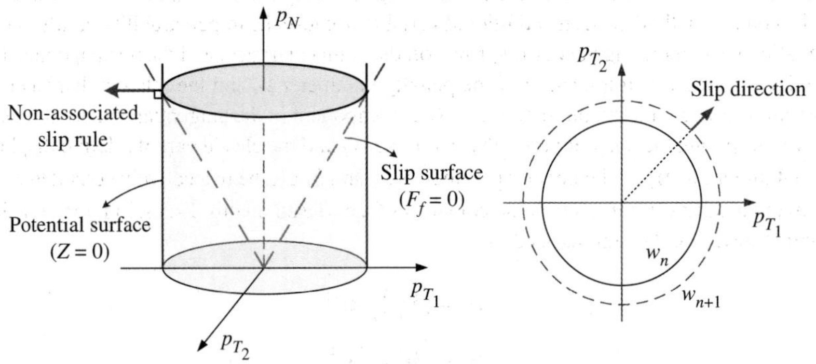

Figure 6.4 The potential surface and non-associated slip rule

where μ_f is the Coulomb friction coefficient defined by $\mu_f = \tan\varphi_f$, with φ_f denoting the interface friction angle. The friction coefficient μ_f is assumed to depend on the adopted state variables, that is, the frictional work parameter w and the normal contact force $\mathbf{p}_N$. If $F_f < 0$, a stick situation occurs, that is, there is no relative movement between two bodies in contact. Conversely, if $F_f > 0$, the slip situation occurs and the shear stress at the interface is constrained to the value $F_f = 0$. The slip criterion is depicted as a generalized cone in Figure 6.3 in the space ($\mathbf{p}_N$, $\mathbf{p}_T$) due to dependence on the state variable $\mathbf{p}_N$.

The direction of slip is governed by an appropriate slip rule, which can be derived from the gradient of a convex potential Z. If the potential Z is replaced by the slip criterion F_f, the slip rule becomes associated (Figure 6.3). In the classical theory of plasticity, the flow rule associated with the standard yield criteria can be employed for relatively large classes of materials. However, in the plasticity theory of friction, the slip rule associated with the friction criterion (6.8) would generate a normal force that may result in the creation of gaps (separation). In order to avoid the slave body separation from the contacting surface, a non-associated slip rule is typically adopted (Curnier, 1984). Hence, the slip potential Z is introduced as a cylinder with the radius of $\|\mathbf{p}_T\|$ for an isotropic frictional contact, and the slip direction is defined as the outward normal to the slip potential Z, as shown in Figure 6.4. Consequently, the slip rule of the tangential plastic deformation $[\![\mathbf{u}_T^p]\!]$ in Eq. (6.5) can be defined as

$$d[\![\mathbf{u}_T^p]\!] = d\gamma \frac{\partial Z}{\partial \mathbf{p}_T} \equiv d\gamma \mathbf{m}_{\Gamma_c} \tag{6.9}$$

where $d\gamma$ is the plastic constant expressing the collinearity of the slip increment with the outward normal to the potential Z and $\mathbf{m}_{\Gamma_c} = \mathbf{p}_T / \|\mathbf{p}_T\|$ is the unit tangential vector.

6.2.3 Continuum Tangent Matrix of Contact Problem

According to the standard arguments of elasto-plasticity, the continuum tangent matrix for the contact problem with nonlinear frictional evolution can be derived using the consistency condition as

$$\left(\frac{\partial F_f(\mathbf{p}_c, w)}{\partial \mathbf{p}_T}\right) d\mathbf{p}_T + \left(\frac{\partial F_f(\mathbf{p}_c, w)}{\partial \mathbf{p}_N}\right) d\mathbf{p}_N + \left(\frac{\partial F_f(\mathbf{p}_c, w)}{\partial w}\right) dw = 0 \tag{6.10}$$

Substituting $d\gamma$ into the constitutive law, a linearized equation can be obtained as

$$d\mathbf{p}_c = \mathbf{D}_f^{ep} d[\![\mathbf{u}]\!] \tag{6.11}$$

where the continuum tangent matrix of contact problem $\mathbf{D}_f^{ep}$ can be defined as

$$\begin{aligned}\mathbf{D}_f^{ep} &= k_N(\mathbf{n}_{\Gamma_c} \otimes \mathbf{n}_{\Gamma_c}) + k_T(\mathbf{I} - \mathbf{m}_{\Gamma_c} \otimes \mathbf{m}_{\Gamma_c} - \mathbf{n}_{\Gamma_c} \otimes \mathbf{n}_{\Gamma_c}) \\ &- k_T \frac{1}{\beta}\left(\frac{\mu_f}{k_T} \| \mathbf{p}_N \|^2 \left(\frac{\partial \mu_f}{\partial w}\right)\right)(\mathbf{m}_{\Gamma_c} \otimes \mathbf{m}_{\Gamma_c}) \\ &- k_N \frac{1}{\beta}\left(\mu_f + \|\mathbf{p}_N\| \left(\frac{\partial \mu_f}{\partial \|\mathbf{p}_N\|}\right)\right)(\mathbf{m}_{\Gamma_c} \otimes \mathbf{n}_{\Gamma_c})\end{aligned} \tag{6.12}$$

where

$$\beta = 1 + \frac{\mu_f}{k_T} \| \mathbf{p}_N \|^2 \left(\frac{\partial \mu_f}{\partial w}\right) \tag{6.13}$$

In Eq. (6.12), the first term of continuum tangent matrix $\mathbf{D}_f^{ep}$ indicates the stiffness in the normal direction to the contact surface. The second term denotes the stick stiffness perpendicular to the sliding direction on the contact surface. The third and fourth terms indicate the adhesion/slip stiffness with hardening, or softening, in the sliding direction, respectively. The forces of friction, tear, and wear are respectively analogous to the stress states, kinematic stress, and isotropic (or equivalent) stress in the theory of plasticity. Since the existing experimental data do not provide enough accurate information for the precise modeling of frictional behavior, it is difficult to obtain appropriate friction factors and thus, modeling of the influence of wear and tear forces to sliding resistance remains only a dream in the numerical simulation. In the case of frictional slip without hardening or softening, Eq. (6.12) can be simplified as

$$\mathbf{D}_f^{ep} = k_N(\mathbf{n}_{\Gamma_c} \otimes \mathbf{n}_{\Gamma_c}) + k_T(\mathbf{I} - \mathbf{m}_{\Gamma_c} \otimes \mathbf{m}_{\Gamma_c} - \mathbf{n}_{\Gamma_c} \otimes \mathbf{n}_{\Gamma_c}) - k_N \mu_f(\mathbf{m}_{\Gamma_c} \otimes \mathbf{n}_{\Gamma_c}) \tag{6.14}$$

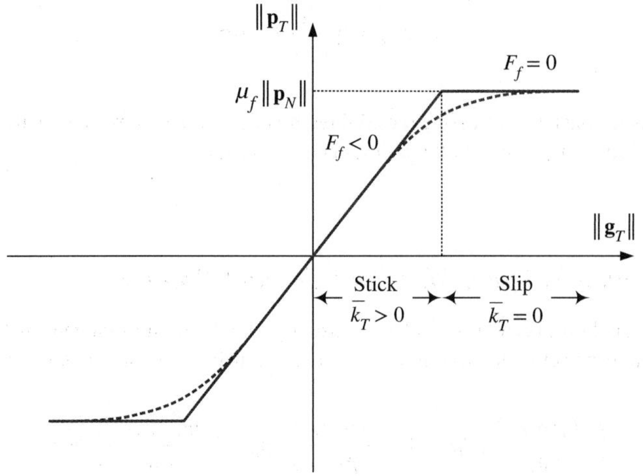

Figure 6.5 Frictional force relationship in stick and slip regions

The resemblance between the plasticity theory of friction and the classical theory of plasticity allows us to utilize a similar numerical elastic-predictor/plastic-corrector scheme for the contact problem, which has been extensively used in the theory of elasto-plasticity.

The continuum tangent matrix $\mathbf{D}_f^{ep}$ is an unsymmetric matrix due to non-associated slip rule adopted to derive the contact constitutive matrix. In order to preserve the symmetry of the numerical formulation, the coupling between the normal and tangential tractions at the contact interface can be neglected. In this way, the problem is artificially decomposed into a pure contact in the normal direction and a frictional resistance in the tangential direction. Hence, the contact constitutive relations can be rewritten to relate the normal and tangential contact tractions to the relative displacement as

$$d\mathbf{p}_N = \left(\mathbf{D}_f\right)_N d[\![\mathbf{u}]\!]$$
$$d\mathbf{p}_T = \left(\bar{\mathbf{D}}_f\right)_T d[\![\mathbf{u}]\!] \tag{6.15}$$

where $\left(\mathbf{D}_f\right)_N = k_N(\mathbf{n}_{\Gamma_c} \otimes \mathbf{n}_{\Gamma_c})$ and $\left(\bar{\mathbf{D}}_f\right)_T = \bar{k}_T(\mathbf{I} - \mathbf{n}_{\Gamma_c} \otimes \mathbf{n}_{\Gamma_c})$, with $\bar{k}_T$ denoting the tangential stiffness parameter. The continuum tangent matrix (6.14) can be simplified as

$$\bar{\mathbf{D}}_f^{ep} = k_N(\mathbf{n}_{\Gamma_c} \otimes \mathbf{n}_{\Gamma_c}) + \bar{k}_T(\mathbf{I} - \mathbf{n}_{\Gamma_c} \otimes \mathbf{n}_{\Gamma_c}) \tag{6.16}$$

in which the stick–slip condition can be obtained by an appropriate variation of $\bar{k}_T$ derived from the stick–slip relationship, as shown in Figure 6.5. In the stick condition, there is no contact movement but the contact friction forces are built up to $\|\mathbf{p}_T\| = \mu_f\|\mathbf{p}_N\|$. In such case, a stick tangential stiffness $\bar{k}_T = k_T^{\text{Stick}} > 0$ is derived directly from the slope of ($\|\mathbf{p}_T\|$, $\|\mathbf{g}_T\|$) curve. On the commencement of movement, the frictional forces remain constant according to $F_f = 0$; and hence, a slip tangential stiffness $\bar{k}_T = k_T^{\text{Slip}} = 0$ is obtained. The predictor–corrector algorithm used in the classical plasticity theory with a return mapping algorithm can be readily implemented, as described in Box 6.1. It is notable that such an approach yields to a smooth transition from the stick to slip, as presented in Figure 6.5.

Box 6.1 The predictor–corrector algorithm for a frictional contact problem

Evaluate the normal traction

$$\mathbf{p}_N^{n+1} = \mathbf{p}_N^n + \left(\mathbf{D}_f^e\right)_N [\Delta \mathbf{u}]^n$$

Compute the elastic (stick) predictor phase

Set $\bar{k}_T = k_T^{\text{Stick}}$ and calculate the trial value of elastic frictional traction

$$\left(\mathbf{p}_T^{n+1}\right)^{\text{trial}} = \mathbf{p}_T^n + \left(\bar{\mathbf{D}}_f\right)_T [\Delta \mathbf{u}]^n$$

Check the current stick–slip condition

IF $F_f = \left\| \left(\mathbf{p}_T^{n+1}\right)^{\text{trial}} \right\| - \mu_f \left\| \mathbf{p}_N^{n+1} \right\| < 0$, THEN accept the predictor as the final value (stick condition).
ELSE go to slip corrector phase (step c),

Calculate the plastic (slip) corrector phase

Correct the trial value of frictional traction and friction stiffness for plastic sliding (slip condition) respectively as

$$\mathbf{p}_T^{n+1} = \mu_f \left\| \mathbf{p}_N^{n+1} \right\| \frac{\Delta \mathbf{g}_T^{n+1}}{\left\| \Delta \mathbf{g}_T^{n+1} \right\|} \text{ and } \bar{k}_T = \frac{\mu_f \left\| \mathbf{p}_N^{n+1} \right\| - \left\| \mathbf{p}_T^{n+1} \right\|}{\left\| \Delta \mathbf{g}_T^{n+1} \right\|}$$

6.3 X-FEM Modeling of the Contact Problem

In order to numerically model the contact friction algorithm described in the preceding section, the X-FEM is employed in which the contact surface is embedded inside the FE mesh by adding appropriate enrichment functions to the standard FE approximation. In the derivation of the variational formulation of FEM solution, it is implicitly assumed that the displacement field is continuous over the domain of problem. However, due to the fact that in the X-FEM technique a discontinuous displacement field is considered, necessary modifications must be applied in the variational formulation of X-FEM solution. To this end, the *Gauss–Green theorem* is presented for discontinuous problems to provide a mathematical description for the variational formulation of the X-FEM solution of contact problems. Based on this theorem, an appropriate term is introduced into the variational formulation to incorporate the contact discontinuity into the displacement field.

6.3.1 The Gauss–Green Theorem for Discontinuous Problems

Consider a two-dimensional (2D) domain Ω with the external boundary Γ in which the outward unit normal vector is denoted by $\mathbf{n}_\Gamma$, as shown in Figure 6.6a. Let the tensor A_{ij} and vector a_i are the field variables of the problem, the domain integration can be partly converted to the boundary integration based on the Gauss–Green theorem as

$$\int_\Omega \frac{\partial A_{ij}}{\partial x_j} a_i \, d\Omega = \int_\Gamma a_i A_{ij} n_{\Gamma_j} \, d\Gamma - \int_\Omega A_{ij} \frac{\partial a_i}{\partial x_j} \, d\Omega \qquad (i,j = 1,2,3) \tag{6.17}$$

Now consider the tensor A_{ij} and vector a_i are discontinuous fields over a boundary Γ_d in the domain Ω, the domain of problem can be divided into two distinct sub-domains Ω^+ and Ω^- with the outer boundaries

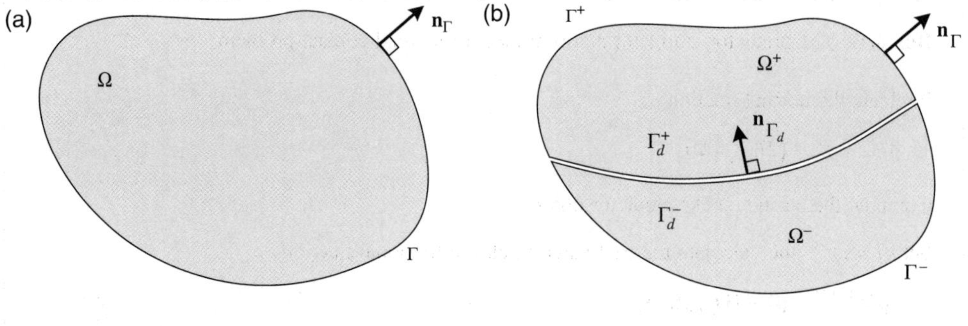

Figure 6.6 A 2D domain Ω with an external boundary Γ: (a) the continuous domain Ω; (b) the discontinuous domain with a discontinuity over the boundary Γ_d

defined as $\Gamma_d^+ \cup \Gamma^+$ and $\Gamma_d^- \cup \Gamma^-$, respectively, such that $\Omega = \Omega^+ \cup \Omega^-$ and $\Gamma = \Gamma^+ \cup \Gamma^-$, as shown in Figure 6.6b. Furthermore, the unit normal vector on Γ_d^- is denoted by $\mathbf{n}_{\Gamma_d}$, that is, the unit normal on Γ_d in the direction of $\Omega^- \to \Omega^+$. The domain integration of left hand side in Eq. (6.17) can be decomposed into two parts as

$$\int_\Omega \frac{\partial A_{ij}}{\partial x_j} a_i \, d\Omega = \int_{\Omega^+} \frac{\partial A_{ij}}{\partial x_j} a_i \, d\Omega + \int_{\Omega^-} \frac{\partial A_{ij}}{\partial x_j} a_i \, d\Omega \tag{6.18}$$

On the basis of Gauss–Green theorem, the right hand side integrals of Eq. (6.18) can be written for the two sub-domains Ω^+ and Ω^- as

$$\int_{\Omega^+} \frac{\partial A_{ij}}{\partial x_j} a_i \, d\Omega = \int_{\Gamma^+} a_i A_{ij} n_{\Gamma_j} \, d\Gamma + \int_{\Gamma_d^+} a_i A_{ij} \left(-n_{\Gamma_{d_j}}\right) d\Gamma - \int_{\Omega^+} A_{ij} \frac{\partial a_i}{\partial x_j} \, d\Omega \tag{6.19}$$

$$\int_{\Omega^-} \frac{\partial A_{ij}}{\partial x_j} a_i \, d\Omega = \int_{\Gamma^-} a_i A_{ij} n_{\Gamma_j} \, d\Gamma + \int_{\Gamma_d^-} a_i A_{ij} n_{\Gamma_{d_j}} \, d\Gamma - \int_{\Omega^-} A_{ij} \frac{\partial a_i}{\partial x_j} \, d\Omega \tag{6.20}$$

Substituting Eqs. (6.19) and (6.20) into (6.18), it yields to

$$\int_\Omega \frac{\partial A_{ij}}{\partial x_j} a_i \, d\Omega = \int_\Gamma a_i A_{ij} n_{\Gamma_j} \, d\Gamma - \int_{\Gamma_d} \left[(a_i A_{ij})^+ - (a_i A_{ij})^-\right] n_{\Gamma_{d_j}} \, d\Gamma - \int_\Omega A_{ij} \frac{\partial a_i}{\partial x_j} \, d\Omega \tag{6.21}$$

in which the value of $(a_i A_{ij})$ on Γ_d^+ and Γ_d^- is respectively denoted by $(a_i A_{ij})^+$ and $(a_i A_{ij})^-$. By defining the jump of $(a_i A_{ij})$ as $[\![a_i A_{ij}]\!] = (a_i A_{ij})^+ - (a_i A_{ij})^-$, the Gauss–Green theorem can, therefore, be expressed for discontinuous problems as

$$\int_\Omega \frac{\partial A_{ij}}{\partial x_j} a_i \, d\Omega = \int_\Gamma a_i A_{ij} n_{\Gamma_j} \, d\Gamma - \int_{\Gamma_d} [\![a_i A_{ij}]\!] \, n_{\Gamma_{d_j}} \, d\Gamma - \int_\Omega A_{ij} \frac{\partial a_i}{\partial x_j} \, d\Omega \tag{6.22}$$

This Gauss–Green theorem for discontinuous problems provides a mathematical basis for the variational formulation of the X-FEM solution in contact problem.

6.3.2 The Weak Form of Governing Equation for a Contact Problem

Consider a 2D body Ω with an internal discontinuous boundary denoted by Γ_d as shown in Figure 6.7, the static equilibrium equation of the body can be written as

$$\nabla \cdot \boldsymbol{\sigma} + \mathbf{b} = 0 \tag{6.23}$$

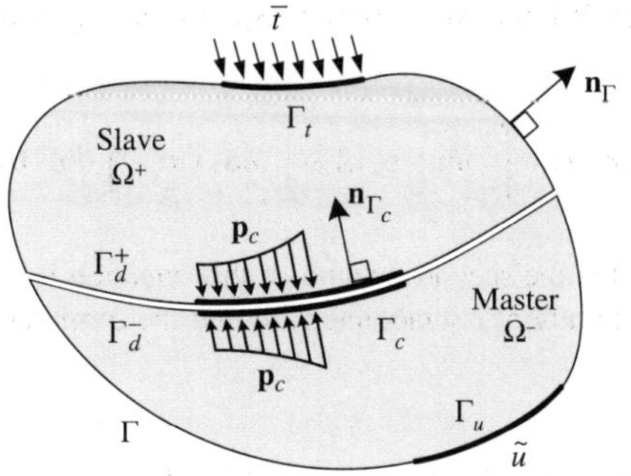

Figure 6.7 Definition of a contact problem. The domain Ω contains the contact interface $\Gamma_c \subset \Gamma_d$

where ∇ is the gradient operator, $\boldsymbol{\sigma}$ is the Cauchy stress tensor that can be related to the strain tensor $\boldsymbol{\varepsilon}$ by an appropriate constitutive relation, and $\mathbf{b}$ is the body force vector.

The essential and natural boundary conditions are defined as follows; the displacement (Dirichlet) boundary condition as $\mathbf{u} = \tilde{\mathbf{u}}$ on $\Gamma_u \subset \Gamma$, the traction (Neumann) boundary condition as $\boldsymbol{\sigma} \cdot \mathbf{n}_\Gamma = \bar{\mathbf{t}}$ on $\Gamma_t \subset \Gamma$, and the contact boundary condition as $\boldsymbol{\sigma} \cdot \mathbf{n}_{\Gamma_c} = \mathbf{p}_c$ on Γ_c in the context of the standard Kuhn–Tucker relations as $g_N \geq 0$, $p_N \leq 0$, and $g_N p_N = 0$ (Figure 6.7). In these boundary conditions, it is assumed that $\Gamma_u \cup \Gamma_t = \Gamma$ and $\Gamma_u \cap \Gamma_t = \emptyset$, and the contact interface $\Gamma_c \subset \Gamma_d$ is known. For simplicity, it is assumed that the remainder of the discontinuity Γ_d experiences an opening mode and is free of traction.

This boundary value problem includes the inequalities related to contact condition, which is called the *Signorini* problem. In the X-FEM solution of contact problem, the variational formulation caused by the contact condition inequalities characterizes the solution of Signorini problem, which is different from the standard variational formulation derived in the continuous continuum mechanics. Due to the inequality constraint on the deformation field, the contact problem has a nonlinear form, even in the case of linear material behavior. Hence, the weak form can be written as an equality assuming that the contact constraint $g_N = 0$ is enforced on the contact interface Γ_c within an each incremental solution.

The notion of slave and master nodes/segments of classical FEM can also be applied in the framework of X-FEM technique, where the interface of bodies in contact is characterized by the discontinuity Γ_c. In such a case, the slave and master nodes/segments pertain to each side of the discontinuity. Consider Ω^+ is assigned to the slave sub-domain and Ω^- to the master sub-domain, the weak form of equilibrium equation can be obtained by multiplication of the virtual displacement field $\delta\mathbf{u}(\boldsymbol{x}, t)$ and the equilibrium equation (6.23) as

$$\int_\Omega \delta\mathbf{u}(\boldsymbol{x},t)\,(\nabla \cdot \boldsymbol{\sigma} + \mathbf{b})\,d\Omega = 0 \tag{6.24}$$

in which $\delta\mathbf{u}(\boldsymbol{x}, t) = \{\delta\mathbf{u} \in C_0 | \delta\mathbf{u} = \mathbf{0}$ on Γ_u and is discontinuous on $\Gamma_c\}$. Applying the Gauss–Green theorem (6.22) for discontinuous problems, Eq. (6.24) can be written as

$$\int_\Omega \nabla\delta\mathbf{u} : \boldsymbol{\sigma}\,d\Omega + \int_{\Gamma_c} [\![\delta\mathbf{u} \,.\, \boldsymbol{\sigma}]\!]\mathbf{n}_{\Gamma_c} d\Gamma - \int_{\Gamma_t} \delta\mathbf{u} \,.\, \bar{\mathbf{t}}\,d\Gamma - \int_\Omega \delta\mathbf{u} \,.\, \mathbf{b}\,d\Omega = 0 \tag{6.25}$$

In this equation, the integration over the boundary Γ is reduced to Γ_t since $\delta\mathbf{u}(\boldsymbol{x}, t)$ is zero on Γ_u. In addition, due to the fact that the tractions are continuous over the contact interface, the second integral of (6.25) can be written as

$$\begin{aligned} \int_{\Gamma_c} [\![\delta\mathbf{u} \,.\, \boldsymbol{\sigma}]\!]\mathbf{n}_{\Gamma_c} d\Gamma &= \int_{\Gamma_c} [(\delta\mathbf{u} \,.\, \boldsymbol{\sigma})^+ - (\delta\mathbf{u} \,.\, \boldsymbol{\sigma})^-]\mathbf{n}_{\Gamma_c} d\Gamma \\ &= \int_{\Gamma_c} (\delta\mathbf{u}^+ - \delta\mathbf{u}^-)\boldsymbol{\sigma} \,.\, \mathbf{n}_{\Gamma_c} d\Gamma = \int_{\Gamma_c} [\![\delta\mathbf{u}]\!]\boldsymbol{\sigma} \,.\, \mathbf{n}_{\Gamma_c} d\Gamma = \int_{\Gamma_c} [\![\delta\mathbf{u}]\!]\,\mathbf{p}_c\,d\Gamma \end{aligned} \tag{6.26}$$

Substituting Eq. (6.26) into (6.25), the weak form of equilibrium equation for contact problem can be finally obtained as

$$\int_{\Omega} \delta \boldsymbol{\varepsilon} : \boldsymbol{\sigma}\, d\Omega + \int_{\Gamma_c} [\![\delta \mathbf{u}]\!] \cdot \mathbf{p}_c\, d\Gamma - \int_{\Gamma_t} \delta \mathbf{u} \,.\, \bar{\mathbf{t}}\, d\Gamma - \int_{\Omega} \delta \mathbf{u} \,.\, \mathbf{b}\, d\Omega = 0 \tag{6.27}$$

It is worthwhile mentioning that the second integral of this equation is derived from the Gauss–Green theorem that takes the discontinuity of contact interface into the formulation.

6.3.3 The Enrichment of Displacement Field

In order to model the discontinuity of contact interface into the X-FEM, the nodal points of elements cut by the contact interface are enriched by the Heaviside jump function $H(\boldsymbol{x})$. The discontinuity of displacement can be modeled by introducing enriched degrees of freedom (DOF) to the nodes associated with the discontinuity. The displacement field of an enriched element $\mathbf{u}(\boldsymbol{x}, t)$ can be approximated by

$$\begin{aligned} \mathbf{u}(\boldsymbol{x},t) &= \sum_{I \in \mathcal{N}^{std}} N_I^{std}(\boldsymbol{x})\, \bar{\mathbf{u}}_I + \sum_{J \in \mathcal{N}^{enr}} N_J^{enr}(\boldsymbol{x})\, \bar{\mathbf{a}}_J \\ &\equiv \mathbf{N}^{std}(\boldsymbol{x})\, \bar{\mathbf{u}} + \mathbf{N}^{enr}(\boldsymbol{x})\, \bar{\mathbf{a}} \end{aligned} \tag{6.28}$$

where $\mathcal{N}^{std}$ is the set of all nodal points and $\mathcal{N}^{enr}$ is the set of enriched nodal points in the domain, $\bar{\mathbf{u}}_I$ is the standard nodal DOF, and $\bar{\mathbf{a}}_J$ is the enriched nodal DOF. In fact, for the set of nodes which their support is cut by the contact surface, the new DOF are introduced to enhance the standard FE approximation. In the earlier relation, $N^{std}(\boldsymbol{x})$ denotes the standard shape function and the enriched shape function $N^{enr}(\boldsymbol{x})$ is defined by $N_J^{enr}(\boldsymbol{x}) = N_J^{std}(\boldsymbol{x})\,(H(\boldsymbol{x}) - H(\boldsymbol{x}_J))$ with $H(\boldsymbol{x})$ denoting the Heaviside jump function introduced as

$$H(\boldsymbol{x}) = \begin{cases} +1 & \varphi(\boldsymbol{x}) \geq 0 \\ 0 & \varphi(\boldsymbol{x}) < 0 \end{cases} \tag{6.29}$$

where $\varphi(\boldsymbol{x})$ is the signed distance function defined based on the absolute value of level set function as

$$\varphi(\boldsymbol{x}) = \min \|\boldsymbol{x} - \boldsymbol{x}^*\| \operatorname{sign}((\boldsymbol{x} - \boldsymbol{x}^*) \cdot \mathbf{n}_{\Gamma_c}) \tag{6.30}$$

where $\boldsymbol{x}^*$ is a point on the contact surface that is the closest to the point $\boldsymbol{x}$, and $\mathbf{n}_{\Gamma_c}$ is the normal vector to the contact surface at point $\boldsymbol{x}^*$. Because of independent displacement fields at both sides of the contact surface, the Heaviside jump function is an appropriate function for simulation of a contact problem. The displacement jump on the contact surface can therefore be written from (6.28) as

$$\begin{aligned} [\![\mathbf{u}]\!] &= [\![\mathbf{N}^{std}(\boldsymbol{x})\, \bar{\mathbf{u}} + \mathbf{N}^{enr}(\boldsymbol{x})\, \bar{\mathbf{a}}]\!] = [\![\mathbf{N}^{std}(\boldsymbol{x})\, \bar{\mathbf{u}}]\!] + [\![\mathbf{N}^{enr}(\boldsymbol{x})\, \bar{\mathbf{a}}]\!] = [\![\mathbf{N}^{enr}(\boldsymbol{x})\, \bar{\mathbf{a}}]\!] \\ &= \mathbf{N}^{std}(\boldsymbol{x})[\![H(\boldsymbol{x}) - H(\boldsymbol{x}_J)]\!]\bar{\mathbf{a}} \equiv \mathbf{N}^{std}(\boldsymbol{x})\, \bar{\mathbf{a}} \end{aligned} \tag{6.31}$$

Accordingly, the strain vector $\boldsymbol{\varepsilon}(\boldsymbol{x}, t)$ corresponding to the approximate displacement field (6.28) can be written in terms of the standard and enriched nodal values as

$$\begin{aligned} \boldsymbol{\varepsilon}(\boldsymbol{x},t) &= \sum_{I \in \mathcal{N}^{std}} B_I^{std}(\boldsymbol{x})\, \bar{\mathbf{u}}_I + \sum_{J \in \mathcal{N}^{enr}} B_J^{enr}(\boldsymbol{x})\, \bar{\mathbf{a}}_J \\ &\equiv \mathbf{B}^{std}(\boldsymbol{x})\, \bar{\mathbf{u}} + \mathbf{B}^{enr}(\boldsymbol{x})\, \bar{\mathbf{a}} \end{aligned} \tag{6.32}$$

where $\mathbf{B}^{std}(\boldsymbol{x}) \equiv \mathbf{L}\mathbf{N}^{std}(\boldsymbol{x})$ involve the spatial derivatives of the standard shape functions and $\mathbf{B}^{enr}(\boldsymbol{x}) \equiv \mathbf{L}\mathbf{N}^{enr}(\boldsymbol{x})$ contain the spatial derivatives of the enriched shape functions associated with the Heaviside function, where $\mathbf{L}$ denotes the matrix differential operator defined as

$$\mathbf{L} = \begin{bmatrix} \partial/\partial x & 0 \\ 0 & \partial/\partial y \\ \partial/\partial y & \partial/\partial x \end{bmatrix} \tag{6.33}$$

Finally, the required variations of the displacement and strain fields $\delta\mathbf{u}(\boldsymbol{x}, t)$ and $\delta\boldsymbol{\varepsilon}(\boldsymbol{x}, t)$ for the weak form of governing equation (6.27) can be respectively introduced in the same approximating space as the displacement and strain fields $\mathbf{u}(\boldsymbol{x}, t)$ and $\boldsymbol{\varepsilon}(\boldsymbol{x}, t)$ defined in (6.28) and (6.32) as

$$\begin{aligned} \delta\mathbf{u}(\boldsymbol{x},t) &= \sum_{I\in\mathcal{N}^{std}} N_I^{std}(\boldsymbol{x})\,\delta\bar{\mathbf{u}}_I + \sum_{J\in\mathcal{N}^{enr}} N_J^{enr}(\boldsymbol{x})\,\delta\bar{\mathbf{a}}_J \\ &\equiv \mathbf{N}^{std}(\boldsymbol{x})\,\delta\bar{\mathbf{u}} + \mathbf{N}^{enr}(\boldsymbol{x})\,\delta\bar{\mathbf{a}} \end{aligned} \tag{6.34}$$

$$\begin{aligned} \delta\boldsymbol{\varepsilon}(\boldsymbol{x},t) &= \sum_{I\in\mathcal{N}^{std}} B_I^{std}(\boldsymbol{x})\,\delta\bar{\mathbf{u}}_I + \sum_{J\in\mathcal{N}^{enr}} B_J^{enr}(\boldsymbol{x})\,\delta\bar{\mathbf{a}}_J \\ &\equiv \mathbf{B}^{std}(\boldsymbol{x})\,\delta\bar{\mathbf{u}} + \mathbf{B}^{enr}(\boldsymbol{x})\,\delta\bar{\mathbf{a}} \end{aligned} \tag{6.35}$$

6.4 Modeling of Contact Constraints via the Penalty Method

In order to implement the contact constraints into the weak form of X-FEM equilibrium equation of the contact problem, various contact constraints techniques can be employed. There are a large variety of approaches proposed for the numerical solution of contact constraints, which are well known methods in optimization theory. Each method has its own advantages and disadvantages. Among them, the most frequently used techniques in the numerical treatment of contact problems are the penalty method, the Lagrange multipliers technique, and the augmented-Lagrange multipliers method. These methods are proposed here into the weak form of X-FEM equilibrium equation (6.27) in the following sections. The Lagrange and augmented-Lagrange multipliers methods are formulated to impose the contact constraints in the normal direction whereas in the tangential direction the contact constitutive law, described by the tangential stiffness of the contact surface, is used to incorporate the frictional forces to distinguish between the stick and slip conditions. In this section, the contact constraint formulations are derived based on the penalty method in a general form of frictional contact so that the frictionless contact can be derived by assuming the frictional stiffness to be zero.

In the penalty method, the penetration between two contacting bodies is introduced where the normal contact force is related to the penetration by using the normal stiffness of contact interface. In fact, this normal stiffness allows the contact surface to slightly overlap, that is, $g_N < 0$; which results in approximately satisfying the contact constraint. The normal stiffness of contact interface is analogous to a penalty parameter that imposes the contact constraint, that is, $g_N = 0$, through the penalty method. In order to derive the X-FEM solution of contact problem, the displacement and strain fields $\delta\mathbf{u}(\boldsymbol{x}, t)$ and $\delta\boldsymbol{\varepsilon}(\boldsymbol{x}, t)$ defined in (6.34) and (6.35) are substituted into the weak form of equilibrium equation (6.27) as

$$\begin{aligned} &\int_{\Omega} \left(\mathbf{B}^{std}\,\delta\bar{\mathbf{u}} + \mathbf{B}^{enr}\,\delta\bar{\mathbf{a}}\right)^T \boldsymbol{\sigma}\,d\Omega + \int_{\Gamma_c} [\![\mathbf{N}^{std}\,\delta\bar{\mathbf{u}} + \mathbf{N}^{enr}\,\delta\bar{\mathbf{a}}]\!]^T \mathbf{p}_c\,d\Gamma \\ &- \int_{\Gamma_t} \left(\mathbf{N}^{std}\,\delta\bar{\mathbf{u}} + \mathbf{N}^{enr}\,\delta\bar{\mathbf{a}}\right)^T \bar{\mathbf{t}}\,d\Gamma - \int_{\Omega} \left(\mathbf{N}^{std}\,\delta\bar{\mathbf{u}} + \mathbf{N}^{enr}\,\delta\bar{\mathbf{a}}\right)^T \mathbf{b}\,d\Omega = 0 \end{aligned} \tag{6.36}$$

The second term of this equation can be written according to (6.31) as

$$\int_{\Gamma_c} \left[\mathbf{N}^{std}\,\delta\bar{\mathbf{u}} + \mathbf{N}^{enr}\,\delta\bar{\mathbf{a}}\right]^T \mathbf{p}_c\, d\Gamma = \int_{\Gamma_c} \delta\bar{\mathbf{a}}^T \left(\mathbf{N}^{std}\right)^T \mathbf{p}_c\, d\Gamma \tag{6.37}$$

Substituting relation (6.37) into (6.36) and rearranging the integral equilibrium equation, it yields to

$$\begin{aligned} &\delta\bar{\mathbf{u}}^T \left\{ \int_\Omega \left(\mathbf{B}^{std}\right)^T \boldsymbol{\sigma}\, d\Omega - \int_{\Gamma_t} \left(\mathbf{N}^{std}\right)^T \bar{\mathbf{t}}\, d\Gamma - \int_\Omega \left(\mathbf{N}^{std}\right)^T \mathbf{b}\, d\Omega \right\} \\ &+ \delta\bar{\mathbf{a}}^T \left\{ \int_\Omega (\mathbf{B}^{enr})^T \boldsymbol{\sigma}\, d\Omega + \int_{\Gamma_c} \left(\mathbf{N}^{std}\right)^T \mathbf{p}_c\, d\Gamma - \int_{\Gamma_t} (\mathbf{N}^{enr})^T \bar{\mathbf{t}}\, d\Gamma - \int_\Omega (\mathbf{N}^{enr})^T \mathbf{b}\, d\Omega \right\} = 0 \end{aligned} \tag{6.38}$$

This equilibrium equation results in the following sets of equations as

$$\begin{aligned} \boldsymbol{\Psi}_1(\bar{\mathbf{u}},\bar{\mathbf{a}}) &\equiv \int_\Omega \left(\mathbf{B}^{std}\right)^T \boldsymbol{\sigma}\, d\Omega - \int_{\Gamma_t} \left(\mathbf{N}^{std}\right)^T \bar{\mathbf{t}}\, d\Gamma - \int_\Omega \left(\mathbf{N}^{std}\right)^T \mathbf{b}\, d\Omega = \mathbf{0} \\ \boldsymbol{\Psi}_2(\bar{\mathbf{u}},\bar{\mathbf{a}}) &\equiv \int_\Omega (\mathbf{B}^{enr})^T \boldsymbol{\sigma}\, d\Omega + \int_{\Gamma_c} \left(\mathbf{N}^{std}\right)^T \mathbf{p}_c\, d\Gamma - \int_{\Gamma_t} (\mathbf{N}^{enr})^T \bar{\mathbf{t}}\, d\Gamma - \int_\Omega (\mathbf{N}^{enr})^T \mathbf{b}\, d\Omega = \mathbf{0} \end{aligned} \tag{6.39}$$

in which

$$\begin{aligned} \mathbf{F}_{\text{int}}^{std} &= \int_\Omega \left(\mathbf{B}^{std}\right)^T \boldsymbol{\sigma}\, d\Omega \\ \mathbf{F}_{\text{ext}}^{std} &= \int_{\Gamma_t} \left(\mathbf{N}^{std}\right)^T \bar{\mathbf{t}}\, d\Gamma + \int_\Omega \left(\mathbf{N}^{std}\right)^T \mathbf{b}\, d\Omega \\ \mathbf{F}_{\text{int}}^{enr} &= \int_\Omega (\mathbf{B}^{enr})^T \boldsymbol{\sigma}\, d\Omega \\ \mathbf{F}_{\text{ext}}^{enr} &= \int_{\Gamma_t} (\mathbf{N}^{enr})^T \bar{\mathbf{t}}\, d\Gamma + \int_\Omega (\mathbf{N}^{enr})^T \mathbf{b}\, d\Omega \\ \mathbf{f}_{\text{int}}^{\text{con}} &= \int_{\Gamma_c} \left(\mathbf{N}^{std}\right)^T \mathbf{p}_c\, d\Gamma \end{aligned} \tag{6.40}$$

in which the integral over the discontinuity Γ_c in load vector $\mathbf{f}_{\text{int}}^{\text{con}}$ emanates from the enriched part that represents the internal force due to contact traction $\mathbf{p}_c$. The nonlinearities in equilibrium equations (6.39) arise from two distinct sources; the nonlinearity in material behavior of the system and the nonlinearity in contact behavior between two bodies. In fact, in the case of linear material behavior of the system, the nonlinearity due to frictional contact behavior can be observed since the contact traction $\mathbf{p}_c$ is dependent nonlinearly on the displacement jump along the contact surface Γ_c. Hence, the Newton–Raphson iterative procedure can be applied to linearize the discretized governing equations (6.39). The linearized system of equations for iteration $i+1$ of time interval $(n, n+1)$ can be obtained by expanding the residual equations (6.39) with the first-order truncated Taylor series as

$$\begin{bmatrix} \mathbf{K}_{uu} & \mathbf{K}_{ua} \\ \mathbf{K}_{au} & \mathbf{K}_{aa} \end{bmatrix}_{n+1}^{i} \begin{Bmatrix} d\bar{\mathbf{u}}_n^i \\ d\bar{\mathbf{a}}_n^i \end{Bmatrix} = -\begin{Bmatrix} \boldsymbol{\Psi}_1 \\ \boldsymbol{\Psi}_2 \end{Bmatrix}_{n+1}^{i} \tag{6.41}$$

where $d\bar{\mathbf{u}}$ and $d\bar{\mathbf{a}}$ denote the increments of the standard and enriched nodal displacements, $\boldsymbol{\Psi}_1 = \mathbf{F}_{\text{int}}^{std} - \mathbf{F}_{\text{ext}}^{std}$ and $\boldsymbol{\Psi}_2 = \mathbf{F}_{\text{int}}^{enr} - \mathbf{F}_{\text{ext}}^{enr} + \mathbf{f}_{\text{int}}^{\text{con}}$ stand for the vectors that can be evaluated from the known initial values, and the components of stiffness matrix can be obtained as

$$
\begin{aligned}
\mathbf{K}_{uu} &= \frac{\partial \boldsymbol{\Psi}_1}{\partial \bar{\mathbf{u}}} = \int_\Omega \left(\mathbf{B}^{std}\right)^T \frac{\partial \boldsymbol{\sigma}}{\partial \bar{\mathbf{u}}} d\Omega \\
\mathbf{K}_{ua} &= \frac{\partial \boldsymbol{\Psi}_1}{\partial \bar{\mathbf{a}}} = \int_\Omega \left(\mathbf{B}^{std}\right)^T \frac{\partial \boldsymbol{\sigma}}{\partial \bar{\mathbf{a}}} d\Omega \\
\mathbf{K}_{au} &= \frac{\partial \boldsymbol{\Psi}_2}{\partial \bar{\mathbf{u}}} = \int_\Omega \left(\mathbf{B}^{enr}\right)^T \frac{\partial \boldsymbol{\sigma}}{\partial \bar{\mathbf{u}}} d\Omega \\
\mathbf{K}_{aa} &= \frac{\partial \boldsymbol{\Psi}_2}{\partial \bar{\mathbf{a}}} = \int_\Omega \left(\mathbf{B}^{enr}\right)^T \frac{\partial \boldsymbol{\sigma}}{\partial \bar{\mathbf{a}}} d\Omega + \int_{\Gamma_c} \left(\mathbf{N}^{std}\right)^T \frac{\partial \mathbf{p}_c}{\partial \bar{\mathbf{a}}} d\Gamma
\end{aligned}
\tag{6.42}
$$

Hence, the stiffness matrix $\mathbf{K}$ is composed of

$$
\mathbf{K} = \begin{bmatrix} \int_\Omega \left(\mathbf{B}^{std}\right)^T \mathbf{D}\,\mathbf{B}^{std}\, d\Omega & \int_\Omega \left(\mathbf{B}^{std}\right)^T \mathbf{D}\,\mathbf{B}^{enr}\, d\Omega \\ \int_\Omega \left(\mathbf{B}^{enr}\right)^T \mathbf{D}\,\mathbf{B}^{std}\, d\Omega & \int_\Omega \left(\mathbf{B}^{enr}\right)^T \mathbf{D}\,\mathbf{B}^{enr}\, d\Omega + \underbrace{\int_{\Gamma_c} \left(\mathbf{N}^{std}\right)^T \mathbf{D}_f^{ep}\, \mathbf{N}^{std}\, d\Gamma}_{\mathbf{K}^{\text{con}}} \end{bmatrix}
\tag{6.43}
$$

in which $\mathbf{K}^{\text{con}}$ and $\mathbf{f}_{\text{int}}^{\text{con}}$ can be respectively interpreted as the contact stiffness matrix and contact force vector that incorporate the contact behavior into the X-FEM formulation. In these relations, the behavior of the bulk material is assumed to be the linear elastic, so its constitutive relation is defined as

$$
d\boldsymbol{\sigma} = \mathbf{D}\, d\boldsymbol{\varepsilon} \equiv \mathbf{D}\mathbf{B}^{std} d\bar{\mathbf{u}} + \mathbf{D}\mathbf{B}^{enr} d\bar{\mathbf{a}} \tag{6.44}
$$

where $\mathbf{D}$ is the tangential constitutive matrix of the bulk material. The nonlinear behavior of contact friction is characterized by the contact constitutive law, that is, Eq. (6.11), relating the contact traction $\mathbf{p}_c$ to the displacement jump along the contact surface Γ_c as

$$
d\mathbf{p}_c = \mathbf{D}_f^{ep}\, d[\![\mathbf{u}]\!] \equiv \mathbf{D}_f^{ep}\, \mathbf{N}^{std}\, d\bar{\mathbf{a}} \tag{6.45}
$$

In order to derive the contact stiffness matrix $\mathbf{K}^{\text{con}}$ and the contact force vector $\mathbf{f}_{\text{int}}^{\text{con}}$, the relevant integrals must be performed over the contact surface Γ_c. For a two-dimensional problem, the interface is first subdivided into a set of one-dimensional (1D) segments, as shown in Figure 6.8. For each segment, the intersections of the contact interface with the elements cut by the discontinuity are determined to construct a set of 1D elements along Γ_c. The Gauss quadrature points are then used for each 1D element in order to

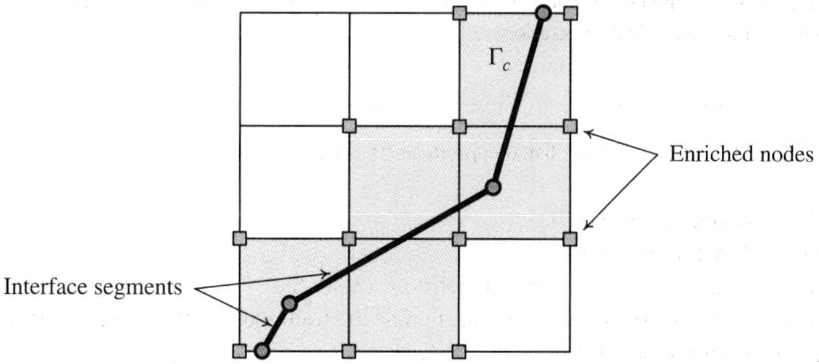

Figure 6.8 Discretization of the contact interface into a set of 1D segments

numerically integrate the integrals of $\mathbf{K}^{\text{con}}$ and $\mathbf{f}_{\text{int}}^{\text{con}}$ over Γ_c, as shown in Figure 6.9. It must be noted that the proposed numerical integration is performed to evaluate the contact stiffness matrix and the contact force vector along the contact surface, and no new DOF are added to the system.

The accuracy of imposing contact constraints in the normal direction strongly depends on the magnitude of penalty parameter k_N; the larger the value of the penalty parameter, the more accurate contact constraints can be achieved. However, a very large value of the penalty parameter results in an ill conditioned stiffness matrix. So, the chosen value of penalty parameter k_N is a key point in the performance of penalty method. Furthermore, in order to preserve the symmetry of the numerical formulation, the continuum tangent matrix can be decomposed into a pure contact in the normal direction and a frictional resistance in the tangential direction replacing $\mathbf{D}_f^{ep}$ by $\bar{\mathbf{D}}_f^{ep}$ defined in (6.16). The solution procedure of the penalty method is presented in Box 6.2. The convergence of the solution can be finally evaluated at each increment using the vector of residual force obtained from the Newton–Raphson procedure by $\eta = \|\langle \boldsymbol{\Psi}_1^T, \boldsymbol{\Psi}_2^T \rangle\| / \|\mathbf{F}_{ext}\| \le \eta_{\text{aim}}$, with η_{aim} denoting the prescribed target percentage error.

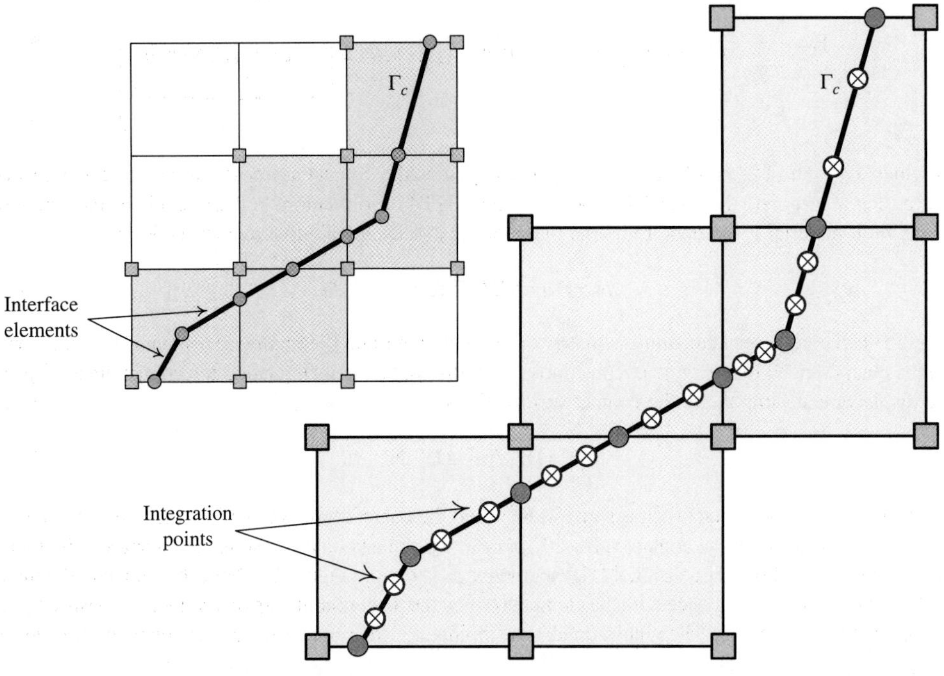

Figure 6.9 Modeling of contact interface in the X-FEM: Generation of interface elements together with the integration points used for numerical integration

Box 6.2 The solution procedure for the penalty method

Initialize the solution; set $(\bar{\mathbf{u}}, \bar{\mathbf{a}}) = \mathbf{0}$,
LOOP over the Newton iterations i
Solve system of Eq. (6.41) (for the first iteration $i = 1$ set the stick condition),
For each Gauss point on the contact interface, update the frictional tractions and determine the state of stick or slip condition according to Box 6.1,
Check for the convergence, IF $\eta = \|\langle \boldsymbol{\Psi}_1^T, \boldsymbol{\Psi}_2^T \rangle\| / \|\mathbf{F}_{ext}\| \le \eta_{\text{aim}}$ then END LOOP,
END LOOP over the Newton iterations.

6.4.1 Modeling of an Elastic Bar with a Discontinuity at Its Center

In order to illustrate the performance of proposed penalty X-FEM formulation described in previous section, an elastic 1D bar with a discontinuity at its center is modeled numerically. The geometry, boundary conditions, and X-FEM mesh of the problem are shown in Figure 6.10. The material property of the member is chosen as $AE = 2 \times 10^4$ kg, and the bar is subjected to a compression load of $p = 200$ kg at the free end. In order to model the discontinuity in the bar, the enriched DOF are assumed at nodal points 2 and 3, and the contact constraint is imposed using the penalty method in the enriched element (2). The Heaviside jump function is defined as

$$H(x) = \begin{cases} +1 & x \geq 3\ell/2 \\ 0 & x < 3\ell/2 \end{cases} \tag{6.46}$$

The numerical integration over Γ_c for this 1D problem is performed using one integration point at the position of discontinuity. The standard and enriched parts of the shape function $\mathbf{N}(x)$, the vector of Cartesian shape function derivatives $\mathbf{B}(x)$, and the stiffness matrix $\mathbf{K}$ for these three elements are as follows

$$\begin{aligned}
&{}^{(1)}\mathbf{N}^{std}_{(x)} = \langle 1 - x/\ell \quad x/\ell \rangle, \quad {}^{(1)}\mathbf{N}^{enr}_{(x)} = (x/\ell)H(x) \\
&{}^{(1)}\mathbf{B}^{std}_{(x)} = \langle -1/\ell \quad 1/\ell \rangle, \quad {}^{(1)}\mathbf{B}^{enr}_{(x)} = (1/\ell)H(x) \\
&{}^{(1)}\mathbf{K} = \begin{bmatrix} \int_\ell (\mathbf{B}^{std})^T AE\, \mathbf{B}^{std} dx & \int_\ell (\mathbf{B}^{std})^T AE\, \mathbf{B}^{enr} dx \\ \int_\ell (\mathbf{B}^{enr})^T AE\, \mathbf{B}^{std} dx & \int_\ell (\mathbf{B}^{enr})^T AE\, \mathbf{B}^{enr} dx \end{bmatrix} = \frac{AE}{\ell} \begin{bmatrix} +1 & -1 & 0 \\ -1 & +1 & 0 \\ 0 & 0 & 0 \end{bmatrix} \begin{matrix} u_1 \\ u_2 \\ a_2 \end{matrix}
\end{aligned} \tag{6.47a}$$

for element (1), whereas for element (2) as

$$\begin{aligned}
&{}^{(2)}\mathbf{N}^{std}_{(x)} = \langle 2 - x/\ell \quad -1 + x/\ell \rangle, \quad {}^{(2)}\mathbf{N}^{enr}_{(x)} = \langle (2 - x/\ell)H(x) \quad (-1 + x/\ell)(H(x) - 1) \rangle \\
&{}^{(2)}\mathbf{B}^{std}_{(x)} = \langle -1/\ell \quad 1/\ell \rangle, \qquad {}^{(2)}\mathbf{B}^{enr}_{(x)} = \langle (-1/\ell)H(x) \quad (1/\ell)(H(x) - 1) \rangle \\
&{}^{(2)}\mathbf{K} = \begin{bmatrix} \int_\ell (\mathbf{B}^{std})^T AE\, \mathbf{B}^{std} dx & \int_\ell (\mathbf{B}^{std})^T AE\, \mathbf{B}^{enr} dx \\ \int_\ell (\mathbf{B}^{enr})^T AE\, \mathbf{B}^{std} dx & \int_\ell (\mathbf{B}^{enr})^T AE\, \mathbf{B}^{enr} dx + \left| (\mathbf{N}^{std})^T k_N\, \mathbf{N}^{std} \right|_{x = 3\ell/2} \end{bmatrix}
\end{aligned}$$

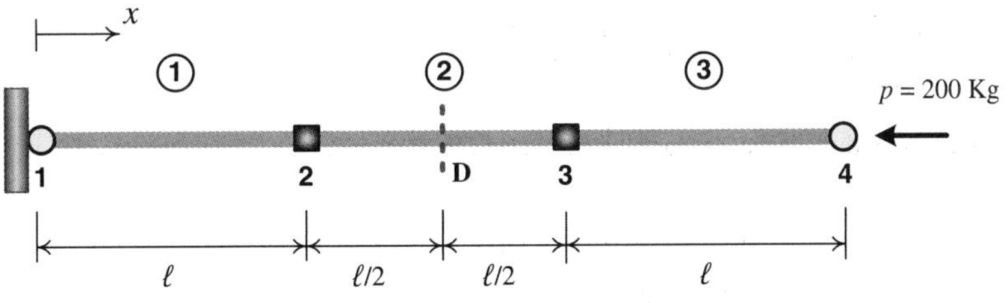

Figure 6.10 An elastic 1D bar with a discontinuity at its center

$$
=\frac{AE}{\ell}\begin{bmatrix} +1 & -1 & +\frac{1}{2} & +\frac{1}{2} \\ -1 & +1 & -\frac{1}{2} & -\frac{1}{2} \\ +\frac{1}{2} & -\frac{1}{2} & \frac{1}{2}+\frac{k_N\ell}{4AE} & \frac{k_N\ell}{4AE} \\ +\frac{1}{2} & -\frac{1}{2} & \frac{k_N\ell}{4AE} & \frac{1}{2}+\frac{k_N\ell}{4AE} \end{bmatrix}\begin{matrix} u_2 \\ u_3 \\ a_2 \\ a_3 \end{matrix} \tag{6.47b}
$$

and for element (3) as

$$
{}^{(3)}\mathbf{N}^{std}_{(x)} = \langle 3-x/\ell \quad -2+x/\ell \rangle, \quad {}^{(3)}\mathbf{N}^{enr}_{(x)} = (3-x/\ell)(H(x)-1)
$$

$$
{}^{(3)}\mathbf{B}^{std}_{(x)} = \langle -1/\ell \quad 1/\ell \rangle, \qquad {}^{(3)}\mathbf{B}^{enr}_{(x)} = (-1/\ell)(H(x)-1)
$$

$$
{}^{(3)}\mathbf{K} = \begin{bmatrix} \int_\ell (\mathbf{B}^{std})^T AE\, \mathbf{B}^{std} dx & \int_\ell (\mathbf{B}^{std})^T AE\, \mathbf{B}^{enr} dx \\ \int_\ell (\mathbf{B}^{enr})^T AE\, \mathbf{B}^{std} dx & \int_\ell (\mathbf{B}^{enr})^T AE\, \mathbf{B}^{enr} dx \end{bmatrix} = \frac{AE}{\ell}\begin{bmatrix} +1 & -1 & 0 \\ -1 & +1 & 0 \\ 0 & 0 & 0 \end{bmatrix}\begin{matrix} u_3 \\ u_4 \\ a_3 \end{matrix} \tag{6.47c}
$$

Thus, the total stiffness matrix can be obtained as

$$
\mathbf{K} = \frac{AE}{\ell}\begin{bmatrix} +1 & -1 & 0 & 0 & 0 & 0 \\ -1 & +2 & -1 & 0 & +\frac{1}{2} & +\frac{1}{2} \\ 0 & -1 & +2 & -1 & -\frac{1}{2} & -\frac{1}{2} \\ 0 & 0 & -1 & +1 & 0 & 0 \\ 0 & +\frac{1}{2} & -\frac{1}{2} & 0 & \frac{1}{2}+\frac{k_N\ell}{4AE} & \frac{k_N\ell}{4AE} \\ 0 & +\frac{1}{2} & -\frac{1}{2} & 0 & \frac{k_N\ell}{4AE} & \frac{1}{2}+\frac{k_N\ell}{4AE} \end{bmatrix}\begin{matrix} u_1 \\ u_2 \\ u_3 \\ u_4 \\ a_2 \\ a_3 \end{matrix} \tag{6.48}
$$

Note that in the absence of frictional behavior, a linear system of equations can be solved, which results in the following standard and enriched nodal displacements as

$$
\begin{aligned}
\bar{\mathbf{u}}^T &= \langle u_1 \quad u_2 \quad u_3 \quad u_4 \rangle = \left\langle 0 \quad -\frac{p\ell}{AE} \quad -\frac{p}{k_N}-\frac{2p\ell}{AE} \quad -\frac{p}{k_N}-\frac{3p\ell}{AE} \right\rangle \\
\bar{\mathbf{a}}^T &= \langle a_2 \quad a_3 \rangle = \left\langle -\frac{p}{k_N} \quad -\frac{p}{k_N} \right\rangle
\end{aligned} \tag{6.49}
$$

The normal gap g_N and the contact traction p_c at point D can be obtained as

$$
\begin{aligned}
g_N &= \left| {}^{(2)}\mathbf{N}^{std}_{(x)}\, \bar{\mathbf{a}} \right|_{x=3\ell/2} = \frac{1}{2}(a_2+a_3) = -\frac{p}{k_N} \\
p_c &= k_N g_N = -p
\end{aligned} \tag{6.50}
$$

Table 6.1 An elastic 1D bar with a discontinuity at its center

k_N (kg/cm)	u_2 (cm)	u_3 (cm)	u_4 (cm)	a_2 (cm)	a_3 (cm)	g_N (cm)
AE/ℓ	−0.1	−0.3	−0.4	−0.1	−0.1	−0.1
$10AE/\ell$	−0.1	−0.210	−0.310	−0.010	−0.010	−0.01
$100AE/\ell$	−0.1	−0.201	−0.301	−0.001	−0.001	−0.001
$1000AE/\ell$	−0.1	−0.2001	−0.3001	−0.0010	−0.0010	−0.0001

The nodal displacement values for different values of k_N.

Obviously, the accuracy of the solution strongly depends on the value of k_N. If $k_N \to \infty$, the values of displacement converge to the exact solution and the normal gap g_N approaches zero. However, the finite values of k_N results in an approximate solution of nodal displacements and contact traction p_c. In Table 6.1, the values of nodal displacements are given for different values of k_N.

6.4.2 *Modeling of an Elastic Plate with a Discontinuity at Its Center*

The next example is of an elastic plate with a discontinuity at its center chosen to illustrate the performance of proposed penalty X-FEM formulation for a 2D problem. The geometry, boundary conditions, and X-FEM mesh of the problem are shown in Figure 6.11. The material properties of the plate are chosen as follows; $E = 2\times 10^6$ kg/cm^2 and $\nu = 0.3$. The plate is in plane strain condition, which is fixed at the left edge and subjected to a compression load $q = 2\times 10^4$ kg/cm^2 at the right edge. In order to model the discontinuity in the plate, the enriched DOF are assumed at nodal points of element (2), and the contact constraints are imposed using the penalty method in this enriched element. The problem is modeled by three 4-noded elements using the standard shape functions $N^{std}_{I(\xi,\eta)} = \frac{1}{4}(1+\xi_I\xi)(1+\eta_I\eta)$ and the enriched shape functions $N^{enr}_{J(\xi,\eta)} = N^{std}_{J(\xi,\eta)}(H(\boldsymbol{x}) - H(\boldsymbol{x}_J))$, with the Heaviside function defined as

$$H(x) = \begin{cases} +1 & x \geq x_c \\ 0 & x < x_c \end{cases} \tag{6.51}$$

The numerical integration over the contact interface is performed using the 1D standard integration rule with two Gauss integration points along the contact interface Γ_c. The stiffness matrices for these three elements can be obtained according to (6.43) as follows

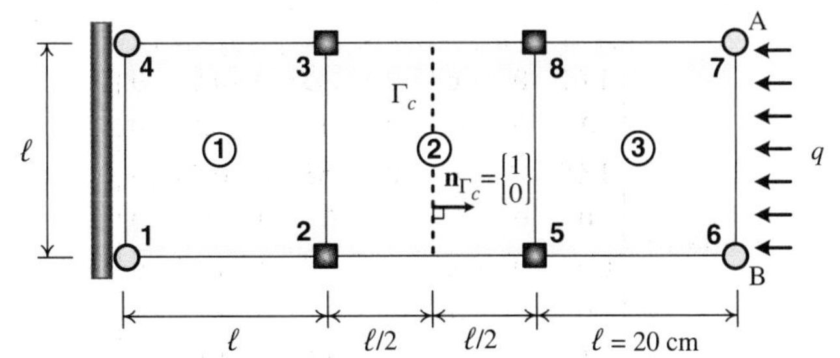

Figure 6.11 An elastic plate with a discontinuity in the middle

$$
{}^{(1)}\mathbf{K} = {}^{(3)}\mathbf{K} = \int_{-1}^{+1}\int_{-1}^{+1} \mathbf{B}_{(\xi,\eta)}^{T} \mathbf{D}\,\mathbf{B}_{(\xi,\eta)} \det[\mathbf{J}] d\xi\, d\eta \equiv \left({}^{(1)}\mathbf{K}_{uu} = {}^{(3)}\mathbf{K}_{uu}\right)
$$

$$
= 1\times 10^{6}
\begin{array}{c}
\begin{array}{cccccccc} u_{x_1} & u_{y_1} & u_{x_2} & u_{y_2} & u_{x_3} & u_{y_3} & u_{x_4} & u_{y_4} \end{array} \\
\begin{bmatrix}
1.154 & 0.481 & -0.769 & 0.096 & -0.577 & -0.481 & 0.192 & -0.096 \\
0.481 & 1.154 & -0.096 & 0.192 & -0.481 & -0.577 & 0.096 & -0.769 \\
-0.769 & -0.096 & 1.154 & -0.481 & 0.192 & 0.096 & -0.577 & 0.481 \\
0.096 & 0.192 & -0.481 & 1.154 & -0.096 & -0.769 & 0.481 & -0.577 \\
-0.577 & -0.481 & 0.192 & -0.096 & 1.154 & 0.481 & -0.769 & 0.096 \\
-0.481 & -0.577 & 0.096 & -0.769 & 0.481 & 1.154 & -0.096 & 0.192 \\
0.192 & 0.096 & -0.577 & 0.481 & -0.769 & -0.096 & 1.154 & -0.481 \\
-0.096 & -0.769 & 0.481 & -0.577 & 0.096 & 0.192 & -0.481 & 1.154
\end{bmatrix}
\end{array}
\begin{array}{c} u_{x_1} \\ u_{y_1} \\ u_{x_2} \\ u_{y_2} \\ u_{x_3} \\ u_{y_3} \\ u_{x_4} \\ u_{y_4} \end{array}
\tag{6.52a}
$$

in which for elements (1) and (3), the stiffness matrices corresponding to enriched DOF are zero, that is, $\mathbf{K}_{ua} = \mathbf{K}_{au} = \mathbf{K}_{aa} = \mathbf{0}$. The stiffness matrix of element (2) that contains the contact interface Γ_c can be obtained using (6.43) from the following two matrices as

$$
{}^{(2)}\overline{\mathbf{K}} = \begin{bmatrix} \overline{\mathbf{K}}_{uu} & \overline{\mathbf{K}}_{ua} \\ \overline{\mathbf{K}}_{au} & \overline{\mathbf{K}}_{aa} \end{bmatrix} = \begin{bmatrix} \int_{-1}^{+1}\int_{-1}^{+1} \mathbf{B}_{(\xi,\eta)}^{std\,T} \mathbf{D}\,\mathbf{B}_{(\xi,\eta)}^{std} \det[\mathbf{J}] d\xi\, d\eta & \int_{-1}^{+1}\int_{-1}^{+1} \mathbf{B}_{(\xi,\eta)}^{std\,T} \mathbf{D}\,\mathbf{B}_{(\xi,\eta)}^{enr} \det[\mathbf{J}] d\xi\, d\eta \\ \int_{-1}^{+1}\int_{-1}^{+1} \mathbf{B}_{(\xi,\eta)}^{enr\,T} \mathbf{D}\,\mathbf{B}_{(\xi,\eta)}^{std} \det[\mathbf{J}] d\xi\, d\eta & \int_{-1}^{+1}\int_{-1}^{+1} \mathbf{B}_{(\xi,\eta)}^{enr\,T} \mathbf{D}\,\mathbf{B}_{(\xi,\eta)}^{enr} \det[\mathbf{J}] d\xi\, d\eta \end{bmatrix}
$$

$$
= 1.0\times 10^{6}
\begin{array}{c}
\begin{array}{cccccccccccccccc} u_{x_2} & u_{y_2} & u_{x_5} & u_{y_5} & u_{x_8} & u_{y_8} & u_{x_3} & u_{y_3} & a_{x_2} & a_{y_2} & a_{x_5} & a_{y_5} & a_{x_8} & a_{y_8} & a_{x_3} & a_{y_3} \end{array} \\
\begin{bmatrix}
1.154 & 0.481 & -0.769 & 0.096 & -0.577 & -0.481 & 0.192 & -0.096 & 0.481 & 0.120 & 0.385 & 0.072 & 0.288 & 0.216 & 0.192 & -0.024 \\
0.481 & 1.154 & -0.096 & 0.192 & -0.481 & -0.577 & 0.096 & -0.769 & 0.120 & 0.240 & 0.168 & -0.096 & 0.264 & 0.288 & 0.024 & -0.048 \\
-0.769 & -0.096 & 1.154 & -0.481 & 0.192 & 0.096 & -0.577 & 0.481 & -0.385 & 0.072 & -0.481 & 0.120 & -0.192 & -0.024 & -0.288 & 0.216 \\
0.096 & 0.192 & -0.481 & 1.154 & -0.096 & -0.769 & 0.481 & -0.577 & 0.168 & 0.096 & 0.120 & -0.240 & 0.024 & 0.048 & 0.264 & -0.288 \\
-0.577 & -0.481 & 0.192 & -0.096 & 1.154 & 0.481 & -0.769 & 0.096 & -0.288 & -0.216 & -0.192 & 0.024 & -0.481 & -0.120 & -0.385 & -0.072 \\
-0.481 & -0.577 & 0.096 & -0.769 & 0.481 & 1.154 & -0.096 & 0.192 & -0.264 & -0.288 & -0.024 & 0.048 & -0.120 & -0.240 & -0.168 & 0.096 \\
0.192 & 0.096 & -0.577 & 0.481 & -0.769 & -0.096 & 1.154 & -0.481 & 0.192 & 0.024 & 0.288 & -0.216 & 0.385 & -0.072 & 0.481 & -0.120 \\
-0.096 & -0.769 & 0.481 & -0.577 & 0.096 & 0.192 & -0.481 & 1.154 & -0.024 & -0.048 & -0.264 & 0.288 & -0.168 & -0.096 & -0.120 & 0.240 \\
0.481 & 0.120 & -0.385 & 0.168 & -0.288 & -0.264 & 0.192 & -0.024 & 0.481 & 0.120 & 0 & 0 & 0 & 0 & 0.192 & -0.024 \\
0.120 & 0.240 & 0.072 & 0.096 & -0.216 & -0.288 & 0.024 & -0.048 & 0.120 & 0.240 & 0 & 0 & 0 & 0 & 0.024 & -0.048 \\
0.385 & 0.168 & -0.481 & 0.120 & -0.192 & -0.024 & 0.288 & -0.264 & 0 & 0 & 0.481 & -0.120 & 0.192 & 0.024 & 0 & 0 \\
0.072 & -0.096 & 0.120 & -0.240 & 0.024 & 0.048 & -0.216 & 0.288 & 0 & 0 & -0.120 & 0.240 & -0.024 & -0.048 & 0 & 0 \\
0.288 & 0.264 & -0.192 & 0.024 & -0.481 & -0.120 & 0.385 & -0.168 & 0 & 0 & 0.192 & -0.024 & 0.481 & 0.120 & 0 & 0 \\
0.216 & 0.288 & -0.024 & 0.048 & -0.120 & -0.240 & -0.072 & -0.096 & 0 & 0 & 0.024 & -0.048 & 0.120 & 0.240 & 0 & 0 \\
0.192 & 0.024 & -0.288 & 0.264 & -0.385 & -0.168 & 0.481 & -0.120 & 0.192 & 0.024 & 0 & 0 & 0 & 0 & 0.481 & -0.120 \\
-0.024 & -0.048 & 0.216 & -0.288 & -0.072 & 0.096 & -0.120 & 0.240 & -0.024 & -0.048 & 0 & 0 & 0 & 0 & -0.120 & 0.240
\end{bmatrix}
\end{array}
\begin{array}{c} u_{x_2} \\ u_{y_2} \\ u_{x_5} \\ u_{y_5} \\ u_{x_8} \\ u_{y_8} \\ u_{x_3} \\ u_{y_3} \\ a_{x_2} \\ a_{y_2} \\ a_{x_5} \\ a_{y_5} \\ a_{x_8} \\ a_{y_8} \\ a_{x_3} \\ a_{y_3} \end{array}
\tag{6.52b}
$$

and the contact stiffness matrix of element (2) for penalty parameter of $k_N = E/\ell$ in the absence of friction can be obtained as

$$
{}^{(2)}\mathbf{K}^{\text{con}} = \int_{\Gamma_c} \left(\mathbf{N}^{std}\right)^T \mathbf{D}_f^{ep}\, \mathbf{N}^{std}\, d\Gamma \equiv \int_{-1}^{+1} \mathbf{N}_{(\eta)}^{std\,T} \begin{bmatrix} k_N & 0 \\ 0 & 0 \end{bmatrix} \mathbf{N}_{(\eta)}^{std} \det[\mathbf{J}] d\eta \Bigg|_{\xi=0}
$$

$$
= 1.0\times 10^{5}
\begin{array}{c}
\begin{array}{cccccccc} a_{x_2} & a_{y_2} & a_{x_5} & a_{y_5} & a_{x_8} & a_{y_8} & a_{x_3} & a_{y_3} \end{array} \\
\begin{bmatrix}
1.67 & 0 & 1.67 & 0 & 0.83 & 0 & 0.83 & 0 \\
0 & 0 & 0 & 0 & 0 & 0 & 0 & 0 \\
1.67 & 0 & 1.67 & 0 & 0.83 & 0 & 0.83 & 0 \\
0 & 0 & 0 & 0 & 0 & 0 & 0 & 0 \\
0.83 & 0 & 0.83 & 0 & 1.67 & 0 & 1.67 & 0 \\
0 & 0 & 0 & 0 & 0 & 0 & 0 & 0 \\
0.83 & 0 & 0.83 & 0 & 1.67 & 0 & 1.67 & 0 \\
0 & 0 & 0 & 0 & 0 & 0 & 0 & 0
\end{bmatrix}
\end{array}
\begin{array}{c} a_{x_2} \\ a_{y_2} \\ a_{x_5} \\ a_{y_5} \\ a_{x_8} \\ a_{y_8} \\ a_{x_3} \\ a_{y_3} \end{array}
\tag{6.52c}
$$

Table 6.2 An elastic plate with a discontinuity in the middle

k_N (kg/cm^3)	$u_A = u_B$ (cm)	$g_N\|_{\Gamma_c}$ (cm)	$p_N\|_{\Gamma_c}$ (kg/cm^2)
E/ℓ	−0.7338	−0.2	-2×10^4
$100E/\ell$	−0.5358	-2×10^{-3}	-2×10^4
$1\times 10^4 E/\ell$	−0.5338	-2×10^{-5}	-2×10^4
$1\times 10^6 E/\ell$	−0.5338	-2×10^{-7}	-2×10^4

The values of nodal displacement, normal gap and contact pressure for different values of k_N.

By assembling the stiffness matrices of elements (1), (2), and (3), a set of linear system of equations can be obtained in the frictionless condition as

$$\begin{bmatrix} {}^{(1)}\mathbf{K} + {}^{(2)}\overline{\mathbf{K}}_{uu} + {}^{(3)}\mathbf{K} & {}^{(2)}\overline{\mathbf{K}}_{ua} \\ {}^{(2)}\overline{\mathbf{K}}_{au} & {}^{(2)}\overline{\mathbf{K}}_{aa} + {}^{(2)}\mathbf{K}^{\text{con}} \end{bmatrix}_{24\times 24} \begin{Bmatrix} \bar{\mathbf{u}}_{16\times 1} \\ \bar{\mathbf{a}}_{8\times 1} \end{Bmatrix} = \begin{Bmatrix} \mathbf{F}_{\text{ext}}^{std} \\ \mathbf{F}_{\text{ext}}^{enr} \end{Bmatrix}_{24\times 1} \tag{6.53}$$

In Table 6.2, the results of nodal displacements at the right edge are given together with the normal gap g_N and contact traction p_c at the position of discontinuity for different values of k_N. Obviously, the normal gap g_N approaches zero when the penalty parameter $k_N \to \infty$ (see Companion Website).

6.5 Modeling of Contact Constraints via the Lagrange Multipliers Method

In the Lagrange multipliers method, the contact constraints are exactly enforced by introducing additional unknowns called the *Lagrange multipliers*. In order to impose the contact constraints into the weak form of X-FEM equilibrium equation, the Lagrange multipliers method is employed. In this manner, the technique is used to impose the contact constraints in the normal direction whereas the stick–slip motion between two bodies is modeled by introducing a tangential penalty parameter $\bar{k}_T$ in the tangential direction, which is analogous to the tangential stiffness of the contact interface. The weak form of equilibrium equation of contact problem has been derived in (6.27); this equation can be similarly obtained based on the principle of minimum potential energy by minimizing the total potential energy of the system Π as

$$\delta\Pi = \int_\Omega \delta\boldsymbol{\varepsilon}^T\boldsymbol{\sigma}\,d\Omega + \int_{\Gamma_c} [\![\delta\mathbf{u}]\!]^T \mathbf{p}_c\,d\Gamma - \int_{\Gamma_t} \delta\mathbf{u}^T\bar{\mathbf{t}}\,d\Gamma - \int_\Omega \delta\mathbf{u}^T\mathbf{b}\,d\Omega = 0 \tag{6.54}$$

Consider the contact constraint is defined by $g_N(\mathbf{u}) = 0$ in the normal direction to the contact interface Γ_c, in which the normal gap function g_N is defined according to (6.2) as $g_N = [\![\mathbf{u}]\!]\cdot\mathbf{n}_{\Gamma_c}$. A *constraint functional* is defined to represent the potential energy of system due to contact constraint as

$$\Pi^{\text{con}} = \int_{\Gamma_c} \lambda_N\, g_N(\mathbf{u})\,d\Gamma \tag{6.55}$$

where λ_N is the new unknown variable analogous to the normal contact force defined on the contact interface Γ_c, called the *Lagrange multiplier*. This *constraint functional* is added to the potential energy of the system Π to define the *Lagrange functional* $\bar{\Pi}$ as

$$\bar{\Pi} = \Pi + \Pi^{\text{con}} \equiv \Pi + \int_{\Gamma_c} \lambda_N\, g_N(\mathbf{u})\,d\Gamma \tag{6.56}$$

By taking the variation from (6.55) to minimize the *Lagrange functional* $\bar{\Pi}$, it leads to

$$\begin{aligned}\delta\bar{\Pi} &= \delta\Pi + \delta\Pi^{\mathrm{con}} \\ &= \int_{\Omega} \delta\boldsymbol{\varepsilon}^T \boldsymbol{\sigma}\, d\Omega + \int_{\Gamma_c} [\![\delta\mathbf{u}]\!]^T \mathbf{p}_c\, d\Gamma - \int_{\Gamma_t} \delta\mathbf{u}^T \bar{\mathbf{t}}\, d\Gamma - \int_{\Omega} \delta\mathbf{u}^T \mathbf{b}\, d\Omega \\ &\quad + \int_{\Gamma_c} \delta\lambda_N\, g_N\, d\Gamma + \int_{\Gamma_c} \lambda_N\, \delta g_N\, d\Gamma = 0\end{aligned} \tag{6.57}$$

in which the contact constitutive model is incorporated into the previous variational formulation through the second integral term on Γ_c, which can be rewritten in terms of the normal and tangential components as

$$\int_{\Gamma_c} [\![\delta\mathbf{u}]\!]^T \mathbf{p}_c\, d\Gamma = \int_{\Gamma_c} g_N\, p_N\, d\Gamma + \int_{\Gamma_c} [\![\delta\mathbf{u}_T]\!]^T \mathbf{p}_T\, d\Gamma \tag{6.58}$$

Since the jump of normal displacement is enforced to be zero on Γ_c, the first term of right hand side vanishes, and Eq. (6.57) can be rewritten as

$$\begin{aligned}&\int_{\Omega} \delta\boldsymbol{\varepsilon}^T \boldsymbol{\sigma}\, d\Omega + \int_{\Gamma_c} [\![\delta\mathbf{u}_T]\!]^T \mathbf{p}_T\, d\Gamma - \int_{\Gamma_t} \delta\mathbf{u}^T \bar{\mathbf{t}}\, d\Gamma - \int_{\Omega} \delta\mathbf{u}^T \mathbf{b}\, d\Omega \\ &\quad + \int_{\Gamma_c} \delta\lambda_N\, g_N\, d\Gamma + \int_{\Gamma_c} \lambda_N\, \delta g_N\, d\Gamma = 0\end{aligned} \tag{6.59}$$

In order to discretize this integral equation in the domain, the Lagrange multiplier λ_N needs to be approximated on Γ_c (Belytschko *et al.*, 2000a). For this purpose, the contact interface is discretized to a set of interface elements on which λ_N is approximated using the standard FE shape functions. Note that for a 2D problem the contact interface is discretized by 1D elements, while in a 3D problem, the contact surface is characterized by 2D elements, in which the integration over Γ_c is performed using the Gauss quadrature points over these interface elements. In Figure 6.12, the procedure for approximation of Lagrange multiplier λ_N along the contact interface Γ_c in the X-FEM Lagrange multipliers method is presented. It must be noted that additional DOF are introduced in this technique associated with the Lagrange multiplier λ_N for the generated interface elements on Γ_c.

The Lagrange multiplier λ_N can be approximated using the shape functions of interface elements $\widehat{\mathbf{N}}$ as $\lambda_N = \widehat{\mathbf{N}}\boldsymbol{\lambda}_N$, with $\boldsymbol{\lambda}_N$ denoting the vector of discretized values of Lagrange multiplier λ_N (Figure 6.12). The

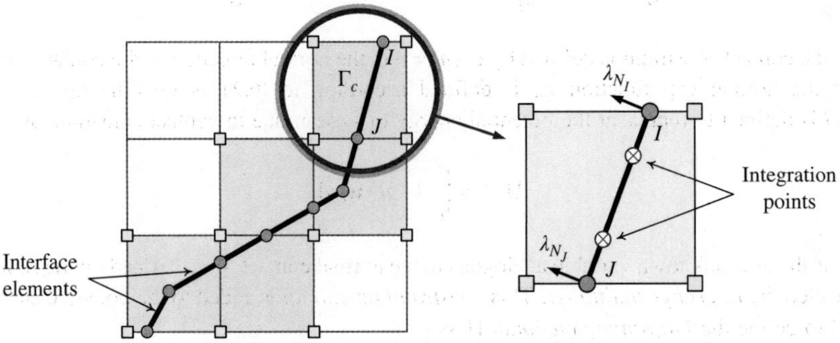

Figure 6.12 Approximation of Lagrange multipliers λ_N along the contact interface Γ_c in the X-FEM Lagrange multipliers method

variation of Lagrange multiplier $\delta\lambda_N$ for the weak form of governing equation (6.59) can be assumed by $\delta\lambda_N = \widehat{\mathbf{N}}\delta\boldsymbol{\lambda}_N$ in the same approximating space as the Lagrange multiplier λ_N. The X-FEM solution of contact problem on the basis of Lagrange multiplier method can therefore be obtained by substituting the displacement and strain fields $\delta\mathbf{u}$ and $\delta\boldsymbol{\varepsilon}$ from (6.34) and (6.35), and λ_N and $\delta\lambda_N$ into the weak form of equilibrium equation (6.59) as

$$\begin{aligned}
&\delta\bar{\mathbf{u}}^T\left\{\int_\Omega(\mathbf{B}^{std})^T\boldsymbol{\sigma}\,d\Omega-\int_{\Gamma_t}(\mathbf{N}^{std})^T\bar{\mathbf{t}}\,d\Gamma-\int_\Omega(\mathbf{N}^{std})^T\mathbf{b}\,d\Omega\right\}\\
&\quad+\delta\bar{\mathbf{a}}^T\left\{\int_\Omega(\mathbf{B}^{enr})^T\boldsymbol{\sigma}\,d\Omega+\int_{\Gamma_c}(\mathbf{N}^{std})^T\mathbf{p}_T\,d\Gamma+\left(\int_{\Gamma_c}(\mathbf{N}^{std})^T\mathbf{n}_{\Gamma_c}\widehat{\mathbf{N}}\,d\Gamma\right)\boldsymbol{\lambda}_N\right.\\
&\quad\left.-\int_{\Gamma_t}(\mathbf{N}^{enr})^T\bar{\mathbf{t}}\,d\Gamma-\int_\Omega(\mathbf{N}^{enr})^T\mathbf{b}\,d\Omega\right\}+\delta\boldsymbol{\lambda}_N^T\left\{\left(\int_{\Gamma_c}\widehat{\mathbf{N}}^T\mathbf{n}_{\Gamma_c}^T\mathbf{N}^{std}\,d\Gamma\right)\bar{\mathbf{a}}\right\}=0
\end{aligned}\tag{6.60}$$

This equilibrium equation can be written for any arbitrary values of $\delta\bar{\mathbf{u}}^T$, $\delta\bar{\mathbf{a}}^T$, and $\delta\boldsymbol{\lambda}_N^T$ in the following sets of equations as

$$\begin{aligned}
&\bar{\boldsymbol{\Psi}}_1\equiv\int_\Omega(\mathbf{B}^{std})^T\boldsymbol{\sigma}\,d\Omega-\int_{\Gamma_t}(\mathbf{N}^{std})^T\bar{\mathbf{t}}\,d\Gamma-\int_\Omega(\mathbf{N}^{std})^T\mathbf{b}\,d\Omega=\mathbf{0}\\
&\bar{\boldsymbol{\Psi}}_2\equiv\int_\Omega(\mathbf{B}^{enr})^T\boldsymbol{\sigma}\,d\Omega+\int_{\Gamma_c}(\mathbf{N}^{std})^T\mathbf{p}_T\,d\Gamma+\left(\int_{\Gamma_c}(\mathbf{N}^{std})^T\mathbf{n}_{\Gamma_c}\widehat{\mathbf{N}}\,d\Gamma\right)\boldsymbol{\lambda}_N\\
&\qquad-\int_{\Gamma_t}(\mathbf{N}^{enr})^T\bar{\mathbf{t}}\,d\Gamma-\int_\Omega(\mathbf{N}^{enr})^T\mathbf{b}\,d\Omega=\mathbf{0}\\
&\bar{\boldsymbol{\Psi}}_3\equiv\left(\int_{\Gamma_c}\widehat{\mathbf{N}}^T\mathbf{n}_{\Gamma_c}^T\mathbf{N}^{std}\,d\Gamma\right)\bar{\mathbf{a}}=\mathbf{0}
\end{aligned}\tag{6.61}$$

in which

$$\begin{aligned}
&\mathbf{F}_{\text{int}}^{std}=\int_\Omega(\mathbf{B}^{std})^T\boldsymbol{\sigma}\,d\Omega\\
&\mathbf{F}_{\text{ext}}^{std}=\int_{\Gamma_t}(\mathbf{N}^{std})^T\bar{\mathbf{t}}\,d\Gamma+\int_\Omega(\mathbf{N}^{std})^T\mathbf{b}\,d\Omega\\
&\mathbf{F}_{\text{int}}^{enr}=\int_\Omega(\mathbf{B}^{enr})^T\boldsymbol{\sigma}\,d\Omega\\
&\mathbf{F}_{\text{ext}}^{enr}=\int_{\Gamma_t}(\mathbf{N}^{enr})^T\bar{\mathbf{t}}\,d\Gamma+\int_\Omega(\mathbf{N}^{enr})^T\mathbf{b}\,d\Omega\\
&\mathbf{f}_{\text{int}}^{\text{con}}=\int_{\Gamma_c}(\mathbf{N}^{std})^T\mathbf{p}_T\,d\Gamma\\
&\mathbf{G}_{\text{int}}^{\text{con}}=\int_{\Gamma_c}(\mathbf{N}^{std})^T\mathbf{n}_{\Gamma_c}\widehat{\mathbf{N}}\,d\Gamma
\end{aligned}\tag{6.62}$$

Since the tangential contact traction $\mathbf{p}_T$ is nonlinearly dependent on the normal contact traction $\boldsymbol{\lambda}_N$ and the displacement jump along the contact surface Γ_c, the Newton–Raphson procedure may be applied to solve the set of nonlinear equations (6.61). The linearized system of equations for iteration $i+1$ of time interval $(n, n+1)$ can be obtained by expanding the residual equations (6.61) with the first-order truncated Taylor series as

$$\begin{bmatrix} \mathbf{K}_{uu} & \mathbf{K}_{ua} & \mathbf{K}_{u\lambda} \\ \mathbf{K}_{au} & \mathbf{K}_{aa} & \mathbf{K}_{a\lambda} \\ \mathbf{K}_{\lambda u} & \mathbf{K}_{\lambda a} & \mathbf{K}_{\lambda\lambda} \end{bmatrix}_{n+1}^{i} \begin{Bmatrix} d\bar{\mathbf{u}} \\ d\bar{\mathbf{a}} \\ d\boldsymbol{\lambda}_N \end{Bmatrix}_{n}^{i} = -\begin{Bmatrix} \bar{\boldsymbol{\Psi}}_1 \\ \bar{\boldsymbol{\Psi}}_2 \\ \bar{\boldsymbol{\Psi}}_3 \end{Bmatrix}_{n+1}^{i} \tag{6.63}$$

in which $\bar{\boldsymbol{\Psi}}_1$, $\bar{\boldsymbol{\Psi}}_2$, and $\bar{\boldsymbol{\Psi}}_3$ stand for the vectors that can be evaluated from the known initial values, and the components of stiffness matrix $\mathbf{K}$ can be obtained as

$$\begin{aligned}
\mathbf{K}_{uu} &= \frac{\partial \bar{\boldsymbol{\Psi}}_1}{\partial \bar{\mathbf{u}}} = \int_\Omega \left(\mathbf{B}^{std}\right)^T \frac{\partial \boldsymbol{\sigma}}{\partial \bar{\mathbf{u}}} d\Omega \\
\mathbf{K}_{ua} &= \frac{\partial \bar{\boldsymbol{\Psi}}_1}{\partial \bar{\mathbf{a}}} = \int_\Omega \left(\mathbf{B}^{std}\right)^T \frac{\partial \boldsymbol{\sigma}}{\partial \bar{\mathbf{a}}} d\Omega \\
\mathbf{K}_{u\lambda} &= \frac{\partial \bar{\boldsymbol{\Psi}}_1}{\partial \boldsymbol{\lambda}_N} = \mathbf{0} \\
\mathbf{K}_{au} &= \frac{\partial \bar{\boldsymbol{\Psi}}_2}{\partial \bar{\mathbf{u}}} = \int_\Omega \left(\mathbf{B}^{enr}\right)^T \frac{\partial \boldsymbol{\sigma}}{\partial \bar{\mathbf{u}}} d\Omega \\
\mathbf{K}_{aa} &= \frac{\partial \bar{\boldsymbol{\Psi}}_2}{\partial \bar{\mathbf{a}}} = \int_\Omega \left(\mathbf{B}^{enr}\right)^T \frac{\partial \boldsymbol{\sigma}}{\partial \bar{\mathbf{a}}} d\Omega + \int_{\Gamma_c} \left(\mathbf{N}^{std}\right)^T \frac{\partial \mathbf{p}_T}{\partial \bar{\mathbf{a}}} d\Gamma \\
\mathbf{K}_{a\lambda} &= \frac{\partial \bar{\boldsymbol{\Psi}}_2}{\partial \boldsymbol{\lambda}_N} = \mathbf{G}_{\text{int}}^{\text{con}} \\
\mathbf{K}_{\lambda u} &= \frac{\partial \bar{\boldsymbol{\Psi}}_3}{\partial \bar{\mathbf{u}}} = \mathbf{0} \\
\mathbf{K}_{\lambda a} &= \frac{\partial \bar{\boldsymbol{\Psi}}_3}{\partial \bar{\mathbf{a}}} = \left(\mathbf{G}_{\text{int}}^{\text{con}}\right)^T \\
\mathbf{K}_{\lambda\lambda} &= \frac{\partial \bar{\boldsymbol{\Psi}}_3}{\partial \boldsymbol{\lambda}_N} = \mathbf{0}
\end{aligned} \tag{6.64}$$

Hence, the stiffness matrix can be defined as

$$\mathbf{K} = \begin{bmatrix} \int_\Omega \left(\mathbf{B}^{std}\right)^T \mathbf{D}\,\mathbf{B}^{std} d\Omega & \int_\Omega \left(\mathbf{B}^{std}\right)^T \mathbf{D}\,\mathbf{B}^{enr} d\Omega & \mathbf{0} \\ \int_\Omega (\mathbf{B}^{enr})^T \mathbf{D}\,\mathbf{B}^{std} d\Omega & \int_\Omega (\mathbf{B}^{enr})^T \mathbf{D}\,\mathbf{B}^{enr} d\Omega + \int_{\Gamma_c} \left(\mathbf{N}^{std}\right)^T \left(\bar{\mathbf{D}}_f\right)_T \mathbf{N}^{std} d\Gamma & \int_{\Gamma_c} \left(\mathbf{N}^{std}\right)^T \mathbf{n}_{\Gamma_c} \widehat{\mathbf{N}}\, d\Gamma \\ \mathbf{0} & \int_{\Gamma_c} \widehat{\mathbf{N}}^T \mathbf{n}_{\Gamma_c}^T \mathbf{N}^{std} d\Gamma & \mathbf{0} \end{bmatrix} \tag{6.65}$$

in which the tangential contact constitutive relation is defined according to (6.15) as

$$d\mathbf{p}_T = \left(\bar{\mathbf{D}}_f\right)_T d[\mathbf{u}] \equiv \left(\bar{\mathbf{D}}_f\right)_T \mathbf{N}^{std} d\bar{\mathbf{a}} \tag{6.66}$$

where $\left(\bar{\mathbf{D}}_f\right)_T = \bar{k}_T(\mathbf{I} - \mathbf{n}_{\Gamma_c} \otimes \mathbf{n}_{\Gamma_c})$, with $\bar{k}_T$ denoting the tangential stiffness of contact surface.

The stick–slip motion between two bodies in contact can be modeled by an appropriate variation of $\bar{k}_T$, as discussed in Section 6.3.2. The algorithm for the Lagrange multipliers method is presented in Box 6.3. It must be noted that the advantage of Lagrange multipliers method in comparison to the penalty method is that the Lagrange multipliers method accurately enforces the constraint $g_N = 0$ on the contact interface Γ_c.

Box 6.3 The solution procedure for the Lagrange multipliers method

Initialize the solution; set $(\bar{\mathbf{u}}, \bar{\mathbf{a}}, \lambda_N) = \mathbf{0}$,
LOOP over the Newton iterations i
Solve system of Eq. (6.63) (for the first iteration $i = 1$ set the stick condition),
For each Gauss point on the contact interface, set $p_N = \lambda_N$, update the frictional tractions and determine the state of stick or slip according to Box 6.1,
Check for the convergence, IF $\eta = \left\| \left\langle \bar{\mathbf{\Psi}}_1^T, \bar{\mathbf{\Psi}}_2^T, \bar{\mathbf{\Psi}}_3^T \right\rangle \right\| / \|\mathbf{F}_{ext}\| \leq \eta_{\text{aim}}$ then END LOOP,
END LOOP over the Newton iterations.

However, the need to set up a nodal point and an element topology on the contact interface is a major disadvantage of the Lagrange multipliers method, which is far more complicated in 3D problems. In addition, the set of equations given in (6.63) is not banded and obtaining an arrangement of unknowns to make the formulation banded is difficult. Moreover, the solution of the system of equations is complicated since there exist zero diagonal terms. The situation is more involved when a large number of additional unknowns associated to the Lagrange multiplier are introduced.

6.5.1 Modeling the Discontinuity in an Elastic Bar

In order to illustrate the performance of proposed contact constraint model in the X-FEM formulation described in previous section, an elastic 1D bar with a discontinuity at its center described in Section 6.4.1 is modeled on the basis of Lagrange multipliers method. The geometry, boundary conditions, and X-FEM mesh of the problem are similar to that given in Figure 6.10. In order to model the discontinuity in an axial bar, the enriched DOF are assumed at points 2 and 3, and the contact constraint is imposed using the Lagrange multipliers method in the enriched element (2). For the proposed 1D problem, the shape function of contact discontinuity is defined as $\widehat{\mathbf{N}} = \langle 1 \rangle$, and the unit normal to the discontinuity as $\mathbf{n}_{\Gamma_c} = \langle 1 \rangle$. The matrix $\mathbf{G}_{\text{int}}^{\text{con}}$ for element (2) can be obtained from (6.64) as

$$ {}^{(2)}\mathbf{G}_{\text{int}}^{\text{con}} = \int_{\Gamma_c} {}^{(2)}\mathbf{N}_{(x)}^{std^T} \mathbf{n}_{\Gamma_c} \widehat{\mathbf{N}} \, d\Gamma = \left. {}^{(2)}\mathbf{N}_{(x)}^{std^T} \right|_{x=3\ell/2} = \left\langle \frac{1}{2} \;\; \frac{1}{2} \right\rangle^T \tag{6.67} $$

Based on the stiffness matrices of elements (1), (2), and (3) derived in Section 6.4.1, the total stiffness matrix can therefore be obtained as

$$ \mathbf{K} = \frac{AE}{\ell} \begin{bmatrix} +1 & -1 & 0 & 0 & 0 & 0 & 0 \\ -1 & +2 & -1 & 0 & +\frac{1}{2} & +\frac{1}{2} & 0 \\ 0 & -1 & +2 & -1 & -\frac{1}{2} & -\frac{1}{2} & 0 \\ 0 & 0 & -1 & +1 & 0 & 0 & 0 \\ 0 & +\frac{1}{2} & -\frac{1}{2} & 0 & +\frac{1}{2} & 0 & \frac{\ell}{2AE} \\ 0 & +\frac{1}{2} & -\frac{1}{2} & 0 & 0 & +\frac{1}{2} & \frac{\ell}{2AE} \\ 0 & 0 & 0 & 0 & \frac{\ell}{2AE} & \frac{\ell}{2AE} & 0 \end{bmatrix} \begin{matrix} u_1 \\ u_2 \\ u_3 \\ u_4 \\ a_2 \\ a_3 \\ \lambda_N \end{matrix} \tag{6.68} $$

Note that in the absence of frictional behavior, a linear system of equations can be solved, which results in the following standard and enriched nodal displacements as

$$\begin{aligned}
\bar{\mathbf{u}}^T &= \langle u_1 \;\; u_2 \;\; u_3 \;\; u_4 \rangle = \left\langle 0 \;\; -\frac{p\ell}{AE} \;\; -\frac{2p\ell}{AE} \;\; -\frac{3p\ell}{AE} \right\rangle \\
\bar{\mathbf{a}}^T &= \langle a_2 \;\; a_3 \rangle = \langle 0 \;\; 0 \rangle \\
\lambda_N &= -p
\end{aligned} \tag{6.69}$$

These results can be compared with those obtained using the penalty method in Section 6.4.1. Obviously, the contact constraint can be exactly enforced using the Lagrange multipliers method, that is, $g_N = \left| {}^{(2)}\mathbf{N}^{std}_{(x)} \,\bar{\mathbf{a}} \right|_{x=3\ell/2} = 0$.

6.5.2 *Modeling the Discontinuity in an Elastic Plate*

In the next example, an elastic plate with a discontinuity in the middle, demonstrated in Section 6.4.2, is chosen to illustrate the performance of Lagrange multipliers method in the framework of X-FEM formulation. The geometry, boundary conditions, and X-FEM mesh of the problem are similar to that given in Figure 6.11. The discontinuity in the plate is modeled by enrichment of the nodal points of element (2), and the contact constraints are imposed using the Lagrange multipliers method in this enriched element. The problem is solved in the frictionless condition. In the X-FEM Lagrange multipliers method, additional DOF are introduced associated with Lagrange multiplier $\boldsymbol{\lambda}_N$ in element (2), in which $\boldsymbol{\lambda}_N$ is approximated on Γ_c using the 1D FE shape functions. In Figure 6.13, the unknown Lagrange multipliers λ_{N_1} and λ_{N_2} are shown on the contact interface Γ_c for element (2) together with the position of integration points used for numerical integration. For this 2D problem, the shape functions of contact interface is defined as $\widehat{\mathbf{N}}(\eta) = \left\langle \frac{1}{2}(1-\eta) \;\; \frac{1}{2}(1+\eta) \right\rangle$, and the unit normal vector to the interface is given by $\mathbf{n}^T_{\Gamma_c} = \langle 1 \;\; 0 \rangle$. The matrix $\mathbf{G}^{\rm con}_{\rm int}$ for element (2) can be obtained from (6.64) using the standard shape functions of a 4-noded element $\mathbf{N}^{std}_{(\xi,\eta)}$ as

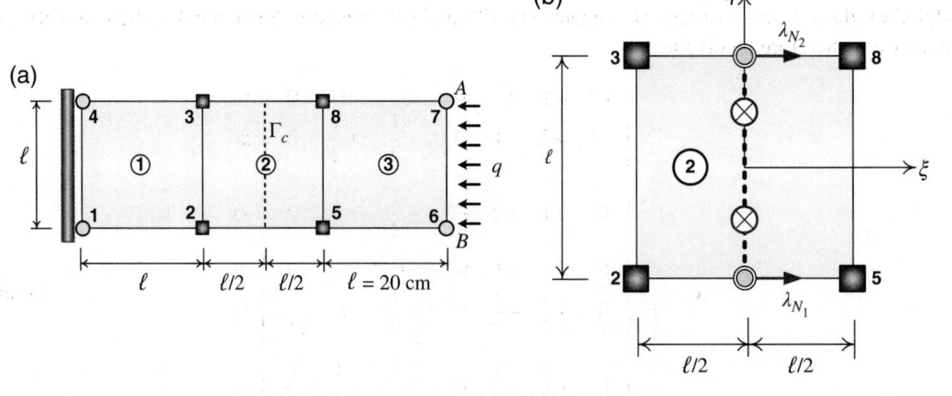

Figure 6.13 An elastic plate with a discontinuity in the middle: (a) problem definition; (b) the unknown Lagrange multipliers λ_{N_1} and λ_{N_2} along the contact interface Γ_c in element (2)

$$
\begin{aligned}
{}^{(2)}\mathbf{G}_{\text{int}}^{\text{con}} &= \int_{\Gamma_c} \left(\mathbf{N}^{std}\right)^T \mathbf{n}_{\Gamma_c} \widehat{\mathbf{N}}\, d\Gamma = \int_{-1}^{+1} \mathbf{N}_{(\xi,\eta)}^{std\,T} \mathbf{n}_{\Gamma_c} \widehat{\mathbf{N}}_{(\eta)} \det[\mathbf{J}] d\eta \Big|_{\xi=0} \\
&= \begin{bmatrix} 3.33 & 0 & 3.33 & 0 & 1.67 & 0 & 1.67 & 0 \\ 1.67 & 0 & 1.67 & 0 & 3.33 & 0 & 3.33 & 0 \end{bmatrix}^T
\end{aligned} \tag{6.70}
$$

On the basis of stiffness matrices of elements (1), (2), and (3) derived in Section 6.4.2, a linear system of equations can be obtained from (6.63) in the case of frictionless condition as

$$
\begin{bmatrix} {}^{(1)}\mathbf{K} + {}^{(2)}\overline{\mathbf{K}}_{uu} + {}^{(3)}\mathbf{K} & {}^{(2)}\overline{\mathbf{K}}_{ua} & \mathbf{0} \\ {}^{(2)}\overline{\mathbf{K}}_{au} & {}^{(2)}\overline{\mathbf{K}}_{aa} & {}^{(2)}\mathbf{G}_{\text{int}}^{\text{con}} \\ \mathbf{0} & \left({}^{(2)}\mathbf{G}_{\text{int}}^{\text{con}}\right)^T & \mathbf{0} \end{bmatrix}_{26\times 26} \begin{Bmatrix} \bar{\mathbf{u}}_{16\times 1} \\ \bar{\mathbf{a}}_{8\times 1} \\ \lambda_{N2\times 1} \end{Bmatrix} = \begin{Bmatrix} \mathbf{F}_{\text{ext}}^{std} \\ \mathbf{F}_{\text{ext}}^{enr} \\ \mathbf{0} \end{Bmatrix}_{26\times 1} \tag{6.71}
$$

The solution of these equations results in the following nodal displacements at points A and B at the right edge, and the Lagrange multipliers λ_{N_1} and λ_{N_2} as

$$
\begin{aligned}
\bar{\mathbf{u}}_{\text{A,B}}^T &= \langle u_{\text{A}} \;\; u_{\text{B}} \rangle = \langle -0.5338 \;\; -0.5338 \rangle \\
\lambda_N^T &= \langle \lambda_{N_1} \;\; \lambda_{N_2} \rangle = \langle -2\times 10^4 \;\; -2\times 10^4 \rangle
\end{aligned} \tag{6.72}
$$

These results can be compared with those obtained using the penalty method in Section 6.4.2 (see Companion Website).

6.6 Modeling of Contact Constraints via the Augmented-Lagrange Multipliers Method

The Lagrange multipliers method described in preceding section accurately enforces the contact constraints on the contact interface Γ_c; however, it makes the algebraic system of equations more complicated by introducing zero diagonal terms. It can be shown that, as the penalty parameter $k_N \to \infty$, the solution of the Lagrange multipliers method can be recovered by the formulation of penalty approach. However, a major problem associated with the numerical treatment of the penalty method is the ill-conditioning that arises from a very large value of penalty parameter added to the stiffness matrix of the system. In order to overcome the ill-conditioning problem in the penalty method and the zero diagonal terms in the Lagrange multipliers method, the augmented-Lagrange multipliers method is introduced based on the combination of these two techniques. This method eliminates the drawbacks of penalty and Lagrange multipliers techniques, and attempts to achieve a predetermined tolerance for the contact constraint through an iterative procedure. The main idea of this technique is to combine the penalty and Lagrange multipliers methods to inherit the advantages of both techniques, that is, decreasing the ill-conditioning of governing equations, and essentially satisfying the contact constraints with finite values of penalty parameters. In such case, the convergence of the solution can be assumed to be independent of the penalty parameters.

The solution procedure of augmented-Lagrange multipliers method consists of a double loop algorithm in which the Lagrange multipliers are kept constant during an outer loop to solve the equilibrium equations of the system in an inner loop. An updated algorithm is then employed within the outer loop to achieve the convergence for the values of Lagrange multipliers. This procedure is also known as the *Uzawa algorithm* (Wriggers, 2006). In this method, it is assumed that the Lagrange multipliers are not primary unknowns, and the values are obtained using an iterative procedure. The *augmented-Lagrange functional* $\widehat{\Pi}$ can be defined by adding the *constraint functional* Π^{con} to the potential energy of the system Π as

$$
\widehat{\Pi} = \Pi + \Pi^{\text{con}} \equiv \Pi + \int_{\Gamma_c} \lambda_N \, g_N(\mathbf{u}) \, d\Gamma \tag{6.73}
$$

in which the Lagrange multiplier value $\bar{\lambda}_N$ is associated with the augmented-Lagrange multipliers method. It must be noted that $\bar{\lambda}_N$ is not primary unknown and is obtained by an iterative procedure. By taking the variation from (6.73) to minimize the *augmented-Lagrange functional* $\widehat{\Pi}$, it results in

$$\begin{aligned}\delta\widehat{\Pi} = & \int_\Omega \delta\boldsymbol{\varepsilon}^T\boldsymbol{\sigma}\,d\Omega + \int_{\Gamma_c}[\delta\mathbf{u}]^T\mathbf{p}_c\,d\Gamma \\ & - \int_{\Gamma_t}\delta\mathbf{u}^T\bar{\mathbf{t}}\,d\Gamma - \int_\Omega \delta\mathbf{u}^T\mathbf{b}\,d\Omega + \int_{\Gamma_c}\bar{\lambda}_N\,\delta g_N\,d\Gamma = 0\end{aligned} \tag{6.74}$$

in which $\bar{\lambda}_N$ is approximated by $\bar{\lambda}_N = \widehat{\mathbf{N}}\bar{\boldsymbol{\lambda}}_N$, where $\bar{\boldsymbol{\lambda}}_N$ denotes the vector of discretized values of Lagrange multiplier $\bar{\lambda}_N$, and $\widehat{\mathbf{N}}$ are the shape functions of interface elements generated by discretization of the contact interface Γ_c (Figure 6.12). The X-FEM formulation of augmented-Lagrange multiplier method can then be obtained by substituting the displacement and strain fields $\delta\mathbf{u}$ and $\delta\boldsymbol{\varepsilon}$ from (6.34) and (6.35) and $\bar{\lambda}_N = \widehat{\mathbf{N}}\bar{\boldsymbol{\lambda}}_N$ into the weak form of equilibrium equation (6.74) as

$$\begin{aligned}& \delta\bar{\mathbf{u}}^T\left\{\int_\Omega(\mathbf{B}^{std})^T\boldsymbol{\sigma}\,d\Omega - \int_{\Gamma_t}(\mathbf{N}^{std})^T\bar{\mathbf{t}}\,d\Gamma - \int_\Omega(\mathbf{N}^{std})^T\mathbf{b}\,d\Omega\right\} \\ & + \delta\bar{\mathbf{a}}^T\left\{\int_\Omega(\mathbf{B}^{enr})^T\boldsymbol{\sigma}\,d\Omega + \int_{\Gamma_c}(\mathbf{N}^{std})^T\mathbf{p}_c\,d\Gamma + \left(\int_{\Gamma_c}(\mathbf{N}^{std})^T\mathbf{n}_{\Gamma_c}\widehat{\mathbf{N}}\,d\Gamma\right)\bar{\boldsymbol{\lambda}}_N\right. \\ & \left. - \int_{\Gamma_t}(\mathbf{N}^{enr})^T\bar{\mathbf{t}}\,d\Gamma - \int_\Omega(\mathbf{N}^{enr})^T\mathbf{b}\,d\Omega\right\} = 0\end{aligned} \tag{6.75}$$

This equilibrium equation can be written for any arbitrary values of $\delta\bar{\mathbf{u}}^T$ and $\delta\bar{\mathbf{a}}^T$ in the following sets of equations as

$$\begin{aligned}\widehat{\boldsymbol{\Psi}}_1 \equiv & \int_\Omega(\mathbf{B}^{std})^T\boldsymbol{\sigma}\,d\Omega - \int_{\Gamma_t}(\mathbf{N}^{std})^T\bar{\mathbf{t}}\,d\Gamma - \int_\Omega(\mathbf{N}^{std})^T\mathbf{b}\,d\Omega = \mathbf{0} \\ \widehat{\boldsymbol{\Psi}}_2 \equiv & \int_\Omega(\mathbf{B}^{enr})^T\boldsymbol{\sigma}\,d\Omega + \int_{\Gamma_c}(\mathbf{N}^{std})^T\mathbf{p}_c\,d\Gamma + \left(\int_{\Gamma_c}(\mathbf{N}^{std})^T\mathbf{n}_{\Gamma_c}\widehat{\mathbf{N}}\,d\Gamma\right)\bar{\boldsymbol{\lambda}}_N \\ & - \int_{\Gamma_t}(\mathbf{N}^{enr})^T\bar{\mathbf{t}}\,d\Gamma - \int_\Omega(\mathbf{N}^{enr})^T\mathbf{b}\,d\Omega = \mathbf{0}\end{aligned} \tag{6.76}$$

In order to solve the set of nonlinear equations (6.76), the Newton–Raphson procedure can be applied to linearize the system of equations for iteration $i+1$ in the time interval $(n, n+1)$ by expanding the residual equations (6.76) with the first-order truncated Taylor series as

$$\begin{bmatrix}\mathbf{K}_{uu} & \mathbf{K}_{ua} \\ \mathbf{K}_{au} & \mathbf{K}_{aa}\end{bmatrix}_{n+1}^i \begin{Bmatrix} d\bar{\mathbf{u}}_n^i \\ d\bar{\mathbf{a}}_n^i\end{Bmatrix} = -\begin{Bmatrix}\boldsymbol{\Psi}_1 \\ \boldsymbol{\Psi}_2\end{Bmatrix}_{n+1}^i \tag{6.77}$$

It must be noted that the Lagrange multipliers $\bar{\boldsymbol{\lambda}}_N^k$ are kept constant during an augmentation loop k to solve the equilibrium equations (6.77) in an inner iteration i. Substituting the components of stiffness matrix from (6.42) into (6.77), this system of equations can be rewritten as

$$\begin{aligned}& \begin{bmatrix}\int_\Omega(\mathbf{B}^{std})^T\mathbf{D}\mathbf{B}^{std}\,d\Omega & \int_\Omega(\mathbf{B}^{std})^T\mathbf{D}\mathbf{B}^{enr}\,d\Omega \\ \int_\Omega(\mathbf{B}^{enr})^T\mathbf{D}\mathbf{B}^{std}\,d\Omega & \int_\Omega(\mathbf{B}^{enr})^T\mathbf{D}\mathbf{B}^{enr}\,d\Omega + \int_{\Gamma_c}(\mathbf{N}^{std})^T\mathbf{D}_f^{ep}\mathbf{N}^{std}\,d\Gamma\end{bmatrix}_{n+1}^i \begin{Bmatrix} d\bar{\mathbf{u}}_n^i \\ d\bar{\mathbf{a}}_n^i\end{Bmatrix} \\ & = \begin{Bmatrix}\mathbf{F}_{\text{ext}}^{std} - \mathbf{F}_{\text{int}}^{std} \\ \mathbf{F}_{\text{ext}}^{enr} - \mathbf{F}_{\text{int}}^{enr} - \mathbf{f}_{\text{int}}^{\text{con}}\end{Bmatrix}_{n+1}^i - \begin{Bmatrix}\mathbf{0} \\ \mathbf{G}_{\text{int}}^{\text{con}}\end{Bmatrix}_{n+1}^i \bar{\boldsymbol{\lambda}}_N^k\end{aligned} \tag{6.78}$$

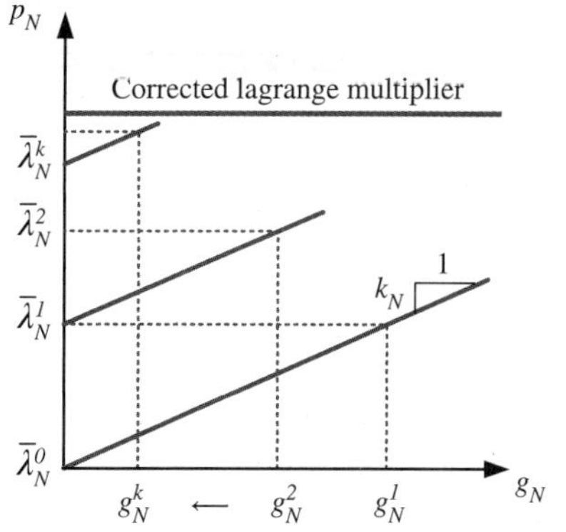

Figure 6.14 The schematic updated procedure of augmented Lagrange multipliers method

in which $\mathbf{F}_{\text{int}}^{std}$, $\mathbf{F}_{\text{ext}}^{std}$, $\mathbf{F}_{\text{int}}^{enr}$, $\mathbf{F}_{\text{ext}}^{enr}$, and $\mathbf{f}_{\text{int}}^{\text{con}}$ are defined in (6.40) and $\mathbf{G}_{\text{int}}^{\text{con}}$ is given in (6.62). In the solution of augmented-Lagrange multipliers method, the outbreak of penetration is first checked. If it is greater than the prescribed tolerance, the augmented-Lagrange rebounding force $\bar{\boldsymbol{\lambda}}_N$ is determined based on an iterative procedure as $\bar{\lambda}_N^{k+1} = \bar{\lambda}_N^k + d\bar{\lambda}_N^k$, where $d\bar{\lambda}_N^k$ denotes the normal force produced by the penetration and the index k is used to compute the Lagrange multipliers over the augmentation loop. The augmented-Lagrange rebounding force is evaluated by $d\bar{\lambda}_N^k = k_N\, g_N^k$, in which $d\bar{\lambda}_N^k = \widehat{\mathbf{N}}\, d\bar{\boldsymbol{\lambda}}_N^k$ and the normal gap function is defined by $g_N^k = [\![\mathbf{u}]\!] \cdot \mathbf{n}_{\Gamma_c}$. In Figure 6.14, the schematic iterative solution of the augmented-Lagrange multipliers method is presented. Obviously, if $g_N^k \to 0$, then $d\bar{\lambda}_N^k$ approaches zero, and the value of $\bar{\lambda}_N^{k+1}$ approaches the correct value. Otherwise, the procedure is repeated using an updated value of $\bar{\lambda}_N^{k+1}$ until the constraints are satisfied within a user defined tolerance. If the convergence is obtained, that is, $d\bar{\lambda}_N^k \approx 0$, the tangential and normal forces are determined at each slave-master point. Finally, the out of balance, or residual forces of contact constraints are evaluated, and the computational algorithm is repeated until the norm of residual forces of the Newton–Raphson iterations in the inner loop and the residual forces of contact constraints along the contact interface are both less than the prescribed tolerance. The augmented Lagrange multipliers algorithm with an inner Newton–Raphson procedure and an outer augmentation iterative solution is shown in Box 6.4.

The convergence of the solution can be achieved based on the global and local error criteria. The global error criterion is on the basis of residual force vector of the Newton–Raphson iteration in an inner loop of the solution defined as

$$\eta_{\text{global}} = \frac{\left\| \left\langle \widehat{\boldsymbol{\Psi}}_1^T, \widehat{\boldsymbol{\Psi}}_2^T \right\rangle \right\|}{\|\mathbf{F}_{ext}\|} \leq \eta_{\text{global}}^{\text{aim}} \tag{6.79}$$

The local error criterion is related to the convergence of augmentation iteration based on the unilateral contact constraint, which is defined by the average penetration of contacting bodies along the contact interface Γ_c as

$$\eta_{\text{local}} = \frac{1}{\ell_c^2} \int_{\Gamma_c} \|\mathbf{g}_N\| \mathrm{d}\Gamma \leq \eta_{\text{local}}^{\text{aim}} \tag{6.80}$$

where ℓ_c denotes the length of contact interface.

Box 6.4 The solution procedure for the augmented-Lagrange multipliers method (Uzawa algorithm)

Initialize the solution; set $(\bar{\mathbf{u}},\bar{\mathbf{a}})=\mathbf{0}$ and $\left.(\bar{\lambda}_N)_n^k\right|_{k=0}=\mathbf{0}$,
LOOP over the augmentation iterations k (local convergence)
LOOP over the Newton iterations i (global convergence)
Solve system of Eq. (6.78) (for the first Newton iteration $i=1$ and the first augmentation iteration $k=1$ set the stick condition),
For each Gauss point on the contact interface: set $p_N=\bar{\lambda}_N+k_Ng_N$, update the frictional tractions and determine the state of stick or slip according to Box 6.1.
Check for the global convergence, IF $\eta^i_{\text{gobal}}\le\eta_{\text{aim}}$ then END LOOP,
END LOOP over the Newton iterations,
LOOP over the interface nodal points
Update the Lagrange multiplier at each nodal point $(\bar{\lambda}_N)_{n+1}^{k+1}=(\bar{\lambda}_N)_{n+1}^k+k_N(g_N)_n^k$,
END LOOP over the interface nodal points,
Check for the local convergence, IF $\eta^k_{\text{local}}\le\eta_{\text{aim}}$ then STOP,
END LOOP over the augmentation iterations.

6.6.1 Modeling an Elastic Bar with a Discontinuity

In order to illustrate the performance of proposed augmented-Lagrange multipliers method in modeling the contact constraints, an elastic 1D bar with a discontinuity at its center described in Section 6.4.1 is modeled on the basis of X-FEM formulation described in previous section. The geometry, boundary conditions, and X-FEM mesh of the problem are similar to that given in Figure 6.10. In order to model the discontinuity in the bar, the enriched DOF are assumed at nodal points 2 and 3, and the contact constraint is imposed using the augmented-Lagrange multipliers method in the enriched element (2). In the absence of frictional behavior, the system of Eq. (6.78) can be solved linearly as

$$\begin{bmatrix}\mathbf{K}_{uu} & \mathbf{K}_{ua}\\ \mathbf{K}_{au} & \mathbf{K}_{aa}\end{bmatrix}\begin{Bmatrix}\bar{\mathbf{u}}\\ \bar{\mathbf{a}}\end{Bmatrix}=\begin{Bmatrix}\mathbf{F}^{std}_{\text{ext}}\\ \mathbf{F}^{enr}_{\text{ext}}\end{Bmatrix}-\begin{Bmatrix}\mathbf{0}\\ \mathbf{G}^{\text{con}}_{\text{int}}\end{Bmatrix}(\bar{\lambda}_N) \tag{6.81}$$

which results in

$$\frac{AE}{\ell}\begin{bmatrix}+1 & -1 & 0 & 0 & 0 & 0\\ -1 & +2 & -1 & 0 & +\frac{1}{2} & +\frac{1}{2}\\ 0 & -1 & +2 & -1 & -\frac{1}{2} & -\frac{1}{2}\\ 0 & 0 & -1 & +1 & 0 & 0\\ 0 & +\frac{1}{2} & -\frac{1}{2} & 0 & \frac{1}{2}+\frac{k_N\ell}{4AE} & \frac{k_N\ell}{4AE}\\ 0 & +\frac{1}{2} & -\frac{1}{2} & 0 & \frac{k_N\ell}{4AE} & \frac{1}{2}+\frac{k_N\ell}{4AE}\end{bmatrix}\begin{Bmatrix}u_1\\ u_2\\ u_3\\ u_4\\ a_2\\ a_3\end{Bmatrix}=\begin{Bmatrix}0\\ 0\\ 0\\ -p\\ 0\\ 0\end{Bmatrix}-\begin{Bmatrix}0\\ 0\\ 0\\ 0\\ +\frac{1}{2}\\ +\frac{1}{2}\end{Bmatrix}(\bar{\lambda}_N) \tag{6.82}$$

Table 6.3 An elastic bar with a discontinuity at its center

k_N (kg/cm)	k	λ_N	u_2 (cm)	u_3 (cm)	u_4 (cm)	a_2 (cm)	a_3 (cm)	Residual force
$0.1AE/\ell$	1	0	−0.1	−1.2	−1.3	−1	−1	1
	2	−200	−0.1	−0.2	−0.3	≅0	≅0	≅0
AE/ℓ	1	0	−0.1	−0.3	−0.4	−0.1	−0.1	−0.1
	2	−200	−0.1	−0.2	−0.3	≅0	≅0	≅0
$10AE/\ell$	1	0	−0.1	−0.21	−0.31	−0.01	−0.01	−0.01
	2	−200	−0.1	−0.2	−0.3	0	0	≅0

The values of nodal displacement, contact pressure and residual force for different values of k_N.

Table 6.4 An elastic plate with a discontinuity in the middle

k_N (kg/cm^3)	k	$\bar{\lambda}_{N_1} = \bar{\lambda}_{N_2}$	$u_A = u_B$	Residual force
E/ℓ	1	0	−0.7338	0.2
	2	-2×10^4	−0.5338	≅0
$10E/\ell$	1	0	−0.5538	−0.02
	2	-2×10^4	−0.5338	≅0
$100E/\ell$	1	0	−0.5358	−0.002
	2	-2×10^4	−0.5338	≅0

The values of nodal displacement, contact pressure and residual force for different values of k_N.

In Table 6.3, the results of standard and enriched nodal displacements $\bar{\mathbf{u}}$ and $\bar{\mathbf{a}}$ are shown together with the augmented-Lagrange rebounding force $\bar{\lambda}_N$ for three values of penalty parameter k_N. In order to solve this system of equations, the augmented-Lagrange multiplier $\bar{\lambda}_N$ is set to zero at the first augmentation loop, and the contact constrain is approximately enforced using the value of penalty parameter. However, the convergence is achieved by only one augmentation loop even for a small value of the penalty parameter $k_N = 0.1AE/\ell$.

6.6.2 *Modeling an Elastic Plate with a Discontinuity*

In the next example, an elastic plate with a discontinuity, demonstrated in Section 6.4.2, is chosen to illustrate the performance of proposed augmented-Lagrange multipliers method in the framework of X-FEM formulation. The geometry, boundary conditions, and X-FEM mesh of the problem are similar to that presented in Figure 6.11. The discontinuity in the plate is modeled by enrichment of the nodal points of element (2), and the contact constraints are imposed using the augmented-Lagrange multipliers method in this enriched element. As depicted in Figure 6.13, the Lagrange multipliers $\bar{\lambda}_{N_1}$ and $\bar{\lambda}_{N_2}$ are chosen as unknowns along the contact interface Γ_c in element (2). In the absence of frictional behavior, the system of Eq. (6.78) can be solved linearly as

$$\begin{bmatrix} {}^{(1)}\mathbf{K} + {}^{(2)}\overline{\mathbf{K}}_{uu} + {}^{(3)}\mathbf{K} & {}^{(2)}\overline{\mathbf{K}}_{ua} \\ {}^{(2)}\overline{\mathbf{K}}_{au} & {}^{(2)}\overline{\mathbf{K}}_{aa} + {}^{(2)}\mathbf{K}^{\text{con}} \end{bmatrix}_{24\times 24} \begin{Bmatrix} \bar{\mathbf{u}}_{16\times 1} \\ \bar{\mathbf{a}}_{8\times 1} \end{Bmatrix} = \begin{Bmatrix} \mathbf{F}^{std}_{\text{ext}} \\ \mathbf{F}^{enr}_{\text{ext}} \end{Bmatrix}_{24\times 1} - \begin{bmatrix} \mathbf{0} \\ {}^{(2)}\mathbf{G}^{\text{con}}_{\text{int}} \end{bmatrix}_{24\times 2} \{\bar{\lambda}_N\}_{2\times 1} \tag{6.83}$$

In Table 6.4, the results of standard and enriched nodal displacements $\bar{\mathbf{u}}$ and $\bar{\mathbf{a}}$ are shown together with the augmented-Lagrange rebounding forces $\bar{\lambda}_{N_1}$ and $\bar{\lambda}_{N_2}$ for three values of penalty parameter k_N. Obviously, the convergence is achieved again after one augmentation loop for different values of penalty parameter (see Companion Website).

6.7 X-FEM Modeling of Large Sliding Contact Problems

In preceding sections, it has been assumed that sliding along the discontinuous contact interface is small enough so that there is no distinction between the deformed and undeformed configuration of two bodies in contact. However, if the contact between two bodies involves a large slide, it is necessary to incorporate the effect of that large slide in the solution. The geometrically nonlinear system of equations can be derived by substituting the X-FEM approximation field into the weak form of governing equations. In previous chapter, the geometric nonlinearity due to large deformations has been discussed in the framework of X-FEM formulation; in this chapter an updated Lagrangian description is employed to model the frictional contact behavior due to large sliding within the context of penalty method. The study is mainly focused on the effect of a large sliding contact problem in the weak form of X-FEM formulation that arises from the surface integral of $\int_{\Gamma_c} [\![\delta \mathbf{u}]\!] \, \mathbf{p}_c \, d\Gamma$.

In small deformation contact problems, both the slave and master segment at a contact point are located at the same position; however, in the case of large slides, the slave and master segment may belong to different elements. Obviously, the assumptions considered for small deformations are not valid for large slides, and proper treatment is necessary to communicate between a part of one cut element and parts belonging to other cut elements in order to take the effect of large sliding contact in the X-FEM formulation. Hence, provision of a geometrical update is required along with a contact search algorithm for the large sliding regime to account for the evolving geometry of contact, and to determine the elements of slave and master segment at each stage of body motion. In order to approximate the displacement jump along the discontinuity in large sliding contact problem, the concept of the *node-to-surface algorithm* is used within the X-FEM technique by indicating the elements of slave and master segment at each contact point. A typical situation of a large sliding contact problem along the contact interface between two bodies is shown in Figure 6.15. The contact interface is denoted by Γ_c and the bodies at both sides of contact interface are indicated by a master and a slave body. Note that the slave and master segment are denoted based on the sign of Heaviside function and the direction of the unit normal on the contact interface. As can be seen from this figure at the initial configuration, the slave and master pair for all contact points are located in the same element. Once sliding begins, the slave–master contact pairs must be determined using a contact search algorithm since each side of slave point belongs to different master element. Based on the

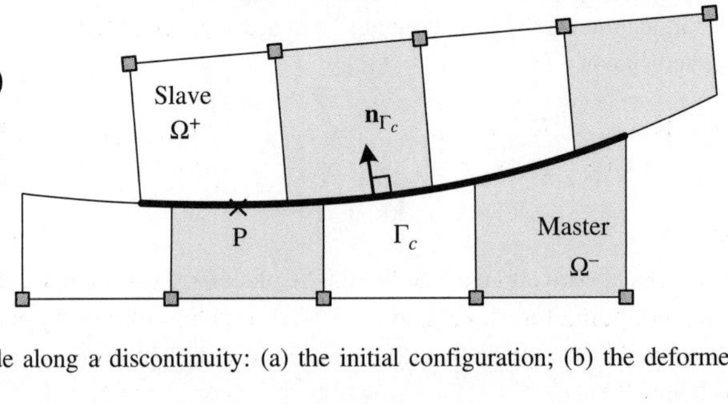

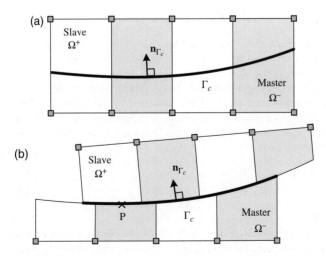

Figure 6.15 Large slide along a discontinuity: (a) the initial configuration; (b) the deformed configuration after sliding

new slave–master contact pairs, the displacement jump can be evaluated along the contact interface Γ_c, and the required contact force vector and contact stiffness matrix can be finally computed.

In order to evaluate the integration over Γ_c, the intersection points of slave sub-elements with the contact interface Γ_c are determined to construct a set of slave segments defining the interface elements, as depicted in Figure 6.16. In this figure, the intersection points of contact interface with the generated master and slave elements are displayed. During the slide, the value of *TSP* is determined based on the position of slave node *S* on the master edge $(m_1 - m_2)$ to indicate the active slave node and master edge. The numerical integration can be performed using the Gauss quadrature points for each interface element along Γ_c. It must be noted that these interface elements are introduced with no DOF and are only used to provide a domain for numerical integration along the contact surface Γ_c. Moreover, a search algorithm can be employed based on the procedure given in Box 6.5 to obtain the associated master element to each integration point, in which each integration point pertains to a slave segment along the contact interface. A set of master segments is also needed in the algorithm that can be generated by considering the intersection points of master sub-elements with the contact interface.

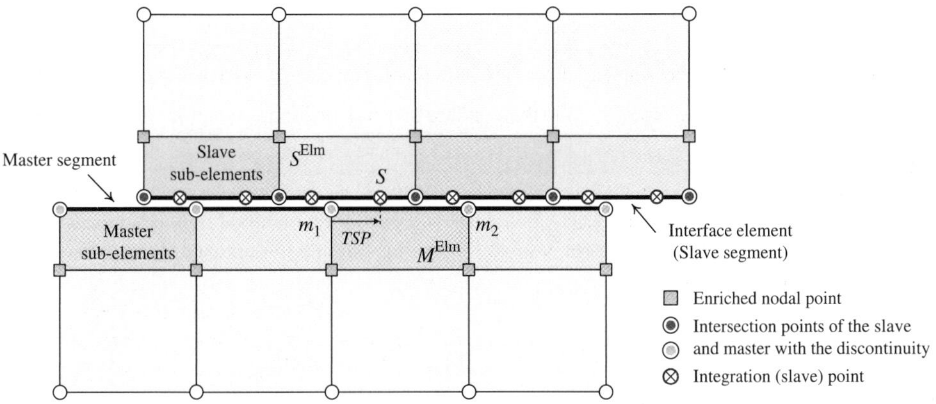

Figure 6.16 Illustration of the contact interface discretization in a large slide

Box 6.5 The contact search algorithm for the large sliding problem

LOOP over the interface elements (slave segments)
LOOP over the integration points (slave points)
LOOP over the master segments
Determine the normalized tangent vector on the master segment as

$$\mathbf{m}_{\Gamma_c} = (\boldsymbol{x}_{m_2} - \boldsymbol{x}_{m_1})/\ell \text{ where } \ell = \|\boldsymbol{x}_{m_2} - \boldsymbol{x}_{m_1}\|$$

Calculate the value of *TSP* based on the position of integration (slave) point on the master segment as (Figure 6.16)

$$TSP = (\boldsymbol{x}_S - \boldsymbol{x}_{m_1}) \cdot \mathbf{m}_{\Gamma_c}$$

IF $0 \le TSP \le 1$, the integration (slave) point is linked to the master segment. Store the information of the slave point and associated master element.
END LOOP on master segments
END LOOP on integration points
END LOOP on interface elements

Consider that the slave point S comes into contact with the master segment $(m_1 - m_2)$, as shown in Figure 6.16; the displacement jump can be evaluated as

$$
\begin{aligned}
[\![\mathbf{u}]\!] &= \left[\mathbf{N}_{(\xi)}^{S\,std}\bar{\mathbf{u}}^S + \mathbf{N}_{(\xi)}^{S\,enr}\bar{\mathbf{a}}^S\right] - \left[\mathbf{N}_{(\eta)}^{m\,std}\bar{\mathbf{u}}^m + \mathbf{N}_{(\eta)}^{m\,enr}\bar{\mathbf{a}}^m\right] \\
&= \left[\mathbf{N}^{S\,std} \quad \mathbf{N}^{S\,enr} \quad -\mathbf{N}^{m\,std} \quad -\mathbf{N}^{m\,enr}\right] \begin{Bmatrix} \bar{\mathbf{u}}^S \\ \bar{\mathbf{a}}^S \\ \bar{\mathbf{u}}^m \\ \bar{\mathbf{a}}^m \end{Bmatrix} \equiv \mathbf{N}^{S-m}\bar{\mathbf{U}}^{\text{con}}
\end{aligned}
\tag{6.84}
$$

in which ξ is the reference coordinate of slave point in the initial slave element S^{Elm}, and η is the reference coordinate of slave point projected on master edge in the initial master element M^{Elm}. In this relation, the shape functions of the slave–master at the contact surface are defined as

$$
\mathbf{N}^{S-m} = \left[\mathbf{N}^{S\,std} \quad \mathbf{N}^{S^{enr}} \quad -\mathbf{N}^{m\,std} \quad -\mathbf{N}^{m^{enr}}\right] \tag{6.85}
$$

where $\mathbf{N}^{S^{std}}$ and $\mathbf{N}^{S^{enr}}$ are the standard and enriched shape functions of slave element, with $N_{J(\boldsymbol{x})}^{S^{enr}} = N_{J(\boldsymbol{x})}^{S^{std}}(H(\boldsymbol{x}) - H(\boldsymbol{x}_J))$ and, $\mathbf{N}^{m\,std}$ and $\mathbf{N}^{m^{enr}}$ are the standard and enriched shape functions of master element, with $N_{J(\boldsymbol{x})}^{m^{enr}} = N_{J(\boldsymbol{x})}^{m^{std}}(H(\boldsymbol{x}) - H(\boldsymbol{x}_J))$.

In order to derive the stiffness matrix corresponding to active slave node and associated master segment, the concept of node-to-segment element is applied into the weak form of equilibrium equation (6.27). By taking the variation from the displacement jump (6.84) and substituting $[\![\delta\mathbf{u}]\!]$ into the integration over contact interface Γ_c, it results in

$$
\int_{\Gamma_c} [\![\delta\mathbf{u}]\!]\,\mathbf{p}_c\,d\Gamma = \left(\delta\bar{\mathbf{U}}^{\text{con}}\right)^T \left(\int_{\Gamma_c} \left(\mathbf{N}^{S-m}\right)^T \mathbf{p}_c\,d\Gamma\right) \tag{6.86}
$$

Thus, the contact load vector can be obtained as

$$
\mathbf{f}_{\text{LS}}^{\text{con}} = \int_{\Gamma_c} \left(\mathbf{N}^{S-m}\right)^T \mathbf{p}_c\,d\Gamma \tag{6.87}
$$

and the contact stiffness matrix can be computed as

$$
\mathbf{K}_{\text{LS}}^{\text{con}} = \int_{\Gamma_c} \left(\mathbf{N}^{S-m}\right)^T \mathbf{D}_f^{ep}\,\mathbf{N}^{S-m}\,d\Gamma \tag{6.88}
$$

in which the components of contact stiffness matrix and contact load vector are respectively defined for the large sliding problem as

$$
\begin{aligned}
\mathbf{K}_{\text{LS}}^{\text{con}} &= \begin{bmatrix}
\mathbf{K}^{S\,std,\,S\,std} & \mathbf{K}^{S\,std,\,S\,enr} & -\mathbf{K}^{S\,std,\,m\,std} & -\mathbf{K}^{S\,std,\,m\,enr} \\
\mathbf{K}^{S\,enr,\,S\,std} & \mathbf{K}^{S\,enr,\,S\,enr} & -\mathbf{K}^{S\,enr,\,m\,std} & -\mathbf{K}^{S\,enr,\,m\,enr} \\
-\mathbf{K}^{m\,std,\,S\,std} & -\mathbf{K}^{m\,std,\,S\,enr} & \mathbf{K}^{m\,std,\,m\,std} & \mathbf{K}^{m\,std,\,m\,enr} \\
-\mathbf{K}^{m\,enr,\,S\,std} & -\mathbf{K}^{m\,enr,\,S\,enr} & \mathbf{K}^{m\,enr,\,m\,std} & \mathbf{K}^{m\,enr,\,m\,enr}
\end{bmatrix} \\
\left(\mathbf{f}_{\text{LS}}^{\text{con}}\right)^T &= \left\langle \mathbf{f}^{S\,std} \quad \mathbf{f}^{S\,enr} \quad -\mathbf{f}^{m\,std} \quad -\mathbf{f}^{m\,enr} \right\rangle
\end{aligned}
\tag{6.89}
$$

where

$$\mathbf{K}^{\alpha^{std},\beta^{std}} = \int_{\Gamma_c} \left(\mathbf{N}^{\alpha^{std}}\right)^T \mathbf{D}_f^{ep}\, \mathbf{N}^{\beta^{std}}\, d\Gamma$$
$$\mathbf{K}^{\alpha^{std},\beta^{enr}} = \int_{\Gamma_c} \left(\mathbf{N}^{\alpha^{std}}\right)^T \mathbf{D}_f^{ep}\, \mathbf{N}^{\beta^{enr}}\, d\Gamma$$
$$\mathbf{K}^{\alpha^{enr},\beta^{std}} = \int_{\Gamma_c} \left(\mathbf{N}^{\alpha^{enr}}\right)^T \mathbf{D}_f^{ep}\, \mathbf{N}^{\beta^{std}}\, d\Gamma \tag{6.90}$$
$$\mathbf{K}^{\alpha^{enr},\beta^{enr}} = \int_{\Gamma_c} \left(\mathbf{N}^{\alpha^{enr}}\right)^T \mathbf{D}_f^{ep}\, \mathbf{N}^{\beta^{enr}}\, d\Gamma$$

and

$$\mathbf{f}^{\alpha^{std}} = \int_{\Gamma_c} \left(\mathbf{N}^{\alpha^{std}}\right)^T \mathbf{p}_c\, d\Gamma$$
$$\mathbf{f}^{\alpha^{enr}} = \int_{\Gamma_c} \left(\mathbf{N}^{\alpha^{enr}}\right)^T \mathbf{p}_c\, d\Gamma \tag{6.91}$$

in which $(\alpha,\ \beta) = (S, m)$. Furthermore, in order to eliminate the coupling between the normal and tangential contact tractions, the continuum tangent matrix can be decomposed into a pure contact in the normal direction and a frictional resistance in the tangential direction by replacing $\mathbf{D}_f^{ep}$ by $\bar{\mathbf{D}}_f^{ep}$ defined in (6.16). Note that the large sliding contact stiffness matrix $\mathbf{K}_{\mathrm{LS}}^{\mathrm{con}}$ and contact force vector $\mathbf{f}_{\mathrm{LS}}^{\mathrm{con}}$ are 32×32 and 32×1, respectively, with the components of the standard and enriched DOF of the slave and master elements given in (6.89), and the vector of unknowns associated with the contact stiffness matrix and force vector is defined as $\left(\bar{\mathbf{U}}^{\mathrm{con}}\right)^T = \left\langle \bar{\mathbf{u}}^S \ \bar{\mathbf{a}}^S \ \bar{\mathbf{u}}^m \ \bar{\mathbf{a}}^m \right\rangle$.

6.7.1 *Large Sliding with Horizontal Material Interfaces*

In order to illustrate the performance of X-FEM technique in modeling of a large sliding contact problem presented in the preceding section, the large sliding contact behavior of a frictionless problem is investigated for a plate with horizontal material interfaces, as shown in Figure 6.17. The X-FEM is employed to simulate the sliding behavior at contact surfaces. The problem is composed of a 2D rectangular plate with the length of 4 m and width of 9 m, as shown in Figure 6.17. The top and bottom edges of plate are subjected to the vertical displacement of 1 mm, while the horizontal displacement of 2 m is applied at the left edge of the middle plate. The X-FEM modeling is performed using three X-FEM meshes, including a coarse mesh of 36 elements and 2 fine meshes of 216 and 324 elements. The deformed configurations are presented in Figure 6.18a for three X-FEM meshes. Also presented in Figure 6.18b are the distributions of stress σ_y contour for various X-FEM meshes. A good agreement can be seen among various X-FEM meshes, particularly the stress distributions obtained by two fine meshes 2 and 3. The results of stress contours obtained here can be compared with those reported by Nistor *et al.* (2009). Finally, the distributions of contact pressure are plotted along the contact interface at different slides, as shown in Figure 6.19. This figure clearly presents the variations of contact pressure along the contact surface during sliding at different steps of movement. This example adequately illustrates the performance of proposed X-FEM technique in modeling of large deformation-large sliding contact problem with horizontal interfaces.

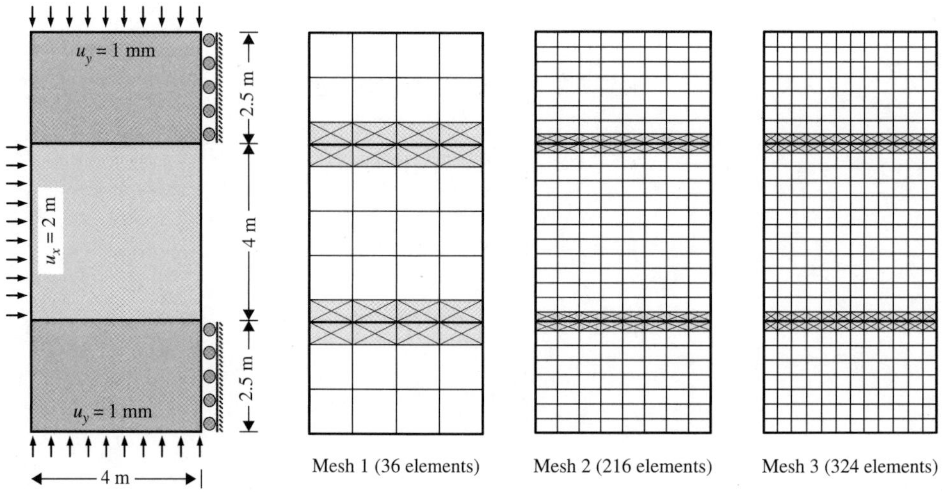

Figure 6.17 Large slide with horizontal material interfaces: Problem definition together with the initial X-FEM meshes

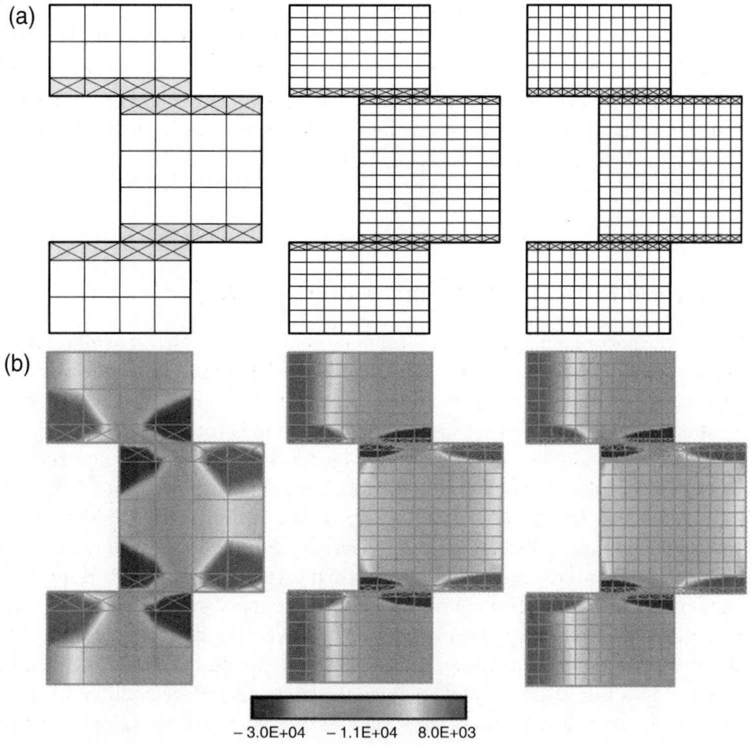

Figure 6.18 Large slide with horizontal material interfaces: (a) the deformed configuration of X-FEM meshes; (b) The distribution of stress σ_y contours for various X-FEM meshes

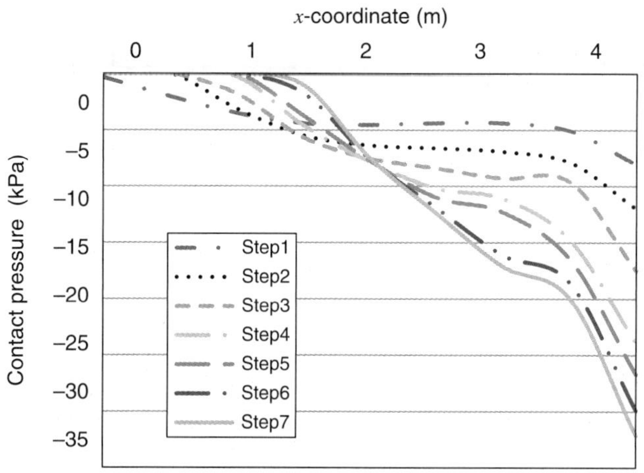

Figure 6.19 The distribution of contact pressure along the contact surface at different slides

6.8 Application of X-FEM Method in Frictional Contact Problems

In order to illustrate the accuracy and versatility of proposed computational algorithm several numerical examples are presented here in modeling frictional contact behavior of the straight, curved, and *S*–shaped discontinuities. All problems are solved in plane-strain condition and the infinitesimal deformation is assumed. The material behavior is considered to be linearly elastic with the Young modulus of $E = 1 \times 10^4$ MPa and Poisson ratio of $\nu = 0.3$. In order to provide more meaningful discussions of the numerical results, a comparison is performed between the X-FEM simulation results and those reported in literature. The numerical integration along the contact interface Γ_c is performed using the Gauss quadrature rule with two integration points at each contact segment. The first example is an elastic square plate with a horizontal interface at its center chosen to illustrate the accuracy and efficiency of augmented Lagrange X-FEM technique in two different cases; first, the effect of normal penalty parameter is studied on the convergence rate of unilateral contact constraints, and then the effect of relative tangential movement is investigated on the convergence rate of proposed technique where the frictional stick-slip behavior occurs. The second example is an elastic square plate with an inclined crack at its center chosen to demonstrate the performance of augmented Lagrange X-FEM formulation for a frictional crack surface where the crack undergoes the tangential movement. The third example is a double-clamped beam with a central crack chosen to indicate the active contact surface along the crack interface, where the contact region is *a priori* unknown and could undergo the opening, or closing mode. This example clearly illustrates how the closing mode of contact region can be determined along a crack surface within the Newton–Raphson iterations until the convergence is achieved. The last example is chosen to demonstrate the performance of augmented Lagrange X-FEM for a challenging problem, where a rectangular block with an *S*–shaped frictional contact interface is modeled under pure shear loading. It is shown that the technique can be efficiently used to achieve the desired local error tolerance of normal contact constraint with finite penalty parameter and reasonable computational effort. Furthermore, the quadratic convergence rate of Newton iterations is obtained for the global solution even in the presence of geometrical and frictional nonlinearities.

6.8.1 An Elastic Square Plate with Horizontal Interface

The first example is of an elastic square plate with a horizontal interface at its center, as shown in Figure 6.20. The plate is 1 × 1 m which is modeled by 75 × 75 structured FE mesh consisting of 5625

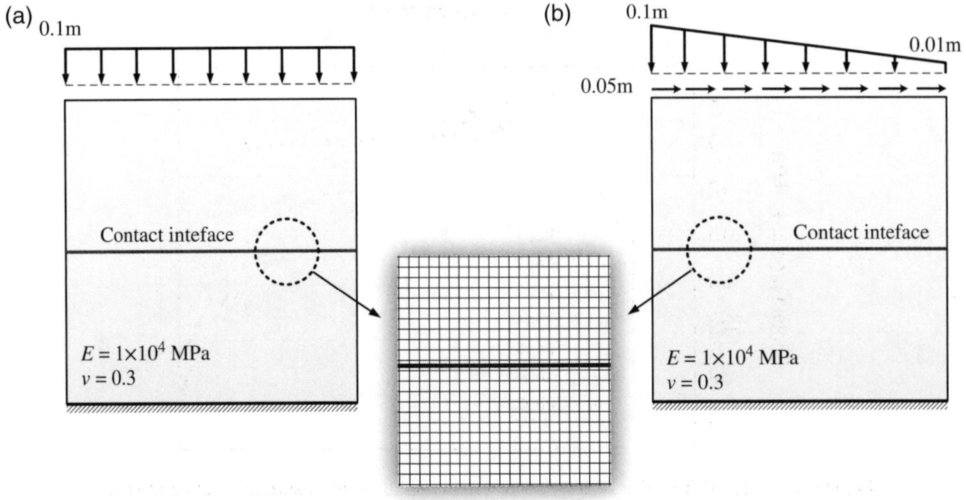

Figure 6.20 An elastic square plate with a horizontal interface: (a) the plate is subjected to a uniform vertical compression; (b) the plate is subjected to a non-uniform vertical displacement and a uniform horizontal displacement

bilinear elements. This example is chosen to illustrate the accuracy and efficiency of augmented Lagrange X-FEM technique in two different cases; that is, imposing the unilateral contact constraint in the normal direction and modeling the frictional stick-slip behavior in the tangential direction along the contact surface. The first case investigates the effect of normal penalty parameter on the convergence rate of augmented Lagrange X-FEM by comparing to the LArge Time INcrement (LATIN) and penalty methods. The second case evaluates the effect of relative tangential movement on the convergence rate of the proposed technique where frictional stick–slip behavior occurs.

6.8.1.1 Imposing the Unilateral Contact Constraint

In order to illustrate the effect of normal penalty parameter k_N on the convergence rate of augmented Lagrange X-FEM, the plate is modeled under a uniform vertical compression of 0.1 m imposed at the top edge while the bottom edge is fixed, as shown in Figure 6.20a. Note that the frictional effects are not generally involved in this particular case since relative sliding does not occur. The results are compared with the LATIN and penalty methods originally proposed by Dolbow *et al.* (2001) and Liu and Borja (2008). In the LATIN method, an iterative strategy is employed to impose the contact constraints at the contact interface. In this method, it is basically assumed that the interface is initially open, and it will then be closed iteratively to satisfy the unilateral contact constraints. In the penalty method, the contact constraint is enforced by considering a normal stiffness for the contact surface that is, in principle, analogous to the normal penalty parameter k_N. The solution of penalty method can be recovered from the solution of augmented Lagrange multipliers method at the first augmentation iteration where the augmented Lagrange multiplier is zero and the normal contact constraint is enforced approximately.

In Table 6.5, the convergence profiles of augmentation iteration are presented for three values of penalty parameter, that is, $k_N^{(1)} = 5\times10^7\,\mathrm{MN/m^3}$, $k_N^{(2)} = 0.1\,k_N^{(1)}$, and $k_N^{(3)} = 0.01\,k_N^{(1)}$. It must be noted that since the contact interface is known *a priori* and the problem does not involve the friction, the set of equations are linear within the augmentation loops, and the inner Newton iteration converges to the machine precision with only one iteration. It can be seen from Table 6.5 that the smaller values of k_N delay the

Table 6.5 The convergence profiles of augmentation iterations for an elastic square plate with a horizontal interface using three values of penalty parameter

Augmentation iterations	$k_N^{(1)} = 5 \times 10^7\,\text{MN/m}^3$	$k_N^{(2)} = 0.1\,k_N^{(1)}$	$k_N^{(3)} = 0.01\,k_N^{(1)}$
0 (penalty solution)	2.312E−05	3.307E−04	2.260E−03
1	5.367E−09	5.345E−07	5.127E−05
2	1.344E−12	1.300E−09	1.175E−06
3	2.229E−15 ≅ 0	1.374E−11	3.248E−08
4		1.628E−12	3.122E−09
5		2.488E−13	1.458E−09
6		3.851E−14	9.074E−10
7		6.001E−15 ≅ 0	5.747E−10
8			3.657E−10
9			2.336E−10
⋮			⋮
16			1.086E−11
17			7.041E−12
18			4.572E−12
⋮			⋮
25			1.079E−13
26			5.439E−14
27			9.657E−15 ≅ 0

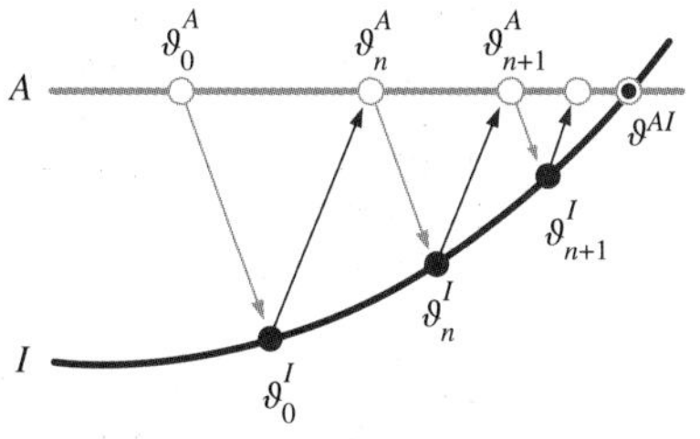

Figure 6.21 The schematic representation of LATIN iterative procedure

convergence of the solution, however, the desired tolerance can be achieved using more augmentation loops. Nevertheless, Table 6.5 clearly presents that the convergence can be obtained independent of the penalty parameter in the augmented Lagrange X-FEM algorithm and the penalty parameter only influences the convergence rate of the solution. It can be seen from this table that as the penalty parameter k_N increases, a more precise contact constraint can be achieved for the penalty method. However, it is well-known that a major problem associated with the numerical treatment of the penalty method is ill-conditioning, which arises when a very large value of penalty parameter is combined with the stiffness matrices of the master and slave bodies. As a result, it can be concluded that if a high precision of contact constraints is desired, the proposed computational method has an advantage over the penalty method since the contact constraints are exactly enforced even if a relatively small penalty parameter is used.

In order to provide more comparative discussions, the LATIN method is also implemented here. The LATIN method, which is called the LArge Time Increment, is an iterative strategy originally proposed by Ladevèze (1999) for solving nonlinear evolution problems. The technique consists of partitioning governing equations into a nonlinear group ***I*** that is local in space and time, and a linear group ***A*** that is global in the spatial variables. This two-step approach is schematically shown in Figure 6.21. In order to obtain the value of ϑ^{AI}, which is located at the intersection of the sets $\boldsymbol{A}$ and $\boldsymbol{I}$, the iterative strategy of LATIN method is applied at each increment of governing equations of global system within an increment $(n, n+1)$ in the following manner. The algorithm begins with an initial value of ϑ_0^A in $\boldsymbol{A}$. The value of ϑ_0^I is then obtained by satisfying the governing equation of set $\boldsymbol{I}$. In the next step, the governing equations of set $\boldsymbol{A}$ is preserved for obtaining the value of ϑ_1^A. This procedure is carried out until the convergence is achieved. In this algorithm, the value of ϑ_{i+1}^A at iteration $i+1$ can be calculated from ϑ_i^A at iteration i in two steps; first ϑ_i^A to ϑ_i^I and then, ϑ_i^I to ϑ_{i+1}^A by satisfying the governing equations of set $\boldsymbol{I}$ and $\boldsymbol{A}$, respectively. The iteration will be stopped when the difference between ϑ_i^I and ϑ_{i+1}^A is below a specified tolerance. It must be noted that going from one set of equation to the other set requires having extra equations in order to achieve the fast convergence. In this approach, the regularization parameter of the LATIN method k affects the convergence rate of iterations; so a proper value of LATIN parameter must be chosen to achieve an optimum convergence rate. Table 6.6 presents the convergence profile of the LATIN method for $k = 1 \times 10^4$ MN/m^3. In this table, the global and local error indicators are obtained according to relations (6.79) and (6.80). Obviously, the rate of convergence exhibited by this method is much slower than the augmented Lagrange X-FEM. It can be expected that in the case of geometrical and frictional nonlinearities where the contact surface is not known *a priori*, the convergence rate of the LATIN method could get

Table 6.6 The convergence profile of LATIN method for an elastic square plate with a horizontal interface using $k = 1 \times 10^4$ MN/m^3

LATIN iterations	Global error η_{global}	Local error η_{local}
0	1.00E+00	0.00E+00
1	2.28E−01	1.00E−01
2	3.94E−01	5.66E−02
3	5.99E−02	1.13E−02
4	8.31E−03	1.32E−03
5	2.06E−03	1.27E−04
6	2.28E−01	1.54E−04
⋮	⋮	⋮
19	2.74E−06	1.07E−07
20	2.06E−06	8.32E−08
⋮	⋮	⋮
49	7.46E−08	2.88E−09
50	7.04E−08	2.72E−09
⋮	⋮	⋮
99	1.05E−08	3.04E−10
100	1.02E−08	2.88E−10

Table 6.7 The total contact reaction for an elastic square plate with a horizontal interface using various techniques

Penalty X-FEM technique $\left(k_N^{(3)}\right)$	−1130.04517
Penalty X-FEM technique $\left(k_N^{(2)}\right)$	−1153.60263
Penalty X-FEM technique $\left(k_N^{(1)}\right)$	−1156.01283
LATIN method	−1156.28121
Augmented Lagrange X-FEM technique	−1156.28125
Standard FEM technique	−1156.28125

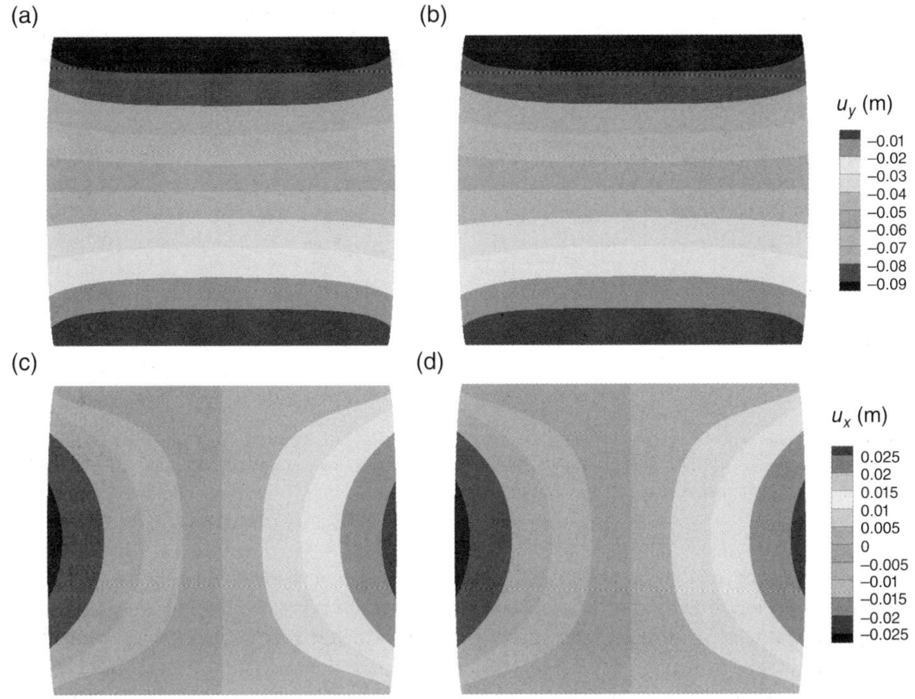

Figure 6.22 The contours of vertical and horizontal displacements for a plate with horizontal interface subjected to a uniform vertical compression: (a and c) the augmented Lagrange X-FEM; (b and d) the standard FEM technique

worse. In Table 6.7, a comparison of the total contact reaction is performed between the penalty, LATIN, and augmented Lagrange X-FEM techniques, and the results are compared with the standard FEM. Since the augmented Lagrange X-FEM exactly enforces the contact constraint, its solution is identical to FEM. Finally, the contours of vertical and horizontal displacements are presented in Figure 6.22 for the augmented Lagrange X-FEM and those of standard FEM solution. Obviously, the displacement contours evaluated by two methods are completely identical.

6.8.1.2 Modeling the Frictional Stick–Slip Behavior

In order to evaluate the effect of frictional stick–slip behavior on the convergence rate of augmented Lagrange X-FEM, the plate is solved under a non-uniform vertical displacement of $\delta_y = 0.09x - 0.1$ m where $0 \le x \le 1$ m, and a uniform horizontal displacement of $\delta_x = 0.05$ m, while the bottom edge is fixed, as shown in Figure 6.20b. In order to investigate the effect of frictional contact behavior along the contact surface, the plate is modeled for two values of the friction coefficient, that is, $\mu = 0.25$ and 0.4, with the stick stiffness of $k_T^{\text{stick}} = 5 \times 10^7\,\text{MN/m}^3$. The contact constraints are imposed using the normal penalty parameter of $k_N = 5 \times 10^7$ MN/m^3. This example is solved in three augmentation iterations to achieve the required accuracy of the global and local error criteria.

In Figure 6.23a, the evolutions of normal contact stress are plotted along the contact interface for the friction coefficients of 0.25 and 0.4. Due to the fact that the normal and tangential components of contact traction are treated in a decoupled manner, the normal contact stress is independent of friction coefficients and the results are identical for two values of friction coefficient. In Figure 6.23b, the evolutions of tangential relative displacement are plotted along the contact interface for two friction coefficients.

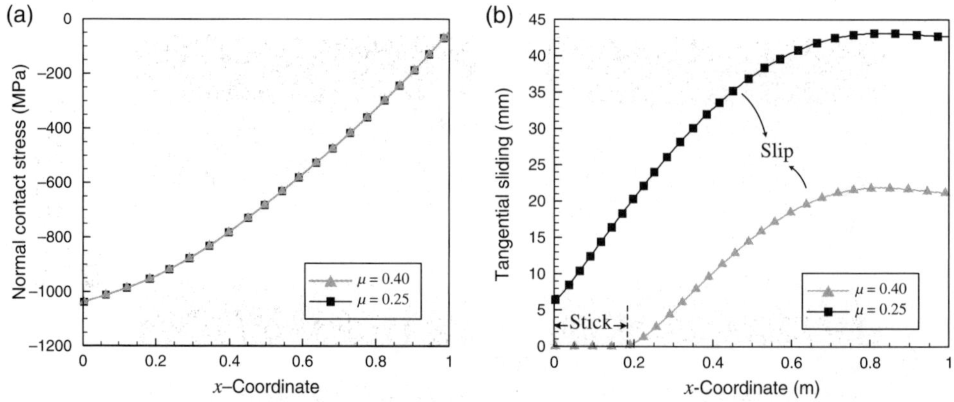

Figure 6.23 A plate with the horizontal interface subjected to a non-uniform vertical displacement and a uniform horizontal displacement: (a) the evolutions of normal contact stress along the contact interface; (b) the evolutions of tangential relative displacement along the contact interface

Obviously, two distinct regions of sticking and slipping can be distinguished for the friction coefficient of 0.4. In the stick region, where the tangential contact stresses drop below the slip limit, the tangential relative displacement consists of the elastic (stick) behavior, which is much smaller than that of the slip region, and can be approximately considered to be zero. It can be observed from Figure 6.23 that the normal contact stress decreases along the contact surface from the left to right edge, and as a result, the tangential contact stress reaches the slip limit along the contact surface where the transition from the stick to slip condition occurs, which leads to a relatively large tangential movement. The slip limit is generally much lower at the friction coefficient 0.25 than that of the friction coefficient 0.4, and the tangential contact stresses readily reach the slip limit over the whole contact surface. Hence, the stick condition cannot occur for $\mu = 0.25$ and the whole contact interface is in a slip condition where a relatively large movement occurs. In order to investigate the convergence rate of the solution for a frictional contact problem, the convergence profile of Newton iterations is plotted in Figure 6.24 within the first augmentation loop for two values of friction coefficient. It must be noted that the frictional traction changes alternatively according to the predictor-corrector contact algorithm until the stick–slip behavior is obtained along the contact surface. Obviously, the accuracy is achieved within eight iterations for $\mu = 0.4$ and nine iterations for $\mu = 0.25$ that indicate the efficiency of the proposed computational algorithm.

6.8.2 A Square Plate with an Inclined Crack

The next example is of an elastic square plate with an inclined crack at its center, as shown in Figure 6.25. The plate is 1 × 1 m that contains a central crack of 0.566 m oriented at $\theta = 45°$ relative to the horizontal direction, in which a uniform vertical displacement of − 0.1 m is applied at the top edge while the bottom edge is restrained. The FE mesh consists of 2500 bilinear elements with 2601 nodal points, and the standard Gauss integration rule is utilized for numerical integration of cut elements. In order to not intersect the crack with the FE nodal points, the crack tips are specified at coordinates of (0.29, 0.28) and (0.72, 0.71) m. The material property is assumed to be linearly elastic with the Young's modulus $E = 1 \times 10^4$ MPa and the Poisson ratio $\nu = 0.3$. The contact constraints are imposed using the normal penalty parameter $k_N = 5 \times 10^7$ MN/m^3. The crack has a frictional behavior with the friction coefficient of $\mu = 0.1$ and stick stiffness of $k_T^{\text{stick}} = 5 \times 10^7\,\text{MN/m}^3$. On the basis of vertical displacement applied at the top edge, the crack may undergo the relative tangential movement, however, it is assumed that the

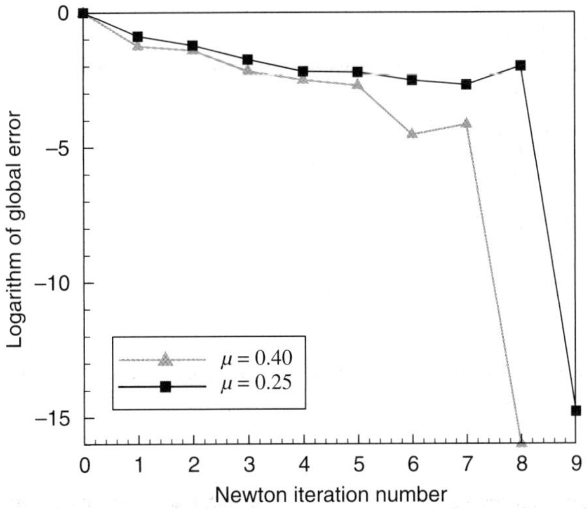

Figure 6.24 Convergence profiles of Newton iterations within the first augmentation loop for a plate with the horizontal interface subjected to a non-uniform vertical displacement and a uniform horizontal displacement for friction coefficients of 0.25 and 0.4

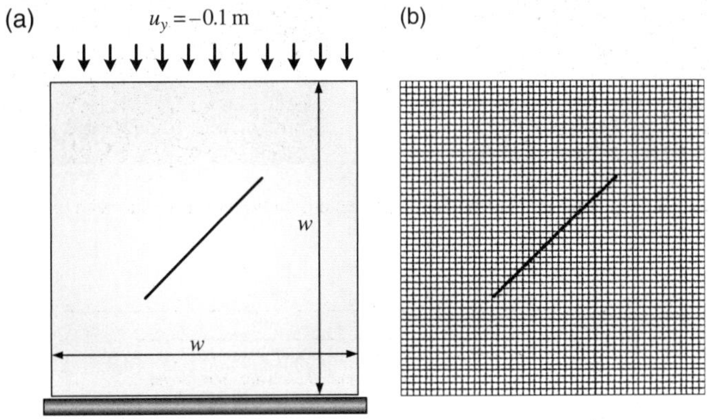

Figure 6.25 An elastic square plate with an inclined crack at its center: (a) Problem definition; (b) The X-FEM mesh

crack tips could not advance. This example is originally modeled by Dolbow *et al.* (2001), and then simulated by Liu and Borja (2008) using 20000 CST elements.

In Figure 6.26, the deformed configuration of X-FEM mesh is shown for the vertical deformation of −0.1 m from the top edge. Obviously, there is a tangential movement along the crack surface. In Figure 6.27, the contours of vertical and horizontal displacements are shown at the final stage of loading, which can be compared with those reported by Liu and Borja (2008). As can be observed, there is a tangential movement along the crack surface. Obviously, the displacement discontinuity can be clearly observed from the displacement contours obtained by the augmented Lagrange X-FEM formulation. Also plotted in Figure 6.28 is the total vertical reaction at different vertical movements. It must be noted that the predicted force–displacement curve is obtained using one step solution for each prescribed vertical displacement. On the other hand, an incremental prescribed solution is

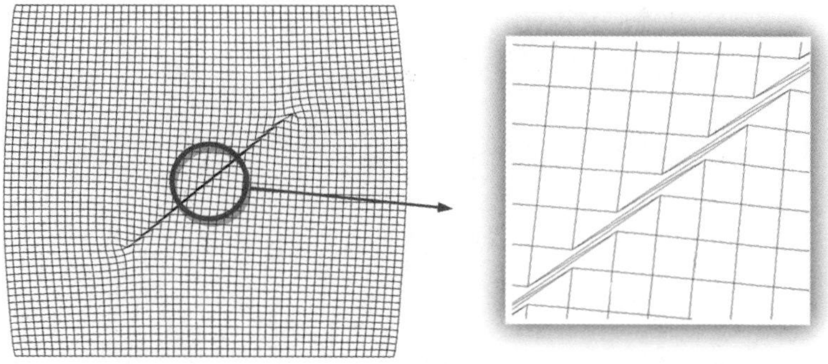

Figure 6.26 The deformed X-FEM mesh of square plate with an inclined crack. A tangential relative displacement is obvious along the contact interface

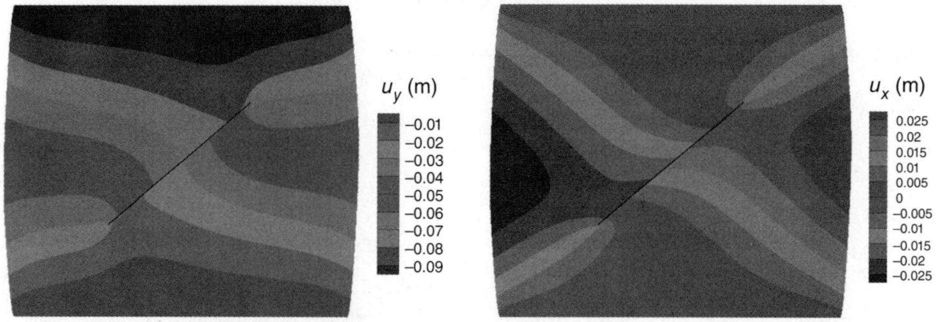

Figure 6.27 The vertical and horizontal displacement contours for a plate with an inclined crack

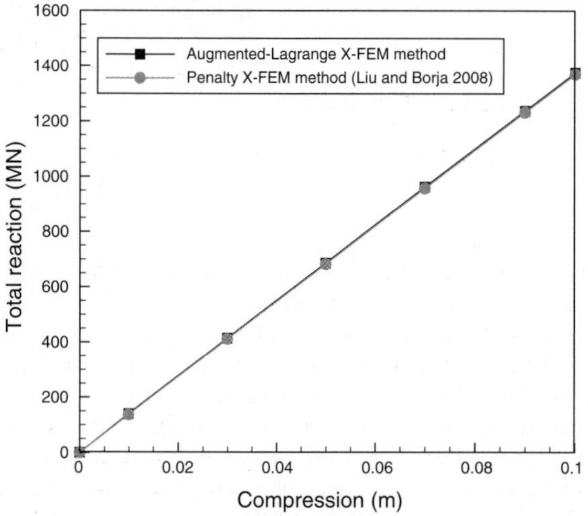

Figure 6.28 The variation of total reaction with vertical deformation for a plate with an inclined crack: A comparison between the augmented Lagrange X-FEM and the penalty approach reported by Liu and Borja (2008)

not employed here in order to perform a direct comparison with the solution of Dolbow *et al.* (2001). In fact, the solution at a given vertical displacement corresponds to that generated in one time step from the starting value of zero, and not a continuation of the solution generated from the previous calculated time step. In this example, three augmentation iterations are used with a maximum number of seven Newton iterations for each augmentation loop to achieve the required accuracy of the global and local error criteria since the frictional traction changes alternatively according to the predictor-corrector contact algorithm described in Box 6.1 until the stick–slip behavior is achieved along the contact surface.

6.8.3 A Double-Clamped Beam with a Central Crack

In the next example, a beam fixed at both ends is modeled under a vertical crack at its center, as shown in Figure 6.29. The beam has a length of $\ell = 10$ m and height of $h = 3$ m, which is subjected to a prescribed horizontal displacement of 0.06 m from the right edge and a vertical force of 240 MN from the top edge distributed over a length of 0.4 m at its center. The beam is divided into two parts due to a vertical crack located at the middle of the beam. The material property of the beam is assumed to be linearly elastic with the Young's modulus of $E = 1 \times 10^4$ MPa and the Poisson ratio of $\nu = 0.3$. It must be noted that frictional effects are not generally relevant to this problem since the discontinuity experiences only opening/closing modes and no relative movement occurs. Hence, a frictionless contact condition is assumed for the crack interface, and a normal penalty parameter $k_N = 5 \times 10^7$ MN/m^3 is employed to impose the contact constraint on the crack interface. In contrast to the previous example where the entire crack was in contact condition, in the current example the crack interface may undergo the opening, or closing mode. In fact, the contact region is unknown *a priori*, and it is important to indicate the active contact region along the crack interface. To this end, the entire crack interface is first discretized by a set of 1D segments, and the integration points are generated along the crack interface. The contact region is determined within the Newton–Raphson iteration loops based on a non-positive value of the normal gap distance g_N calculated at the integration points along the crack interface.

In order to investigate the performance of Heaviside enrichment function in the proposed augmented Lagrange X-FEM, four X-FEM meshes consisting of 250, 816, 1782, and 3780 bilinear elements are employed, as shown in Figure 6.30. In Figure 6.31, the convergence profiles of Newton iterations are presented within the first augmentation loop for various FE meshes. Obviously, the active contact surface changes alternatively within the Newton iterations until the convergence is achieved. Although the number of iterations increases by refining the X-FEM meshes, the convergence is obtained within three augmentation loops with a maximum number of 12 Newton iterations at each augmentation loop for

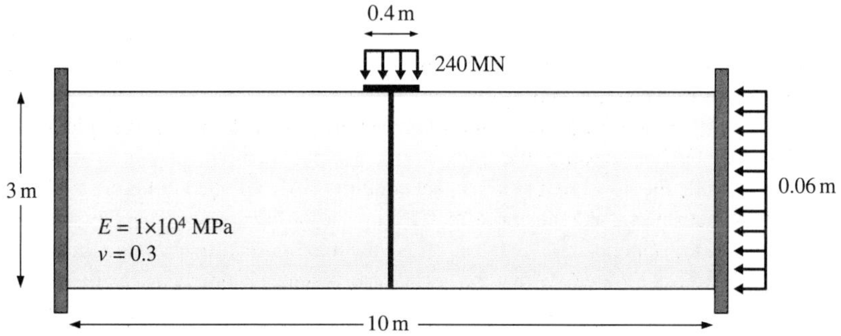

Figure 6.29 A double-clamped beam with a central crack subjected to a prescribed horizontal displacement of 0.06 m from the right edge and a vertical force of 240 MN from the top edge distributed over a length of 0.4 m at its center

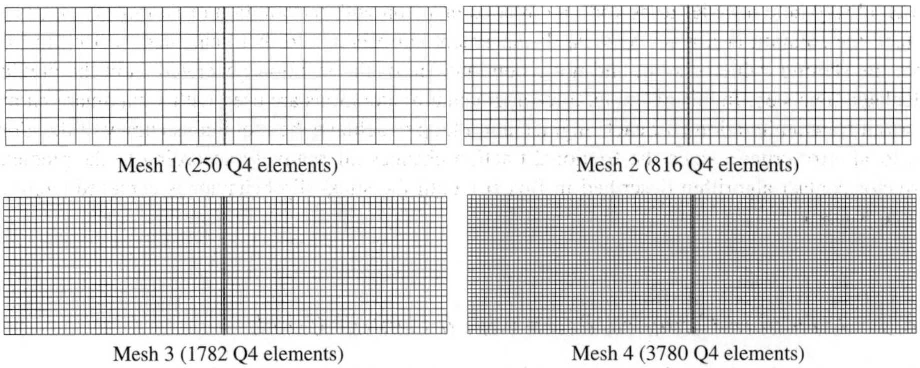

Figure 6.30 Modeling of a double-clamped beam with a central crack using four X-FEM meshes

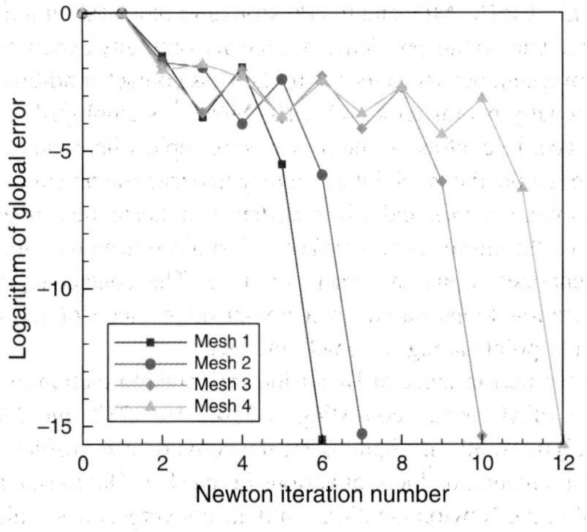

Figure 6.31 Convergence profiles of Newton iterations within the first augmentation loop for a double-clamped beam with a central crack using various X-FEM meshes

all 4 meshes. It must be noted that the number of Newton iterations increases slightly as the mesh is refined because the number of integration points increases along the contact interface that makes the initial Newton solution far from the converged solution.

In Figure 6.32a, the variations of normal gap distance along the crack interface are plotted for various FE meshes. Obviously, the lower part of the crack interface experiences an opening mode where the normal traction is zero, while the upper part is in contact condition. Also plotted in Figure 6.32b are the variations of normal contact stress along the crack interface for various FE meshes. Clearly, the compression traction is developed in the upper part of the beam, where the maximum normal contact stress occurs at the top edge. It can be seen from Figure 6.32 that the normal gap distance is nearly insensitive to mesh refinement, whereas the normal contact stress is substantially influenced by the mesh refinement, particularly near the top edge of the beam where the maximum normal contact stress occurs. Nevertheless, the convergence of the solution can be observed in this figure by refining the X-FEM meshes. Finally, the distributions of stress σ_x contour are presented in Figure 6.33 for four X-FEM meshes. The stress σ_x contours

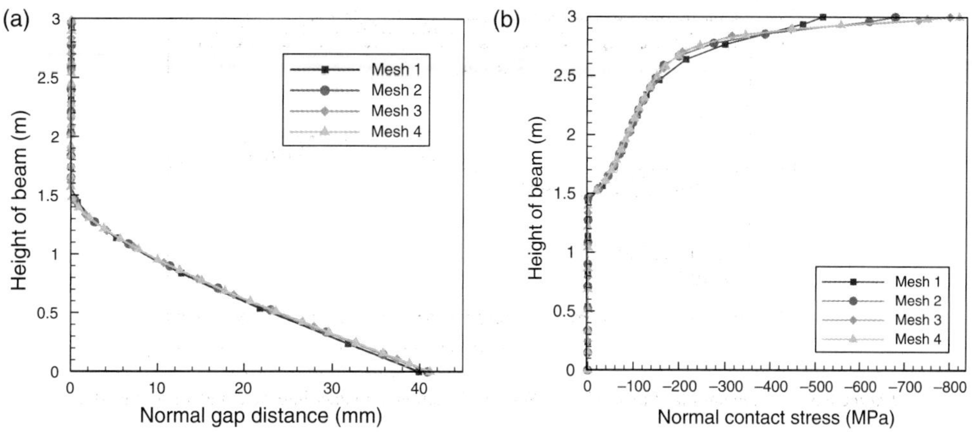

Figure 6.32 A double-clamped beam with a central crack: (a) the variations of normal gap distance; (b) the variations of normal contact stress along the crack interface for various X-FEM meshes

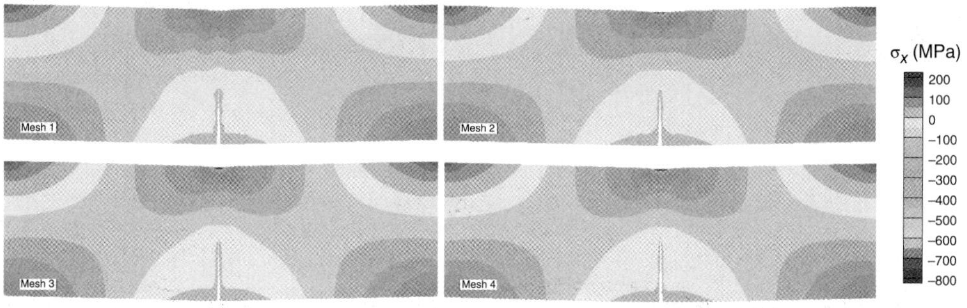

Figure 6.33 The distributions of stress σ_x contour for a double-clamped beam with a central crack using four X-FEM meshes

clearly present that the closing mode occurs in the upper region of the crack interface while the lower region is in opening mode.

6.8.4 A Rectangular Block with an S–Shaped Frictional Contact Interface

The last example is chosen to demonstrate the performance of proposed computational algorithm for a challenging problem, where a rectangular block with an *S*–shaped frictional contact interface is modeled under shear loading, as shown in Figure 6.34. The wavy discontinuities in geological problems, such as natural joints, faults, flaws, and bedding planes, are generally subjected to frictional contact condition. An appropriate modeling of such frictional discontinuities is of great importance in computational geosciences in order to efficiently assess the behavior of the cracked media. In this example, an *S*–shaped (sine-wave) frictional contact interface with an amplitude of 0.01 m embedded in a rectangular block of 0.2 m wide and 0.1 m high is modeled under pure shear loading by applying a horizontal displacement of 3 mm at the top edge while the bottom edge is fixed. The crack has a frictional behavior with the friction coefficient of $\mu = 0.25$ and the stick stiffness of $k_T^{\text{stick}} = 1 \times 10^6\,\text{MN/m}^3$.

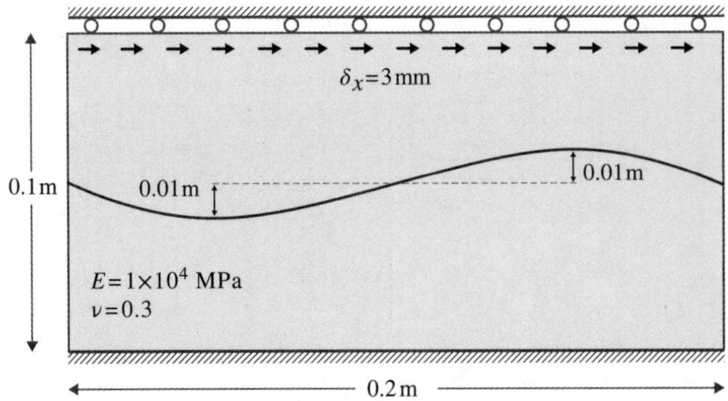

Figure 6.34 A rectangular block with an *S*–shaped frictional contact interface under shear loading

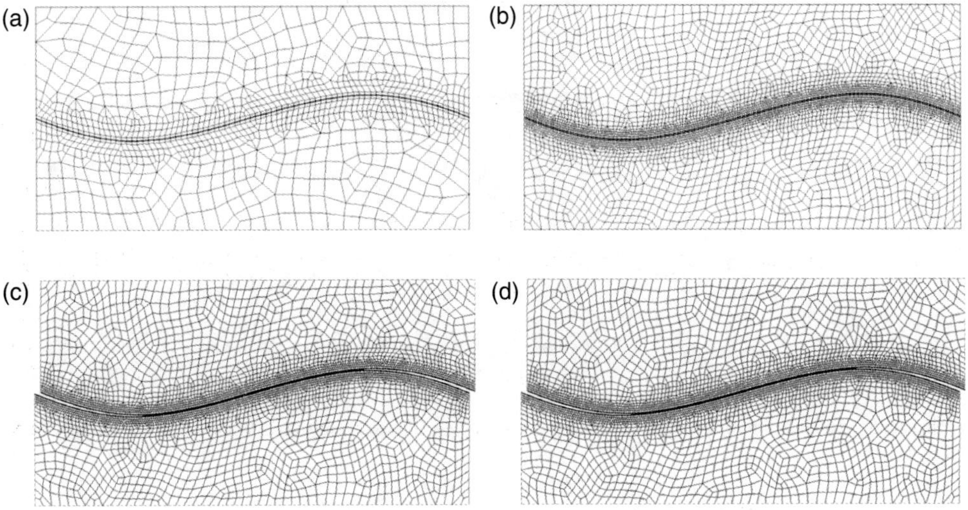

Figure 6.35 Modeling of a rectangular block with an *S*–shaped frictional contact interface using two X-FEM meshes: (a) the "coarse" mesh 1 with 1016 Q4 elements; (b) the "fine" mesh 2 with 4376 Q4 elements; (c) the deformed "coarse" X-FEM mesh; (d) the deformed "fine" X-FEM mesh

The contact constraints are imposed using the normal penalty parameter of $k_N = 1 \times 10^5$ MN/m^3 with an error tolerance of 10^{-5} for the convergence of augmentation iterations. The problem is modeled using two X-FEM meshes, that is, a "coarse" mesh of 1016 and a "fine" mesh of 4376 bilinear elements, as shown in Figure 6.35a and b. The wavy contact interface is characterized by a set of interface segments obtained from the intersections of contact interface with the X-FEM mesh. The deformed configurations of X-FEM meshes are presented in Figure 6.35c and d, in which an anti-symmetric deformed shape can be observed with respect to the contact interface. Despite the presence of geometrical and frictional nonlinearities, the profiles of Newton iterations plotted in Figure 6.36 exhibit a stable convergence of global error at the first augmentation loop for both X-FEM meshes. In this example, the convergence is obtained within seven

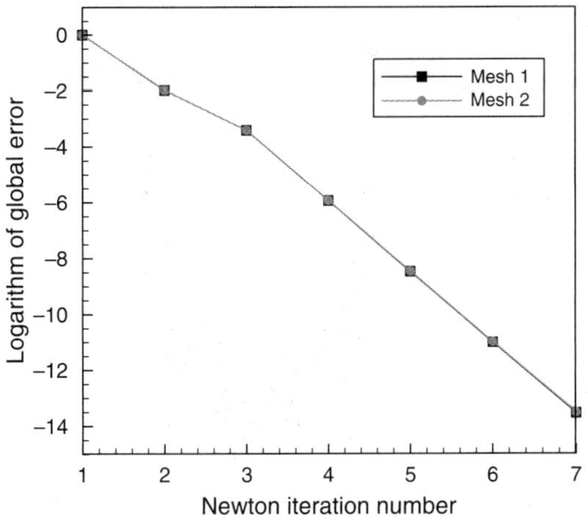

Figure 6.36 Convergence profiles of Newton iterations within the first augmentation loop for a rectangular block with an *S*–shaped frictional contact interface

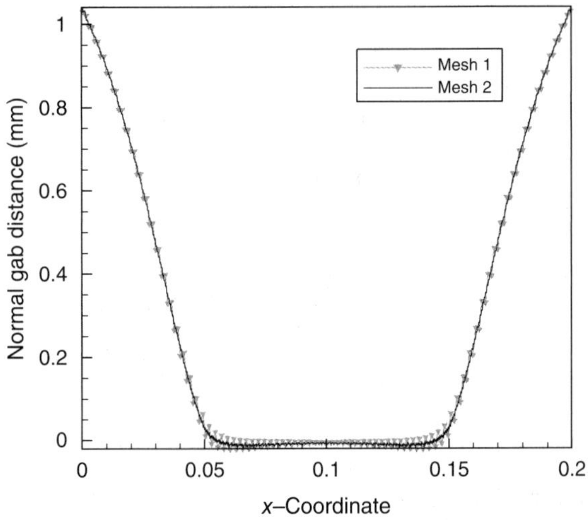

Figure 6.37 The evolutions of normal gap distance along the contact interface for a rectangular block with an *S*–shaped frictional contact interface using two X-FEM meshes

augmentation loops with a maximum number of seven Newton iterations at each augmentation loop for both X-FEM meshes.

In Figure 6.37, the evolutions of normal gap distance along the contact interface are plotted for two X-FEM meshes. Obviously, the opening modes with traction free surfaces appear in roughly one-quarter of contact interface at the left and right regions, while the remaining zone at the middle of contact surface is in the frictional contact condition. Clearly, the oscillations of normal gap distance can be observed using the coarse X-FEM mesh, which is a result of non-smooth contact approximation, and can be removed by employing the fine X-FEM mesh. A comparison of the normal contact stress is performed in Figure 6.38

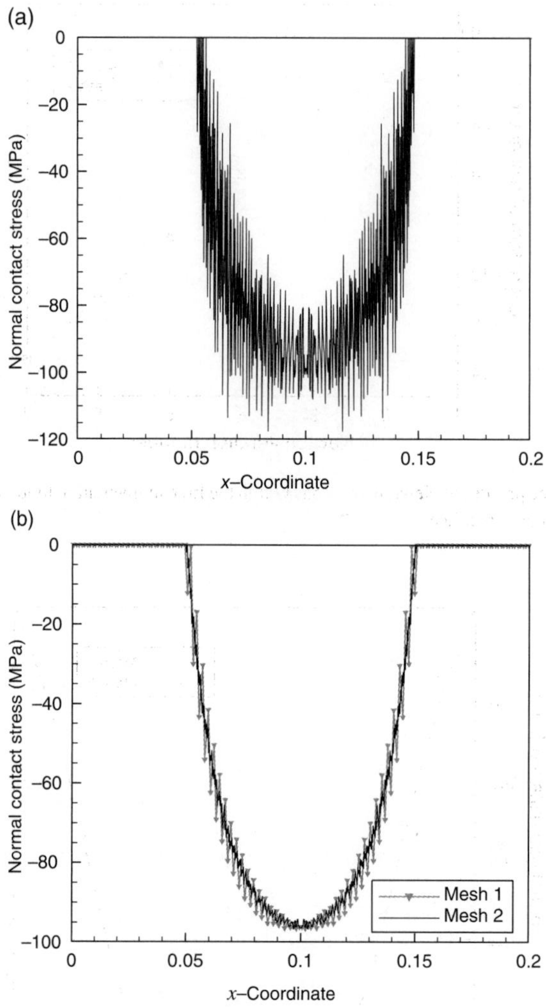

Figure 6.38 The evolutions of normal contact stress along the contact interface for a rectangular block with an *S*–shaped frictional contact interface: (a) the penalty X-FEM using "fine" X-FEM mesh; (b) the augmented Lagrange X-FEM using the "coarse" and "fine" X-FEM meshes

between the penalty method and augmented Lagrange X-FEM technique. In Figure 6.38a, the evolution of normal contact stress is plotted using the penalty X-FEM technique for the fine X-FEM mesh. This curve is obtained using the normal penalty parameter of $k_N = 8 \times 10^6$ MN/m^3 to acquire a similar local convergence error of $\eta_{\text{local}}^{\text{aim}} = 10^{-5}$, as considered in the augmented Lagrange X-FEM technique. Comparing the result of normal contact stress obtained from the penalty X-FEM technique with those reported in Figure 6.38b using the augmented Lagrange X-FEM technique for the "coarse" and "fine" X-FEM meshes illustrate that the penalty X-FEM technique has great oscillations. This figure clearly presents the efficiency and robustness of the augmented Lagrange X-FEM technique compared to the penalty X-FEM technique. Finally, the contours of vertical and horizontal displacements are presented in Figure 6.39

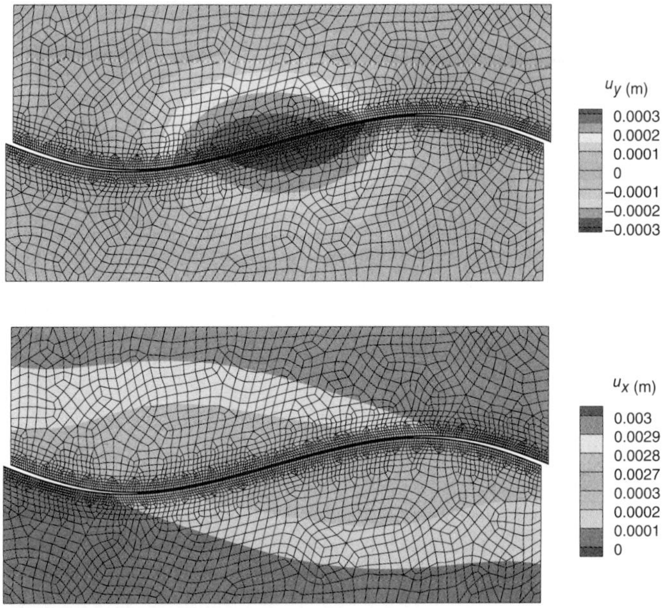

Figure 6.39 The vertical and horizontal displacement contours for a rectangular block with an *S*–shaped frictional contact interface using "fine" X-FEM mesh

using the augmented Lagrange X-FEM technique for the fine X-FEM mesh. This example adequately illustrates the performance of augmented Lagrange X-FEM technique in modeling frictional contact interfaces with complex geometries.

7

Linear Fracture Mechanics with the X-FEM Technique

7.1 Introduction

History has witnessed numerous counterexamples of structural failure due to the existence of cracks. Buildings, dams, bridges, airplanes, railroads, and many other constructions have been faced with this phenomenon that has a lot of detrimental financial and human effects in the world. Many solutions have been formulated in order to decrease the effects of cracks and prevent their appearance in structures. The understanding and analysis of the mixed-mode fracture is an important subject in fracture mechanics because material flaws, or pre-cracks, can have an impact with respect to the loading applied to the component. The analysis of cracked structures is an important aspect of structural safety assessment. In general, analytical solutions to crack problems can only be established for simple cases. Most realistic problems encountered in practical engineering applications require complex solutions that can be obtained by computational techniques. The finite element method (FEM) plays an important role in computational fracture mechanics. However, the most difficult part with this technique is the proper modeling of strong discontinuities in displacements caused by the growth of crack. This particular difficulty constitutes a main challenge for many researchers to propose a solution strategy within the context of FEM. Historically, the numerical methods proposed for simulation of crack growth can be classified into three basic categories (Tabeie and Wu, 2003); including the smeared crack approach, the discrete crack propagation using the nodal release approach, and the discrete crack propagation using the delete-and-fill remeshing method.

The so-called smeared crack approach was first introduced by Rashid (1968) for fracture behavior of prestressed concrete pressure vessels. In this technique, the crack was not represented explicitly, but modeled as a region of material degradation by modifying the material constitutive equations. To model the crack numerically, the stiffness of elements that satisfy a failure criterion are reduced to small values. In this approach, there is no need for remeshing the process at each step of crack growth. However, it has some disadvantages, since the state variables are computed at the integration points that are near the center of elements. In fact, the values of these variables are smaller than those at the crack tip so that the applied stress causing fracture is not accurate (Marzougui, 1998). Moreover, the effective loss of material at the crack tip may change the crack tip geometry significantly. The crack trajectories obtained by this approach are mesh dependent and are highly sensitive to mesh size. For those materials where the crack tip remains relatively distinct, such as metals, the crack tip cannot be modeled by this procedure since large values of stresses can be produced in the elements that are near the crack front. The strain energy near the crack tip is not estimated well, despite accurate overall displacement (Koenke *et al.*, 1998). Furthermore, if the crack does not propagate parallel to the direction of element edges, stress-locking may occur.

Extended Finite Element Method: Theory and Applications, First Edition. Amir R. Khoei.

A number of mesh update procedures have been used to simulate crack geometry and displacement discontinuities in a continuum explicitly. In these approaches, the representation of crack as a moving boundary is performed by updating the finite element (FE) model at each time step of crack propagation. Since the boundary update and mesh regeneration procedure seems to be a difficult and expensive task, some researchers have proposed the inter-element crack propagation scheme, which allows the propagation of crack only along the edges of the elements in the FE mesh (Ngo and Scordelis, 1967). Discrete crack propagation using the nodal release method is one of these techniques used for simulation (Xu and Needleman, 1994). In this method, two element edges, initially constrained to identical displacements, are allowed to separate by releasing the constraints and nodal forces that hold the elements together. A new, crack free surface is therefore generated at the crack interface. The nodal release approach has been proven to be very robust and easy to implement, even in commercial FE codes, and shown as a popular technique for modeling fracture procedure. However, the technique is greatly mesh dependent. For problems in which the crack path is uncertain, a very fine mesh with considerable numerical effort has to be applied. So, an alternative approach has been performed for crack growth simulation on the basis of discrete crack propagation via the delete-and-fill remeshing method. The technique was first introduced by Wawrzynek and Ingraffea (1980) in simulation of crack propagation problems irrespective of the existing mesh. In this method, a group of elements in a region around the expected new crack tip is first deleted. After the crack is extended, the local domain is refilled with new elements for the new crack tip. A similar technique was also developed by Swenson (1985), in which a partial remeshing was only performed in a small area around the new crack segment.

The adaptive mesh strategy was employed in simulation of crack propagation based on the recovery of gradients using the analytical solution of crack tip fields. The adaptive remeshing technique enables us to estimate the error of FE discretization and adapt the mesh in such a way that the error remains in a prescribed range. In this manner, the FE model is adaptively modified until the structure is fully cracked or the prescribed loading level is reached. In order to improve the rate of convergence of FE solution and the accuracy of stress intensity factors (SIFs), the collapsed quarter point singular elements proposed by Barsoum (1976) and Henshell and Shaw (1975) are used at the crack tips. The well-known Zienkiewicz-Zhu (1992) error estimator with a recovery procedure based on the modified superconvergent patch recovery (SPR) technique (Khoei, Azadi, and Moslemi, 2008c) is used and provides more accurate error estimation for fracture problems. Several modifications must be taken into the recovery process when using the standard SPR technique in conjunction with singular elements (Moslemi and Khoei, 2009; Azadi and Khoei, 2011). The implementation of a polynomial with the same order of shape function for smoothing procedure cannot describe the behavior of stresses in the vicinity of a crack tip. Although the derivatives of singular element shape functions represent an appropriate order of singularity due to singular mapping between the natural and parametric coordinates, the polynomial shape functions are not able to represent such a feature and therefore the recovery process loses its efficiency near the crack tip. To overcome this problem, the modified SPR technique was employed to incorporate the analytical solution of crack tip fields into the recovery process as the smoothing function at the crack tip region (Khoei *et al.*, 2013). Considerable improvements in the accuracy of the error estimator were observed using this modified patch recovery procedure.

Despite of capability of remeshing method to model the crack growth, various techniques have been developed to model the crack propagation without remeshing; these meshless methods do not use remeshing and discretize the problem by nodal points (Belytschko, Lu, and Gu, 1994); the arbitrary local mesh replacement method is a moving mesh technique and divides the problem into two distinct meshes; one that surrounds the advancing crack front and moves with crack, and the other that fills the entire domain (Rashid, 1998); and the strong discontinuity technique that is used to model the crack implicitly by a jump in the displacement field without affecting the neighboring elements by using the additional degrees of freedom (DOF) into the FE model (Simo, Oliver, and Armero, 1993; Garikipati, 1996). One of the most popular techniques in modeling crack propagation that does not need a remeshing process is the extended finite element method (X-FEM), in which the effect of crack on the surrounding strain–displacement field is embedded in the local approximant within element, leaving the geometry and mesh unaltered. In this method, the displacement field is enriched near the crack by incorporating both discontinuous fields and the near tip asymptotic fields through the partition of unity method (PUM) (Babuška and Melenk, 1997).

The X-FEM technique was first developed by Dolbow (1999), Belytschko and Black (1999), and Moës, Dolbow, and Belytschko (1999) to model cracks, voids, and inhomogeneities. They proposed appropriate additional terms in FE approximation based on the PUM that allows for the entire crack geometry to be modeled independently of the mesh, and completely avoids the need to remesh as the crack grows. A methodology that constructs the enriched approximation based on the interaction of the discontinuous geometric features with the mesh was developed by Daux *et al.* (2000) in modeling crack discontinuities. The implementation of the X-FEM into three-dimensional crack modeling was presented by Sukumar *et al.* (2000) using the notion of partition of unity (PU). Belytschko *et al.* (2001) developed a technique in modeling arbitrary discontinuities in FE, in which the discontinuous approximation was constructed in terms of a signed distance function and the level sets were used to update the position of the discontinuities. Dolbow, Moës, and Belytschko (2001) proposed special treatments in numerical simulation of crack growth, in which the frictional contact was modeled in the framework of X-FEM technique and an iterative procedure, called the " LATIN method" was applied in the analysis of contact faces of crack. In this method, the nonlinear behavior of contact was incorporated into the linear global equations by an iterative scheme to resolve the nonlinear boundary value problem. The combination of this method with X-FEM was enhanced the solution procedure of contact analysis, efficiently. An algorithm which couples the level set method with the X-FEM was proposed by Stolarska *et al.* (2001), in which a discontinuous function based on the Heaviside step function was employed in modeling two-dimensional (2D) linear elastic crack tip displacement fields. A methodology for modeling crack discontinuities was presented by Sukumar and Prévost (2003) with implementation of X-FEM in isotropic and bimaterial media. The technique was implemented into three-dimensional fatigue crack propagation simulation of multiple coplanar cracks by Sukumar, Chopp, and Moran (2003a) and Chopp and Sukumar (2003). They combined the X-FEM to the fast marching method using the PUM to model the entire crack geometry, including one or more cracks, by a single signed distance (level set) function. An enriched FEM with arbitrary discontinuities in space–time was presented by Chessa and Belytschko (2004). They modeled discontinuities by the X-FEM with a local PU enrichment to introduce discontinuities along a moving hyper-surface. In this chapter, the X-FEM technique is presented to model the crack propagation in the linear elastic fracture mechanics (LEFM). The basis of LEFM is first introduced by defining the stress and displacement distributions around the crack tip and the SIFs for different loading modes. The governing equation of a solid fractured body is then derived in the framework of X-FEM. The procedure of crack growth simulation using the X-FEM technique is illustrated, and the method is employed to model the curved crack with higher order elements, and the crack growth simulation for bimaterial interface cracks in complex composite components.

7.2 The Basis of LEFM

Fracture mechanics originally concentrated on the elastic material behavior where Hook's law was obeyed. A number of experiments and theories were presented by Orowan (1948), Irwin (1957), and Barenblatt (1962) to illustrate the behavior of cracked domain in linear elastic materials. Irwin (1960) and Shih and Hutchinson (1976) extended the concept of LEFM to nonlinear behavior of materials such as plastic solids. A general description of the fractured materials is presented in Figure 7.1a, where a tension load is applied to an infinite plate that contains a crack with the length of $2a$ at the center of the plate. The fractured domain causes a singularity in the stress field at the crack tip region for the elastic material. This singularity is depicted in Figure 7.1b where the stress reaches an infinite value. However, in plastic materials, a plastic zone occurs at the crack tip that causes the stress to reach a finite value equal to yield stress of the material.

There are a number of microscopic and macroscopic studies performed by researchers to investigate the fracture behavior of the material. From the microscopic point of view, the crack can be propagated if the

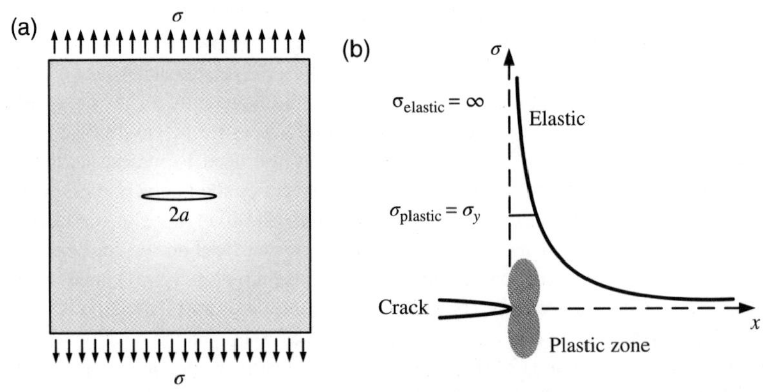

Figure 7.1 (a) A crack of length $2a$ in an infinite plate subjected to a uniform tensile stress σ; (b) the distribution of normal stress ahead of a crack

potential energy of atoms exceeds the bound energy existed between two adjacent atoms. Consider that x_0 stands for the equilibrium space between two atoms subjected to the tension force, the stress value that needs to overcome the atomic bound can be obtained by (Anderson, 1994):

$$\sigma_c = \sqrt{\frac{E\gamma_s}{x_0}} \tag{7.1}$$

where E is the Young module of the material and γ_s is the surface energy per unit area.

From another point of view, fracture of the material can be studied on the macroscopic level, in which the crack growth is modeled in the framework of continuum mechanics. In what follows, the basis of LEFM, including the energy release rate, the evaluation of stress and displacement distributions around the crack tip, and the SIFs in different modes of loading are introduced.

7.2.1 Energy Balance in Crack Propagation

The concept of energy balance in fracture mechanics was first presented by Griffith in 1920. According to this theory, the minimum potential energy of a cracked body can be achieved, when the crack grows and the system stands in equilibrium condition. Consider the total energy of the system can be decomposed into two parts as

$$\mathcal{E} = \mathcal{P} + \mathcal{W}_s \tag{7.2}$$

where $\mathcal{E}$ is the total energy of the system, $\mathcal{P}$ is the potential energy produced by the internal and external forces, and $\mathcal{W}_s$ is the work done to propagate the crack. By taking the derivation from this equation with respect to the increased value of crack surface $d\mathcal{A}$, it leads to

$$-\frac{d\mathcal{P}}{d\mathcal{A}} = \frac{d\mathcal{W}_s}{d\mathcal{A}} \tag{7.3}$$

which denotes that the amount of energy needed to increase the crack length to $d\mathcal{A}$ is equal to the variation of the dropped potential energy with respect to $d\mathcal{A}$. Irwin, Kies, and Smith (1958) introduced the concept of energy release rate $\mathcal{G}$ based on the rate of change in potential energy with respect to the crack area as

$\mathcal{G} = -\partial\mathcal{P}/\partial\mathcal{A}$. Consider a wide plate with the crack length $2a$ and unit thickness, as shown in Figure 7.1, the potential energy of the plate was derived by Inglis (1913) as $\mathcal{P} = \mathcal{P}_0 - \pi\sigma^2 a^2/E$, where $\mathcal{P}_0$ is the potential energy of an uncracked plate, the energy release rate can therefore be obtained as

$$\mathcal{G} = \frac{\pi\sigma^2 a^2}{E} \tag{7.4}$$

in which the energy release rate is a function of the stress imposed on both sides of the crack. As can be physically observed, the crack propagation happens by increasing the tensile stress σ. The amount of stress for crack growth is denoted by σ_c and the critical value of energy release rate obtained from (7.4) is indicated by $\mathcal{G}_c$ that is a measure of the crack toughness of the material, and is related to the length of fracture and the material properties of the domain.

7.2.2 Displacement and Stress Fields at the Crack Tip Area

The importance of stress singularity at the crack tip area was presented by Westergard (1939) and Williams (1957) by evaluation of the displacement and stress fields around the crack tip area. In order to describe the behavior of crack at this region, the loading applied to the fractured body is resolved into three modes, called the opening mode (mode I), sliding or shearing mode (mode II), and tearing mode (mode III), as shown in Figure 7.2. Williams (1957) used the Airy stress function to capture the singularity at the crack tip area using a polar coordinate system (r, θ) as

$$\Phi = r^{\lambda+1}\left(c_1 \sin(\lambda+1)\hat{\theta} + c_2 \cos(\lambda+1)\hat{\theta} + c_3 \sin(\lambda-1)\hat{\theta} + c_4 \sin(\lambda-1)\hat{\theta}\right) \tag{7.5}$$

where c_i are the coefficients and $\hat{\theta}$ is depicted in Figure 7.3.

Substituting the Airy stress function (7.5) in the equilibrium equation of the system, that is, $\nabla^2\nabla^2\Phi = 0$ in the absence of body forces, applying the traction-free boundary conditions at the crack faces, and neglecting the higher order terms, the displacement fields in the vicinity of crack tip can be obtained for mode I loading as

$$\begin{aligned} u_x &= \frac{K_I(1+\nu)}{E}\sqrt{\frac{r}{2\pi}}\cos\frac{\theta}{2}\left(\kappa - 1 + 2\sin^2\frac{\theta}{2}\right) \\ u_y &= \frac{K_I(1+\nu)}{E}\sqrt{\frac{r}{2\pi}}\sin\frac{\theta}{2}\left(\kappa + 1 - 2\cos^2\frac{\theta}{2}\right) \\ u_z &= 0 \end{aligned} \tag{7.6}$$

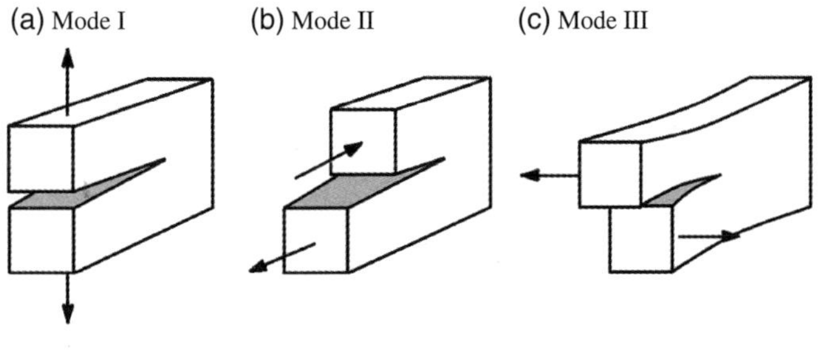

Figure 7.2 The three basic modes of crack extension: (a) Opening mode I; (b) Sliding mode II; (c) Tearing mode III

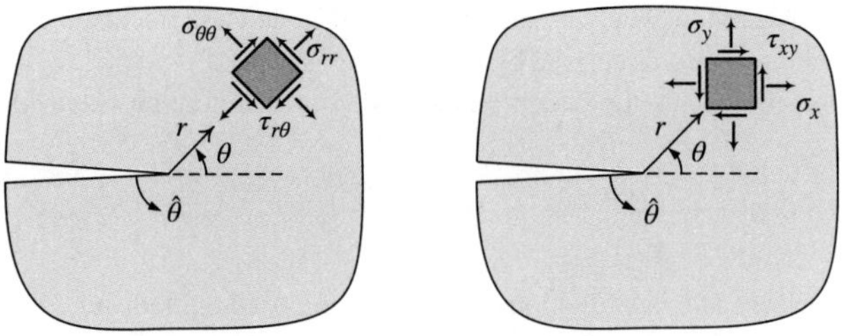

Figure 7.3 The polar coordinate system associated with the crack tip and stress notation

and the stress fields as

$$\begin{aligned}
&\sigma_x = \frac{K_I}{\sqrt{2\pi r}}\cos\frac{\theta}{2}\left(1-\sin\frac{\theta}{2}\sin\frac{3\theta}{2}\right), \quad && \tau_{xy} = \frac{K_I}{\sqrt{2\pi r}}\sin\frac{\theta}{2}\cos\frac{\theta}{2}\cos\frac{3\theta}{2} \\
&\sigma_y = \frac{K_I}{\sqrt{2\pi r}}\cos\frac{\theta}{2}\left(1+\sin\frac{\theta}{2}\sin\frac{3\theta}{2}\right), \quad && \tau_{yz} = 0 \\
&\sigma_z = \begin{cases} \nu\left(\sigma_x+\sigma_y\right) & \text{plane strain} \\ 0 & \text{plane stress} \end{cases}, \quad && \tau_{zx} = 0
\end{aligned} \tag{7.7}$$

In mode II loading, the displacement fields around the crack tip area are obtained as

$$\begin{aligned}
&u_x = \frac{K_{II}(1+\nu)}{E}\sqrt{\frac{r}{2\pi}}\sin\frac{\theta}{2}\left(\kappa+1+2\cos^2\frac{\theta}{2}\right) \\
&u_y = -\frac{K_{II}(1+\nu)}{E}\sqrt{\frac{r}{2\pi}}\cos\frac{\theta}{2}\left(\kappa-1-2\sin^2\frac{\theta}{2}\right) \\
&u_z = 0
\end{aligned} \tag{7.8}$$

and the stress fields as

$$\begin{aligned}
&\sigma_x = -\frac{K_{II}}{\sqrt{2\pi r}}\sin\frac{\theta}{2}\left(2+\cos\frac{\theta}{2}\cos\frac{3\theta}{2}\right), \quad && \tau_{xy} = \frac{K_{II}}{\sqrt{2\pi r}}\cos\frac{\theta}{2}\left(1-\sin\frac{\theta}{2}\sin\frac{3\theta}{2}\right) \\
&\sigma_y = \frac{K_{II}}{\sqrt{2\pi r}}\sin\frac{\theta}{2}\cos\frac{\theta}{2}\cos\frac{3\theta}{2}, \quad && \tau_{yz} = 0 \\
&\sigma_z = \nu\left(\sigma_x+\sigma_y\right), \quad && \tau_{zx} = 0
\end{aligned} \tag{7.9}$$

Finally, in mode III loading, the displacement fields in the vicinity of crack tip are as follows

$$\begin{aligned}
&u_x = 0 \\
&u_y = 0 \\
&u_z = \frac{K_{III}(1+\nu)}{E}\sqrt{\frac{r}{2\pi}}\sin\frac{\theta}{2}
\end{aligned} \tag{7.10}$$

and the stress fields as

$$\begin{aligned}
&\sigma_x = 0\ , \quad \tau_{xy} = 0 \\
&\sigma_y = 0\ , \quad \tau_{yz} = \frac{K_{III}}{\sqrt{2\pi r}}\cos\frac{\theta}{2} \\
&\sigma_z = 0\ , \quad \tau_{zx} = -\frac{K_{III}}{\sqrt{2\pi r}}\sin\frac{\theta}{2}
\end{aligned} \tag{7.11}$$

in which ν is the Poisson ratio and the constant κ is defined as $\kappa = (3-\nu)/(1+\nu)$ for plane stress and $\kappa = 3-4\nu$ for plane strain problems. In these relations, K_I, K_{II}, and K_{III} are the SIFs in modes I, II, and III, respectively, defined as

$$\begin{aligned}
K_I &= \lim_{\substack{r\to 0\\ \theta=0}} \sigma_y\sqrt{2\pi r} \\
K_{II} &= \lim_{\substack{r\to 0\\ \theta=0}} \sigma_{xy}\sqrt{2\pi r} \\
K_{III} &= \lim_{\substack{r\to 0\\ \theta=0}} \sigma_{yz}\sqrt{2\pi r}
\end{aligned} \tag{7.12}$$

It can be seen from relations (7.7) and (7.9) that for the 2D mixed-mode problems the stresses σ_x, σ_y, and τ_{xy} are singular near the crack tip, when $r \to 0$. Finally, the total stress and displacement fields for mixed-mode loading can be obtained as

$$\begin{aligned}
\sigma_{ij}^{\text{total}} &= \sigma_{ij}^{K_I} + \sigma_{ij}^{K_{II}} + \sigma_{ij}^{K_{III}} \\
u_i^{\text{total}} &= u_i^{K_I} + u_i^{K_{II}} + u_i^{K_{III}}
\end{aligned} \tag{7.13}$$

7.2.3 The SIFs

In the linear elastic behavior assumption (LEFM), the stress, strain, and displacement fields can be determined by employing the concept of the SIFs near the crack tip region. It is therefore important to accurately evaluate the SIFs for the FE analysis of LEFM. A number of techniques have been presented in literature for extracting SIFs, including: the displacement correlation method, the virtual crack extension method, the modified crack closure integral, and the *J*–integral method. Basically, the computational algorithms developed to evaluate the SIFs can be categorized into two groups; the "direct" approach and the "energy" approach. The direct approach correlates the SIFs with the FEM results directly, while the energy approach is based on the computation of energy release rate. In general, the energy approaches are more accurate than the direct procedures; however, the direct approaches are more popular and are usually used to verify the results of energy approaches, since their expressions are simple.

The boundary integral and domain integral methods are the most convenient and accurate techniques proposed to compute the SIFs in mixed-mode conditions on the basis of interaction energy integrals. Eshelby (1974) introduced a number of contour integrals based on the theorem of energy conservation, which were path independent. The *J*–integral technique was originally defined by Cherepanow (1967) and Rice (1968) to calculate the energy release rate in crack problems using a local crack tip coordinate system (x_1, x_2) as

$$J = \int_\Gamma \left(w\,\delta_{1j} - \sigma_{ij}\frac{\partial u_i}{\partial x_1} \right) n_j\, d\Gamma \tag{7.14}$$

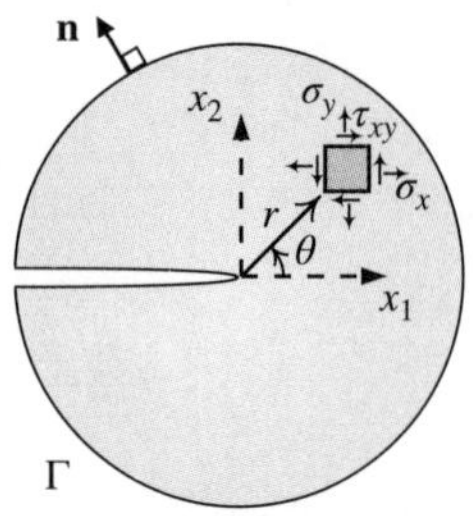

Figure 7.4 A simple circular path for the contour integral around the crack tip

where $\mathcal{W}$ is the strain energy density defined as $\mathcal{W} = \frac{1}{2}\sigma_{ij}\varepsilon_{ij}$ with σ_{ij} denoting the stress tensor, u_i is the displacement field, n_j is the unit outward normal vector to the contour integral Γ, δ is the Kronecker delta, and $T_i = \sigma_{ij}n_j$ is the traction on the contour integral Γ (Figure 7.4).

The contour J–integral is path independent with no body forces and tractions on the crack faces. It is based on the small deformation, elastic material behavior, and quasi-static isothermal conditions. It has been shown that the J–integral defined in relation (7.14) is the variation of potential energy for an infinitesimal virtual crack extension and therefore for linear elastic material, the energy release rate is equal to the value of J–integral (Rice, 1968).

Consider two states of a cracked body, that is, the present state (1) called by $\left(u_i^{(1)}, \varepsilon_{ij}^{(1)}, \sigma_{ij}^{(1)}\right)$ and an auxiliary state (2) denoted by $\left(u_i^{(2)}, \varepsilon_{ij}^{(2)}, \sigma_{ij}^{(2)}\right)$, the J–integral for the sum of states (1) and (2) can be written according to (7.14) as

$$J^{(1+2)} = \int_\Gamma \left(\frac{1}{2}\left(\sigma_{ij}^{(1)} + \sigma_{ij}^{(2)}\right)\left(\varepsilon_{ij}^{(1)} + \varepsilon_{ij}^{(2)}\right)\delta_{1j} - \left(\sigma_{ij}^{(1)} + \sigma_{ij}^{(2)}\right)\frac{\partial}{\partial x_1}\left(u_i^{(1)} + u_i^{(2)}\right)\right) n_j \, d\Gamma \tag{7.15}$$

Expanding this expression results in

$$J^{(1+2)} = J^{(1)} + J^{(2)} + I^{(1,2)} \tag{7.16}$$

where $I^{(1,2)}$ is called the interaction integral for states (1) and (2) defined as

$$I^{(1,2)} = \int_\Gamma \left(\mathcal{W}^{(1,2)}\delta_{1j} - \sigma_{ij}^{(1)}\frac{\partial u_i^{(2)}}{\partial x_1} - \sigma_{ij}^{(2)}\frac{\partial u_i^{(1)}}{\partial x_1}\right) n_j \, d\Gamma \tag{7.17}$$

where $\mathcal{W}^{(1,2)}$ is the interaction strain energy defined by

$$\mathcal{W}^{(1,2)} = \sigma_{ij}^{(1)}\varepsilon_{ij}^{(2)} = \sigma_{ij}^{(2)}\varepsilon_{ij}^{(1)} \tag{7.18}$$

The energy release rate for 2D mixed-mode crack problems can be generally defined based on the modes I and II SIFs K_I and K_{II} as (Rice, 1968)

$$J \equiv \mathcal{G} = \frac{1}{E'}\left(K_I^2 + K_{II}^2\right) \tag{7.19}$$

in which $E' = E$ for plane stress and $E' = E/(1 - \nu^2)$ for plane strain problems. The J–integral definition in (7.19) can be written for the combined states (1) and (2) as

$$J^{(1+2)} = J^{(1)} + J^{(2)} + \frac{2}{E'}\left(K_I^{(1)}K_I^{(2)} + K_{II}^{(2)}K_{II}^{(1)}\right) \tag{7.20}$$

Comparing relation (7.16) with (7.20), the interaction integration $I^{(1,2)}$ can be obtained as

$$I^{(1,2)} = \frac{2}{E'}\left(K_I^{(1)}K_I^{(2)} + K_{II}^{(2)}K_{II}^{(1)}\right) \tag{7.21}$$

If the auxiliary state (2) is assumed as the pure mode I asymptotic fields, that is, $K_I^{(2)} = 1$ and $K_{II}^{(2)} = 0$, the mode I SIF $K_I^{(1)}$ can be stated based on the contour integral $I^{(1)}_{\text{mode } I}$ as

$$K_I^{(1)} = \frac{E'}{2} I^{(1)}_{\text{mode } I} \tag{7.22}$$

and if the auxiliary state (2) is assumed as the pure mode II asymptotic fields, that is, $K_I^{(2)} = 0$ and $K_{II}^{(2)} = 1$, the mode II SIF $K_{II}^{(1)}$ can be obtained as

$$K_{II}^{(1)} = \frac{E'}{2} I^{(1)}_{\text{mode } II} \tag{7.23}$$

The contour *J*–integral (7.14) can be directly evaluated along a contour of FE mesh. This contour can be usually evaluated by passing through the element Gauss integration points, where the stresses are computed more accurately. However, the practical implementation of this technique rarely exhibits path independence and the result becomes mesh dependent. Li, Shih, and Needleman (1985) performed the computation of *J*–integral by transforming the contour integral to an equivalent area integral. This technique is simple to implement in an FE code and the numerical implementation presents a great precision and its independence to the surface of integration. The area form of *J*–integral is defined as

$$J = \int_A \left(\sigma_{ij}\frac{\partial u_i}{\partial x_1} - w\,\delta_{1j}\right)\frac{\partial q}{\partial x_j}\,dA \tag{7.24}$$

where q is a weighting function defined over the domain of integration. The domain of integration must be selected in such a way that it is firstly sufficient close to the crack tip for complex crack pattern, secondly it must be simple to implement in a fully automatic simulation procedure, and finally it needs to be consistent with the geometry and boundary limitations in complex boundaries and multiple crack problems (Figure 7.5). The function q has the value of unity at the crack tip and vanishes on an outer prescribed contour.

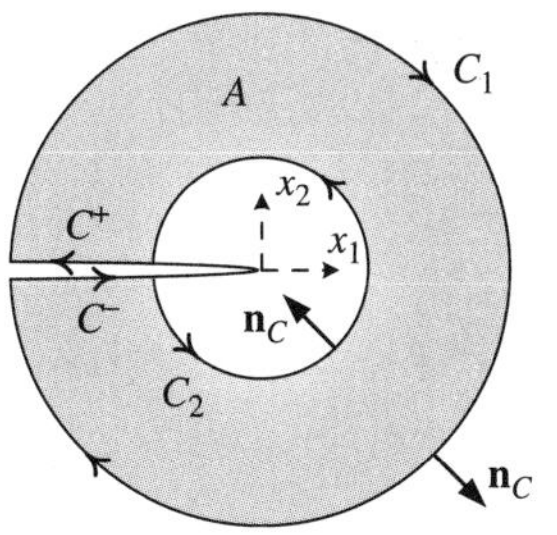

Figure 7.5 The *J*–integral domain for computation of mixed-mode stress intensity factors

Based on the area J–integral definition given in (7.24), the new interaction integration can be defined as

$$I^{(1,2)} = \int_A \left(-w^{(1,2)}\delta_{1j} + \sigma_{ij}^{(1)}\frac{\partial u_i^{(2)}}{\partial x_1} + \sigma_{ij}^{(2)}\frac{\partial u_i^{(1)}}{\partial x_1} \right)\frac{\partial q}{\partial x_j} dA \tag{7.25}$$

The modes I and II SIFs K_I and K_{II} can be finally obtained according to (7.22) and (7.23) by assuming the auxiliary state (2) as a pure mode I and/or a pure mode II asymptotic fields, by evaluating the displacement and stress fields at the crack tip area, and by substituting these variables into relation (7.25).

7.3 Governing Equations of a Cracked Body

In order to derive the governing equation of a solid fractured domain using the X-FEM, consider a 2D domain Ω crossed by the discontinuity Γ_d, as shown in Figure 7.6. The strong form of the equilibrium equation of the cracked body can be expressed as

$$\nabla \cdot \boldsymbol{\sigma} + \mathbf{b} = 0 \tag{7.26}$$

where ∇ is the gradient operator, $\boldsymbol{\sigma}$ is the Cauchy stress tensor, and $\mathbf{b}$ is the body force loading applied to the system. It is assumed that the material has a linear elastic behavior, with the constitutive relation defined as $\boldsymbol{\sigma} = \mathbf{D}\boldsymbol{\varepsilon}$, where $\mathbf{D}$ is the tangential constitutive matrix of the material.

The essential and natural boundary conditions are respectively defined as $\mathbf{u} = \tilde{\mathbf{u}}$ on Γ_u, $\boldsymbol{\sigma} \cdot \mathbf{n}_\Gamma = \bar{\mathbf{t}}$ on Γ_t, and $\boldsymbol{\sigma} \cdot \mathbf{n}_{\Gamma_d} = 0$ on Γ_d, in which $\mathbf{n}_\Gamma$ is the outward unit normal vector to the external boundary Γ, and $\mathbf{n}_{\Gamma_d}$ is the unit normal vector to the discontinuity Γ_d. In addition, $\tilde{\mathbf{u}}$ is the prescribed displacement at the boundary Γ_u, and $\bar{\mathbf{t}}$ is the prescribed traction at the boundary Γ_t.

In the standard FE formulation with conventional elements, in which the polynomial based shape functions are used, it is known that the FE solution initially converges by refining the mesh, but eventually diverges from the exact solution. This is because of the fact that the polynomial shape functions cannot basically capture the theoretical singular behavior predicted for the stress and strain fields. In order to overcome this deficiency, the quarter-point singular element enhancement has been introduced at the crack tip region in the standard FEM, in which the mid-side node of the quadratic order isoparametric elements is moved to its position at one-quarter of the way from crack tip node to the far end of the element. This node repositioning introduces a singularity in the displacement field of element when mapping between the parametric and Cartesian coordinate systems. The implementation of the collapsed quarter-point singular element can improve the frequently observed low convergence rate of the standard FE solution for crack problems. However, in the X-FEM, the discontinuities are not considered in the mesh generation operation and the implementation of the collapsed quarter-point singular element cannot be

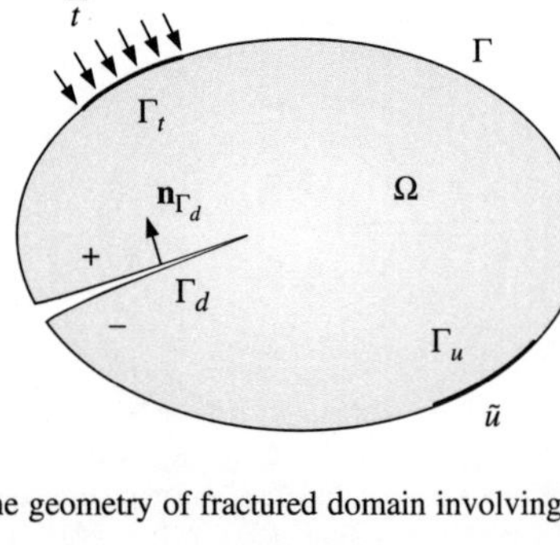

Figure 7.6 The geometry of fractured domain involving a discontinuity Γ_d

applied simply. In such case, special functions, which depend on the nature of discontinuity, must be included into the FE approximation at the crack tip region where the solution shows a singular behavior. In the X-FEM, the asymptotic crack tip functions can be incorporated from the analytical solution to accurately evaluate the values of stresses at the crack tip region. It must be noted that the asymptotic crack tip functions are employed in the elements that contain the crack tip in order to improve the crack tip fields.

7.3.1 The Enrichment of Displacement Field

In order to model the discontinuity of crack interface in the X-FEM, the nodal points of elements cut by the discontinuity are enhanced using the enrichment function $\psi(\boldsymbol{x})$. The enhanced nodal points take additional DOF based on the position of discontinuity in the domain (Figure 7.7). The displacement field of an enriched element can be approximated by

$$\mathbf{u}(\boldsymbol{x},t) = \sum_{I \in \mathcal{N}} N_I(\boldsymbol{x})\,\bar{\mathbf{u}}_I + \sum_{J \in \mathcal{N}^{enr}} N_J(\boldsymbol{x})\,\psi(\boldsymbol{x})\,\bar{\mathbf{a}}_J \tag{7.27}$$

where $\mathcal{N}$ is the set of all nodal points and $\mathcal{N}^{enr}$ is the set of enriched nodal points in the domain, $\bar{\mathbf{u}}_I$ is the standard nodal DOF, $\bar{\mathbf{a}}_J$ is the enriched nodal DOF, $\psi(\boldsymbol{x})$ is the enrichment function, and $N(\boldsymbol{x})$ denotes the standard shape function. It can be highlighted from relation (7.27) that the enhanced shape functions in the X-FEM can be divided into the standard and enriched parts as

$$\mathbf{N}^{enh} = [\mathbf{N}(\boldsymbol{x}),\;\; \mathbf{N}(\boldsymbol{x})\,\psi(\boldsymbol{x})] \tag{7.28}$$

Considering the enriched approximation of displacement field defined by Eq. (7.27), the value of $\mathbf{u}(\boldsymbol{x})$ on an enriched node K in set $\mathcal{N}^{enr}$ can be written as

$$\mathbf{u}(\boldsymbol{x}_K) = \bar{\mathbf{u}}_K + \psi(\boldsymbol{x}_K)\,\bar{\mathbf{a}}_K \tag{7.29}$$

Since $\psi(\boldsymbol{x}_K)$ is not necessarily zero, this expression is not equal to the real nodal value $\bar{\mathbf{u}}_K$. Hence, the enriched displacement field can be corrected to

$$\mathbf{u}(\boldsymbol{x},t) = \sum_{I \in \mathcal{N}} N_I(\boldsymbol{x})\,\bar{\mathbf{u}}_I + \sum_{J \in \mathcal{N}^{enr}} N_J(\boldsymbol{x})\,(\psi(\boldsymbol{x}) - \psi(\boldsymbol{x}_J))\,\bar{\mathbf{a}}_J \tag{7.30}$$

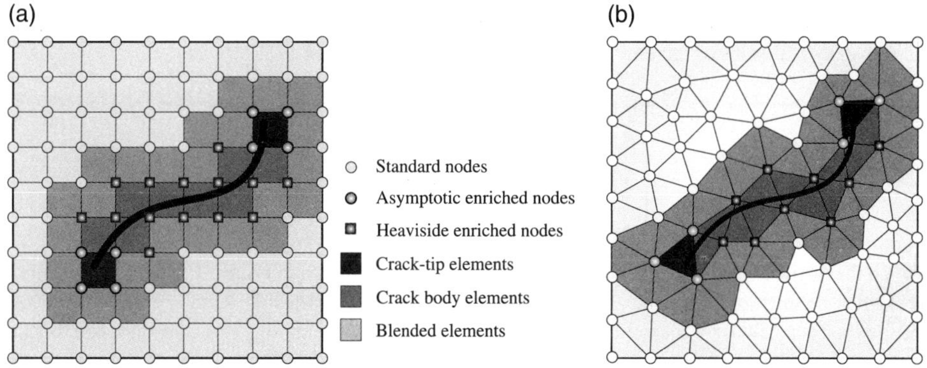

Figure 7.7 The X-FEM modeling of a cracked body with the Heaviside and crack tip asymptotic enrichment functions: (a) Quadrilateral mesh; (b) Triangular mesh

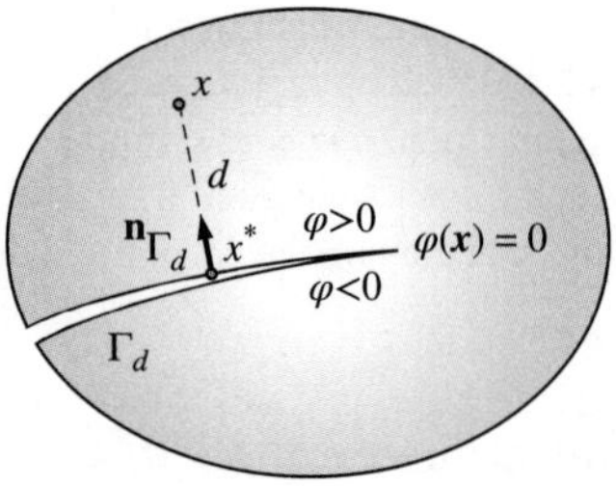

Figure 7.8 The signed distance function

in which the new definition in last term of relation (7.30) results in the expected property, that is, $\mathbf{u}(\boldsymbol{x}_K) = \bar{\mathbf{u}}_K$. In this relation, $\psi(\boldsymbol{x}_J)$ is the value of enriched function at J^{th} enriched nodal point. Hence, the enhanced shape functions (7.28) can be rewritten as

$$\mathbf{N}^{enh} = [\mathbf{N}(\boldsymbol{x}), \mathbf{N}(\boldsymbol{x})\,(\psi(\boldsymbol{x}) - \psi(\boldsymbol{x}_I))] \tag{7.31}$$

The choice of enrichment function $\psi(\boldsymbol{x})$ is dependent on the position of discontinuity in the domain. Generally, there are two types of enrichment functions used for the cracked body in an isotropic LEFM problem; the Heaviside jump function and the asymptotic crack tip function. In order to model the discontinuity due to different displacement fields on either sides of the crack, the Heaviside jump function is used as

$$H(\boldsymbol{x}) = \begin{cases} +1 & \varphi(\boldsymbol{x}) \geq 0 \\ -1 & \varphi(\boldsymbol{x}) < 0 \end{cases} \tag{7.32}$$

where $\varphi(\boldsymbol{x})$ is the signed distance function defined based on the absolute value of level set function as

$$\varphi(\boldsymbol{x}) = \min \|\boldsymbol{x} - \boldsymbol{x}^*\| \operatorname{sign}((\boldsymbol{x} - \boldsymbol{x}^*) \cdot \mathbf{n}_{\Gamma_d}) \tag{7.33}$$

where $\boldsymbol{x}^*$ is a point on the discontinuity which has the closest distance from the point $\boldsymbol{x}$, and $\mathbf{n}_{\Gamma_d}$ is the normal vector to the contact interface at point $\boldsymbol{x}^*$, as shown in Figure 7.8. Because of independent displacement fields at both sides of the crack, the Heaviside jump function is an appropriate function for simulation of the cracked body.

In order to model the displacement field at the crack tip region, the asymptotic functions are extracted from the analytical solution given in relations (7.6), (7.8), and (7.10) as

$$\begin{aligned} \mathcal{B}(r,\theta) &= \{\mathcal{B}_1, \mathcal{B}_2, \mathcal{B}_3, \mathcal{B}_4\} \\ &= \left\{ \sqrt{r}\sin\frac{\theta}{2}, \sqrt{r}\cos\frac{\theta}{2}, \sqrt{r}\sin\frac{\theta}{2}\sin\theta, \sqrt{r}\cos\frac{\theta}{2}\sin\theta \right\} \end{aligned} \tag{7.34}$$

in which the distributions of these four asymptotic functions are shown in Figure 7.9. Obviously, only the first asymptotic function represents the discontinuity near the tip on both sides of the crack while the other three functions can be used to ameliorate the accuracy of the approximation. Thus, the enriched displacement field (7.30) for the cracked element can be rewritten by employing the crack tip asymptotic functions (7.34) as

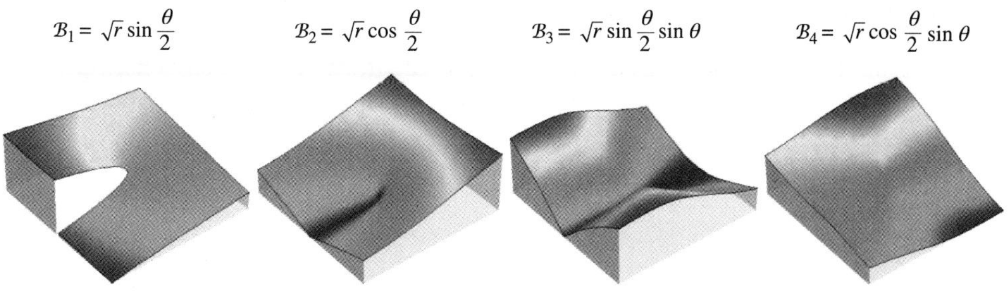

Figure 7.9 Schematic view of asymptotic crack tip functions

$$\begin{aligned}\mathbf{u}(\boldsymbol{x},t) = &\sum_{I\in\mathcal{N}} N_I(\boldsymbol{x})\,\bar{\mathbf{u}}_I + \sum_{J\in\mathcal{N}^{dis}} N_J(\boldsymbol{x})\,(H(\boldsymbol{x})-H(\boldsymbol{x}_J))\,\bar{\mathbf{d}}_J \\ &+ \sum_{K\in\mathcal{N}^{tip}} N_K(\boldsymbol{x})\sum_{\alpha=1}^{4}(\mathcal{B}_\alpha(\boldsymbol{x})-\mathcal{B}_\alpha(\boldsymbol{x}_K))\,\bar{\mathbf{b}}_{\alpha K}\end{aligned} \tag{7.35}$$

where $\mathcal{N}$ is the set of all nodal points, $\mathcal{N}^{dis}$ is the set of enriched nodes whose support is bisected by the crack, and $\mathcal{N}^{tip}$ is the set of nodes which contain the crack tip in the support of their shape functions enriched by the asymptotic functions. In this relation, $\bar{\mathbf{u}}_I$ are the unknown standard nodal DOF at I^{th} node, $\bar{\mathbf{d}}_J$ are the unknown enriched nodal DOF associated with the Heaviside enrichment function at node J, and $\bar{\mathbf{b}}_{\alpha K}$ are the additional enriched nodal DOF associated with the asymptotic functions at node K.

Finally, the extended FE approximation of the displacement field (7.35) can be summarized as

$$\begin{aligned}\mathbf{u}(\boldsymbol{x},t) &= \mathbf{N}^{std}(\boldsymbol{x})\,\bar{\mathbf{u}}(t) + \mathbf{N}^{Hev}(\boldsymbol{x})\,\bar{\mathbf{d}}(t) + \mathbf{N}^{tip}(\boldsymbol{x})\,\bar{\mathbf{b}}(t) \\ &\equiv \mathbf{N}^{std}(\boldsymbol{x})\,\bar{\mathbf{u}}(t) + \mathbf{N}^{enr}(\boldsymbol{x})\,\bar{\mathbf{a}}(t)\end{aligned} \tag{7.36}$$

where $\mathbf{N}^{std}(\boldsymbol{x}) \equiv \mathbf{N}(\boldsymbol{x})$ are the matrix of standard FE shape functions and $\mathbf{N}^{enr} = [\mathbf{N}^{Hev}(\boldsymbol{x}),\ \mathbf{N}^{tip}(\boldsymbol{x})]$ are referred to as the matrix of enriched shape functions associated with the Heaviside enrichment function and the crack tip asymptotic functions.

Accordingly, the strain vector corresponding to the approximate displacement field (7.36) can be written in terms of the standard and enriched nodal values as

$$\begin{aligned}\boldsymbol{\varepsilon}(\boldsymbol{x},t) &= \mathbf{B}^{std}(\boldsymbol{x})\,\bar{\mathbf{u}}(t) + \mathbf{B}^{Hev}(\boldsymbol{x})\,\bar{\mathbf{d}}(t) + \mathbf{B}^{tip}(\boldsymbol{x})\,\bar{\mathbf{b}}(t) \\ &\equiv \mathbf{B}^{std}(\boldsymbol{x})\,\bar{\mathbf{u}}(t) + \mathbf{B}^{enr}(\boldsymbol{x})\,\bar{\mathbf{a}}(t)\end{aligned} \tag{7.37}$$

where $\mathbf{B}^{std}(\boldsymbol{x}) \equiv \mathbf{L}\,\mathbf{N}^{std}(\boldsymbol{x})$ involve the spatial derivatives of the standard shape functions and $\mathbf{B}^{enr}(\boldsymbol{x}) \equiv \mathbf{L}\,\mathbf{N}^{enr}(\boldsymbol{x})$ contain the spatial derivatives of the enriched shape functions associated with the Heaviside enrichment function as $\mathbf{B}^{Hev}(\boldsymbol{x}) \equiv \mathbf{L}\,\mathbf{N}^{Hev}(\boldsymbol{x})$ and the crack tip asymptotic functions as $\mathbf{B}^{tip}(\boldsymbol{x}) \equiv \mathbf{L}\,\mathbf{N}^{tip}(\boldsymbol{x})$, with $\mathbf{L}$ denoting the matrix differential operator defined as

$$\mathbf{L} = \begin{bmatrix} \partial/\partial x & 0 \\ 0 & \partial/\partial y \\ \partial/\partial y & \partial/\partial x \end{bmatrix} \tag{7.38}$$

7.3.2 Discretization of Governing Equations

The weak form of the governing equation (7.26) can be derived based on the Galerkin discretization technique by integrating the product of the equilibrium equation multiplied by admissible test function over the domain. The test function $\delta\mathbf{u}(\boldsymbol{x}, t)$ can be considered in the same approximating space as the displacement field $\mathbf{u}(\boldsymbol{x}, t)$ defined in (7.36) as

$$\delta\mathbf{u}(\boldsymbol{x},t) = \mathbf{N}^{std}(\boldsymbol{x})\,\delta\bar{\mathbf{u}}(t) + \mathbf{N}^{Hev}(\boldsymbol{x})\,\delta\bar{\mathbf{d}}(t) + \mathbf{N}^{tip}(\boldsymbol{x})\,\delta\bar{\mathbf{b}}(t) \tag{7.39}$$

Applying the Galerkin procedure, the weak form of the equilibrium equation (7.26) can be written as

$$\int_{\Omega} \delta\mathbf{u}(\boldsymbol{x},t)(\nabla\cdot\boldsymbol{\sigma} + \mathbf{b})\,d\Omega = 0 \tag{7.40}$$

Applying the Divergence theorem, imposing the natural boundary conditions, and satisfying the traction-free boundary condition on the discontinuity surface, the integral equation (7.40) can be obtained as

$$\int_{\Omega} \nabla\delta\mathbf{u} : \boldsymbol{\sigma}\, d\Omega + \int_{\Gamma_d} [\![\delta\mathbf{u}]\!]\, \boldsymbol{\sigma}\cdot\mathbf{n}_{\Gamma_d}\, d\Gamma = \int_{\Gamma_t} \delta\mathbf{u}\cdot\bar{\mathbf{t}}\, d\Gamma + \int_{\Omega} \delta\mathbf{u}\cdot\mathbf{b}\, d\Omega \tag{7.41}$$

where the symbol $[\![\ ]\!]$ denotes the jump across the crack discontinuity, which is the difference between the corresponding values at the two crack faces, that is, $[\![\Xi]\!] = \Xi^+ - \Xi^-$. It must be noted that the second integral in the left hand side of Eq. (7.41) on the discontinuity Γ_d can be eliminated from the integral equation assigning the positive and negative sides to Γ_d and imposing the traction free boundary condition on the crack faces, that is, $\boldsymbol{\sigma}\cdot\mathbf{n}_{\Gamma_d} = 0$ on Γ_d, as

$$\begin{aligned} -\int_{\Gamma_d^+} \delta\mathbf{u}\left(\boldsymbol{\sigma}\cdot\mathbf{n}_{\Gamma_d^+}\right) d\Gamma - \int_{\Gamma_d^-} \delta\mathbf{u}\left(\boldsymbol{\sigma}\cdot\mathbf{n}_{\Gamma_d^-}\right) d\Gamma \\ = \int_{\Gamma_d} (\delta\mathbf{u}^+ - \delta\mathbf{u}^-)\boldsymbol{\sigma}\cdot\mathbf{n}_{\Gamma_d}\, d\Gamma = \int_{\Gamma_d} [\![\delta\mathbf{u}]\!]\boldsymbol{\sigma}\cdot\mathbf{n}_{\Gamma_d}\, d\Gamma = 0 \end{aligned} \tag{7.42}$$

where $\mathbf{n}_{\Gamma_d^+}$ and $\mathbf{n}_{\Gamma_d^-}$ are the unit normal vectors directed to Ω^- and Ω^+, respectively, as depicted in Figure 7.6, and the superscripts + and – above Γ_d represent the two sides of the discontinuity, in which $\mathbf{n}_{\Gamma_d}$ is defined as $\mathbf{n}_{\Gamma_d} = \mathbf{n}_{\Gamma_d^-} = -\mathbf{n}_{\Gamma_d^+}$.

The integral equation (7.41) can therefore be written as

$$\int_{\Omega} \nabla\delta\mathbf{u} : \boldsymbol{\sigma}\, d\Omega = \int_{\Gamma_t} \delta\mathbf{u}\cdot\bar{\mathbf{t}}\, d\Gamma + \int_{\Omega} \delta\mathbf{u}\cdot\mathbf{b}\, d\Omega \tag{7.43}$$

Applying the extended FE discretization method and using the enhanced test function $\delta\mathbf{u}(\boldsymbol{x}, t)$ defined in (7.39), the earlier equation results in

$$\int_{\Omega} \bar{\mathbf{B}}^T \boldsymbol{\sigma}\, d\Omega = \int_{\Gamma_t} \bar{\mathbf{N}}^T \bar{\mathbf{t}}\, d\Gamma + \int_{\Omega} \bar{\mathbf{N}}^T \mathbf{b}\, d\Omega \tag{7.44}$$

where $\bar{\mathbf{B}} = \left[\mathbf{B}^{std}, \mathbf{B}^{Hev}, \mathbf{B}^{tip}\right]$ and $\bar{\mathbf{N}} = \left[\mathbf{N}^{std}, \mathbf{N}^{Hev}, \mathbf{N}^{tip}\right]$. The discrete system of equations can be obtained from (7.44) as $\mathbf{K}\bar{\mathbf{U}} - \mathbf{F} = 0$, where $\bar{\mathbf{U}}^T = \left[\bar{\mathbf{u}}^T, \bar{\mathbf{d}}^T, \bar{\mathbf{b}}^T\right]$ is the vector of unknowns at the nodal points, $\mathbf{K}$ is the total stiffness matrix, and $\mathbf{F}$ is the external force vector. The discrete system of equations can be finally obtained as

$$\begin{bmatrix} \mathbf{K}_{uu} & \mathbf{K}_{ud} & \mathbf{K}_{ub} \\ \mathbf{K}_{du} & \mathbf{K}_{dd} & \mathbf{K}_{db} \\ \mathbf{K}_{bu} & \mathbf{K}_{bd} & \mathbf{K}_{bb} \end{bmatrix} \begin{Bmatrix} \bar{\mathbf{u}} \\ \bar{\mathbf{d}} \\ \bar{\mathbf{b}} \end{Bmatrix} = \begin{Bmatrix} \mathbf{F}_u \\ \mathbf{F}_d \\ \mathbf{F}_b \end{Bmatrix} \tag{7.45}$$

where

$$\mathbf{K} = \begin{bmatrix} \int_\Omega (\mathbf{B}^{std})^T \mathbf{D} \mathbf{B}^{std} d\Omega & \int_\Omega (\mathbf{B}^{std})^T \mathbf{D} \mathbf{B}^{Hev} d\Omega & \int_\Omega (\mathbf{B}^{std})^T \mathbf{D} \mathbf{B}^{tip} d\Omega \\ \int_\Omega (\mathbf{B}^{Hev})^T \mathbf{D} \mathbf{B}^{std} d\Omega & \int_\Omega (\mathbf{B}^{Hev})^T \mathbf{D} \mathbf{B}^{Hev} d\Omega & \int_\Omega (\mathbf{B}^{Hev})^T \mathbf{D} \mathbf{B}^{tip} d\Omega \\ \int_\Omega (\mathbf{B}^{tip})^T \mathbf{D} \mathbf{B}^{std} d\Omega & \int_\Omega (\mathbf{B}^{tip})^T \mathbf{D} \mathbf{B}^{Hev} d\Omega & \int_\Omega (\mathbf{B}^{tip})^T \mathbf{D} \mathbf{B}^{tip} d\Omega \end{bmatrix} \tag{7.46}$$

and

$$\mathbf{F} = \begin{Bmatrix} \int_{\Gamma_t} (\mathbf{N}^{std})^T \bar{\mathbf{t}}\, d\Gamma + \int_\Omega (\mathbf{N}^{std})^T \mathbf{b}\, d\Omega \\ \int_{\Gamma_t} (\mathbf{N}^{Hev})^T \bar{\mathbf{t}}\, d\Gamma + \int_\Omega (\mathbf{N}^{Hev})^T \mathbf{b}\, d\Omega \\ \int_{\Gamma_t} (\mathbf{N}^{tip})^T \bar{\mathbf{t}}\, d\Gamma + \int_\Omega (\mathbf{N}^{tip})^T \mathbf{b}\, d\Omega \end{Bmatrix} \tag{7.47}$$

where $\mathbf{N}_I^{std} = N_I\,\mathbf{I}$, $\mathbf{N}_I^{Hev} = N_I(H(\boldsymbol{x}) - H(\boldsymbol{x}_I))\,\mathbf{I}$, and $\mathbf{N}_{\alpha I}^{tip} = N_I(\mathcal{B}_\alpha(\boldsymbol{x}) - \mathcal{B}_\alpha(\boldsymbol{x}_I))\,\mathbf{I}$, with $\mathbf{I}$ denoting a 2×2 identity matrix, and the matrices $\mathbf{B}^{std}$, $\mathbf{B}^{Hev}$, and $\mathbf{B}^{tip}$ can be defined for the X-FEM formulation using the enrichment displacement field for node I as

$$\mathbf{B}_I^{std} = \begin{bmatrix} \partial N_I/\partial x & 0 \\ 0 & \partial N_I/\partial y \\ \partial N_I/\partial y & \partial N_I/\partial x \end{bmatrix} \tag{7.48a}$$

$$\mathbf{B}_I^{Hev} = \begin{bmatrix} \partial(N_I(H(\boldsymbol{x}) - H(\boldsymbol{x}_I)))/\partial x & 0 \\ 0 & \partial(N_I(H(\boldsymbol{x}) - H(\boldsymbol{x}_I)))/\partial y \\ \partial(N_I(H(\boldsymbol{x}) - H(\boldsymbol{x}_I)))/\partial y & \partial(N_I(H(\boldsymbol{x}) - H(\boldsymbol{x}_I)))/\partial x \end{bmatrix} \tag{7.48b}$$

$$\mathbf{B}_{\alpha I}^{tip} = \begin{bmatrix} \partial(N_I(\mathcal{B}_\alpha(\boldsymbol{x}) - \mathcal{B}_\alpha(\boldsymbol{x}_I)))/\partial x & 0 \\ 0 & \partial(N_I(\mathcal{B}_\alpha(\boldsymbol{x}) - \mathcal{B}_\alpha(\boldsymbol{x}_I)))/\partial y \\ \partial(N_I(\mathcal{B}_\alpha(\boldsymbol{x}) - \mathcal{B}_\alpha(\boldsymbol{x}_I)))/\partial y & \partial(N_I(\mathcal{B}_\alpha(\boldsymbol{x}) - \mathcal{B}_\alpha(\boldsymbol{x}_I)))/\partial x \end{bmatrix}_{(\alpha=1,2,3,4)} \tag{7.48c}$$

where the derivative of Heaviside enrichment function in (7.48b) can be simply obtained as

$$\frac{\partial}{\partial \boldsymbol{x}}(N_I(H(\boldsymbol{x}) - H(\boldsymbol{x}_I))) = \frac{\partial N_I}{\partial \boldsymbol{x}}(H(\boldsymbol{x}) - H(\boldsymbol{x}_I)) + N_I \frac{\partial H(\boldsymbol{x})}{\partial \boldsymbol{x}} \tag{7.49}$$

and the derivative of crack tip asymptotic functions in (7.48c) can be performed as

$$\frac{\partial}{\partial \boldsymbol{x}}(N_I(\mathcal{B}_\alpha(\boldsymbol{x}) - \mathcal{B}_\alpha(\boldsymbol{x}_I))) = \frac{\partial N_I}{\partial \boldsymbol{x}}(\mathcal{B}_\alpha(\boldsymbol{x}) - \mathcal{B}_\alpha(\boldsymbol{x}_I)) + N_I \frac{\partial \mathcal{B}_\alpha(\boldsymbol{x})}{\partial \boldsymbol{x}} \tag{7.50}$$

in which $\partial \mathcal{B}_\alpha(\boldsymbol{x})/\partial \boldsymbol{x}$ can be first obtained using a transformation between the polar and Cartesian coordinates in a local Cartesian coordinate system (x_1, x_2) as

$$\begin{aligned}\frac{\partial \mathcal{B}_\alpha}{\partial x_1} &= \frac{\partial \mathcal{B}_\alpha}{\partial r}\frac{\partial r}{\partial x_1} + \frac{\partial \mathcal{B}_\alpha}{\partial \theta}\frac{\partial \theta}{\partial x_1} \\ \frac{\partial \mathcal{B}_\alpha}{\partial x_2} &= \frac{\partial \mathcal{B}_\alpha}{\partial r}\frac{\partial r}{\partial x_2} + \frac{\partial \mathcal{B}_\alpha}{\partial \theta}\frac{\partial \theta}{\partial x_2}\end{aligned} \tag{7.51}$$

where $\partial r/\partial x_1 = \cos\theta$, $\partial r/\partial x_2 = \sin\theta$, $\partial \theta/\partial x_1 = -\frac{1}{r}\sin\theta$, and $\partial \theta/\partial x_2 = \frac{1}{r}\cos\theta$. Hence, the local derivatives of the crack tip asymptotic functions $\mathcal{B}(r,\theta)$ defined in (7.34) can be obtained using relations (7.51) as

$$\begin{aligned}
&\frac{\partial \mathcal{B}_1}{\partial x_1} = -\frac{1}{2\sqrt{r}}\sin\frac{\theta}{2}, &&\frac{\partial \mathcal{B}_1}{\partial x_2} = \frac{1}{2\sqrt{r}}\cos\frac{\theta}{2} \\
&\frac{\partial \mathcal{B}_2}{\partial x_1} = \frac{1}{2\sqrt{r}}\cos\frac{\theta}{2}, &&\frac{\partial \mathcal{B}_2}{\partial x_2} = \frac{1}{2\sqrt{r}}\sin\frac{\theta}{2} \\
&\frac{\partial \mathcal{B}_3}{\partial x_1} = -\frac{1}{2\sqrt{r}}\sin\frac{3\theta}{2}\sin\theta, &&\frac{\partial \mathcal{B}_3}{\partial x_2} = \frac{1}{2\sqrt{r}}\left(\sin\frac{\theta}{2} + \sin\frac{3\theta}{2}\cos\theta\right) \\
&\frac{\partial \mathcal{B}_4}{\partial x_1} = -\frac{1}{2\sqrt{r}}\cos\frac{3\theta}{2}\sin\theta, &&\frac{\partial \mathcal{B}_4}{\partial x_2} = \frac{1}{2\sqrt{r}}\left(\cos\frac{\theta}{2} + \cos\frac{3\theta}{2}\cos\theta\right)
\end{aligned} \tag{7.52}$$

Finally, the derivatives of crack tip asymptotic functions with respect to the global coordinate system (x, y) can be obtained by

$$\begin{aligned}\frac{\partial \mathcal{B}_\alpha}{\partial x} &= \frac{\partial \mathcal{B}_\alpha}{\partial x_1}\cos\alpha - \frac{\partial \mathcal{B}_\alpha}{\partial x_2}\sin\alpha \\ \frac{\partial \mathcal{B}_\alpha}{\partial y} &= \frac{\partial \mathcal{B}_\alpha}{\partial x_1}\sin\alpha + \frac{\partial \mathcal{B}_\alpha}{\partial x_2}\cos\alpha\end{aligned} \tag{7.53}$$

in which α indicates the angle of crack with respect to the global coordinate system, as shown in Figure 7.10.

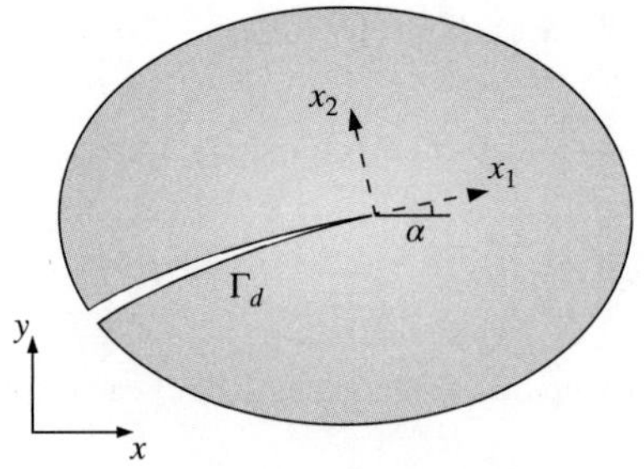

Figure 7.10 The local and global coordinate systems at the crack tip

7.4 Mixed-Mode Crack Propagation Criteria

In crack propagation problems, there are two main requirements at each time step. It must be first determined whether the crack propagates or not; and, if so, in which direction it propagates. Based on these two requirements, two criteria must be utilized; one for crack propagation and the other for crack kinking. The crack propagation criteria may be a function of the SIFs, the strain energy release rate, the strain energy density, and so on. The crack kinking criteria determines the direction of the crack and can be determined based on the fracture toughness of brittle material, which is usually measured in a pure mode I loading conditions noted by K_{IC}. For a general mixed-mode case, a criterion is needed to determine the angle of incipient propagation with respect to crack direction, and a critical combination of SIFs that leads to crack propagation. The crack kinking criteria can be generally classified into two categories for brittle materials. The first group is based on the local fields at the crack tip following a local approach, such as the maximum circumferential stress criterion, or the minimum strain energy density criterion. The second group is based on the energy distribution throughout the cracked part, following a global approach, such as the maximum strain energy release rate criterion. In these two criteria, the accuracy of computed direction is directly dependent on the accuracy of the numerical computation of local or energetic parameters. In what follows, three of the most widely used criteria for the mixed-mode crack growth, that is, the maximum circumferential tensile stress criterion, the minimum strain energy density criterion, and the maximum energy release rate, are briefly presented.

7.4.1 The Maximum Circumferential Tensile Stress Criterion

The maximum circumferential tensile stress theory was first presented by Erdogan and Sih (1963) based on the state of stress near the crack tip. Based on this theory, the crack propagates at the crack tip in a radial direction on the plane perpendicular to the direction of maximum tension, when $(\sigma_\theta)_{\max}$ reaches a critical material constant. In this case, the hoop stress reaches its maximum value on the plane of zero shear stress. Assuming that the size of plastic zone at the crack tip is negligible, the singular term solutions of stress at the crack tip can be used to determine the crack propagation angle, where the shear stress becomes zero. Considering the mixed-mode loading conditions, the asymptotic crack tip circumferential and shear stress can be obtained using relations (7.7) and (7.9) by transforming the stress values to a polar coordinate system as

$$\begin{Bmatrix} \sigma_{\theta\theta} \\ \sigma_{r\theta} \end{Bmatrix} = \frac{K_I}{4\sqrt{2\pi r}} \begin{Bmatrix} 3\cos\frac{\theta}{2} + \cos\frac{3\theta}{2} \\ \sin\frac{\theta}{2} + \sin\frac{3\theta}{2} \end{Bmatrix} + \frac{K_{II}}{4\sqrt{2\pi r}} \begin{Bmatrix} -3\sin\frac{\theta}{2} - 3\sin\frac{3\theta}{2} \\ \cos\frac{\theta}{2} + 3\cos\frac{3\theta}{2} \end{Bmatrix} \tag{7.54}$$

The critical angle of crack propagation θ_c can be determined by setting the shear stress $\sigma_{r\theta}$ to zero, which leads to the following equation

$$\frac{1}{\sqrt{2\pi r}}\cos\frac{\theta}{2}\left[\frac{1}{2}K_I\sin\theta + \frac{1}{2}K_{II}(3\cos\theta - 1)\right] = 0 \tag{7.55}$$

The solution of this equation results in the crack propagation angle θ_c that can be expressed using the angle between the line of crack and the crack growth direction, with the positive value defined in an anti-clockwise direction, as

$$\theta_c = 2\arctan\frac{1}{4}\left(\frac{K_I}{K_{II}} \pm \sqrt{\left(\frac{K_I}{K_{II}}\right)^2 + 8}\right) \tag{7.56}$$

in which $K_{II}=0$ results in a pure mode I condition, that is, $\theta_c=0$. If $K_{II}>0$, the crack growth direction $\theta_c<0$, and if $K_{II}<0$, the crack growth direction $\theta_c>0$. An efficient expression of the critical angle of crack propagation can also be given as

$$\theta_c = 2\arctan\left[\frac{-2K_{II}/K_I}{1+\sqrt{1+8(K_{II}/K_I)^2}}\right] \tag{7.57}$$

7.4.2 *The Minimum Strain Energy Density Criterion*

The minimum strain energy density criterion was first introduced by Sih (1974). Based on this criterion, the crack propagates in the direction, where the material has its minimum strength. In this method, the critical angle of crack propagation θ_c is computed in the direction that the strain energy density at a critical distance reaches a minimum value. The strain energy density is defined as

$$S = rw = \frac{r}{4\mu}\left[\frac{\kappa+1}{4}(\sigma_x+\sigma_y)^2 - 2\left(\sigma_x\sigma_y-\tau_{xy}\right)^2\right] \tag{7.58}$$

where μ is the shear module. For plane stress problem $\kappa=(3-\nu)/(1+\nu)$ and plane strain problem $\kappa=3-4\nu$. Substituting the near tip stress values given in (7.7) and (7.9) into this expression, after some manipulations it results in

$$\frac{8\mu}{(\kappa-1)}\left[a_{11}\left(\frac{K_I}{K_{Ic}}\right)^2 + 2a_{12}\left(\frac{K_IK_{II}}{K_{Ic}}\right) + a_{22}\left(\frac{K_{II}}{K_{Ic}}\right)^2\right] = 1 \tag{7.59}$$

where K_{Ic} is the critical value of SIF, and the constants a_{11}, a_{12}, and a_{22} are defined as

$$\begin{aligned} a_{11} &= \frac{1}{16\mu}[(1+\cos\theta)(\kappa-\cos\theta)] \\ a_{12} &= \frac{1}{16\mu}\sin\theta(2\cos\theta-\kappa+1) \\ a_{22} &= \frac{1}{16\mu}[(\kappa+1)(1-\cos\theta)+(1+\cos\theta)(3\cos\theta-1)] \end{aligned} \tag{7.60}$$

The critical value of crack propagation angle θ_c can be obtained by minimizing relation (7.59) with respect to θ.

7.4.3 *The Maximum Energy Release Rate*

Erdogan and Sih (1963) proposed the Griffith theory as a valid criterion for crack growth, and showed that the crack propagates in the direction where the elastic energy release rate per unit crack extension becomes a maximum value. In this case, the crack grows when the energy release reaches a critical value. For the SIFs of a major crack with the infinitesimal kink at an angle θ, Hussain, Pu, and Underwood (1974) evaluated $K_I(\theta)$ and $K_{II}(\theta)$ in terms of the SIFs of the original crack K_I and K_{II} (Figure 7.11) as

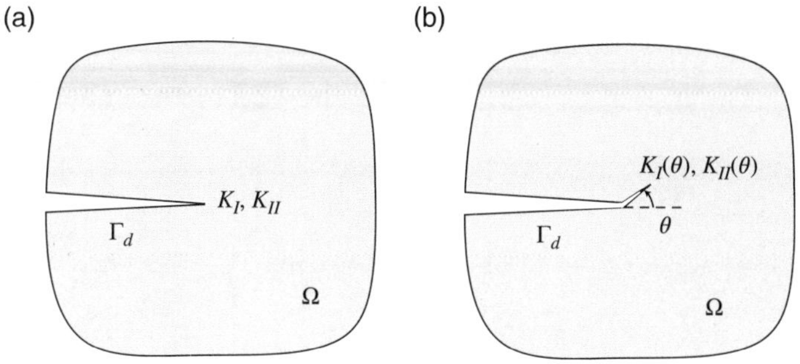

Figure 7.11 The crack propagation with infinitesimal kink at an angle θ: (a) the initial crack; (b) the kinked crack

$$K_I(\theta) = g(\theta)\left(K_I\cos\theta + \frac{3}{2}K_{II}\sin\theta\right)$$
$$K_{II}(\theta) = g(\theta)\left(K_I\cos\theta - \frac{3}{2}K_{II}\sin\theta\right) \tag{7.61}$$

where $g(\theta)$ is defined as

$$g(\theta) = \left(\frac{4}{3+\cos^2\theta}\right)\left(\frac{1-\theta/\pi}{1+\theta/\pi}\right)^{\frac{\theta}{2\pi}} \tag{7.62}$$

Substituting $K_I(\theta)$ and $K_{II}(\theta)$ into the Irwin's generalized expression, the energy release rate for the linear crack growth at the kinked angle θ can be written according to (7.19) as

$$\mathcal{G}(\theta) = \frac{1}{E'}\left(K_I^2(\theta) + K_{II}^2(\theta)\right) \tag{7.63}$$

Substituting relations (7.61) into (7.63), it results in

$$\mathcal{G}(\theta) = \frac{1}{4E'}g^2(\theta)\left[\left(1+3\cos^2\theta\right)K_I^2 + 8\sin\theta\cos\theta K_I K_{II} + \left(9-5\cos^2\theta\right)K_{II}^2\right] \tag{7.64}$$

The angle of propagation can be obtained by maximizing $\mathcal{G}(\theta)$, that is,

$$\frac{d\mathcal{G}}{d\theta} = 0 \quad \text{and} \quad \frac{d^2\mathcal{G}}{d^2\theta} < 0 \tag{7.65}$$

Substituting $K_{II} = 0$ and $\mathcal{G}_{cr} = (K_I)_{cr}^2/E'$ in Eq. (7.65), it results in the relationship between two modes of fracture.

7.5 Crack Growth Simulation with X-FEM

The crack growth simulation is one of the most cumbersome tasks in the FE framework. In the standard FEM, since the crack evolves through the FE mesh, the solution algorithm must be updated at each crack growth by modifying the FE mesh in new configuration. However, in the extended FE modeling, the crack

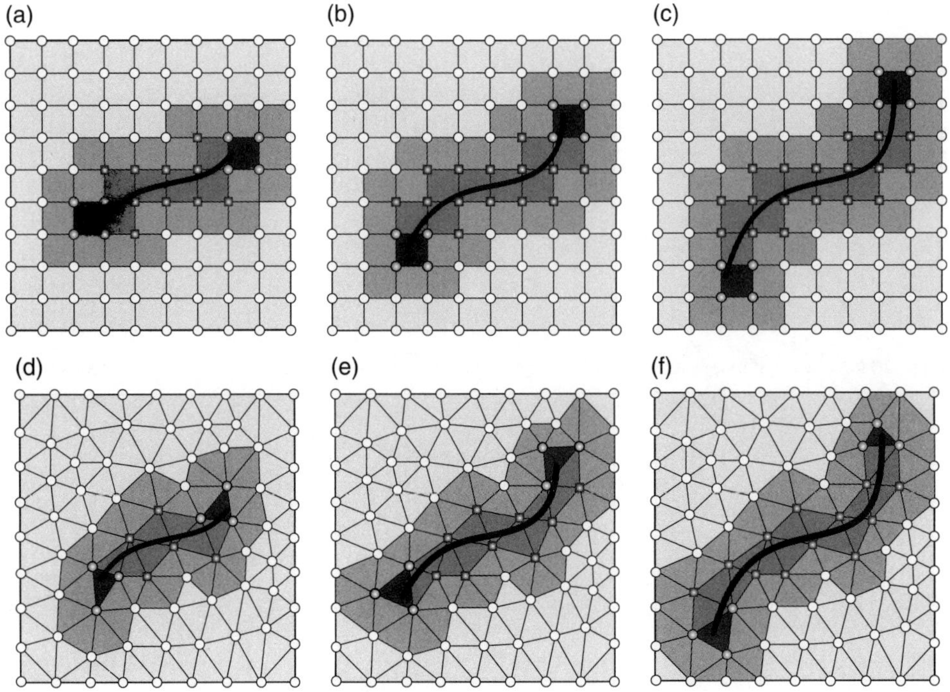

Figure 7.12 The enriched nodal points during crack propagation: The squared nodes are enriched by the Heaviside function whereas the circled nodes are enriched by the crack tip asymptotic functions: (a–c) Quadrilateral meshes; (d–f) Triangular meshes

geometry is modeled independent of the FE mesh by enriching the nodal points of elements that are intersected by the crack during crack propagation. In crack growth simulation, the crack propagates with a predefined value of Δl if the crack propagation criterion is satisfied. There are basically two different techniques proposed for the enrichment of the nodal points at crack tip area based on the " crack tip enrichment element" and the " fixed enrichment area". In the crack tip enrichment element scheme, the enrichment is performed according to the current position of the crack tip at each time step. According to Figure 7.12, the nodal points of elements that are cut by the crack are enriched by the Heaviside step function, as indicated with the squared points; while the crack tip asymptotic functions are used to enrich the crack tip nodal points, as shown with the circled points. Based on these enrichments, the simulation is performed to obtain the stress and displacement fields at the crack tip region, in order to indicate its position in the next step of crack growth. If the new crack tip position is in the area of former element, no update is necessary for the enriched elements; however, if the new crack segment crosses the next element, the enrichment of nodal points must be updated, as shown in Figure 7.12. The simulation can then be carried out according to the new enriched elements based on the new configuration of crack propagation. It must be noted that this method needs a high accurate recognition function to diagnose the position of new crack tip elements.

In the "fixed enrichment area" scheme, an area of enrichment is assumed at the crack tip region in order to improve the convergence rate in the crack growth simulation. It was shown by Laborde *et al.* (2005) that in the standard crack tip enrichment element scheme, which was used to enrich only the nodal points of the tip element, the support of the enriched functions vanishes when the element size goes to zero. In fact, decreasing the element size at the crack tip area with a single enriched tip-element has a significant effect on the convergence rate of the solution. In order to overcome this problem, a fixed enrichment area is

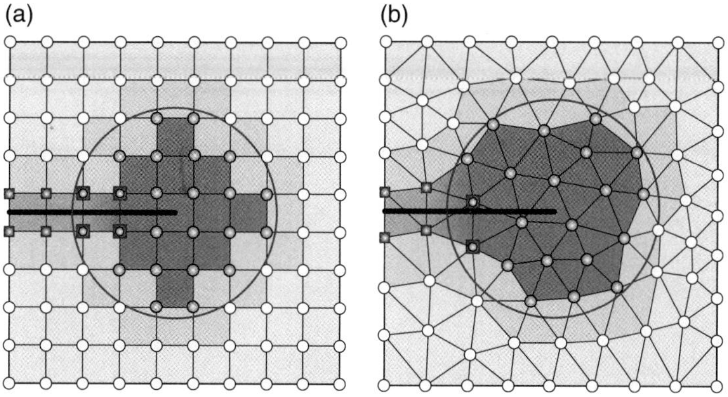

Figure 7.13 The fixed enrichment area scheme used at the crack tip region to enhance the nodal points inserted in the circle by crack tip branch functions: (a) Quadrilateral mesh; (b) Triangular mesh

assumed at the crack tip region independent of the element mesh size to enhance the elements of this specific area. Based on this technique, a virtual circle is considered around the crack tip at position x_{tip} with the radius R at each step. The radius of this circle is assumed to be equal of 0.1^{th} of the domain dimension. All nodal points inserted in this circle are enriched by the crack tip branch functions, as shown in Figure 7.13; these nodal points are indicated with the circled points. The Heaviside jump function is yet used to enhance the nodal points of elements intersected by the crack and are outside the circle on both sides of the crack, as displayed by the squared points. Of course, there are some nodal points in this approach that are enhanced using both the crack tip and Heaviside jump functions, as indicated in Figure 7.13.

According to the "fixed enrichment area" scheme, the enriched displacement field (7.35) can be rewritten as

$$\begin{aligned}\mathbf{u}(\boldsymbol{x},t) = & \sum_{I\in\mathcal{N}} N_I(\boldsymbol{x})\,\bar{\mathbf{u}}_I + \sum_{J\in\mathcal{N}^{dis}} N_J(\boldsymbol{x})\,(H(\boldsymbol{x})-H(\boldsymbol{x}_J))\,\bar{\mathbf{d}}_J \\ & + \sum_{K\in\mathcal{N}^{Fixed}} N_K(\boldsymbol{x}) \sum_{\alpha=1}^{4} (\mathcal{B}_\alpha(\boldsymbol{x})-\mathcal{B}_\alpha(\boldsymbol{x}_K))\,\bar{\mathbf{b}}_{\alpha K}\end{aligned} \tag{7.66}$$

in which $\mathcal{N}^{Fixed}$ is the set of nodes included in a virtual circle around the crack tip with the radius R, and enriched by the crack tip asymptotic functions.

7.5.1 Numerical Integration Scheme

The derivation of the weak form of equilibrium equation (7.43) has been performed for a cracked body in Section 7.3.2 using the Divergence theorem, however, this theorem must be applied on the domain with a sufficiently regular displacement field that does not contain discontinuities, or singularities. If the discontinuity is not properly taken into account when the integration is performed, this may lead to poor numerical results, or even to a non-invertible set of equations. Hence, the crack discontinuity must be taken as an internal boundary of the domain of element, and the integration must be carried out by dividing the domain of element into two sub-domains in order to perform the integration on these two sub-domains. In a continuous domain, the Gaussian integration method can be applied with a specific number of Gauss integration points over the domain of element. However, the numerical integration of

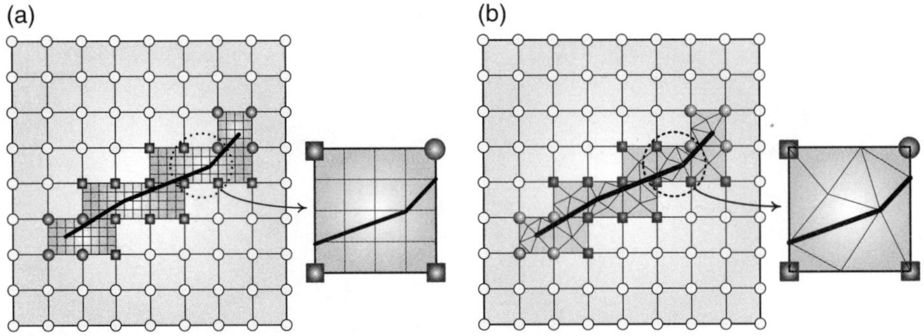

Figure 7.14 The numerical integration of discontinuity in X-FEM: (a) the rectangular sub-divisions algorithm; (b) the triangular sub-divisions scheme

the X-FEM formulations involves two difficulties; the discontinuity along the crack and the singularity at the crack tip. In order to implement the standard Gaussian integration method in the X-FEM formulation, the element cut by the crack surface must be first divided into several sub-polygons, and the numerical integration is then performed over each sub-polygon. There are basically two techniques proposed for the numerical integration of discontinuity, as shown in Figure 7.14. The first method is based on rectangular sub-divisions, as shown in Figure 7.14a, in which the crack does not coincide with the rectangular edges that causes some approximation in numerical simulation, however, a reasonable result can be achieved by increasing the number of sub-divisions. In the second technique, the discontinuity is divided into sub-triangles, as shown in Figure 7.14b, which leads to more accurate results than the previous method. It must be noted that because of the singularity at crack tip element, a more triangular sub-divisions are necessary in the tip element. Generally, a high degree of accuracy can be achieved in both techniques by increasing the number of sub-divisions but is not optimal for computational cost.

An alternative approach was presented by Laborde *et al.* (2005) for the numerical integration of crack tip elements, which can improve the rate of convergence in the numerical simulation. Consider the singular term in the integration of stiffness matrix (7.46), that is,

$$\mathbf{K}_{bb} = \int_{\Omega} \left(\mathbf{B}^{tip}\right)^T \mathbf{D}\,\mathbf{B}^{tip} d\Omega \tag{7.67}$$

This singularity occurs in the integration of the derivative of crack tip asymptotic functions, that is, $\partial\mathcal{B}(r,\theta)/\partial\mathbf{x}$, over the domain of tip-elements. It is obvious that integration (7.67) in polar coordinates will remove the $1/\sqrt{r}$ singularity of $\mathcal{B}(r,\theta)$. It was shown by Laborde *et al.* (2005) that a polar integration approach on the basis of a geometric transformation from a square to a triangle for crack tip functions, as shown in Figure 7.15, improves the results with a very low number of integration points when using the sub-triangles at the crack tip element. Based on this geometric transformation, it is possible to build a quadrature rule on the triangle from a quadrature rule on a unit square. It has been observed that the polar integration approach with 25 Gauss points is enough for the most accurate convergence test. A polar integration technique was also presented by Béchet *et al.* (2005), which also uses the non-singular characteristic of the polar form of the integrals. In this method, the real triangular interpolation cell is mapped on a reference quadrilateral cell, which can transform singular functions into regular ones. It was shown that this mapping can be efficiently used to integrate the singular functions with few sampling points.

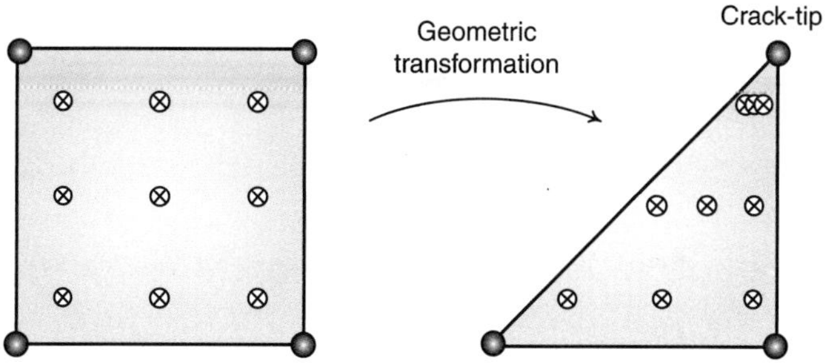

Figure 7.15 A polar integration approach based on the geometric transformation from a square into a triangle for crack tip functions

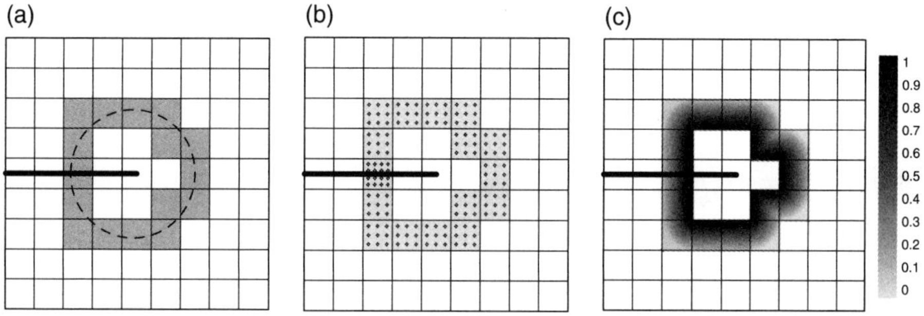

Figure 7.16 Computation of the area *J*–integral: (a) Elements selected around the crack tip; (b) Gauss integration points used for evaluation of the *J*–integral; (c) The distribution of weighting function q over the elements

7.5.2 Numerical Integration of Contour J–Integral

It was stated in Section 7.2.3 that the SIFs are the most important fracture parameters used for the determination of mixed-mode near tip stress, strain, and displacement fields in LEFM. In order to evaluate the SIFs, the area *J*–integral is defined in relation (7.25), which can be computed over an area of the FE mesh. In X-FEM modeling, the area *J*–integral is calculated by assuming a virtual circle with a specific radius around the crack tip, and the integration is performed over the elements crossed by this circle, as shown in Figure 7.16. For this purpose, the X-FEM equilibrium equation (7.44) is solved based on the enriched displacement field (7.66) to obtain the displacement, strain, and stress fields of state (1), that is, $u_i^{(1)}$, $\varepsilon_{ij}^{(1)}$, and $\sigma_{ij}^{(1)}$; and the values of displacement, strain, and stress fields are then transferred from the global to local crack tip coordinate system (x_1, x_2) by using an appropriate transformation.

The distribution of weighting function q in relation (7.25) can be obtained within an element using the standard FE interpolation as

$$q = \sum_{I=1}^{\mathcal{N}^{elem}} N_I(\boldsymbol{x})\, q_I \tag{7.68}$$

where $\mathcal{N}^{elem}$ is the number of nodes of an element and q_I are the nodal values of q. The derivation of weighting function can be obtained as

$$\frac{\partial q}{\partial \mathbf{x}} = \sum_{I=1}^{\mathcal{N}^{elem}} \frac{\partial N_I}{\partial \mathbf{x}} q_I \tag{7.69}$$

A transformation is then performed to convert $\partial q/\partial \mathbf{x}$ to the local crack tip coordinate $\partial q/\partial x_j (j = 1, 2)$.

In order to evaluate the displacement, strain, and stress fields of an auxiliary state (2), that is, $u_i^{(2)}$, $\varepsilon_{ij}^{(2)}$, and $\sigma_{ij}^{(2)}$, a pure mode I asymptotic fields, that is, $K_I^{(2)} = 1$ and $K_{II}^{(2)} = 0$, is assumed to obtain the contour integral $I^{(1)}$ from relation (7.25). Similarly, by assuming the pure mode II asymptotic fields, that is, $K_I^{(2)} = 0$ and $K_{II}^{(2)} = 1$, the contour integral $I^{(2)}$ can be obtained from (7.25). The contour integral (7.25) can be numerically evaluated using the Gaussian integration rule over an element as

$$I^{(1,2)} = \sum_{m=1}^{\mathcal{N}^{Gauss}} \left\{ \left(-W^{(1,2)} \delta_{1j} + \sigma_{ij}^{(1)} \frac{\partial u_i^{(2)}}{\partial x_1} + \sigma_{ij}^{(2)} \frac{\partial u_i^{(1)}}{\partial x_1} \right) \frac{\partial q}{\partial x_j} \right\}_m w_m \det \mathbf{J} \tag{7.70}$$

where $\mathcal{N}^{Gauss}$ is the number of integration points of an element, w_m is the Gauss weighting factor, and $\mathbf{J}$ is the well-known Jacobian matrix. Finally, substituting the contour integrals $I^{(1)}$ and $I^{(2)}$ from (7.70) into relations (7.22) and (7.23), it results in the modes I and II SIFs $K_I^{(1)}$ and $K_{II}^{(1)}$, respectively.

7.6 Application of X-FEM in Linear Fracture Mechanics

In order to demonstrate the performance of X-FEM together with the implemented fracture technique, several examples are analyzed numerically. The first two examples are chosen to illustrate the accuracy of computational algorithm for the mode I fracture by comparison of the X-FEM model with analytical solution for the SIFs and fracture parameters in a double cantilever beam (DCB) and an uniaxial tensile stress. The effect of tip enrichment at the crack tip element, and the effect of integration radius on the evaluation of *J*–integral are investigated in X-FEM models. The third example demonstrates the robustness of X-FEM computational model in a mixed-mode fracture analysis of an infinite plate with an inclined crack. The numerical simulation is compared with the theoretical solution by evaluation of the SIFs for modes I and II fractures at different crack angles. The last two examples are chosen to demonstrate the robustness of X-FEM technique in simulation of crack propagation in plates with pre-existing cracks and complex geometries. The maximum circumferential stress criterion is employed to determine the crack growth direction, and the computational algorithm presented in Section 7.5 is performed to obtain the crack trajectory at different load steps for these complex geometries.

7.6.1 X-FEM Modeling of a DCB

The first example is chosen to illustrate the performance of proposed X-FEM formulation for modeling a DCB, as shown in Figure 7.17. This problem was modeled by Belytschko *et al.* (1996) using a meshless method, and by Huang, Sukumar, and Prévost (2003b) employing an X-FEM technique. The beam is in plane strain condition, which is fixed at the left edge and subjected to the prescribed displacement $u_0 = 0.5$ mm at the top and bottom corners of the right edge. The crack has a length of 30 cm which is located in the middle of the beam. The geometry, boundary conditions and X-FEM mesh of the problem are shown in Figure 7.17. Obviously, the pure mode I stress field can be observed in this example. The material properties of the plate is chosen as follows; $E = 2 \times 10^6$ kg/cm^2 and $\nu = 0.3$. In order to model the crack

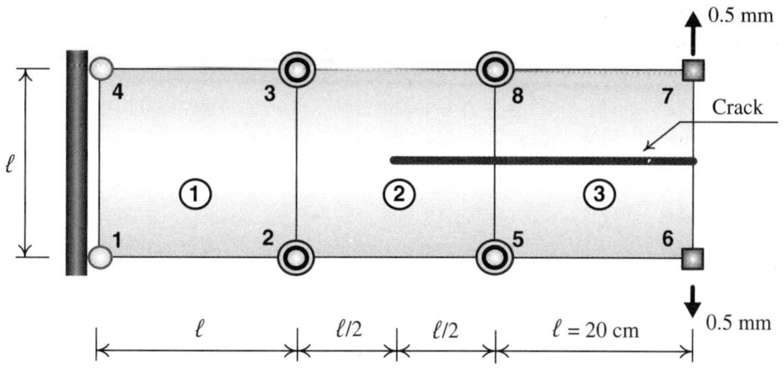

Figure 7.17 A double cantilever beam

in the beam, the Heaviside jump function (7.32) is used for the crack discontinuity and the asymptotic functions (7.34) for the crack tip. The beam is modeled with three 4-noded elements where the standard shape functions $N^{std}_{I(\xi,\eta)} = \frac{1}{4}(1+\xi_I\xi)(1+\eta_I\eta)$ are used at nodes 1 and 4, the enriched Heaviside functions $N^{enr}_J(\boldsymbol{x}) = N^{std}_J(\boldsymbol{x})(H(\boldsymbol{x})-H(\boldsymbol{x}_J))$ are applied at nodes 6 and 7, and the enriched asymptotic tip functions $N^{enr}_K(\boldsymbol{x}) = N^{std}_K(\boldsymbol{x})(\mathcal{B}_\alpha(\boldsymbol{x})-\mathcal{B}_\alpha(\boldsymbol{x}_K))$ are employed at nodes 2, 3, 5, and 8. The numerical integration is performed employing the rectangular sub-grids method, where each element is divided into 8×8 sub-grids with four Gauss points at each sub-rectangle. The stiffness matrices for these three elements are obtained according to (7.46) for the standard nodal points 1 and 4, the enriched nodal points 6 and 7 enhanced by the Heaviside jump function, and the enriched nodal points 2, 3, 5, and 8 enhanced by four asymptotic crack tip functions; for example, the stiffness sub-matrix of node 7 is obtained according to the Heaviside jump function as

$$\mathbf{K}^{\boxed{7}} = 1\times 10^6 \begin{bmatrix} 1.154 & 0.481 & -0.481 & -0.240 \\ 0.481 & 1.154 & -0.240 & -0.962 \\ -0.481 & -0.240 & 0.962 & 0.481 \\ -0.240 & -0.962 & 0.481 & 1.923 \end{bmatrix} \begin{matrix} u_{x_7} \\ u_{y_7} \\ a_{x_7} \\ a_{y_7} \end{matrix} \tag{7.71a}$$

The stiffness sub-matrix of node 8 corresponding to element (3) is obtained according to the asymptotic crack tip functions as

$${}^{(3)}\mathbf{K}^{\boxed{8}} = 1\times 10^6 \begin{bmatrix} 1.154 & -0.481 & 1.715 & -0.752 & -0.586 & 0.006 & -0.886 & -0.052 & 0.600 & -0.087 \\ -0.481 & 1.154 & -0.856 & 3.184 & 0.092 & -0.325 & 0.106 & -0.417 & -0.156 & 0.417 \\ 1.715 & -0.856 & 13.871 & -5.198 & -2.050 & 0.718 & -3.537 & 1.216 & 1.561 & -0.610 \\ -0.752 & 3.184 & -5.198 & 26.465 & 0.679 & -5.280 & 1.135 & -9.091 & -0.753 & 3.713 \\ -0.586 & 0.092 & -2.050 & 0.679 & 1.098 & -0.169 & 1.822 & -0.235 & -0.843 & 0.223 \\ 0.006 & -0.325 & 0.718 & -5.280 & -0.169 & 2.052 & -0.246 & 3.583 & 0.234 & -1.222 \\ -0.886 & 0.106 & -3.537 & 1.135 & 1.822 & -0.246 & 3.075 & -0.350 & -1.317 & 0.322 \\ -0.052 & -0.417 & 1.216 & -9.091 & -0.235 & 3.583 & -0.350 & 6.348 & 0.330 & -1.973 \\ 0.600 & -0.156 & 1.561 & -0.753 & -0.843 & 0.234 & -1.317 & 0.330 & 0.869 & -0.282 \\ -0.087 & 0.417 & -0.610 & 3.713 & 0.223 & -1.222 & 0.322 & -1.973 & -0.282 & 1.180 \end{bmatrix} \begin{matrix} u_{x_8} \\ u_{y_8} \\ b^{(1)}_{x_8} \\ b^{(1)}_{y_8} \\ b^{(2)}_{x_8} \\ b^{(2)}_{y_8} \\ b^{(3)}_{x_8} \\ b^{(3)}_{y_8} \\ b^{(4)}_{x_8} \\ b^{(4)}_{y_8} \end{matrix} \tag{7.71b}$$

and the stiffness sub-matrix of node 8 corresponding to element (2) is obtained according to the asymptotic crack tip functions as

$$
{}^{(2)}\mathbf{K}^{\boxed{8}} = 1\times 10^{6}\begin{bmatrix}
1.154 & 0.481 & 1.568 & 0.544 & -0.444 & 0.12 & -0.847 & 0.016 & 0.378 & -0.337 \\
0.481 & 1.154 & 0.575 & 2.668 & -0.114 & 0.175 & -0.165 & -0.271 & -0.055 & -0.358 \\
1.568 & 0.575 & 12.432 & 4.896 & -0.824 & -1.69 & -3.114 & -3.125 & 0.134 & 0.537 \\
0.544 & 2.668 & 4.896 & 17.018 & -1.359 & -3.322 & -2.444 & -6.366 & 0.465 & 3.07 \\
-0.444 & -0.114 & -0.824 & -1.359 & 2.639 & -0.459 & 1.648 & -0.253 & -1.869 & 0.939 \\
0.12 & 0.175 & -1.69 & -3.322 & -0.459 & 3.921 & -0.214 & 4.699 & 0.909 & -2.729 \\
-0.847 & -0.165 & -3.114 & -2.444 & 1.648 & -0.214 & 4.063 & -0.01 & -1.386 & 0.645 \\
0.016 & -0.271 & -3.125 & -6.366 & -0.253 & 4.699 & -0.01 & 7.896 & 0.685 & -2.804 \\
0.378 & -0.055 & 0.134 & 0.465 & -1.869 & 0.909 & -1.386 & 0.685 & 3.081 & -0.859 \\
-0.337 & -0.358 & 0.537 & 3.07 & 0.939 & -2.729 & 0.645 & -2.804 & -0.859 & 4.694
\end{bmatrix}
\begin{matrix} u_{x_8} \\ u_{y_8} \\ b_{x_8}^{(1)} \\ b_{y_8}^{(1)} \\ b_{x_8}^{(2)} \\ b_{y_8}^{(2)} \\ b_{x_8}^{(3)} \\ b_{y_8}^{(3)} \\ b_{x_8}^{(4)} \\ b_{y_8}^{(4)} \end{matrix}
\tag{7.71c}
$$

Hence, the total stiffness sub-matrix of node 8 can be obtained as

$$
\mathbf{K}^{\boxed{8}} = 1\times 10^{6}\begin{bmatrix}
2.308 & 0.000 & 3.283 & -0.208 & -1.030 & 0.126 & -1.733 & -0.036 & 0.979 & -0.424 \\
0.000 & 2.308 & -0.282 & 5.852 & -0.022 & -0.150 & -0.060 & -0.687 & -0.211 & 0.059 \\
3.283 & -0.282 & 26.303 & -0.302 & -2.874 & -0.972 & -6.651 & -1.910 & 1.695 & -0.073 \\
-0.208 & 5.852 & -0.302 & 43.483 & -0.680 & -8.601 & -1.310 & -15.456 & -0.289 & 6.783 \\
-1.030 & -0.022 & -2.874 & -0.680 & 3.736 & -0.628 & 3.470 & -0.488 & -2.712 & 1.162 \\
0.126 & -0.150 & -0.972 & -8.601 & -0.628 & 5.973 & -0.460 & 8.282 & 1.142 & -3.951 \\
-1.733 & -0.060 & -6.651 & -1.310 & 3.470 & -0.460 & 7.139 & -0.360 & -2.703 & 0.966 \\
-0.036 & -0.687 & -1.910 & -15.456 & -0.488 & 8.282 & -0.360 & 14.244 & 1.015 & -4.777 \\
0.979 & -0.211 & 1.695 & -0.289 & -2.712 & 1.142 & -2.703 & 1.015 & 3.950 & -1.141 \\
-0.424 & 0.059 & -0.073 & 6.783 & 1.162 & -3.951 & 0.966 & -4.777 & -1.141 & 5.874
\end{bmatrix}
\begin{matrix} u_{x_8} \\ u_{y_8} \\ b_{x_8}^{(1)} \\ b_{y_8}^{(1)} \\ b_{x_8}^{(2)} \\ b_{y_8}^{(2)} \\ b_{x_8}^{(3)} \\ b_{y_8}^{(3)} \\ b_{x_8}^{(4)} \\ b_{y_8}^{(4)} \end{matrix}
\tag{7.71d}
$$

By assembling the stiffness matrices of these three elements and solving the set of linear system X-FEM equations, the standard and enriched nodal displacements 6 and 7 can be obtained as

$$
\begin{aligned}
&\qquad\qquad\quad\ \ u_{x_6} \qquad\ \ u_{y_6} \qquad\ \ u_{x_7} \qquad u_{y_7} \\
&\bar{\mathbf{U}}^{T} = 1\times 10^{-3}\langle -8.17 \quad -50.00 \quad -8.17 \quad 50.00\rangle \\
&\qquad\qquad\quad\ \ a_{x_6} \qquad\ \ a_{y_6} \qquad\ \ a_{x_7} \qquad a_{y_7} \\
&\bar{\mathbf{a}}^{T} = 1\times 10^{-3}\langle 15.69 \quad 49.88 \quad -15.69 \quad 49.88\rangle
\end{aligned}
\tag{7.72a}
$$

The jump of displacement at the crack mouth can be obtained from $[\![\mathbf{u}]\!] = 2\mathbf{N}\bar{\mathbf{a}}$ as

$$
[\![\mathbf{u}]\!]_{\mathrm{CM}} = 2\begin{bmatrix} N_6 & 0 & N_7 & 0 \\ 0 & N_6 & 0 & N_7 \end{bmatrix}_{x=60,y=10}\bar{\mathbf{a}} = 2\begin{bmatrix} 0.5 & 0 & 0.5 & 0 \\ 0 & 0.5 & 0 & 0.5 \end{bmatrix}\bar{\mathbf{a}} = \begin{Bmatrix} 0 \\ 99.76\times 10^{-3} \end{Bmatrix}
\tag{7.72b}
$$

The normal gap g_N and the tangential gap g_T at the crack mouth can be obtained as

$$
\begin{aligned}
g_N &= \mathbf{n}_{\Gamma_d}[\![\mathbf{u}]\!] = 99.76\times 10^{-3}\ \mathrm{m} \\
g_T &= \mathbf{t}_{\Gamma_d}[\![\mathbf{u}]\!] = 0
\end{aligned}
\tag{7.72c}
$$

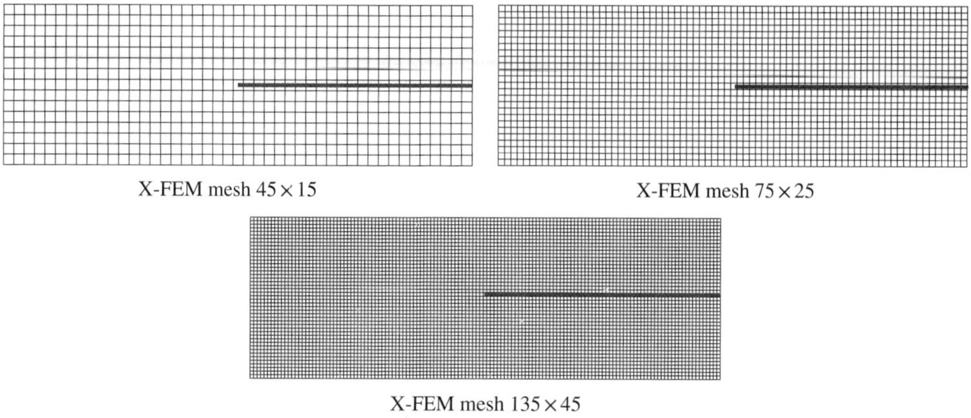

Figure 7.18 The X-FEM meshes of a double cantilever beam

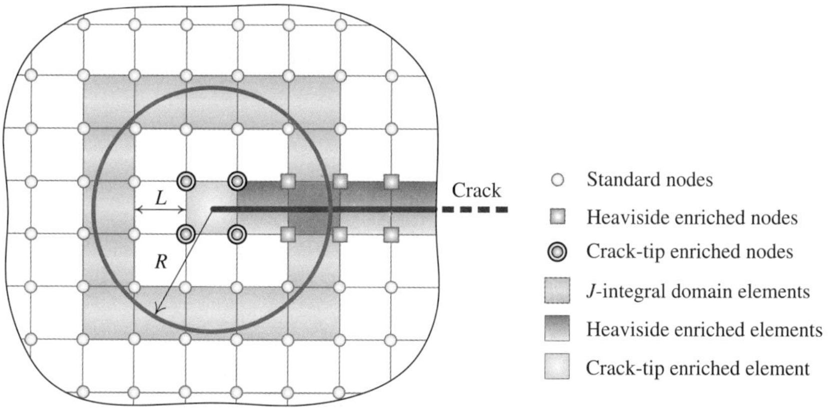

Figure 7.19 Computation of the area *J*–integral by assuming a virtual circle of radius *R* around the crack tip: The integration is performed over the elements crossed by this circle

In order to illustrate the accuracy of computational algorithm for the mode I fracture, the X-FEM analysis is performed using three structured meshes with the refinements of 45 × 15, 75 × 25, and 135 × 45, as shown in Figure 7.18. The effect of integration radius on the evaluation of *J*–integral and the mode I SIF are investigated for three X-FEM models. The numerical integration is performed using 8 × 8 rectangular sub-grids with four Gauss points at each sub-rectangle for the elements cut by the crack, and 16 Gauss integration points for elements in the domain of *J*–integral. The computation of *J*–integral is performed by assuming a virtual circle of radius *R* around the crack tip, and the numerical integration is carried out over the elements crossed by this circle, as shown in Figure 7.19. Table 7.1 presents the values of *J*–integral and SIF K_I for various ratios of R/L, where L is the element size, using three X-FEM meshes. Obviously, the values of *J*–integral are independent of the radius of integration for $R/L \geq 3$.

The standard and enriched displacement values for the top and bottom nodes of crack mouth element are evaluated for the X-FEM mesh of 135 × 45 as

Table 7.1 A double cantilever beam: The values of J–integral and stress intensity factor K_I for various ratios of R/L using three X-FEM meshes

X-FEM meshes	R/L	1.0	2.0	3.0	4.0	5.0
45 × 15	J–integral	2.981	2.286	2.119	2.117	2.115
	K_I	2559.729	2241.703	2158.025	2157.079	2156.142
75 × 25	J–integral	2.921	2.285	2.114	2.113	2.112
	K_I	2533.527	2240.865	2155.333	2154.904	2154.240
135 × 45	J–integral	2.869	2.274	2.101	2.101	2.100
	K_I	2510.949	2235.567	2148.986	2148.936	2148.523

$$\begin{aligned} &\qquad\qquad\quad u_x^{\text{bot}} \quad\;\; u_y^{\text{bot}} \quad\; u_x^{\text{top}} \quad u_y^{\text{top}} \\ \bar{\mathbf{U}}^T &= 1\times10^{-3}\langle 9.12 \quad -48.17 \quad 9.12 \quad 48.17\rangle \\ &\qquad\qquad\quad a_x^{\text{bot}} \quad\;\; a_y^{\text{bot}} \quad\; a_x^{\text{top}} \quad a_y^{\text{top}} \\ \bar{\mathbf{a}}^T &= 1\times10^{-3}\langle 0.43 \quad 48.17 \quad -0.43 \quad 48.17\rangle \end{aligned} \tag{7.73a}$$

The jump of displacement at the crack mouth is obtained as

$$[\![\mathbf{u}]\!]_{\text{CM}} = 2\begin{bmatrix} N^{\text{bot}} & 0 & N^{\text{top}} & 0 \\ 0 & N^{\text{bot}} & 0 & N^{\text{top}} \end{bmatrix}_{x=60,y=10} \bar{\mathbf{a}} = 2\begin{bmatrix} 0.5 & 0 & 0.5 & 0 \\ 0 & 0.5 & 0 & 0.5 \end{bmatrix}\bar{\mathbf{a}} = \begin{Bmatrix} 0 \\ 96.34\times10^{-3} \end{Bmatrix} \tag{7.73b}$$

and the normal gap g_N and the tangential gap g_T at the crack mouth are evaluated as

$$\begin{aligned} g_N &= \mathbf{n}_{\Gamma_d}[\![\mathbf{u}]\!] = 96.34\times10^{-3}\ \text{m} \\ g_T &= \mathbf{t}_{\Gamma_d}[\![\mathbf{u}]\!] = 0 \end{aligned} \tag{7.73c}$$

Finally, the distribution of stress contours σ_x, σ_y, and τ_{xy} are presented in Figure 7.20 over the beam (see Companion Website).

7.6.2 An Infinite Plate with a Finite Crack in Tension

The second example is of an infinite plate with a finite crack at its center subjected to uniform far-field tension of $\sigma = 2000$ kg/cm^2 in mode I loading. The material properties are as follows $E = 2.1\times10^6$ kg/cm^2 and $\nu = 0.3$, and the plate is in plane stress condition. In order to perform the far-field condition of an infinite plate, the ratio of plate width w to crack length a is chosen to be 50. In Figure 7.21a, the problem definition is presented together with the geometry and boundary conditions of the plate. The analytical solution is available for this plate, where $\sigma_{22} = \sigma$, $\sigma_{11} = 0$,and $\sigma_{12} = 0$. The stress components are defined as

$$\begin{aligned} \sigma_{11} &= \sigma\left[\text{Re}\left(\frac{z}{\sqrt{z^2-a^2}}\right) - x_2\text{Im}\left(\frac{1}{\sqrt{z^2-a^2}} - \frac{z^2}{(z^2-a^2)^{3/2}}\right)\right] - \sigma \\ \sigma_{22} &= \sigma\left[\text{Re}\left(\frac{z}{\sqrt{z^2-a^2}}\right) + x_2\text{Im}\left(\frac{1}{\sqrt{z^2-a^2}} - \frac{z^2}{(z^2-a^2)^{3/2}}\right)\right] \\ \sigma_{12} &= -\sigma x_2\text{Re}\left(\frac{1}{\sqrt{z^2-a^2}} - \frac{z^2}{(z^2-a^2)^{3/2}}\right) \end{aligned} \tag{7.74}$$

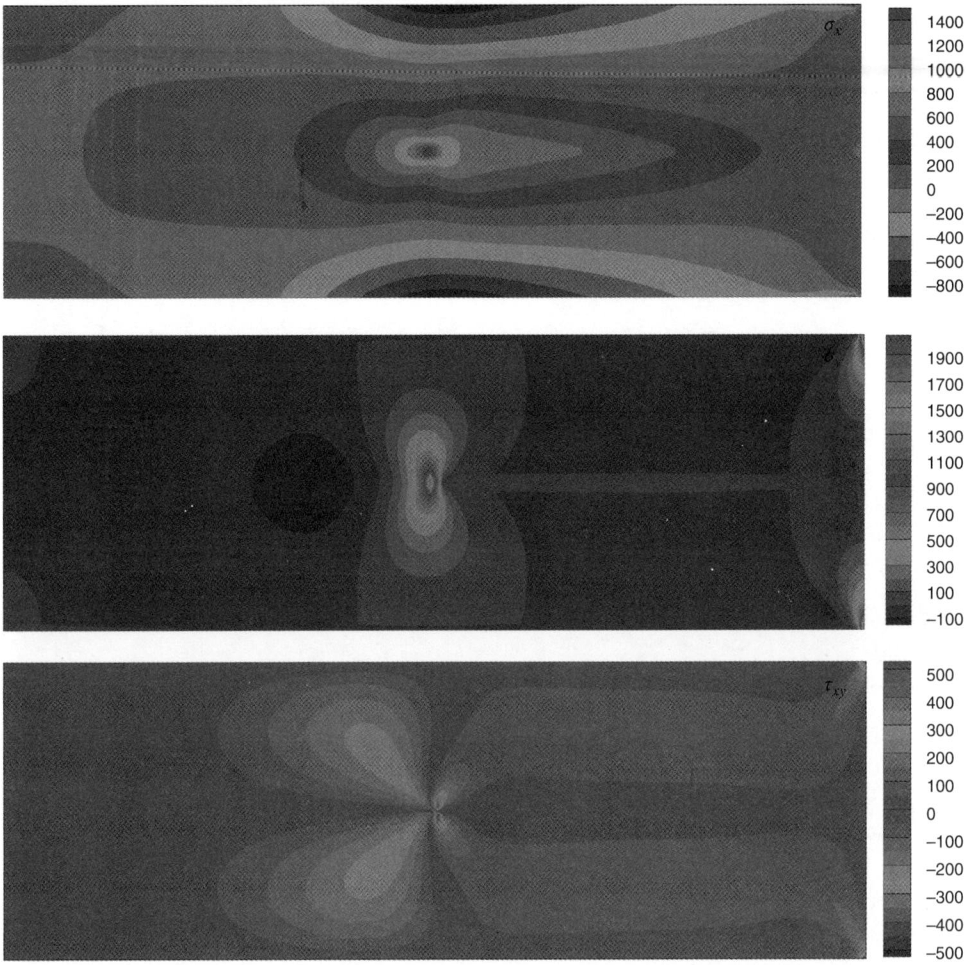

Figure 7.20 The distribution of stress contours σ_x, σ_y, and τ_{xy} for a double cantilever beam using the X-FEM mesh of 135×45

where $z = x_1 + ix_2$. The mode I SIF for this problem is equal to $K_I = \sigma\sqrt{\pi a}$ (Broek, 1986).

A structured FE mesh with a fine grid in the cracked region and a coarse grid in other parts of the domain is used, as shown in Figure 7.21b, to reduce the computational cost of the analysis and provide the required accuracy. In the X-FEM analysis carried out here, the stiffness matrices corresponding to a standard nodal point, a nodal point enriched by the Heaviside jump function, and a nodal point enriched by four asymptotic crack tip functions are computed according to (7.46) as

$$\mathbf{K} = \begin{array}{c} \begin{matrix} u_x & \quad u_y \end{matrix} \\ \begin{bmatrix} 1038462 & 375000 \\ 375000 & 1038462 \end{bmatrix} \end{array} \begin{array}{c} \\ u_x \\ u_y \end{array} \tag{7.75a}$$

for a nodal point of a standard element, whereas for a nodal point of an element cut by the crack surface and enriched by the Heaviside function as

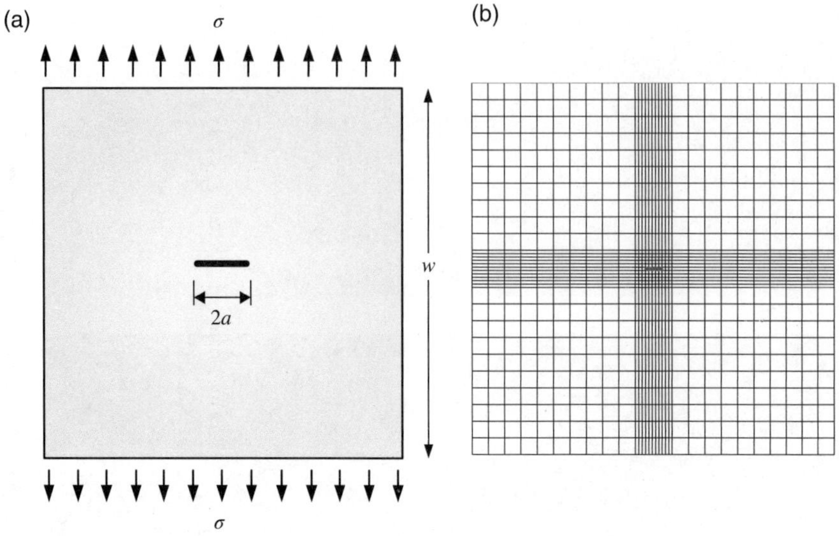

Figure 7.21 An infinite plate with a finite crack at its center: (a) Problem definition; (b) The X-FEM mesh

$$\mathbf{K}=\begin{array}{c} \begin{array}{cccc} u_x & u_y & d_x & d_y \end{array} \\ \begin{bmatrix} 1038462 & 375000 & 461539 & 187500 \\ 375000 & 1038462 & 187500 & 836539 \\ 461539 & 187500 & 923077 & 375000 \\ 187500 & 836539 & 375000 & 1673077 \end{bmatrix} \end{array} \begin{array}{c} u_x \\ u_y \\ d_x \\ d_y \end{array} \tag{7.75b}$$

and for a nodal point of the crack tip element enriched by asymptotic tip functions as

$$\mathbf{K}=\begin{array}{c} \begin{array}{cccccccccc} u_x & u_y & b_{1x} & b_{1y} & b_{2x} & b_{2y} & b_{3x} & b_{3y} & b_{4x} & b_{4y} \end{array} \\ \begin{bmatrix} 1038462 & 375000 & -135236 & -23972 & -79459 & 27261 & -85765 & -4582 & -195723 & -30218 \\ 375000 & 1038461 & -8339 & -62160 & 22002 & -111717 & 14002 & 52753 & -17047 & -116604 \\ -135236 & -8339 & 203764 & -18456 & -8177 & 14809 & 112937 & -38447 & 76237 & 4750 \\ -23972 & -62160 & -18456 & 257556 & 12101 & 83812 & -39102 & 152102 & 524 & 213959 \\ -79459 & 22002 & -8177 & 12101 & 170346 & 8640 & -35625 & 4311 & 67397 & 47344 \\ 27261 & -111717 & 14809 & 83812 & 8640 & 155586 & 4431 & 86115 & 46174 & 96197 \\ -85765 & 14002 & 112937 & -39102 & -35625 & 4431 & 187683 & -23256 & 59420 & -507 \\ -4582 & 52753 & -38447 & 152102 & 4311 & 86115 & -23256 & 260934 & -4546 & 106352 \\ -195723 & -17047 & 76237 & 524 & 67397 & 46174 & 59420 & -4546 & 233380 & 26645 \\ -30218 & -116604 & 4750 & 213959 & 47344 & 96197 & -507 & 106352 & 26645 & 314838 \end{bmatrix} \end{array} \begin{array}{c} u_x \\ u_y \\ b_{1x} \\ b_{1y} \\ b_{2x} \\ b_{2y} \\ b_{3x} \\ b_{3y} \\ b_{4x} \\ b_{4y} \end{array} \tag{7.75c}$$

It must be noted that the values of these matrices are valid for nodal points of all square elements at the cracked area. In all three matrices, the first two rows and columns correspond to the standard DOF in $x-$ and $y-$ directions and the other terms are related to the enriched DOF in $x-$ and $y-$ directions. In Figure 7.22, the distribution of stress contours σ_x, σ_y, and τ_{xy} are presented over the cracked body. In order to investigate the accuracy of the computed SIF K_I, the error of interaction integral is evaluated for various DOF. Table 7.2 presents the normalized values of SIF with respect to the analytical solution and the interaction integral error for various ratios of a/s, where $2a$ indicates the crack size and $2s$ denotes the

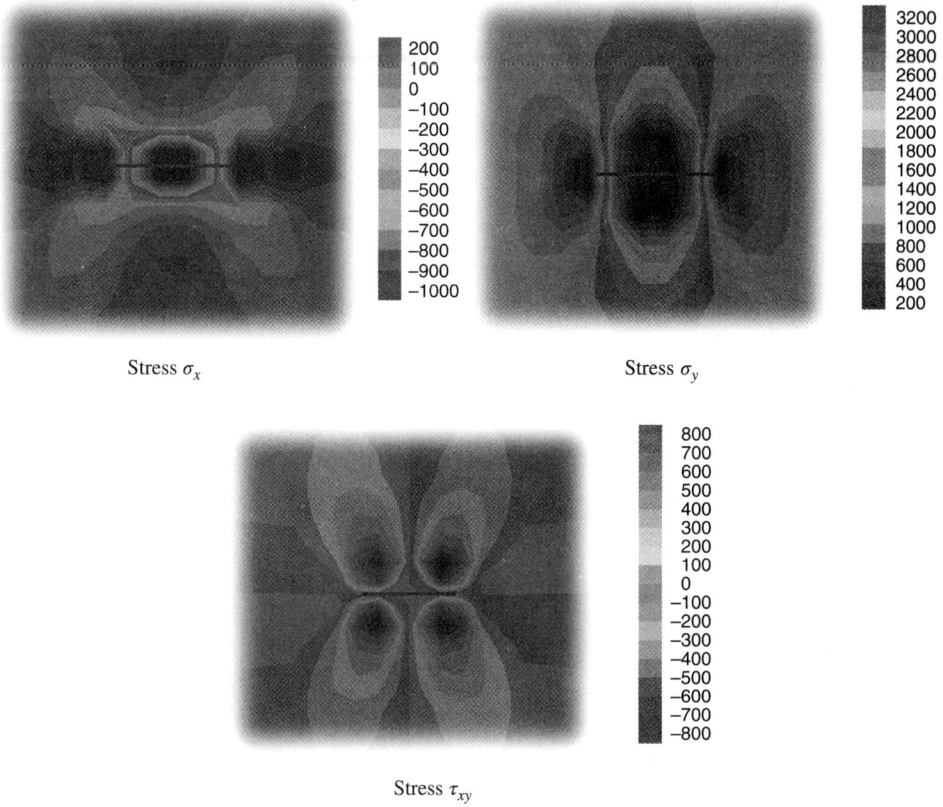

Figure 7.22 The distribution of stress contours σ_x, σ_y, and τ_{xy} for an infinite plate with a finite crack at its center (For color details, please see color plate section)

Table 7.2 An infinite plate with a finite crack in its center: The normalized values of stress intensity factor (SIF) with respect to analytical solution and the interaction integral error for various ratios of a/s

Degrees of freedom	2120	3616	5512	7808	10153
Ratio a/s	2	4	6	8	10
Normalized SIF	0.986	0.992	0.993	0.994	0.994
Error of SIF (%)	1.395	0.825	0.667	0.595	0.557

element size at the cracked area. Obviously, the error of SIF obtained by interaction integral method decreases by increasing the ratio of a/s, where the mesh of cracked domain becomes finer. In order to investigate the significance of crack tip enrichment, the variations of SIF K_I with the ratio of a/s are plotted in Figure 7.23a for two cases; when the asymptotic enrichment functions are employed and in the case that no enrichment is applied at the tip element. It can be clearly seen from the figure that the tip enrichment functions increase the accuracy of SIF at the crack tip region. Finally, the effect of integration radius on the evaluation of J–integral is shown in Figure 7.23b for the ratio of $a/s = 4$. In this figure, the J–integral error is obtained for different ratios of r/s, where r denotes the radius of integration. It shows that the value of J–integral is not dependent on the radius of integration, except for the values of $r/s \leq 2$.

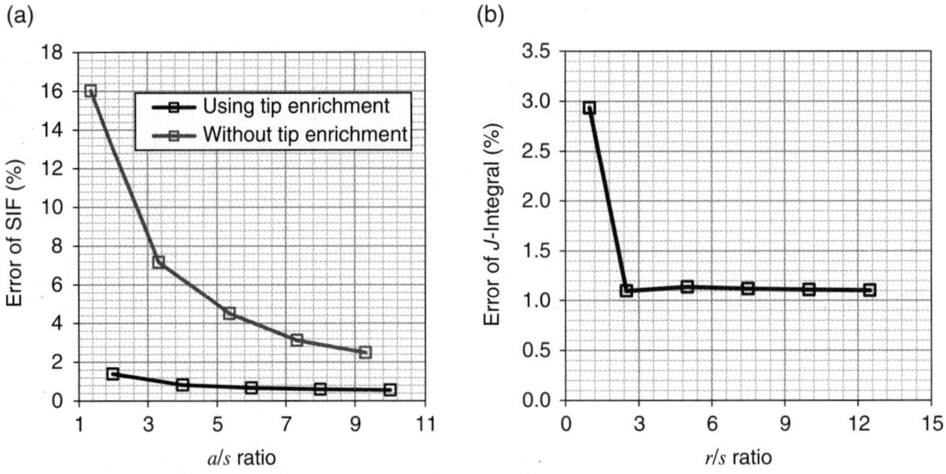

Figure 7.23 An infinite plate with a finite crack at its center: (a) The error of stress intensity factor K_I with the ratio of a/s; (b) The variations of J–integral error with the ratio of r/s

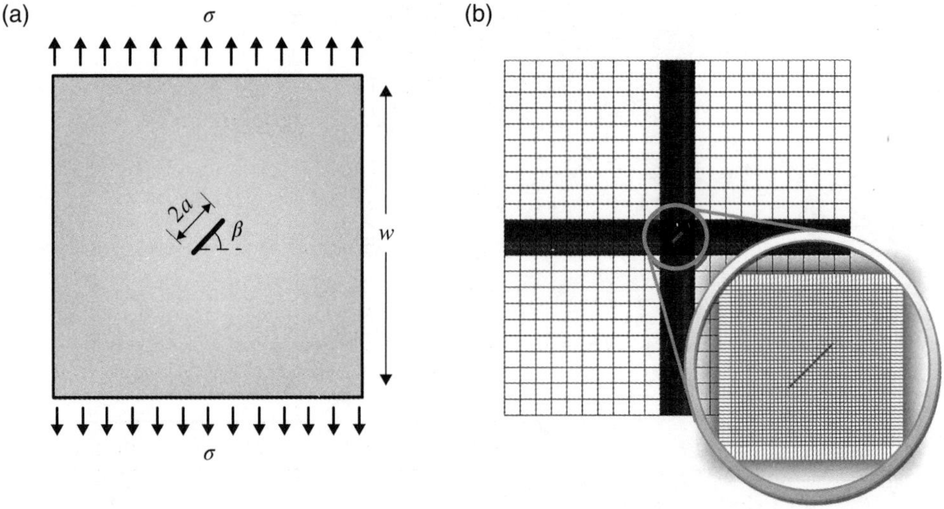

Figure 7.24 An infinite plate with an inclined crack: (a) Problem definition; (b) The X-FEM mesh

7.6.3 An Infinite Plate with an Inclined Crack

The third example is chosen to illustrate the performance of X-FEM technique in mixed-mode loading for an infinite plate with an inclined crack subjected to uniform far-field tension stress. The material properties of the plate are similar to the previous example. The geometry and boundary conditions of the plate are shown in Figure 7.24a. The ratio of plate width to crack size and the ratio of crack size to element size at the cracked region are chosen as $w/a = 50$ and $a/s = 8$, respectively. The plate is in plane stress condition and the far-field tension stress is equal to $\sigma = 2000$ kg/cm^2. The analytical solution is available for this plate with the SIFs defined for modes I and II fracture as (Broek, 1986)

$$K_I = \sigma \cos^2\beta \sqrt{\pi a}$$
$$K_{II} = \sigma \sin\beta \cos\beta \sqrt{\pi a} \quad (7.76)$$

where β is the crack angle with the horizontal axis.

The X-FEM analysis is performed with a structured FE mesh, as shown in Figure 7.24b. In Table 7.3, a comparison is performed for the modes I and II SIFs between the numerical simulation and those obtained from the analytical solution at different crack angles β. Also plotted in Figure 7.25 are the evolutions of SIFs K_I and K_{II} at various crack angles for the numerical and theoretical solutions. Obviously, the relative error is almost negligible and a remarkable agreement can be seen between the numerical results and theoretical values. Finally, the distribution of stress contours σ_x, σ_y, and τ_{xy} are presented over the cracked body in Figure 7.26.

Table 7.3 An infinite plate with an inclined crack: A comparison of modes I and II stress intensity factors between the numerical and analytical solutions at different crack angles β

	K_I (kg/cm$^{3/2}$)			K_{II} (kg/cm$^{3/2}$)		
Crack angle β	Numerical	Analytical	Error (%)	Numerical	Analytical	Error (%)
0	4983.4	5013.3	0.595	0.02	0.00	—
10	4832.5	4862.1	0.609	853.4	857.3	0.460
20	4401.6	4426.8	0.569	1603.1	1611.2	0.506
30	3738.2	3759.9	0.578	2160.0	2170.8	0.497
40	2924.5	2941.9	0.593	2457.2	2468.5	0.460
50	2059.2	2071.4	0.589	2457.5	2468.5	0.448
60	1246.0	1253.3	0.583	2160.3	2170.8	0.482
70	583.0	586.4	0.593	1603.4	1611.2	0.487
80	150.1	151.2	0.686	853.0	857.3	0.506
90	0	0	0	0	0	—

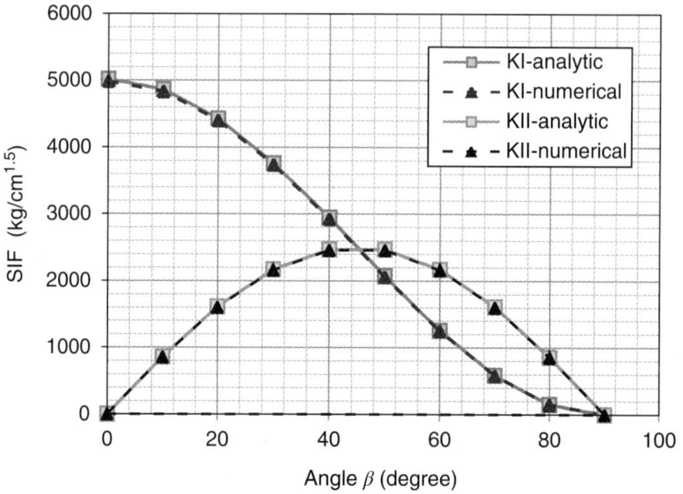

Figure 7.25 The evolutions of stress intensity factors K_I and K_{II} at various crack angles: A comparison between the numerical and analytical solutions

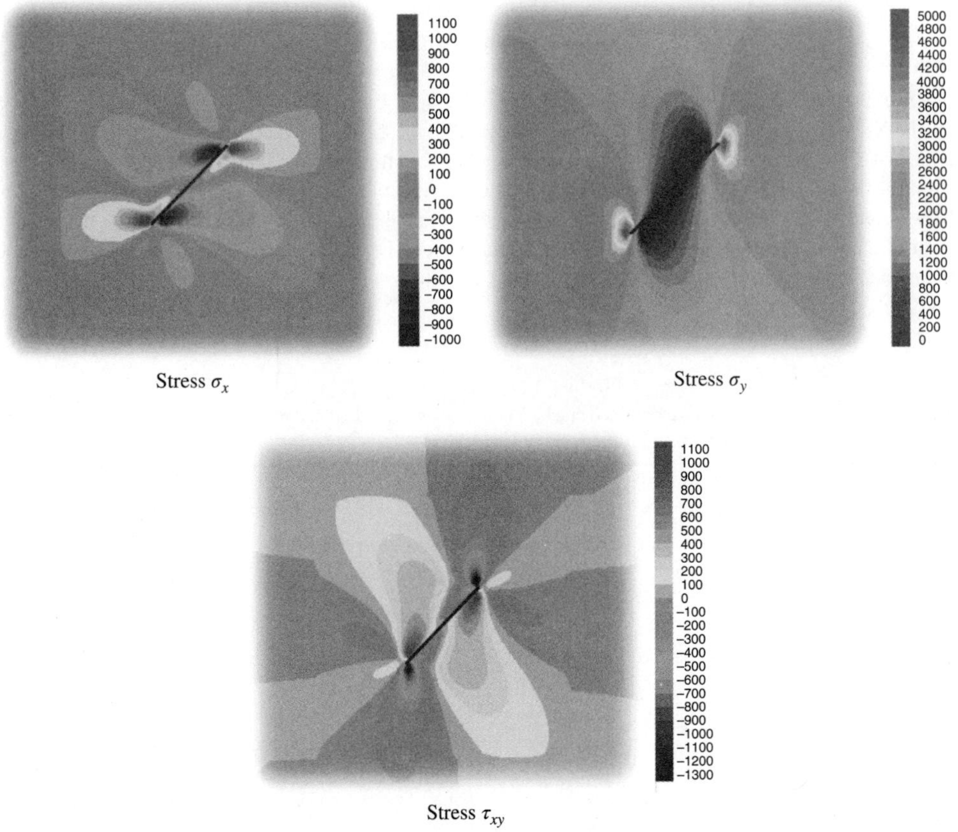

Figure 7.26 The distribution of stress contours σ_x, σ_y, and τ_{xy} for an infinite plate with an inclined crack (For color details, please see color plate section)

7.6.4 A Plate with Two Holes and Multiple Cracks

The next example presents a rectangular plate with two off-center circular holes subjected to a prescribed displacement at one edge, as shown in Figure 7.27. Two pre-cracked parts are considered beyond the holes, which can be propagated with the same length since they are symmetric. This example was modeled by Khoei, Azadi, and Moslemi (2008c) to show the applicability of their remeshing technique in multiple cracks and complex boundaries. The plate is in a plane strain condition and a prescribed displacement of 0.1 mm is applied in 20 loading steps. The material properties are as follows; $E = 2 \times 10^5$ N/mm^2, $\nu = 0.3$, and $K_{IC} = 1500$ N/mm$^{3/2}$. Both cracks have a length of 1 mm at the initial configuration and are oriented horizontally. The primary objective of this example is to investigate the effect of X-FEM technique on the prediction of crack trajectory. It must be noted that, although the geometry and boundary conditions of problem are symmetric, if the FE discretization is not symmetric, the calculated SIFs will not be exactly the same for both crack tips (Khoei, Azadi, and Moslemi, 2008c). This produces a difficulty into the simulation procedure, if a coarse FEM mesh is used. In this case, the left crack grows sooner than the right one and consequently, no crack growth occurs on the right hand side. By increasing the load, the right crack begins to grow, and the final configuration of crack propagation will not be symmetric. However, in the X-FEM numerical simulation proposed here, a symmetric X-FEM mesh can prevent such a drawback and the final pattern of the crack growth becomes symmetric, as shown in Figure 7.28. In this

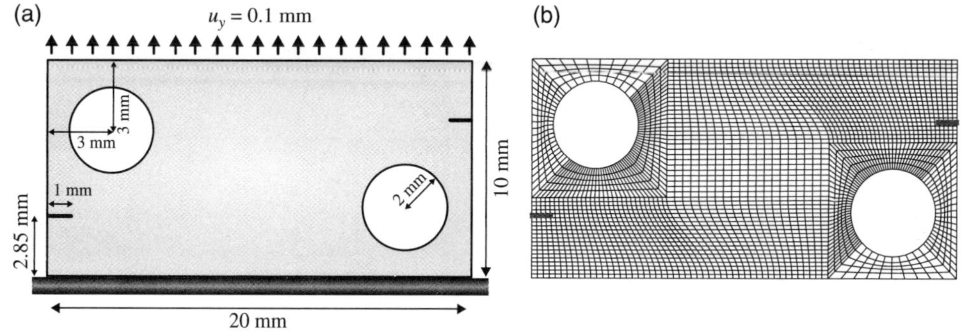

Figure 7.27 A rectangular plate with two circular holes: (a) Problem definition; (b) The X-FEM mesh

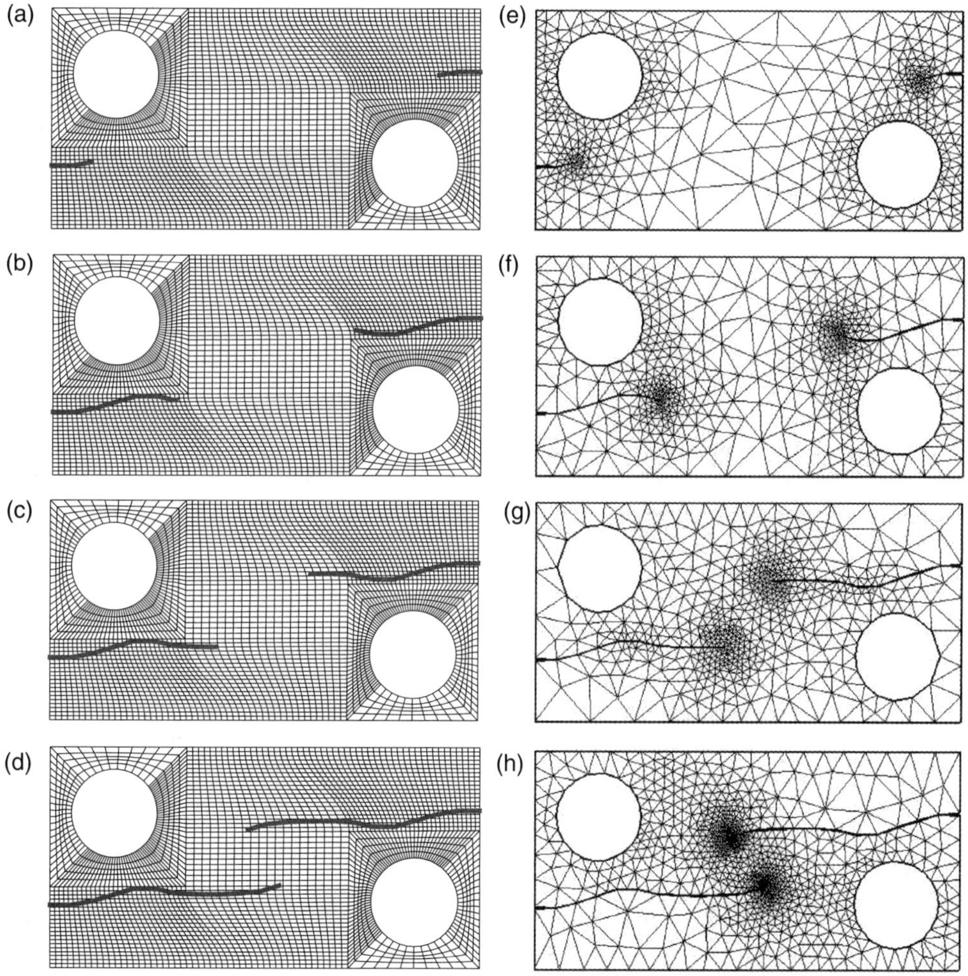

Figure 7.28 The crack trajectory for a plate with two holes during crack growth: (a–d) the X-FEM technique; (e–h) the adaptive FEM

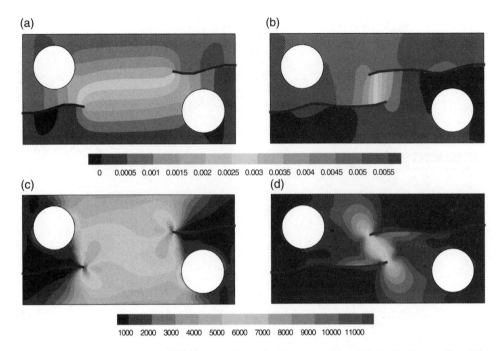

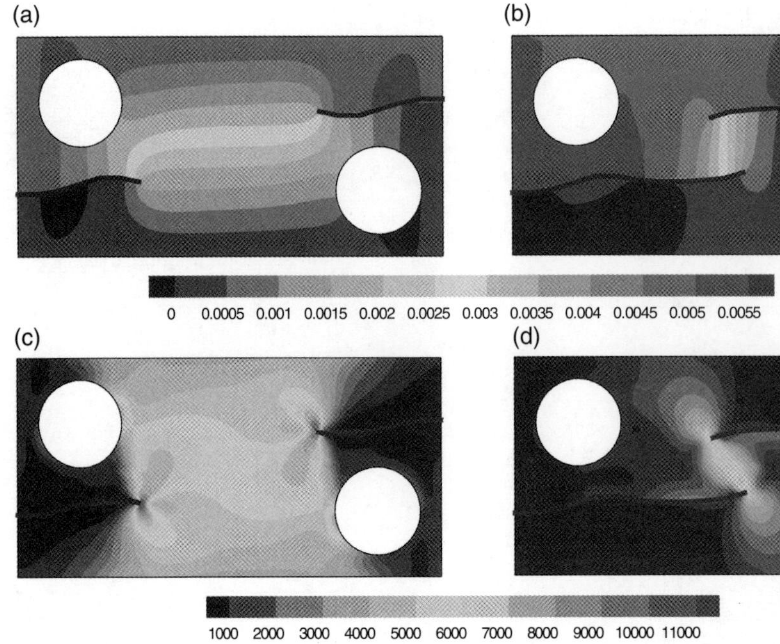

Figure 7.29 (a and b) The contours of displacement in $y-$ direction (cm) and (c and d) the distribution of von-Mises stress (kg/cm^2) at the half and final loading steps of crack propagation (For color details, please see color plate section)

figure, the X-FEM mesh configurations are shown at different crack growths, and the crack trajectory is compared with the adaptive mesh refinements reported by Khoei, Azadi, and Moslemi (2008c). It can be observed that the X-FEM can be efficiently used to capture the interaction of the two cracks. In Figure 7.29, the contours of displacement in $y-$ direction and the distributions of von-Mises stress are presented at the half and final loading steps of crack propagation. The variation of vertical reaction is plotted with crack length in Figure 7.30a. Obviously, during the crack propagation, a drop occurs in the force–crack length curve corresponding to a reduction in the stiffness of the structure. A complete agreement can be observed between the X-FEM and adaptive remeshing techniques. In Figure 7.30b, the variation of mode I SIF is plotted at various crack lengths.

This example with different boundary conditions is solved once again to illustrate the performance of X-FEM technique in modeling a rectangular plate with cracks emanating from two holes, as shown in Figure 7.31. This example was also solved by Khoei, Azadi, and Moslemi (2008c) using an adaptive mesh refinement strategy. The plate is in plane strain condition and a prescribed displacement tensile loading is applied in 20 steps. In the initial configuration, both cracks have a length of 1 mm and are oriented with the angle of 45 °. The plate is made of 4340 steel with linear elastic properties as follows; $E = 2 \times 10^5$ N/mm^2, $\nu = 0.3$, and $K_{IC} = 1300$ N/mm$^{3/2}$. The crack growth simulation is shown in Figure 7.32 based on the X-FEM technique, which is in good agreement with those reported by Khoei, Azadi, and Moslemi (2008c) using an adaptive remeshing strategy. In Figure 7.33, the contours of displacement in $y-$ direction and the distributions of von-Mises stress are presented at the half and final loading steps of crack propagation. In Figure 7.34a, the variation of vertical reaction is plotted with crack length. Complete agreement can be observed between the X-FEM and adaptive remeshing techniques. Finally, the mode I SIF is plotted in Figure 7.34b for different crack lengths.

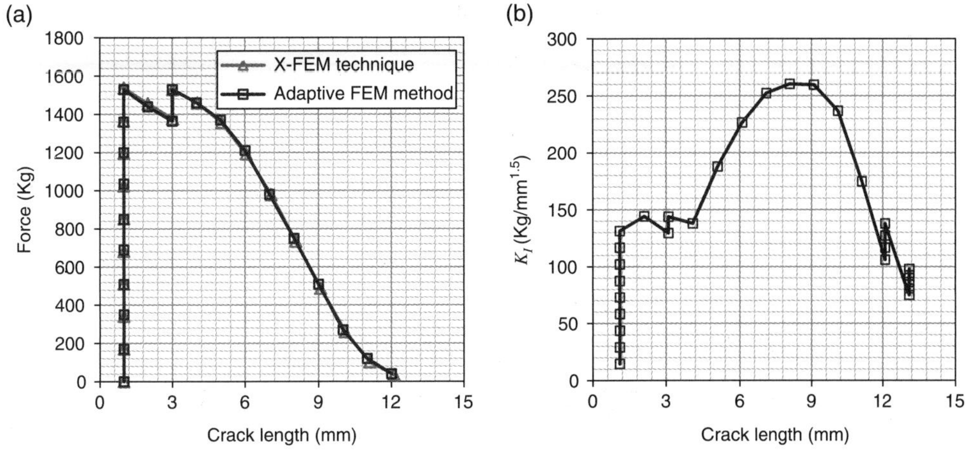

Figure 7.30 (a) The variation of vertical reaction with crack length: (b) the variation of mode I stress intensity factor with crack length

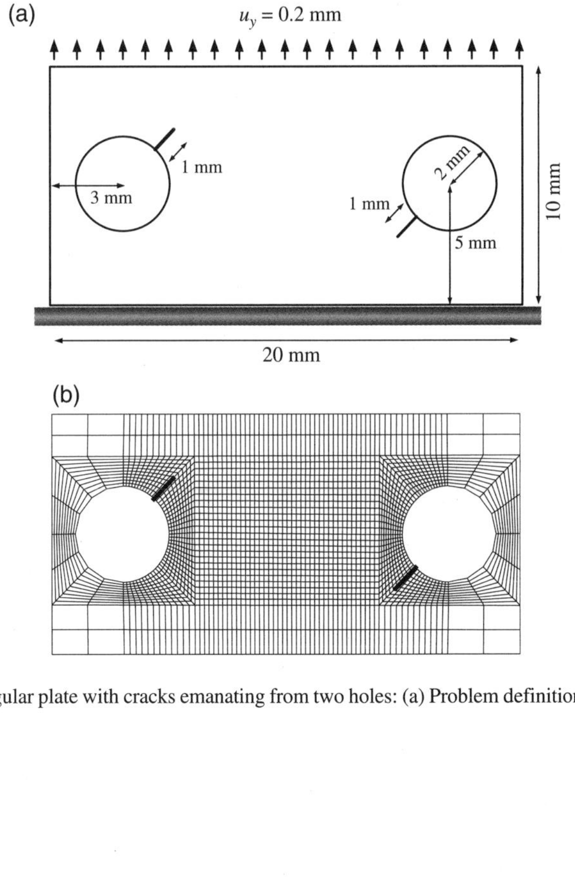

Figure 7.31 A rectangular plate with cracks emanating from two holes: (a) Problem definition; (b) The X-FEM mesh

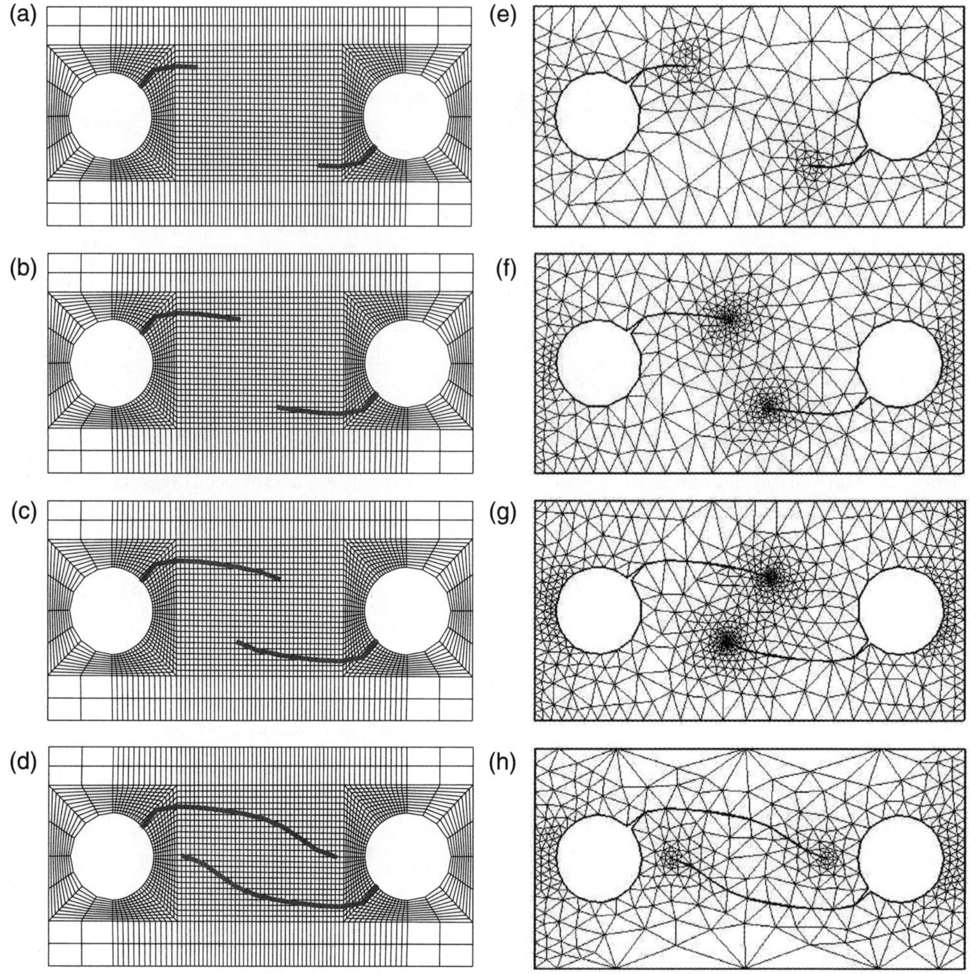

Figure 7.32 The crack trajectory for a plate with two holes during crack growth: (a–d) the X-FEM technique; (e–h) the adaptive FEM

7.7 Curved Crack Modeling with X-FEM

The simulation of straight crack in a fractured domain has been presented in preceding sections based on the linear triangular and quadrilateral elements. In order to model the curved crack in a fractured domain, the higher order elements can be employed to discretize the X-FEM formulation. In Chapter 2, it was illustrated that the higher order elements are a good candidate for curved crack simulation than the linear elements that are capable of representing the curved discontinuity, however, the extension to a *higher order* X-FEM requires the study of several necessities, including the convergence issue, the numerical integration, and the blending problem. The use of higher order elements in X-FEM was originally proposed by Wells and Sluys (2001) and Wells, Sluys, and de Borst (2002) using 6-noded triangular elements, however, no convergence studies were reported. The curved crack modeling with higher order elements was presented by Stazi *et al.* (2003) using different order of shape functions for the standard and

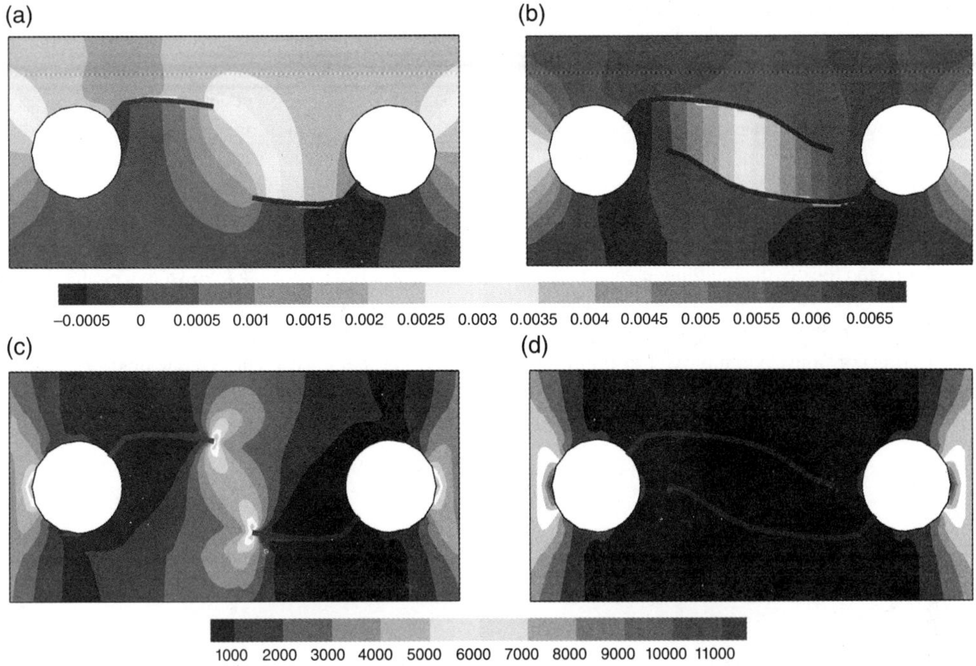

Figure 7.33 (a and b) The contours of displacement in y – direction (cm) and (c and d) the distribution of von-Mises stress (kg/cm^2) at the half and final loading steps of crack propagation (For color details, please see color plate section)

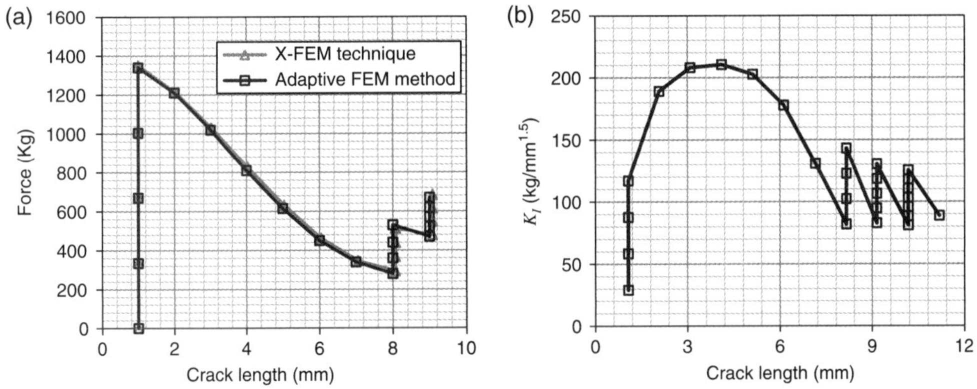

Figure 7.34 (a) The variation of vertical reaction with crack length; (b) the variation of mode I stress intensity factor with crack length

enriched parts. Chessa, Wang, and Belytschko (2003) highlighted that the higher order elements has a negative effect on the convergence rate of the X-FEM solution in weak discontinuities, however, it can be efficiently employed for strong discontinuities using the Heaviside enrichment function. Laborde *et al.* (2005) investigated the rate of convergence for higher order X-FEM, and highlighted that the asymptotic tip functions must be employed in a whole zone around the crack tip. Moreover, a numerical integration rule was proposed based on the polar integration algorithm that was numerically efficient. Byfut and Schröder (2012) proposed the higher order Lagrange-type partition of unity functions, and shown that the optimal convergence rates can be obtained in the X-FEM. However, the method severely increases the condition numbers of the linear system of equations, so they proposed to use a PU with lower-order in the vicinity of crack tips and higher order away from crack tips.

Basically, in order to obtain an optimal accuracy for the singular problem of elastic fracture mechanics, several requirements must be considered; firstly the crack tip enrichment must be applied in a fixed area around the crack tip, secondly an advanced integration technique is required for the element containing the crack tip, and thirdly the blending elements must be treated appropriately. It was shown in Chapter 4 that the higher order shape functions must not be employed as a PU support in the enriched parts. In fact, the quadratic or higher order polynomials can be used as the standard shape functions, whereas the linear shape function must be employed to build the PU. Hence, the enriched displacement field (7.35) can be written for the curved crack simulation as

$$\begin{aligned}\mathbf{u}(\boldsymbol{x},t) = \sum_{I\in\mathcal{N}} N_I(\boldsymbol{x})\,\bar{\mathbf{u}}_I &+ \sum_{J\in\mathcal{N}^{dis}} \widehat{N}_J(\boldsymbol{x})\,(H(\boldsymbol{x})-H(\boldsymbol{x}_J))\,\bar{\mathbf{d}}_J \\ &+ \sum_{K\in\mathcal{N}^{tip}} \widehat{N}_K(\boldsymbol{x}) \sum_{\alpha=1}^{4} (\mathcal{B}_\alpha(\boldsymbol{x})-\mathcal{B}_\alpha(\boldsymbol{x}_K))\,\bar{\mathbf{b}}_{\alpha K}\end{aligned} \tag{7.77}$$

in which $N_I(\boldsymbol{x})$ are the quadratic shape functions whereas $\widehat{N}_J(\boldsymbol{x})$ and $\widehat{N}_K(\boldsymbol{x})$ are the linear shape functions that construct the PU in the enriched parts. Figure 7.35 presents the implementation of higher order elements in a curved crack problem. It must be noted that the implementation of different order interpolants is crucial for the standard and enriched shape functions in achieving the optimal convergence rate. The reason can be related to the presence of elements located in the blending sub-domain in which only some nodal points of these elements are enriched. Since these elements are partially enriched, the enriched nodes of these elements do not form a PU, as mentioned earlier. In fact, the blending elements play an important role in the approximation properties and convergence of the enriched scheme.

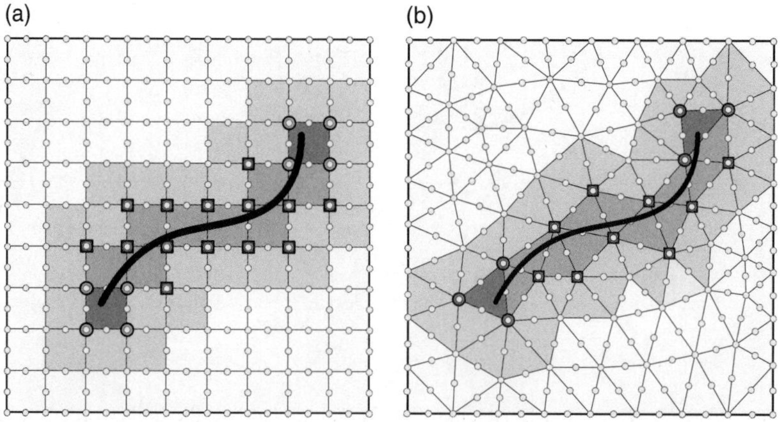

Figure 7.35 The X-FEM modeling of a cracked body with higher order elements: (a) quadrilateral mesh; (b) triangular mesh

7.7.1 Modeling a Curved Center Crack in an Infinite Plate

In order to illustrate the performance of X-FEM formulation with higher order elements, a curved center crack in an infinite plate is modeled numerically. The analytical solution is available for this problem as reported by Gdoutos (1993). The numerical simulation was also performed by Stazi *et al.* (2003) based on the X-FEM technique. According to the enriched displacement field (7.77), the displacement approximation includes the standard and enriched parts, where the higher order shape functions must be carefully taken into the computation. In this example, the 9-noded quadratic element is employed for the standard part and the 4-noded rectangular element for a PU.

Consider a square plate of 40 × 40 m subjected to tensile stress $\sigma = 1.0$ Pa from the top and bottom with a curved crack at its center, as shown in Figure 7.36, where the radius of circular arc is $R = 8.5$ m and the curved crack angle is $\beta = 56.145\,^{\circ}$. The analytical SIFs for this problem are given in modes I and II as (Gdoutos, 1993)

$$K_I = \frac{\sigma}{2}\left(\pi R \sin\frac{\beta}{2}\right)^{1/2}\left[\frac{\left(1-\sin^2\frac{\beta}{4}\cos^2\frac{\beta}{4}\right)\cos\frac{\beta}{4}}{1+\sin^2\frac{\beta}{4}} + \cos\frac{3\beta}{4}\right]$$

$$K_{II} = \frac{\sigma}{2}\left(\pi R \sin\frac{\beta}{2}\right)^{1/2}\left[\frac{\left(1-\sin^2\frac{\beta}{4}\cos^2\frac{\beta}{4}\right)\sin\frac{\beta}{4}}{1+\sin^2\frac{\beta}{4}} + \sin\frac{3\beta}{4}\right] \qquad (7.78)$$

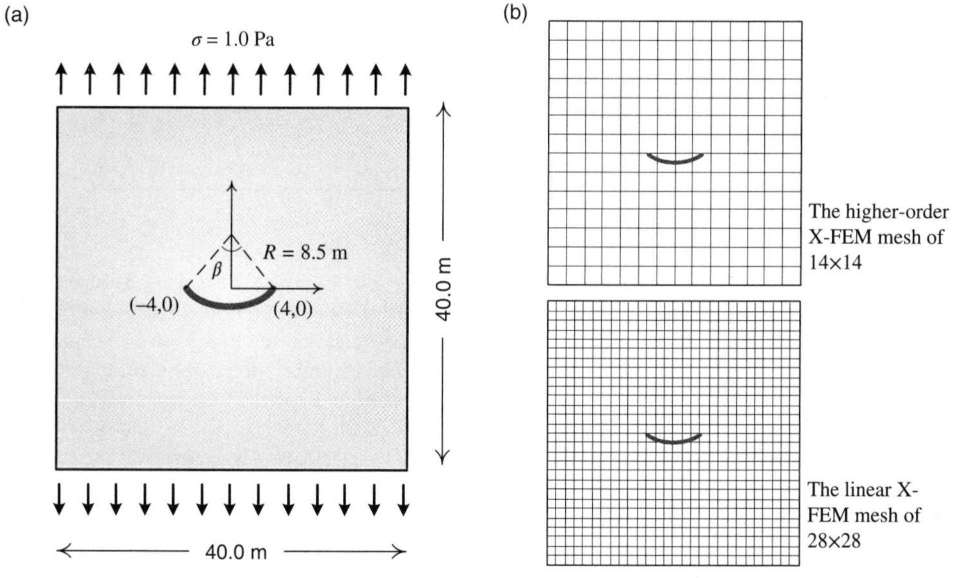

Figure 7.36 A curved center crack in an infinite plate: (a) Problem definition; (b) The X-FEM meshes of 14 × 14 higher-order elements and 28 × 28 linear elements

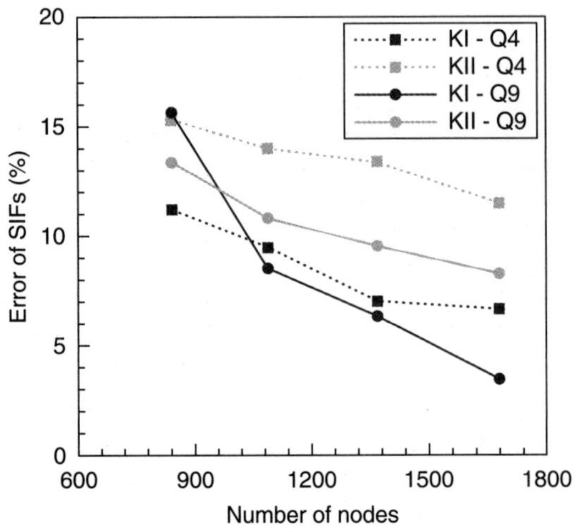

Figure 7.37 The evolutions of SIF errors with the number of nodal points for modes I and II fracture using linear and higher-order X-FEM meshes

Table 7.4 The values of stress intensity factors (SIFs) K_I and K_{II} together with the error of SIFs for a curved center crack in an infinite plate using linear and higher-order X-FEM meshes

		Numerical analysis		Error of SIFs (%)	
X-FEM meshes		K_I	K_{II}	K_I	K_{II}
Linear meshes	28 × 28	3.17	1.33	11.23	15.29
	32 × 32	3.12	1.35	9.47	14.01
	36 × 36	3.05	1.36	7.02	13.38
	40 × 40	3.04	1.39	6.67	11.46
Higher order meshes	14 × 14	3.30	1.78	15.62	13.38
	16 × 16	3.09	1.74	8.53	10.83
	18 × 18	3.03	1.72	6.35	9.55
	20 × 20	2.75	1.70	3.42	8.28

in which the analytical SIFs are obtained as $K_I = 2.848$ and $K_{II} = 1.571$. The X-FEM analysis is performed for four higher-order structured meshes with the refinements of 14 × 14, 16 × 16, 18 × 18, and 20 × 20, where the higher order Q9 elements are used for the standard part and the linear Q4 elements for the PU. In order to perform a comparison between the linear and higher order X-FEM solutions, the X-FEM analyses are also carried out using the linear standard and enriched elements with the same DOF, that is, four structured meshes with refinements of 28 × 28, 32 × 32, 36 × 36, and 40 × 40, and the results are compared to those obtained from higher order elements. The numerical integration is performed using nine Gauss points for higher order standard elements, four Gauss points for linear standard elements, and the rectangular sub-grids of 5 × 5 with four Gauss points at each sub-grid for enriched elements. In Figure 7.37, the variations of error of the SIFs K_I and K_{II} are plotted with the number of nodal points for modes I and II fracture using the linear and higher-order X-FEM meshes. In Table 7.4, the values of SIFs K_I and K_{II} are evaluated together with the error of SIFs for a curved center crack in an infinite plate using linear and higher-order X-FEM meshes. Obviously, the error of SIFs K_I and K_{II}

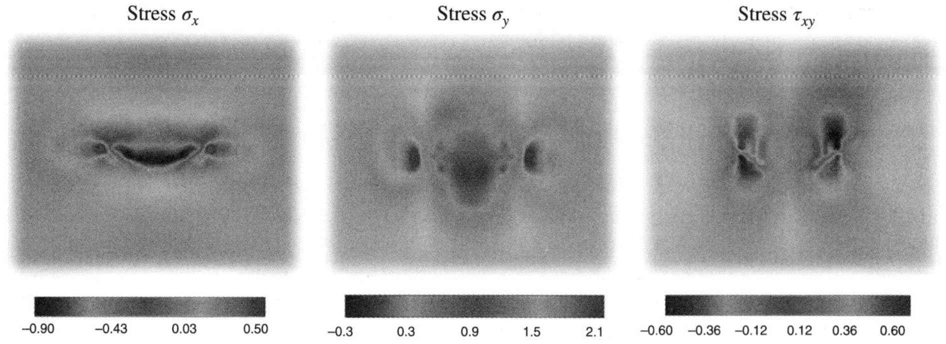

Figure 7.38 The distribution of stress contours σ_x, σ_y, and τ_{xy} for a curved center crack in an infinite plate (For color details, please see color plate section)

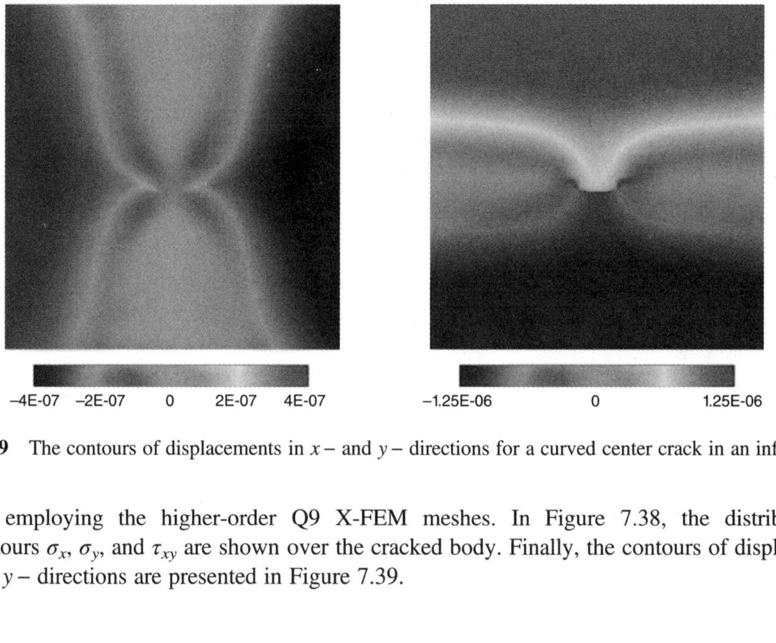

Figure 7.39 The contours of displacements in $x-$ and $y-$ directions for a curved center crack in an infinite plate

decreases employing the higher-order Q9 X-FEM meshes. In Figure 7.38, the distribution of stress contours σ_x, σ_y, and τ_{xy} are shown over the cracked body. Finally, the contours of displacements in $x-$ and $y-$ directions are presented in Figure 7.39.

7.8 X-FEM Modeling of a Bimaterial Interface Crack

The multi-functional and multi-layered material systems have attracted considerable attention in mechanical, aerospace, and biomedical applications. The overall mechanical behavior and response of layered systems are directly dependent on the mechanical properties and fracture behavior of the interfaces. The presence of weak fiber–matrix interfaces in composites provides the preferential crack paths that enhance the overall fracture toughness of the composite. Thus, the development of a robust technique to characterize the crack driving force and interfacial toughness in bimaterial systems can lead to a better understanding of the role and influence of the mismatch in properties and their effects on crack growth. In fact, an accurate evaluation of SIFs is essential for the prediction of failure and crack growth rate in these multi-layered structures. The analytical solution of bimaterial interfacial cracks was originally presented by Williams (1959), and then extended by Rice and Sih (1965) and Rice (1988) to describe an interpretation for the complex SIFs on interfacial crack mechanics.

In order to evaluate the SIFs of bimaterial interface cracks in complex composite components, the FEM has been used extensively by researchers. However, the fracture analysis of bimaterial systems by the standard FEM is a formidable task. Even aligning a mesh so that the fiber–matrix interfaces correspond to element interfaces, which is necessary in the standard FEM, is quite difficult for complex composites. The purpose of this section is to model the bimaterial interfacial cracks using a partition of unity-based enrichment method (PUM). It aims to apply the X-FEM in the analysis of cracks that lie at the interface of two elastically homogeneous isotropic materials. The implementation of X-FEM in the fracture analysis of composites and layered structures were presented by Nagashima, Omoto, and Tani (2003), Sukumar *et al.* (2004), and Liu, Xiao, and Karihaloo (2004). In these studies, only the crack tip enrichments were employed in the computations, and the FE mesh was aligned with the material interfaces. The X-FEM was also proposed by Hettich and Ramm (2006) and Huynh and Belytschko (2009) to model the failure in material interfaces, in which both material interfaces and singularities were taken into the X-FEM analysis, consequently, no mesh restriction was imposed.

7.8.1 The Interfacial Fracture Mechanics

Consider a bimaterial interface crack located along the interface between two semi-infinite planes, as shown in Figure 7.40. The plane above the crack is denoted by material (1) with the Young's modulus E_1 and Poisson ratio ν_1, and the plane below the crack is indicated by material (2) with the Young's modulus E_2 and Poisson ratio ν_2. For this type of problem, the SIFs K_I and K_{II} are inseparable in the vicinity of interface crack tip, and are not similar to those of standard mode I and mode II fracture defined in isotropic media. In such case, the SIF is defined in a complex form on the basis of LEFM as (Wiliams, 1959)

$$\mathbf{K} = K_I + i K_{II} \tag{7.79}$$

where $i = \sqrt{-1}$. In addition, the energy release rate of bimaterial interface crack can be related to the SIF as (Malyshev and Salganik, 1965)

$$\mathcal{G} = \frac{1}{E^*} \frac{|\mathbf{K}|^2}{\cosh^2(\pi \varepsilon)} \tag{7.80}$$

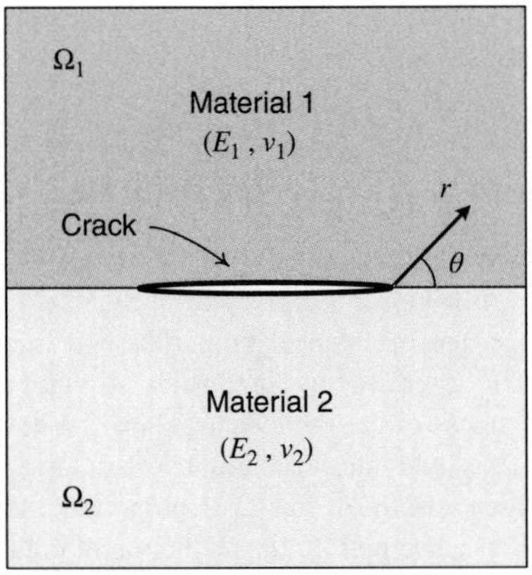

Figure 7.40 A bimaterial interface crack along the interface between two semi-infinite plates

where $|\mathbf{K}|^2 = K_I^2 + K_{II}^2$ and $2/E^* = 1/\bar{E}_1 + 1/\bar{E}_2$, in which $\bar{E}_i = E_i$ for plane stress and $\bar{E}_i = E_i/\left(1-\nu_i^2\right)$ for plane strain problems.

The near tip displacement fields in the upper-half domain, that is, $0 \le \theta \le \pi$, was introduced by Rice (1988) in the polar coordinates (r, θ) as

$$u_j(r,\theta) = \frac{1}{2\mu_1}\sqrt{\frac{r}{2\pi}}\left(\mathrm{Re}\left[\mathbf{K}\,r^{i\varepsilon}\right]\tilde{u}_j^{\mathrm{I}}(\theta,\varepsilon,\nu_1) + \mathrm{Im}\left[\mathbf{K}\,r^{i\varepsilon}\right]\tilde{u}_j^{\mathrm{II}}(\theta,\varepsilon,\nu_1)\right) \quad (j=1,2) \tag{7.81}$$

where ε is a bimaterial constant and $r^{i\varepsilon} = e^{i\varepsilon \log r} \equiv \cos(\varepsilon \log r) + i\,\sin(\varepsilon \log r)$. In this relation, $\mathrm{Re}[\Xi]$ and $\mathrm{Im}[\Xi]$ denote the real and imaginary parts of a complex number Ξ, and the angular functions $\tilde{u}_j^{\mathrm{I}}$ and $\tilde{u}_j^{\mathrm{II}}$ in the asymptotic displacement fields (7.81) are defined as

$$\begin{aligned}
\tilde{u}_1^{\mathrm{I}} &= A\left[-e^{2\varepsilon(\pi-\theta)}\left(\cos\frac{\theta}{2} + 2\varepsilon\sin\frac{\theta}{2}\right) + \kappa_1\left(\cos\frac{\theta}{2} - 2\varepsilon\sin\frac{\theta}{2}\right) + (1+4\varepsilon^2)\sin\frac{\theta}{2}\sin\theta\right]\\
\tilde{u}_1^{\mathrm{II}} &= A\left[e^{2\varepsilon(\pi-\theta)}\left(\sin\frac{\theta}{2} - 2\varepsilon\cos\frac{\theta}{2}\right) + \kappa_1\left(\sin\frac{\theta}{2} + 2\varepsilon\cos\frac{\theta}{2}\right) + (1+4\varepsilon^2)\cos\frac{\theta}{2}\sin\theta\right]\\
\tilde{u}_2^{\mathrm{I}} &= A\left[e^{2\varepsilon(\pi-\theta)}\left(\sin\frac{\theta}{2} - 2\varepsilon\cos\frac{\theta}{2}\right) + \kappa_1\left(\sin\frac{\theta}{2} + 2\varepsilon\cos\frac{\theta}{2}\right) - (1+4\varepsilon^2)\cos\frac{\theta}{2}\sin\theta\right]\\
\tilde{u}_2^{\mathrm{II}} &= A\left[e^{2\varepsilon(\pi-\theta)}\left(\cos\frac{\theta}{2} + 2\varepsilon\sin\frac{\theta}{2}\right) - \kappa_1\left(\cos\frac{\theta}{2} - 2\varepsilon\sin\frac{\theta}{2}\right) + (1+4\varepsilon^2)\sin\frac{\theta}{2}\sin\theta\right]
\end{aligned} \tag{7.82}$$

where $A = e^{-\varepsilon(\pi-\theta)}/[(1+4\varepsilon^2)\cosh(\pi\,\varepsilon)]$. In these relations, the polar coordinates system (r, θ) is defined at the right crack tip, as shown in Figure 7.40, and the bimaterial constant ε is defined as (Dundurs, 1969)

$$\varepsilon = \frac{1}{2\pi}\log\left(\frac{1-\beta}{1+\beta}\right) \quad \text{with} \quad \beta = \frac{\mu_1(\kappa_2-1) - \mu_2(\kappa_1-1)}{\mu_1(\kappa_2+1) + \mu_2(\kappa_1+1)} \tag{7.83}$$

in which the constant κ_i is defined as $\kappa_i = (3-\nu_i)/(1+\nu_i)$ for plane stress and $\kappa_i = 3 - 4\nu_i$ for plane strain problems, and the parameters μ_i, ν_i, and κ_i denote the material type $i(i=1,2)$. In order to obtain the near tip displacement fields in the lower-half domain, the variable $\varepsilon\pi$ in (7.82) can be replaced by $-\varepsilon\pi$. In Figure 7.41, the angular functions $\tilde{u}_j^{\mathrm{I}}$ and $\tilde{u}_j^{\mathrm{II}}$ are plotted for $E_1/E_2 = 0.1$ and $\nu_1 = \nu_2 = 0.3$ in two cases of the bimaterial constant ε, that is, $\varepsilon = 0$ (no mismatch) and $\varepsilon = 0.076$, in the plane strain condition.

7.8.2 The Enrichment of the Displacement Field

In crack simulation of isotropic linear elasticity, the Heaviside jump function and the 2D asymptotic crack tip displacement fields are used to account for the crack discontinuity. This enables the domain to be modeled by FE without explicitly meshing the crack surfaces. For the bimaterial interface crack, the analytical solution of near tip displacement fields is introduced in relation (7.81). In order to model the interface crack within the X-FEM framework, the Heaviside enrichment function $H(\boldsymbol{x})$ is used to model the crack surface, and the crack tip asymptotic functions $\mathcal{B}(r,\theta)$ are employed to model the crack tip for an interface

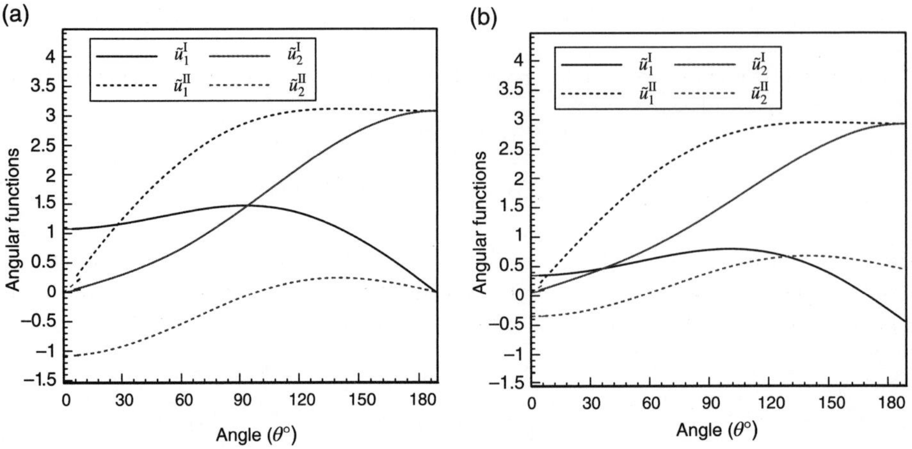

Figure 7.41 Angular functions $\tilde{u}_j^{\mathrm{I}}$ and $\tilde{u}_j^{\mathrm{II}}$ in the asymptotic displacement fields for $E_1/E_2 = 0.1$ and $\nu_1 = \nu_2 = 0.3$ at two bimaterial constants: (a) $\varepsilon = 0$ (no mismatch); (b) $\varepsilon = 0.076$

crack. Based on relations (7.82), the crack tip asymptotic functions for a bimaterial interfacial crack can be defined using the following 12 functions as

$$
\begin{aligned}
\mathcal{B}(r,\theta) &= \{\mathcal{B}_1, \mathcal{B}_2, \ldots, \mathcal{B}_{12}\} \\
&= \Bigg\{ \sqrt{r}\cos(\varepsilon\log r)e^{-\varepsilon\theta}\sin\frac{\theta}{2},\ \sqrt{r}\cos(\varepsilon\log r)e^{-\varepsilon\theta}\cos\frac{\theta}{2}, \\
&\quad \sqrt{r}\cos(\varepsilon\log r)e^{+\varepsilon\theta}\sin\frac{\theta}{2},\ \sqrt{r}\cos(\varepsilon\log r)e^{+\varepsilon\theta}\cos\frac{\theta}{2}, \\
&\quad \sqrt{r}\cos(\varepsilon\log r)e^{+\varepsilon\theta}\sin\frac{\theta}{2}\sin\theta,\ \sqrt{r}\cos(\varepsilon\log r)e^{+\varepsilon\theta}\cos\frac{\theta}{2}\sin\theta, \\
&\quad \sqrt{r}\sin(\varepsilon\log r)e^{-\varepsilon\theta}\sin\frac{\theta}{2},\ \sqrt{r}\sin(\varepsilon\log r)e^{-\varepsilon\theta}\cos\frac{\theta}{2}, \\
&\quad \sqrt{r}\sin(\varepsilon\log r)e^{+\varepsilon\theta}\sin\frac{\theta}{2},\ \sqrt{r}\sin(\varepsilon\log r)e^{+\varepsilon\theta}\cos\frac{\theta}{2}, \\
&\quad \sqrt{r}\sin(\varepsilon\log r)e^{+\varepsilon\theta}\sin\frac{\theta}{2}\sin\theta,\ \sqrt{r}\sin(\varepsilon\log r)e^{+\varepsilon\theta}\cos\frac{\theta}{2}\sin\theta \Bigg\}
\end{aligned}
\tag{7.84}
$$

If the bimaterial constant ε is set to zero, these tip asymptotic functions result in the span of near tip enrichment functions used for isotropic material in (7.34). The enriched displacement field (7.35) can therefore be written for the bimaterial interface crack as

$$
\begin{aligned}
\mathbf{u}(\boldsymbol{x},t) &= \sum_{I\in\mathcal{N}} N_I(\boldsymbol{x})\,\bar{\mathbf{u}}_I + \sum_{J\in\mathcal{N}^{dis}} N_J(\boldsymbol{x})\,(H(\boldsymbol{x})-H(\boldsymbol{x}_J))\,\bar{\mathbf{d}}_J \\
&\quad + \sum_{K\in\mathcal{N}^{tip}} N_K(\boldsymbol{x}) \sum_{\alpha=1}^{12} (\mathcal{B}_\alpha(\boldsymbol{x}) - \mathcal{B}_\alpha(\boldsymbol{x}_K))\,\bar{\mathbf{b}}_{\alpha K}
\end{aligned}
\tag{7.85}
$$

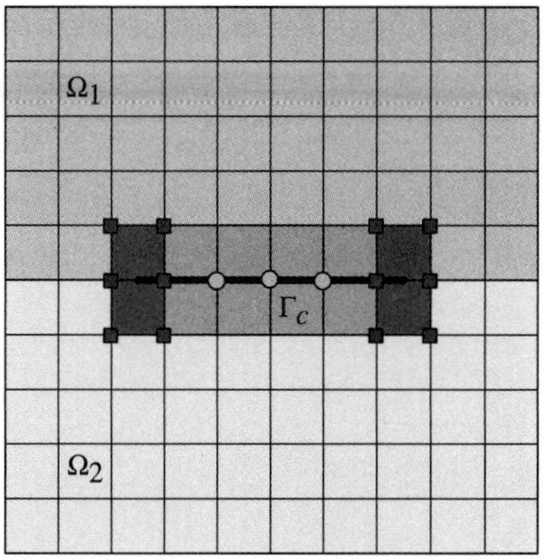

Figure 7.42 The enriched nodal points for a bimaterial interface crack, where the circled nodes enriched by the Heaviside function and the near tip functions by the squared nodes

in which 12 asymptotic functions are used in the last term of relation (7.85) to enrich the nodal points of the crack tip element. In Figure 7.42, the enriched nodal points for a bimaterial interface crack are presented, where the circled nodes are enriched by the Heaviside function, and the squared nodes are enriched with the near tip functions given in (7.84). Since the edges of the FE mesh coincide with the interface crack, all nodal points that are enriched by the Heaviside function lie on the crack surface. It must be noted that if the other nodal points of the same element are also enriched, it then leads to the ill-conditioned stiffness matrix, as discussed in Section 2.7.2. In the case that the crack has an arbitrary orientation with respect to the FE mesh, the partitioning algorithm needs to be employed.

In order to perform the numerical integration of the weak form of equilibrium equation, it is necessary to use the higher-order Gauss quadrature rules for those elements enriched by the Heaviside and crack tip asymptotic functions. For this purpose, the crack tip elements are split into sub-triangles with 12 Gauss integration points in each sub-triangle. It was shown by Sukumar *et al.* (2004) that the use of 6 integration points leads to an almost singular ill-conditioned stiffness matrix; and the 12 integration points rule are sufficient to obtain a well-conditioned matrix for an accurate solution of the discrete system.

In order to compute the SIFs in the mixed-mode fracture given in (7.79), the *J*–integral method can be applied for the bimaterial interface crack. Based on the *J*–integral method described in Section 7.2.3, the interaction integral can be related to the SIFs as

$$I^{(1,2)} = \frac{2}{E^* \cosh^2(\pi \varepsilon)} \left(K_I^{(1)} K_I^{(2)} + K_{II}^{(2)} K_{II}^{(1)} \right) \tag{7.86}$$

where, by selecting the auxiliary state (2) as the pure mode I asymptotic fields, that is, $K_I^{(2)} = 1$ and $K_{II}^{(2)} = 0$, the mode I SIF K_I can be obtained; and in an analogous manner the mode II SIF K_{II} can be determined as

$$K_I = \frac{E^* \cosh^2(\pi \varepsilon)}{2} I_{\text{mode } I} \quad \text{and} \quad K_{II} = \frac{E^* \cosh^2(\pi \varepsilon)}{2} I_{\text{mode } II} \tag{7.87}$$

This interaction integral is a well-established technique that can be used to determine the mixed-mode SIFs in 2D interfacial fracture computations.

7.8.3 Modeling of a Center Crack in an Infinite Bimaterial Plate

In order to illustrate the performance of X-FEM technique in modeling the bimaterial interface crack, an infinite bimaterial plate with a center-crack located along the interface between two materials is modeled, as shown in Figure 7.43; and the results of SIFs are compared with the analytical solution for the pure tension far-field loading. The analytical solution of this problem subjected to the far-field tension stress σ was obtained by Rice and Sih (1965), where the SIFs K_I and K_{II} are defined as

$$\mathbf{K} = K_I + i K_{II} = \sigma (1 + 2i\varepsilon)\sqrt{\pi a}(2a)^{-i\varepsilon} \tag{7.88}$$

where $2a$ is the crack length, and the tensile stress σ is imposed on the upper and lower edges of the plate. The material parameters used for numerical simulation are as follows; $E_1/E_2 = 22$, $\nu_1 = 0.2571$,

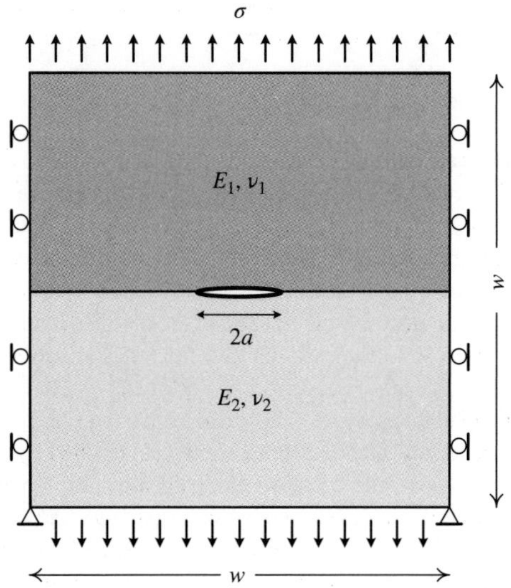

Figure 7.43 An infinite bimaterial plate with a center crack along the interface: The geometry and boundary conditions

Table 7.5 An infinite bimaterial plate with a center-crack: The normalized stress intensity factors for two X-FEM meshes at $w/a = 40$

X-FEM mesh	a/s	r_d	K_I/K_0	K_{II}/K_0
200 × 200	5	2	1.011	−0.1154
		3	1.009	−0.1132
		4	1.010	−0.1126
400 × 400	10	2	1.011	−0.1145
		3	1.010	−0.1121
		4	1.010	−0.1115
		5	1.010	−0.1113
		6	1.010	−0.1113

and $\nu_2 = 0.3$; and the problem is solved in the plane strain condition. The exact values of SIFs K_I and K_{II} can be obtained from (7.88) as

$$\frac{K_I}{K_0} = 1.008, \quad \frac{K_{II}}{K_0} = -0.1097 \tag{7.89}$$

in which the factor K_0 is defined by $K_0 = \sigma\sqrt{\pi a}$ to normalize the SIF. The X-FEM modeling is performed for two structured meshes of 200×200 and 400×400. The numerical results are obtained for the ratio of plate width to crack size $w/a = 40$ and the normalized SIFs are given in Table 7.5. In this table, the ratio of crack size to element size is chosen by a/s. In addition, a domain radius of $r_{tip} = r_d\, s$ is defined to compute the interaction integral (7.25), where r_d is a scalar multiple.

8

Cohesive Crack Growth with the X-FEM Technique

8.1 Introduction

Linear elastic fracture mechanics (LEFM) has proven to be a useful tool for solving fracture problems when the nonlinear zone ahead of the crack tip is negligible. This is not always the case and for quasi-brittle materials, such as geomaterials and concrete, the size of the nonlinear zone at the crack front due to plasticity or micro-cracking is not negligible in comparison with other dimensions of the crack geometry (Bazant and Planas, 1998). Moreover, the LEFM assumption is quite restrictive for certain types of failure in materials like structural steels, though such materials can be prone to brittle fracture but in some cases show a quasi-brittle or a ductile behavior for which plasticity plays an important role during fracture. The cohesive model is employed to describe the nonlinear fracture processes developing in the area in front of the crack tip called the fracture process zone (FPZ) where the energy dissipation takes place.

In the cohesive model, it is assumed that the near tip FPZ is lumped into the crack line, unlike the LEFM in which the fracture processes are considered to occur at the crack tip. The basic assumption is the formation of the process zone, where the material, albeit damaged, is still able to transfer stresses (Figure 8.1). From a meso-scale point of view, a material like concrete consists of mortar matrix and aggregates. The micro-cracks are initiated close to the interface between the mortar matrix and aggregate, and a macroscopic crack is created from an assemblage of micro-cracks. The heterogeneous structure of concrete leads to phenomena such as crack bridging, where two parallel cracks are connected by an aggregate. This allows the cohesive forces to be transferred through an existing crack and enables the model to describe materials that exhibit a strain-softening type behavior. The point separating the stress-free area, that is, the real crack, from the process zone, is called the real crack tip, while the point separating the process zone from the uncracked material is referred to as the fictitious crack tip. Moreover, at the end of the fictitious crack, the ultimate stress is equal to the value of tensile strength. Thus, no singularities arise in the state of stress. In other words, the cohesive crack model allows abandoning the singularity of the crack tip stress field, an unrealistic assumption of LEFM. In the cohesive crack model, the nonlinear behavior of the material in the FPZ is described using a cohesive constitutive relation. The cohesive constitutive relation represents the failure characteristics of the material and characterizes the separation process.

The simplest cohesive constitutive relation is one where in the FPZ the cohesive surface traction is a function of the displacement jump, while in the uncracked zone the behavior of material is linear-elastic. Such a cohesive constitutive relation incorporates the strength f_t and the work of separation (or fracture energy) E_f into the model. Figure 8.2 depicts various cohesive constitutive relations in terms of the

Extended Finite Element Method: Theory and Applications, First Edition. Amir R. Khoei.

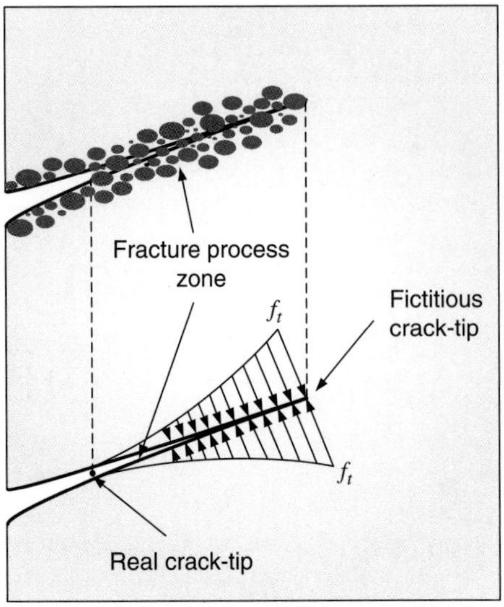

Figure 8.1 The fracture process zone

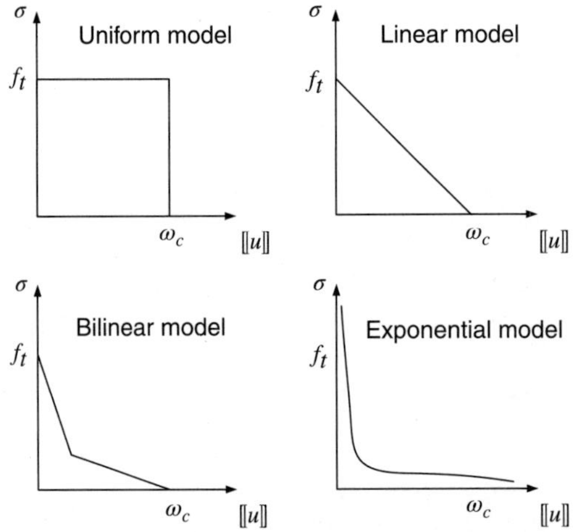

Figure 8.2 Different cohesive laws in terms of the normalized traction and opening displacement jump

normalized traction and opening displacement jump. The cohesive crack model is able to describe materials that exhibit a strain-softening type behavior. The area under the closing stress versus crack opening displacement curve represents the fracture energy E_f, which is assumed as a material property. The concept of cohesive fracture was originally introduced by Barenblatt (1959, 1962) and Dugdale (1960) to describe the near tip nonlinear processes taking place in brittle and ductile materials, respectively. In both models, the crack was divided into two parts; the first part is the crack surface, which is stress free, and the second part is a cohesive zone, which is introduced ahead of the existing crack tip and loaded by cohesive stresses. Barenblatt (1962) investigated the fracture of brittle materials and proposed several assumptions about cohesive stresses; first, the extension of the cohesive zone is constant for a given material independent of the global load and second, it is small compared to other dimensions. The stresses in the cohesive zone follow a prescribed distribution, which can be specified for a given material but independent of the global loading conditions. Dugdale (1960) introduced a finite stress as the yield stress for plane stress problems, where the crack opening stresses can be much higher than the equivalent stress in a multi-axial stress state. The relation between the work expended in the cohesive zone and that in the crack tip field is typically such that the stress singularity vanishes and near tip stresses are finite. Barenblatt also presented that the cohesive forces essentially have effect only on the displacement field close to the edge of the crack and not on those in the main part of the crack.

The cohesive models of Barenblatt and Dugdales were proposed by Bilby, Cottrell, and Swinden (1963), Willis (1967), Rice (1968), and Wnuk (1974) to analyze the strain concentration in notches and cracks. An extension of the Dugdale–Barenblatt plastic crack tip model was presented by Hillerborg, Modéer, and Petersson (1976) to simulate cohesive crack growth in concrete in the concept of a fictitious crack model, which relates normal stress to normal crack opening. They introduced the concept of fracture energy into the cohesive crack model and proposed a number of traction–displacement relationships for concrete. In fact, the development of an appropriate numerical simulation of cohesive zone model into the finite element method (FEM) was originally proposed by Hillerborg, Modéer, and Petersson (1976) to investigate the localized failure of concrete bending beam with the crack opening displacement related equilibrium forces. In this model, the crack was assumed to propagate when the stress at the crack tip reaches the tensile strength. The stresses applied on the crack surfaces decrease with the increase in crack opening and do not drop to zero suddenly. This fact makes the crack close smoothly and no singularity remains at the crack tip at the onset of crack propagation, which makes the stress intensity factor (SIF) in mode I (K_I) vanish. There are some similarities among the Barenblatt, Dugdale, and Hillerborg models; the crack tip faces close smoothly (the SIF vanishes at the crack-tip in mode I propagation) and the FPZ is of negligible thickness. On the other hand, the closing stresses in the FPZ are constant only in Dugdale model, while the size of this zone is constant and small in comparison with the length of main crack in Barenblatt model.

There are various numerical methods developed in literature to model cohesive fracture problems, including the boundary element method (BEM), meshless techniques, the FEM, and the extended-FEM technique. The BEM has been used extensively in fracture mechanics since it only needs to mesh the boundary of the domain that results in a smaller stiffness matrix compared to the FEM. Different BEM techniques have been developed by researchers in cohesive fracture mechanics; including a multi-domain BEM by Cen and Maier (1992) in modeling the bifurcation and instability in fracture of cohesive-softening structures, a symmetric Galerkin BEM by Maier, Novati, and Cen (1993) for quasi-brittle-fracture with frictional contact, a dual BEM by Saleh and Aliabadi (1995) in modeling the mode I and mixed-mode cohesive crack propagation in concrete, a single-domain dual boundary element approach by Yang and Ravi-Chandar (1998) in simulation of the cohesive crack growth, and the multi-zone BEM by Chen *et al.* (1999) for cohesive crack growth. There are also various meshless techniques used by researchers to model cohesive fracture problems, such as the Element Free Galerkin (EFG) method by Belytschko (2000b) for mixed-mode dynamic crack propagation in concrete, and an extended Element-Free Galerkin technique using the local partition of unity method (PUM) by Rabczuk and Zi (2007) for cohesive crack growth.

The FEM has been extensively employed to model cohesive crack propagation and can be generally categorized into two approaches. The first approach is based on the inter-element algorithm, in which the cohesive constitutive model is applied between elements and remeshing is necessary when the crack path is not known in advance. This approach has been used for both brittle and ductile materials by various researchers; including a cohesive zone model by Xu and Needleman (1994) where cohesive elements are inserted into the finite element mesh, a stress-based extrinsic cohesive law by Camacho and Ortiz (1996) where a new surface is adaptively created by duplicating nodes which were previously bonded, and a rate-dependent cohesive crack model by Bazant and Li (1997) and Xu, Siegmund, and Ramani (2003) in FE (finite element) modeling of cohesive crack growth. Carpinteri *et al.* (2003) employed the inter-element approach to analyze crack stability in elastic-softening materials like concrete, in which the value of principal stress was taken as a crack propagation criterion in the cohesive crack model. A bilinear cohesive zone model was adopted by Song, Paulino, and Buttlar (2006b) to efficiently reduce the artificial compliance inherent in the intrinsic cohesive law and to model mixed-mode crack propagation in asphalt concrete. The inter-element approach was also applied in hydraulic fracturing of poroelastic media; such as the cohesive interface element by Khoei, Barani, and Mofid (2011) and Barani, Khoei, and Mofid (2011) in the FE analysis of fracture propagation, and the zero-thickness cohesive element by Sarris and Papanastasiou (2012) and Carrier and Granet (2012) in simulation of hydraulic fracturing. The second approach is based on the intra-element algorithm, which essentially consists of enriching the continuous displacement modes of the standard finite elements with additional discontinuous displacements, devised for capturing the physical discontinuity, that is, fractures, cracks, and so on. This approach is generally employed in the concept of PUM to include the enrichment functions and known as the Extended Finite Element Method (X-FEM).

The X-FEM technique is developed to model the discontinuity in the displacement field along the crack path, wherever the crack is located with respect to the mesh. The technique was originally proposed by Dolbow, Moës, and Belytschko (2001) in modeling the crack growth with frictional contact. The technique was also presented by Wells and Sluys (2001) and Moës and Belytschko (2002a) based on the PU property of FEs for cohesive crack problems. A cohesive segment method was introduced by Remmers, de Borst, and Needleman (2003, 2008) to model the crack nucleation and discontinuous crack growth, where the segments are added to the finite element mesh by using the partition of unity property. Li and Ghosh (2006) developed an extended Voronoi cell finite element model to predict the multiple cohesive crack interaction and propagation in brittle materials. The X-FEM was employed by Unger, Eckardt, and Könke (2007) for discrete crack simulation of concrete using an adaptive crack growth algorithm. A methodology was proposed by Réthoré, de Borst, and Abellan (2008) to insert discontinuities, such as cracks, faults, or shear bands, in an unsaturated porous medium, in which a two-scale X-FEM model was developed for fluid flow in a deforming, unsaturated, and progressively fracturing porous medium. The X-FEM was developed by Mohammadnejad and Khoei (2013b, c) in modeling the hydraulic fracture propagation of multi-phase porous media in conjunction with the cohesive crack model. In this chapter, the X-FEM technique is utilized to discretize the weak form of governing equations of a cracked body combined with the cohesive crack model by introducing the enriched DOF to the nodes associated with the discontinuity. Various cohesive crack growths are demonstrated in the framework of extended-FEM based on the stress criterion, the SIF criterion, and the cohesive segments method.

8.2 Governing Equations of a Cracked Body

A two-dimensional domain Ω crossed by a discontinuity Γ_d is considered, as shown in Figure 8.3. The strong form of the equilibrium equation of a body can be expressed as

$$\nabla\cdot\boldsymbol{\sigma}+\mathbf{b}=0 \tag{8.1}$$

where ∇ is the gradient operator, $\boldsymbol{\sigma}$ is the Cauchy stress tensor, and $\mathbf{b}$ is the body force vector. The behavior of the bulk material is assumed to be the linear elastic, so its constitutive relation is defined by $\boldsymbol{\sigma}=\mathbf{D}\boldsymbol{\varepsilon}$, where $\mathbf{D}$ is the tangential constitutive matrix of the bulk material. The essential and natural boundary conditions are respectively as follows

$$\begin{aligned}\mathbf{u}&=\tilde{\mathbf{u}} && \text{on} \quad \Gamma_u\\ \boldsymbol{\sigma}\cdot\mathbf{n}_\Gamma&=\bar{\mathbf{t}} && \text{on} \quad \Gamma_t\\ \boldsymbol{\sigma}\cdot\mathbf{n}_{\Gamma_d}&=\mathbf{t}_d && \text{on} \quad \Gamma_d\end{aligned} \tag{8.2}$$

where $\mathbf{n}_\Gamma$ is the outward unit normal vector to the external boundary Γ, and $\mathbf{n}_{\Gamma_d}$ is the unit normal vector to the discontinuity Γ_d pointing to Ω^+. Moreover, $\tilde{\mathbf{u}}$ is the prescribed displacement at the boundary Γ_u, $\bar{\mathbf{t}}$ is the

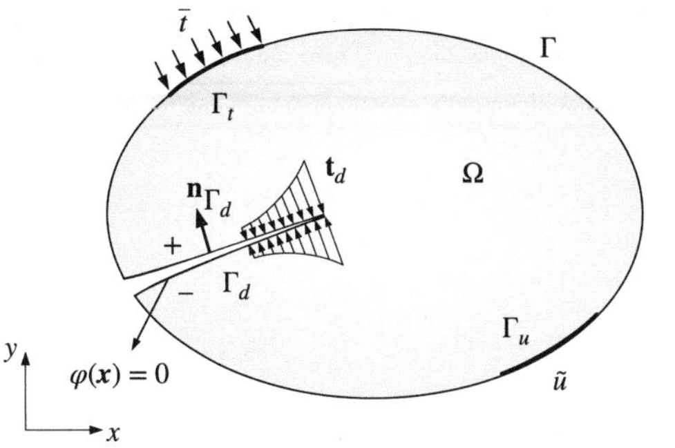

Figure 8.3 A two-dimensional domain involving a discontinuity Γ_d

prescribed traction at the boundary Γ_t, and $\mathbf{t}_d$ is the cohesive traction transferred across Γ_d, which is related to the displacement jump at the discontinuity using the cohesive law. It is noted that in imposing the boundary conditions distinction is made between the crack mouth and the internal boundary of the crack. Conditions at the crack mouth are imposed as the essential or natural boundary conditions on the external boundary of the body, while boundary conditions on the crack are the direct consequence of the governing equations.

The nonlinear behavior of the fracturing material in the cohesive zone is characterized by a traction-separation law relating the cohesive traction to the relative displacement as

$$\mathbf{t}_d = \mathbf{t}_d([\![\mathbf{u}]\!]) \tag{8.3}$$

where $\mathbf{t}_d$ is the cohesive traction transmitted across the FPZ and $[\![\mathbf{u}]\!]$ is defined as the relative displacement vector across the discontinuity, the difference between the displacement vectors at the two faces of the discontinuity. Indeed, the symbol $[\![\ ]\!]$ denotes the jump across the discontinuity, which is further discussed later. The cohesive crack concept implies that the cohesive tractions on either side of the crack are identical. In quasi-brittle materials, as soon as the failure limit of the material is exceeded, the cohesive zone develops in which the material begins to fail and exhibits a softening behavior. The softening induced by the material failure is simulated using a softening cohesive law. This implies that the cohesive traction transferred across the cohesive zone is made of a decaying function of the relative displacement. It is noted that in quasi-brittle materials the cohesive zone undergoes softening due to the occurrence of micro-cracks.

The linearization of the cohesive relation (8.3) results in the following differential form of the traction-separation law as

$$d\mathbf{t}_d = \mathbf{T}d[\![\mathbf{u}]\!] \tag{8.4}$$

where $\mathbf{T}$ represents the tangential modulus matrix of the discontinuity, which is obtained from the following relation

$$\mathbf{T} = \frac{\partial \mathbf{t}_d}{\partial [\![\mathbf{u}]\!]} \tag{8.5}$$

In order to develop a relation for the material tangent matrix $\mathbf{T}$, it is first formulated in the local orthogonal coordinate system, constructed from the tangential and normal unit vectors to the discontinuity, $\mathbf{t}_{\Gamma_d}$ and $\mathbf{n}_{\Gamma_d}$, with the orientation $\mathbf{t}_{\Gamma_d}$ taken such that the unit vector perpendicular to the plane of the two-dimensional medium forms a right-handed system. Then, it is rotated into the global coordinate system in the following form

$$\mathbf{T} = \mathbf{A}^T\mathbf{T}'\mathbf{A} \tag{8.6}$$

where **A** is the rotation matrix, applied to perform the transformation to the global coordinate system, and $\mathbf{T}'$ is related to the local coordinate system, obtained by differentiating the tangential and normal components of the cohesive traction with respect to the displacement jump in the tangential and normal directions as

$$\mathbf{T}' = \begin{bmatrix} \dfrac{\partial t_s}{\partial [\![u_s]\!]} & \dfrac{\partial t_s}{\partial [\![u_n]\!]} \\ \dfrac{\partial t_n}{\partial [\![u_s]\!]} & \dfrac{\partial t_n}{\partial [\![u_n]\!]} \end{bmatrix} \tag{8.7}$$

where t_s and t_n are the shear and normal cohesive tractions, respectively, and $[\![u_s]\!]$ and $[\![u_n]\!]$ denote the crack sliding and the crack opening displacement, respectively. The components of the cohesive traction and displacement jump tangent and normal to the crack are determined by decomposing the related vector with respect to the local orthogonal basis. The explicit form for $\mathbf{T}'$ is obtained considering the adopted cohesive law.

In the case of the occurrence of mode I failure, the relation given in Eq. (8.6) is simply substituted by the following relation in terms of the cohesive traction and the relative displacement in the normal direction to the crack as

$$\mathbf{T} = \mathbf{n}_{\Gamma_d} \frac{\partial t_n}{\partial [\![u_n]\!]} \mathbf{n}_{\Gamma_d}^T \tag{8.8}$$

whose detailed expression depends on the constitutive model applied at the discontinuity. It is noted that in this case the shear cohesive traction acting tangent to the crack and the shear relative displacement in the tangential direction with respect to the crack are zero. That is, this failure mode only involves the normal cohesive traction and the crack opening.

8.2.1 The Enrichment of Displacement Field

The X-FEM is an efficient method to solve problems containing a discontinuity. By exploiting the PU property of FE shape functions, the displacement discontinuity can be directly modeled by introducing enriched DOF to the nodes associated with the discontinuity. The displacement field can be decomposed into two parts; a continuous and a discontinuous part, defined as

$$\mathbf{u}(\boldsymbol{x},t) = \hat{\mathbf{u}}(\boldsymbol{x},t) + H_{\Gamma_d}(\boldsymbol{x})[\![\mathbf{u}(\boldsymbol{x},t)]\!] \tag{8.9}$$

where $\hat{\mathbf{u}}$ and $[\![\mathbf{u}]\!]$ are continuous functions, and H_{Γ_d} is the Heaviside function centered on the discontinuity, defined as

$$H_{\Gamma_d}(\boldsymbol{x}) = \begin{cases} 0 & \text{if } \boldsymbol{x} \in \Omega^- \\ 1 & \text{if } \boldsymbol{x} \in \Omega^+ \end{cases} \tag{8.10}$$

in which for the discontinuity Γ_d, it can be stated $\Omega^+ \cup \Omega^- = \Omega$. As was pointed out in the previous section, the notation $[\![\Xi]\!] = \Xi^+ - \Xi^-$ represents the difference between the corresponding values at the two crack faces. The components of displacement jump at the discontinuity are given by $[\![\mathbf{u}]\!]_{x\in\Gamma_d}$. The strain field for a body crossed by a discontinuity is given by taking the gradient of Eq. (8.9) as

$$\boldsymbol{\varepsilon}(\boldsymbol{x},t) = \nabla^s \hat{\mathbf{u}}(\boldsymbol{x},t) + H_{\Gamma_d} \nabla^s [\![\mathbf{u}(\boldsymbol{x},t)]\!] + \delta_{\Gamma_d}([\![\mathbf{u}(\boldsymbol{x},t)]\!] \otimes \mathbf{n}_{\Gamma_d})^s \tag{8.11}$$

where the superscript s denotes the symmetric part of a differential operator, δ_{Γ_d} is the Dirac-delta distribution centered on the discontinuity and $\mathbf{n}_{\Gamma_d}$ is the unit normal vector to the discontinuity.

Due to the jump in the displacement field, the Dirac-delta distribution arises in the strain field. Since the Dirac-delta distribution is unbounded, it is physically meaningless, but it is useful for inserting the strain field into the virtual work equation.

The extended FE approximation of the displacement field can be written as

$$\begin{aligned}\mathbf{u}(\boldsymbol{x},t) &= \mathbf{N}(\boldsymbol{x})\bar{\mathbf{u}}(t) + H_{\Gamma_d}\mathbf{N}(\boldsymbol{x})\bar{\mathbf{a}}(t)\\ &\equiv \mathbf{N}^{std}(\boldsymbol{x})\bar{\mathbf{u}}(t) + \mathbf{N}^{enr}(\boldsymbol{x})\bar{\mathbf{a}}(t)\end{aligned} \tag{8.12}$$

where $\bar{\mathbf{u}}$ and $\bar{\mathbf{a}}$ represent the standard and enriched nodal DOF, respectively, $\mathbf{N}^{std}(\boldsymbol{x}) \equiv \mathbf{N}(\boldsymbol{x})$ are the standard FE shape functions, $\mathbf{N}^{enr}(\boldsymbol{x}) = H_{\Gamma_d}\mathbf{N}(\boldsymbol{x})$ are referred to as the matrix of the enriched shape functions, and H_{Γ_d} is the Heaviside step function. The standard displacement DOF represent the continuous part of the displacement field, while the enriched DOF stand for the displacement jump across the discontinuity Γ_d defined by $[\![\mathbf{u}]\!] = \mathbf{N}(\boldsymbol{x})\bar{\mathbf{a}}(t)$. Accordingly, the strain vector corresponding to the approximate displacement field $\mathbf{u}(\boldsymbol{x}, t)$ can be written in terms of the standard and enriched nodal values as

$$\boldsymbol{\varepsilon}(\boldsymbol{x},t) = \mathbf{B}^{std}(\boldsymbol{x})\bar{\mathbf{u}}(t) + \mathbf{B}^{enr}(\boldsymbol{x})\bar{\mathbf{a}}(t) \tag{8.13}$$

where the standard and enriched matrices $\mathbf{B}^{std}$ and $\mathbf{B}^{enr}$ involve the spatial derivatives of the standard and enriched shape functions, respectively, defined as $\mathbf{B}^{std}(\boldsymbol{x}) = \mathbf{L}\mathbf{N}(\boldsymbol{x})$ and $\mathbf{B}^{enr}(\boldsymbol{x}) = \mathbf{L}\mathbf{N}^{enr}(\boldsymbol{x})$, and $\mathbf{L}$ is the matrix differential operator expressed as

$$\mathbf{L} = \begin{bmatrix}\partial/\partial x & 0\\ 0 & \partial/\partial y\\ \partial/\partial y & \partial/\partial x\end{bmatrix} \tag{8.14}$$

8.2.2 Discretization of Governing Equations

The weak form of the governing partial differential equation (8.1) can be derived based on the Galerkin discretization technique by integrating the product of the equilibrium equation multiplied by admissible test function over the analyzed domain, applying the Divergence theorem, imposing the natural boundary conditions, and satisfying the boundary conditions on the discontinuity. The test function is considered in the same approximating space as the displacement field $\mathbf{u}(\boldsymbol{x},t)$ in Eq. (8.9) as

$$\delta\mathbf{u}(\boldsymbol{x},t) = \delta\hat{\mathbf{u}}(\boldsymbol{x},t) + H_{\Gamma_d}(\boldsymbol{x})[\![\delta\mathbf{u}(\boldsymbol{x},t)]\!] \tag{8.15}$$

where $\delta\hat{\mathbf{u}}$ and $[\![\delta\mathbf{u}]\!]$ denote the standard and enriched parts of the test function approximation, respectively. Applying the Galerkin procedure, the weak form of the equilibrium equation (8.1) can be written as

$$\int_\Omega (\delta\hat{\mathbf{u}} + H_{\Gamma_d}[\![\delta\mathbf{u}]\!])(\nabla\cdot\boldsymbol{\sigma} + \mathbf{b})d\Omega = 0 \tag{8.16}$$

Applying the Divergence theorem and using the admissible test function $\delta\mathbf{u}(\boldsymbol{x}, t)$ that has been decomposed into two independent admissible test functions $\delta\hat{\mathbf{u}}$ and $[\![\delta\mathbf{u}]\!]$, the following set of equations can be obtained as

$$\begin{aligned}&\int_\Omega \nabla\delta\hat{\mathbf{u}} : \boldsymbol{\sigma} d\Omega = \int_{\Gamma_t} \delta\hat{\mathbf{u}}\cdot\bar{\mathbf{t}} d\Gamma + \int_\Omega \delta\hat{\mathbf{u}}\cdot\mathbf{b} d\Omega\\ &\int_{\Omega^+} \nabla[\![\delta\mathbf{u}]\!] : \boldsymbol{\sigma} d\Omega + \int_{\Gamma_d} [\![\delta\mathbf{u}]\!]\cdot\mathbf{t}_d d\Gamma = \int_{\Gamma_t^+} [\![\delta\mathbf{u}]\!]\cdot\bar{\mathbf{t}} d\Gamma + \int_{\Omega^+} [\![\delta\mathbf{u}]\!]\cdot\mathbf{b} d\Omega\end{aligned} \tag{8.17}$$

It must be noted that in the second integral equation of (8.17), H_{Γ_d} has been eliminated by transferring the domain of integration from Ω to Ω^+. Moreover, the integral on the discontinuity Γ_d in the second integral equation (8.17) comes from the following derivation assigning the positive and negative sides to Γ_d and imposing the boundary condition on the discontinuity, that is, $\sigma\cdot\mathbf{n}_{\Gamma_d}=\mathbf{t}_d$ on Γ_d given in Eq. (8.2),

$$-\int_{\Gamma_d^+}\delta\mathbf{u}\cdot\left(\boldsymbol{\sigma}\cdot\mathbf{n}_{\Gamma_d^+}\right)d\Gamma-\int_{\Gamma_d^-}\delta\mathbf{u}\cdot\left(\boldsymbol{\sigma}\cdot\mathbf{n}_{\Gamma_d^-}\right)d\Gamma=\int_{\Gamma_d}(\delta\mathbf{u}^+-\delta\mathbf{u}^-)\cdot(\boldsymbol{\sigma}\cdot\mathbf{n}_{\Gamma_d})d\Gamma=\int_{\Gamma_d}[\delta\mathbf{u}]\cdot\mathbf{t}_d d\Gamma \tag{8.18}$$

where $\mathbf{n}_{\Gamma_d^+}$ and $\mathbf{n}_{\Gamma_d^-}$ are the unit normal vectors directed to Ω^- and Ω^+, respectively, as shown in Figure 8.1, and the superscripts + and – above Γ_d represent the two sides of the discontinuity. It should be mentioned that $\mathbf{n}_{\Gamma_d}$ has been taken such that $\mathbf{n}_{\Gamma_d}=\mathbf{n}_{\Gamma_d^-}=-\mathbf{n}_{\Gamma_d^+}$. It is manifest from this equation that the test function space allows for functions discontinuous on Γ_d.

In the framework of the X-FEM, Eq. (8.17) can be discretized as

$$\begin{aligned}&\int_{\Omega}\mathbf{B}^T\boldsymbol{\sigma}d\Omega=\int_{\Gamma_t}\mathbf{N}^T\bar{\mathbf{t}}d\Gamma+\int_{\Omega}\mathbf{N}^T\mathbf{b}d\Omega\\&\int_{\Omega^+}\mathbf{B}^T\boldsymbol{\sigma}d\Omega+\int_{\Gamma_d}\mathbf{N}^T\mathbf{t}_d d\Gamma=\int_{\Gamma_t^+}\mathbf{N}^T\bar{\mathbf{t}}d\Gamma+\int_{\Omega^+}\mathbf{N}^T\mathbf{b}d\Omega\end{aligned} \tag{8.19}$$

where $\mathbf{B}$ denotes the strain-deformation matrix. Because of the nonlinearity of the system of equations, the Newton–Raphson iterative procedure is employed to linearize the discretized governing equations (8.19). The linearized system of equations can be obtained as

$$\begin{bmatrix}\mathbf{K}_{uu} & \mathbf{K}_{ua}\\ \mathbf{K}_{au} & \mathbf{K}_{aa}\end{bmatrix}\begin{Bmatrix}d\bar{\mathbf{u}}\\ d\bar{\mathbf{a}}\end{Bmatrix}=\begin{Bmatrix}\mathbf{f}_u^{\text{ext}}\\ \mathbf{f}_a^{\text{ext}}\end{Bmatrix}-\begin{Bmatrix}\mathbf{f}_u^{\text{int}}\\ \mathbf{f}_a^{\text{int}}\end{Bmatrix} \tag{8.20}$$

in which $d\bar{\mathbf{u}}$ and $d\bar{\mathbf{a}}$ denote the increments of the standard and enriched nodal displacements, and the stiffness matrix is composed of

$$\begin{aligned}&\mathbf{K}_{uu}=\int_{\Omega}\mathbf{B}^T\mathbf{D}\mathbf{B}d\Omega\\&\mathbf{K}_{ua}=\mathbf{K}_{au}^T=\int_{\Omega^+}\mathbf{B}^T\mathbf{D}\mathbf{B}d\Omega\\&\mathbf{K}_{aa}=\int_{\Omega^+}\mathbf{B}^T\mathbf{D}\mathbf{B}d\Omega+\int_{\Gamma_d}\mathbf{N}^T\mathbf{T}\,\mathbf{N}d\Gamma\end{aligned} \tag{8.21}$$

and the internal and external force vectors are defined as

$$\begin{aligned}&\mathbf{f}_u^{\text{int}}=\int_{\Omega}\mathbf{B}^T\boldsymbol{\sigma}d\Omega\\&\mathbf{f}_a^{\text{int}}=\int_{\Omega^+}\mathbf{B}^T\boldsymbol{\sigma}d\Omega+\int_{\Gamma_d}\mathbf{N}^T\mathbf{t}_d d\Gamma\\&\mathbf{f}_u^{\text{ext}}=\int_{\Gamma_t}\mathbf{N}^T\bar{\mathbf{t}}d\Gamma+\int_{\Omega}\mathbf{N}^T\mathbf{b}d\Omega\\&\mathbf{f}_a^{\text{ext}}=\int_{\Gamma_t^+}\mathbf{N}^T\bar{\mathbf{t}}d\Gamma+\int_{\Omega^+}\mathbf{N}^T\mathbf{b}d\Omega\end{aligned} \tag{8.22}$$

8.3 Cohesive Crack Growth Based on the Stress Criterion

A method for modeling cohesive cracks was developed by Wells and Sluys (2001), which allows the introduction of the displacement jump into the classical FEM based on the PU property of FE shape functions. In this model, the discontinuity can propagate through an unstructured FE mesh independently of the mesh structure and the crack path is insensitive to the element size. Moreover, there is no restriction on the type of underlying FEs that can be used and displacement jumps are continuous across element boundaries and the interpolation of the jump can be of any polynomial order, which is not the case for all EAS-based (enhanced assumed strains) approaches. A crucial advantage of this method over conventional interface elements is that deformations at the discontinuity are purely inelastic. The consequence is that there is no need for a dummy elastic stiffness, which enhances the robustness of the numerical procedure. The displacement jump across a crack is represented by extra nodal DOF at existing nodes by using discontinuous shape functions. Functions with displacement jumps are added selectively to the support of individual nodes to model a propagating discontinuity. A cohesive crack model is used for the simulation of discontinuity in quasi-brittle materials, with tractions as functions of crack opening acting on discontinuity surfaces. The formulation begins by considering the kinematics of a body crossed by a discontinuity. The FE formulation is developed by considering first the interpolation of the displacement field and then inserting the discontinuous displacement field into the virtual work equation. The underlying polynomial basis is enriched with the discontinuous Heaviside jump function in the displacement field. Based on this enrichment function, the crack tip can propagate element by element.

A computational modeling of cohesive crack in quasi-brittle materials was performed by Mergheim, Kuhl, and Steinmann (2005). In this model, the crack is allowed to propagate freely through the elements, and the crack path is not limited to inter-element boundaries. In the elements which are intersected by the discontinuity additional displacement DOF are introduced at the existing nodes. Therefore, two independent copies of the standard basis functions are used. One set is put to zero on one side of the discontinuity, while it takes its usual values on the opposite side, and vice versa for the other set. In order to model the inelastic material behavior at the discontinuity, a discrete damage-type constitutive law is applied, which is formulated in terms of the displacement jumps and cohesive tractions. Xu and Yuan (2009, 2011) introduced the cohesive zone model with a threshold combined with the X-FEM for simulating mixed-mode cracks. In conventional cohesive zone models the traction-separation law starts from zero load, so that the model cannot be applied to predict mixed-mode cracking. The computations on both brittle and ductile materials predict that the contribution of the shear stress is negligible for the considered cracked specimens. Thus, the crack initiation and propagation under mixed-mode loading conditions can be characterized by normal stress dominated cohesive zone model, and the crack growth direction can be governed by a nonlocal maximum principal stress criterion. However, the shear stress becomes important for uncracked specimens to yield the correct crack initiation angle.

8.3.1 Cohesive Constitutive Law

In order to describe the nonlinear fracture processes occurring ahead of the crack tip, the cohesive crack model is used. This model is an appropriate alternative when the size of the FPZ is not negligible in comparison with the crack length, which is a commonly seen feature for cracks in quasi-brittle materials, such as geomaterials and concrete. In the cohesive crack model, it is assumed that the near tip FPZ is lumped into the crack line, unlike LEFM in which the fracture processes are assumed to take place at the crack tip. Moreover, the cohesive crack model avoids the singular stress field at the crack tip, a physically meaningless characteristic of LEFM. In the cohesive crack model, the nonlinear behavior of the material in the FPZ is modeled using a cohesive constitutive relation to describe the near tip nonlinear processes taking place in brittle and ductile materials.

If a certain failure criterion is met, the discontinuity is introduced into the underlying finite element mesh. A discrete damage-type constitutive model is applied at the discontinuity to describe the inelastic behavior of the quasi-brittle material. The constitutive model relates the traction vector $\mathbf{t}_d$ to the jump in the displacement field $[\![\mathbf{u}]\!]$ (Eq. 8.3). In the direction normal to the discontinuity, an exponential softening model is assumed, and in the tangential direction, a constant shear stiffness is adopted. Thus, the normal and shear components of the traction vector are obtained as

$$
\begin{aligned}
t_n &= f_t \exp\left(-\frac{f_t}{E_f} [\![u_n]\!] \right) \\
t_s &= G [\![u_s]\!]
\end{aligned}
\tag{8.23}
$$

in which $\mathbf{t}_d = t_n \mathbf{n}_{\Gamma_d} + t_s \mathbf{t}_{\Gamma_d}$, with $\mathbf{n}_{\Gamma_d}$ and $\mathbf{t}_{\Gamma_d}$ denoting the unit vector of normal and tangential directions, respectively. In these relations, f_t is the tensile strength of the material, E_f is the fracture energy, and G is the shear stiffness. Due to the constant shear stiffness, the tangential stiffness matrix preserves its symmetry.

Because of the applied constitutive law at the discontinuity, the system of equations becomes nonlinear and has to be solved iteratively. The Newton–Raphson iterative procedure is performed to linearize the weak form of the governing equations. The consistent linearization of the adopted constitutive relation at the discontinuity results in the tangential stiffness matrix defined as

$$
\mathbf{T} = -\frac{f_t^2}{E_f} \exp\left(-\frac{f_t}{E_f} [\![u_n]\!] \right) \mathbf{n}_{\Gamma_d} \otimes \mathbf{n}_{\Gamma_d} + G \mathbf{t}_{\Gamma_d} \otimes \mathbf{t}_{\Gamma_d}
\tag{8.24}
$$

where $\mathbf{T}$ represents the tangent stiffness of the traction-separation law at the discontinuity. It can be observed that the tangent stiffness $\mathbf{T}$ retains its symmetry due to the constant shear stiffness in the tangential direction.

8.3.2 Crack Growth Criterion and Crack Growth Direction

In order to describe the propagation of a discontinuity, it is required that a failure criterion be introduced, and the direction in which the discontinuity propagates be determined. Moreover, an adequate integration scheme for integrating over the intersected elements and along the discontinuity must be presented. The constitutive model applied to the discontinuity is based upon the cohesive crack concept, in which the inelastic deformation around the crack tip is compressed onto a crack line and represented as traction forces acting on a crack. Away from the discontinuity, the elastic behavior is assumed for the medium. To obtain the traction–separation relationship, a loading function is defined. If the loading function is positive, loading (opening) occurs at a discontinuity; otherwise, unloading (closing) occurs.

A discontinuity is incorporated into an element when a certain crack growth criterion is satisfied. Once debonding occurs, the cohesive tractions transmitted across the discontinuity are determined by means of the traction–separation law. During the calculation, the maximum principal tensile stress is checked at all integration points in the element ahead of the tip of the discontinuity at the end of a load increment. If the maximum principal tensile stress at any of the integration points in the element ahead of the discontinuity tip attains the tensile strength of the material f_t, the discontinuity is introduced through the entire element. The discontinuity is inserted as a straight line within the element and is enforced to be geometrically continuous. Since, within the cohesive crack framework energy is dissipated upon crack opening rather than upon crack extension (as with the linear elastic fracture), the numerical result is not particularly sensitive to exactly when a discontinuity is extended or the length of the discontinuity extension. It is possible (and is inevitable upon mesh refinement) that a discontinuity propagates through more than one element at the end of a load increment. The introduction of a discontinuity through an entire element can lead to stress jumps upon introduction of a discontinuity, although experience indicates that this has no influence on the robustness of the algorithm, and upon mesh refinement the stress jump approaches zero. Unlike the continuum-type models, in this model a crack must propagate from a discrete point. Therefore, discontinuities can be initiated in two ways, the first by choosing a point before the calculation and the second by performing an elastic loading and checking where the principal stresses are greatest. In this way, the method is more closely related to linear elastic fracture simulations where a crack must propagate from a point. Discontinuities are extended only at the end of a load increment in order to preserve the quadratic convergence rate of the full Newton–Raphson solution procedure.

The most important consideration when extending a discontinuity is that the correct direction is chosen. Since the tip of the discontinuity is not located at a point where the stress state is known accurately (such as conventional Gauss points), the local stress field cannot be relied upon to accurately yield the correct normal vector to a discontinuity. To overcome this, the averaged stress tensor or the so-called non-local stress tensor is used at the discontinuity tip to obtain the principal stress direction and to determine the right direction of the discontinuity extension. The discontinuity is extended in the direction perpendicular to the maximum non-local principal stress direction. According to Wells and Sluys (2001), the non-local stress tensor is calculated as a weighted average of stresses using a Gaussian weight function defined by

$$w(r)=\frac{1}{(2\pi)^{3/2}\ell^{3}}\exp\left(-\frac{r^{2}}{2\ell^{2}}\right) \tag{8.25}$$

where w is the weight function, ℓ determines how quickly the weight function decays away from the discontinuity tip and r is the distance of a point from the discontinuity tip. It is emphasized that this does not imply any non-locality in the model, but is a method of smoothing stresses in order to accurately determine the principal stress direction. The parameter ℓ is taken as approximately three times the typical element size. It is noted that using a non-local measure for determining the direction of discontinuity propagation leads to a more reliable prediction of the crack path.

According to Mergheim, Kuhl, and Steinmann (2005), the non-local stress tensor is calculated as the weighted average of the stresses at the n_G Gauss points within an interaction radius around the tip of the discontinuity. More precisely, the non-local stress tensor results from the sum of the local stresses at the n_G Gauss points, weighted with w_i and the associated area A_i as

$$\bar{\boldsymbol{\sigma}}=\left(\sum_{i=1}^{n_G} w_i A_i\right)^{-1}\sum_{i=1}^{n_G}\boldsymbol{\sigma}_i w_i A_i \tag{8.26}$$

where the weighted Gauss function is defined by

$$w(r)=\frac{1}{(2\pi)^{1/2}\ell}\exp\left(-\frac{r^{2}}{4\ell^{2}}\right) \tag{8.27}$$

where r is the distance of the Gauss point to the tip of the discontinuity, and ℓ determines the decline of w with respect to r.

The jump in the displacement field at the tip of the discontinuity must be equal to zero. In order to enforce this condition, the nodes belonging to the element edge on which the tip of the discontinuity lies are not enriched. The path of the discontinuity and the nodes that are enriched are depicted in Figure 8.4. Since the enrichment functions are multiplied by the shape functions of a particular node, the enhanced basis at a particular node has an influence only over the support of that node. Therefore, the Heaviside function is added only to the enhanced basis of nodes whose support is crossed by a discontinuity. Another condition that must be satisfied is that the displacement jump at the crack tip be zero. To ensure this, the nodes on the element boundary touched by the discontinuity tip are not enhanced. When a discontinuity propagates into the next element, all nodes behind the crack tip are enhanced. It is noted that since only a relatively small number of the total number of nodes need to be enhanced, the procedure is computationally efficient.

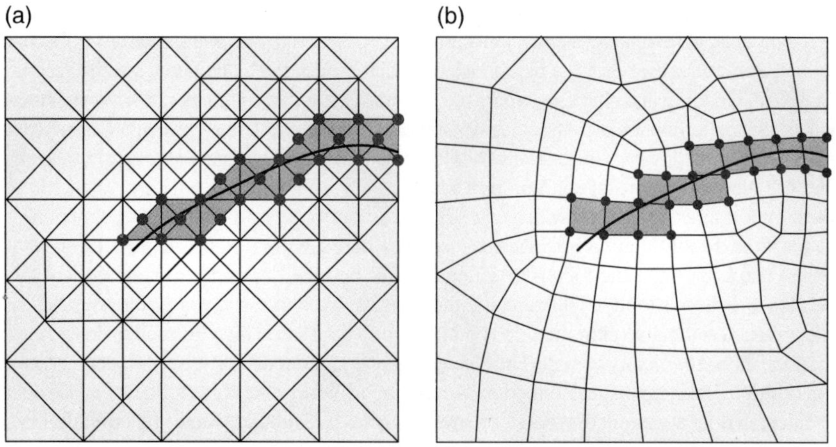

Figure 8.4 The path of discontinuity in a structured (a) and an unstructured (b) X-FEM mesh where the circled nodes are enriched with the jump function

8.3.3 Numerical Integration Scheme

The most important requirement when using discontinuous functions (non-standard shape functions) is that the strain field is adequately integrated on both sides of the discontinuity. Failure to integrate on both sides of a discontinuity results in a linearly dependent system of equations since the Heaviside function cannot be distinguished from a constant function. Often when a discontinuity crosses an element, the conventional Gauss integration scheme is not sufficient to ensure that the shape functions remain linearly independent. That is, for the integration of the intersected elements, the conventional Gauss quadrature scheme is not valid. The integration over the domain is performed element by element. Elements which are not crossed by a discontinuity are integrated by the standard Gauss quadrature. For the elements cut by the crack, the integration is performed on each side of the crack separately. When an element is crossed by a discontinuity, the domains on either side of the discontinuity are triangulated into sub-domains. Within each triangular sub-domain, the standard Gauss quadrature is applied. In the elements close to the crack tip, a higher order quadrature rule is used for the case of branch function enrichments. For integration over the cohesive zone, the cohesive zone is geometrically represented by a sequence of 1D segments. The integration is thus performed by looping over these segments. In addition to Gauss points in the bulk of the element, two Gauss points are set on each segment of the discontinuity in order to integrate the traction forces.

If the numerical procedure is implemented using the 6-node triangle as the underlying FE mesh, the proposed integration scheme requires 23 Gauss points per element, which at first seems excessive. However, since only elements crossed by a discontinuity require the modified integration scheme, the computational burden is small. In the bulk of the element, the proposed scheme over-integrates the stress field. The scheme is adopted for the maximum flexibility since it may be desirable to add enhancement functions other than the Heaviside jump (e.g., the near tip enrichment functions for the LEFM). It is undesirable that the integration scheme be modified for every different set of enhancement functions since the computational cost of over integrating a small number of elements is negligible.

8.4 Cohesive Crack Growth Based on the SIF Criterion

An X-FEM was developed by Moës and Belytschko (2002a) to model the cohesive crack growth in a uniform FE mesh. Since displacement discontinuities do not need to conform to the element edges, and no remeshing is required as the crack propagates, the cohesive crack can propagate arbitrarily with respect to the underlying FE mesh. In this method, the cohesive crack growth is governed by requiring the SIF at the mathematical crack tip to vanish. This energetic method avoids the evaluation of the stress field at the mathematical crack tip, whose accuracy is questionable in the finite element context. Zi and Belytschko (2003) employed an X-FEM for the static cohesive crack propagation and developed a new formulation for elements containing the crack tip. This method gives the possibility for crack growth without remeshing. Moreover, the crack path is independent of the underlying FE mesh. The proposed strategy treats the entire crack with only one type of enrichment function. That is, all elements that are cut by the crack, including the elements containing the crack tip, are enriched with the sign function. Thus, not only the inner blending does not occur, but also blending to outside the enriched sub-domain does not happen. This implies that the PU holds in the entire domain. Although the branch function is not used in the presented formulation, the crack tip can be located anywhere within the element. This scheme is applied to 3-node linear triangular elements and 6-node quadratic triangular elements. The parent domain of the partially cracked element, that is, the tip element, is divided into two parts; the cracked and the uncracked parts. Only the cracked part is enriched with the sign function. In order to guarantee the smooth closing of the cohesive crack tip, the stress component normal to the crack tip is imposed to be equal to the tensile strength of the material. The equilibrium equation of the bulk material and the traction condition at the crack tip are solved simultaneously using the Newton–Raphson iterative scheme to obtain the nodal displacements.

8.4.1 *The Enrichment of Displacement Field*

The approximation of the displacement field is accomplished by the X-FEM, which allows the cohesive crack to propagate arbitrarily with respect to the underlying FE mesh. In the formulation derived by Moës and Belytschko (2002a), the extended FE approximation of the displacement field is written as

$$\mathbf{u}^h(\boldsymbol{x}) = \sum_{I \in \mathcal{N}} N_{uI}(\boldsymbol{x})\bar{\mathbf{u}}_I + \sum_{I \in \mathcal{N}^{enr}} N_{uI}(\boldsymbol{x}) H_{\Gamma_d}(\boldsymbol{x})\bar{\mathbf{a}}_I + \sum_{I \in \mathcal{N}^{tip}} N_{uI}(\boldsymbol{x}) \left(\sum_J F_J(\boldsymbol{x})\bar{\mathbf{b}}_I^J \right) \tag{8.28}$$

That is, the classical FE approximation of the displacement field is enriched with two additional terms. The first enrichment function in Eq. (8.28) is the jump function. The nodal set $\mathcal{N}^{enr}$ is the set of enriched nodes which is defined as the set of nodes in the mesh whose support, that is, the support of their nodal shape function, is bisected by the crack. The second additional term involves a set of branch functions in order to incorporate the near tip displacement field. The branch functions are chosen based on the asymptotic solution of the displacement field around the tip of the discontinuity, that is,

$$\{F_J(r,\theta)\} = \left\{ \sqrt{r}\sin\frac{\theta}{2},\ \sqrt{r}\cos\frac{\theta}{2},\ \sqrt{r}\sin\frac{\theta}{2}\sin\theta,\ \sqrt{r}\cos\frac{\theta}{2}\sin\theta, \right\} \tag{8.29}$$

where (r,θ) are the local polar coordinates at the crack tip. However, these functions do not produce the non-singular stress field at the tip of the cohesive crack. Hence, the non-singular branch functions of the following form can be considered within the X-FEM framework as

$$\{F_J(r,\theta)\} = \left\{ r\sin\frac{\theta}{2},\quad \text{or}\ r^{3/2}\sin\frac{\theta}{2},\quad \text{or}\ r^2\sin\frac{\theta}{2} \right\} \tag{8.30}$$

The nodes whose support, that is, the support of their nodal shape function, contains the cohesive tip are enriched with the branch functions. Figure 8.5 displays the nodes enriched with the branch functions and those enriched with the jump function. These enrichment functions give the ability to deal with the crack tips within elements. That is, the crack tip can be located anywhere within an element.

For elements which are bisected by the crack, a discontinuous function must be used. In the scheme proposed by Zi and Belytschko (2003), the sign function enrichment is used instead of the Heaviside step

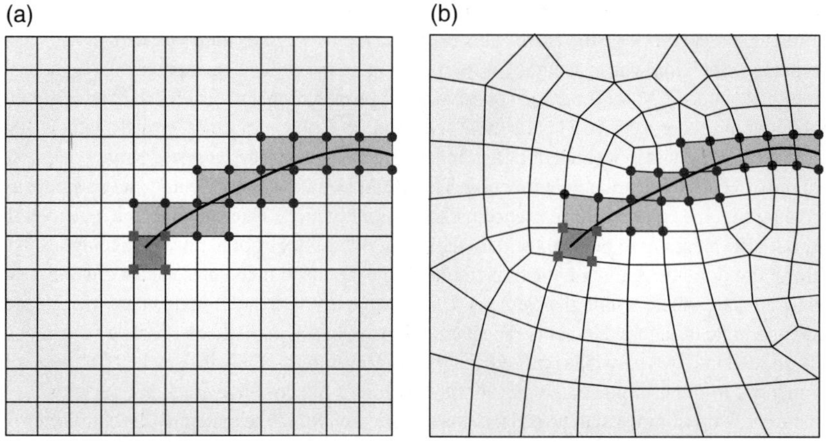

Figure 8.5 The path of discontinuity in a structured (a) and an unstructured (b) X-FEM mesh: The circled nodes are enriched with the jump function, while the squared nodes are enriched with the asymptotic tip functions

function enrichment because of its appealing symmetry. It is noted that the numerical discretization of these two enrichments is identical. If a crack propagates element by element, that is, if the crack tip always lies on an edge of an element, a successful element can be devised by using only the sign function enrichment without other enrichment functions. When a crack grows smoothly, the crack tip can lie within an element. In this case, the sign function enrichment cannot be used solely for elements partially cut by the crack, that is, the tip elements. This is because the sign function enrichment does not yield the appropriate displacement field for the tip element.

Branch functions of the type $r^m \sin \theta/2$ where $m = 1.0$, 1.5, 2.0, and so on, have been proposed for the enrichment of the tip element. However, when branch functions are used in conjunction with the step function, the PU property will not hold any longer in the elements surrounding the tip element. The enrichment of these elements is a local PU and it must be blended to the rest of the domain for an optimal performance because the branch functions do not vanish at the edges of the tip element. It is noted that the branch functions are not piecewise constant functions like the step function. In order to avoid the difficulties associated with the branch functions, a new enrichment is developed for the tip element.

Consider the two-dimensional domain Ω with an external boundary Γ containing a cohesive crack Γ_d, as shown in Figure 8.3. The crack is described implicitly by the level set function $\varphi(\boldsymbol{x}) = 0$. Depending on the cohesive law, a cohesive traction $\mathbf{t}_d$ is transmitted across the crack surfaces. The displacement field $\mathbf{u}$ can be decomposed additively into a continuous part $\mathbf{u}_{\text{con}}$ and a discontinuous part $\mathbf{u}_{\text{dis}}$ as

$$\mathbf{u}(\boldsymbol{x}) = \mathbf{u}_{\text{con}}(\boldsymbol{x}) + \mathbf{u}_{\text{dis}}(\boldsymbol{x}) \tag{8.31}$$

The continuous part of the displacement field is approximated by the standard piecewise polynomial shape functions as

$$\mathbf{u}_{\text{con}}(\boldsymbol{x}) = \sum\nolimits_{I \in \mathcal{N}} N_I(\boldsymbol{x}) \bar{\mathbf{u}}_I \tag{8.32}$$

where $\mathcal{N}$ is the set of all nodes in the finite element mesh and $\bar{\mathbf{u}}_I$ are the standard nodal displacement DOF. The discontinuous part of the displacement field can be approximated by

$$\mathbf{u}_{\text{dis}}(\boldsymbol{x}) = \sum\nolimits_{I \in \mathcal{N}^{enr}} N_I(\boldsymbol{x}) \psi_I(\boldsymbol{x}) \bar{\mathbf{a}}_I \tag{8.33}$$

where $\mathcal{N}^{enr}$ is the set of nodes of the elements cut by the crack, $\psi_I(\boldsymbol{x})$ is the enrichment function and $\bar{\mathbf{a}}_I$ are the enriched nodal DOF.

The standard level set function is the signed distance function. The absolute value of the signed distance function at any point is defined as the minimum distance to the crack line. The choice of the sign of the signed distance function does not change the numerical results. When the sign of the signed distance function is altered, the enriched nodal DOF change the sign as well. Therefore, the sign of the signed distance function can be chosen arbitrarily as long as the sign is consistent along the crack. The signed distance function at any point in the enriched sub-domain can be interpolated by means of the FE shape functions as

$$\varphi(\boldsymbol{x}) = \sum\nolimits_{I \in \mathcal{N}^{enr}} N_I(\boldsymbol{x})\varphi_I \tag{8.34}$$

in which φ_I are the nodal values of the signed distance function. The unit normal vector $\mathbf{n}$ in the positive direction is easily determined by $\mathbf{n} = \partial\varphi/\partial\boldsymbol{x}$ (Figure 8.3).

For elements completely cut by the crack, the enrichment function is constructed as follows

$$\psi_I(\boldsymbol{x}) = \text{sign}(\varphi(\boldsymbol{x})) - \text{sign}(\varphi(\boldsymbol{x}_I)) \tag{8.35}$$

That is, the sign function is shifted by its value at enriched nodes, that is, sign($\varphi(\boldsymbol{x}_I)$). Otherwise, the enriched displacement field does not vanish outside the enriched subdomain. This shift does not change the approximating space with respect to the unshifted enrichment function, but it simplifies the numerical implementation. This is because the resulting enrichment function vanishes in the sub-domain containing elements not cut by the crack.

For an element containing a crack tip, consider that the crack passes through side $\overline{23}$ of a 3-node linear triangular element with node numbers 1, 2, and 3, as shown in Figure 8.6a. The crack is assumed to be straight within the tip element. Let the direction of the crack be such that it intersects side $\overline{12}$ of the tip element. For compatibility condition, the enrichment should vanish on the two edges $\overline{12}$ and $\overline{13}$ of the tip element that are not cut by the crack. Moreover, it should be continuous across the edge $\overline{23}$ that is cut by the crack. In order to satisfy these conditions, only node 3 of the tip element is enriched. The discontinuous displacement field in this element can be obtained as

$$\mathbf{u}_{\text{dis}} = \boldsymbol{\xi}_3^* \, \psi_3(\boldsymbol{\xi}^*)\,\bar{\mathbf{a}}_3 \tag{8.36}$$

where $\boldsymbol{\xi}^* = \{\xi_1^*, \xi_2^*, \xi_3^*\}$ are the area (triangular) coordinates of the shaded region 23P in Figure 8.6b, that is, the element formed by sides $\overline{23}$, $\overline{3\text{P}}$, and $\overline{\text{P}2}$, where P is the intersection point of the line joining node 2 to the crack tip and side $\overline{31}$. The shaded parent area coordinates are related by $\xi_3^* = 1 - \xi_1^* - \xi_2^*$ and $\psi_3(\boldsymbol{\xi}^*) = \text{sign}(\varphi(\boldsymbol{\xi}^*)) - \text{sign}(\varphi(\boldsymbol{\xi}_3^*))$. The relation between $\boldsymbol{\xi}^*$ and $\boldsymbol{\xi}$ is defined by

$$\xi_1^* = \frac{\xi_1}{\xi_{1\text{P}}}, \quad \xi_2^* = \xi_2 \tag{8.37}$$

where $\xi_{1\text{P}}$ is the area coordinate of point P.

In general, the discontinuous part of the displacement field at the crack tip elements can be approximated as

$$\mathbf{u}_{\text{dis}} = \sum_I \boldsymbol{\xi}_I^* \, \psi_I(\boldsymbol{\xi}^*)\,\bar{\mathbf{a}}_I \tag{8.38}$$

in which the nodal DOF $\bar{\mathbf{a}}_I$ that are associated with nodes belonging to the edge toward where the crack is propagating are restrained.

The enriched part of the displacement field vanishes on the boundary of the enriched sub-domain. Therefore, only the elements in the enriched sub-domain need a special treatment to model the crack.

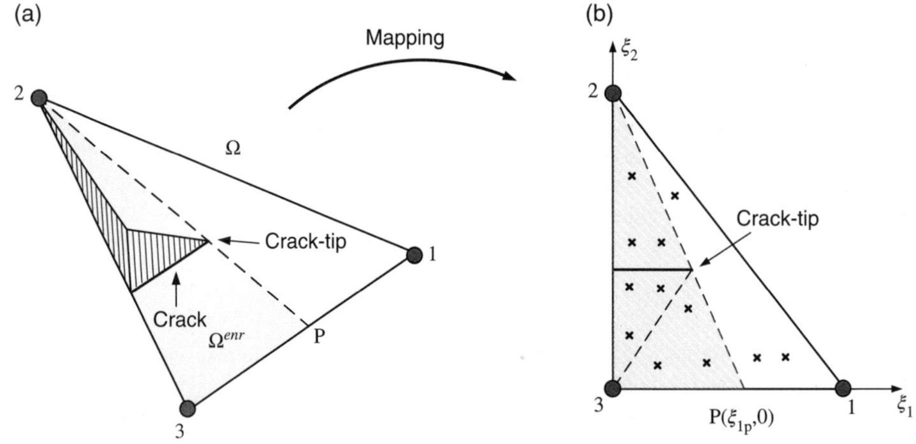

Figure 8.6 A 3-noded linear triangular tip element

Moreover, all elements in the enriched sub-domain have enrichments of the same type. Therefore, contrary to enriched methods in which branch functions are used together with step function, there is no inner blending between different enrichment functions. Furthermore, since the enrichment vanishes on the boundary of the enriched sub-domain, blending to outside the enriched sub-domain does not occur. Thus, although this is a local PU, it is indistinguishable from a global PU.

8.4.2 The Condition for Smooth Crack Closing

In the cohesive crack model, the stress singularity at the crack tip is avoided. Thus, the SIF at the crack tip should vanish, which implies that the crack closes smoothly. This condition is referred to as the zero SIF condition. It is assumed that the crack propagates under the mode I loading condition, so only the mode I SIF is taken into account

$$K_{I_{\text{tip}}} = 0 \tag{8.39}$$

where $K_{I_{\text{tip}}}$ is the mode I SIF at the crack tip. In the FEM, the SIF at the crack tip is calculated by means of the domain integration technique. The zero SIF at the crack tip implies that the stress at the crack tip should be finite. The zero SIF condition is equivalent to the condition that the stress component in the normal direction to the crack be equal to the tensile strength of the material. This condition is referred to as the stress condition, that is,

$$\mathbf{n}^T \boldsymbol{\sigma}_{\text{tip}}\, \mathbf{n} = f_t \tag{8.40}$$

where $\boldsymbol{\sigma}_{\text{tip}}$ is the stress at the crack tip and f_t is the tensile strength of the material. The zero SIF condition and the stress condition can be used interchangeably, but the stress condition (8.40) is simpler to implement.

8.4.3 Crack Growth Criterion and Crack Growth Direction

The location of the cohesive zone is characterized by two tips; a mathematical or fictitious crack tip and a physical or real crack tip, as discussed in Figure 8.1. At the mathematical tip, the crack opening is zero and the cohesive traction is equal to the cohesive strength of the material, whereas at the physical tip the cohesive traction is zero and the crack opening equals the critical crack opening. Before any loading is applied to the specimen, these two tips coincide. As the loading is imposed, the cohesive zone develops.

In cohesive crack growth, snap-back may occur. The snap-back branch can be captured numerically only if the loading process is controlled by a monotonically increasing function of the crack length. If the crack length itself is taken as an increasing function, it is called a crack length control scheme. In this scheme, a scalar load factor is defined such that the traction at the crack tip is equal to the cohesive strength of the material. In this way, it is ensured that the crack tip is indeed the mathematical tip. If the loading is too small, the cohesive zone will close before the crack tip, and if the loading is too high, the traction at the crack tip will be lower than the cohesive strength. In the FE framework, where the cohesive crack is directly discretized by the element edges, the value of the traction at the tip may be obtained from the nodal forces at the tip, or using an elastic "dummy" (penalty) interface stiffness and calculating the stress at the integration points close to the tip. The shortcoming of this approach is that it involves evaluating quantities near the crack tip, whose accuracy is questionable in the finite element framework. Instead of evaluating pointwise quantities, a method based on the energetic considerations is adopted. In this method, the scalar load factor is obtained such that the mode I SIF vanishes at the mathematical crack-tip, that is, $K_I = 0$. Indeed, the cohesive fracture mechanics avoids the non-physical, singular stress field at the crack tip. The evaluation of the SIF is carried out with a domain integral. It is noted that, since the bulk of the material is linearly elastic, the SIF is the sum of the SIF caused by the external loading and that caused by the cohesive forces, that is, $K_I = K_I^{ext} + K_I^{coh} = 0$. Note also that the SIF caused by the external loading K_I^{ext} is linear in the load factor, while the SIF caused by the cohesive forces K_I^{coh} is nonlinear in the load factor.

The direction of crack propagation is determined based on the LEFM criterion. In the LEFM theory at least two criteria can be used to predict the trajectory of the cohesive crack; the principle of local symmetry and the maximum hoop stress criterion. In the principle of local symmetry, the crack growth direction is obtained so that the second SIF vanishes at the new tip that is therefore an implicit criterion. In order to implement the technique, the direction of crack propagation is iterated until the second SIF is close to zero at the crack tip. It is noted that these iterations do not need any remeshing because in the X-FEM framework the mesh does not require to conform to the crack. On the other hand, the maximum hoop stress criterion is an explicit criterion. In this criterion, the crack growth direction is normal to the maximum hoop stress direction. The maximum hoop stress criterion and the principle of local symmetry are comparable for an infinitesimal crack growth.

Applying the maximum hoop stress criterion, the angle of crack propagation with respect to the tangent of crack line at the tip can be obtained in the terms of mode I and II external SIFs as

$$\theta = 2\arctan\frac{1}{4}\left(K_{\mathrm{I}}^{ext}/K_{\mathrm{II}}^{ext} \pm \sqrt{\left(K_{\mathrm{I}}^{ext}/K_{\mathrm{II}}^{ext}\right)^2 + 8}\right) \tag{8.41}$$

in which the sign is chosen such that the hoop stress is positive. It is noted that the ratio of the SIFs appearing in Eq. (8.41) is not dependent on the load factor. It should be mentioned that in this scheme the cohesive crack growth direction is determined based upon the LEFM, while the global load–deflection curve for the structure is determined based on the cohesive crack model. This assumption is in agreement with experiments on concrete in which it was observed that the size effect in concrete affects the load–deflection curves whereas the crack path is much less sensitive to the size of structure.

The cohesive crack growth simulation can then be implemented as follows; the direction of crack growth is determined using Eq. (8.41) and a predefined crack length is added to the crack tip. An initial guess for the load factor is also provided. In the case that no crack exists initially, the first segment initiates at a point where the tensile strength of the material is reached. The mode I SIF is calculated at the crack tip. If the SIF is zero within a given tolerance, the crack propagates from the new crack tip. Otherwise, the load factor is updated. This algorithm continues until the SIF vanishes within the given tolerance. If, on the last 1D segment away from the crack tip the crack opening at all integration points exceeds the critical opening, the 1D segment is removed from the cohesive zone. That is, the cohesive zone is replaced by the fully damaged zone.

8.5 Cohesive Crack Growth Based on the Cohesive Segments Method

A numerical method was presented by Remmers, de Borst, and Needleman (2003, 2008) for cohesive crack growth in which the crack is not regarded as a single discontinuity that propagates continuously. Instead, the crack is represented by a collection of overlapping cohesive segments with a finite length. These cohesive segments are incorporated into the underlying FE mesh as discontinuities in the displacement field by exploiting the PU property of FE shape functions. A combination of overlapping cohesive segments can behave as a continuous cohesive zone. Furthermore, since the cohesive segments can be inserted at any arbitrary positions and with arbitrary orientations, the cohesive segments method can be used for complex crack patterns, including the simulation of crack nucleation at multiple locations, followed by the growth and coalescence, or branching of an existing crack. The cohesive segments can be inserted at arbitrary locations and directions and are not dependent on the underlying mesh structure. The evolution of the debonding of the cohesive segments is governed by a cohesive law. The cohesive segments method permits the simulation of crack nucleation as well as the discontinuous crack propagation, irrespective of the structure of the FE mesh, and it provides a promising tool for modeling complex fracture behavior. In this approach, a single element can be crossed by multiple cohesive segments, each with its own enriched DOF.

The cohesive segments method was presented by Remmers, de Borst, and Needleman (2003, 2008) within the extended FE framework that allows for the simulation of the nucleation, growth, coalescence, and branching of multiple cracks in solids. In fact, the important characteristic of the cohesive segments method is the possible emergence of multiple cohesive cracks in a domain. In this method, displacement discontinuities are introduced as jumps in the displacement field by employing the PU property of FE shape functions. The magnitude of the displacement jumps is governed by the cohesive constitutive relation employed. In the proposed method, the cohesive segments are only inserted when they are needed, contrary to some models in which the cohesive segments must be inserted at the potential inter-element boundaries. Since the crack growth direction does not depend on the mesh, the new cohesive segments can nucleate with arbitrary orientations and the existing ones can propagate in arbitrary directions when a critical condition is met.

8.5.1 *The Enrichment of Displacement Field*

The significant characteristic of the cohesive segments method is the possible emergence of multiple cracks in a domain. A domain Ω with boundary Γ is considered, as shown in Figure 8.7. The domain includes m discontinuities $\Gamma_{d,j}$, where $j = 1, 2, \ldots, m$. Each discontinuity splits the domain into two sub-domains, which are denoted by by Ω_j^- and Ω_j^+. For all discontinuities it can be stated $\Omega_j^+ \cup \Omega_j^- = \Omega$, where $j = 1, 2, \ldots, m$.

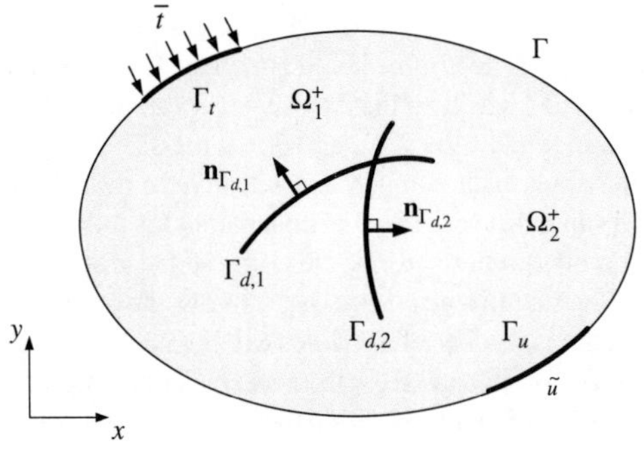

Figure 8.7 The domain Ω crossed by two discontinuities $\Gamma_{d,1}$ and $\Gamma_{d,2}$

The displacement field can be defined based on an extension of Eq. (8.9), composed of a continuous regular displacement field $\hat{\mathbf{u}}$ plus m additional continuous displacement fields $[\![\mathbf{u}_j]\!]$ as follows

$$\mathbf{u}(\boldsymbol{x},t)=\hat{\mathbf{u}}(\boldsymbol{x},t)+\sum_{j=1}^{m}H_{\Gamma_{d,j}}(\boldsymbol{x})\,[\![\mathbf{u}_j(\boldsymbol{x},t)]\!] \tag{8.42}$$

where $\boldsymbol{x}$ denotes the position of a material point, t is the time, and $H_{\Gamma_{d,j}}$ is the Heaviside step function, defined as

$$H_{\Gamma_{d,j}}(\boldsymbol{x})=\begin{cases}0 & \text{if } \boldsymbol{x}\in\Omega_j^-\\ 1 & \text{if } \boldsymbol{x}\in\Omega_j^+\end{cases} \tag{8.43}$$

The strain field can be obtained by taking the derivative of the displacement field (8.42) as

$$\boldsymbol{\varepsilon}(\boldsymbol{x},t)=\nabla^s\hat{\mathbf{u}}(\boldsymbol{x},t)+\sum_{j=1}^{m}H_{\Gamma_{d,j}}(\boldsymbol{x})\nabla^s\,[\![\mathbf{u}_j(\boldsymbol{x},t)]\!] \tag{8.44}$$

in which a superscript s denotes the symmetric part of a differential operator. It is noted that the strain field is not defined at the discontinuities $\Gamma_{d,j}$. The magnitude of the displacement jump at the discontinuities can be obtained by $[\![\mathbf{u}_j]\!]$ according to Eq. (8.42). Hence, the discrete form of the displacement field can be written using Eq. (8.42) as

$$\mathbf{u}(\boldsymbol{x},t)=\mathbf{N}(\boldsymbol{x})\bar{\mathbf{u}}(t)+\sum_{j=1}^{m}H_{\Gamma_{d,j}}(\boldsymbol{x})\,\mathbf{N}(\boldsymbol{x})\,\bar{\mathbf{a}}_j(t) \tag{8.45}$$

where the vector $\bar{\mathbf{u}}$ contains the standard nodal degrees of freedom and $\bar{\mathbf{a}}_j$ contains the enriched nodal degrees of freedom associated with the discontinuity $\Gamma_{d,j}$. The matrix $\mathbf{N}(\boldsymbol{x})$ contains the conventional shape functions. The discrete form of the strain field can be derived by differentiating equation (8.45) as

$$\boldsymbol{\varepsilon}(\boldsymbol{x},t)=\mathbf{B}^{std}(\boldsymbol{x})\bar{\mathbf{u}}(t)+\sum_{j=1}^{m}\mathbf{B}_j^{enr}(\boldsymbol{x})\,\bar{\mathbf{a}}_j(t) \tag{8.46}$$

where $\mathbf{B}^{std}(\boldsymbol{x})=\mathbf{L}\,\mathbf{N}(\boldsymbol{x})$ contains the spatial derivatives of the standard shape functions and $\mathbf{B}_j^{enr}(\boldsymbol{x})=\mathbf{L}\,\mathbf{N}_j^{enr}(\boldsymbol{x})$ involves the spatial derivatives of the enriched shape functions, where $\mathbf{N}_j^{enr}(\boldsymbol{x})=H_{\Gamma_{d,j}}\mathbf{N}(\boldsymbol{x})$ is referred to as the matrix of the enriched shape functions for the discontinuity $\Gamma_{d,j}$, and $\mathbf{L}$ is a differential operator matrix that is given in relation (8.14) for two-dimensional problem. Finally, the discretized displacement jump at the discontinuity $\Gamma_{d,j}$ can be obtained according to Eq. (8.45) by $[\![\mathbf{u}_j]\!]=\mathbf{N}(\boldsymbol{x})\bar{\mathbf{a}}_j$.

8.5.2 *Cohesive Constitutive Law*

The heterogeneity of the material in heterogeneous quasi-brittle materials such as concrete, that is, the presence of particles of different size and stiffness, causes a complex stress field in which new cracks can nucleate and existing cracks can grow and branch. The smeared cohesive zone models are not able to capture the processes of crack initiation, crack growth, coalescence, and crack branching properly. This is because essential characteristics of these processes are lost in the averaging technique. On the contrary, the cohesive segments method can simulate these physical processes. In the cohesive segments method, the cohesive segments are not considered at the beginning of the analysis. This implies that an initially rigid cohesive constitutive relation must be employed, contrary to the methods in which the cohesive interface elements are incorporated into inter-element boundaries beforehand, and accordingly, cohesive constitutive relations with an initial elastic branch is used. In the cohesive segments framework, when decohesion initiates, a cohesive segment is inserted through the integration point. The cohesive segment is extended throughout the element to which the integration point belongs and into the neighboring elements. The evolution of the cohesive segment is governed by a cohesive constitutive relation. Thus, it is required to specify a criterion for the initiation of a new cohesive segment, a criterion for the direction of the added cohesive segment, and a cohesive constitutive relation which governs the evolution of the cohesive segment.

Consider the normal (mode I) decohesion, the initiation of decohesion occurs when the maximum tensile principal stress at an integration point within an element reaches the cohesive strength of the material f_t. The orientation of the added cohesive segment is normal to the maximum tensile principal stress direction. The exponential form of the cohesive constitutive relation is employed by

$$t_n = f_t \exp\left(-\frac{f_t}{E_f} k\right) \tag{8.47}$$

where t_n is the normal cohesive traction transmitted across the cohesive segment, E_f is the cohesive fracture energy, and k is a specified cohesive parameter. Obviously, in mode I decohesion the shear cohesive traction transmitted across the cohesive segment vanishes. In the extended FE framework, the cohesive segments are inserted into the underlying FE mesh by exploiting the PU property of FE shape functions. In this way, the orientation of the added cohesive segments is not connected to any direction associated with the FE discretization.

The stress rate in the bulk material $\dot{\boldsymbol{\sigma}}$ is a function of the strain rate $\dot{\boldsymbol{\varepsilon}}$ as

$$\dot{\boldsymbol{\sigma}} = \mathbf{D}\dot{\boldsymbol{\varepsilon}} = \mathbf{D}\left(\mathbf{B}(\boldsymbol{x})\dot{\bar{\mathbf{u}}} + \sum_{j=1}^{m} \mathbf{B}_j^{enr}(\boldsymbol{x})\dot{\bar{\mathbf{a}}}_j\right) \tag{8.48}$$

where $\mathbf{D}$ is the tangential stiffness matrix of the bulk material surrounding the discontinuity. The traction rate $\dot{\mathbf{t}}_{d_j}$ at the jth discontinuity is expressed in terms of the corresponding enriched nodal velocities $\dot{\bar{\mathbf{a}}}_j$ as

$$\dot{\mathbf{t}}_{d_j} = \mathbf{T}[\![\dot{\mathbf{u}}_j]\!] = \mathbf{T}\mathbf{N}(\boldsymbol{x})\dot{\bar{\mathbf{a}}}_j \tag{8.49}$$

where $\mathbf{T}$ is the tangential stiffness matrix of the traction-separation law at the discontinuity. The cohesive constitutive relation is defined in the local coordinate system which is aligned with the discontinuity.

8.5.3 Crack Growth Criterion and Its Direction for Continuous Crack Propagation

The presence of a crack can be divided into three stages; the nucleation, the growth, and finally the coalescence with other cracks. In all cases, a similar criterion is used to determine when and in which direction a cohesive segment is created, or extended. In order to arrive at a consistent numerical formulation, this fracture criterion is related to the cohesive constitutive relation that governs the debonding process of a cohesive segment. When the stress state in an integration point in the bulk material is such that the cohesive strength of the material is attained, a new cohesive segment is added. The cohesive segment, which is assumed to be straight, crosses this integration point and is extended into the neighboring elements until it touches the boundary of the patch of the neighboring elements, as shown in Figure 8.8. The patch of the neighboring elements is composed of all elements that share one of the nodes of the central element that contains the integration point at which the criterion was violated. The nodes that

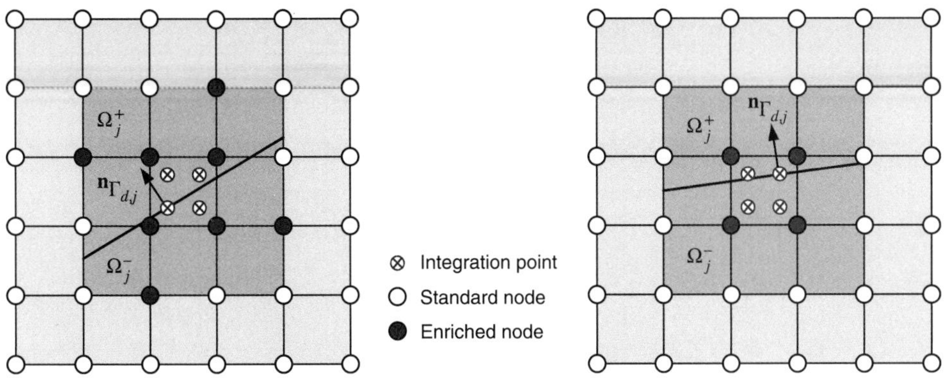

Figure 8.8 The nucleation of a new cohesive segment in the bulk material

support these outer boundaries are not enriched in order to ensure a zero crack opening at the tips of the new cohesive segment. It is noted that a new segment does not immediately reduce the high stresses in the bulk material in the vicinity of the new cohesive segment. Depending on the fracture toughness of the material and the magnitude of the applied load, the stresses in the vicinity of the new cohesive segment will only slowly diminish away. Even, especially in the dynamic simulations, the stress state can locally increase for a short period of time. To prevent new cohesive segments from being nucleated in the same position in a short period of time, an element and its neighboring elements can be the source of a new cohesive segment only once in a simulation.

The growth of a cohesive segment is carried out as follows; when the stress state at the tip of the cohesive segment reaches the cohesive strength of the material at a specified direction, the cohesive segment is extended into the next element in that direction until it touches the boundary of that element. Since the stress state at the tip of the cohesive segment is not known exactly, the stress state is estimated by calculating the average stress state in the vicinity of the tip using a Gauss averaging technique. In order to predict the direction of the crack extension accurately the averaging length must be chosen three times the characteristic element length in the mesh around the tip. In this way, the stress state at the tip of the cohesive segment is underestimated, and in general, the cohesive segment is extended somewhat too late. Thus, a discrepancy is created because the fracture criterion is employed to determine both the nucleation of a new cohesive segment as well as its extension. The nucleation of a new cohesive segment is based on the stress state in a single integration point, while the extension of an existing cohesive segment is based on an average stress state that is normally too low. Hence, the nucleation of a new cohesive segment is preferred to the extension of an existing one, which will hinder the quantitative analysis of fracture problems. In transient simulations, this problem is even more pronounced. For this reason, two alternative approaches are presented.

In the first approach, different stress states are used to determine when and in which direction a cohesive segment is extended. In this approach, the stress state in the nearest integration point to the tip of the cohesive segment is monitored. This stress state is in general a good representative of the peak stress at the tip. When this stress state violates the crack growth criterion, the direction of the crack growth is determined based upon the averaged stress state. This averaged stress state is computed on the basis of a number of integration points within a certain radius of the tip. It is noted that the averaged stress state can be significantly smaller than the peak stress at the tip. The advantage of this approach is that the cohesive segment is extended at the correct load step and that it is extended in the right direction. However, this approach is unreliable in situations where mode transitions become an important factor. In the second approach, both the creation and the extension of a cohesive segment are based on an averaged stress state. A predefined average length is used for both cases. In order to reduce the computational effort, a new

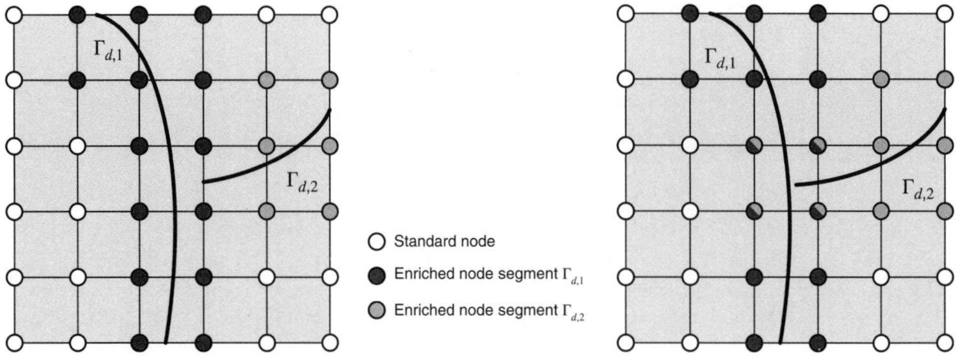

Figure 8.9 Merging of two cohesive segments, where segment $\Gamma_{d,2}$ is approaching segment $\Gamma_{d,1}$

cohesive segment can only nucleate at one point for each element, for example, the geometric center of the element. In both cases, the average stress state is used to determine the instance as well as the direction of cohesive segment creation, or extension. Obviously, the critical condition for adding a cohesive segment will occur with some delay, but the creation of a new cohesive segment is no longer favored over the extension of an existing one.

An important characteristic of the cohesive segments method is the possibility to have multiple, interacting cohesive segments. To do this, each cohesive segment is supported by a unique set of additional degrees of freedom. When two cohesive segments meet within a single element, the nodes of that element are enriched twice. Merging of two cohesive segments is modeled as follows; consider the case shown in Figure 8.9a, in which segment $\Gamma_{d,2}$ is approaching another segment $\Gamma_{d,1}$. When the fracture criterion at the tip of segment $\Gamma_{d,2}$ is violated, the segment is extended accordingly. Segment $\Gamma_{d,2}$ is only extended until it touches segment $\Gamma_{d,1}$, as shown in Figure 8.9b. Since segment $\Gamma_{d,1}$ forms a free edge in the material, there is no new tip for segment $\Gamma_{d,2}$. Consequently, all nodes of the corresponding element are enriched in order to support displacement discontinuity of segment $\Gamma_{d,2}$. Crack branching is considered as follows; when the fracture criterion in an integration point in the vicinity of an existing cohesive segment $\Gamma_{d,1}$ is satisfied, that is, the effective traction in the integration point exceeds the cohesive strength of the material, a new cohesive segment $\Gamma_{d,2}$ is created. This new cohesive segment is extended until it touches either the boundary of its neighboring elements, or the existing cohesive segment, as shown in Figure 8.10. In the latter case, the new cohesive segment only has a single tip.

The criterion for adding a new cohesive segment or extending the existing one is based on a stress-based formulation proposed by Camacho and Ortiz (1996). From a stress $\boldsymbol{\sigma}$, the normal and shear tractions t_n and t_s can be computed along an axis η, which is rotated by an angle θ with respect to the x-axis shown in Figure 8.11 as

$$t_n = \mathbf{n}^T \boldsymbol{\sigma}_{\text{tip}}\, \mathbf{n}, \quad t_s = \mathbf{s}^T \boldsymbol{\sigma}_{\text{tip}}\, \mathbf{n} \tag{8.50}$$

in which $\mathbf{n}$ and $\mathbf{s}$ are respectively the unit normal and tangent vectors to the axis η, defined as

$$\mathbf{n} = [-\sin\theta,\ \cos\theta]^T, \quad \mathbf{s} = [\cos\theta,\ \sin\theta]^T \tag{8.51}$$

The normal and shear tractions t_n and t_s are used to construct an effective traction t_{eff}, which is a function of the angle θ of the η -axis as

$$t_{eff} = \sqrt{\langle t_n \rangle^2 + \frac{1}{\beta} t_s^2} \tag{8.52}$$

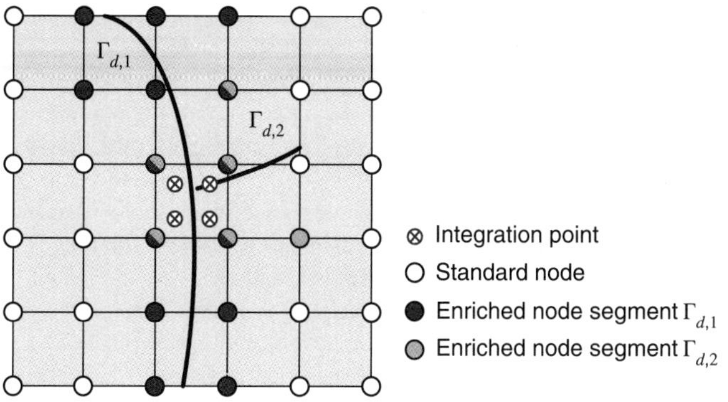

Figure 8.10 Crack branching by creating a new cohesive segment $\Gamma_{d,2}$ when the fracture criterion in an integration point close to cohesive segment $\Gamma_{d,1}$ is satisfied

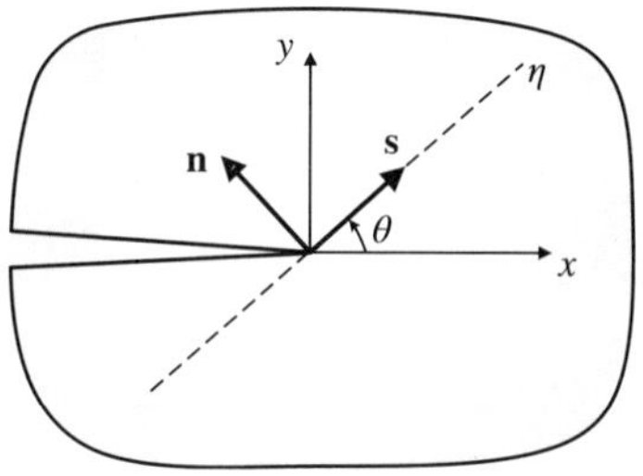

Figure 8.11 The axis η rotated by an angle θ with respect to the x-axis

where the operator $\Xi = \frac{1}{2}(\Xi + |\Xi|)$ is the McCauley bracket, and parameter β is a scaling factor that scales the contribution of the shear stress (Ortiz and Pandolfi, 1999). For a given crack tip stress $\boldsymbol{\sigma}_{\text{tip}}$, the effective traction t_{eff} is determined for all orientations θ of the η-axis. The angle for which the effective traction exceeds the cohesive strength of the material is the direction in which the crack propagates. This angle can be obtained by solving a minimization problem. It can be demonstrated that for mode I crack propagation, this direction is identical to the direction perpendicular to the principal stress direction. However, in this formulation, a distinction is made between a tensile and a compressive stress. In order to avoid sudden jumps in stresses when a cohesive segment is inserted, the constitutive behavior of the cohesive segment is related to the stress state in the bulk material that violates the fracture criterion and reaches the cohesive strength of the material. The initial tractions in the irreversible cohesive relation are taken to be equal to the normal and shear tractions in the bulk material at the exact moment of nucleation.

8.5.4 Crack Growth Criterion and Its Direction for Discontinuous Crack Propagation

A procedure is implemented using the FE meshes composed of 4-noded quadrilateral elements. A new cohesive segment is added when the maximum principal tensile stress at an integration point within an element exceeds the cohesive strength of the material f_t. The added cohesive segment passes through the integration point where the cohesive strength has been attained, and extends through the entire element and into the neighboring elements until it touches the boundary of a patch of elements, as shown in

Figure 8.8. This figure displays the elements that belong to the patch of elements that is influenced by the cohesive segment. The orientation of the cohesive segment is taken to be normal to the direction of the maximum principal tensile stress at the integration point where the cohesive strength of the material f_t has been reached. Since the cohesive segment is taken to be a straight line, the normal vector to the cohesive segment $\mathbf{n}_{\Gamma_{d,j}}$ is constant along the patch of elements. The magnitude of the displacement jump of the cohesive segment is governed by the set of enriched DOF that are added to the nodes belonging to the central element of the patch. Since the nodes at the boundary of the patch of elements are not enriched, the displacement jump at the boundary of the patch of elements vanishes. In this way, it is guaranteed that a zero crack opening at the tip of the cohesive segment is satisfied.

An important characteristic of the cohesive segments approach is the possibility of having overlapping, interacting, and intersecting cohesive segments. In this approach, a new cohesive segment $\Gamma_{d,2}$ can be added next to an existing cohesive segment $\Gamma_{d,1}$, as depicted in Figure 8.12a. The displacement jump of the new cohesive segment $\Gamma_{d,2}$ is supported by an additional set of enriched DOF, which is added to the nodes of the central element of the new patch of elements. This new set of enriched DOF has no relation to the enriched DOF of the existing cohesive segment $\Gamma_{d,1}$. The hatched elements that belong

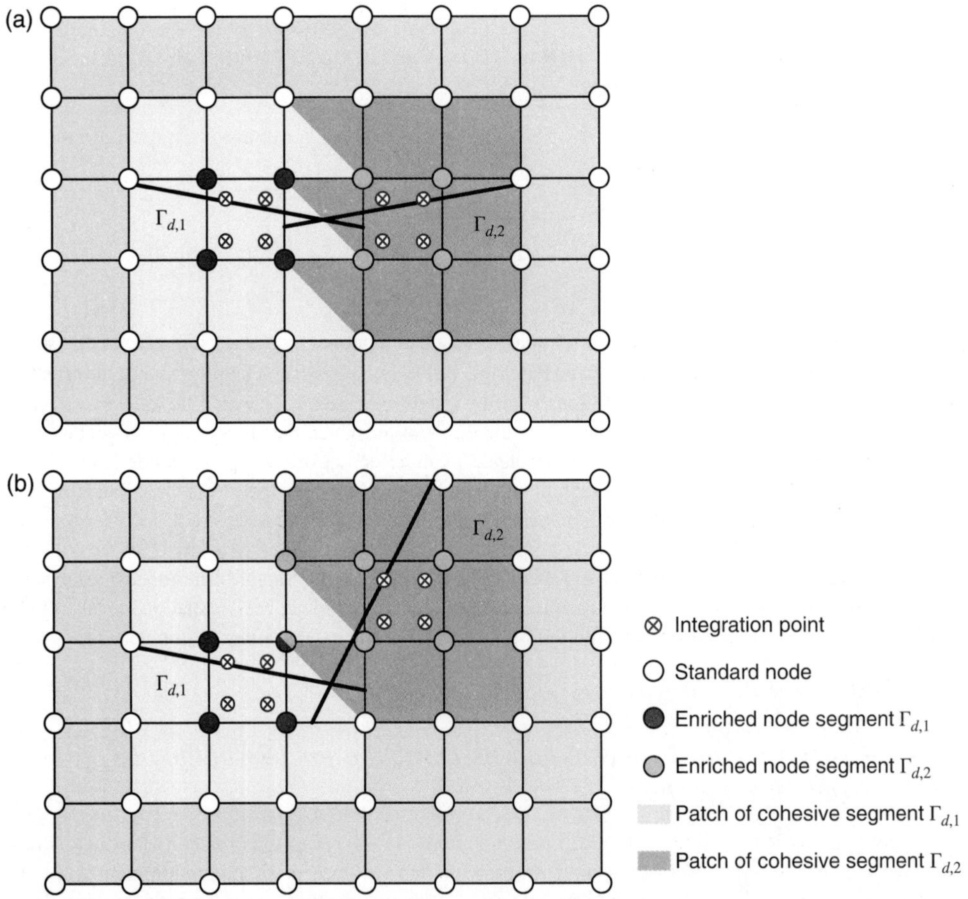

Figure 8.12 The interaction of two cohesive segments: A new cohesive segment $\Gamma_{d,2}$ is added next to an existing cohesive segment $\Gamma_{d,1}$

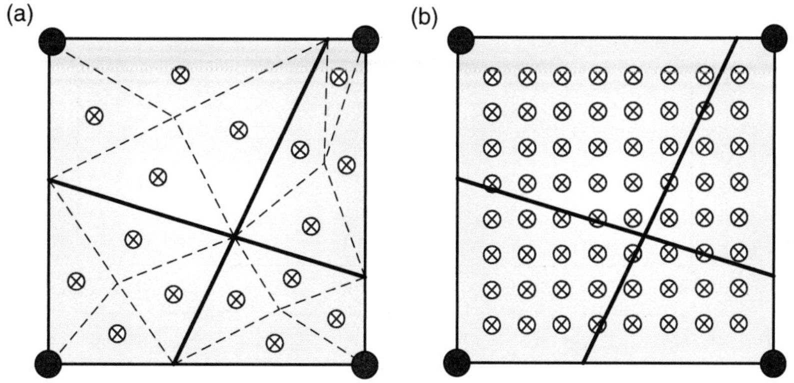

Figure 8.13 The numerical integration of an element cut by two cohesive segments: (a) The triangular partitioning method, (b) The standard Gauss integration points

to both patches of cohesive segments have the sets of enriched DOF of both cohesive segments, although each set of enriched DOF has been added to a different set of nodes. Another possible situation is shown in Figure 8.12b where two cohesive segments intersect to form a cohesive zone with a sharp bend. In this case, the patches of two cohesive segments overlap in such a way that one node contains enriched DOF related to both cohesive segments. It is noted that these configurations act as a single cohesive zone on a global level.

8.5.5 Numerical Integration Scheme

Special care should be taken to the numerical integration of elements that contain one or more cohesive segments. Since the displacement field is only piecewise continuous within these elements, the conventional integration scheme is no longer applicable. One alternative approach is to divide the element crossed by one or more cohesive segments into sub-elements, as shown in Figure 8.13a. Since the displacement field is continuous within each sub-element, these sub-elements can be integrated numerically using the conventional Gauss integration scheme with one integration point. The second alternative technique is to integrate the elements cut by cohesive segments using a large number of fixed integration points, as shown in Figure 8.13b. Although the error with this integration scheme is greater than that associated with the first alternative algorithm, the effect on the global accuracy of the computation is generally negligible.

8.6 Application of X-FEM Method in Cohesive Crack Growth

8.6.1 A Three-Point Bending Beam with Symmetric Edge Crack

The first example is a simply supported beam, which is loaded by an imposed displacement at the center of the beam on the top edge, as shown in Figure 8.14. The problem is chosen to illustrate the performance of a cohesive X-FEM technique for a beam with symmetric edge crack. The material properties of the beam are as follows; a Young's modulus of $E = 100$ MPa, Poisson ratio of $\nu = 0$, the tensile strength of material $f_t = 1.0$ MPa, and the fracture energy $E_f = 0.1$ N/mm. A prescribed displacement is gradually exerted to the center of top edge of the beam, increasing until the failure of the beam. The top row of the elements of the beam are prevented from cracking since if a crack propagates through the entire beam, the system of

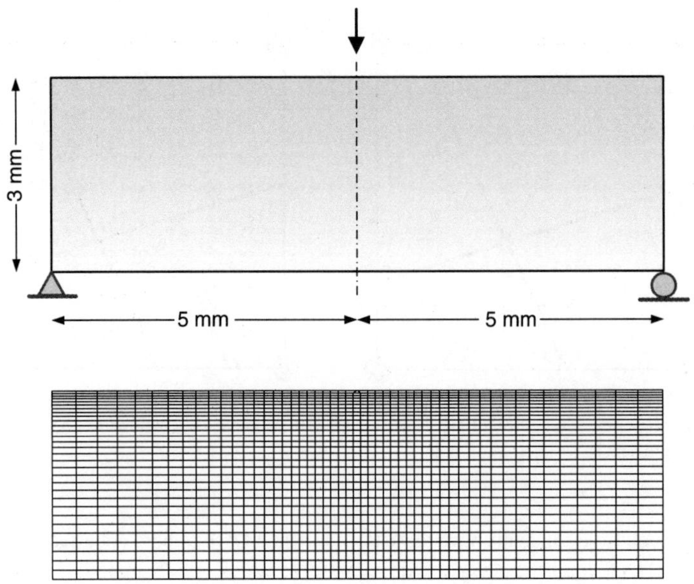

Figure 8.14 A three-point bending beam with symmetric edge crack: Geometry, boundary conditions, and X-FEM mesh

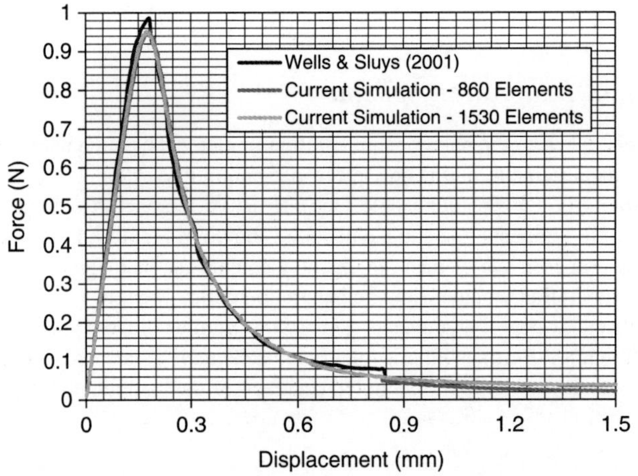

Figure 8.15 The load-displacement curve for a three-point bending beam

equations becomes singular. The crack is initiated at the center of the beam on the bottom edge and propagates toward the upper edge.

The beam is discretized using a "coarse" and a "fine" mesh with 860 and 1530 bilinear quadrilateral elements, respectively. The coarse FE mesh is presented in Figure 8.14. The load-displacement curve of the three-point bending beam is plotted in Figure 8.15. The predicted result is in a good agreement with that reported by Wells and Sluys (2001). In Table 8.1, the values of cohesive stiffness matrix and cohesive force vector are given for the element of bottom edge at different steps of crack growth. The contours of stress in the x-direction are depicted in Figure 8.16 at various displacements where the crack tip

Table 8.1 A three-point bending beam with symmetric edge crack

Step	$\mathbf{K}^{coh}$	$\mathbf{f}^{coh}$
73	$\begin{bmatrix} -180.04 & -180.04 \\ -180.04 & -180.04 \end{bmatrix}$	$\begin{bmatrix} -5.40D-2 \\ -5.40D-2 \end{bmatrix}$
82	$\begin{bmatrix} -176.93 & -176.93 & -88.81 & -88.81 \\ -176.93 & -176.93 & -88.81 & -88.81 \\ -88.81 & -88.81 & -178.31 & -178.31 \\ -88.81 & -88.81 & -178.31 & -178.31 \end{bmatrix}$	$\begin{bmatrix} -5.31D-2 \\ -5.31D-2 \\ -5.34D-2 \\ -5.34D-2 \end{bmatrix}$
90	$\begin{bmatrix} -172.90 & -172.90 & -86.98 & -86.98 \\ -172.90 & -172.90 & -86.98 & -86.98 \\ -86.98 & -86.98 & -175.05 & -175.05 \\ -86.98 & -86.98 & -175.05 & -175.05 \end{bmatrix}$	$\begin{bmatrix} -5.19D-2 \\ -5.19D-2 \\ -5.24D-2 \\ -5.24D-2 \end{bmatrix}$
99	$\begin{bmatrix} -160.76 & -160.76 & -81.18 & -81.18 \\ -160.76 & -160.76 & -81.18 & -81.18 \\ -81.18 & -81.18 & -163.99 & -163.99 \\ -81.18 & -81.18 & -163.99 & -163.99 \end{bmatrix}$	$\begin{bmatrix} -4.83D-2 \\ -4.83D-2 \\ -4.90D-2 \\ -4.90D-2 \end{bmatrix}$
107	$\begin{bmatrix} -148.18 & -148.18 & -75.10 & -75.10 \\ -148.18 & -148.18 & -75.10 & -75.10 \\ -75.10 & -75.10 & -152.22 & -152.22 \\ -75.10 & -75.10 & -152.22 & -152.22 \end{bmatrix}$	$\begin{bmatrix} -4.44D-2 \\ -4.44D-2 \\ -4.56D-2 \\ -4.56D-2 \end{bmatrix}$

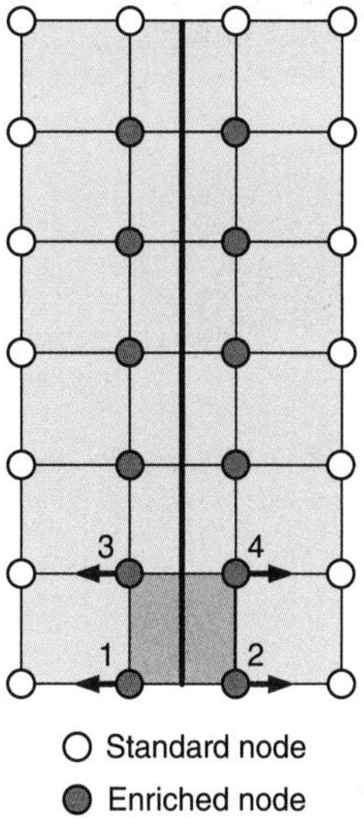

The cohesive stiffness matrices and cohesive force vectors for the element of bottom edge at different steps of crack growth.

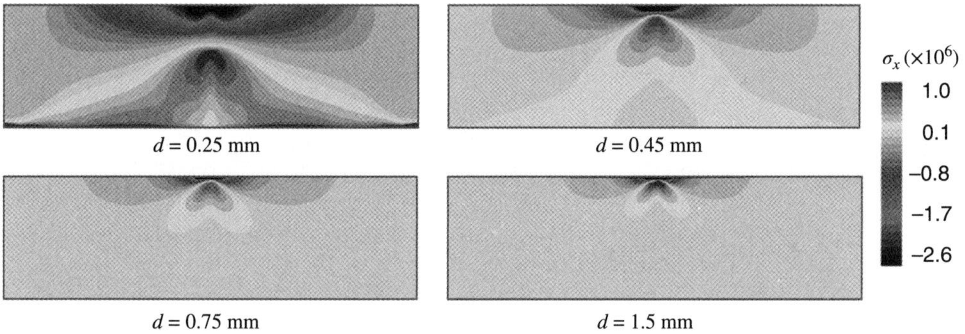

Figure 8.16 The distribution of stress contour in the *x*-direction at different steps of crack growth (For color details, please see color plate section)

approaches the upper boundary of the beam. Also presented in Figure 8.17 are the contours of horizontal displacement at various vertical deformations. Moreover, the variation of predicted cohesive traction at the crack mouth with the crack mouth opening displacement (CMOD) is depicted in Figure 8.18.

8.6.2 A Plate with an Edge Crack under Impact Velocity

The next example is a plate with an edge crack under impact velocity at the top and bottom edges, as shown in Figure 8.19. Consider a square block with the sides of 3 mm long, which is loaded under tension

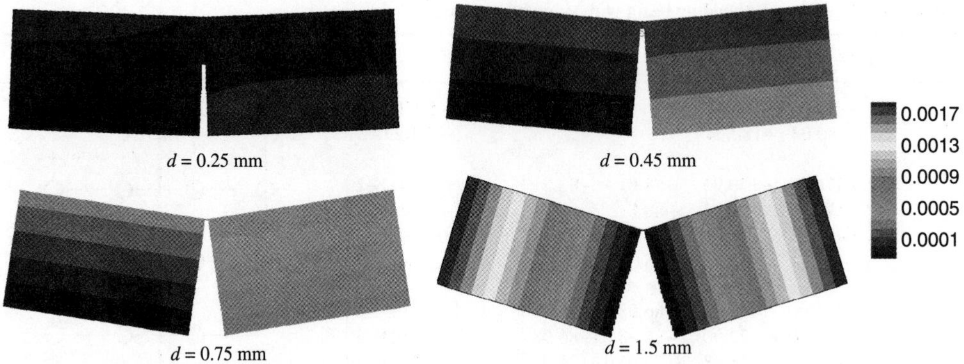

Figure 8.17 The distribution of displacement contour in x-direction at different steps of crack growth (For color details, please see color plate section)

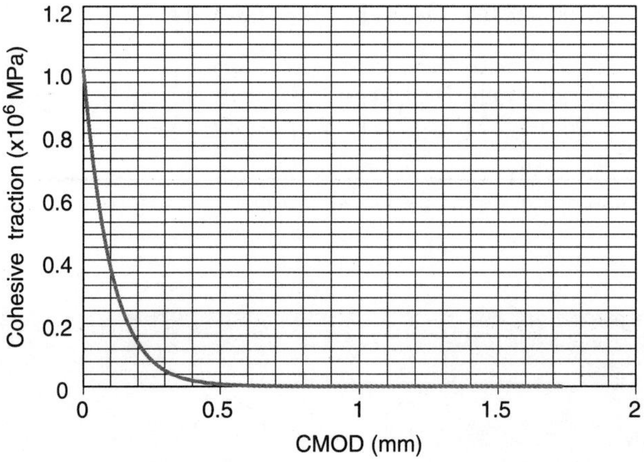

Figure 8.18 The cohesive traction at the crack mouth versus crack mouth opening displacement

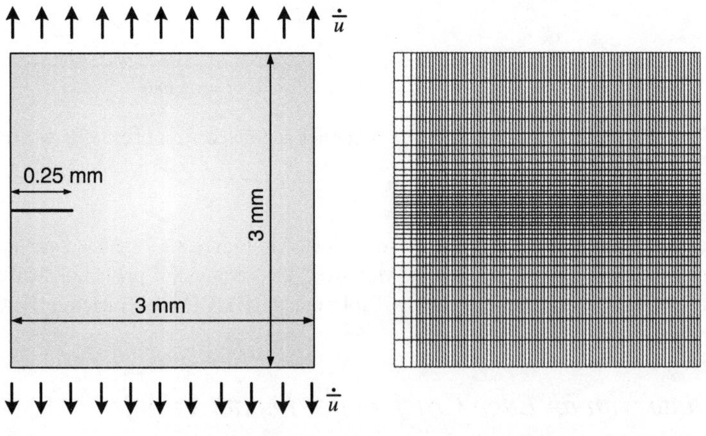

Figure 8.19 A plate with an edge crack: The geometry, boundary conditions, and X-FEM mesh

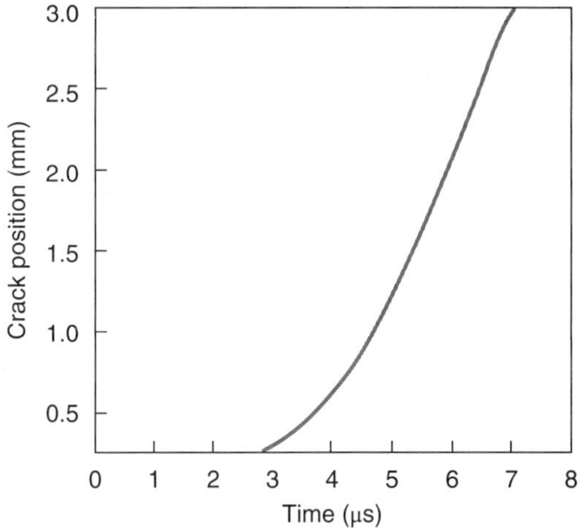

Figure 8.20 The position of crack tip with time

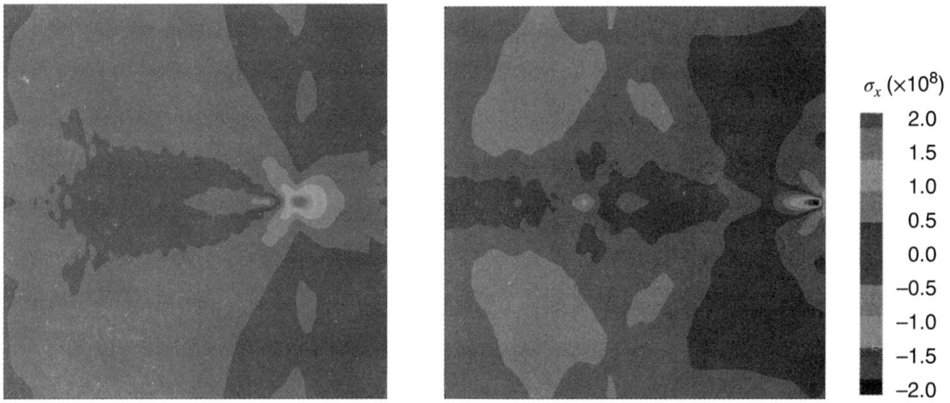

Crack-tip at 2.4 mm Crack-tip at 3.0 mm

Figure 8.21 The distribution of stress contour in the *x*-direction at different steps of crack growth (For color details, please see color plate section)

by two pulse loads applied at the top and bottom edges of the plate. These pulses are considered as the prescribed velocities increased linearly to a constant value $v = 6$ m/s. This magnitude is reached after a rise time $t_r = 10^{-7}$ s. The square plate contains an edge crack of length 0.25 mm. The material properties of the block are as follows; Young's modulus of $E = 3.24$ GPa, Poisson ratio of $\nu = 0.35$, and density of $\rho = 1190\ \text{kg/m}^3$. The linear softening cohesive law is employed to model nonlinear fracture processes in the cohesive zone, in which the tensile strength of the material is set to $f_t = 10^8\ \text{N/m}^2$ and the cohesive fracture energy is taken as $E_c = 700\ \text{N/m}$. Due to the symmetry of the problem, the crack propagates in a straight line along the *x*-axis. The edge crack is modeled as a traction free cohesive crack. For the numerical simulation, the plate is modeled with a FE mesh composed of 3220 bilinear quadrilateral elements, as depicted in Figure 8.19. The position of crack tip is shown in Figure 8.20 as a function of the time. The predicted result agrees well with the result reported by Réthoré, de Borst, and Abellan (2008).

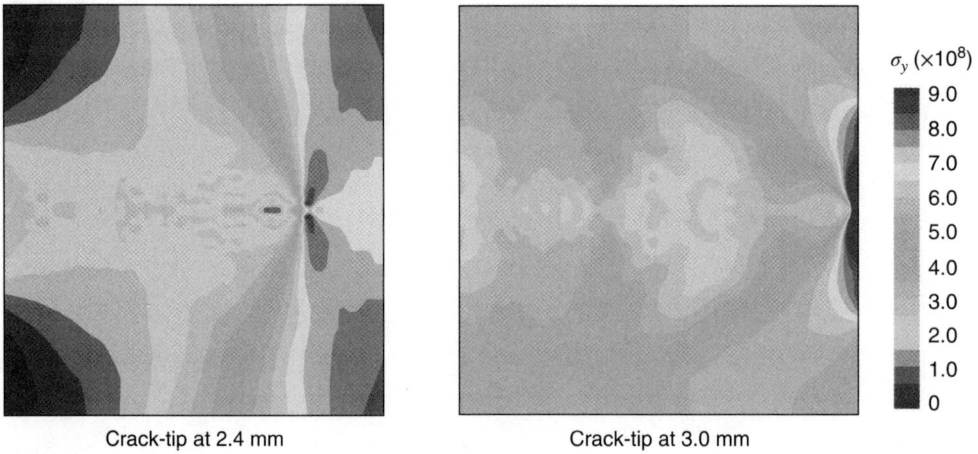

Figure 8.22 The distribution of stress contour in the y-direction at different steps of crack growth (For color details, please see color plate section)

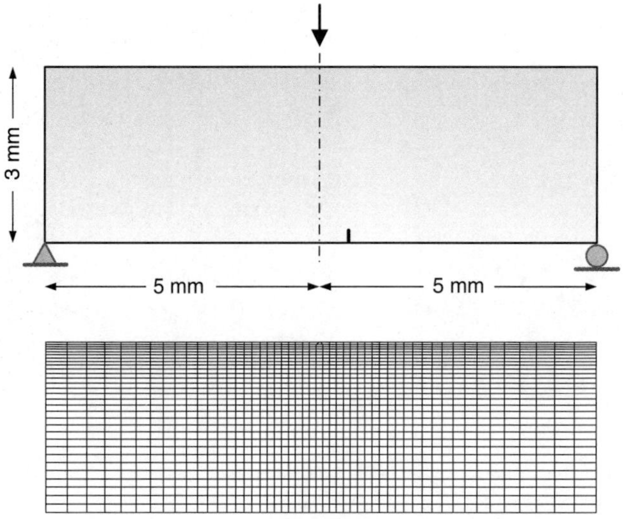

Figure 8.23 A three-point bending beam with an eccentric crack: The geometry, boundary conditions, and X-FEM mesh

The contours of stress in x- and y-directions are displayed in Figures 8.21 and 8.22 at different steps of crack growth when the crack tip approaches the right hand side of the plate.

8.6.3 A Three-Point Bending Beam with an Eccentric Crack

In the last example, the X-FEM cohesive crack simulation is performed for the beam in the first example, in which the crack is forced to initiate eccentrically. The geometry, boundary conditions, and material properties are similar to the first example, as shown in Figure 8.23. The eccentric crack leads to the mixed

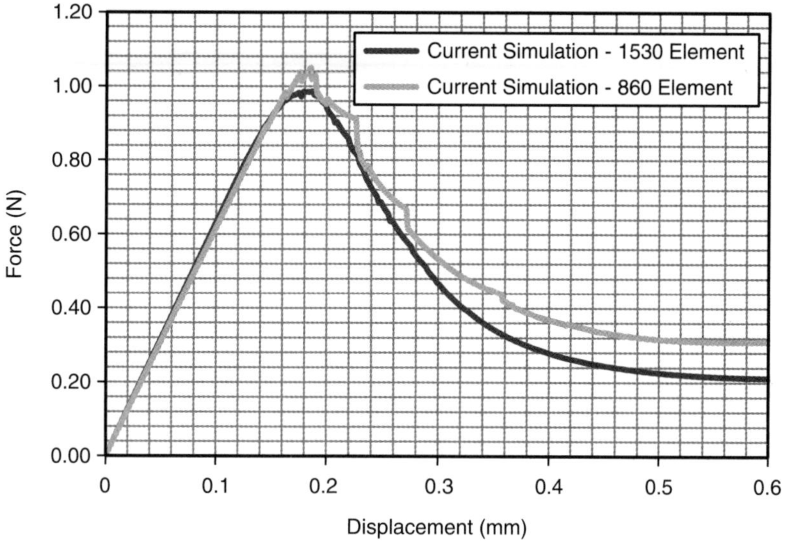

Figure 8.24 The load-displacement curve for a three-point bending beam

Table 8.2 A three-point bending beam with an eccentric crack

Step	$\mathbf{K}^{coh}$	$\mathbf{f}^{coh}$
67	$\begin{bmatrix} -19.20 & -0.61 & -98.28 & -3.12 \\ -0.61 & -0.01 & -3.12 & -0.09 \\ -98.28 & -3.12 & -504.02 & -16.03 \\ -3.12 & -0.09 & -16.03 & -0.51 \end{bmatrix}$	$\begin{bmatrix} -1.80D-2 \\ -5.72D-4 \\ -9.00D-2 \\ -2.86D-3 \end{bmatrix}$
77	$\begin{bmatrix} -19.01 & -0.60 & -97.31 & -3.09 & -51.23 & -1.63 & -10.82 & -0.34 \\ -0.60 & -0.01 & -3.09 & -0.09 & -1.63 & -0.05 & -0.34 & -0.01 \\ -97.31 & -3.09 & -498.94 & -15.87 & -244.20 & -7.77 & -51.23 & -1.60 \\ -3.09 & -0.09 & -15.87 & -0.50 & -7.77 & -0.24 & -1.63 & -0.05 \\ -51.23 & -1.63 & -244.20 & -7.77 & -477.86 & -15.20 & -107.62 & -3.42 \\ -1.63 & -0.05 & -7.77 & -0.24 & -15.20 & -0.48 & -3.42 & -0.10 \\ -10.82 & -0.34 & -51.23 & -1.63 & -107.62 & -3.42 & -24.27 & -0.77 \\ -0.34 & -0.01 & -0.05 & -0.05 & -3.42 & -0.10 & -0.77 & -0.02 \end{bmatrix}$	$\begin{bmatrix} -1.78D-2 \\ -5.67D-4 \\ -8.92D-2 \\ -2.83D-3 \\ -8.81D-2 \\ -2.80D-3 \\ -1.94D-2 \\ -6.17D-4 \end{bmatrix}$
85	$\begin{bmatrix} -18.40 & -0.58 & -94.16 & -2.99 & -49.73 & -1.58 & -10.50 & -0.33 \\ -0.58 & -0.01 & -2.99 & -0.09 & -1.58 & -0.05 & -0.33 & -0.01 \\ -94.16 & -2.99 & -482.76 & -15.36 & -236.97 & -7.54 & -49.73 & -1.58 \\ -2.99 & -0.09 & -15.36 & -0.48 & -7.54 & -0.24 & -1.58 & -0.05 \\ -49.73 & -1.58 & -236.97 & -7.54 & -465.13 & -14.80 & -104.76 & -3.33 \\ -1.58 & -0.05 & -7.54 & -0.24 & -14.80 & -0.47 & -3.33 & -0.10 \\ -10.50 & -0.33 & -49.73 & -1.58 & -104.76 & -3.33 & -23.63 & -0.75 \\ -0.33 & -0.01 & -1.58 & -0.05 & -3.33 & -0.10 & -0.75 & -0.02 \end{bmatrix}$	$\begin{bmatrix} -1.72D-2 \\ -5.50D-4 \\ -8.64D-2 \\ -2.75D-3 \\ -8.57D-2 \\ -2.72D-3 \\ -1.88D-2 \\ -6.00D-4 \end{bmatrix}$

(continued overleaf)

Table 8.2 (*continued*)

Step	$\mathbf{K}^{coh}$	$\mathbf{f}^{coh}$
92	$\begin{bmatrix} -17.46 & -0.55 & -89.35 & -2.84 & -47.33 & -1.50 & -10.01 & -0.31 \\ -0.55 & -0.01 & 2.84 & -0.09 & -1.50 & -0.04 & -0.31 & -0.01 \\ -89.35 & 2.84 & -458.07 & -14.57 & -225.47 & -7.17 & -47.33 & -1.50 \\ -2.84 & -0.09 & -14.57 & -0.46 & -7.17 & -0.22 & -1.50 & -0.04 \\ -47.33 & -1.50 & -225.47 & -7.17 & -443.82 & -14.12 & -99.96 & -3.18 \\ -1.50 & -0.04 & -7.17 & -0.22 & -14.12 & -0.44 & -3.18 & -010 \\ -10.01 & -0.31 & -47.33 & -1.50 & -99.96 & -3.18 & -22.55 & -0.71 \\ -0.31 & -0.01 & -1.50 & -0.04 & -3.18 & -010 & -0.71 & -0.02 \end{bmatrix}$	$\begin{bmatrix} -1.64D-2 \\ -5.22D-4 \\ -8.20D-2 \\ -2.61D-3 \\ -8.17D-2 \\ -2.60D-3 \\ -1.79D-2 \\ -5.72D-4 \end{bmatrix}$

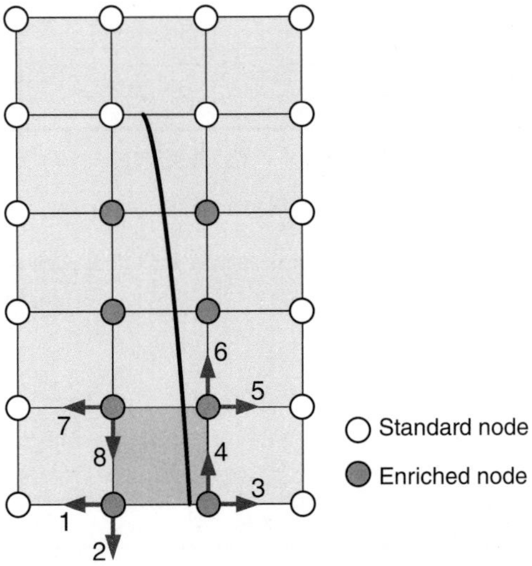

The cohesive stiffness matrices and cohesive force vectors for the element of bottom edge at different steps of crack growth.

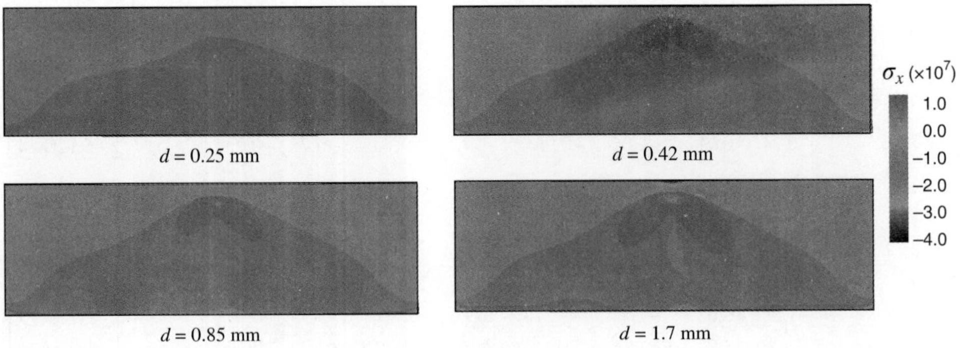

Figure 8.25 The distribution of stress contour in the *x*-direction at different steps of crack growth (For color details, please see color plate section)

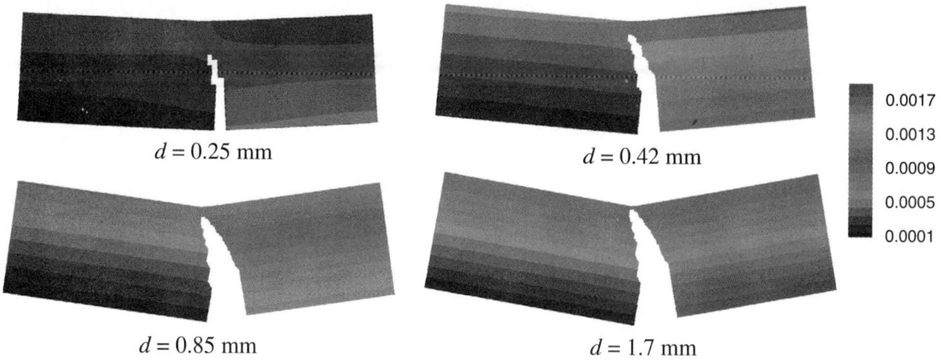

Figure 8.26 The distribution of displacement contour in the x-direction at different steps of crack growth (For color details, please see color plate section)

combination of mode I and II crack behavior. The beam is discretized using a "coarse" and a "fine" mesh with 860 and 1530 bilinear quadrilateral elements, respectively. The load-displacement curves of the three-point bending beam are plotted in Figure 8.24 for two meshes. In Table 8.2, the values of cohesive stiffness matrix and cohesive force vector are given for the element of bottom edge at different steps of crack growth. The contours of stress in x- and y-directions are displayed in Figures 8.25 and 8.26 at different steps of crack growth when the crack-tip approaches the upper boundary of the beam.

9

Ductile Fracture Mechanics with a Damage-Plasticity Model in X-FEM

9.1 Introduction

Crack initiation and propagation phenomena in structural and mechanical failures have led to disastrous events during recent decades; especially in the aerospace and nuclear industries. In many engineering problems, failures are subjected to cyclic loading and deformations that may occur at a high rate, such as earthquake loading, high pressure turbine discs in airplane engines, and so on. Low cycle fatigue and ductile fracture are two common failure mechanisms observed in these loading conditions. These phenomena are highly nonlinear and complex in nature; plastic yielding may eventually occur, accumulation of which in consecutive cycles leads to complete degradation of the material. These catastrophic events forced the researchers to study the main causes of such events. The most important reason for such issues was found to be crack propagation due to pre-existing defects in the structure. Initiating a crack in an engineering part can be considered to be a lifetime estimation criterion but leads to an inefficient design. In a proper economical design, it is necessary to consider the full load-bearing capacity of the material, which means the load-bearing capacity of the part including the crack. So, having a precise understanding of the crack behavior in mechanical parts can be beneficial for design purposes and from an operational design considerations point of view.

Fracture mechanics can be generally divided into two main categories, namely brittle and ductile material behaviors. In brittle fracture mechanics, a small plastic deformation occurs at the crack tip region, in which the assumptions of linear elastic fracture mechanics (LEFM) are valid. In contrast, ductile fracture mechanics is preceded by large plastic deformation, in which the plastic energy dissipation cannot be negligible at the crack-tip area. The fracture of ductile materials is the consequence of a progressive damage process where a considerable plastic deformation usually precedes the ultimate failure. Ductile fracture occurs in a stable manner where extra loading is needed for the crack propagation. Ductile fracture models are usually referred to as plastic models and include several behavioral models; of which two of the more popular are the cohesive-zone and damage-plasticity models. In the cohesive-zone model, crack influence on material behavior is restricted to a limited portion of the body; while in the damage-plasticity model, the damage parameter is defined to represent crack initiation and propagation phenomena as part of mechanical failures. The numerical prediction of damage evolution and crack propagation can be described by the means of continuum damage mechanics (CDM).

Ductile fracture presents the process of fracture from the micro-structural point of view that can be explained in three stages; at the first step, the micro-voids and micro-cracks are formed in different parts

Extended Finite Element Method: Theory and Applications, First Edition. Amir R. Khoei.

of the material, the growth and coalescence of these micro-voids leads to the formation of macro-cracks at the second step, and the propagation of crack causes the failure of structure at the last step. Experimental investigations demonstrate that most fracture processes are nonlinear in nature and occur at a region around the crack tip called the fracture process zone. In fact, the ductile fracture is initiated by cracking of inclusion particles or separation at the interfaces of metal and inclusions. The voids grow about these defects as the deformation increases, and the fracture occurs when these adjacent voids coalescence and form macro-cracks (Dragon, 1985; Tai, 1990). From the macro-mechanical point of view, two important factors are involved as the driving forces of ductile fracture; the plastic deformation and the hydrostatic stress. The relative importance of these factors depends on the type of loading; for example, in tensile loading, both the plastic deformation and hydrostatic stress are important but in the case of shear loading, the plastic deformation becomes dominant.

Introducing a powerful numerical tool that is able to model the component behavior from the beginning of loading to the final failure stage can significantly decrease production costs while improving the quality. There are various computational algorithms proposed in literature in modeling the fracture of ductile materials. The first approach is based on the continuous algorithm, in which the fracture process zone is appropriately modeled but the cracks, micro-voids, and flaws are not discretely considered and their effects are introduced to continua by changing the overall material properties using phenomenological models, such as softening plasticity or damage plasticity models. This kind of modeling predicts the correct behavior of the structure only up to the onset of fracture. This approach is prone to localization and ill-posedness issues that need suitable regularization techniques. Since the pioneering studies of McClintock (1968) and Rice and Tracy (1969), there have been two main techniques for modeling these phenomena in metals. The first model is based on the porous metal plasticity, in which material degradation is modeled implicitly with softening plasticity; such as the famous model proposed by Gurson (1977) and Tvergaard and Needleman (1984). The second model is based on the CDM that was introduced by Kachanov (1986) and Lemaitre and Desmorat (2005); in which the material degradation, void nucleation, and coalescence are modeled through a definition of damage as a thermo-dynamic state variable. However, these approaches are prone to localization and ill-posedness issues that need suitable regularization techniques. The best way to overcome these drawbacks is to introduce discontinuities into the media.

The second approach is based on the discontinuous algorithm, in which the crack discontinuity is modeled into the displacement field while the fracture process zone is assumed to be small compared to the size of structure, or it is concentrated in a small region, such as the cohesive fracture zone, where its effect on the overall behavior of system is neglected (Sancho *et al.*, 2006). The adaptive finite element method (FEM) is one of the most popular techniques that has been extensively used for modeling fracture mechanics problems; as examples, Mediavilla, Peerlings, and Geers (2006), Moslemi and Khoei (2009), and Khoei *et al.* (2013) employed the adaptive FEM strategy to model complex 2D and 3D components under different loading conditions. However, in this technique the mesh must be confined to the crack body, and crack propagation requires a mesh generation step, which is both time consuming and costly. An alternative technique that alleviates these problems is based on the partition of unity method (PUM) (Babuska and Melenk, 1997), in which the X-FEM method (extended FEM) is the most popular one (Belytschko and Black, 1999). Obviously, a combination of these two procedures can be implemented as an alternative technique for capturing the fracture process zone at the crack tip region by employing a numerical algorithm to model the crack discontinuity using a damage-plasticity model.

In this chapter, the X-FEM technique is presented for the crack growth simulation of ductile fracture problems. Since local damage models lead to ill-posedness in the governing differential equation, a non-local damage model is proposed that involves the solution of a Helmholtz type equation in conjunction with the X-FEM equilibrium equation. A non-local version of the Lemaitre damage-plasticity model (Lemaitre and Desmorat, 2005) is employed to capture the fracture process zone. The corrected X-FEM method originally proposed by Fries (2008) is implemented to overcome the blending issues and obtain the optimum convergence rate in ductile fracture problems. An updated Lagrangian X-FEM formulation is used to efficiently model the large deformation crack propagation and avoid the data transfer that spoils the accuracy at the fracture process zone. In order to simulate the process of failure and crack propagation in dynamic and cyclic loading conditions, the damage–plasticity model is then incorporated into the dynamic large deformation X-FEM formulation. An explicit central difference method is employed to model the dynamic effects, in which the method is efficiency enhanced with mass lumping and numerical damping. In order to prevent locking issues and to decrease computational costs, the reduced integration method with hourglassing control is employed. Since the loading can be reversed, the contact behavior between the crack faces is taken into account by employing the crack closure effects. The constitutive material behavior is modeled using the Lemaitre-damage plasticity model that uses von-Mises plasticity with an isotropic-kinematic hardening rule combined with a scalar damage variable. Moreover, the issues relating to large strains, crack propagation criteria, and crack direction detection are discussed.

9.2 Large FE Deformation Formulation

Consider a cracked body Ω with the surface Γ subjected to the body force loading $\mathbf{b}$ in Ω, the prescribed traction $\mathbf{t}$, and displacement $\mathbf{u}_0$ on Γ_t and Γ_u, respectively, and the traction free cracked surface Γ_c, as shown in Figure 9.1. The governing equilibrium equation of the system can be written as

$$\nabla \cdot \mathbf{P} + \mathbf{b} = 0 \tag{9.1}$$

where ∇ is the gradient operator, and σ is the Cauchy stress tensor defined on the current configuration. The boundary conditions are defined as

$$\begin{aligned} \boldsymbol{\sigma} \cdot \mathbf{n}|_{\Gamma_t} &= \mathbf{t} \\ \boldsymbol{\sigma} \cdot \mathbf{n}|_{\Gamma_c} &= \mathbf{0} \\ \mathbf{u}|_{\Gamma_u} &= \mathbf{u}_0 \end{aligned} \tag{9.2}$$

In ductile fracture mechanics, the nonlinearities arise from two distinct sources; constitutive nonlinearities and geometric nonlinearities, due to material behavior and large deformations. Consider the state variables are known at step n, the goal is to obtain the variables at step $n+1$, as shown in Figure 9.2. Since the configuration of step $n+1$ is not known, the first Piola–Kirchhof (PK1) stress is used to define the equilibrium equation on the available configuration of step n. This is the core idea of an updated Lagrangian framework proposed here; Eq. (9.1) can be therefore rewritten as

$$\nabla_n \cdot \mathbf{P} + \mathbf{b} = 0 \tag{9.3}$$

where ∇_n is the gradient operator with respect to $\boldsymbol{x}_n$, and $\mathbf{P}$ is the first Piola–Kirchhof stress tensor of current configuration with respect to step n configuration and $\boldsymbol{x}_n$ is the vector of coordinates in the current

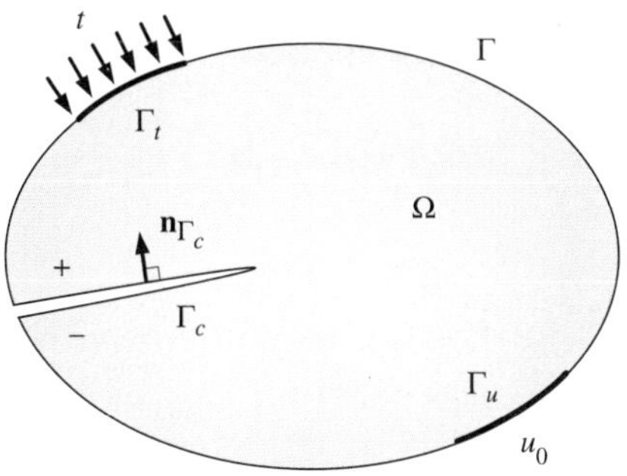

Figure 9.1 Problem definition of a cracked body in domain Ω

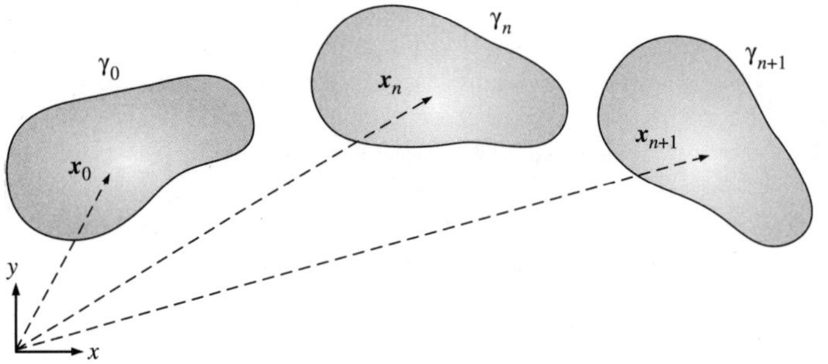

Figure 9.2 The Lagrangian framework: The initial configuration γ_0, the known current configuration γ_n, and the unknown incremental configuration γ_{n+1}

configuration at step n. The weak form of this equation can be obtained by multiplying the test function $\delta\mathbf{u}$ and integrating over the domain as

$$\int_\Omega \delta\mathbf{u}(\nabla_n \cdot \mathbf{P} + \mathbf{b} = \mathbf{0})d\Omega = 0 \quad \text{for all admissible} \quad \delta\mathbf{u} \in \mathcal{H}_0^1(\Omega) \tag{9.4}$$

where $\mathcal{H}_0^1(\Omega)$ is the first Hilbert space with generalized zero boundary conditions. Applying the Gaussian theorem, it yields

$$\int_\Omega \nabla_n \delta\mathbf{u} : \mathbf{P}\, d\Omega = \int_{\Gamma_t} \delta\mathbf{u} \cdot \bar{\mathbf{t}}\, d\Gamma + \int_\Omega \delta\mathbf{u} \cdot \mathbf{b}\, d\Omega \quad \text{for all admissible} \quad \delta\mathbf{u} \in \mathcal{H}_0^1(\Omega) \tag{9.5}$$

where $\bar{\mathbf{t}}$ is the traction relating to the first Piola–Kirchhof stress. Since $\boldsymbol{x}_{n+1} = \boldsymbol{x}_n + \mathbf{u}$, the corresponding deformation gradient is defined by $\mathbf{F}_n = \mathbf{I} + \partial\mathbf{u}/\partial\boldsymbol{x}_n$, and the variation of deformation gradient is given by $\delta\mathbf{F}_n = \partial\delta\mathbf{u}/\partial\boldsymbol{x}_n$. Substitution of the above expressions into Eq. (9.5) yields

$$\int_\Omega \delta\mathbf{F}_n^T \mathbf{P}\, d\Omega = \int_{\Gamma_t} \delta\mathbf{u}^T \bar{\mathbf{t}}\, d\Gamma + \int_\Omega \delta\mathbf{u}^T \mathbf{b}\, d\Omega \tag{9.6}$$

In order to obtain the symmetric matrices, the second Piola–Kirchhof (PK2) stress $\mathbf{S}$ can be employed using $\mathbf{P} = \mathbf{F}_n\, \mathbf{S}$, hence

$$\int_\Omega \delta\mathbf{F}_n^T \mathbf{F}_n \mathbf{S}\, d\Omega = \int_{\Gamma_t} \delta\mathbf{u}^T \bar{\mathbf{t}}\, d\Gamma + \int_\Omega \delta\mathbf{u}^T \mathbf{b}\, d\Omega \tag{9.7}$$

in which the second Piola–Kirchhof stress is work conjugate stress to the Green strain $\boldsymbol{E}^G$ defined as

$$\boldsymbol{E}^G = \frac{1}{2}\left(\mathbf{F}_n^T \mathbf{F}_n - \mathbf{I}\right) \tag{9.8}$$

The variational form of this relation can be written as

$$\delta\boldsymbol{E}^G = \frac{1}{2}\left(\delta\mathbf{F}_n^T \mathbf{F}_n + \mathbf{F}_n^T \delta\mathbf{F}_n\right) \tag{9.9}$$

Substituting relation (9.9) into Eq. (9.7) results in

$$\int_{\Omega} \delta E^G \mathbf{S}\, d\Omega = \int_{\Gamma_t} \delta \mathbf{u}^T \bar{\mathbf{t}}\, d\Gamma + \int_{\Omega} \delta \mathbf{u}^T \mathbf{b}\, d\Omega \tag{9.10}$$

Applying the FE Galerkin discretization, the independent approximations of $\mathbf{u}$ can be defined as $\mathbf{u} = \mathbf{N}\bar{\mathbf{u}}$, with $\mathbf{N}$ denoting the matrix of shape functions and $\bar{\mathbf{u}}$ is the vector of corresponding nodal degrees of freedom. Based on the Green strain definition, the strain–displacement relationship can be defined in terms of linear and nonlinear components as

$$E^G = E_L + E_{NL} = \mathbf{B}_L \bar{\mathbf{u}} + \frac{1}{2} A_{\theta(\bar{\mathbf{u}})} \mathbf{G} \bar{\mathbf{u}} \tag{9.11}$$

where $A_{\theta(\bar{\mathbf{u}})}$ is a function of $\bar{\mathbf{u}}$ defined as

$$A_{\theta(\bar{\mathbf{u}})} = \begin{bmatrix} \frac{\partial u}{\partial x_n} & \frac{\partial v}{\partial x_n} & 0 & 0 \\ 0 & 0 & \frac{\partial u}{\partial y_n} & \frac{\partial v}{\partial y_n} \\ \frac{\partial u}{\partial y_n} & \frac{\partial v}{\partial y_n} & \frac{\partial u}{\partial x_n} & \frac{\partial v}{\partial x_n} \end{bmatrix} \tag{9.12}$$

and $\mathbf{B}_L$ is the standard matrix of shape function gradients defined for node I in linear FEM as

$$\mathbf{B}_L^I = \begin{bmatrix} \frac{\partial N_I}{\partial x_n} & 0 \\ 0 & \frac{\partial N_I}{\partial y_n} \\ \frac{\partial N_I}{\partial y_n} & \frac{\partial N_I}{\partial x_n} \end{bmatrix} \tag{9.13}$$

and $\mathbf{G}$ is the matrix of shape function derivatives defined as

$$\mathbf{G}^I = \begin{bmatrix} \frac{\partial N_I}{\partial x_n} & 0 \\ 0 & \frac{\partial N_I}{\partial x_n} \\ \frac{\partial N_I}{\partial y_n} & 0 \\ 0 & \frac{\partial N_I}{\partial y_n} \end{bmatrix} \tag{9.14}$$

Assuming $\mathbf{B}_{NL} = A_\theta \mathbf{G}$ and taking the variation from (9.11), it results in $dE^G = (\mathbf{B}_L + \mathbf{B}_{NL}) d\bar{\mathbf{u}}$. Thus, the FE formulation (9.10) can be obtained as

$$\psi(\bar{\mathbf{u}}) = \int_{\Omega} \bar{\mathbf{B}}^T \mathbf{S}\, d\Omega - \int_{\Gamma_t} \mathbf{N}^T \bar{\mathbf{t}}\, d\Gamma - \int_{\Omega} \mathbf{N}^T \mathbf{b}\, d\Omega = 0 \tag{9.15}$$

where $\boldsymbol{\psi}(\bar{\mathbf{u}})$ is the residual vector and $\bar{\mathbf{B}} = \mathbf{B}_L + \mathbf{B}_{NL}$. This nonlinear system of equations must be solved for $\bar{\mathbf{u}}$. The Newton–Raphson method is used to linearize Eq. (9.15) as

$$\boldsymbol{\psi}_{n+1}^{i+1}(\bar{\mathbf{u}}) \simeq \boldsymbol{\psi}_{n+1}^{i}(\bar{\mathbf{u}}) + d\boldsymbol{\psi} = 0 \tag{9.16}$$

where i refers to the i th iteration of the Newton–Raphson method and $n+1$ refers to step $n+1$. In equation (9.16), $d\boldsymbol{\psi}$ is defined as

$$d\boldsymbol{\psi}(\bar{\mathbf{u}}) = \int_\Omega d\bar{\mathbf{B}}^T \mathbf{S}\, d\Omega + \int_\Omega \bar{\mathbf{B}}^T d\mathbf{S}\, d\Omega \tag{9.17}$$

Consider the hypoelasto-plastic constitutive relation as $\mathbf{S} = \mathbf{C}^{ep}\, E^G$, in which the tensor $\mathbf{C}^{ep}$ can be obtained from the constitutive equations. The incremental form of constitutive relation can be written as $d\mathbf{S} = \mathbf{C}^{ep}\, dE^G$. Substituting this relation into (9.17), it yields to the tangent stiffness matrix $\mathbf{K}_T$ that consists of the geometric and material parts defined as (Khoei, 2005)

$$d\boldsymbol{\psi}(\bar{\mathbf{u}}) = \mathbf{K}_T d\bar{\mathbf{u}} = \left(\underbrace{\int_\Omega \mathbf{G}^T \mathbf{M}_S\, \mathbf{G}\, d\Omega}_{\text{Geometric stiffness}} + \underbrace{\int_\Omega \bar{\mathbf{B}}^T \mathbf{C}^{ep}\, \bar{\mathbf{B}}\, d\Omega}_{\text{Material stiffness}} \right) d\bar{\mathbf{u}} \tag{9.18}$$

where the matrix $\mathbf{M}_S$ is defined as

$$\mathbf{M}_S = \begin{bmatrix} S_{xx}\,\mathbf{I}_{2\times 2} & S_{xy}\,\mathbf{I}_{2\times 2} \\ sym. & S_{yy}\,\mathbf{I}_{2\times 2} \end{bmatrix} \tag{9.19}$$

where $\mathbf{I}$ is the identity matrix. Substituting relation (9.18) into (9.16), the incremental nodal displacements can be obtained in an iterative procedure as

$$\mathbf{K}_T\, d\bar{\mathbf{u}} = -\boldsymbol{\psi}_{n+1}^{i} \tag{9.20}$$

In order to solve the system of Eq. (9.20), the residual vector $\boldsymbol{\psi}_{n+1}^{i}$ is set to load increment $\Delta\mathbf{f}$ at the first iteration of step $n+1$. The incremental nodal displacements $d\bar{\mathbf{u}}$ can be obtained from (9.20) using the updated values of stresses and strains obtained from the constitutive relation. For the next iteration, the residual vector $\boldsymbol{\psi}_{n+1}^{i}$ is obtained from (9.15) and the incremental displacement vector $d\bar{\mathbf{u}}$ from (9.20). This iterative procedure continues until the norm of residual vector becomes less than the prescribed tolerance. It must be noted that the stiffness matrix $\mathbf{K}_T$ must be updated at each iteration in the Newton–Raphson method. If the convergence of solution is obtained, the nodal coordinates are updated and the second Piola–Kirchhof stresses are transferred to the Cauchy stresses at the new configuration by $\boldsymbol{\sigma} = J^{-1}\mathbf{F}_n \mathbf{S}\, \mathbf{F}_n^T$, where J is the determinant of $\mathbf{F}_n$.

9.3 Modified X-FEM Formulation

The X-FEM was originally proposed by Belytschko and Black (1999). The method has been successfully applied to elastic and cohesive fracture problems (Moës and Belytschko, 2002a; Areias and Belytschko, 2005a; Asferg, Poulsen, and Nielsen, 2007) with a great advantage over the standard FEM, in which the mesh does not need to be confined to the crack body and hence the remeshing strategy is removed in crack propagation. In this technique, the discontinuities are taken into account by adding appropriate functions into the standard approximation spaces through a PUM, where the singularities and high gradients can be achieved by an optimal convergence rate (Stazi *et al.*, 2003; Sukumar and Prévost, 2003; Huang, Sukumar, and Prévost, 2003b). The PUM allows the inclusion of prior knowledge of the problem to the FEM space and provides the ability to construct the FEM space of any desired regularity (Melenk and Babuška, 1996).

A PU on a domain $\Omega \in \mathbb{R}^2$ is a set of functions $\{\Phi_i\} \in C^m(\Omega)$ such that

$$\sum_i \Phi_i(x) = 1 \qquad \forall x \in \Omega \tag{9.21}$$

where $C^m(\Omega)$ is the set of m times continuous functions defined on domain Ω. Hence, each desired function $f \in \mathcal{H}^1(\Omega)$ can be expressed as

$$f(x) = \sum_i f_i(x)\,\Phi_i \tag{9.22}$$

where $\mathcal{H}^1(\Omega)$ is the first Hilbert space function. This property allows to locally approximate the function $f(x)$ with functions, which are tailored to the problem under consideration. It is well-known that the FEM shape functions $N_i(x)$ constitute the PU. Based on this property the X-FEM displacement field for the elastic fracture problem is defined as

$$\mathbf{u}(\boldsymbol{x}) = \underbrace{\sum_{i \in \mathcal{N}^{std}} N_i(\boldsymbol{x})\,\bar{\mathbf{u}}_i}_{\text{standard part}} + \underbrace{\sum_{i \in \mathcal{N}^{enr}_{Hev}} N_i(\boldsymbol{x}) H(\boldsymbol{x})\,\bar{\mathbf{a}}_i + \sum_{i \in \mathcal{N}^{enr}_{Tip}} \sum_{j=1}^{4} N_i(\boldsymbol{x}) A_j(\boldsymbol{x})\,\bar{\mathbf{b}}_i^j}_{\text{enriched part}} \tag{9.23}$$

where

$$H(\boldsymbol{x}) = \begin{cases} -1 & \boldsymbol{x} > 0 \\ +1 & \boldsymbol{x} < 0 \end{cases} \tag{9.24}$$

and

$$\{A_j(\boldsymbol{x})\} = \left\{ \sqrt{r}\sin\frac{\theta}{2}, \sqrt{r}\cos\frac{\theta}{2}, \sqrt{r}\sin\frac{\theta}{2}\sin\theta, \sqrt{r}\cos\frac{\theta}{2}\sin\theta \right\} \tag{9.25}$$

in which the displacement field is made of the standard and enriched parts. The standard part represents the standard FEM displacement field, and the enriched part is constructed from two terms; the first term displays the Heaviside enrichment function $H(\boldsymbol{x})$ used to represent the crack body discontinuity, and the second term illustrates the asymptotic enrichment functions $A_j(\boldsymbol{x})$, chosen to describe the singularity of the stress field at the crack tip area. The four asymptotic functions (9.25) are derived from the LEFM analytical solution around the crack tip. It was shown by Legrain, Moës, and Verron (2005) that the first enrichment function, that is, $\sqrt{r}\sin\frac{\theta}{2}$, is sufficient for the large deformation fracture problems. In these relations, $N_i(\boldsymbol{x})$ is the shape function of node i; $\bar{\mathbf{u}}_i$, $\bar{\mathbf{a}}_i$ and $\bar{\mathbf{b}}_i^j$ are the corresponding degrees of freedom for the standard, Heaviside and asymptotic enrichments; and r and θ are the polar coordinates with respect to the crack-tip under consideration. As shown in Figure 9.3, $\mathcal{N}^{enr}_{Hev}$ are the nodes of all elements which are cut by the crack and enriched by the Heaviside function, and $\mathcal{N}^{enr}_{Tip}$ are the nodes of crack tip elements that are enriched by the asymptotic functions. Consequently, there are four types of elements; the standard elements with no enriched nodes, the crack body elements whose nodes are enriched by the Heaviside

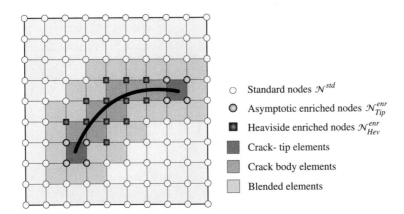

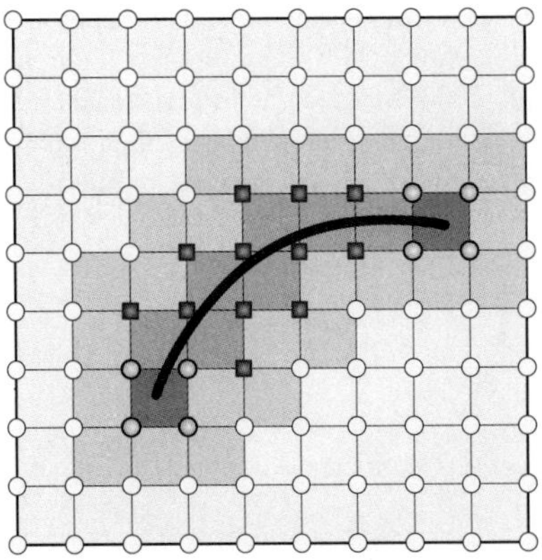

Figure 9.3 The crack body elements together with the crack tip elements and blended elements along with corresponding enriched nodes in the X-FEM technique

function, the crack tip elements whose nodes are enriched by the asymptotic functions, and the blended elements with nodes that have different kinds of enrichments.

In practice a shifted version of relation (9.23) is generally used, in which the nodal displacements take the actual value of $\bar{\mathbf{u}}_i$ and the effect of enrichment is alleviated. This formulation has several advantages, for instance the Dirichlet boundary conditions can be applied without any modification compared to the standard FEM. The shifted X-FEM displacement field is defined as

$$\mathbf{u}(\boldsymbol{x}) = \sum_{i\in\mathcal{N}^{std}} N_i(\boldsymbol{x})\,\bar{\mathbf{u}}_i + \sum_{i\in\mathcal{N}_{Hev}^{enr}} N_i(\boldsymbol{x})(H(\boldsymbol{x}) - H(\boldsymbol{x}_i))\,\bar{\mathbf{a}}_i + \sum_{i\in\mathcal{N}_{Tip}^{enr}} \sum_{j=1}^{4} N_i(\boldsymbol{x})\left(A_j(\boldsymbol{x}) - A_j(\boldsymbol{x}_i)\right)\bar{\mathbf{b}}_i^j \tag{9.26}$$

It was shown by Laborde *et al.* (2005) that in the LEFM problems with Q4 standard elements, the crack body and crack tip elements represent a convergence rate of $O(h)$ in the energy norm, however, the blending elements reduce the convergence rate to $O(\sqrt{h})$. The existence of blending layer is crucial in C_0 continuity and these elements introduce parasitic terms into approximation space which leads to the suboptimal convergence rate. Various techniques have been introduced by researchers to overcome the blending issues, including: the enhanced strain method by Chessa, Wang, and Belytschko (2003) and Gracie, Wang, and Belytschko (2008b), the cutoff function by Chahine, Laborde, and Renard (2008), the blending weight function by Ventura, Gracie, and Belytschko (2009), and the discontinuous Galerkin method by Shen and Lew (2010a). In this study, a modified X-FEM technique is employed for blending elements, in which the enrichment is assumed to be geometric (Béchet *et al.*, 2005). The geometric enrichment means that all elements within a predefined area with a constant radius around the crack tip are enriched. In the modified X-FEM method, in addition to those nodal points enriched by the standard X-FEM, all nodes of the blending elements are also enriched. According to Figure 9.4, $\mathcal{N}_{Tip}^{enr}$ are the set of nodes of elements that are fully prescribed in the enrichment radius, and $\mathcal{N}_{Tip}^{*enr}$ consists of the set of $\mathcal{N}_{Tip}^{enr}$ and the nodes of blending elements which are enriched by asymptotic functions. $\mathcal{N}_{Hev}^{enr}$ is the set of nodes of elements that are cut by the crack and are not in the set of $\mathcal{N}_{Tip}^{enr}$.

In order to alleviate the negative effect of parasitic terms in blending elements, the asymptotic enrichment functions are modified by application of a cutoff function $R(\boldsymbol{x})$, which is produced from the shape functions of the nodes in $\mathcal{N}_{Tip}^{enr}$ as

$$A_j^{\text{Mod}}(\boldsymbol{x}) = R(\boldsymbol{x})A_j(\boldsymbol{x}) \tag{9.27}$$

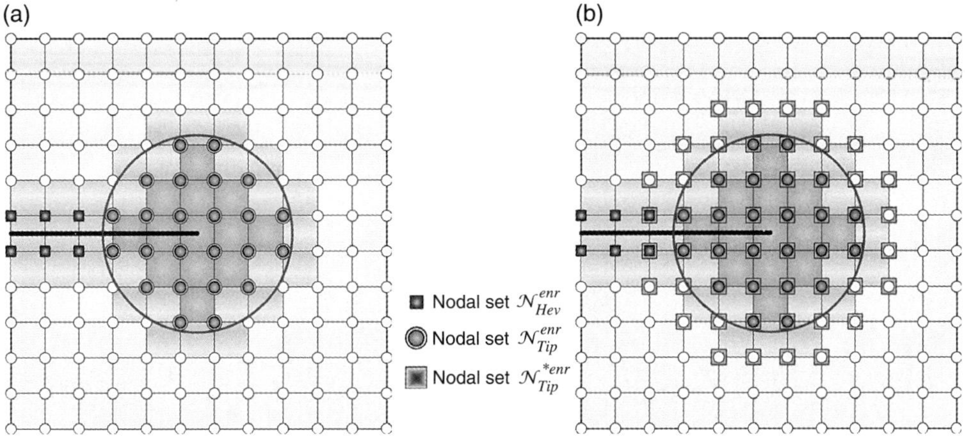

Figure 9.4 The geometric enrichment strategy: (a) The standard X-FEM; (b) The modified X-FEM

where

$$R(\boldsymbol{x}) = \sum_{i \in \mathcal{N}^{enr}_{Tip}} N_i(\boldsymbol{x}) \tag{9.28}$$

in which the cutoff function $R(\boldsymbol{x})$ is zero at the standard elements and varies continuously to unity in the fully enriched elements. In this method, the partition of unity exists in the whole domain. Hence, the modified X-FEM displacement field can be defined as

$$\mathbf{u}(\boldsymbol{x}) = \sum_{i \in \mathcal{N}^{std}} N_i(\boldsymbol{x})\,\bar{\mathbf{u}}_i + \sum_{i \in \mathcal{N}^{enr}_{Hev}} N_i(\boldsymbol{x})\bar{H}(\boldsymbol{x})\,\bar{\mathbf{a}}_i + \sum_{i \in \mathcal{N}^{enr}_{Tip}} \sum_{j=1}^{4} N_i(\boldsymbol{x})\,R(\boldsymbol{x})\bar{A}_j(\boldsymbol{x})\,\bar{\mathbf{b}}^j_i \tag{9.29}$$

where $\bar{H}(\boldsymbol{x}) = H(\boldsymbol{x}) - H(\boldsymbol{x}_i)$ and $\bar{A}_j(\boldsymbol{x}) = A_j(\boldsymbol{x}) - A_j(\boldsymbol{x}_i)$.

9.4 Large X-FEM Deformation Formulation

In order to model the ductile fracture mechanics in large deformations, the Lagrangian formulation is employed here in the framework of the X-FEM technique (Khoei, Biabanaki, and Anahid, 2009a). Since the crack tip enrichment functions described in preceding section are defined for the small deformation problem, the implementation of these asymptotic functions with the crack tip polar coordinates r and θ on the current configuration is absolutely meaningless. Hence, the enrichment functions are defined on the initial (*undeformed*) configuration and an appropriate mapping is used to transfer the values of enrichment functions on the current configuration in an updated Lagrangian framework, as shown in Figure 9.5, which is tailored to the X-FEM.

In the standard FEM, the map between the parent and current element is a straightforward manner that can be given by a bilinear relation for the Q4 element, as shown in Figure 9.6, by

$$\boldsymbol{x}_0 = \sum_{i=1}^{4} N_i(\xi,\eta)\boldsymbol{x}_{i0} \quad \text{and} \quad \boldsymbol{x} = \sum_{i=1}^{4} N_i(\xi,\eta)\boldsymbol{x}_i \tag{9.30}$$

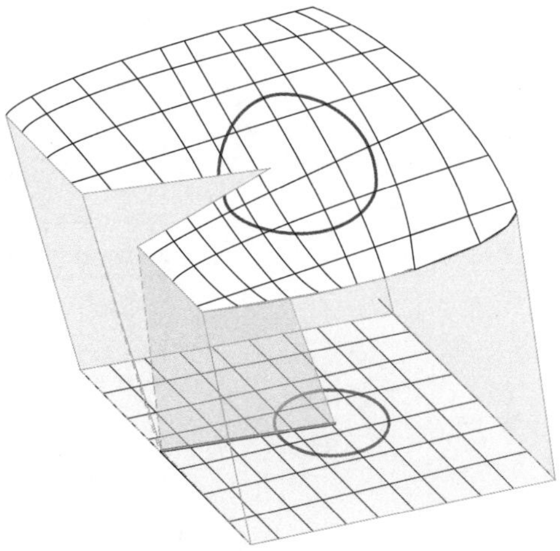

Figure 9.5 The mapping of enrichment functions from the initial (*undeformed*) configuration to the current configuration

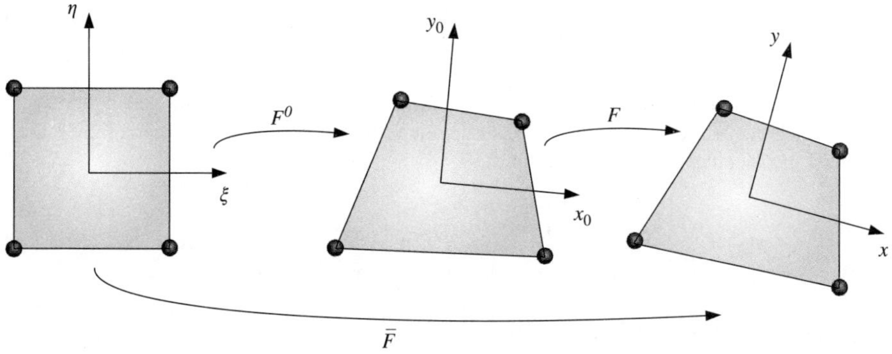

Figure 9.6 The map between the parent and current element for the Q4 element; $\mathbf{F}^0$ is the map between the parent element and initial configuration, $\mathbf{F}$ is the map between initial and current configurations called the deformation gradient, and $\bar{\mathbf{F}}$ is the map between the parent element and current configuration

where $\boldsymbol{x}_0, \boldsymbol{x}_{i0} \in \mathbb{R}^2$ are the initial position and the initial *nodal* position vectors, respectively; and $\boldsymbol{x}, \boldsymbol{x}_i \in \mathbb{R}^2$ are the current configuration counterparts. The Jacobian matrices for the initial and current configurations are defined as

$$\mathbf{J}^0 = \begin{bmatrix} \dfrac{\partial x_0}{\partial \xi} & \dfrac{\partial y_0}{\partial \xi} \\ \dfrac{\partial x_0}{\partial \eta} & \dfrac{\partial y_0}{\partial \eta} \end{bmatrix} = \begin{bmatrix} \sum_i \dfrac{\partial N_i}{\partial \xi} x_{i0} & \sum_i \dfrac{\partial N_i}{\partial \xi} y_{i0} \\ \sum_i \dfrac{\partial N_i}{\partial \eta} x_{i0} & \sum_i \dfrac{\partial N_i}{\partial \eta} y_{i0} \end{bmatrix} \tag{9.31a}$$

$$\mathbf{J} = \begin{bmatrix} \dfrac{\partial x}{\partial \xi} & \dfrac{\partial y}{\partial \xi} \\ \dfrac{\partial x}{\partial \eta} & \dfrac{\partial y}{\partial \eta} \end{bmatrix} = \begin{bmatrix} \sum_i \dfrac{\partial N_i}{\partial \xi} x_i & \sum_i \dfrac{\partial N_i}{\partial \xi} y_i \\ \sum_i \dfrac{\partial N_i}{\partial \eta} x_i & \sum_i \dfrac{\partial N_i}{\partial \eta} y_i \end{bmatrix} \tag{9.31b}$$

where $\mathbf{J}^0$ is the Jacobian matrix between the parent and initial elements, and $\mathbf{J}$ is the Jacobian matrix between the parent and current elements. Hence, the mapping between the parent, initial, and current configurations can be performed as

$$\mathbf{F}^0 = \begin{bmatrix} \dfrac{\partial \xi}{\partial x_0} & \dfrac{\partial \eta}{\partial x_0} \\ \dfrac{\partial \xi}{\partial y_0} & \dfrac{\partial \eta}{\partial y_0} \end{bmatrix} = \frac{1}{J_{11}^0 J_{22}^0 - J_{12}^0 J_{21}^0} \begin{bmatrix} J_{22}^0 & -J_{12}^0 \\ -J_{21}^0 & J_{11}^0 \end{bmatrix} \tag{9.32a}$$

$$\bar{\mathbf{F}} = \begin{bmatrix} \dfrac{\partial \xi}{\partial x} & \dfrac{\partial \eta}{\partial x} \\ \dfrac{\partial \xi}{\partial y} & \dfrac{\partial \eta}{\partial y} \end{bmatrix} = \frac{1}{J_{11} J_{22} - J_{12} J_{21}} \begin{bmatrix} J_{22} & -J_{12} \\ -J_{21} & J_{11} \end{bmatrix} \tag{9.32b}$$

$$\mathbf{F} = \begin{bmatrix} \dfrac{\partial x}{\partial x_0} & \dfrac{\partial x}{\partial y_0} \\ \dfrac{\partial y}{\partial x_0} & \dfrac{\partial y}{\partial y_0} \end{bmatrix} = \begin{bmatrix} J_{11}F_{11}^0 + J_{21}F_{12}^0 & J_{11}F_{21}^0 + J_{21}F_{22}^0 \\ J_{12}F_{11}^0 + J_{22}F_{12}^0 & J_{12}F_{21}^0 + J_{22}F_{22}^0 \end{bmatrix} \tag{9.32c}$$

In the X-FEM formulation, these mappings must be modified according to the corresponding displacement field. Since the enrichments are added to the approximation space, the current position vector is defined by $\boldsymbol{x} = \boldsymbol{x}_0 + \mathbf{u}$, where $\mathbf{u}$ is the total displacement vector and $\boldsymbol{x}_0$ is defined in (9.30). Based on the X-FEM displacement field (9.29), the current position vector can be obtained as

$$\boldsymbol{x} = \sum_{i \in \mathcal{N}^{std}} N_i(\xi,\eta)\boldsymbol{x}_i + \sum_{i \in \mathcal{N}^{enr}_{Hev}} N_i(\xi,\eta)\bar{H}(\boldsymbol{x})\,\bar{\mathbf{a}}_i + \sum_{i \in \mathcal{N}^{*enr}_{Tip}} \sum_{j=1}^{4} N_i(\xi,\eta)\bar{A}_j(\boldsymbol{x}_0)\,R(\boldsymbol{x}_0)\,\bar{\mathbf{b}}_i^j \tag{9.33}$$

in which the crack tip enrichment functions A_j are defined on the initial configuration $\boldsymbol{x}_0$. The components of matrix $\mathbf{J}$ can therefore be written as

$$\begin{aligned} J_{11} = \frac{\partial x}{\partial \xi} = & \sum_{i \in \mathcal{N}^{std}} \frac{\partial N_i}{\partial \xi} x_i + \sum_{i \in \mathcal{N}^{enr}_{Hev}} \frac{\partial N_i}{\partial \xi} \bar{H}(\boldsymbol{x})\, a_i^x + \sum_{i \in \mathcal{N}^{*enr}_{Tip}} \sum_{j=1}^{4} \frac{\partial N_i}{\partial \xi} \bar{A}_j(\boldsymbol{x}_0)\, R(\boldsymbol{x}_0)\, b_i^{j_x} \\ & + \sum_{i \in \mathcal{N}^{*enr}_{Tip}} \sum_{j=1}^{4} N_i \left(\frac{\partial \bar{A}_j(\boldsymbol{x}_0)}{\partial x_0} J_{11}^0 R(\boldsymbol{x}_0) + \bar{A}_j(\boldsymbol{x}_0) \frac{\partial R(\boldsymbol{x}_0)}{\partial x_0} J_{11}^0 \right. \\ & \left. + \frac{\partial \bar{A}_j(\boldsymbol{x}_0)}{\partial y_0} J_{12}^0 R(\boldsymbol{x}_0) + \bar{A}_j(\boldsymbol{x}_0) \frac{\partial R(\boldsymbol{x}_0)}{\partial y_0} J_{12}^0 \right) b_i^{j_x} \end{aligned} \tag{9.34a}$$

$$\begin{aligned} J_{12} = \frac{\partial y}{\partial \xi} = & \sum_{i \in \mathcal{N}^{std}} \frac{\partial N_i}{\partial \xi} y_i + \sum_{i \in \mathcal{N}^{enr}_{Hev}} \frac{\partial N_i}{\partial \xi} \bar{H}(\boldsymbol{x})\, a_i^y + \sum_{i \in \mathcal{N}^{*enr}_{Tip}} \sum_{j=1}^{4} \frac{\partial N_i}{\partial \xi} \bar{A}_j(\boldsymbol{x}_0)\, R(\boldsymbol{x}_0)\, b_i^{j_y} \\ & + \sum_{i \in \mathcal{N}^{*enr}_{Tip}} \sum_{j=1}^{4} N_i \left(\frac{\partial \bar{A}_j(\boldsymbol{x}_0)}{\partial x_0} J_{11}^0 R(\boldsymbol{x}_0) + \bar{A}_j(\boldsymbol{x}_0) \frac{\partial R(\boldsymbol{x}_0)}{\partial x_0} J_{11}^0 \right. \\ & \left. + \frac{\partial \bar{A}_j(\boldsymbol{x}_0)}{\partial y_0} J_{12}^0 R(\boldsymbol{x}_0) + \bar{A}_j(\boldsymbol{x}_0) \frac{\partial R(\boldsymbol{x}_0)}{\partial y_0} J_{12}^0 \right) b_i^{j_y} \end{aligned} \tag{9.34b}$$

$$J_{21}=\frac{\partial x}{\partial \eta}=\sum_{i\in\mathcal{N}^{std}}\frac{\partial N_i}{\partial \eta}x_i+\sum_{i\in\mathcal{N}^{enr}_{Hev}}\frac{\partial N_i}{\partial \eta}\bar{H}(\boldsymbol{x})\,a_i^x+\sum_{i\in\mathcal{N}^{*enr}_{Tip}}\sum_{j=1}^{4}\frac{\partial N_i}{\partial \eta}\bar{A}_j(\boldsymbol{x}_0)R(\boldsymbol{x}_0)\,b_i^{j_x}$$
$$+\sum_{i\in\mathcal{N}^{*enr}_{Tip}}\sum_{j=1}^{4}N_i\left(\frac{\partial \bar{A}_j(\boldsymbol{x}_0)}{\partial x_0}J_{21}^0R(\boldsymbol{x}_0)+\bar{A}_j(\boldsymbol{x}_0)\frac{\partial R(\boldsymbol{x}_0)}{\partial x_0}J_{21}^0\right.\tag{9.34c}$$
$$\left.+\frac{\partial \bar{A}_j(\boldsymbol{x}_0)}{\partial y_0}J_{22}^0R(\boldsymbol{x}_0)+\bar{A}_j(\boldsymbol{x}_0)\frac{\partial R(\boldsymbol{x}_0)}{\partial y_0}J_{22}^0\right)b_i^{j_x}$$

$$J_{22}=\frac{\partial y}{\partial \eta}=\sum_{i\in\mathcal{N}^{std}}\frac{\partial N_i}{\partial \eta}y_i+\sum_{i\in\mathcal{N}^{enr}_{Hev}}\frac{\partial N_i}{\partial \eta}\bar{H}(\boldsymbol{x})\,a_i^y+\sum_{i\in\mathcal{N}^{*enr}_{Tip}}\sum_{j=1}^{4}\frac{\partial N_i}{\partial \eta}\bar{A}_j(\boldsymbol{x}_0)\,R(\boldsymbol{x}_0)\,b_i^{j_y}$$
$$+\sum_{i\in\mathcal{N}^{*enr}_{Tip}}\sum_{j=1}^{4}N_i\left(\frac{\partial \bar{A}_j(\boldsymbol{x}_0)}{\partial x_0}J_{21}^0R(\boldsymbol{x}_0)+\bar{A}_j(\boldsymbol{x}_0)\frac{\partial R(\boldsymbol{x}_0)}{\partial x_0}J_{21}^0\right.\tag{9.34d}$$
$$\left.+\frac{\partial \bar{A}_j(\boldsymbol{x}_0)}{\partial y_0}J_{22}^0R(\boldsymbol{x}_0)+\bar{A}_j(\boldsymbol{x}_0)\frac{\partial R(\boldsymbol{x}_0)}{\partial y_0}J_{22}^0\right)b_i^{j_y}$$

in which the mapping between the parent, initial, and current configurations, that is, $\mathbf{F}^0$, $\bar{\mathbf{F}}$ and $\mathbf{F}$ defined in (9.32) can be obtained using the Jacobian matrix given in relations (9.34). In an updated Lagrangian formulation, the deformation gradient between step n and $n+1$ can be defined by $\boldsymbol{F}_n=\partial \boldsymbol{x}_{n+1}/\partial \boldsymbol{x}_n$ as

$$\mathbf{F}_n=\begin{bmatrix}\dfrac{\partial x_{n+1}}{\partial x_n} & \dfrac{\partial x_{n+1}}{\partial y_n}\\[2mm] \dfrac{\partial y_{n+1}}{\partial x_n} & \dfrac{\partial y_{n+1}}{\partial y_n}\end{bmatrix}=\frac{1}{\det(\mathbf{J}_n)}\begin{bmatrix}J_{11}^{n+1}J_{22}^{n}-J_{21}^{n+1}J_{12}^{n} & -J_{11}^{n+1}J_{21}^{n}+J_{21}^{n+1}J_{11}^{n}\\ J_{12}^{n+1}J_{22}^{n}-J_{22}^{n+1}J_{12}^{n} & -J_{12}^{n+1}J_{21}^{n}+J_{22}^{n+1}J_{11}^{n}\end{bmatrix}\tag{9.35}$$

In the same manner, the matrices $\mathbf{B}_L$, $\mathbf{B}_{NL}$, and $\mathbf{G}$ given in Eq. (9.18) can be defined for the X-FEM formulation using the enrichment displacement field. The introduction of enrichment functions into these matrices results in three distinct parts, including the standard, Heaviside, and tip asymptotic enrichments, which are defined for node I as

$$\left(\mathbf{B}_L^{Std}\right)_I=\begin{bmatrix}\dfrac{\partial N_I}{\partial x_n} & 0\\[2mm] 0 & \dfrac{\partial N_I}{\partial y_n}\\[2mm] \dfrac{\partial N_I}{\partial y_n} & \dfrac{\partial N_I}{\partial x_n}\end{bmatrix}\tag{9.36a}$$

$$\left(\mathbf{B}_L^{Hev}\right)_I=\begin{bmatrix}\dfrac{\partial}{\partial x_n}(N_I\bar{H}(\boldsymbol{x})) & 0\\[2mm] 0 & \dfrac{\partial}{\partial y_n}(N_I\bar{H}(\boldsymbol{x}))\\[2mm] \dfrac{\partial}{\partial y_n}(N_I\bar{H}(\boldsymbol{x})) & \dfrac{\partial}{\partial x_n}(N_I\bar{H}(\boldsymbol{x}))\end{bmatrix}\tag{9.36b}$$

$$\left(\mathbf{B}_L^{Tip}\right)_I = \begin{bmatrix} \frac{\partial}{\partial x_n}\left(N_I\bar{A}_j(\boldsymbol{x}_0)R(\boldsymbol{x}_0)\right) & 0 \\ 0 & \frac{\partial}{\partial y_n}\left(N_I\bar{A}_j(\boldsymbol{x}_0)R(\boldsymbol{x}_0)\right) \\ \frac{\partial}{\partial y_n}\left(N_I\bar{A}_j(\boldsymbol{x}_0)R(\boldsymbol{x}_0)\right) & \frac{\partial}{\partial x_n}\left(N_I\bar{A}_j(\boldsymbol{x}_0)R(\boldsymbol{x}_0)\right) \end{bmatrix} \tag{9.36c}$$

in which the derivatives are defined at the current configuration as $\frac{\partial \Xi}{\partial x_n} = \frac{\partial \Xi}{\partial \xi}\bar{F}_{11}^n + \frac{\partial \Xi}{\partial \eta}\bar{F}_{12}^n$ and $\frac{\partial \Xi}{\partial y_n} = \frac{\partial \Xi}{\partial \xi}\bar{F}_{21}^n + \frac{\partial \Xi}{\partial \eta}\bar{F}_{22}^n$, and are computed for the crack tip enrichments according to the initial configuration as $\frac{\partial \Xi}{\partial x_n} = \frac{\partial \Xi}{\partial x_0}F_{11}^{-1} + \frac{\partial \Xi}{\partial y_0}F_{12}^{-1}$ and $\frac{\partial \Xi}{\partial y_n} = \frac{\partial \Xi}{\partial x_0}F_{21}^{-1} + \frac{\partial \Xi}{\partial y_0}F_{22}^{-1}$.

Similarly, the nonlinear parts of matrix $\mathbf{B}$ can be obtained using $\mathbf{B}_{NL} = \boldsymbol{A}_\theta \mathbf{G}$ as

$$\left(\mathbf{B}_{NL}^{Std}\right)_I = \begin{bmatrix} \frac{\partial u}{\partial x_n}\frac{\partial N_I}{\partial x_n} & \frac{\partial v}{\partial x_n}\frac{\partial N_I}{\partial x_n} \\ \frac{\partial u}{\partial y_n}\frac{\partial N_I}{\partial y_n} & \frac{\partial v}{\partial y_n}\frac{\partial N_I}{\partial y_n} \\ \frac{\partial u}{\partial x_n}\frac{\partial N_I}{\partial y_n} + \frac{\partial u}{\partial y_n}\frac{\partial N_I}{\partial x_n} & \frac{\partial v}{\partial x_n}\frac{\partial N_I}{\partial y_n} + \frac{\partial v}{\partial y_n}\frac{\partial N_I}{\partial x_n} \end{bmatrix} \tag{9.37a}$$

$$\left(\mathbf{B}_{NL}^{Hev}\right)_I = \begin{bmatrix} \frac{\partial u}{\partial x_n}\frac{\partial}{\partial x_n}\left(N_I\bar{H}(\boldsymbol{x})\right) & \frac{\partial v}{\partial x_n}\frac{\partial}{\partial x_n}\left(N_I\bar{H}(\boldsymbol{x})\right) \\ \frac{\partial u}{\partial y_n}\frac{\partial}{\partial y_n}\left(N_I\bar{H}(\boldsymbol{x})\right) & \frac{\partial v}{\partial y_n}\frac{\partial}{\partial y_n}\left(N_I\bar{H}(\boldsymbol{x})\right) \\ \frac{\partial u}{\partial x_n}\frac{\partial}{\partial y_n}\left(N_I\bar{H}(\boldsymbol{x})\right) + \frac{\partial u}{\partial y_n}\frac{\partial}{\partial x_n}\left(N_I\bar{H}(\boldsymbol{x})\right) & \frac{\partial v}{\partial x_n}\frac{\partial}{\partial y_n}\left(N_I\bar{H}(\boldsymbol{x})\right) + \frac{\partial v}{\partial y_n}\frac{\partial}{\partial x_n}\left(N_I\bar{H}(\boldsymbol{x})\right) \end{bmatrix} \tag{9.37b}$$

$$\left(\mathbf{B}_{NL}^{Tip}\right)_I = \begin{bmatrix} \frac{\partial u}{\partial x_n}\frac{\partial}{\partial x_n}\left(N_I\bar{A}_j(\boldsymbol{x}_0)R(\boldsymbol{x}_0)\right) & \frac{\partial v}{\partial x_n}\frac{\partial}{\partial x_n}\left(N_I\bar{A}_j(\boldsymbol{x}_0)R(\boldsymbol{x}_0)\right) \\ \frac{\partial u}{\partial y_n}\frac{\partial}{\partial y_n}\left(N_I\bar{A}_j(\boldsymbol{x}_0)R(\boldsymbol{x}_0)\right) & \frac{\partial v}{\partial y_n}\frac{\partial}{\partial y_n}\left(N_I\bar{A}_j(\boldsymbol{x}_0)R(\boldsymbol{x}_0)\right) \\ \frac{\partial u}{\partial x_n}\frac{\partial}{\partial y_n}\left(N_I\bar{A}_j(\boldsymbol{x}_0)R(\boldsymbol{x}_0)\right) + \frac{\partial u}{\partial y_n}\frac{\partial}{\partial x_n}\left(N_I\bar{A}_j(\boldsymbol{x}_0)R(\boldsymbol{x}_0)\right) & \frac{\partial v}{\partial x_n}\frac{\partial}{\partial y_n}\left(N_I\bar{A}_j(\boldsymbol{x}_0)R(\boldsymbol{x}_0)\right) + \frac{\partial v}{\partial y_n}\frac{\partial}{\partial x_n}\left(N_I\bar{A}_j(\boldsymbol{x}_0)R(\boldsymbol{x}_0)\right) \end{bmatrix} \tag{9.37c}$$

and the standard, Heaviside, and tip asymptotic enrichments of the matrix $\mathbf{G}$ are defined as

$$\mathbf{G}_I^{Std} = \begin{bmatrix} \frac{\partial N_I}{\partial x_n} & 0 \\ 0 & \frac{\partial N_I}{\partial x_n} \\ \frac{\partial N_I}{\partial y_n} & 0 \\ 0 & \frac{\partial N_I}{\partial y_n} \end{bmatrix} \tag{9.38a}$$

$$\mathbf{G}_I^{Hev} = \begin{bmatrix} \frac{\partial}{\partial x_n}(N_I\bar{H}(\boldsymbol{x})) & 0 \\ 0 & \frac{\partial}{\partial x_n}(N_I\bar{H}(\boldsymbol{x})) \\ \frac{\partial}{\partial y_n}(N_I\bar{H}(\boldsymbol{x})) & 0 \\ 0 & \frac{\partial}{\partial y_n}(N_I\bar{H}(\boldsymbol{x})) \end{bmatrix} \tag{9.38b}$$

$$\mathbf{G}_I^{Tip} = \begin{bmatrix} \frac{\partial}{\partial x_n}(N_I\bar{A}_j(\boldsymbol{x}_0)R(\boldsymbol{x}_0)) & 0 \\ 0 & \frac{\partial}{\partial x_n}(N_I\bar{A}_j(\boldsymbol{x}_0)R(\boldsymbol{x}_0)) \\ \frac{\partial}{\partial y_n}(N_I\bar{A}_j(\boldsymbol{x}_0)R(\boldsymbol{x}_0)) & 0 \\ 0 & \frac{\partial}{\partial y_n}(N_I\bar{A}_j(\boldsymbol{x}_0)R(\boldsymbol{x}_0)) \end{bmatrix} \tag{9.38c}$$

9.5 The Damage–Plasticity Model

In order to model the nonlinear behavior of ductile fracture mechanics, particularly at the fracture process zone, the damage-plasticity model is utilized by the means of CDM. A hypoelasto-plastic constitutive model is employed in the framework of large X-FEM deformation formulation described in preceding section. In the CDM, the damage is a state parameter that provides a macroscopic measure of the micro-structural defects in the material, for example, the micro-cracks and micro-voids (Kachanov, 1986; Voyiadjis and Kattan, 2005). The damage variable D is defined based on the ratio of damaged surface to the virgin one with respect to a plane that passes through the point under consideration as

$$D = \frac{S_\Phi}{S} \tag{9.39}$$

where S_Φ is the area of micro-cracks and intersections and S is the total area of cross section, as shown in Figure 9.7. Because of an anisotropic behavior of the damage, the CDM is generally defined by using a

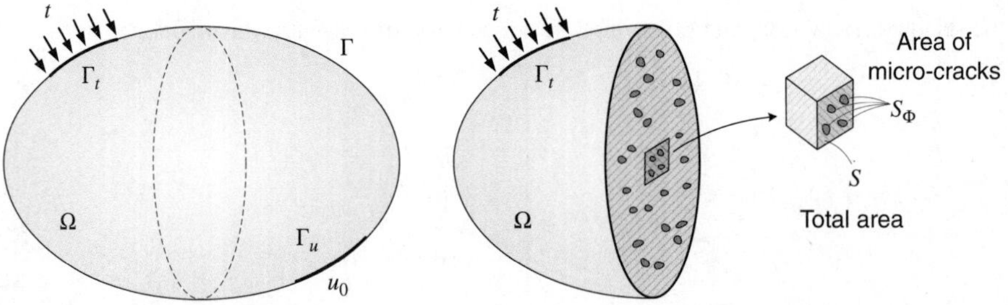

Figure 9.7 A damaged element presenting the areas S and S_Φ

tensor field. However, it is assumed here that the micro-cracks and cavities are distributed uniformly in all directions, and an isotropic damage variable is employed based on the Lemaitre simplified damage plasticitymodel (LSDP). In this model, the behavior of damaged material is defined based on the principle of strain equivalence, in which the strain behavior of a damaged material is obtained by the constitutive equations of an undamaged material, and the effective stress is employed instead of the conventional stress field. Hence, the effective stress tensor is related to the true stress tensor as

$$\bar{\boldsymbol{\sigma}} = \frac{\boldsymbol{\sigma}}{1 - D} \tag{9.40}$$

where $\boldsymbol{\sigma}$ and $\bar{\boldsymbol{\sigma}}$ are the conventional stress and effective stress tensors, respectively. Note that the cyclic plasticity characteristics, such as the Bauschinger effect, can be modeled through the introduction of the Armstrong–Fredrick nonlinear kinematic hardening model, where the original formulation is modified to account for the damage.

Based on thermo-dynamics principles, three requirements are necessary for a material constitutive model; the first includes the definition of state variables which completely determine the thermodynamic state at a given point. The second involves the definition of a proper state potential in which the thermo-elasticity law and the state forces conjugated to state variables are addressed, and the last is an energy dissipation potential proposed to obtain the evolution laws for the state variables. By neglecting the thermal effects, the state variables of the model can be expressed as the elastic strain tensor $\boldsymbol{\varepsilon}^e$, the damage variable D, the equivalent plastic strain $\varepsilon^p_{eq} = \sqrt{\frac{2}{3}\varepsilon^P_{ij}\varepsilon^P_{ij}}$, the isotropic hardening parameter r, and the kinematic hardening parameter $\boldsymbol{\alpha}$, which is a second order tensor and is used for dynamic–cyclic loading. Based on the hypothesis of decoupling between elasticity-damage and plastic hardening the state potential can be defined as

$$\Psi = \Psi_{ED} + \Psi_P \tag{9.41}$$

where Ψ_{ED} and Ψ_P are the elastic-damage and plastic hardening contributions to the Helmholtz free energy, respectively; these parameters are defined as

$$\begin{aligned} \rho\Psi_{ED} &= \frac{1}{2}\boldsymbol{\varepsilon}^e(1-D)\mathbf{C}\,\boldsymbol{\varepsilon}^e \\ \rho\Psi_P &= \int_0^r R(r)\,dr + \frac{1}{2}a(1-D)\boldsymbol{\alpha}:\boldsymbol{\alpha} \end{aligned} \tag{9.42}$$

where ρ is the density, $\mathbf{C}$ is an undamaged elasticity tensor, $\boldsymbol{\varepsilon}^e$ is the elastic strain tensor, $R(r)$ is an isotropic hardening force defined as a function of hardening parameter r, and a is a linear kinematic hardening material parameter. The corresponding state forces of the state variables are defined as

$$\begin{aligned} \boldsymbol{\sigma} &= \rho\frac{\partial\Psi}{\partial\boldsymbol{\varepsilon}^e} = (1-D)\mathbf{C}\,\boldsymbol{\varepsilon}^e \\ -Y &= \rho\frac{\partial\Psi}{\partial D} = \frac{\sigma_{eq}^2}{2E(1-D)^2}\left[\frac{2}{3}(1+\nu) + 3(1-2\nu)\left(\frac{p}{\sigma_{eq}}\right)^2\right] \\ R(r) &= \rho\frac{\partial\Psi}{\partial r} = H\,r + \left(\sigma_{y\infty} - \sigma_{y0}\right)\left(1 - e^{-mr}\right) \\ \boldsymbol{\chi} &= \rho\frac{\partial\Psi}{\partial\boldsymbol{\alpha}} = a(1-D)\boldsymbol{\alpha} \end{aligned} \tag{9.43}$$

where Y is the damage energy release rate, σ_{eq} is the von-Mises equivalent stress, E is the Young's modulus, ν is the Poisson ratio, p is the pressure, H is the linear isotropic hardening parameter, $\sigma_{y\infty}$ is the saturated yield stress, σ_{y0} is the initial yield stress of the material, m is a material constant, and $\boldsymbol{\chi}$ is the back stress tensor that is a second order deviatoric tensor.

The energy dissipation potential can be decomposed into the plastic, damage, and nonlinear kinematic hardening terms as

$$\mathcal{E} = \mathcal{E}_P + \mathcal{E}_D + \mathcal{E}_\chi \tag{9.44}$$

On the basis of an associated von-Mises plasticity with the hardening rule $R(r)$, the plastic part of energy dissipation potential $\mathcal{E}_P$ can be defined as

$$\mathcal{E}_P = \sqrt{\frac{3}{2}} \left\| \frac{\boldsymbol{\sigma}_{dev}}{1-D} - \boldsymbol{\chi} \right\| - (\sigma_{y0} + R(r)) \tag{9.45}$$

where $\boldsymbol{\sigma}_{dev}$ is the deivatoric part of stress tensor, $\| \ \|$ is the norm of tensor, and σ_{y0} is the initial yield stress. In the LSDP model, the damage part of dissipation potential $\mathcal{E}_D$ is defined as

$$\mathcal{E}_D = \frac{z}{(1+s)(1-D)} \left(\frac{Y}{z} \right)^{s+1} \tag{9.46}$$

where s and z are material parameters. Based on the Armstrong–Frederick nonlinear kinematic hardening the dissipation potential $\mathcal{E}_\chi$ is defined as

$$\mathcal{E}_\chi = \frac{b}{2(1-D)a} \boldsymbol{\chi} : \boldsymbol{\chi} \tag{9.47}$$

where a and b are the linear and nonlinear kinematic hardening constants. Hence, the plastic flow rule and the evolution laws for state variables D, r, and $\boldsymbol{\alpha}$, can be obtained as

$$\begin{aligned}
\dot{\boldsymbol{\varepsilon}}^p &= \dot{\lambda} \frac{\partial \mathcal{E}_P}{\partial \boldsymbol{\sigma}} = \frac{\dot{\lambda}}{1-D} \mathbf{n}_g \\
\dot{D} &= -\dot{\lambda} \frac{\partial \mathcal{E}_D}{\partial Y} = \frac{\dot{\lambda}}{1-D} \left(-\frac{Y}{z} \right)^s \\
\dot{r} &= -\dot{\lambda} \frac{\partial \mathcal{E}_P}{\partial R} = \dot{\lambda} \\
\dot{\boldsymbol{\alpha}} &= -\dot{\lambda} \frac{\partial \mathcal{E}}{\partial \boldsymbol{\chi}} = \frac{\dot{\lambda}}{1-D} \left(\mathbf{n}_g - \frac{b}{a} \boldsymbol{\chi} \right)
\end{aligned} \tag{9.48}$$

where $\dot{\lambda}$ is the consistency parameter that is governed by the Kuhn–Tucker loading/unloading criteria, that is, $\dot{\lambda} \geq 0$, $\mathcal{E}_P \leq 0$, and $\dot{\lambda} \mathcal{E}_P = 0$, and the flow vector $\mathbf{n}_g$ is defined as

$$\mathbf{n}_g = \sqrt{\frac{3}{2}} \left(\frac{\boldsymbol{\sigma}_{dev}}{1-D} - \boldsymbol{\chi} \right) \Big/ \left\| \frac{\boldsymbol{\sigma}_{dev}}{1-D} - \boldsymbol{\chi} \right\| \tag{9.49}$$

The experimental observations illustrate that there is no damage evolution until a certain accumulated or irreversible plastic strain reaches a threshold value ${}^{Thr}\varepsilon_{eq}^p$ that is dependent on both the material and loading type. This is because the damage initiation is related to the amount of energy needed for the growth of defects, which is known as the stored energy threshold of the material w_s (Lemaitre and Desmorat, 2005). For isotropic and nonlinear kinematic hardening plasticity, the stored energy w_s can be defined as

$$w_s = \int_0^t (R(r)\,Z(r)\,\dot{r} + \boldsymbol{\chi} : \dot{\boldsymbol{\alpha}})dt \tag{9.50}$$

in which $Z(r) = \frac{A}{m} r^{\frac{1-m}{m}}$ is an isotropic hardening correction function, where A and m are material parameters. Consider a proportionality factor P_D that accounts for the nonlinear damage growth, the damage evolution in Eq. (9.48) can be rewritten as

$$\dot{D} = H\left(\varepsilon_{eq}^p - {}^{Thr}\varepsilon_{eq}^p\right) \frac{\dot{\lambda}}{1-D} P_D \left(-\frac{Y}{z}\right)^s \tag{9.51}$$

where H is the Heaviside function, and P_D is a function of damage variable defined for the linear and nonlinear material behavior as

$$P_D = \begin{cases} 1 & \text{Linear} \\ \dfrac{3\left(1 - \tanh^2(3)(2D-1)^2\right)}{\tanh(3)\left({}^{Cr}\varepsilon_{eq}^p - {}^{Thr}\varepsilon_{eq}^p\right)} & \text{Nonlinear} \end{cases} \tag{9.52}$$

where ${}^{Cr}\varepsilon_{eq}^P$ is the critical plastic strain defined as a material parameter. It can be observed from Eqs. (9.43) and (9.48) that the plastic deformation and hydrostatic pressure are the two driving forces of damage. In CDM, the damage parameter presents itself in degradation of mechanical properties, that is, the elasticity modulus and yield stress, and produces some kind of softening effects, which can produce desired effects in modeling the fracture process zone.

These constitutive equations are coupled and must be solved as a unique system of equations simultaneously. In this study, an uncoupled solution is performed, in which the plasticity variables are first obtained based on an elastic predictor–plastic corrector scheme at each iteration of the Newton–Raphson procedure. The damage is then calculated explicitly as a post-process output of the plasticity solution at the end of each loading step by using a discretized form of the damage evolution law as

$$D_{n+1} = D_n + P_D(D_n)\left(-\frac{Y_n}{z}\right)^s \Delta\varepsilon_{eq}^p \tag{9.53}$$

where $\Delta\varepsilon_{eq}^p$ is an incremental equivalent plastic strain. Based on the definition of effective stress $\bar{\boldsymbol{\sigma}}$ given in (9.40), the computed damage variable represents the stress softening behavior. In some engineering applications, the damage is completely decoupled from the formulation and is used as an indicator of material degradation. In this situation, relation (9.40) can be rewritten as

$$\bar{\boldsymbol{\sigma}} = \frac{\boldsymbol{\sigma}}{1 - D_{\text{index}} D} \tag{9.54}$$

where D_{index} is an integer with the value of unit if the damage is coupled with the stress and zero if it is decoupled.

A similar integer may be used to account for the crack closure effects in compression during the cyclic loading behavior. In fact, in the damage plasticity model it is assumed that the material degradation is similar in both tension and compression; this assumption leads to a shorter material life prediction. Observations show that the micro defects are partially closed in compression and hence the area which effectively carries the load and the stiffness, may partially be recovered. This issue can be modeled by introduction of the micro-crack closure parameter h that affects the elastic stress–strain relationship and the damage growth rate as follows

$$\begin{aligned} \sigma &= \rho\frac{\partial\Psi}{\partial\boldsymbol{\varepsilon}^e} = (1-hD)\mathbf{C}\,\boldsymbol{\varepsilon}^e \\ -Y &= \rho\frac{\partial\Psi}{\partial D} = \frac{h\sigma_{eq}^2}{2E(1-hD)^2}\left[\frac{2}{3}(1+\nu)+3(1-2\nu)\left(\frac{p}{\sigma_{eq}}\right)^2\right] \end{aligned} \tag{9.55}$$

An important issue for the crack closure effects in compression is to identify the compression and tension states. Lemaitre and Desmorat (2005) decomposed the stress tensor into the compressive and tensile parts based on an Eigenvalue decomposition, and applied appropriate h to each part. However, the decomposition is expensive and it results in a nonlinear stress–strain relationship that requires an iterative solution even for an elastic material. In this study, the volumetric strain and as a result the pressure is employed to determine the compressive or tensile state. Since the volumetric strain is known before the return mapping process in plasticity algorithm, no extra iteration is needed. In such case, for a positive value of volumetric strain the micro-crack closure parameter is taken as $h = 1$ and for a negative value as $h = 0.2$.

9.6 The Nonlocal Gradient Damage Plasticity

The introduction of damage into constitutive equations results in material softening behavior that leads to a mesh dependent problem. In fact, damage localizes in a vanishing volume while the surrounding material remains unaffected. This localization phenomenon contradicts the smoothness of damage field that is physically unacceptable. Mathematically, the mesh dependent problem is due to the fact that the governing PDE (partial differential equation) loses its ellipticity and becomes ill-posed (Peerlings *et al.*, 2002). This pathological localization effect can be avoided by using regularization methods, such as the nonlocal, or gradient damage methods (Jirasek, 1998), and the rate dependent viscose models (Pedersen, Simone, and Sluys, 2008). In this study, a non-local version of the LSDP is utilized, which was originally proposed by Mediavilla, Peerlings, and Geers (2006) and Cesar, Areias, and Zheng (2006). In this technique, the damage is computed using a nonlocal equivalent plastic strain variable $\bar{\varepsilon}^p_{eq}$ that is obtained from the local equivalent plastic strain ε^p_{eq} by solving a scalar Helmholtz type PDE equation on the initial (*undeformed*) configuration as

$$\bar{\varepsilon}^p_{eq} - \ell_c^2\,\nabla_0^2\bar{\varepsilon}^p_{eq} = \varepsilon^p_{eq} \tag{9.56}$$

where ∇_0^2 is the Laplacian operator in which the subscript 0 denoting the initial configuration, and ℓ_c is the internal characteristic length that implicitly sets the width of the localization band. This PDE equation is subject to the Neumann type boundary conditions, that is,

$$\nabla_0\bar{\varepsilon}^p_{eq}\,\mathbf{n}_0 = 0 \tag{9.57}$$

where $\mathbf{n}_0$ is the unit normal to the boundary surface Γ_0. It must be noted that the PDE Eq. (9.56) should be solved simultaneously with the governing equilibrium equation of the system (9.3), that is,

$$\begin{cases} \nabla_n \cdot \mathbf{P} + \mathbf{b} = 0 & (9.3-\text{repeated}) \\ \bar{\varepsilon}_{eq}^p - \ell_c^2 \, \nabla_0^2 \bar{\varepsilon}_{eq}^p = \varepsilon_{eq}^p & (9.56-\text{repeated}) \end{cases}$$

with the following boundary conditions

$$\left.\begin{aligned} \boldsymbol{\sigma} \cdot \mathbf{n}|_{\Gamma_t} &= \mathbf{t} \\ \mathbf{u}|_{\Gamma_u} &= \mathbf{u}_0 \end{aligned}\right. \quad (9.2-\text{repeated})$$

$$\nabla_0 \bar{\varepsilon}_{eq}^p \, \mathbf{n}_0 = 0 \quad (9.57-\text{repeated})$$

where $\mathbf{t}$ is the prescribed traction on the boundary Γ_t and $\mathbf{u}_0$ is the prescribed displacement on the boundary Γ_u with $\Gamma_u \cup \Gamma_t = \Gamma$.

The strategy for the solution of this coupled system of equations is to use an operator-split technique. This approach reduces the computational costs and can easily be applied into a local damage-plasticity program. In this method, the coupled system of equations is split into two uncoupled equations; the first is an equilibrium equation with the fixed non-local $\bar{\varepsilon}_{eq}^p$ and the second is the non-local damage Eq. (9.56) on the initial configuration as

$$\underbrace{\begin{bmatrix} \nabla_n \cdot \mathbf{P} + \mathbf{b} = 0 \\ \bar{\varepsilon}_{eq}^p - \ell_c^2 \, \nabla_0^2 \bar{\varepsilon}_{eq}^p = \varepsilon_{eq}^p \end{bmatrix}}_{\text{Coupled problem}} \cong \underbrace{\begin{bmatrix} \nabla_n \cdot \mathbf{P} + \mathbf{b} = 0 \\ \end{bmatrix}}_{\substack{\text{Equilibrium phase} \\ (\text{fixed damage})}} + \underbrace{\begin{bmatrix} \\ \bar{\varepsilon}_{eq}^p - \ell_c^2 \, \nabla_0^2 \bar{\varepsilon}_{eq}^p = \varepsilon_{eq}^p \end{bmatrix}}_{\text{Non-local phase(fixed configuration)}} \quad (9.58)$$

In order to solve the uncoupled system of Eq. (9.58) numerically, consider that the position vector $\boldsymbol{x}_n$, the non-local equivalent plastic strain $\left(\bar{\varepsilon}_{eq}^p\right)_n$ and the non-local damage variable $\bar{D}_n$ are known at step n. To obtain the corresponding unknown values at step $n+1$, the backward Euler technique is employed to the first part of Eq. (9.58), that is, the equilibrium phase, with a fixed damage value D_n. If the convergence of Newton–Raphson scheme is obtained, the position vector $\boldsymbol{x}_{n+1}$ and the local equivalent plastic strain $\left(\varepsilon_{eq}^p\right)_{n+1}$ are computed. The second part of Eq. (9.58), that is, the non-local phase, is then solved on the initial configuration to obtain the non-local equivalent plastic strain $\left(\bar{\varepsilon}_{eq}^p\right)_{n+1}$. By substituting the non-local equivalent plastic strain increment $\Delta\bar{\varepsilon}_{eq}^p$ into its local counterpart in Eq. (9.53), the non-local damage variable $\bar{D}_{n+1}$ can be obtained at step $n+1$. It must be noted that since a coupled system is approximated with two uncoupled systems there may be some inconsistencies in the solution, which can be overcome by taking a small value of increment.

9.7 Ductile Fracture with X-FEM Plasticity Model

In this section, the performance of X-FEM plasticity model is investigated by studying the convergence rate of solution in ductile fracture problem; and the implementation of X-FEM damage-plasticity model will be described in the next section. There are a few research works proposed in literature for X-FEM modeling of plasticity problems; including: the crack propagation in plastic fracture mechanics by Elguedj, Gravouil, and Combescure (2006) and Prabel *et al.* (2007); the plasticity of pressure-sensitive materials by Khoei, Shamloo, and Azami (2006b); the plasticity of frictional contact on arbitrary interfaces by Khoei and Nikbakht (2007); the plasticity of large deformation problems by Khoei, Anahid, and Shahim (2008a) and Khoei, Biabanaki, and Anahid (2008d); and strain localization in the higher-order Cosserat plasticity model by Khoei and Karimi (2008). In these studies, it was shown that a nearly optimal convergence rate can be obtained for the X-FEM plasticity model with the level-set enrichment function. In plastic fracture mechanics, appropriate crack tip enrichment functions were utilized based on the works of Hutchinson (1967) and Rice and Rosengren (1968) in the framework of a confined plasticity model, in which plastic deformation was restricted to the crack tip area. In fact, the applied stress is assumed to be sufficiently small to ensure that the plastic zone is small compared to the crack length.

Consider the Ramberg–Osgood power-law hardening rule; the following relationship can be written between the uniaxial stress and strain for HRR (Hutchinson–Rice–Rosengren) materials as

$$\frac{\varepsilon}{\varepsilon_y} = \frac{\sigma}{\sigma_y} + \alpha\left(\frac{\sigma}{\sigma_y}\right)^n \tag{9.59}$$

where $\varepsilon_y = \sigma_y/E$ is the yield strain and, n and α are material parameters. For these types of materials, an approximate analytical solution can be obtained at the onset of crack tip for pure mode I fracture without the solution of a fourth order partial differential equation. An analytical crack tip solution was obtained by Hutchinson (1967) for various material parameters, and its truncated Fourier expansion was then used by Elguedj, Gravouil, and Combescure (2007) to obtain appropriate asymptotic functions as

$$A_j = r^{\frac{1}{n+1}}\left\{\sin\frac{\theta}{2}, \cos\frac{\theta}{2}, \sin\frac{\theta}{2}\sin\theta, \cos\frac{\theta}{2}\sin\theta, \sin\frac{\theta}{2}\sin 3\theta, \cos\frac{\theta}{2}\sin 3\theta\right\} \tag{9.60}$$

It must be noted that these enrichment bases are only valid for HRR materials with confined plasticity. In the case of linear hardening rule, the first four terms of these functions are similar to the LEFM enrichment functions, however, for other types of materials, the elastic enrichment functions can be used approximately.

In order to investigate the convergence rate of X-FEM plasticity model, an edge crack problem is analyzed numerically, which is subjected to mode I and II loading conditions, and the results are compared with the LEFM enrichment functions. The geometry and boundary conditions of the problem are presented in Figure 9.8. The plate has a thickness of 0.1 mm in the plane stress condition. The material parameters are given in Table 9.1. For linear fracture analysis, the exact crack tip solution is used as given by Broek (1986), and it is applied to the prescribed displacements at the boundaries of plate as

$$\text{Mode I}\begin{cases} u(r,\theta) = \dfrac{K_I}{\mu}\sqrt{\dfrac{r}{2\pi}}\cos\dfrac{\theta}{2}\left(1-2\nu+\sin^2\dfrac{\theta}{2}\right) \\ v(r,\theta) = \dfrac{K_I}{\mu}\sqrt{\dfrac{r}{2\pi}}\sin\dfrac{\theta}{2}\left(2-2\nu+\cos^2\dfrac{\theta}{2}\right) \end{cases} \tag{9.61a}$$

$$\text{Mode II}\begin{cases} u(r,\theta) = \dfrac{K_{II}}{\mu}\sqrt{\dfrac{r}{2\pi}}\cos\dfrac{\theta}{2}\left(-1+2\nu+\sin^2\dfrac{\theta}{2}\right) \\ v(r,\theta) = \dfrac{K_{II}}{\mu}\sqrt{\dfrac{r}{2\pi}}\sin\dfrac{\theta}{2}\left(2-2\nu-\cos^2\dfrac{\theta}{2}\right) \end{cases} \tag{9.61b}$$

For the fracture analysis, a very fine FEM mesh with 90,000 DOF (degrees of freedom) is used as the plastic reference solution with similar boundary conditions. The reference FEM analysis is performed using the quadratic 8-noded Q8 elements with 3×3 Gauss quadrature points. The X-FEM plasticity model

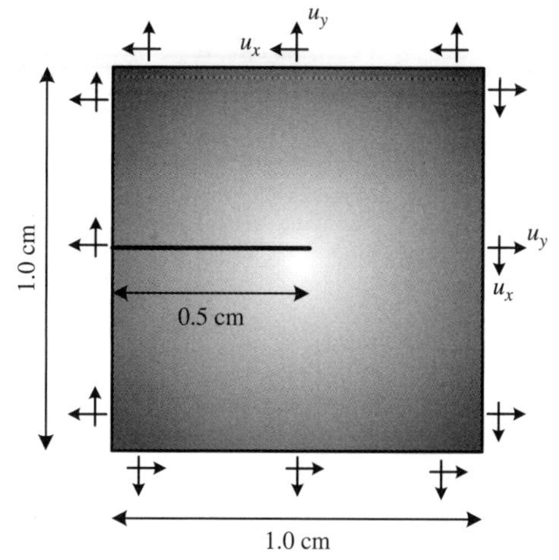

Figure 9.8 The edge crack problem subjected to modes I and II loading conditions: The geometry and boundary conditions

Table 9.1 Material properties of the edge crack problem

Property	Value
Young's modulusy E(GPa)	200
Poisson ratio ν	0.3
Hardening law $R(r)$ (MPa)	$620 + 2000\ r$

is performed by standard Q4 elements with the uniform meshes of $11 \times 11, 21 \times 21, 31 \times 31, 41 \times 41$, and 51×51 elements. The numerical integration of standard elements is carried out with four Gauss points, and the rectangular subdivisions of 15×15 with four Gauss points for each sub-rectangle are used for enriched elements. The enrichment radius at the crack-tip area is set to 2 mm.

The assessment of results illustrates that the implementation of modified X-FEM is crucial in X-FEM plasticity analysis. On the other hand, the standard X-FEM model results in the large value of error at the crack tip area with a non-smooth stress distribution at this region, as shown in Figure 9.9a. Consequently, the extracted information from the fracture process zone is not valid with the standard X-FEM method and, results in the poor crack growth detection and deviated growth direction. In addition, the radius of crack tip enrichment must be large enough to avoid the previously mentioned difficulties, as shown in Figure 9.9b. It was observed that a minimum of two/three layers of elements must be employed for the enrichment of crack tip area, as shown in Figure 9.9c. It must be noted that a larger value of enrichment radius at the crack tip area causes the high condition number and increases the number of DOF that leads to great computational costs. Thus, the radius must be chosen so that it strikes a balance between the accuracy, condition number, and computational costs. In Figures 9.10a and b, the L_2 –error norms of displacement are plotted for linear and plastic fracture mechanics. Obviously, the convergence rates of displacement error are 1.83 and 1.72 for modes I and II of the elastic solution, respectively, which are close to the optimal rate of convergence. However, the convergence rates reduce to 1.1 and 1.23 for modes I and II in the plastic fracture solution, respectively. It can be seen from these graphs that although the error of mode II is higher than mode I, the convergence rate is closer to optimal rate. It can be highlighted that the nonlinearity of plastic behavior has a deteriorating effect on the convergence of solution. Also, plotted in Figures 9.10c and d are the errors of energy norm for modes I and II of linear and plastic

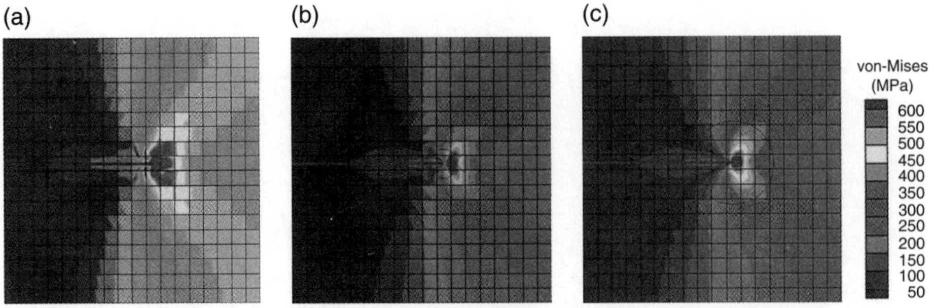

Figure 9.9 The von-Mises stress (MPa) contours for an edge crack problem: (a) the standard X-FEM; (b) the modified X-FEM with one-layer enriched elements; (c) the modified X-FEM with two-layer enriched elements (For color details, please see color plate section)

fracture mechanics. Obviously, the convergence rates of energy error for the elastic fracture solution are 0.95 and 0.96 for modes I and II, respectively, which are very close to the optimal rate of unity. However, the sub-optimal convergence rates are obtained for the plastic solution with the rates of 0.43 and 0.34 for modes I and II, respectively. Finally, the L_2–error norms of von-Mises stress are plotted in Figures 9.10e and for linear and plastic fracture mechanics. This error is equivalent to the $\mathcal{H}^1$–error norm of displacement for the elastic solution. Since the elastic energy cannot be used meaningfully for the plastic solution, the error norm of von-Mises stress is applied here. Obviously, the convergence rates of von-Mises stress error in the elastic fracture solution are 0.95 and 0.96 for modes I and II, however, the sub-optimal convergence rates of 0.5 and 0.58 are obtained for modes I and II of the plastic solution.

In order to study the size effect of fracture plastic zone on the convergence rate of solution, the problem is solved for mode I at various loading amplitudes. For this purpose, the prescribed displacements are applied at different stress intensity factors, that is, 1500, 2000, and 2500 $kg/cm^{3/2}$. In fact, a large value of SIFs results in a large prescribed displacement that yields to a great fracture plastic zone. In Figure 9.11, the L_2–error norms of displacement, the errors of energy norm, and the L_2–error norms of von-Mises stress are plotted for mode I of the linear and plastic fracture mechanics. Obviously for all error norms, the value of error increases and the rate of convergence decreases by increasing the size of plastic zone. These results highlight that, although the modified X-FEM method notably improves the quality of stress distribution at the crack tip area, the optimal convergence rate cannot be obtained for the plastic fracture analysis, and only a sub-optimal convergence rate can be achieved for the X-FEM plasticity model. Thus, a technique that improves the convergence rate of plastic solution remains an open question.

9.8 Ductile Fracture with X-FEM Non-Local Damage-Plasticity Model

9.8.1 Crack Initiation and Crack Growth Direction

In fracture mechanics there are two important factors that need to be considered for crack growth simulation; crack initiation criteria and crack propagation direction. In linear fracture mechanics, various criteria have been proposed for the mixed-mode fracture based on the critical stress intensity factor and fracture toughness, including: the minimum strain energy density criterion (Sih, 1974), the maximum hoop stress criterion (Sih, 1974) and the maximum energy release rate (Hussain, et al., 1974). Since these criteria are generally based on the overall state of the structure, they cannot accurately predict the behavior of fracture process zone and the local nonlinearity at the crack tip region. In this study, a robust technique is presented on the basis of damage state that provides a local crack growth criterion at the fracture process zone. In this model, the damage parameter D_{CP} is computed at the crack-tip area for each loading step

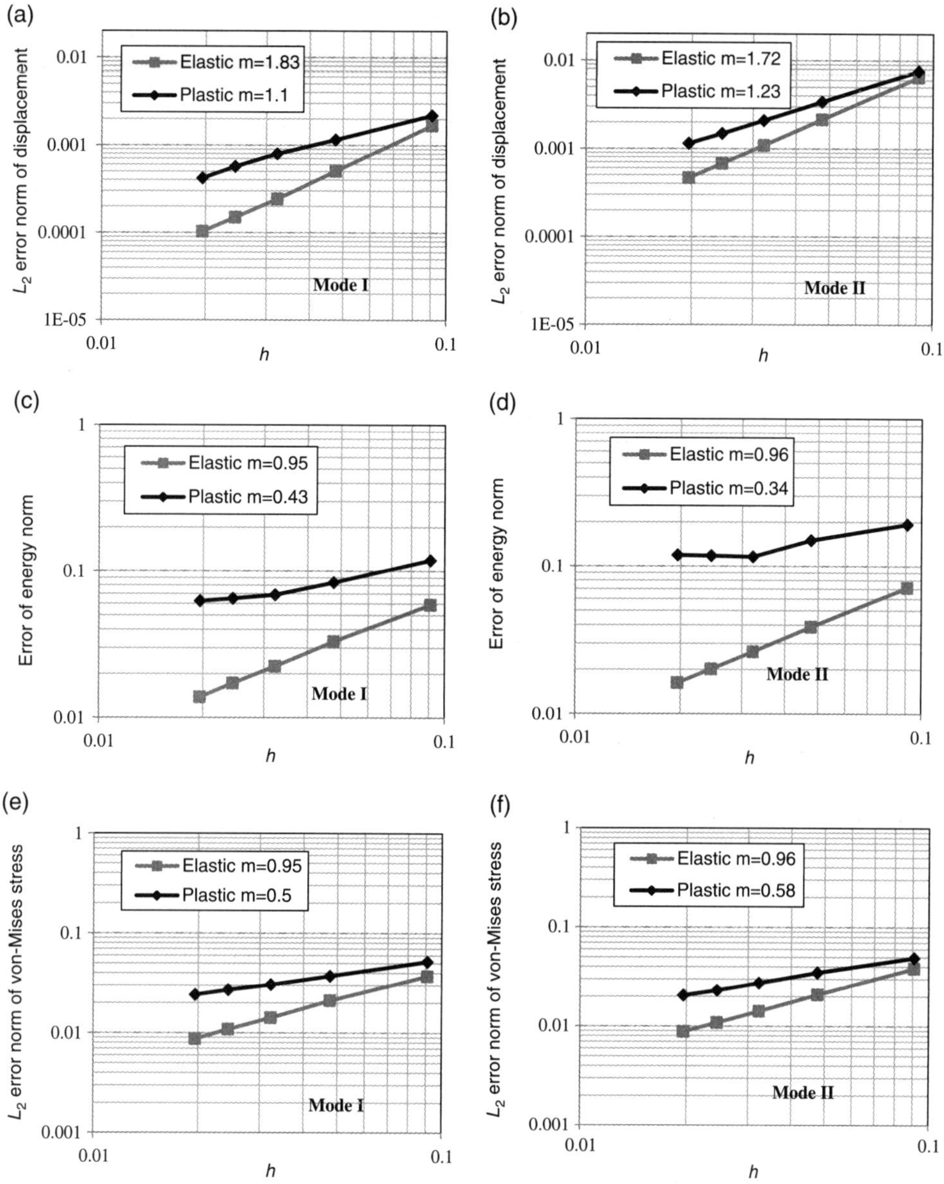

Figure 9.10 An edge crack problem: (a and b) The L_2 error norm of displacement for modes I and II; (c and d) The error of energy norm for modes I and II; (e and f) The L_2 error norm of von-Mises stress for modes I and II

called the damage calculation point (DCP), and its value is compared with a critical damage value D_{cr}; if $D_{CP} \geq D_{cr}$, the crack propagation happens in the structure. Since the crack growth simulation is basically performed in a discontinuous manner, a more reliable value of damage is used to reduce the influence of local damage variations based on the weighted average value of damage at the crack tip area. The size of this area can be related to the internal characteristic length of material ℓ_c that provides good information

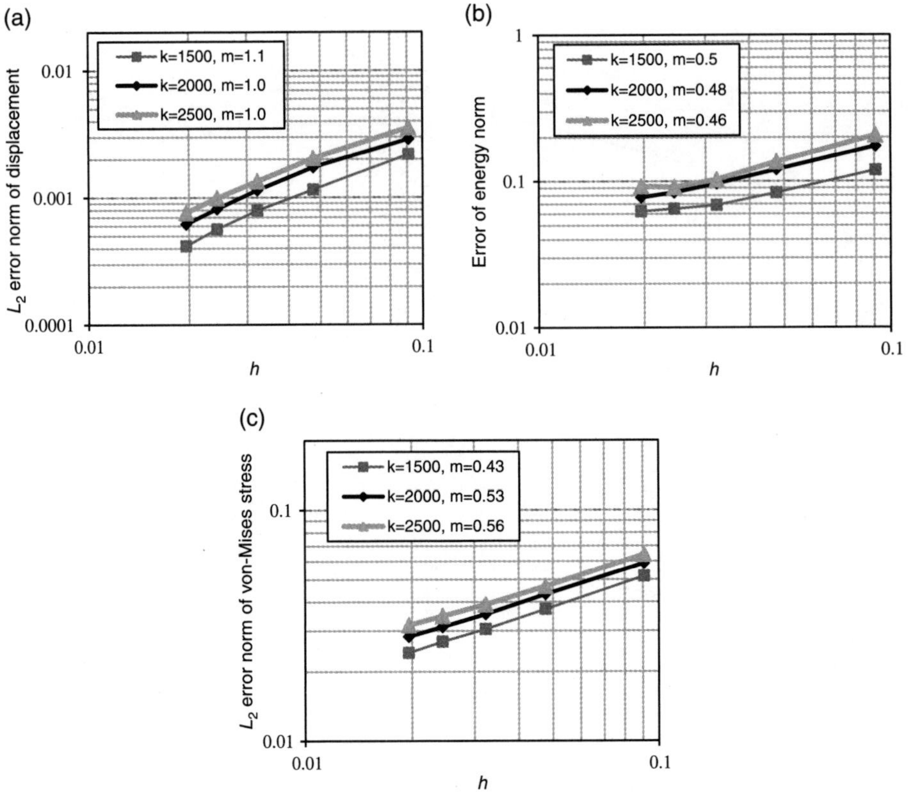

Figure 9.11 The size effect of fracture plastic zone on the convergence rate for an edge crack problem: (a) The L_2 error norm of displacement; (b) The error of energy norm; (c) The L_2 error norm of von-Mises stress

about the size of material particles. The weighted average value of damage is obtained at the DCPs using the Gauss quadrature points of a half circle with the radius of ℓ_c from the crack tip at various angles of θ_{CP} in the range of $-90 \le \theta_{CP} \le 90$, as shown in Figure 9.12. The weighting function is chosen based on a Gaussian distribution, which decreases monotonically away from the crack tip. The value of damage at each DCP is calculated using the weighted average value of damage at all neighboring Gauss points by

$$D_{CP} = \left(\sum_j w_j\right)^{-1} \left(\sum_j D_j w_j\right) \quad \text{with} \quad w(\boldsymbol{x},\boldsymbol{y}) = \frac{1}{(2\pi)^{1.5}\ell_c^3} \exp\left(-\frac{|\boldsymbol{x}-\boldsymbol{y}|^2}{\ell_c^2}\right) \tag{9.62}$$

where D_{CP} is the damage at the DCP and, D_j and w_j are the damage parameter and weighting factor of Gauss point j, respectively, $\boldsymbol{x}$ is the coordinates of Gauss point, and $\boldsymbol{y}$ is the coordinates of crack tip. In this relation, the closer a Gauss point is to the DCP the higher its effect on the damage value. It must be noted that the weighted average value of damage parameter and the crack growth direction are computed on the initial (*undeformed*) configuration in the large X-FEM deformation analysis.

The orientation of crack propagation can be obtained in the direction of maximum damage value. In fact, the damage value can be determined by first identifying the element that contains the DCP, and then employing the FEM shape functions to damage nodal values. However, it was observed here that the resulting value of damage is not satisfactory; since in this strategy all neighboring Gauss points have equal

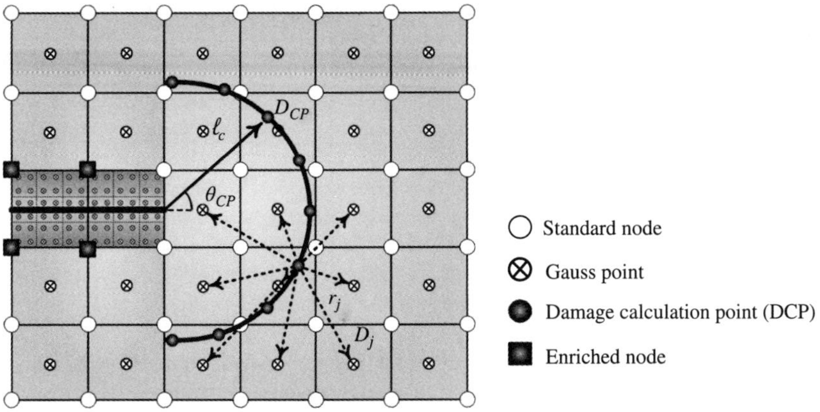

Figure 9.12 The crack initiation and crack growth direction: The DCP at the distance ℓ_c from the crack tip at various angles θ_{CP} using a weighted summation of the damage value of the neighboring Gauss points D_j

effect on the damage value. For this purpose, the damage parameter D_{CP} is computed for all calculation points at each loading step, and their values are compared with a critical damage value D_{cr}; if $D_{CP} \geq D_{cr}$ the crack propagates in the direction θ_{CP} of the DCP with the maximum damage value, as shown in Figure 9.12. This criterion can be properly used for the shear dominated problems, but it cannot provide the proper crack growth direction in mode I fracture. In fact, for a specimen under tensile loading the crack propagates in a straight line while the maximum damage occurs at $\theta = \pm 70^{\circ}$. Hence, the resulting crack propagation profile produces the zig-zag shape. Mediavilla, Peerlings, and Geers (2006) proposed a similar criterion for determination of the crack growth direction in ductile fracture propagation. However, for the crack initiation criterion, they compared the value of damage at the crack-tip with D_{cr} that causes the mesh dependence results, and different crack growth directions can be obtained using different values of crack segment length.

9.8.2 Crack Growth with a Null Step Analysis

In linear fracture mechanics, X-FEM analysis can be performed independently for each step of loading. In fact, during the crack propagation simulation the geometry of problem changes by adding the new crack segment into the X-FEM mesh. The new boundary interfaces are defined independent of the X-FEM mesh based on the geometry of body, and X-FEM analysis can be performed from the beginning of loading. However, in plastic fracture mechanics, the material behavior is dependent on the history of loading and the analysis must be carried out incrementally. So it is important to transfer the material state variables from the previous geometry to the new one in an efficient manner in order not to spoil the stability of the solution. One of the main advantages of X-FEM is that the technique does not need remeshing as the crack propagates. However, because of the element sub-divisions and the changes of enrichment status during the crack propagation, the global stiffness matrix must be updated at each increment. In this case, the data transfer process is needed during crack growth simulation that may cause the error and inconsistency at the fracture process zone and surrounding region.

Two main concerns usually encountered in the data transfer process are the consistency and equilibrium. After transferring the state variables from the previous geometry to new one, the equilibrium condition between the internal and external forces cannot be usually met. Furthermore, the transferred variables cannot satisfy the constitutive equations. Without overcoming these drawbacks, further computations produce serious stability and convergence problems. The inconsistency problem is mainly due to employing the linear transfer operator while the stress–strain relation is nonlinear. In order to remove such inconsistency, the stress is controlled at each Gauss integration point, and if it lies outside the yield surface, a multiplier is applied to bring the stress to the yield condition. This leads to a slight reduction in the internal force and so a drop can be observed in the force-displacement curve. On the other side, the lack of equilibrium is primarily because of numerical discrepancies occurred during the data transfer process. To address this issue, a *null step* without any changes in loading and boundary conditions can be performed to reach the equilibrium in a number of iterations. In *null step* analysis, the constraints on material behavior can be released and an elastic behavior can be assumed. In this manner although the equilibrium can be reached in two or three iterations, the stresses again lie outside the yield surface and it is then necessary to expand the yield surface in order not to disturb the consistency condition. Expanding the yield surface leads to an artificial increase in the material strength and as a result, a rise can be observed in the force-displacement curve.

In order to avoid the error and inconsistency at the fracture process zone, Elguedj, Gravouil, and Combescure (2007) proposed a sub-rectangular integration strategy in the plastic fracture analysis, as shown in Figure 9.13a. In this algorithm, each quadrilateral element is divided into several regular sub-rectangles with four Gauss points for each sub-rectangle. A predefined area of two layers of elements is modeled around the crack tip by rectangular sub-divisions, and the crack growth segment is chosen smaller than the radius of rectangulations. In this case, when the crack propagates, the new crack tip falls within an area that was rectangulated beforehand and the data transfer is not needed in the crack tip area. It must be noted that the data transfer process is required for those new elements located within the predefined area around the new crack tip, as shown in Figure 9.13b, however, these elements are far from the old fracture process zone that caused the crack growth initiation, and hence the data transfer process does not create a serious problem.

In the X-FEM crack growth simulation, the data transfer process only needs to be applied to state variables obtained at the Gauss points of old geometry. Data transferring can be performed using a simple linear operator in two stages. At the first step, the state variables are transferred from the old Gauss points to nodal points, as shown in Figure 9.14a, by the following integral equation over each element area A as

$$\boldsymbol{\Lambda}_N^{\text{old}} = \left(\int_A \mathbf{N}^T\mathbf{N}\,dA\right)^{-1}\int_A \mathbf{N}^T\boldsymbol{\Lambda}_G^{\text{old}}\,dA \tag{9.63}$$

where $\boldsymbol{\Lambda}_N^{\text{old}}$ are the desired nodal values of state variable $\boldsymbol{\Lambda}$ in previous geometry, $\boldsymbol{\Lambda}_G^{\text{old}}$ are the calculated values of state variable at the old Gauss points, and $\mathbf{N}$ is the matrix of shape functions, in which for node j of a Q4 element it is defined as $\mathbf{N}_j = N_j\mathbf{I}_{m\times m}$, with m denoting the size of $\boldsymbol{\Lambda}_G^{\text{old}}$. In the second stage, the

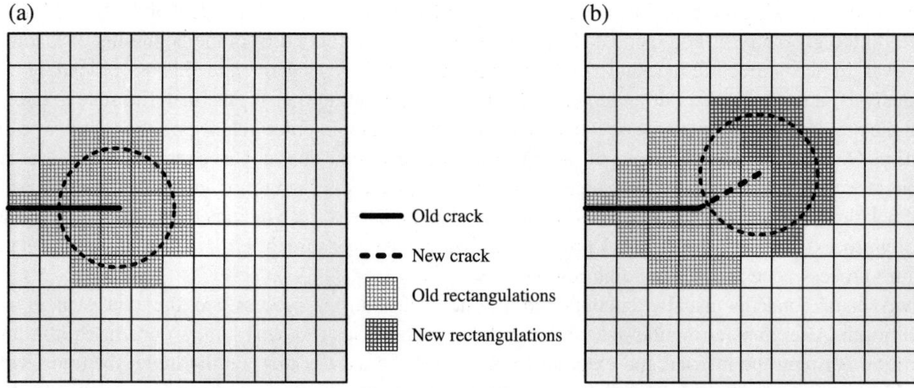

Figure 9.13 The integration strategy using rectangular sub-divisions in a predefined area of two layers of elements around the crack-tip: (a) before crack growth; (b) after crack growth

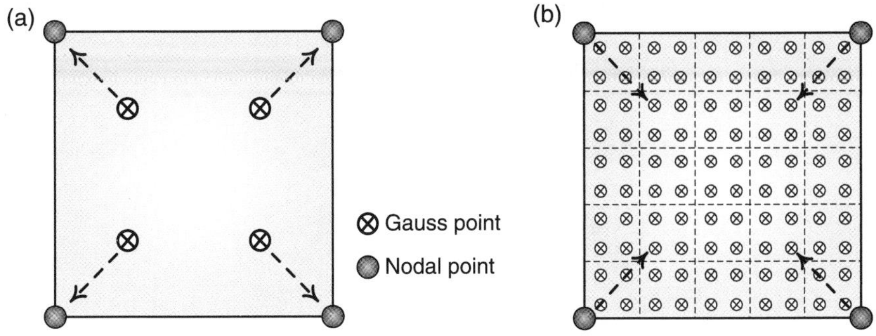

Figure 9.14 The data transfer process: (a) Data transferring from the Gauss points to nodal points in old geometry; (b) Data transferring from the nodal points of old geometry to the Gauss points of new geometry

shape functions of Q4 element are used to transfer the nodal values of old geometry $\mathbf{\Lambda}_N^{\text{old}}$ to the Gauss points of new geometry $\mathbf{\Lambda}_G^{\text{new}}$, as shown in Figure 9.14b by

$$\mathbf{\Lambda}_G^{\text{new}}(\xi,\eta) = \sum_i N_i(\xi,\eta)\mathbf{\Lambda}_N^{\text{old}} \tag{9.64}$$

It must be noted that, despite the nonlinear behavior of state variables, that is, stresses, strains, damage parameter, and so on, the linear transfer operators are used in relations (9.63) and (9.64) to transfer the state variables from the Gauss points of the old geometry to the new one. In this case, the equilibrium and constitutive equations cannot be satisfied and the inconsistency can be caused in the solution. In order to overcome the inconsistency problem a reduced set of variables are used instead of all state variables, and the other variables are obtained by satisfying the equilibrium and constitutive equations. In this study, the stress tensor $\boldsymbol{\sigma}$, the damage parameter D, and the hardening parameter r are transferred from the old to new geometry. Although this strategy reduces the inconsistency of the solution, the transferred stress tensors, which were in the plastic state before data transfer, may not be on the corresponding yield surface, that is, $\mathcal{E}_P > 0$. In addition, the transferred stress tensors may not satisfy the overall equilibrium of the structure.

The equilibrium of the system can be restored by using an artificial *null step*, in which no new loading is employed to the structure, and the equilibrium between the internal and external forces is satisfied with a number of iterations in the Newton–Raphson procedure. Since the *null step* only eliminates the unwanted unbalances produced by the data transfer process, which are not the result of physical deformation, an *elastic* constitutive behavior is assumed in this step that improves the convergence and stability of the solution. It must be noted that the implementation of material constitutive behavior, that is, the damage-plasticity model, in the *null step* analysis may decrease the convergence rate of the solution and even shows divergence. By satisfying the equilibrium of the system, the problem of inconsistency between the yield surface and the stress values obtained from the *null step* can be resolved by scaling the yield stress $\sigma_{y0} + R(r)$ to the equivalent effective stress $\bar{\sigma}_{eq}$. In this algorithm, the new value of hardening parameter r_{new} can be obtained by solving the following nonlinear equation as

$$\bar{\sigma}_{eq} = \sigma_{y0} + R(r_{new}) \tag{9.65}$$

9.8.3 *Crack Growth with a Relaxation Phase Analysis*

In the crack growth simulation, when the new crack segment is added to the X-FEM mesh, the values of stresses that were obtained at the old crack tip must be vanished to comply with the new boundary conditions. Since there is no continuous influence of the damage on the stresses, large stress

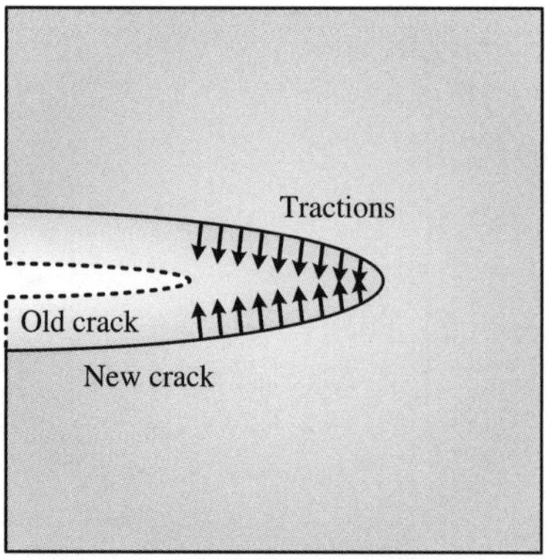

Figure 9.15 The unbalanced tractions at the new crack segment that must be removed by using a *relaxation phase* analysis

redistributions may take place at the crack tip area and a considerable amount of energy may be released during crack propagation. If this is not taken properly into the computations, these stress redistributions may have a detrimental effect on the stability of solution. In order to remove these unbalanced tractions from the new crack segment, as shown in Figure 9.15, which were in the self-equilibrium condition at the old crack tip, a *relaxation phase* is implemented in the new boundary condition. It must be noted that these tractions cannot be removed in a single step analysis, since a large stress redistribution can be occurred, and as a result the system may not be able to restore the equilibrium and the solution may show divergence. Thus, the complete removal of tractions must be performed in several steps called *relaxation steps* with no external loading. Note that the *relaxation steps* cannot be performed as elastic, and the material constitutive model based on the damage-plasticity model must be applied, since the crack opening represents a physical change. The number of *relaxation steps* is related to the maximum allowable damage and the crack growth length. In fact, the high value of damage reduces the stress value at the fracture process zone and results in a small stress redistribution at the crack tip area. Also, the smaller crack segment causes the lower stress redistribution and fewer number of *relaxation steps*.

The value of unbalanced tractions at the crack faces can be obtained using the total internal force of the system. In the case of prescribed displacement algorithm, the internal force of the system is zero at the end of previous step in old geometry. However, by adding the new crack segment and changing the enrichments of the system, a non-zero value of the internal force will be obtained that is equal to the tractions of the new crack faces as

$$\mathbf{f}^{\text{Traction}} = \int_{\Omega^*} \bar{\mathbf{B}}^T \mathbf{S}\, d\Omega \tag{9.66}$$

where $\mathbf{f}^{\text{Traction}}$ is the vector of tractions at the crack faces and Ω^* is the domain with new enrichments. The most important issue in the crack growth simulation is the change of enrichments at the crack tip area, which has been carried out in two stages. At the first step, it is assumed that no enrichments are applied to the elements of fracture process zone during rectangulations, data transfer, and elastic *null step*, since the new crack segment is still closed. At the second stage, the new enrichments are applied during the *relaxation phase* where the values of new enriched DOF are set to zero, and as a result the crack opening happens due to the stress redistribution of *relaxation phase*. Figure 9.16 describes the flowchart of required sequences proposed in modeling the ductile fracture mechanics using the non-local damage plasticity with the X-FEM technique.

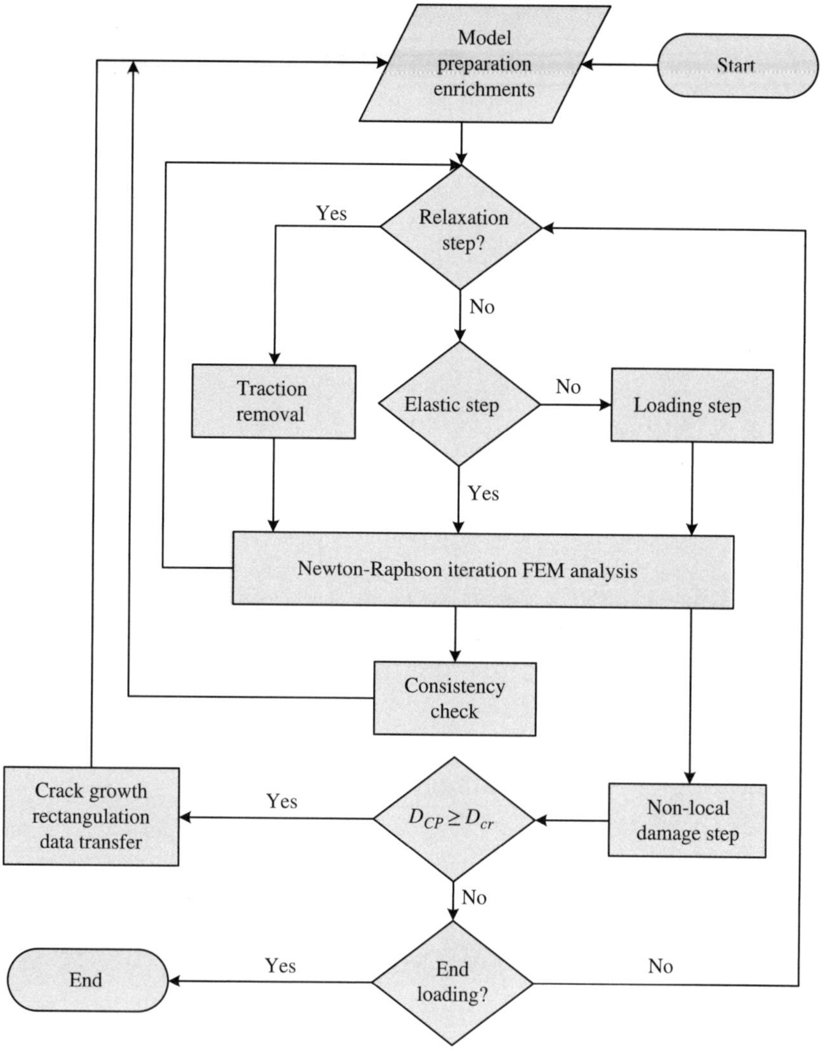

Figure 9.16 The flowchart of required sequences proposed in modeling the ductile fracture mechanics using a non-local damage plasticity with the X-FEM technique

9.8.4 Locking Issues in Crack Growth Modeling

In the X-FEM technique, the implementation of quadrilateral Q4 element is very popular; however, it is well known that this type of element is subjected to the volumetric locking in incompressible, or nearly incompressible materials, where the J_2 plasticity is used. In ductile fracture mechanics, the volumetric locking is probable and has a considerable effect on the damage growth and crack growth pattern. There are a number of remedies for the volumetric locking problem, such as the implementation of higher order elements, the mixed and hybrid formulations, the reduced and selective integrations, and so on. It has been shown that the mixed formulation is efficiently used to solve the volumetric locking; however, it increases the number of DOF and computational costs. The reduced and selective integration methods are equivalent to the mixed formulation, and provide the necessary singularity of the constraint part of the stiffness matrix that avoids volumetric locking.

In this study, a selective integration strategy is employed to eliminate the volumetric locking in the X-FEM damage plasticity model. In this technique, the actual volume changes at the Gauss points are replaced by the average volume change of the element to reduce the order of integration in selected terms and modify the stress–strain relation. Consider the incremental Green strain tensor $\Delta \boldsymbol{E}^G$ defined in relation 9.8, the modified strain increment $\Delta \bar{\boldsymbol{E}}$ can be defined as

$$\Delta \bar{\boldsymbol{E}} = \Delta \boldsymbol{E}^G + \frac{1}{2}\mathbf{I}\left(\text{trace}\left(\overline{\Delta \boldsymbol{E}}\right) - \text{trace}\left(\Delta \boldsymbol{E}^G\right)\right) \tag{9.67}$$

where $\overline{\Delta \boldsymbol{E}}$ is the average strain increment of the element defined as

$$\overline{\Delta \boldsymbol{E}} = \frac{1}{\Omega^e}\int_{\Omega^e} \Delta \boldsymbol{E}\, d\Omega \tag{9.68}$$

On the basis of modified strain increment $\overline{\Delta \boldsymbol{E}}$ obtained from relation (9.68) at each Gauss point, the constitutive equations described in Section 9.5 are used to calculate the corresponding stress value. The internal force vector is then obtained based on the summation of two parts, that is, the deviatoric and hydrostatic components. In fact, the deviatoric stress is integrated using four Gauss points and the uniform pressure is integrated by a single Gauss point. This selective integration procedure is only employed for the standard elements. The enriched elements are less prone to volumetric locking, since the enriched DOF provide more flexibility in the deformation of element, and hence eliminate the volumetric locking problem.

9.9 Application of X-FEM Damage-Plasticity Model

In order to demonstrate the performance of proposed computational algorithm, three examples are analyzed numerically. The first example is of a necking problem chosen to illustrate the efficiency and accuracy of proposed model in large deformation with the nonlocal damage model. The next two examples are chosen to demonstrate the robustness of X-FEM damage plasticity model in simulation of modes I and II crack propagation for a compact tension (CT) test and a more complex geometry of the double-notched specimen.

9.9.1 *The Necking Problem*

The first example is of a necking problem, which is a well–known benchmark test in large deformation solid mechanics. This example is chosen to illustrate the performance of proposed computational model due to implementation of a large deformation formulation, the volumetric locking strategy, and a non-local damage-plasticity model. A plate with the width of 1 cm and length of 5 cm is subjected to the uniaxial extension, as shown in Figure 9.17a. The geometry and boundary conditions of the problem are shown in this figure. The material parameters are given in Table 9.2. The plate is in plane strain condition, which is modeled using the 5×20 quadrilateral Q4 FE mesh, and is subjected to 1 cm vertical displacement at the top edge in 100 steps. The numerical simulation is first performed using the von-Mises plasticity model without the damage effect. In Figures 9.17b–d, the von-Mises stress contours are presented for the small deformation formulation, the updated Lagrangin large deformation formulation, and the large deformation formulation with the volumetric locking scheme, respectively. Also plotted in Figure 9.18 are the variations with displacement of vertical force for three different cases. As can be observed from Figures 9.17 to 9.18, softening behavior does not happen in the small deformation

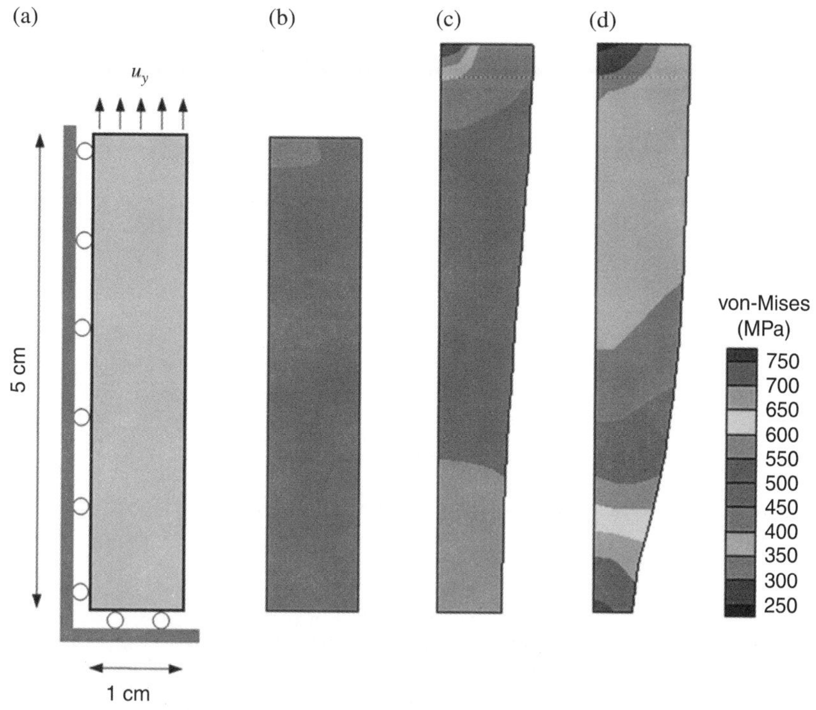

Figure 9.17 The necking problem: (a) the geometry and boundary conditions; the von-Mises stress contours for (b) the small deformation formulation; (c) the large deformation formulation; (d) the large deformation formulation with the volumetric locking scheme

Table 9.2 Material properties of necking problem

Property	Value
Young's modulus E(GPa)	180
Poisson ratio ν	0.28
Hardening law $R(r)$ (MPa)	$443+250\,\varepsilon_p^{0.35}$
Damage parameter z (MPa)	3.5
Damage parameter s	0
Critical damage D_{cr}	0.85

formulation, however, large deformation formulation captures geometrical softening, and the necking phenomenon can be clearly captured when the volumetric locking strategy is taken into computational modeling, as shown in Figure 9.17d.

In order to describe the performance of the damage plasticity model, the problem is also modeled using the local and non-local damage models for four different meshes, as shown in Figure 9.19. In the case of the non-local damage plasticity model, the material characteristic length is assumed to be $\ell_c = 1$ mm, which is relatively high compared to the real characteristic length of material to avoid the computational cost of using very fine meshes. In Figure 9.19, the distributions of damage contours are shown for the local and non-local damage models at four different meshes. Clearly the mesh dependency is obvious in the local damage model and the zone of shear band is confined to the size of element, while the non-local damage model preserves a smooth behavior in the deformed shape of various FE meshes.

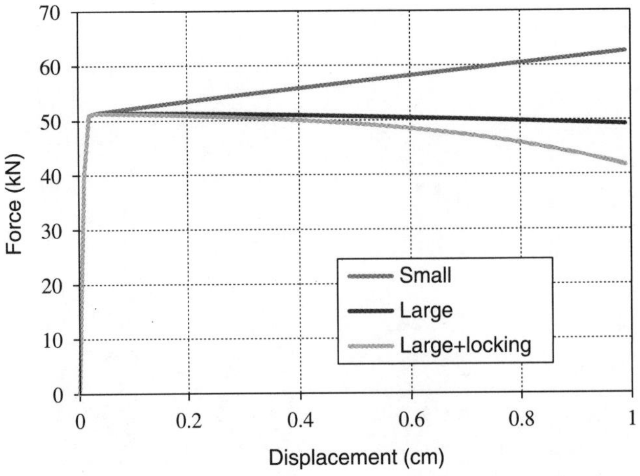

Figure 9.18 The variations with displacement of vertical force for three different formulations

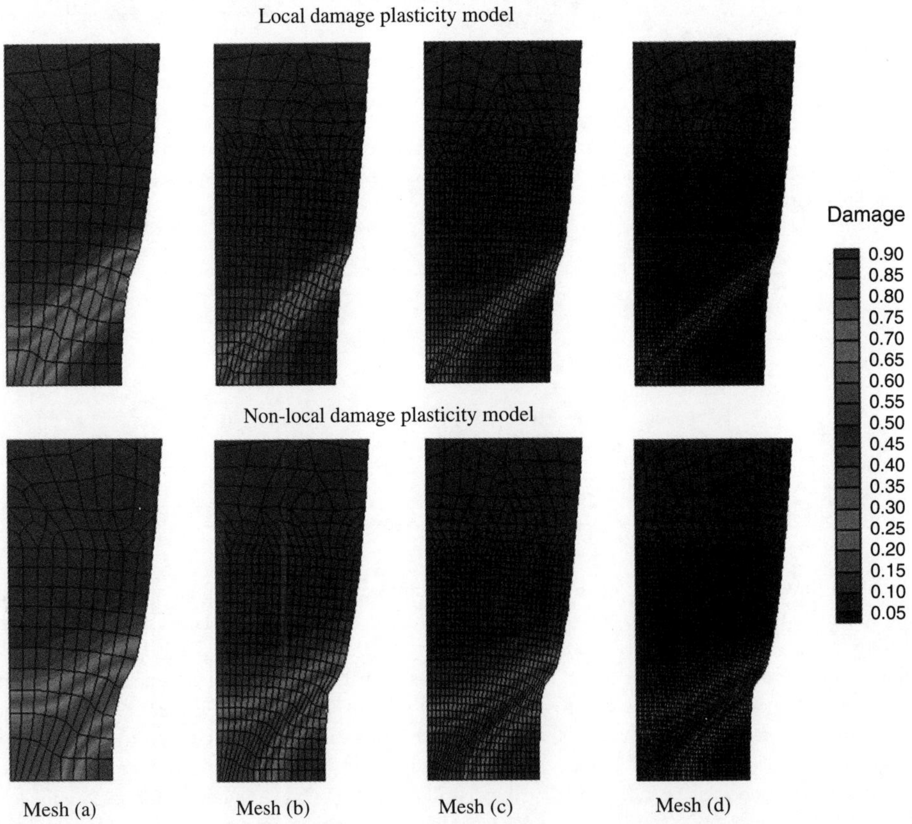

Figure 9.19 A comparison of damage contours for four different meshes between the local and non-local damage-plasticity models at the vertical displacement of 0.5 cm (For color details, please see color plate section)

9.9.2 *The CT Test*

The second example is the well-known CT test chosen to illustrate the performance of X-FEM damage plasticity model in mode I crack propagation. The outlook of the problem is shown in Figure 9.20, where the dimensions are $w = 152.4$ mm and $a = 61$ mm. The specimen is subjected to a prescribed displacement of 1 cm in 200 steps. Two rigid pins are considered in the upper and lower holes, in which the prescribed displacements are applied at the midpoint of these rigid pins. The specimen is made of Al-2024 T3 with material parameters given in Table 9.3. The damage is considered to be decoupled from the plasticity model, that is, $D_{\text{index}} = 0$, and a linear damage law is assumed as $P_D = 1$. The initial crack length is 1 mm and the crack growth segment is 3 mm. The weighted average value of damage is calculated using the Gauss points of a half circle with the radius of $\ell_c = 0.5$ mm around the crack-tip.

In Figure 9.21, the distributions of von-Mises and damage contours are shown at different loading steps, in which the circle indicates the enrichment radius and the gray line presents the crack growth. Since the damage is decoupled from the plasticity model, the softening behavior is due to crack propagation. As can be expected, the maximum damage value is concentrated at the crack tip region; although some values of damage can be seen around the rigid pins in the upper and lower holes where the loadings are applied, and in the compressive region at the right hand side of the plate. In Figure 9.22, the predicted vertical force versus displacement is plotted for the CT test, and the result is compared with that reported by Mediavilla, Peerlings, and Geers (2006) using an adaptive FE remeshing strategy. Obviously, it can be seen from the force-displacement curves that there is a little difference between the results of X-FEM and adaptive remeshing techniques in the softening part of the diagram before the onset of fracture. The only noticeable difference consists in the bumps that occur at each step of crack growth, which are larger for X-FEM analysis since the larger crack growth segment is employed here. In fact, the larger jumps in the

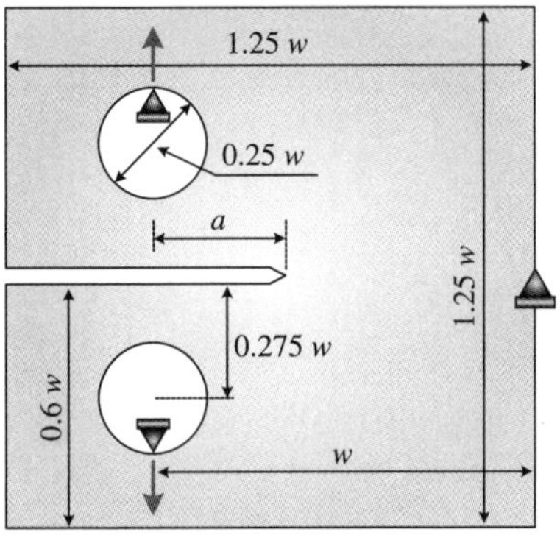

Figure 9.20 The CT test: The geometry and boundary conditions

Table 9.3 Material properties of the Al-2024 T3

Property	Value
Young's modulus E(GPa)	72.6
Poisson ratio ν	0.32
Hardening law σ_y(MPa)	$345 + 250\,\varepsilon_P^{0.35}$
Damage parameter z (MPa)	1.7
Damage parameter s	1
Critical damage D_{cr}	0.85

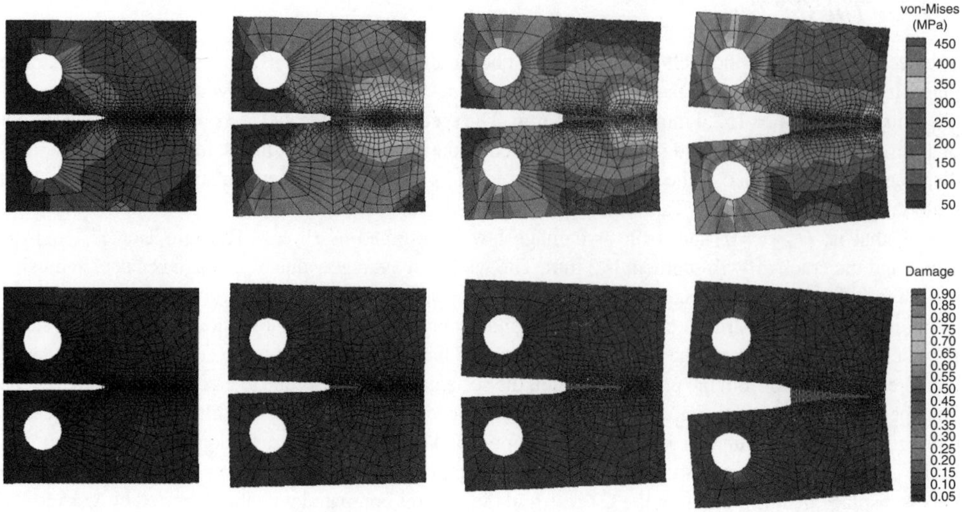

Step 5 Step 50 Step 150 Step 275

Figure 9.21 The distributions of von-Mises stress and damage contours at different crack growths for the CT test (For color details, please see color plate section)

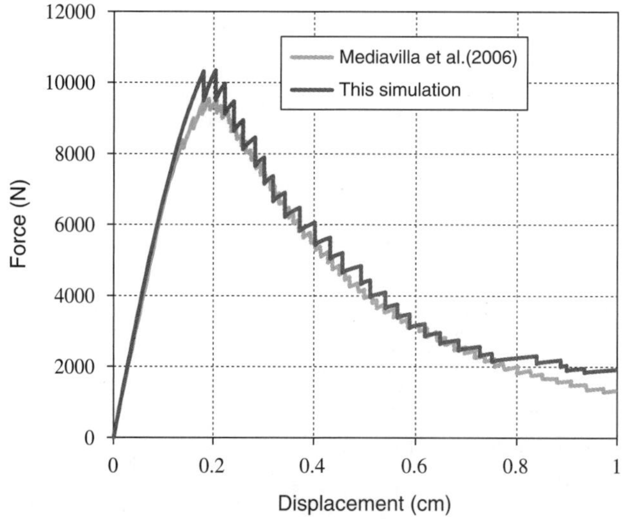

Figure 9.22 The predicted vertical force versus displacement for the CT test: A comparison between the X-FEM technique and an adaptive FE strategy

X-FEM model are the consequence of the combined effects of large transfer errors and large crack segments, which result in a drop in the force-displacement curve. In order to obtain a smooth force-displacement curve, a fine mesh with a small crack segment length can be used, however, it must be noted that there is a combination of various factors that play important roles in the simulation of ductile fracture mechanics, including: critical damage value, radius of calculated damage area, enrichment radius at the crack tip region, and crack segment length.

The critical damage value D_{cr} has been reported in the literature for various materials, as given by Lemaitre and Desmorat (2005). For instance, the value of $D_{cr} = 0.23$ is recommended for Al-2024, however, it was observed here that this value leads to early crack growth and as a result, an unacceptable force-displacement curve. For this reason, a critical damage value of $D_{cr} = 1.0$ was necessary to obtain a reasonable result. Of course, the higher value of critical damage may prevent the crack growth and results in an excessive material degradation. In order to prevent such circumstances, the crack is propagated based on the crack initiation criterion where there are some Gauss points in the weighted damage area that have the maximum value of damage, and the weighted average value of damage at the crack tip area remains constant for two successive steps. In addition, the radius size of calculated damage plays an important role in the evaluation of damage variable at the crack tip region that must be chosen as a function of the material characteristic length obtained from the experiments. Furthermore, the enrichment radius must be chosen so that the enrichment circle contains at least two/three layers of elements; hence it depends on the local mesh size at the crack tip region. The larger crack segments result in higher stress redistribution and larger bumps in the force-displacement curve. Finally, the number of *relaxation steps* must be increased by increasing the crack segment length.

9.9.3 The Double-Notched Specimen

The last example is chosen to illustrate the performance of proposed computational model for a blanking process, in which a plate with two asymmetrically placed rounded notches is subjected to tensile loading. This complex geometry is modeled numerically to demonstrate the robustness of X-FEM damage-plasticity model in the simulation of mode II crack propagation for a double-notched specimen. The geometry and boundary conditions of the problem are shown in Figure 9.23. The vertical displacements are imposed on the top and left boundaries, while the bottom and right edges remain fixed. The specimen is subjected to a prescribed displacement of 1 mm in y-direction in 200 steps. The specimen is in plane strain condition and is made of a steel with material parameters given in Table 9.2. The non-local damage-plasticity model is employed with the material characteristic length of $\ell_c = 1$ mm. A nonlinear damage law is assumed based on the nonlinear damage function defined in relation 9.52. The radius of calculated damage area is taken equal to the material characteristic length, that is, $\ell_c = 1$ mm. The plate is modeled using the quadrilateral Q4 elements with the selective reduced integration algorithm.

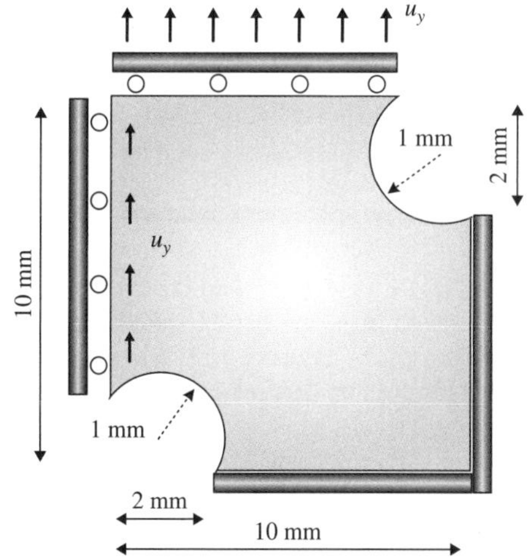

Figure 9.23 A double-notched specimen: The geometry and boundary conditions

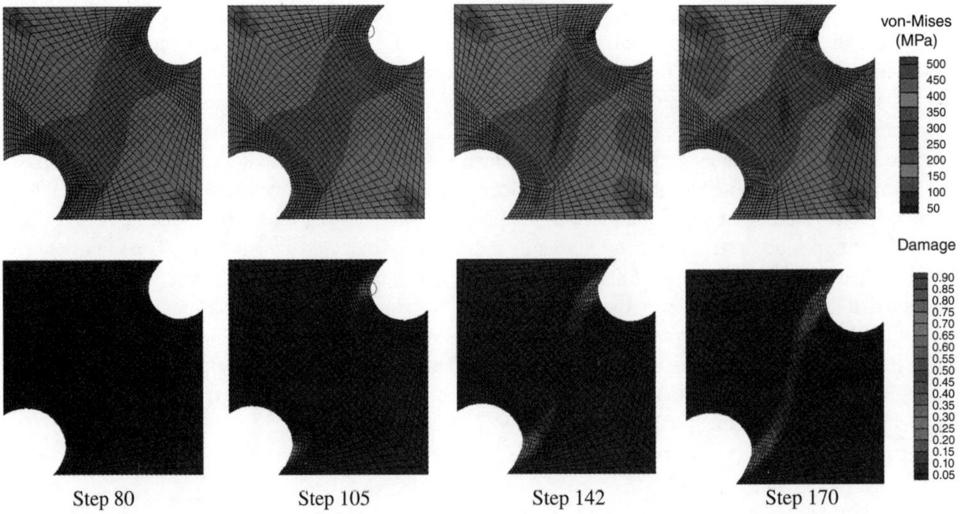

Figure 9.24 The distributions of von-Mises stress and damage contours at different crack growths for a double-notched specimen (For color details, please see color plate section)

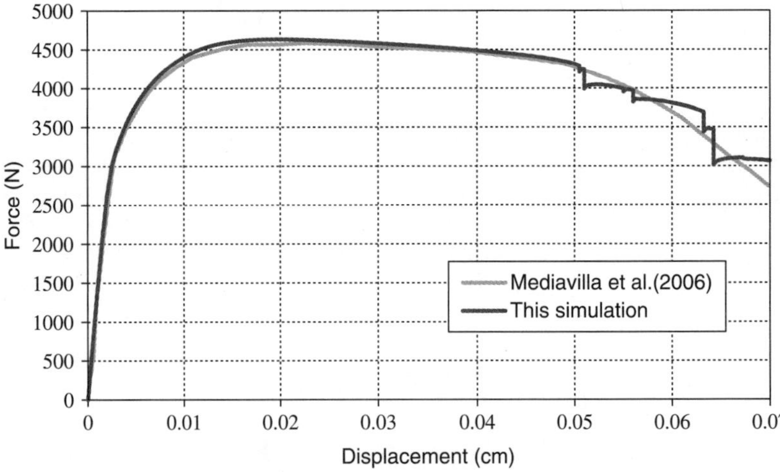

Figure 9.25 The predicted vertical force versus displacement for a double-notched specimen: A comparison between the X-FEM technique and an adaptive FE strategy

In Figure 9.24, the distributions of von-Mises and damage contours are shown at different loading steps, in which the circle indicates the enrichment radius at the crack tip area. The tensile loading causes the development of plastic shear band localization between the two notches and leads to the maximum damage values at the notches. Obviously, a localization band is formed in the middle of plate where the crack propagation takes place. An initial crack is automatically induced in the plate where the damage at a Gauss point reaches the critical damage value, and a new crack segment is introduced from the boundary to that Gauss point. The crack is then propagated in the direction of maximum damage on the basis of a semi-circular algorithm around the crack tip, described in Figure 9.12. In fact, the maximum damage value obtained at the two notches enforces the nucleation of two cracks which run first from the top notch, then from the bottom notch, and finally towards the center of the specimen. In Figure 9.25, the predicted force versus displacement is plotted for the double-notched specimen, and the result is compared with that reported by Mediavilla, Peerlings, and Geers (2006) using an adaptive FE remeshing strategy. Similar to previous example, the only difference between two techniques consists in the bumps that occur at each step of crack growth. As noted earlier, because of the large crack growth segment employed here, the larger force jumps at each increment of crack growth can be seen in X-FEM analysis. It must be noted that because of the material and geometrical nonlinearities, the shear band localization and the volumetric locking issues, this problem is highly nonlinear; hence a line-search method with the full Newton–Raphson procedure is used to improve the stability of the solution, particularly in the *relaxation steps* where the solution can easily diverge.

9.10 Dynamic Large X-FEM Deformation Formulation

In many engineering applications, the useful life of components is a crucial item in the design process; the non-scheduled stops for the maintenance owing to the unpredicted failure, may incur serious economic consequences. The situation becomes even more complicated when the deformation of solids occurs under dynamic loading, either low or high strain rate. This is the case in blast shields and shaped charges, where the liner is deformed by the strain rate that varies radially due to the propagation of pressure loading generated by the explosive detonation. It can also happen in the case of an earthquake; the cyclic loading overloads the reeling of pipelines and the high pressure turbine discs in aero-engines, where the engineering components are subjected to cyclically varying loading conditions. The material can eventually be forced to undergo the cyclic plastic flow, and the accumulation of the number of cycles is a source of damage that, sooner or later, leads to material failure. In this section, the X-FEM is developed for large deformation–large strain ductile fracture problems, which is coupled with the Lemaitre damage-plasticity model based on an isotropic-kinematic hardening rule, in order to perform the crack propagation simulation under dynamic and cyclic loadings. An explicit central difference technique is employed in conjunction with mass lumping, reduced integration, and numerical damping to model the dynamic effects. The crack behavior under cyclic loading is modeled using the contact friction algorithm within the X-FEM governing equations, and the crack closure effect is incorporated into the damage plasticity model.

Consider a cracked body ${}^0\Omega$ bounded externally by surface ${}^0\Gamma$ in reference configuration ${}^0\gamma$. The crack faces are Γ_c^+ and Γ_c^- corresponding to the either sides of the crack. Under the loading and boundary conditions, the body deforms and occupies the new configuration ${}^t\gamma$. The motion of a particle is described at the reference position ${}^0\boldsymbol{x}$ and the current position ${}^t\boldsymbol{x}$ at time t with the mapping ${}^t\boldsymbol{x} = \Phi({}^0\boldsymbol{x}, t)$. The equation of motion of the current configuration at time t can be expressed as

$$ {}^t\rho\,{}^t\ddot{\mathbf{u}} = \nabla\cdot{}^t\boldsymbol{\sigma} + {}^t\mathbf{b} \quad \text{in} \quad \Omega \tag{9.69} $$

which is subjected to the boundary and initial conditions as

$$ \begin{aligned} {}^t\mathbf{u}({}^0\boldsymbol{x},t) &= {}^t\mathbf{u}_d({}^0\boldsymbol{x},t) && \text{on } \Gamma_u \\ {}^t\boldsymbol{\sigma}\cdot\mathbf{n} &= {}^t\mathbf{t} && \text{on } \Gamma_t \\ {}^t\boldsymbol{\sigma}\cdot\mathbf{n}_c &= {}^t\mathbf{t}_c && \text{on } \Gamma^- \\ {}^t\boldsymbol{\sigma}\cdot(-\mathbf{n}_c) &= -{}^t\mathbf{t}_c && \text{on } \Gamma^+ \\ {}^0\mathbf{u}({}^0\boldsymbol{x},0) &= \mathbf{u}_0({}^0\boldsymbol{x}) && \text{in } \Omega \\ \dot{\mathbf{u}}({}^0\boldsymbol{x},0) &= \dot{\mathbf{u}}_0({}^0\boldsymbol{x}) && \text{in } \Omega \end{aligned} \tag{9.70} $$

where ${}^t\rho$ is the density, ${}^t\mathbf{u}$ is the displacement, ${}^t\boldsymbol{\sigma}$ is the Cauchy stress tensor, ${}^t\mathbf{b}$ is the body force vector, ${}^t\mathbf{t}$ is surface traction, $\mathbf{n}$ is the normal to body surface. Here, (∇) stands for the divergence operator; that is, the derivatives with respect to the time are shown with dotted symbols. Note that the third and fourth relations in Eq. (9.70) are implemented only when the crack faces are in contact otherwise the crack faces are considered traction free.

Equation (9.69) is a nonlinear equilibrium equation in nature; one way to handle it is to use a dynamic explicit formulation that is fully compatible with a step-by-step incremental Lagrangian approach. In such case, equation of motion is considered at time step t and all variables are determined at time step $t + \Delta t$. Since ${}^{t+\Delta t}\boldsymbol{x} = {}^{0}\boldsymbol{x} + {}^{t+\Delta t}\mathbf{u}$ and ${}^{t}\boldsymbol{x} = {}^{0}\boldsymbol{x} + {}^{t}\mathbf{u}$, it can be written as ${}^{t+\Delta t}\boldsymbol{x} = {}^{t}\boldsymbol{x} + {}_{t}\mathbf{u}$ and ${}_{t}\mathbf{u} = {}^{t+\Delta t}\mathbf{u} - {}^{t}\mathbf{u}$. Applying the virtual work principle to the configuration at time t and employing the discontinuous version of Gauss-Green Divergence theorem, it leads to

$$\int_{{}^{t}\Omega} \delta\mathbf{u}^{T\,t}\rho\,{}^{t}\ddot{\mathbf{u}}\,d^{t}\Omega = -\int_{{}^{t}\Omega} \delta_{t}\boldsymbol{\varepsilon}^{T\,t}\boldsymbol{\sigma}\,d^{t}\Omega - \int_{\Gamma_C^-} [\![\delta\mathbf{u}^{T}]\!]\,{}^{t}\mathbf{t}_{c}\,d^{t}\Gamma + \int_{{}^{t}\Omega} \delta\mathbf{u}^{T\,t}\mathbf{b}\,d^{t}\Omega + \int_{{}^{t}\Gamma_t} \delta\mathbf{u}^{T\,t}\mathbf{t}\,d^{t}\Gamma \tag{9.71}$$

where $\delta\mathbf{u}$ is the virtual displacement imposed at time t, and $\delta_t\boldsymbol{\varepsilon} = \delta(\nabla^{s}{}_{t}\mathbf{u})$ is the corresponding virtual strain and ∇^{s} is the symmetric gradient operator. The second term in the right hand side of the equation is the contribution of contact force, here $[\![\delta\mathbf{u}]\!]$ is the jump of $\delta\mathbf{u}$ across the discontinuity and ${}^{t}\mathbf{t}_{c} = {}^{t}\boldsymbol{\sigma}\cdot\mathbf{n}_{c}^{-}$ is corresponding contact traction. This formulation may seem like small deformation formulation, but here the derivatives are taken with respect to coordinates at time t and all integrals are calculated with respect to the configuration at this time.

9.10.1 The Dynamic X-FEM Discretization

In order to derive the dynamic X-FEM formulation, the crack body and crack tip singularities are modeled using the Heaviside and crack tip asymptotic functions, respectively. Hence, the modified X-FEM displacement field is defined according to relation (9.29) as

$$\mathbf{u}(\boldsymbol{x},t) = \sum_{i\in\mathcal{N}^{std}} N_i({}^{0}\boldsymbol{x})\,\bar{\mathbf{u}}_i(t) + \sum_{i\in\mathcal{N}^{enr}_{Hev}} N_i({}^{t}\boldsymbol{x})\,\overline{H}({}^{0}\boldsymbol{x})\,\bar{\mathbf{a}}_i(t) + \sum_{i\in\mathcal{N}^{*enr}_{Tip}}\sum_{j=1}^{4} N_i({}^{t}\boldsymbol{x})\,R({}^{t}\boldsymbol{x})\,\overline{A}_j({}^{0}\boldsymbol{x})\,\bar{\mathbf{b}}_i^{j}(t) \tag{9.72}$$

where $\overline{H}({}^{0}\boldsymbol{x}) = H({}^{0}\boldsymbol{x}) - H({}^{0}\boldsymbol{x}_i)$ and $\overline{A}_j({}^{0}\boldsymbol{x}) = A_j({}^{0}\boldsymbol{x}) - A_j({}^{0}\boldsymbol{x}_i)$ are accordingly the shifted Heaviside and asymptotic functions; $R({}^{t}\boldsymbol{x})$ is the cutoff function that varies between 0 and 1 in blending layer with the 0 value at the standard elements and the unit value at the fully enriched elements; $\bar{\mathbf{u}}_i$, $\bar{\mathbf{a}}_i$ and $\bar{\mathbf{b}}_i^{j}$ are the corresponding standard, Heaviside, and crack tip nodal DOF. In a simplified form, this relation can be written as $\mathbf{u} = \mathbf{N}\bar{\mathbf{u}}$, where $\bar{\mathbf{u}}^{T} = \left[\bar{\mathbf{u}}_i^{T}, \bar{\mathbf{a}}_i^{T}, \left(\bar{\mathbf{b}}_i^{j}\right)^{T}\right]$ is the set of nodal values. Note that since these enrichment functions are tailored to small deformation problems, they must be defined on the reference configuration ${}^{0}\gamma$ for large deformation problems.

In order to derive the discretized matrix form of the equation of motion, substitute relation (9.72) into (9.71), it yields to

$$\mathbf{M}\,{}^{t}\ddot{\bar{\mathbf{u}}} = {}^{t}\mathbf{R} - {}^{t}\mathbf{F} - {}^{t}\mathbf{G} \tag{9.73}$$

where ${}^{t}\ddot{\bar{\mathbf{u}}}$ is the nodal acceleration vector, and $\mathbf{M}$, ${}^{t}\mathbf{R}$, ${}^{t}\mathbf{F}$, and ${}^{t}\mathbf{G}$ are defined as

$$\begin{aligned}
\mathbf{M} &= \int_{{}^{t}\Omega} {}^{t}\rho\,\mathbf{N}^{T}\mathbf{N}\,d^{t}\Omega && \text{(inerital term)}\\
{}^{t}\mathbf{R} &= \int_{{}^{t}\Gamma_t} \mathbf{N}^{T\,t}\mathbf{t}\,d^{t}\Gamma + \int_{{}^{t}\Omega} \mathbf{N}^{T\,t}\mathbf{b}\,d^{t}\Omega && \text{(external force term)}\\
{}^{t}\mathbf{F} &= \int_{{}^{t}\Omega}^{t} \mathbf{B}_L^{T\,t}\boldsymbol{\sigma}\,d^{t}\Omega && \text{(internal force term)}\\
{}^{t}\mathbf{G} &= \int_{\Gamma_c^-} [\![\mathbf{N}^{T}]\!]\,{}^{t}\mathbf{t}_c\,d^{t}\Gamma && \text{(crack contact force term)}
\end{aligned} \tag{9.74}$$

Considering the 2D plane stress/strain, or an axisymmetric condition, the matrices $\mathbf{N}$, $[\![\mathbf{N}]\!]$ and $\mathbf{B}_L$ for each node I can be defined as

$$\mathbf{N}_I = \left[\mathbf{N}_I^{std}, \mathbf{N}_I^{Hev}, \mathbf{N}_I^{Tip}\right] \tag{9.75a}$$

$$[\![\mathbf{N}_I]\!] = \left[0_{2\times 2}, 2\mathbf{N}_I^{std}, 2\sqrt{r}\,R\left({}^t\boldsymbol{x}\right)\mathbf{N}_I^{std}\right] \tag{9.75b}$$

$$(\mathbf{B}_L)_I = \left[\left(\mathbf{B}_L^{std}\right)_I, \left(\mathbf{B}_L^{Hev}\right)_I, \left(\mathbf{B}_L^{Tip}\right)_I\right] \tag{9.75c}$$

where

$$\mathbf{N}_I^{std} = \begin{bmatrix} N_I & 0 \\ 0 & N_I \end{bmatrix} \tag{9.76a}$$

$$\mathbf{N}_I^{Hev} = \begin{bmatrix} N_I\overline{H}\left({}^t\boldsymbol{x}\right) & 0 \\ 0 & N_I\overline{H}\left({}^t\boldsymbol{x}\right) \end{bmatrix} \tag{9.76b}$$

$$\mathbf{N}_I^{Tip} = \begin{bmatrix} N_I R\left({}^t\boldsymbol{x}\right)\overline{A}_j\left({}^0\boldsymbol{x}\right) & 0 \\ 0 & N_I R\left({}^t\boldsymbol{x}\right)\overline{A}_j\left({}^0\boldsymbol{x}\right) \end{bmatrix} \tag{9.76c}$$

and

$$\left(\mathbf{B}_L^{std}\right)_I = \begin{bmatrix} \dfrac{\partial N_I}{\partial^t x_1} & 0 \\ 0 & \dfrac{\partial N_I}{\partial^t x_2} \\ \dfrac{\partial N_I}{\partial^t x_2} & \dfrac{\partial N_I}{\partial^t x_1} \\ \cdots & \cdots \\ \dfrac{N_I}{{}^t x_1} & 0 \end{bmatrix} \tag{9.77a}$$

$$\left(\mathbf{B}_L^{Hev}\right)_I = \begin{bmatrix} \dfrac{\partial}{\partial^t x_1}\left(N_I\overline{H}\left({}^0\boldsymbol{x}\right)\right) & 0 \\ 0 & \dfrac{\partial}{\partial^t x_2}\left(N_I\overline{H}\left({}^0\boldsymbol{x}\right)\right) \\ \dfrac{\partial}{\partial^t x_2}\left(N_I\overline{H}\left({}^0\boldsymbol{x}\right)\right) & \dfrac{\partial}{\partial^t x_1}\left(N_I\overline{H}\left({}^0\boldsymbol{x}\right)\right) \\ \cdots & \cdots \\ \dfrac{N_I\overline{H}\left({}^0\boldsymbol{x}\right)}{{}^t x_1} & 0 \end{bmatrix} \tag{9.77b}$$

$$\left(\mathbf{B}_L^{Tip}\right)_I = \begin{bmatrix} \frac{\partial}{\partial^t x_1}\left(N_I R({}^t\boldsymbol{x})\bar{A}_j({}^0\boldsymbol{x})\right) & 0 \\ 0 & \frac{\partial}{\partial^t x_2}\left(N_I R({}^t\boldsymbol{x})\bar{A}_j({}^0\boldsymbol{x})\right) \\ \frac{\partial}{\partial^t x_2}\left(N_I R({}^t\boldsymbol{x})\bar{A}_j({}^0\boldsymbol{x})\right) & \frac{\partial}{\partial^t x_1}\left(N_I R({}^t\boldsymbol{x})\bar{A}_j({}^0\boldsymbol{x})\right) \\ \cdots & \cdots \\ \frac{N_I R({}^t\boldsymbol{x})\bar{A}_j({}^0\boldsymbol{x})}{{}^t x_1} & 0 \end{bmatrix} \tag{9.77c}$$

This formulation provides a general form with high quality stress field around the crack tip, but in the presence of singular terms the calculation of integrals in Eq. (9.74) requires a high number of Gauss points; in addition the enrichment of a predefined area around the crack tip increases the number of Gauss points dramatically. As will be discussed in the next section, the main process overhead in an explicit dynamic formulation is due to the calculation of integrals (9.74) in the Gauss integration points, so reducing the number of Gauss points dramatically enhances the performance of solution. Thus, tip enrichment is not considered in this section and only the crack body is modeled with the Heaviside function. This can also be justified by the fact that the crack tip blunts in ductile materials and crack tip singularity is decreased, hence the need for enrichment is reduced.

9.10.2 The Large Strain Model

The Lagrangian formulation presented in previous section, accounts for large deformations, however, in many problems the strains are also large; for example, in low cycle fatigue problems the strains at the crack-tip area are in order of 200%. Constitutive modeling of elasto-plastic materials in the finite deformation range has received considerable attention over the past decades. The most constitutive models in large strain formulations are based on the hypo-elastic stress–strain rules, which are based on an additive decomposition of strain tensor that are conceptually simple. However, such formulations do not assume a stored energy function and generally lead to non-zero dissipation in a closed cycle loading. In addition, the hypo-elastic material descriptions utilize the rate form quantities that need to be integrated in time, and as a result lead to numerical errors even if the numerical objectivity is preserved. Alternative constitutive models in large strain formulations are based on the hyper-elastic laws that assume stored energy functions and do not have the drawbacks mentioned earlier. In this study, the hyper-elastic large strain formulation proposed by Eterovic and Bathe (1990) is implemented, which is a direct extension of infinitesimal elasto-plasticity formulation. Based on this method, the product decomposition of deformation gradient can be written as

$$\mathbf{F} = \mathbf{F}^e\,\mathbf{F}^p \tag{9.78}$$

where $\mathbf{F}^e$ and $\mathbf{F}^p$ are the elastic and plastic parts of deformation gradient tensor $\mathbf{F}$. Applying the Hencky strain, which is conceptually an extension of the true strain in uniaxial loading into the multi-axial case, this can be defined as

$$\boldsymbol{E}^e = \ln\left(\mathbf{F}^e\mathbf{F}^{eT}\right) = \frac{1}{2}\ln(\mathbf{B}^e) \tag{9.79}$$

where $\mathbf{B}$ is the left Cauchy–Green tensor and ln(.) is the tensorial logarithm function. The presence of the Hencky stored energy function results in a linear relationship between the Hencky elastic strain and the Kirchoff stress $\boldsymbol{\tau}$ as

Table 9.4 The stress updating procedure in large strain model

Given ${}_t\bar{\mathbf{u}}$, compute the incremental deformation gradient ${}^{\Delta t}\mathbf{F} = \mathbf{I} + \nabla_t \bar{\mathbf{u}}$ and the current deformation gradient ${}^{t+\Delta t}\mathbf{F} = {}^{\Delta t}\mathbf{F}\,{}^{t}\mathbf{F}$,
Compute the elastic trial state as
${}^{t}\mathbf{B}^e = \exp\left(2\,{}^{t}\boldsymbol{E}^e\right)$
${}^{t+\Delta t}\mathbf{B}^e_{trial} = {}^{\Delta t}\mathbf{F}\,{}^{t}\mathbf{B}^e \left({}^{\Delta t}\mathbf{F}\right)^T$
${}^{t+\Delta t}\mathbf{E}^e_{trial} = \frac{1}{2}\ln\left({}^{t+\Delta t}\mathbf{B}^e_{trial}\right)$
where exp(.) is the tensorial exponential function,
Perform the standard infinitesimal return mapping by using ${}^{t+\Delta t}\boldsymbol{E}^e_{trial}$ instead of ${}^{t+\Delta t}\boldsymbol{\varepsilon}^e_{trial}$,
Calculate the internal force vector by determination of the Cauchy stress from the Kirchoff stress by
${}^{t+\Delta t}\boldsymbol{\sigma} = \det({}^{t+\Delta t}\mathbf{F})^{-1}\,{}^{t+\Delta t}\boldsymbol{\tau}$

$$\rho\Psi_{ED} = \frac{1}{2}\boldsymbol{E}^e(1-D)\mathbf{C}\boldsymbol{E}^e$$

$$\boldsymbol{\tau} = \rho\frac{\partial\Psi_{ED}}{\partial\boldsymbol{E}^e} = (1-D)\mathbf{C}\boldsymbol{E}^e \tag{9.80}$$

Neglecting the details of derivation of the method, the procedure for stress updating in a large strain model is described in Table 9.4, where the updated stress–strain state is obtained at time step $t + \Delta t$ from the incremental displacement ${}_t\bar{\mathbf{u}} = {}^{t+\Delta t}\bar{\mathbf{u}} - {}^{t}\bar{\mathbf{u}}$ using Eq. (9.73).

9.10.3 The Contact Friction Model

In the case of cyclic and dynamic loading, cracks may undergo the opening or closing mode; in addition the friction plays an important role in the case of sliding. Hence, it is important to model the frictional contact behavior between crack faces. In essence, the frictional contact is similar to an elasto-plasticity theory where several requirements must be taken into account (Khoei and Nikbakht, 2007); these requirements are the *stick law* that describes the stress state when the bodies in contact have no relative displacements, the *stick–slip law* that describes the relationship between the stress and stick–slip (elasto-plastic) conditions, the *slip criterion*, which is a yield criterion that indicates the stress level at which relative slip motion occurs, and the *slip rule* that is a flow rule to describe a relationship between stress and slip motion. Consider the positive side of the crack as the master and the negative side as the slave. For each point with coordinates $\boldsymbol{x}$ on the crack face Γ_c, the normal gap function $\mathbf{g}_N$ is defined as

$$\mathbf{g}_N(\boldsymbol{x}) = g_N(\boldsymbol{x})\,\mathbf{n}_c \quad \text{and} \quad g_N(\boldsymbol{x}) = \mathbf{n}_c^T[\![\mathbf{u}]\!] \tag{9.81}$$

and the impenetrability condition in normal direction is expressed by the Kuhn–Tucker rule, that is, $g_N \geq 0$, $t_{cN} \leq 0$, and $g_N\, t_{cN} = 0$, where t_{cN}, is the normal contact traction on the crack faces. There are various techniques proposed to impose the contact constraints between two bodies, such as the penalty, Lagrange multiplier, and augmented Lagrange methods. The penalty method is a popular technique among researchers since it is simple to implement and adds no extra DOF to the problem. However, the method may lead to locking and conditioning issues in the case of high penalty parameters and causes the contact surface penetration. Applying the penalty method, the normal contact traction $\mathbf{t}_{cN}$ can be defined as a function of normal gap as

$$\mathbf{t}_{cN}(\boldsymbol{x}) = t_{cN}(\boldsymbol{x})\,\mathbf{n}_c \quad \text{and} \quad t_{cN}(\boldsymbol{x}) = \kappa_N\, g_N(\boldsymbol{x}) \tag{9.82}$$

where κ_N is the penalty parameter in the normal direction. In a similar manner, the tangential gap function can be defined as

$$\mathbf{g}_T(\boldsymbol{x}) = g_T(\boldsymbol{x})\,\mathbf{m}_c \quad \text{and} \quad g_T(\boldsymbol{x}) = \mathbf{m}_c[\![\mathbf{u}]\!] \tag{9.83}$$

where $\mathbf{m}_c$ is the unit tangential vector. The tangential gap can be decomposed into the elastic (stick) and the plastic (slip) parts as $\mathbf{g}_T(\boldsymbol{x}) = g_T^e(\boldsymbol{x}) + g_T^p(\boldsymbol{x})$; hence, the stick law can be defined as

$$\mathbf{t}_{cT}(\boldsymbol{x}) = t_{cT}(\boldsymbol{x})\,\mathbf{m}_c \quad \text{and} \quad t_{cT}(\boldsymbol{x}) = \kappa_T\, g_T^e(\boldsymbol{x}) \tag{9.84}$$

where κ_T is the penalty parameter in the tangential direction. Considering the classical Coulomb frictional theory, the slip criterion can be written in the absence of cohesion as

$$F = t_{cT} + \mu\, t_{cN} \begin{cases} <0 & \text{stick} \\ =0 & \text{slip} \end{cases} \tag{9.85}$$

where μ is the friction coefficient. Considering an associated slip rule, it leads to

$$\dot{\mathbf{g}}_T^p = \dot{\gamma}\frac{\partial F}{\partial \mathbf{t}_{cT}} = \dot{\gamma}\,\mathbf{m}_c \tag{9.86}$$

where $\dot{\gamma}$ is a slip proportionality factor that can be defined according to the Kuhn–Tucker rule on the basis of stick and slip conditions, that is, $\dot{\gamma} \geq 0$, $F \leq 0$ and $\dot{\gamma}F = 0$. Considering the rate form of Eq. (9.83) and using (9.84) and (9.86), it yields to

$$\dot{\mathbf{t}}_{cT} = \kappa_T(\dot{\mathbf{g}}_T - \dot{\gamma}\,\mathbf{m}_c) \tag{9.87}$$

For the stick condition $\dot{\gamma} = 0$ and $\dot{\mathbf{t}}_{cT} = \kappa_T\, \dot{g}_T\, \mathbf{m}_c$, however, for the slip condition the plastic consistency condition holds as

$$\dot{F} = \left(\frac{\partial F}{\partial \mathbf{t}_{cT}}\right)^T \dot{\mathbf{t}}_{cT} + \left(\frac{\partial F}{\partial \mathbf{t}_{cN}}\right)^T \dot{\mathbf{t}}_{cN} = 0 \tag{9.88}$$

The plastic proportionality factor $\dot{\gamma}$ can be obtained by substituting Eqs. (9.82) and (9.87) into the consistency Eq. (9.88) as

$$\kappa_T \mathbf{m}_c^T(\dot{\mathbf{g}}_T - \dot{\gamma}\,\mathbf{m}_c) + \mu \mathbf{n}_c^T \kappa_N\, \dot{\mathbf{g}}_N = 0 \quad \text{or} \quad \dot{\gamma} = \frac{1}{\kappa_T}(\kappa_T \dot{g}_T + \mu\, \kappa_N \dot{g}_N) \tag{9.89}$$

Substitution (9.89) into (9.87), the tangential traction can be obtained as

$$\dot{\mathbf{t}}_{cT} = -\mu\, \kappa_N \dot{g}_N \mathbf{m}_c \tag{9.90}$$

Thus, the total contact traction can be defined as $\mathbf{t}_c = \mathbf{t}_{cT} + \mathbf{t}_{cN}$, and the crack contact force can be finally obtained from the last integral of Eq. (9.74).

9.11 The Time Domain Discretization: The Dynamic Explicit Central Difference Method

The equation of motion (9.73) is a system of second-order differential equations that must be integrated in time. In this study, the central difference method is employed, which is an explicit direct time integration method with the second-order accuracy. In comparison to the dynamic implicit method, the dynamic explicit method has several advantages; it is simple, needs a low storage space, does not need to solve the nonlinear system of equations iteratively, and it does not require the stiffness matrix assembly. The dynamic explicit method is the most suitable technique for transient (short time) problems. The method can be used for quasi-static analysis of problems with highly nonlinear behavior that usually have convergence issues. The main disadvantage of the method is that it is conditionally stable. The stability criterion enforces the small time step Δt particularly for fine and distorted meshes. In this method, the set of discretized equations in time and space can be written as

$$\begin{aligned} \mathbf{M}\,{}^{t+\Delta t}\bar{\mathbf{u}} &= \Delta t^2\left({}^t\mathbf{R}-{}^t\mathbf{F}-{}^t\mathbf{G}\right)+\mathbf{M}\left(2\,{}^t\bar{\mathbf{u}}-{}^{t-\Delta t}\bar{\mathbf{u}}\right) \\ {}^t\ddot{\bar{\mathbf{u}}} &= \frac{1}{\Delta t^2}\left({}^{t+\Delta t}\bar{\mathbf{u}}-2\,{}^t\bar{\mathbf{u}}+{}^{t-\Delta t}\bar{\mathbf{u}}\right) \\ {}^t\dot{\bar{\mathbf{u}}} &= \frac{1}{2\Delta t}\left({}^{t+\Delta t}\bar{\mathbf{u}}-{}^{t-\Delta t}\bar{\mathbf{u}}\right) \end{aligned} \tag{9.91}$$

The method is not self-starting, hence, it needs two conditions for the start of the solution as

$${}^{-\Delta t}\bar{\mathbf{u}} = {}^0\bar{\mathbf{u}} - \Delta t\,{}^0\dot{\bar{\mathbf{u}}} + \frac{1}{2}\Delta t^2\,{}^0\ddot{\bar{\mathbf{u}}} \qquad \text{with} \qquad {}^0\ddot{\bar{\mathbf{u}}} = \mathbf{M}^{-1}\left({}^0\mathbf{R}-{}^0\mathbf{F}-{}^0\mathbf{G}\right) \tag{9.92}$$

The stability limit Δt_c is approximately equal to the time of an elastic dilatational (pressure) wave to cross the smallest dimension of the model $\ell_{\min}$, that is,

$$\Delta t_c = \frac{\ell_{\min}}{c_d} \qquad \text{with} \qquad c_d = \sqrt{\frac{\lambda+2\mu}{\rho}} \tag{9.93}$$

where c_d is the speed of dilatational wave and, λ and μ are the Lame's constants. In Eq. (9.91), if an appropriate lumped mass matrix is used the method can be performed efficiently. In the X-FEM, the conventional FEM lumping techniques, such as the row summation scheme, cannot be used since the critical time step tends to zero as the discontinuity approaches the element boundaries. Belytschko *et al.* (2003b) proposed an implicit-explicit time integration scheme, in which the enriched elements are integrated using an unconditionally stable implicit method and the standard elements are integrated using an explicit method. De Borst, Réthoré, and Abellan (2006) employed the conventional explicit X-FEM method, however, the algorithm prevents a crack discontinuity to cross an element boundary in the vicinity of a supporting node. Menouillard *et al.* (2006) introduced a lumped mass matrix for the enriched elements that enables the use of a pure explicit formulation in the X-FEM. It has been shown that the method provides a lower bound for the critical time step regardless of the location of the discontinuity. In this method, the components of the lumped mass matrix for the enriched DOF are obtained as

$$\overline{M}_{ii} = \frac{m}{NA}\int_{\Omega^e}\psi^2\,d\Omega \tag{9.94}$$

where Ω^e is the area of the corresponding element for which the mass matrix is calculated, m is the mass of element, N is the number of nodes, and ψ is the corresponding enrichment function. The row summation technique is then employed for the rest of the structure as

$$\overline{M}_{ii} = \sum_j M_{ij} \tag{9.95}$$

where M_{ij} are the components of the consistent mass matrix obtained from the first integral of Eq. (9.74). Menouillard *et al.* (2008) proposed a method that provides a critical time step equal to the corresponding FEM time step, but the resulting mass matrix is the block diagonal which increases the computational costs.

An important factor concerning the time integration is the introduction of a numerical damping into the computational algorithm. The artificial bulk viscosity has long been used in explicit methods as a damping source that can be expressed as (Johnson and Beissel, 2001)

$$Q_{BV} = \begin{cases} C_L \rho c_d h |\dot{\varepsilon}_v| + C_Q \rho h^2 \dot{\varepsilon}_v^2 & \text{if } \dot{\varepsilon}_v < 0 \\ 0 & \text{if } \dot{\varepsilon}_v \geq 0 \end{cases} \tag{9.96}$$

where Q_{BV} is an artificial bulk viscosity stress, C_L and C_Q are the linear and quadratic coefficients which are usually taken as 0.06 and 1.2, respectively, $\dot{\varepsilon}_v$ is the rate of volumetric strain change, and h is a characteristic length of the element. Note that Q_{BV} has only a numerical effect; it is added to the stress tensor in the calculation of internal force, and is not included in the real stresses of the material points. As it can be seen from Eq. (9.96), the numerical damping is made of a linear and a quadratic term. The role of the linear term is to damp out the Gibbs oscillations behind the shock fronts, and the quadratic term smears the shocks over several elements and prevents the element collapse under high velocity gradients. Figure 9.26 presents the effect of linear and quadratic terms on a shock wave front in a one-dimensional problem.

In the equation of motion (9.73), the computational cost is mainly due to the calculation of internal force term ${}^t\mathbf{F}$ which must be performed numerically in the Gauss integration points. Reducing the number of Gauss points not only reduces the calculation cost of the integrals but also reduces the number of solutions required for the expensive material constitutive equations. For this purpose, the reduced integration method that is the most popular techniques in an explicit formulation is employed using only one

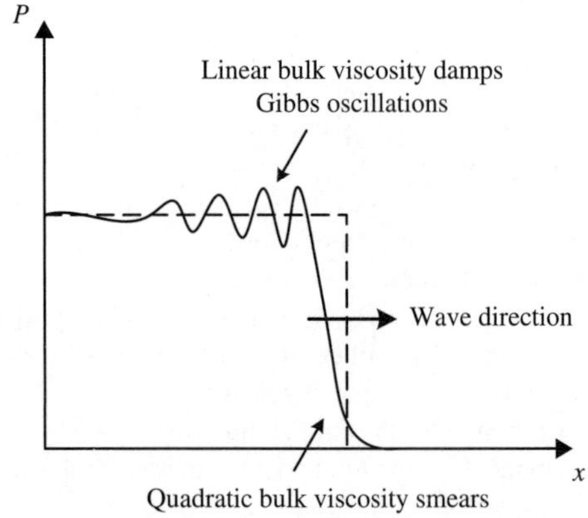

Figure 9.26 The role of linear and quadratic bulk viscosity in the propagation of a shock wave in a one-dimensional problem

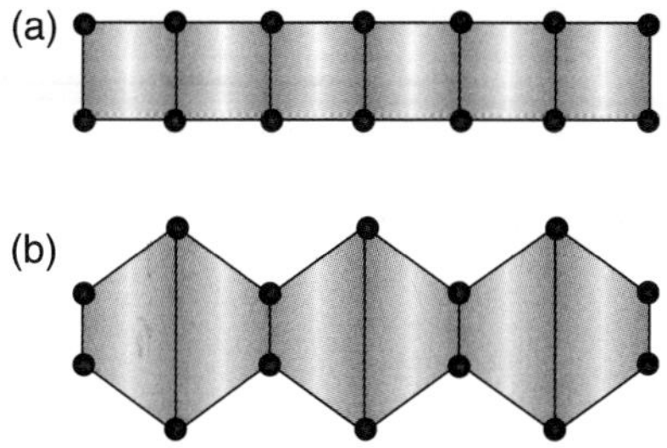

Figure 9.27 The hourglassing phenomenon: (a) the undeformed mesh; (b) the deformed mesh due to hourglassing phenomenon

integration point for a Q4 bilinear element. The method is also a remedy for the locking issue in lower-order elements of incompressible and nearly incompressible media. However, the reduced integration method results in the singular element stiffness matrix and extra zero energy modes called the hourglass modes. These zero energy modes do not affect the strain and stress fields but can arouse the displacements significantly (Figure 9.27). The concentrated forces and restrained boundary conditions are more susceptible to excite these modes. Various techniques have been proposed to control the hourglass phenomenon; among them are the artificial damping and artificial stiffness methods introduced by Flanagan and Belytschko (1981). In these two methods, the hourglass shape vector $\boldsymbol{\Upsilon}$ is first constructed as

$$\boldsymbol{\Upsilon}_I = \frac{1}{4A}\begin{bmatrix} x_2(y_3-y_4)+x_3(y_4-y_2)+x_4(y_2-y_3) \\ x_3(y_1-y_4)+x_4(y_3-y_1)+x_1(y_4-y_3) \\ x_4(y_1-y_2)+x_1(y_2-y_4)+x_2(y_4-y_1) \\ x_1(y_3-y_2)+x_2(y_1-y_3)+x_3(y_2-y_1) \end{bmatrix} \tag{9.97}$$

where x_i and y_i are the coordinates of node i. The subscript I represents x and y directions. The anti-hourglass forces $\mathbf{f}^{HG}$ are then computed as

$$\mathbf{f}_{iI}^{HG} = \frac{1}{2} Q_i \boldsymbol{\Upsilon}_I \tag{9.98}$$

in which for the viscose damping method, Q_i is defined as

$$\dot{Q}_i = \varepsilon t \sqrt{\frac{\rho(\lambda+2\mu)B_{jk}B_{jk}}{2}}\dot{q}_i \tag{9.99}$$

whereas for the stiffness method, Q_i is given by

$$\dot{Q}_i = \kappa t \frac{(\lambda+2\mu)B_{jk}B_{jk}}{2A}\dot{q}_i \tag{9.100}$$

where ε and κ are two constants that dependent on the problem, and $\dot{q}_i$ are the hourglass velocities defined as

$$\dot{q}_i = \frac{1}{2}\dot{u}_{iI}\boldsymbol{\Upsilon}_I \tag{9.101}$$

and

$$B_{iI} = \frac{1}{2}\begin{bmatrix} y_2-y_4 & y_3-y_1 & y_4-y_2 & y_1-y_3 \\ x_4-x_2 & x_1-x_3 & x_2-x_4 & x_3-x_1 \end{bmatrix} \tag{9.102}$$

In comparison to the artificial stiffness method, the artificial damping method needs less storage space and does not affect the low frequency modes of the structure even with a great damping. However, the method does not suppress the hourglassing issue completely and the permanent mesh distortion occurs; in addition the introduction of damping decreases the critical time step in an explicit dynamic method. The stiffness method with a small stiffness ratio ensures the mesh stability while it does not change the solution notably. In this study, the stiffness hourglass control method is only applied to the standard elements since the enriched elements are fully integrated.

9.12 Implementation of Dynamic X-FEM Damage-Plasticity Model

In order to perform the crack propagation simulation under dynamic and cyclic loadings, the dynamic X-FEM formulation described in the preceding sections is employed in conjunction with the damage-plasticity model. The technique is performed on the basis of damage state to provide a local crack growth criterion at the fracture process zone. The damage is calculated at the DCPs that are placed at the distant ℓ_c from the crack-tip at various angles of θ_{CP} in the range of $-90 \le \theta_{CP} \le 90$, as shown in Figure 9.12. The value of damage at each DCP is calculated using a weighted average value of damage at all neighboring Gauss points by

$$D_{CP} = \left(\sum_j \frac{1}{r_j^2} \right)^{-1} \left(\sum_j \frac{D_j}{r_j^2} \right) \tag{9.103}$$

where D_{CP} is the damage at the DCP and, D_j and r_j are the damage and the distance from a calculation point at a nearby Gauss point, respectively. The crack propagation simulation is then followed by adding a new crack segment when the crack initiation criterion is met. In this stage, the new crack segments are inactive; while the element sub-divisions are carried out and the data transfer process is performed. The next stage will be followed by updating the nodal enrichments and activating the new crack segments for further crack propagation simulation, as shown in Figure 9.28. Since the crack tip enrichment is not taken into account, it is assumed that the crack tip intersects the edge of the last element. Due to

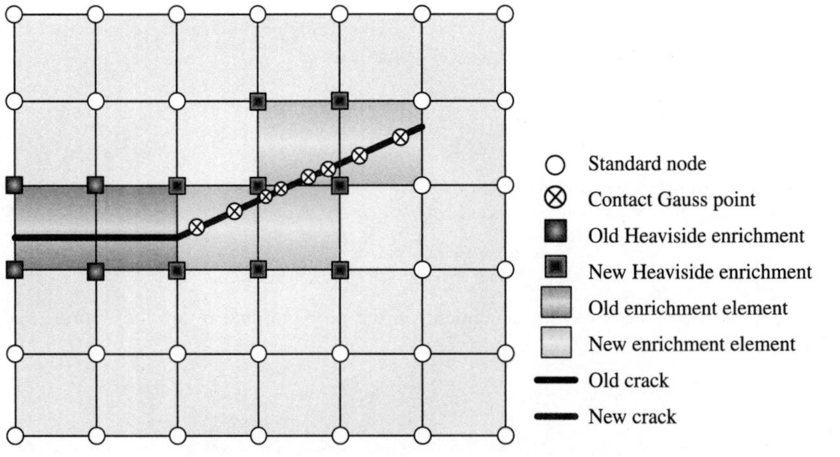

Figure 9.28 Crack propagation strategy: The nodes of elements cut by the new crack segment are enriched using the Heaviside enrichment, and for an element cut by the crack tip, the nodes of an element edge intersected by the crack tip are not enriched. The contact Gauss points are added to the crack interface at each sub-segment

discontinuous displacement field, the new cracked elements need a special numerical integration rule. The rectangular subdivision strategy is employed here by partitioning the parent element into uniform structured sub-rectangles, where the Gauss points of each sub-rectangle are used for numerical integration. Since in the produced enriched elements, the element subdivisions and hence the positions of Gauss points are updated, the variables of old Gauss points are transferred into new Gauss points. The data transfer process is performed in a similar manner to DCP damage values. For each new enriched element, all neighboring elements including the enriched element are selected in a set called $\Im$ (Figure 9.12). The transferred data at each new Gauss point is obtained using a weighted average of the data in the old Gauss points of the set $\Im$ as

$$\widehat{\boldsymbol{A}} = \left(\sum_j \frac{1}{r_j^2}\right)^{-1} \left(\sum_j \frac{\boldsymbol{A}_j}{r_j^2}\right) \qquad j \in \Im \tag{9.104}$$

where $\widehat{\boldsymbol{A}}$ is the transferred data into the new Gauss point, $\boldsymbol{A}_j$ is the data in the old Gauss point j, and r_j is the distance between the new Gauss point and the old Gauss point j. It must be noted that the transfer operator employed in Eq. (9.104) does not necessarily retain the nonlinear relationship that exists among the original state variables, that is, stresses, strains and damage parameter, in the transferred state variables. Hence, the equation of motion and constitutive equations cannot be satisfied and the inconsistency can be caused in the solution. In order to overcome the inconsistency problem a reduced set of variables are used instead of all state variables, and the other variables are obtained by satisfying the equation of motion and constitutive equations. In this study, the Cauchy stress tensor $\boldsymbol{\sigma}$, the back stress tensor $\boldsymbol{\chi}$, the damage parameter D, and the hardening parameter r are transferred from the old Gauss points to the new ones. Although this strategy reduces the inconsistency of the solution, the transferred stress tensors, which were in the plastic state before data transferring, may not be on the corresponding yield surface, that is, $\mathcal{E}_P > 0$. In addition, the transferred data may not satisfy the equation of motion of the structure.

The equation of motion of the system can be restored by using an artificial *null step*, in which no new loading is employed to the structure, and the system is allowed to obtain an equilibrium in one single time step Δt. Since the null step only eliminates the unwanted unbalances produced by the data transfer process, which are not the result of physical deformation, an elastic constitutive behavior is assumed at this step. The problem of inconsistency between the yield surface and the stress values obtained from the null step can be resolved by scaling the yield stress $\sigma_{y0} + R(r)$ to the equivalent effective stress $\bar{\sigma}_{eq} = \sqrt{\frac{3}{2}}\left\|\boldsymbol{\sigma}_{dev}/(1-D) - \boldsymbol{\chi}\right\|$. In fact, the new value of hardening parameter r_{new} can be obtained using the solution of nonlinear equation of $\bar{\sigma}_{eq} = \sigma_{y0} + R(r_{new})$. It must be noted that in the static analysis of ductile fracture problem, a *relaxation* phase is generally employed in the new boundary condition of crack propagation simulation, as described in Section 9.8.3, in which the extra tractions at the new crack surface are removed with a few relaxation increments that were in a self-equilibrium condition at the old crack tip. However, this step is not necessary in a dynamic simulation since the work produced by these extra tractions is automatically transferred into the kinematic energy at the elements of the crack tip region. In physical reality, a portion of this work is transferred into heat and sound that are neglected here.

In order to numerically integrate the crack contact forces, the contact Gauss points are employed along the crack segment with two Gauss points at each sub-segment, as shown in Figure 9.28. An important issue in crack propagation simulation is to update the enrichments of an element cut by the new crack segment. If a new crack segment is added, the Heaviside enrichment function is applied at all nodes of an element cut by the crack interface. The only exception is the element that contains the crack tip, in which the nodes of an element edge intersected by the crack tip are not enriched (Figure 9.28). This enrichment strategy introduces a singularity behavior at the crack-tip region. In fact, if all nodes of this element are enriched the two sides of the cracked element move like two rigid body objects and no stress concentration occurs. In such case, the enriched DOF corresponding to the nodes of an element edge intersected by the crack tip of the new crack segment are set to zero.

Finally, in order to illustrate the robustness of proposed dynamic X-FEM damage-plasticity algorithm, several examples are analyzed numerically. For all numerical examples, the bilinear 2D elements with a reduced integration rule are employed; the enriched elements are subdivided into 3 × 3 sub-rectangles with four Gauss points at each sub-rectangle. The hourglassing stiffness factor κ is set to 0.01; the linear and quadratic numerical damping factors are used as 0.06 and 1.2, respectively. The penalty parameters are chosen as $\kappa_N = \kappa_T = 100E$ and the friction coefficient is set to $\mu = 0.1$.

9.12.1 A Plate with an Inclined Crack

The first example is chosen to illustrate the performance of dynamic large X-FEM deformation with frictional contact behavior. The problem is a plate of 1 × 1 m^2 with a stationary 45° inclined crack at its center. The geometry, boundary conditions, and the X-FEM mesh of the problem are shown in Figure 9.29. Since the deformation of the plate is large, a special mesh configuration is employed to avoid excessive element distortions. The plate is in plane strain condition with the elastic material properties given as $\rho = 7800$ kg/m^3, $E = 113$ GPa, and $\nu = 0.3$. The plate is subjected to a sinusoidal vertical displacement with amplitude of 0.1 m and frequency of 100 Hz. This problem was solved by Liu and Borja (2008) under a vertical ramp loading in the static condition. In Figure 9.30, the distributions of vertical and

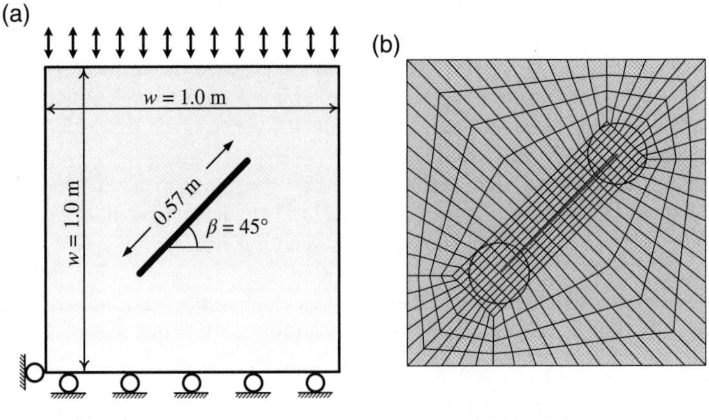

Figure 9.29 A plate with a 45° inclined crack at its center: (a) the geometry and boundary conditions; (b) the X-FEM mesh

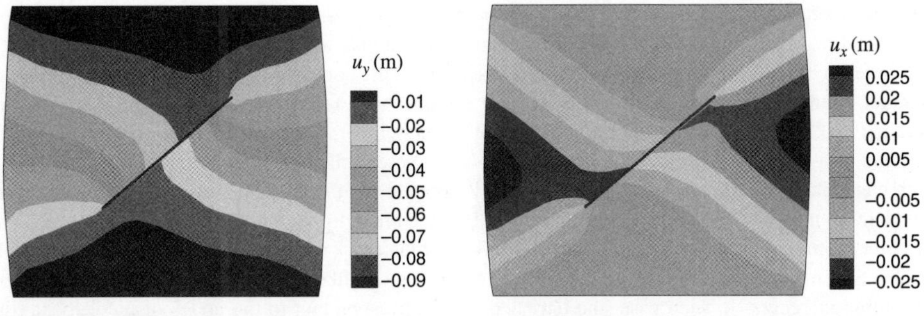

Figure 9.30 The vertical and horizontal displacement contours for a plate with an inclined crack at $u = -0.1$ m

horizontal displacement contours are shown for the dynamic analysis at $u = -0.1$ m that can be compared with those reported in Chapter 6 using a static analysis. The contours present a good agreement between the static and dynamic analyses. In Figure 9.31, the variations of normal traction force with the vertical displacement are plotted for two computational models. Obviously, the contact between two crack edges in compression increases the stiffness of the plate. In Figures 9.32a and b, the distributions of von-Mises stress contour are shown at the displacements of $u_y = \pm 0.1$ m when the contact is in opening mode. Also presented in Figure 9.32c is the von-Mises stress contour at the displacement of $u_y = -0.1$ m when the contact is in closing mode. It is interesting to highlight that when the crack is in opening mode with no contact forces, the mode I loading condition is dominant, however, when the crack is in closing mode with sliding behavior, the mode II loading condition is dominant. Moreover, the crack edges open extensively at the displacement of $u_y = +0.1$ m where a large deformation can be observed, however, an excessive mesh distortion occurs at the displacement of $u_y = -0.1$ m. The main issue in closing mode with an excessive mesh distortion is that the value of Jacobian becomes almost 0 and the value of critical time step reduces.

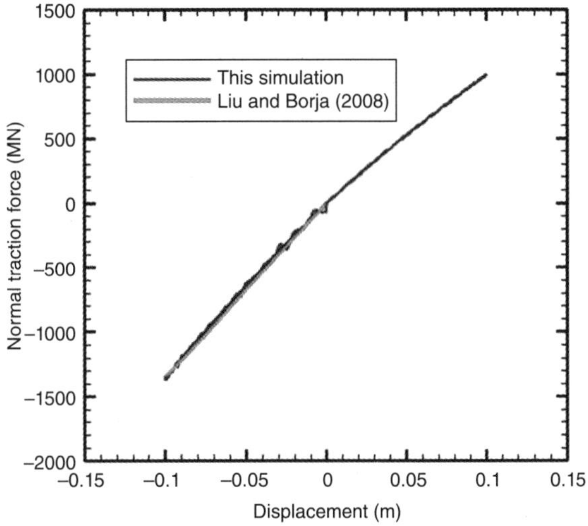

Figure 9.31 The variation of normal traction force with the vertical displacement: A comparison between the current numerical simulation and that reported by Liu and Borja (2008)

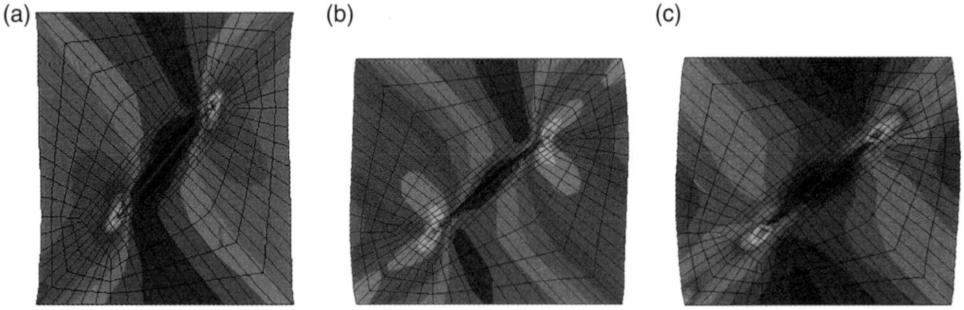

Figure 9.32 The von-Mises stress contours (MPa) at the displacement of: (a) $u = +0.1$ m in opening mode; (b) $u = -0.1$ m in opening mode; (c) $u = -0.1$ m in closing mode (For color details, please see color plate section)

9.12.2 The Low Cycle Fatigue Test

The second example is chosen to validate the accuracy of proposed damage-plasticity model for the crack growth simulation in a benchmark problem of the low cycle fatigue test. The test was performed both experimentally and numerically by Khan *et al.* (2010) on the 2024-T351 aluminum alloy. The alloy is widely used in aircraft manufacturing and presents a minimum ductility at room temperature. The material properties of aluminum alloy are given in Table 9.5. In the first part of this example, the numerical simulation of low cycle fatigue experimental test is performed on a damage low cycle specimen (DLC) and the results are compared with those of experiments. The experimental test presents the behavior of material under uniaxial stress state. A symmetric, cyclic displacement controlled loading is applied to the specimen. The numerical simulation is performed on a single axisymmetric $1 \times 1\ \text{mm}^2$ element under a triangular wave displacement with the compressive to tensile load ratio of $R = -1$, the amplitude of 0.01 m and the frequency of 100 Hz. The ratio of kinematic energy to total energy is less than 1 % and the loading is considered as a quasi-static condition. In Figure 9.33a, the numerical cyclic stress–strain diagram is plotted, where the crack closure effect can be observed. In this figure, the effect of damage on the tensile behavior of the specimen is obvious. In Figure 9.33b, the variations of peak stress with the number of cyclic loading are plotted for the current numerical simulation and those of experimental and numerical results reported by Khan *et al.* (2010). This figure clearly illustrates that the numerical simulation is in a good agreement with the experimental data. It is worth noting that since the von-Mises yield criterion is employed here, the difference between the tension and compression behavior cannot be captured.

In the second part of this example, the low cycle fatigue test is performed on a round notched bar (RNB), as shown in Figure 9.34. The geometry of the specimen and corresponding X-FEM mesh are depicted in this figure; the notch is circumferential with the radius of 2 mm and the length of specimen is 35 mm. The loading is applied using a triangular wave displacement controlled algorithm with the amplitude of 0.1 mm and the frequency of 200 Hz. The specimen is modeled using axisymmetric X-FEM elements. Due to a lack of accuracy of the material parameter calibration, the crack growth simulation of a RNB is carried out using the damage parameters of $w_D = 0.5$ and $r = 0.5$, and the crack segment growth is set to 0.055 mm. In Figure 9.35a, the cyclic force-displacement diagram is plotted for a RNB. Obviously the numerical simulation captures the overall behavior of the specimen both in tension and compression; once again, the effect of damage crack closure is obvious in this figure. In Figure 9.35b, the variations of normal stress with the time are plotted for the X-FEM analysis and those of experiments reported by Khan *et al.* (2010); it shows a good agreement between the numerical and experimental results. Note that the jumps observed in the predicted diagram are due to crack propagation steps.

Table 9.5 Material properties of the Al-2024 alloy

Property		Value
Dynamic	ρ (kg/m^3)	2780
Elasticity	E(GPa)	70
	ν	0.3
Plasticity	σ_y (MPa)	284
	σ_∞ (MPa)	434
	ξ	4
	a(MPa)	11300
	b	80
Damage	m	3.9
	A	0.0113
	w_D(MJ/m^3)	0.897
	r (MJ/m^3)	1.3
	S	1

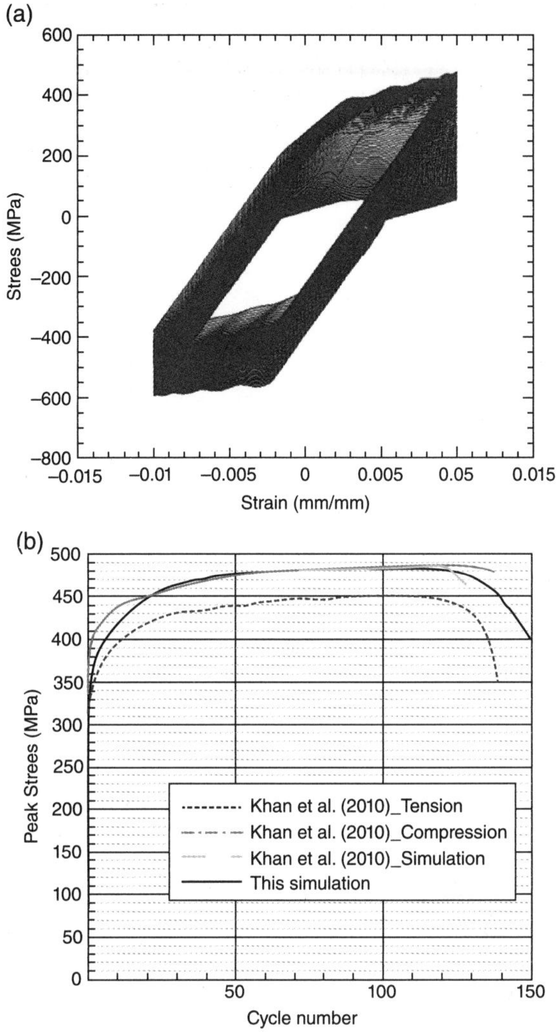

Figure 9.33 The low cycle fatigue test: (a) the numerical simulation of cyclic stress–strain diagram; (b) the variation of peak stress with the number cycle. A comparison between the current numerical simulation and those reported by Khan *et al.* (2010)

Moreover, the simulation is performed here with a higher frequency than the experiment, so the time axis was scaled to that of the experiment. In Figure 9.36, the distributions of damage contours are shown at different stages of crack propagation. Clearly, the crack initiates from the notch root and propagates towards the center of the specimen which coincides with experimental observations.

9.12.3 The Cyclic CT Test

In the next example, the cyclic behavior of TP304 stainless steel is investigated for the CT test under mode I loading regime. The material is widely used in piping systems, and is chosen here as a

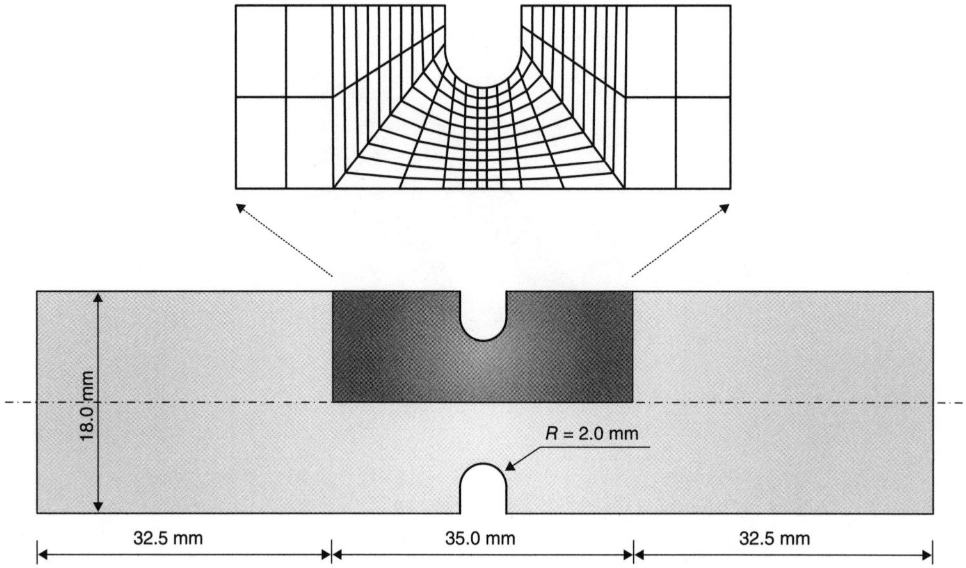

Figure 9.34 The low cycle fatigue test on a RNB: (a) the geometry of a RNB with a circumferential notch of radius 2 mm; (b) the X-FEM mesh

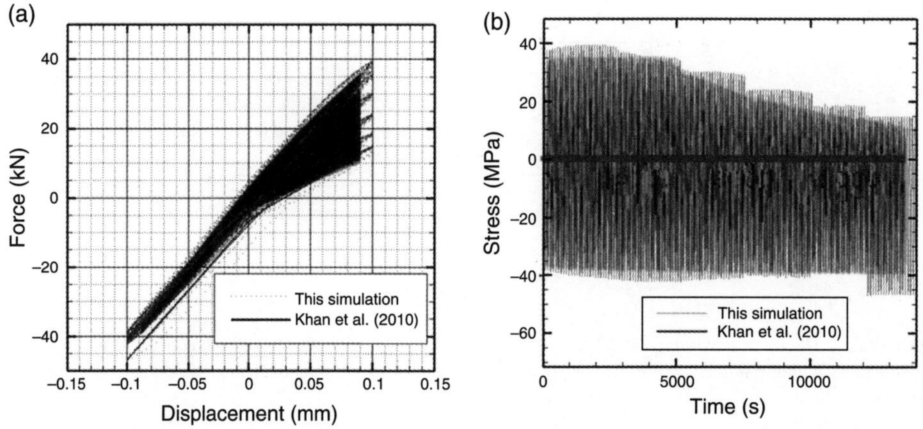

Figure 9.35 A comparison between the current numerical simulation and those of experiments reported by Khan *et al.* (2010) for a RNB specimen: (a) the cyclic force-displacement diagram; (b) the variation of normal stress with the time

benchmark problem. Rudland *et al.* (1996) performed extensive experiments on steel pipes in nuclear industries to investigate the material behavior under sever loading and high temperature conditions. The material properties of TP304 steel at the temperature of 288 °C are given in Table 9.6. The rate-independent elasto-plastic parameters are calibrated by available monotonic experimental data, and the damage parameters are calibrated through trial and error. The geometry and boundary conditions of the CT test are presented in Figure 9.37. In the experimental test, the loading is applied as a

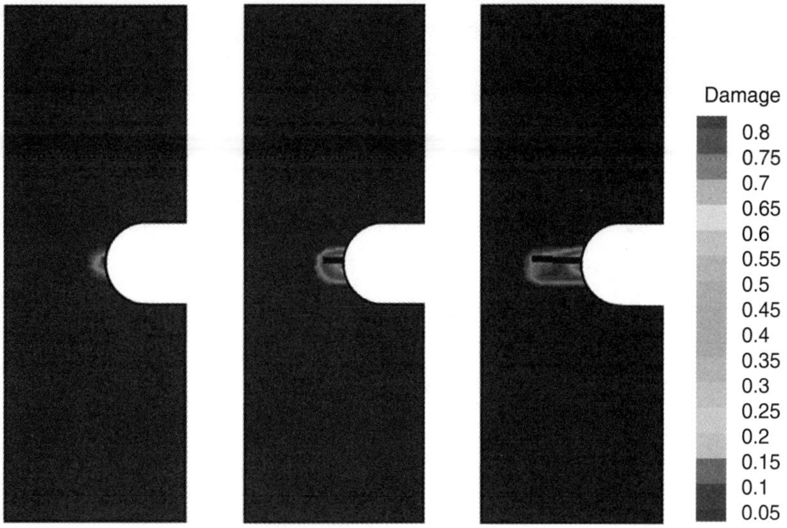

Figure 9.36 The distributions of damage contour for a RNB specimen at $t = 0.14$, 0.3, and 0.7 s

Table 9.6 Material properties of TP304 steel at 288 °C

Property		Value
Dynamic	ρ (kg/m^3)	7800
Elasticity	E(GPa)	60
	ν	0.32
Plasticity	σ_y (MPa)	138
	σ_∞ (MPa)	406
	ξ	8
	a(MPa)	4000
	b	100
Damage	m	—
	A	—
	w_D(MJ/m^3)	0.0
	r (MJ/m^3)	1.7
	S	1

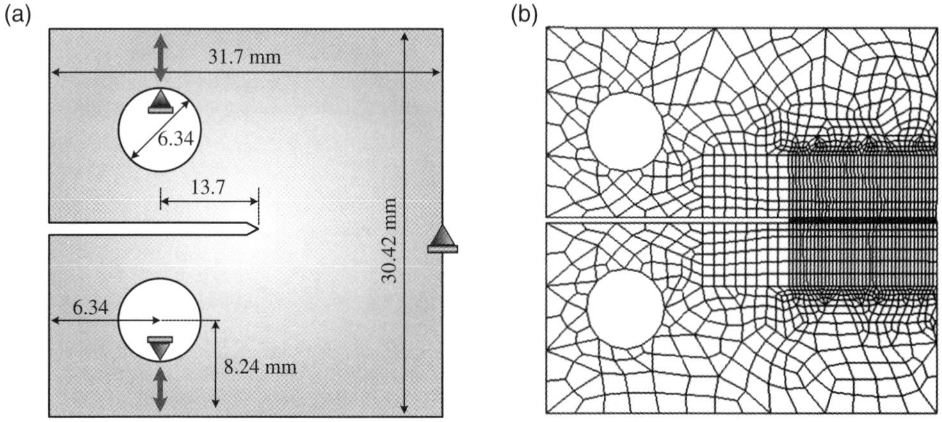

Figure 9.37 The CT test under the mode I loading condition: (a) the geometry and boundary conditions; (b) the X-FEM mesh

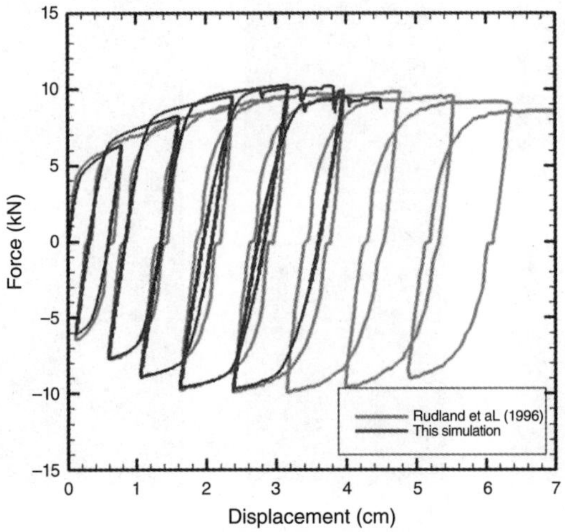

Figure 9.38 The cyclic force-displacement diagram for the CT test: A comparison between the current numerical simulation and that reported by Rudland *et al.* (1996)

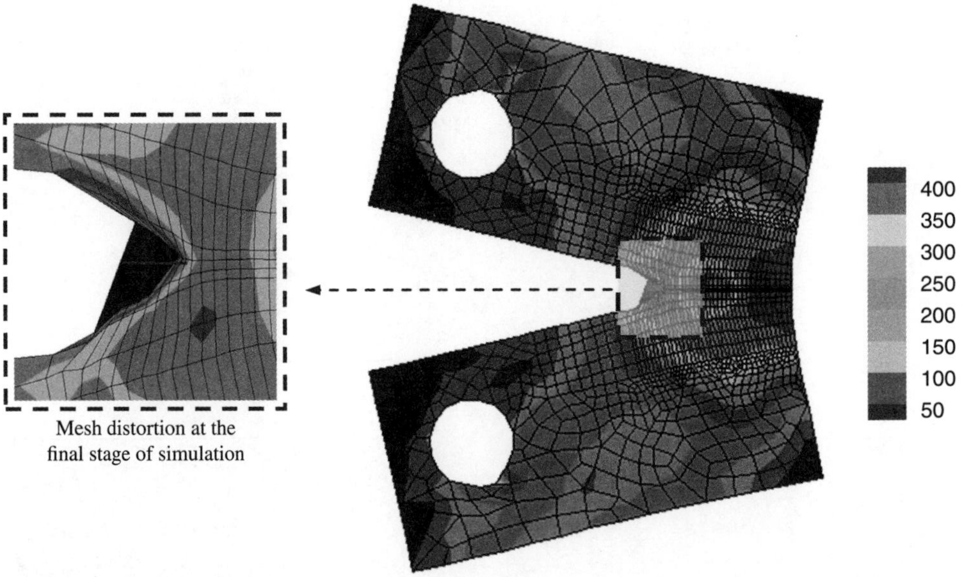

Figure 9.39 The von-Mises stress contour (MPa) for the CT test at the final stage of simulation: An excessive mesh distortion can be observed at the zoomed crack tip region (For color details, please see color plate section)

displacement control at two rigid pins placed in the two holes of the plate. In the numerical simulation, the pins are not explicitly modeled and the loading is applied by restraining the nodal points at the contact surface between the pins and the plate. A special loading strategy is applied in which at each cycle the specimen is first loaded up to $n \times 0.8$ mm, where n is the number of cycles; the corresponding maximum tensile reaction force is then determined; and the specimen is subsequently loaded in the reverse direction until the compressive reaction force becomes equal to the maximum tensile force. In Figure 9.38, a comparison of the cyclic force-displacement diagram is performed between the current numerical simulation and that obtained from the experimental test. Obviously there is a good agreement between the numerical and experimental results. Note that the jumps in the numerical simulation curve are due to crack propagation steps. Since the problem is highly nonlinear and the deformations are large, an excessive mesh distortion prevents the numerical simulation proceeds further. In order to continue the numerical simulation a remeshing step needs to be performed. In Figure 9.39, the contour of von-Mises stress is shown at the final step of simulation. Obviously, the damage is localized in a highly nonlinear region with large strains that results in the mesh distortion. It must be noted that the technique is able to predict the complete process of crack initiation up to the final failure stage, however, an excessive mesh distortion may stop the numerical simulation. In fact, although the technique removes the need for remeshing in crack propagation simulation, it may need the mesh generation due to mesh distortion.

9.12.4 The Double Notched Specimen in Cyclic Loading

The last example is chosen to demonstrate the capability of the proposed computational algorithm in a complex geometry with mixed-mode crack propagation. The example is a plate with two asymmetrically placed notches on the top and bottom, as discussed in Section 9.9.3. The geometry and boundary conditions of the problem are shown in Figure 9.23. The plate is in the plane strain condition; it is made of lead free solder with material properties given in Table 9.7. In the first part of this example, the specimen is subjected to a prescribed monotonic displacement of 1 mm in the y-direction. In Figure 9.40, the distributions of von-Mises stress and damage contours are shown at different stages of loading. As can be seen from this figure after a few load steps, a localization zone appears in the specimen that finally leads to the crack initiation from the top notch; it is then followed by emanating the crack from the bottom notch. The two cracks grow in the direction of the localization zone until complete failure occurs. A similar failure mechanism was observed in Section 9.9.3, in which different material properties were employed using a static analysis. In the second part of this example, the specimen is subjected to cyclic loading condition; the time history of the prescribed cyclic displacement u_y is shown in Figure 9.41a. In Figure 9.41b, the

Table 9.7 Material properties of lead free solder

Property		Value
Dynamic	ρ (kg/m^3)	7800
Elasticity	E(GPa)	30
	ν	0.36
Plasticity	σ_y (MPa)	40
	σ_∞ (MPa)	60
	ξ	8
	a(MPa)	17000
	b	2000
Damage	m	—
	A	—
	w_D(MJ/m^3)	0.0
	r (MJ/m^3)	5.6
	S	1

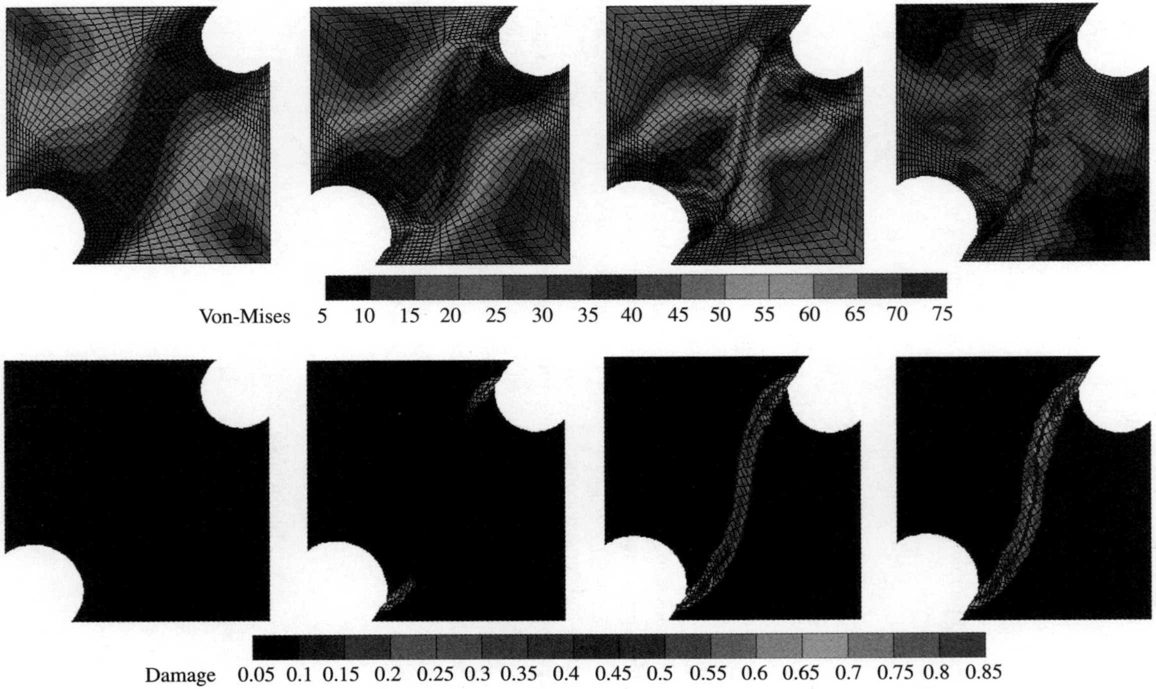

Figure 9.40 A double notched specimen under monotonic loading: The distributions of von-Mises stress (MPa) and damage contours at different stages of crack propagation t = 0.02, 0.04, 0.076, and 0.097 s (For color details, please see color plate section)

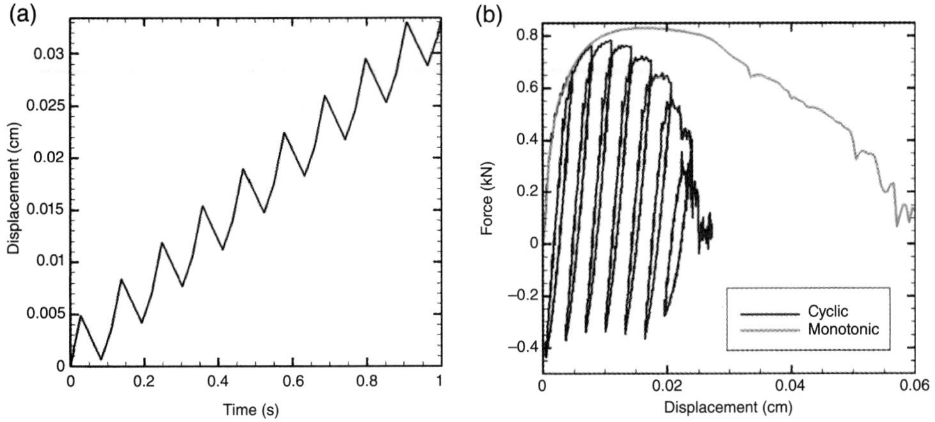

Figure 9.41 A double notched specimen: (a) the prescribed cyclic displacement time history; (b) the force-displacement diagram. A comparison between the monotonic and cyclic loadings

force-displacement diagrams are shown for monotonic and cyclic loadings. As can be observed from this figure, cyclic loading reduces the ductility of the specimen after the second cycle that results in early failure of the specimen.

10

X-FEM Modeling of Saturated/ Semi-Saturated Porous Media

10.1 Introduction

Porous media are a composition of particles of different sizes and shapes constituting the so-called solid skeleton, together with at least one fluid medium filling the remaining pores. The mechanical behavior of the fully or partially saturated porous medium is generally governed by the interaction of the solid skeleton with the pore fluid (Figure 10.1). Thereby, coupled problems of fluid flow and deformation of the solid skeleton are typically encountered if the consolidation of a soil under a constant applied load is considered. Except in some special loading conditions, the complete prediction of behavior of the solid material deformation interacting with a fluid flow is achieved by a complete solution of coupled governing equations of porous media. This interaction is much important in dynamic loading with low permeability values, such as earthquake loading of the dams and engineering soil structures. Basically, different kinematical and mechanical variables are considered for each phase of porous media, in which the strain and stress components are decomposed for each phase. In fact, porous media are described as multiphase media with separate velocities and stresses for each phase; so the stress acting on the solid skeleton is usually referred to as effective stress, and the hydrostatic stress acting on the fluid phase is denoted by the excess pore fluid pressure.

The flow problem in deforming porous media has been a topic of interest to research workers for a considerable time. A complete review of the history of development of porous media theories was reported by de Boer (1996). In order to present an introductory review of developed theories, the theory of Terzaghi can be considered the first study in deformable porous media, which has led to the one-dimensional consolidation theory on the basis of effective stress concept (Terzaghi, 1943). This theory was extended to three-dimensional cases by Biot (1941, 1956, 1962). Modern mixture theories were developed based on the concept of volume fractions by Morland (1972), Goodman and Cowin (1979), Sampaio and Williams (1979), and Bowen (1980, 1982). Averaging theories were developed by Whitaker (1977) and Hassanizadeh and Gary (1979a, b, 1980). The Biot theory was extended to three-phase conditions with pore air as the third phase by Fredlund and Morgenstern (1977) and Chang and Duncan (1983). The theory was developed by De Boer and Kowalski (1983) on material nonlinearity behavior of the soil skeleton in the Terzaghi–Biot framework. A generalized incremental form that includes large deformation and nonlinear material behavior was derived by Zienkiewicz *et al.* (1990a) for liquefaction analysis of soil structures. A simple extension on two-phase formulation of porous media considering air pressure to be constant and equal to atmospheric pressure was proposed by Zienkiewicz *et al.* (1990b). Coupled formulations that involve the air and water phases in soils were proposed by Alonso, Gens, and Josa (1990), Schrefler and Zhan

Extended Finite Element Method: Theory and Applications, First Edition. Amir R. Khoei.

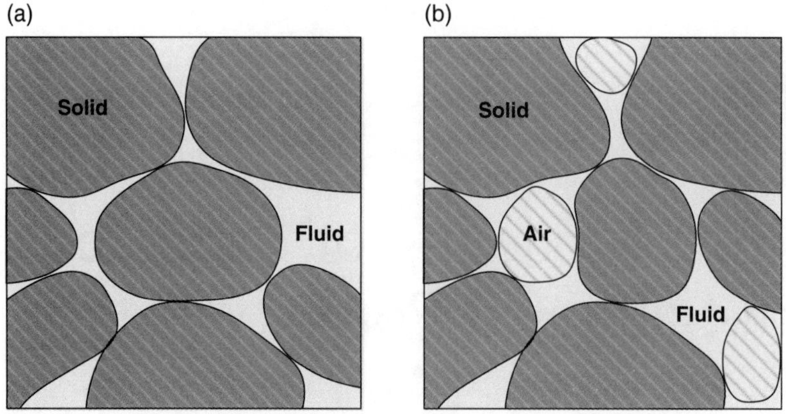

Figure 10.1 A schematic view of the interaction of the solid skeleton with the pore fluids in a deformable porous medium: (a) A saturated porous medium with one fluid in pores of granular solid (water); (b) A semi-saturated porous medium with two fluids in pores of granular solid (water and air)

(1993), and Gawin, Baggio, and Schrefler (1995). There are several recent mathematical models proposed in the literature for fully and partially saturated porous materials; they are generally based on typical simplifying assumptions, such as the rigid soil skeleton with no solid deformation (Wu and Forsyth, 2001), the static gas phase with null gas flow and uniform gas pressure equal to the atmospheric pressure (Sheng *et al.*, 2003), the omission of phase transitions (Schrefler and Scotta, 2001; Oettl, Stark, and Hofstetter, 2004), and the quasi-static condition (Stelzer and Hofstetter, 2005). A detailed literature review on porous media can be found in Lewis and Schrefler (1998) and Zienkiewicz *et al.* (1999).

In spite of the importance of the subject, flow in *damaged* porous media has received little attention. The presence of damage, such as cracks, faults, shear bands, can noticeably change the physical behavior of porous medium. In fact, the physics of the flow within such discontinuities is different from the flow of the interstitial fluid in the deforming bulk material. These differences in physics affect the flow pattern and the deformations in the vicinity of the discontinuity; the local differences in flow characteristics even influence the flow and deformations in the entire body of interest, as the porosity on this scale is altered. The problem of modeling fluid-driven fractures has been addressed with the pioneering work of Khristianovic and Zheltov (1955) and Geertsma and de Klerk (1969), who developed the so-called KGD model for plane strain hydraulic fracture. Early research work provided analytical solutions for the hydraulic fracturing problem (Geertsma and Haafkens, 1979; Spence and Sharp, 1985; Garagash and Detournay, 2000; Detournay, 2004; Lecampion and Detournay, 2007). These solutions usually suffer from the limitations of the analytical models. There are also a number of finite element (FE) approaches in the literature that have been proposed for the simulation of fracture in porous media; Boone and Ingraffea (1990) presented a numerical procedure for the simulation of hydraulically-driven fractures in poroelastic materials combining the finite element method (FEM) with the finite difference method, Simoni and Secchi (2003) and Secchi, Simoni, and Schrefler (2007) modeled the hydraulic cohesive crack growth using the FEM with mesh adaptation, Segura and Carol (2008a) proposed a hydro-mechanical formulation for geomaterials with pre-existing discontinuities based on the FEM with zero-thickness interface elements, Lobao *et al.* (2010) presented the FE modeling of hydro-fracture flow in porous media using zero-thickness interface elements, Khoei, Barani, and Mofid (2011) and Barani, Khoei, and Mofid (2011) performed the FE analysis of fracture propagation using the cohesive interface elements, Sarris and Papanastasiou (2012) modeled hydraulic fracturing in poroelastic media using the FEM with cohesive elements, and Carrier and Granet (2012) developed a zero-thickness cohesive element for the simulation of hydraulic fracturing in poroelastic media.

In classical FEM, the crack path is restricted to the inter-element boundaries, which usually encounters the problem of mesh dependency (Camacho and Ortiz, 1996; Ortiz and Pandolfi, 1999). To overcome the sensitivity to the mesh generated and to avoid the preferred directions when extending the crack, mesh adaptivity is required (Khoei, Azadi, and Moslemi, 2008c; Khoei *et al.*, 2009b; Moslemi and Khoei, 2009; Azadi and Khoei, 2011), which makes crack growth simulation a computationally expensive and cumbersome process. The difficulties confronted in the conventional FEM are handled by locally enriching the conventional FE approximation with an additional function through the concept of the partition of unity (PU), which was introduced in the pioneering work of Melenk and Babuska (1996). This idea was exploited to set up the framework of the extended FEM (X-FEM) by Dolbow (1999), Belytschko and Black (1999), and Moës, Dolbow, and Belytschko (1999). By appropriately selecting the enrichment function and enriching specific nodal points, the enriched approximation would be capable of directly capturing the local property in the solution. The X-FEM is used as a powerful and accurate approach to model discontinuities, which are independent of the FE mesh topology. The technique has been successfully applied to problems exhibiting discontinuities and inhomogeneities, such as cracks, holes, or material interfaces. The method allows modeling of the entire crack geometry independently of the mesh, and completely avoids the need to regenerate the mesh as the crack grows. The X-FEM was used to model the crack growth and arbitrary discontinuities by enriching the discontinuous approximation in terms of a signed distance and level sets functions (Daux *et al.*, 2000; Belytschko *et al.*, 2001; Sukumar *et al.*, 2001). The technique was applied in linear crack problems, including: the crack growth with frictional contact by Dolbow, Moës, and Belytschko (2001), cohesive crack propagation by Moës and Belytschko (2002a), stationary and growing cracks by Ventura, Budyn, and Belytschko (2003), and three-dimensional crack propagation by Areias and Belytschko (2005a). The other application area of the X-FEM includes the fluid flow in porous media that has drawn considerable attention; the shear band evolution in fluid saturated porous media was modeled by Réthoré, de Borst, and Abellan (2007a), the fluid flow in fractured fully saturated systems and in fracturing unsaturated systems with passive gas phase was treated by Réthoré, de Borst, and Abellan (2007b, 2008), the strong discontinuities in partially saturated porous media with passive gas phase was considered by Callari, Armero, and Abati (2010), the fully and partially saturated porous media with material interfaces were modeled by Khoei and Haghighat (2011) and Mohamadnejad and Khoei (2013a), and the hydro-mechanical modeling of cohesive crack propagation in saturated and multi-phase porous media was carried out by Mohamadnejad and Khoei (2013b, c).

The aim of this chapter is to model deformable porous media with weak and strong discontinuities using an enriched FE model based on the X-FEM technique. The mass balance equation of the fluid phase is applied together with the momentum balance of bulk and fluid phases to obtain the fully coupled set of equations in the framework of u–p formulation. The X-FEM is applied to the governing equations of porous media in an updated Lagrangian framework, followed by a generalized Newmark scheme used for the time domain discretization. For the numerical solution, the unconditionally stable direct time-stepping procedure is applied to resolve the resulting system of strongly coupled nonlinear algebraic equations using the Newton–Raphson iterative algorithm. The weak and strong discontinuities are modeled by additional degrees of freedom (DOF) for both the solid and fluid phases. The X-FEM is used in modeling weak and strong discontinuities, in which the process is accomplished by partitioning the elements cut by the interface for integration of the domain of elements. Several numerical examples are presented to demonstrate the capability and efficiency of the X-FEM in the fully coupled simulation of weak and strong discontinuities in deformable porous media.

10.1.1 Governing Equations of Deformable Porous Media

Porous medium refers to material that consists of solid grains known as the solid skeleton, and at least one fluid phase, such as water or gas that flows through the solid grains. Basically, two strategies have been used to describe the porous media as a continuum body, one is the *mixture theory* integrated by the concept of volume fractions, and the other is the *averaging theory* that describes the porous media from a micromechanical point of view. In this study, the mixture theory of porous media, based on the Biot theory, is employed to describe the behavior of deformable porous media. In most applications, pores are filled partially with water and partially with moist air. One of the most important concepts in the mixture theory is the degree of saturation for non-solid phases in a representative elementary volume (REV). The degree of saturation of each fluid phase in a REV can be defined by the ratio of the volume of phase i, dV_i, to the total volume of pores dV_v as

$$S_i = \frac{dV_i}{dV_v} = dV_i \Big/ \sum_{j=1}^{NSP} dV_j \tag{10.1}$$

where NSP is the number of non-solid phases. It is obvious that for two-phase fluid flow $S_w + S_g = 1$, with w and g denoting the water and gas phases, respectively. In a fully saturated porous media with only one pore fluid, the degree of saturation of non-solid phase is defined as $S_w = 1$. In addition, the porosity of solid grains, or simply the porosity of media, is defined as the ratio of pore space to the total volume, that is, $n = dV_v/dV_s$. Thus, the density of total mixture can be defined for the soil-fluid domain as

$$\rho = n\rho_v + (1-n)\rho_s \tag{10.2}$$

where ρ_v is the density of fluid phase that fills the pores, and ρ_s is the density of solid grains. In a fully saturated media with water as the pore fluid phase, it results in $\rho = n\rho_w + (1-n)\rho_s$.

The stress at each point can be decomposed into the effective stress which acts between the solid grains and controls their loading capacity and deformation, and the average pore pressure of fluid phases that participates in the load bearing of the total soil sample. Hence, the total stress tensor can be defined as $\boldsymbol{\sigma} = \boldsymbol{\sigma}' - \alpha p\,\mathbf{I}$, with $\boldsymbol{\sigma}'$ denoting the effective stress acting between solid grains, and p is an average pressure of fluid phases. Parameter α is a corrective coefficient of pore pressure effect on solid grains, in which for isotropic materials can be computed by $\alpha = 1 - K_T/K_S$, where K_T and K_S are the bulk modulus of soil sample and solid grains, respectively, and for most soils are given as $\alpha \approx 1$. The average pore pressure p can be defined as

$$p = \chi_w p_w + \chi_g p_g \tag{10.3}$$

where χ_w and χ_g are coefficients that depend on the water and gas interfaces with solid grains and can be approximated by S_w and S_g. In a fully saturated media, the average pressure p is defined by the total pressure of fluid.

In order to derive the governing equations of porous media, the Biot theory is employed in an updated Lagrangian framework. Consider a two-phase media, in which the fluid phase is partially saturated under the condition of $p_g = 0$, and the convective terms are neglected. The motion of total mixture is defined by the motion of solid–fluid mixture $u_i(\boldsymbol{x}, t)$ and the relative motion of fluid with respect to the mixture $w_i(\boldsymbol{x}, t)$. The governing equations are the total solid–fluid mixture, the linear momentum balance, and the continuity equation for each phase. The linear momentum balance of solid–fluid mixture can be written by neglecting the relative acceleration of fluid phase with respect to the solid phase, that is, $\ddot{w}_i \approx 0$, as

$$\nabla \cdot \boldsymbol{\sigma} - \rho\,\ddot{\mathbf{u}} + \rho\,\mathbf{b} = 0 \tag{10.4}$$

where $\ddot{\mathbf{u}}$ is the acceleration vector of the solid phase, $\mathbf{b}$ is the body force vector, ρ is the average density of the mixture, and the symbol ∇ denotes the vector gradient operator.

The conservation of linear momentum for the fluid flow results in the generalized Darcy equation. Assuming negligibility of the relative acceleration term of the fluid, the momentum balance of the fluid phase, or the Darcy relation for the pore fluid flow through the porous medium surrounding the fracture can be written as

$$-\nabla p-\mathbf{R}-\rho_f\,\ddot{\mathbf{u}}+\rho_f\,\mathbf{b}=0 \tag{10.5}$$

where **R** denotes the viscous drag force defined by the Darcy seepage law as $\dot{\mathbf{w}}=\boldsymbol{k}_f\,\mathbf{R}$, with $\boldsymbol{k}_f$ denoting the permeability matrix of the porous medium with respect to the pore fluid in which its dimension is defined as $(length)^3(time)/(mass)$.

Finally, the continuity equation of fluid phase can be written as

$$\nabla\cdot\dot{\mathbf{w}}+\alpha\,\nabla\cdot\dot{\mathbf{u}}+\frac{1}{Q}\dot{p}=0 \tag{10.6}$$

where $1/Q\equiv(\alpha-n)/K_s+n/K_f$, in which K_s and K_f are the bulk moduli of the solid and fluid phases, respectively. Equations (10.4)–(10.6) are the coupled governing equations of porous saturated media with variable set (u_i, w_i, p). Performing some mathematical manipulations, the set of equations can be modified as $u-p$, the formulation of porous media is more appropriate for modeling the saturated porous material and is much more appropriate for loading conditions up to earthquake frequencies (Zienkiewicz *et al.*, 1999). Combining Eqs. (10.5) and (10.6) results in

$$\nabla\cdot\left[\boldsymbol{k}_f\left(-\nabla p-\rho_f\,\ddot{\mathbf{u}}+\rho_f\,\mathbf{b}\right)\right]+\alpha\,\nabla\cdot\dot{\mathbf{u}}+\frac{1}{Q}\dot{p}=0 \tag{10.7}$$

Equations (10.4) and (10.7) are complemented by appropriate kinematical relation and the stress-strain constitutive equation. The kinematic equation is defined by the incremental strain–displacement relationship, and the stress-strain constitutive equation is defined by $d\boldsymbol{\sigma}'=\mathbf{D}\,d\boldsymbol{\varepsilon}$, with **D** denoting the forth order tangential stiffness matrix computed using an appropriate constitutive law.

The governing equations (10.4) and (10.7) can be solved using the essential and natural boundary conditions. The solid-phase boundary conditions are $\mathbf{u}=\bar{\mathbf{u}}$ on Γ_u and $\mathbf{t}=\boldsymbol{\sigma}\cdot\mathbf{n}_\Gamma\equiv\bar{\mathbf{t}}$ on Γ_t, with $\mathbf{n}_\Gamma$ denoting the unit outward normal vector to the external boundary Γ, in which $\Gamma=\Gamma_t\cup\Gamma_u$ is the boundary of domain. The fluid-phase boundary conditions are $\mathbf{p}=\bar{\mathbf{p}}$ on Γ_p and $\dot{\mathbf{w}}\cdot\mathbf{n}_\Gamma=\bar{q}$ on Γ_w, with $\bar{q}$ denoting the prescribed outflow of the fluid imposed on Γ_w with $\Gamma=\Gamma_p\cup\Gamma_w$, as shown in Figure 10.2. Moreover, the internal boundary condition is defined as $\boldsymbol{\sigma}\cdot\mathbf{n}_{\Gamma_d}=\bar{\mathbf{t}}_d$ on the discontinuity Γ_d, in which $\mathbf{n}_{\Gamma_d}$ is the unit normal vector to Γ_d pointing to Ω^+ with $\mathbf{n}_{\Gamma_d}=\mathbf{n}_{\Gamma_d-}=-\mathbf{n}_{\Gamma_d+}$. In a domain with the weak discontinuity such as a material interface, $\bar{\mathbf{t}}_d$ is the traction transferred across the discontinuity Γ_d.

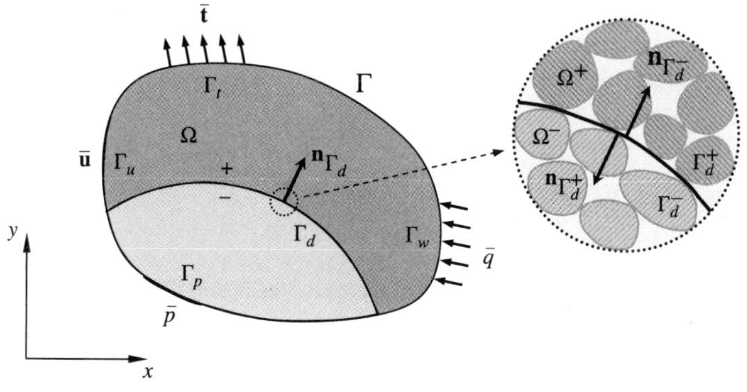

Figure 10.2 The boundary conditions of a saturated porous medium Ω with a material discontinuity Γ_d

10.2 The X-FEM Formulation of Deformable Porous Media with Weak Discontinuities

In order to numerically model a deformable porous media with a weak discontinuity, the X-FEM is employed by adding appropriate enrichment functions to the standard FE approximation. It must be noted that in the derivation of X-FEM formulation, it is implicitly assumed that the displacement and pressure fields are continuous over the domain of problem, however, necessary modifications must be applied in the variational formulation of X-FEM solution to model the discontinuous displacement and pressure fields across the discontinuity. In Chapter 2, the *Divergence theorem* is presented for discontinuous problems to provide a mathematical description for the variational formulation of the X-FEM solution, and an appropriate term is obtained to incorporate the discontinuity into the X-FEM formulation. Based on the *Divergence theorem*, the integration over the domain Ω can be converted to the integration over the boundary Γ for a discontinuous function $\mathbf{F}$ as

$$\int_\Omega \operatorname{div}\mathbf{F}\,d\Omega=\int_\Gamma \mathbf{F}\cdot\mathbf{n}_\Gamma\,d\Gamma-\int_{\Gamma_d}[\![\mathbf{F}]\!]\cdot\mathbf{n}_{\Gamma_d}\,d\Gamma \qquad (2.51-\text{repeated})$$

where $[\![\mathbf{F}]\!]=\mathbf{F}^+-\mathbf{F}^-$ represents the jump of function $\mathbf{F}$, with $\mathbf{F}^+$ and $\mathbf{F}^-$ denoting the values of $\mathbf{F}$ on Γ_d^+ and Γ_d^-, respectively.

In order to derive the weak form of governing equations (10.4) and (10.7), the trial functions $\mathbf{u}(\boldsymbol{x},t)$ and $p(\boldsymbol{x},t)$ are required to satisfy all essential boundary conditions and to be smooth enough to define the derivatives of equations. In addition, the test functions $\delta\mathbf{u}(\boldsymbol{x},t)$ and $\delta p(\boldsymbol{x},t)$ are required to be smooth enough to be vanished on the prescribed strong boundary conditions. To obtain the weak form of governing equations, the test functions $\delta\mathbf{u}(\boldsymbol{x},t)$ and $\delta p(\boldsymbol{x},t)$ are respectively multiplied by Eqs. (10.4) and (10.7) and integrated over the domain Ω as

$$\begin{aligned}&\int_\Omega \delta\mathbf{u}(\nabla\cdot\boldsymbol{\sigma}-\rho\,\ddot{\mathbf{u}}+\rho\,\mathbf{b})d\Omega=0\\&\int_\Omega \delta p\left(\nabla\cdot\left[\boldsymbol{k}_f\left(-\nabla p-\rho_f\,\ddot{\mathbf{u}}+\rho_f\,\mathbf{b}\right)\right]+\alpha\,\nabla\cdot\dot{\mathbf{u}}+\frac{1}{Q}\dot{p}\right)d\Omega=0\end{aligned} \qquad (10.8)$$

Expanding these integral equations and using the Divergence theorem (2.51), it leads to the following weak form of governing equations

$$\int_\Omega \nabla\delta\mathbf{u}:\boldsymbol{\sigma}\,d\Omega+\int_\Omega \delta\mathbf{u}\cdot\rho\,\ddot{\mathbf{u}}\,d\Omega+\int_{\Gamma_d}[\![\delta\mathbf{u}]\!]\boldsymbol{\sigma}\cdot\mathbf{n}_{\Gamma_d}d\Gamma-\int_\Omega \delta\mathbf{u}\cdot\rho\,\mathbf{b}\,d\Omega-\int_{\Gamma_t}\delta\mathbf{u}\cdot\bar{\mathbf{t}}\,d\Gamma=0 \qquad (10.9a)$$

$$\begin{aligned}&\int_\Omega \nabla\delta p\,\boldsymbol{k}_f\,\nabla p\,d\Omega+\int_\Omega \nabla\delta p\,\boldsymbol{k}_f\cdot\rho_f\,\ddot{\mathbf{u}}\,d\Omega-\int_{\Gamma_d}\delta p[\![\dot{\mathbf{w}}]\!]\cdot\mathbf{n}_{\Gamma_d}d\Gamma\\&+\int_\Omega \delta p\,\alpha\,\nabla\cdot\dot{\mathbf{u}}\,d\Omega+\int_\Omega \delta p\frac{1}{Q}\dot{p}\,d\Omega-\int_\Omega \nabla\delta p\,\boldsymbol{k}_f\cdot\rho_f\,\mathbf{b}\,d\Omega+\int_{\Gamma_w}\delta p(\dot{\bar{\mathbf{w}}}\cdot\mathbf{n}_\Gamma)d\Gamma=0\end{aligned} \qquad (10.9b)$$

It must be noted that because of the traction continuity across the material interface, the following integral on the weak discontinuity disappears in the weak form of the equilibrium equation (10.9a), that is,

$$\int_{\Gamma_d}[\![\delta\mathbf{u}]\!]\cdot\boldsymbol{\sigma}\cdot\mathbf{n}_{\Gamma_d}\,d\Gamma=\int_{\Gamma_d}[\![\delta\mathbf{u}]\!]\cdot\bar{\mathbf{t}}_d\,d\Gamma=\int_{\Gamma_d}(\delta\mathbf{u}^+-\delta\mathbf{u}^-)\cdot\bar{\mathbf{t}}_d\,d\Gamma=0 \qquad (10.10)$$

Similarly, on account of the pore fluid flow continuity across the material interface, the corresponding integral on the weak discontinuity disappears in the weak form of the flow continuity equation (10.9b), that is,

$$\int_{\Gamma_d} \delta p[\![\dot{\mathbf{w}}]\!]\cdot \mathbf{n}_{\Gamma_d} d\Gamma = \int_{\Gamma_d} \delta p(\dot{\mathbf{w}}^+ - \dot{\mathbf{w}}^-)\cdot \mathbf{n}_{\Gamma_d} d\Gamma = 0 \tag{10.11}$$

Thus, the weak form of governing equations (10.9) can be simplified to

$$\int_\Omega \nabla\delta\mathbf{u} : \boldsymbol{\sigma}\, d\Omega + \int_\Omega \delta\mathbf{u}\cdot\rho\,\ddot{\mathbf{u}}\, d\Omega - \int_\Omega \delta\mathbf{u}\cdot\rho\,\mathbf{b}\, d\Omega - \int_{\Gamma_t} \delta\mathbf{u}\cdot\bar{\mathbf{t}}\, d\Gamma = 0 \tag{10.12a}$$

$$\begin{aligned}\int_\Omega \nabla\delta p\, \boldsymbol{k}_f\, \nabla p\, d\Omega &+ \int_\Omega \nabla\delta p\, \boldsymbol{k}_f\cdot\rho_f\,\ddot{\mathbf{u}}\, d\Omega + \int_\Omega \delta p\, \alpha\, \nabla\cdot\dot{\mathbf{u}}\, d\Omega \\ &+ \int_\Omega \delta p \frac{1}{Q}\dot{p}\, d\Omega - \int_\Omega \nabla\delta p\, \boldsymbol{k}_f\cdot\rho_f\,\mathbf{b}\, d\Omega + \int_{\Gamma_w} \delta p(\dot{\mathbf{w}}\cdot\mathbf{n}_\Gamma)\, d\Gamma = 0\end{aligned} \tag{10.12b}$$

10.2.1 Approximation of Displacement and Pressure Fields

In order to solve the integral equations (10.12), the X-FEM is employed here to obtain the values of $\mathbf{u}(\boldsymbol{x}, t)$ and $p(\boldsymbol{x}, t)$. To account for the strain jump across the material interface, it is required that the displacement field be continuous, while its gradient be discontinuous on Γ_d. In order to provide the discontinuity in the displacement derivatives across the material interface, the displacement field is approximated by locally enriching the classical FE space. A key feature of this work is the introduction of an enrichment function that models the discontinuity in the derivatives of the solution in the classical FE approximation complemented by the addition of enriched DOF pertinent to the enrichment function. By choosing an appropriate enrichment, the local enrichment strategy allows for the incorporation of the strain discontinuity across the material interface into the solution of the problem. Various enrichment functions can be used to enhance the displacement approximation field. The level set method is generally used to model the material interfaces, and is proposed here to enhance the FEM approximation of displacement field for problems containing weak discontinuities, such as material interfaces.

In the X-FEM, the approximation is based upon an additive decomposition into a standard part and an enriched part. Hence, the approximation field is approximated as a linear combination of the standard and enriched shape functions. The enriched approximation of the X-FEM for the displacement field $\mathbf{u}(\boldsymbol{x}, t)$ with multiple material interfaces can be written as

$$\begin{aligned}\mathbf{u}(\boldsymbol{x},t) \approx \mathbf{u}^h(\boldsymbol{x},t) &= \sum_{I\in\mathcal{N}} N_{uI}(\boldsymbol{x})\,\bar{\mathbf{u}}_I(t) + \sum_{j\in\mathcal{M}^{dis}}\left(\sum_{J\in\mathcal{N}^{dis}} N_{uJ}(\boldsymbol{x})\,(\psi(\boldsymbol{x})-\psi(\boldsymbol{x}_J))\,\bar{\mathbf{d}}_J(t)\right) \\ &+ \sum_{k\in\mathcal{M}^{jun}}\left(\sum_{K\in\mathcal{N}^{jun}} N_{uK}(\boldsymbol{x})\,(J(\boldsymbol{x})-J(\boldsymbol{x}_K))\,\bar{\mathbf{b}}_K(t)\right)\end{aligned} \tag{10.13}$$

where $\mathcal{M}^{dis}$ is the number of material interfaces including the main interfaces and the secondary interfaces, and $\mathcal{M}^{jun}$ is the number of junctions, as shown in Figure 10.3. In relation (10.13), $\mathcal{N}$ is the set of all nodal points and $\bar{\mathbf{u}}_I(t)$ are the corresponding standard nodal displacements; $\mathcal{N}^{dis}$ is the subset of nodes to enrich for the $j^{\text{th}} \in \mathcal{M}^{dis}$ material interface and $\bar{\mathbf{d}}_J(t)$ are the corresponding additional

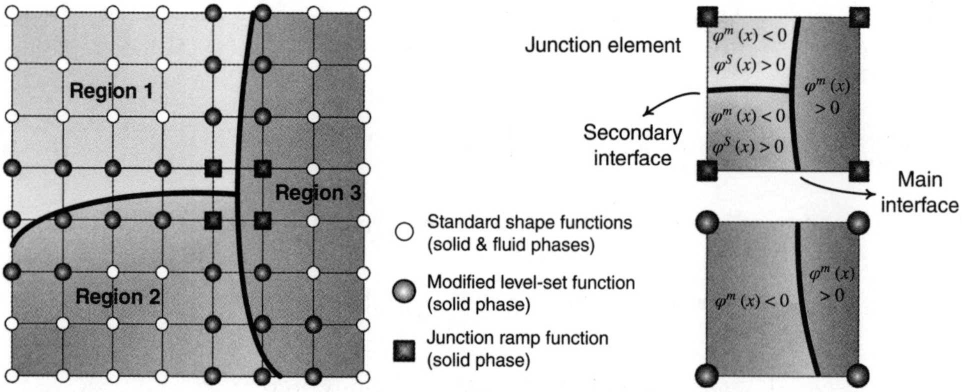

Figure 10.3 The enriched nodal points of the displacement field in porous media with different material regions

DOF; and $\mathcal{N}^{jun}$ is the subset of nodes to enrich for the $k^{\text{th}} \in \mathcal{M}^{jun}$ junction and $\bar{\mathbf{b}}_K$ are the corresponding additional DOF. In relation (10.13), $N_u(\boldsymbol{x})$ denotes the standard displacement shape functions, $\psi(\boldsymbol{x})$ is the enriched shape function associated with the modified level set function for elements cut by one material interface, and $J(\boldsymbol{x})$ is the enriched shape function associated with the junction *ramp* function for elements cut by two material interfaces.

The enriched shape function based on the modified level set function $\psi(\boldsymbol{x})$ is defined as

$$\psi(\boldsymbol{x}) = \psi_{\text{ridge}}(\varphi(\boldsymbol{x})) = \sum_{I \in \mathcal{N}^{dis}} N_{uI}(\boldsymbol{x})|\varphi_I| - \left| \sum_{I \in \mathcal{N}^{dis}} N_{uI}(\boldsymbol{x})\, \varphi_I \right| \tag{10.14}$$

where φ_I are the nodal values of the level set function $\varphi(\boldsymbol{x})$, which is defined on the basis of a signed distance function as

$$\varphi(\boldsymbol{x}) = \min \|\boldsymbol{x} - \boldsymbol{x}^*\| \operatorname{sign}((\boldsymbol{x} - \boldsymbol{x}^*) \cdot \mathbf{n}_{\Gamma_d}) \tag{10.15}$$

where $\boldsymbol{x}^*$ is a point on the discontinuity that is the closest to the point $\boldsymbol{x}$, and $\mathbf{n}_{\Gamma_d}$ is the normal vector to the discontinuity interface at point $\boldsymbol{x}^*$. It can be observed from the definition (10.14) that this enrichment function has a ridge centered on the interpolated interface and vanishes in elements not containing the material interface (Figure 10.4a). This enrichment function is used since the resulting enrichment function is zero in the blending elements, and the unwanted terms appearing in the approximating space of the blending elements are avoided. It must be noted that this enrichment is only required in the elements where the discontinuity on the gradient of displacement needs to be captured.

In order to enrich the element cut by two material interfaces, as depicted in Figure 10.3, the junction *ramp* function $J(\boldsymbol{x})$ is employed. Consider the main (major) interface is denoted by m and the secondary (minor) interface by S, where the approaching interface S is arrested by the main interface m. The junction *ramp* function $J(\boldsymbol{x})$ is defined as

$$J(\boldsymbol{x}) = J_{\text{ramp}}(\varphi(\boldsymbol{x})) = \begin{cases} |\varphi^m(\boldsymbol{x})||\varphi^S(\boldsymbol{x})| & \text{if } \varphi^m(\boldsymbol{x}) \le 0 \\ |\varphi^m(\boldsymbol{x})| & \text{if } \varphi^m(\boldsymbol{x}) > 0 \end{cases} \tag{10.16}$$

where $|\varphi^s(\boldsymbol{x})|$ is the signed distance function of the secondary (minor) interface that joins the main (major) interface given by the signed distance function $|\varphi^m(\boldsymbol{x})|$. Obviously, the junction enrichment function (10.16)

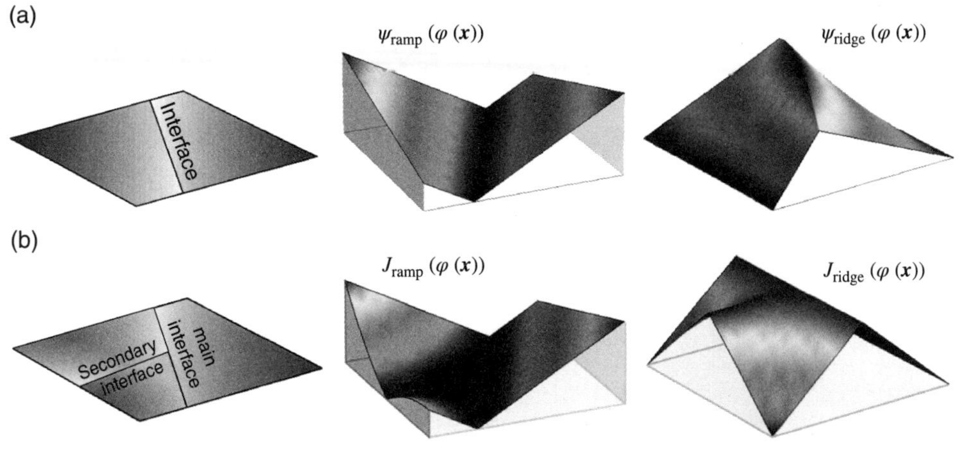

Figure 10.4 The enrichment functions of displacement field in porous media: (a) the level set ramp function together with the *ridged* level set function; (b) the junction ramp function along with the *ridged* junction ramp function

provides the continuity in the displacement field, while there are discontinuities in the displacement derivatives across the main (major) and secondary (minor) material interfaces. The schematic views of the junction *ramp* function along with the *ridged* junction ramp function are shown in Figure 10.4b.

In order to approximate the pressure field, the standard shape functions are employed for the fluid pressure. Since the material interface does not impose any discontinuity in the fluid flow, the fluid pressure must not be enriched to ensure the fluid flow continuity across the material interface. Hence, the fluid pressure field $p(\boldsymbol{x}, t)$ is approximated using the classical FEM as

$$p(\boldsymbol{x},t) \approx p^h(\boldsymbol{x},t) = \sum_{I \in \mathcal{N}} N_{pI}(\boldsymbol{x}) \bar{p}_I(t) \tag{10.17}$$

where $\bar{p}_I(t)$ is the unknown standard nodal pressure at node I, and $N_p(\boldsymbol{x})$ are the standard pressure shape functions.

Finally, the FE approximation of the enriched displacement field (10.13) and the standard pressure field (10.17) can be symbolically written in the following form

$$\mathbf{u}^h(\boldsymbol{x},t) = \mathbf{N}_u^{std}(\boldsymbol{x})\,\bar{\mathbf{u}}(t) + \mathbf{N}_u^{enr}(\boldsymbol{x})\,\bar{\mathbf{a}}(t) \tag{10.18a}$$

$$p^h(\boldsymbol{x},t) = \mathbf{N}_p^{std}(\boldsymbol{x})\,\bar{\mathbf{p}}(t) \tag{10.18b}$$

where $\mathbf{N}_u^{std}(\boldsymbol{x})$ is the matrix of standard displacement shape functions, $\mathbf{N}_u^{enr}(\boldsymbol{x})$ is the matrix of enriched shape functions associated with the modified level set and junction *ramp* functions, $\bar{\mathbf{u}}(t)$ is the vector of standard displacement DOF, and $\bar{\mathbf{a}}(t)$ is the vector of enriched displacement DOF associated with the enriched DOF $\bar{\mathbf{d}}_J(t)$ and $\bar{\mathbf{b}}_K(t)$. Moreover, $\mathbf{N}_p^{std}(\boldsymbol{x})$ is the matrix of standard pressure shape functions, and $\bar{\mathbf{p}}(t)$ is the vector of pressure nodal DOF.

Accordingly, the strain vector corresponding to the approximate displacement field (10.18a) can be written in terms of the standard and enriched nodal values as

$$\boldsymbol{\varepsilon}^h(\boldsymbol{x},t) = \mathbf{B}_u^{std}(\boldsymbol{x})\,\bar{\mathbf{u}}(t) + \mathbf{B}_u^{enr}(\boldsymbol{x})\,\bar{\mathbf{a}}(t) \tag{10.19}$$

where $\mathbf{B}_u^{std}(\boldsymbol{x}) \equiv \mathbf{L}\,\mathbf{N}_u^{std}(\boldsymbol{x})$ involve the spatial derivatives of the standard shape functions, $\mathbf{B}_u^{enr}(\boldsymbol{x}) \equiv \mathbf{L}\,\mathbf{N}_u^{enr}(\boldsymbol{x})$ contain the spatial derivatives of the enriched shape functions associated with the modified level set and junction *ramp* functions, with **L** denoting the matrix differential operator defined by

$$\mathbf{L} = \begin{bmatrix} \partial/\partial x & 0 \\ 0 & \partial/\partial y \\ \partial/\partial y & \partial/\partial x \end{bmatrix} \tag{10.20}$$

10.2.2 The X-FEM Spatial Discretization

The discrete form of integral equations (10.12) can be obtained in the framework of X-FEM formulation using the test and trial functions of the enriched displacement and standard pressure fields. Applying the trial functions of enriched displacement and standard pressure fields (10.18) together with the test functions of displacement and pressure fields $\delta\mathbf{u}(\boldsymbol{x}, t)$ and $\delta p(\boldsymbol{x}, t)$, which are respectively defined in the same approximating space as the approximate displacement and fluid pressure fields as

$$\delta\mathbf{u}^h(\boldsymbol{x}, t) = \mathbf{N}_u^{std}(\boldsymbol{x})\,\delta\bar{\mathbf{u}}(t) + \mathbf{N}_u^{enr}(\boldsymbol{x})\,\delta\bar{\mathbf{a}}(t) \tag{10.21a}$$

$$\delta p^h(\boldsymbol{x}, t) = \mathbf{N}_p^{std}(\boldsymbol{x})\,\delta\bar{\mathbf{p}}(t) \tag{10.21b}$$

and satisfying the necessity that the weak form should hold for all kinematically admissible test functions, the discretized form of integral equations (10.12) can therefore be obtained based on the Bubnov–Galerkin technique by substituting the trial and test functions (10.18) and (10.21) into the weak form of governing equations (10.12) and considering the arbitrariness of displacement and pressure variations as

$$\begin{gathered} \begin{pmatrix} \mathbf{M}_{uu} & \mathbf{M}_{ua} \\ \mathbf{M}_{au} & \mathbf{M}_{aa} \end{pmatrix} \begin{Bmatrix} \ddot{\bar{\mathbf{u}}} \\ \ddot{\bar{\mathbf{a}}} \end{Bmatrix} + \begin{pmatrix} \mathbf{K}_{uu} & \mathbf{K}_{ua} \\ \mathbf{K}_{au} & \mathbf{K}_{aa} \end{pmatrix} \begin{Bmatrix} \bar{\mathbf{u}} \\ \bar{\mathbf{a}} \end{Bmatrix} - \begin{pmatrix} \mathbf{Q}_{up} \\ \mathbf{Q}_{ap} \end{pmatrix} \bar{\mathbf{p}} - \begin{Bmatrix} \mathbf{f}_u^{\text{ext}} \\ \mathbf{f}_a^{\text{ext}} \end{Bmatrix} = \mathbf{0} \\ \langle \mathbf{C}_{pu} \;\; \mathbf{C}_{pa} \rangle \begin{Bmatrix} \ddot{\bar{\mathbf{u}}} \\ \ddot{\bar{\mathbf{a}}} \end{Bmatrix} + \left\langle \mathbf{Q}_{up}^T \;\; \mathbf{Q}_{ap}^T \right\rangle \begin{Bmatrix} \dot{\bar{\mathbf{u}}} \\ \dot{\bar{\mathbf{a}}} \end{Bmatrix} + \mathbf{H}_{pp}\,\bar{\mathbf{p}} + \mathbf{S}_{pp}\,\dot{\bar{\mathbf{p}}} - \mathbf{q}_p^{\text{ext}} = \mathbf{0} \end{gathered} \tag{10.22}$$

or

$$\begin{gathered} \mathbf{M}_{uu}\ddot{\bar{\mathbf{u}}} + \mathbf{M}_{ua}\ddot{\bar{\mathbf{a}}} + \mathbf{K}_{uu}\bar{\mathbf{u}} + \mathbf{K}_{ua}\bar{\mathbf{a}} - \mathbf{Q}_{up}\bar{\mathbf{p}} - \mathbf{f}_u^{\text{ext}} = \mathbf{0} \\ \mathbf{M}_{au}\ddot{\bar{\mathbf{u}}} + \mathbf{M}_{aa}\ddot{\bar{\mathbf{a}}} + \mathbf{K}_{au}\bar{\mathbf{u}} + \mathbf{K}_{aa}\bar{\mathbf{a}} - \mathbf{Q}_{ap}\bar{\mathbf{p}} - \mathbf{f}_a^{\text{ext}} = \mathbf{0} \\ \mathbf{C}_{pu}\ddot{\bar{\mathbf{u}}} + \mathbf{C}_{pa}\ddot{\bar{\mathbf{a}}} + \mathbf{Q}_{up}^T\dot{\bar{\mathbf{u}}} + \mathbf{Q}_{ap}^T\dot{\bar{\mathbf{a}}} + \mathbf{H}_{pp}\,\bar{\mathbf{p}} + \mathbf{S}_{pp}\,\dot{\bar{\mathbf{p}}} - \mathbf{q}_p^{\text{ext}} = \mathbf{0} \end{gathered} \tag{10.23}$$

where the mass matrix **M**, stiffness matrix **K**, coupling matrix **Q**, inertial matrix **C**, permeability matrix **H**, compressibility matrix **S**, and external force vectors $\mathbf{f}^{\text{ext}}$ and $\mathbf{q}^{\text{ext}}$ are defined as

$$\begin{aligned} \mathbf{M}_{\alpha\beta} &= \int_\Omega (\mathbf{N}_u^\alpha)^T \rho\, \mathbf{N}_u^\beta \, d\Omega \\ \mathbf{K}_{\alpha\beta} &= \int_\Omega (\mathbf{B}_u^\alpha)^T \mathbf{D}\, \mathbf{B}_u^\beta \, d\Omega \\ \mathbf{Q}_{\alpha p} &= \int_\Omega (\mathbf{B}_u^\alpha)^T \alpha\, \mathbf{m}\, \mathbf{N}_p^{std} \, d\Omega \\ \mathbf{f}_\alpha^{\text{ext}} &= \int_\Omega (\mathbf{N}_u^\alpha)^T \rho\, \mathbf{b}\, d\Omega + \int_{\Gamma_t} (\mathbf{N}_u^\alpha)^T \bar{\mathbf{t}}\, d\Gamma \end{aligned} \tag{10.24}$$

and

$$
\begin{aligned}
\mathbf{C}_{p\beta} &= \int_{\Omega} \left(\nabla \mathbf{N}_p^{std}\right)^T \boldsymbol{k}_f \cdot \rho_f \, \mathbf{N}_u^{\beta} \, d\Omega \\
\mathbf{H}_{pp} &= \int_{\Omega} \left(\nabla \mathbf{N}_p^{std}\right)^T \boldsymbol{k}_f \left(\nabla \mathbf{N}_p^{std}\right) d\Omega \\
\mathbf{S}_{pp} &= \int_{\Omega} \left(\mathbf{N}_p^{std}\right)^T \frac{1}{Q} \mathbf{N}_p^{std} \, d\Omega \\
\mathbf{q}_p^{\text{ext}} &= \int_{\Omega} \left(\nabla \mathbf{N}_p^{std}\right)^T \boldsymbol{k}_f \cdot \rho_f \, \mathbf{b} \, d\Omega - \int_{\Gamma_w} \left(\mathbf{N}_p^{std}\right)^T \bar{q} \, d\Gamma
\end{aligned}
\tag{10.25}
$$

where $\mathbf{N}_u^{\alpha}$, $\mathbf{N}_u^{\beta}(\alpha,\beta) \in (std, enr)$ denote the "standard" and "enriched" shape functions of displacement field and, $\mathbf{N}_p^{std}$ denote the "standard" shape functions of pressure field. In these definitions, $\mathbf{m}$ is the vector of delta Dirac function defined as $\mathbf{m} = \{1 \;\; 1 \;\; 0\}^T$.

Finally, the X-FEM discretization equations (10.22) can be rewritten as

$$
\begin{aligned}
&\mathbf{M}\ddot{\bar{\mathbf{U}}} + \mathbf{K}\bar{\mathbf{U}} - \mathbf{Q}\bar{\mathbf{p}} - \mathbf{f}^{\text{ext}} = \mathbf{0} \\
&\mathbf{C}\ddot{\bar{\mathbf{U}}} + \mathbf{Q}^T\dot{\bar{\mathbf{U}}} + \mathbf{H}\bar{\mathbf{p}} + \mathbf{S}\dot{\bar{\mathbf{p}}} - \mathbf{q}^{\text{ext}} = \mathbf{0}
\end{aligned}
\tag{10.26}
$$

where $\bar{\mathbf{U}} = \langle \bar{\mathbf{u}}, \bar{\mathbf{a}} \rangle$ are the complete set of the standard and enriched DOF of displacement field, respectively.

10.2.3 The Time Domain Discretization and Solution Procedure

In order to complete the numerical solution of X-FEM equations, it is necessary to integrate the time differential equations (10.26) in time. The set of equations is discretized in the time domain following the line of the well-known Newmark scheme, in which it is supposed that the system of equations is satisfied at each discrete time step. Thus, the problem must be solved at a series of sequential discrete time steps. To advance the solution in time, the link between the successive values of the unknown field variables at time t_{n+1} and the known field variables at time t_n is established by applying the minimum order of the generalized Newmark scheme required considering the highest order of the time derivatives in the differential equations. Hence, the generalized Newmark GN22 method is employed for the displacement field $\bar{\mathbf{U}}$ and GN11 method for the pressure field $\bar{\mathbf{p}}$. The link between the successive values of the unknown field variables at time $t_{n+1} = t_n + \Delta t$ and the known field variables at time t_n can be established as

$$
\begin{aligned}
\ddot{\bar{\mathbf{U}}}_{n+1} &= \frac{1}{\beta \Delta t^2}(\bar{\mathbf{U}}_{n+1} - \bar{\mathbf{U}}_n) - \frac{1}{\beta \Delta t}\dot{\bar{\mathbf{U}}}_n - \left(\frac{1}{2\beta} - 1\right)\ddot{\bar{\mathbf{U}}}_n \\
\dot{\bar{\mathbf{U}}}_{n+1} &= \frac{\gamma}{\beta \Delta t}(\bar{\mathbf{U}}_{n+1} - \bar{\mathbf{U}}_n) - \left(\frac{\gamma}{\beta} - 1\right)\dot{\bar{\mathbf{U}}}_n - \Delta t\left(\frac{\gamma}{2\beta} - 1\right)\ddot{\bar{\mathbf{U}}}_n
\end{aligned}
\tag{10.27a}
$$

and

$$
\dot{\bar{\mathbf{p}}}_{n+1} = \frac{1}{\theta \Delta t}(\bar{\mathbf{p}}_{n+1} - \bar{\mathbf{p}}_n) - \left(\frac{1}{\theta} - 1\right)\dot{\bar{\mathbf{p}}}_n \tag{10.27b}
$$

where β, γ, and θ are Newmark parameters, which are usually chosen in the range of $[0-1]$. However, for unconditionally stability of the time integration procedure, it is required that $\gamma \geq 0.5, \beta \geq 0.25(0.5+\gamma)^2$ and $\theta \geq 0.5$. In these relations, $\bar{\mathbf{U}}_n$, $\dot{\bar{\mathbf{U}}}_n$, and $\ddot{\bar{\mathbf{U}}}_n$ denote the known values of displacement, velocity, and acceleration of the standard and enriched DOF at time t_n, and $\bar{\mathbf{p}}_n$ and $\dot{\bar{\mathbf{p}}}_n$ are the known values of pressure and gradient of pressure of the standard DOFs at time t_n. Substituting relations (10.27) into the spatial discrete equations (10.26), the following nonlinear equation can be achieved as

$$\begin{aligned}
\mathbf{\Psi}_{\mathbf{U}_{n+1}} &= \frac{1}{\beta\,\Delta t^2}\mathbf{M}\,\bar{\mathbf{U}}_{n+1} + \mathbf{K}\,\bar{\mathbf{U}}_{n+1} - \mathbf{Q}\,\bar{\mathbf{p}}_{n+1} - \mathbf{G}_{\mathbf{U}_{n+1}} = \mathbf{0} \\
\mathbf{\Psi}_{\mathbf{p}_{n+1}} &= \frac{1}{\beta\,\Delta t^2}\mathbf{C}\,\bar{\mathbf{U}}_{n+1} + \frac{\gamma}{\beta\,\Delta t}\mathbf{Q}^T\,\bar{\mathbf{U}}_{n+1} + \mathbf{H}\,\bar{\mathbf{p}}_{n+1} + \frac{1}{\theta\,\Delta t}\mathbf{S}\,\bar{\mathbf{p}}_{n+1} - \mathbf{G}_{\mathbf{p}_{n+1}} = \mathbf{0}
\end{aligned} \tag{10.28}$$

where $\mathbf{G}_{\mathbf{U}_{n+1}}$ and $\mathbf{G}_{\mathbf{p}_{n+1}}$ are the vector of known values at time t_n defined as

$$\begin{aligned}
\mathbf{G}_{\mathbf{U}_{n+1}} &= \mathbf{f}^{\text{ext}}_{\mathbf{U}_{n+1}} + \mathbf{M}\left(\frac{1}{\beta\,\Delta t^2}\bar{\mathbf{U}}_n + \frac{1}{\beta\,\Delta t}\dot{\bar{\mathbf{U}}}_n + \left(\frac{1}{2\beta}-1\right)\ddot{\bar{\mathbf{U}}}_n\right) \\
\mathbf{G}_{\mathbf{p}_{n+1}} &= \mathbf{q}^{\text{ext}}_{\mathbf{p}_{n+1}} + \mathbf{C}\left(\frac{1}{\beta\,\Delta t^2}\bar{\mathbf{U}}_n + \frac{1}{\beta\,\Delta t}\dot{\bar{\mathbf{U}}}_n + \left(\frac{1}{2\beta}-1\right)\ddot{\bar{\mathbf{U}}}_n\right) \\
&\quad + \mathbf{Q}^T\left(\frac{\gamma}{\beta\,\Delta t}\bar{\mathbf{U}}_n + \left(\frac{\gamma}{\beta}-1\right)\dot{\bar{\mathbf{U}}}_n + \Delta t\left(\frac{\gamma}{2\beta}-1\right)\ddot{\bar{\mathbf{U}}}_n\right) + \mathbf{S}\left(\frac{1}{\theta\,\Delta t}\bar{\mathbf{p}}_n + \left(\frac{1}{\theta}-1\right)\dot{\bar{\mathbf{p}}}_n\right)
\end{aligned} \tag{10.29}$$

The set of nonlinear equations (10.28) can be solved using an appropriate approach, such as the Newton–Raphson iterative algorithm to linearize the nonlinear system of equations, that is, $\mathbf{\Psi}_{n+1} = \mathbf{0}$. By expanding equations (10.28) with the first-order truncated Taylor series as $\mathbf{\Psi}^{i+1}_{n+1} = \mathbf{\Psi}^{i}_{n+1} + (\partial\mathbf{\Psi}/\partial\mathbf{X})^{i}_{n+1} d\mathbf{X}^{i}_{n}$, where the vector of nodal unknowns are $\mathbf{X}^T = \langle \bar{\mathbf{u}}^T, \bar{\mathbf{a}}^T, \bar{\mathbf{p}}^T \rangle$, the following linear approximation can be obtained as

$$\begin{Bmatrix} \mathbf{\Psi}^{i+1}_{\mathbf{U}_{n+1}} \\ \mathbf{\Psi}^{i+1}_{\mathbf{p}_{n+1}} \end{Bmatrix} = \begin{Bmatrix} \mathbf{\Psi}^{i}_{\mathbf{U}_{n+1}} \\ \mathbf{\Psi}^{i}_{\mathbf{p}_{n+1}} \end{Bmatrix} + \begin{bmatrix} \dfrac{\partial \mathbf{\Psi}_{\mathbf{U}}}{\partial \bar{\mathbf{U}}} & \dfrac{\partial \mathbf{\Psi}_{\mathbf{U}}}{\partial \bar{\mathbf{p}}} \\ \dfrac{\partial \mathbf{\Psi}_{\mathbf{p}}}{\partial \bar{\mathbf{U}}} & \dfrac{\partial \mathbf{\Psi}_{\mathbf{p}}}{\partial \bar{\mathbf{p}}} \end{bmatrix}^{i}_{n+1} \begin{Bmatrix} d\bar{\mathbf{U}}^{i}_{n} \\ d\bar{\mathbf{p}}^{i}_{n} \end{Bmatrix} = \mathbf{0} \tag{10.30}$$

where the Jacobian matrix $\mathbf{J}$ in Eq. (10.30) can be obtained as

$$\mathbf{J} = \begin{bmatrix} \dfrac{\partial \mathbf{\Psi}_{\mathbf{U}}}{\partial \bar{\mathbf{U}}} & \dfrac{\partial \mathbf{\Psi}_{\mathbf{U}}}{\partial \bar{\mathbf{p}}} \\ \dfrac{\partial \mathbf{\Psi}_{\mathbf{p}}}{\partial \bar{\mathbf{U}}} & \dfrac{\partial \mathbf{\Psi}_{\mathbf{p}}}{\partial \bar{\mathbf{p}}} \end{bmatrix} = \begin{bmatrix} \dfrac{1}{\beta\,\Delta t^2}\mathbf{M} + \mathbf{K} & -\mathbf{Q} \\ \dfrac{1}{\beta\,\Delta t^2}\mathbf{C} + \dfrac{\gamma}{\beta\,\Delta t}\mathbf{Q}^T & \mathbf{H} + \dfrac{1}{\theta\,\Delta t}\mathbf{S} \end{bmatrix} \tag{10.31}$$

or

$$\mathbf{J} = \begin{bmatrix} \dfrac{1}{\beta\,\Delta t^2}\mathbf{M}_{uu} + \mathbf{K}_{uu} & \dfrac{1}{\beta\,\Delta t^2}\mathbf{M}_{ua} + \mathbf{K}_{ua} & -\mathbf{Q}_{up} \\ \dfrac{1}{\beta\,\Delta t^2}\mathbf{M}_{au} + \mathbf{K}_{au} & \dfrac{1}{\beta\,\Delta t^2}\mathbf{M}_{aa} + \mathbf{K}_{aa} & -\mathbf{Q}_{ap} \\ \dfrac{1}{\beta\,\Delta t^2}\mathbf{C}_{pu} + \dfrac{\gamma}{\beta\,\Delta t}\mathbf{Q}^T_{up} & \dfrac{1}{\beta\,\Delta t^2}\mathbf{C}_{pa} + \dfrac{\gamma}{\beta\,\Delta t}\mathbf{Q}^T_{ap} & \mathbf{H}_{pp} + \dfrac{1}{\theta\,\Delta t}\mathbf{S}_{pp} \end{bmatrix} \tag{10.32}$$

In Eq. (10.32), the Jacobian matrix needs to be evaluated at each iteration i of time step $\Delta t = t_{n+1} - t_n$. In fact, at each time step a linearized system of equations is solved until the iteration convergence is achieved. Furthermore, it can be seen from (10.32) that the Jacobian matrix is a non-symmetric matrix, so it is advantageous to make the Jacobian matrix symmetric to save the core storage and computational cost. For this purpose, the first and second rows of the Jacobian matrix are multiplied by the scalar value $-\frac{\gamma}{\beta\,\Delta t}$. In addition, the inertial matrices $\mathbf{C}_{pu}$ and $\mathbf{C}_{pa}$ are omitted from the Jacobian matrix, since the contribution of the dynamic seepage forcing terms to the solution is negligible compared to the other terms. By the imposition of the previously-mentioned simplifications, a symmetric approximation to the Jacobian matrix is obtained as follows

$$\mathbf{J} = \begin{bmatrix} -\frac{\gamma}{\beta\,\Delta t}\left(\frac{1}{\beta\,\Delta t^2}\mathbf{M}_{uu} + \mathbf{K}_{uu}\right) & -\frac{\gamma}{\beta\,\Delta t}\left(\frac{1}{\beta\,\Delta t^2}\mathbf{M}_{ua} + \mathbf{K}_{ua}\right) & \frac{\gamma}{\beta\,\Delta t}\mathbf{Q}_{up} \\ -\frac{\gamma}{\beta\,\Delta t}\left(\frac{1}{\beta\,\Delta t^2}\mathbf{M}_{au} + \mathbf{K}_{au}\right) & -\frac{\gamma}{\beta\,\Delta t}\left(\frac{1}{\beta\,\Delta t^2}\mathbf{M}_{aa} + \mathbf{K}_{aa}\right) & \frac{\gamma}{\beta\,\Delta t}\mathbf{Q}_{ap} \\ \frac{\gamma}{\beta\,\Delta t}\mathbf{Q}_{up}^T & \frac{\gamma}{\beta\,\Delta t}\mathbf{Q}_{ap}^T & \mathbf{H}_{pp} + \frac{1}{\theta\,\Delta t}\mathbf{S}_{pp} \end{bmatrix} \tag{10.33}$$

10.2.4 Numerical Integration Scheme

For the elements cut by the interface, the standard Gauss quadrature points are insufficient for numerical integration, and may not adequately integrate the interface boundary. Thus, it is necessary to modify the element quadrature points to accurately evaluate the contribution to the weak form for both sides of interface. In the standard FEM method, the numerical integration can be performed by discretizing the domain into a number of elements and employing the standard Gauss integration points over each element. In the X-FEM method, the elements located on interface boundary can be partitioned by triangular sub-domains with the boundaries aligned with the material interface. The numerical integration of enriched elements can be performed over these new sub-triangles using the standard Gauss quadrature points. It is important that the Gauss points of sub-triangles are only used for the numerical integration of the elements cut by the interface and no new DOF are added to the system.

The proposed computational model presented in preceding sections is capable of considering any number of material interfaces for an enriched element. In this case, it is only required to apply an efficient technique for obtaining the elements located on material interfaces and the material properties of corresponding integration points. In the literature, it was suggested to use the sign of distance function in order to obtain the material properties of each integration point; however, it is not simple for an element cut by various interfaces. In this study, an efficient technique is developed based on the Gauss point position in *polygon routines*, which can be simply used to determine the material property of each Gauss point in an enriched element with several material interfaces. It must be noted that the position of each sample point is computed in the local coordinates of the element, so the requirement is not to update its position in global coordinates. In order to obtain the local coordinates of each sample point of triangular sub-domain, the following nonlinear equations must be solved using the global coordinates as

$$\begin{aligned} f_1(\xi,\eta) &= x_p - \sum_i N_i(\xi,\eta)\bar{x}_i = 0 \\ f_2(\xi,\eta) &= y_p - \sum_i N_i(\xi,\eta)\bar{y}_i = 0 \end{aligned} \tag{10.34}$$

where (x_p, y_p) are the global coordinates of the sample point. This nonlinear equation can be solved using the Newton–Raphson procedure as

$$\left\{ \begin{array}{c} \xi^{k+1} \\ \eta^{k+1} \end{array} \right\} = \left\{ \begin{array}{c} \xi^{k} \\ \eta^{k} \end{array} \right\} - \begin{bmatrix} \dfrac{\partial f_1^k}{\partial \xi} & \dfrac{\partial f_1^k}{\partial \eta} \\ \dfrac{\partial f_2^k}{\partial \xi} & \dfrac{\partial f_2^k}{\partial \eta} \end{bmatrix}^{-1} \left\{ \begin{array}{c} f_1(\xi^k, \eta^k) \\ f_2(\xi^k, \eta^k) \end{array} \right\} \tag{10.35}$$

10.3 Application of the X-FEM Method in Deformable Porous Media with Arbitrary Interfaces

In order to illustrate the accuracy and versatility of the X-FEM in deformable porous media with arbitrary interfaces, two numerical examples of an elastic soil column and an elastic foundation are presented. Both examples are chosen to assess the accuracy of the X-FEM model in the dynamic analysis of fully saturated media. The examples are solved using both FEM and X-FEM techniques, and the results are compared. In order to perform a real comparison, the number of elements in FEM and X-FEM meshes is taken to be almost equal. In both examples, a uniform X-FEM mesh is used independent of the shape of material interfaces, while the FEM mesh is conformed to the geometry of interface. Both examples are simulated by a plane strain representation and the convergence tolerance is set to 10^{-8}.

10.3.1 An Elastic Soil Column

The first example is a fully saturated soil column subjected to a surface step loading, which is solved by the X-FEM and the results are compared with those of FEM model. The column has a width of $w = 1.0$ m and a height of $H = 30$ m, which is subjected to a surface step loading of 1 kN/m^2 applied in 0.1 s with a fixed pore pressure of $p = 0$ at the top of column. The geometry and boundary conditions of soil column is shown in Figure 10.5, and the material properties are summarized in Table 10.1. The problem includes

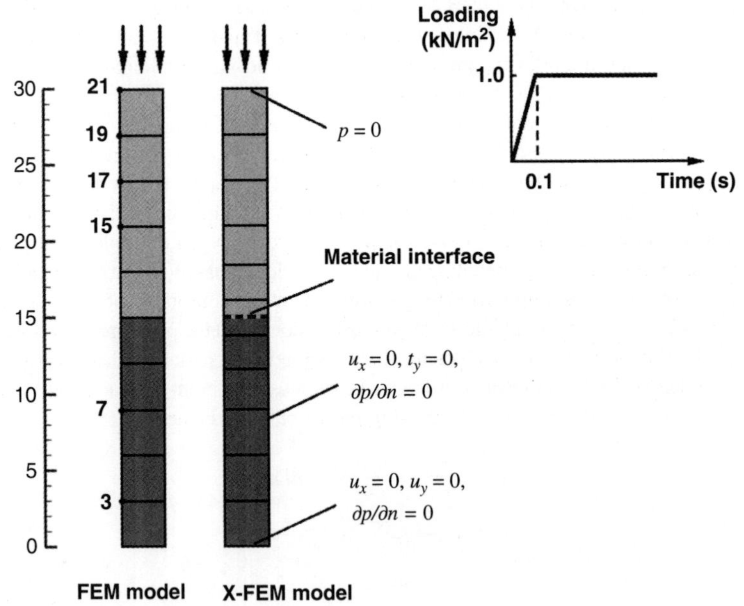

Figure 10.5 Soil column subjected to a surface step loading: The geometry and boundary condition together with the X-FEM and FEM meshes

Table 10.1 Material properties of a soil column

Material properties	Region 1	Region 2
Young's modulus E (Pa)	3×10^7	6×10^7
Poisson ratio ν	0.2	0.2
Solid density ρ_s (kg/m^3)	2000	2000
Fluid density ρ_f (kg/m^3)	1000	1000
Bulk modulus of fluid K_f (Pa)	2.1×10^9	2.1×10^9
Bulk modulus of solid K_s (Pa)	1.0×10^{20}	1.0×10^{20}
Porosity n	0.3	0.3
Permeability k (m^{3}s/kg)	1.02×10^{-8}	1.02×10^{-9}

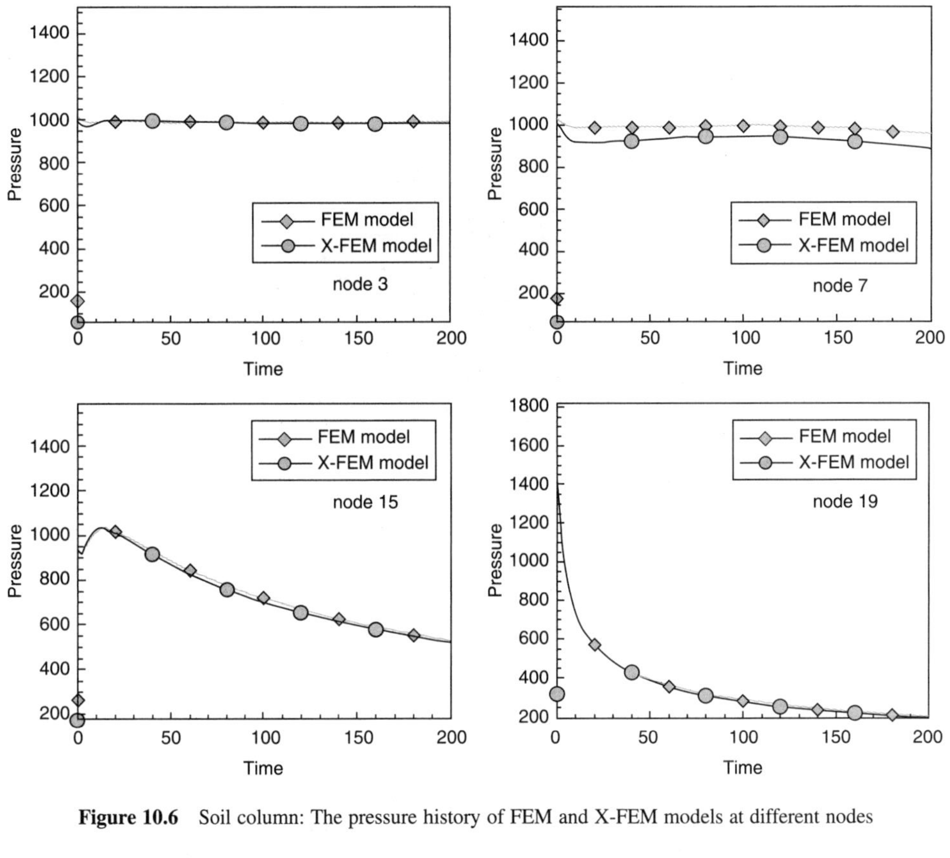

Figure 10.6 Soil column: The pressure history of FEM and X-FEM models at different nodes

two regions with a horizontal material interface. The mesh consists of ten 4-noded quadrilateral elements for both displacement and pressure fields. The generalized Newmark GN22 method is employed for the standard and enriched displacement DOF and the GN11 method for the standard DOF of pressure. The time step is set to 0.05 s. The FEM and X-FEM meshes are shown in Figure 10.5. The variations with time of pore pressure are plotted in Figure 10.6 at four points of column, that is, points 3, 7, 15, and 19. As can be seen from this figure, the proposed dynamic X-FEM simulation is in good agreement with those obtained by FEM. Also plotted in Figure 10.7 are the variations with time of the vertical displacement at points 17 and 21. Reasonable agreement can be observed between the FEM and X-FEM models.

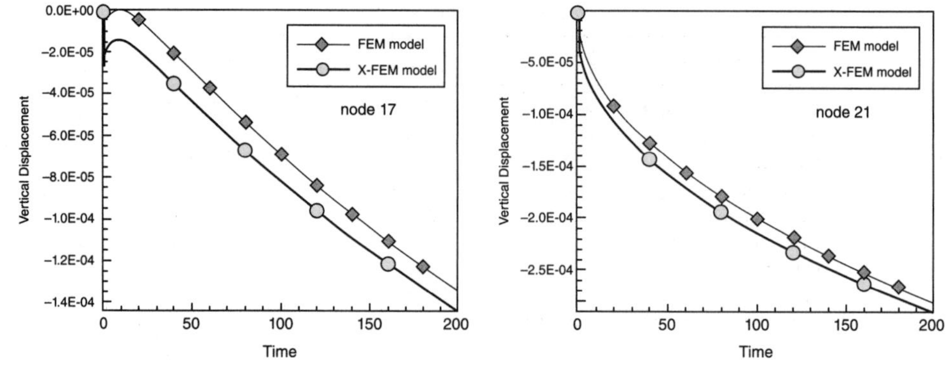

Figure 10.7 Soil column: The displacement history of FEM and X-FEM models at different nodes

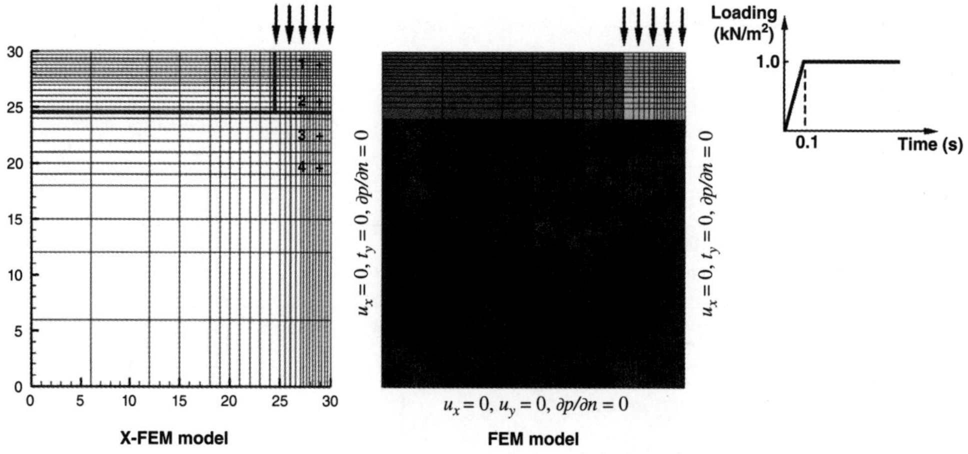

Figure 10.8 Elastic foundation subjected to a surface step loading: The geometry and boundary condition together with the X-FEM and FEM meshes

10.3.2 An Elastic Foundation

The next example is an elastic foundation subjected to a surface step loading with three internal sub-domains. The foundation problem is one of the most popular examples used by various researchers to capture the shear band localization (Khoei *et al.*, 2007). In this study, a fully saturated soil foundation with a step loading of 350 kN/m^2 is applied in 0.1 s and, a fixed pore pressure of $p = 0$ is assumed at the top of foundation. The foundation is 30×30 m with the geometry and boundary conditions shown in Figure 10.8. The material properties of foundation for three regions are given in Table 10.2. The 4-noded quadrilateral elements are used for both displacement and pressure fields in the FEM and X-FEM analyses. The time step is set to 0.01 s. In a FE simulation, the FE mesh is conformed to the boundary of material interfaces, while in X-FEM analysis the interfaces pass through the elements. In the X-FEM, a partitioning procedure is used to generate the sub-triangles. For those elements cut by the interfaces, the Gauss points of sub-triangles are then employed to evaluate the stiffness, mass,

Table 10.2 Material properties of an elastic foundation

Material properties	Region 1	Region 2	Region 3
Young's modulus E (Pa)	20×10^6	40×10^6	100×10^6
Poisson ratio ν	0.2	0.2	0.2
Solid density ρ_s (kg/m^3)	2000	2000	2000
Fluid density ρ_f (kg/m^3)	1000	1000	1000
Bulk modulus of fluid K_f (Pa)	2.1×10^9	2.1×10^9	2.1×10^9
Bulk modulus of solid K_s (Pa)	1.0×10^{20}	1.0×10^{20}	1.0×10^{20}
Porosity n	0.25	0.3	0.35
Permeability k (m^{3}s/kg)	1.0×10^{-8}	5.0×10^{-8}	2.0×10^{-7}

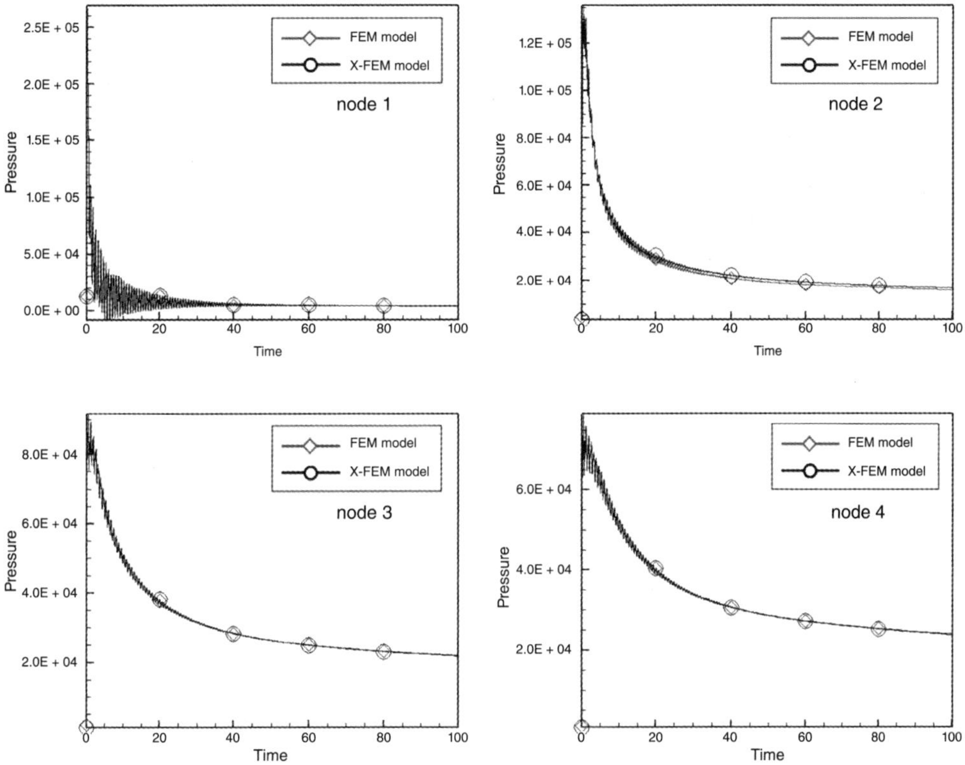

Figure 10.9 Elastic foundation: The pressure history of FEM and X-FEM models at different nodes

compressibility, and permeability matrices. The variations with time of pore pressure are plotted in Figure 10.9 at four points of foundation. Remarkable agreement can be seen between the X-FEM and FEM dynamic simulations. In Figure 10.10, the variations with time of the vertical displacement are plotted at different points, which are in good agreement with those obtained by the FEM analyses. Finally, the distributions of pressure contour are shown in Figure 10.11 at the time step of 10.0 s using the FEM and X-FEM models. A good agreement can be observed between two techniques.

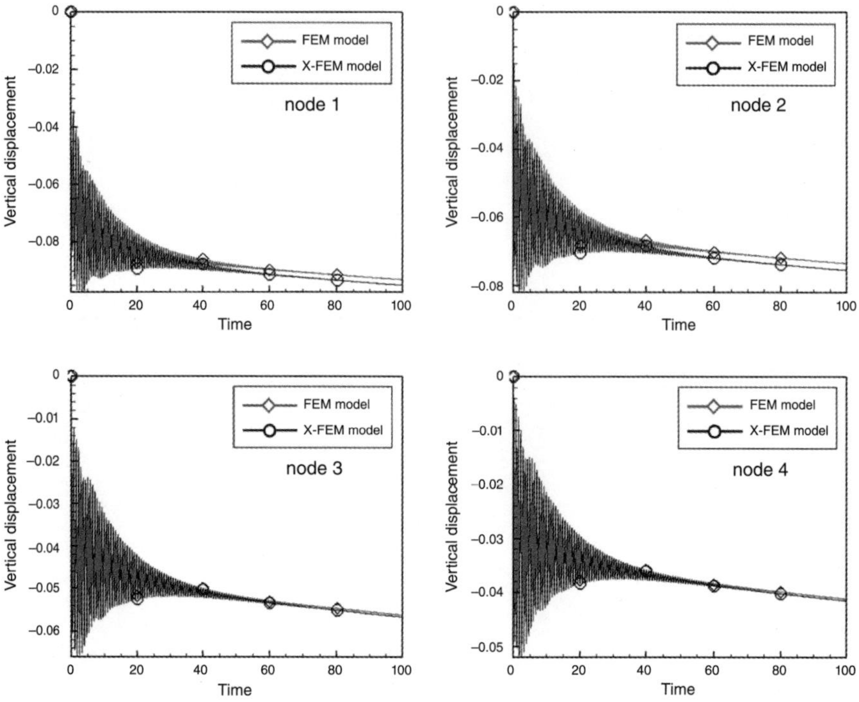

Figure 10.10 Elastic foundation: The displacement history of FEM and X-FEM models at different nodes

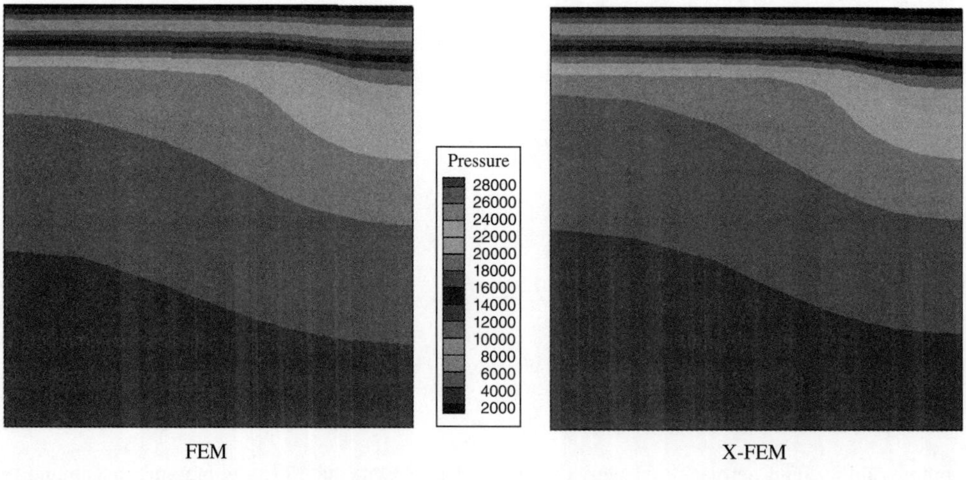

Figure 10.11 Elastic foundation: The distribution of pressure contours using the FEM and X-FEM models at $t = 10$ s

10.4 Modeling Hydraulic Fracture Propagation in Deformable Porous Media

The problem of modeling hydraulic fracture propagation has attracted considerable attention since the early 1950s. Interest in hydraulic fracturing is mainly due to its practical applications in a broad range of engineering areas. Hydraulic fracturing is the most commonly used stimulation technique for reservoirs. This technique is widely used in the petroleum industry to enhance the recovery of hydrocarbons, such as gas and oil, from underground reservoirs. The other important application of hydraulic fracturing includes toxic or radioactive waste disposal. In hydraulic fracturing, a viscous fluid is injected into the fracture, which pressurizes the fracture faces (Figure 10.12). With increasing internal pressure, the fracture propagation criterion is met ahead of the fracture tip, leading to hydraulic fracture propagation. In fact, the hydraulic fracturing process involves pumping of a viscous fluid into the underground formation at a high enough injection rate to fracture the formation and drive the fracture hydraulically. This section aims to illustrate a fully coupled numerical model on the basis of the X-FEM technique for the hydro-mechanical analysis of hydraulic fracture propagation in porous media.

In order to perform a realistic modeling of the hydraulically-driven fractures, the medium surrounding the fracture is treated as a porous medium, and the coupling between various physical phenomena is taken into account. The involved physical phenomena that govern the propagation of the hydraulic fracture in porous media include: the flow of the fracturing fluid within the fracture, the flow of the pore fluid through the surrounding porous medium, the deformation of the surrounding porous medium, the leak-off of the fracturing fluid from the fracture into the surrounding porous medium, and hydraulic fracture propagation. Challenging difficulties in the numerical modeling of fluid-driven fractures in permeable porous media emanate from hydro-mechanical coupling between partial differential equations governing fluid flow within the fracture, the pore fluid flow in the porous medium surrounding the fracture, and solid deformation. This coupling results in a system of fully coupled nonlinear equations, which must be solved using an iterative solution procedure. Such modeling allows one to thoroughly simulate the hydraulic fracturing process and to predict hydraulic fracture response to both fluid and formation properties.

There are several classical models presented in the literature for the problem of a fracture driven by injection of a viscous fluid. The essential differences between these models concern the modeling of near tip processes, which may significantly affect hydraulic fracture propagation. Generally, analytical solutions are based on the restrictive assumption that a hydraulic fracture is completely filled with the injected fluid. That is, the solution is obtained assuming that the fluid front coincides with the fracture tip. In analytical solutions, which account for the existence of the fluid lag, the fluid pressure in the lag region is

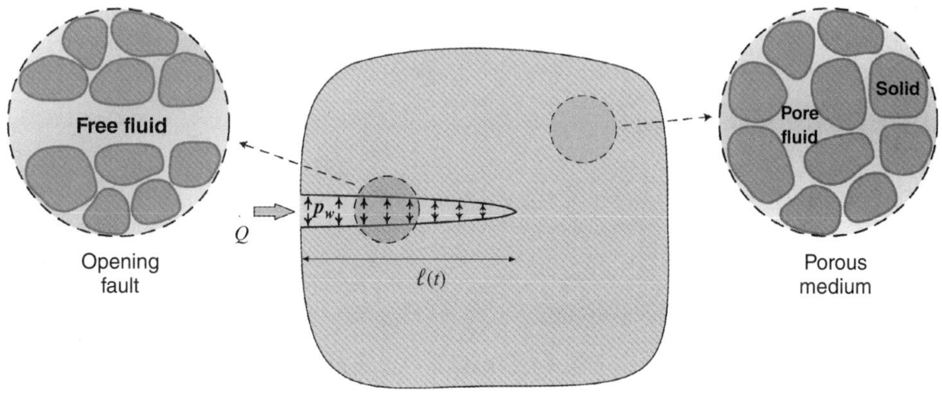

Figure 10.12 Hydraulic fracturing process: The fluid flow in a fractured deformable porous medium

assumed to be known, while the presented numerical model allows for the development of *a priori* unknown lag between the fracturing fluid front and the fracture tip. Moreover, the analytical solutions are either restricted to the limiting condition of zero fluid leak-off or incorporate the fluid leak-off from the fracture by the application of the Carter leak-off model (Carter, 1957). This model assumes that the fluid leak-off is one-dimensional, which contrasts with the more general case where the fluid leak-off exhibits a two-dimensional flow pattern. In practice, when the hydraulic fracture is internally pressurized by injecting the fracturing fluid, the fluid may leak into the medium surrounding the fracture depending on the permeability of the medium. The proposed model calculates the leak-off of the fracturing fluid through the fracture faces into the surrounding medium on the basis of the fully coupled poroelastic analysis. Research works show that the injection pressure measured during the hydraulic fracturing treatment is higher than the one predicted by the classical models (Carrier and Granet, 2012; Sarris and Papanastasiou, 2012). The discrepancy between model predictions and field measurements is due to the fact that commonly used models consider the medium surrounding the fracture to be a solid continuum and consequently neglect pore pressure development and fluid flow in pore spaces of the surrounding medium. The proposed numerical model accounts for the poroelastic effects. It shows that poroelasticity can have a strong influence on hydraulic fracturing behavior.

In this section, the X-FEM is employed to model the hydraulic fracturing process in saturated two phase porous media. The mass balance equation of fluid phase is applied together with the momentum balance of bulk and fluid phases to obtain the fully coupled set of equations in the framework of *u–p* formulation. The fluid flow within the fracture is modeled using Darcy's Law, in which the fracture permeability is assumed according to the well-known cubic law. The spatial domain discritization is performed based on the X-FEM, and the time domain discretization is carried out using the generalized Newmark scheme. In order to incorporate the fracture opening and the fluid exchange between the fracture and the surrounding porous medium, several modifications are applied into the X-FEM formulation. In this regard, the displacement jump requires that the displacement field be discontinuous across the fracture. In addition, the fluid exchange implies that the fluid flow, which is governed by the Darcy velocity of the fluid, in the normal direction to the fracture be discontinuous. Since the Darcy velocity is related to the fluid pressure gradient by means of Darcy's Law, the normal gradient of the fluid pressure must be discontinuous across the fracture. In the context of the X-FEM, the discontinuity in the displacement field is modeled by enhancing the standard piecewise polynomial basis with a discontinuous enrichment function; and the discontinuity in the fluid flow normal to the fracture is modeled by enhancing the standard FE approximation of the pressure field with the commonly used enrichment function for weak discontinuities. For the numerical solution, the unconditionally stable direct time-stepping procedure is applied to resolve the resulting system of strongly coupled nonlinear algebraic equations using the Newton–Raphson iterative algorithm. As a result, the solid displacement and fluid pressure fields are obtained simultaneously together with the fracture length. Furthermore, the fluid leak-off is obtained as a part of the solution process without introducing any simplifying assumption. Finally, several numerical examples are presented to demonstrate the capability and the efficiency of the developed model in the fully coupled simulation of hydraulically-driven fractures in deformable porous media.

10.4.1 Governing Equations of a Fractured Porous Medium

Modeling of the hydraulic fracture propagating in the porous medium involves the coupling of various physical phenomena, including deformation of the solid skeleton, pore fluid flow through the porous medium surrounding the fracture, fluid flow within the fracture, fluid exchange between the fracture and surrounding porous medium, and propagation of the hydraulic fracture. The partial differential equations governing hydraulic fracture propagation in the porous medium consist of the equilibrium equation for the whole mixture and the continuity equation for fluid flow inside the fracture and through the surrounding porous medium. Darcy's Law with constant intrinsic permeability is assumed to hold for pore fluid flow in the porous medium surrounding the fracture. Assuming that the flow of the fracturing fluid within the fracture is steady, Darcy's Law is valid for fluid flow within the fracture. However, in contrast to the porous medium, the intrinsic permeability within the fracture cannot be assumed to be constant. Thus, Poiseuille's or cubic law is employed to define the intrinsic permeability within the fracture. The validity of cubic law has been shown in the case of the laminar flow of the Newtonian viscous fluid through fractures with smooth, parallel walls (Witherspoon *et al.*, 1980). The deviation from the ideal parallel faces condition causes an apparent decrease in fluid flow through the fracture. This effect can be taken into account by incorporating a coefficient into the cubic law. In fact, this coefficient considers the influence of fracture roughness and fracture opening variation on fluid flow.

The governing equations of a deformable porous medium was presented for the solid–fluid mixture in the framework of Biot theory in Section 10.2 as

$$\nabla\cdot\boldsymbol{\sigma}-\rho\,\ddot{\mathbf{u}}+\rho\,\mathbf{b}=0 \qquad (10.4-\text{repeated})$$

$$\nabla\cdot\left[\boldsymbol{k}_f\left(-\nabla p-\rho_f\,\ddot{\mathbf{u}}+\rho_f\,\mathbf{b}\right)\right]+\alpha\,\nabla\cdot\dot{\mathbf{u}}+\frac{1}{Q}\dot{p}=0 \qquad (10.7-\text{repeated})$$

These governing equations can be solved for a fractured porous medium based on the essential and natural boundary conditions. The solid-phase boundary conditions are $\mathbf{u}=\bar{\mathbf{u}}$ on Γ_u and $\mathbf{t}=\boldsymbol{\sigma}\cdot\mathbf{n}_\Gamma\equiv\bar{\mathbf{t}}$ on Γ_t. The fluid-phase boundary conditions are $\mathbf{p}=\bar{\mathbf{p}}$ on Γ_p and $\dot{\mathbf{w}}\cdot\mathbf{n}_\Gamma=\bar{q}$ as the prescribed outflow rate of the fluid imposed on Γ_w. From the physics of the problem, the existence of a discontinuity Γ_d in the porous medium leads to the mechanical and mass transfer couplings between fracture and the porous medium surrounding the fracture. Mechanical coupling arises from the fluid pressure exerted on the fracture faces by injecting a viscous fluid into the fracture. Mass transfer coupling emanates from the fluid exchange between the fracture and the surrounding permeable porous medium. Hence, the essential and natural boundary conditions, which hold on the complementary parts of the external boundary of the body, compete with the following boundary conditions on the discontinuity as $\boldsymbol{\sigma}\cdot\mathbf{n}_{\Gamma_d}=-p\,\mathbf{n}_{\Gamma_d}$ and $[\![\dot{\mathbf{w}}]\!]\cdot\mathbf{n}_{\Gamma_d}=\bar{q}_d$ on Γ_d, where $\bar{q}_d$ is the leakage flux of the fluid along the fracture toward the surrounding porous medium, which implies that there exists a discontinuity in the normal fluid flow across Γ_d, and $\mathbf{n}_{\Gamma_d}$ is the unit normal vector to the discontinuity Γ_d pointing to Ω^+ with $\mathbf{n}_{\Gamma_d}=\mathbf{n}_{\Gamma_{d^-}}=-\mathbf{n}_{\Gamma_{d^+}}$, as shown in Figure 10.13. The notation $[\![\Xi]\!]=\Xi^+-\Xi^-$ represents the difference between the corresponding values at the two crack faces. It is considered that fluid pressure has the same value at both faces of the crack and pressure continuity from the crack faces to the surrounding porous medium holds along the crack boundary. It is noted that in imposing the

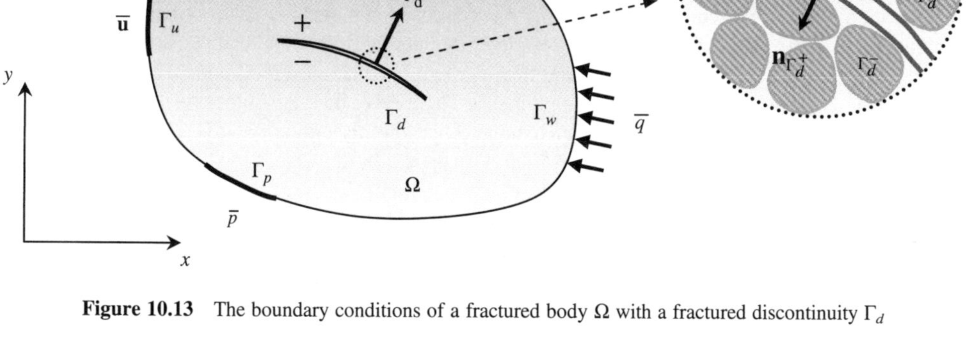

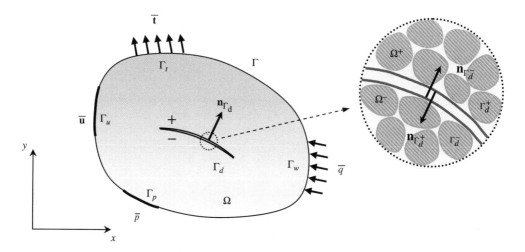

Figure 10.13 The boundary conditions of a fractured body Ω with a fractured discontinuity Γ_d

boundary conditions a distinction is made between the crack mouth and internal boundary of the crack. Conditions at the crack mouth are imposed as essential or natural boundary conditions on the external boundary of the body, while boundary conditions on the crack are the direct consequence of the governing equations.

In order to model the fluid flow through the discontinuity, the continuity equation for the fluid flow within the fracture can be written according to Eq. (10.7) as

$$\nabla \cdot \dot{\mathbf{w}} + \alpha \nabla \cdot \dot{\mathbf{u}} + \frac{1}{K_f}\dot{p} = 0 \tag{10.36}$$

where $\dot{\mathbf{w}}$ is the Darcy velocity vector of the fracturing fluid and is in fact the relative velocity of the fracturing fluid with respect to the solid phase defined as

$$\dot{\mathbf{w}} = k_{f_d}\left(-\nabla p - \rho_f\,\ddot{\mathbf{u}} + \rho_f\,\mathbf{b}\right) \tag{10.37}$$

where k_{f_d} is the fracture permeability with respect to the fracturing fluid. For a viscous fluid flow with Newtonian rheology, the fracture permeability is estimated using the cubic law as

$$k_{f_d} = \frac{1}{\kappa}\frac{w^2}{12\mu_f} \tag{10.38}$$

where $w = 2h$ is the fracture opening, μ_f denotes the dynamic viscosity of the fluid, and κ is a coefficient varying from 1.04 to 1.65. The parameter κ accounts for the effect of the deviation from the ideal parallel faces conditions. In this relation, $w^2/12\kappa$ is the intrinsic permeability of the fracture.

The existence of fracture besides the longitudinal conductivity may represent an obstacle for fluid flow in the transversal direction because of the potential drop due to the transition from a pore system into an open channel and back into a pore system. Thus, defining the flow potential Φ equal to $\Phi = p + \rho_f(\ddot{u}_i - b_i)\,x_i$ with x_i denoting the distance from a datum in ith direction, a jump is admitted in the total flow potential field across the fracture related to transverse fluid flux q_t traveling normal to the discontinuity. Considering a discrete version of Darcy's law, in which the total flow potential drop plays the role of total flow potential gradient, the transverse fluid flux is linked to the difference of hydraulic potentials between the two surfaces defining the discontinuity by $q_t = k_t(\Phi^- - \Phi^+)$, where k_t is a transverse hydraulic permeability coefficient and the superscripts − and + stand for each side of the discontinuity.

10.4.2 The Weak Formulation of a Fractured Porous Medium

In order to derive the X-FEM formulation for a deformable porous media with a strong discontinuity, the Divergence theorem for discontinuous problems was employed into the integral formulation of governing equations (10.4) and (10.7). The weak form of governing partial differential equations was obtained by integrating the product of the equilibrium and flow continuity equations (10.4) and (10.7) multiplied by admissible test functions over the domain, applying the Divergence theorem, imposing the natural boundary conditions, and satisfying the boundary conditions on the discontinuity. It is noteworthy that the mechanical and mass transfer coupling terms naturally appear in the weak form of the equilibrium and flow continuity equations as a result of the presence of discontinuity in the porous medium. In this manner, it is ensured that the coupling between the discontinuity and the continuum medium surrounding the discontinuity is met in the weak form. The weak form of governing equations was finally obtained multiplying the test functions $\delta\mathbf{u}(\boldsymbol{x}, t)$ and $\delta p(\boldsymbol{x}, t)$ by Eqs. (10.4) and (10.7) and integrating over the domain Ω as

$$\int_{\Omega} \nabla \delta \mathbf{u} : \boldsymbol{\sigma}\, d\Omega + \int_{\Omega} \delta \mathbf{u} \cdot \rho \ddot{\mathbf{u}}\, d\Omega + \int_{\Gamma_d} [\![\delta \mathbf{u}]\!] \boldsymbol{\sigma} \cdot \mathbf{n}_{\Gamma_d} d\Gamma - \int_{\Omega} \delta \mathbf{u} \cdot \rho\, \mathbf{b}\, d\Omega - \int_{\Gamma_t} \delta \mathbf{u} \cdot \bar{\mathbf{t}}\, d\Gamma = 0 \quad (10.9\text{a} - \text{repeated})$$

$$\begin{aligned} &\int_{\Omega} \nabla \delta p\, \mathbf{k}_f\, \nabla p\, d\Omega + \int_{\Omega} \nabla \delta p\, \mathbf{k}_f \cdot \rho_f\, \ddot{\mathbf{u}}\, d\Omega - \int_{\Gamma_d} \delta p [\![\dot{\mathbf{w}}]\!] \cdot \mathbf{n}_{\Gamma_d} d\Gamma \\ &+ \int_{\Omega} \delta p\, \alpha\, \nabla \cdot \dot{\mathbf{u}}\, d\Omega + \int_{\Omega} \delta p \frac{1}{Q} \dot{p}\, d\Omega - \int_{\Omega} \nabla \delta p\, \mathbf{k}_f \cdot \rho_f\, \mathbf{b}\, d\Omega + \int_{\Gamma_w} \delta p (\dot{\mathbf{w}} \cdot \mathbf{n}_{\Gamma}) d\Gamma = 0 \end{aligned} \quad (10.9\text{b} - \text{repeated})$$

in which the mechanical coupling term in the weak form of Eq. (10.9a) is computed by assigning the positive and negative sides to Γ_d and imposing the internal boundary condition on the discontinuity as

$$\begin{aligned} -\int_{\Gamma_d^+} \delta \mathbf{u} \cdot \boldsymbol{\sigma} \cdot \mathbf{n}_{\Gamma_d^+} d\Gamma - \int_{\Gamma_d^-} \delta \mathbf{u} \cdot \boldsymbol{\sigma} \cdot \mathbf{n}_{\Gamma_d^-} d\Gamma &= \int_{\Gamma_d} (\delta \mathbf{u}^+ - \delta \mathbf{u}^-) \cdot \boldsymbol{\sigma} \cdot \mathbf{n}_{\Gamma_d} d\Gamma \\ &= \int_{\Gamma_d} [\![\delta \mathbf{u}]\!] (\boldsymbol{\sigma} \cdot \mathbf{n}_{\Gamma_d}) d\Gamma = \int_{\Gamma_d} [\![\delta \mathbf{u}]\!] (-p \cdot \mathbf{n}_{\Gamma_d}) d\Gamma \end{aligned} \quad (10.39)$$

where $\mathbf{n}_{\Gamma_d^+}$ and $\mathbf{n}_{\Gamma_d^-}$ are the unit normal vectors directed to Ω^- and Ω^+, respectively, as clarified in Figure 10.13, and the superscripts + and – above Γ_d represent the two sides of the discontinuity. It should be mentioned that $\mathbf{n}_{\Gamma_d}$ has been taken such that $\mathbf{n}_{\Gamma_d} = \mathbf{n}_{\Gamma_d^-} = -\mathbf{n}_{\Gamma_d^+}$.

In addition, the mass transfer coupling term in the weak form of the continuity equation (10.9b) is obtained by imposing the boundary condition on the discontinuity as

$$\begin{aligned} &-\int_{\Gamma_d^+} \delta p \left(\dot{\mathbf{w}} \cdot \mathbf{n}_{\Gamma_d^+} \right) d\Gamma - \int_{\Gamma_d^-} \delta p \left(\dot{\mathbf{w}} \cdot \mathbf{n}_{\Gamma_d^-} \right) d\Gamma \\ &= \int_{\Gamma_d} \delta p (\dot{\mathbf{w}}^+ - \dot{\mathbf{w}}^-) \cdot \mathbf{n}_{\Gamma_d} d\Gamma = \int_{\Gamma_d} \delta p [\![\dot{\mathbf{w}}]\!] \cdot \mathbf{n}_{\Gamma_d} d\Gamma = \int_{\Gamma_d} \delta p\, \bar{q}_d\, d\Gamma \end{aligned} \quad (10.40)$$

in which the term $[\![\dot{\mathbf{w}}]\!] \cdot \mathbf{n}_{\Gamma_d} = \bar{q}_d$ corresponds to fluid flow from the crack normal to the cavity. In fact, the mass transfer coupling term implies that the Darcy velocity of the fluid normal to the fracture is discontinuous on Γ_d.

In order to derive a relation for the mass transfer coupling term emerging in the weak form of the flow continuity equation (10.9b) within the porous medium surrounding the fracture, the flow continuity of fluid inside the fracture is employed using Eq. (10.36). The weak form of flow continuity equation within the fracture can be obtained multiplying the test functions $\delta p(\boldsymbol{x}, t)$ by Eq. (10.36) and integrating over the domain of the discontinuity Ω' (Figure 10.14) as

$$\int_{\Omega'} \delta p \left(\nabla \cdot \dot{\mathbf{w}} + \alpha\, \nabla \cdot \dot{\mathbf{u}} + \frac{1}{K_f} \dot{p} \right) d\Omega = 0 \quad (10.41)$$

or

$$\int_{\Omega'} \delta p \left(\nabla \cdot \left[k_{f_d} \left(-\nabla p - \rho_f\, \ddot{\mathbf{u}} + \rho_f\, \mathbf{b} \right) \right] + \alpha\, \nabla \cdot \dot{\mathbf{u}} + \frac{1}{K_f} \dot{p} \right) d\Omega = 0 \quad (10.42)$$

The implementation of the Divergence theorem for discontinuous problems and the incorporation of the Darcy relation lead to the weak form of continuity equation of flow within the fracture as

$$\begin{aligned} &\int_{\Omega'} \nabla \delta p\, k_{f_d}\, \nabla p\, d\Omega + \int_{\Omega'} \nabla \delta p\, k_{f_d}\, \rho_f\, \ddot{\mathbf{u}}\, d\Omega - \int_{\Gamma_d} \delta p [\![\dot{\mathbf{w}}]\!] \cdot \mathbf{n}_{\Gamma_d} d\Gamma \\ &+ \int_{\Omega'} \delta p\, \alpha\, \nabla \cdot \dot{\mathbf{u}}\, d\Omega + \int_{\Omega'} \delta p \frac{1}{K_f} \dot{p}\, d\Omega - \int_{\Omega} \nabla \delta p\, k_{f_d}\, \rho_f\, \mathbf{b}\, d\Omega = 0 \end{aligned} \quad (10.43)$$

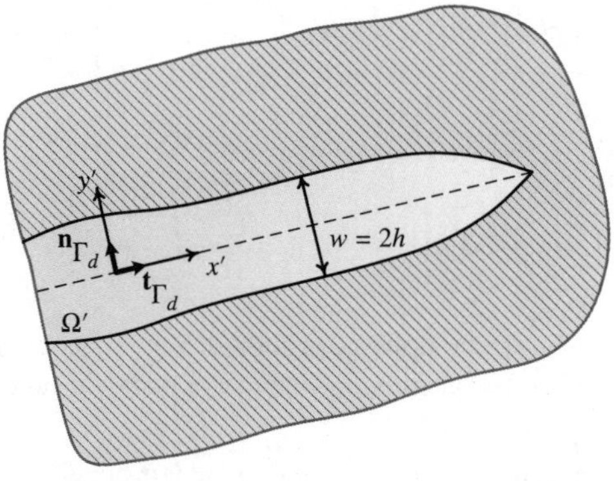

Figure 10.14 The geometry of the fluid flow inside the fracture

Hence, the mass transfer coupling term in the weak form of the flow continuity equation (10.9b) can be obtained from the weak form of the flow continuity equation (10.43) within the fracture as

$$\int_{\Gamma_d} \delta p [\dot{\mathbf{w}}] \cdot \mathbf{n}_{\Gamma_d} d\Gamma = -\underbrace{\int_{\Omega'} \nabla \delta p \, k_{f_d} \, \nabla p \, d\Omega}_{\text{integral (I)}} - \underbrace{\int_{\Omega'} \nabla \delta p \, k_{f_d} \, \rho_f \, \ddot{\mathbf{u}} \, d\Omega}_{\text{integral (II)}}$$
$$-\underbrace{\int_{\Omega'} \delta p \, \alpha \, \nabla \cdot \dot{\mathbf{u}} \, d\Omega}_{\text{integral (III)}} - \underbrace{\int_{\Omega'} \delta p \frac{1}{K_f} \dot{p} \, d\Omega}_{\text{integral (IV)}} + \underbrace{\int_{\Omega} \nabla \delta p \, k_{f_d} \, \rho_f \, \mathbf{b} \, d\Omega}_{\text{integral (V)}} \tag{10.44}$$

In order to evaluate the mass transfer coupling term in Eq. (10.44), the integral terms (I)–(V) are calculated in the local Cartesian coordinate system (x', y'), where the directions of x' and y' are aligned with the tangent and normal unit vectors to the discontinuity, $\mathbf{t}_{\Gamma_d}$ and $\mathbf{n}_{\Gamma_d}$ respectively, as shown in Figure 10.14. The integrals over the domain of discontinuity Ω' are performed in the local coordinate system, in which the integral over the cross section of the discontinuity is evaluated by ignoring the variation of fluid pressure across the discontinuity width since the width of the discontinuity $2h$ is insignificant relative to its length. As a result, the fluid pressure as well as its corresponding test function is assumed to be uniform over the cross section of the discontinuity. Hence, the derivation of integrals (I)–(V) in Eq. (10.44) can be obtained as

$$\underbrace{\int_{\Omega'} \nabla \delta p \, k_{f_d} \, \nabla p \, d\Omega}_{\text{integral (I)}} = \int_{\Gamma_d} \int_{-h}^{+h} \nabla \delta p \, k_{f_d} \nabla p \, dy' \, d\Gamma$$
$$= \int_{\Gamma_d} \int_{-h}^{+h} k_{f_d} \left(\frac{\partial \delta p}{\partial x'} \frac{\partial p}{\partial x'} + \frac{\partial \delta p}{\partial y'} \frac{\partial p}{\partial y'} \right) dy' \, d\Gamma = \int_{\Gamma_d} k_{f_d} (2h) \frac{\partial \delta p}{\partial x'} \frac{\partial p}{\partial x'} d\Gamma \tag{10.45a}$$

$$\underbrace{\int_{\Omega'} \nabla \delta p \, k_{f_d} \, \rho_f \, \ddot{\mathbf{u}} \, d\Omega}_{\text{integral (II)}} = \int_{\Gamma_d} \int_{-h}^{+h} \nabla \delta p \, k_{f_d} \, \rho_f \, \ddot{\mathbf{u}} \, dy' \, d\Gamma$$
$$= \int_{\Gamma_d} \int_{-h}^{+h} k_{f_d} \, \rho_f \left(\frac{\partial \delta p}{\partial x'} \ddot{u}_{x'} + \frac{\partial \delta p}{\partial y'} \ddot{u}_{y'} \right) dy' \, d\Gamma \tag{10.45b}$$
$$= \int_{\Gamma_d} k_{f_d} \, \rho_f \, (2h) \frac{\partial \delta p}{\partial x'} \frac{1}{2} \left(\ddot{u}_{x'}|_{+h} + \ddot{u}_{x'}|_{-h} \right) d\Gamma = \int_{\Gamma_d} k_{f_d} \, \rho_f (2h) \frac{\partial \delta p}{\partial x'} \langle \ddot{u}_{x'} \rangle d\Gamma$$

$$\underbrace{\int_{\Omega'} \delta p\, \alpha\, \nabla \cdot \dot{\mathbf{u}}\, d\Omega}_{\text{integral (III)}} = \int_{\Gamma_d} \int_{-h}^{+h} \delta p\, \alpha\, \nabla \cdot \dot{\mathbf{u}}\, dy'\, d\Gamma = \int_{\Gamma_d} \int_{-h}^{+h} \delta p\, \alpha \left(\frac{\partial \dot{u}_{x'}}{\partial x'} + \frac{\partial \dot{u}_{y'}}{\partial y'} \right) dy'\, d\Gamma$$

$$= \int_{\Gamma_d} \delta p\, \alpha\, (2h) \frac{1}{2} \left(\left. \frac{\partial \dot{u}_{x'}}{\partial x'} \right|_{+h} + \left. \frac{\partial \dot{u}_{x'}}{\partial x'} \right|_{-h} \right) d\Gamma + \int_{\Gamma_d} \delta p\, \alpha \left(\dot{u}_{y'}|_{+h} - \dot{u}_{y'}|_{-h} \right) d\Gamma \tag{10.45c}$$

$$= \int_{\Gamma_d} \delta p\, \alpha\, (2h) \left\langle \frac{\partial \dot{u}_{x'}}{\partial x'} \right\rangle d\Gamma + \int_{\Gamma_d} \delta p\, \alpha\, [\![\dot{u}_{y'}]\!]\, d\Gamma$$

$$\underbrace{\int_{\Omega'} \delta p \frac{1}{K_f} \dot{p}\, d\Omega}_{\text{integral (IV)}} = \int_{\Gamma_d} \int_{-h}^{+h} \delta p \frac{1}{K_f} \dot{p}\, dy'\, d\Gamma = \int_{\Gamma_d} \delta p\, (2h) \frac{1}{K_f} \dot{p}\, d\Gamma \tag{10.45d}$$

$$\underbrace{\int_{\Omega} \nabla \delta p\, k_{f_d}\, \rho_f\, \mathbf{b}\, d\Omega}_{\text{integral (V)}} = \int_{\Gamma_d} k_{f_d}\, \rho_f\, (2h) \frac{\partial \delta p}{\partial x'} b_{x'}\, d\Gamma \tag{10.45e}$$

In these equations, $\dot{u}_{x'}$ and $\dot{u}_{y'}$ are the components of the solid velocity vector projected on the longitudinal and transversal axes, respectively. It is assumed that the velocity component of the solid phase in the longitudinal direction $\dot{u}_{x'}$ varies linearly with y' over the width of the discontinuity $2h$, so its derivative with respect to x' also varies linearly with y'. Similar to that assumed for the solid velocity in the tangential direction of the discontinuity, the solid acceleration in the tangential direction $\ddot{u}_{x'}$ also varies linearly with y'. As noted earlier, the fluid pressure and its corresponding test function are assumed to be uniform across the width of the discontinuity, so their derivatives in the tangential direction do not vary with y', and their derivatives in the normal direction vanish. Furthermore, the tangential derivatives in the previous equations are analytically integrated over the width of the discontinuity $2h$, in which $\langle \Xi \rangle = (\Xi^+ + \Xi^-)/2$ is defined as the average of corresponding values at the discontinuity faces.

Substituting the constituents of Eq. (10.44) and rearranging it, the mass transfer coupling term appeared in the weak form of the flow continuity equation (10.9b) in the porous medium can be obtained as

$$\begin{aligned} \int_{\Gamma_d} \delta p\, [\![\dot{\mathbf{w}}]\!] \cdot \mathbf{n}_{\Gamma_d} d\Gamma = & -\int_{\Gamma_d} k_{f_d} (2h) \frac{\partial \delta p}{\partial x'} \frac{\partial p}{\partial x'} d\Gamma - \int_{\Gamma_d} k_{f_d}\, \rho_f (2h) \frac{\partial \delta p}{\partial x'} \langle \ddot{u}_{x'} \rangle d\Gamma \\ & -\int_{\Gamma_d} \delta p\, \alpha\, (2h) \left\langle \frac{\partial \dot{u}_{x'}}{\partial x'} \right\rangle d\Gamma - \int_{\Gamma_d} \delta p\, \alpha\, [\![\dot{u}_{y'}]\!]\, d\Gamma \\ & -\int_{\Gamma_d} \delta p\, (2h) \frac{1}{K_f} \dot{p}\, d\Gamma + \int_{\Gamma_d} k_{f_d}\, \rho_f\, (2h) \frac{\partial \delta p}{\partial x'} b_{x'}\, d\Gamma \end{aligned} \tag{10.46}$$

This equation presents an expression for the leak-off from the crack into the porous media. Since the flow within the fracture is generally assumed to be incompressible, the integral (IV) can be removed from Eq. (10.46). In order to model the flow within the fracture, a viscous Newtonian fluid flow is assumed between the crack faces, in which the fracture permeability is defined according to (10.38) based on the cubic law as $k_{f_d} = (2h)^2 / 12\kappa\, \mu_f$. It must be noted that this relation is only valid if the fluid regime inside the cavity is laminar.

10.5 The X-FEM Formulation of Deformable Porous Media with Strong Discontinuities

In this section, the weak form of the equilibrium and flow continuity equations of the porous medium obtained in the preceding section (Section 10.4) is used to derive the discrete form of governing equations by discretization the governing equations first in the spatial domain employing the X-FEM and then in the time domain applying the generalized Newmark scheme. The resulting system of fully coupled non-linear equations is finally solved using the Newton–Raphson procedure. In order to discretize the integral equations (10.9), the X-FEM is employed in the spatial domain to obtain the values of $\mathbf{u}(\boldsymbol{x}, t)$ and $p(\boldsymbol{x}, t)$. Furthermore, to integrate the X-FEM differential equations in time, the generalized Newmark GN22 scheme is employed for the displacement field and the GN11 method for the pressure field.

10.5.1 Approximation of the Displacement and Pressure Fields

In the X-FEM method, the conventional FEM shape functions are enriched using proper enrichment functions based on the type of discontinuity. The displacement and pressure fields are enriched based on the analytical solutions to make the approximations capable of tracking the discontinuities. In order to capture the hydro-mechanical coupling associated with the tractions acting on the crack edges and the fluid leak-off from the fracture into the domain, the displacement, and fluid pressure fields are enhanced using appropriate enrichment functions. To describe the displacement jump across the fracture, the displacement field is assumed to be discontinuous over the crack edges. In addition, to present the fluid flow jump normal to the fracture, the fluid pressure field is assumed to be continuous, while its gradient normal to the fracture is discontinuous over the crack edges. Thus, the displacement discontinuity over the crack edges is modeled using the Heaviside and crack tip asymptotic functions, and the fluid pressure is modeled using the modified level set function to capture the discontinuity on the gradient of fluid pressure normal to the fracture.

The enriched approximation of the X-FEM for the displacement field $\mathbf{u}(\boldsymbol{x}, t)$ can be written as

$$\begin{aligned}\mathbf{u}(\boldsymbol{x},t)\approx\mathbf{u}^h(\boldsymbol{x},t)&=\sum_{I\in\mathcal{N}}N_{uI}(\boldsymbol{x})\,\bar{\mathbf{u}}_I(t)+\sum_{J\in\mathcal{N}^{dis}}N_{uJ}(\boldsymbol{x})\,(H(\varphi(\boldsymbol{x}))-H(\varphi(\boldsymbol{x}_J)))\,\bar{\mathbf{a}}_J(t)\\&+\sum_{K\in\mathcal{N}^{tip}}N_{uK}(\boldsymbol{x})\sum_{\alpha=1}^{4}(\mathcal{B}_\alpha(\boldsymbol{x})-\mathcal{B}_\alpha(\boldsymbol{x}_K))\,\bar{\mathbf{b}}_{\alpha K}(t)\end{aligned} \tag{10.47}$$

where $\mathcal{N}$ is the set of all nodal points, $\mathcal{N}^{dis}$ is the set of enriched nodes whose support is bisected by the crack, and $\mathcal{N}^{tip}$ is the set of nodes which contain the crack-tip in the support of their shape functions enriched by the asymptotic functions. In the previous relation, $\bar{\mathbf{u}}_I(t)$ are the unknown standard nodal displacements at node I, vector $\bar{\mathbf{a}}_J(t)$ are the unknown enriched nodal DOF associated with the Heaviside enrichment function at node J, and $\bar{\mathbf{b}}_{\alpha K}(t)$ are the additional enriched nodal DOF associated with the asymptotic functions at node K. In relation 10.47, $N_u(\boldsymbol{x})$ denote the standard displacement shape functions, $H(\varphi(\boldsymbol{x}))$ is the Heaviside jump function used to model the discontinuity due to different displacement fields on either sides of the crack, and $\mathcal{B}_\alpha(\boldsymbol{x})$ are the asymptotic functions extracted from the analytical solution and used to model the displacement field at the crack tip region. The asymptotic tip functions $\mathcal{B}_\alpha(\boldsymbol{x})$ and the Heaviside jump function $H(\varphi(\boldsymbol{x}))$ are defined as

$$\begin{aligned}\mathcal{B}(r,\theta)&=\{\mathcal{B}_1,\mathcal{B}_2,\mathcal{B}_3,\mathcal{B}_4\}\\&=\left\{\sqrt{r}\sin\frac{\theta}{2},\sqrt{r}\cos\frac{\theta}{2},\sqrt{r}\sin\frac{\theta}{2}\sin\theta,\sqrt{r}\cos\frac{\theta}{2}\sin\theta\right\}\end{aligned} \tag{10.48}$$

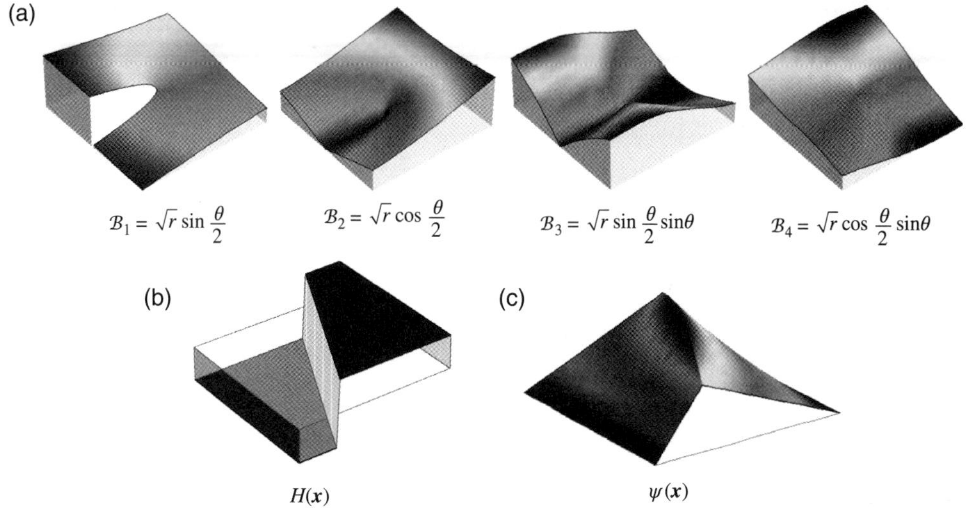

Figure 10.15 The enrichment functions of displacement and pressure fields in a fractured porous medium: (a) the crack tip asymptotic functions; (b) the Heaviside function; (c) the modified level-set function

and

$$H(\varphi(\boldsymbol{x})) = \begin{cases} +1 & \varphi(\boldsymbol{x}) \geq 0 \\ 0 & \varphi(\boldsymbol{x}) < 0 \end{cases} \tag{10.49}$$

where $\varphi(\boldsymbol{x})$ is the signed distance function defined in (10.15). It can be observed from the definition (10.49) that the Heaviside function exhibits a unit jump across the discontinuity with independent displacement fields at both sides of the crack, so it is an appropriate function for simulation of the cracked domain. The distributions of four crack-tip asymptotic functions together with the Heaviside step function are shown in Figure 10.15a and b.

The enriched approximation of the X-FEM for the pressure field $p(\boldsymbol{x}, t)$ can be written as

$$p(\boldsymbol{x},t) \approx p^h(\boldsymbol{x},t) = \sum_{I \in \mathcal{N}} N_{pI}(\boldsymbol{x})\bar{p}_I(t) + \sum_{J \in \mathcal{N}^{dis}} N_{pJ}(\boldsymbol{x})(\psi(\boldsymbol{x}) - \psi(\boldsymbol{x}_J))\bar{c}_J(t) \tag{10.50}$$

where $\bar{p}_I(t)$ is the unknown standard nodal pressure at node I, and $\bar{c}_J(t)$ is the unknown enriched nodal DOF associated with the modified level set function at node J. In this relation, $N_p(\boldsymbol{x})$ are the standard pressure shape functions, and $\psi(\boldsymbol{x})$ is the modified level set function defined as

$$\psi(\boldsymbol{x}) = \sum_{I \in \mathcal{N}^{dis}} N_{pI}(\boldsymbol{x})|\varphi_I| - \left| \sum_{I \in \mathcal{N}^{dis}} N_{pI}(\boldsymbol{x})\,\varphi_I \right| \tag{10.51}$$

where φ_I are the nodal values of the level set function. This enrichment function has a ridge centered on the interpolated interface and vanishes in elements not containing the material interface. The distribution of the modified level set function is shown in Figure 10.15c. It must be noted that this enrichment is only required in the elements where the discontinuity on the gradient of fluid pressure needs to be captured. In relation (10.51), the enrichment function is continuous across the fracture, while its gradient is

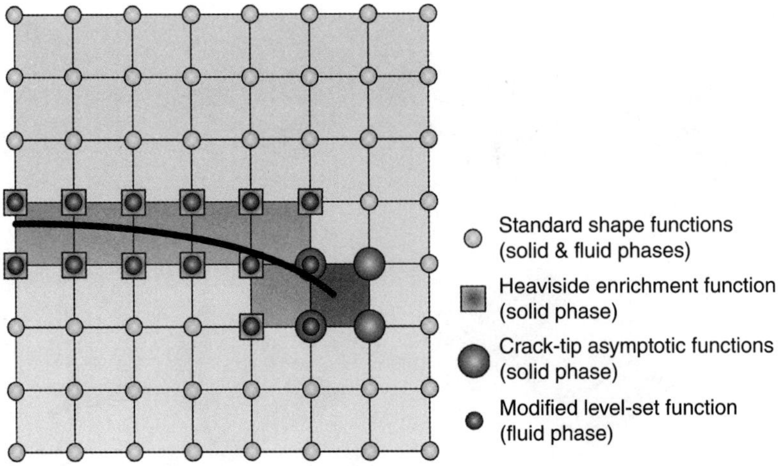

Figure 10.16 The enriched nodal points of the displacement and pressure fields in porous media

discontinuous in the normal direction to the discontinuity. Possessing this desirable property, the chosen enrichment function enables the approximate pressure field to be discontinuous in its normal derivative across the discontinuity, accounting for the leak-off of the fluid from the discontinuity. It is important that the leakage flux vanishes at the crack tip, which can be assured by requiring that the nodes on the element edge with which the crack-tip coincides not be enriched. In Figure 10.16, the enriched nodal points of the displacement and pressure fields are presented for a discontinuous domain according to the Heaviside jump function, asymptotic tip functions, and the modified level-set function.

Finally, the enriched FE approximation of the displacement and pressure fields (10.47) and (10.50) can be symbolically written in the following form

$$\mathbf{u}^h(\boldsymbol{x},t)=\mathbf{N}_u^{std}(\boldsymbol{x})\bar{\mathbf{u}}(t)+\mathbf{N}_u^{Hev}(\boldsymbol{x})\bar{\mathbf{a}}(t)+\mathbf{N}_u^{tip}(\boldsymbol{x})\bar{\mathbf{b}}(t) \tag{10.52}$$

$$p^h(\boldsymbol{x},t)=\mathbf{N}_p^{std}(\boldsymbol{x})\bar{\mathbf{p}}(t)+\mathbf{N}_p^{abs}(\boldsymbol{x})\bar{\mathbf{c}}(t) \tag{10.53}$$

where $\mathbf{N}_u^{std}(\boldsymbol{x})$ is the matrix of displacement shape functions that include the standard piecewise polynomial shape functions, $\mathbf{N}_u^{Hev}(\boldsymbol{x})$ is the matrix of enriched shape functions associated with the Heaviside function, $\mathbf{N}_u^{tip}(\boldsymbol{x})$ is the matrix of enriched shape functions associated with the asymptotic tip functions, $\bar{\mathbf{u}}(t)$ is the vector of standard displacement DOF, $\bar{\mathbf{a}}(t)$ is the vector of enriched displacement DOF associated with the Heaviside function, and $\bar{\mathbf{b}}(t)$ is the vector of enriched displacement DOF associated with the asymptotic tip functions. In relation (10.53), $\mathbf{N}_p^{std}(\boldsymbol{x})$ is the matrix of pressure shape functions, $\mathbf{N}_p^{abs}(\boldsymbol{x})$ is the matrix of enriched shape functions associated with the modified level set function, $\bar{\mathbf{p}}(t)$ is the vector of pressure nodal DOF, and $\bar{\mathbf{c}}(t)$ is the vector of enriched nodal DOF associated with the modified level set function.

Accordingly, the strain vector corresponding to the approximate displacement field (10.52) can be written in terms of the standard and enriched nodal values as

$$\boldsymbol{\varepsilon}^h(\boldsymbol{x},t)=\mathbf{B}_u^{std}(\boldsymbol{x})\bar{\mathbf{u}}(t)+\mathbf{B}_u^{Hev}(\boldsymbol{x})\bar{\mathbf{a}}(t)+\mathbf{B}_u^{tip}(\boldsymbol{x})\bar{\mathbf{b}}(t) \tag{10.54}$$

where $\mathbf{B}_u^{std}(\boldsymbol{x})\equiv\mathbf{L}\,\mathbf{N}_u^{std}(\boldsymbol{x})$ involve the spatial derivatives of the standard shape functions, $\mathbf{B}_u^{Hev}(\boldsymbol{x})\equiv\mathbf{L}\,\mathbf{N}_u^{Hev}(\boldsymbol{x})$ contain the spatial derivatives of the enriched shape functions associated with the Heaviside enrichment function, and $\mathbf{B}_u^{tip}(\boldsymbol{x})\equiv\mathbf{L}\,\mathbf{N}_u^{tip}(\boldsymbol{x})$ contain the spatial derivatives of the enriched shape functions associated with the crack-tip asymptotic functions, where $\mathbf{L}$ is defined in (10.20).

10.5.2 The X-FEM Spatial Discretization

The discrete form of integral equations (10.9) in the framework of X-FEM formulation can be obtained using the test and trial functions of the enriched displacement and pressure fields. Applying the trial functions of enriched displacement and pressure fields (10.52) and (10.53) together with the test functions of displacement and pressure fields $\delta\mathbf{u}(\boldsymbol{x}, t)$ and $\delta p(\boldsymbol{x}, t)$, which are respectively defined in the same enriched approximating space as the approximate displacement and fluid pressure fields as

$$\delta\mathbf{u}^h(\boldsymbol{x},t)=\mathbf{N}_u^{std}(\boldsymbol{x})\,\delta\bar{\mathbf{u}}(t)+\mathbf{N}_u^{Hev}(\boldsymbol{x})\,\delta\bar{\mathbf{a}}(t)+\mathbf{N}_u^{tip}(\boldsymbol{x})\,\delta\bar{\mathbf{b}}(t) \tag{10.55}$$

$$\delta p^h(\boldsymbol{x},t)=\mathbf{N}_p^{std}(\boldsymbol{x})\,\delta\bar{\mathbf{p}}(t)+\mathbf{N}_p^{abs}(\boldsymbol{x})\,\delta\bar{\mathbf{c}}(t) \tag{10.56}$$

and satisfying the necessity that the weak form should hold for all kinematically admissible test functions, the discretized form of integral equations (10.9) can therefore be obtained according to the Bubnov–Galerkin technique as

$$\begin{pmatrix}\mathbf{M}_{uu} & \mathbf{M}_{ua} & \mathbf{M}_{ub}\\ \mathbf{M}_{au} & \mathbf{M}_{aa} & \mathbf{M}_{ab}\\ \mathbf{M}_{bu} & \mathbf{M}_{ba} & \mathbf{M}_{bb}\end{pmatrix}\begin{Bmatrix}\ddot{\bar{\mathbf{u}}}\\ \ddot{\bar{\mathbf{a}}}\\ \ddot{\bar{\mathbf{b}}}\end{Bmatrix}+\begin{pmatrix}\mathbf{K}_{uu} & \mathbf{K}_{ua} & \mathbf{K}_{ub}\\ \mathbf{K}_{au} & \mathbf{K}_{aa} & \mathbf{K}_{ab}\\ \mathbf{K}_{bu} & \mathbf{K}_{ba} & \mathbf{K}_{bb}\end{pmatrix}\begin{Bmatrix}\bar{\mathbf{u}}\\ \bar{\mathbf{a}}\\ \bar{\mathbf{b}}\end{Bmatrix}-\begin{pmatrix}\mathbf{Q}_{up} & \mathbf{Q}_{uc}\\ \mathbf{Q}_{ap} & \mathbf{Q}_{ac}\\ \mathbf{Q}_{bp} & \mathbf{Q}_{bc}\end{pmatrix}\begin{Bmatrix}\bar{\mathbf{p}}\\ \bar{\mathbf{c}}\end{Bmatrix}+\begin{Bmatrix}\mathbf{f}_u^{\text{int}}\\ \mathbf{f}_a^{\text{int}}\\ \mathbf{f}_b^{\text{int}}\end{Bmatrix}-\begin{Bmatrix}\mathbf{f}_u^{\text{ext}}\\ \mathbf{f}_a^{\text{ext}}\\ \mathbf{f}_b^{\text{ext}}\end{Bmatrix}=\mathbf{0}$$

$$\begin{pmatrix}\mathbf{C}_{pu} & \mathbf{C}_{pa} & \mathbf{C}_{pb}\\ \mathbf{C}_{cu} & \mathbf{C}_{ca} & \mathbf{C}_{cb}\end{pmatrix}\begin{Bmatrix}\ddot{\bar{\mathbf{u}}}\\ \ddot{\bar{\mathbf{a}}}\\ \ddot{\bar{\mathbf{b}}}\end{Bmatrix}+\begin{pmatrix}\mathbf{Q}_{up}^T & \mathbf{Q}_{ap}^T & \mathbf{Q}_{bp}^T\\ \mathbf{Q}_{uc}^T & \mathbf{Q}_{ac}^T & \mathbf{Q}_{bc}^T\end{pmatrix}\begin{Bmatrix}\dot{\bar{\mathbf{u}}}\\ \dot{\bar{\mathbf{a}}}\\ \dot{\bar{\mathbf{b}}}\end{Bmatrix}+\begin{pmatrix}\mathbf{H}_{pp} & \mathbf{H}_{pc}\\ \mathbf{H}_{cp} & \mathbf{H}_{cc}\end{pmatrix}\begin{Bmatrix}\bar{\mathbf{p}}\\ \bar{\mathbf{c}}\end{Bmatrix}+$$

$$\begin{pmatrix}\mathbf{S}_{pp} & \mathbf{S}_{pc}\\ \mathbf{S}_{cp} & \mathbf{S}_{cc}\end{pmatrix}\begin{Bmatrix}\dot{\bar{\mathbf{p}}}\\ \dot{\bar{\mathbf{c}}}\end{Bmatrix}-\begin{Bmatrix}\mathbf{q}_p^{\text{int}}\\ \mathbf{q}_c^{\text{int}}\end{Bmatrix}-\begin{Bmatrix}\mathbf{q}_p^{\text{ext}}\\ \mathbf{q}_c^{\text{ext}}\end{Bmatrix}=\mathbf{0} \tag{10.57}$$

where the mass matrix $\mathbf{M}$, stiffness matrix $\mathbf{K}$, coupling matrix $\mathbf{Q}$, permeability matrix $\mathbf{H}$, compressibility matrix $\mathbf{S}$, and external force vectors $\mathbf{f}^{\text{ext}}$ and $\mathbf{q}^{\text{ext}}$ are defined as

$$\begin{aligned}
\mathbf{M}_{\alpha\beta}&=\int_\Omega(\mathbf{N}_u^\alpha)^T\rho\,\mathbf{N}_u^\beta\,d\Omega\\
\mathbf{K}_{\alpha\beta}&=\int_\Omega(\mathbf{B}_u^\alpha)^T\mathbf{D}\,\mathbf{B}_u^\beta\,d\Omega\\
\mathbf{Q}_{\alpha\gamma}&=\int_\Omega(\mathbf{B}_u^\alpha)^T\alpha\,\mathbf{m}\,\mathbf{N}_p^\gamma\,d\Omega\\
\mathbf{f}_\alpha^{\text{ext}}&=\int_\Omega(\mathbf{N}_u^\alpha)^T\rho\,\mathbf{b}\,d\Omega+\int_{\Gamma_t}(\mathbf{N}_u^\alpha)^T\bar{\mathbf{t}}\,d\Gamma
\end{aligned} \tag{10.58}$$

and

$$\begin{aligned}
\mathbf{C}_{\delta\beta}&=\int_\Omega\left(\nabla\mathbf{N}_p^\delta\right)^T\boldsymbol{k}_f\cdot\rho_f\,\mathbf{N}_u^\beta\,d\Omega\\
\mathbf{H}_{\delta\gamma}&=\int_\Omega\left(\nabla\mathbf{N}_p^\delta\right)^T\boldsymbol{k}_f\left(\nabla\mathbf{N}_p^\gamma\right)d\Omega\\
\mathbf{S}_{\delta\gamma}&=\int_\Omega\left(\mathbf{N}_p^\delta\right)^T\frac{1}{Q}\mathbf{N}_p^\gamma\,d\Omega\\
\mathbf{q}_\delta^{\text{ext}}&=\int_\Omega\left(\nabla\mathbf{N}_p^\delta\right)^T\boldsymbol{k}_f\cdot\rho_f\,\mathbf{b}\,d\Omega-\int_{\Gamma_w}\left(\mathbf{N}_p^\delta\right)^T\bar{q}\,d\Gamma
\end{aligned} \tag{10.59}$$

where $(\alpha, \beta) \in (std,\ Hev,\ tip)$ denote the "standard", "Heaviside", and "asymptotic tip" functions of displacement field and, $(\delta, \gamma) \in (std,\ abs)$ denote the "standard" and "modified level set" functions of pressure field. In these definitions, $\mathbf{m}$ is the vector of delta Dirac function defined as $\mathbf{m} = \{1\ \ 1\ \ 0\}^T$.

In order to discretize the mechanical and mass transfer couplings between the fracture and the porous medium in integral equations (10.9), the interfacial force vector $\mathbf{f}_\alpha^{\text{int}}$ due to the fluid pressure exerted on the fracture faces by injecting a viscous fluid into the fracture, and the flux vectors $\mathbf{q}_\delta^{\text{int}}$ due to the fluid exchange between the fracture and the surrounding permeable porous medium in Eq. (10.57) can be obtained as

$$\mathbf{f}_\alpha^{\text{int}} = \int_{\Gamma_d} [\![\mathbf{N}_u^\alpha]\!]^T \boldsymbol{\sigma} \cdot \mathbf{n}_{\Gamma_d} d\Gamma = -\int_{\Gamma_d} [\![\mathbf{N}_u^\alpha]\!]^T p \cdot \mathbf{n}_{\Gamma_d} d\Gamma \tag{10.60}$$

where $\alpha \in (std,\ Hev,\ tip)$, and

$$\begin{aligned}
\mathbf{q}_\delta^{\text{int}} &= \int_{\Gamma_d} \left(\mathbf{N}_p^\delta\right)^T [\dot{\mathbf{w}}] \cdot \mathbf{n}_{\Gamma_d} d\Gamma = \int_{\Gamma_d} \left(\mathbf{N}_p^\delta\right)^T \bar{q}_d \, d\Gamma \\
&= -\int_{\Gamma_d} \left(\nabla \mathbf{N}_p^\delta\right)^T \cdot \mathbf{t}_{\Gamma_d} k_{fd} (2h) \nabla p \cdot \mathbf{t}_{\Gamma_d} d\Gamma - \int_{\Gamma_d} \left(\nabla \mathbf{N}_p^\delta\right)^T \cdot \mathbf{t}_{\Gamma_d} k_{fd} \rho_f (2h) \langle \ddot{\mathbf{u}} \rangle \cdot \mathbf{t}_{\Gamma_d} d\Gamma \\
&\quad - \int_{\Gamma_d} \left(\mathbf{N}_p^\delta\right)^T \alpha (2h) \, \mathbf{t}_{\Gamma_d} \cdot \langle \nabla \dot{\mathbf{u}} \rangle \cdot \mathbf{t}_{\Gamma_d} d\Gamma - \int_{\Gamma_d} \left(\mathbf{N}_p^\delta\right)^T \alpha \, [\dot{\mathbf{u}}] \cdot \mathbf{n}_{\Gamma_d} d\Gamma \\
&\quad - \int_{\Gamma_d} \left(\mathbf{N}_p^\delta\right)^T (2h) \frac{1}{K_f} \dot{p} \, d\Gamma + \int_{\Gamma_d} \left(\nabla \mathbf{N}_p^\delta\right)^T \cdot \mathbf{t}_{\Gamma_d} k_{fd} \rho_f (2h) \, \mathbf{b} \cdot \mathbf{t}_{\Gamma_d} d\Gamma
\end{aligned} \tag{10.61}$$

where $\delta \in (std,\ abs)$. Finally, the X-FEM discretization equations (10.57) can be rewritten as

$$\begin{aligned}
&\mathbf{M}\ddot{\bar{\mathbb{U}}} + \mathbf{K}\bar{\mathbb{U}} - \mathbf{Q}\bar{\mathbb{P}} + \mathbf{f}_{\mathbb{U}}^{\text{int}} - \mathbf{f}_{\mathbb{U}}^{\text{ext}} = \mathbf{0} \\
&\mathbf{C}\ddot{\bar{\mathbb{U}}} + \mathbf{Q}^T \dot{\bar{\mathbb{U}}} + \mathbf{H}\bar{\mathbb{P}} + \mathbf{S}\dot{\bar{\mathbb{P}}} - \mathbf{q}_{\mathbb{P}}^{\text{int}} - \mathbf{q}_{\mathbb{P}}^{\text{ext}} = \mathbf{0}
\end{aligned} \tag{10.62}$$

where $\bar{\mathbb{U}} = \langle \bar{\mathbf{u}}, \bar{\mathbf{a}}, \bar{\mathbf{b}} \rangle$ and $\bar{\mathbb{P}} = \langle \bar{\mathbf{p}}, \bar{\mathbf{c}} \rangle$ are the complete set of the standard and enriched DOF of displacement and pressure fields, respectively.

10.5.3 The Time Domain Discretization and Solution Procedure

In order to complete the numerical solution of FE equations, it is necessary to integrate the time differential equations (10.62) in time. The well-known generalized Newmark GN22 scheme is employed for the displacement field $\bar{\mathbb{U}}$ and the GN11 method for the pressure field $\bar{\mathbb{P}}$. The link between the successive values of the unknown field variables at time $t_{n+1} = t_n + \Delta t$ and the known field variables at time t_n can be established as

$$\begin{aligned}
\ddot{\bar{\mathbb{U}}}_{n+1} &= \frac{1}{\beta \Delta t^2} \left(\bar{\mathbb{U}}_{n+1} - \bar{\mathbb{U}}_n\right) - \frac{1}{\beta \Delta t} \dot{\bar{\mathbb{U}}}_n - \left(\frac{1}{2\beta} - 1\right) \ddot{\bar{\mathbb{U}}}_n \\
\dot{\bar{\mathbb{U}}}_{n+1} &= \frac{\gamma}{\beta \Delta t} \left(\bar{\mathbb{U}}_{n+1} - \bar{\mathbb{U}}_n\right) - \left(\frac{\gamma}{\beta} - 1\right) \dot{\bar{\mathbb{U}}}_n - \Delta t \left(\frac{\gamma}{2\beta} - 1\right) \ddot{\bar{\mathbb{U}}}_n
\end{aligned} \tag{10.63a}$$

and

$$\dot{\bar{\mathbb{P}}}_{n+1} = \frac{1}{\theta\,\Delta t}(\bar{\mathbb{P}}_{n+1} - \bar{\mathbb{P}}_n) - \left(\frac{1}{\theta} - 1\right)\dot{\bar{\mathbb{P}}}_n \tag{10.63b}$$

where β, γ, and θ are parameters, which are usually chosen in the range of 0–1. However, for unconditionally stability of the time integration procedure, it is required that $\gamma \geq 0.5$, $\beta \geq 0.25(0.5+\gamma)^2$, and $\theta \geq 0.5$. In these relations, $\bar{\mathbb{U}}_n$, $\dot{\bar{\mathbb{U}}}_n$, and $\ddot{\bar{\mathbb{U}}}_n$ denote the known values of displacement, velocity, and acceleration of the standard and enriched DOF at time t_n, and $\bar{\mathbb{P}}_n$, and $\dot{\bar{\mathbb{P}}}_n$ are the known values of pressure and gradient of pressure of the standard and enriched DOF at time t_n. Substituting relations (10.63a) and (10.63b) into the spatial discrete equations (10.62), the following nonlinear equation can be achieved as

$$\begin{aligned} \mathbf{\Psi}_{\mathbb{U}_{n+1}} &= \frac{1}{\beta\,\Delta t^2}\mathbf{M}\,\bar{\mathbb{U}}_{n+1} + \mathbf{K}\,\bar{\mathbb{U}}_{n+1} - \mathbf{Q}\,\bar{\mathbb{P}}_{n+1} + \mathbf{f}^{\text{int}}_{\mathbb{U}_{n+1}} - \mathbf{G}_{\mathbb{U}_{n+1}} = \mathbf{0} \\ \mathbf{\Psi}_{\mathbb{P}_{n+1}} &= \frac{1}{\beta\,\Delta t^2}\mathbf{C}\,\bar{\mathbb{U}}_{n+1} + \frac{\gamma}{\beta\,\Delta t}\mathbf{Q}^T\,\bar{\mathbb{U}}_{n+1} + \mathbf{H}\,\bar{\mathbb{P}}_{n+1} + \frac{1}{\theta\,\Delta t}\mathbf{S}\,\bar{\mathbb{P}}_{n+1} - \mathbf{q}^{\text{int}}_{\mathbb{P}_{n+1}} - \mathbf{G}_{\mathbb{P}_{n+1}} = \mathbf{0} \end{aligned} \tag{10.64}$$

where $\mathbf{G}_{\mathbb{U}_{n+1}}$ and $\mathbf{G}_{\mathbb{P}_{n+1}}$ are the vector of known values at time t_n defined as

$$\begin{aligned} \mathbf{G}_{\mathbb{U}_{n+1}} &= \mathbf{f}^{\text{ext}}_{\mathbb{U}_{n+1}} + \mathbf{M}\left(\frac{1}{\beta\,\Delta t^2}\bar{\mathbb{U}}_n + \frac{1}{\beta\,\Delta t}\dot{\bar{\mathbb{U}}}_n + \left(\frac{1}{2\beta} - 1\right)\ddot{\bar{\mathbb{U}}}_n\right) \\ \mathbf{G}_{\mathbb{P}_{n+1}} &= \mathbf{q}^{\text{ext}}_{\mathbb{P}_{n+1}} + \mathbf{C}\left(\frac{1}{\beta\,\Delta t^2}\bar{\mathbb{U}}_n + \frac{1}{\beta\,\Delta t}\dot{\bar{\mathbb{U}}}_n + \left(\frac{1}{2\beta} - 1\right)\ddot{\bar{\mathbb{U}}}_n\right) \\ &\quad + \mathbf{Q}^T\left(\frac{\gamma}{\beta\,\Delta t}\bar{\mathbb{U}}_n + \left(\frac{\gamma}{\beta} - 1\right)\dot{\bar{\mathbb{U}}}_n + \Delta t\left(\frac{\gamma}{2\beta} - 1\right)\ddot{\bar{\mathbb{U}}}_n\right) + \mathbf{S}\left(\frac{1}{\theta\,\Delta t}\bar{\mathbb{P}}_n + \left(\frac{1}{\theta} - 1\right)\dot{\bar{\mathbb{P}}}_n\right) \end{aligned} \tag{10.65}$$

The set of nonlinear equations (10.64) can be solved using an appropriate approach, such as the Newton–Raphson iterative algorithm to linearize the nonlinear system of equations. By expanding equations (10.64) with the first-order truncated Taylor series, the following linear approximation can be obtained as

$$\begin{Bmatrix} \mathbf{\Psi}^{i+1}_{\mathbb{U}_{n+1}} \\ \mathbf{\Psi}^{i+1}_{\mathbb{P}_{n+1}} \end{Bmatrix} = \begin{Bmatrix} \mathbf{\Psi}^{i}_{\mathbb{U}_{n+1}} \\ \mathbf{\Psi}^{i}_{\mathbb{P}_{n+1}} \end{Bmatrix} + \begin{bmatrix} \dfrac{\partial \mathbf{\Psi}_{\mathbb{U}}}{\partial \bar{\mathbb{U}}} & \dfrac{\partial \mathbf{\Psi}_{\mathbb{U}}}{\partial \bar{\mathbb{P}}} \\ \dfrac{\partial \mathbf{\Psi}_{\mathbb{P}}}{\partial \bar{\mathbb{U}}} & \dfrac{\partial \mathbf{\Psi}_{\mathbb{P}}}{\partial \bar{\mathbb{P}}} \end{bmatrix}^{i}_{n+1} \begin{Bmatrix} d\bar{\mathbb{U}}^{i}_{n} \\ d\bar{\mathbb{P}}^{i}_{n} \end{Bmatrix} = \mathbf{0} \tag{10.66}$$

where the Jacobian matrix $\mathbf{J}$ in Eq. (10.66) is defined as

$$\mathbf{J} = \begin{bmatrix} \dfrac{\partial \mathbf{\Psi}_{\mathbb{U}}}{\partial \bar{\mathbb{U}}} & \dfrac{\partial \mathbf{\Psi}_{\mathbb{U}}}{\partial \bar{\mathbb{P}}} \\ \dfrac{\partial \mathbf{\Psi}_{\mathbb{P}}}{\partial \bar{\mathbb{U}}} & \dfrac{\partial \mathbf{\Psi}_{\mathbb{P}}}{\partial \bar{\mathbb{P}}} \end{bmatrix} = \begin{bmatrix} \dfrac{1}{\beta\,\Delta t^2}\mathbf{M} + \mathbf{K} + \dfrac{\partial \mathbf{f}^{\text{int}}_{\mathbb{U}}}{\partial \bar{\mathbb{U}}} & -\mathbf{Q} + \dfrac{\partial \mathbf{f}^{\text{int}}_{\mathbb{U}}}{\partial \bar{\mathbb{P}}} \\ \dfrac{1}{\beta\,\Delta t^2}\mathbf{C} + \dfrac{\gamma}{\beta\,\Delta t}\mathbf{Q}^T - \dfrac{\partial \mathbf{q}^{\text{int}}_{\mathbb{P}}}{\partial \bar{\mathbb{U}}} & \mathbf{H} + \dfrac{1}{\theta\,\Delta t}\mathbf{S} - \dfrac{\partial \mathbf{q}^{\text{int}}_{\mathbb{P}}}{\partial \bar{\mathbb{P}}} \end{bmatrix} \tag{10.67}$$

or

$$\mathbf{J}=\begin{bmatrix} \frac{1}{\beta\Delta t^2}\mathbf{M}_{uu}+\mathbf{K}_{uu}+\frac{\partial\mathbf{f}_u^{\text{int}}}{\partial\bar{\mathbf{u}}} & \frac{1}{\beta\Delta t^2}\mathbf{M}_{ua}+\mathbf{K}_{ua}+\frac{\partial\mathbf{f}_u^{\text{int}}}{\partial\bar{\mathbf{a}}} & \frac{1}{\beta\Delta t^2}\mathbf{M}_{ub}+\mathbf{K}_{ub}+\frac{\partial\mathbf{f}_u^{\text{int}}}{\partial\bar{\mathbf{b}}} & -\mathbf{Q}_{up}+\frac{\partial\mathbf{f}_u^{\text{int}}}{\partial\bar{\mathbf{p}}} & -\mathbf{Q}_{uc}+\frac{\partial\mathbf{f}_u^{\text{int}}}{\partial\bar{\mathbf{c}}} \\ \frac{1}{\beta\Delta t^2}\mathbf{M}_{au}+\mathbf{K}_{au}+\frac{\partial\mathbf{f}_a^{\text{int}}}{\partial\bar{\mathbf{u}}} & \frac{1}{\beta\Delta t^2}\mathbf{M}_{aa}+\mathbf{K}_{aa}+\frac{\partial\mathbf{f}_a^{\text{int}}}{\partial\bar{\mathbf{a}}} & \frac{1}{\beta\Delta t^2}\mathbf{M}_{ab}+\mathbf{K}_{ab}+\frac{\partial\mathbf{f}_a^{\text{int}}}{\partial\bar{\mathbf{b}}} & -\mathbf{Q}_{ap}+\frac{\partial\mathbf{f}_a^{\text{int}}}{\partial\bar{\mathbf{p}}} & -\mathbf{Q}_{ac}+\frac{\partial\mathbf{f}_a^{\text{int}}}{\partial\bar{\mathbf{c}}} \\ \frac{1}{\beta\Delta t^2}\mathbf{M}_{bu}+\mathbf{K}_{bu}+\frac{\partial\mathbf{f}_b^{\text{int}}}{\partial\bar{\mathbf{u}}} & \frac{1}{\beta\Delta t^2}\mathbf{M}_{ba}+\mathbf{K}_{ba}+\frac{\partial\mathbf{f}_b^{\text{int}}}{\partial\bar{\mathbf{a}}} & \frac{1}{\beta\Delta t^2}\mathbf{M}_{bb}+\mathbf{K}_{bb}+\frac{\partial\mathbf{f}_b^{\text{int}}}{\partial\bar{\mathbf{b}}} & -\mathbf{Q}_{bp}+\frac{\partial\mathbf{f}_b^{\text{int}}}{\partial\bar{\mathbf{p}}} & -\mathbf{Q}_{bc}+\frac{\partial\mathbf{f}_b^{\text{int}}}{\partial\bar{\mathbf{c}}} \\ \frac{1}{\beta\Delta t^2}\mathbf{C}_{pu}+\frac{\gamma}{\beta\Delta t}\mathbf{Q}_{pu}^T-\frac{\partial\mathbf{q}_p^{\text{int}}}{\partial\bar{\mathbf{u}}} & \frac{1}{\beta\Delta t^2}\mathbf{C}_{pa}+\frac{\gamma}{\beta\Delta t}\mathbf{Q}_{pa}^T-\frac{\partial\mathbf{q}_p^{\text{int}}}{\partial\bar{\mathbf{a}}} & \frac{1}{\beta\Delta t^2}\mathbf{C}_{pb}+\frac{\gamma}{\beta\Delta t}\mathbf{Q}_{pb}^T-\frac{\partial\mathbf{q}_p^{\text{int}}}{\partial\bar{\mathbf{b}}} & \mathbf{H}_{pp}+\frac{1}{\theta\Delta t}\mathbf{S}_{pp}-\frac{\partial\mathbf{q}_p^{\text{int}}}{\partial\bar{\mathbf{p}}} & \mathbf{H}_{pc}+\frac{1}{\theta\Delta t}\mathbf{S}_{pc}-\frac{\partial\mathbf{q}_p^{\text{int}}}{\partial\bar{\mathbf{c}}} \\ \frac{1}{\beta\Delta t^2}\mathbf{C}_{cu}+\frac{\gamma}{\beta\Delta t}\mathbf{Q}_{cu}^T-\frac{\partial\mathbf{q}_c^{\text{int}}}{\partial\bar{\mathbf{u}}} & \frac{1}{\beta\Delta t^2}\mathbf{C}_{ca}+\frac{\gamma}{\beta\Delta t}\mathbf{Q}_{ca}^T-\frac{\partial\mathbf{q}_c^{\text{int}}}{\partial\bar{\mathbf{a}}} & \frac{1}{\beta\Delta t^2}\mathbf{C}_{cb}+\frac{\gamma}{\beta\Delta t}\mathbf{Q}_{cb}^T-\frac{\partial\mathbf{q}_c^{\text{int}}}{\partial\bar{\mathbf{b}}} & \mathbf{H}_{cp}+\frac{1}{\theta\Delta t}\mathbf{S}_{cp}-\frac{\partial\mathbf{q}_c^{\text{int}}}{\partial\bar{\mathbf{p}}} & \mathbf{H}_{cc}+\frac{1}{\theta\Delta t}\mathbf{S}_{cc}-\frac{\partial\mathbf{q}_c^{\text{int}}}{\partial\bar{\mathbf{c}}} \end{bmatrix} \tag{10.68}$$

In Eq. (10.66), the Jacobian matrix needs to be evaluated at each iteration i of time step $\Delta t = t_{n+1} - t_n$. In fact, at each time step a linearized system of equations is solved until the iteration convergence is achieved. Furthermore, it can be seen from (10.68) that the Jacobian matrix is a non-symmetric matrix, so it is advantageous to make the Jacobian matrix symmetric to save the core storage and computational cost. For this purpose, the first three rows of the Jacobian matrix are multiplied by the scalar value $-\dfrac{\gamma}{\beta\Delta t}$. In addition, the inertial matrices $\mathbf{C}_{\delta\beta}$ are omitted from the Jacobian matrix since the contribution of dynamic seepage forcing terms to the solution is negligible compared to the other terms.

Moreover, it can be seen form (10.60) that $\mathbf{f}_u^{\text{int}} = 0$ since $[\![\mathbf{N}_u^{std}]\!] = 0$, as a result the derivatives of $\mathbf{f}_u^{\text{int}}$ with respect to the standard and enriched pressure variables are zero, that is, $\partial\mathbf{f}_u^{\text{int}}/\partial\bar{\mathbf{p}} = 0$ and $\partial\mathbf{f}_u^{\text{int}}/\partial\bar{\mathbf{c}} = 0$. In addition, it can be obtained from (10.60) that

$$\frac{\partial\mathbf{f}_{\mathbb{U}}^{\text{int}}}{\partial\bar{\mathbb{U}}} = \begin{bmatrix} \frac{\partial\mathbf{f}_u^{\text{int}}}{\partial\bar{\mathbf{u}}} & \frac{\partial\mathbf{f}_u^{\text{int}}}{\partial\bar{\mathbf{a}}} & \frac{\partial\mathbf{f}_u^{\text{int}}}{\partial\bar{\mathbf{b}}} \\ \frac{\partial\mathbf{f}_a^{\text{int}}}{\partial\bar{\mathbf{u}}} & \frac{\partial\mathbf{f}_a^{\text{int}}}{\partial\bar{\mathbf{a}}} & \frac{\partial\mathbf{f}_a^{\text{int}}}{\partial\bar{\mathbf{b}}} \\ \frac{\partial\mathbf{f}_b^{\text{int}}}{\partial\bar{\mathbf{u}}} & \frac{\partial\mathbf{f}_b^{\text{int}}}{\partial\bar{\mathbf{a}}} & \frac{\partial\mathbf{f}_b^{\text{int}}}{\partial\bar{\mathbf{b}}} \end{bmatrix} = \mathbf{0} \tag{10.69}$$

Hence, in order to make the Jacobian matrix symmetric the derivatives of interfacial flux vectors, that is, $\partial\mathbf{q}_p^{\text{int}}/\partial\bar{\mathbf{u}}$ and $\partial\mathbf{q}_c^{\text{int}}/\partial\bar{\mathbf{u}}$, are eliminated from the first column of the Jacobian matrix. Also, to retain the symmetry of the Jacobian matrix, it is assumed that $\partial\mathbf{f}_{\mathbb{U}}^{\text{int}}/\partial\bar{\mathbb{P}} = \frac{\beta\Delta t}{\gamma}\left(\partial\mathbf{q}_{\mathbb{P}}^{\text{int}}/\partial\bar{\mathbb{U}}\right)^T$, that is,

$$\begin{bmatrix} \frac{\partial\mathbf{f}_u^{\text{int}}}{\partial\bar{\mathbf{p}}} & \frac{\partial\mathbf{f}_u^{\text{int}}}{\partial\bar{\mathbf{c}}} \\ \frac{\partial\mathbf{f}_a^{\text{int}}}{\partial\bar{\mathbf{p}}} & \frac{\partial\mathbf{f}_a^{\text{int}}}{\partial\bar{\mathbf{c}}} \\ \frac{\partial\mathbf{f}_b^{\text{int}}}{\partial\bar{\mathbf{p}}} & \frac{\partial\mathbf{f}_b^{\text{int}}}{\partial\bar{\mathbf{c}}} \end{bmatrix} = \frac{\beta\Delta t}{\gamma}\begin{bmatrix} \frac{\partial\mathbf{q}_p^{\text{int}}}{\partial\bar{\mathbf{u}}} & \frac{\partial\mathbf{q}_p^{\text{int}}}{\partial\bar{\mathbf{a}}} & \frac{\partial\mathbf{q}_p^{\text{int}}}{\partial\bar{\mathbf{b}}} \\ \frac{\partial\mathbf{q}_c^{\text{int}}}{\partial\bar{\mathbf{u}}} & \frac{\partial\mathbf{q}_c^{\text{int}}}{\partial\bar{\mathbf{a}}} & \frac{\partial\mathbf{q}_c^{\text{int}}}{\partial\bar{\mathbf{b}}} \end{bmatrix}^T \tag{10.70a}$$

where

$$\frac{\partial\mathbf{f}_a^{\text{int}}}{\partial\bar{\mathbf{p}}_\delta} = -\int_{\Gamma_d} [\![\mathbf{N}_u^\alpha]\!]^T \mathbf{n}_{\Gamma_d} \mathbf{N}_p^\delta \, d\Gamma \tag{10.70b}$$

In addition, it is assumed that $\partial \mathbf{q}_c^{\text{int}} / \partial \bar{\mathbf{p}} = \left(\partial \mathbf{q}_p^{\text{int}} / \partial \bar{\mathbf{c}} \right)^T$, thus

$$\frac{\partial \mathbf{q}_{\mathbb{P}}^{\text{int}}}{\partial \bar{\mathbb{P}}} = \begin{bmatrix} \dfrac{\partial \mathbf{q}_p^{\text{int}}}{\partial \bar{\mathbf{p}}} & \dfrac{\partial \mathbf{q}_p^{\text{int}}}{\partial \bar{\mathbf{c}}} \\ \left(\dfrac{\partial \mathbf{q}_p^{\text{int}}}{\partial \bar{\mathbf{c}}} \right)^T & \dfrac{\partial \mathbf{q}_c^{\text{int}}}{\partial \bar{\mathbf{c}}} \end{bmatrix} \tag{10.71a}$$

where

$$\frac{\partial \mathbf{q}_\delta^{\text{int}}}{\partial \bar{\mathbf{p}}_\gamma} = -\int_{\Gamma_d} \left(\nabla \mathbf{N}_p^\delta \right)^T \cdot \mathbf{t}_{\Gamma_d} k_{f_d} (2h) \, \mathbf{t}_{\Gamma_d}^T \cdot \nabla \mathbf{N}_p^\gamma \, d\Gamma - \frac{1}{\theta \Delta t} \int_{\Gamma_d} \left(\mathbf{N}_p^\delta \right)^T (2h) \frac{1}{K_f} \mathbf{N}_p^\gamma \, d\Gamma \tag{10.71b}$$

Based on the previously-mentioned simplifications, a symmetric Jacobian matrix can be obtained as

$$\mathbf{J} = \begin{bmatrix} -\dfrac{\gamma}{\beta \Delta t} \left(\dfrac{1}{\beta \Delta t^2} \mathbf{M}_{uu} + \mathbf{K}_{uu} \right) & -\dfrac{\gamma}{\beta \Delta t} \left(\dfrac{1}{\beta \Delta t^2} \mathbf{M}_{ua} + \mathbf{K}_{ua} \right) & -\dfrac{\gamma}{\beta \Delta t} \left(\dfrac{1}{\beta \Delta t^2} \mathbf{M}_{ub} + \mathbf{K}_{ub} \right) & \dfrac{\gamma}{\beta \Delta t} \mathbf{Q}_{up} & \dfrac{\gamma}{\beta \Delta t} \mathbf{Q}_{uc} \\ -\dfrac{\gamma}{\beta \Delta t} \left(\dfrac{1}{\beta \Delta t^2} \mathbf{M}_{au} + \mathbf{K}_{au} \right) & -\dfrac{\gamma}{\beta \Delta t} \left(\dfrac{1}{\beta \Delta t^2} \mathbf{M}_{aa} + \mathbf{K}_{aa} \right) & -\dfrac{\gamma}{\beta \Delta t} \left(\dfrac{1}{\beta \Delta t^2} \mathbf{M}_{ab} + \mathbf{K}_{ab} \right) & \dfrac{\gamma}{\beta \Delta t} \left(\mathbf{Q}_{ap} - \dfrac{\partial \mathbf{f}_a^{\text{int}}}{\partial \bar{\mathbf{p}}} \right) & \dfrac{\gamma}{\beta \Delta t} \left(\mathbf{Q}_{ac} - \dfrac{\partial \mathbf{f}_a^{\text{int}}}{\partial \bar{\mathbf{c}}} \right) \\ -\dfrac{\gamma}{\beta \Delta t} \left(\dfrac{1}{\beta \Delta t^2} \mathbf{M}_{bu} + \mathbf{K}_{bu} \right) & -\dfrac{\gamma}{\beta \Delta t} \left(\dfrac{1}{\beta \Delta t^2} \mathbf{M}_{ba} + \mathbf{K}_{ba} \right) & -\dfrac{\gamma}{\beta \Delta t} \left(\dfrac{1}{\beta \Delta t^2} \mathbf{M}_{bb} + \mathbf{K}_{bb} \right) & \dfrac{\gamma}{\beta \Delta t} \left(\mathbf{Q}_{bp} - \dfrac{\partial \mathbf{f}_b^{\text{int}}}{\partial \bar{\mathbf{p}}} \right) & \dfrac{\gamma}{\beta \Delta t} \left(\mathbf{Q}_{bc} - \dfrac{\partial \mathbf{f}_b^{\text{int}}}{\partial \bar{\mathbf{c}}} \right) \\ \dfrac{\gamma}{\beta \Delta t} \mathbf{Q}_{pu}^T & \dfrac{\gamma}{\beta \Delta t} \left(\mathbf{Q}_{pa}^T - \dfrac{\partial \mathbf{f}_a^{\text{int}}}{\partial \bar{\mathbf{p}}} \right)^T & \dfrac{\gamma}{\beta \Delta t} \left(\mathbf{Q}_{pb}^T - \dfrac{\partial \mathbf{f}_b^{\text{int}}}{\partial \bar{\mathbf{p}}} \right)^T & \mathbf{H}_{pp} + \dfrac{1}{\theta \Delta t} \mathbf{S}_{pp} - \dfrac{\partial \mathbf{q}_p^{\text{int}}}{\partial \bar{\mathbf{p}}} & \mathbf{H}_{pc} + \dfrac{1}{\theta \Delta t} \mathbf{S}_{pc} - \dfrac{\partial \mathbf{q}_p^{\text{int}}}{\partial \bar{\mathbf{c}}} \\ \dfrac{\gamma}{\beta \Delta t} \mathbf{Q}_{cu}^T & \dfrac{\gamma}{\beta \Delta t} \left(\mathbf{Q}_{ca}^T - \dfrac{\partial \mathbf{f}_a^{\text{int}}}{\partial \bar{\mathbf{c}}} \right)^T & \dfrac{\gamma}{\beta \Delta t} \left(\mathbf{Q}_{cb}^T - \dfrac{\partial \mathbf{f}_b^{\text{int}}}{\partial \bar{\mathbf{c}}} \right)^T & \mathbf{H}_{cp} + \dfrac{1}{\theta \Delta t} \mathbf{S}_{cp} - \left(\dfrac{\partial \mathbf{q}_p^{\text{int}}}{\partial \bar{\mathbf{c}}} \right)^T & \mathbf{H}_{cc} + \dfrac{1}{\theta \Delta t} \mathbf{S}_{cc} - \dfrac{\partial \mathbf{q}_c^{\text{int}}}{\partial \bar{\mathbf{c}}} \end{bmatrix} \tag{10.72}$$

From the computational aspect, one main difficulty encountered when implementing the X-FEM deals with the numerical integration in the presence of the weak or strong discontinuity. In the conventional FEM, the standard shape functions are in terms of the polynomial order and the Gauss integration rule can be used efficiently to evaluate the integral terms. However, in the X-FEM the enhanced shape functions are not smooth over an enriched element due to the presence of weak or strong discontinuity inside of the element. Hence, the standard Gauss quadrature rule cannot be used if the element is crossed by a discontinuity, and the discontinuous nature of the solution requires a special treatment to accurately integrate the non-smooth enrichment functions appearing in the enriched part of the X-FEM approximation. In order to carry out the numerical integration in X-FEM, two approaches were introduced in Chapter 2 for the numerical integration of an enriched element based on the triangular partitioning method and the rectangular sub-grids method. In this study, the triangular partitioning method is employed in which the element bisected by the interface, is divided into sub-triangles and the Gauss integration rule is performed over each sub-triangle. In this technique, the numerical integration of elements bisected by the discontinuity is performed separately on each sub-element into which the element is divided. In practice, in order to be able to perform the integration over these sub-elements on either side of the discontinuity, each sub-element is partitioned into triangles over which the standard Gauss quadrature is applied. Moreover, a number of Gauss points must be set along the one-dimensional segments of the discontinuity delimited with the element edges to integrate the mechanical and mass transfer coupling terms along the discontinuity.

10.6 Alternative Approaches to Fluid Flow Simulation within the Fracture

In preceding sections, the fluid flow within the fracture was modeled by discretizing the flow continuity equation of fluid inside the fracture. For this purpose, the hydro-mechanical coupling associated with the tractions acting on the crack edges and the fluid leak-off from the fracture into the domain was obtained by discretizing the weak form of flow continuity equation of fluid flow inside the fracture in order to compute the interfacial force vector $\mathbf{f}_{\alpha}^{\text{int}}$ due to the fluid pressure exerted on the fracture faces, and the flux vectors $\mathbf{q}_{\delta}^{\text{int}}$ due to the fluid exchange between the fracture and the surrounding permeable porous medium in Eq. (10.57). There are basically various techniques proposed by researchers in the literature to simulate the fluid flow through the fracture with simplified computational algorithms. These approaches are generally employed for modeling hydraulic fracturing through impermeable porous media, in which the fluid pressure behaves as an interfacial force on the fracture faces with no direct effect on the fluid pressure of the porous medium surrounding the fracture. These simplifications are mostly valid for rapid crack propagation in an impermeable porous media and widely used in hydraulic fracturing, which is known as the *displacement discontinuity method* (Dong and de Pater, 2001; Akulich and Zvyagin, 2008; Zhang and Ghassemi, 2011). In this study, two alternative computational algorithms are employed to compute the interfacial forces due to fluid pressure exerted on the fracture faces based on a "partitioned solution algorithm" and a "time-dependent constant pressure algorithm" that are mostly applicable to impermeable media, and the results are compared with the coupling X-FEM model described in Section 10.5.

10.6.1 A Partitioned Solution Algorithm for Interfacial Pressure

A procedure for numerical approximation of hydraulically-driven fracture propagation in a poroelastic material was proposed by Boone and Ingraffea (1990). The method was based on a partitioned solution procedure used to solve the FE approximation of poroelasticity equations in conjunction with a finite difference approximation for modeling the fluid flow along the fracture. In this method, several assumptions were incorporated in the numerical procedure; firstly the crack tip is assumed to close smoothly: that is a distinctive feature of the Dugdale-Barenblatt fracture model, secondly there is a coupling between the fluid flow in the fracture and the surrounding porous medium, and finally the fluid flow along the fracture does not reach the crack tip so a region may exist near it that is void of fluid.

In this study, the partitioned solution procedure proposed by Boone and Ingraffea (1990) is employed as an alternative solution to compute the fluid pressure exerted on the fracture faces of an impermeable porous media. In this simplified method, it is assumed that the flow is laminar, the fluid is incompressible, and it accounts for the time-dependent rate of crack opening. Hence, the fluid mass conservation equation can be stated as

$$\frac{\partial q}{\partial \bar{x}} - \frac{\partial w}{\partial t} + q_1 = 0 \tag{10.73}$$

where w is the fracture opening ($w = 2h$), and q_1 is the fluid leak-off into the surrounding porous medium, which is equal to zero for the case of impermeable porous media. In equation (10.73), q is the flow rate along the fracture length $\bar{x}$ defined as

$$q = -\frac{w^3}{12\mu_f}\frac{\partial p}{\partial \bar{x}} \tag{10.74}$$

where μ_f is the viscosity of fluid within the fracture and p is the fluid pressure along the fracture. It must be noted that this approach can be used for the case of permeable porous medium by employing some experimental relations for the fluid leak-off q_1.

In order to solve the fluid mass conservation equation (10.73), the finite difference algorithm is employed using a predetermined value of fracture length, where the fracture is in equilibrium between the *in situ* stress and the pressure distribution in the fracture. The crack interface allows the fracture to open when subjected to an internal pressure. A control-volume approach is proposed to model the fluid flow within the fracture. Basically, an iterative procedure is used to solve for the length and opening of the crack. For this purpose, a number of control points are determined along the crack interface and Eq. (10.73) is satisfied for each control point. The set of equilibrium equations is finally solved over all control points along the crack interface, as shown in the flowchart of Figure 10.17a. In this algorithm, the crack opening w is obtained from the X-FEM solution so that the unknown value from Eq. (10.73) is the fluid pressure p. Assembling the set of equations for the individual control volumes constructed over the control point results in a tridiagonal matrix. This fluid flow model is first constructed for the complete length of the predetermined fracture path. The control volumes are given a nominal, small initial width so that the equations are well-posed. The boundary conditions for each time step are specified as a fixed pressure at the crack mouth and zero flow at the other end of the fracture. The fluid within the fracture is assumed to be incompressible so that there can be no flow past the tip of the fracture, even though the zero-flow condition is imposed at the end of the row of control volumes. Generally, it would be desirable to be able to impose a flow rate as a boundary condition at the crack mouth. However, the imposition of a

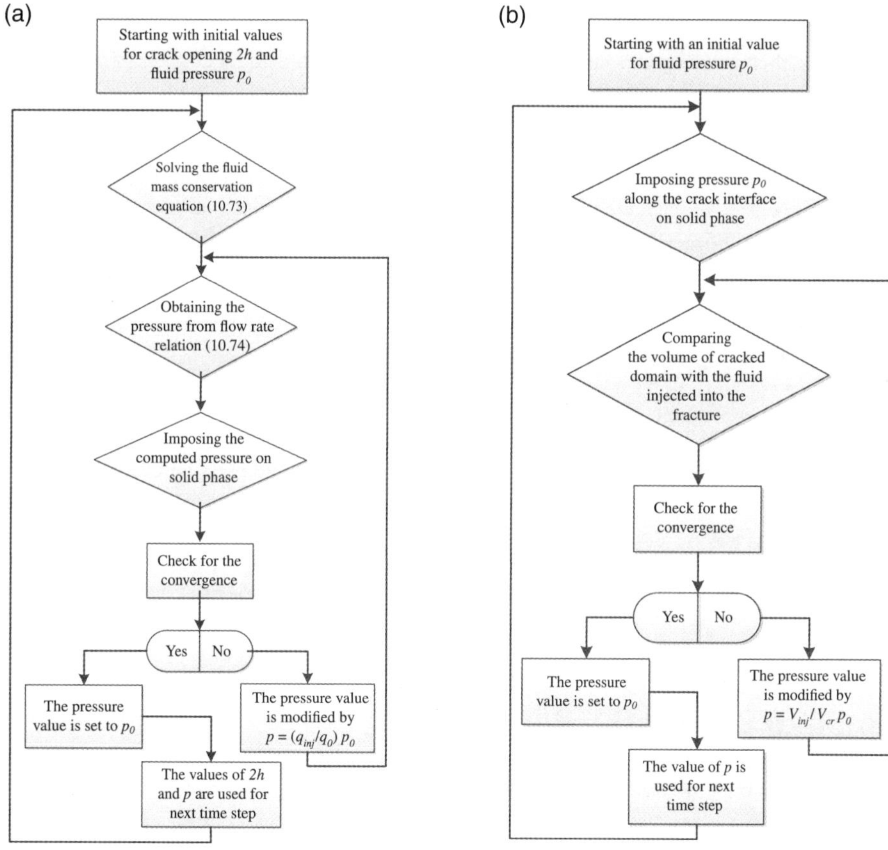

Figure 10.17 The flowchart of the solution: (a) the partitioned solution algorithm; (b) the time-dependent constant pressure algorithm

Neumann boundary condition as the crack mouth pressure (CMP) results in an ill-posed set of equations. In order to impose this boundary condition in conjunction with the partitioned analysis procedure, it is necessary to impose a pressure boundary condition elsewhere along the fracture. However, the pressure along the fracture is not known and it cannot be imposed at the crack tip owing to the existence of a singularity in the pressure at this point.

It must be noted that there must be a one-to-one correspondence between the control volumes and the elements bisected by the crack along the fracture interface in the X-FEM solution. This ensures that fluid mass is conserved across the boundary. Basically, two situations may occur for coupling along the fracture depending on whether the fracture faces are assumed permeable, or impermeable. For impermeable faces, the fluid pressures in the fracture are converted to equivalent external loads on the fracture faces. For permeable faces, the fluid pressure in the fracture must be applied as a pore pressure and a total stress boundary condition. It is also necessary to obtain a solution where the fluid leak-off from the fracture is consistent with the flow across the interface in the X-FEM solution. It must be noted that the proposed finite difference algorithm is not unconditionally stable and the convergence can be generally achieved using a large number of iterations.

10.6.2 A Time-Dependent Constant Pressure Algorithm

In hydraulic fracturing, an essential ingredient in establishing a formulation for hydraulically-driven fracture propagating in the porous medium is to perform a relationship between the injection rate, fluid leak-off, fracture width, fracture length, and the total volume pumped into the formation. This formulation was comprehensively derived in Section 10.5 by discretizing the continuity equation of fluid flow inside the fracture in the framework of X-FEM formulation and evaluating the tractions acting on the crack edges and the fluid leak-off from the fracture into the porous medium based on Eqs. (10.60) and (10.61). An alternative technique was proposed in the previous section on the basis of a "partitioned solution algorithm" in the framework of the finite difference method by satisfying the fluid mass conservation equation over a number of control points along the fracture interface. Although the method was presented for the case of impermeable porous media, it can easily be extended to the permeable porous media using some experimental relations for the fluid leak-off behavior. In this section, a simplified approach is presented based on a "time-dependent constant pressure algorithm" to evaluate the traction forces acting on the fracture edges for the case of impermeable porous media.

It has been observed from the solution of hydraulic fracturing in porous media that the pressure distribution in the fracture is almost constant along the crack interface. Geertsma and de Klerk (1969) developed a method based on the concept of equilibrium fracture propagation, and derived a simplified pressure distribution in the direction of fracture propagation. An analytical solution was obtained by Spence and Sharp (1985), in which the internal flow was modeled by lubrication theory that results in a nonlinear partial differential equation connecting the pressure to the cavity shape, and the solution was studied for a special case of the constant pressure distribution. The idea of constant pressure distribution along the fracture for the case of impermeable porous media has been widely used in various numerical implementation of hydraulic fracturing due to its simplicity. It must be noted that the only measurable quantities that are directly related to the fracture propagation process are the total volume of fluid injected into the fractured domain and the time to accomplish this process. Moreover, considering the case of no leak-off of fracture fluid into the surrounding porous medium, it implies that all fluid injected into the fracture must be used to propagate the fracture, both in the width and the length.

In this study, the idea of constant pressure distribution along the crack is modified to be able to model the propagation of fluid-driven fracture in time. In fact, the modification is performed because of some limitations on the "partitioned solution algorithm" discussed in previous section, due to imposition of the Neumann boundary condition as a flow rate at the crack mouth and the low convergence rate of the solution. Basically, two types of boundary conditions can be employed in hydraulic fracturing problems; the CMP as an essential boundary condition, in which no modification is needed and the imposed pressure can be used everywhere along the fracture during the solution; and the flow rate at the crack mouth as a natural boundary condition, in which an auxiliary assumption is used to solve the fluid mass conservation equation (10.73). By injecting the fluid flow into the fracture with no leak-off into the surrounding porous medium, the volume of cracked domain must be equal to the total volume of fluid injected into the fractured zone. This results in a decrease of fluid pressure through time as the crack propagates and reaches almost a steady value when the crack tip proceeds far away from the injection point, as has been already observed in hydraulic fracturing problems. The flowchart of the solution is presented in Figure 10.17b for the time-dependent constant pressure algorithm. The computational algorithm starts with an initial value of pressure along the fracture for each time step; the value of fracture opening is then obtained from the X-FEM solution by imposing the predetermined pressure on the fracture faces of the porous medium as external tractions; the solution follows by comparing the volume of cracked domain with the total volume of fluid injected into the fractured zone; the value of pressure is then modified based on the ratio of the volume of cracked domain to the volume of fluid injected into the domain. The process continues until the convergence is achieved. It has been shown through the numerical simulation results that the proposed technique results in reasonable accuracy and a great numerical stability; the number of iterations and the rate of convergence are quite satisfying proving the soundness of proposed computational algorithm.

10.7 Application of the X-FEM Method in Hydraulic Fracture Propagation of Saturated Porous Media

In order to illustrate a part of the wide range of problems that can be solved by the proposed approach and to validate the performance of the computational algorithm in modeling of the hydraulic fracturing problem, several numerical examples are solved. The first example demonstrates the robustness of X-FEM in mixed-mode fracture analysis of an infinite saturated porous media with an inclined crack. The numerical simulation is compared with an available numerical modeling reported in literature. The next two examples are chosen to demonstrate the robustness of X-FEM technique in simulation of hydraulically-driven fracture propagation in an infinite poroelastic medium and a gravity dam under hydrostatic pressure. In order to model the hydraulically-driven fracture propagating in X-FEM, crack geometry is modeled independent of the FE mesh by enriching the nodal points of elements that are intersected by the crack during crack propagation. In crack growth simulation, the crack propagates with a predefined value of crack length (CL), if the crack propagation criterion is satisfied. In X-FEM, the Heaviside and crack tip enrichments are used according to the current position of the crack tip at each time step. Based on these enrichments, the simulation is performed to obtain the stress and displacement fields at the crack tip region, in order to indicate the crack tip position in the next step of crack growth. If the new crack tip position is in the area of former element, no update is necessary for the enriched elements; however, if the new crack segment crosses the next element, the enrichment of nodal points must be updated. The simulation can then be carried out according to the new enriched elements based on the new configuration of crack propagation. It must be noted that this method needs a high accurate recognition function to diagnose the position of new crack tip elements.

In crack propagation problems, there are two main requirements at each time step. It must be first determined whether the crack propagates or not; and if so, in which direction. Based on these two requirements, two criteria must be utilized; one for crack propagation and the other for crack kinking. In this study, two different enrichment strategies are employed for crack growth simulation. In the first case, the asymptotic enrichment functions are employed at the crack tip element to model the singularity at the tip of discontinuity, the Heaviside enrichment function to model the displacement jump across the fracture, and the modified level set function to model the discontinuity on the gradient of fluid pressure normal to the fracture. While in the second case, the crack tip enrichment functions are removed from the X-FEM simulation, and the Heaviside enrichment function and the modified level set function are used only to model the displacement and pressure fields, respectively.

In order to model the crack propagation in the framework of linear fracture mechanics, a criterion can be defined based on a function of the stress intensity factors, the strain energy release rate, the strain energy density, and so on. The crack kinking criteria determines the direction of the crack and can be determined based on the fracture toughness of brittle material, which is usually measured in a pure mode I loading conditions noted by K_{IC}. For a general mixed-mode case, a criterion is needed to determine the angle of incipient propagation with respect to crack direction, and a critical combination of stress intensity factors that leads to crack propagation. There are various criteria which have been proposed by researchers for the mixed mode crack propagation, including the maximum energy release rate, the minimum strain energy density criteria, the maximum circumferential tensile stress, and so on. In this study, the maximum circumferential tensile stress is employed, where the hoop stress reaches its maximum value on the plane of zero shear stress. The crack propagation angle θ_0 is expressed by using the angle between the line of crack and the crack growth direction, with the positive value defined in an anti-clockwise direction, as

$$\theta_0 = 2\arctan\left(\frac{K_I}{4K_{II}} \pm \frac{1}{4}\sqrt{\left(\frac{K_I}{K_{II}}\right)^2 + 8}\right) \tag{10.75}$$

in which the sign is chosen such that the hoop stress is positive. To initiate crack propagation, it is required that the maximum circumferential tensile stress σ_θ reaches a critical value. However, when the plastic zone size cannot be ignored, it is necessary to use the stress state at a material dependent finite distance from the crack tip.

In order to perform the crack growth simulation with no crack tip enrichments, the maximum principal tensile stress is checked at all integration points in the element ahead of the tip of the discontinuity at the end of a load increment. If the maximum principal tensile stress at any of the integration points in the element ahead of the crack tip reaches the tensile strength of the material, the discontinuity is introduced through the entire element. The discontinuity is inserted as a straight line within the element and is enforced to be geometrically continuous. In this model, since a crack propagates from a discrete point, the discontinuity can be handled in two ways, the first by choosing a point before the calculation, and the second by performing an elastic loading and checking where the principal stresses are greatest. It must be noted that the discontinuity is extended only at the end of a load increment in order to preserve the quadratic convergence rate of the full Newton–Raphson solution procedure. The most important ingredient when extending a discontinuity is that the correct direction is chosen. Since the tip of the discontinuity is not located at a point where the stress state is known accurately, such as conventional Gauss points, the local stress field cannot be relied upon to accurately yield the correct normal vector to a discontinuity. To overcome this, the averaged stress tensor or the so-called non-local stress tensor is used at the crack tip to obtain the principal stress direction and to determine the right direction of the discontinuity extension. The discontinuity is extended in the direction perpendicular to the maximum non-local principal stress direction. It must be noted that the jump in the displacement field at the tip of the discontinuity must be equal to zero. In order to enforce this condition, the nodes belonging to the element edge on which the tip of the discontinuity lies are not enriched. Since the enrichment functions are multiplied by the shape functions of a particular node, the enhanced basis at a particular node has an influence only over the support of that node. Therefore, the Heaviside function is added only to the enhanced basis of nodes whose support is crossed by a discontinuity. Another condition that must be satisfied is that the displacement jump at the crack tip be zero. To ensure this, the nodes on the element boundary touched by the crack tip are not enhanced. When a discontinuity propagates into the next element, all nodes behind the crack tip are enhanced.

10.7.1 An Infinite Saturated Porous Medium with an Inclined Crack

The first example is chosen to illustrate the performance of X-FEM model in hydro-mechanical analysis of an inclined crack within an infinite saturated porous media, as shown in Figure 10.18. This example was

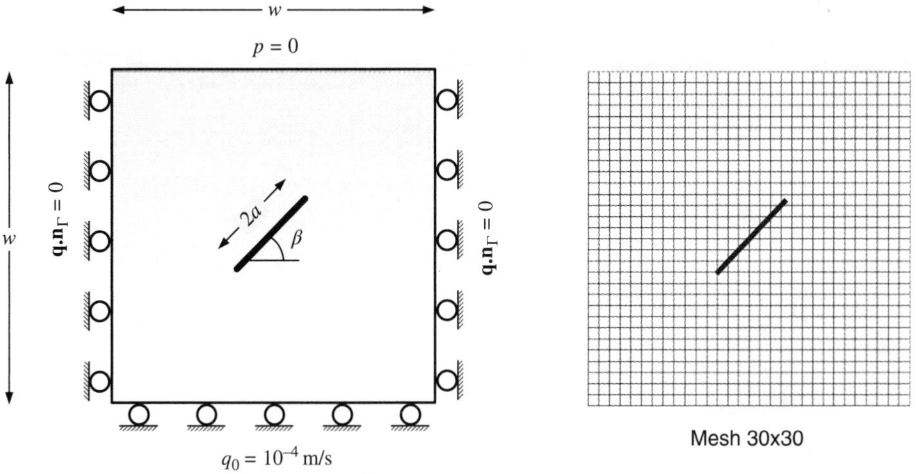

Figure 10.18 An infinite saturated porous medium with an inclined crack: Problem definition and the X-FEM mesh

Table 10.3 Material properties of an infinite saturated porous medium with an inclined crack

Young's modulus	$E = 9$ GPa
Poisson ratio	$\nu = 0.4$
Biot constant	$\alpha = 1$
Porosity	$n = 0.3$
Solid phase density	$\rho_s = 2000$ kg/m^3
Water density	$\rho_w = 1000$ kg/m^3
Bulk modulus of solid phase	$K_s = 1.0 \times 10^{27}$ GPa
Bulk modulus of water	$K_w = 1.0 \times 10^{27}$ GPa
Permeability	$k = 1.0 \times 10^{-9}$ m^3/N s
Viscosity of water	$\mu_w = 1 \times 10^{-3}$ Pa s

originally proposed by Réthoré, de Borst, and Abellan (2007b) to present their X-FEM formulation for modeling fluid flow in a fractured porous medium and is used here for comparison. A square-shaped fractured domain of 10 × 10 m, with an inclined crack of length 2 m at its center, is modeled in a plane strain condition, which is subjected to a normal fluid flux of $q_0 = 10^{-4}$ m/s at the bottom surface while the top surface is imposed as a drained condition. The geometry, boundary conditions, and the position of the fault are shown in Figure 10.18. The left and right edges are assumed to have an undrained boundary condition. The material properties of the soil are given in Table 10.3. The X-FEM analysis is performed using a quadrilateral structured FE mesh of 30 × 30 at different crack angles of $\beta = 15$, 30, and 45°. The problem is solved for a total period of 10 s in 75 time steps.

In Figure 10.19a, the distribution of vertical displacement contour is shown for the fracture angle of $\beta = 30°$ at time step $t = 10$ s. Obviously, the imposed fluid flux at the bottom of the porous medium increases the fluid pressure through the domain and as a result inside the fracture, so the crack opens and the fluid flows inside the fracture. In Figure 10.19b, the contour of pressure gradient is shown in the normal direction to the fracture at time step $t = 10$ s. Clearly, a jump can be observed at both tips of the discontinuity. It is obvious that the fluid flows inside the crack from the left crack tip to the right one because of the imposed fluid flux at the bottom edge. A maximum pressure gradient can be seen at the bottom left of the crack tip

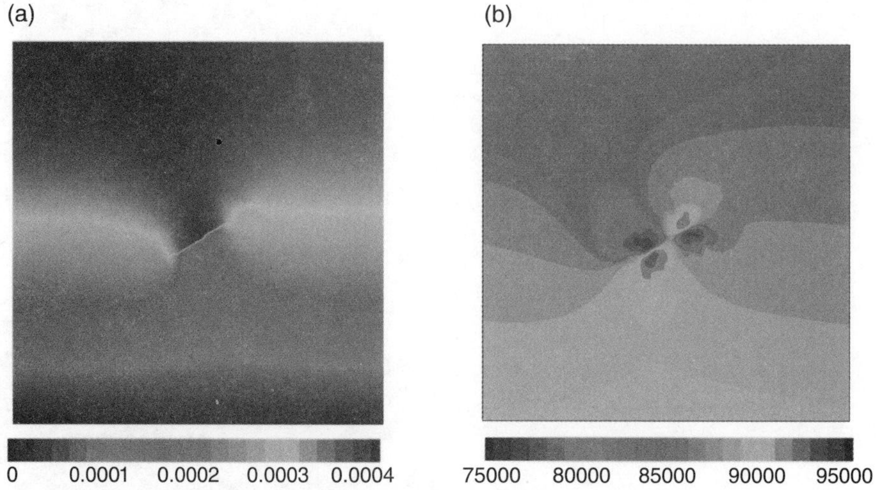

Figure 10.19 An infinite saturated porous medium with an inclined crack at the crack angle of $\beta = 30^o$: (a) the distribution of vertical displacement contour; (b) the contour of pressure gradient normal to discontinuity at the time step of $t = 10$ s (For color details, please see color plate section)

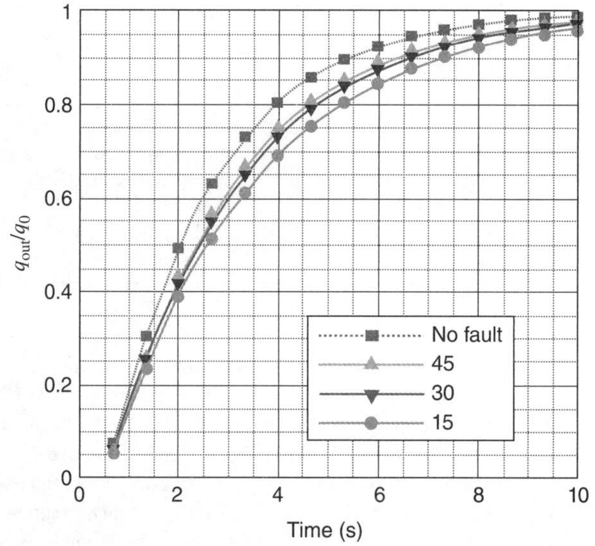

Figure 10.20 The ratio of out-flow to in-flow (q_{out}/q_0) for a fractured porous media at different crack angles

and a minimum value at the top left, showing that a considerable amount of fluid flows through the fracture. In contrast, the minimum pressure gradient at the bottom right and its maximum value at the top right of the crack tip cause the flow of fluid from the fracture into the porous medium. In Figure 10.20, the evolutions of the ratio of out-flow to in-flow (imposed fluid flux), that is, q_{out}/q_0, are plotted for different fracture angles, and the results are compared with that obtained without a fracture. It can be clearly observed that the presence of the fracture affects the flow of fluid from the bottom to the top of the domain, described in terms of the ratio of out-flow to in-flow in this figure. In fact, a part of the fluid can be stored inside the fracture that causes the fluid flow to decrease for the lower fracture angle. Similar results were reported by Réthoré, de Borst, and Abellan (2007b) that demonstrates a good performance of proposed computational algorithm for modeling the fluid flow through a fractured porous medium.

10.7.2 Hydraulic Fracture Propagation in an Infinite Poroelastic Medium

The next example illustrates the performance of proposed X-FEM model for the simulation of hydraulically-driven fracture propagation in infinitely porous media. An analytical solution was obtained for this example by Spence and Sharp (1985) and Emerman, Turcotte, and Spence (1986) and used here for comparison. The example was also modeled by Boone and Ingraffea (1990) to present their staggered procedure, in which a FEM was applied for the mechanical problem and a finite difference method for flow analysis through the fracture. The hydraulic fracturing problem was modeled using the adaptive FEM strategy to study the static and dynamic behavior of fractured domain in saturated porous media (Schrefler, Secchi, and Simoni, 2006; Secchi, Simoni, and Schrefler, 2007; Barani, Khoei, and Mofid, 2011; Khoei, Barani, and Mofid, 2011). In this study, hydraulic fracture propagation is modeled based on the proposed X-FEM technique to overcome the expensive and cumbersome computational costs encountered during the mesh generation process in crack growth simulation. The problem is solve using the coupling X-FEM computational algorithm described in Section 10.5, and the results are compared with the simplified computational models of a partitioned solution algorithm and a time-dependent constant pressure algorithm described in Section 10.6.

The hydraulic fracture problem demonstrates the injection of fluid into a borehole at the constant rate of Q that causes the fracture to advance into the porous medium. Due to the symmetry of problem, the circular borehole is solved for one-half of the specimen containing a tip, from which the crack enucleates. In Figure 10.21, the geometry and boundary condition of problem are presented together with the X-FEM mesh. A finite element mesh of 2420 quadrilateral elements is employed where the size of elements around the borehole is assumed to be 0.05×0.05 m. The mesh size is fine enough to represent accurately the distribution of the fluid pressure in the direction of the propagating fracture and to guarantee the mesh independence and numerical convergence of the solution. The material properties of poroelastic medium are given in Table 10.4. An initial crack of length 0.05 m is assumed at the borehole where the fluid injection is imposed, and a constant flow rate of 0.0001 m^2/s is applied at the crack mouth. The crack propagates in the normal direction to the maximum principal tensile stress when the principle effective stress at the crack tip reaches the ultimate tensile strength of the material 1.0 MPa. In order to evaluate the accuracy of proposed computational algorithm, the mass matrix is set to zero. The problem is solved for 10 s with the time increment chosen as 0.01 s to achieve the results independent of the temporal discretization.

In the limiting case where there is no fluid leak-off and under the simplifying assumption that the fracturing fluid is incompressible and the surrounding porous medium is impermeable, analytical solutions were reported by Geertsma and Klerk (1969) and Spence and Sharp (1985) in terms of the crack mouth opening displacement (CMOD), crack mouth pressure (CMP) and crack length (CL) as

$$\begin{aligned}
\text{CMOD} &= A\left(\frac{\mu(1-\nu)Q^3}{G}\right)^{1/6} t^{1/3} \\
\text{CMP} &= B\left(\frac{G^3 Q\mu}{(1-\nu)^3 L^2}\right)^{1/4} + S \\
\text{CL} &= C\left(\frac{GQ^3}{\mu(1-\nu)}\right)^{1/6} t^{2/3}
\end{aligned} \tag{10.76}$$

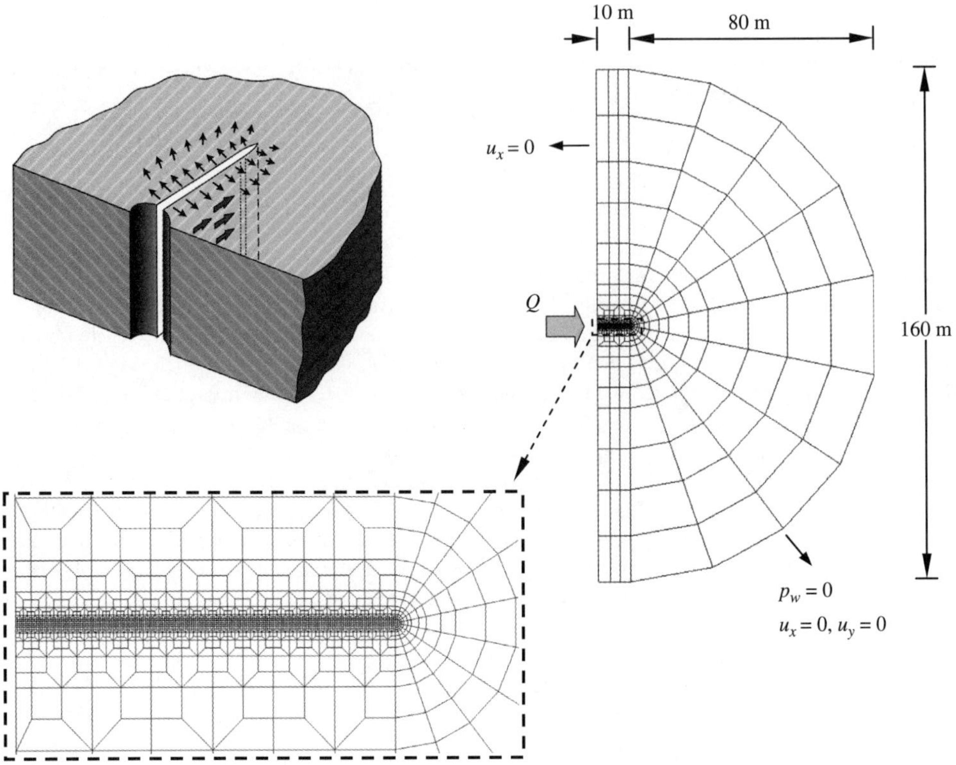

Figure 10.21 A hydraulically-driven fracture propagation in an infinite porous media: A schematic illustration of the problem along with the geometry, boundary conditions, and X-FEM mesh

Table 10.4 Material properties of hydraulic fracture propagation in an infinite poroelastic medium

Shear modulus	$G = 6$ GPa
Drained Poisson ratio	$\nu = 0.2$
Undrained Poisson ratio	$\nu_u = 0.33$
Biot constant	$\alpha = 1$
Porosity	$n = 0.19$
Solid phase density	$\rho_s = 2000$ kg/m^3
Water density	$\rho_w = 1000$ kg/m^3
Bulk modulus of solid phase	$K_s = 36.0$ GPa
Bulk modulus of water	$K_w = 3.0$ GPa
Permeability	$k = 6.0 \times 10^{-12}$ m^3/N s
Viscosity of water	$\mu_w = 1 \times 10^{-3}$ Pa s
Tensile strength	$\sigma_{ult} = 1$ MPa

where S is the *in situ* stress normal to the crack growth direction, G is the shear modulus, μ is the fluid viscosity, t is the time, and parameters A, B, and C are constants given in Table 10.5 based on the solutions of Geertsma and de Klerk (1969) and Spence and Sharp (1985).

Table 10.5 Constants *A*, *B*, and *C* given by Geertsma and de Klerk (1969) and Spence and Sharp (1985)

Parameter	Spence and Sharp (1985)	Geertsma and de Klerk (1969)
A	2.14	1.87
B	1.97	1.38
C	0.65	0.68

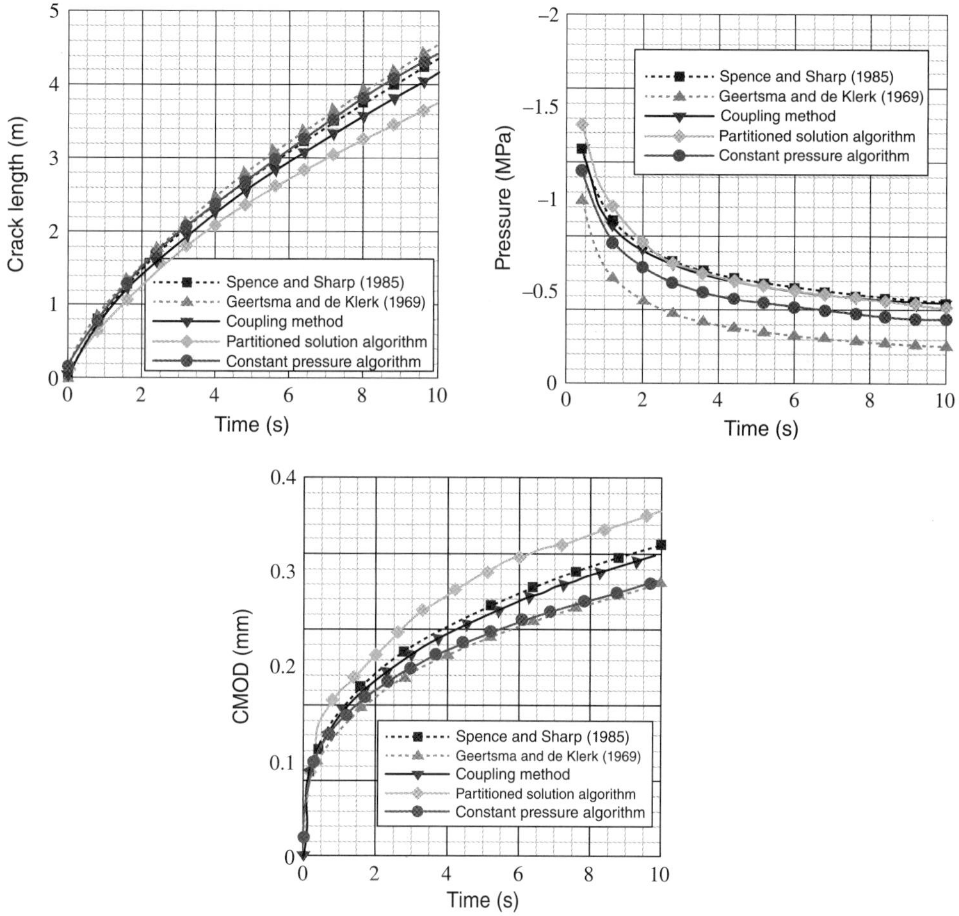

Figure 10.22 The variations with time of the CL, CMP, and CMOD: A comparison between the coupling X-FEM, the X-FEM based on the partitioned solution algorithm, the X-FEM based on the time-dependent constant pressure algorithm, and those of analytical solutions

In Figure 10.22, the variations with time of the CL, CMP, and CMOD are plotted using the coupling X-FEM method, the X-FEM method based on the "partitioned solution algorithm", and the X-FEM method based on the "time-dependent constant pressure algorithm", and the results are compared with those of analytical solutions reported by Geertsma and de Klerk (1969) and Spence and Sharp (1985). Obviously, there is good agreement between the results of coupling X-FEM method and the analytical

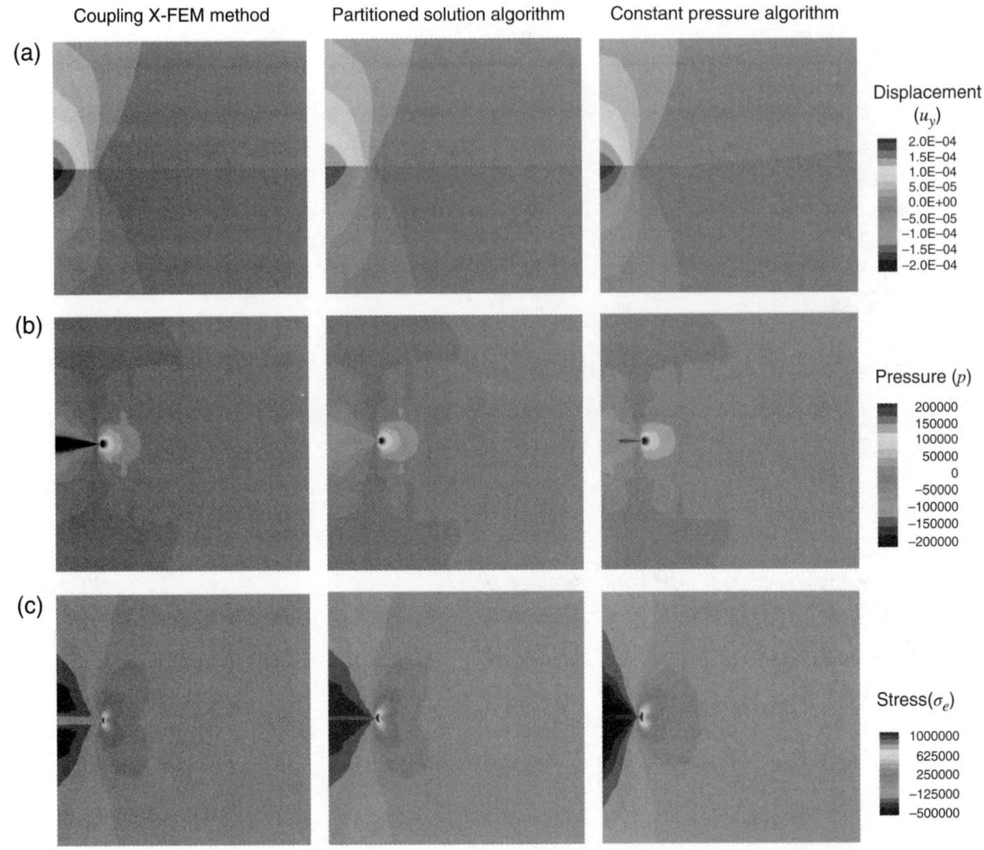

Figure 10.23 The contours of: (a) vertical displacement; (b) fluid pressure; and (c) maximum principal stress using the coupling X-FEM, the X-FEM based on the partitioned solution algorithm, and the X-FEM based on the time-dependent constant pressure algorithm (For color details, please see color plate section)

solution obtained by Spence and Sharp (1985). It can be seen that the results of time-dependent constant pressure algorithm are close to those of the analytical solution obtained by Geertsma and de Klerk (1969); particularly the profiles of the CL and CMP. It must be noted that the convergence is achieved in this method with a maximum number of three iterations for each time step. However, the results of partitioned solution algorithm present a noticeable difference from the two analytical solutions, as can be seen from the profiles of the CL and CMOD. Moreover, a low convergence rate was observed in the solution of partitioned solution algorithm. Hence, it can be highlighted that the time-dependent constant pressure algorithm proposed here can be considered an alternative simplified algorithm for the case of impermeable medium. Finally, the contours of vertical displacement, fluid pressure, and maximum principal stress are shown in Figure 10.23 at the time step $t = 10$ s for three computational algorithms; that is, the coupling X-FEM method, the X-FEM method based on the partitioned solution algorithm, and the X-FEM method based on the time-dependent constant pressure algorithm. Obviously, the overall performances of all three computational models are acceptable.

10.7.3 Hydraulic Fracturing in a Concrete Gravity Dam

The last example is chosen to illustrate the performance of proposed X-FEM method for the challenging problem of a concrete gravity dam under hydrostatic pressure due to water pressure in the reservoir and the

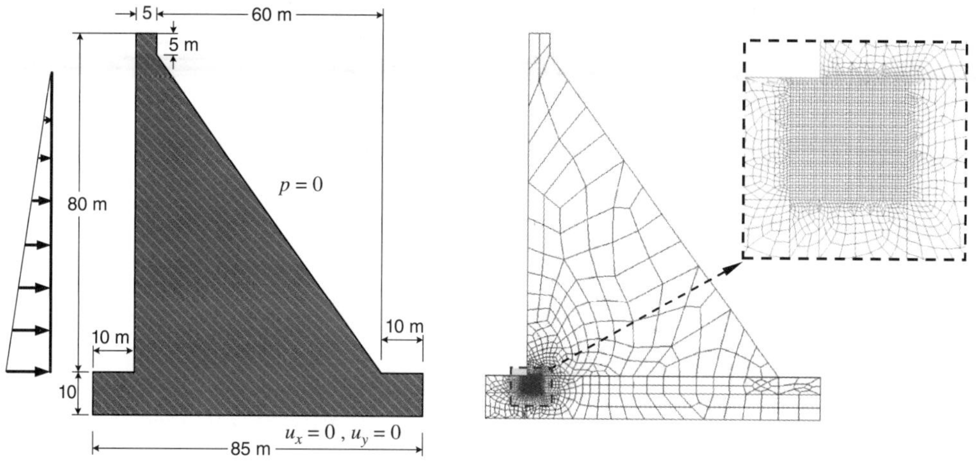

Figure 10.24 A concrete gravity dam under hydrostatic pressure: The geometry, boundary condition, and X-FEM mesh

Table 10.6 Material properties of the concrete gravity dam

Elasticity modulus	$E = 24$ GPa
Poisson ratio	$\nu = 0.15$
Biot constant	$\alpha = 1$
Porosity	$n = 0.19$
Density	$\rho_s = 2400\ \text{kg/m}^3$
Water density	$\rho_w = 1000\ \text{kg/m}^3$
Bulk modulus of solid phase	$K_s = 36.0$ GPa
Bulk modulus of water	$K_w = 3.0$ GPa
Permeability	$k = 10^{-18}\ \text{m}^3/\text{N s}$
Viscosity of water	$\mu_w = 1 \times 10^{-3}$ Pa s
Tensile strength	$\sigma_{ult} = 1.5$ MPa

dam's self-weight. The dam geometry is similar to the ICOLD (International Commission On Large Dams) benchmark exercise A2 (*Proceedings of 5th ICOLD International Benchmark Workshop on Numerical Analysis of Dams*, 1999). This practical example was modeled by Schrefler, Secchi, and Simoni (2006) using a quasi-static cohesive fracture analysis, and Khoei, Barani, and Mofid (2011) using a dynamic analysis of cohesive fracture propagation. In the simulation presented here, the coupling X-FEM method is employed to evaluate the pattern of crack growth in the dam concrete foundation, and the results are compared with the constant pressure algorithm. The geometry, boundary condition, and X-FEM mesh are shown in Figure 10.24. The material parameters for numerical simulation are given in Table 10.6. The zero initial pore pressure is assumed for the concrete dam. The problem is solved for the total time of 0.7 s with the time step chosen as 0.002 s. The dam is modeled under the hydrostatic pressure of the reservoir, where the level of water is set to 70 m. The crack is automatically induced in the dam using the maximum tensile effective stress criterion, and propagated in a direction perpendicular to the maximum tensile effective stress.

In Figure 10.25, the contours of maximum principal stress are presented at time step $t = 0.7$ s, where the fracture length is equal to 3.5 m using the coupling X-FEM method and the X-FEM method based on the constant pressure algorithm. It must be noted that in the constant pressure algorithm proposed here, a pressure boundary condition is assumed at the crack mouth, so the analysis is carried out using a constant water

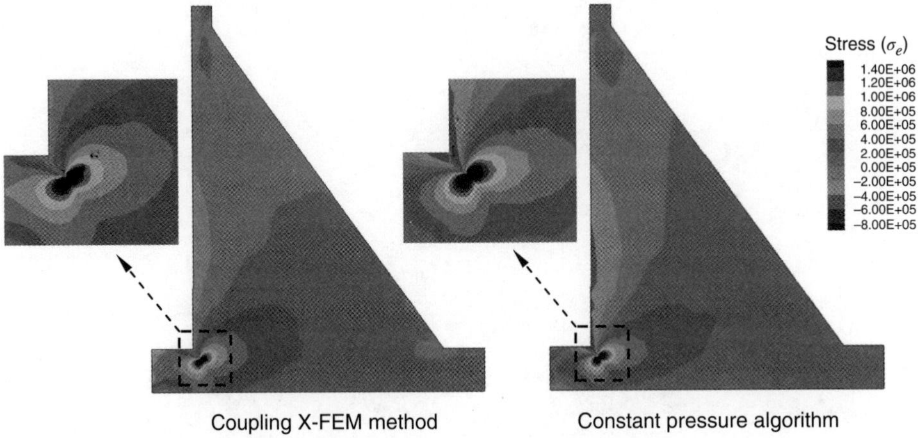

Figure 10.25 The contours of maximum principal stress using the coupling X-FEM method and the X-FEM based on the time-dependent constant pressure algorithm (For color details, please see color plate section)

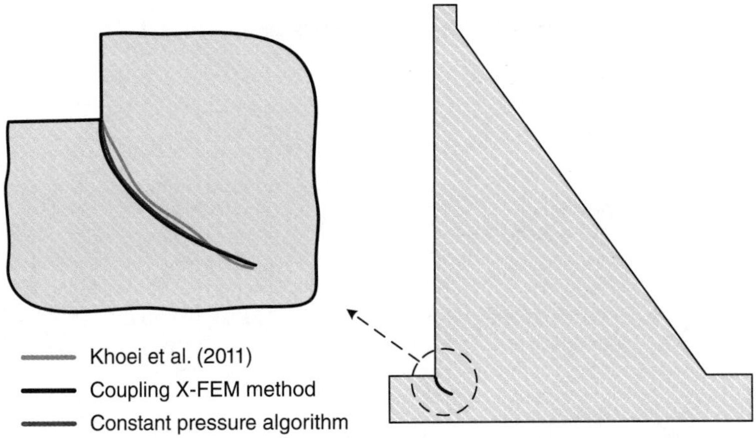

Figure 10.26 The patterns of crack growth in a concrete dam using the coupling X-FEM and the constant pressure algorithm

pressure acting on crack faces along the fracture interface equal to the level of water in the reservoir. Obviously, the similar stress contours can be seen between two computational models, which are in a good agreement with those reported by Schrefler, Secchi, and Simoni (2006) and Khoei, Barani, and Mofid (2011) using an adaptive mesh refinement technique. In Figure 10.26, the patterns of crack growth in concrete dam are depicted for the coupling X-FEM method and the constant pressure algorithm that show a good agreement with that reported by Khoei, Barani, and Mofid (2011). In Figure 10.27, the variations with time of the CL and CMOD are plotted for the coupling X-FEM method and the constant pressure algorithm, and the results are compared with those reported by Khoei, Barani, and Mofid (2011). Clearly, a reasonable agreement can be observed between the X-FEM computational models and those obtained by an adaptive FEM. It can be concluded that while the proposed coupling X-FEM can be used for modeling hydraulic fracture propagation in both permeable and impermeable domains, the time-dependent constant

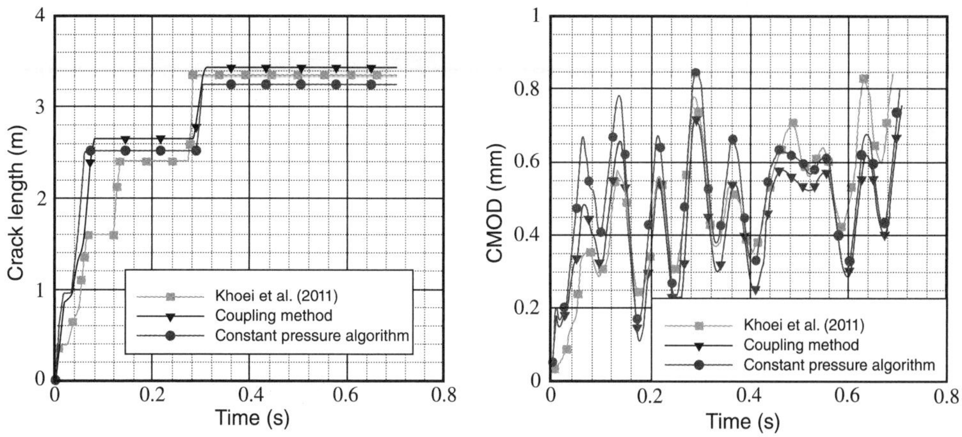

Figure 10.27 The variations with time of the CL and CMOD: A comparison between the coupling X-FEM and the constant pressure algorithm

pressure algorithm can efficiently be applied as a simple computational algorithm in the framework of X-FEM formulation of deformable porous media in the case of an impermeable medium.

10.8 X-FEM Modeling of Contact Behavior in Fractured Porous Media

One of the most advanced issues in the deformable porous medium is the discontinuity because of fracture in the medium. Fractures in porous media are very often in unsaturated soils or rocks, and are sometimes available in saturated clays. Such discontinuities in the fractured porous medium are prone to opening and/or closing modes that result in the fluid flow within the fracture, or contact behavior at the crack edges. In fact, the cyclic loading, such as in an earthquake, in porous media may result in an opening and/or closure along the fracture interface. Thus, modeling the contact behavior in a saturated porous medium is an important task and must be taken precisely into the computation. Contact problems in continuum mechanics have been modeled based on various computational techniques in the FE framework (see Chapter 6). In geomechanical problems, contact behavior has been studied by various researchers (Dyskin and Caballero, 2009; Zhang and Ghassemi, 2011), in which the standard contact elements were used to transfer the internal compression forces along fracture interfaces. In this section, the contact condition in a fractured porous medium is modeled using the X-FEM, and the effect of contact on the fluid phase is employed by considering no leak-off from/into the porous medium.

10.8.1 Contact Behavior in a Fractured Medium

Modeling the mechanical behavior of a fracture in a porous medium is an important task since an opening or a closing mode can be occurred along the fracture that results in the fluid flow within it, or contact behavior at the crack edges. The contact behavior in a fractured porous medium produces additional complexity through the solution of governing equations (10.57) that needs to satisfy the inequalities of contact conditions for both solid and fluid phases. The contact constraints in the solid phase are imposed to prevent the penetration of contacting bodies along the fracture interface, while the contact constraints in fluid phase are imposed to consider no leak-off from/into the porous medium. The objective of a mathematical theory of the contact constitutive law is to provide a theoretical description of motion at the interface of bodies in contact. As described in Chapter 6, the plasticity theory of friction can be achieved by an analogy between plastic and frictional phenomena. In order to formulate such a theory of friction several requirements have to be considered. These requirements, which are similar to the requirements considered in the theory of elasto-plasticity, are as follows; the stick (or adhesion) law, the stick-slip law, the slip criterion, the slip rule, and the wear-tear rule.

The gap function at the contact interface can be defined based on the relative displacement of two bodies at two sides of the discontinuity Γ_d as $[\![\mathbf{u}]\!] = \mathbf{u}^+ - \mathbf{u}^-$, where $\mathbf{u}^+$ and $\mathbf{u}^-$ are the values of $\mathbf{u}$ on two sides of Γ_d. The normal and tangential components of gap function can be defined as $g_N = [\![\mathbf{u}]\!] \cdot \mathbf{n}_{\Gamma_d}$ and $g_T = [\![\mathbf{u}]\!] \cdot \mathbf{m}_{\Gamma_d}$, respectively, with $\mathbf{n}_{\Gamma_d}$ and $\mathbf{m}_{\Gamma_d}$ denoting the normal and tangential unit vectors at the discontinuity. The contact conditions are imposed on the surface of the discontinuity Γ_d in the normal direction through the standard *Kuhn–Tucker* relations, defined by $g_N \geq 0$, $t_c^N \leq 0$, and $g_N t_c^N = 0$, where $t_c^N = t_c \cdot \mathbf{n}_{\Gamma_d}$ is the normal contact stress with t_c denoting the contact traction acting on the discontinuity Γ_d. In the tangent direction, the normal traction in conjunction with the friction between the fracture faces result in the tangential traction t_c^T that leads to the stick-slip behavior. The stick–slip behavior can be determined using a slip criterion F defined based on Coulomb's Law as $|t_c^T| - \mu_f |t_c^N| - c_f = 0$, with μ_f and c_f denoting the contact friction coefficient and the cohesion friction, respectively. The tangential relative sliding is defined by employing a slip rule as $dg_T = d\lambda(\partial Z/\partial t_c^T)$, where $d\lambda$ is the plastic constant and the potential Z is defined by $Z = |t_c^T|$.

10.8.2 X-FEM Formulation of Contact along the Fracture

In order to apply the discretized governing equations (10.57) in the case of fracture closing condition due to contact behavior between the fracture faces, the contact constraints are imposed in the definition of interfacial force vector $\mathbf{f}_\alpha^{\text{int}}$ given in (10.60) to prevent the penetration of contacting bodies along the fracture, and in the definition of flux vector $\mathbf{q}_\delta^{\text{int}}$ given in (10.61) to prevent the leak-off from/into the porous medium. The contact constraints in the solid phase can be incorporated into the force vector $\mathbf{f}_\alpha^{\text{int}}$ by imposing the internal boundary condition $t_c = \boldsymbol{\sigma} \cdot \mathbf{n}_{\Gamma_d}$ on the discontinuity as

$$\mathbf{f}_\alpha^{\text{int}} = \int_{\Gamma_d} [\![\mathbf{N}_u^\alpha]\!]^T \boldsymbol{\sigma} \cdot \mathbf{n}_{\Gamma_d} d\Gamma = \int_{\Gamma_d} [\![\mathbf{N}_u^\alpha]\!]^T t_c \, d\Gamma \qquad \alpha \in (std, Hev, tip) \equiv (\bar{\mathbf{u}}, \bar{\mathbf{a}}, \bar{\mathbf{b}}) \tag{10.77}$$

Moreover, the contact constraints in the fluid phase are incorporated into the flux vector $\mathbf{q}_\delta^{\text{int}}$ by imposing the boundary condition on the discontinuity Γ_d as

$$\begin{aligned} \mathbf{q}_\delta^{\text{int}} &= \int_{\Gamma_d} \left(\mathbf{N}_p^\delta\right)^T [\![\dot{\mathbf{w}}]\!] \cdot \mathbf{n}_{\Gamma_d} d\Gamma = \int_{\Gamma_d} \left(\mathbf{N}_p^\delta\right)^T \bar{q}_d \, d\Gamma \\ &= \int_{\Gamma_d} \left(\mathbf{N}_p^\delta\right)^T k_f \left[\!\left[\nabla \mathbf{N}_p^\delta\right]\!\right] \cdot \mathbf{n}_{\Gamma_d} p \, d\Gamma \qquad \delta \in (std, abs) \equiv (\bar{\mathbf{p}}, \bar{\mathbf{c}}) \end{aligned} \tag{10.78}$$

in which an injection equal to $\bar{q}_d = [\![\dot{\mathbf{w}}]\!] \cdot \mathbf{n}_{\Gamma_d}$ is imposed in opposite direction to prevent the leak-off from/into the porous medium at the contact interface.

In order to evaluate the derivatives of $\mathbf{f}_\alpha^{\text{int}}$ and $\mathbf{q}_\delta^{\text{int}}$ with respect to $\mathbb{U}$ and $\mathbb{P}$ in the Jacobian matrix (10.67), the force and flux vectors are employed in the closing condition of the fracture according to (10.77) and (10.78) as

$$\frac{\partial \mathbf{f}_\alpha^{\text{int}}}{\partial \bar{\mathbb{U}}} = \int_{\Gamma_d} [\![\mathbf{N}_u^\alpha]\!]^T \frac{\partial t_c}{\partial \bar{\mathbb{U}}} d\Gamma \qquad \alpha \in (std, Hev, tip) \equiv (\bar{\mathbf{u}}, \bar{\mathbf{a}}, \bar{\mathbf{b}}) \tag{10.79a}$$

$$\frac{\partial \mathbf{f}_{\alpha}^{\text{int}}}{\partial \bar{\mathbb{P}}} = \frac{\partial \mathbf{q}_{\delta}^{\text{int}}}{\partial \bar{\mathbb{U}}} = \mathbf{0} \qquad \delta \in (std, abs) \equiv (\bar{\mathbf{p}}, \bar{\mathbf{c}}) \tag{10.79b}$$

$$\frac{\partial \mathbf{q}_{\delta}^{\text{int}}}{\partial \bar{\mathbb{P}}} = \int_{\Gamma_d} \left(\mathbf{N}_p^{\delta}\right)^T k_f \left[\!\left[\nabla \mathbf{N}_p^{\delta}\right]\!\right] \cdot \mathbf{n}_{\Gamma_d} \, d\Gamma \tag{10.79c}$$

Obviously, it can be seen from (10.79a) that an additional term must be applied into the stiffness matrix **K** given in the set of Eq. (10.57) when the fracture is in contact condition. In this case, the enriched stiffness matrix $\mathbf{K}_{aa}$ is modified as

$$\widehat{\mathbf{K}}_{aa} = \mathbf{K}_{aa} + \mathbf{K}^{\text{con}} = \int_{\Omega} \left(\mathbf{B}_u^{enr}\right)^T \mathbf{D}\, \mathbf{B}_u^{enr} \, d\Omega + \int_{\Gamma_d} \left(\mathbf{N}_u^{std}\right)^T \mathbf{D}^{\text{con}} \, \mathbf{N}_u^{std} \, d\Gamma \tag{10.80}$$

in which $\mathbf{K}^{\text{con}}$ is used based on the penalty method to incorporate the contact behavior into the X-FEM formulation. In this definition, $\mathbf{D}^{\text{con}}$ is the tangent matrix of contact problem relating the contact traction $\boldsymbol{t}_c$ to the displacement jump along the contact interface Γ_d as $dt_c = \mathbf{D}^{\text{con}}\, d[\![\mathbf{u}]\!]$. The contact tangent matrix $\mathbf{D}^{\text{con}}$ is basically an unsymmetric matrix due to non-associated slip rule adopted to derive the contact constitutive matrix; however, in order to preserve the symmetry of the numerical formulation, the coupling between the normal and tangential tractions at the contact interface can be neglected. In this way, the problem can be decomposed into a pure contact in the normal direction and a frictional resistance in the tangential direction as $dt_c^N = k_N\, d[\![\mathbf{u}_N]\!]$ and $dt_c^T = k_T\, d[\![\mathbf{u}_T]\!]$, where the penalty parameters k_N and k_T are considered as the normal and tangential stiffness constants at the contact interface.

10.8.3 Consolidation of a Porous Block with a Vertical Discontinuity

In order to illustrate the robustness of proposed X-FEM contact computational algorithm in comparison with those reported in literature based on the conventional FEM in conjunction with the zero thickness elements, the consolidation of a porous block with a vertical discontinuity is solved using the X-FEM method. This example is chosen to demonstrate the opening and closing of the fracture due to injection of the fluid into the fracture, and the contact behavior between fracture faces. This problem was modeled originally by Ng and Small (1977) and then by Segura and Carol (2008b) to present the performance of their zero thickness elements in a finite element analysis. A comparison is performed between the results of proposed X-FEM model and those reported in literature to identify the effect of fracture in a deformable porous medium and to illustrate the performance of coupling between the porous medium and the discontinuity aperture. The problem includes a square porous block of 1 × 1 m with a vertical discontinuity at the middle of porous medium, as shown in Figure 10.28. A linear elastic behavior is assumed for the block with Young's modulus of $E = 1.0$ MPa and Poisson ratio of $\nu = 0.2$ (Table 10.7). An isotropic material behavior is assumed with the permeability of $k_f = 1.157 \times 10^{-9}$ m^3/N s (equivalent to 1 m/day). The left and right boundaries are restrained in the horizontal direction, while the lower boundary is restrained in both directions. A distributed uniform loading of $w = 10$ kPa is exerted together with the zero fluid pressure at the upper edge. A structured mesh of 11 × 11 elements with 144 nodes is utilized. In order to highlight the effect of the fracture from a hydraulic point of view, a high longitudinal permeability is assumed through the discontinuity in comparison with the permeability of porous medium. The permeability of the fracture is assumed based on the cubic law, which is the aperture-dependent permeability.

The initial boundary conditions are similar to a typical consolidation problem, so an increase of the pore fluid pressure can be observed over the whole domain at the beginning of the analysis. However, by increasing the load at the upper edge, the porous medium starts evacuating the fluid from the upper edge where the fluid pressure is zero. The excess fluid pressure reduces from the upper edge to the lower edge and as a result the effective stress decreases; this results in the compaction of porous medium and the

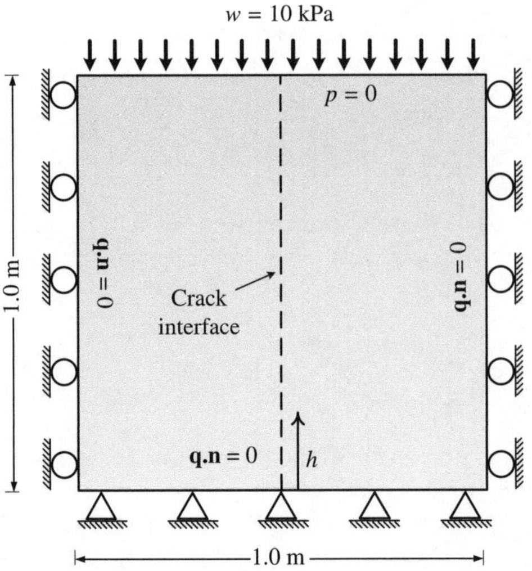

Figure 10.28 The consolidation of a porous block with a vertical discontinuity: Problem definition

Table 10.7 Material properties for the consolidation of a porous block with a vertical discontinuity

Elasticity modulus	$E = 1$ MPa
Poisson ratio	$\nu = 0.2$
Biot constant	$\alpha = 1$
Porosity	$n = 0.3$
Density	$\rho_s = 2000\ \text{kg/m}^3$
Water density	$\rho_w = 1000\ \text{kg/m}^3$
Bulk modulus of water	$K_w = 3.0$ GPa
Permeability	$k = 1.157 \times 10^{-9}\text{m}^3/\text{N s}$
Viscosity of water	$\mu_w = 1 \times 10^{-3}$ Pa s

closure of discontinuity towards the lower edge. This closure first appears in the upper part and then propagates toward the lower part; as a result it causes the closure of the fracture on the upper part at the early stages of loading that prevents the evacuation of the fluid through the fracture. Because of a preferential path for the fluid flow in the lower part, it leads to an increase in fluid pressure at the upper part and a decrease in fluid pressure in the lower part. In fact, the initially fast consolidation becomes slower after the upper part of the fracture closes, consolidating at the same rate as the porous medium with no discontinuity. It is noteworthy to highlight that during the initial stages of loading, the X-FEM modeling based on the coupled hydromechanical formulation is activated to model the opening of the fracture due to fluid flow through the fracture. However, by increasing the load, an iterative strategy is used to indicate the contact region on the upper part and the opening zone in the lower part of the fracture. It can be observed from the simulation that at the final stages of loading, the whole fracture is in contact behavior and a coupled hydromechanical X-FEM analysis based on the contact constraints model is utilized to evaluate the excess fluid pressure over the domain. In Figure 10.29, the evolutions of excess fluid pressure are plotted along the interface at different time steps, that is, $T = 0.0007$, 0.0021, 0.0035, and 0.007 day. A comparison is performed between the X-FEM model with the penalty and LATIN (LArge Time Increment) methods and those reported by Segura and Carol (2008b). Obviously, a reasonable agreement can be observed between the proposed X-FEM model and those obtained by the conventional FEM method with zero thickness elements. In Figures 10.30 and 10.31, the distributions of excess pore water pressure

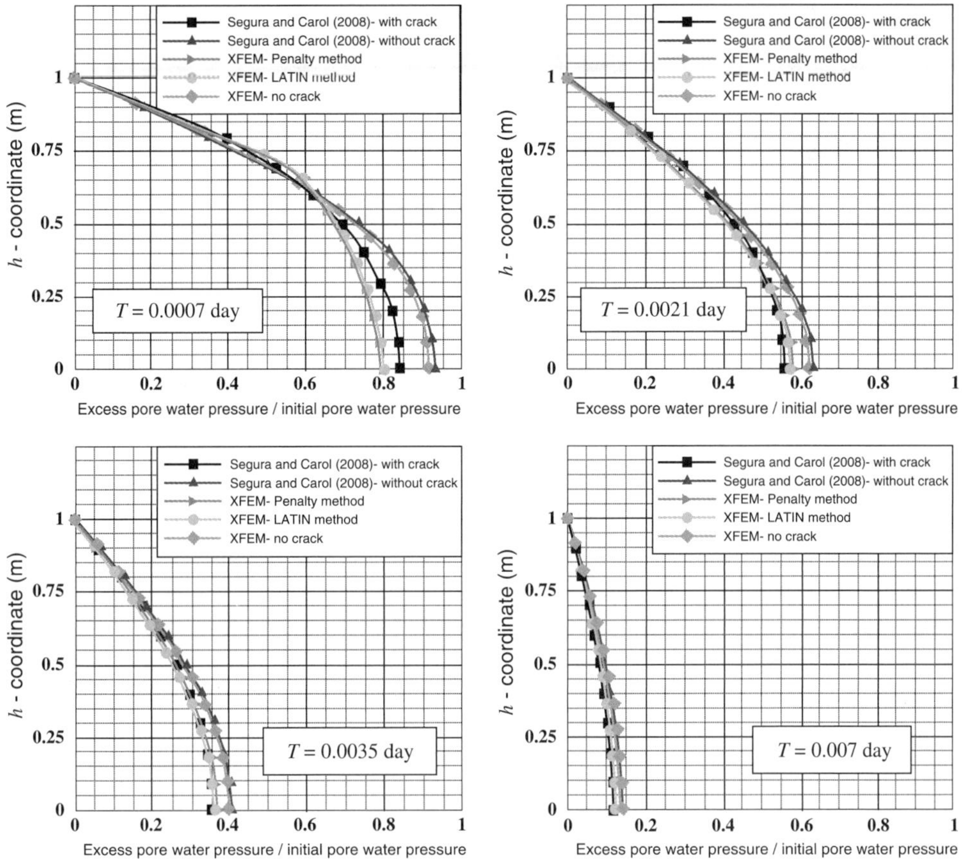

Figure 10.29 The fluid pressure distribution along the interface at different time steps: A comparison between the X-FEM model with the penalty and LATIN methods and those reported in literature

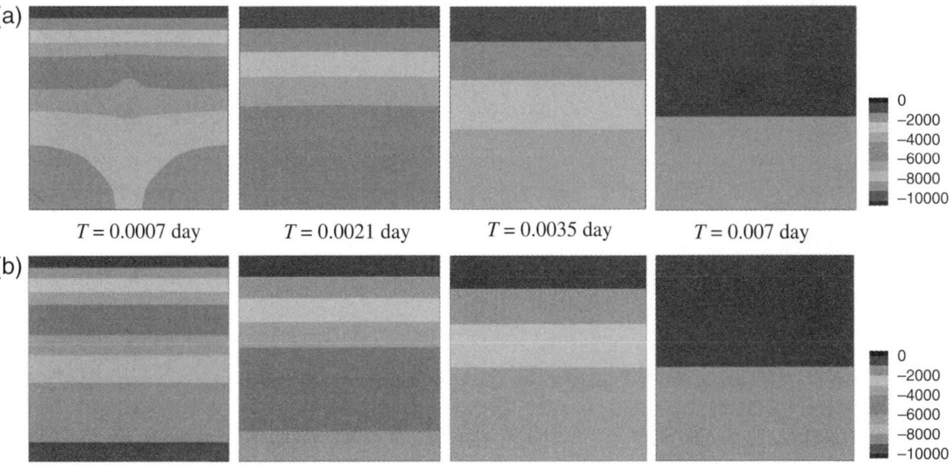

Figure 10.30 The distributions of excess pore water pressure contours at various time steps: (a) the X-FEM model with the penalty method; (b) the FEM with no crack (For color details, please see color plate section)

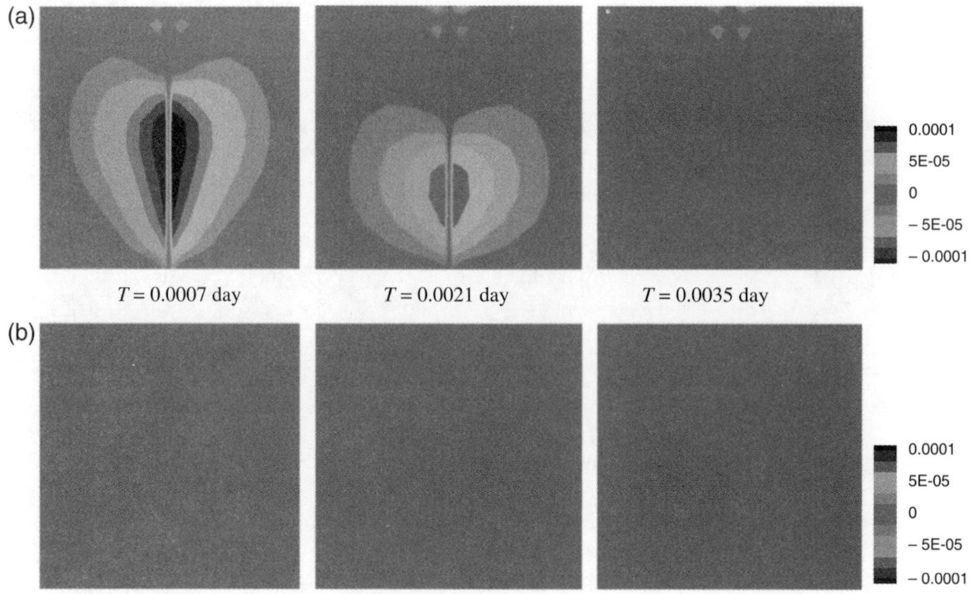

Figure 10.31 The distributions of horizontal displacement contours at various time steps: (a) the X-FEM model with the penalty method; (b) the FEM with no crack (For color details, please see color plate section)

and horizontal displacement contours are presented respectively at different time steps. It can be observed from these figures that the excess pore water pressure reduces quickly from the upper edge to the lower edge along the fracture at the early stages of loading. However, after the closure of the fracture due to consolidation of porous medium, the initially fast consolidation becomes slower.

11

Hydraulic Fracturing in Multi-Phase Porous Media with X-FEM

11.1 Introduction

The present chapter focuses on the hydromechanical modeling of two-phase fluid flow in deforming partially saturated porous media containing propagating cohesive cracks, which has practical applications in a broad range of engineering areas. In the literature, the topic of fluid flow in fractured/fracturing porous media has been dealt with in different ways; Boone and Ingraffea (1990) presented a numerical procedure for the simulation of hydraulically driven fracture propagation in poroelastic materials combining the finite element method (FEM) with the finite difference method; Schrefler, Secchi, and Simoni (2006) and Secchi, Simoni, and Schrefler (2007) modeled the hydraulic cohesive crack growth in fully saturated porous media using the FEM with mesh adaptation; Segura and Carol (2008a) proposed a hydromechanical formulation for fully saturated geomaterials with pre-existing discontinuities based on the FEM with zero-thickness interface elements; Khoei, Barani, and Mofid (2011) and Barani, Khoei, and Mofid (2011) presented the dynamic analysis of cohesive fracture propagation in fully saturated and partially saturated porous media with a passive gas phase, respectively. Recently, the fluid flow in fractured fully saturated porous media and fracturing unsaturated porous media with passive gas phase was presented by Réthoré, De Borst, and Abellan (2007b, 2008) using the extended FEM, which is now extended to three-phase porous media. The three-phase numerical model developed here is based upon the mechanics of deformable porous media on the basis of the generalization of the Biot theory (1941) in conjunction with cohesive fracture mechanics, which provides a suitable framework for describing coupled hydromechanical and fracture mechanisms occurring in fracturing, multi-phase porous media. In such multi-phase systems, the coupling between the flow of the wetting and non-wetting phases in the pore spaces of the continuous porous medium and the discontinuity, the deformation of the solid phase, the fluid exchange between the discontinuity and the surrounding porous medium, and the possible development of the discontinuity across which the cohesive tractions are transmitted, is usually strong, which demands the fully coupled treatment of the problem. In the formulation presented herein, all these components are brought together to thoroughly simulate the deforming, partially saturated porous medium behavior in the presence of geomechanical discontinuities, thus exhibiting fluid flow, deformation, and fracture processes properly.

In the approaches based on the standard FEM, the crack path is restricted to the inter-element boundaries, suffering from the problem of mesh dependency (Xu and Needleman, 1994; Camacho and Ortiz, 1996; Ortiz and Pandolfi, 1999), or successive remeshing is carried out to overcome the sensitivity to the mesh generated and avoid the preferred directions when the crack is propagating (Khoei, Azadi, and Moslemi, 2008c; Khoei *et al.*, 2009b; Moslemi and Khoei, 2009; Azadi and Khoei, 2011), which makes crack growth simulation a computationally expensive and cumbersome process. The difficulties

Extended Finite Element Method: Theory and Applications, First Edition. Amir R. Khoei.

confronted in the standard FEM are handled by locally enriching the conventional finite element (FE) approximation with an additional function through the concept of the partition of unity (PU), which was introduced in the pioneering work of Melenk and Babuska (1996). This idea was exploited to set up the frame of the extended finite element method (X-FEM) by Belytschko and Black (1999) and Moës, Dolbow, and Belytschko (1999). Indeed, the X-FEM approximation relies on the PU property of FE shape functions for the incorporation of local enrichments into the classical FE basis. By appropriately selecting the enrichment function and enriching specific nodal points through the addition of extra degrees of freedom (DOF) relevant to the chosen enrichment function to these nodes, the enriched approximation would be capable of directly capturing the local property in the solution (Daux *et al.*, 2000; Belytschko *et al.*, 2001; Sukumar *et al.*, 2001). On the basis of X-FEM technique, the evolving cohesive crack is simulated independently of the underlying FE mesh and without continuous remeshing of the domain as the crack grows. Since the FE mesh does not need to conform to the crack geometry, and thus the need for the costly mesh regeneration and data transfer between two successive meshes is eliminated, the X-FEM facilitates modeling of the propagating crack. The technique has been applied in crack problems, including: crack growth with frictional contact by Dolbow, Moës, and Belytschko (2001), cohesive crack growth by Wells and Sluys (2001) and Moës and Belytschko (2002a), stationary and growing cracks by Ventura, Budyn, and Belytschko (2003), and dynamic cohesive crack propagation by Remmers, de Borst, and Needleman (2008). The X-FEM has also been employed in elasto-plastic problems, including: plastic fracture mechanics by Elguedj, Gravouil, and Combescure (2006, 2007), plasticity of frictional contact by Khoei and Nikbakht (2007), localization phenomenon in higher-order Cosserat theory by Khoei and Karimi (2008), and porous media with arbitrary interfaces by Khoei and Haghighat (2011).

In fracturing partially saturated porous media, crack growth occurs as the progressive decay of the cohesive tractions transferred across the fracture process zone and the imposition of the mean pore pressure onto the crack faces by means of the pore fluids within the crack. The tractions acting on the fracture faces give rise to mechanical coupling between the fracture and porous medium surrounding it. Besides, the flux of the two fluid phases through the fracture borders leads to the mass transfer coupling, which is a subject of great interest in hydraulic fracturing. In numerical modeling, in order to provide for the displacement jump across the crack borders and mass transfer between the crack and the surrounding porous medium, some requirements must be met. Crack opening requires that the displacement field be discontinuous across the crack. In addition, the mass transfer implies that the fluid flow, which is governed by the Darcy velocity of the pore fluid, in the normal direction to the crack be discontinuous. Taking into account the Darcy relation which relates the Darcy velocity to the pressure gradient, it can be concluded that the normal gradient of the fluid pressure must also be discontinuous across the crack. In the context of the extended FEM, the discontinuity in the displacement field and fluid flow normal to the crack along the crack trajectory is modeled by inserting a discontinuous enrichment function and an enrichment function with discontinuous normal derivative, respectively, in the corresponding standard approximating space. In an enriched formulation of hydraulic fracture propagation, the standard FE approximation of the displacement field is enriched by incorporating the sign function, and for the water pressure and the capillary pressure approximation, a modified definition of the commonly used distance function is introduced for the enrichment function. The local enrichment of the standard approximating space enables the consequent enriched approximation to capture the special characteristics of the solution in the local parts of the domain where the field variables exhibit discontinuity or discontinuous normal derivative.

To arrive at the discrete equations, the X-FEM is utilized to discretize the weak form of the governing equations, that is, the linear momentum balance equation and flow continuity equations, in spatial domain along with the generalized Newmark scheme for time domain discretization. For the numerical solution, the unconditionally stable direct time stepping procedure is applied to resolve the resulting system of strongly coupled nonlinear algebraic equations using the Newton–Raphson iterative procedure. In order to illustrate the performance of proposed computational algorithm, a numerical analysis of convergence is performed for different local enrichment methods that model the weak discontinuities in the displacement field across the bimaterial boundary. For this purpose, a convergence study of various approaches in the X-FEM framework for treating weak discontinuities is investigated, including the standard X-FEM with the abs-enrichment function, the standard X-FEM with the weighted abs-enrichment function, the standard X-FEM with the modified abs-enrichment function, and the corrected X-FEM. It is well-known that the standard X-FEM approximation with the abs-enrichment function leads to a sub-optimal convergence rate and reduces the accuracy of the solution because of problems that emanate from the parasitic terms arising in the approximating space of the blending elements; that is, elements where only some of the nodes are enriched. The cause of the appearance of these parasitic terms is that the enriched part of the X-FEM approximation is not spanned by the standard basis functions in the blending elements. It is proved that, in order to achieve an optimal convergence rate, it is necessary to eliminate the parasitic terms appearing in the approximating space of the blending elements (see Chapter 4). Finally, a practical example is presented to demonstrate the capability of proposed computational model in hydraulic fracture problems. Apart from illustrating the considerable influence of the hydromechanical coupling between the continuum and the discontinuity on the computational results, numerical simulations confirm the intense contribution of the propagating discontinuity to the deformability, permeability, and flow characteristics of the multi-phase porous system. Besides, it has been shown that the flow of the gas phase and, hence, the development of the gas pressure disregarded in the simplified model based on the passive gas phase assumption, can be as significant as those of the water phase. This emphasizes the importance of the inclusion of the gaseous phase in the approach.

11.2 The Physical Model of Multi-Phase Porous Media

The partially saturated porous medium surrounding the fracture is modeled as a multi-phase system where wetting and non-wetting pore fluids exist simultaneously in the void spaces of the solid skeleton. In the presented physical model, immiscibility is supposed for the two-phase fluid flow through the porous medium. That is, it is assumed that there is no phase change and no mass transfer between the two porous fluids. It is also assumed that the multi-phase system remains under isothermal conditions (Lewis and Schrefler, 1998).

A proper choice of primary variables is a crucial step in efficiently modeling of the highly nonlinear problem of multi-phase flow in porous media. It is an indisputable fact that the choice of primary variables not only impacts the computational performance of the computer code and the conditioning of the Jacobian matrix and hence the number of iterations, but may also determine the feasibility of a numerical modeling. In what follows, the equations describing the problem are written in terms of the solid skeleton displacements, the wetting phase pressure, and the capillary pressure. The choice of the capillary pressure as the third primary variable is preferable to the water saturation from a numerical point of view. The option of the water saturation as the independent variable introduces $\partial p_c/\partial S_w$ into the formulation, while the formulation based on the selection of the capillary pressure as the third variable introduces $\partial S_w/\partial p_c$ into the equations. In the water saturation formulation, the steepest or infinite slope of a capillary curve at both ends of dry and wet conditions causes the infinity of $\partial p_c/\partial S_w$ if saturation is near the residual or unity. Hence, the numerical simulation is prone to convergence problem and instability. This problem may become more distinct if it is intended to preserve symmetry in the Jacobian matrix viewing the fact that the gas phase mass balance equation must be wholly multiplied by $\partial p_c/\partial S_w$. In contrast, in the capillary pressure formulation at the limits of fully saturated and dry conditions $\partial S_w/\partial p_c$ tends to zero, which assures the better numerical performance of the computer code. Furthermore, in the water saturation formulation the off-diagonal terms of the Jacobian matrix which make it non-symmetric are more compared with the capillary formulation. Thus, further approximation is necessary if it is intended to render the Jacobian matrix symmetric.

The pores of the solid skeleton in the partially saturated porous medium are assumed to be filled up partly with water (w) as the wetting phase and partly with gas (g) as the non-wetting phase. Thus, the degrees of saturation of the liquid phase S_w and the gaseous phase S_g always sum to unity, that is, $S_w + S_g = 1$. The degree of saturation is specified through the experimentally determined function of the capillary pressure ($S_w = S_w(p_c)$). The capillary pressure between the two fluid phases is defined as the difference between the water pressure p_w and the gas pressure $p_g\left(p_c = p_g - p_w\right)$, which plays a key role in determining the state of the multi-phase system.

In the mechanics of partially saturated porous media, the stress relation is derived by introducing the concept of the modified effective stress to account for the compressibility of the solid grains, that is,

$$\boldsymbol{\sigma}'' = \boldsymbol{\sigma} + \alpha \mathbf{m} p \tag{11.1}$$

where $\boldsymbol{\sigma}$ is the total stress vector, $\boldsymbol{\sigma}''$ is the modified effective stress vector, $\mathbf{m}$ is the identity vector defined as $[1 \;\; 1 \;\; 0]^{\mathrm{T}}$ for the two-dimensional case, p denotes the mean pore pressure applied by the porous fluids on the solid skeleton, which is given by the averaging technique $p = S_w p_w + S_g p_g$, and α is the Biot constant defined as $\alpha = 1 - K_T/K_S \leq 1$, with K_T and K_S denoting the bulk moduli of the porous medium and the solid grains, respectively. According to the experimental observations, the modified effective stress governs the major deformation of the solid skeleton, so the mechanical behavior of the partially saturated material is characterized in its term. To take account of the material nonlinearity in the continuum medium surrounding the crack, the constitutive equation of the solid phase relating the modified effective stress to the total strain is expressed by means of an incrementally linear modified effective stress-strain relationship as

$$d\boldsymbol{\sigma}'' = \mathbf{D}\, d\boldsymbol{\varepsilon} \tag{11.2}$$

where $\mathbf{D}$ represents the tangential constitutive matrix of the continuum defined by a suitable constitutive law. Throughout this chapter, the stress is regarded as tension positive, while the pore fluid pressure is regarded as compression positive.

In order to describe the nonlinear fracture processes developing ahead of the crack tip, the cohesive crack model is used. This model is an appropriate alternative when the size of the fracture process zone at the crack front is not negligible in comparison with the crack length (Bazant and Planas, 1998), which is a commonly seen feature for cracks in quasi-brittle materials, such as geomaterials and concrete. In this model, it is assumed that the near tip fracture process zone is lumped into the crack line, unlike the linear elastic fracture mechanics (LEFM) in which the fracture processes are considered to occur at the crack tip. Moreover, the cohesive crack model allows abandoning the singularity of the crack tip stress field, an unrealistic characteristic of LEFM. In the cohesive crack model, the nonlinear behavior of the material in the fracture process zone is described using a cohesive constitutive relation. The concept of cohesive crack was originally introduced by Barenblatt (1959, 1969) and Dugdale (1960) to describe the near tip nonlinear processes taking place in brittle and ductile materials, respectively. Afterwards, the cohesive crack concept was extended to simulate the cohesive crack development in concrete by Hillerborg, Modéer, and Petersson (1976).

The nonlinear behavior of the fracturing material in the cohesive zone is governed by a traction-separation law relating the cohesive tractions to the relative displacements, that is, $\mathbf{t}_d = \mathbf{t}_d([\![\mathbf{u}]\!])$, where $\mathbf{t}_d$ is the cohesive traction transmitted across the fracture process zone and $[\![\mathbf{u}]\!]$ is defined as the relative displacement vector at the discontinuity, the difference in the displacement vector between the two faces of the discontinuity. Indeed, the symbol $[\![\;]\!]$ denotes the jump across the discontinuity. In quasi-brittle materials, as soon as the failure limit of the material is exceeded, the cohesive zone develops in which the material begins to fail and exhibits a softening behavior. The softening induced by the material failure is simulated using a softening cohesive law. This implies that the cohesive traction transferred across the cohesive zone is made a decaying function of the relative displacement.

The linearization of the cohesive relation results in the differential form of $d\mathbf{t}_d = \mathbf{T}\, d[\![\mathbf{u}]\!]$, in which $\mathbf{T}$ represents the tangential modulus matrix of the discontinuity to be used in the iterative solution procedure, obtained from the following relation

$$\mathbf{T} = \frac{\partial \mathbf{t}_d}{\partial [\![\mathbf{u}]\!]} \tag{11.3}$$

In order to develop a relation for the material tangent matrix $\mathbf{T}$, it is first formulated in the local orthogonal coordinate system, constructed from the tangential and normal unit vectors to the discontinuity, $\mathbf{t}_{\Gamma_d}$ and $\mathbf{n}_{\Gamma_d}$, with the orientation of $\mathbf{t}_{\Gamma_d}$ taken such that the unit vector perpendicular to the plane of the two-dimensional medium forms a right-handed system. Then, it is rotated into the global coordinate system by $\mathbf{T}=\mathbf{A}^T\mathbf{T}'\mathbf{A}$, in which $\mathbf{A}$ is the rotation matrix, applied to perform the transformation to the global coordinate system, and $\mathbf{T}'$ is related to the local coordinate system, obtained by differentiating the tangential and normal components of the cohesive traction with respect to the displacement jump in the tangential and normal directions as

$$\mathbf{T}'=\begin{bmatrix} \dfrac{\partial t_s}{\partial [\![u_s]\!]} & \dfrac{\partial t_s}{\partial [\![u_n]\!]} \\ \dfrac{\partial t_n}{\partial [\![u_s]\!]} & \dfrac{\partial t_n}{\partial [\![u_n]\!]} \end{bmatrix} \tag{11.4}$$

where t_s and t_n are the shear and the normal cohesive traction, respectively, and $[\![u_s]\!]$ and $[\![u_n]\!]$ denote the crack sliding and the crack opening displacement, respectively. The components of the cohesive traction and displacement jump tangential and normal to the crack are determined by decomposing the related vector with respect to the local orthonormal basis. The explicit form of $\mathbf{T}'$ can be obtained using the cohesive law (see Chapter 8).

In the case of the occurrence of mode I failure, the relation given in (11.3) is simply substituted by a relation in terms of the cohesive traction and relative displacement in the direction normal to the crack as $\mathbf{T}=\mathbf{n}_{\Gamma_d}(\partial t_n/\partial[\![u_n]\!])\,\mathbf{n}_{\Gamma_d}^T$, whose detailed expression depends on the constitutive model applied at the discontinuity. It is noted that in this case the shear cohesive traction acting tangent to the crack and the shear relative displacement in the tangential direction with respect to the crack are zero. That is, this failure mode only involves the normal cohesive traction and the crack opening.

11.3 Governing Equations of Multi-Phase Porous Medium

In order to derive the partial differential equations governing the solid skeleton deformation and the wetting and the non-wetting pore fluid flow through the partially saturated porous medium surrounding the fracture, the balance equations of linear momentum and mass are employed. The governing equations, composed of the equilibrium equation for the whole mixture and the continuity equation of flow for each porous fluid, constitute the basis of the multi-phase formulation. These equations are complemented by the mechanical constitutive law of the solid phase together with the constitutive relationships between the hydraulic properties of the pore fluid phases. It is assumed that the pore fluid flow through the fracturing porous medium is based on Darcy's law. In what follows, the equations specifying the problem are written in terms of the displacement of the solid phase, the pressure of the wetting phase and the capillary pressure (Khoei and Mohammadnejad, 2011).

In the low frequency range, the relative acceleration of the fluid phases with respect to the solid phase is negligible when compared to the acceleration of the solid phase (Zienkiewicz and Shiomi, 1984). With taking this into consideration, the linear momentum balance equation for the multi-phase porous medium can be written as

$$\nabla\cdot\boldsymbol{\sigma}+\rho\,\mathbf{b}-\rho\,\ddot{\mathbf{u}}=0 \tag{11.5}$$

where $\ddot{\mathbf{u}}$ is the acceleration vector of the solid phase, $\mathbf{b}$ is the body force vector, ρ is the average density of the multi-phase system defined as $\rho=(1-n)\rho_s+n(S_w\rho_w+S_g\rho_g)$, in which n stands for the porosity of the porous medium and ρ_s, ρ_w and ρ_g are the densities of the solid phase, porous water, and gas, respectively, and the symbol ∇ denotes the vector gradient operator.

The continuity equations for the flow of wetting and non-wetting phase fluids through the deforming, isothermal porous medium can be written as

$$\begin{aligned}&\frac{1}{Q_{ww}}\dot{p}_w+\frac{1}{Q_{wc}}\dot{p}_c+\alpha\nabla\cdot\dot{\mathbf{u}}+\nabla\cdot\dot{\mathbf{w}}_w+\nabla\cdot\dot{\mathbf{w}}_g=0\\&\frac{1}{Q_{cw}}\dot{p}_w+\frac{1}{Q_{cc}}\dot{p}_c+\alpha S_g\nabla\cdot\dot{\mathbf{u}}+\nabla\cdot\dot{\mathbf{w}}_g=0\end{aligned} \tag{11.6}$$

where $\dot{\mathbf{u}}$ is the solid velocity vector, $\dot{\mathbf{w}}_w$ and $\dot{\mathbf{w}}_g$ are the Darcy velocity vectors of two flowing fluids, and the superposed dot denotes the material time derivative. The compressibility coefficients are defined as

$$\begin{aligned}&\frac{1}{Q_{ww}}=\frac{\alpha-n}{K_s}+\frac{nS_w}{K_w}+\frac{nS_g}{K_g}\\&\frac{1}{Q_{wc}}=\frac{\alpha-n}{K_s}\left((1-S_w)-p_c\frac{\partial S_w}{\partial p_c}\right)+\frac{nS_g}{K_g}\\&\frac{1}{Q_{cw}}=\frac{(\alpha-n)S_g}{K_s}+\frac{nS_g}{K_g}\\&\frac{1}{Q_{cc}}=\frac{(\alpha-n)S_g}{K_s}\left((1-S_w)-p_c\frac{\partial S_w}{\partial p_c}\right)-n\frac{\partial S_w}{\partial p_c}+\frac{nS_g}{K_g}\end{aligned} \tag{11.7}$$

in which K_w and K_g are the bulk moduli of the porous fluids.

The linear momentum balance equation for each fluid phase results in the generalized Darcy equation for multi-phase flow. Assuming negligibility of the relative acceleration term of the fluids, identical to what was supposed previously, the Darcy relation for pore fluid flow can be written as

$$\dot{\mathbf{w}}_\alpha=\mathbf{k}_\alpha[-\nabla p_\alpha+\rho_\alpha(\mathbf{b}-\ddot{\mathbf{u}})] \qquad \alpha=w,g \tag{11.8}$$

where $\mathbf{k}_w$ and $\mathbf{k}_g$ are the permeability matrices of the porous medium to the pore fluids, which are generally evaluated by the following expression

$$\mathbf{k}_\alpha=\mathbf{k}\frac{k_{r\alpha}}{\mu_\alpha} \qquad \alpha=w,g \tag{11.9}$$

in which $\mathbf{k}$ denotes the intrinsic permeability matrix of the porous medium, which is simply replaced by a scalar value k for the isotropic medium, $k_{r\alpha}$ is the relative permeability coefficient of the fluid phase, which is related to the degree of saturation through an experimental function known from the laboratory $(0 \le k_{r\alpha}(S_\alpha) \le 1)$, and μ_α denotes the dynamic viscosity of the fluid phase.

Darcy's Law with constant intrinsic permeability is assumed to hold for the pore fluid flow in the porous medium surrounding the fracture. This assumption is not valid for the fluid flow in the fracture. The permeability inside the fracture, that is, the fully damaged zone and the micro-cracked zone, is strongly influenced by the change in the pore spaces of the solid skeleton as a result of cracking and micro-cracking processes. Consequently, the effect of variation in the micro-structure of the cracked porous material inside the fracture on the permeability of this zone must be accounted. To this end, the pore fluid flow within the fracture is modeled by means of Darcy's law with porosity dependent permeability, in which the dependence of the fracture permeability on the porosity is incorporated into the formulation via the coefficient k_{n_d}, that is,

$$\mathbf{k}_{\alpha d} = \mathbf{k}\frac{k_{n_d} k_{r\alpha}}{\mu_\alpha} \qquad \alpha = w, g \tag{11.10}$$

According to Meschke and Grasberger (2003), the following relation is assigned to k_{n_d} as

$$k_{n_d}(n_d) = 10^{\delta_{n_d}}, \qquad \delta_{n_d} = \frac{6(n_d - n)}{0.3 - 0.4n} \tag{11.11}$$

where n_d and n are the current and the initial porosity of the fracture material, respectively. Therefore, $(n_d - n)$ denotes the change of porosity in the fracture. An explicit relation for n_d is introduced in the next section.

11.4 The X-FEM Formulation of Multi-Phase Porous Media with Weak Discontinuities

In order to develop the weak form of the governing equations of two-phase fluid flow in deformable partially saturated porous media containing weak discontinuities, such as material interfaces, via the extended FEM, a two-dimensional domain Ω bounded by the boundary Γ is considered, as depicted in Figure 11.1, in which the domain contains a material interface Γ_d. It must be noted that in the case where a material interface exists in the porous medium, the fluid flow is not affected by the material interface. That is, the fluid flow is continuous across the material interface. This implies that the water pressure and capillary pressure fields are smooth and need not be enriched. The presence of the material interface in the porous medium only results in a weak discontinuity in the displacement field. In the context of the X-FEM, the discontinuity in the strain field is modeled by inserting an enrichment function with discontinuous derivatives in the standard approximating space. The local enrichment of the standard approximating space enables the consequent enriched approximation to capture the discontinuity in the displacement derivatives.

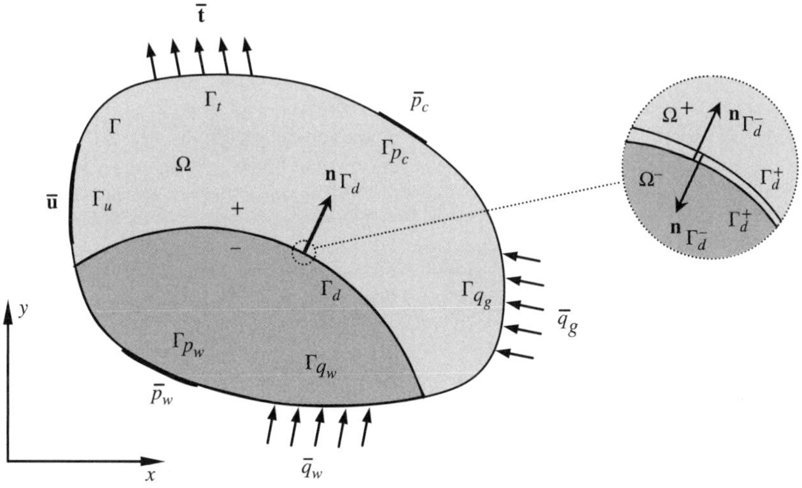

Figure 11.1 Boundary conditions of the body Ω involving a weak discontinuity Γ_d

The initial boundary conditions, specifying the displacement and the velocity field of the solid phase as well as the water pressure and the capillary pressure field of the fluid phases at time $t=0$ in the whole analyzed domain, are acquired through a preliminary static analysis before the dynamic analysis is launched to ensure the satisfaction of the equilibrium and flow continuity equations at the beginning of the dynamic analysis. The essential boundary conditions are imposed on the external boundary by using the prescribed primary variables as $\mathbf{u}=\bar{\mathbf{u}}$ on Γ_u, $p_w=\bar{p}_w$ on Γ_{p_w} and $p_c=\bar{p}_c$ on Γ_{p_c}. The natural boundary conditions are imposed on the external boundary by using the prescribed traction and volume out-fluxes as $\boldsymbol{\sigma}\cdot\mathbf{n}_\Gamma=\bar{\mathbf{t}}$ on Γ_t, $\dot{\mathbf{w}}_w\cdot\mathbf{n}_\Gamma=\bar{q}_w$ on Γ_{q_w} and $\dot{\mathbf{w}}_g\cdot\mathbf{n}_\Gamma=\bar{q}_g$ on Γ_{q_g}, where $\bar{\mathbf{t}}$ is the prescribed traction applied on the boundary Γ_t and, $\bar{q}_w$ and $\bar{q}_g$ are the prescribed out-flow rates of the porous fluids imposed on the permeable boundaries Γ_{q_w} and Γ_{q_g}, respectively, and $\mathbf{n}_\Gamma$ is the unit outward normal vector to the external boundary Γ, in which the usual conditions $\Gamma_u\cup\Gamma_t=\Gamma$, $\Gamma_{p_w}\cup\Gamma_{q_w}=\Gamma$ and $\Gamma_{p_c}\cup\Gamma_{q_g}=\Gamma$ hold.

The essential and natural boundary conditions, which hold on the complementary parts of the external boundary of the body, complete with the continuity condition on the internal boundary between two different materials. The traction and pore fluid flow continuity must be satisfied across the material interface Γ_d as $[\![\boldsymbol{\sigma}]\!]\cdot\mathbf{n}_{\Gamma_d}=0$, $[\![\dot{\mathbf{w}}_w]\!]\cdot\mathbf{n}_{\Gamma_d}=0$ and $[\![\dot{\mathbf{w}}_g]\!]\cdot\mathbf{n}_{\Gamma_d}=0$, where $\mathbf{n}_{\Gamma_d}$ is the unit normal vector to the weak discontinuity Γ_d pointing to Ω^+. The notation $[\![\Xi]\!]=\Xi^+-\Xi^-$ represents the difference between the corresponding values at the material interface. The interface conditions defined here imply that there is no damage along the internal boundary. It must be noted that the presence of damage, such as the crack, cavity, or diaphragm, at the material interface affects the physical behavior of the fluid flow. In these cases, the fluid flow is no longer continuous across the material interface, leading to a strong or weak discontinuity in the pressure field, as discussed in Section 11.6.

The weak form of the governing partial differential equations of multi-phase porous media with weak discontinuities can be derived by integrating the product of the equilibrium and flow continuity equations (11.5) and (11.6) multiplied by admissible test functions over the analyzed domain, applying the Divergence theorem, imposing the natural boundary conditions, and satisfying the continuity conditions on the material interface. The weak form of the equilibrium equation for the multi-phase system is obtained as

$$\int_\Omega \nabla^s\delta\mathbf{u}:\boldsymbol{\sigma}\,d\Omega+\int_\Omega \rho\,\delta\mathbf{u}\cdot\ddot{\mathbf{u}}\,d\Omega=\int_{\Gamma_t}\delta\mathbf{u}\cdot\bar{\mathbf{t}}\,d\Gamma+\int_\Omega \rho\,\delta\mathbf{u}\cdot\mathbf{b}\,d\Omega \tag{11.12}$$

which must hold for any kinematically admissible test function for the solid phase displacement $\delta\mathbf{u}$, satisfying the homogenized essential boundary condition, that is, vanishing on part of the external boundary with prescribed displacements. In (11.12), ∇^s denotes the symmetric part of the gradient operator. The test function space allows for continuous functions with discontinuous derivatives on Γ_d. This also holds for the set of kinematically admissible displacement fields, in which the displacement solution is required for the initial boundary value problem, usually a subset of a Sobolev space. It must be noted that because of the traction continuity across the material interface, the following integral on the weak discontinuity disappears in the weak form of the equilibrium equation (11.12), that is,

$$-\int_{\Gamma_d^+}\delta\mathbf{u}\cdot\left(\boldsymbol{\sigma}\cdot\mathbf{n}_{\Gamma_d^+}\right)d\Gamma-\int_{\Gamma_d^-}\delta\mathbf{u}\cdot\left(\boldsymbol{\sigma}\cdot\mathbf{n}_{\Gamma_d^-}\right)d\Gamma=\int_{\Gamma_d}\delta\mathbf{u}\cdot\left([\![\boldsymbol{\sigma}]\!]\cdot\mathbf{n}_{\Gamma_d}\right)d\Gamma \tag{11.13}$$

where $\mathbf{n}_{\Gamma_d^+}$ and $\mathbf{n}_{\Gamma_d^-}$ are the unit normal vectors directed to Ω^- and Ω^+, respectively, as shown in Figure 11.1, and the superscripts + and – above Γ_d represent the two sides of the weak discontinuity. It must be noted that $\mathbf{n}_{\Gamma_d}$ has been taken such that $\mathbf{n}_{\Gamma_d}=\mathbf{n}_{\Gamma_d^-}=-\mathbf{n}_{\Gamma_d^+}$.

Incorporating Darcy's Law, the weak form of the continuity equation of flow for each fluid phase can be obtained as

$$\begin{aligned}
&\int_\Omega \delta p_w \frac{1}{Q_{ww}} \dot{p}_w \, d\Omega + \int_\Omega \delta p_w \frac{1}{Q_{wc}} \dot{p}_c \, d\Omega + \int_\Omega \delta p_w \alpha \, \nabla \cdot \dot{\mathbf{u}} \, d\Omega + \int_\Omega (k_w + k_g) \, \nabla \delta p_w \cdot \nabla p_w \, d\Omega \\
&+ \int_\Omega k_g \, \nabla \delta p_w \cdot \nabla p_c \, d\Omega + \int_\Omega \left(k_w \rho_w + k_g \rho_g\right) \nabla \delta p_w \cdot \ddot{\mathbf{u}} \, d\Omega \\
&= \int_\Omega \left(k_w \rho_w + k_g \rho_g\right) \nabla \delta p_w \cdot \mathbf{b} \, d\Omega - \int_{\Gamma_{qw}} \delta p_w \, \bar{q}_w \, d\Gamma \\
&- \int_{\Gamma_{qg} \cap \Gamma_{qw}} \delta p_w \, \bar{q}_g \, d\Gamma - \int_{\Gamma_{pc} \cap \Gamma_{qw}} \delta p_w \dot{\mathbf{w}}_g \cdot \mathbf{n}_\Gamma \, d\Gamma
\end{aligned} \tag{11.14}$$

and

$$\begin{aligned}
&\int_\Omega \delta p_c \frac{1}{Q_{cw}} \dot{p}_w \, d\Omega + \int_\Omega \delta p_c \frac{1}{Q_{cc}} \dot{p}_c \, d\Omega + \int_\Omega \delta p_c \, \alpha \left(1 - S_w\right) \nabla \cdot \dot{\mathbf{u}} \, d\Omega \\
&+ \int_\Omega k_g \nabla \delta p_c \cdot \nabla p_w \, d\Omega + \int_\Omega k_g \nabla \delta p_c \cdot \nabla p_c \, d\Omega + \int_\Omega k_g \rho_g \nabla \delta p_c \cdot \ddot{\mathbf{u}} \, d\Omega \\
&= \int_\Omega k_g \rho_g \nabla \delta p_c \cdot \mathbf{b} \, d\Omega - \int_{\Gamma_{qg}} \delta p_c \, \bar{q}_g \, d\Gamma
\end{aligned} \tag{11.15}$$

which must hold for any kinematically admissible test function for the wetting phase pressure δp_w and the capillary pressure δp_c, respectively, each vanishing on the boundary portion where the corresponding essential boundary condition is imposed. On account of the pore fluid flow continuity across the material interface, the following integrals on the weak discontinuity disappear in the weak form of the flow continuity equation of the wetting and non-wetting pore fluids, (11.14) and (11.15), respectively, that is,

$$\begin{aligned}
&-\int_{\Gamma_d^+} \delta p_w \left(\dot{\mathbf{w}}_w \cdot \mathbf{n}_{\Gamma_d^+}\right) d\Gamma - \int_{\Gamma_d^-} \delta p_w \left(\dot{\mathbf{w}}_w \cdot \mathbf{n}_{\Gamma_d^-}\right) d\Gamma - \int_{\Gamma_d^+} \delta p_w \left(\dot{\mathbf{w}}_g \cdot \mathbf{n}_{\Gamma_d^+}\right) d\Gamma - \int_{\Gamma_d^-} \delta p_w \left(\dot{\mathbf{w}}_g \cdot \mathbf{n}_{\Gamma_d^-}\right) d\Gamma \\
&= \int_{\Gamma_d} \delta p_w \left(\llbracket \dot{\mathbf{w}}_w \rrbracket \cdot \mathbf{n}_{\Gamma_d} + \llbracket \dot{\mathbf{w}}_g \rrbracket \cdot \mathbf{n}_{\Gamma_d}\right) d\Gamma
\end{aligned} \tag{11.16}$$

and

$$-\int_{\Gamma_d^+} \delta p_c \left(\dot{\mathbf{w}}_g \cdot \mathbf{n}_{\Gamma_d^+}\right) d\Gamma - \int_{\Gamma_d^-} \delta p_c \left(\dot{\mathbf{w}}_g \cdot \mathbf{n}_{\Gamma_d^-}\right) d\Gamma = \int_{\Gamma_d} \delta p_c \left(\llbracket \dot{\mathbf{w}}_g \rrbracket \cdot \mathbf{n}_{\Gamma_d}\right) d\Gamma \tag{11.17}$$

11.4.1 Approximation of the Primary Variables

In order to account for the strain jump across the material interface, it is required that the displacement field be continuous, while its gradient be discontinuous on Γ_d, referred to as the weak discontinuity. Viewing this, so as to provide for the inclusion of the discontinuity in the displacement derivatives across the material interface, the displacement field is approximated by locally enriching the classical FE space. The key feature of this work is the introduction of an enrichment function that models the discontinuity in the derivatives of the solution in the classical FE approximation complemented by the addition of enriched DOF pertinent to the enrichment function. This is achieved by exploiting the PU property of FE shape functions, leading to the enrichment localized to the desired areas. For the success of the enriched approximation, it is vital that the enrichment function be chosen properly. By choosing an appropriate enrichment, the local enrichment strategy allows for the incorporation of the strain discontinuity across the material interface into the solution of the problem. In the X-FEM, the approximation is based upon an additive decomposition into a standard part and an enriched part. That is, the considered field is approximated as a linear combination of the standard and enriched shape functions. Following this, the X-FEM approximation of the displacement field $\mathbf{u}(\boldsymbol{x}, t)$ can be written as

$$\mathbf{u}^h(\boldsymbol{x},t) = \sum_{I\in\mathcal{N}} N_{uI}(\boldsymbol{x})\ \mathbf{u}_I(t) + \sum_{I\in\mathcal{N}^{enr}} N_{uI}(\boldsymbol{x})(\psi(\boldsymbol{x})-\psi(\boldsymbol{x}_I))\,\tilde{\mathbf{u}}_I(t) \tag{11.18}$$

where $N_{uI}(\boldsymbol{x})$ is the standard FE shape function of node I, the nodal set $\mathcal{N}$ is the set of all nodes in the mesh, the nodal set $\mathcal{N}^{enr}$ is the set of enriched nodes defined as the set of nodes in the mesh whose support, that is the support of their nodal shape function, is bisected by the material interface, and $\mathbf{u}_I(t)$ are the standard displacement DOF of node I. The nodes in the enriched nodal set $\mathcal{N}^{enr}$ hold additional displacement DOF $\tilde{\mathbf{u}}_I(t)$ associated with the so-called shifted enrichment function $(\psi(\boldsymbol{x}) - \psi(\boldsymbol{x}_I))$, in which $\psi(\boldsymbol{x}_I)$ is the value of $\psi(\boldsymbol{x})$ at the enriched node I. The enrichment function $\psi(\boldsymbol{x})$ is taken as the distance function, the frequently used enrichment function for weak discontinuities, defined as the absolute value of the signed distance function, that is,

$$\psi(\boldsymbol{x}) = |\varphi(\boldsymbol{x})| \tag{11.19}$$

This enrichment function is referred to as the abs-enrichment function. The abs-enrichment function is continuous, whereas its gradient is discontinuous across Γ_d, which is illustrated in Figure 11.2a for the one-dimensional case. Possessing this desirable property, the chosen enrichment function enables the approximate displacement field to be discontinuous in its derivatives across Γ_d. Thus, the strain field discontinuity across the material interface can be obtained. However, using the abs-enrichment function in the standard X-FEM approximation (11.18) may significantly degrade both the accuracy and the overall

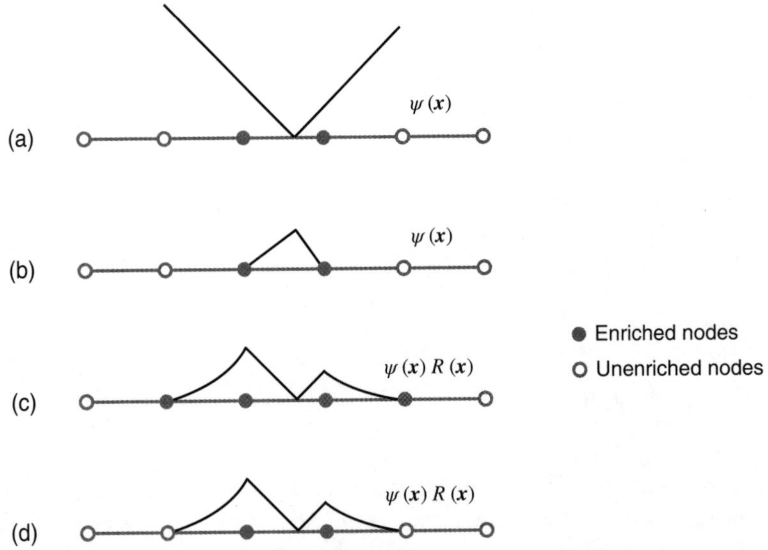

Figure 11.2 Schematic description of the enrichment function and enriched nodes for the one-dimensional case: (a) the standard X-FEM with the abs-enrichment function; (b) the standard X-FEM with the modified abs-enrichment function; (c) the corrected X-FEM; (d) the standard X-FEM with the weighted abs-enrichment function

convergence rate of the solution. The reason for the decrease in the accuracy and the sub-optimal convergence rate is that this choice of enrichment leads to problems in the blending elements. As mentioned in Chapter 4, blending elements are the partially enriched elements that blend the enriched sub-domain, that is, the sub-domain formed by the union of elements with all their nodes being enriched, with the unenriched one where the elements do not have their nodes enriched. The fact that the PU property is not satisfied in the blending elements not only causes the enrichment function not to be represented exactly in the blending elements, but also introduces unwanted terms in the approximation that cannot be compensated for by the standard part of the X-FEM approximation. In other words, the enriched part of the X-FEM approximation cannot be spanned by the standard basis functions in the blending elements that result in the unwanted terms in the approximation. In what follows, some enrichment strategies for weak discontinuities are described that alleviate the errors and improve the convergence rate as compared with the standard X-FEM with the abs-enrichment function.

In the standard X-FEM, the displacement field can be approximated using a modification of the abs-enrichment function suggested by Moës *et al.* (2003) as

$$\mathbf{u}^h(\boldsymbol{x},t)=\sum_{I\in\mathcal{N}}N_{uI}(\boldsymbol{x})\;\mathbf{u}_I(t)+\sum_{I\in\mathcal{N}^{enr}}N_{uI}(\boldsymbol{x})\;\psi(\boldsymbol{x})\,\tilde{\mathbf{u}}_I(t) \tag{11.20}$$

in which $\psi(\boldsymbol{x})$ is given by

$$\psi(\boldsymbol{x})=\sum_{I\in\mathcal{N}^{enr}}N_{uI}(\boldsymbol{x})\;|\varphi_I|-\left|\sum_{I\in\mathcal{N}^{enr}}N_{uI}(\boldsymbol{x})\,\varphi_I\right| \tag{11.21}$$

where φ_I are the nodal values of the level set function. The modified abs-enrichment function is a ridge centered on the interpolated interface and vanishes in elements not containing the material interface, as shown schematically in Figure 11.2b. It is noted that the enrichment is only required in the sub-domain where the local property in the solution is to be captured. In the surrounding sub-domain where the usual smoothness condition is satisfied, the classical finite element approximation is capable of approximating the solution closely. Regarding that the standard abs-enrichment function is modified such that the resulting enrichment function is zero in the blending elements, the unwanted terms appearing in the approximating space of the blending elements are avoided. This choice of the enrichment function is illustrated to give more accurate numerical results than the standard abs-enrichment function for a given discretization and to converge optimally.

In the corrected X-FEM formulation, proposed by Fries (2008), some modifications are introduced to the standard X-FEM formulation as

$$\mathbf{u}^h(\boldsymbol{x},t)=\sum_{I\in\mathcal{N}}N_{uI}(\boldsymbol{x})\;\mathbf{u}_I(t)+\sum_{I\in\mathcal{N}'^{\,enr}}N_{uI}(\boldsymbol{x})(\psi(\boldsymbol{x})-\psi(\boldsymbol{x}_I))R(\boldsymbol{x})\,\tilde{\mathbf{u}}_I(t) \tag{11.22}$$

where the nodal set $\mathcal{N}'^{\,enr}$ is the set of nodes of the bisected and blending elements in which the enriched DOF are added. Obviously, $\mathcal{N}^{enr}\subset\mathcal{N}'^{\,enr}$. That is, the nodes which are enriched in the corrected X-FEM are slightly increased compared with the standard X-FEM, leading to more enriched DOF. $\tilde{\mathbf{u}}_I(t)$ are the enriched displacement DOF corresponding to the shifted enrichment function $(\psi(\boldsymbol{x})-\psi(\boldsymbol{x}_I))R(\boldsymbol{x})$ in which $\psi(\boldsymbol{x})$ is the standard abs-enrichment function defined in (11.19) and $R(\boldsymbol{x})$ is a weight function with compact support that localizes the enrichment function given by

$$R(\boldsymbol{x})=\sum_{I\in\mathcal{N}^{enr}}N_{uI}(\boldsymbol{x}) \tag{11.23}$$

As stated earlier, nodes in $\mathcal{N}^{enr}$ have their support bisected by the material interface. It is noted that the multiplication with $R(\boldsymbol{x})$ causes the weighted abs-enrichment function to vary continuously between the standard abs-enrichment function and zero in the elements whose some nodes are in $\mathcal{N}^{enr}$, reproducing the standard abs-enrichment function in the elements whose all nodes are in $\mathcal{N}^{enr}$, as illustrated in Figure 11.2c for the one-dimensional case. This is because $R(\boldsymbol{x})$ serves as a ramp function smoothly decreasing from unity to zero in the elements with some of their nodes in $\mathcal{N}^{enr}$, surrounded by the elements with all of their nodes in $\mathcal{N}^{enr}$ in which the weight function is a constant function equal to unity and the elements with none of their nodes in $\mathcal{N}^{enr}$ in which the weight function has zero value. It is shown that the corrected X-FEM provides higher accuracy and convergence rate than the standard X-FEM with the abs-enrichment function. The theoretical validation of the corrected X-FEM was presented by Shibanuma and Utsunomiya (2011).

Both of the X-FEM approximations given in (11.20) and (11.22) are based upon the modification of the enrichment function, such that it vanishes beyond the enriched sub-domain, where the elements have all their nodes enriched. Thus, the enrichment function can be reproduced exactly in the entire domain in contrast to the standard X-FEM approximation with the abs-enrichment, which is not able to recover the enrichment function in the blending elements due to the lack of the PU property. More importantly, no unwanted terms are introduced in the approximating space. Preventing the unwanted terms associated with the enrichment, these approaches yield to both the improved accuracy and the optimal rate of convergence.

In the context of the standard X-FEM, the classical FE approximation of the displacement field can be enriched by incorporating a modified definition of the commonly used abs-enrichment function as

$$\mathbf{u}^h(\boldsymbol{x},t) = \sum_{I\in\mathcal{N}} N_{uI}(\boldsymbol{x})\, \mathbf{u}_I(t) + \sum_{I\in\mathcal{N}^{enr}} N_{uI}(\boldsymbol{x})(\psi(\boldsymbol{x})-\psi(\boldsymbol{x}_I))R(\boldsymbol{x})\,\tilde{\mathbf{u}}_I(t) \tag{11.24}$$

in which the weighted abs-enrichment function declines smoothly in the blending sub-domain, surrounding the enriched one, and vanishes at the boundary with the unenriched sub-domain contrary to the standard abs-enrichment function that increases with distance from the interface. The enriched formulation (11.24) has the form of the enrichment function in common with the corrected formulation (11.22), however, the nodes chosen for enrichment conform with those of the standard X-FEM that differs from the nodes introduced to be enriched in the corrected X-FEM, as clarified in Figure 11.2d. The weighting of the standard abs-enrichment function results in a better accuracy as compared with the standard abs-enrichment function.

The standard level set function is the signed distance function to the interface defined as

$$\varphi(\boldsymbol{x}) = \|\boldsymbol{x}-\boldsymbol{x}_{\Gamma_d}\|\operatorname{sign}((\boldsymbol{x}-\boldsymbol{x}_{\Gamma_d})\cdot\mathbf{n}_{\Gamma_d}) \tag{11.25}$$

where $\boldsymbol{x}_{\Gamma_d}$ is the closest point projection of $\boldsymbol{x}$ onto the material interface, and $\|\ \|$ denotes the Euclidean norm; accordingly, $\|\boldsymbol{x}-\boldsymbol{x}_{\Gamma_d}\|$ specifies the distance of point $\boldsymbol{x}$ to the material interface. Readily, it can be found from this definition that the sign on one side of the interface is opposite to that of the other side. Through this definition, the interface can be represented implicitly as the zero iso-contour of the level set function. It can be shown that the norm of the gradient of the signed distance function is equal to unity, that is, $\|\nabla\varphi\| = 1$. From this property, it is clear that the gradient of the signed distance function at the weak discontinuity is indeed the unit normal $\mathbf{n}_{\Gamma_d}$ oriented to Ω^+, the φ -positive domain. That is, the following equality holds for the signed distance function at the weak discontinuity $\nabla\varphi = \mathbf{n}_{\Gamma_d}$.

The level set function $\varphi(\boldsymbol{x})$ can be interpolated by the standard FE shape functions as

$$\varphi^h(\boldsymbol{x}) = \sum_{I\in\mathcal{N}} N_{uI}(\boldsymbol{x})\,\varphi_I \tag{11.26}$$

where φ_I is the value of the level set function at node I, which is indeed the signed distance of node I to the material interface, that is, $\varphi_I = \varphi(\boldsymbol{x}_I)$.

The enriched FE approximation of the displacement field can be symbolically rewritten in the following form

$$\mathbf{u}^h(\boldsymbol{x},t) = \mathbf{N}_u(\boldsymbol{x})\ \mathbf{U}(t) + \mathbf{N}_u^{enr}(\boldsymbol{x})\ \tilde{\mathbf{U}}(t) \tag{11.27}$$

which consists of the standard and enriched parts. In relation (11.27), $\mathbf{N}_u(\boldsymbol{x})$ is the matrix of the standard shape functions, including the standard piecewise polynomial shape functions, $\mathbf{N}_u^{enr}(\boldsymbol{x})$ is referred to as the matrix of the enriched shape functions, $\mathbf{U}(t)$ is the vector of the standard displacement DOF, and $\tilde{\mathbf{U}}(t)$ is the vector of the enriched DOF. It must be noted that the standard and enriched nodal shape functions construct the basis for the enriched approximating space, where the approximate solution is required. Accordingly, the strain vector corresponding to the approximate displacement field $\mathbf{u}^h(\boldsymbol{x}, t)$ can be written in terms of the standard and enriched nodal values as

$$\boldsymbol{\varepsilon}^h(\boldsymbol{x},t) = \mathbf{B}(\boldsymbol{x})\ \mathbf{U}(t) + \mathbf{B}^{enr}(\boldsymbol{x})\ \tilde{\mathbf{U}}(t) \tag{11.28}$$

where the standard and enriched matrices $\mathbf{B}(\boldsymbol{x})$ and $\mathbf{B}^{enr}(\boldsymbol{x})$ involve the spatial derivatives of the standard and enriched shape functions, respectively, in which $\mathbf{B}(\boldsymbol{x}) = \mathbf{L}\ \mathbf{N}_u(\boldsymbol{x})$ and $\mathbf{B}^{enr}(\boldsymbol{x}) = \mathbf{L}\,\mathbf{N}_u^{enr}(\boldsymbol{x})$ with $\mathbf{L}$ denoting the matrix differential operator expressed as

$$\mathbf{L} = \begin{bmatrix} \partial/\partial x & 0 \\ 0 & \partial/\partial y \\ \partial/\partial y & \partial/\partial x \end{bmatrix} \tag{11.29}$$

Since the material interface does not impose any discontinuity in the fluid flow, the water pressure as well as the capillary pressure should not be enriched to ensure the fluid flow continuity across the material interface. Hence, both the water pressure and capillary pressure fields $p_\alpha(\boldsymbol{x}, t)$ are approximated using the standard FEM as

$$p_\alpha^h(\boldsymbol{x},t) = \sum_{I\in\mathcal{N}} N_{p_\alpha I}(\boldsymbol{x})\, p_{\alpha I}(t) \qquad \alpha = w, c \tag{11.30}$$

where $N_{p_\alpha I}(\boldsymbol{x})$ are the standard FE shape functions and $p_{\alpha I}(t)$ are the standard pressure DOF. Likewise, the standard FE approximation of the water pressure as well as the capillary pressure in Eq. (11.30) can be rewritten as

$$p_\alpha^h(\boldsymbol{x},t) = \mathbf{N}_{p_\alpha}(\boldsymbol{x})\, \mathbf{P}_\alpha(t) \qquad \alpha = w, c \tag{11.31}$$

where $\mathbf{N}_{p_\alpha}(\boldsymbol{x})$ is the matrix of the standard FE shape functions and $\mathbf{P}_\alpha(t)$ is the vector of the standard nodal DOF.

11.4.2 Discretization of Equilibrium and Flow Continuity Equations

In this section, the weak form of the equilibrium and flow continuity equations (11.12), (11.14), and (11.15) is rendered in a discrete form by means of discretization firstly in space employing the X-FEM and then in time applying the generalized Newmark scheme. The resulting system of fully coupled non-linear algebraic equations is directly solved using the direct solution procedure combined with the Newton–Raphson iterative process. As a result, the main unknowns are obtained simultaneously, leading to the full solution of the problem.

In order to discretize the weak form of the equilibrium and flow continuity equations, the test functions for the displacement $\delta\mathbf{u}$, water pressure δp_w, and capillary pressure δp_c are considered in the same approximating space following the Bubnov–Galerkin technique, that is, the finite-dimensional subspace spanned by the linearly independent basis functions, as the approximate displacement, water pressure, and capillary pressure fields, respectively. In other words, they both are formed from the linear combination of the same basis functions. Inserting the FE approximations into the weak form of the governing partial differential equations (11.12), (11.14), and (11.15) and satisfying the necessity that the weak form should hold for all kinematically admissible test functions, the discretized form of the equations defining the multiphase problem with weak discontinuities can be obtained as

$$\begin{aligned}
&\mathbf{M}_{uu}\ddot{\mathbf{U}} + \mathbf{M}_{u\tilde{u}}\ddot{\tilde{\mathbf{U}}} + \int_{\Omega} \mathbf{B}^T \boldsymbol{\sigma}''\, d\Omega - \mathbf{Q}_{uw}\mathbf{P}_w - \mathbf{Q}_{uc}\mathbf{P}_c = \mathbf{F}_u^{\text{ext}} \\
&\mathbf{M}_{u\tilde{u}}^T\ddot{\mathbf{U}} + \mathbf{M}_{\tilde{u}\tilde{u}}\ddot{\tilde{\mathbf{U}}} + \int_{\Omega^{enr}} (\mathbf{B}^{enr})^T \boldsymbol{\sigma}''\, d\Omega - \mathbf{Q}_{\tilde{u}w}\mathbf{P}_w - \mathbf{Q}_{\tilde{u}c}\mathbf{P}_c = \mathbf{F}_{\tilde{u}}^{\text{ext}} \\
&\mathbf{M}_{wu}\ddot{\mathbf{U}} + \mathbf{M}_{w\tilde{u}}\ddot{\tilde{\mathbf{U}}} + \mathbf{Q}_{uw}^T\dot{\mathbf{U}} + \mathbf{Q}_{\tilde{u}w}^T\dot{\tilde{\mathbf{U}}} + \mathbf{C}_{ww}\dot{\mathbf{P}}_w + \mathbf{C}_{wc}\dot{\mathbf{P}}_c + \mathbf{H}_{ww}\mathbf{P}_w + \mathbf{H}_{wc}\mathbf{P}_c = \mathbf{F}_w^{\text{ext}} \\
&\mathbf{M}_{cu}\ddot{\mathbf{U}} + \mathbf{M}_{c\tilde{u}}\ddot{\tilde{\mathbf{U}}} + \mathbf{Q}_{uc}^T\dot{\mathbf{U}} + \mathbf{Q}_{\tilde{u}c}^T\dot{\tilde{\mathbf{U}}} + \mathbf{C}_{cw}\dot{\mathbf{P}}_w + \mathbf{C}_{cc}\dot{\mathbf{P}}_c + \mathbf{H}_{wc}^T\mathbf{P}_w + \mathbf{H}_{cc}\mathbf{P}_c = \mathbf{F}_c^{\text{ext}}
\end{aligned} \tag{11.32}$$

where the coefficient matrices in the system of discretized governing equations (11.32) are defined as

$$\begin{aligned}
\mathbf{M}_{uu} &= \int_{\Omega} \mathbf{N}_u^T \rho \mathbf{N}_u \, d\Omega \\
\mathbf{M}_{u\tilde{u}} &= \int_{\Omega^{enr}} \mathbf{N}_u^T \rho \mathbf{N}_u^{enr} \, d\Omega \\
\mathbf{M}_{\tilde{u}\tilde{u}} &= \int_{\Omega^{enr}} (\mathbf{N}_u^{enr})^T \rho \mathbf{N}_u^{enr} \, d\Omega \\
\mathbf{M}_{wu} &= \int_{\Omega} \nabla\mathbf{N}_{p_w}^T (k_w\rho_w + k_g\rho_g)\mathbf{N}_u \, d\Omega \\
\mathbf{M}_{w\tilde{u}} &= \int_{\Omega^{enr}} \nabla\mathbf{N}_{p_w}^T (k_w\rho_w + k_g\rho_g)\mathbf{N}_u^{enr} \, d\Omega \\
\mathbf{M}_{cu} &= \int_{\Omega} \nabla\mathbf{N}_{p_c}^T k_g\rho_g \mathbf{N}_u \, d\Omega \\
\mathbf{M}_{c\tilde{u}} &= \int_{\Omega^{enr}} \nabla\mathbf{N}_{p_c}^T k_g\rho_g \mathbf{N}_u^{enr} \, d\Omega \\
\mathbf{Q}_{uw} &= \int_{\Omega} \mathbf{B}^T \alpha \mathbf{m} \mathbf{N}_{p_w} \, d\Omega \\
\mathbf{Q}_{\tilde{u}w} &= \int_{\Omega^{enr}} (\mathbf{B}^{enr})^T \alpha \mathbf{m} \mathbf{N}_{p_w} \, d\Omega \\
\mathbf{Q}_{uc} &= \int_{\Omega} \mathbf{B}^T \alpha (1 - S_w) \mathbf{m} \mathbf{N}_{p_c} \, d\Omega \\
\mathbf{Q}_{\tilde{u}c} &= \int_{\Omega^{enr}} (\mathbf{B}^{enr})^T \alpha (1 - S_w) \mathbf{m} \mathbf{N}_{p_c} \, d\Omega \\
\mathbf{C}_{ww} &= \int_{\Omega} \mathbf{N}_{p_w}^T \frac{1}{Q_{ww}} \mathbf{N}_{p_w} \, d\Omega \\
\mathbf{C}_{cc} &= \int_{\Omega} \mathbf{N}_{p_c}^T \frac{1}{Q_{cc}} \mathbf{N}_{p_c} \, d\Omega
\end{aligned}$$

$$\mathbf{C}_{wc} = \int_{\Omega} \mathbf{N}_{p_w}^T \frac{1}{Q_{wc}} \mathbf{N}_{p_c}\, d\Omega$$

$$\mathbf{C}_{cw} = \int_{\Omega} \mathbf{N}_{p_c}^T \frac{1}{Q_{cw}} \mathbf{N}_{p_w}\, d\Omega$$

$$\mathbf{H}_{ww} = \int_{\Omega} \nabla \mathbf{N}_{p_w}^T \left(k_w + k_g\right) \nabla \mathbf{N}_{p_w}\, d\Omega$$

$$\mathbf{H}_{cc} = \int_{\Omega} \nabla \mathbf{N}_{p_c}^T\, k_g\, \nabla \mathbf{N}_{p_c}\, d\Omega$$

$$\mathbf{H}_{wc} = \int_{\Omega} \nabla \mathbf{N}_{p_w}^T\, k_g\, \nabla \mathbf{N}_{p_c}\, d\Omega \tag{11.33}$$

and the external force and flux vectors are specified as

$$\begin{aligned}
\mathbf{F}_u^{\text{ext}} &= \int_{\Omega} \mathbf{N}_u^T \rho\, \mathbf{b}\, d\Omega + \int_{\Gamma_t} \mathbf{N}_u^T \bar{\mathbf{t}}\, d\Gamma \\
\mathbf{F}_{\tilde{u}}^{\text{ext}} &= \int_{\Omega^{enr}} \left(\mathbf{N}_u^{enr}\right)^T \rho\, \mathbf{b}\, d\Omega + \int_{\Gamma_t^{enr}} \left(\mathbf{N}_u^{enr}\right)^T \bar{\mathbf{t}}\, d\Gamma \\
\mathbf{F}_w^{\text{ext}} &= \int_{\Omega} \nabla \mathbf{N}_{p_w}^T \left(k_w \rho_w + k_g \rho_g\right) \mathbf{b}\, d\Omega - \int_{\Gamma_{qw}} \mathbf{N}_{p_w}^T \bar{q}_w\, d\Gamma - \int_{\Gamma_{qg} \cap \Gamma_{qw}} \mathbf{N}_{p_w}^T \bar{q}_g\, d\Gamma - \int_{\Gamma_{pc} \cap \Gamma_{qw}} \mathbf{N}_{p_w}^T \dot{\mathbf{w}}_g \cdot \mathbf{n}_\Gamma\, d\Gamma \\
\mathbf{F}_c^{\text{ext}} &= \int_{\Omega} \nabla \mathbf{N}_{p_c}^T k_g \rho_g \mathbf{b}\, d\Omega - \int_{\Gamma_{qg}} \mathbf{N}_{p_c}^T \bar{q}_g\, d\Gamma
\end{aligned} \tag{11.34}$$

The system of discretized governing equations (11.32) can then be discretized in the time domain following the line of the well-known Newmark scheme, in which it is supposed that the system of equations is satisfied at each discrete time step. Thus, the problem must be solved at a series of sequential discrete time steps. To advance the solution in time, the link between the successive values of the unknown field variables at time t_{n+1} and the known field variables at time t_n is established by applying the minimum order of the generalized Newmark scheme required considering the highest order of the time derivatives in the differential equations; that is, through the application of GN22 and GN11 to the displacement and pressure variables, respectively, as

$$\begin{aligned}
\ddot{\mathbf{U}}_{n+1} &= a_0 \left(\mathbf{U}_{n+1} - \mathbf{U}_n\right) - a_2 \dot{\mathbf{U}}_n - a_4 \ddot{\mathbf{U}}_n \\
\dot{\mathbf{U}}_{n+1} &= a_1 \left(\mathbf{U}_{n+1} - \mathbf{U}_n\right) - a_3 \dot{\mathbf{U}}_n - a_5 \ddot{\mathbf{U}}_n \\
\ddot{\tilde{\mathbf{U}}}_{n+1} &= a_0 \left(\tilde{\mathbf{U}}_{n+1} - \tilde{\mathbf{U}}_n\right) - a_2 \dot{\tilde{\mathbf{U}}}_n - a_4 \ddot{\tilde{\mathbf{U}}}_n \\
\dot{\tilde{\mathbf{U}}}_{n+1} &= a_1 \left(\tilde{\mathbf{U}}_{n+1} - \tilde{\mathbf{U}}_n\right) - a_3 \dot{\tilde{\mathbf{U}}}_n - a_5 \ddot{\tilde{\mathbf{U}}}_n \\
\dot{\mathbf{P}}_{\alpha_{n+1}} &= a_1' \left(\mathbf{P}_{\alpha_{n+1}} - \mathbf{P}_{\alpha_n}\right) - a_3' \dot{\mathbf{P}}_{\alpha_n} \qquad \alpha = w, c
\end{aligned} \tag{11.35}$$

where $a_0 = 1/\beta\Delta t^2$, $a_1 = \gamma/\beta\Delta t$, $a_2 = 1/\beta\Delta t$, $a_3 = \gamma/\beta - 1$, $a_4 = 1/2\beta - 1$, $a_5 = \Delta t(\gamma/2\beta - 1)$, $a_1' = 1/\theta\Delta t$ and $a_3' = 1/\theta - 1$. In these relations, $\Delta t = t_{n+1} - t_n$ is the time increment and, γ and θ are the Newmark parameters. To guarantee the unconditional stability of the time integration procedure, the Newmark parameters must be chosen such that $\gamma \geq 0.5$, $\beta \geq 0.25(0.5 + \gamma)^2$ and $\theta \geq 0.5$.

11.4.3 Solution Procedure of Discretized Equilibrium Equations

In order to solve the system of fully coupled nonlinear algebraic equations at each time step, the direct solution procedure is employed. In this numerical strategy, the Newton–Raphson scheme is implemented to linearize the whole system of nonlinear equations. So as to obtain the nodal unknown values at any specified time t_{n+1}, the discrete system of equations is solved at the given time step applying the Newton–Raphson iterative algorithm to its residual form $\boldsymbol{\Psi}_{n+1} = \mathbf{0}$. By expanding the residual equations with the first-order truncated Taylor series, the following linear approximation for the nonlinear system can be obtained as

$$\boldsymbol{\Psi}_{n+1}^{i+1} = \begin{Bmatrix} \boldsymbol{\Psi}_{u_{n+1}}^{i+1} \\ \boldsymbol{\Psi}_{\tilde{u}_{n+1}}^{i+1} \\ \boldsymbol{\Psi}_{w_{n+1}}^{i+1} \\ \boldsymbol{\Psi}_{c_{n+1}}^{i+1} \end{Bmatrix} = \begin{Bmatrix} \boldsymbol{\Psi}_{u_{n+1}}^{i} \\ \boldsymbol{\Psi}_{\tilde{u}_{n+1}}^{i} \\ \boldsymbol{\Psi}_{w_{n+1}}^{i} \\ \boldsymbol{\Psi}_{c_{n+1}}^{i} \end{Bmatrix} + \mathbf{J} \begin{Bmatrix} d\mathbf{U}_n^i \\ d\tilde{\mathbf{U}}_n^i \\ d\mathbf{P}_{w_n}^i \\ d\mathbf{P}_{c_n}^i \end{Bmatrix} = 0 \tag{11.36}$$

where $\mathbf{J}$ is the well-known Jacobian matrix. By obtaining the solution of the linearized system of equations (11.36), that is, the increment of the standard and enriched nodal DOF, the corresponding nodal unknowns are subsequently attained through the following incremental relation

$$\mathbf{X}_{n+1}^{i+1} = \mathbf{X}_{n+1}^{i} + d\mathbf{X}_n^i \tag{11.37}$$

in which $\mathbf{X}$ represents the vector of nodal unknowns, $\mathbf{X}^T = \begin{bmatrix} \mathbf{U}^T & \tilde{\mathbf{U}}^T & \mathbf{P}_w^T & \mathbf{P}_c^T \end{bmatrix}$. This relation implies that the vector of nodal unknowns is improved at each iteration as the iterative algorithm proceeds. The Jacobian matrix is defined as $\partial \boldsymbol{\Psi}_{n+1}^i / \partial \mathbf{X}_{n+1}^i$, which can be approximated as follows

$$\mathbf{J} = \begin{bmatrix} a_0\mathbf{M}_{uu} + \int_\Omega \mathbf{B}^T\mathbf{D}\mathbf{B}\,d\Omega & a_0\mathbf{M}_{u\tilde{u}} + \int_{\Omega^{enr}} \mathbf{B}^T\mathbf{D}\mathbf{B}^{enr}\,d\Omega & -\mathbf{Q}_{uw} & -\mathbf{Q}_{uc} \\ a_0\mathbf{M}_{u\tilde{u}}^T + \int_{\Omega^{enr}} (\mathbf{B}^{enr})^T\mathbf{D}\mathbf{B}\,d\Omega & a_0\mathbf{M}_{\tilde{u}\tilde{u}} + \int_{\Omega^{enr}} (\mathbf{B}^{enr})^T\mathbf{D}\mathbf{B}^{enr}\,d\Omega & -\mathbf{Q}_{\tilde{u}w} & -\mathbf{Q}_{\tilde{u}c} \\ a_0\mathbf{M}_{wu} + a_1\mathbf{Q}_{uw}^T & a_0\mathbf{M}_{w\tilde{u}} + a_1\mathbf{Q}_{\tilde{u}w}^T & a_1'\mathbf{C}_{ww} + \mathbf{H}_{ww} & a_1'\mathbf{C}_{wc} + \mathbf{H}_{wc} \\ a_0\mathbf{M}_{cu} + a_1\mathbf{Q}_{uc}^T & a_0\mathbf{M}_{c\tilde{u}} + a_1\mathbf{Q}_{\tilde{u}c}^T & a_1'\mathbf{C}_{cw} + \mathbf{H}_{wc}^T & a_1'\mathbf{C}_{cc} + \mathbf{H}_{cc} \end{bmatrix} \tag{11.38}$$

where the index $(i, n+1)$ is assumed for the matrices, inferring that the Jacobian matrix $\mathbf{J}$ as well as the residual vector $\boldsymbol{\Psi}$ is updated at each iteration. Hence, within each time step a sequence of linearized system of equations is solved to attain a solution satisfying the residual equation. This iterative process continues until the iteration convergence is achieved. That is, the residual vector vanishes within the given tolerance.

Evidently, the formulation results in the non-symmetric Jacobian matrix notwithstanding the choice of the constitutive model applied in the medium. It is worthwhile rendering the Jacobian matrix symmetric from the computational point of view so as to save on core storage and computational effort. For this purpose, the first and second rows of the Jacobian matrix are multiplied by the scalar factor $-a_1$ to symmetrize it relative to the coefficients. In addition, since the contribution of the dynamic seepage forcing terms to the solution is negligible compared to the other terms, the inertial matrices $\mathbf{M}_{wu}$, $\mathbf{M}_{w\tilde{u}}$, $\mathbf{M}_{cu}$ and $\mathbf{M}_{c\tilde{u}}$ are omitted from the Jacobian matrix to obtain a symmetric matrix. Furthermore, it seems rational to replace the compressibility matrix $\mathbf{C}_{wc}$ by $\mathbf{C}_{cw}^T$ in the Jacobian matrix by comparison between the expressions of the compressibility coefficients $1/Q_{wc}$ and $1/Q_{cw}$. By the imposition of the previously-mentioned simplifications, a symmetric approximation to the Jacobian matrix is obtained as follows, provided that the tangential constitutive matrix of the medium is itself symmetric

$$\mathbf{J}=\begin{bmatrix} -a_1\left(a_0\mathbf{M}_{uu}+\int_{\Omega}\mathbf{B}^T\mathbf{D}\,\mathbf{B}\,d\Omega\right) & -a_1\left(a_0\mathbf{M}_{u\tilde{u}}+\int_{\Omega^{enr}}\mathbf{B}^T\mathbf{D}\,\mathbf{B}^{enr}\,d\Omega\right) & a_1\mathbf{Q}_{uw} & a_1\mathbf{Q}_{uc} \\ -a_1\left(a_0\mathbf{M}_{u\tilde{u}}^T+\int_{\Omega^{enr}}(\mathbf{B}^{enr})^T\mathbf{D}\,\mathbf{B}\,d\Omega\right) & -a_1\left(a_0\mathbf{M}_{\tilde{u}\tilde{u}}+\int_{\Omega^{enr}}(\mathbf{B}^{enr})^T\mathbf{D}\,\mathbf{B}^{enr}\,d\Omega\right) & a_1\mathbf{Q}_{\tilde{u}w} & a_1\mathbf{Q}_{\tilde{u}c} \\ a_1\mathbf{Q}_{uw}^T & a_1\mathbf{Q}_{\tilde{u}w}^T & a_1'\,\mathbf{C}_{ww}+\mathbf{H}_{ww} & a_1'\,\mathbf{C}_{cw}^T+\mathbf{H}_{wc} \\ a_1\mathbf{Q}_{uc}^T & a_1\mathbf{Q}_{\tilde{u}c}^T & a_1'\,\mathbf{C}_{cw}+\mathbf{H}_{wc}^T & a_1'\,\mathbf{C}_{cc}+\mathbf{H}_{cc} \end{bmatrix} \tag{11.39}$$

From the computational aspect, one main difficulty encountered when implementing the X-FEM deals with the numerical integration in the presence of weak discontinuities. The discontinuous nature of the solution derivatives in this case requires a special treatment to accurately integrate the functions possessing a non-smooth character over the integration domain. To fulfill the usual requirements of the integration that involves the non-smooth enrichment function appearing in the enriched part of the X-FEM approximation, a common approach is that the numerical integration of the elements bisected by the material interface be done separately on each sub-element into which the element is divided. It must be noted that the material properties differ in the two sub-elements on either side of the interface. In practice, in order to be able to perform the integration over these sub-domains, each sub-domain is partitioned into triangles to which the standard Gauss quadrature rule is applied. The remaining elements that are not cut by the interface are numerically integrated employing the conventional Gauss integration scheme.

11.5 Application of X-FEM Method in Multi-Phase Porous Media with Arbitrary Interfaces

In this section, a numerical example is presented that includes a material interface to assess the performance of the standard X-FEM with the abs-enrichment, the weighted abs-enrichment, and the modified abs-enrichment, as well as the corrected X-FEM for the weak discontinuity by comparing the computed approximation error and convergence rate of the solution. It is illustrated that the standard X-FEM with the modified abs-enrichment function and the corrected X-FEM have similar accuracy and convergence, and they both achieve the optimal rate of convergence. Furthermore, it is shown that although the error in the L_2 norm is reduced for a given discretization in the case of standard X-FEM with the weighted abs-enrichment function in comparison with the standard abs-enrichment function, the optimal rate of convergence is not reached. This degradation of the convergence rate is attributed to the parasitic terms produced in the approximating space of the blending elements due to the lack of the PU property. It is interesting to note that the enriched approximation of the weakly discontinuous displacement field using different enrichment strategies mentioned before not only affects the convergence properties of the displacement field itself, as expected, but also affects the convergence properties of the fluid variables, which are approximated using the classical FEM.

The relative L_2 norm of the error in the displacement field over the entire domain Ω at time $t=\bar{t}$ introduced by the X-FEM approximation is computed as

$$\|e_{\mathbf{u}}\|_{L_2(\Omega)}=\left(\int_{\Omega}\|\mathbf{u}(\boldsymbol{x},\bar{t})-\mathbf{u}^h(\boldsymbol{x},\bar{t})\|^2 d\Omega\right)^{\frac{1}{2}}\Bigg/\left(\int_{\Omega}\|\mathbf{u}(\boldsymbol{x},\bar{t})\|^2 d\Omega\right)^{\frac{1}{2}} \tag{11.40}$$

where $\mathbf{u}^h(\boldsymbol{x},\bar{t})$ is the approximate displacement field obtained by the X-FEM and $\mathbf{u}(\boldsymbol{x},\bar{t})$ is the numerical reference solution. It is noted that the FE solution computed on a very fine mesh conforming to the material interface is taken as the numerical reference solution. The relative error in the FE approximation of the pressure field over the entire domain Ω at time $t=\bar{t}$ measured in the L_2 norm is computed as

$$\left\|e_{p_\alpha}\right\|_{L_2(\Omega)} = \left(\int_\Omega \left|p_\alpha(\boldsymbol{x},\bar{t}) - p_\alpha^h(\boldsymbol{x},\bar{t})\right|^2 d\Omega\right)^{\frac{1}{2}} \Big/ \left(\int_\Omega \left|p_\alpha(\boldsymbol{x},\bar{t})\right|^2 d\Omega\right)^{\frac{1}{2}} \qquad \alpha = w, c \tag{11.41}$$

where $p_\alpha^h(\boldsymbol{x},\bar{t})$ is the approximate pressure field obtained using the classical FEM and $p_\alpha(\boldsymbol{x},\bar{t})$ is the FE reference solution.

The example involves a partially saturated consolidation of a bimaterial soil column due to the change of the pore water pressure at the upper boundary, whose geometry and boundary conditions are displayed in Figure 11.3. This example is presented to assess the influence of the blending problems on the approximation error and the convergence rate of the solution. As shown in Figure 11.3, a vertical soil column of 1 m depth and 0.05 m width is subjected to an external compressive surface load of 1000 N/m^2 at the top. Initially, the partially saturated condition is supposed for the soil column in which the initial water saturation is set to 0.52, and the initial water pressure is imposed to be – 280 kN/m^2, as considered in the original homogeneous example (Lewis and Rahman, 1999). Subsequently, the pore water pressure at the top surface is changed immediately from – 280 to – 420 kN/m^2 while it is exposed to air pressure.

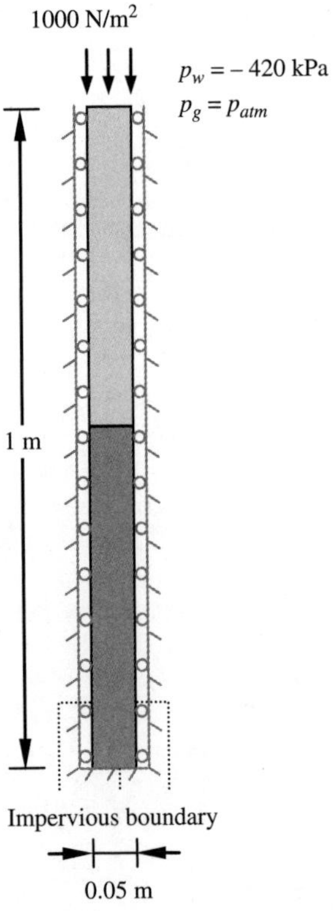

Figure 11.3 The geometry and boundary conditions of the problem of the partially saturated consolidation of a bimaterial soil column

The boundary conditions of the problem are as follows; the bottom and lateral surfaces of the soil column are impermeable to both water and air. That is, the top surface is the only drained boundary of the soil column. At the top boundary, the water pressure is set to – 420 kN/m^2 and the air pressure remains at the initial atmospheric pressure. The horizontal and vertical displacements are restrained at the bottom, and at the lateral surfaces the condition of constrained horizontal displacement is applied. The soil column is assumed to be elastically deformable. The material properties of the partially saturated soil are listed in Table 11.1. There is a discontinuity in the Young's modulus across the material interface located at 0.515625 m of the bottom face. The Young's modulus is set to 6 MPa at the top of the interface and 3 MPa at the bottom.

In order to describe the flow of the immiscible fluids through the porous material, the constitutive relations for the water saturation as well as the water and air relative permeabilities are assumed based on that proposed by Brooks and Corey (1966) as

$$\begin{aligned} S_w &= S_{rw} + (1 - S_{rw})\left(\frac{p_b}{p_c}\right)^{\lambda} \\ k_{rw} &= S_e^{(2+3\lambda)/\lambda} \\ k_{rg} &= (1 - S_e)^2\left(1 - S_e^{(2+\lambda)/\lambda}\right) \end{aligned} \tag{11.42}$$

where the residual water saturation $S_{rw} = 0.3966$, the bubbling pressure $p_b = 225$ kN/m^2, the pore size distribution index $\lambda = 3$, and the effective water saturation S_e is defined as

$$S_e = \frac{S_w - S_{rw}}{1 - S_{rw}} \tag{11.43}$$

In the numerical study of convergence, the computational domain is uniformly discretized with equally sized quadrilateral elements. As a result, a series of meshes composed of 20, 80, 320 and 1280 elements with bilinear shape functions are used in the simulation. To obtain the numerical reference solution, the FE analysis is performed on a very fine mesh aligning with the material interface which consists of 5120 bilinear elements. Figure 11.4 displays the numerical reference solution for the computed distributions of the relative water pressure, capillary pressure, water saturation, and vertical displacement along the soil column at times $t = 0.5$ and 2 days. As expected, there is a weak discontinuity in the displacement solution at the bimaterial boundary due to the discontinuity in the material properties.

Table 11.1 Material properties of the soil column consolidation

Poisson ratio	$\nu = 0.4$
Biot constant	$\alpha = 1$
Porosity	$n = 0.3$
Solid phase density	$\rho_s = 2000$ kg/m^3
Water density	$\rho_w = 1000$ kg/m^3
Air density	$\rho_g = 1.22$ kg/m^3
Bulk modulus of solid phase	$K_S = 0.14 \times 10^{10}$ Pa
Bulk modulus of water	$K_w = 0.43 \times 10^{13}$ Pa
Bulk modulus of air	$K_g = 0.1 \times 10^6$ Pa
Intrinsic permeability	$k = 0.46 \times 10^{-11}$ m^2
Dynamic viscosity of water	$\mu_w = 1 \times 10^{-3}$ Pa s
Dynamic viscosity of air	$\mu_g = 1 \times 10^{-3}$ Pa s
Gravitational acceleration	$g = 9.806$ m/s^2
Atmospheric pressure	$p_{\text{atm}} = 101.8967$ kPa

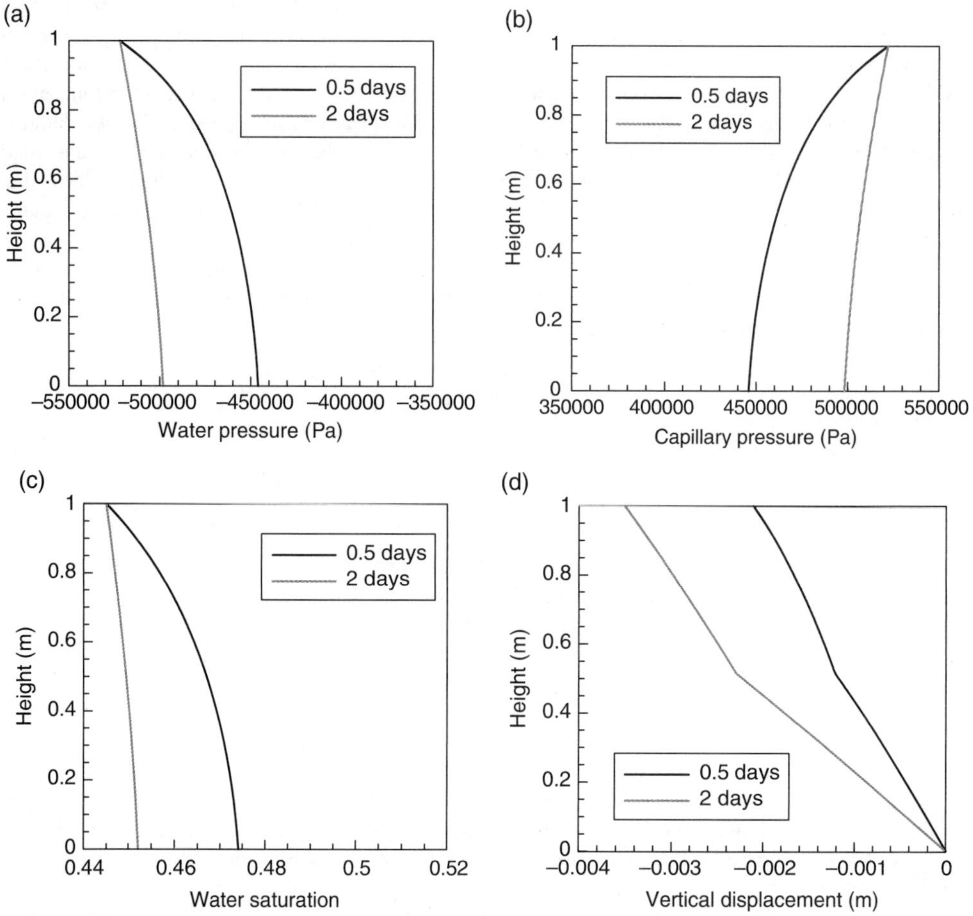

Figure 11.4 Numerical reference solution for the partially saturated consolidation of a bimaterial soil column: The variations with the column height of (a) relative water pressure (Pa), (b) capillary pressure (Pa), (c) water saturation, and (d) vertical displacement (m)

Figure 11.5a indicates the comparison between the convergence plots of the water pressure relative to the element size at $t = 0.5$ days. The capillary pressure and water saturation show almost the same behavior, so their convergence plots are not presented. From these curves, it can be observed that all the enriched approaches nearly exhibit the same convergence behavior. More precisely, the convergence rate in the L_2 norm of the approximation error remains to $O(h^2)$ notwithstanding the choice of the local enrichment strategy utilized to approximate the displacement field. This means that the optimal rate of convergence is maintained independently of the enrichment strategy employed. At time $t = 2$ days, the convergence rate of the standard X-FEM with the abs-enrichment function and the weighted abs-enrichment function is lowered from order h^2, yet the latter attains a convergence rate close to the optimal one, as observed in Figure 11.5b. Moreover, both the standard X-FEM with the modified abs-enrichment function and the corrected X-FEM lead to the similar accuracy and convergence rate, which are superior to the other two methods. It can be noticed that for the coarse mesh with 20 elements the error level is almost identical

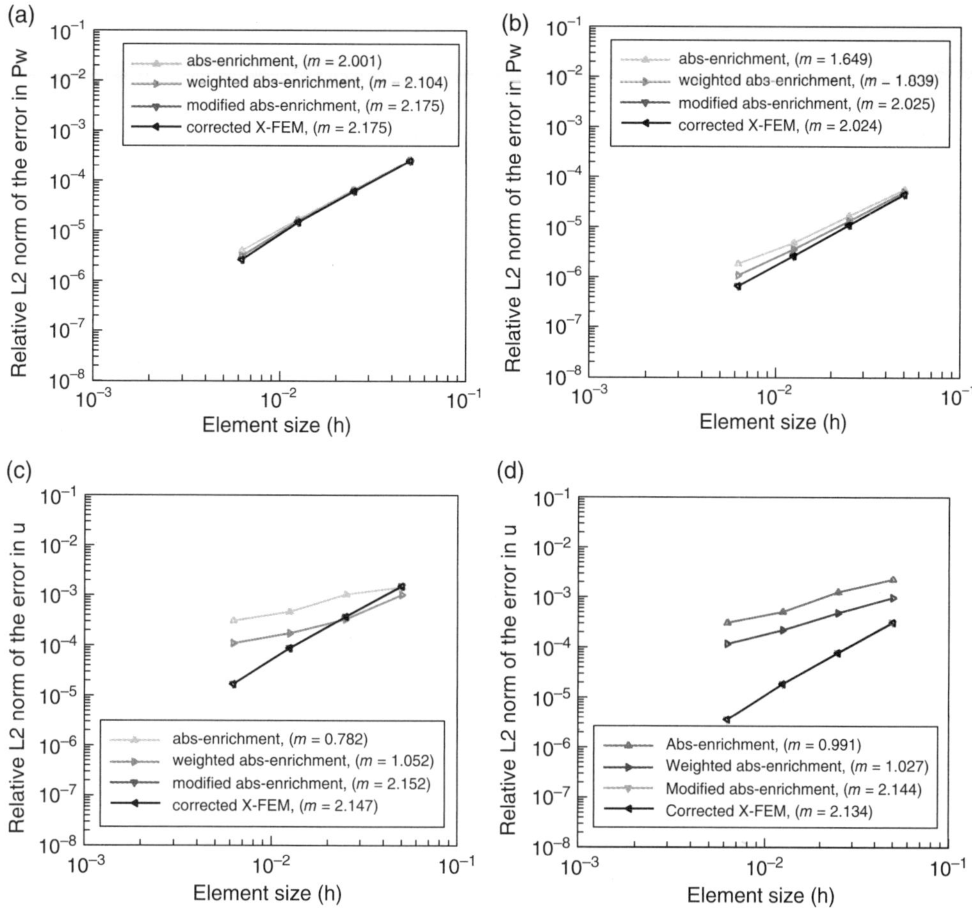

Figure 11.5 Comparison of the convergence rates for the partially saturated consolidation of a bimaterial soil column: The relative L_2 error norm of (a–b) water pressure at $t = 0.5$ and 2 days, (c–d) vertical displacement at $t = 0.5$ and 2 days

for all the enriched approaches considered. Figures 11.5c and d compare the convergence rate and the error in the approximation of the displacement field at times 0.5 and 2 days, respectively. The standard X-FEM with the modified abs-enrichment function converges at the optimal rate, while the convergence rate of the standard X-FEM with the abs-enrichment function and the weighted abs-enrichment function is sub-optimal. In addition, the numerical results show the same rate of convergence as the standard X-FEM with the modified abs-enrichment function for the corrected X-FEM, which is optimal. For the given mesh, the standard X-FEM with the abs-enrichment function shows the poorest accuracy. It can also be seen that the weighted abs-enrichment function results in the lower approximation error than the commonly used abs-enrichment function, but it is less effective in the convergence rate. In summary, the comparison of the convergence plots makes obvious that the blending elements are crucial for the efficiency and performance of the local enrichment approaches.

11.6 The X-FEM Formulation for Hydraulic Fracturing in Multi-Phase Porous Media

The hydraulic fracture propagation occurs as the progressive decay of the cohesive tractions transferred across the fracture process zone and the imposition of the fluid pressure onto the fracture faces by means of the fracturing fluid within the fracture. The tractions acting on the fracture faces give rise to the mechanical coupling, and the fluid leak-off through the fracture faces leads to the mass transfer coupling between the fracture and the porous medium surrounding the fracture. In order to provide a proper numerical modeling for the fracture opening and the fluid exchange between the fracture and the surrounding porous medium, some requirements must be taken into the computation. The displacement jump requires that the displacement field be discontinuous across the fracture. The fluid exchange implies that the fluid flow, which is governed by the Darcy velocity of the fluid, in the normal direction to the fracture must be discontinuous. Furthermore, since the Darcy velocity is related to the fluid pressure gradient by means of Darcy's Law, the normal gradient of the fluid pressure must be discontinuous across the fracture. In the context of the extended FEM, the discontinuity in the displacement field is modeled by enhancing the standard piecewise polynomial basis with a discontinuous enrichment function. In order to model the discontinuity in the fluid flow normal to the fracture, the standard FE approximation of the water pressure and the capillary pressure fields are enriched by incorporating a modified definition of the commonly used enrichment function for weak discontinuities.

For the resolution of the initial boundary value problem, the knowledge of the values of the field variables associated with the initial and boundary conditions is necessary. In order to develop the governing equations, the two-dimensional domain Ω bounded by the boundary Γ is considered, as depicted in Figure 11.6, in which the domain contains the geomechanical discontinuity Γ_d. The essential boundary conditions are imposed on the external boundary as the prescribed primary variables with $\mathbf{u}=\bar{\mathbf{u}}$ on Γ_u, $p_w=\bar{p}_w$ on Γ_{p_w} and $p_c=\bar{p}_c$ on Γ_{p_c}. The natural boundary conditions are also imposed on the external boundary as the prescribed traction and volume out-fluxes as $\boldsymbol{\sigma}\cdot\mathbf{n}_\Gamma=\bar{\mathbf{t}}$ on Γ_t, $\dot{\mathbf{w}}_w\cdot\mathbf{n}_\Gamma=\bar{q}_w$ on Γ_{q_w} and $\dot{\mathbf{w}}_g\cdot\mathbf{n}_\Gamma=\bar{q}_g$ on Γ_{q_g}, where $\bar{\mathbf{t}}$ is the prescribed traction applied on the boundary Γ_t and, $\bar{q}_w$ and $\bar{q}_g$ are the prescribed out-flow rates of the porous fluids imposed on the permeable boundaries Γ_{q_w} and Γ_{q_g}, respectively, and $\mathbf{n}_\Gamma$ is the unit outward normal vector to the external boundary Γ.

From the physics of the problem, the existence of the fracture in the multi-phase porous medium leads to the mechanical and mass transfer couplings between the fracture and the porous medium surrounding the fracture. The mechanical coupling emanates from the developing crack still transferring stress through the fracture process zone that is actually the micro-cracked zone and the mean pore pressure exerted on the crack faces by means of the pore fluids within the crack. The mass transfer coupling originates from the flux of the wetting and non-wetting pore fluids flowing through the borders of the crack.

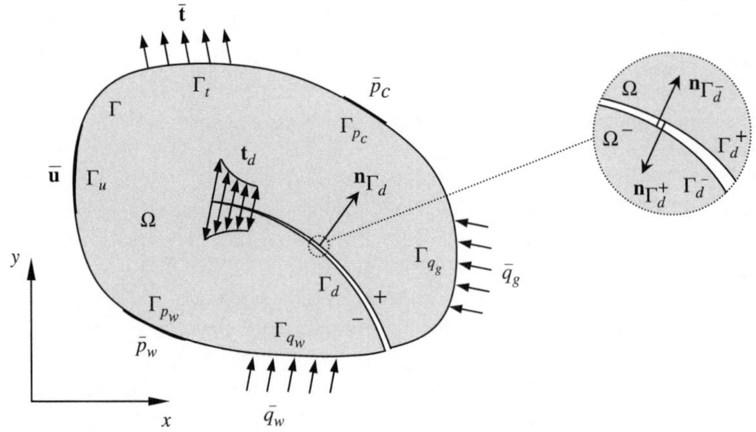

Figure 11.6 Boundary conditions of the body Ω involving the geomechanical discontinuity Γ_d

The essential and natural boundary conditions, which hold on the complementary parts of the external boundary of the body, complete with the boundary conditions on the discontinuity. In conformity with what was mentioned earlier, the following conditions should be fulfilled on the discontinuity Γ_d as $\boldsymbol{\sigma}\cdot\mathbf{n}_{\Gamma_d}=\mathbf{t}_d-\alpha p\,\mathbf{n}_{\Gamma_d}$, $[\![\dot{\mathbf{w}}_w]\!]\cdot\mathbf{n}_{\Gamma_d}=q_{wd}$ and $[\![\dot{\mathbf{w}}_g]\!]\cdot\mathbf{n}_{\Gamma_d}=q_{gd}$, where $\mathbf{t}_d$ is the cohesive traction acting at the fracture process zone, q_{wd} and q_{gd} are the leakage fluxes of the two porous fluids along the fracture, that is, the fully damaged zone and the fracture process zone, toward the surrounding porous medium, which implies that a discontinuity exists in the normal flow of the porous fluids across Γ_d, and $\mathbf{n}_{\Gamma_d}$ is the unit normal vector to the discontinuity Γ_d pointing to Ω^+. As already pointed out earlier, the notation $[\![\Xi]\!]=\Xi^+-\Xi^-$ represents the difference between the corresponding values at the two fracture faces. The cohesive crack concept implies that the cohesive tractions on either side of the crack are identical. Furthermore, it is assumed that the fluid pressure has the same values at both faces of the crack and pressure continuity from the crack faces to the surrounding medium holds along the crack boundary.

The weak form of the governing partial differential equations for the hydraulic fracture propagation in multi-phase porous media can be derived by integrating the product of the equilibrium and flow continuity equations (11.5) and (11.6) multiplied by admissible test functions over the domain, applying the Divergence theorem, imposing the natural boundary conditions, and satisfying the boundary conditions on the discontinuity. It is noteworthy that the mechanical and mass transfer coupling terms naturally appear in the weak form of the equilibrium and flow continuity equations on account of the presence of the geomechanical discontinuity in the medium. In this manner, it is ensured that the coupling between the discontinuity and the continuum medium surrounding the discontinuity is met in the weak form. The weak form of the equilibrium equation for the multi-phase system can be written as

$$\int_\Omega \nabla^s\delta\mathbf{u}:\boldsymbol{\sigma}\,d\Omega+\int_\Omega \rho\,\delta\mathbf{u}\cdot\ddot{\mathbf{u}}\,d\Omega+\int_{\Gamma_d}[\![\delta\mathbf{u}]\!]\cdot(\mathbf{t}_d-\alpha p\,\mathbf{n}_{\Gamma_d})\,d\Gamma=\int_{\Gamma_t}\delta\mathbf{u}\cdot\bar{\mathbf{t}}\,d\Gamma+\int_\Omega \rho\,\delta\mathbf{u}\cdot\mathbf{b}\,d\Omega \tag{11.44}$$

which holds for any admissible test function for the solid phase displacement $\delta\mathbf{u}$. The mechanical coupling term, which involves the integral on the discontinuity, comes from the following derivation assigning the positive and negative sides to Γ_d and imposing the boundary condition on the discontinuity

$$\begin{aligned}-\int_{\Gamma_d^+}\delta\mathbf{u}\cdot\left(\boldsymbol{\sigma}\cdot\mathbf{n}_{\Gamma_d^+}\right)d\Gamma-\int_{\Gamma_d^-}\delta\mathbf{u}\cdot\left(\boldsymbol{\sigma}\cdot\mathbf{n}_{\Gamma_d^-}\right)d\Gamma&=\int_{\Gamma_d}(\delta\mathbf{u}^+-\delta\mathbf{u}^-)\cdot(\boldsymbol{\sigma}\cdot\mathbf{n}_{\Gamma_d})\,d\Gamma\\ =\int_{\Gamma_d}[\![\delta\mathbf{u}]\!]\cdot(\boldsymbol{\sigma}\cdot\mathbf{n}_{\Gamma_d})\,d\Gamma&=\int_{\Gamma_d}[\![\delta\mathbf{u}]\!]\cdot(\mathbf{t}_d-\alpha p\,\mathbf{n}_{\Gamma_d})\,d\Gamma\end{aligned} \tag{11.45}$$

where $\mathbf{n}_{\Gamma_d^+}$ and $\mathbf{n}_{\Gamma_d^-}$ are the unit normal vectors directed to Ω^- and Ω^+, respectively, as shown in Figure 11.6, and $\mathbf{n}_{\Gamma_d}$ has been taken such that we have $\mathbf{n}_{\Gamma_d}=\mathbf{n}_{\Gamma_d^-}=-\mathbf{n}_{\Gamma_d^+}$. It is manifest from the mechanical coupling term that the test function space allows for functions discontinuous on Γ_d. This also holds for the set of kinematically admissible displacement fields, in which the actual displacement field is required, usually a subset of a specific Sobolev space (Babuška and Rosenzweig, 1972; Grisvard, 1985).

Incorporating Darcy's Law, the weak form of the continuity equation of flow for each of the fluid phases can be written as

$$\begin{aligned}&\int_\Omega \delta p_w\frac{1}{Q_{ww}}\dot{p}_w\,d\Omega+\int_\Omega \delta p_w\frac{1}{Q_{wc}}\dot{p}_c\,d\Omega+\int_\Omega \delta p_w\alpha\,\nabla\cdot\dot{\mathbf{u}}\,d\Omega+\int_\Omega (k_w+k_g)\,\nabla\delta p_w\cdot\nabla p_w\,d\Omega\\ &+\int_\Omega k_g\,\nabla\delta p_w\cdot\nabla p_c\,d\Omega+\int_\Omega (k_w\rho_w+k_g\rho_g)\,\nabla\delta p_w\cdot\ddot{\mathbf{u}}\,d\Omega-\int_{\Gamma_d}\delta p_w(q_{wd}+q_{gd})\,d\Gamma\\ &=\int_\Omega (k_w\rho_w+k_g\rho_g)\,\nabla\delta p_w\cdot\mathbf{b}\,d\Omega-\int_{\Gamma_{qw}}\delta p_w\,\bar{q}_w\,d\Gamma\\ &-\int_{\Gamma_{qg}\cap\Gamma_{qw}}\delta p_w\,\bar{q}_g\,d\Gamma-\int_{\Gamma_{pc}\cap\Gamma_{qw}}\delta p_w\dot{\mathbf{w}}_g\cdot\mathbf{n}_\Gamma\,d\Gamma\end{aligned} \tag{11.46}$$

and

$$\int_{\Omega}\delta p_c\frac{1}{Q_{cw}}\dot{p}_w\,d\Omega+\int_{\Omega}\delta p_c\frac{1}{Q_{cc}}\dot{p}_c\,d\Omega+\int_{\Omega}\delta p_c\,\alpha\,(1-S_w)\,\nabla\cdot\dot{\mathbf{u}}\,d\Omega$$
$$+\int_{\Omega}k_g\nabla\delta p_c\cdot\nabla p_w\,d\Omega+\int_{\Omega}k_g\nabla\delta p_c\cdot\nabla p_c\,d\Omega+\int_{\Omega}k_g\rho_g\nabla\delta p_c\cdot\ddot{\mathbf{u}}\,d\Omega-\int_{\Gamma_d}\delta p_c\,q_{gd}\,d\Gamma \tag{11.47}$$
$$=\int_{\Omega}k_g\rho_g\nabla\delta p_c\cdot\mathbf{b}\,d\Omega-\int_{\Gamma_{qg}}\delta p_c\,\bar{q}_g\,d\Gamma$$

which hold for any admissible test functions of the wetting phase pressure δp_w and the capillary pressure δp_c, respectively. The mass transfer coupling terms in the two weak forms here have resulted from the following sequence of statements and the imposition of the boundary conditions on the discontinuity

$$-\int_{\Gamma_d^+}\delta p_w\left(\dot{\mathbf{w}}_w\cdot\mathbf{n}_{\Gamma_d^+}\right)d\Gamma-\int_{\Gamma_d^-}\delta p_w\left(\dot{\mathbf{w}}_w\cdot\mathbf{n}_{\Gamma_d^-}\right)d\Gamma-\int_{\Gamma_d^+}\delta p_w\left(\dot{\mathbf{w}}_g\cdot\mathbf{n}_{\Gamma_d^+}\right)d\Gamma-\int_{\Gamma_d^-}\delta p_w\left(\dot{\mathbf{w}}_g\cdot\mathbf{n}_{\Gamma_d^-}\right)d\Gamma$$
$$=\int_{\Gamma_d}\delta p_w\left(\left(\dot{\mathbf{w}}_w^+-\dot{\mathbf{w}}_w^-\right)\cdot\mathbf{n}_{\Gamma_d}+\left(\dot{\mathbf{w}}_g^+-\dot{\mathbf{w}}_g^-\right)\cdot\mathbf{n}_{\Gamma_d}\right)d\Gamma \tag{11.48}$$
$$=\int_{\Gamma_d}\delta p_w\left([\![\dot{\mathbf{w}}_w]\!]\cdot\mathbf{n}_{\Gamma_d}+[\![\dot{\mathbf{w}}_g]\!]\cdot\mathbf{n}_{\Gamma_d}\right)d\Gamma=\int_{\Gamma_d}\delta p_w\left(q_{wd}+q_{gd}\right)d\Gamma$$

and

$$-\int_{\Gamma_d^+}\delta p_c\left(\dot{\mathbf{w}}_g\cdot\mathbf{n}_{\Gamma_d^+}\right)d\Gamma-\int_{\Gamma_d^-}\delta p_c\left(\dot{\mathbf{w}}_g\cdot\mathbf{n}_{\Gamma_d^-}\right)d\Gamma=\int_{\Gamma_d}\delta p_c\left(\dot{\mathbf{w}}_g^+-\dot{\mathbf{w}}_g^-\right)\cdot\mathbf{n}_{\Gamma_d}\,d\Gamma$$
$$=\int_{\Gamma_d}\delta p_c\,[\![\dot{\mathbf{w}}_g]\!]\cdot\mathbf{n}_{\Gamma_d}\,d\Gamma=\int_{\Gamma_d}\delta p_c\,q_{gd}\,d\Gamma \tag{11.49}$$

The mass transfer coupling terms imply that the Darcy velocity of the flowing fluids normal to the fracture is discontinuous across the fracture. According to the Darcy law, which relates the Darcy velocity to the corresponding fluid pressure gradient, it necessitates that the gradient of the fluid pressure in the normal direction to the fracture be also discontinuous across the fracture. Thus, the set of kinematically admissible pressure fields, to which the actual pressure field belongs, as well as the test function space should embody continuous functions with discontinuous normal derivative on Γ_d.

In order to arrive at a relation for the mass transfer coupling terms emerging in the weak form of the continuity equations of flow in the porous medium surrounding the fracture, the flow continuity equation for each flowing fluid inside the fracture is considered, in which the impact of cracking and therefore the change in porosity on the permeability of this damaged porous zone is taken into account. It is noticed that the fluid flow continuity equation at the discontinuity is identical to that of the continuum porous medium. The application of the Divergence theorem and the incorporation of the Darcy relation lead to the weak form of the continuity equation for the flow of the wetting phase inside the discontinuity

$$\int_{\Omega'}\delta p_w\frac{1}{Q_{ww}}\dot{p}_w\,d\Omega'+\int_{\Omega'}\delta p_w\frac{1}{Q_{wc}}\dot{p}_c\,d\Omega'+\int_{\Omega'}\delta p_w\alpha\,\nabla\cdot\dot{\mathbf{u}}\,d\Omega'+\int_{\Omega'}\left(k_{wd}+k_{gd}\right)\nabla\delta p_w\cdot\nabla p_w\,d\Omega'$$
$$+\int_{\Omega'}k_{gd}\nabla\delta p_w\cdot\nabla p_c\,d\Omega'+\int_{\Omega'}\left(k_{wd}\rho_w+k_{gd}\rho_g\right)\nabla\delta p_w\cdot\ddot{\mathbf{u}}\,d\Omega'+\int_{\Gamma_d}\delta p_w\left(q_{wd}+q_{gd}\right)d\Gamma \tag{11.50}$$
$$=\int_{\Omega'}\left(k_{wd}\rho_w+k_{gd}\rho_g\right)\nabla\delta p_w\cdot\mathbf{b}\,d\Omega'$$

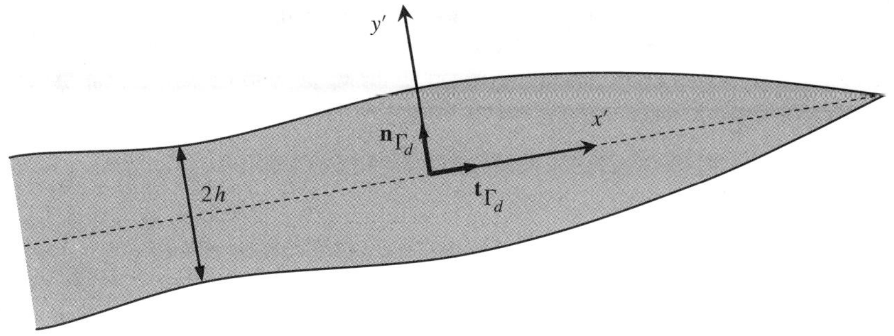

Figure 11.7 The geometry of the geomechanical discontinuity

where Ω' is the domain of the discontinuity, as shown in Figure 11.7. In what follows, the formulations are presented in the local Cartesian coordinate system (x', y') where x' and y' denote the local Cartesian coordinates whose directions are aligned with the tangent and normal unit vectors to the discontinuity, $\mathbf{t}_{\Gamma_d}$ and $\mathbf{n}_{\Gamma_d}$, respectively, as displayed in Figure 11.7. The integrals over the domain of the discontinuity are performed in the local coordinate system. The integral over the cross section of the discontinuity is first evaluated. Regarding that the width of the discontinuity $2h$ is insignificant relative to its length, the variation of the fluid pressure across the discontinuity width is ignored. That is, the fluid pressure as well as the corresponding test function is assumed to be uniform over the cross section of the discontinuity. Hence, the first two integrals in Eq. (11.50) can be rewritten as

$$\begin{aligned}
&\int_{\Omega'} \delta p_w \frac{1}{Q_{ww}} \dot{p}_w \, d\Omega' + \int_{\Omega'} \delta p_w \frac{1}{Q_{wc}} \dot{p}_c \, d\Omega' \\
&= \int_{\Gamma_d} \int_{-h}^{+h} \delta p_w \frac{1}{Q_{ww}} \dot{p}_w \, dy' d\Gamma + \int_{\Gamma_d} \int_{-h}^{+h} \delta p_w \frac{1}{Q_{wc}} \dot{p}_c \, dy' d\Gamma \\
&= \int_{\Gamma_d} \delta p_w \, 2h \frac{1}{Q_{ww}} \dot{p}_w \, d\Gamma + \int_{\Gamma_d} \delta p_w \, 2h \frac{1}{Q_{wc}} \dot{p}_c \, d\Gamma
\end{aligned} \tag{11.51}$$

By applying the definition of the Divergence in the local coordinate system, the following expression can be obtained for the third integral of Eq. (11.50) as

$$\int_{\Omega'} \delta p_w \, \alpha \nabla \cdot \dot{\mathbf{u}} \, d\Omega' = \int_{\Gamma_d} \int_{-h}^{+h} \delta p_w \, \alpha \nabla \cdot \dot{\mathbf{u}} \, dy' d\Gamma = \int_{\Gamma_d} \int_{-h}^{+h} \delta p_w \, \alpha \left(\frac{\partial \dot{u}_{x'}}{\partial x'} + \frac{\partial \dot{u}_{y'}}{\partial y'} \right) dy' d\Gamma \tag{11.52}$$

where $\dot{u}_{x'}$ and $\dot{u}_{y'}$ are the components of the solid velocity vector projected on the longitudinal and transversal axes, respectively. It is supposed that the velocity component of the solid phase in the longitudinal direction varies linearly with y' over the discontinuity width. In consequence, its derivative with respect to x' also varies linearly with y'. Taking this into account, the first contribution on the right-hand side of Eq. (11.52), which involves the tangential derivative, as well as the contribution which includes the normal derivative can be integrated analytically. Thus, specifying $\langle \Xi \rangle = (\Xi^+ + \Xi^-)/2$ as the average of the corresponding values at the discontinuity faces, Eq. (11.52) can be obtained as

$$\begin{aligned}
\int_{\Omega'} \delta p_w \, \alpha \nabla \cdot \dot{\mathbf{u}} \, d\Omega' &= \int_{\Gamma_d} \delta p_w \, 2h \, \alpha \frac{1}{2} \left(\frac{\partial \dot{u}_{x'}}{\partial x'}(h) + \frac{\partial \dot{u}_{x'}}{\partial x'}(-h) \right) d\Gamma + \int_{\Gamma_d} \delta p_w \, \alpha \left(\dot{u}_{y'}(h) - \dot{u}_{y'}(-h) \right) d\Gamma \\
&= \int_{\Gamma_d} \delta p_w \, 2h \, \alpha \left\langle \frac{\partial \dot{u}_{x'}}{\partial x'} \right\rangle d\Gamma + \int_{\Gamma_d} \delta p_w \, \alpha [\![\dot{u}_{y'}]\!] \, d\Gamma
\end{aligned} \tag{11.53}$$

Since the fluid pressure is supposed to be uniform across the width of the discontinuity, its derivative in the tangential direction does not vary with y', and its derivative in the normal direction vanishes. This is also the case for the corresponding test function. In view of all these, the fourth integral in Eq. (11.50) can be solved analytically as

$$\begin{aligned}\int_{\Omega'}(k_{wd}+k_{gd})\,\nabla\delta p_w\cdot\nabla p_w\,d\Omega' &= \int_{\Gamma_d}\int_{-h}^{+h}(k_{wd}+k_{gd})\,\nabla\delta p_w\cdot\nabla p_w\,dy'd\Gamma \\ &= \int_{\Gamma_d}\int_{-h}^{+h}(k_{wd}+k_{gd})\left(\frac{\partial\delta p_w}{\partial x'}\frac{\partial p_w}{\partial x'}+\frac{\partial\delta p_w}{\partial y'}\frac{\partial p_w}{\partial y'}\right)dy'd\Gamma \\ &= \int_{\Gamma_d}2h\,(k_{wd}+k_{gd})\,\frac{\partial\delta p_w}{\partial x'}\frac{\partial p_w}{\partial x'}d\Gamma\end{aligned}\tag{11.54}$$

Under the same considerations as for the fourth integral, the following analytical solution can be obtained for the fifth integral in Eq. (11.50) as

$$\int_{\Omega'}k_{gd}\nabla\delta p_w\cdot\nabla p_c\,d\Omega' = \int_{\Gamma_d}2h\,k_{gd}\,\frac{\partial\delta p_w}{\partial x'}\frac{\partial p_c}{\partial x'}d\Gamma\tag{11.55}$$

In agreement with the assumption made for the solid velocity in the direction tangent to the discontinuity, the solid acceleration in the tangential direction also varies linearly with y'. As mentioned earlier, the derivative of the test function with respect to x' is constant over the discontinuity width, and its derivative with respect to y' has a zero value. Thus, the next integral of Eq. (11.50) can be obtained as

$$\begin{aligned}\int_{\Omega'}(k_{wd}\rho_w+k_{gd}\rho_g)\,\nabla\delta p_w\cdot\ddot{\mathbf{u}}\,d\Omega' &= \int_{\Gamma_d}\int_{-h}^{+h}(k_{wd}\rho_w+k_{gd}\rho_g)\,\nabla\delta p_w\cdot\ddot{\mathbf{u}}\,dy'\,d\Gamma \\ &= \int_{\Gamma_d}\int_{-h}^{+h}(k_{wd}\rho_w+k_{gd}\rho_g)\left(\frac{\partial\delta p_w}{\partial x'}\ddot{u}_{x'}+\frac{\partial\delta p_w}{\partial y'}\ddot{u}_{y'}\right)dy'\,d\Gamma \\ &= \int_{\Gamma_d}2h\,(k_{wd}\rho_w+k_{gd}\rho_g)\,\frac{\partial\delta p_w}{\partial x'}\langle\ddot{u}_{x'}\rangle d\Gamma\end{aligned}\tag{11.56}$$

Applying the dot product and regarding that the body force that is generally taken as the gravity acceleration is constant over the entire domain and so the discontinuity width, the solution of the remaining integral over the domain of the discontinuity can be directly obtained as

$$\int_{\Omega'}(k_{wd}\rho_w+k_{gd}\rho_g)\,\nabla\delta p_w\cdot\mathbf{b}\,d\Omega' = \int_{\Gamma_d}2h\,(k_{wd}\rho_w+k_{gd}\rho_g)\,\frac{\partial\delta p_w}{\partial x'}b_{x'}\,d\Gamma\tag{11.57}$$

By substituting the constituents of equation (11.50) and rearranging the equation, the mass transfer coupling term appearing in the weak form of the water flow continuity equation of the continuum porous medium can be achieved as

$$\begin{aligned}\int_{\Gamma_d}\delta p_w\,(q_{wd}+q_{gd})\,d\Gamma = &-\int_{\Gamma_d}\delta p_w\,2h\frac{1}{Q_{ww}}\dot{p}_w\,d\Gamma-\int_{\Gamma_d}\delta p_w\,2h\frac{1}{Q_{wc}}\dot{p}_c\,d\Gamma \\ &-\int_{\Gamma_d}\delta p_w\,2h\,\alpha\left\langle\frac{\partial\dot{u}_{x'}}{\partial x'}\right\rangle d\Gamma-\int_{\Gamma_d}\delta p_w\,\alpha\,[\![\dot{u}_{y'}]\!]\,d\Gamma \\ &-\int_{\Gamma_d}\frac{\partial\delta p_w}{\partial x'}2h\,(k_{wd}+k_{gd})\,\frac{\partial p_w}{\partial x'}d\Gamma-\int_{\Gamma_d}\frac{\partial\delta p_w}{\partial x'}2h\,k_{gd}\,\frac{\partial p_c}{\partial x'}d\Gamma \\ &-\int_{\Gamma_d}\frac{\partial\delta p_w}{\partial x'}2h\,(k_{wd}\rho_w+k_{gd}\rho_g)\langle\ddot{u}_{x'}\rangle d\Gamma+\int_{\Gamma_d}\frac{\partial\delta p_w}{\partial x'}2h\,(k_{wd}\rho_w+k_{gd}\rho_g)\,b_{x'}\,d\Gamma\end{aligned}\tag{11.58}$$

In a similar manner, the weak form of the non-wetting fluid flow continuity equation of the discontinuity yields the relevant mass transfer coupling term as

$$\begin{aligned}
\int_{\Gamma_d} \delta p_c\, q_{gd}\, d\Gamma = & -\int_{\Gamma_d} \delta p_c\, 2h \frac{1}{Q_{cw}} \dot{p}_w\, d\Gamma - \int_{\Gamma_d} \delta p_c\, 2h \frac{1}{Q_{cc}} \dot{p}_c\, d\Gamma \\
& -\int_{\Gamma_d} \delta p_c\, 2h\, \alpha (1 - S_w) \left\langle \frac{\partial \dot{u}_{x'}}{\partial x'} \right\rangle d\Gamma - \int_{\Gamma_d} \delta p_c\, \alpha (1 - S_w) [\![\dot{u}_{y'}]\!]\, d\Gamma \\
& -\int_{\Gamma_d} \frac{\partial \delta p_c}{\partial x'} 2h\, k_{gd} \frac{\partial p_w}{\partial x'} d\Gamma - \int_{\Gamma_d} \frac{\partial \delta p_c}{\partial x'} 2h\, k_{gd} \frac{\partial p_c}{\partial x'} d\Gamma \\
& -\int_{\Gamma_d} \frac{\partial \delta p_c}{\partial x'} 2h\, k_{gd}\, \rho_g \langle \ddot{u}_{x'} \rangle d\Gamma + \int_{\Gamma_d} \frac{\partial \delta p_c}{\partial x'} 2h\, k_{gd}\, \rho_g\, b_{x'}\, d\Gamma
\end{aligned} \tag{11.59}$$

The test results indicate that the permeability is dependent on the porosity, which varies with the solid skeleton deformation. In the porous medium surrounding the fracture this effect can be neglected, but in the fracture the change in porosity induced by decohesion and cracking of the solid skeleton is significant and its effect should be considered. An analytical relation for the porosity within the discontinuity is derived from the conservation of mass for the solid phase. By taking the Biot coefficient inside the discontinuity equal to unity, the mass balance equation for the solid phase can be written as

$$\frac{\dot{n}_d}{1 - n_d} = \nabla \cdot \dot{\mathbf{u}} \tag{11.60}$$

Integrating the mass balance equation over the discontinuity width, assuming that porosity and thus its time derivative are constant across the discontinuity cross section, using the result obtained in Eq. (11.53), and finally dividing the whole result by the discontinuity width $2h$, it follows that

$$\frac{\dot{n}_d}{1 - n_d} = \left\langle \frac{\partial \dot{u}_{x'}}{\partial x'} \right\rangle + \frac{[\![\dot{u}_{y'}]\!]}{[\![u_{y'}]\!]} \tag{11.61}$$

It must be noted that the width of the discontinuity is in fact equal to the normal component of the displacement jump vector at the discontinuity, that is, $2h = [\![u_{y'}]\!]$. By taking the integral of expression (11.61) over the typical time interval from t_n to t_{n+1}, after some manipulations an explicit relation for the porosity inside the discontinuity at the current time step can be obtained as a function of its value at the previous time step and the current and previous values of the crack opening and longitudinal strain as

$$n_d(t_{n+1}) = 1 - (1 - n_d(t_n)) \frac{[\![u_{y'}(t_n)]\!]}{[\![u_{y'}(t_{n+1})]\!]} \exp\left(-\left\langle \frac{\partial u_{x'}}{\partial x'}(t_{n+1}) \right\rangle + \left\langle \frac{\partial u_{x'}}{\partial x'}(t_n) \right\rangle \right) \tag{11.62}$$

11.7 Discretization of Multi-Phase Governing Equations with Strong Discontinuities

In order to derive the discretization system of the governing equations of multi-phase porous media described in previous section, the weak form of the equilibrium and flow continuity equations is rendered in a discrete form by means of discretization firstly in space employing the extended FEM and then in time applying the generalized Newmark scheme. The resulting system of fully coupled nonlinear algebraic equations is directly solved using the direct solution procedure combined with the Newton–Raphson iterative process. As a result, the main unknowns are obtained simultaneously, leading to the full solution of the problem. In order to capture the hydromechanical coupling associated with the tractions acting on the fracture and the leakage flux of the two fluid phases along the fracture, the displacement, water pressure, and capillary pressure fields must satisfy some requirements. To account for the displacement jump across the fracture, the displacement field must be discontinuous on Γ_d, referred to as the strong discontinuity. In addition, to take into account each fluid flow jump normal to the fracture, it is required that the water pressure and also the capillary pressure field be continuous, while their corresponding gradient normal to the fracture should be discontinuous on Γ_d, which is called the weak discontinuity. Viewing all these, so as to provide for the inclusion of the discontinuity in the displacement field and in the normal derivative of the water pressure and capillary pressure fields across the crack line, the corresponding field is approximated by locally enriching the classical FE space. The key feature of this work is the introduction of an enrichment function that models the discontinuity in the solution, or in the derivatives of the solution in the classical FE approximation complemented by the addition of enriched DOF pertinent to the enrichment function. This is achieved by exploiting the PU property of FE shape functions, leading to the enrichment localized to the desired areas. For the success of the enriched approximation, it is vital that the enrichment function be chosen properly. In this way, the local enrichment strategy allows for the incorporation of the discontinuity in the displacement field and fluid flux across the crack faces into the solution of the problem.

In the X-FEM, the displacement approximation as well as the water pressure and capillary pressure approximations is based upon an additive decomposition into a standard part and an enriched part. Following this, the X-FEM approximation of the displacement field $\mathbf{u}(\boldsymbol{x}, t)$ is written as

$$\mathbf{u}^h(\boldsymbol{x},t)=\sum_{I\in\mathcal{N}} N_{uI}(\boldsymbol{x})\ \mathbf{u}_I(t)+\sum_{I\in\mathcal{N}^{enr}} N_{uI}(\boldsymbol{x})\frac{1}{2}(H_{\Gamma_d}(\boldsymbol{x})-H_{\Gamma_d}(\boldsymbol{x}_I))\,\tilde{\mathbf{u}}_I(t) \tag{11.63}$$

where $N_{uI}(\boldsymbol{x})$ is the standard FE shape function of node I, the nodal set $\mathcal{N}$ is the set of all nodes in the domain and the nodal set $\mathcal{N}^{enr}$ is the set of enriched nodes whose support is bisected by the discontinuity. To ensure that the displacement jump is zero at the discontinuity tip, the nodes belonging to the element edge on which the discontinuity tip lies are not enriched. In relation (11.63), $\mathbf{u}_I(t)$ are the standard displacement DOF at node I, and the enriched nodes hold additional displacement DOF $\tilde{\mathbf{u}}_I(t)$ associated with the shifted enrichment function $\frac{1}{2}(H_{\Gamma_d}(\boldsymbol{x})-H_{\Gamma_d}(\boldsymbol{x}_I))$. The discontinuous function $H_{\Gamma_d}(\boldsymbol{x})$ is taken as the sign function centered on the line of the discontinuity Γ_d as

$$H_{\Gamma_d}(\boldsymbol{x})=\operatorname{sign}(\varphi(\boldsymbol{x})) \tag{11.64}$$

in which $\varphi(\boldsymbol{x})$ is the standard level set function defined in (11.25). $H_{\Gamma_d}(\boldsymbol{x}_I)$ is the value of $H_{\Gamma_d}(\boldsymbol{x})$ at the enriched node I. It follows from the earlier definition that the enrichment function exhibits a unit jump across the discontinuity, so the displacement discontinuity across Γ_d can be obtained. Moreover, the shifted enrichment function vanishes in all elements not containing the discontinuity.

Symbolically, the enriched FE approximation of the displacement field in Eq. (11.63) can be written in the following form

$$\mathbf{u}^h(\boldsymbol{x},t)=\mathbf{N}_u(\boldsymbol{x})\ \mathbf{U}(t)+\mathbf{N}_u^{enr}(\boldsymbol{x})\ \tilde{\mathbf{U}}(t) \tag{11.65}$$

where $\mathbf{N}_u(\boldsymbol{x})$ is the matrix of the standard shape functions, $\mathbf{N}_u^{enr}(\boldsymbol{x})$ is the matrix of the enriched shape functions, $\mathbf{U}(t)$ is the vector of the standard displacement DOF, and $\tilde{\mathbf{U}}(t)$ is the vector of the enriched displacement DOF. Accordingly, the strain vector corresponding to the approximate displacement field $\mathbf{u}^h(\boldsymbol{x}, t)$ can be written in terms of the standard and enriched nodal values as

$$\boldsymbol{\varepsilon}^h(\boldsymbol{x},t) = \mathbf{B}(\boldsymbol{x})\ \mathbf{U}(t) + \mathbf{B}^{enr}(\boldsymbol{x})\ \tilde{\mathbf{U}}(t) \tag{11.66}$$

where $\mathbf{B}(\boldsymbol{x}) = \mathbf{L}\ \mathbf{N}_u(\boldsymbol{x})$ and $\mathbf{B}^{enr}(\boldsymbol{x}) = \mathbf{L}\,\mathbf{N}_u^{enr}(\boldsymbol{x})$ are the standard and enriched matrices that involve spatial derivatives of the standard and enriched shape functions, respectively, with $\mathbf{L}$ denoting the matrix differential operator defined in (11.29).

Similarly, the water pressure and capillary pressure $p_\alpha(\boldsymbol{x}, t)$, $(\alpha = w, c)$ are approximated using a linear combination of the standard and enriched shape functions as

$$p_\alpha^h(\boldsymbol{x},t) = \sum_{I\in\mathcal{N}} N_{p_\alpha I}(\boldsymbol{x})\, p_{\alpha I}(t) + \sum_{I\in\mathcal{N}^{enr}} N_{p_\alpha I}(\boldsymbol{x})\big(D_{\Gamma_d}(\boldsymbol{x}) - D_{\Gamma_d}(\boldsymbol{x}_I)\big)R(\boldsymbol{x})\,\tilde{p}_{\alpha I}(t) \qquad \alpha = w, c \tag{11.67}$$

where $N_{p_\alpha I}(\boldsymbol{x})$, $(\alpha = w, c)$ are the standard FE shape functions. Nodes in $\mathcal{N}^{enr}$, to which enriched DOF are added, have their support bisected by the discontinuity. It is essential that the leakage flux vanish at the discontinuity tip. This is assured by requiring that the nodes on the element edge with which the discontinuity tip coincides not be enriched. $p_{\alpha I}(t)$, $(\alpha = w, c)$ are the standard pressure DOF and $\tilde{p}_{\alpha I}(t)$ are the enriched pressure DOF corresponding to the shifted enrichment function $(D_{\Gamma_d}(\boldsymbol{x}) - D_{\Gamma_d}(\boldsymbol{x}_I))R(\boldsymbol{x})$. In the above relation, $D_{\Gamma_d}(\boldsymbol{x})$ is the distance function, the frequently used enrichment function for weak discontinuities, defined as the absolute value of the signed distance function as $D_{\Gamma_d}(\boldsymbol{x}) = |\varphi(\boldsymbol{x})|$, which is indeed the distance to the discontinuity. The distance function is continuous, whereas its gradient in the normal direction to the discontinuity is discontinuous across Γ_d, which can be simply observed by taking its normal derivative leading to the sign function, that is, $\nabla D_{\Gamma_d} \cdot \mathbf{n}_{\Gamma_d} = H_{\Gamma_d}$. Possessing this desirable property, the chosen enrichment function enables the approximate water pressure and capillary pressure fields to be discontinuous in their normal derivative across the discontinuity, accounting for the leakage flux of the pore fluids into the continuum medium surrounding the discontinuity. In relation (11.67), $R(x)$ is a weight function with compact support defined as

$$R(\boldsymbol{x}) = \sum_{I\in\mathcal{N}^{enr}} N_{p_\alpha I}(\boldsymbol{x}) \qquad \alpha = w, c \tag{11.68}$$

As noted earlier from relation (11.23), the multiplication with $R(\boldsymbol{x})$ causes the weighted enrichment function to vary continuously between the standard enrichment function $(D_{\Gamma_d}(\boldsymbol{x}) - D_{\Gamma_d}(\boldsymbol{x}_I))$ and zero in the elements whose some nodes are in the enriched nodal set, reproducing the standard enrichment function in the elements whose all nodes are in the enriched nodal set. This choice of the enrichment function shows a better accuracy and improves the convergence rate as compared with the standard enrichment function. Likewise, the enriched FE approximation of the water pressure and capillary pressure fields in Eq. (11.67) can be rewritten as

$$p_\alpha^h(\boldsymbol{x},t) = \mathbf{N}_{p_\alpha}(\boldsymbol{x})\ \mathbf{P}_\alpha(t) + \mathbf{N}_{p_\alpha}^{enr}(\boldsymbol{x})\tilde{\mathbf{P}}_\alpha(t) \qquad \alpha = w, c \tag{11.69}$$

where $\mathbf{N}_{p_\alpha}(\boldsymbol{x})$ is the matrix of the standard shape functions, $\mathbf{N}_{p_\alpha}^{enr}(\boldsymbol{x})$ is the matrix of the enriched shape functions, $\mathbf{P}_\alpha(t)$ is the vector of the standard nodal DOF, and $\tilde{\mathbf{P}}_\alpha(t)$ is the vector of the enriched nodal DOF.

Following the Bubnov–Galerkin technique, the test functions for the displacement $\delta\mathbf{u}$, water pressure δp_w and capillary pressure δp_c in the discrete weak form are considered in the same enriched approximating space, that is, the finite-dimensional subspace spanned by the linearly independent basis functions, as the approximate displacement, water pressure, and capillary pressure fields, respectively. Inserting the enriched FE approximations into the weak form of the governing partial differential equations (11.44), (11.46), and (11.47) and satisfying the necessity that the weak form should hold for all admissible test functions, the discretized form of the equations defining the multi-phase problem can be obtained as

$$
\begin{aligned}
&\mathbf{M}_{uu}\ddot{\mathbf{U}}+\mathbf{M}_{u\tilde{u}}\ddot{\tilde{\mathbf{U}}}+\int_{\Omega}\mathbf{B}^{T}\boldsymbol{\sigma}''\,d\Omega-\mathbf{Q}_{uw}\mathbf{P}_{w}-\mathbf{Q}_{u\tilde{w}}\tilde{\mathbf{P}}_{w}-\mathbf{Q}_{uc}\mathbf{P}_{c}-\mathbf{Q}_{u\tilde{c}}\tilde{\mathbf{P}}_{c}=\mathbf{F}_{u}^{\text{ext}}\\
&\mathbf{M}_{\tilde{u}u}^{T}\ddot{\mathbf{U}}+\mathbf{M}_{\tilde{u}\tilde{u}}\ddot{\tilde{\mathbf{U}}}+\int_{\Omega^{enr}}(\mathbf{B}^{enr})^{T}\boldsymbol{\sigma}''\,d\Omega-\mathbf{Q}_{\tilde{u}w}\mathbf{P}_{w}-\mathbf{Q}_{\tilde{u}\tilde{w}}\tilde{\mathbf{P}}_{w}-\mathbf{Q}_{\tilde{u}c}\mathbf{P}_{c}-\mathbf{Q}_{\tilde{u}\tilde{c}}\tilde{\mathbf{P}}_{c}+\mathbf{F}_{\tilde{u}}^{\text{int}}=\mathbf{F}_{\tilde{u}}^{\text{ext}}\\
&\mathbf{M}_{wu}\ddot{\mathbf{U}}+\mathbf{M}_{w\tilde{u}}\ddot{\tilde{\mathbf{U}}}+\mathbf{Q}_{uw}^{T}\dot{\mathbf{U}}+\mathbf{Q}_{\tilde{u}w}^{T}\dot{\tilde{\mathbf{U}}}+\mathbf{C}_{ww}\dot{\mathbf{P}}_{w}+\mathbf{C}_{w\tilde{w}}\dot{\tilde{\mathbf{P}}}_{w}+\mathbf{C}_{wc}\dot{\mathbf{P}}_{c}+\mathbf{C}_{w\tilde{c}}\dot{\tilde{\mathbf{P}}}_{c}\\
&\qquad+\mathbf{H}_{ww}\mathbf{P}_{w}+\mathbf{H}_{w\tilde{w}}\tilde{\mathbf{P}}_{w}+\mathbf{H}_{wc}\mathbf{P}_{c}+\mathbf{H}_{w\tilde{c}}\tilde{\mathbf{P}}_{c}-\mathbf{F}_{w}^{\text{int}}=\mathbf{F}_{w}^{\text{ext}}\\
&\mathbf{M}_{\tilde{w}u}\ddot{\mathbf{U}}+\mathbf{M}_{\tilde{w}\tilde{u}}\ddot{\tilde{\mathbf{U}}}+\mathbf{Q}_{u\tilde{w}}^{T}\dot{\mathbf{U}}+\mathbf{Q}_{\tilde{u}\tilde{w}}^{T}\dot{\tilde{\mathbf{U}}}+\mathbf{C}_{w\tilde{w}}^{T}\dot{\mathbf{P}}_{w}+\mathbf{C}_{\tilde{w}\tilde{w}}\dot{\tilde{\mathbf{P}}}_{w}+\mathbf{C}_{\tilde{w}c}\dot{\mathbf{P}}_{c}+\mathbf{C}_{\tilde{w}\tilde{c}}\dot{\tilde{\mathbf{P}}}_{c}\\
&\qquad+\mathbf{H}_{w\tilde{w}}^{T}\mathbf{P}_{w}+\mathbf{H}_{\tilde{w}\tilde{w}}\tilde{\mathbf{P}}_{w}+\mathbf{H}_{\tilde{w}c}\mathbf{P}_{c}+\mathbf{H}_{\tilde{w}\tilde{c}}\tilde{\mathbf{P}}_{c}-\mathbf{F}_{\tilde{w}}^{\text{int}}=\mathbf{F}_{\tilde{w}}^{\text{ext}}\\
&\mathbf{M}_{cu}\ddot{\mathbf{U}}+\mathbf{M}_{c\tilde{u}}\ddot{\tilde{\mathbf{U}}}+\mathbf{Q}_{uc}^{T}\dot{\mathbf{U}}+\mathbf{Q}_{\tilde{u}c}^{T}\dot{\tilde{\mathbf{U}}}+\mathbf{C}_{cw}\dot{\mathbf{P}}_{w}+\mathbf{C}_{c\tilde{w}}\dot{\tilde{\mathbf{P}}}_{w}+\mathbf{C}_{cc}\dot{\mathbf{P}}_{c}+\mathbf{C}_{c\tilde{c}}\dot{\tilde{\mathbf{P}}}_{c}+\mathbf{H}_{wc}^{T}\mathbf{P}_{w}\\
&\qquad+\mathbf{H}_{\tilde{w}c}^{T}\tilde{\mathbf{P}}_{w}+\mathbf{H}_{cc}\mathbf{P}_{c}+\mathbf{H}_{c\tilde{c}}\tilde{\mathbf{P}}_{c}-\mathbf{F}_{c}^{\text{int}}=\mathbf{F}_{c}^{\text{ext}}\\
&\mathbf{M}_{\tilde{c}u}\ddot{\mathbf{U}}+\mathbf{M}_{\tilde{c}\tilde{u}}\ddot{\tilde{\mathbf{U}}}+\mathbf{Q}_{u\tilde{c}}^{T}\dot{\mathbf{U}}+\mathbf{Q}_{\tilde{u}\tilde{c}}^{T}\dot{\tilde{\mathbf{U}}}+\mathbf{C}_{\tilde{c}w}\dot{\mathbf{P}}_{w}+\mathbf{C}_{\tilde{c}\tilde{w}}\dot{\tilde{\mathbf{P}}}_{w}+\mathbf{C}_{c\tilde{c}}^{T}\dot{\mathbf{P}}_{c}+\mathbf{C}_{\tilde{c}\tilde{c}}\dot{\tilde{\mathbf{P}}}_{c}\\
&\qquad+\mathbf{H}_{w\tilde{c}}^{T}\mathbf{P}_{w}+\mathbf{H}_{\tilde{w}\tilde{c}}^{T}\tilde{\mathbf{P}}_{w}+\mathbf{H}_{c\tilde{c}}^{T}\mathbf{P}_{c}+\mathbf{H}_{\tilde{c}\tilde{c}}\tilde{\mathbf{P}}_{c}-\mathbf{F}_{\tilde{c}}^{\text{int}}=\mathbf{F}_{\tilde{c}}^{\text{ext}}
\end{aligned}
\tag{11.70}
$$

where the coefficient matrices of the discretized governing equations (11.70) are defined as

$$\mathbf{M}_{uu}=\int_{\Omega}\mathbf{N}_{u}^{T}\rho\,\mathbf{N}_{u}\,d\Omega \qquad \mathbf{M}_{u\tilde{u}}=\int_{\Omega^{enr}}\mathbf{N}_{u}^{T}\rho\,\mathbf{N}_{u}^{enr}\,d\Omega$$

$$\mathbf{M}_{\tilde{u}\tilde{u}}=\int_{\Omega^{enr}}(\mathbf{N}_{u}^{enr})^{T}\rho\,\mathbf{N}_{u}^{enr}\,d\Omega \qquad \mathbf{M}_{wu}=\int_{\Omega}\nabla\mathbf{N}_{p_w}^{T}(k_w\rho_w+k_g\rho_g)\mathbf{N}_{u}\,d\Omega$$

$$\mathbf{M}_{w\tilde{u}}=\int_{\Omega^{enr}}\nabla\mathbf{N}_{p_w}^{T}(k_w\rho_w+k_g\rho_g)\mathbf{N}_{u}^{enr}\,d\Omega \qquad \mathbf{M}_{\tilde{w}u}=\int_{\Omega^{enr}}\left(\nabla\mathbf{N}_{p_w}^{enr}\right)^{T}(k_w\rho_w+k_g\rho_g)\mathbf{N}_{u}\,d\Omega$$

$$\mathbf{M}_{\tilde{w}\tilde{u}}=\int_{\Omega^{enr}}\left(\nabla\mathbf{N}_{p_w}^{enr}\right)^{T}(k_w\rho_w+k_g\rho_g)\mathbf{N}_{u}^{enr}\,d\Omega \qquad \mathbf{M}_{cu}=\int_{\Omega}\nabla\mathbf{N}_{p_c}^{T}k_g\rho_g\,\mathbf{N}_{u}\,d\Omega$$

$$\mathbf{M}_{c\tilde{u}}=\int_{\Omega^{enr}}\nabla\mathbf{N}_{p_c}^{T}k_g\rho_g\,\mathbf{N}_{u}^{enr}\,d\Omega \qquad \mathbf{M}_{\tilde{c}u}=\int_{\Omega^{enr}}\left(\nabla\mathbf{N}_{p_c}^{enr}\right)^{T}k_g\rho_g\,\mathbf{N}_{u}\,d\Omega$$

$$\mathbf{M}_{\tilde{c}\tilde{u}}=\int_{\Omega^{enr}}\left(\nabla\mathbf{N}_{p_c}^{enr}\right)^{T}k_g\rho_g\,\mathbf{N}_{u}^{enr}\,d\Omega \qquad \mathbf{Q}_{uw}=\int_{\Omega}\mathbf{B}^{T}\alpha\,\mathbf{m}\,\mathbf{N}_{p_w}\,d\Omega$$

$$\mathbf{Q}_{u\tilde{w}}=\int_{\Omega^{enr}}\mathbf{B}^{T}\alpha\,\mathbf{m}\,\mathbf{N}_{p_w}^{enr}\,d\Omega \qquad \mathbf{Q}_{\tilde{u}w}=\int_{\Omega^{enr}}(\mathbf{B}^{enr})^{T}\alpha\,\mathbf{m}\,\mathbf{N}_{p_w}\,d\Omega$$

$$\mathbf{Q}_{\tilde{u}\tilde{w}}=\int_{\Omega^{enr}}(\mathbf{B}^{enr})^{T}\alpha\,\mathbf{m}\,\mathbf{N}_{p_w}^{enr}\,d\Omega \qquad \mathbf{Q}_{uc}=\int_{\Omega}\mathbf{B}^{T}\alpha(1-S_w)\,\mathbf{m}\,\mathbf{N}_{p_c}\,d\Omega$$

$$\mathbf{Q}_{u\tilde{c}}=\int_{\Omega^{enr}}\mathbf{B}^{T}\alpha(1-S_w)\,\mathbf{m}\,\mathbf{N}_{p_c}^{enr}\,d\Omega \qquad \mathbf{Q}_{\tilde{u}c}=\int_{\Omega^{enr}}(\mathbf{B}^{enr})^{T}\alpha(1-S_w)\,\mathbf{m}\,\mathbf{N}_{p_c}\,d\Omega$$

$$\mathbf{Q}_{\tilde{u}\tilde{c}}=\int_{\Omega^{enr}}(\mathbf{B}^{enr})^{T}\alpha(1-S_w)\,\mathbf{m}\,\mathbf{N}_{p_c}^{enr}\,d\Omega \qquad \mathbf{C}_{ww}=\int_{\Omega}\mathbf{N}_{p_w}^{T}\frac{1}{Q_{ww}}\mathbf{N}_{p_w}\,d\Omega$$

$$\mathbf{C}_{w\tilde{w}}=\int_{\Omega^{enr}}\mathbf{N}_{p_w}^{T}\frac{1}{Q_{ww}}\mathbf{N}_{p_w}^{enr}\,d\Omega \qquad \mathbf{C}_{\tilde{w}\tilde{w}}=\int_{\Omega^{enr}}\left(\mathbf{N}_{p_w}^{enr}\right)^{T}\frac{1}{Q_{ww}}\mathbf{N}_{p_w}^{enr}\,d\Omega$$

$$\mathbf{C}_{cc}=\int_{\Omega}\mathbf{N}_{p_c}^{T}\frac{1}{Q_{cc}}\mathbf{N}_{p_c}\,d\Omega$$

$$\mathbf{C}_{\tilde{c}\tilde{c}} = \int_{\Omega^{enr}} \left(\mathbf{N}_{p_c}^{enr}\right)^T \frac{1}{Q_{cc}} \mathbf{N}_{p_c}^{enr}\, d\Omega \qquad \mathbf{C}_{c\tilde{c}} = \int_{\Omega^{enr}} \mathbf{N}_{p_c}^T \frac{1}{Q_{cc}} \mathbf{N}_{p_c}^{enr}\, d\Omega$$

$$\mathbf{C}_{w\tilde{c}} = \int_{\Omega^{enr}} \mathbf{N}_{p_w}^T \frac{1}{Q_{wc}} \mathbf{N}_{p_c}^{enr}\, d\Omega \qquad \mathbf{C}_{wc} = \int_{\Omega} \mathbf{N}_{p_w}^T \frac{1}{Q_{wc}} \mathbf{N}_{p_c}\, d\Omega$$

$$\mathbf{C}_{\tilde{w}\tilde{c}} = \int_{\Omega^{enr}} \left(\mathbf{N}_{p_w}^{enr}\right)^T \frac{1}{Q_{wc}} \mathbf{N}_{p_c}^{enr}\, d\Omega \qquad \mathbf{C}_{\tilde{w}c} = \int_{\Omega^{enr}} \left(\mathbf{N}_{p_w}^{enr}\right)^T \frac{1}{Q_{wc}} \mathbf{N}_{p_c}\, d\Omega$$

$$\mathbf{C}_{c\tilde{w}} = \int_{\Omega^{enr}} \mathbf{N}_{p_c}^T \frac{1}{Q_{cw}} \mathbf{N}_{p_w}^{enr}\, d\Omega \qquad \mathbf{C}_{cw} = \int_{\Omega} \mathbf{N}_{p_c}^T \frac{1}{Q_{cw}} \mathbf{N}_{p_w}\, d\Omega$$

$$\mathbf{C}_{\tilde{c}\tilde{w}} = \int_{\Omega^{enr}} \left(\mathbf{N}_{p_c}^{enr}\right)^T \frac{1}{Q_{cw}} \mathbf{N}_{p_w}^{enr}\, d\Omega \qquad \mathbf{C}_{\tilde{c}w} = \int_{\Omega^{enr}} \left(\mathbf{N}_{p_c}^{enr}\right)^T \frac{1}{Q_{cw}} \mathbf{N}_{p_w}\, d\Omega \qquad (11.71)$$

$$\mathbf{H}_{w\tilde{w}} = \int_{\Omega^{enr}} \nabla\mathbf{N}_{p_w}^T (k_w + k_g) \nabla\mathbf{N}_{p_w}^{enr}\, d\Omega \qquad \mathbf{H}_{ww} = \int_{\Omega} \nabla\mathbf{N}_{p_w}^T (k_w + k_g) \nabla\mathbf{N}_{p_w}\, d\Omega$$

$$\mathbf{H}_{cc} = \int_{\Omega} \nabla\mathbf{N}_{p_c}^T k_g \nabla\mathbf{N}_{p_c}\, d\Omega \qquad \mathbf{H}_{\tilde{w}\tilde{w}} = \int_{\Omega^{enr}} \left(\nabla\mathbf{N}_{p_w}^{enr}\right)^T (k_w + k_g) \nabla\mathbf{N}_{p_w}^{enr}\, d\Omega$$

$$\mathbf{H}_{\tilde{c}\tilde{c}} = \int_{\Omega^{enr}} \left(\nabla\mathbf{N}_{p_c}^{enr}\right)^T k_g \nabla\mathbf{N}_{p_c}^{enr}\, d\Omega \qquad \mathbf{H}_{c\tilde{c}} = \int_{\Omega^{enr}} \nabla\mathbf{N}_{p_c}^T k_g \nabla\mathbf{N}_{p_c}^{enr}\, d\Omega$$

$$\mathbf{H}_{w\tilde{c}} = \int_{\Omega^{enr}} \nabla\mathbf{N}_{p_w}^T k_g \nabla\mathbf{N}_{p_c}^{enr}\, d\Omega \qquad \mathbf{H}_{wc} = \int_{\Omega} \nabla\mathbf{N}_{p_w}^T k_g \nabla\mathbf{N}_{p_c}\, d\Omega$$

$$\mathbf{H}_{\tilde{w}\tilde{c}} = \int_{\Omega^{enr}} \left(\nabla\mathbf{N}_{p_w}^{enr}\right)^T k_g \nabla\mathbf{N}_{p_c}^{enr}\, d\Omega \qquad \mathbf{H}_{\tilde{w}c} = \int_{\Omega^{enr}} \left(\nabla\mathbf{N}_{p_w}^{enr}\right)^T k_g \nabla\mathbf{N}_{p_c}\, d\Omega$$

Moreover, the external force and flux vectors are specified as

$$\mathbf{F}_u^{\text{ext}} = \int_{\Omega} \mathbf{N}_u^T \rho\, \mathbf{b}\, d\Omega + \int_{\Gamma_t} \mathbf{N}_u^T \bar{\mathbf{t}}\, d\Gamma$$

$$\mathbf{F}_{\tilde{u}}^{\text{ext}} = \int_{\Omega^{enr}} (\mathbf{N}_u^{enr})^T \rho\, \mathbf{b}\, d\Omega + \int_{\Gamma_t^{enr}} (\mathbf{N}_u^{enr})^T \bar{\mathbf{t}}\, d\Gamma$$

$$\mathbf{F}_w^{\text{ext}} = \int_{\Omega} \nabla\mathbf{N}_{p_w}^T (k_w\rho_w + k_g\rho_g)\mathbf{b}\, d\Omega - \int_{\Gamma_{qw}} \mathbf{N}_{p_w}^T \bar{q}_w\, d\Gamma - \int_{\Gamma_{qg}\cap\Gamma_{qw}} \mathbf{N}_{p_w}^T \bar{q}_g\, d\Gamma - \int_{\Gamma_{pc}\cap\Gamma_{qw}} \mathbf{N}_{p_w}^T \dot{\mathbf{w}}_g \cdot \mathbf{n}_\Gamma\, d\Gamma$$

$$\mathbf{F}_{\tilde{w}}^{\text{ext}} = \int_{\Omega^{enr}} \left(\nabla\mathbf{N}_{p_w}^{enr}\right)^T (k_w\rho_w + k_g\rho_g)\mathbf{b}\, d\Omega - \int_{\Gamma_{qw}^{enr}} \left(\mathbf{N}_{p_w}^{enr}\right)^T \bar{q}_w\, d\Gamma$$
$$- \int_{\Gamma_{qg}^{enr}\cap\Gamma_{qw}^{enr}} \left(\mathbf{N}_{p_w}^{enr}\right)^T \bar{q}_g\, d\Gamma - \int_{\Gamma_{pc}^{enr}\cap\Gamma_{qw}^{enr}} \left(\mathbf{N}_{p_w}^{enr}\right)^T \dot{\mathbf{w}}_g \cdot \mathbf{n}_\Gamma\, d\Gamma$$

$$\mathbf{F}_c^{\text{ext}} = \int_{\Omega} \nabla\mathbf{N}_{p_c}^T k_g\rho_g\, \mathbf{b}\, d\Omega - \int_{\Gamma_{qg}} \mathbf{N}_{p_c}^T \bar{q}_g\, d\Gamma$$

$$\mathbf{F}_{\tilde{c}}^{\text{ext}} = \int_{\Omega^{enr}} \left(\nabla\mathbf{N}_{p_c}^{enr}\right)^T k_g\rho_g\, \mathbf{b}\, d\Omega - \int_{\Gamma_{qg}^{enr}} \left(\mathbf{N}_{p_c}^{enr}\right)^T \bar{q}_g\, d\Gamma \qquad (11.72)$$

and the interfacial force and flux vectors are specified as

$$\mathbf{F}_{\tilde{u}}^{\text{int}} = \int_{\Gamma_d} \mathbf{N}_u^T (\mathbf{t}_d - \alpha p\, \mathbf{n}_{\Gamma_d})\, d\Gamma = \int_{\Gamma_d} \mathbf{N}_u^T \mathbf{t}_d\, d\Gamma - \int_{\Gamma_d} \mathbf{N}_u^T \alpha\, \mathbf{n}_{\Gamma_d} p_w\, d\Gamma - \int_{\Gamma_d} \mathbf{N}_u^T \alpha (1 - S_w) \mathbf{n}_{\Gamma_d} p_c\, d\Gamma$$

$$\mathbf{F}_{w}^{\text{int}} = \int_{\Gamma_d} \mathbf{N}_{p_w}^T (q_{wd} + q_{gd})\, d\Gamma = -\int_{\Gamma_d} \mathbf{N}_{p_w}^T 2h \frac{1}{Q_{ww}} \dot{p}_w\, d\Gamma - \int_{\Gamma_d} \mathbf{N}_{p_w}^T 2h \frac{1}{Q_{wc}} \dot{p}_c\, d\Gamma$$

$$- \int_{\Gamma_d} \mathbf{N}_{p_w}^T 2h\, \alpha\, \mathbf{t}_{\Gamma_d} \cdot \langle \nabla \dot{\mathbf{u}} \rangle \cdot \mathbf{t}_{\Gamma_d}\, d\Gamma - \int_{\Gamma_d} \mathbf{N}_{p_w}^T \alpha [\![\dot{\mathbf{u}}]\!] \cdot \mathbf{n}_{\Gamma_d}\, d\Gamma$$

$$- \int_{\Gamma_d} \nabla \mathbf{N}_{p_w}^T \mathbf{t}_{\Gamma_d} 2h (k_{wd} + k_{gd}) \nabla p_w \cdot \mathbf{t}_{\Gamma_d}\, d\Gamma - \int_{\Gamma_d} \nabla \mathbf{N}_{p_w}^T \mathbf{t}_{\Gamma_d} 2h\, k_{gd} \nabla p_c \cdot \mathbf{t}_{\Gamma_d}\, d\Gamma$$

$$- \int_{\Gamma_d} \nabla \mathbf{N}_{p_w}^T \mathbf{t}_{\Gamma_d} 2h (k_{wd}\rho_w + k_{gd}\rho_g) \langle \ddot{\mathbf{u}} \rangle \cdot \mathbf{t}_{\Gamma_d}\, d\Gamma + \int_{\Gamma_d} \nabla \mathbf{N}_{p_w}^T \mathbf{t}_{\Gamma_d} 2h (k_{wd}\rho_w + k_{gd}\rho_g) \mathbf{b} \cdot \mathbf{t}_{\Gamma_d}\, d\Gamma$$

$$\mathbf{F}_{\tilde{w}}^{\text{int}} = \int_{\Gamma_d} \left(\mathbf{N}_{p_w}^{enr}\right)^T (q_{wd} + q_{gd})\, d\Gamma = -\int_{\Gamma_d} \left(\mathbf{N}_{p_w}^{enr}\right)^T 2h \frac{1}{Q_{ww}} \dot{p}_w\, d\Gamma - \int_{\Gamma_d} \left(\mathbf{N}_{p_w}^{enr}\right)^T 2h \frac{1}{Q_{wc}} \dot{p}_c\, d\Gamma$$

$$- \int_{\Gamma_d} \left(\mathbf{N}_{p_w}^{enr}\right)^T 2h\, \alpha\, \mathbf{t}_{\Gamma_d} \cdot \langle \nabla \dot{\mathbf{u}} \rangle \cdot \mathbf{t}_{\Gamma_d}\, d\Gamma - \int_{\Gamma_d} \left(\mathbf{N}_{p_w}^{enr}\right)^T \alpha [\![\dot{\mathbf{u}}]\!] \cdot \mathbf{n}_{\Gamma_d}\, d\Gamma$$

$$- \int_{\Gamma_d} \left(\nabla \mathbf{N}_{p_w}^{enr}\right)^T \mathbf{t}_{\Gamma_d} 2h (k_{wd} + k_{gd}) \nabla p_w \cdot \mathbf{t}_{\Gamma_d}\, d\Gamma - \int_{\Gamma_d} \left(\nabla \mathbf{N}_{p_w}^{enr}\right)^T \mathbf{t}_{\Gamma_d} 2h\, k_{gd} \nabla p_c \cdot \mathbf{t}_{\Gamma_d}\, d\Gamma$$

$$- \int_{\Gamma_d} \left(\nabla \mathbf{N}_{p_w}^{enr}\right)^T \mathbf{t}_{\Gamma_d} 2h (k_{wd}\rho_w + k_{gd}\rho_g) \langle \ddot{\mathbf{u}} \rangle \cdot \mathbf{t}_{\Gamma_d}\, d\Gamma + \int_{\Gamma_d} \left(\nabla \mathbf{N}_{p_w}^{enr}\right)^T \mathbf{t}_{\Gamma_d} 2h (k_{wd}\rho_w + k_{gd}\rho_g) \mathbf{b} \cdot \mathbf{t}_{\Gamma_d}\, d\Gamma$$

$$\mathbf{F}_{c}^{\text{int}} = \int_{\Gamma_d} \mathbf{N}_{p_c}^T q_{gd}\, d\Gamma = -\int_{\Gamma_d} \mathbf{N}_{p_c}^T 2h \frac{1}{Q_{cw}} \dot{p}_w\, d\Gamma - \int_{\Gamma_d} \mathbf{N}_{p_c}^T 2h \frac{1}{Q_{cc}} \dot{p}_c\, d\Gamma$$

$$- \int_{\Gamma_d} \mathbf{N}_{p_c}^T 2h\, \alpha (1 - S_w)\, \mathbf{t}_{\Gamma_d} \cdot \langle \nabla \dot{\mathbf{u}} \rangle \cdot \mathbf{t}_{\Gamma_d}\, d\Gamma - \int_{\Gamma_d} \mathbf{N}_{p_c}^T \alpha (1 - S_w) [\![\dot{\mathbf{u}}]\!] \cdot \mathbf{n}_{\Gamma_d}\, d\Gamma$$

$$- \int_{\Gamma_d} \nabla \mathbf{N}_{p_c}^T \mathbf{t}_{\Gamma_d} 2h\, k_{gd} \nabla p_w \cdot \mathbf{t}_{\Gamma_d}\, d\Gamma - \int_{\Gamma_d} \nabla \mathbf{N}_{p_c}^T \mathbf{t}_{\Gamma_d} 2h\, k_{gd} \nabla p_c \cdot \mathbf{t}_{\Gamma_d}\, d\Gamma$$

$$- \int_{\Gamma_d} \nabla \mathbf{N}_{p_c}^T \mathbf{t}_{\Gamma_d} 2h\, k_{gd}\rho_g \langle \ddot{\mathbf{u}} \rangle \cdot \mathbf{t}_{\Gamma_d}\, d\Gamma + \int_{\Gamma_d} \nabla \mathbf{N}_{p_c}^T \mathbf{t}_{\Gamma_d} 2h\, k_{gd}\rho_g \mathbf{b} \cdot \mathbf{t}_{\Gamma_d}\, d\Gamma$$

$$\mathbf{F}_{\tilde{c}}^{\text{int}} = \int_{\Gamma_d} \left(\mathbf{N}_{p_c}^{enr}\right)^T q_{gd}\, d\Gamma = -\int_{\Gamma_d} \left(\mathbf{N}_{p_c}^{enr}\right)^T 2h \frac{1}{Q_{cw}} \dot{p}_w\, d\Gamma - \int_{\Gamma_d} \left(\mathbf{N}_{p_c}^{enr}\right)^T 2h \frac{1}{Q_{cc}} \dot{p}_c\, d\Gamma$$

$$- \int_{\Gamma_d} \left(\mathbf{N}_{p_c}^{enr}\right)^T 2h\, \alpha (1 - S_w)\, \mathbf{t}_{\Gamma_d} \cdot \langle \nabla \dot{\mathbf{u}} \rangle \cdot \mathbf{t}_{\Gamma_d}\, d\Gamma - \int_{\Gamma_d} \left(\mathbf{N}_{p_c}^{enr}\right)^T \alpha (1 - S_w) [\![\dot{\mathbf{u}}]\!] \cdot \mathbf{n}_{\Gamma_d}\, d\Gamma$$

$$- \int_{\Gamma_d} \left(\nabla \mathbf{N}_{p_c}^{enr}\right)^T \mathbf{t}_{\Gamma_d} 2h\, k_{gd} \nabla p_w \cdot \mathbf{t}_{\Gamma_d}\, d\Gamma - \int_{\Gamma_d} \left(\nabla \mathbf{N}_{p_c}^{enr}\right)^T \mathbf{t}_{\Gamma_d} 2h\, k_{gd} \nabla p_c \cdot \mathbf{t}_{\Gamma_d}\, d\Gamma$$

$$- \int_{\Gamma_d} \left(\nabla \mathbf{N}_{p_c}^{enr}\right)^T \mathbf{t}_{\Gamma_d} 2h\, k_{gd}\rho_g \langle \ddot{\mathbf{u}} \rangle \cdot \mathbf{t}_{\Gamma_d}\, d\Gamma + \int_{\Gamma_d} \left(\nabla \mathbf{N}_{p_c}^{enr}\right)^T \mathbf{t}_{\Gamma_d} 2h\, k_{gd}\rho_g \mathbf{b} \cdot \mathbf{t}_{\Gamma_d}\, d\Gamma \tag{11.73}$$

The system of discretized governing equations (11.70) can now be discretized in the time domain following the line of the well-known Newmark scheme, in which it is supposed that the system of equations is satisfied at each discrete time step. Thus, the problem is solved at a series of sequential discrete time steps. To advance the solution in time, the link between the successive values of the unknown field variables at time t_{n+1} and the known field variables at time t_n is established by applying the minimum order of the generalized Newmark scheme considering the highest order of the time derivatives in the differential equations; that is, through the application of GN22 and GN11 to the displacement and pressure variables, respectively, as

$$\begin{aligned}
&\ddot{\mathbf{U}}_{n+1} = a_0\left(\mathbf{U}_{n+1}-\mathbf{U}_n\right)-a_2\,\dot{\mathbf{U}}_n-a_4\,\ddot{\mathbf{U}}_n \\
&\dot{\mathbf{U}}_{n+1} = a_1\left(\mathbf{U}_{n+1}-\mathbf{U}_n\right)-a_3\,\dot{\mathbf{U}}_n-a_5\,\ddot{\mathbf{U}}_n \\
&\ddot{\tilde{\mathbf{U}}}_{n+1} = a_0\left(\tilde{\mathbf{U}}_{n+1}-\tilde{\mathbf{U}}_n\right)-a_2\,\dot{\tilde{\mathbf{U}}}_n-a_4\,\ddot{\tilde{\mathbf{U}}}_n \\
&\dot{\tilde{\mathbf{U}}}_{n+1} = a_1\left(\tilde{\mathbf{U}}_{n+1}-\tilde{\mathbf{U}}_n\right)-a_3\,\dot{\tilde{\mathbf{U}}}_n-a_5\,\ddot{\tilde{\mathbf{U}}}_n \\
&\dot{\mathbf{P}}_{\alpha_{n+1}} = a'_1\left(\mathbf{P}_{\alpha_{n+1}}-\mathbf{P}_{\alpha_n}\right)-a'_3\,\dot{\mathbf{P}}_{\alpha_n} \qquad \alpha = w, c \\
&\dot{\tilde{\mathbf{P}}}_{\alpha_{n+1}} = a'_1\left(\tilde{\mathbf{P}}_{\alpha_{n+1}}-\tilde{\mathbf{P}}_{\alpha_n}\right)-a'_3\,\dot{\tilde{\mathbf{P}}}_{\alpha_n} \qquad \alpha = w, c
\end{aligned} \tag{11.74}$$

where a_0, a_1, a_2, a_3, a_4, a_5, a'_1, and a'_3 are given in Section 11.4.2, and β, γ and θ are the Newmark parameters.

11.8 Solution Procedure for Fully Coupled Nonlinear Equations

In order to solve the system of coupled nonlinear equations at each time step, the Newton–Raphson scheme is implemented to linearize the whole system of nonlinear equations. To obtain the nodal unknown values at time t_{n+1}, the discrete system of equations is solved at the given time step applying the Newton–Raphson iterative algorithm to its residual form, $\boldsymbol{\Psi}_{n+1} = \mathbf{0}$. By expanding the residual equations with the first-order truncated Taylor series, the following linear approximation for the nonlinear system can be obtained as

$$\boldsymbol{\Psi}^{i+1}_{n+1} = \begin{Bmatrix} \boldsymbol{\Psi}^{i+1}_{u_{n+1}} \\ \boldsymbol{\Psi}^{i+1}_{\tilde{u}_{n+1}} \\ \boldsymbol{\Psi}^{i+1}_{w_{n+1}} \\ \boldsymbol{\Psi}^{i+1}_{\tilde{w}_{n+1}} \\ \boldsymbol{\Psi}^{i+1}_{c_{n+1}} \\ \boldsymbol{\Psi}^{i+1}_{\tilde{c}_{n+1}} \end{Bmatrix} = \begin{Bmatrix} \boldsymbol{\Psi}^{i}_{u_{n+1}} \\ \boldsymbol{\Psi}^{i}_{\tilde{u}_{n+1}} \\ \boldsymbol{\Psi}^{i}_{w_{n+1}} \\ \boldsymbol{\Psi}^{i}_{\tilde{w}_{n+1}} \\ \boldsymbol{\Psi}^{i}_{c_{n+1}} \\ \boldsymbol{\Psi}^{i}_{\tilde{c}_{n+1}} \end{Bmatrix} + \mathbf{J} \begin{Bmatrix} d\mathbf{U}^{i}_{n} \\ d\tilde{\mathbf{U}}^{i}_{n} \\ d\mathbf{P}^{i}_{w_n} \\ d\tilde{\mathbf{P}}^{i}_{w_n} \\ d\mathbf{P}^{i}_{c_n} \\ d\tilde{\mathbf{P}}^{i}_{c_n} \end{Bmatrix} = \mathbf{0} \tag{11.75}$$

where $\mathbf{J}$ is the Jacobian matrix. The solution of the linearized system of equations (11.75) results in the increment of the standard and enriched nodal DOF, and the corresponding nodal unknowns are subsequently attained by

$$\mathbf{X}^{i+1}_{n+1} = \mathbf{X}^{i}_{n+1} + d\mathbf{X}^{i}_{n} \tag{11.76}$$

where $\mathbf{X}$ represents the vector of nodal unknowns, $\mathbf{X}^T = \left[\mathbf{U}^T \ \tilde{\mathbf{U}}^T \ \mathbf{P}^T_w \ \tilde{\mathbf{P}}^T_w \ \mathbf{P}^T_c \ \tilde{\mathbf{P}}^T_c\right]$. The Jacobian matrix is defined as $\partial\boldsymbol{\Psi}^{i}_{n+1}/\partial\mathbf{X}^{i}_{n+1}$, which can be approximated as follows

$$\mathbf{J}=\left[\begin{array}{cccccc}
a_0\mathbf{M}_{uu}+\int_{\Omega}\mathbf{B}^T\mathbf{D}\,\mathbf{B}\,d\Omega & a_0\mathbf{M}_{u\tilde{u}}+\int_{\Omega^{enr}}\mathbf{B}^T\mathbf{D}\,\mathbf{B}^{enr}\,d\Omega & -\mathbf{Q}_{uw} & -\mathbf{Q}_{u\tilde{w}} & -\mathbf{Q}_{uc} & -\mathbf{Q}_{u\tilde{c}} \\
a_0\mathbf{M}_{u\tilde{u}}^T+\int_{\Omega^{enr}}(\mathbf{B}^{enr})^T\mathbf{D}\,\mathbf{B}\,d\Omega & a_0\mathbf{M}_{\tilde{u}\tilde{u}}+\int_{\Omega^{enr}}(\mathbf{B}^{enr})^T\mathbf{D}\,\mathbf{B}^{enr}\,d\Omega+\frac{\partial\mathbf{F}_{\tilde{u}}^{\text{int}}}{\partial\tilde{\mathbf{U}}} & -\mathbf{Q}_{\tilde{u}w}+\frac{\partial\mathbf{F}_{\tilde{u}}^{\text{int}}}{\partial\mathbf{P}_w} & -\mathbf{Q}_{\tilde{u}\tilde{w}}+\frac{\partial\mathbf{F}_{\tilde{u}}^{\text{int}}}{\partial\tilde{\mathbf{P}}_w} & -\mathbf{Q}_{\tilde{u}c}+\frac{\partial\mathbf{F}_{\tilde{u}}^{\text{int}}}{\partial\mathbf{P}_c} & -\mathbf{Q}_{\tilde{u}\tilde{c}}+\frac{\partial\mathbf{F}_{\tilde{u}}^{\text{int}}}{\partial\tilde{\mathbf{P}}_c} \\
a_0\mathbf{M}_{wu}+a_1\mathbf{Q}_{uw}^T-\frac{\partial\mathbf{F}_{w}^{\text{int}}}{\partial\mathbf{U}} & a_0\mathbf{M}_{w\tilde{u}}+a_1\mathbf{Q}_{\tilde{u}w}^T-\frac{\partial\mathbf{F}_{w}^{\text{int}}}{\partial\tilde{\mathbf{U}}} & a_1'\mathbf{C}_{ww}+\mathbf{H}_{ww}-\frac{\partial\mathbf{F}_{w}^{\text{int}}}{\partial\mathbf{P}_w} & a_1'\mathbf{C}_{w\tilde{w}}+\mathbf{H}_{w\tilde{w}}-\frac{\partial\mathbf{F}_{w}^{\text{int}}}{\partial\tilde{\mathbf{P}}_w} & a_1'\mathbf{C}_{wc}+\mathbf{H}_{wc}-\frac{\partial\mathbf{F}_{w}^{\text{int}}}{\partial\mathbf{P}_c} & a_1'\mathbf{C}_{w\tilde{c}}+\mathbf{H}_{w\tilde{c}}-\frac{\partial\mathbf{F}_{w}^{\text{int}}}{\partial\tilde{\mathbf{P}}_c} \\
a_0\mathbf{M}_{\tilde{w}u}+a_1\mathbf{Q}_{u\tilde{w}}^T-\frac{\partial\mathbf{F}_{\tilde{w}}^{\text{int}}}{\partial\mathbf{U}} & a_0\mathbf{M}_{\tilde{w}\tilde{u}}+a_1\mathbf{Q}_{\tilde{u}\tilde{w}}^T-\frac{\partial\mathbf{F}_{\tilde{w}}^{\text{int}}}{\partial\tilde{\mathbf{U}}} & a_1'\mathbf{C}_{w\tilde{w}}^T+\mathbf{H}_{w\tilde{w}}^T-\frac{\partial\mathbf{F}_{\tilde{w}}^{\text{int}}}{\partial\mathbf{P}_w} & a_1'\mathbf{C}_{\tilde{w}\tilde{w}}+\mathbf{H}_{\tilde{w}\tilde{w}}-\frac{\partial\mathbf{F}_{\tilde{w}}^{\text{int}}}{\partial\tilde{\mathbf{P}}_w} & a_1'\mathbf{C}_{\tilde{w}c}+\mathbf{H}_{\tilde{w}c}-\frac{\partial\mathbf{F}_{\tilde{w}}^{\text{int}}}{\partial\mathbf{P}_c} & a_1'\mathbf{C}_{\tilde{w}\tilde{c}}+\mathbf{H}_{\tilde{w}\tilde{c}}-\frac{\partial\mathbf{F}_{\tilde{w}}^{\text{int}}}{\partial\tilde{\mathbf{P}}_c} \\
a_0\mathbf{M}_{cu}+a_1\mathbf{Q}_{uc}^T-\frac{\partial\mathbf{F}_{c}^{\text{int}}}{\partial\mathbf{U}} & a_0\mathbf{M}_{c\tilde{u}}+a_1\mathbf{Q}_{\tilde{u}c}^T-\frac{\partial\mathbf{F}_{c}^{\text{int}}}{\partial\tilde{\mathbf{U}}} & a_1'\mathbf{C}_{cw}+\mathbf{H}_{wc}^T-\frac{\partial\mathbf{F}_{c}^{\text{int}}}{\partial\mathbf{P}_w} & a_1'\mathbf{C}_{c\tilde{w}}+\mathbf{H}_{\tilde{w}c}^T-\frac{\partial\mathbf{F}_{c}^{\text{int}}}{\partial\tilde{\mathbf{P}}_w} & a_1'\mathbf{C}_{cc}+\mathbf{H}_{cc}-\frac{\partial\mathbf{F}_{c}^{\text{int}}}{\partial\mathbf{P}_c} & a_1'\mathbf{C}_{c\tilde{c}}+\mathbf{H}_{c\tilde{c}}-\frac{\partial\mathbf{F}_{c}^{\text{int}}}{\partial\tilde{\mathbf{P}}_c} \\
a_0\mathbf{M}_{\tilde{c}u}+a_1\mathbf{Q}_{u\tilde{c}}^T-\frac{\partial\mathbf{F}_{\tilde{c}}^{\text{int}}}{\partial\mathbf{U}} & a_0\mathbf{M}_{\tilde{c}\tilde{u}}+a_1\mathbf{Q}_{\tilde{u}\tilde{c}}^T-\frac{\partial\mathbf{F}_{\tilde{c}}^{\text{int}}}{\partial\tilde{\mathbf{U}}} & a_1'\mathbf{C}_{\tilde{c}w}+\mathbf{H}_{w\tilde{c}}^T-\frac{\partial\mathbf{F}_{\tilde{c}}^{\text{int}}}{\partial\mathbf{P}_w} & a_1'\mathbf{C}_{\tilde{c}\tilde{w}}+\mathbf{H}_{\tilde{w}\tilde{c}}^T-\frac{\partial\mathbf{F}_{\tilde{c}}^{\text{int}}}{\partial\tilde{\mathbf{P}}_w} & a_1'\mathbf{C}_{c\tilde{c}}^T+\mathbf{H}_{c\tilde{c}}^T-\frac{\partial\mathbf{F}_{\tilde{c}}^{\text{int}}}{\partial\mathbf{P}_c} & a_1'\mathbf{C}_{\tilde{c}\tilde{c}}+\mathbf{H}_{\tilde{c}\tilde{c}}-\frac{\partial\mathbf{F}_{\tilde{c}}^{\text{int}}}{\partial\tilde{\mathbf{P}}_c}
\end{array}\right] \tag{11.77}$$

where the index $(i, n+1)$ displays that the Jacobian matrix $\mathbf{J}$ as well as the residual vector $\boldsymbol{\Psi}$ is updated at each iteration. As a result, a sequence of linearized system of equations is solved to yield a solution satisfying the residual equation within each time step. This iterative process continues until the iteration convergence is achieved.

The Jacobian matrix (11.77) results in a non-symmetric matrix notwithstanding the choice of the constitutive models applied in the continuum and at the discontinuity. It is worthwhile rendering the Jacobian matrix symmetric from the computational point of view so as to save the core storage and computational cost. For this purpose, the first and second rows of the Jacobian matrix are multiplied by the scalar factor $-a_1$ to symmetrize it relative to the coefficients. Furthermore, since the contribution of the dynamic seepage forcing terms to the solution is negligible compared to the other terms, the inertial matrices $\mathbf{M}_{wu}$, $\mathbf{M}_{w\tilde{u}}$, $\mathbf{M}_{\tilde{w}u}$, $\mathbf{M}_{\tilde{w}\tilde{u}}$, $\mathbf{M}_{cu}$, $\mathbf{M}_{c\tilde{u}}$, $\mathbf{M}_{\tilde{c}u}$, and $\mathbf{M}_{\tilde{c}\tilde{u}}$, are omitted from the Jacobian matrix to obtain a symmetric matrix. In addition, by comparing the expressions of the compressibility coefficients $1/Q_{wc}$ and $1/Q_{cw}$, it seems that the compressibility matrices $\mathbf{C}_{wc}$, $\mathbf{C}_{w\tilde{c}}$, $\mathbf{C}_{\tilde{w}c}$, and $\mathbf{C}_{\tilde{w}\tilde{c}}$ can be replaced by $\mathbf{C}_{cw}^T$, $\mathbf{C}_{\tilde{c}w}^T$, $\mathbf{C}_{c\tilde{w}}^T$, and $\mathbf{C}_{\tilde{c}\tilde{w}}^T$, respectively. Concerning the terms that involve the partial derivatives of the interfacial flux vectors, $\partial \mathbf{F}_w^{\text{int}}/\partial \mathbf{U}$, $\partial \mathbf{F}_{\tilde{w}}^{\text{int}}/\partial \mathbf{U}$, $\partial \mathbf{F}_c^{\text{int}}/\partial \mathbf{U}$, and $\partial \mathbf{F}_{\tilde{c}}^{\text{int}}/\partial \mathbf{U}$, are eliminated from the Jacobian matrix to prevent the non-symmetric case. As a matter of fact, the terms $\partial \mathbf{F}_w^{\text{int}}/\partial \tilde{\mathbf{U}}$, $\partial \mathbf{F}_{\tilde{w}}^{\text{int}}/\partial \tilde{\mathbf{U}}$, $\partial \mathbf{F}_c^{\text{int}}/\partial \tilde{\mathbf{U}}$, and $\partial \mathbf{F}_{\tilde{c}}^{\text{int}}/\partial \tilde{\mathbf{U}}$ are equal to $a_1\left(\partial \mathbf{F}_{\tilde{u}}^{\text{int}}/\partial \mathbf{P}_w\right)^T$, $a_1\left(\partial \mathbf{F}_{\tilde{u}}^{\text{int}}/\partial \tilde{\mathbf{P}}_w\right)^T$, $a_1\left(\partial \mathbf{F}_{\tilde{u}}^{\text{int}}/\partial \mathbf{P}_c\right)^T$, and $a_1\left(\partial \mathbf{F}_{\tilde{u}}^{\text{int}}/\partial \tilde{\mathbf{P}}_c\right)^T$, respectively, which retains symmetry in the Jacobian matrix. Again, for the same reason as mentioned for the compressibility matrices, $\partial \mathbf{F}_w^{\text{int}}/\partial \mathbf{P}_c$, $\partial \mathbf{F}_w^{\text{int}}/\partial \tilde{\mathbf{P}}_c$, $\partial \mathbf{F}_{\tilde{w}}^{\text{int}}/\partial \mathbf{P}_c$, and $\partial \mathbf{F}_{\tilde{w}}^{\text{int}}/\partial \tilde{\mathbf{P}}_c$ are substituted by $\left(\partial \mathbf{F}_c^{\text{int}}/\partial \mathbf{P}_w\right)^T$, $\left(\partial \mathbf{F}_{\tilde{c}}^{\text{int}}/\partial \mathbf{P}_w\right)^T$, $\left(\partial \mathbf{F}_c^{\text{int}}/\partial \tilde{\mathbf{P}}_w\right)^T$, and $\left(\partial \mathbf{F}_{\tilde{c}}^{\text{int}}/\partial \tilde{\mathbf{P}}_w\right)^T$, respectively, to restore the Jacobian matrix symmetry. Applying the previously mentioned simplifications results in a symmetric approximation of the Jacobian matrix provided that the tangential constitutive matrices of the cohesive crack and the medium surrounding the crack are, themselves, symmetric. Thus, the simplified Jacobian matrix (11.77) can be written as

$$
\mathbf{J} = \begin{bmatrix}
-a_1\left(a_0\mathbf{M}_{uu} + \int_\Omega \mathbf{B}^T\mathbf{D}\,\mathbf{B}\,d\Omega\right) & -a_1\left(a_0\mathbf{M}_{u\tilde{u}} + \int_{\Omega^{enr}} \mathbf{B}^T\mathbf{D}\,\mathbf{B}^{enr}\,d\Omega\right) & a_1\mathbf{Q}_{uw} & a_1\mathbf{Q}_{u\tilde{w}} & a_1\mathbf{Q}_{uc} & a_1\mathbf{Q}_{u\tilde{c}} \\
-a_1\left(a_0\mathbf{M}_{u\tilde{u}}^T + \int_{\Omega^{enr}} (\mathbf{B}^{enr})^T\mathbf{D}\,\mathbf{B}\,d\Omega\right) & -a_1\left(a_0\mathbf{M}_{\tilde{u}\tilde{u}} + \int_{\Omega^{enr}} (\mathbf{B}^{enr})^T\mathbf{D}\,\mathbf{B}^{enr}\,d\Omega + \dfrac{\partial \mathbf{F}_{\tilde{u}}^{\text{int}}}{\partial \tilde{\mathbf{U}}}\right) & a_1\mathbf{Q}_{\tilde{u}w} - a_1\dfrac{\partial \mathbf{F}_{\tilde{u}}^{\text{int}}}{\partial \mathbf{P}_w} & a_1\mathbf{Q}_{\tilde{u}\tilde{w}} - a_1\dfrac{\partial \mathbf{F}_{\tilde{u}}^{\text{int}}}{\partial \tilde{\mathbf{P}}_w} & a_1\mathbf{Q}_{\tilde{u}c} - a_1\dfrac{\partial \mathbf{F}_{\tilde{u}}^{\text{int}}}{\partial \mathbf{P}_c} & a_1\mathbf{Q}_{\tilde{u}\tilde{c}} - a_1\dfrac{\partial \mathbf{F}_{\tilde{u}}^{\text{int}}}{\partial \tilde{\mathbf{P}}_c} \\
a_1\mathbf{Q}_{uw}^T & a_1\mathbf{Q}_{\tilde{u}w}^T - a_1\left(\dfrac{\partial \mathbf{F}_{\tilde{u}}^{\text{int}}}{\partial \mathbf{P}_w}\right)^T & a_1'\mathbf{C}_{ww} + \mathbf{H}_{ww} - \dfrac{\partial \mathbf{F}_w^{\text{int}}}{\partial \mathbf{P}_w} & a_1'\mathbf{C}_{w\tilde{w}} + \mathbf{H}_{w\tilde{w}} - \dfrac{\partial \mathbf{F}_w^{\text{int}}}{\partial \tilde{\mathbf{P}}_w} & a_1'\mathbf{C}_{cw}^T + \mathbf{H}_{wc} - \left(\dfrac{\partial \mathbf{F}_c^{\text{int}}}{\partial \mathbf{P}_w}\right)^T & a_1'\mathbf{C}_{\tilde{c}w}^T + \mathbf{H}_{w\tilde{c}} - \left(\dfrac{\partial \mathbf{F}_{\tilde{c}}^{\text{int}}}{\partial \mathbf{P}_w}\right)^T \\
a_1\mathbf{Q}_{u\tilde{w}}^T & a_1\mathbf{Q}_{\tilde{u}\tilde{w}}^T - a_1\left(\dfrac{\partial \mathbf{F}_{\tilde{u}}^{\text{int}}}{\partial \tilde{\mathbf{P}}_w}\right)^T & a_1'\mathbf{C}_{w\tilde{w}}^T + \mathbf{H}_{w\tilde{w}}^T - \left(\dfrac{\partial \mathbf{F}_w^{\text{int}}}{\partial \tilde{\mathbf{P}}_w}\right)^T & a_1'\mathbf{C}_{\tilde{w}\tilde{w}} + \mathbf{H}_{\tilde{w}\tilde{w}} - \dfrac{\partial \mathbf{F}_{\tilde{w}}^{\text{int}}}{\partial \tilde{\mathbf{P}}_w} & a_1'\mathbf{C}_{c\tilde{w}}^T + \mathbf{H}_{\tilde{w}c} - \left(\dfrac{\partial \mathbf{F}_c^{\text{int}}}{\partial \tilde{\mathbf{P}}_w}\right)^T & a_1'\mathbf{C}_{\tilde{c}\tilde{w}}^T + \mathbf{H}_{\tilde{w}\tilde{c}} - \left(\dfrac{\partial \mathbf{F}_{\tilde{c}}^{\text{int}}}{\partial \tilde{\mathbf{P}}_w}\right)^T \\
a_1\mathbf{Q}_{uc}^T & a_1\mathbf{Q}_{\tilde{u}c}^T - a_1\left(\dfrac{\partial \mathbf{F}_{\tilde{u}}^{\text{int}}}{\partial \mathbf{P}_c}\right)^T & a_1'\mathbf{C}_{cw} + \mathbf{H}_{wc}^T - \dfrac{\partial \mathbf{F}_c^{\text{int}}}{\partial \mathbf{P}_w} & a_1'\mathbf{C}_{c\tilde{w}} + \mathbf{H}_{\tilde{w}c}^T - \dfrac{\partial \mathbf{F}_c^{\text{int}}}{\partial \tilde{\mathbf{P}}_w} & a_1'\mathbf{C}_{cc} + \mathbf{H}_{cc} - \dfrac{\partial \mathbf{F}_c^{\text{int}}}{\partial \mathbf{P}_c} & a_1'\mathbf{C}_{c\tilde{c}} + \mathbf{H}_{c\tilde{c}} - \dfrac{\partial \mathbf{F}_c^{\text{int}}}{\partial \tilde{\mathbf{P}}_c} \\
a_1\mathbf{Q}_{u\tilde{c}}^T & a_1\mathbf{Q}_{\tilde{u}\tilde{c}}^T - a_1\left(\dfrac{\partial \mathbf{F}_{\tilde{u}}^{\text{int}}}{\partial \tilde{\mathbf{P}}_c}\right)^T & a_1'\mathbf{C}_{\tilde{c}w} + \mathbf{H}_{w\tilde{c}}^T - \dfrac{\partial \mathbf{F}_{\tilde{c}}^{\text{int}}}{\partial \mathbf{P}_w} & a_1'\mathbf{C}_{\tilde{c}\tilde{w}} + \mathbf{H}_{\tilde{w}\tilde{c}}^T - \dfrac{\partial \mathbf{F}_{\tilde{c}}^{\text{int}}}{\partial \tilde{\mathbf{P}}_w} & a_1'\mathbf{C}_{c\tilde{c}}^T + \mathbf{H}_{c\tilde{c}}^T - \left(\dfrac{\partial \mathbf{F}_c^{\text{int}}}{\partial \tilde{\mathbf{P}}_c}\right)^T & a_1'\mathbf{C}_{\tilde{c}\tilde{c}} + \mathbf{H}_{\tilde{c}\tilde{c}} - \dfrac{\partial \mathbf{F}_{\tilde{c}}^{\text{int}}}{\partial \tilde{\mathbf{P}}_c}
\end{bmatrix}
\tag{11.78}
$$

in which the partial derivatives of the interfacial force and flux vectors are given by

$$
\begin{aligned}
&\frac{\partial \mathbf{F}_{\tilde{u}}^{\text{int}}}{\partial \tilde{\mathbf{U}}} = \int_{\Gamma_d} \mathbf{N}_u^T \, \mathbf{T} \, \mathbf{N}_u \, d\Gamma \\
&\frac{\partial \mathbf{F}_{\tilde{u}}^{\text{int}}}{\partial \mathbf{P}_w} = -\int_{\Gamma_d} \mathbf{N}_u^T \alpha \, \mathbf{n}_{\Gamma_d} \mathbf{N}_{p_w} \, d\Gamma \\
&\frac{\partial \mathbf{F}_{\tilde{u}}^{\text{int}}}{\partial \tilde{\mathbf{P}}_w} = -\int_{\Gamma_d} \mathbf{N}_u^T \alpha \, \mathbf{n}_{\Gamma_d} \mathbf{N}_{p_w}^{enr} \, d\Gamma \\
&\frac{\partial \mathbf{F}_{\tilde{u}}^{\text{int}}}{\partial \mathbf{P}_c} = -\int_{\Gamma_d} \mathbf{N}_u^T \alpha (1-S_w) \mathbf{n}_{\Gamma_d} \mathbf{N}_{p_c} \, d\Gamma \\
&\frac{\partial \mathbf{F}_{\tilde{u}}^{\text{int}}}{\partial \tilde{\mathbf{P}}_c} = -\int_{\Gamma_d} \mathbf{N}_u^T \alpha (1-S_w) \mathbf{n}_{\Gamma_d} \mathbf{N}_{p_c}^{enr} \, d\Gamma \\
&\frac{\partial \mathbf{F}_{w}^{\text{int}}}{\partial \mathbf{P}_w} = -a_1' \int_{\Gamma_d} \mathbf{N}_{p_w}^T 2h \frac{1}{Q_{ww}} \mathbf{N}_{p_w} \, d\Gamma - \int_{\Gamma_d} \nabla \mathbf{N}_{p_w}^T \mathbf{t}_{\Gamma_d} 2h \left(k_{wd} + k_{gd}\right) \mathbf{t}_{\Gamma_d}^T \nabla \mathbf{N}_{p_w} \, d\Gamma \\
&\frac{\partial \mathbf{F}_{w}^{\text{int}}}{\partial \tilde{\mathbf{P}}_w} = \left(\frac{\partial \mathbf{F}_{\tilde{w}}^{\text{int}}}{\partial \mathbf{P}_w}\right)^T = -a_1' \int_{\Gamma_d} \mathbf{N}_{p_w}^T 2h \frac{1}{Q_{ww}} \mathbf{N}_{p_w}^{enr} \, d\Gamma - \int_{\Gamma_d} \nabla \mathbf{N}_{p_w}^T \mathbf{t}_{\Gamma_d} 2h \left(k_{wd} + k_{gd}\right) \mathbf{t}_{\Gamma_d}^T \nabla \mathbf{N}_{p_w}^{enr} \, d\Gamma \\
&\frac{\partial \mathbf{F}_{\tilde{w}}^{\text{int}}}{\partial \tilde{\mathbf{P}}_w} = -a_1' \int_{\Gamma_d} \left(\mathbf{N}_{p_w}^{enr}\right)^T 2h \frac{1}{Q_{ww}} \mathbf{N}_{p_w}^{enr} \, d\Gamma - \int_{\Gamma_d} \left(\nabla \mathbf{N}_{p_w}^{enr}\right)^T \mathbf{t}_{\Gamma_d} 2h \left(k_{wd} + k_{gd}\right) \mathbf{t}_{\Gamma_d}^T \nabla \mathbf{N}_{p_w}^{enr} \, d\Gamma \qquad (11.79) \\
&\frac{\partial \mathbf{F}_{c}^{\text{int}}}{\partial \mathbf{P}_c} = -a_1' \int_{\Gamma_d} \mathbf{N}_{p_c}^T 2h \frac{1}{Q_{cc}} \mathbf{N}_{p_c} \, d\Gamma - \int_{\Gamma_d} \nabla \mathbf{N}_{p_c}^T \mathbf{t}_{\Gamma_d} 2h \, k_{gd} \, \mathbf{t}_{\Gamma_d}^T \nabla \mathbf{N}_{p_c} \, d\Gamma \\
&\frac{\partial \mathbf{F}_{c}^{\text{int}}}{\partial \tilde{\mathbf{P}}_c} = \left(\frac{\partial \mathbf{F}_{\tilde{c}}^{\text{int}}}{\partial \mathbf{P}_c}\right)^T = -a_1' \int_{\Gamma_d} \mathbf{N}_{p_c}^T 2h \frac{1}{Q_{cc}} \mathbf{N}_{p_c}^{enr} \, d\Gamma - \int_{\Gamma_d} \nabla \mathbf{N}_{p_c}^T \mathbf{t}_{\Gamma_d} 2h \, k_{gd} \, \mathbf{t}_{\Gamma_d}^T \nabla \mathbf{N}_{p_c}^{enr} \, d\Gamma \\
&\frac{\partial \mathbf{F}_{\tilde{c}}^{\text{int}}}{\partial \tilde{\mathbf{P}}_c} = -a_1' \int_{\Gamma_d} \left(\mathbf{N}_{p_c}^{enr}\right)^T 2h \frac{1}{Q_{cc}} \mathbf{N}_{p_c}^{enr} \, d\Gamma - \int_{\Gamma_d} \left(\nabla \mathbf{N}_{p_c}^{enr}\right)^T \mathbf{t}_{\Gamma_d} 2h \, k_{gd} \, \mathbf{t}_{\Gamma_d}^T \nabla \mathbf{N}_{p_c}^{enr} \, d\Gamma \\
&\frac{\partial \mathbf{F}_{c}^{\text{int}}}{\partial \mathbf{P}_w} = -a_1' \int_{\Gamma_d} \mathbf{N}_{p_c}^T 2h \frac{1}{Q_{cw}} \mathbf{N}_{p_w} \, d\Gamma - \int_{\Gamma_d} \nabla \mathbf{N}_{p_c}^T \mathbf{t}_{\Gamma_d} 2h \, k_{gd} \, \mathbf{t}_{\Gamma_d}^T \nabla \mathbf{N}_{p_w} \, d\Gamma \\
&\frac{\partial \mathbf{F}_{c}^{\text{int}}}{\partial \tilde{\mathbf{P}}_w} = -a_1' \int_{\Gamma_d} \mathbf{N}_{p_c}^T 2h \frac{1}{Q_{cw}} \mathbf{N}_{p_w}^{enr} \, d\Gamma - \int_{\Gamma_d} \nabla \mathbf{N}_{p_c}^T \mathbf{t}_{\Gamma_d} 2h \, k_{gd} \, \mathbf{t}_{\Gamma_d}^T \nabla \mathbf{N}_{p_w}^{enr} \, d\Gamma \\
&\frac{\partial \mathbf{F}_{\tilde{c}}^{\text{int}}}{\partial \mathbf{P}_w} = -a_1' \int_{\Gamma_d} \left(\mathbf{N}_{p_c}^{enr}\right)^T 2h \frac{1}{Q_{cw}} \mathbf{N}_{p_w} \, d\Gamma - \int_{\Gamma_d} \left(\nabla \mathbf{N}_{p_c}^{enr}\right)^T \mathbf{t}_{\Gamma_d} 2h \, k_{gd} \, \mathbf{t}_{\Gamma_d}^T \nabla \mathbf{N}_{p_w} \, d\Gamma \\
&\frac{\partial \mathbf{F}_{\tilde{c}}^{\text{int}}}{\partial \tilde{\mathbf{P}}_w} = -a_1' \int_{\Gamma_d} \left(\mathbf{N}_{p_c}^{enr}\right)^T 2h \frac{1}{Q_{cw}} \mathbf{N}_{p_w}^{enr} \, d\Gamma - \int_{\Gamma_d} \left(\nabla \mathbf{N}_{p_c}^{enr}\right)^T \mathbf{t}_{\Gamma_d} 2h \, k_{gd} \, \mathbf{t}_{\Gamma_d}^T \nabla \mathbf{N}_{p_w}^{enr} \, d\Gamma
\end{aligned}
$$

11.9 Computational Notes in Hydraulic Fracture Modeling

In order to perform the numerical integration in the presence of strong and weak discontinuities in the X-FEM analysis, a special treatment is needed to accurately integrate the functions possessing a non-smooth character over the integration domain. To fulfill the usual requirements of the integration that involves the non-smooth enrichment functions appearing in the enriched part of the approximation, a common approach is that the numerical integration of the elements bisected by the discontinuity be performed separately on each sub-element into which the element is divided. In practice, in order to be able to perform the integration over these sub-elements on either side of the discontinuity, each sub-element is partitioned into triangles in which the standard Gauss quadrature rule is enforced. The remaining elements which are not cut by the discontinuity are numerically integrated employing the conventional Gauss integration scheme. Moreover, for the purpose of integrating the coupling terms along the one-dimensional segments of the discontinuity delimited with the element edges, the standard Gauss quadrature is applied. For the integration purpose, it is required that a number of Gauss points be set along the discontinuity segments to integrate tractions and leakage fluxes on the discontinuity additional to the Gauss points positioned within the elements not cut by the discontinuity and on both sides of the bisected elements to integrate hydromechanical fields on the continuum domain surrounding the discontinuity.

Concerning the crack growth simulation, some issues involved in implementing the crack growth with the X-FEM are discussed. These issues may considerably affect the accuracy of the solution and the computational efficiency of the crack growth modeling. Simulating a propagating crack necessitates a criterion that governs the crack evolution. Under mixed-mode loading condition, the maximum equivalent traction at all Gauss points in the element ahead of the crack tip is checked against the cohesive strength of the material σ_c. If the maximum equivalent traction at any of the Gauss points in the element ahead of the crack tip reaches the cohesive strength, the crack is extended into that element until it touches one of the element edges. In other words, the discontinuity to which the cohesive law is applied is inserted in the element ahead of the crack tip when the condition for crack extension is fulfilled. This procedure is repeated until the criterion for crack growth is not satisfied anymore. It is therefore possible that the crack propagates through several elements at each time step. In this manner, the crack grows such that the crack tip always lays on an edge of an element in accordance with the adopted enrichment functions, otherwise the chosen enrichment functions do not yield the acceptable approximation matching the case which allows the crack tip to lie within an element. The equivalent traction t_{eq} is computed as (Ortiz and Pandolfi, 1999)

$$t_{eq}(\theta) = \sqrt{\beta^{-2} t_s^2 + t_n^2} \tag{11.80}$$

where t_s and t_n are the shear and normal tractions along an axis which constructs an angle θ with the axis x. The shear and normal tractions are computed as $t_s = \mathbf{t} \cdot \boldsymbol{\sigma}'' \cdot \mathbf{n}$ and $t_n = \mathbf{n} \cdot \boldsymbol{\sigma}'' \cdot \mathbf{n}$, respectively, with $\mathbf{t}$ and $\mathbf{n}$ denoting the unit tangent and normal vectors to the axis. The parameter β indicates the ratio between the shear and the normal strength of the material.

Due to the intense variation of the stress field in the vicinity of the crack tip, the local stress field is not reliable for the prediction of the crack growth direction. Thus, the direction in which the crack propagates is determined on the basis of a non-local measure of the stress field (Dolbow, Moës, and Belytschko, 2001). When the aforementioned fracture criterion is met, the discontinuity is extended along the direction in which the equivalent traction computed from the non-local stress tensor at the crack tip is at a maximum. It is evident that for the mode I failure, it yields the direction perpendicular to the direction of the maximum principal non-local stress. The non-local stress tensor at the crack tip is obtained by weighted averaging of the stresses within an interaction radius around the crack tip as

$$\boldsymbol{\sigma}''_{tip} = \left(\int_A w \, dA \right)^{-1} \left(\int_A \boldsymbol{\sigma}'' w \, dA \right) \tag{11.81}$$

in which w is the Gaussian weight function that smoothens the stress field at the tip neighborhood defined as

$$w(r) = \frac{1}{(2\pi)^{3/2}\ell^3} \exp\left(-\frac{r^2}{2\ell^2}\right) \tag{11.82}$$

where r is the distance to the crack tip and the parameter ℓ determines the decline of the weight function away from the crack tip. According to Wells and Sluys (2001), the parameter ℓ was taken equal to three times the typical element size in the mesh around the crack tip; it was suggested by Dias-da-Costa *et al.* (2010) a value of approximately 1 % of the Hillerborg et al.'s characteristic length (1976).

In case that the fracture occurs under mode I loading condition, the crack growth simulation can be implemented as follows; under the assumption of pure opening mode, the crack propagates along a straight line. Thus, since the direction of crack propagation is known in advance, the partitioning is carried out within the elements that are to be bisected by the crack as the crack propagates at the onset of the dynamic analysis. This avoids the need for transferring history variables, the issue that arises in the integration scheme adapted when a discontinuity is introduced within an element. To be consistent with the cohesive fracture mechanics, the crack growth criterion is extracted from the cohesive crack concept that characterizes the debonding of the solid skeleton. Thus, in order to determine when the crack propagates, the calculation of the tractions at the Gauss points within the FEs is not needed; instead, the tractions are calculated directly at the Gauss points along the potential crack. Consequently, the Gauss integration points which need to be set on the discontinuity segments to integrate coupling terms over the discontinuity are placed beforehand. The stress field at these Gauss points, located on the trajectory of the potential crack, can be calculated correctly, so it can be relied upon to predict the instance of crack extension. It is emphasized that the potential crack is fully coherent prior to the satisfaction of the crack growth condition. In this way, the modified effective stress component in the direction normal to the crack line, computed by $\mathbf{n}_{\Gamma_d} \cdot \boldsymbol{\sigma}'' \cdot \mathbf{n}_{\Gamma_d}$, is checked against a threshold for the cohesive crack extension at the Gauss points lying on the potential crack path. This normal stress is identical to the maximum principal tensile stress, and in harmony with this the shear stress is zero along the crack growth direction. This means that the crack propagates in perpendicular direction to the maximum principal stress direction as expected for mode I crack propagation. The checking algorithm is carried out element by element beginning from the one ahead of the crack tip. If on the considered element the normal stress to the crack line at all Gauss points on the potential crack trajectory attains the cohesive strength of the material σ_c, the discontinuity is introduced within the element. This procedure continues until the aforementioned stress condition for insertion of a discontinuity is violated. In this manner, the existing discontinuity can propagate into more than one element at each time step. This approach for mode I fracture leads to the same results as the one mentioned previously for the general case of mixed-mode fracture upon mesh refinement. However, this strategy makes it possible to relieve the stress evaluation at the crack tip whose accuracy is questionable regarding the fact that the stress state only at Gauss points is exactly known. Moreover, this strategy avoids the data transfer between the old and new Gauss points which is inevitable when the element partitioning is done for numerical integration. It is well-known that in transferring history variables, there is no guarantee that the transferred quantities satisfy the constitutive equations together with the governing equations simultaneously. Therefore, this approach makes the mode I crack growth implementation practically very efficient. It is noted that in both of these approaches the length of crack extension, one main concern when extending a discontinuity, is obtained during the analysis rather than set in advance. In the X-FEM context, once decohesion of the solid skeleton is initiated in an element, enriched nodes, to which enriched DOF are added, are introduced into the element to impose the discontinuity within the element. Hence, during the crack propagation modeling, some unenriched nodal points become enriched as the crack enters their support. It is recalled that no enrichments are added to the nodes on the element edge touched by the crack tip.

In order to assure the convergence of the solution and to obtain mesh independent results, the distribution of tractions in the cohesive zone must be exhibited properly. Therefore, it is necessary that the cohesive zone be discretized by a sufficient number of elements so as to accurately resolve the cohesive traction distribution and to arrive at a converged solution. In the X-FEM context, a minimum of about two elements in the cohesive zone was suggested by Moës and Belytschko (2002a). Under the plane strain conditions, the size of the cohesive zone for quasi-brittle materials can be estimated as

$$\ell_{coh} = \frac{E}{1-\nu^2}\frac{G_c}{\sigma_c^2} \tag{11.83}$$

which is identified as the characteristic length of the material introduced by Hillerborg, Modéer, and Petersson (1976). In this relation, E is the Young's modulus, ν is the Poisson ratio, σ_c is the cohesive strength of the material, and G_c is the cohesive fracture energy, the area under the softening curve.

11.10 Application of the X-FEM Method to Hydraulic Fracture Propagation of Multi-Phase Porous Media

In order to illustrate a part of the wide range of problems that can be modeled by the proposed approach, and to validate the performance of the computational algorithm in dynamic modeling of coupled hydro-mechanical problems, an example is solved based on the aforementioned numerical model. This example involves the propagation of an edge crack in a square plate subjected to tensile loading, which illustrates the substantial effect of the hydromechanical coupling between the crack and the surrounding porous medium on the results and the significance of the use of the three-phase model in simulating the hydro-mechanical behavior of partially saturated porous media including cracks. The numerical simulation of this problem has been carried out previously by Réthoré, De Borst, and Abellan (2008) using the passive gas phase assumption. In this study, the effect of the gas flow is considered further through the three-phase simulation.

The square plate with the edge crack of length 0.05 m lying along its symmetry axis is simulated under the plane strain condition. The geometry and loading of the plate are displayed in Figure 11.8. As shown in

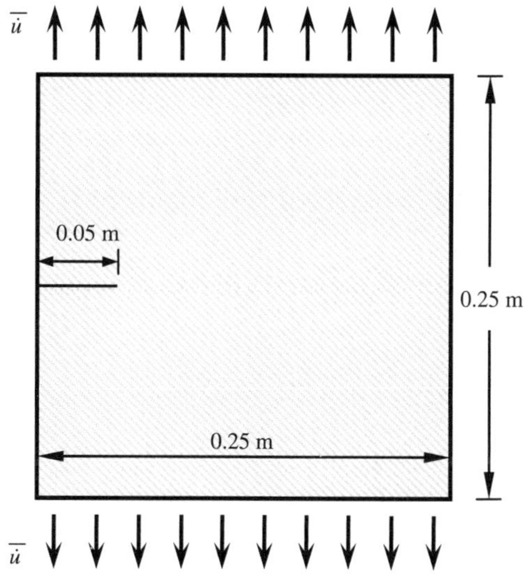

Figure 11.8 The geometry and loading of a square plate of partially saturated porous medium with an edge crack

Table 11.2 Material properties of the partially saturated porous medium with an edge crack

Young's modulus	$E = 25.85$ GPa
Poisson ratio	$\nu = 0.18$
Biot constant	$\alpha = 1$
Initial porosity	$n = 0.2$
Solid phase density	$\rho_s = 2000$ kg/m^3
Water density	$\rho_w = 1000$ kg/m^3
Air density	$\rho_g = 1.2$ kg/m^3
Bulk modulus of solid phase	$K_S = 13.46$ GPa
Bulk modulus of water	$K_w = 0.2$ GPa
Bulk modulus of air	$K_g = 0.1 \times 10^{-3}$ GPa
Intrinsic permeability	$k = 2.78 \times 10^{-21}$ m^2
Dynamic viscosity of water	$\mu_w = 5 \times 10^{-4}$ Pa s
Dynamic viscosity of air	$\mu_g = 5 \times 10^{-6}$ Pa s
Atmospheric pressure	$p_{atm} = 0$ Pa

this figure, the length of sides of the plate are all 0.25 m. The plate is loaded in tension by two uniform vertical velocities with magnitude $\bar{u} = 2.35 \times 10^{-2}$ μm/s applied in opposite directions to the top and bottom edges of the plate. The given equal and opposite normal velocities are imposed as prescribed velocities on the top and bottom boundaries of the plate. It is assumed that the plate has impermeable boundaries to both fluids. That is, the undrained boundary condition is enforced at all faces of the plate. Initially, the fully saturated condition is supposed in the entire domain in which the initial water pressure and gas pressure are set equal to the atmospheric pressure. The linear elastic constitutive law is assumed for the medium surrounding the crack, and in the cohesive zone the linear softening cohesive law is applied. Because of the symmetry, the cohesive crack propagation occurs under the pure opening mode. The material properties of the partially saturated porous medium considered in this numerical example are listed in Table 11.2. For fracture analysis, the cohesive fracture parameters of the material are set as follows; the cohesive strength $\sigma_c = 2.7$ MPa and the cohesive fracture energy $G_c = 95$ N/m. Regarding the adopted traction-separation law, the critical crack opening displacement at which the crack has fully developed and the cohesive traction transferred across the crack has reduced to zero, is obtained as $2G_c/\sigma_c$. In numerical modeling, once the normal displacement jump exceeds the critical opening, the normal cohesive traction is set to zero. It is noticed that the edge crack is modeled as a traction free cohesive crack.

In order to describe the flow of the immiscible fluids through the porous material, the constitutive relations for the water saturation as well as the water and gas relative permeabilities are assumed on the basis of the van Genuchten–Mualem model (Van Genuchten, 1980; Dury, Fischer, and Schulin, 1999)

$$
\begin{aligned}
S_w &= S_{rw} + (1 - S_{rw})\left[1 + \left(p_c/p_{ref}\right)^{1/(1-m)}\right]^{-m} \\
k_{rw} &= S_e^{\,1/2}\left[1 - \left(1 - S_e^{\,1/m}\right)^m\right]^2 \\
k_{rg} &= (1 - S_e)^{1/2}\left[1 - S_e^{\,1/m}\right]^{2m}
\end{aligned}
\tag{11.84}
$$

in which the residual water saturation $S_{rw} = 0$, the empirical curve-fitting parameter $m = 0.4396$, the reference pressure $p_{ref} = 18.6$ MPa, and the effective water saturation S_e is defined in relation (11.43).

The numerical analysis of the plate is performed employing the three-phase model as well as the passive gas phase assumption to obtain its response against the tensile loading. In the so-called passive gas phase assumption, it is supposed that the flow of gas is negligible, and the gas phase remains constantly at the atmospheric pressure in the partially saturated zone. That is, the constant atmospheric pressure is

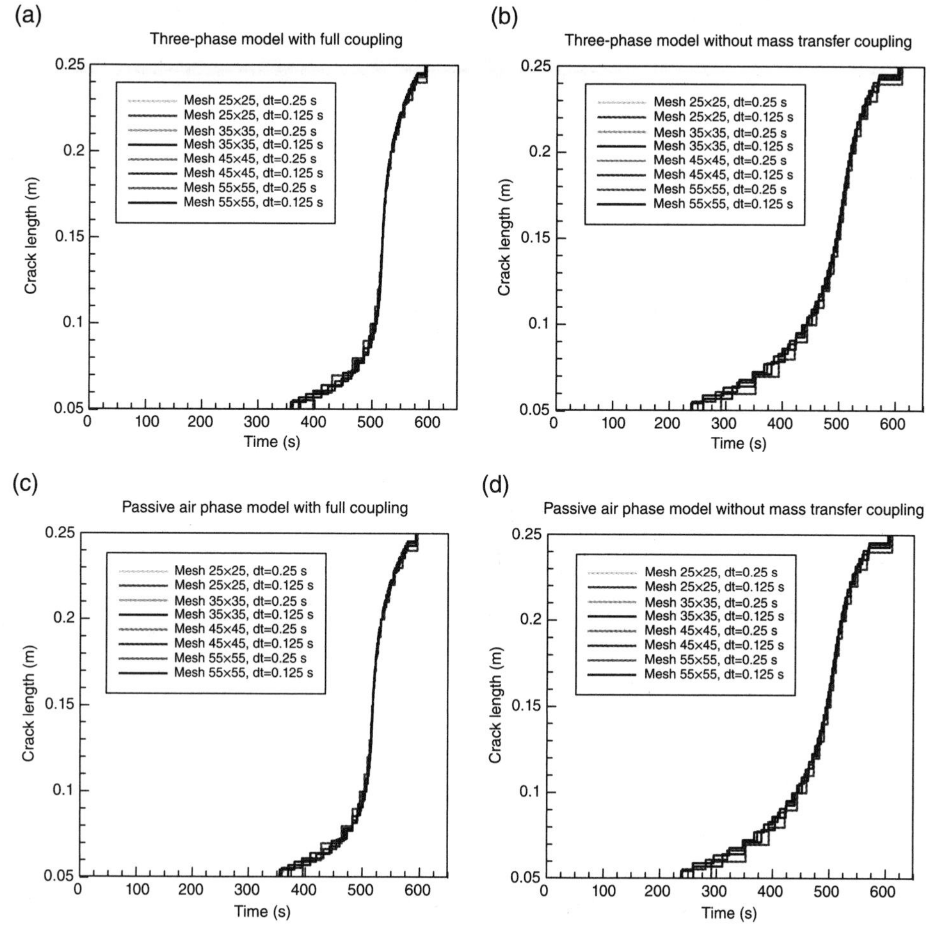

Figure 11.9 Crack length time histories: (a) three-phase model with full coupling; (b) three-phase model without mass transfer coupling; (c) passive air phase model with full coupling; (d) passive air phase model without mass transfer coupling

assigned to the gas pressure, and therefore the continuity equation of gas flow is ignored. In this modeling, a comparison is made between the numerical results obtained considering the full coupling, that is, the mechanical and the mass transfer coupling between the crack and the surrounding porous medium, and disregarding the mass transfer coupling. In the latter case, the interfacial flux vectors (i.e. $\mathbf{F}_w^{int}$, $\mathbf{F}_{\tilde{w}}^{int}$, $\mathbf{F}_c^{int}$ and $\mathbf{F}_{\tilde{c}}^{int}$) in the system of equations (11.70) are omitted. Subsequently, the water pressure and capillary pressure fields need not be enriched any longer, so their corresponding normal derivative remains continuous across the crack. In this case, the crack is not identified as a discontinuity in the fluid flow normal to the crack. This implies that in the case without the mass transfer coupling term, no distinction is made between the flow of the pore fluids in the crack and in the porous medium surrounding the crack.

For the chosen material properties, relation (11.83) gives the cohesive zone length of 0.348 m. Since the specimen size is too small compared to the estimated cohesive zone size, the suggested criterion in the previous section is not appropriate for the choice of the mesh size. In this numerical study, the plate is

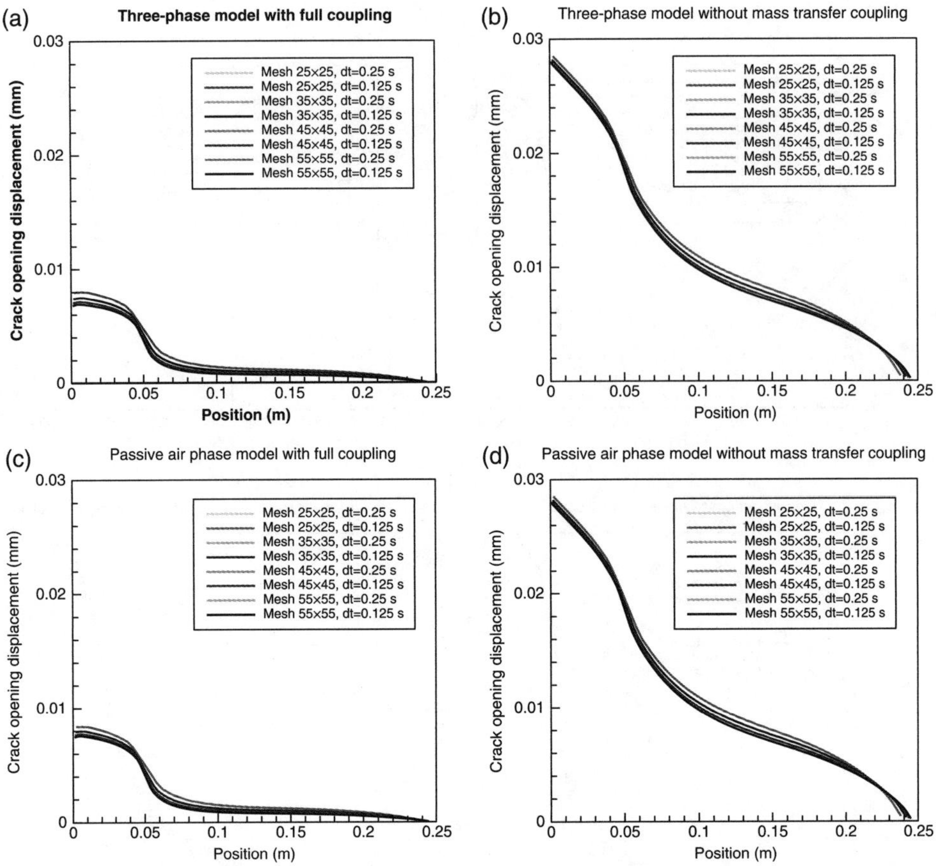

Figure 11.10 Crack opening displacement curves: (a) three-phase model with full coupling; (b) three-phase model without mass transfer coupling; (c) passive air phase model with full coupling; (d) passive air phase model without mass transfer coupling

modeled using FE meshes of various sizes that uniformly discretize the domain with 25×25, 35×35, 45×45, and 55×55 bilinear quadrilateral elements. In the analysis, two different time increments are used, $\Delta t = 0.25$ s and $\Delta t = 0.125$ s. The simulation is performed with several spatial and temporal discretizations to examine the convergence of the solution. The numerical analysis continues until the crack tip gets to the right hand side of the plate. Figure 11.9 compares the position of the crack tip as a function of time for different meshes and time increments. It can be observed from this figure that the crack length time histories for various meshes are in good agreement with each other, and with increasing refinement the results converge quickly. The time histories of the crack length clearly indicate that the results are entirely independent of the temporal discretization and coinciding for both time increments. It can also be found that crack tip velocity is roughly the same in all cases. The crack initiation occurs relatively earlier for the finest mesh. However, the difference between the crack initiation times becomes negligible with decreasing element size, showing that the converged solution has been achieved. The influence of the combined spatial and temporal discretizations is further investigated by comparing the crack opening displacement curves shown in Figure 11.10 for the time step before the crack tip reaches the right hand side of the plate. As can be seen from this figure, the crack opening displacement curves show minor

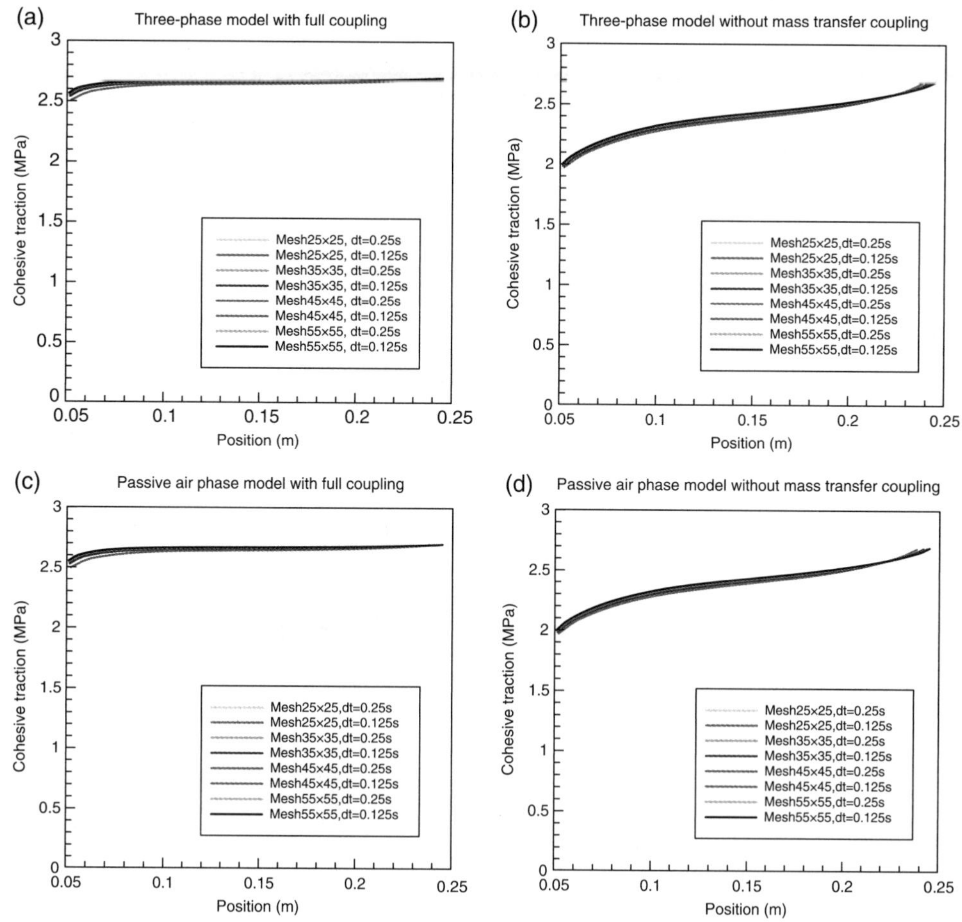

Figure 11.11 Cohesive traction profiles: (a) three-phase model with full coupling; (b) three-phase model without mass transfer coupling; (c) passive air phase model with full coupling; (d) passive air phase model without mass transfer coupling

differences as the mesh is refined, and for the two finest meshes the results are quite close to each other, which indicates the solution convergence. Consequently, further mesh refinement does not impact the results. Moreover, it is obvious that both of the time increments yield identical results. This means that the results are not sensitive to the time increment. The corresponding cohesive traction profiles are compared in Figure 11.11. In agreement with the crack opening displacement curves, the cohesive traction profiles converge quickly as the mesh size is decreased. The achieved results show that the considered meshes are fine enough to resolve the details of the cohesive zone. Again, the results are not affected by the variation in the time domain discretization.

In the following, only the simulation results with the most refined mesh are displayed. The results are presented for the time step before the crack propagates through the whole plate, that is, $t = 602.5$ s for the cases without the mass transfer coupling, $t = 590.75$ s for the three-phase model with full coupling, and $t = 591.5$ s for the passive air phase model with full coupling. Figure 11.12 exhibits the contours of the water pressure for different simulations. The comparison of these contours makes obvious that the influence of the mass transfer coupling on the results is strong. As can be seen from the contours,

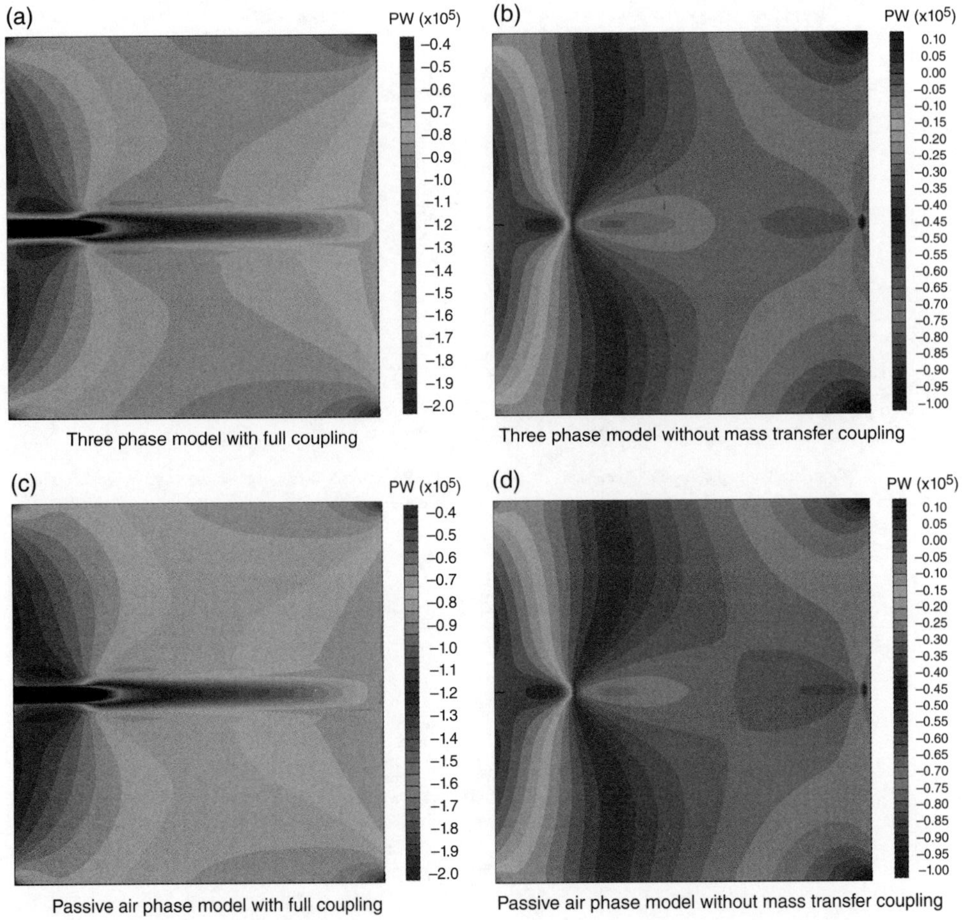

Figure 11.12 Water pressure (Pa) contours: (a) three-phase model with full coupling; (b) three-phase model without mass transfer coupling; (c) passive air phase model with full coupling; (d) passive air phase model without mass transfer coupling (For color details, please see color plate section)

incorporating the mass transfer coupling term into the formulation results in high negative water pressures concentrated in the vicinity of the crack, which implies that the pore water is drawn into the crack. These effects can be also distinguished in the contours of the gas pressure shown in Figure 11.13, which result from the three-phase model. As observed from this figure, the negative pressures are greater in the case with full coupling than those without the mass transfer coupling. Moreover, it can be noticed that allowing for the interfacial flux along the crack leads to the considerable decrease of the gas pressure in the area surrounding the crack. This causes the pore gas to flow toward the crack. The gas pressure contours reveal that the values of the gas pressure, ignored in the model based on the assumption of the passive gas phase, can be as large as those of the water pressure. The impact of the incorporation of the mass transfer coupling between the discontinuity and the continuum medium on the results can further be evidenced by comparing the contours given in Figures 11.14 and 11.15 representing the norm of the water pressure and gas pressure gradients, respectively. In accordance with what was observed before, pressure gradients with high values develop in the zone around the crack due to the mass transfer coupling. It also appears that

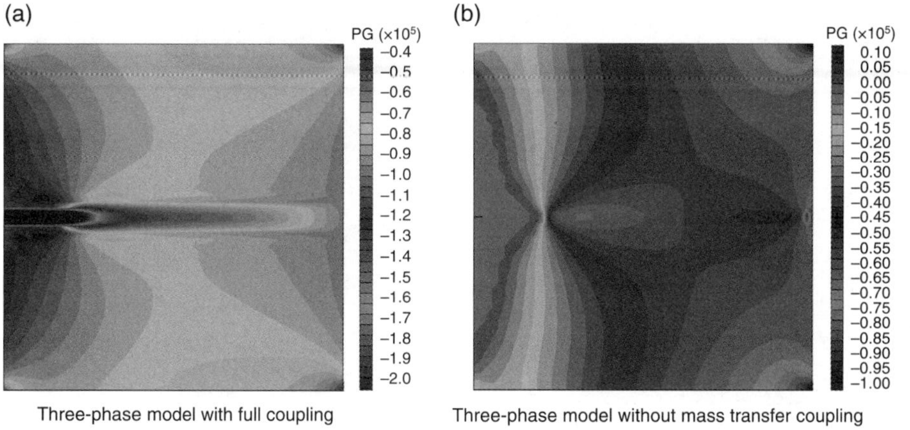

Figure 11.13 Gas pressure (Pa) contours: (a) three-phase model with full coupling; (b) three-phase model without mass transfer coupling (For color details, please see color plate section)

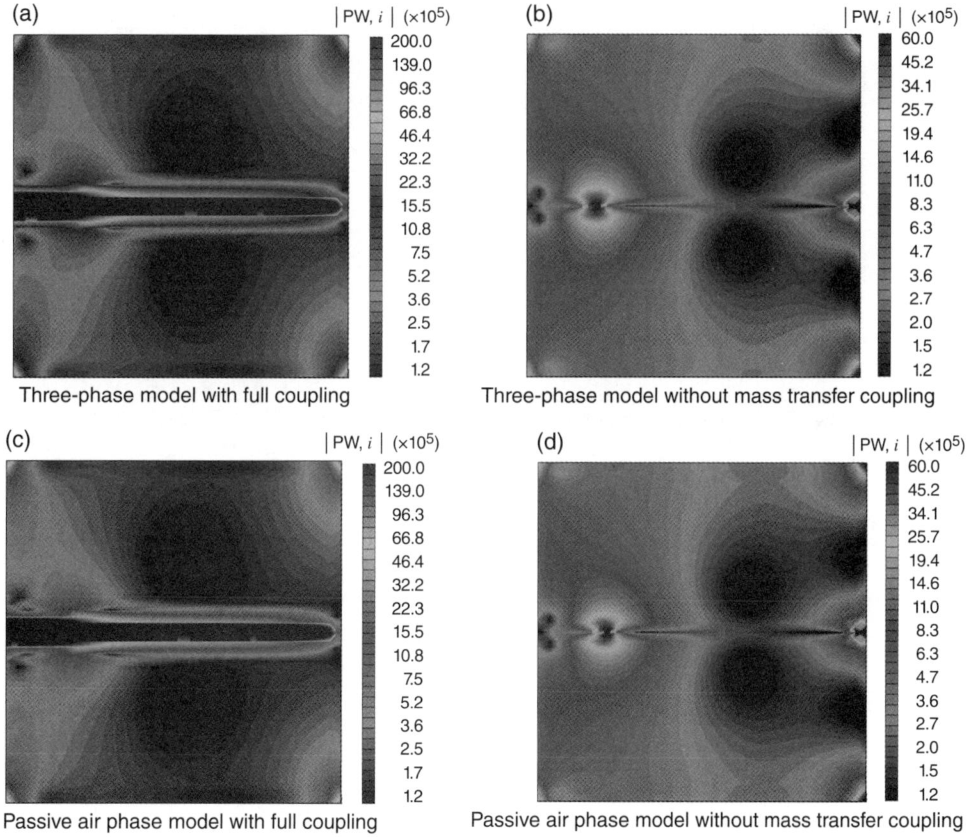

Figure 11.14 Norm of the water pressure gradient (Pa/m) contours (logarithmic scale): (a) three-phase model with full coupling; (b) three-phase model without mass transfer coupling; (c) passive air phase model with full coupling; (d) passive air phase model without mass transfer coupling (For color details, please see color plate section)

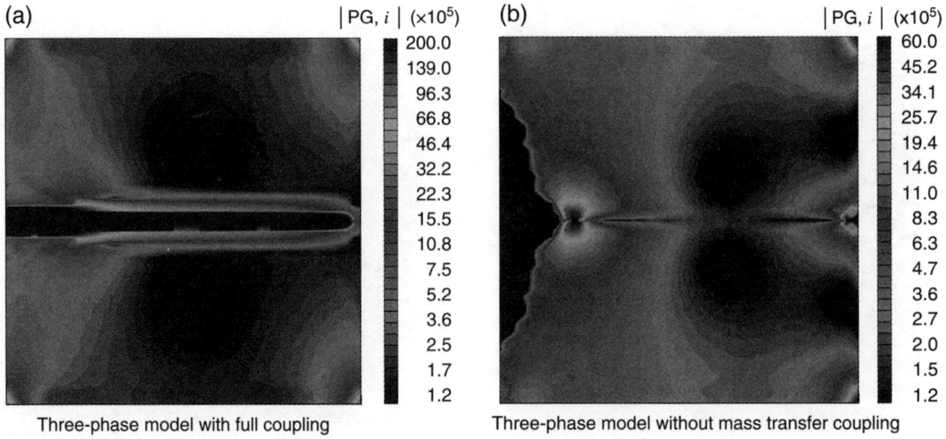

Figure 11.15 Norm of the gas pressure gradient (Pa/m) contours (logarithmic scale): (a) three-phase model with full coupling; (b) three-phase model without mass transfer coupling (For color details, please see color plate section)

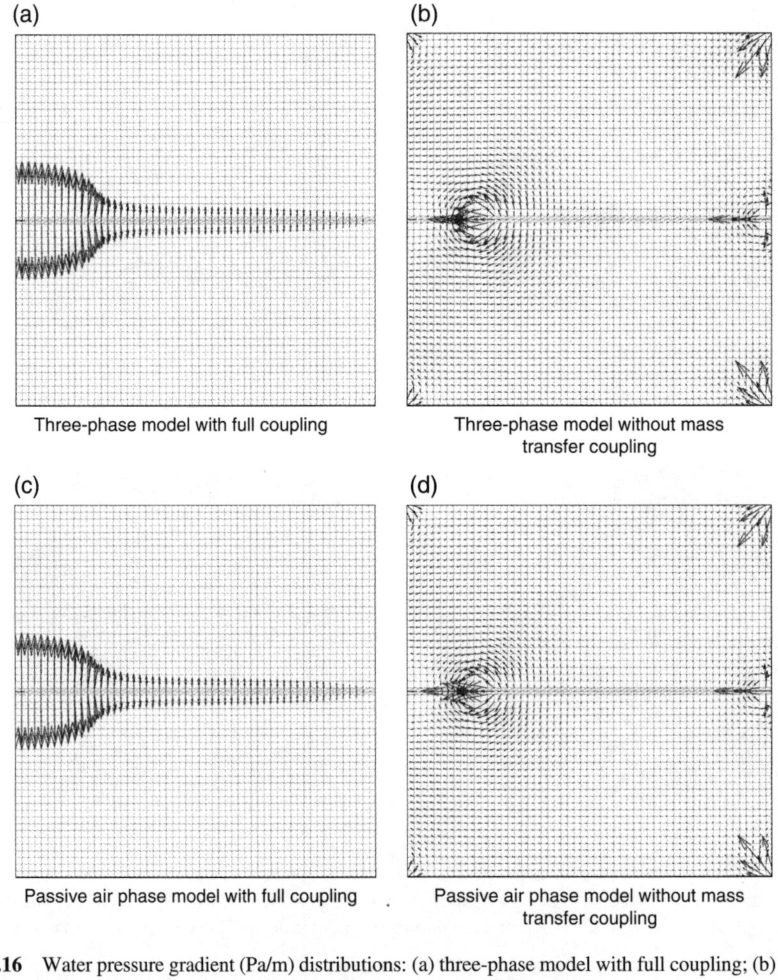

Figure 11.16 Water pressure gradient (Pa/m) distributions: (a) three-phase model with full coupling; (b) three-phase model without mass transfer coupling; (c) passive air phase model with full coupling; (d) passive air phase model without mass transfer coupling

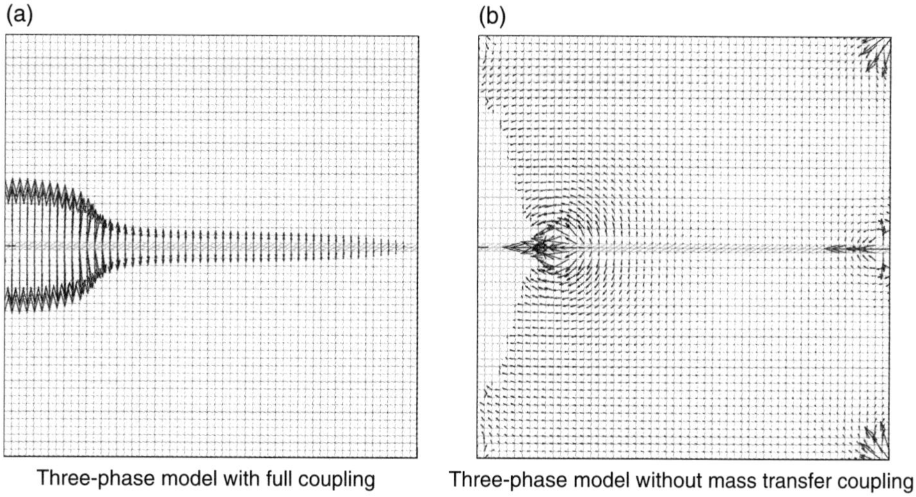

Figure 11.17 Gas pressure gradient (Pa/m) distributions: (a) three-phase model with full coupling; (b) three-phase model without mass transfer coupling

the simulation in which all primary variables are enriched, results in much higher values of the pressure gradient compared with those obtained without the water pressure and the capillary pressure enrichment. Figure 11.16 displays the distribution of the water pressure gradient in the plate. This figure also characterizes the distribution of the Darcy velocity of the pore water with negative sign. It is observable that there is a noticeable discrepancy between the results obtained from the simulation with full coupling at the discontinuity and those obtained from the simulation without mass transfer coupling. Accordingly, it is confirmed that mass transfer coupling can have a significant effect on the flow of pore fluids throughout the fracturing porous medium. From the results obtained by inserting both the mechanical and mass transfer coupling terms, it can be found that the flux of water through the crack borders is toward the crack compatible with what was concluded earlier, and its amount is strongly dependent on the crack opening displacement. This is also the case for the gaseous phase. As can be observed from the gas pressure gradient distribution for the case with full coupling indicated in Figure 11.17, the direction of the gaseous phase flow is oriented to the crack with a magnitude varying with the normal displacement jump across the crack. Furthermore, it can be discerned from Figures 11.16 and 11.17 when the enriched DOF corresponding to the fluid variables are also activated the main flow happens in the region close to the crack. Obviously, this area matches the area where high pressure gradients are dominant.

In sum, it can be concluded that the omission of the mass transfer coupling terms can make significant differences between the calculated results of the two simulations. In addition, the calculations indicate that the results related to the gas phase are quantitatively comparable with those related to the water phase. This reflects the need for the inclusion of the flow continuity equation of the gaseous phase, disregarded in the model in which the passive gas phase assumption is made, that is, the application of the three-phase model.

12

Thermo-Hydro-Mechanical Modeling of Porous Media with X-FEM

12.1 Introduction

Thermo-Hydro-Mechanical (THM) modeling in porous media is one of the most important subjects in geotechnical and environmental engineering. The coupled THM behavior in porous media has been a subject of great interest in many engineering disciplines. Many attempts have been made to develop numerical prediction capabilities associated with topics such as the movement of pollutant plumes, gas injection, energy storage, permafrost and frozen soil engineering, safety assessment of waste repositories for radioactive waste and spent nuclear fuel, and geothermal energy extraction. The effects of heat transfer from emplaced radio-active waste to the surrounding rockmass are matters of great concern in nuclear waste disposal. The decomposition of carbohydrates, fats, and proteins generates heat in sanitary landfills that influences the migrations of fluid, heat, and gas. The environmental contamination from such migrations is a major concern due to its potential hazards to public safety. These and many other geo-engineering cases are associated with the problem of heat transfer and its effects on rocks/soils that require comprehensive understanding of the THM behavior, such as temperature distribution, deformation behavior, and so on, for the safe design of the ground structure.

There are various mathematical formulations proposed by researchers for THM model of porous saturated–unsaturated media in literature. Coupled THM processes are commonly simulated using the theories of porous media. The first theory was proposed by Terzaghi (1943) for one-dimensional consolidation of soils, followed later by Biot (1941, 1956) for isothermal consolidation of elastic porous media that was a phenomenological approach of poroelasticity, and then the mixture theory was described by Morland (1972), Bowen (1980) and others. The non-isothermal consolidation of deformable porous media has been the basis of modern coupled THM models, using either the averaging approach as proposed first by Hassanizadeh and Gray (1979a, b, 1980) and Achanta, Cushman, and Okos (1994), or an extension of Biot phenomenological approach with a thermal component by de Boer (1998). A fully coupled numerical model to simulate the slow transient phenomena involving the heat and mass transfer in deforming porous media was developed by Gawin and Klemm (1994) and Gawin, Baggio, and Schrefler (1995), in which the heat transfer was taken through the conduction and convection into the model. A model in terms of displacements, temperature, capillary pressure, and gas pressure was proposed by Schrefler, Zhan, and Simoni (1995) and Gawin and Schrefler (1996) in partially saturated deformable porous medium, where the effects of temperature on capillary pressure were investigated for drying process of partially saturated porous media. A finite element (FE) formulation of multi-phase fluid flow

Extended Finite Element Method: Theory and Applications, First Edition. Amir R. Khoei.

and heat transfer within a deforming porous medium was presented by Vaziri (1996) in terms of displacement, pore pressure, and temperature, and its application was demonstrated in one-dimensional thermal saturated soil layer. A fully coupled THM model was applied by Neaupane and Yamabe (2001) to describe the nonlinear behavior of freezing and thawing of rock. A general governing equation was proposed by Rutqvist *et al.* (2001) for coupled THM process in saturated and unsaturated geologic formations. An object-oriented FE analysis was performed by Wang and Kolditz (2007) in THM problems of porous media. A combined THM–damage model was presented by Gatmiri and Arson (2008) on the basis of a suction-based heat, moisture transfer, and skeleton deformation equations for unsaturated media. A thermal conductivity model of a three-phase mixture of gas, water, and solid was developed by Chen, Zhou, and Jing (2009) and Tong, Jing, and Zimmerman (2009, 2010) for simulation of THM processes of geological porous media by combining the effects of solid mineral composition, temperature, liquid saturation degree, porosity, and pressure on the effective thermal conductivity of porous media. The THM model was proposed by Dumont *et al.* (2010) for unsaturated soils, in which the effective stress is extended to unsaturated soils by introducing the capillary pressure based on a micro-structural model and taking the effects of desaturation and thermal softening phenomenon into the model.

Modeling of discontinuity with finite element method (FEM) in fractured/fracturing porous media has been attractive to researchers. One of the earlier research works in THM modeling of two-phase fractured media was illustrated by Noorishad, Tsang, and Witherspoon (1984), where a numerical approach was given for the saturated fractured porous rocks. Boone and Ingraffea (1990) presented a numerical procedure for the simulation of hydraulically-driven fracture propagation in poroelastic materials combining the FEM with the finite difference method. A cohesive segments method was proposed by Remmers, de Borst, and Needleman (2003) for the simulation of crack growth, where the cohesive segments were inserted into FEs as discontinuities in the displacement field by exploiting the partition of unity (PU) property. Schrefler *et al.* (2006) and Secchi, Simoni, and Schrefler (2007) modeled the hydraulic cohesive crack growth problem in fully saturated porous media using the FEM with mesh adaptation. Crack propagation simulation was performed by Radi and Loret (2007) for an elastic isotropic fluid-saturated porous media at an intersonic constant speed. Hoteit and Firoozabadi (2008) proposed a numerical procedure for incompressible fluid flow in fractured porous media based on the combination of finite difference, finite volume, and FEMs. Segura and Carol (2008a) proposed a hydromechanical formulation for fully saturated geomaterials with pre-existing discontinuities based on the FEM with zero-thickness interface elements. The cohesive segment method was employed by Remmers, de Borst, and Needleman (2008) in dynamic analysis of the nucleation, growth, and coalescence of multiple cracks in brittle solids. Khoei, Barani, and Mofid (2011) and Barani, Khoei, and Mofid (2011) presented a dynamic analysis of cohesive fracture propagation in saturated and partially saturated porous media. The importance of the cohesive zone model in a fluid driven fracture was studied by Sarris and Papanastasiou (2011), and it was shown that the crack mouth opening has larger value in the case of elastic-softening cohesive model compared to the rigid softening model. A numerical model based on the fully coupled approach was presented by Carrier and Granet (2012) for the hydraulic fracturing of permeable medium, where four limiting propagation regimes were assumed.

In modeling of discontinuity based on the standard FEM, the discontinuity is restricted to the inter-element boundaries suffering from the mesh dependency. In such cases, successive remeshing must be carried out to overcome sensitivity to the FEM mesh that makes the computation an expensive and cumbersome process. The difficulties confronted in the standard FEM are handled by locally enriching the conventional FE approximation with an additional function through the concept of the PU, which was introduced in the pioneering work of Melenk and Babuska (1996). The idea was exploited to set up the frame of the extended finite element method (X-FEM) by Belytschko and Black (1999) and Moës, Dolbow, and Belytschko (1999). Indeed, the X-FEM approximation relies on the PU property of FE shape functions for the incorporation of local enrichments into the classical FE basis. By appropriately selecting the enrichment function and enriching specific nodal points through the addition of extra degrees of freedom (DOF) relevant to the chosen enrichment function to these nodes, the enriched approximation would be capable of directly capturing the local property in the solution. The X-FEM was originally applied in linear fracture mechanics problems (Daux *et al.*, 2000; Dolbow, Moës, and Belytschko, 2001; Ventura, Budyn, and Belytschko, 2003; Arcias and Belytschko, 2005a), and then extended to the cohesive and plastic fracture mechanics applications (Elguedj, Gravouil, and Combescure, 2006, 2007; Khoei and Nikbakht, 2007; Khoei and Karimi, 2008; Broumand and Khoei, 2013, 2014). The X-FEM technique was employed in coupled problems by its application in multi-phase porous media. The technique was proposed by de Borst, Réthoré, and Abellan (2006) and Réthoré, de Borst, and Abellan (2007b) for the fluid flow in fractured porous media. It was proposed by QingWen, YuWen, and TianTang (2009) in modeling hydraulic fracturing in concrete by imposing a constant pressure value along the crack faces. The technique was applied by Lecamipon (2009) in hydraulic fracture problems using special crack tip functions in the presence of internal pressure inside the crack. The X-FEM was employed in modeling arbitrary interfaces by Khoei and Haghighat (2011) in the nonlinear dynamic analysis of deformable porous media. It was proposed by Khoei, Moallemi, and Haghighat (2012) in THM modeling of impermeable discontinuities in saturated porous media. The technique was employed by Mohamadnejad and Khoei (2013a, b, c) in hydromechanical modeling of deformable, progressively fracturing porous media interacting with the flow of two immiscible, compressible wetting and non-wetting pore fluids.

In this chapter, the X-FEM technique is presented in THM modeling of impermeable discontinuities in saturated porous media. In order to derive the THM governing equations, the momentum equilibrium equation, mass balance equation, and the energy conservation relation are applied. The spatial discretization of governing equations of THM porous media is performed by the X-FEM technique, and followed by the generalized Newmark scheme for the time domain discretization. The displacement, pressure, and temperature discontinuities are defined by enriching the displacement field using the Heaviside and crack tip asymptotic functions, and the pressure and temperature fields using the Heaviside and appropriate asymptotic functions. Finally, numerical examples are analyzed to demonstrate the performance and capability of proposed computational algorithm in modeling of impermeable discontinuity in THM behavior of porous soils.

12.2 THM Governing Equations of Saturated Porous Media

The THM processes in geological porous media are governed by the momentum, mass, and energy conservation laws of the continuum mechanics. In the literature, both the extended Biot phenomenological approach with thermal components and the averaging approach of the mixture theory are commonly used to establish the governing partial differential equations, in which the former is more suited to macroscopic description whereas the latter is more suitable for understanding the microscopic behavior of the porous media. In this study, the Biot theory is employed to derive the governing equations of THM behavior in saturated medium, which is coupled with the heat transfer analysis.

The effective stress is an essential concept in the deformation of porous media, which can be defined by $\boldsymbol{\sigma}' = \boldsymbol{\sigma} + \alpha p\mathbf{I}$, where $\boldsymbol{\sigma}'$ and $\boldsymbol{\sigma}$ are the effective stress and the total stress, respectively, $\mathbf{I}$ is the identity tensor and p is the pore pressure. In this relation, α is the Biot coefficient related to the material properties defined by $\alpha = 1 - K_T/K_S$, where K_T and K_S are the bulk modulus of porous media and its solid grains, respectively, which is assumed to be $\alpha = 1$ for most soils. By neglecting the acceleration of fluid particles with respect to the solids, the linear momentum balance of solid-fluid mixture can be written as

$$\nabla \cdot \boldsymbol{\sigma} - \rho\,\ddot{\mathbf{u}} + \rho\,\mathbf{b} = 0 \tag{12.1}$$

where the symbol ∇ denotes the vector gradient operator, $\ddot{\mathbf{u}}$ is the acceleration of mixture, $\mathbf{b}$ is the gravitational acceleration, and ρ is the total density of mixture defined by $\rho = n\,\rho_f + (1-n)\rho_s$, with ρ_f and ρ_s denoting the density of fluid and solid phases, respectively, and n is the porosity of mixture defined as the ratio of pore space to the total volume in a representative elementary volume.

In order to derive the mass balance equation of the mixture, the density of mixture is applied to each phase of the domain as

$$\frac{\partial \rho_{pI}}{\partial t}+\nabla\cdot\left(\rho_{pI}\mathbf{v}_{pI}\right)=0 \tag{12.2}$$

where $\mathbf{v}_{pI}$ denotes the absolute velocity of phase I in the bulk. Considering the solid-fluid mixture as a homogenous domain, Eq. (12.2) can be rewritten as

$$\frac{1-n}{\rho_s}\frac{\partial \rho_s}{\partial t}+\nabla\cdot\mathbf{v}_s+\frac{n}{\rho_f}\frac{\partial \rho_f}{\partial t}+n\,\nabla\cdot\mathbf{w}_f=0 \tag{12.3}$$

where $\mathbf{w}_f$ is the relative velocity of the fluid to the solid phase. In this equation, the variation of fluid density with respect to time is derived by assuming the fluid density ρ_f as a function of the temperature and pressure defined by $\rho_f=\rho_{f0}\exp\left[-\beta_f T+1/K_f(p-p_0)\right]$ as described by Fernandez (1972), where β_f is the volumetric thermal expansion coefficient of fluid, T is the temperature, and K_f is the bulk module of the fluid phase. Considering the solid density as a function of the temperature, pressure, and the first invariant of effective stress, the mass balance equation of the mixture can be obtained as

$$\alpha\,\nabla\cdot\mathbf{v}_s+n\,\nabla\cdot\mathbf{w}_f+\left(\frac{\alpha-n}{K_s}+\frac{n}{K_f}\right)\dot{p}-\left((\alpha-n)\,\beta_s+n\,\beta_f\right)\dot{T}=0 \tag{12.4}$$

Applying Darcy's Law to Eq. (12.4), the fluid velocity $\mathbf{w}_f$ can be expressed in term of the pressure gradient for an isotropic porous medium as $\mathbf{w}_f=(\boldsymbol{k}_f/n)(-\nabla p+\rho_f\,\mathbf{b})$, and Eq. (12.4) can be written as

$$\alpha\,\nabla\cdot\mathbf{v}_s+\nabla\cdot\left[\boldsymbol{k}_f\left(-\nabla p+\rho_f\,\mathbf{b}\right)\right]+\frac{\dot{p}}{Q}-\beta\,\dot{T}=0 \tag{12.5}$$

where $\boldsymbol{k}_f$ is the permeability matrix of the media, β is the thermal expansion coefficient of the bulk defined as $\beta=(\alpha-n)\,\beta_s+n\,\beta_f$ and $1/Q=(\alpha-n)/K_s+n/K_f$.

In order to derive the THM formulation, the heat transfer formulation is incorporated into the governing equations of porous saturated media using the energy conserving equation (enthalpy balance) for each phase. The governing equation of heat conduction is derived for a continuous medium from the principle of conservation of heat energy over an arbitrary fixed volume. Based on this principle, the heat increase rate of the system is equal to the summation of heat conduction and heat generation rate in a fixed volume. Applying the Fourier law of heat conduction, the energy balance equation can be written as

$$(\rho c)_{pI}\frac{\partial T}{\partial t}+(\rho c)_{pI}\mathbf{v}_{pI}\,\nabla T-\nabla\cdot\left(\mathbb{k}_{pI}\,\nabla T\right)=0 \tag{12.6}$$

where c denotes the heat capacity, and $\mathbb{k}_{pI}$ is the heat conductivity matrix of phase I. Multiplying the energy Eq. (12.6) by its porosity for each phase, neglecting the velocity of solid phase, and using Darcy's Law for the fluid velocity, the energy equation for the mixture can be obtained as (Lewis and Schrefler, 1998)

$$(\rho c)_{avg}\frac{\partial T}{\partial t}+\left(\rho_f\,c_f\left[\boldsymbol{k}_f\left(-\nabla p+\rho\,\mathbf{b}\right)\right]\right)\nabla T-\nabla(\mathbb{k}\;\nabla T)=0 \tag{12.7}$$

where $(\rho c)_{avg}=(1-n)\,(\rho c)_s+n\,(\rho c)_f$ and $\mathbb{k}=(1-n)\,\mathbb{k}_s+n\,\mathbb{k}_f$. The second term of Eq. (12.7) implies the effect of fluid flow on the heat transfer as a convection term.

12.3 Discontinuities in a THM Medium

The singularity in a discontinuous porous media can be caused due to the thermal and pressure loading in the vicinity of singular points. Since the governing equation of fluid flow in porous media are similar to the heat transfer equation, the treatment of thermal field near the singular points is assumed to be similar to the fluid phase. By neglecting the effect of transient terms in the heat transfer equation at the vicinity of singular points, Eq. (12.7) can be transformed to $\nabla^2 T = 0$ in the absence of heat source, which is also valid near the discontinuity. In order to solve the heat transfer equation $\nabla^2 T = 0$, the boundary conditions must be applied at the discontinuity edges, as shown in Figure 12.1. The boundary conditions near the singular point p is considered as one of the following cases

$$T(x,y) = 0 \quad (x,y) \in \Gamma_1, \Gamma_2 \quad \text{Dirichlet BC} \tag{12.8}$$

$$\frac{\partial T(x,y)}{\partial n} = 0 \quad (x,y) \in \Gamma_1, \Gamma_2 \quad \text{Neumann BC} \tag{12.9}$$

$$\begin{aligned} T(x,y) = 0 \quad (x,y) \in \Gamma_1 \\ \frac{\partial T(x,y)}{\partial n} = 0 \quad (x,y) \in \Gamma_2 \end{aligned} \quad \text{Dirichlet \& Neumann BCs} \tag{12.10}$$

These boundary conditions can be used to solve the heat transfer equation near the singularity. The solution of the differential equation is given as (Yosibash, 1995)

$$T(r,\alpha) = \sum_{n=1}^{\infty} b_n r^{\frac{n\pi}{\beta}} \sqrt{\frac{2}{\beta}} \sin\left(\frac{n\pi\alpha}{\beta}\right) \quad \text{Dirichlet BC} \tag{12.11}$$

$$T(r,\alpha) = \sum_{n=0}^{\infty} b_n r^{\frac{n\pi}{\beta}} \cos\left(\frac{n\pi\alpha}{\beta}\right) \quad \text{Neumann BC} \tag{12.12}$$

$$T(r,\alpha) = \sum_{n=1}^{\infty} b_n r^{\frac{n\pi}{2\beta}} \sqrt{\frac{2}{\beta}} \sin\left(\frac{n\pi\alpha}{2\beta}\right) \quad \text{Dirichlet and Neumann BCs} \tag{12.13}$$

where $0 \le \alpha \le 2\pi$. In the case of thermal discontinuity, the thermal flux on the discontinuity Γ_1 and Γ_2 is equal to zero, so the Neumann boundary condition must be applied on discontinuity faces and relation (12.12) is used as the near tip singular solution (Duflot, 2008; Zamani and Eslami, 2010).

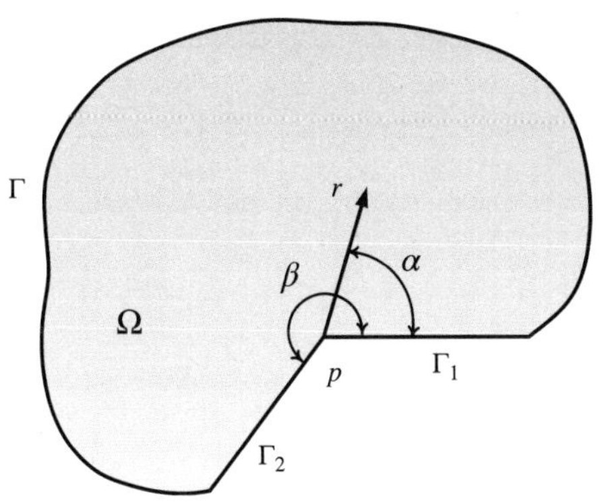

Figure 12.1 Geometry of singularity: The domain Ω contains a re-entrant corner with an internal angle β

The same procedure can be applied to obtain the pressure distribution in the vicinity of singular point. Considering the Darcy equation as $\mathbf{w}_f = (\boldsymbol{k}_f/n)(-\nabla p + \rho_f\,\mathbf{b})$ and the continuity equation as $\nabla.\,\mathbf{w}_f = 0$, the pressure differential equation can be obtained in the absence of gravity forces as

$$\nabla^2 p = 0 \tag{12.14}$$

This equation is similar to the partial differential equation derived for the steady-state thermal condition. In the case of impermeable discontinuity, the leak-off from the discontinuity faces is equal to zero. Thus, the solution of partial differential Eq. (12.14) at the singular point p can be obtained similar to relation (12.12) for the impermeable boundary condition as

$$P(r,\alpha) = \sum_{n=0}^{\infty} b_n\, r^{\frac{n\pi}{\beta}} \cos\left(\frac{n\pi\alpha}{\beta}\right) \tag{12.15}$$

and the fluid flow and thermal flux can be obtained by taking the derivation from (12.12) or (12.15) with respect to r and inserting $\beta = 2\pi$ as

$$q(r,\alpha) = \sum_{n=0}^{\infty} c_n\, r^{\frac{n}{2}-1} \cos\left(\frac{n\alpha}{2}\right)\hat{e}_r - \sum_{n=0}^{\infty} c_n\, r^{\frac{n}{2}-1} \sin\left(\frac{n\alpha}{2}\right)\hat{e}_\alpha \tag{12.16}$$

This relation presents the singularity of fluid and thermal fluxes in the vicinity of point p.

12.4 The X-FEM Formulation of THM Governing Equations

In order to numerically model a THM porous medium with impermeable discontinuities, the X-FEM is employed by adding appropriate enrichment functions to the standard FE approximation. It must be noted that in the derivation of X-FEM formulation, it is implicitly assumed that the displacement, pressure, and temperature fields are continuous over the domain of problem, however, necessary modifications are applied in the variational formulation of the X-FEM solution to model the discontinuous displacement, pressure, and temperature fields across the discontinuity. In order to derive the weak form of governing Eqs. (12.1), (12.5), and (12.7), the trial functions $\mathbf{u}(\boldsymbol{x}, t)$, $p(\boldsymbol{x}, t)$, $T(\boldsymbol{x}, t)$ and the test functions $\delta\mathbf{u}(\boldsymbol{x}, t)$, $\delta p(\boldsymbol{x}, t)$ and $\delta T(\boldsymbol{x}, t)$ are required to be smooth enough in order to satisfy all essential boundary conditions and define the derivatives of equations. Furthermore, the test functions $\delta\mathbf{u}(\boldsymbol{x}, t)$, $\delta p(\boldsymbol{x}, t)$ and $\delta T(\boldsymbol{x}, t)$ must be vanished on the prescribed strong boundary conditions. To obtain the weak form of governing equations, the test functions $\delta\mathbf{u}(\boldsymbol{x}, t)$, $\delta p(\boldsymbol{x}, t)$, and $\delta T(\boldsymbol{x}, t)$ are multiplied by Eqs. (12.1), (12.5), and (12.7), respectively, and integrated over the domain Ω as

$$\int_\Omega \delta\mathbf{u}(\nabla\cdot\boldsymbol{\sigma} - \rho\,\ddot{\mathbf{u}} + \rho\,\mathbf{b})\,d\Omega = 0 \tag{12.17a}$$

$$\int_\Omega \delta p\left(\alpha\,\nabla\cdot\mathbf{v}_s + \nabla\cdot\left[\boldsymbol{k}_f\left(-\nabla p + \rho_f\,\mathbf{b}\right)\right] + \frac{\dot{p}}{Q} - \beta\,\dot{T}\right)d\Omega = 0 \tag{12.17b}$$

$$\int_\Omega \delta T\left((\rho c)_{avg}\dot{T} + \left(\rho_f c_f\left[\boldsymbol{k}_f\left(-\nabla p + \rho\,\mathbf{b}\right)\right]\right)\nabla T - \nabla(\boldsymbol{k}\;\nabla T)\right)d\Omega = 0 \tag{12.17c}$$

Expanding the integral equations (12.17) and applying the Divergence theorem, leads to the following weak form of governing equations

$$\int_\Omega \delta\mathbf{u}\cdot\rho\,\ddot{\mathbf{u}}\,d\Omega + \int_\Omega \nabla\delta\mathbf{u} : \boldsymbol{\sigma}\,d\Omega - \int_\Omega \delta\mathbf{u}\cdot\rho\,\mathbf{b}\,d\Omega - \int_{\Gamma_t} \delta\mathbf{u}\cdot\bar{\mathbf{t}}\,d\Gamma = 0 \tag{12.18a}$$

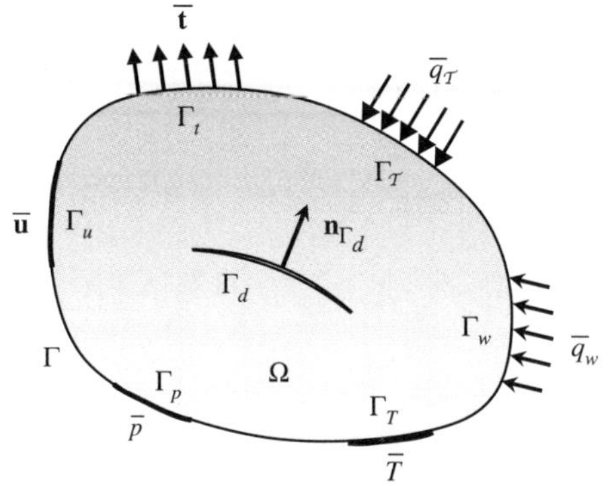

Figure 12.2 Geometry of discontinues domain and its boundary conditions

$$\begin{aligned}&\int_\Omega \delta p\, \alpha\, \nabla \cdot \mathbf{v}_s \, d\Omega + \int_\Omega \nabla \delta p\, \boldsymbol{k}_f \nabla p\, d\Omega + \int_\Omega \delta p \frac{1}{Q} \dot{p}\, d\Omega - \int_\Omega \delta p\, \beta\, \dot{T}\, d\Omega \\ &\quad - \int_\Omega \nabla \delta p\, \boldsymbol{k}_f \left(\rho_f \mathbf{b}\right) d\Omega + \int_{\Gamma_w} \delta p \left(\mathbf{w}_f \cdot \mathbf{n}_\Gamma\right) d\Gamma = 0\end{aligned} \tag{12.18b}$$

$$\begin{aligned}&\int_\Omega \delta T\, (\rho c)_{avg} \dot{T}\, d\Omega + \int_\Omega \delta T \left(\rho_f c_f \left[\boldsymbol{k}_f(-\nabla p + \rho\, \mathbf{b})\right]\right) \nabla T\, d\Omega \\ &\quad + \int_\Omega \nabla \delta T\, \boldsymbol{k}\, \nabla T\, d\Omega - \int_{\Gamma_T} \delta T \left(\bar{q}_T\right) d\Gamma = 0\end{aligned} \tag{12.18c}$$

in which the essential and natural boundary conditions are as follows; the solid-phase boundary conditions are $\mathbf{u} = \bar{\mathbf{u}}$ on Γ_u and $\mathbf{t} = \boldsymbol{\sigma} \cdot \mathbf{n}_\Gamma \equiv \bar{\mathbf{t}}$ on Γ_t, with $\mathbf{n}_\Gamma$ denoting the unit outward normal vector to the external boundary Γ; the fluid-phase boundary conditions are $\mathbf{p} = \bar{\mathbf{p}}$ on Γ_p and $\mathbf{w}_f \cdot \mathbf{n}_\Gamma = \bar{q}_w$ on Γ_w, with $\bar{q}_w$ denoting the prescribed outflow of the fluid imposed on Γ_w; and the thermal boundary conditions are $T = \bar{T}$ on Γ_T and $\bar{q}_T$ as the prescribed thermal flux imposed on Γ_T, as shown in Figure 12.2. It must be noted that the total stress $\boldsymbol{\sigma}$ in the integral Eq. (12.18a) must be replaced by $\boldsymbol{\sigma} = \boldsymbol{\sigma}' - \alpha p\, \mathbf{I}$, in which the effective stress $\boldsymbol{\sigma}'$ can be related to the total strain $\boldsymbol{\varepsilon}$ and the thermal strain $\boldsymbol{\varepsilon}^t$ in the constitutive relation of $\boldsymbol{\sigma}' = \mathbf{D}\, (\boldsymbol{\varepsilon} - \boldsymbol{\varepsilon}^t)$, that is,

$$\int_\Omega \nabla \delta \mathbf{u} : \boldsymbol{\sigma}\, d\Omega = \int_\Omega \nabla \delta \mathbf{u} : \mathbf{D}\left(\boldsymbol{\varepsilon} - \boldsymbol{\varepsilon}^t\right) d\Omega - \int_\Omega \nabla \delta \mathbf{u} : (\alpha p\, \mathbf{I})\, d\Omega \tag{12.19}$$

where $\mathbf{D}$ is the material property of solid skeleton. In what follows, the spatial and time domain discretization of THM integral equations (12.18) are derived using the X-FEM and generalized Newmark approaches.

12.4.1 Approximation of Displacement, Pressure, and Temperature Fields

In order to solve the integral equations (12.18), the X-FEM is employed for the spatial discretization. To achieve this aim, the displacement, pressure, and temperature fields are enriched using the analytical solution to make the approximations capable of tracking the discontinuities. Various enrichment functions can be used to enhance the approximation of displacement, pressure, and temperature fields. In the THM modeling of impermeable discontinuities proposed here, the displacement field is enriched by the Heaviside and crack tip asymptotic functions, and the pressure and temperature fields are enriched by the Heaviside and appropriate asymptotic functions. The enriched approximation of the extended FEM for the field variable $\varphi(\boldsymbol{x}, t)$ can be approximated using the enhanced trial function $\varphi^h(\boldsymbol{x}, t)$ as

$$\begin{aligned}\varphi(\boldsymbol{x},t) \approx \varphi^h(\boldsymbol{x},t) &= \sum_{I\in\mathcal{N}} N_I(\boldsymbol{x})\,\bar{\varphi}_I(t) + \sum_{J\in\mathcal{N}^{dis}} N_J(\boldsymbol{x})\,\psi(\boldsymbol{x})\,\bar{a}_J(t)\\ &+ \sum_{K\in\mathcal{N}^{tip}} \sum_{\alpha=1}^{\mathcal{M}^{tip}} N_K(\boldsymbol{x})\,\mathcal{B}_\alpha(\boldsymbol{x})\,\bar{b}_{\alpha K}(t)\end{aligned} \tag{12.20}$$

where $\mathcal{N}$ is the set of all nodal points, $N_I(\boldsymbol{x})$ is the standard shape function, $\psi(\boldsymbol{x})$ denotes the Heaviside enrichment function used for strong discontinuities, $\mathcal{B}_\alpha(\boldsymbol{x})$ indicates the additional asymptotic enrichment functions depending on the tip singularity of the displacement, pressure, or temperature field, and $\mathcal{M}^{tip}$ is the required number of asymptotic solution used as the enrichment functions. In order to achieve the Kronecker property of standard FEM shape functions, that is, $\varphi(\boldsymbol{x}_J) = \varphi_J$ at the nodal point J, Eq. (12.20) can be rewritten as

$$\begin{aligned}\varphi(\boldsymbol{x},t) \approx \varphi^h(\boldsymbol{x},t) &= \sum_{I\in\mathcal{N}} N_I(\boldsymbol{x})\,\bar{\varphi}_I(t) + \sum_{J\in\mathcal{N}^{dis}} N_J(\boldsymbol{x})\,(\psi(\boldsymbol{x}) - \psi(\boldsymbol{x}_J))\,\bar{a}_J(t)\\ &+ \sum_{K\in\mathcal{N}^{tip}} \sum_{\alpha=1}^{\mathcal{M}^{tip}} N_K(\boldsymbol{x})\,(\mathcal{B}_\alpha(\boldsymbol{x}) - \mathcal{B}_\alpha(\boldsymbol{x}_K))\,\bar{b}_{\alpha K}(t)\end{aligned} \tag{12.21}$$

in which $\psi(\boldsymbol{x})$ is defined for the strong discontinuity based on the Heaviside step function $H(\boldsymbol{x})$ as

$$\psi(\boldsymbol{x}) = H(\boldsymbol{x}) = \begin{cases} +1 & \text{if } (\boldsymbol{x}-\boldsymbol{x}^*)\cdot\mathbf{n}_d \geq 0\\ 0 & \text{otherwise}\end{cases} \tag{12.22}$$

where $\boldsymbol{x}^*$ is the point on the discontinuity which has the closest distance from the point $\boldsymbol{x}$, and $\mathbf{n}_{\Gamma_d}$ is the unit normal vector to the discontinuity at point $\boldsymbol{x}^*$, as shown in Figure 12.2. In Eq. (12.21), $\mathcal{B}_\alpha(\boldsymbol{x})$ is the tip asymptotic functions, which is defined based on the analytical solutions of displacement, pressure, and temperature fields at the vicinity of singularity as

$$\begin{aligned}\mathcal{B}_u(r,\theta) &= \{\mathcal{B}_{u1}, \mathcal{B}_{u2}, \mathcal{B}_{u3}, \mathcal{B}_{u4}\}\\ &= \left\{\sqrt{r}\sin\frac{\theta}{2}, \sqrt{r}\cos\frac{\theta}{2}, \sqrt{r}\sin\frac{\theta}{2}\sin\theta, \sqrt{r}\cos\frac{\theta}{2}\sin\theta\right\}\end{aligned} \tag{12.23a}$$

$$\mathcal{B}_p(r,\theta) = \sqrt{r}\sin\frac{\theta}{2} \tag{12.23b}$$

$$\mathcal{B}_T(r,\theta) = \sqrt{r}\sin\frac{\theta}{2} \tag{12.23c}$$

where θ is in the range of $[-\pi, \pi]$. It must be noted that the value of θ is computed by using the local tip coordinate and obtained by inserting $\theta = \alpha - \pi$. The discrete form of integral equations (12.18) can be obtained in the X-FEM using the test and trial functions for the displacement, pressure, and temperature fields. Applying the enriched field (12.21), the trial functions $\mathbf{u}(\boldsymbol{x}, t)$, $p(\boldsymbol{x}, t)$, and $T(\boldsymbol{x}, t)$ can be defined as

$$\begin{aligned}\mathbf{u}(\boldsymbol{x},t) \approx \mathbf{u}^h(\boldsymbol{x},t) &= \sum_{I\in\mathcal{N}} N_{uI}(\boldsymbol{x})\,\bar{\mathbf{u}}_I(t) + \sum_{J\in\mathcal{N}^{dis}} N_{uJ}(\boldsymbol{x})\,(H(\boldsymbol{x}) - H(\boldsymbol{x}_J))\,\bar{\mathbf{a}}_J(t)\\ &+ \sum_{K\in\mathcal{N}^{tip}} \sum_{\alpha=1}^{4} N_{uK}(\boldsymbol{x})\,(\mathcal{B}_{u\alpha}(\boldsymbol{x}) - \mathcal{B}_{u\alpha}(\boldsymbol{x}_K))\,\bar{\mathbf{b}}_{\alpha K}(t)\end{aligned} \tag{12.24a}$$

$$p(\boldsymbol{x},t) \approx p^h(\boldsymbol{x},t) = \sum_{I \in \mathcal{N}} N_{pI}(\boldsymbol{x})\,\bar{\mathbf{p}}_I(t) + \sum_{J \in \mathcal{N}^{dis}} N_{pJ}(\boldsymbol{x})\,(H(\boldsymbol{x}) - H(\boldsymbol{x}_J))\,\bar{\mathbf{c}}_J(t) + \sum_{K \in \mathcal{N}^{tip}} N_{pK}(\boldsymbol{x})\,(\mathcal{B}_p(\boldsymbol{x}) - \mathcal{B}_p(\boldsymbol{x}_K))\,\bar{\mathbf{d}}_K(t) \tag{12.24b}$$

$$T(\boldsymbol{x},t) \approx T^h(\boldsymbol{x},t) = \sum_{I \in \mathcal{N}} N_{TI}(\boldsymbol{x})\,\bar{\mathbf{T}}_I(t) + \sum_{J \in \mathcal{N}^{dis}} N_{TJ}(\boldsymbol{x})\,(H(\boldsymbol{x}) - H(\boldsymbol{x}_J))\,\bar{\mathbf{e}}_J(t) + \sum_{K \in \mathcal{N}^{tip}} N_{TK}(\boldsymbol{x})\,(\mathcal{B}_T(\boldsymbol{x}) - \mathcal{B}_T(\boldsymbol{x}_K))\,\bar{\mathbf{f}}_K(t) \tag{12.24c}$$

in which variables $\bar{\mathbf{a}}_J$, $\bar{\mathbf{b}}_{\alpha K}$, $\bar{\mathbf{c}}_J$, $\bar{\mathbf{d}}_K$, $\bar{\mathbf{e}}_J$, and $\bar{\mathbf{f}}_K$ are the additional DOF according to additional enrichment functions, and N_u, N_p, and N_T are the standard shape functions of displacement, pressure, and temperature fields. The enriched FE approximation of the displacement, pressure, and temperature fields (12.24) can be symbolically written in the following form

$$\mathbf{u}^h(\boldsymbol{x},t) = \mathbf{N}_u^{std}(\boldsymbol{x})\,\bar{\mathbf{u}}(t) + \mathbf{N}_u^{Hev}(\boldsymbol{x})\,\bar{\mathbf{a}}(t) + \mathbf{N}_u^{tip}(\boldsymbol{x})\,\bar{\mathbf{b}}(t) \tag{12.25a}$$

$$p^h(\boldsymbol{x},t) = \mathbf{N}_p^{std}(\boldsymbol{x})\,\bar{\mathbf{p}}(t) + \mathbf{N}_p^{Hev}(\boldsymbol{x})\,\bar{\mathbf{c}}(t) + \mathbf{N}_p^{tip}(\boldsymbol{x})\,\bar{\mathbf{d}}(t) \tag{12.25b}$$

$$T^h(\boldsymbol{x},t) = \mathbf{N}_T^{std}(\boldsymbol{x})\,\bar{\mathbf{T}}(t) + \mathbf{N}_T^{Hev}(\boldsymbol{x})\,\bar{\mathbf{e}}(t) + \mathbf{N}_T^{tip}(\boldsymbol{x})\,\bar{\mathbf{f}}(t) \tag{12.25c}$$

where $\mathbf{N}_u^{std}(\boldsymbol{x})$ is the matrix of standard displacement shape functions, $\mathbf{N}_u^{Hev}(\boldsymbol{x})$ and $\mathbf{N}_u^{tip}(\boldsymbol{x})$ are the matrices of enriched displacement shape functions associated with the Heaviside and asymptotic tip functions; $\mathbf{N}_p^{std}(\boldsymbol{x})$ is the matrix of standard pressure shape functions, $\mathbf{N}_p^{Hev}(\boldsymbol{x})$ and $\mathbf{N}_p^{tip}(\boldsymbol{x})$ are the matrices of enriched pressure shape functions associated with the Heaviside and asymptotic tip functions; $\mathbf{N}_T^{std}(\boldsymbol{x})$ is the matrix of standard temperature shape functions, and $\mathbf{N}_T^{Hev}(\boldsymbol{x})$ and $\mathbf{N}_T^{tip}(\boldsymbol{x})$ are the matrices of enriched temperature shape functions associated with the Heaviside and asymptotic tip functions.

Accordingly, the strain vector corresponding to the approximate displacement field (12.25a) can be written in terms of the standard and enriched nodal values as

$$\boldsymbol{\varepsilon}^h(\boldsymbol{x},t) = \mathbf{B}_u^{std}(\boldsymbol{x})\,\bar{\mathbf{u}}(t) + \mathbf{B}_u^{Hev}(\boldsymbol{x})\,\bar{\mathbf{a}}(t) + \mathbf{B}_u^{tip}(\boldsymbol{x})\,\bar{\mathbf{b}}(t) \tag{12.26}$$

where $\mathbf{B}_u^{std}(\boldsymbol{x}) \equiv \mathbf{L}\,\mathbf{N}_u^{std}(\boldsymbol{x})$ involve the spatial derivatives of the standard shape functions, $\mathbf{B}_u^{Hev}(\boldsymbol{x}) \equiv \mathbf{L}\,\mathbf{N}_u^{Hev}(\boldsymbol{x})$ contain the spatial derivatives of the enriched shape functions associated with the Heaviside enrichment function, and $\mathbf{B}_u^{tip}(\boldsymbol{x}) \equiv \mathbf{L}\,\mathbf{N}_u^{tip}(\boldsymbol{x})$ contain the spatial derivatives of the enriched shape functions associated with the tip asymptotic functions, where $\mathbf{L}$ is defined as

$$\mathbf{L} = \begin{bmatrix} \partial/\partial x & 0 \\ 0 & \partial/\partial y \\ \partial/\partial y & \partial/\partial x \end{bmatrix} \tag{12.27}$$

12.4.2 The X-FEM Spatial Discretization

The discrete form of integral equations (12.18) in the framework of X-FEM formulation can be obtained using the test and trial functions of the enriched displacement, pressure, and temperature fields. Applying the trial functions of enriched displacement, pressure, and temperature fields (12.24) together with the test functions of these approximation fields $\delta\mathbf{u}(\boldsymbol{x},t)$, $\delta p(\boldsymbol{x},t)$, and $\delta T(\boldsymbol{x},t)$, which are respectively defined in the same enriched approximating space as

$$\begin{aligned}\delta\mathbf{u}(\boldsymbol{x},t)\approx\delta\mathbf{u}^h(\boldsymbol{x},t)&=\sum_{I\in\mathcal{N}}N_{uI}(\boldsymbol{x})\,\delta\bar{\mathbf{u}}_I(t)+\sum_{J\in\mathcal{N}^{dis}}N_{uJ}(\boldsymbol{x})\,(H(\boldsymbol{x})-H(\boldsymbol{x}_J))\,\delta\bar{\mathbf{a}}_J(t)\\&+\sum_{K\in\mathcal{N}^{tip}}\sum_{\alpha=1}^{4}N_{uK}(\boldsymbol{x})\,(\mathcal{B}_{u\alpha}(\boldsymbol{x})-\mathcal{B}_{u\alpha}(\boldsymbol{x}_K))\,\delta\bar{\mathbf{b}}_{\alpha K}(t)\end{aligned} \tag{12.28a}$$

$$\begin{aligned}\delta p(\boldsymbol{x},t)\approx\delta p^h(\boldsymbol{x},t)&=\sum_{I\in\mathcal{N}}N_{pI}(\boldsymbol{x})\,\delta\bar{\mathbf{p}}_I(t)+\sum_{J\in\mathcal{N}^{dis}}N_{pJ}(\boldsymbol{x})\,(H(\boldsymbol{x})-H(\boldsymbol{x}_J))\,\delta\bar{\mathbf{c}}_J(t)\\&+\sum_{K\in\mathcal{N}^{tip}}N_{pK}(\boldsymbol{x})\,(\mathcal{B}_p(\boldsymbol{x})-\mathcal{B}_p(\boldsymbol{x}_K))\,\delta\bar{\mathbf{d}}_K(t)\end{aligned} \tag{12.28b}$$

$$\begin{aligned}\delta T(\boldsymbol{x},t)\approx\delta T^h(\boldsymbol{x},t)&=\sum_{I\in\mathcal{N}}N_{TI}(\boldsymbol{x})\,\delta\bar{\mathbf{T}}_I(t)+\sum_{J\in\mathcal{N}^{dis}}N_{TJ}(\boldsymbol{x})\,(H(\boldsymbol{x})-H(\boldsymbol{x}_J))\,\delta\bar{\mathbf{e}}_J(t)\\&+\sum_{K\in\mathcal{N}^{tip}}N_{TK}(\boldsymbol{x})\,(\mathcal{B}_T(\boldsymbol{x})-\mathcal{B}_T(\boldsymbol{x}_K))\,\delta\bar{\mathbf{f}}_K(t)\end{aligned} \tag{12.28c}$$

or in the symbolic form as

$$\delta\mathbf{u}^h(\boldsymbol{x},t)=\mathbf{N}_u^{std}(\boldsymbol{x})\,\delta\bar{\mathbf{u}}(t)+\mathbf{N}_u^{Hev}(\boldsymbol{x})\,\delta\bar{\mathbf{a}}(t)+\mathbf{N}_u^{tip}(\boldsymbol{x})\,\delta\bar{\mathbf{b}}(t) \tag{12.29a}$$

$$\delta p^h(\boldsymbol{x},t)=\mathbf{N}_p^{std}(\boldsymbol{x})\,\delta\bar{\mathbf{p}}(t)+\mathbf{N}_p^{Hev}(\boldsymbol{x})\,\delta\bar{\mathbf{c}}(t)+\mathbf{N}_p^{tip}(\boldsymbol{x})\,\delta\bar{\mathbf{d}}(t) \tag{12.29b}$$

$$\delta T^h(\boldsymbol{x},t)=\mathbf{N}_T^{std}(\boldsymbol{x})\,\delta\bar{\mathbf{T}}(t)+\mathbf{N}_T^{Hev}(\boldsymbol{x})\,\delta\bar{\mathbf{e}}(t)+\mathbf{N}_T^{tip}(\boldsymbol{x})\,\delta\bar{\mathbf{f}}(t) \tag{12.29c}$$

The discretized form of integral equations (12.18) can therefore be obtained according to the Bubnov–Galerkin technique as

$$\begin{aligned}&\begin{pmatrix}\mathbf{M}_{uu}&\mathbf{M}_{ua}&\mathbf{M}_{ub}\\\mathbf{M}_{au}&\mathbf{M}_{aa}&\mathbf{M}_{ab}\\\mathbf{M}_{bu}&\mathbf{M}_{ba}&\mathbf{M}_{bb}\end{pmatrix}\begin{Bmatrix}\ddot{\bar{\mathbf{u}}}\\\ddot{\bar{\mathbf{a}}}\\\ddot{\bar{\mathbf{b}}}\end{Bmatrix}+\begin{pmatrix}\mathbf{K}_{uu}&\mathbf{K}_{ua}&\mathbf{K}_{ub}\\\mathbf{K}_{au}&\mathbf{K}_{aa}&\mathbf{K}_{ab}\\\mathbf{K}_{bu}&\mathbf{K}_{ba}&\mathbf{K}_{bb}\end{pmatrix}\begin{Bmatrix}\bar{\mathbf{u}}\\\bar{\mathbf{a}}\\\bar{\mathbf{b}}\end{Bmatrix}-\begin{pmatrix}\mathbf{Q}_{up}&\mathbf{Q}_{uc}&\mathbf{Q}_{ud}\\\mathbf{Q}_{ap}&\mathbf{Q}_{ac}&\mathbf{Q}_{ad}\\\mathbf{Q}_{bp}&\mathbf{Q}_{bc}&\mathbf{Q}_{bd}\end{pmatrix}\begin{Bmatrix}\bar{\mathbf{p}}\\\bar{\mathbf{c}}\\\bar{\mathbf{d}}\end{Bmatrix}\\&-\begin{pmatrix}\mathbf{W}_{ut}&\mathbf{W}_{ue}&\mathbf{W}_{uf}\\\mathbf{W}_{at}&\mathbf{W}_{ae}&\mathbf{W}_{af}\\\mathbf{W}_{bt}&\mathbf{W}_{be}&\mathbf{W}_{bf}\end{pmatrix}\begin{Bmatrix}\bar{\mathbf{T}}\\\bar{\mathbf{e}}\\\bar{\mathbf{f}}\end{Bmatrix}-\begin{Bmatrix}\mathbf{f}_u^{\text{ext}}\\\mathbf{f}_a^{\text{ext}}\\\mathbf{f}_b^{\text{ext}}\end{Bmatrix}=\mathbf{0}\\&\begin{pmatrix}\mathbf{Q}_{up}^T&\mathbf{Q}_{ap}^T&\mathbf{Q}_{bp}^T\\\mathbf{Q}_{uc}^T&\mathbf{Q}_{ac}^T&\mathbf{Q}_{bc}^T\\\mathbf{Q}_{ud}^T&\mathbf{Q}_{ad}^T&\mathbf{Q}_{bd}^T\end{pmatrix}\begin{Bmatrix}\dot{\bar{\mathbf{u}}}\\\dot{\bar{\mathbf{a}}}\\\dot{\bar{\mathbf{b}}}\end{Bmatrix}+\begin{pmatrix}\mathbf{H}_{pp}&\mathbf{H}_{pc}&\mathbf{H}_{pd}\\\mathbf{H}_{cp}&\mathbf{H}_{cc}&\mathbf{H}_{cd}\\\mathbf{H}_{dp}&\mathbf{H}_{dc}&\mathbf{H}_{dd}\end{pmatrix}\begin{Bmatrix}\bar{\mathbf{p}}\\\bar{\mathbf{c}}\\\bar{\mathbf{d}}\end{Bmatrix}+\begin{pmatrix}\mathbf{S}_{pp}&\mathbf{S}_{pc}&\mathbf{S}_{pd}\\\mathbf{S}_{cp}&\mathbf{S}_{cc}&\mathbf{S}_{cd}\\\mathbf{S}_{dp}&\mathbf{S}_{dc}&\mathbf{S}_{dd}\end{pmatrix}\begin{Bmatrix}\dot{\bar{\mathbf{p}}}\\\dot{\bar{\mathbf{c}}}\\\dot{\bar{\mathbf{d}}}\end{Bmatrix}\\&+\begin{pmatrix}\mathbf{R}_{pt}&\mathbf{R}_{pe}&\mathbf{R}_{pf}\\\mathbf{R}_{ct}&\mathbf{R}_{ce}&\mathbf{R}_{cf}\\\mathbf{R}_{dt}&\mathbf{R}_{de}&\mathbf{R}_{df}\end{pmatrix}\begin{Bmatrix}\dot{\bar{\mathbf{T}}}\\\dot{\bar{\mathbf{e}}}\\\dot{\bar{\mathbf{f}}}\end{Bmatrix}-\begin{Bmatrix}\mathbf{q}_p^{\text{ext}}\\\mathbf{q}_c^{\text{ext}}\\\mathbf{q}_d^{\text{ext}}\end{Bmatrix}=\mathbf{0}\\&\begin{pmatrix}\mathbf{L}_{tt}&\mathbf{L}_{te}&\mathbf{L}_{tf}\\\mathbf{L}_{et}&\mathbf{L}_{ee}&\mathbf{L}_{ef}\\\mathbf{L}_{ft}&\mathbf{L}_{fe}&\mathbf{L}_{ff}\end{pmatrix}\begin{Bmatrix}\dot{\bar{\mathbf{T}}}\\\dot{\bar{\mathbf{e}}}\\\dot{\bar{\mathbf{f}}}\end{Bmatrix}+\begin{pmatrix}\mathbf{C}_{tt}&\mathbf{C}_{te}&\mathbf{C}_{tf}\\\mathbf{C}_{et}&\mathbf{C}_{ee}&\mathbf{C}_{ef}\\\mathbf{C}_{ft}&\mathbf{C}_{fe}&\mathbf{C}_{ff}\end{pmatrix}\begin{Bmatrix}\bar{\mathbf{T}}\\\bar{\mathbf{e}}\\\bar{\mathbf{f}}\end{Bmatrix}-\begin{Bmatrix}\mathbf{g}_t^{\text{ext}}\\\mathbf{g}_e^{\text{ext}}\\\mathbf{g}_f^{\text{ext}}\end{Bmatrix}=\mathbf{0}\end{aligned} \tag{12.30}$$

where the matrices **M**, **K**, **Q**, **W**, **H**, **S**, **R**, **L**, **C** and external force vectors **f**, **q**, and **g** are defined as

$$\begin{aligned}
\mathbf{M}_{\alpha\beta} &= \int_{\Omega} (\mathbf{N}_u^{\alpha})^T \rho \mathbf{N}_u^{\beta}\, d\Omega \\
\mathbf{K}_{\alpha\beta} &= \int_{\Omega} (\mathbf{B}_u^{\alpha})^T \mathbf{D} \mathbf{B}_u^{\beta}\, d\Omega \\
\mathbf{Q}_{\alpha\gamma} &= \int_{\Omega} (\mathbf{B}_u^{\alpha})^T \alpha \mathbf{m} \mathbf{N}_p^{\gamma}\, d\Omega \\
\mathbf{W}_{\alpha\kappa} &= \int_{\Omega} (\mathbf{B}_u^{\alpha})^T \mathbf{D} \frac{1}{3}\beta \mathbf{m} \mathbf{N}_T^{\kappa}\, d\Omega \\
\mathbf{f}_{\alpha}^{\text{ext}} &= \int_{\Omega} (\mathbf{N}_u^{\alpha})^T \rho \mathbf{b}\, d\Omega + \int_{\Gamma_t} (\mathbf{N}_u^{\alpha})^T \bar{\mathbf{t}}\, d\Gamma
\end{aligned} \tag{12.31}$$

and

$$\begin{aligned}
\mathbf{H}_{\delta\gamma} &= \int_{\Omega} \left(\nabla \mathbf{N}_p^{\delta}\right)^T \boldsymbol{k}_f \left(\nabla \mathbf{N}_p^{\gamma}\right) d\Omega \\
\mathbf{S}_{\delta\gamma} &= \int_{\Omega} \left(\mathbf{N}_p^{\delta}\right)^T \frac{1}{Q} \mathbf{N}_p^{\gamma}\, d\Omega \\
\mathbf{R}_{\delta\kappa} &= \int_{\Omega} \left(\mathbf{N}_p^{\delta}\right)^T \beta\, \mathbf{N}_T^{\kappa}\, d\Omega \\
\mathbf{q}_{\delta}^{\text{ext}} &= \int_{\Omega} \left(\nabla \mathbf{N}_p^{\delta}\right)^T \boldsymbol{k}_f \rho_f \mathbf{b}\, d\Omega - \int_{\Gamma_w} \left(\mathbf{N}_p^{\delta}\right)^T \bar{q}_w\, d\Gamma
\end{aligned} \tag{12.32}$$

$$\begin{aligned}
\mathbf{L}_{\lambda\kappa} &= \int_{\Omega} (\mathbf{N}_T^{\lambda})^T (\rho c)_{avg} \mathbf{N}_T^{\kappa}\, d\Omega \\
\mathbf{C}_{\lambda\kappa} &= \int_{\Omega} (\mathbf{N}_T^{\lambda})^T \left(\rho_f c_f \left[\boldsymbol{k}_f(-\nabla p + \rho \mathbf{b})\right]\right) \nabla \mathbf{N}_T^{\kappa}\, d\Omega + \int_{\Omega} (\nabla \mathbf{N}_T^{\lambda})^T \boldsymbol{\kappa} \nabla \mathbf{N}_T^{\kappa}\, d\Omega \\
\mathbf{g}_{\lambda}^{\text{ext}} &= \int_{\Gamma_T} (\mathbf{N}_T^{\lambda})^T \bar{q}_T\, d\Gamma
\end{aligned} \tag{12.33}$$

in which $(\alpha, \beta) \in (u^{std}, a^{Hev}, b^{tip})$ denote the "standard", "Heaviside", and "asymptotic tip" functions of the displacement field, $(\delta, \gamma) \in (p^{std}, c^{Hev}, d^{tip})$ denote the "standard", "Heaviside", and "asymptotic tip" functions of the pressure field, and $(\lambda, \kappa) \in (T^{std}, e^{Hev}, f^{tip})$ denote the "standard", "Heaviside", and "asymptotic tip" functions of the temperature field. In these definitions, **m** is the vector of delta Dirac function defined as $\mathbf{m} = \{1\ \ 1\ \ 0\}^T$.

Finally, the X-FEM discretization equations (12.30) can be rewritten as

$$\begin{aligned}
&\mathbf{M}\ddot{\bar{\mathbb{U}}} + \mathbf{K}\bar{\mathbb{U}} - \mathbf{Q}\bar{\mathbb{P}} - \mathbf{W}\bar{\mathbb{T}} - \mathbf{f}^{\text{ext}} = \mathbf{0} \\
&\mathbf{Q}^T\dot{\bar{\mathbb{U}}} + \mathbf{H}\bar{\mathbb{P}} + \mathbf{S}\dot{\bar{\mathbb{P}}} + \mathbf{R}\dot{\bar{\mathbb{T}}} - \mathbf{q}^{\text{ext}} = \mathbf{0} \\
&\mathbf{L}\dot{\bar{\mathbb{T}}} + \mathbf{C}\bar{\mathbb{T}} - \mathbf{g}^{\text{ext}} = \mathbf{0}
\end{aligned} \tag{12.34}$$

where $\bar{\mathbb{U}}^T = \left\langle \bar{\mathbf{u}}^T, \bar{\mathbf{a}}^T, \bar{\mathbf{b}}^T \right\rangle$, $\bar{\mathbb{P}}^T = \left\langle \bar{\mathbf{p}}^T, \bar{\mathbf{c}}^T, \bar{\mathbf{d}}^T \right\rangle$ and $\bar{\mathbb{T}}^T = \left\langle \bar{\mathbf{T}}^T, \bar{\mathbf{e}}^T, \bar{\mathbf{f}}^T \right\rangle$ are the complete set of the standard and enriched DOF of displacement, pressure, and temperature fields, respectively.

12.4.3 The Time Domain Discretization

In order to complete the numerical solution of X-FEM equations, it is necessary to integrate the differential equations (12.34) in time. The set of equations is discretized in the time domain following the line of the well-known Newmark scheme, in which it is supposed that the system of equations is satisfied at each discrete time step. In this manner, the problem is solved at a series of sequential discrete time steps. To advance the solution in time, the link between the successive values of the unknown field variables at time t_{n+1} and the known field variables at time t_n is established by applying the minimum order of the generalized Newmark scheme required considering the highest order of the time derivatives in the differential equations. The Newmark GN22 method is employed for the displacement field $\bar{\mathbb{U}}$, and the GN11 method for the pressure field $\bar{\mathbb{P}}$ and the temperature field $\bar{\mathbb{T}}$ as

$$\begin{aligned}
\ddot{\bar{\mathbb{U}}}^{t+\Delta t} &= \ddot{\bar{\mathbb{U}}}^{t} + \Delta\ddot{\bar{\mathbb{U}}}^{t} \\
\dot{\bar{\mathbb{U}}}^{t+\Delta t} &= \dot{\bar{\mathbb{U}}}^{t} + \ddot{\bar{\mathbb{U}}}^{t}\Delta t + \beta_1 \Delta\ddot{\bar{\mathbb{U}}}^{t}\Delta t \\
\bar{\mathbb{U}}^{t+\Delta t} &= \bar{\mathbb{U}}^{t} + \dot{\bar{\mathbb{U}}}^{t}\Delta t + \frac{1}{2}\ddot{\bar{\mathbb{U}}}^{t}\Delta t^2 + \frac{1}{2}\beta_2 \Delta\ddot{\bar{\mathbb{U}}}^{t}\Delta t^2
\end{aligned} \tag{12.35}$$

and

$$\begin{aligned}
\dot{\bar{\mathbb{P}}}^{t+\Delta t} &= \dot{\bar{\mathbb{P}}}^{t} + \Delta\dot{\bar{\mathbb{P}}}^{t} \\
\bar{\mathbb{P}}^{t+\Delta t} &= \bar{\mathbb{P}}^{t} + \dot{\bar{\mathbb{P}}}^{t}\Delta t + \bar{\beta}\Delta\dot{\bar{\mathbb{P}}}^{t}\Delta t
\end{aligned} \tag{12.36}$$

and

$$\begin{aligned}
\dot{\bar{\mathbb{T}}}^{t+\Delta t} &= \dot{\bar{\mathbb{T}}}^{t} + \Delta\dot{\bar{\mathbb{T}}}^{t} \\
\bar{\mathbb{T}}^{t+\Delta t} &= \bar{\mathbb{T}}^{t} + \dot{\bar{\mathbb{T}}}^{t}\Delta t + \bar{\gamma}\Delta\dot{\bar{\mathbb{T}}}^{t}\Delta t
\end{aligned} \tag{12.37}$$

where $\beta_1, \beta_2, \bar{\beta}$, and $\bar{\gamma}$ are parameters, which are chosen in the range of 0–1. However, for unconditionally stability of the algorithm, it is required that $\beta_1 \geq \beta_2 \geq \frac{1}{2}$, $\bar{\beta} \geq \frac{1}{2}$, and $\bar{\gamma} \geq \frac{1}{2}$. In these relations, $\bar{\mathbb{U}}^{t}$, $\dot{\bar{\mathbb{U}}}^{t}$, and $\ddot{\bar{\mathbb{U}}}^{t}$ denote the values of displacement, velocity, and acceleration of the standard and enriched DOF at time t, $\bar{\mathbb{P}}^{t}$, and $\dot{\bar{\mathbb{P}}}^{t}$ are the values of pressure and gradient of pressure of the standard and enriched DOF at time t, and $\bar{\mathbb{T}}^{t}$ and $\dot{\bar{\mathbb{T}}}^{t}$ are the values of temperature and gradient of temperature of the standard and enriched DOF at time t. Substituting relations (12.35), (12.36), and (12.37) into the space-discrete equations (12.34), the following nonlinear equation can be achieved as

$$\begin{pmatrix}
\mathbf{M} + \frac{1}{2}\beta_2 \Delta t^2 \mathbf{K} & -\bar{\beta}\Delta t\mathbf{Q} & -\bar{\gamma}\,\Delta t\mathbf{W} \\
\beta_1\,\Delta t\mathbf{Q}^T & \mathbf{S} + \bar{\beta}\Delta t\mathbf{H} & \mathbf{R} \\
\mathbf{0} & \mathbf{0} & \mathbf{L} + \bar{\gamma}\,\Delta t\mathbf{C}
\end{pmatrix}
\begin{Bmatrix} \Delta\ddot{\bar{\mathbb{U}}}^{t} \\ \Delta\dot{\bar{\mathbb{P}}}^{t} \\ \Delta\dot{\bar{\mathbb{T}}}^{t} \end{Bmatrix}
=
\begin{Bmatrix} \mathbb{G}^{(1)} \\ \mathbb{G}^{(2)} \\ \mathbb{G}^{(3)} \end{Bmatrix} \tag{12.38}$$

where the right-hand side of equation (12.38) denotes the vector of known values at time t defined as

$$\begin{aligned}
\mathbb{G}^{(1)} &= {}^{t+\Delta t}\mathbf{f}^{\text{ext}} - \mathbf{M}\ddot{\bar{\mathbb{U}}}^{t} - \mathbf{K}\left(\bar{\mathbb{U}}^{t} + \dot{\bar{\mathbb{U}}}^{t}\Delta t + \frac{1}{2}\ddot{\bar{\mathbb{U}}}^{t}\Delta t^2\right) + \mathbf{Q}\left(\bar{\mathbb{P}}^{t} + \dot{\bar{\mathbb{P}}}^{t}\Delta t\right) + \mathbf{W}\left(\bar{\mathbb{T}}^{t} + \dot{\bar{\mathbb{T}}}^{t}\Delta t\right) \\
\mathbb{G}^{(2)} &= {}^{t+\Delta t}\mathbf{q}^{\text{ext}} - \mathbf{Q}^T\left(\dot{\bar{\mathbb{U}}}^{t} + \ddot{\bar{\mathbb{U}}}^{t}\Delta t\right) - \mathbf{H}\left(\bar{\mathbb{P}}^{t} + \dot{\bar{\mathbb{P}}}^{t}\Delta t\right) - \mathbf{S}\dot{\bar{\mathbb{P}}}^{t} - \mathbf{R}\dot{\bar{\mathbb{T}}}^{t} \\
\mathbb{G}^{(3)} &= {}^{t+\Delta t}\mathbf{g}^{\text{ext}} - \mathbf{L}\,\dot{\bar{\mathbb{T}}}^{t} - \mathbf{C}\left(\bar{\mathbb{T}}^{t} + \dot{\bar{\mathbb{T}}}^{t}\Delta t\right)
\end{aligned} \tag{12.39}$$

The set of nonlinear equations (12.38) can be solved using an appropriate approach, such as the Newton–Raphson procedure.

12.5 Application of the X-FEM Method to THM Behavior of Porous Media

In order to illustrate the accuracy and versatility of the extended FEM in THM modeling of deformable porous media, several numerical examples are presented. The first two examples are chosen to illustrate the robustness and accuracy of computational algorithm for two benchmark problems; the first example illustrates the accuracy of the X-FEM model in the heat transfer analysis of a plate with an inclined crack; and the second example deals with the thermo-mechanical analysis of a plate with an edge crack to verify the stress intensity factor obtained from the numerical analysis with that reported by the analytical solution. In the third example, the X-FEM technique is performed in the hydromechanical analysis of an impermeable discontinuity in the saturated porous media. Finally, the last example is chosen to demonstrate the performance of proposed computational algorithm for the THM modeling of deformable porous media in a challenging problem, where an inclined fault is modeled in a porous medium with various fault angles, and the case of randomly generated faults in a saturated porous medium.

12.5.1 A Plate with an Inclined Crack in Thermal Loading

The first example is of a plate with an inclined crack subjected to thermal loading, as shown in Figure 12.3. This example was modeled by Duflot (2008) and Zamani, Gracie, and Eslami (2010) to present their X-FEM formulation in the thermo-elastic fracture analysis. A rectangular plate of 2×1 m is modeled with

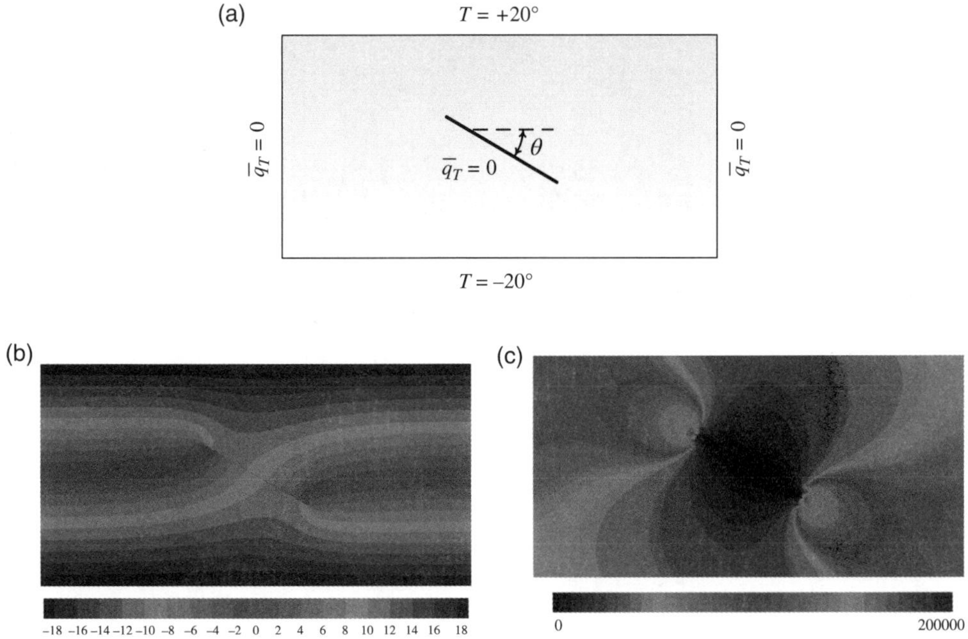

Figure 12.3 A plate with an inclined crack: (a) geometry and boundary conditions, (b) the distribution of temperature, (c) the distribution of heat flux (For color details, please see color plate section)

the crack length of 0.6 m located at the center of plate with the angle of $\theta = 30^{\circ}$. The prescribed temperatures of +20 and −20 °C are implied on the upper and lower edges of the plate, respectively. It is assumed that the normal flux from the left and right edges of the plate is zero, and the heat conductivity is equal to 837w/m°C. A structured uniform mesh of 61 × 31 is used for the X-FEM analysis. In Figure 12.3b, the distribution of temperature contour is shown for the steady state condition. The temperature contour is in complete agreement with that reported by Duflot (2008) and Zamani, Gracie, and Eslami (2010). Obviously, the Heaviside enrichment function causes the jump in the thermal field at the crack faces. In Figure 12.3c, the heat flux distribution is shown in the normal direction to crack, that is, $-\mathbf{k}\,\nabla T\,\mathbf{n}_{\Gamma_d}$, where the singularity can be seen at both crack tips. Clearly, a good agreement can be seen between the proposed model and that reported by Zamani, Gracie, and Eslami (2010).

12.5.2 A Plate with an Edge Crack in Thermal Loading

The second example is a plate with an edge crack under thermal loading, as shown in Figure 12.4a. This example is chosen to illustrate the accuracy of stress intensity factor obtained by the proposed X-FEM modeling of thermo-mechanical analysis. A rectangular plate of 0.5 × 20 m is assumed with an edge crack of 0.25 m at the mid-edge of plate. The prescribed temperatures of −50 and +50 °C are imposed at the left and right edges of the plate, respectively. A structured uniform X-FEM mesh of 20 × 80 is employed for the thermo-mechanical analysis in the plane strain condition. The material properties of the plate are given in Table 12.1. The coupled thermo-elasticity analysis is performed for 100 s. In Figure 12.4b–d, the distribution of temperature contour is shown at the end of simulation together with the displacement contours in *x*- and *y*-directions. According to the thermal boundary conditions, a contraction can be observed on the lower and upper edges of the crack. In order to illustrate the accuracy of proposed computational

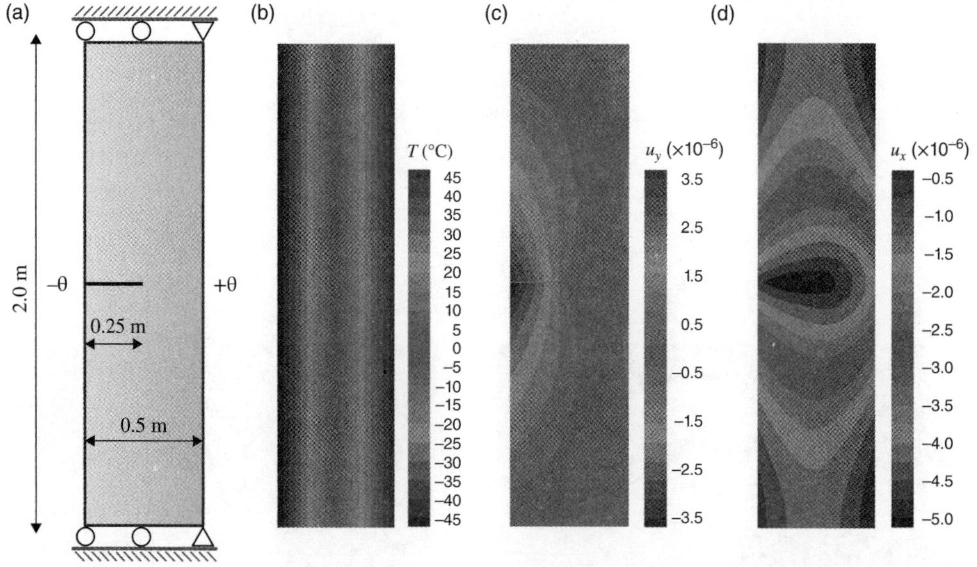

Figure 12.4 A plate with an edge crack in thermal loading: (a) the geometry and boundary conditions, (b) the distribution of temperature, (c) the displacement contour in *y*-direction, (d) the displacement contour in the *x*-direction (For color details, please see color plate section)

Table 12.1 Material parameters for a plate with an edge crack in thermal loading

Young's modulus	9×10^6 (kPa)
Poisson ratio	0.3
Solid density	2×10^3 (kg/m^3)
Thermal conductivity	1×10^3 (W/m °C)
Volumetric thermal expansion coefficient	3×10^{-7} (1/°C)

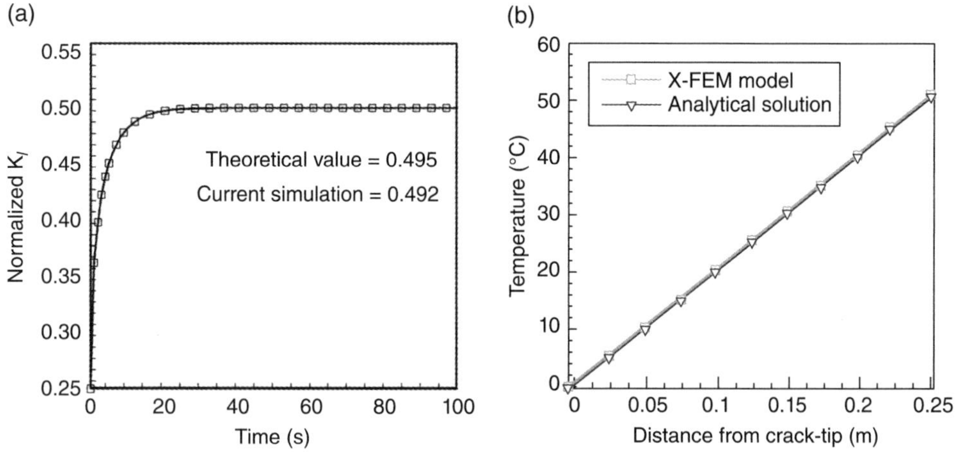

Figure 12.5 A comparison between the numerical result and analytical solution for a plate with an edge crack in thermal loading: (a) the variation with time of normalized stress intensity factor, (b) the variation of temperature with the distance from crack tip

algorithm, the normalized stress intensity factor (SIF) is compared with the analytical solution in Figure 12.5a, in which the normalized SIF is obtained as

$$K_I^{\text{Normalized}} = \frac{3K_I}{\frac{E}{1-\nu}\beta\,\theta_0\sqrt{\pi a}} \tag{12.40}$$

in which E is the elasticity module, a is the crack length, and θ_0 is the absolute value of the prescribed temperature applied at each edge of the plate. It can be seen from Figure 12.5a that the steady state condition occurs after 20 s, where the normalized SIF value is equal to 0.492. Obviously, a good agreement can be seen between the computed value of normalized SIF (0.492) and the analytical value (0.495) reported by Duflot (2008). In Figure 12.5b, a comparison is performed between the temperature obtained along the horizontal direction and that computed from the analytical solution using $\theta(x) = (2x/w)\,\theta_0$, where $\theta(x)$ is the temperature distribution in the horizontal direction at the position of x from the crack tip, and w is the width of the plate. A complete agreement can be seen between the numerical result and analytical solution.

In order to investigate the importance of crack tip enrichment, the variations with time of the temperature gradient in x-direction and the normal stress σ_y are plotted in Figure 12.6 at the crack tip region with and without the tip enrichment function. Obviously, no difference can be observed in the heat flux curve between the two cases, as shown in Figure 12.6a. The main reason is that the thermal streamlines are in the

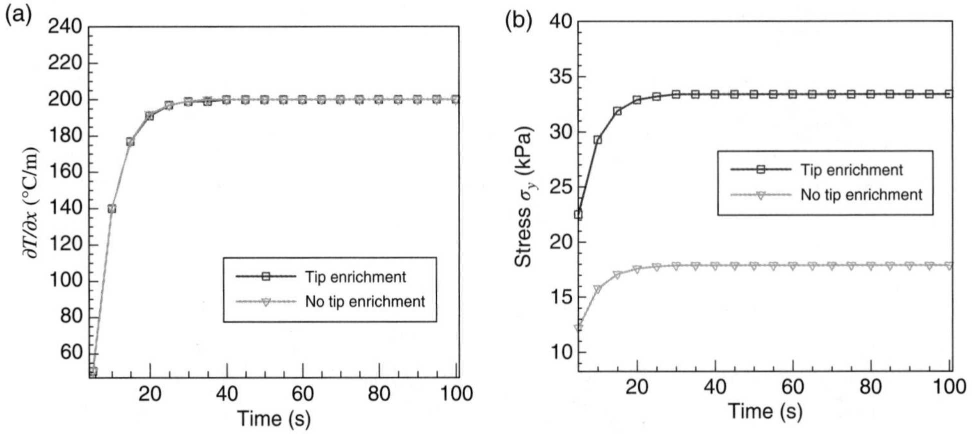

Figure 12.6 The effect of crack tip enrichment on: (a) the temperature gradient in x-direction, (b) the normal stress σ_y

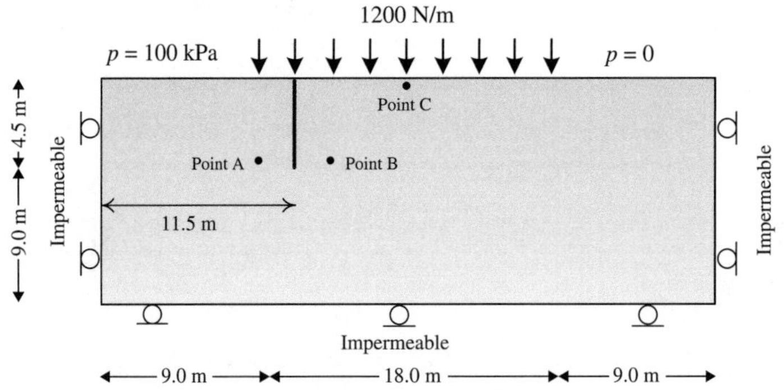

Figure 12.7 An impermeable discontinuity in a saturated porous medium: The geometry and boundary conditions

direction of crack edges, and the discontinuity cannot affect the heat flux paths. As a result, the enrichment DOF $\bar{\mathbf{f}}_K$, defined in Eq. (12.24c), has zero value and the results of temperature gradient between the two cases are identical. However, due to the increase of crack mouth opening, the tip enrichment function has significant affect in the normal stress σ_y, as shown in Figure 12.6b.

12.5.3 An Impermeable Discontinuity in Saturated Porous Media

In the next example, the performance of X-FEM model is presented in the hydromechanical modeling of an impermeable discontinuity in fully saturated porous media, as shown in Figure 12.7. In dam engineering problems, sheet pile is commonly used under the dam to decrease the pore pressure and hydraulic gradient downstream of the dam. In this example, the sheet-pile is considered an impermeable discontinuity in the hydromechanical analysis of saturated porous media. The geometry and boundary conditions of the problem are shown in Figure 12.7. A prescribed pressure of 100 kPa is assumed upstream, and zero pressure downstream. The soil is subjected to the distributed dam weight on the upper surface with the

Table 12.2 Material parameters for an impermeable discontinuity in saturated porous media

Young's modulus	9×10^6 (kPa)
Poisson ratio	0.4
Solid density	2×10^3 (kg/m^3)
Fluid density	1×10^3 (kg/m^3)
Porosity	0.3
Bulk module of fluid	2×10^3 (MPa)
Bulk module of solid	1×10^{14} (MPa)
Permeability	1×10^{-9} (m^2/Pa s)

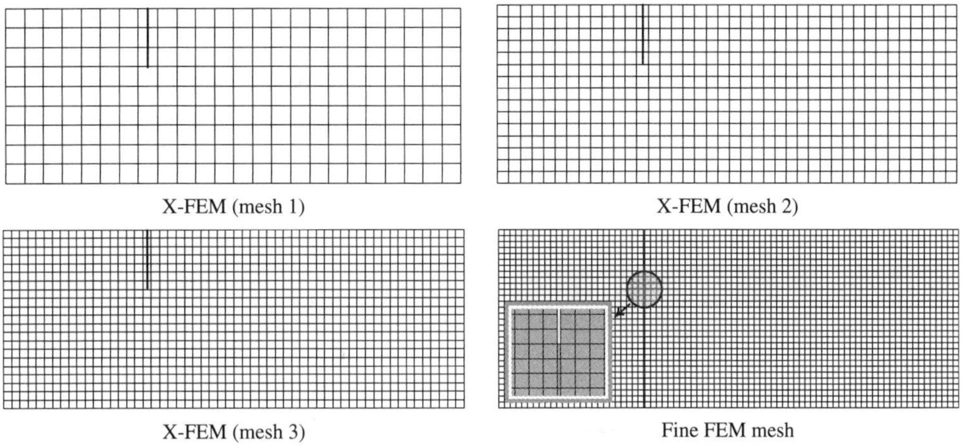

Figure 12.8 Modeling of an impermeable discontinuity in saturated porous media using various X-FEM meshes

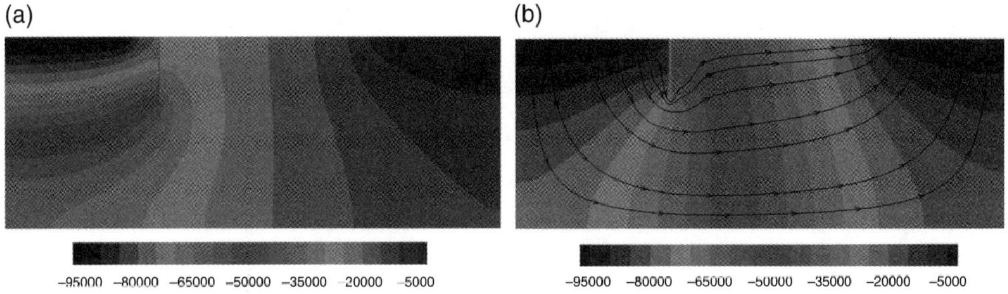

Figure 12.9 The distribution of pressure contours at time steps (a) $t = 10$ and (b) 110 s (For color details, please see color plate section)

material properties given in Table 12.2. In order to investigate the effect of mesh size in X-FEM analysis, three X-FEM meshes are employed and the results are compared with a very fine FE mesh, as shown in Figure 12.8. The analysis is performed for 110 s with the time increment of 0.1 s using the full Newton–Raphson procedure.

In Figure 12.9, the distributions of pressure contour are shown in time steps of $t = 10$ and 110 s. In addition, the fluid flow streamlines are plotted in Figure 12.9b at the end of simulation to illustrate the flow paths in the soil saturated medium. Clearly, the presence of impermeable discontinuity can be seen in the pressure contours around the sheet-pile. In Figure 12.10a, the variations with time of the pressure are plotted for various X-FEM meshes at three points A, B, and C, given in Table 12.3. In this figure, the evolutions of pressure are compared with the FEM technique that illustrates a remarkable agreement between the X-FEM and FEM approaches. In Figure 12.10b, the variations with time of the vertical

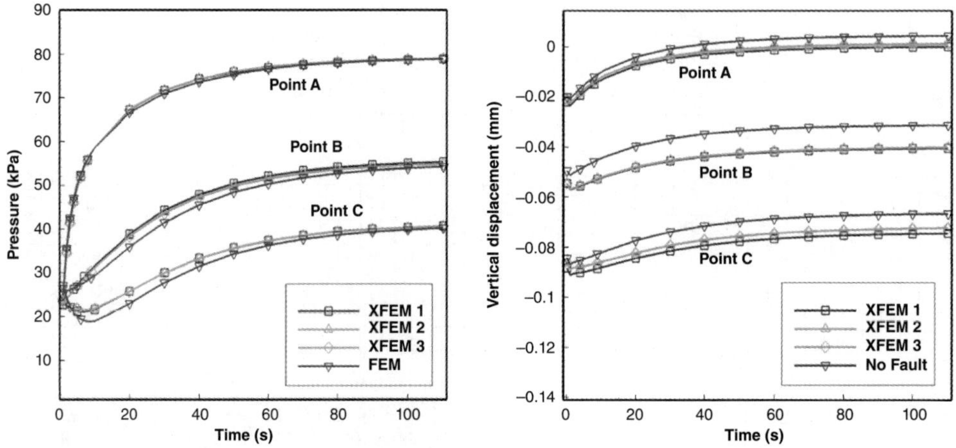

Figure 12.10 The variations with time of the pressure and vertical displacement at points A, B, and C

Table 12.3 Position of the points

	x (m)	y (m)
Point A	9.0	9.0
Point B	13.5	9.0
Point C	18.0	13.5

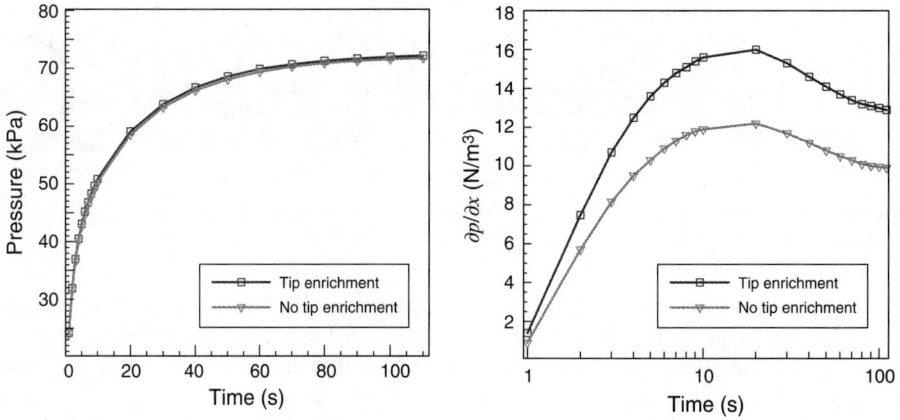

Figure 12.11 The effect of crack tip enrichment on the pressure and the gradient of pressure at the normal direction to crack

displacement are plotted for three X-FEM meshes at selected points. Obviously, the convergence can be seen among various X-FEM meshes when the size of elements is reduced. Also plotted in this figure are the variations with time of the vertical displacement for three points of porous medium, where there is no fault in the domain. Clearly, the displacement curves display the influence of sheet pile in reducing the uplift under the dam. In order to represent the importance of tip enrichment, the variations with time of the pressure and the gradient of pressure are plotted in Figure 12.11 at the vicinity of a singular point with and without the asymptotic enrichment function. Obviously, there is no difference in the curve of pressure between the two cases, as shown in Figure 12.11a, however, the tip enrichment function has significant affect in the gradient of pressure, as shown in Figure 12.11b.

12.5.4 An Inclined Fault in Porous Media

The next example refers to the X-FEM modeling of THM analysis of an inclined fault in the saturated porous media, as shown in Figure 12.12. A square-shaped fractured domain of 10 × 10 m is modeled in the plane strain condition, which is subjected to the prescribed temperature of 100 °C and a normal fluid flux of $q = 10^{-4}$ m/s at the bottom surface while the top surface is imposed as a drained condition with zero pressure and temperature. The geometry, boundary conditions and the position of the fault are shown in Figure 12.12 together with the uniform X-FEM mesh used in the THM analysis. The fault is assumed as an impermeable and adiabatic discontinuity with the length of 2 m located at the center of the domain. The material properties of the soil are given in Table 12.4. The problem is modeled for various fault angles $\beta = 0$, 30°, 45°, 60°, and 90° to investigate the influence of pressure and temperature discontinuities over the domain. The simulation is performed for 10^5 s using the full Newton–Raphson method.

In Figure 12.13, the distributions of pressure and temperature contours are shown at the end of simulation for $\beta = 0$ and 45°. Obviously, the pressure and temperature discontinuities can be seen along the fault, where the Heaviside enrichment function is applied in the thermal and pressure fields at the discontinuity faces. In Figure 12.14, the variations with time of the pressure are plotted for different points given

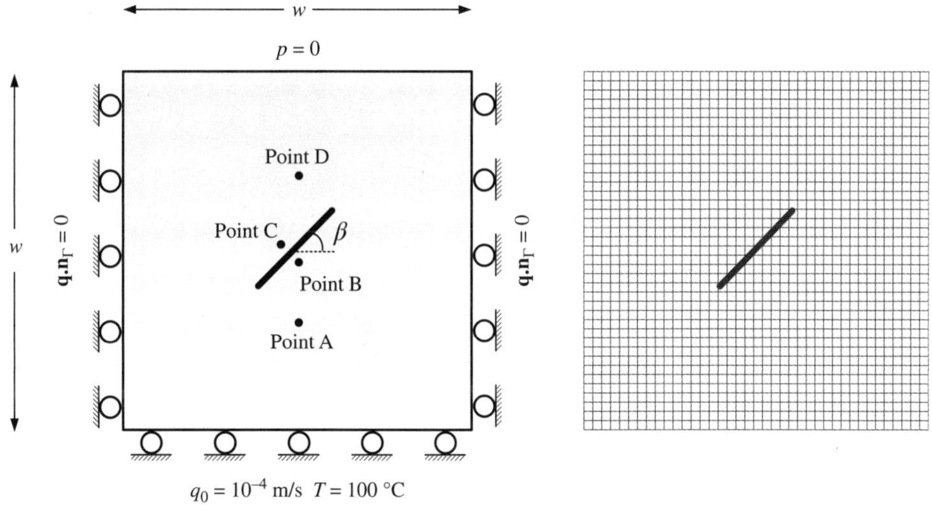

Figure 12.12 An inclined fault in porous media: The geometry, boundary conditions, and X-FEM mesh

Table 12.4 Material parameters for an inclined fault in porous media

Young's modulus	6×10^3 (kPa)
Poisson ratio	0.4
Solid density	2×10^3 (kg/m^3)
Fluid density	1×10^3 (kg/m^3)
Porosity	0.3
Bulk module of fluid	2×10^3 (MPa)
Bulk module of solid	1×10^{14} (MPa)
Permeability	1.1×10^{-9} (m^2/Pa s)
Thermal conductivity of the bulk	837 (W/m °C)
Solid specific heat	878 (J/kg °C)
Fluid specific heat	4184 (J/kg °C)
Volumetric thermal expansion coefficient	9.0×10^{-8} (1/°C)

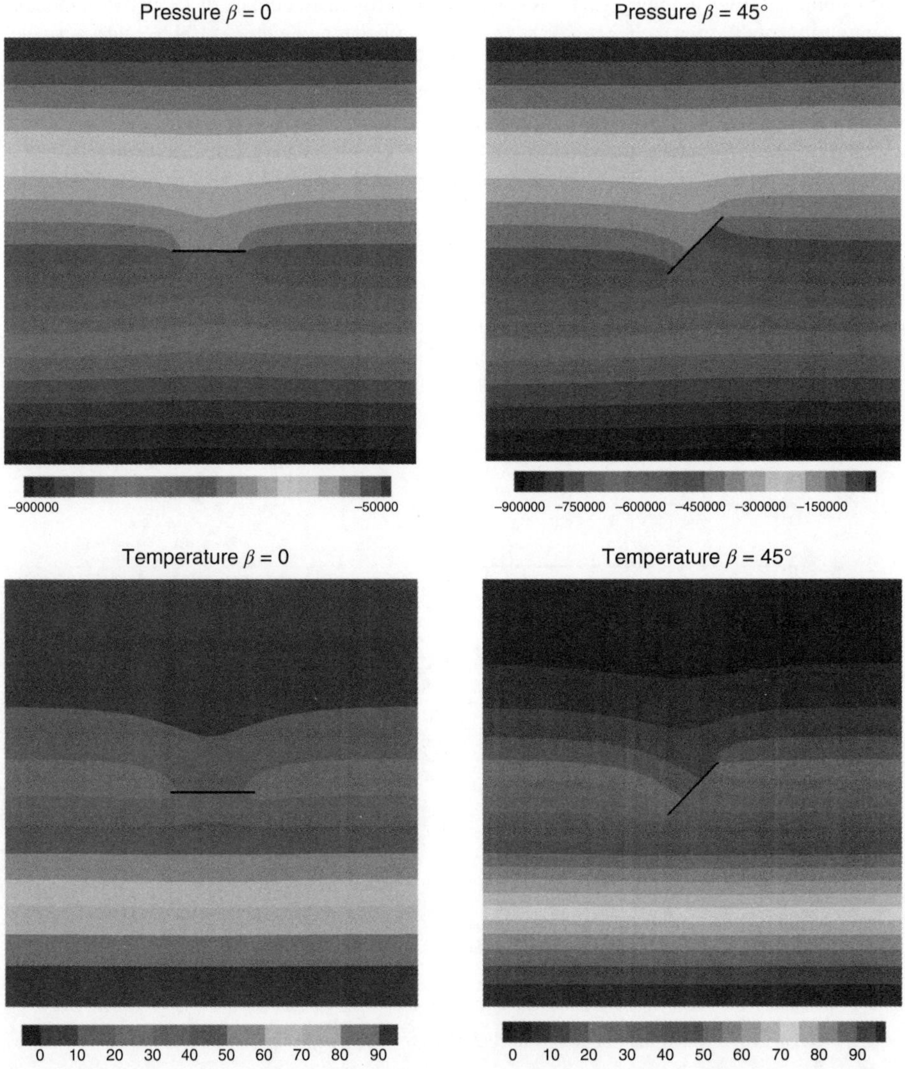

Figure 12.13 The distributions of pressure and temperature contours at the end of simulation for $\beta = 0$ and $\beta = 45°$ (For color details, please see color plate section)

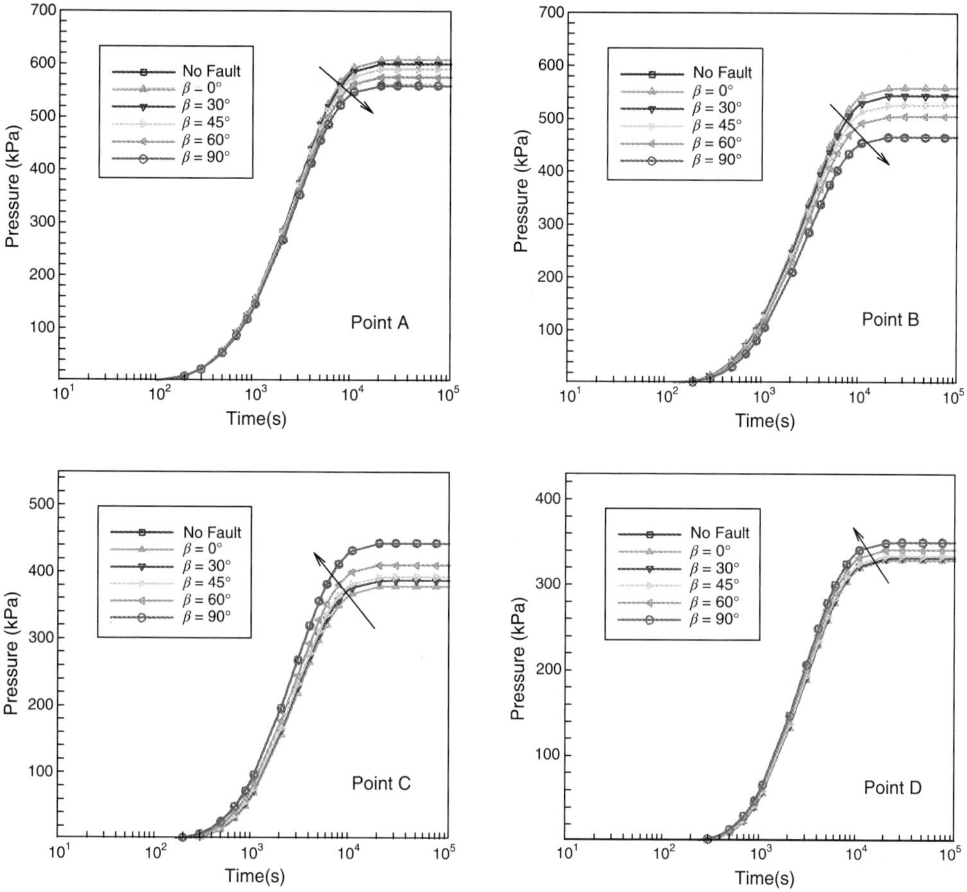

Figure 12.14 The variations with time of the pressure for various fault angles at different points

Table 12.5 Position of the points

	x (m)	y (m)
Point A	5.13	3.85
Point B	5.13	4.87
Point C	4.87	5.13
Point D	5.13	6.15

in Table 12.5, at various fault angles. It can be seen that the pressure increases in the domain due to the imposed fluid flux at the bottom surface, however, the fault causes the pressure discontinuity and prevents the fluid from flowing directly through the domain. As a result, the variation of pressure decreases at points A and B (below the fault) when the fault angle increases, as shown in Figures 12.14a and b. Conversely, the variation of pressure increases at points C and D (above the fault) when the fault angle increases, as shown in Figures 12.14c and d. As it can be seen, the fluid can move freely toward the

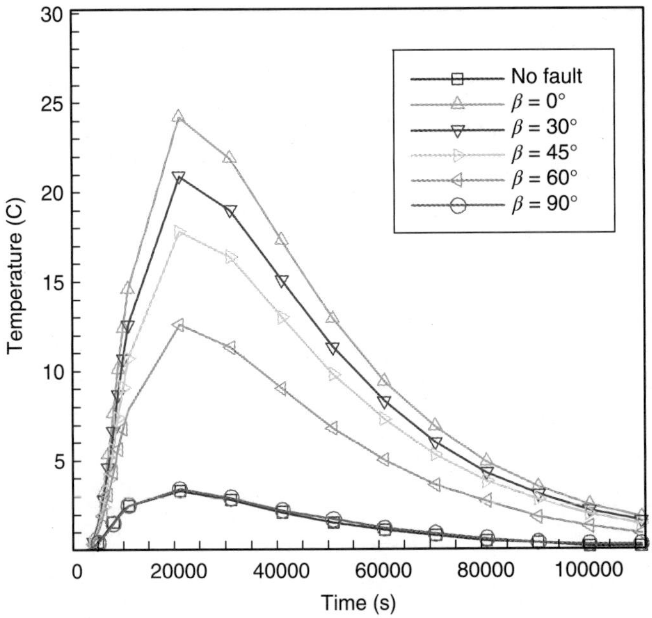

Figure 12.15 The variations with time of the temperature difference between points B and C for various fault angles

top surface in the case of vertical fault, because the fluid streamlines do not cross the discontinuity to affect their paths.

In Figure 12.15, the variations with time of the temperature difference between points B and C (at both sides of the fault) are plotted for various fault angles. Obviously, at the initial time steps, that is, $t < 2 \times 10^4$ s, the temperature difference between points B and C increases to its maximum value, and then decreases because of the imposed temperature at the bottom surface. As expected, the variation of temperature difference (between points B and C) decreases when the fault angle increases. Obviously, in the case of vertical fault, the curve of $\beta = 90°$ is identical to that obtained from the domain with no fault. Moreover, the effect of crack tip enrichment is investigated in Figure 12.16 for pressure, temperature, and their gradients at the crack tip region in the case of $\beta = 45°$. In Figures 12.16a and b, the variations with time of the pressure and pressure gradient are plotted at the crack tip region with and without the pressure enrichment function. Also plotted in Figures 12.16c and d are the variations with time of the temperature and temperature gradient at the crack tip region, with and without the temperature enrichment function. Obviously, the tip enrichment functions have significant affect in the gradient of pressure and temperature, as shown in Figures 12.16b and d.

Finally, in order to demonstrate the performance of the proposed computational algorithm for THM modeling of deformable porous media in a realistic problem, the saturated porous medium is modeled with a number of randomly generated faults. Most of the *in situ* soils contain numerous impermeable materials, such as the hard stones distributed randomly in the media. In this case, the capability of proposed technique is presented for a set of six randomly generated faults in a saturated porous medium, as shown in Figure 12.17. The geometry, boundary conditions, and the material parameters are considered to be similar to the previous case. The faults are assumed as impermeable and adiabatic discontinuities with different lengths and angles randomly distributed within the domain. The simulation is performed for 4×10^4 s, and the distribution of temperature and pressure contours are shown

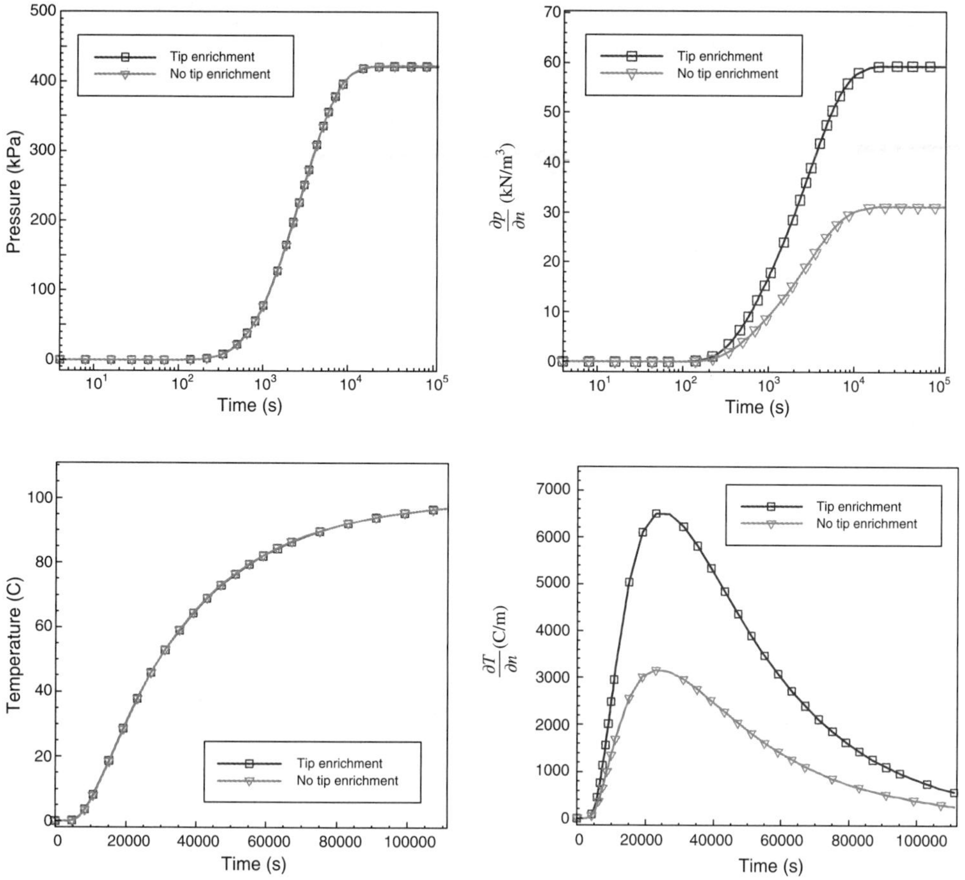

Figure 12.16 The effect of crack tip enrichment on the pressure and temperature and their gradients at the normal direction to the fault in the case of $\beta = 45°$

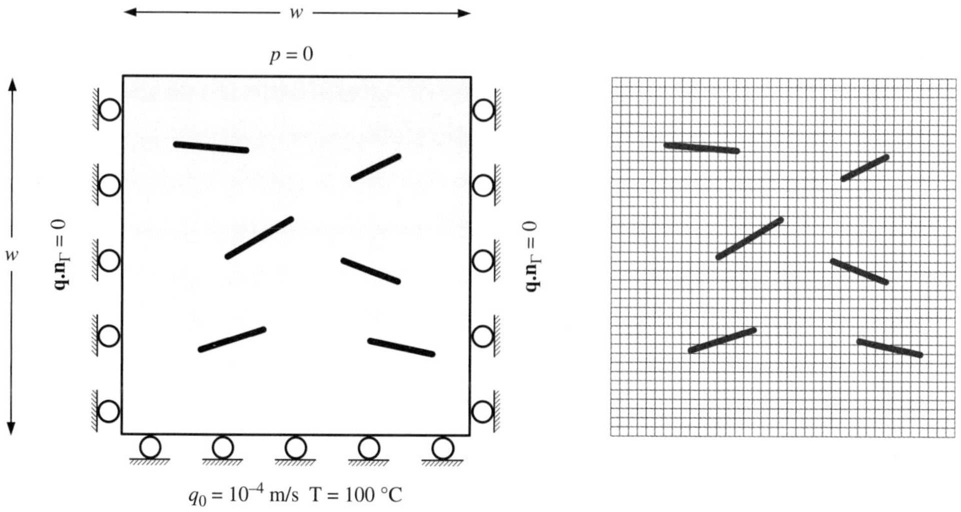

Figure 12.17 Randomly generated faults in saturated porous media: The geometry, boundary conditions, and X-FEM mesh

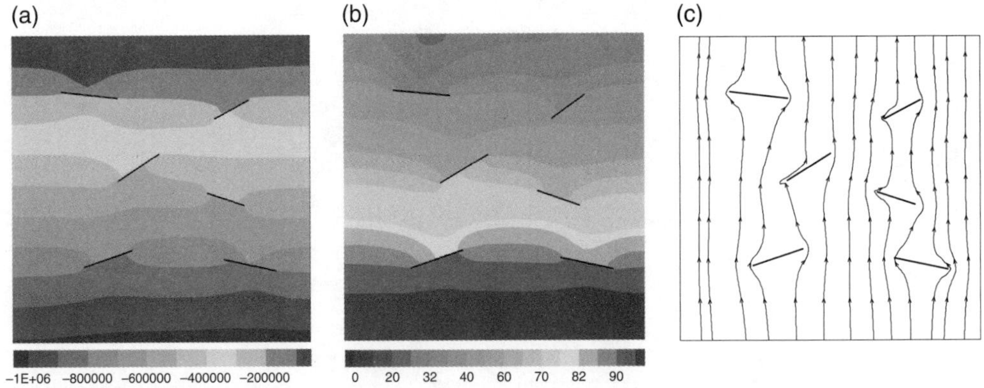

Figure 12.18 Randomly generated faults in saturated porous media: (a) the distribution of pressure contour, (b) the distribution of temperature contour, (c) the streamlines of the pressure (For color details, please see color plate section)

in Figure 12.18 at the end of simulation. This figure clearly presents the influence of discontinuities on the distribution of temperature and pressure through the domain. Also presented in this figure is the pressure streamlines for six randomly generated faults. Obviously, the impermeable–adiabatic behavior of the faults prevents the fluid flow and the heat flux through the discontinuities in the domain. This example clearly presents the capability of the X-FEM technique in the case of distinct enrichments in the THM media.

References

Abbas S, Alizada A, Fries TP. The XFEM for high-gradient solutions in convection-dominated problems. *International Journal for Numerical Methods in Engineering* 2010; **82**: 1044–1072.

Abdelaziz Y, Hamouine A. A survey of the extended finite element. *Computers and Structures* 2008; **86**: 1141–1151.

Achanta S, Cushman JH, Okos MR. On multicomponent, multiphase thermo-mechanics with interfaces. *International Journal of Engineering Science* 1994; **32**: 1717–1738.

Adalsteinsson D, Sethian JA. A fast level set method for propagating interfaces. *Journal of Computational Physics* 1995; **118**: 269–277.

Adalsteinsson D, Sethian JA. The fast construction of extension velocities in level set methods. *Journal of Computational Physics* 1999; **148**: 2–22.

Akulich AV, Zvyagin AV. Interaction between hydraulic and natural fractures. *Fluid Dynamics* 2008; **43**: 428–435.

Alonso EE, Gens A, Josa A. A constitutive model for partially saturated soils. *Géotechnique* 1990; **40**: 405–430.

Amiri F, Anitescu C, Arroyo M, *et al.* XLME interpolants, a seamless bridge between XFEM and enriched meshless methods. *Computational Mechanics* 2014; **53**: 45–57.

Anahid M, Khoei AR. New development in extended finite element modeling of large elasto-plastic deformations. *International Journal for Numerical Methods in Engineering* 2008; **75**: 1133–1171.

Anahid M, Khoei AR. Modeling of moving boundaries in large plasticity deformations via an enriched arbitrary Lagrangian–Eulerian FE method. Scientia Iranica, Transaction A. *Journal of Civil Engineering* 2010; **17**: 141–160.

Anderson TL. *Fracture Mechanics, Fundamentals and Applications*, CRC Press. Boca Raton, FL. 1994.

Aragon AM, Duarte CA, Geubelle PH. Generalized finite element enrichment functions for discontinuous gradient fields. *International Journal for Numerical Methods in Engineering* 2010; **82**: 242–268.

Areias PMA, Belytschko T. Analysis of three-dimensional crack initiation and propagation using the extended finite element method. *International Journal for Numerical Methods in Engineering* 2005a; **63**: 760–788.

Areias PMA, Belytschko T. Non-linear analysis of shells with arbitrary evolving cracks using XFEM. *International Journal for Numerical Methods in Engineering* 2005b; **62**: 384–415.

Areias PMA, Belytschko T. A comment on the article: a finite element method for simulation of strong and weak discontinuities in solid mechanics by Hanbso A, Hansbo P. [Computer Methods in Applied Mechanics and Engineering. 193 (2004)]. *Computer Methods in Applied Mechanics and Engineering*. 2006a; **195**: 1275–1276.

Areias PMA, Belytschko T. Two-scale shear band evolution by local partition of unity. *International Journal for Numerical Methods in Engineering* 2006b; **66**: 878–910.

Armero F, Love E. An arbitrary Lagrangian–Eulerian finite element method for finite strain plasticity. *International Journal for Numerical Methods in Engineering* 2003; **57**: 471–508.

Arnold DN, Brezzi F, CockBurn B, Marini LD. Unified analysis of discontinuous Galerkin methods for elliptic problems. *SIAM Journal on Numerical Analysis* 2002; **39**: 1749–1779.

Asadpoure A Mohammadi S. Developing new enrichment functions for crack simulation in orthotropic media by the extended finite element method. *International Journal for Numerical Methods in Engineering* 2007; **69**: 2150–2172.

Asadpoure A, Mohammadi S, Vafai A. Crack analysis in orthotropic media using the extended finite element method. *Thin-Walled Structures* 2006; **44**: 1031–1038.

Extended Finite Element Method: Theory and Applications, First Edition. Amir R. Khoei.

Asferg JL, Poulsen PN, Nielsen Lo. A consistent partly cracked XFEM element for cohesive crack growth. *International Journal for Numerical Methods in Engineering* 2007; **72**: 464–485.

Aubertin P, Réthoré J, de Borst R. A coupled molecular dynamics and extended finite element method for dynamic crack propagation. *International Journal for Numerical Methods in Engineering* 2010; **81**: 72–88.

Ayhan AO. Stress intensity factors for three-dimensional cracks in functionally graded materials using enriched finite elements. *International Journal of Solids and Structures* 2007; **44**: 8579–8599.

Ayhan AO. Simulation of three-dimensional fatigue crack propagation using enriched finite elements. *Computers and Structures* 2011; **89**: 801–812.

Ayhan AO, Nied HF. Stress intensity factors for three-dimensional surface cracks using enriched finite elements. *International Journal for Numerical Methods in Engineering* 2002; **54**: 899–921.

Azadi H, Khoei AR. Numerical simulation of multiple crack growth in brittle materials with adaptive remeshing. *International Journal for Numerical Methods in Engineering* 2011; **85**: 1017–1048.

Babuška I, Melenk JM. The partition of unity method. *International Journal for Numerical Methods in Engineering* 1997; **40**: 727–758.

Babuška I, Rosenzweig M. A finite element scheme for domains with corners. *Numerische Mathematik* 1972; **20**: 1–21.

Bachene M, Tiberkak R, Rechak S. Vibration analysis of cracked plates using the extended finite element method. *Archive of Applied Mechanics* 2009a; **79**: 249–262.

Bachene M, Tiberkak R, Rechak S, *et al.* Enriched finite element for modal analysis of cracked plates. In: *Damage and Fracture Mechanics*. Boukharaouba T, Elboujdaini M, Pluvinage G (eds), Heidelberg. Springer. 2009b; 463–471.

Baietto MC, Pierres E, Gravouil A, *et al.* Fretting fatigue crack growth simulation based on a combined experimental and XFEM strategy. *International Journal of Fatigue* 2013; **47**: 31–43.

Barani OR, Khoei AR, Mofid M. Modeling of cohesive crack growth in partially saturated porous media: a study on the permeability of cohesive fracture. *International Journal of Fracture* 2011; **167**: 15–31.

Barenblatt GI. The formation of equilibrium cracks during brittle fracture: general ideas and hypotheses. Axially-symmetric cracks. *Journal of Applied Mathematics and Mechanics* 1959; **23**: 622–636.

Barenblatt GI. The mathematical theory of equilibrium of cracks in brittle fracture. *Advances in Applied Mechanics* 1962; **7**: 55–129.

Barrett R, Berry M, Chan TF, *et al. Templates for the solution of linear systems: building blocks for iterative methods*. Philadelphia, PA, Society for Industrial and Applied Mathematics. 1994.

Barsoum RS. On the use of isoparametric finite elements in linear fracture mechanics. *International Journal for Numerical Methods in Engineering* 1976; **10**: 25–37.

Bassi F, Rebay S. A high order accurate discontinuous finite element method for the numerical solution of the compressible Navier–Stokes equations. *Journal of Computational Physics* 1997; **131**: 267–279.

Baydoun M, Fries TP. Crack propagation criteria in three dimensions using the XFEM and an explicit-implicit crack description. *International Journal of Fracture* 2012; **178**: 51–70.

Bayesteh H, Mohammadi S. XFEM fracture analysis of shells: the effect of crack tip enrichments. *Computational Materials Science* 2011; **50**: 2793–2813.

Bayoumi HN, Gadala MS. A complete finite element treatment for the fully coupled implicit ALE formulation. *Computational Mechanics* 2004; **33**: 435–452.

Bazant ZP, Li Y.-N. Cohesive crack model with rate-dependent opening and viscoelasticity: I. Mathematical model and scaling. *International Journal of Fracture* 1997; **86**: 247–265.

Bazant ZP, Planas J. *Fracture and Size Effect in Concrete and Other Quasibrittle Materials*. Boca Raton, FL, CRC Press. 1998.

Béchet E, Minnebol H, Moës N, Burgardt B. Improved implementation and robustness study of the X-FEM for stress analysis around cracks. *International Journal for Numerical Methods in Engineering* 2005; **64**: 1033–1056.

Béchet E, Moës N, Wohlmuth B. A stable Lagrange multiplier space for stiff interface conditions within the extended finite element method. *International Journal for Numerical Methods in Engineering* 2009a; **78**: 931–954.

Béchet E, Scherzer M, Kuna M. Application of the X-FEM to the fracture of piezoelectric materials. *International Journal for Numerical Methods in Engineering* 2009b; **77**: 1535–1565.

Becker R, Burman E, Hansbo P. A Nitsche extended finite element method for incompressible elasticity with discontinuous modulus of elasticity. *Computer Methods in Applied Mechanics and Engineering* 2009; **198**: 3352–3360.

Belytschko T, Black T. Elastic crack growth in finite elements with minimal remeshing. *International Journal for Numerical Methods in Engineering* 1999; **45**: 601–620.

Belytschko T, Black T, Moës N, *et al.* Structured extended finite element methods of solids defined by implicit surfaces. *International Journal for Numerical Methods in Engineering* 2003a; **56**: 609–635.

Belytschko T, Chen H, Xu J, Zi G. Dynamic crack propagation based on loss of hyperbolicity and a new discontinuous enrichment. *International Journal for Numerical Methods in Engineering* 2003b; **58**: 1873–1905.

Belytschko T, Fish J, Englemann BE. A finite element with embedded localization zones. *Computer Methods in Applied Mechanics and Engineering* 1988; **70**: 59–89.

Belytschko T, Gracie R. On XFEM applications to dislocations and interfaces. *International Journal of Plasticity* 2007; **23**: 1721–1738.

Belytschko T, Gracie R, Ventura G. A review of extended/generalized finite element methods for material modeling. *Modelling and Simulation in Materials Science and Engineering* 2009; **17**: 043001 DOI:10.1088/0965-0393/17/4/043001.

Belytschko T, Krongauz Y, Organ D, *et al.* Meshless methods: an overview and recent developments. *Computer Methods in Applied Mechanics and Engineering* 1996; **139**: 3–47.

Belytschko T, Liu WK, Moran B. *Nonlinear Finite Elements for Continua and Structures*. New York: Wiley. 2000a.

Belytschko T, Lu YY, Gu L. Element free Galerkin methods. *International Journal for Numerical Methods in Engineering* 1994; **37**: 229–256.

Belytschko T, Moës N, Usui S, Parimi C. Arbitrary discontinuities in finite elements. *International Journal for Numerical Methods in Engineering* 2001; **50**: 993–1013.

Belytschko T, Organ D, Gerlach C. Element-free Galerkin methods for dynamic fracture in concrete. *Computer Methods in Applied Mechanics and Engineering* 2000b; **187**: 385–399.

Belytschko T, Tabbara M. H-adaptive finite element methods for dynamic problems with emphasis on localization. *International Journal for Numerical Methods in Engineering* 1993; **36**: 4245–4265.

Belytschko T, Xiao SP, Parimi C. Topology optimization with implicit functions and regularization. *International Journal for Numerical Methods in Engineering* 2003c; **57**: 1177–1196.

Benson DJ. An efficient, accurate, simple ALE method for nonlinear FE programs. *Computer Methods in Applied Mechanics and Engineering* 1989; **72**: 305–350.

Benvenuti E. A regularized XFEM framework for embedded cohesive interfaces. *Computer Methods in Applied Mechanics Engineering* 2008; **197**: 4367–4378.

Benvenuti E, Tralli A. Simulation of finite-width process zone in concrete-like materials by means of a regularized extended finite element model. *Computational Mechanics* 2012; **50**: 479–497.

Benvenuti E, Tralli A, Ventura G. A regularized XFEM model for the transition from continuous to discontinuous displacements. *International Journal for Numerical Methods in Engineering* 2008; **74**: 911–944.

Benvenuti E, Ventura G, Ponara N. Finite element quadrature of regularized discontinuous and singular level set functions in 3D problems. *Algorithms* 2012a; **5**: 529–544.

Benvenuti E, Vitarelli O, Tralli A. Delamination of FRP-reinforced concrete by means of an extended finite element formulation. *Composites Part B: Engineering* 2012b; **43**: 3258–3269.

Beshara FBA. Smeared crack analysis for reinforced concrete structures under blast-type loading. *Engineering Fracture Mechanics* 1993; **45**: 119–140.

Bhargava RR, Sharma K. A study of finite size effects on cracked 2D piezoelectric media using extended finite element method. *Computational Materials Science* 2011; **50**: 1834–1845.

Bhargava RR, Sharma K. Generalized crack-tip enrichment functions for X-FEM simulation in piezoelectric media. *Journal of Information and Operations Management* 2012; **3**: 166–169.

Bhattacharya S, Singh IV, Mishra BK. Fatigue-life estimation of functionally graded materials using XFEM. *Engineering with Computers* 2012; **29**: 427–448.

Bhattacharya S, Singh IV, Mishra BK, Bui TQ. Fatigue crack growth simulations of interfacial cracks in bi-layered FGMs using XFEM. *Computational Mechanics* 2013; **52**: 799–814.

Bhushan B, *Principles and Applications of Tribology*. Chichester: Wiley, 1999.

Bilby BA, Cottrell AH, Swinden KH. The spread of plastic yield from a notch. *Proceedings of the Royal Society of London A* 1963; **272**: 304–314.

Biot MA. General theory of three dimensional consolidation. *Journal of Applied Physics* 1941; **12**: 155–164.

Biot MA. Theory of propagation of elastic waves in a fluid saturated porous solid. *Acoustical Society of America* 1956; **28**: 168–191.

Biot MA. Mechanics of deformation and acoustic propagation in porous media. *Journal of Applied Physics* 1962; **33**: 1482–1498.

Blanpied M. *Faulting, Friction and Earthquake Mechanics*. Birkhauser. 1994.

de Boer R, Kowalski SJ. A plasticity theory for fluid saturated porous media. *International Journal of Engineering Science* 1983; **21**: 11–16.

de Boer R. Highlights in the historical development of the porous media theory: toward a consistent macroscopic theory. *Applied Mechanics Reviews* 1996; **69**: 201–262.

de Boer R. The thermodynamic structure and constitutive equations for fluid-saturated compressible and incompressible elastic porous solids. *International Journal of Solids and Structures* 1998; **35**: 4557–4573.

Bonfils N, Chevaugeon N, Moës N. Treating volumetric inequality constraint in a continuum media with a coupled X-FEM/level-set strategy. *Computer Methods in Applied Mechanics and Engineering* 2012; **205**: 16–28.

Boone TJ, Ingraffea AR. A numerical procedure for simulation of hydraulically-driven fracture propagation in poroelastic media. *International Journal for Numerical and Analytical Methods in Geomechanics* 1990; **14**: 27–47.

Bordas SPA, Conley JG, Moran B, *et al.* A simulation-based design paradigm for complex cast components. *Engineering with Computers* 2007a; **23**: 25–37.

Bordas SPA, Legay A. *X-FEM mini-course*. EPFL, Lausanne. 2005.

Bordas SPA, Moran B. Enriched finite elements and level sets for damage tolerance assessment of complex structures. *Engineering Fracture Mechanics* 2006; **73**: 1176–1201.

Bordas SPA, Natarajan S, Kerfriden P, *et al.* On the performance of strain smoothing for quadratic and enriched finite element approximations (XFEM/GFEM/PUFEM). *International Journal for Numerical Methods in Engineering* 2011; **86**: 637–666.

Bordas SPA, Nguyen PV, Dunant C, *et al.* An extended finite element library. *International Journal for Numerical Methods in Engineering* 2007b; **71**: 703–732.

Bordas SPA, Rabczuk T, Hung NX, *et al.* Strain smoothing in FEM and XFEM. *Computers and Structures* 2010; **88**: 1419–1443.

de Borst R. Simulation of strain localization: a reappraisal of the Cosserat continuum. *Engineering Computation* 1991; **8**: 317–332.

de Borst R, Réthoré J, Abellan MA. A numerical approach for arbitrary cracks in a fluid-saturated medium. *Archive of Applied Mechanics* 2006; **75**: 595–606.

Bouhala L, Shao Q, Koutsawa Y, *et al.* An XFEM crack-tip enrichment for a crack terminating at a bi-material interface. *Engineering Fracture Mechanics* 2013; **102**: 51–64.

Bowen RM. Incompressible porous media models by use of the theory of mixtures. *International Journal of Engineering Science* 1980; **18**: 1129–1148.

Bowen RM. Compressible porous media models by use of theories of mixtures. *International Journal of Engineering Science* 1982; **20**: 697–735.

Brezzi F, Manzini G, Marini D, *et al. Discontinuous Finite Elements for Diffusion Problems*. Genova: Accademia di Scienze e Lettere. 1999; 197–217.

Broek D. *Elementary Engineering Fracture Mechanics*, Hingham, MA: Kluwer Academic Publishers. 1986.

Brooks RN, Corey AT. Properties of porous media affecting fluid flow. *Journal of Irrigation and Drainage Engineering, ASCE* 1966; **92**: 61–68.

Broumand P, Khoei AR. The extended finite element method for large deformation ductile fracture problems with a non-local damage–plasticity model. *Engineering Fracture Mechanics* 2013; 112–113: 97–125.

Broumand P, Khoei AR. A large deformation X-FEM method coupled with a damage–plasticity model for dynamic and cyclic crack propagation. *International Journal of Solids and Structures*. 2014.

Budyn E, Zi G, Moës N, Belytschko T. A method for multiple crack growth in brittle materials without remeshing. *International Journal for Numerical Methods in Engineering* 2004; **61**: 1741–1770.

Bui TQ, Zhang C. Extended finite element simulation of stationary dynamic cracks in piezoelectric solids under impact loading. *Computational Materials Science* 2012; **62**: 243–257.

Byfut A, Schröder A. Hp-adaptive extended finite element method. *International Journal for Numerical Methods in Engineering* 2012; **89**: 1392–1418.

Callari C, Armero F, Abati A. Strong discontinuities in partially saturated poroplastic solids. *Computer Methods in Applied Mechanics and Engineering* 2010; **199**: 1513–1535.

Camacho GT, Ortiz M. Computational modeling of impact damage in brittle materials. *International Journal of Solids Structures* 1996; **33**: 2899–2938.

Campilho RDSG, Banea MD, Chaves FJP, da Silva LFM. eXtended finite element method for fracture characterization of adhesive joints in pure mode I. *Computational Material Science* 2011a; **50**: 1543–1549.

Campilho RDSG, Banea MD, Pinto AMG, *et al.* Strength prediction of single- and double-lap joints by standard and extended finite element modelling. *International Journal of Adhesion and Adhesives* 2011b; **31**: 363–372.

Carpinteri A, Cornetti P, Barpi F, Valente S Cohesive crack model description of ductile to brittle size-scale transition: dimensional analysis vs. renormalization group theory. *Engineering Fracture Mechanics* 2003; **70**: 1809–1839.

Carrier B, Granet S. Numerical modeling of hydraulic fracture problem in permeable medium using cohesive zone model. *Engineering Fracture Mechanics* 2012; **79**: 312–328.

Carter RD. Optimum fluid characteristics for fracture extension. In: *API Drilling and Production Operations*. Howard CC, Fast CR (Eds). Tulsa: API. 1957; 261–270.

Cen Z, Maier G. Bifurcations and instabilities in fracture of cohesive-softening structures: a boundary element analysis. *Fatigue & Fracture of Engineering Materials & Structures* 1992; **15**: 911–928.

Cesar de Sa JMA, Areias PMA, Zheng C. Damage modelling in metal forming problems using an implicit non-local gradient model. *Computer Methods in Applied Mechanics and Engineering*. 2006; **195**: 6646–6660.

Chahine E, Laborde P, Renard Y. A quasi-optimal convergence result for fracture mechanics with XFEM. *Comptes Rendus Mathematique* 2006; **342**: 527–532.

Chahine E, Laborde P, Renard Y. A non-conformal extended finite element approach: integral matching XFEM. *Applied Numerical Mathematics* 2011; **61**: 322–343.

Chahine E, Laborde P, Renard Y. Crack tip enrichment in the XFEM using a cutoff function. *International Journal for Numerical Methods in Engineering* 2008; **75**: 629–646.

Chang CS, Duncan JM. Consolidation analysis for partly saturated clay by using an elastic–plastic effective stress–strain model. *International Journal for Numerical and Analytical Methods in Geomechanics* 1983; **7**: 39–55.

Chatzi EN, Hiriyur B, Waisman H, Smyth AW. Experimental application and enhancement of the XFEM-GA algorithm for the detection of flaws in structures. *Computers and Structures* 2011; **89**: 556–570.

Chen L, Rabczuk T, Bordas SPA, *et al.* Extended finite element method with edge-based strain smoothing (ESm-XFEM) for linear elastic crack growth. *Computer Methods in Applied Mechanics and Engineering* 2012; **209**: 250–265.

Chen T, Wang B, Cen Z, Wu Z. A symmetric Galerkin multi-zone boundary element method for cohesive crack growth. *Engineering Fracture Mechanics* 1999; **63**: 591–609.

Chen Y, Zhou C, Jing L. Modeling coupled THM processes of geological porous media with multiphase flow: theory and validation against laboratory and field scale experiments. *Computers and Geotechnics* 2009; **36**: 1308–1329.

Cheng, K.W. and Fries, T.P. (2009) A systematic study of different XFEM-formulations with respect to higher-order accuracy for arbitrarily curved weak discontinuities. International Conference on Extended Finite Element Methods, Aachen, Germany.

Cheng KW, Fries TP. XFEM with hanging nodes for two-phase incompressible flow. *Computer Methods in Applied Mechanics and Engineering* 2012; **245**: 290–312.

Cherepanov GP. Crack propagation in continuous media. *USSR Journal of Applied Mathematics and Mechanics* 1967; **31**: 503–512.

Chessa J, Belytschko T. An enriched finite element method and level sets for axisymmetric two-phase flow with surface tension. *International Journal for Numerical Methods in Engineering* 2003a; **58**: 2041–2064.

Chessa J, Belytschko T. An extended finite element method for two-phase fluids. *Journal of Applied Mechanics* 2003b; **70**: 10–17.

Chessa J, Belytschko T. Arbitrary discontinuities in space-time finite elements by level sets and X-FEM. *International Journal for Numerical Methods in Engineering* 2004; **61**: 2595–2614.

Chessa J, Smolinski P, Belytschko T. The extended finite element method (XFEM) for solidification problems. *International Journal for Numerical Methods in Engineering* 2002; **53**: 1959–1977.

Chessa J, Wang H, Belytschko T. On the construction of blending elements for local partition of unity enriched finite elements. *International Journal for Numerical Methods in Engineering* 2003; **57**: 1015–1038.

Choi YJ, Hulsen MA, Meijer HEH. An extended finite element method for the simulation of particulate viscoelastic flows. *Journal of Non-Newtonian Fluid Mechanics* 2010; **165**: 607–624.

Choi YJ, Hulsen MA, Meijer HEH. Simulation of the flow of a viscoelastic fluid around a stationary cylinder using an extended finite element method. *Computers & Fluids* 2012; **57**: 183–194.

Chopp DL, Sukumar N. Fatigue crack propagation of multiple coplanar cracks with the coupled extended finite element/fast marching method. *International Journal of Engineering Science* 2003; **41**: 845–869.

Combescure A, Gravouil A, Gregoire D, Rethore J. X-FEM a good candidate for energy conservation in simulation of brittle dynamic crack propagation. *Computer Methods in Applied Mechanics and Engineering* 2008; **197**: 309–318.

Comi C, Mariani S. Extended finite element simulation of quasi-brittle fracture in functionally graded materials. *Computer Methods in Applied Mechanics and Engineering* 2007; **196**: 4013–4026.

Cosimo A, Fachinotti V, Cardona A. An enrichment scheme for solidification problems. *Computational Mechanics* 2013; **52**: 17–35.

Court S, Fournié M, Lozinski A. A fictitious domain approach for the Stokes problem based on the extended finite element method. *International Journal for Numerical Methods in Fluids* 2014; **74**: 73–99.

Crisfield MA, Alfano G. Adaptive hierarchical enrichment for delamination fracture using a decohesive zone model. *International Journal for Numerical Methods in Engineering* 2002; **54**: 1369–1390.

Curiel Sosa JL, Karapurath N. Delamination modelling of GLARE using the extended finite element method. *Composites Science and Technology* 2012; **72**: 788–791.

Curnier A. A theory of friction. *International Journal of Solids and Structures* 1984; **20**: 637–647.

Daneshyar A, Mohammadi S. Strong tangential discontinuity modeling of shear bands using the extended finite element method. *Computational Mechanics* 2013; **52**: 1023–1038.

Daux C, Moës N, Dolbow J, *et al.* Arbitrary branched and intersecting cracks with the extended finite element method. *International Journal for Numerical Methods in Engineering* 2000; **48**: 1741–1760.

Deb D, Das KC. Extended finite element method for the analysis of discontinuities in rock masses. *Geotechnical and Geological Engineering* 2010; **28**: 643–659.

Detournay E. Propagation regimes of fluid-driven fractures in impermeable rocks. *International Journal of Geomechanics* 2004; **4**: 35–45.

Dhia HB, Jamond O. On the use of XFEM within the Arlequin framework for the simulation of crack propagation. *Computer Methods in Applied Mechanics and Engineering* 2010; **199**: 1403–1414.

Dias-da-Costa D, Alfaiate J, Sluys LJ, Júlio E. A comparative study on the modelling of discontinuous fracture by means of enriched nodal and element techniques and interface elements. *International Journal of Fracture* 2010; **161**: 97–119.

Dolbow, J.E. (1999) An extended finite element method with discontinuous enrichment for applied mechanics. PhD dissertation, Theoretical and Applied Mechanics, Northwestern University.

Dolbow JE, Devan A. Enrichment of enhanced assumed strain approximations for representing strong discontinuities: addressing volumetric incompressibility and the discontinuous patch test. *International Journal for Numerical Methods in Engineering* 2004; **59**: 47–67.

Dolbow JE, Moës N, Belytschko, T. Discontinuous enrichment in finite elements with a partition of unity method. *Finite Elements in Analysis and Design* 2000a; **36**: 235–260.

Dolbow JE, Moës N, Belytschko T. Modeling fracture in Mindlin-Reissner plates with the extended finite element method. *International Journal of Solids and Structures* 2000b; **37**: 7161–7183.

Dolbow JE, Moës N, Belytschko T. An extended finite element method for modeling crack growth with frictional contact. *Computer Methods in Applied Mechanics and Engineering* 2001; **190**: 6825–6846.

Dolbow JE, Mosso S, Robbins J, Voth T. Coupling volume-of-fluid based interface reconstructions with the extended finite element method. *Computer Methods in Applied Mechanics and Engineering* 2008; **197**: 439–447.

Dong CY, de Pater CJ. Numerical implementation of displacement discontinuity method and its application in hydraulic fracturing. *Computer Methods in Applied Mechanics and Engineering* 2001; **191**: 745–760.

Dragon A. Plasticity and ductile fracture damage: study of void growth in metals. *Engineering Fracture Mechanics* 1985; **21**: 875–885.

Dréau K, Chevaugeon N, Moës N. Studied X-FEM enrichment to handle material interfaces with higher order finite element. *Computer Methods in Applied Mechanics and Engineering* 2010; **199**: 1922–1936.

Duarte CAM, Oden JT. Hp clouds – An hp meshless method. *Numerical Methods for Partial Differential Equations* 1996; **12**: 673–705.

Duddu R, Chopp DL, Voorhees P, Moran B. Diffusional evolution of precipitates in elastic media using the extended finite element and the level set methods. *Journal of Computational Physics* 2011; **230**: 1249–1264.

Duflot M. A meshless method with enriched weight functions for three-dimensional crack propagation. *International Journal for Numerical Methods in Engineering* 2006; **65**: 1970–2006.

Duflot M. The extended finite element method in thermoelastic fracture mechanics. *International Journal for Numerical Methods in Engineering* 2008; **74**: 827–847.

Duflot M, Hung ND. A meshless method with enriched weight functions for fatigue crack growth. *International Journal for Numerical Methods in Engineering* 2004; **59**: 1945–1961.

Dugdale DS. Yielding of steel sheets containing slits. *Journal of the Mechanics and Physics of Solids* 1960; **8**: 100–108.

Dumont M, Taibi S, Fleureau JM, *et al.* Modelling the effect of temperature on unsaturated soil behaviour. *Comptes Rendus Geoscience* 2010; **342**: 892–900.

Dumstorff P, Meschke G. Crack propagation criteria in the framework of X-FEM-based structural analyses. *International Journal for Numerical and Analytical Methods in Geomechanics* 2007; **31**: 239–259.

Dundurs J. Edge-bounded dissimilar orthogonal elastic wedges. *Journal of Applied Mechanics* 1969; **36**: 650–652.

Dunant C, Nguyen P, Belgasmia M, *et al.* Architecture trade-offs of including a mesher in an object-oriented extended finite element code. *European Journal of Computational Mechanics* 2007; **16**: 237–258.

Dury O, Fischer U, Schulin R. A comparison of relative nonwetting-phase permeability models. *Water Resources Research* 1999; **35**: 1481–1493.

Dvorkin EN, Cuitiño AM, Gioia G. Finite elements with displacement interpolated embedded localization lines insensitive to mesh size and distortions. *International Journal for Numerical Methods in Engineering* 1990; **30**: 541–564.

Dyskin AV, Caballero A. Orthogonal crack approaching an interface. *Engineering Fracture Mechanics* 2009; **76**: 2476–2485.

Ebrahimi SH, Mohammadi S, Asadpoure A. An extended finite element (XFEM) approach for crack analysis in composite media. *International Journal of Civil Engineering* 2008; **6**: 198–207.

Edke MS, Chang KH. Shape optimization for 2-D mixed-mode fracture using Extended FEM (XFEM) and Level Set Method (LSM). *Structural and Multidisciplinary Optimization* 2011; **44**: 165–181.
Elguedj T, Gravouil A, Combescure A. Appropriate extended functions for X-FEM simulation of plastic fracture mechanics. *Computer Methods in Applied Mechanics and Engineering* 2006; **195**: 501–515.
Elguedj T, Gravouil A, Combescure A. A mixed augmented Lagrangian-extended finite element method for modelling elastic-plastic fatigue crack growth with unilateral contact. *International Journal for Numerical Methods in Engineering* 2007; **71**: 1569–1597.
Elguedj T, Gravouil A, Maigre H. An explicit dynamics extended finite element method. Part 1: mass lumping for arbitrary enrichment functions. *Computer Methods in Applied Mechanics and Engineering* 2009; **198**: 2297–2317.
Emerman SH, Turcotte DL and Spence DA. Transport of magma and hydrothermal solutions by laminar and turbulent fluid fracture. *Physics of the Earth and Planetary Interiors* 1986; **41**: 249–259.
Erdoghan F, Sih GC. On the extension of plates under plane loading and transverse shear. *ASME Journal of Basic Engineering* 1963; **4**: 519–527.
Eshelby JD. The Calculation of energy release rates. In: Sih GC, Van Elst HC, Broek D. (eds), *Prospects of Fracture Mechanics*, Noordhoff International. 1974; 69–84.
Esna Ashari S, Mohammadi S. Fracture analysis of FRP-reinforced beams by orthotropic XFEM. *Journal of Composite Materials* 2012; **46**: 1367–1389.
Esser P, Grande J, Reusken A. An extended finite element method applied to levitated droplet problems. *International Journal for Numerical Methods in Engineering* 2010; **84**: 757–773.
Eterovic AL, Bathe KJ. A hyperelastic-based large strain elasto-plastic constitutive formulation with combined isotropic-kinematic hardening using the logarithmic stress and strain measures. *International Journal for Numerical Methods in Engineering*. 1990; **30**: 1099–1114.
Fagerstörm M, Larsson R. A thermo-mechanical cohesive zone formulation for ductile fracture. *Journal of the Mechanics and Physics of Solids* 2008; **56**: 3037–3058.
Fan X, Zhang W, Wang T, *et al.* Investigation on periodic cracking of elastic film/substrate system by the extended finite element method. *Applied Surface Science* 2011; **257**: 6718–6724.
Fan X, Zhang W, Wang T, Sun Q. The effect of thermally grown oxide on multiple surface cracking in air plasma sprayed thermal barrier coating system. *Surface and Coatings Technology* 2012; **208**: 7–13.
Farhat C, Harari I, Franca LP. The discontinuous enrichment method. *Computer Methods in Applied Mechanics and Engineering* 2001; **190**: 6455–6479.
Farhat C, Roux F. A method of finite element tearing and interconnecting and its parallel solution algorithm. *International Journal for Numerical Methods in Engineering* 1991; **32**: 1205–1227.
Feerick EM, Liu X, McGarry P. Anisotropic mode-dependent damage of cortical bone using the extended finite element method. *Journal of the Mechanical Behavior of Biomedical Materials* 2013; **20**: 77–89.
Fernandez, R.T. (1972) Natural convection from cylinders buried in porous media. PhD thesis. University of California.
Ferrié E, Buffière JY, Ludwig W, *et al.* Fatigue crack propagation: in situ visualization using X-ray microtomography and 3D simulation using the extended finite element method. *Acta Materialia* 2006; **54**: 1111–1122.
Flanagan DP, Belytschko T. A uniform strain hexahedron and quadrilateral with orthogonal hourglass control. *International Journal for Numerical Methods in Engineering*. 1981; **17**: 679–706.
Fleck NA, Hutchinson JW. A phenomenological theory of strain gradient plasticity. *Journal of Mechanics and Physics of Solids* 1993; **41**: 1825–1857.
Fleming M, Chu YA, Moran B, Belytschko T. Enriched element-free Galerkin methods for singular fields. *International Journal for Numerical Methods in Engineering* 1997; **40**: 1483–1504.
Fredlund DG, Morgenstern NR. Stress state variables for unsaturated soils. *Journal of the Geotechnical Engineering Division, ASCE* 1977; **103**: 447–466.
Fries TP. A corrected XFEM approximation without problems in blending elements. *International Journal for Numerical Methods in Engineering* 2008; **75**: 503–532.
Fries TP. The intrinsic XFEM for two-fluid flows. *International Journal for Numerical Methods in Fluids* 2009; **60**: 437–471.
Fries TP, Baydoun M. Crack propagation with the extended finite element method and a hybrid explicit-implicit crack description. *International Journal for Numerical Methods in Engineering* 2012; **89**: 1527–1558.
Fries TP, Belytschko T. The intrinsic XFEM: a method for arbitrary discontinuities without additional unknowns. *International Journal for Numerical Methods in Engineering* 2006; **68**: 1358–1385.
Fries TP, Belytschko T. The extended/generalized finite element method: an overview of the method and its applications. *International Journal for Numerical Methods in Engineering* 2010; **84**: 253–304.

Fries TP, Zilian A. On time integration in the XFEM. *International Journal for Numerical Methods in Engineering* 2009; **79**: 69–93.

Gadala MS, Wang J. ALE formulation and its application in solid mechanics. *Computer Methods in Applied Mechanics and Engineering* 1998; **167**: 33–55.

Garagash D, Detournay E. The tip region of a fluid-driven fracture in an elastic medium. *Journal of Applied Mechanics* 2000; **67**: 183–192.

Garikipati, K. (1996) On strong discontinuities in inelastic solids and their numerical simulations. PhD thesis. Stanford University.

Gasser TC, Holzapfel GA. 3D crack propagation in unreinforced concrete: a two-step algorithm for tracking 3D crack paths. *Computer Methods in Applied Mechanics and Engineering* 2005a; **195**: 5198–5219.

Gasser TC, Holzapfel GA. Modeling 3D crack propagation in unreinforced concrete using PUFEM. *Computer Methods in Applied Mechanics and Engineering* 2005b; **194**: 2859–2896.

Gatmiri B, Arson C. θ–STOCK, a powerful tool of thermohydromechanical behaviour and damage modeling of unsaturated porous media. *Computers and Geotechnics*. 2008; **35**: 890–915.

Gawin D, Baggio P, Schrefler BA. Coupled heat, water and gas flow in deformable porous media. *International Journal for Numerical Methods in Fluids* 1995; **20**: 969–987.

Gawin D, Klemm P. Coupled heat and moisture transfer with phase change in porous building materials. *Architecture and Civil Engineering* 1994; **40**: 89–104.

Gawin D, Schrefler BA. Thermo-hydro-mechanical analysis of partially saturated porous materials. *Engineering Computations* 1996; **13**: 113–143.

Gdoutos E. *Fracture Mechanics*. Boston, MA: Academics Publisher. 1993.

Geertsma J, de Klerk F. A rapid method of predicting width and extent of hydraulically induced fractures. *Journal of Petroleum Technology* 1969; **21**: 1571–1581.

Geertsma J, Haafkens R. A comparison of the theories for predicting width and extent of vertical hydraulically induced fractures. *Journal of Energy Resources Technology* 1979; **101**: 8–19.

Gerstenberger A, Wall WA. An extended finite element method/Lagrange multiplier based approach for fluid-structure interaction. *Computer Methods in Applied Mechanics and Engineering* 2008a; **197**: 1699–1714.

Gerstenberger A, Wall WA. Enhancement of fixed-grid methods towards complex fluid-structure interaction applications. *International Journal for Numerical Methods in Fluids* 2008b; **57**: 1227–1248.

Ghoneim A, Hunedy J, Ojo OA. An interface-enriched extended finite element-level set simulation of solutal melting of additive powder particles during transient liquid phase bonding. *Metallurgical and Materials Transactions A* 2013; **44**: 1139–1151.

Ghosh S, Raju S. R-S adapted arbitrary Lagrangian–Eulerian finite element method of metal-forming with strain localization. *International Journal for Numerical Methods in Engineering* 1996; **39**: 3247–3272.

Giner E, Navarro C, Sabsabi M, *et al.* Fretting fatigue life prediction using the extended finite element method. *International Journal of Mechanical Sciences* 2011; **53**: 217–225.

Giner E, Sukumar N, Denia FD, Fuenmayor FJ. Extended finite element method for fretting fatigue crack propagation. *International Journal of Solids and Structures* 2008a; **45**: 5675–5687.

Giner E, Sukumar N, Fuenmayor FJ, Vercher A. Singularity enrichment for complete sliding contact using the partition of unity finite element method. *International Journal for Numerical Methods in Engineering* 2008b; **76**: 1402–1418.

Golewski GL, Golewski P, Sadowski T. Numerical modelling crack propagation under Mode II fracture in plain concretes containing siliceous fly-ash additive using XFEM method. *Computational Materials Science* 2012; **62**: 75–78.

González-Albuixech VF, Giner E, Tarancón JE, *et al.* Convergence of domain integrals for stress intensity factor extraction in 2-D curved cracks problems with the extended finite element method. *International Journal for Numerical Methods in Engineering* 2013a; **94**: 740–757.

González-Albuixech VF, Giner E, Tarancón JE, *et al.* Domain integral formulation for 3-D curved and non-planar cracks with the extended finite element method. *Computer Methods in Applied Mechanics and Engineering* 2013b; **264**: 129–144.

Goodman MA, Cowin SC. A continuum theory for granular materials. *Archive for Rational Mechanics and Analysis* 1979; **44**: 249–266.

Gordeliy E, Peirce A. Coupling schemes for modeling hydraulic fracture propagation using the XFEM. *Computer Methods in Applied Mechanics and Engineering* 2013; **253**: 305–322.

Gracie R, Belytschko T. Concurrently coupled atomistic and XFEM models for dislocations and cracks. *International Journal for Numerical Methods in Engineering* 2009; **78**: 354–378.

Gracie R, Craig JR. Modelling well leakage in multilayer aquifer systems using the extended finite element method. *Finite Elements in Analysis and Design* 2010; **46**: 504–513.

Gracie R, Oswald J, Belytschko T. On a new extended finite element method for dislocations: core enrichment and nonlinear formulation. *Journal of the Mechanics and Physics of Solids* 2008a; **56**: 200–214.

Gracie R, Ventura G, Belytschko T. A new fast finite element method for dislocations based on interior discontinuities. *International Journal for Numerical Methods in Engineering* 2007; **69**: 423–441.

Gracie R, Wang H, Belytschko T. Blending in the extended finite element method by discontinuous Galerkin and assumed strain methods. *International Journal for Numerical Methods in Engineering* 2008b; **74**: 1645–1669.

Gravouil A, Elguedj T, Maigre H. An explicit dynamics extended finite element method. Part 2: element-by-element stable-explicit/explicit dynamic scheme. *Computer Methods in Applied Mechanics and Engineering* 2009; **198**: 2318–2328.

Gravouil A, Moës N, Belytschko T. Non-planar 3D crack growth by the extended finite element and level sets. Part II: level set update. *International Journal for Numerical Methods in Engineering* 2002; **53**: 2569–2586.

Griffith AA. The Phenomena of rupture and flow in solids. *Philosophical Transactions A* 1920; **221**: 163–198.

Grisvard P. *Elliptic Problems in Nonsmooth Domains.* Boston, MA, Pitman Publishing. 1985.

Groß S, Reusken A. An extended pressure finite element space for two-phase incompressible flows with surface tension. *Journal of Computational Physics* 2007; **224**: 40–58.

Guidault PA, Allix O, Champaney L, Cornuault C. A multiscale extended finite element method for crack propagation. *Computer Methods in Applied Mechanics and Engineering* 2008; **197**: 381–399.

Gurson AL. Continuum theory of ductile rupture by void nucleation and growth: part I, yield criteria and flow rules for porous ductile media. *Journal of Engineering Materials and Technology* 1977; **99**, 2–15.

Haber RB. A mixed Eulerian-Lagrangian displacement model for large-deformation analysis in solid mechanics. *Computer Methods in Applied Mechanics and Engineering* 1984; **43**: 277–292.

Hansbo A, Hansbo P. A finite element method for the simulation of strong and weak discontinuities in solid mechanics. *Computer Methods in Applied Mechanics and Engineering* 2004; **193**: 3523–3540.

Hassanizadeh M, Gray WG. General conservation equations for multiphase systems: 1. *Averaging procedure, Advances in Water Resources* 1979a; **2**: 131–144.

Hassanizadeh M, Gray WG. General conservation equations for multiphase systems: 2. Mass, momenta, energy and entropy equations. *Advances in Water Resources* 1979b; **2**: 191–203.

Hassanizadeh M, Gray WG. General conservation equations for multiphase systems: 3. Constitutive theory for porous media flow. *Advances in Water Resources* 1980; **3**: 25–40.

Hattori G, Rojas-Díaz R, Sáez A, *et al.* New anisotropic crack-tip enrichment functions for the extended finite element method. *Computational Mechanics* 2012; **50**: 591–601.

Henshell RD, Shaw KG. Crack-tip finite elements are unnecessary. *International Journal for Numerical Methods in Engineering* 1975; **9**: 495–507.

Herrmann LR. Laplacian-isoparametric grid generation scheme. *Journal of the Engineering Mechanics Division of the American Society of Civil Engineers* 1976; **102**: 749–756.

Hettich T, Ramm E. Interface material failure modeled by the extended finite-element method and level sets. *Computer Methods in Applied Mechanics and Engineering* 2006; **195**: 4753–4767.

Hillerborg A, Modéer M, Petersson PE. Analysis of crack formation and crack growth in concrete by means of fracture mechanics and finite elements. *Cement and Concrete Research* 1976; **6**: 163–168.

Hiriyur B, Waisman H, Deodatis G. Uncertainty quantification in homogenization of heterogeneous microstructures modeled by XFEM. *International Journal for Numerical Methods in Engineering* 2011; **88**: 257–278.

Holl M, Rogge T, Loehnert S, *et al.* 3D multiscale crack propagation using the XFEM applied to a gas turbine blade. *Computational Mechanics* 2014; **53**: 173–188.

Hosseini SS, Bayesteh H, Mohammadi S. Thermo-mechanical XFEM crack propagation analysis of functionally graded materials. *Materials Science and Engineering: A* 2013; **561**: 285–302.

Hoteit H, Firoozabadi A. An efficient numerical model for incompressible two-phase flow in fractured media. *Advances in Water Resources* 2008; **31**: 891–905.

Huang H, Long TA, Wan J, Brown WP. On the use of enriched finite element method to model subsurface features in porous media flow problems. *Computational Geosciences* 2011; **15**: 721–736.

Huang R, Prévost JH, Huang ZY, Suo Z. Channel-cracking of thin films with the extended finite element method. *Engineering Fracture Mechanics* 2003a; **70**: 2513–2526.

Huang R, Sukumar N, Prévost JH. Modeling quasi-static crack growth with the extended finite element method Part II: numerical applications. *International Journal of Solids and Structures* 2003b; **40**: 7539–7552.

Hussain M., Pu S, Underwood J. Strain energy release rate for a crack under combined mode I and mode II. *American Society for Testing and Materials* 1974; **560**: 2–28.

Hutchinson JW. Singular behaviour at the end of a tensile crack in a hardening material. *Journal of the Mechanics and Physics of Solids* 1967; **16**: 13–31.

Huynh DBP, Belytschko T. The extended finite element method for fracture in composite materials. *International Journal for Numerical Methods in Engineering* 2009; **77**: 214–239.

Inglis CE. Stresses in a plate due to the presence of crack and sharp corners. *Transactions of the Institute of Naval Architects* 1913; **55**: 219–241.

Irwin GR. Analysis of stress and strain near the end of crack traversing a plate. *Journal of Applied Mechanics* 1957; **24**: 361–364.

Irwin, G.R. (1960) Plastic zone near a crack and fracture toughness. Proceedings of the 7th Sagamore Research Conference, New York, August 1960, **4**, pp. 63–76.

Irwin GR, Kies GA, Smith HL. Fracture strength relative to the onset and arrest of crack propagation. *Proceedings of ASTM.* 1958; **58**: 640–657.

Ji H, Chopp D, Dolbow JE. A hybrid extended finite element/level set method for modeling phase transformations. *International Journal for Numerical Methods in Engineering* 2002; **54**: 1209–1233.

Ji H, Dolbow JE. On strategies for enforcing interfacial constraints and evaluating jump conditions with the extended finite element method. *International Journal for Numerical Methods in Engineering* 2004; **61**: 2508–2535.

Jiang Y, Tay TE, Chen L, Sun XS. An edge-based smoothed XFEM for fracture in composite materials. *International Journal of Fracture* 2013; **179**: 179–199.

Jirasck M. Nonlocal models for damage and fracture: comparison of approaches. *International Journal of Solids and Structures* 1998; **35**: 4133–4145.

Johnson GR, Beissel SR. Damping algorithms and effects for explicit dynamics computations. *International Journal of Impact Engineering.* 2001; **25**: 911–925.

Jung J, Jeong C, Taciroglu E. Identification of a scatterer embedded in elastic heterogeneous media using dynamic XFEM. *Computer Methods in Applied Mechanics and Engineering* 2013; **259**: 50–63.

Kachanov LM. *Introduction to Continuum Damage Mechanics.* Dordrecht: Martinus Nijhoff Publishers. 1986.

Karihaloo BL, Xiao QZ. Modeling of stationary and growing cracks in FE framework without remeshing: a state-of-the-art review. *Computers and Structures* 2003; **81**: 119–129.

Kästner M, Haasemann G, Ulbricht V. Multiscale XFEM-modeling and simulation of the inelastic material behavior of textile-reinforced polymers. *International Journal for Numerical Methods in Engineering* 2011; **86**: 477–498.

Kästner M, Müller S, Goldmann J, *et al.* Higher-order extended FEM for weak discontinuities – level set representation, quadrature and application to magneto-mechanical problems. *International Journal for Numerical Methods in Engineering* 2013; **93**: 1403–1424.

Kennedy JM, Belytschko T. Theory and application of a finite element method for arbitrary Lagrangian–Eulerian fluids and structures. *Nuclear Engineering and Design* 1981; **68**: 119–146.

Khan S, Vyshnevskyy A, Mosler J. Low cycle lifetime assessment of Al2024 alloy. *International Journal of Fatigue.* 2010; **32**: 1270–1277.

Khoei AR. *Computational Plasticity in Powder Forming Processes.* Elsevier, London 2005.

Khoei AR, Anahid M, Shahim K. An extended arbitrary Lagrangian-Eulerian finite element method for large deformation of solid mechanics. *Finite Elements in Analysis and Design* 2008a; **44**: 401–416.

Khoei AR, Anahid M, Shahim K, DorMohammadi H. Arbitrary Lagrangian-Eulerian method in plasticity of pressure-sensitive material with reference to powder forming processes. *Computational Mechanics* 2008b; **42**: 13–38.

Khoei AR, Azadi H, Moslemi H. Modeling of crack propagation via an automatic adaptive mesh refinement based on modified SPR technique. *Engineering Fracture Mechanics* 2008c; **75**: 2921–2945.

Khoei AR, Azami AR, Anahid M, Lewis RW. A three-invariant hardening plasticity for numerical simulation of powder forming processes via the arbitrary Lagrangian-Eulerian FE model. *International Journal for Numerical Methods in Engineering* 2006a; **66**: 843–877.

Khoei AR, Barani OR, Mofid M. Modeling of dynamic cohesive fracture propagation in porous saturated media. *International Journal for Numerical and Analytical Methods in Geomechanics* 2011; **35**: 1160–1184.

Khoei AR, Biabanaki SOR, Anahid M. Extended finite element method for three-dimensional large plasticity deformations on arbitrary interfaces. *Computer Methods in Applied Mechanics and Engineering* 2008d; **197**: 1100–1114.

Khoei AR, Biabanaki SOR, Anahid M. A Lagrangian-extended finite-element method in modeling large-plasticity deformations and contact problems. *International Journal of Mechanical Sciences* 2009a; **51**: 384–401.

Khoei AR, Eghbalian M, Moslemi H, Azadi H. Crack growth modelling via 3D automatic adaptive mesh refinement based on modified–SPR technique. *Applied Mathematical Modeling* 2013; **37**: 357–383.

Khoei AR, Gharehbaghi SA. Modeling of localized plastic deformation via the adaptive mesh refinement. *International Journal of Nonlinear Science and Numerical Simulation* 2003; **4**: 31–46.

Khoei AR, Gharehbaghi SA, Tabarraie AR, Riahi A. Error estimation, adaptivity and data transfer in enriched plasticity continua to analysis of shear band localization. *Applied Mathematical Modelling* 2007; **31**: 983–1000.

Khoei AR, Haghighat E. Extended finite element modeling of deformable porous media with arbitrary interfaces. *Applied Mathematical Modelling* 2011; **35**: 5426–5441.

Khoei AR, Hirmand M, *Vahab M*. An augmented Lagrangian contact formulation for frictional discontinuities with the extended finite element method. *Engineering Fracture Mechanics*. 2014a.

Khoei AR, Karimi K. An enriched-FEM model for simulation of localization phenomenon in Cosserat continuum theory. *Computational Materials Science* 2008; **44**: 733–749.

Khoei AR, Lewis RW. Adaptive finite element remeshing in a large deformation analysis of metal powder forming. *International Journal for Numerical Methods in Engineering* 1999; **45**: 801–820.

Khoei AR, Lewis RW, Zienkiewicz OC. Application of the finite element method for localized failure analysis in dynamic loading. *Finite Elements in Analysis and Design* 1997; **27**: 121–131.

Khoei AR, Moallemi S, Haghighat E. Thermo-hydro-mechanical modeling of impermeable discontinuity in saturated porous media with X-FEM technique. *Engineering Fracture Mechanics* 2012; **96**: 701–723.

Khoei AR, Mohammadnejad T. Numerical modeling of multiphase fluid flow in deforming porous media: a comparison between two- and three-phase models for seismic analysis of earth and rockfill dams. *Computers and Geotechnics* 2011; **38**: 142–166.

Khoei AR, Moslemi H, Ardakany KM, *et al*. Modeling of cohesive crack growth using an adaptive mesh refinement via the modified–SPR technique. *International Journal of Fracture* 2009b; **159**: 21–41.

Khoei AR, Mousavi ST. Modeling of large deformation–Large sliding contact via the penalty X-FEM technique. *Computational Materials Science* 2010; **48**: 471–480.

Khoei AR, Nikbakht M. Contact friction modeling with the extended finite element method (X-FEM). *Journal of Materials Processing Technology* 2006; **177**: 58–62.

Khoei AR, Nikbakht M. An enriched finite element algorithm for numerical computation of contact friction problems. *International Journal of Mechanical Sciences* 2007; **49**: 183–199.

Khoei AR, Shamloo A, Azami AR. Extended finite element method in plasticity forming of powder compaction with contact friction. *International Journal of Solids and Structures* 2006b; **43**: 5421–5448.

Khoei AR, Vahab M. A numerical contact algorithm in saturated porous media with the extended finite element method. *Computational Mechanics*. 2014; DOI: 10.1007/s00466-014-1041-1.

Khoei AR, Vahab M, Haghighat E, Moallemi S. A mesh-independent finite element formulation for modeling crack growth in saturated porous media based on an enriched–FEM technique. *International Journal of Fracture*. 2014b; DOI: 10.1007/s10704-014-9948-2.

Khristianovic SA, Zheltov YP. Formation of vertical fractures by means of highly viscous liquid. *Fourth World Petroleum Congress* 1955; **2**: 579–586.

Kim TY, Dolbow J, Laursen T. A mortared finite element method for frictional contact on arbitrary interfaces. *Computational Mechanics* 2007; **39**: 223–236.

Koenke C, Harte R, Kratzig WB, Rosenstein O. On adaptive remeshing techniques for crack simulation problems. *Engineering Computations* 1998; **15**: 74–88.

Kreissl S, Maute K. Level set based fluid topology optimization using the extended finite element method. *Structural and Multidisciplinary Optimization* 2012; **46**: 311–326.

Kroon M. Dynamic steady-state analysis of crack propagation in rubber-like solids using an extended finite element method. *Computational Mechanics* 2012; **49**: 73–86.

Laborde P, Pommier J, Renard Y, Salaün M. High-order extended finite element method for cracked domains. *International Journal for Numerical Methods in Engineering* 2005; **64**: 354–381.

Ladevèze P. *Non-linear Computational Structural Mechanics*, Springer, New York. 1999.

Lancaster P, Salkauskas K. Surfaces generated by moving least squares methods. *Mathematics of Computation* 1981; **37**: 141–158.

Larsson R, Fagerström M. A framework for fracture modelling based on the material forces concept with XFEM kinematics. *International Journal for Numerical Methods in Engineering* 2005; **62**: 1763–1788.

Larsson R, Mediavilla J, Fagerström M. Dynamic fracture modeling in shell structures based on XFEM. *International Journal for Numerical Methods in Engineering* 2011; **86**: 499–527

Lasry J, Pommier J, Renard Y, Salaün M. Extended finite element methods for thin cracked plates with Kirchhoff–Love theory. *International Journal for Numerical Methods in Engineering* 2010; **84**: 1115–1138.

Lasry J, Renard Y, Salaün M. Stress intensity factors computation for bending plates with extended finite element method. *International Journal for Numerical Methods in Engineering* 2012; **91**: 909–928.

Laursen TA, Puso MA, Sanders J. Mortar contact formulations for deformable-deformable contact: past contributions and new extensions for enriched and embedded interface formulations. *Computer Methods in Applied Mechanics and Engineering* 2012; **205**: 3–15.

Laursen TA, Simo JC. A continuum–based finite element formulation for the implicit solution of multibody, large deformation frictional contact problems. *International Journal for Numerical Methods in Engineering* 1993; **36**: 3451–3485.

Lecampion B. An extended finite element method for hydraulic fracture problems. *Communications in Numerical Methods in Engineering* 2009; **25**: 121–133.

Lecampion B, Detournay E. An implicit algorithm for the propagation of a hydraulic fracture with a fluid lag. *Computer Methods in Applied Mechanics and Engineering* 2007; **196**: 4863–4880.

Lee P, Yang R, Maute K. An extended finite element method for the analysis of submicron heat transfer phenomena. *Multiscale methods in Computational Mechanics* 2011; **55**: 195–212.

Lee SH, Song JH, Yoon YC, *et al.* Combined extended and superimposed finite element method for cracks. *International Journal for Numerical Methods in Engineering* 2004; **59**: 1119–1136.

Legay A. An extended finite element method approach for structural-acoustic problems involving immersed structures at arbitrary positions. *International Journal for Numerical Methods in Engineering* 2013; **93**: 376–399.

Legay A, Chessa J, Belytschko T. An Eulerian-Lagrangian method for fluid-structure interaction based on level sets. *Computer Methods in Applied Mechanics and Engineering* 2006; **195**: 2070–2087.

Legay A, Wang HW, Belytschko T. Strong and weak arbitrary discontinuities in spectral finite elements. *International Journal for Numerical Methods in Engineering* 2005; **64**: 991–1008.

Legrain G. A NURBS enhanced extended finite element approach for unfitted CAD analysis. *Computational Mechanics* 2013; **52**: 913–929.

Legrain G, Allais R, Cartraud P. On the use of the extended finite element method with quadtree/octree meshes. *International Journal for Numerical Methods in Engineering* 2011; **86**: 717–743.

Legrain G, Moës N, Huerta A. Stability of incompressible formulations enriched with X-FEM. *Computer Methods in Applied Mechanics and Engineering* 2008; **197**: 1835–1849.

Legrain G, Moës N, Verron E. Stress analysis around crack tips in finite strain problems using the extended finite element method. *International Journal for Numerical Methods in Engineering* 2005; **63**: 290–314.

Lemaitre J, Desmorat R. *Engineering Damage Mechanics*. New York: Springer. 2005.

Lewis RW, Rahman NA. Finite element modeling of multiphase immiscible flow in deforming porous media for subsurface systems. *Computers and Geotechnics* 1999; **24**: 41–63.

Lewis RW, Schrefler BA. *The Finite Element Method in the Static and Dynamic Deformation and Consolidation of Porous Media*. Wiley, New York. 1998.

Li FZ, Shih CF, Needleman A. A comparison of methods for calculating energy release rates. *Engineering Fracture Mechanics* 1985; **21**: 405–421.

Li J, Zhang XB, Recho N. Stress singularities near the tip of a two-dimensional notch formed from several elastic anisotropic materials. *International Journal of Fracture* 2001; **107**: 379–395.

Li L, Wang MY, Wei P. XFEM schemes for level set based structural optimization. *Frontiers of Mechanical Engineering* 2012; **7**: 335–356.

Li S, Ghosh S. Multiple cohesive crack growth in brittle materials by the extended Voronoi cell finite element model. *International Journal of Fracture* 2006; **141**: 373–393.

Liang J, Huang R, Prévost JH, Suo Z. Evolving crack patterns in thin films with the extended finite element method. *International Journal of Solids and Structures* 2003; **40**: 2343–2354.

Liao JH, Zhuang Z. A consistent projection-based SUPG/PSPG XFEM for incompressible two-phase flows. *Acta Mechanica Sinica* 2012; **28**: 1309–1322.

Liu F, Borja RI. A contact algorithm for frictional crack propagation with the extended finite element method. *International Journal for Numerical Methods in Engineering* 2008; **76**: 1489–1512.

Liu F, Borja RI. An extended finite element framework for slow-rate frictional faulting with bulk plasticity and variable friction. *International Journal for Numerical and Analytical Methods in Geomechanics* 2009; **33**: 1535–1560.

Liu F, Borja RI. Finite deformation formulation for embedded frictional crack with the extended finite element method. *International Journal for Numerical Methods in Engineering* 2010a; **82**: 773–804.

Liu F, Borja RI. Stabilized low-order finite elements for frictional contact with the extended finite element method. *Computer Methods in Applied Mechanics and Engineering* 2010b; **199**: 2456–2471.

Liu F, Borja RI. Extended finite element framework for fault rupture dynamics including bulk plasticity. *International Journal for Numerical and Analytical Methods in Geomechanics* 2013; **37**: 3087–3111.

Liu PF, Zhang BJ, Zheng JY. Finite element analysis of plastic collapse and crack behavior of steel pressure vessels and piping using XFEM. *Journal of Failure Analysis and Prevention* 2012; **12**: 707–718.

Liu WK, Belytschko T. Chang H. An Arbitrary Lagrangian-Eulerian finite element method for path-dependent materials. *Computer Methods in Applied Mechanics and Engineering* 1986; **58**: 227–245.

Liu WK, Chang H, Chen JS, *et al.* Arbitrary Lagrangian-Eulerian Petrov-Galerkin finite elements for nonlinear continua. *Computer Methods in Applied Mechanics and Engineering* 1988; **68**: 259–310.

Liu XY, Xiao QZ, Karihaloo BL. XFEM for direct evaluation of mixed mode SIFs in homogeneous and bi-materials. *International Journal for Numerical Methods in Engineering* 2004; **59**: 1103–1118.

Liu ZL, Menouillard T, Belytschko T. An XFEM/Spectral element method for dynamic crack propagation. *International Journal of Fracture* 2011; **169**: 183–198.

Lobao MC, Eve R, Owen DRJ, de Souza Neto EA. Modelling of hydro-fracture flow in porous media. *Engineering Computations* 2010; **27**: 129–154.

Loehnert S, Belytschko T. Crack shielding and amplification due to multiple microcracks interacting with a macrocrack. *International Journal of Fracture* 2007; **145**: 1–8.

Loehnert S, Mueller-Hoeppe DS, Wriggers P. 3D corrected XFEM approach and extension to finite deformation theory. *International Journal for Numerical Methods in Engineering* 2011; **86**: 431–452.

Loehnert S, Prange C, Wriggers P. Error controlled adaptive multiscale XFEM simulation of cracks. *International Journal of Fracture* 2012; **178**: 147–156.

Macri M, Littlefield A. Enrichment based multiscale modeling for thermo-stress analysis of heterogeneous material. *International Journal for Numerical Methods in Engineering* 2013; **93**: 1147–1169.

Maier G, Novati G, Cen Z. Symmetric Galerkin boundary element method for quasi-brittle-fracture and frictional contact problems. *Computational Mechanics* 1993; **13**: 74–89.

Malyshev BM, Salganik RL. The strength of adhesive joints using the theory of cracks. *International Journal of Fracture* 1965; **1**: 114–128.

Mariani S, Perego U. Extended finite element method for quasi-brittle fracture. *International Journal for Numerical Methods in Engineering* 2003; **58**: 103–126.

Mariano PM, Stazi FL. Strain localization due to crack-microcrack interactions: X-FEM for a multifield approach. *Computer Methods in Applied Mechanics and Engineering* 2004; **193**: 5035–5062.

Marzougui, D. (1998) Implementation of a fracture failure model to a 3D non-linear dynamic finite element code (DYNA3D). PhD thesis. The George Washington University.

Massimi P, Tezaur R, Farhat C. A discontinuous enrichment method for three-dimensional multiscale harmonic wave propagation problems in multi-fluid and fluid-solid media. *International Journal for Numerical Methods in Engineering* 2008; **76**: 400–425.

Mayer UM, Gerstenberger A, Wall WA. Interface handling for three-dimensional higher-order XFEM-computations in fluid-structure interaction. *International Journal for Numerical Methods in Engineering* 2009; **79**: 846–869.

Mayer UM, Popp A, Gerstenberger A, Wall WA. 3D fluid-structure-contact interaction based on a combined XFEM FSI and dual mortar contact approach. *Computational Mechanics* 2010; **46**: 53–67.

McClintock FA. A criterion for ductile fracture by the growth of holes. *Journal of Applied Mechanics* 1968; **35**: 363–371.

Mediavilla J, Peerlings RHJ, Geers MGD. Discrete crack modeling of ductile fracture driven by non-local softening plasticity. *International Journal for Numerical Methods in Engineering* 2006; **66**: 661–688.

Melenk JM, Babuška I. The partition of unity finite element method: basic theory and applications. *Computer Methods in Applied Mechanics and Engineering* 1996; **139**: 289–314.

Menk A, Bordas SPA. Numerically determined enrichment functions for the extended finite element method and applications to bi-material anisotropic fracture and polycrystals. *International Journal for Numerical Methods in Engineering* 2010; **83**: 805–828.

Menk A, Bordas SPA. A robust preconditioning technique for the extended finite element method. *International Journal for Numerical Methods in Engineering* 2011; **85**: 1609–1632.

Menouillard T, Belytschko T. Dynamic fracture with meshfree enriched XFEM. *Acta Mechanica* 2010a; **213**: 53–69.

Menouillard T, Belytschko T. Smoothed nodal forces for improved dynamic crack propagation modeling in XFEM. *International Journal for Numerical Methods in Engineering* 2010b; **84**: 47–72.

Menouillard T, Réthoré J, Combescure A, Bung H. Efficient explicit time stepping for the extended finite element method (X-FEM). *International Journal for Numerical Methods in Engineering* 2006; **68**: 911–939.

Menouillard T, Réthoré J, Moës N, *et al.* Mass lumping strategies for X-FEM explicit dynamics: application to crack propagation. *International Journal for Numerical Methods in Engineering* 2008; **74**: 447–474.

Menouillard T, Song JH, Duan Q, Belytschko T. Time dependent crack tip enrichment for dynamic crack propagation. *International Journal of Fracture* 2010; **162**: 33–49.

Mergheim J, Kuhl E, Steinmann P. A finite element method for the computational modeling of cohesive cracks. *International Journal for Numerical Methods in Engineering* 2005; **63**: 276–289.

Merle R, Dolbow J. Solving thermal and phase change problems with the extended finite element method. *Computational Mechanics* 2002; **28**: 339–350.

Meschke G, Dumstorff P. Energy-based modeling of cohesive and cohesionless cracks via X-FEM. *Computer Methods in Applied Mechanics and Engineering* 2007; **196**: 2338–2357.

Meschke G, Grasberger S. Numerical modeling of coupled hydromechanical degradation of cementitious materials. *Journal of Engineering Mechanics* 2003; **129**: 383–392.

Michlik P, Berndt C. Image-based extended finite element modeling of thermal barrier coatings. *Surface and Coatings Technology* 2006; **201**: 2369–2380.

Minnebo H. Three-dimensional integration strategies of singular functions introduced by the XFEM in the LEFM. *International Journal for Numerical Methods in Engineering* 2012; **92**: 1117–1138.

Moës N, Béchet E, Tourbier M. Imposing Dirichlet boundary conditions in the extended finite element method. *International Journal for Numerical Methods in Engineering* 2006; **67**: 1641–1669.

Moës N, Belytschko T. Extended finite element method for cohesive crack growth. *Engineering Fracture Mechanics* 2002a; **69**: 813–833.

Moës N, Belytschko T. X-FEM, de nouvelles frontières pour les éléments finis. *Revue Européenne des Eléments* 2002b; **11**: 305–318.

Moës N, Cloirec M, Cartraud P, Remacle JF. A computational approach to handle complex microstructure geometries. *Computer Methods in Applied Mechanics and Engineering* 2003; **192**: 3163–3177.

Moës N, Dolbow J, Belytschko T. A finite element method for crack growth without remeshing. *International Journal for Numerical Methods in Engineering* 1999; **46**: 131–150.

Moës N, Gravouil A, Belytschko T. Non-planar 3D crack growth by the extended finite element and level sets, Part I: mechanical model. *International Journal for Numerical Methods in Engineering* 2002; **53**: 2549–2568.

Mohammadi S. *Extended Finite Element Method for Fracture Analysis of Structures*. Malden, MA: Blackwell Publishing, 2008.

Mohammadi S. *XFEM: Fracture Analysis of Composites*. Chichester: John Wiley & Sons, Ltd., 2012.

Mohammadnejad T, Khoei AR. An extended finite element method for fluid flow in partially saturated porous media with weak discontinuities; the convergence analysis of local enrichment strategies. *Computational Mechanics* 2013a; **51**: 327–345.

Mohammadnejad T, Khoei AR. An extended finite element method for hydraulic fracture propagation in deformable porous media with the cohesive crack model. *Finite Elements in Analysis and Design* 2013b; **73**: 77–95.

Mohammadnejad T, Khoei AR. Hydro-mechanical modeling of cohesive crack propagation in multiphase porous media using the extended finite element method. *International Journal for Numerical and Analytical Methods in Geomechanics* 2013c; **37**: 1247–1279.

Morland LW. A simple constitutive theory for simple saturated porous solids. *Journal of Geophysical Research* 1972; **77**: 890–900.

Moslemi H, Khoei AR. 3D adaptive finite element modeling of non-planar curved crack growth using the weighted superconvergent patch recovery method. *Engineering Fracture Mechanics* 2009; **76**: 1703–1728.

Motamedi D, Mohammadi S. Dynamic analysis of fixed cracks in composites by the extended finite element method. *Engineering Fracture Mechanics* 2010a; **77**: 3373–3393.

Motamedi D, Mohammadi S. Dynamic crack propagation analysis of orthotropic media by the extended finite element method. *International Journal of Fracture* 2010b; **161**: 21–39.

Motamedi D, Mohammadi S. Fracture analysis of composites by time independent moving-crack orthotropic XFEM. *International Journal of Mechanical Sciences* 2012; **54**: 20–37.

Mougaard JF, Poulsen PN, Nielsen LO. A partly and fully cracked triangular XFEM element for modeling cohesive fracture. *International Journal for Numerical Methods in Engineering* 2011; **85**: 1667–1686.

Mougaard JF, Poulsen PN, Nielsen LO. Complete tangent stiffness for extended finite element method by including crack growth parameters. *International Journal for Numerical Methods in Engineering* 2013; **95**: 33–45.

Mousavi SE, Grinspun E, Sukumar N. Harmonic enrichment functions: a unified treatment of multiple, intersecting and branched cracks in the extended finite element method. *International Journal for Numerical Methods in Engineering* 2011a; **85**: 1306–1322.

Mousavi SE, Grinspun E, Sukumar N. Higher-order extended finite elements with harmonic enrichment functions for complex crack problems. *International Journal for Numerical Methods in Engineering* 2011b; **86**: 560–574.

Mousavi SE, Sukumar N. Generalized Gaussian quadrature rules for discontinuities and crack singularities in the extended finite element method. *Computer Methods in Applied Mechanics and Engineering* 2010a; **199**: 3237–3249.

Mousavi SE, Xiao H, Sukumar N. Generalized Gaussian quadrature rules on arbitrary polygons, *International Journal for Numerical Methods in Engineering* 2010b; **82**: 99–113.

Mueller-Hoeppe DS, Wriggers P, Loehnert S. Crack face contact for a hexahedral-based XFEM formulation. *Computational Mechanics* 2012; **49**: 725–734.

Muhlhaus HB, Ainfantis E. A variational principle for gradient plasticity. *International Journal of Solids and Structures* 1991; **28**: 845–857.

Nagashima T, Omoto Y, Tani S. Stress intensity factor analysis of interface cracks using X-FEM. *International Journal for Numerical Methods in Engineering* 2003; **56**: 1151–1173.

Natarjan, S. (2011) Enriched finite element methods: advances and applications. PhD thesis. Institute of Mechanics and Advanced Materials Theoretical and Computational Mechanics. Cardiff University.

Natarajan S, Baiz PM, Bordas S, *et al.* Natural frequencies of cracked functionally graded material plates by the extended finite element method. *Composite Structures* 2011; **93**: 3082–3092.

Natarajan S, Bordas SPA, Mahapatra DR. Numerical integration over arbitrary polygonal domains based on Schwarz–Christoffel conformal mapping. *International Journal for Numerical Methods in Engineering* 2009; **80**: 103–134.

Neaupane KM, Yamabe T. A fully coupled thermo-hydro-mechanical nonlinear model for a frozen medium. *Computers and Geotechnics* 2001; **28**: 613–637.

Ng KLA, Small JC. Behavior of joints and interfaces subjected to water pressure. *Computers and Geotechnics* 1997; **20**: 71–93.

Ngo D, Scordelis A. Finite element analysis of reinforced concrete beams. *American Concrete Institute* 1967; **64**: 152–163.

Nguyen-Vinh H, Bakar I, Msekh MA, *et al.* Extended finite element method for dynamic fracture of piezo-electric materials. *Engineering Fracture Mechanics* 2012; **92**: 19–31.

Nicaise S, Renard Y, Chahine E. Optimal convergence analysis for the extended finite element method. *International Journal for Numerical Methods in Engineering* 2011; **86**: 528–548.

Nistor I, Guiton MLE, Massin P, *et al.* An X-FEM approach for large sliding contact along discontinuities. *International Journal for Numerical Methods in Engineering* 2009; **78**: 1407–1435.

Nistor I, Pantale O, Caperaa S. Numerical implementation of the extended finite element method for dynamic crack analysis. *Advances in Engineering Software* 2008; **39**: 573–587.

Noorishad J, Tsang CF, Witherspoon PA. Coupled thermal-hydraulic-mechanical phenomena in saturated fractured porous rock. *Journal of Geophysical Research: Solid Earth* 1984; **89**: 365–373.

Nouy A, Clément A. Extended stochastic finite element method for the numerical simulation of heterogeneous materials with random material interfaces. *International Journal for Numerical Methods in Engineering* 2010; **83**: 1312–1344.

Oden J, Martins J. Models and computational methods for dynamic friction phenomena. *Computer Methods in Applied Mechanics and Engineering* 1985; **52**: 527–634.

Oettl G, Stark RF, Hofstetter G. Numerical simulation of geotechnical problems based on a multi-phase finite element approach. *Computers and Geotechnics* 2004; **31**: 643–664.

Olesen JF, Poulsen PN. An embedded crack in a constant strain triangle utilizing extended finite element concepts. *Computers and Structures* 2013; **117**: 1–9.

Orowan E. Fracture and strength of solids. *Reports on Progress in Physics* 1948; **12**: 185–232.

Ortiz M, Leroy Y, Needleman A. A finite element method for localized failure analysis. *Computer Methods in Applied Mechanics and Engineering* 1987; **61**: 189–214.

Ortiz M, Pandolfi A. Finite-deformation irreversible cohesive elements for three-dimensional crack-propagation analysis. *International Journal for Numerical Methods in Engineering* 1999; **44**: 1267–1282.

Osher S, Fedkiw RP. Level set methods: an overview and some recent results. *Journal of Computational Physics* 2001; **169**: 463–502.

Osher S, Sethian JA. Fronts propagating with curvature dependent speed: algorithms based on Hamilton-Jacobi formulations. *Journal of Computational Physics* 1988; **79**: 12–49.

Oswald J, Gracie R, Khare R, Belytschko T. An extended finite element method for dislocations in complex geometries: thin films and nanotubes. *Computer Methods in Applied Mechanics and Engineering* 2009; **198**: 1872–1886.

Oswald J, Wintersberger E, Bauer G, Belytschko T. A higher-order extended finite element method for dislocation energetics in strained layers and epitaxial islands. *International Journal for Numerical Methods in Engineering* 2011; **85**: 920–938.

Panetier J, Ladeveze P, Chamoin L. Strict and effective bounds in goal-oriented error estimation applied to fracture mechanics problems solved with XFEM. *International Journal for Numerical Methods in Engineering* 2010; **81**: 671–700.

Park K, Pereira JP, Duarte CA, Paulino GH. Integration of singular enrichment functions in the generalized/extended finite element method for three-dimensional problems. *International Journal for Numerical Methods in Engineering* 2009; **78**: 1220–1257.

Pastor M, Peraire J, Zienkiewicz OC. Adaptive remeshing for shear band localization problem. *Archive of Applied Mechanics* 1991; **61**: 30–39.

Pathak H, Singh A, Singh IV. Numerical simulation of bi-material interfacial cracks using EFGM and XFEM. *International Journal of Mechanics and Materials in Design* 2012; **8**: 9–36.

Pathak H, Singh A, Singh IV, Yadav SK. A simple and efficient XFEM approach for 3D cracks simulations. *International Journal of Fracture* 2013; **181**: 189–208.

Patzak B, Jirasek M. Process zone resolution by extended finite elements. *Engineering Fracture Mechanics* 2003; **70**: 957–977.

Pedersen RR, Simone A, Sluys LJ. An analysis of dynamic fracture in concrete with a continuum visco-elastic visco-plastic damage model. *Engineering Fracture Mechanics* 2008; **75**: 3782–3805.

Peerlings RHJ, de Borst R, Brekelmans W, Geers M. Localisation issues in local and nonlocal continuum approaches to fracture. *European Journal of Mechanics A/Solids* 2002; **21**: 175–189.

Perić D, Owen DRJ. Computational model for 3D contact problems with friction based on the penalty method. *International Journal for Numerical Methods in Engineering* 1992; **35**: 1289–1309.

Pierrès E, Baietto MC, Gravouil A. A two-scale extended finite element method for modeling 3D crack growth with interfacial contact. *Computer Methods in Applied Mechanics and Engineering* 2010; **199**: 1165–1177.

Planas J, Elices M. Asymptotic analysis of a cohesive crack: 1. Theoretical background. *International Journal of Fracture* 1992; **55**: 153–177.

Planas J, Elices M. Asymptotic analysis of a cohesive crack: 2. Influence of the softening curve. *International Journal of Fracture* 1993; **64**: 221–237.

Pommier S, Gravouil A, Combescure A, Moës N. *Extended Finite Element Method for Crack Propagation*. Hoboken, NJ: John Wiley & Sons, Inc., 2011.

Pourmodheji R, Mashayekhi M. Improvement of the extended finite element method for ductile crack growth. *Materials Science and Engineering A* 2012; **551**: 255–271.

Prabel B, Combescure A, Gravouil A, Marie S. Level set X-FEM non-matching meshes: application to dynamic crack propagation in elastic-plastic media. *International Journal for Numerical Methods in Engineering* 2007; **69**: 1553–1569.

Prange C, Loehnert S, Wriggers P. Error estimation for crack simulations using the XFEM. *International Journal for Numerical Methods in Engineering* 2012; **91**: 1459–1474.

Proceedings of 5th ICOLD International Benchmark Workshop on Numerical Analysis of Dams, Denver, CO. 1999.

QingWen R, YuWen D, TianTang YU. Numerical modeling of concrete hydraulic fracturing with extended finite element method. *Science in China Series E: Technological Sciences* 2009; **52**: 559–565.

Rabczuk T, Belytschko T. Cracking particles: a simplified meshfree method for arbitrary evolving cracks. *International Journal for Numerical Methods in Engineering* 2004; **61**: 2316–2343.

Rabczuk T, Bordas S, Zi, G. A three dimensional meshfree method for continuous multiple-crack initiation, propagation and junction in statics and dynamics. *Computational Mechanics* 2007; **40**: 473–495.

Rabczuk T, Bordas S, Zi G. On three-dimensional modelling of crack growth using partition of unity methods. *Computers and Structures* 2010; **88**: 1391–1411.

Rabczuk T, Wall WA. *Extended finite element and meshfree methods*. Technical University of Munich, Munich, 2006.

Rabczuk T, Zi G. A meshfree method based on the local partition of unity for cohesive cracks. *Computational Mechanics* 2007; **39**: 43–60.

Rabczuk T, Zi G, Gerstenberger A, Wall WA. A new crack tip element for the phantomnode method with arbitrary cohesive cracks. *International Journal for Numerical Methods in Engineering* 2008; **75**: 577–599.

Rabinovich D, Givoli D, Vigdergauz S. XFEM-based crack detection scheme using a genetic algorithm. *International Journal for Numerical Methods in Engineering* 2007; **71**: 1051–1080.

Rabinovich D, Givoli D, Vigdergauz S. Crack identification by "arrival time" using XFEM and a genetic algorithm. *International Journal for Numerical Methods in Engineering* 2009; **77**: 337–359.

Radi E, Loret B. Mode II intersonic crack propagation in poroelastic media. *International Journal of Fracture* 2007; **147**: 235–267.

Rao BN, Rahman S. An enriched meshless method for non-linear fracture mechanics. *International Journal for Numerical Methods in Engineering* 2004; **50**: 197–223.

Rashid MM. The arbitrary local mesh replacement method: an alternative to remeshing for crack propagation analysis. *Computer Methods in Applied Mechanics and Engineering* 1998; **154**: 133–150.

Rashid YR. Analysis of prestressed concrete reactor vessels. *Nuclear Engineering and Design* 1968; **7**: 334–344.

Reed WH, Hill TR. *Triangular mesh methods for the neutron transport equation. Technical Report LA-UR-73-479*, Los Alamos Scientific Laboratory, Los Alamos, 1973.

Remmers JJC, de Borst R, Needleman A. A cohesive segments method for the simulation of crack growth. *Computational Mechanics* 2003; **31**: 69–77.

Remmers JJC, de Borst R, Needleman A. The simulation of dynamic crack propagation using the cohesive segments method. *Journal of the Mechanics and Physics of Solids* 2008; **56**: 70–92.

Réthoré J, de Borst R, Abellan MA. A discrete model for the dynamic propagation of shear bands in a fluid-saturated medium. *International Journal for Numerical and Analytical Methods in Geomechanics* 2007a; **31**: 347–370.

Réthoré J, de Borst R, Abellan MA. A two-scale approach for fluid flow in fractured porous media. *International Journal for Numerical Methods in Engineering* 2007b; **71**: 780–800.

Réthoré J, de Borst R, Abellan MA. A two-scale model for fluid flow in an unsaturated porous medium with cohesive cracks. *Computational Mechanics* 2008; **42**: 227–238.

Réthoré J, Gravouil A, Combescure A. A combined space-time extended finite element method. *International Journal for Numerical Methods in Engineering* 2005a: **64**: 260–284.

Réthoré J, Gravouil A, Combescure A. An energy-conserving scheme for dynamic crack growth using the extended finite element method. *International Journal for Numerical Methods in Engineering* 2005b; **63**: 631–659.

Réthoré J, Hild F, Roux S. Shear-band capturing using a multiscale extended digital image correlation technique. *Computer Methods in Applied Mechanics and Engineering* 2007c; **196**: 5016–5030.

Réthoré, J., Limodin, N., Buffière, J.Y., *et al.* (2012) Three-dimensional analysis of fatigue crack propagation using X-Ray tomography, digital volume correlation and extended finite element simulations. *Procedia IUTAM* **4**, pp. 151–158.

Rice JR. A path independent integral and the approximate analysis of strain concentrations by notches and cracks. *Journal of Applied Mechanics* 1968; **35**: 379–386.

Rice JR. Elastic fracture mechanics concepts for interfacial cracks. *Journal of Applied Mechanics* 1988; **55**: 98–103.

Rice JR, Rosengren GF. Plane strain deformation near a crack tip in a power law hardening material. *Journal of the Mechanics and Physics of Solids* 1968; **16**: 1–12.

Rice JR, Sih GC. Plane problems of crack in dissimilar media. *Journal of Applied Mechanics* 1965; **32**: 418–423.

Rice JR, Tracey DM. On the ductile enlargement of voids in triaxial stress fields. *Journal of the Mechanics and Physics of Solids* 1969; **17**, 201–217.

Richardson CL, Hegemann J, Sifakis E, *et al.* An XFEM method for modeling geometrically elaborate crack propagation in brittle materials. *International Journal for Numerical Methods in Engineering* 2011; **88**: 1042–1065.

Rochus V, Van Miegroet L, Rixen DJ, Duysinx P. Electrostatic simulation using XFEM for conductor and dielectric interfaces. *International Journal for Numerical Methods in Engineering* 2011; **85**: 1207–1226.

Ródenas JJ, González-Estrada OA, Díez P, Fuenmayor P. Accurate recovery-based upper error bounds for the extended finite element framework. *Computer Methods in Applied Mechanics and Engineering* 2010; **199**: 2607–2621.

Ródenas JJ, González-Estrada OA, Fuenmayor FJ, Chinesta F. Enhanced error estimator based on a nearly equilibrated moving least squares recovery technique for FEM and XFEM. *Computational Mechanics* 2013; **52**: 321–344.

Ródenas JJ, González-Estrada OA, Tarancón JE, Fuenmayor FJ. A recovery-type error estimator for the extended finite element method based on *singular + smooth* stress field splitting. *International Journal for Numerical Methods in Engineering* 2008; **76**: 545–571.

Rodriguez-Ferran A, Casadei F, Huerta A. ALE stress update for transient and quasistatic processes. *International Journal for Numerical Methods in Engineering* 1998; **43**: 241–262.

Rodriguez-Ferran A, Perez-Foguet A, Huerta A. Arbitrary Lagrangian–Eulerian (ALE) formulation for hyperelastoplasticity. *International Journal for Numerical Methods in Engineering* 2002; **53**: 1831–1851.

Rojas-Díaz R, Sukumar N, Sáez A, García-Sánchez F. Fracture in magnetoelectroelastic materials using the extended finite element method. *International Journal for Numerical Methods in Engineering* 2011; **88**: 1238–1259.

Rudland D, Brust F, Wilkowski G. The effect of cyclic and dynamic loading on the fracture resistance of nuclear piping steels. *Technical Report*. 1996.

Rüter M, Gerasimov T, Stein E. Goal-oriented explicit residual-type error estimates in XFEM. *Computational Mechanics* 2013; **52**: 361–376.

Rutqvist J, Börgesson L, Chijimatsu M, *et al.* Thermohydromechanics of partially saturated geological media: governing equations and formulation of four finite element models. *International Journal of Rock Mechanics and Mining Science* 2001; **38**: 105–127.

Saleh AL, Aliabadi MH. Crack growth analysis in concrete using boundary element method. *Engineering Fracture Mechanics* 1995; **51**: 533–545.

Samaniego E, Belytschko T. Continuum-discontinuum modelling of shear bands. *International Journal for Numerical Methods in Engineering* 2005; **62**: 1857–1872.
Sampaio R, Williams WO. Thermodynamics of diffusing mixtures. *Journal de Mecanique* 1979; **18**: 19–45.
Sanborn SE, Prévost JH. Frictional slip plane growth by localization detection and the extended finite element method (XFEM). *International Journal for Numerical and Analytical Methods in Geomechanics* 2011; **35**: 1278–1298.
Sancho JM, Planas J, Galvez JC, *et al.* An embedded cohesive crack model for finite element analysis of mixed mode fracture of concrete. *Fatigue & Fracture of Engineering Materials & Structures.* 2006; **29**: 1056–1065.
Sancho JM, Planas J, Cendon DA, *et al.* An embedded crack model for finite element analysis of concrete fracture. *Engineering Fracture Mechanics* 2007; **74**: 75–86.
Sarhangi Fard A, Hulsen MA, Anderson PD. Extended finite element method for viscous flow inside complex three-dimensional geometries with moving internal boundaries. *International Journal for Numerical Methods in Fluids* 2012a; **70**: 775–792.
Sarhangi Fard A, Hulsen MA, Meijer HEH, *et al.* Adaptive non-conformal mesh refinement and extended finite element method for viscous flow inside complex moving geometries. *International Journal for Numerical Methods in Fluids* 2012b; **68**: 1031–1052.
Sarris E, Papanastasiou P. The influence of the cohesive process zone in hydraulic fracturing modeling. *International Journal of Fracture* 2011; **167**: 33–45.
Sarris E, Papanastasiou P. Modeling of hydraulic fracturing in a poroelastic cohesive formation. *International Journal of Geomechanics* 2012; **12**: 160–167.
Sauerland H, Fries TP. The extended finite element method for two-phase and free-surface flows: a systematic study. *Journal of Computational Physics* 2011; **230**: 3369–3390.
Sauerland H, Fries TP. The stable XFEM for two-phase flows. *Computers and Fluids* 2013; **87**: 41–49.
Schrefler BA, Scotta R. A fully coupled dynamic model for two-phase fluid flow in deformable porous media. *Computer Methods in Applied Mechanics and Engineering* 2001; **190**: 3223–3246.
Schrefler BA, Secchi S and Simoni L. On adaptive refinement techniques in multi-field problems including cohesive fracture. *Computer Methods in Applied Mechanics and Engineering* 2006; **195**: 444–461.
Schrefler BA, Zhan XY. A fully coupled model for water flow and airflow in deformable porous media. *Water Resources Research* 1993; **29**: 155–167.
Schrefler BA, Zhan XY, Simoni L. A coupled model for water flow, air flow and heat flow in deformable porous media. *International Journal of Numerical Methods for Heat and Fluid Flow* 1995; **5**: 531–547.
Seabra MRR, Sustaric P, Cesar de Sa JMA, *Rodic T.* Damage driven crack initiation and propagation in ductile metals using XFEM. *Computational Mechanics.* 2013; **52**: 161–179.
Secchi S, Simoni L, Schrefler BS. Mesh adaptation and transfer schemes for discrete fracture propagation in porous materials. *International Journal for Numerical and Analytical Methods in Geomechanics* 2007; **31**: 331–345.
Segura JM, Carol I. Coupled HM analysis using zero-thickness interface elements with double nodes. Part I: theoretical model. *International Journal for Numerical and Analytical Methods in Geomechanics* 2008a; **32**: 2083–2101.
Segura JM, Carol I. Coupled HM analysis using zero-thickness interface elements with double nodes. Part II: verification and application. *International Journal for Numerical and Analytical Methods in Geomechanics* 2008b; **32**: 2103–2123.
Sethian JA. Curvature and the evolution of fronts. *Communication of Mathematical Physics* 1985; **101**: 487–499.
Sethian JA. A marching level set method for monotonically advancing fronts. *Proceeding of the National Academy of Sciences* 1996; **93**: 1591–1595.
Sethian JA. *Level Set Methods and Fast Marching Methods: Evolving Interfaces in Computational Geometry, Fluid Mechanics, Computer Vision, and Material Science.* Cambridge: Cambridge University Press. 1999.
Shahmiri S, Gerstenberger A, Wolfgang AW. An XFEM-based embedding mesh technique for incompressible viscous flows. *International Journal for Numerical Methods in Engineering* 2011; **65**: 166–190.
Shamloo A, Azami AR, Khoei AR. Modeling of pressure-sensitive materials using a cap plasticity theory in extended finite element method. *Journal of Materials Processing Technology* 2005; **164**: 1248–1257.
Shen Y, Lew A. An optimally convergent discontinuous Galerkin-based extended finite element method for fracture mechanics. *International Journal for Numerical Methods in Engineering* 2010a; **82**: 716–755.
Shen Y, Lew A. Stability and convergence proofs for a discontinuous-Galerkin-based extended finite element method for fracture mechanics. *Computer Methods in Applied Mechanics and Engineering* 2010b; **199**: 2360–2382.
Sheng D, Sloan SW, Gens A, Smith DW. Finite element formulation and algorithms for unsaturated soils. Part I: theory. *International Journal for Numerical and Analytical Methods in Geomechanics* 2003; **27**: 745–765.
Shi J, Chopp D, Lua J, *et al.* Abaqus implementation of extended finite element method using a level set representation for three-dimensional fatigue crack growth and life predictions. *Engineering Fracture Mechanics* 2010; **77**: 2840–2863.

Shibanuma K, Utsunomiya T. Reformulation of XFEM based on PUFEM for solving problem caused by blending elements. *Finite Elements in Analysis and Design* 2009; **45**: 806–816.

Shibanuma K, Utsunomiya T. Evaluation on reproduction of priori knowledge in XFEM. *Finite Element in Analysis and Design* 2011; **47**: 424–433.

Shih CF, Hutchinson JW. Fully plastic solution and large-scale yielding estimates for plane stress crack problems. *Journal of Engineering Materials and Technology* 1976; **98**: 289–295.

Sih, GC. Strain energy factors applied to mixed mode crack problems. *International Journal of Fracture* 1974; **10**: 305–321.

Simo JC, Laursen T. An augmented Lagrangian treatment of contact problems involving friction. *Computers and Structures* 1992; **42**: 97–116.

Simo JC, Oliver J, Armero F. An analysis of strong discontinuities induced by strain softening in rate-independent inelastic solids. *Computational Mechanics* 1993; **12**: 277–296.

Simone A. Partition of unity-based discontinuous elements for interface phenomena: computational issues. *Communications in Numerical Methods in Engineering* 2004; **20**: 465–478.

Simone A, Duarte CA, Giessen EVD. A generalized finite element method for polycrystals with discontinuous grain boundaries. *International Journal for Numerical Methods in Engineering* 2006; **67**: 1122–1145.

Simoni L, Secchi S. Cohesive fracture mechanics for a multi-phase porous medium. *Engineering Computations* 2003; **20**: 675–698.

Singh IV, Mishra BK, Bhattacharya S. XFEM simulation of cracks, holes and inclusions in functionally graded materials. *International Journal of Mechanics and Materials in Design* 2011; **7**: 199–218.

Singh IV, Mishra BK, Bhattacharya S, Patil RU. The numerical simulation of fatigue crack growth using extended finite element method. *International Journal of Fatigue* 2012; **36**: 109–119.

Skrzypczak T. Sharp interface numerical modeling of solidification process of pure metals. *Archives of Metallurgy and Materials* 2012; **57**: 1189–1199.

Smith, B.G. (2008) The extended finite element method for special problems with moving interfaces. PhD thesis. Northwestern University.

Smith BG, Vaughan BL, Chopp DL. The extended finite element method for boundary layer problems in biofilm growth. *Communications in Applied Mathematics and Computational Science* 2007; **2**: 35–56.

Song JH, Areias PMA, Belytschko T. A method for dynamic crack and shear band propagation with phantom nodes. *International Journal for Numerical Methods in Engineering* 2006a; **67**: 868–893.

Song SH, Paulino GH, Buttlar WG. A bilinear cohesive zone model tailored for fracture of asphalt concrete considering viscoelastic bulk material. *Engineering Fracture Mechanics* 2006b; **73**: 2829–2848.

Spence DA, Sharp P. Self-similar solutions for elastohydrodynamic cavity flow. *Proceedings of the Royal Society of London* 1985; **400**: 289–313.

Stazi FL, Budyn E, Chessa J, Belytschko T. An extended finite element method with higher-order elements for curved cracks. *Computational Mechanics* 2003; **31**: 38–48.

Stelzer R, Hofstetter G. Adaptive finite element analysis of multi-phase problems in geotechnics. *Computers and Geotechnics* 2005; **32**: 458–481.

Stolarska M, Chopp DL. Modeling thermal fatigue cracking in integrated circuits by level sets and the extended finite element method. *International Journal of Engineering Science* 2003; **41**: 2381–2410.

Stolarska M, Chopp DL, Moës N, Belytschko T. Modeling of crack growth by level sets in the extended finite element method. *International Journal for Numerical Methods in Engineering* 2001; **51**: 943–960.

Strang G, Fix G. *An Analysis of the Finite Element Method.* Engle-wood Cliffs, NJ, Prentice-Hall. 1973.

Strouboulis T, Babuška I, Copps K. The design and analysis of the generalized finite element method. *Computer Methods in Applied Mechanics and Engineering* 2000; **181**: 43–69.

Strouboulis T, Copps K, Babuška I. The generalized finite element method. *Computer Methods in Applied Mechanics and Engineering* 2001; **190**: 4081–4193.

Sukumar N, Chopp DL, Béchet EB, Moës N. Three-dimensional non-planar crack growth by a coupled extended finite element and fast marching method. *International Journal for Numerical Methods in Engineering* 2008; **76**: 727–748.

Sukumar N, Chopp DL, Moës N, Belytschko T. Modeling holes and inclusions by level sets in the extended finite-element method. *Computer Methods in Applied Mechanics and Engineering* 2001; **190**: 6183–6200.

Sukumar N, Chopp DL, Moran B. Extended finite element method and fast marching method for three-dimensional fatigue crack propagation. *Engineering Fracture Mechanics* 2003a; **70**: 29–48.

Sukumar N, Huang ZY, Prévost JH, Suo Z. Partition of unity enrichment for bimaterial interface cracks. *International Journal for Numerical Methods in Engineering* 2004; **59**: 1075–1102.

Sukumar N, Moës N, Moran B, Belytschko T. Extended finite element method for three-dimensional crack modeling. *International Journal for Numerical Methods in Engineering* 2000; **48**: 1549–1570.

Sukumar N, Prévost JH. Modeling quasi-static crack growth with the extended finite element method. Part I: computer implementation. *International Journal of Solids and Structures* 2003; **40**: 7513–7537.

Sukumar N, Srolovitz DJ. Finite element-based model for crack propagation in polycrystalline materials. *Computational and Applied Mathematics* 2004; **23**: 363–380.

Sukumar N, Srolovitz DJ, Baker TJ, Prévost JH. Brittle fracture in polycrystalline microstructures with the extended finite element method. *International Journal for Numerical Methods in Engineering* 2003b; **56**: 2015–2037.

Swenson, D.V. (1985) Modeling mixed-mode dynamic crack propagation using finite elements. PhD thesis. Cornell University.

Sy A, Renard Y. A fictitious domain approach for structural optimization with a coupling between shape and topological gradient. *Far East Journal of Mathematical Sciences* 2010; **47**: 33–50.

Szabó B, Babuška I. *Finite Element Analysis*. John Wiley & Sons, Inc, New York. 1991.

Tabarraei A, Sukumar N. Extended finite element method on polygonal and quadtree meshes. *Computer Methods in Applied Mechanics and Engineering* 2008; **197**: 425–438.

Tabiei A, Wu J. Development of the DYNA3D simulation code with automated fracture procedure for brick elements. *International Journal for Numerical Methods in Engineering* 2003; **57**: 1979–2006.

Tai HW. Plastic damage and ductile fracture in mild steels. *Engineering Fracture Mechanics* 1990; **36**: 853–880.

Tarancon JE, Vercher A, Giner E, Fuenmayor FJ. Enhanced blending elements for XFEM applied to linear elastic fracture mechanics. *International Journal for Numerical Methods in Engineering* 2009; **77**: 126–148.

Terzaghi K. *Theoretical Soil Mechanics*. New York, John Wiley & Sons, Inc. 1943.

Tong FG, Jing LR, Zimmerman RW. An effective thermal conductivity model of geological porous media for coupled thermo-hydro-mechanical systems with multiphase flow. *International Journal of Rock Mechanics and Mining Science* 2009; **46**: 1358–1369.

Tong FG, Jing LR, Zimmerman RW. A fully coupled thermo-hydro-mechanical model for simulating multiphase flow, deformation and heat transfer in buffer material and rock masses. *International Journal of Rock Mechanics and Mining Science* 2010; **47**: 205–217.

Tran AB, Yvonnet J, He QC, *et al.* A multiple level set approach to prevent numerical artefacts in complex microstructures with nearby inclusions within XFEM. *International Journal for Numerical Methods in Engineering* 2011; **85**: 1436–1459.

Tran TQN, Lee HP, Lim SP. Modelling porous structures by penalty approach in the extended finite element method. *Computer Methods in Biomechanics and Biomedical Engineering* 2013; **16**: 347–357.

Triantafyllidis N, Ainfantis E. A gradient approach to localization of deformation, I. Hyperelastic materials, *Journal of Elasticity* 1986; **16**: 225–237.

Tvergaard V, Needleman A. Analysis of the cup-cone fracture in a round tensile bar. *ACTA Metallurgica* 1984; **32**, 157–169.

Uchibori A, Ohshima H. Numerical analysis of melting/solidification phenomena using a moving boundary analysis method X-FEM. In: Revankar ST (ed.) *Advances in Nuclear Fuel*. 2012; **3**: 53–73.

Unger JF, Eckardt S, Könke C. Modelling of cohesive crack growth in concrete structures with the extended finite element method. *Computer Methods in Applied Mechanics Engineering* 2007; **196**: 4087–4100.

Vajragupta N, Uthaisangsuk V, Schmaling B, *et al.* A micromechanical damage simulation of dual phase steels using XFEM. *Computational Materials Science* 2012; **54**: 271–279.

Van der Bos F, Gravemeier V. Numerical simulation of premixed combustion using an enriched finite element method. *Journal of Computational Physics* 2009; **228**: 3605–3624.

Van Genuchten MT. A closed-form equation for predicting the hydraulic conductivity of unsaturated soils. *Soil Science Society of America Journal* 1980; **44**: 892–898.

Vaziri HH. Theory and application of a fully coupled thermo-hydro-mechanical finite element model. *Computers and Structures* 1996; **61**: 131–146.

Ventura G. On the elimination of quadrature subcells for discontinuous functions in the extended finite-element method. *International Journal for Numerical Methods in Engineering* 2006; **66**: 761–795.

Ventura G. Domain and boundary quadrature for enrichment functions in the extended finite element method. *International Journal of Material Forming* 2008; **1**: 1135–1138.

Ventura G, Budyn E, Belytschko T. Vector level sets for description of propagating cracks in finite elements. *International Journal for Numerical Methods in Engineering* 2003; **58**: 1571–1592.

Ventura G, Gracie R, Belytschko T. Fast integration and weight function blending in the extended finite element method. *International Journal for Numerical Methods in Engineering* 2009; **77**: 1–29.

Ventura G, Moran B, Belytschko T. Dislocations by partition of unity. *International Journal for Numerical Methods in Engineering* 2005; **62**: 1463–1487.

Ventura G, Xu JX, Belytschko T. A vector level set method and new discontinuity approximations for crack growth by EFG. *International Journal for Numerical Methods in Engineering* 2002; **54**: 923–944.

Verhoosel CV, Remmers JJC, Gutiérrez MA. A partition of unity-based multiscale approach for modeling fracture in piezoelectric ceramics. *International Journal for Numerical Methods in Engineering* 2010; **82**: 966–994.

Vidal Y, Villon P, Huerta A. Locking in the incompressible limit: pseudo-divergence-free element free Galerkin. *Communications in Numerical Methods in Engineering* 2003; **19**: 725–735.

Vitali E, Benson DJ. An extended finite element formulation for contact in multi-material arbitrary Lagrangian–Eulerian calculations. *International Journal for Numerical Methods in Engineering* 2006; **67**: 1420–1444.

Vitali E, Benson DJ. Kinetic friction for multi-material arbitrary Lagrangian Eulerian extended finite element formulations. *Computational Mechanics* 2009; **43**: 847–857.

Volterra V. Sur l'quilibre des corps lastiques multiplement connexes. *Annales Scientifiques de l'cole Normale Suprieure Sr. 3*. 1907; **24**: 401–517.

Voyiadjis GZ, Kattan PI. *Damage Mechanics*. Boca Raton, FL: Taylor and Francis. 2005.

Wagner GJ, Moës N, Liu WK, Belytschko T. The extended finite element method for rigid particles in Stokes flow. *International Journal for Numerical Methods in Engineering* 2001; **51**: 293–313.

Wall WA, Gerstenberger A, Küttler U, Mayer UM. An XFEM based fixed-grid approach for 3D fluid-structure interaction. *Fluid Structure Interaction II* 2010; **73**: 327–349.

Wang H, Belytschko T. Fluid-structure interaction by the discontinuous-Galerkin method for large deformations. *International Journal for Numerical Methods in Engineering* 2009; **77**: 30–49.

Wang H, Chessa J, Liu WK, Belytschko T. The immersed/fictitious element method for fluid-structure interaction: volumetric consistency, compressibility and thin members. *International Journal for Numerical Methods in Engineering* 2008; **74**: 32–55.

Wang H, Zhang C, Yang L, You Z. Study on the rubber-modified asphalt mixtures' cracking propagation using the extended finite element method. *Construction and Building Materials* 2013; **47**: 223–230.

Wang J, Gadala MS. Formulation and survey of ALE method in nonlinear solid mechanics. *Finite Elements in Analysis and Design* 1997; **24**: 253–269.

Wang W, Kolditz O. Object-oriented finite element analysis of thermo-hydro-mechanical (THM) problems in porous media. *International Journal for Numerical Methods in Engineering* 2007; **69**: 162–201.

Wang Z, Ma L, Wu L, Yu H. Numerical simulation of crack growth in brittle matrix of particle reinforced composites using the XFEM technique. *Acta Mechanica Solida Sinica* 2012; **25**: 9–21.

Watanabe N, Wang W, Taron J, *et al.* Lower-dimensional interface elements with local enrichment: application to coupled hydro-mechanical problems in discretely fractured porous media. *International Journal for Numerical Methods in Engineering* 2012; **90**: 1010–1034.

Wawrzynek, P.A. and Ingraffea, A.R. (1991) Discrete modeling of crack propagation: Theoretical aspects and implementation issues in two and three dimensions. *Report.* Department of Structural Engineering, Cornell University.

Wei P, Wang MY, Xing X. A study on X-FEM in continuum structural optimization using a level set model. *Computer-Aided Design* 2010; **42**: 708–719.

Wells GN, Sluys LJ. A new method for modelling cohesive cracks using finite elements. *International Journal for Numerical Methods in Engineering* 2001; **50**: 2667–2682.

Wells GN, Sluys LJ, de Borst R. Simulating the propagation of displacement discontinuities in a regularized strain-softening medium. *International Journal for Numerical Methods in Engineering* 2002; **53**: 1235–1256.

Westergard HM. Bearing pressure and cracks. *Journal of Applied Mechanics* 1939; **6**: 49–53.

Whitaker S. Simultaneous heat, mass and momentum transfer in porous media; a theory of drying. *Advances in Heat Transfer* 1977; **13**: 119–203.

Williams ML. Stress singularities resulting from various boundary conditions in angular corners of plates in extension. *Journal of Applied Mechanics* 1952; **19**: 526–528.

Williams ML. On the stress distribution at the base of a stationary crack. *Journal of Applied Mechanics* 1957; **24**: 109–114.

Williams ML. The stress around a fault or crack in dissimilar media. *Bulletin of the Seismology Society of America* 1959; **49**: 199–204.

Willis JR. A comparison of the fracture criteria of Griffith and Barenblatt. *Journal of the Mechanics and Physics of Solids* 1967; **15**: 151–162.

Witherspoon PA, Wang JSY, Iwai K, Gale JE. Validity of cubic law for fluid flow in a deformable rock fracture. *Water Resources Research* 1980; **16**: 1016–1024.

Wnuk MP. Quasi-static extension of a tensile crack contained in viscoelastic–plastic solid. *Journal of Applied Mechanics* 1974; **41**: 234–242.

Wriggers P. *Computational Contact Mechanics*. New York: Springer, 2006.

Wu, S., Hoxha, D., Belayachi, N., and Do, D.P. (2010) Modeling mechanical behavior of geomaterials by the extended finite-element method. 5th International Conference on Multiscale Materials Modeling. Freiburg, October 2010.

Wu YS, Forsyth PA. On the selection of primary variables in numerical formulation for modeling multiphase flow in porous media. *Journal of Contaminant Hydrology* 2001; **48**: 277–304.

Wyart E, Coulon D, Duflot M, *et al.* A substructured FE-shell/XFE-3D method for crack analysis in thin-walled structures. *International Journal for Numerical Methods in Engineering* 2007; **72**: 757–779.

Wyart E, Coulon D, Pardoen T, *et al.* Application of the substructured finite element/extended finite element method (S-FE/XFE) to the analysis of cracks in aircraft thin walled structures. *Engineering Fracture Mechanics* 2009; **76**: 44–58.

Xiao QZ, Karihaloo BL. Improving the accuracy of XFEM crack tip fields using higher order quadrature and statically admissible stress recovery. *International Journal for Numerical Methods in Engineering* 2006; **66**: 1378–1410.

Xiao QZ, Karihaloo BL. Implementation of hybrid crack element on a general finite element mesh and in combination with XFEM. *Computer Methods in Applied Mechanics and Engineering* 2007; **196**: 1864–1873.

Xiao QZ, Karihaloo BL, Liu XY. Incremental-secant modulus iteration scheme and stress recovery for simulating cracking process in quasi-brittle materials using XFEM. *International Journal for Numerical Methods in Engineering* 2007; **69**: 2606–2635.

Xu C, Siegmund T, Ramani K. Rate-dependent crack growth in adhesives I. Modeling approach. *International Journal of Adhesion and Adhesives* 2003; **23**: 9–13.

Xu J, Lee CK, Tan KH. A two-dimensional co-rotational Timoshenko beam element with XFEM formulation. *Computational Mechanics* 2012; **49**: 667–683.

Xu J, Lee CK, Tan KH. An XFEM frame for plate elements in yield line analyses. *International Journal for Numerical Methods in Engineering* 2013a; **96**: 150–175.

Xu J, Lee CK, Tan KH. An XFEM plate element for high gradient zones resulted from yield lines. *International Journal for Numerical Methods in Engineering* 2013b; **93**: 1314–1344.

Xu J, Li Y, Chen X, *et al.* Characteristics of windshield cracking upon low-speed impact: numerical simulation based on the extended finite element method. *Computational Materials Science* 2010; **48**: 582–588.

Xu XP, Needleman A. Numerical simulations of fast crack growth in brittle solids. *Journal of the Mechanics and Physics of Solids* 1994; **42**: 1397–1434.

Xu Y, Yuan H. Computational analysis of mixed-mode fatigue crack growth in quasi-brittle materials using extended finite element methods. *Engineering Fracture Mechanics* 2009; **76**: 165–181.

Xu Y, Yuan H. Applications of normal stress dominated cohesive zone models for mixed-mode crack simulation based on extended finite element methods. *Engineering Fracture Mechanics* 2011; **78**: 544–558.

Yamada T, Kikuchi F. An arbitrary Lagrangian-Eulerian finite element method for incompressible hyperelasticity. *Computer Methods in Applied Mechanics and Engineering* 1993; **102**: 149–177.

Yan Y, Park SH. An extended finite element method for modeling near-interfacial crack propagation in a layered structure. *International Journal of Solids and Structures* 2008; **45**: 4756–4765.

Yang B, Ravi-Chandar K. A single-domain dual-boundary-element formulation incorporating a cohesive zone model for elastostatic cracks. *International Journal of Fracture* 1998; **93**: 115–144.

Ye C, Shi J, Cheng GJ. An extended finite element method (XFEM) study on the effect of reinforcing particles on the crack propagation behavior in a metal-matrix composite. *International Journal of Fatigue* 2012; **44**: 151–156.

Yosibash Z. Numerical thermo-elastic analysis of singularities in two-dimensions. *International Journal of Fracture* 1995; **74**: 341–361.

Yu TT, Gong ZW. Numerical simulation of temperature field in heterogeneous material with the XFEM. *Archives of Civil and Mechanical Engineering* 2013; **13**: 199–208.

Yvonnet J, He QC, Zhu QZ, Shao JF. A general and efficient computational procedure for modelling the Kapitza thermal resistance based on XFEM. *Computational Materials Science* 2011; **50**: 1220–1224.

Yvonnet J, Le Quang H, He QC. An XFEM/level set approach to modelling surface/interface effects and to computing the size-dependent effective properties of nanocomposites. *Computational Mechanics* 2008; **42**: 119–131.

Zabaras N, Ganapathysubramanian B, Tan L. Modelling dendritic solidification with melt convection using the extended finite element method. *Journal of Computational Physics* 2006; **218**: 200–227.

Zak AR, Williams ML. Crack point singularities at a bimaterial interface. *Journal of Applied Mechanics* 1963; **30**: 142–143.

Zamani A, Eslami MR. Implementation of the extended finite element method for dynamic thermoelastic fracture initiation. *International Journal of Solids and Structures* 2010; **47**: 1392–1404.

Zamani A, Gracie R, Eslami MR. Higher order tip enrichment of extended finite element method in thermoelasticity. *Computational Mechanics* 2010; **46**: 851–866.

Zamani A, Gracie R, Eslami MR. Cohesive and non-cohesive fracture by higher-order enrichment of XFEM. *International Journal for Numerical Methods in Engineering* 2012; **90**: 452–483.

Zhang HH, Li LX. Modeling inclusion problems in viscoelastic materials with the extended finite element method. *Finite Elements in Analysis and Design* 2009; **45**: 721–729.

Zhang HH, Rong G, Li LX. Numerical study on deformations in a cracked viscoelastic body with the extended finite element method. *Engineering Analysis with Boundary Elements* 2010; **34**: 619–624.

Zhang S, Wang G, Yu X. Seismic cracking analysis of concrete gravity dams with initial cracks using the extended finite element method. *Engineering Structures* 2013; **56**: 528–543.

Zhang Z, Ghassemi A. Simulation of hydraulic fracture propagation near a natural fracture using virtual multidimensional internal bonds. *International Journal for Numerical and Analytical Methods in Geomechanics* 2011; **35**: 480–495.

Zhao X, Bordas SPA, Qu J. A hybrid smoothed extended finite element/level set method for modeling equilibrium shapes of nano-inhomogeneities. *Computational Mechanics* 2013; **52**: 1417–1428.

Zhou J, Qi L. Treatment of discontinuous interface in liquid-solid forming with extended finite element method. *Transactions of Nonferrous Metals Society of China* 2010; **20**: 911–915.

Zhu QZ. On enrichment functions in the extended finite element method. *International Journal for Numerical Methods in Engineering* 2012; **91**: 186-217.

Zi G, Belytschko T. New crack-tip elements for XFEM and applications to cohesive cracks. *International Journal for Numerical Methods in Engineering* 2003; **57**: 2221–2240.

Zi G, Song JH, Budyn E, *et al.* A method for growing multiple cracks without remeshing and its application to fatigue crack growth. *Modeling and Simulation in Materials Science and Engineering* 2004; **12**: 901–915.

Zienkiewicz OC, Chan AHC, Pastor M, *et al.* Static and dynamic behavior of soils; A rational approach to quantitative solution. I. Fully saturated problems. *Proceedings of the Royal Society of London.* 1990a; **429**: 285–309.

Zienkiewicz OC, Chan AHC, Pastor M, *et al. Computational Geomechanics with Special Reference to Earthquake Engineering*. John Wiley & Sons, Inc, New York. 1999.

Zienkiewicz OC, Huang M, Pastor M. Localization problems in plasticity using finite elements with adaptive remeshing. *International Journal for Numerical and Analytical Methods in Geomechanics* 1995; **19**: 127–148.

Zienkiewicz OC, Shiomi T. Dynamic behaviour of saturated porous media; the generalized Biot formulation and its numerical solution. *International Journal for Numerical and Analytical Methods in Geomechanics* 1984; **8**: 71–96.

Zienkiewicz OC, Xie YM, Schrefler BA, *et al.* Static and dynamic behavior of soils; A rational approach to quantitative solution. II. Semi-saturated problems. *Proceedings of the Royal Society of London.* 1990b; **429**: 311–321.

Zienkiewicz OC, Zhu JZ. The Superconvergent patch recovery (SPR) and a posteriori error estimates. Part 2: error estimates and adaptivity. *International Journal for Numerical Methods in Engineering* 1992; **33**: 1365–1382.

Zilian A, Fries TP. A localized mixed-hybrid method for imposing interfacial constraints in the extended finite element method (XFEM). *International Journal for Numerical Methods in Engineering* 2009; **79**: 733–752.

Zilian A, Legay A. The enriched space-time finite element method (EST) for simultaneous solution of fluid-structure interaction. *International Journal for Numerical Methods in Engineering* 2008; **75**: 305–334.

Zilian A, Netuzhylov H. Hybridized enriched space-time finite element method for analysis of thin-walled structures immersed in generalized Newtonian fluids. *Computers and Structures* 2010; **88**: 1265–1277.

Zunino P. Analysis of backward Euler/extended finite element discretization of parabolic problems with moving interfaces. *Computer Methods in Applied Mechanics and Engineering* 2013; **258**: 152–165.

Index

Extended Finite Element Method: Theory and Applications, First Edition. Amir R. Khoei.

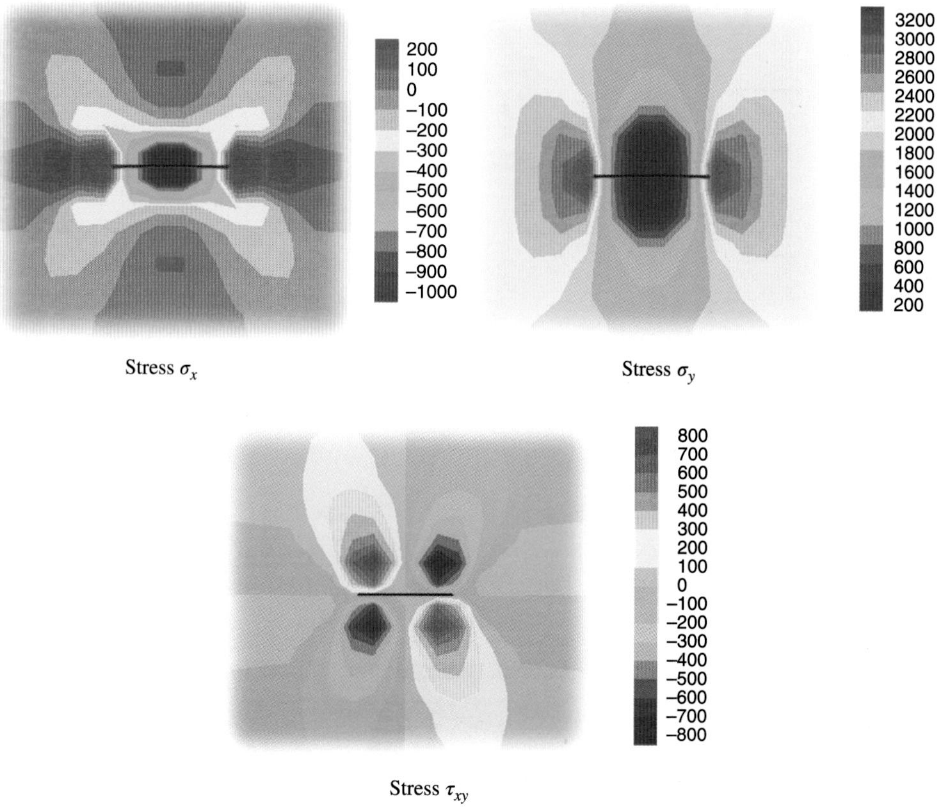

Figure 7.22 The distribution of stress contours σ_x, σ_y, and τ_{xy} for an infinite plate with a finite crack at its center

Extended Finite Element Method: Theory and Applications, First Edition. Amir R. Khoei.

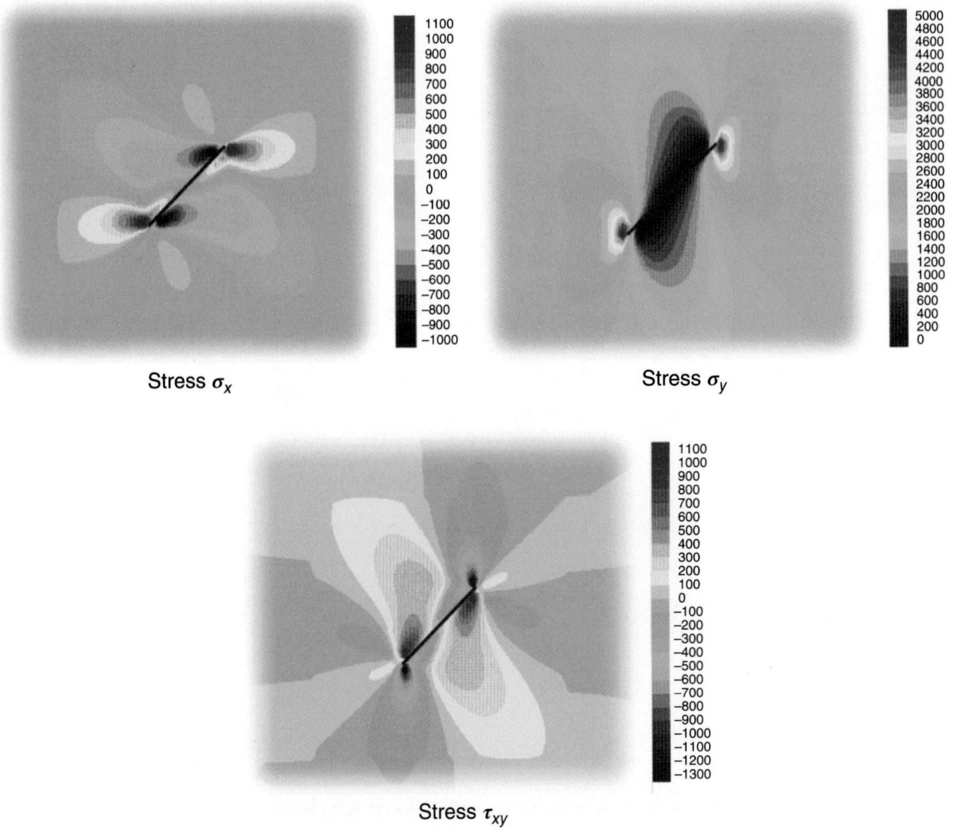

Figure 7.26 The distribution of stress contours σ_x, σ_y, and τ_{xy} for an infinite plate with an inclined crack

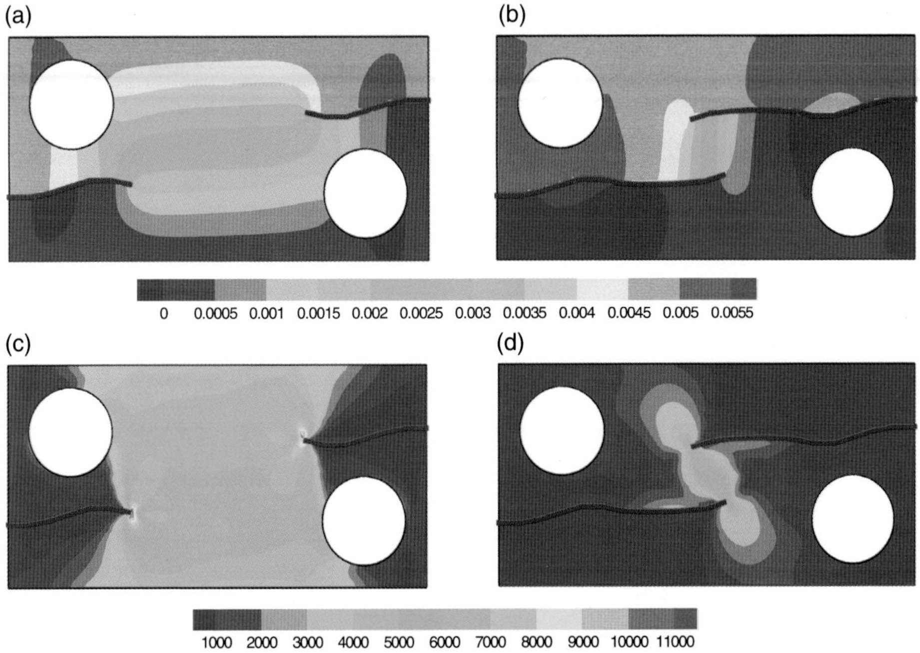

Figure 7.29 (a and b) The contours of displacement in y – direction (cm) and (c and d) the distribution of von-Mises stress (kg/cm^2) at the half and final loading steps of crack propagation

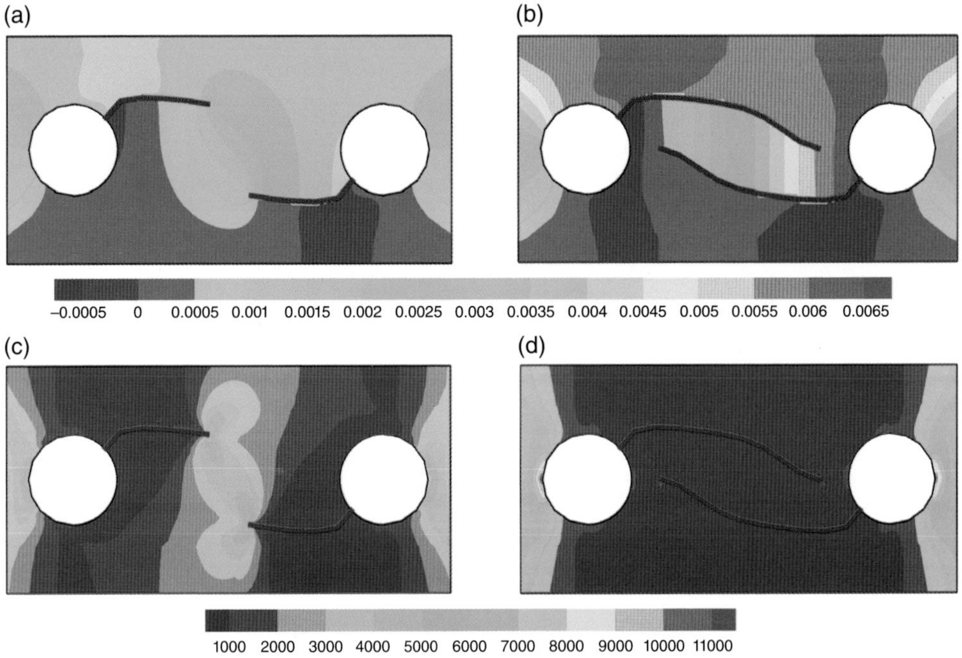

Figure 7.33 (a and b) The contours of displacement in y – direction (cm) and (c and d) the distribution of von-Mises stress (kg/cm^2) at the half and final loading steps of crack propagation

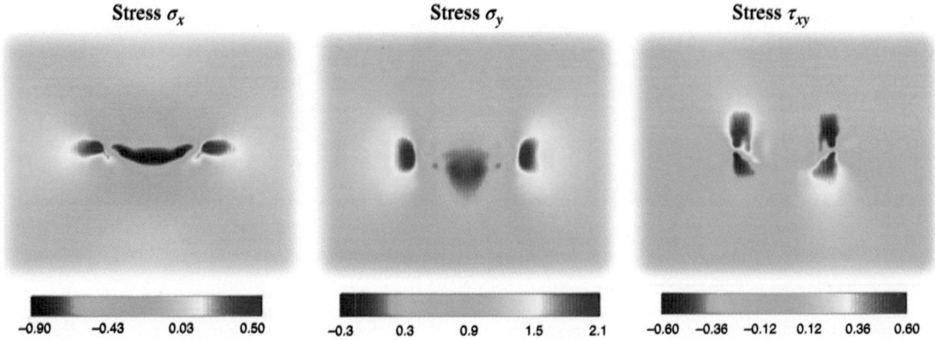

Figure 7.38 The distribution of stress contours σ_x, σ_y, and τ_{xy} for a curved center crack in an infinite plate

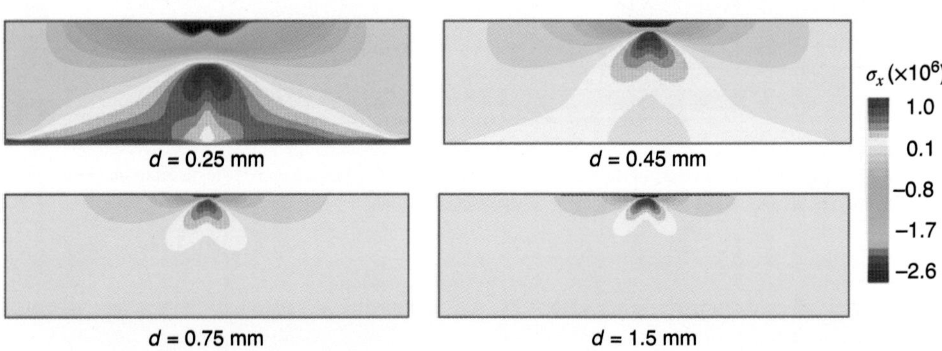

Figure 8.16 The distribution of stress contour in the x-direction at different steps of crack growth

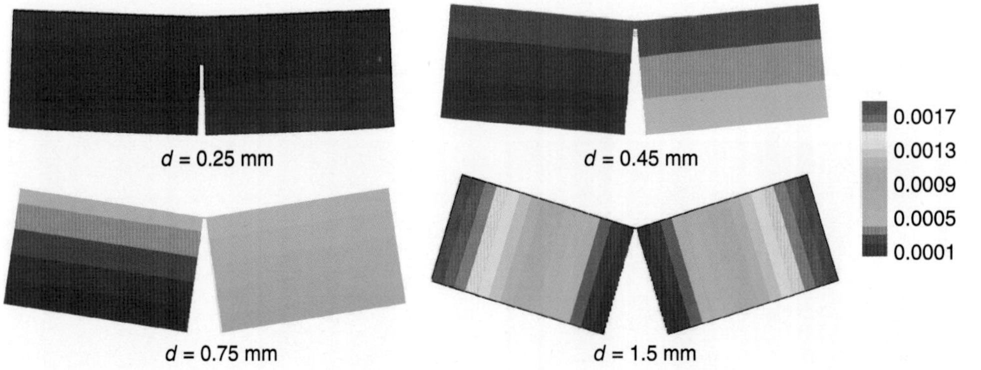

Figure 8.17 The distribution of displacement contour in x-direction at different steps of crack growth

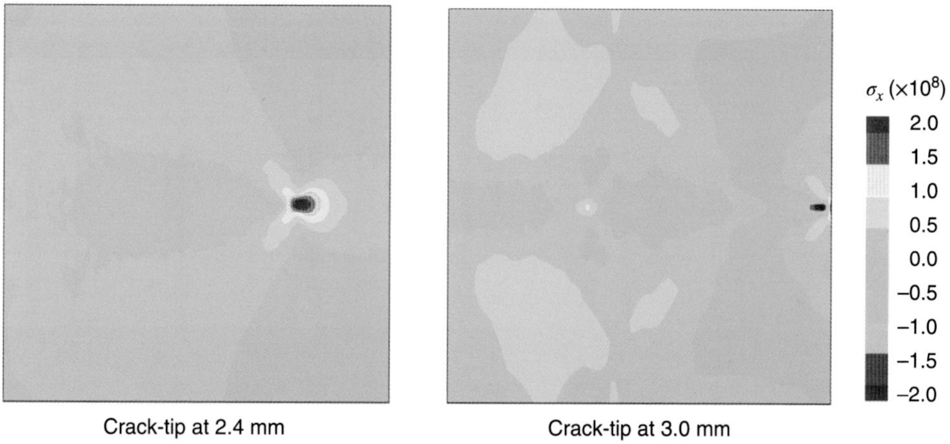

Figure 8.21 The distribution of stress contour in the x-direction at different steps of crack growth

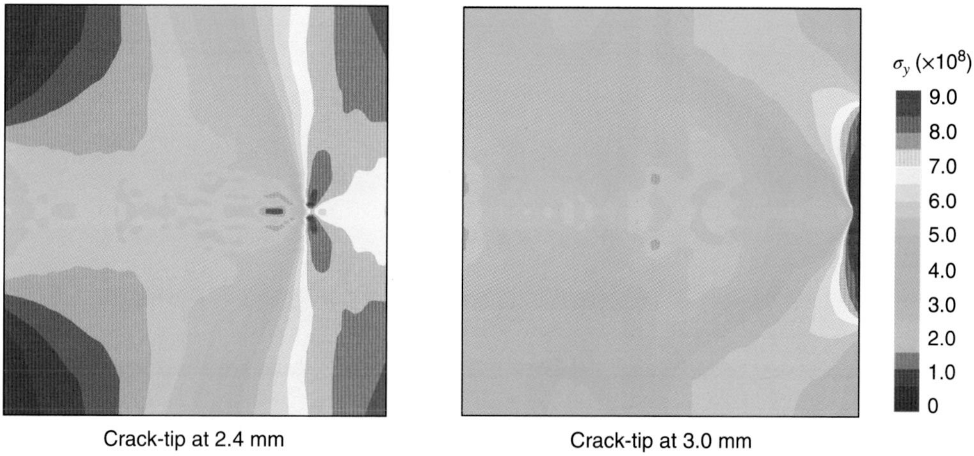

Figure 8.22 The distribution of stress contour in the y-direction at different steps of crack growth

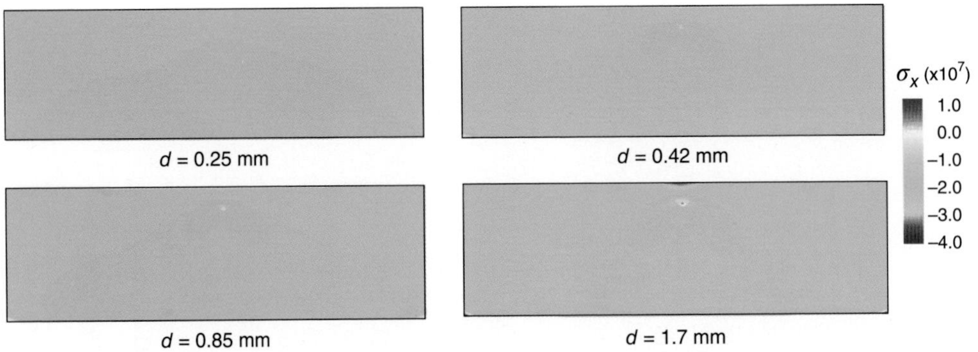

Figure 8.25 The distribution of stress contour in the x-direction at different steps of crack growth

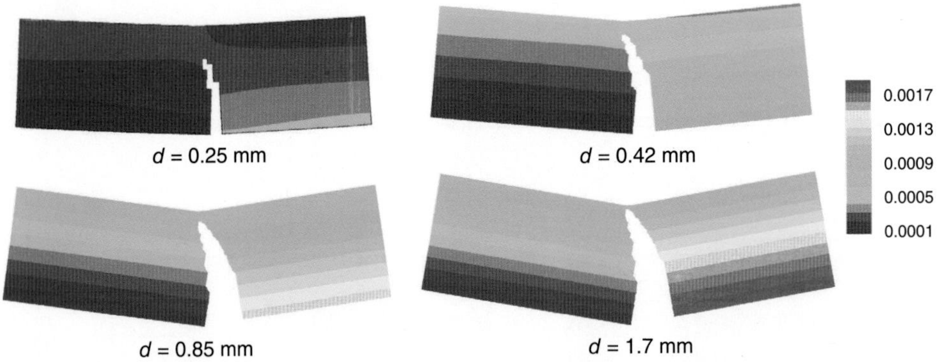

Figure 8.26 The distribution of displacement contour in the x-direction at different steps of crack growth

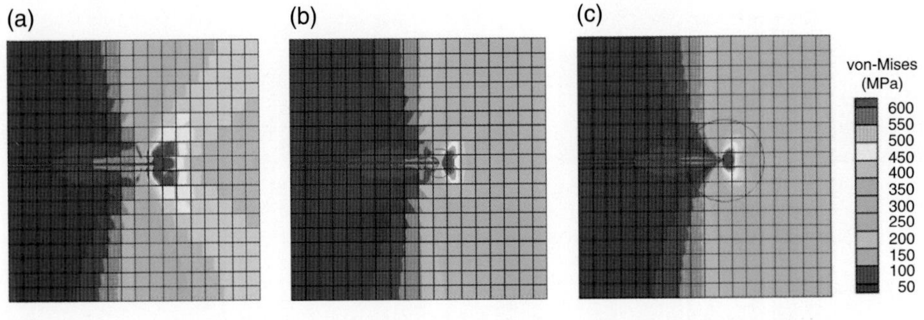

Figure 9.9 The von-Mises stress (MPa) contours for an edge crack problem: (a) the standard X-FEM; (b) the modified X-FEM with one-layer enriched elements; (c) the modified X-FEM with two-layer enriched elements

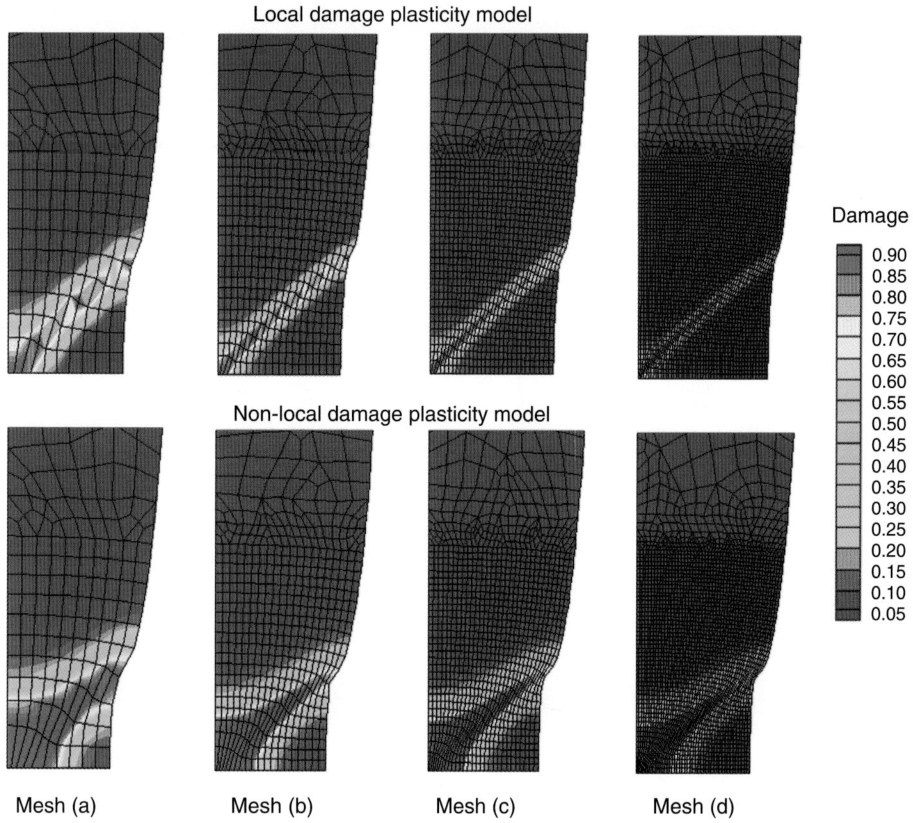

Figure 9.19 A comparison of damage contours for four different meshes between the local and non-local damage-plasticity models at the vertical displacement of 0.5 cm

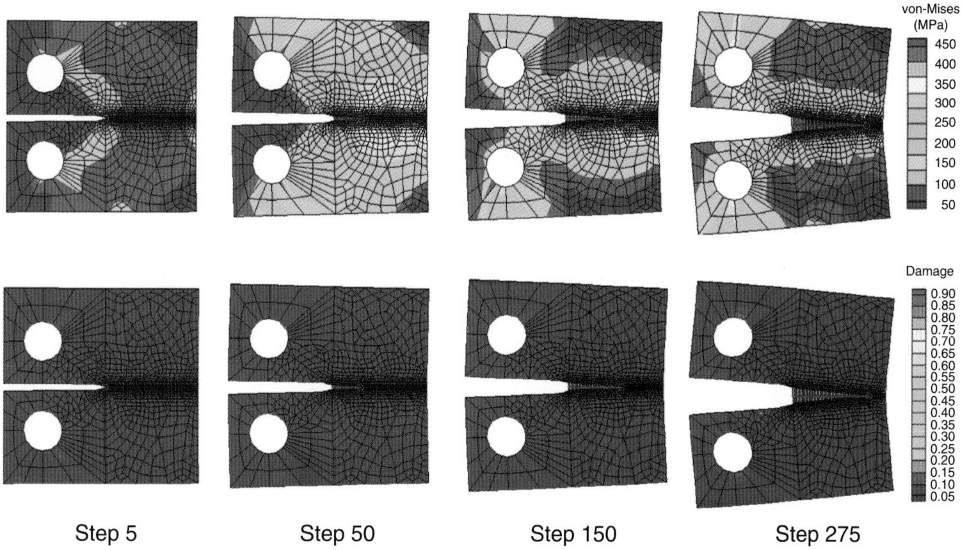

Figure 9.21 The distributions of von-Mises stress and damage contours at different crack growths for the CT test

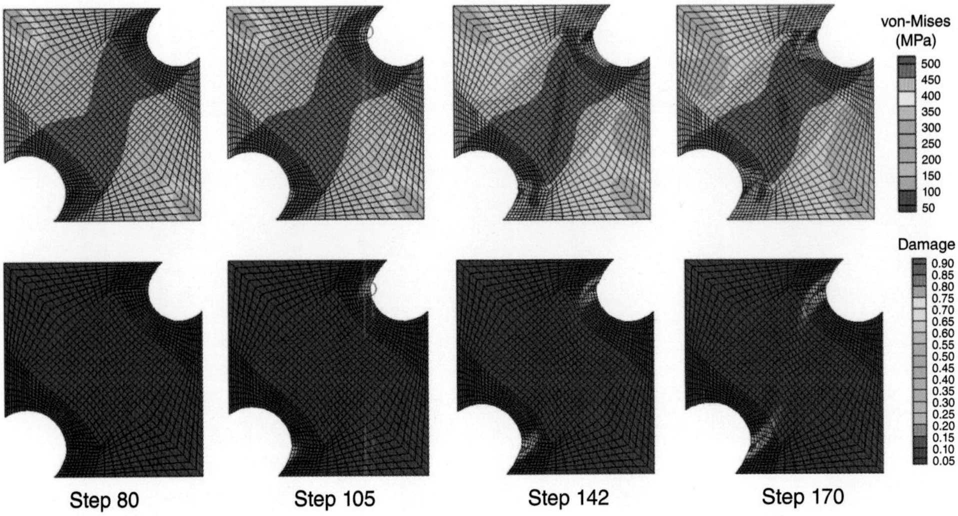

Figure 9.24 The distributions of von-Mises stress and damage contours at different crack growths for a double-notched specimen

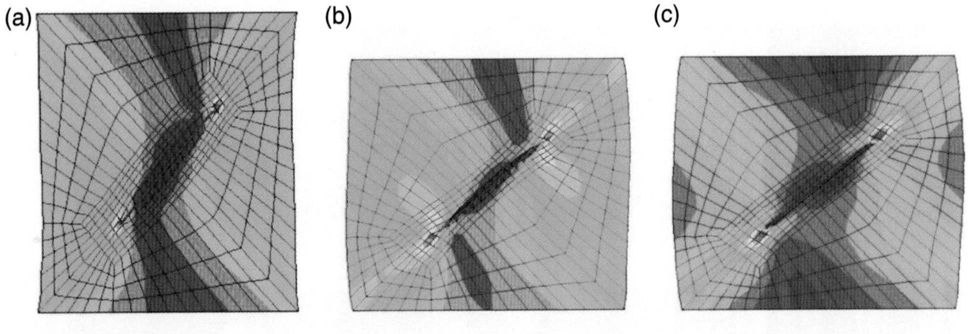

Figure 9.32 The von-Mises stress contours (MPa) at the displacement of: (a) $u = +0.1$ m in opening mode; (b) $u = -0.1$ m in opening mode; (c) $u = -0.1$ m in closing mode

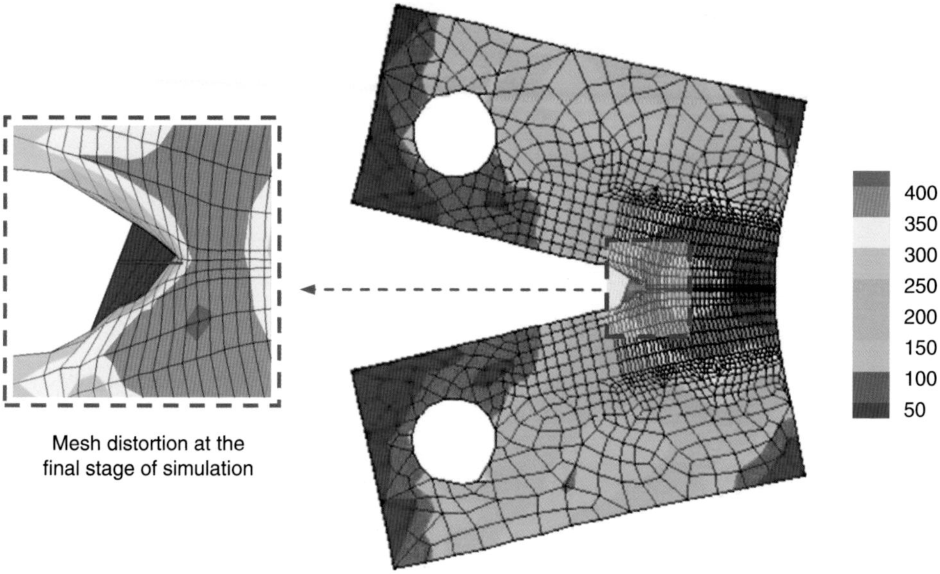

Figure 9.39 The von-Mises stress contour (MPa) for the CT test at the final stage of simulation: An excessive mesh distortion can be observed at the zoomed crack tip region

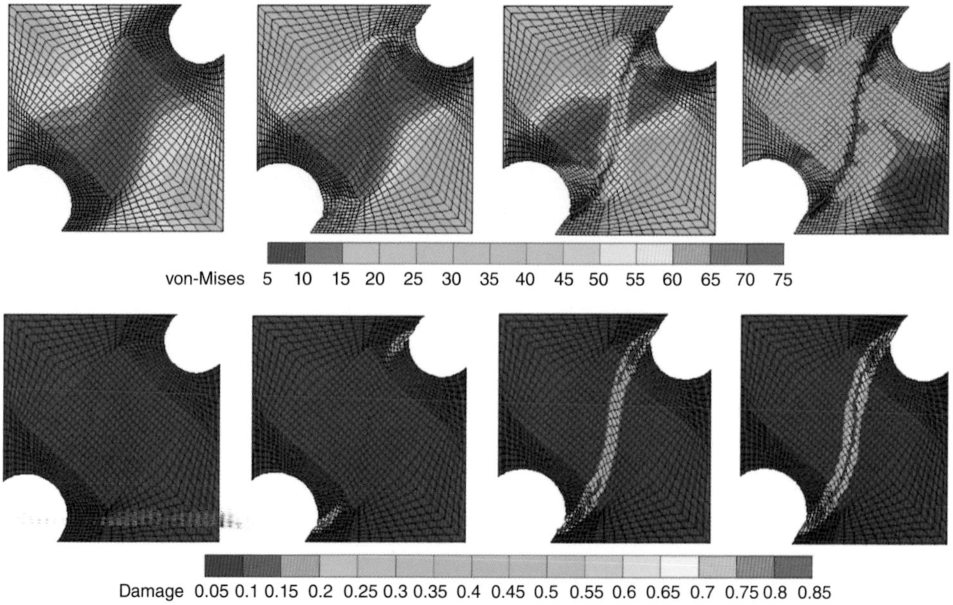

Figure 9.40 A double notched specimen under monotonic loading: The distributions of von-Mises stress (MPa) and damage contours at different stages of crack propagation $t = 0.02$, 0.04, 0.076, and 0.097 s

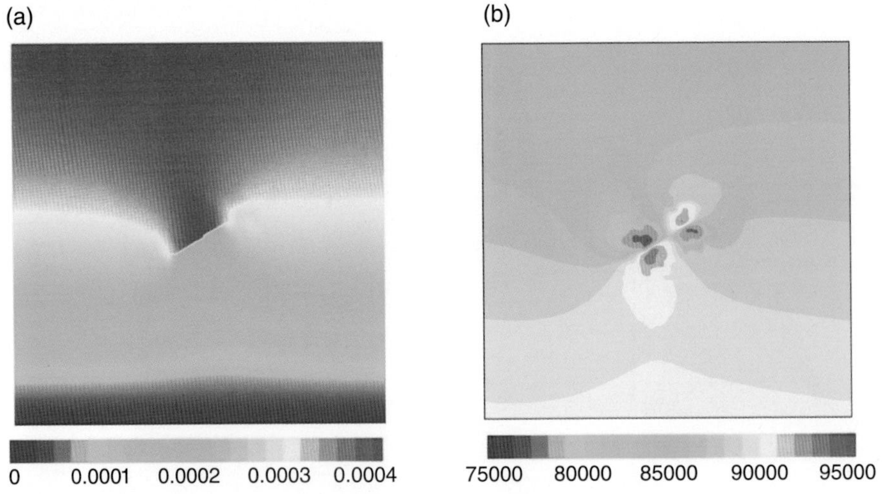

Figure 10.19 An infinite saturated porous medium with an inclined crack at the crack angle of $\beta = 30^{\circ}$: (a) the distribution of vertical displacement contour; (b) the contour of pressure gradient normal to discontinuity at the time step of $t = 10$ s

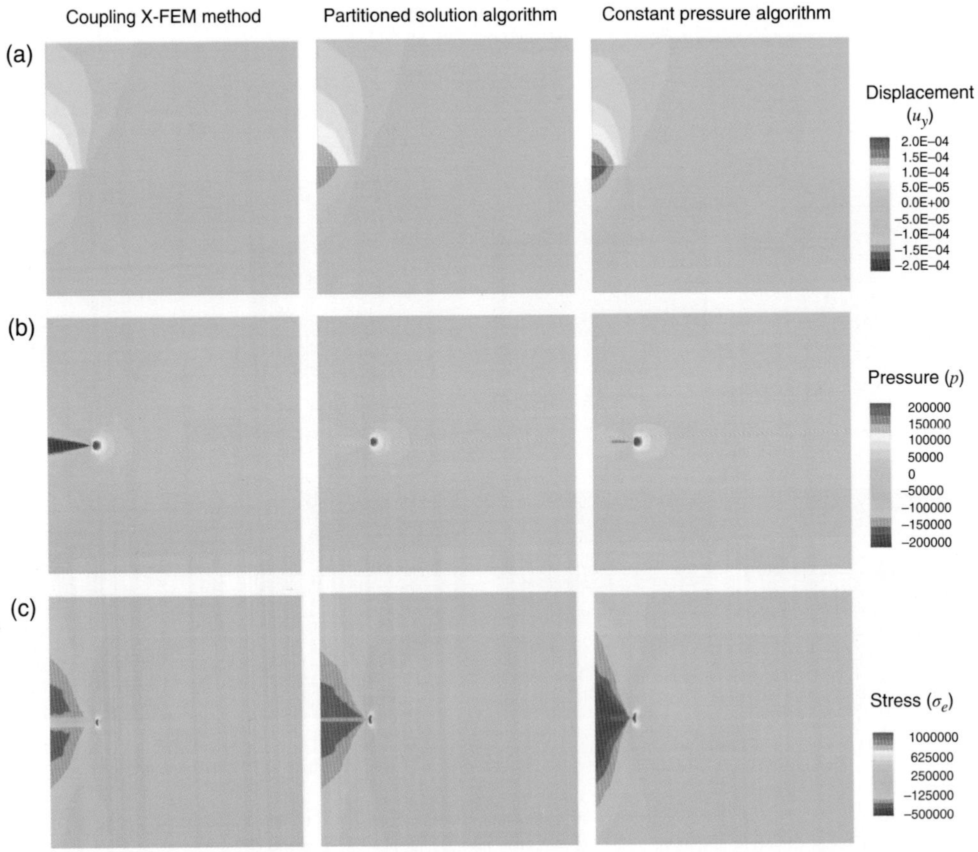

Figure 10.23 The contours of: (a) vertical displacement; (b) fluid pressure; and (c) maximum principal stress using the coupling X-FEM, the X-FEM based on the partitioned solution algorithm, and the X-FEM based on the time-dependent constant pressure algorithm

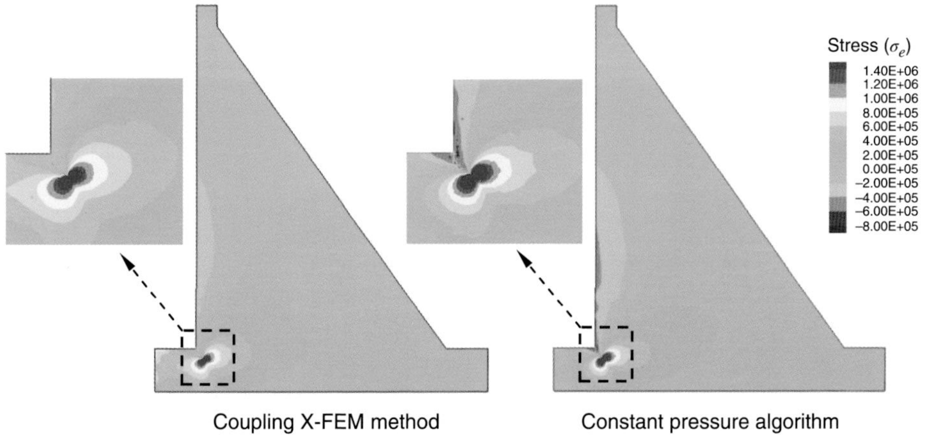

Figure 10.25 The contours of maximum principal stress using the coupling X-FEM method and the X-FEM method based on the time-dependent constant pressure algorithm

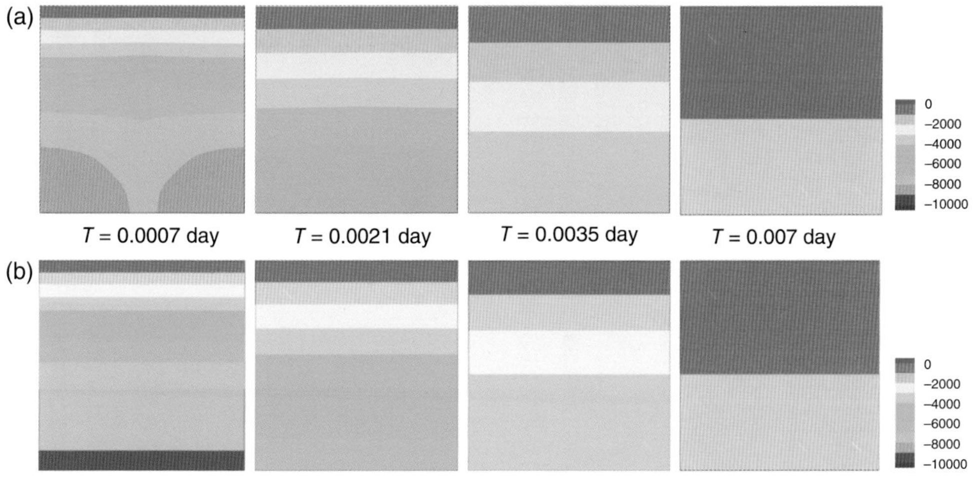

Figure 10.30 The distributions of excess pore water pressure contours at various time steps: (a) the X-FEM model with the penalty method; (b) the FEM with no crack

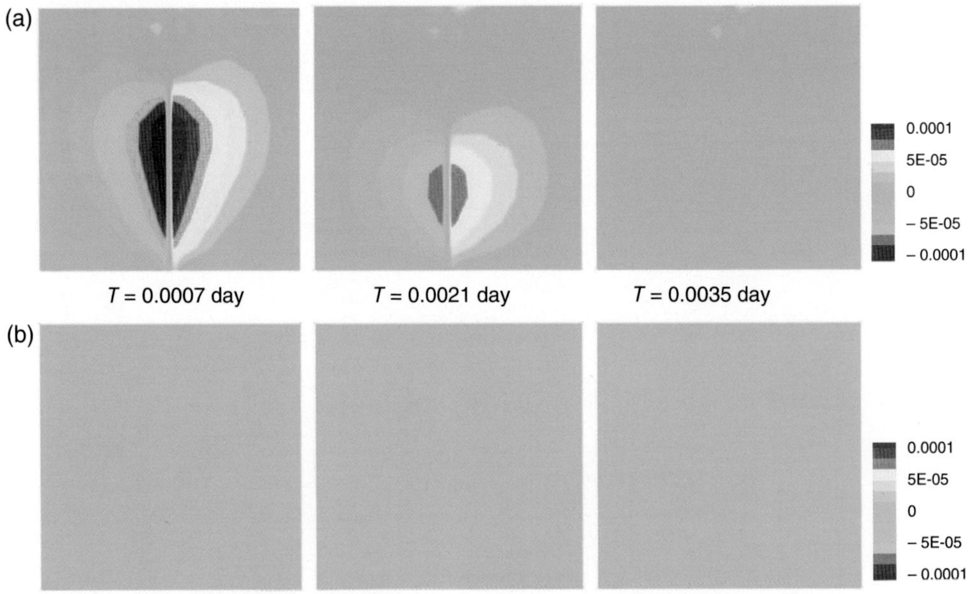

Figure 10.31 The distributions of horizontal displacement contours at various time steps: (a) the X-FEM model with the penalty method; (b) the FEM with no crack

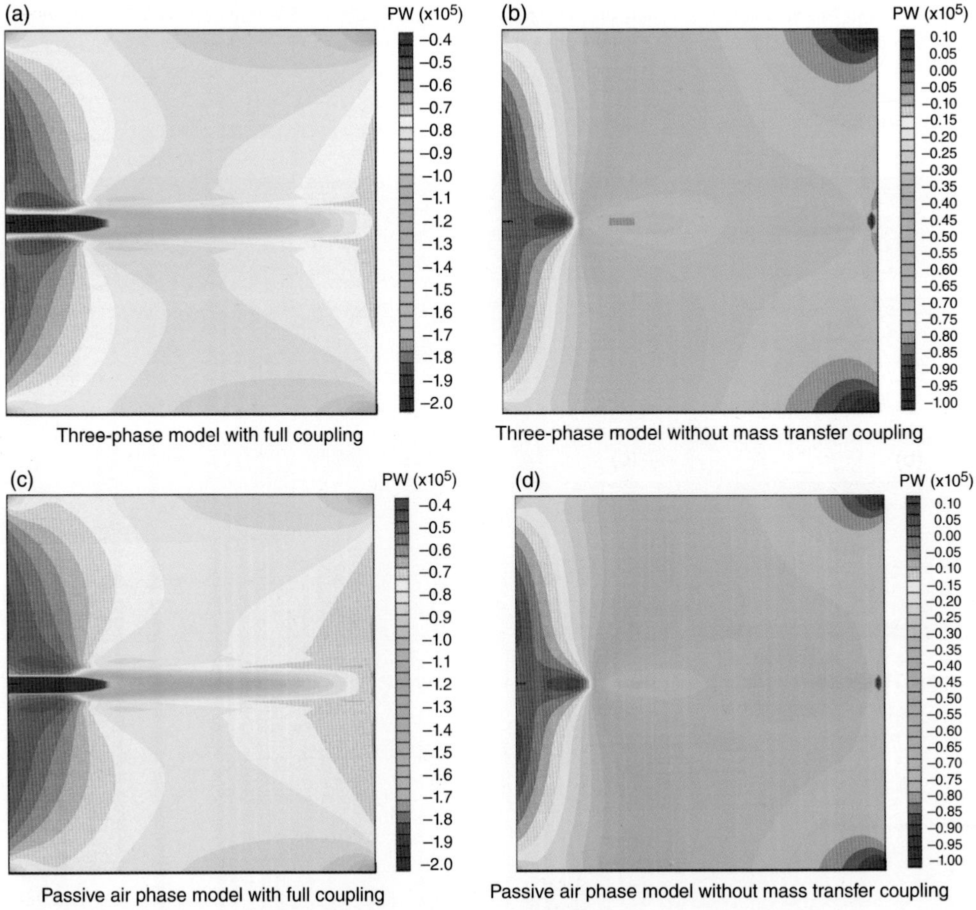

Figure 11.12 Water pressure (Pa) contours; (a) three-phase model with full coupling, (b) three-phase model without mass transfer coupling, (c) passive air phase model with full coupling, (d) passive air phase model without mass transfer coupling

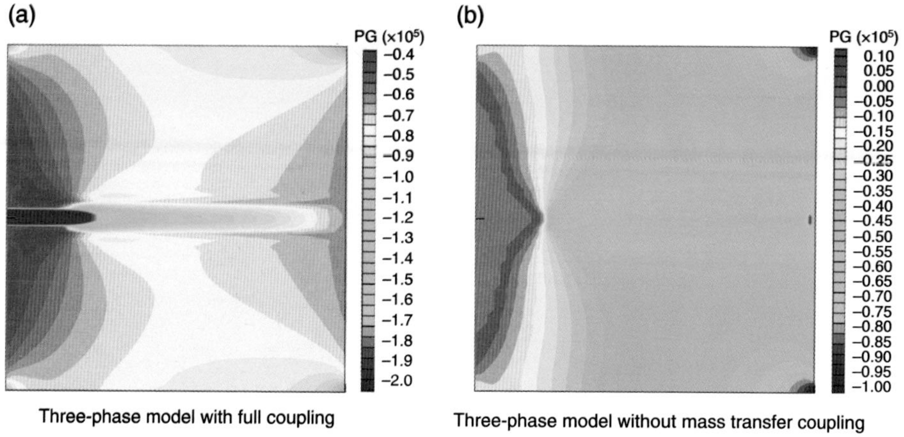

Figure 11.13 Gas pressure (Pa) contours; (a) three-phase model with full coupling, (b) three-phase model without mass transfer coupling

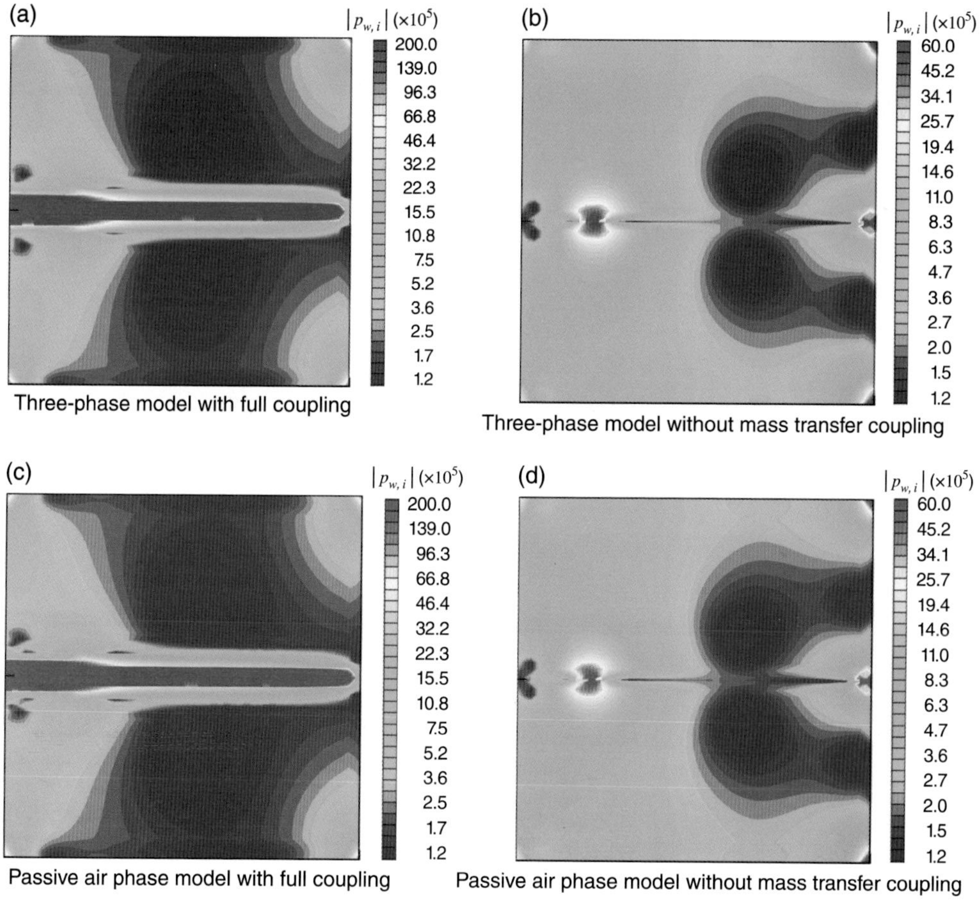

Figure 11.14 Norm of the water pressure gradient (Pa/m) contours (logarithmic scale); (a) three-phase model with full coupling, (b) three-phase model without mass transfer coupling, (c) passive air phase model with full coupling, (d) passive air phase model without mass transfer coupling

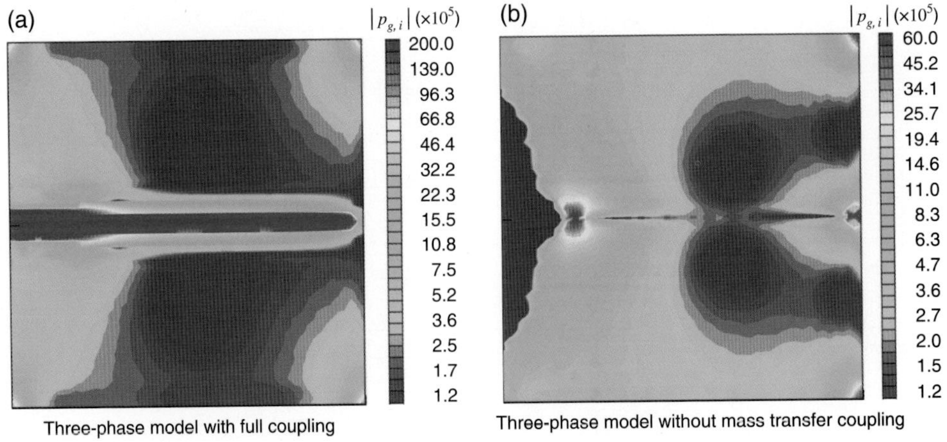

Figure 11.15 Norm of the gas pressure gradient (Pa/m) contours (logarithmic scale); (a) three-phase model with full coupling, (b) three-phase model without mass transfer coupling

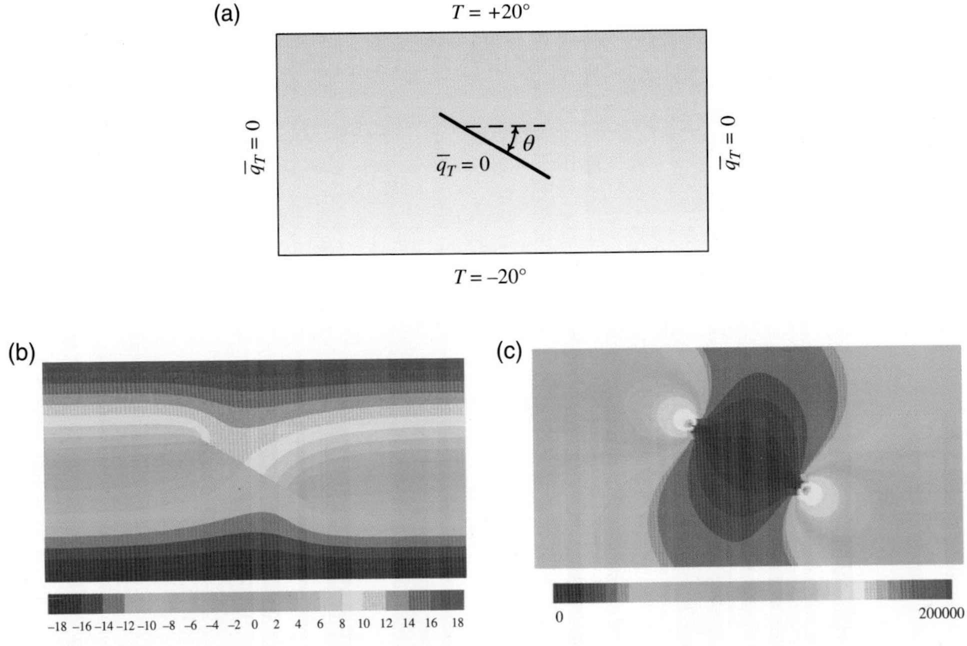

Figure 12.3 A plate with an inclined crack: (a) geometry and boundary conditions, (b) the distribution of temperature, (c) the distribution of heat flux

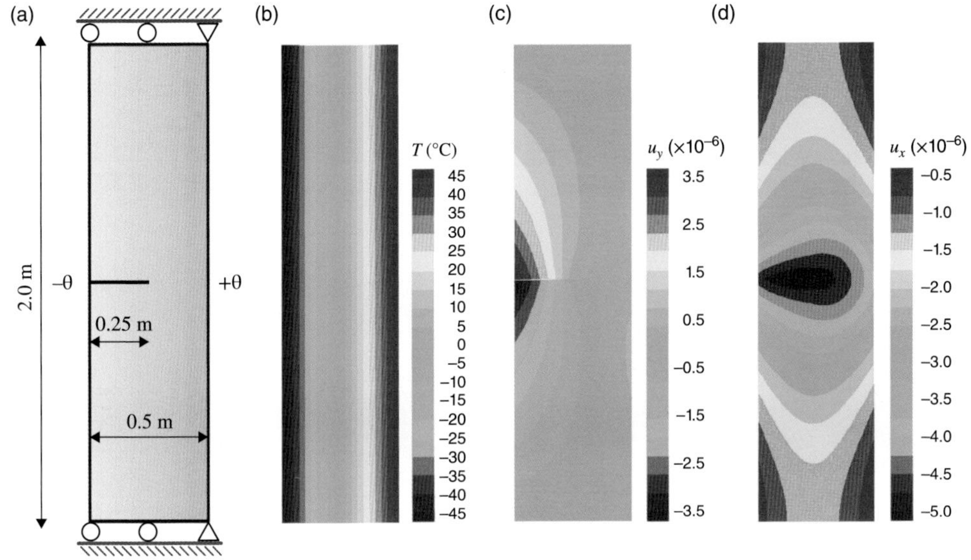

Figure 12.4 A plate with an edge crack in thermal loading: (a) the geometry and boundary conditions, (b) the distribution of temperature, (c) the displacement contour in y-direction, (d) the displacement contour in the x-direction

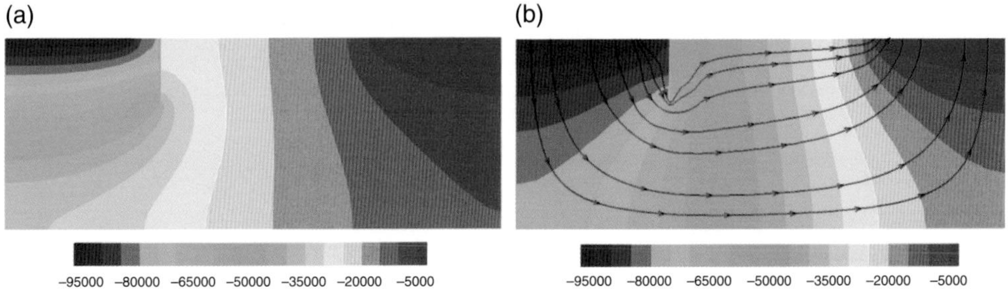

Figure 12.9 The distribution of pressure contours at time steps (a) $t = 10$ and (b) 110 s

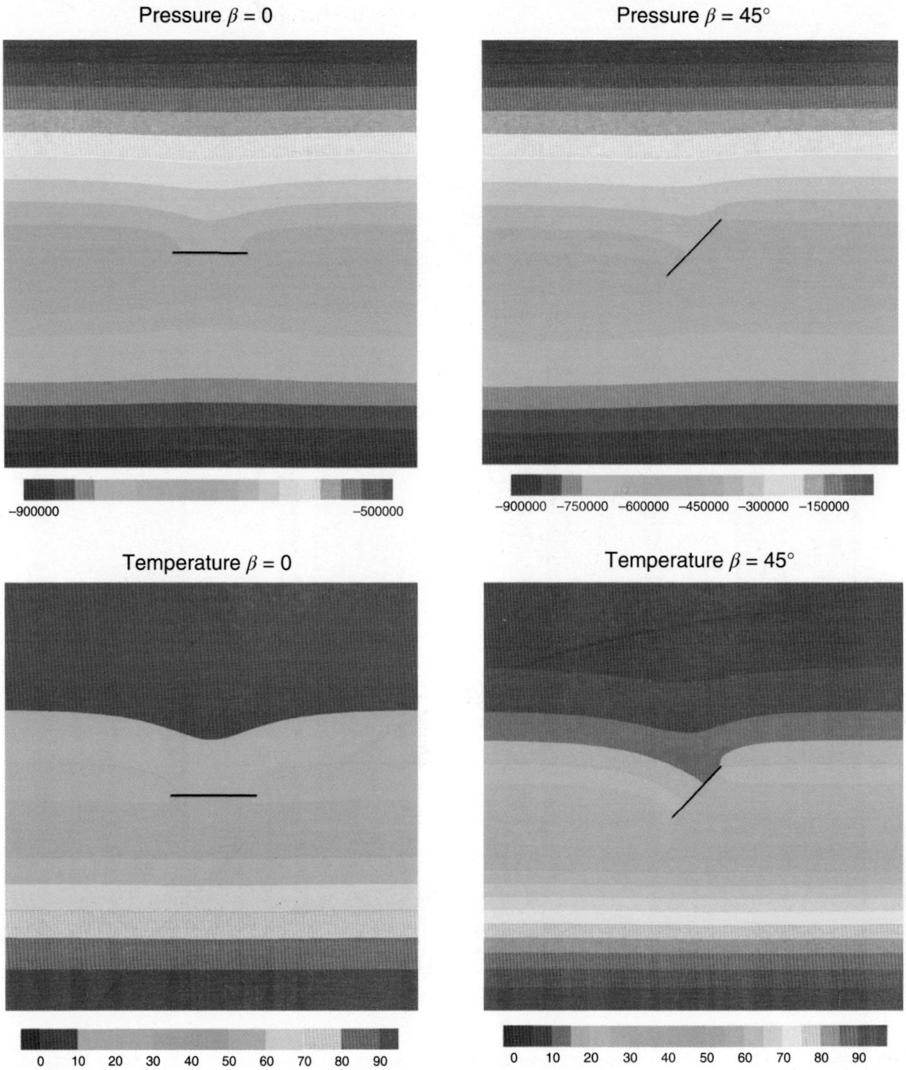

Figure 12.13 The distributions of pressure and temperature contours at the end of simulation for $\beta = 0$ and $\beta = 45^{\circ}$

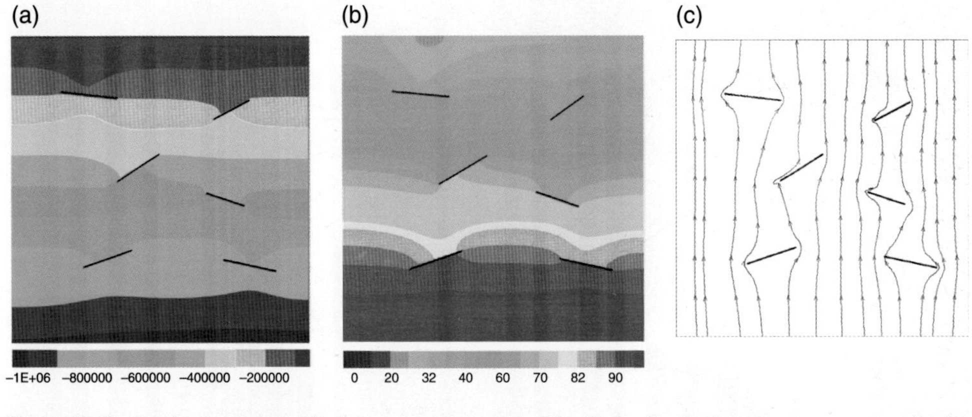

Figure 12.18 Randomly generated faults in saturated porous media: (a) the distribution of pressure contour, (b) the distribution of temperature contour, (c) the streamlines of the pressure